Foundations of Signal Processing

This comprehensive and engaging textbook introduces the basic principles and techniques of signal processing, from the fundamental ideas of signals and systems theory to real-world applications.

- Introduces students to the powerful foundations of modern signal processing, including the basic geometry of Hilbert space, the mathematics of Fourier transforms, and essentials of sampling, interpolation, approximation, and compression.
- Discusses issues in real-world use of these tools such as effects of truncation and quantization, limitations on localization, and computational costs.
- Includes over 160 homework problems and over 220 worked examples, specifically designed to test and expand students' understanding of the fundamentals of signal processing.
- Accompanied by extensive online materials designed to aid learning, including Mathematica® resources and interactive demonstrations.

Martin Vetterli is a Professor of Computer and Communication Sciences at the École Polytechnique Fédérale de Lausanne, and the President of the Swiss National Science Foundation. He has formerly held positions at Columbia University and the University of California, Berkeley, and has received the IEEE Signal Processing Society Technical Achievement Award (2001) and Society Award (2010). He is a Fellow of the ACM, EURASIP, and the IEEE, and is a Thomson Reuters Highly Cited Researcher in Engineering.

Jelena Kovačević is the David Edward Schramm Professor and Head of Electrical and Computer Engineering, and a Professor of Biomedical Engineering, at Carnegie Mellon University. She has been awarded the Belgrade October Prize (1986), the E. I. Jury Award (1991) from Columbia University, and the 2010 Philip L. Dowd Fellowship at Carnegie Mellon University. She is a former Editor-in-Chief of *IEEE Transactions on Image Processing* and a Fellow of the IEEE and EURASIP.

Vivek K Goyal is an Assistant Professor of Electrical and Computer Engineering at Boston University, and a former Esther and Harold E. Edgerton Associate Professor of Electrical Engineering at the Massachusetts Institute of Technology. He has been awarded the IEEE Signal Processing Society Magazine Award (2002) and the Eliahu Jury Award (1998) from the University of California, Berkeley, for outstanding achievement in systems, communications, control, and signal processing. He is a Fellow of the IEEE.

"*Foundations of Signal Processing* by Martin Vetterli, Jelena Kovacevic, and Vivek K Goyal lives up to its title by providing a thorough tour of the subject matter based on selected tools from real analysis which allow sufficient generality to develop the foundations of the classical Fourier methods along with modern wavelet approaches. A key distinction of the book is the use of Hilbert space ideas to provide a geometric interpretation and intuition that enhances both the classic and modern approaches by providing a unified view of their similarities and relative merits in the most important special cases. Many of the specific examples of signal processing considered can be viewed as examples of projections onto subspaces, which yield immediate properties and descriptions from the underlying fundamentals. The development is both pedagogically and theoretically sound, proceeding from the underlying mathematics through discrete-time systems to the more complicated continuous time systems into a wonderfully general and enlightening treatment of sampling and interpolation operations connecting discrete and continuous time. All of the important signal classes are considered and their basic properties and interrelations developed and summarized. The book then develops important topics not ordinarily found in signal processing texts – the accuracy of approximations involving the truncation of series expansions and the quantization of series coefficients, and the localization of signals in the time-frequency plane.

The completeness of the book results in a lengthy volume of roughly 800 pages, but it is easy to navigate to extract portions of interest while saving the many byways and special topics for future reference. The chapter introductions are particularly good at setting the stage in a simple but informative context and then sketching the details to come for the remainder of the chapter. Each chapter closes with a 'Chapter at a glance' section highlighting the primary ideas and results. The book will be a welcome addition to the library of students, practitioners, and researchers in signal processing for learning, reviewing, and referencing the broad array of tools and properties now available to analyze, synthesize, and understand signal processing systems."

Robert M. Gray, Stanford University and Boston University

"Finally a wonderful and accessible book for teaching modern signal processing to undergraduate students."

Stéphane Mallat, École Normale Supérieure

"This is a major book about a serious subject – the combination of engineering and mathematics that goes into modern signal processing: discrete time, continuous time, sampling, filtering, and compression. The theory is beautiful and the applications are so important and widespread."

Gil Strang, Massachusetts Institute of Technology

"This book (FSP) and its companion (FWSP) bring a refreshing new, and comprehensive approach to teaching the fundamentals of signal processing, from analysis and decompositions, to multi-scale representations, approximations, and many other aspects that have a tremendous impact in modern information technology. Whereas classical texts were usually written for students in electrical or communication engineering programs, FSP and FWSP start from basic concepts in algebra and geometry, with the benefit of being easily accessible to a much broader set of readers, and also help those readers develop strong abstract reasoning and intuition about signals and processing operators. A must-read!"

Rico Malvar, Microsoft Research

"This is a wonderful book that connects together all the elements of modern signal processing. From functional analysis and probability theory, to linear algebra and computational methods, it's all here and seamlessly integrated, along with a summary of history and developments in the field. A real tour-de-force, and a must-have on every signal processor's shelf!"

Robert D. Nowak, University of Wisconsin–Madison

"Most introductory signal processing textbooks focus on classical transforms, and study how these can be used. Instead, Foundations of Signal Processing encourages readers to think of signals first. It develops a 'signal-centric' view, one that focuses on signals, their representation and approximation, through the introduction of signal spaces. Unlike most entry-level signal processing texts, this general view, which can be applied to many different signal classes, is introduced right at the beginning. From this, starting from basic concepts, and placing an emphasis on intuition, this book develops mathematical tools that give the readers a fresh perspective on classical results, while providing them with the tools to understand many state-of-the-art signal representation techniques."

Antonio Ortega, University of Southern California

"Foundations of Signal Processing by Vetterli, Kovačević, and Goyal, is a pleasure to read. Drawing on the authors' rich experience of research and teaching of signal processing and signal representations, it provides an intellectually cohesive and modern view of the subject from the geometric point of view of vector spaces. Emphasizing Hilbert spaces, where fine technicalities can be relegated to backstage, this textbook strikes an excellent balance between intuition and mathematical rigor, that will appeal to both undergraduate and graduate engineering students. The last two chapters, on sampling and interpolation, and on localization and uncertainty, take full advantage of the machinery developed in the previous chapters to present these two very important topics of modern signal processing, that previously were only found in specialized monographs. The explanations of advanced topics are exceptionally lucid, exposing the reader to the ideas and thought processes behind the results and their derivation. Students will learn not only a substantial body of knowledge and techniques, but also why things work, at a deep level, which will equip them for independent further reading and research. I look forward to using this text in my own teaching."

Yoram Bresler, University of Illinois at Urbana-Champaign

Foundations of Signal Processing

MARTIN VETTERLI

École Polytechnique Fédérale de Lausanne, Switzerland
University of California, Berkeley, USA

JELENA KOVAČEVIĆ

Carnegie Mellon University, USA

VIVEK K GOYAL

Boston University, USA
Massachusetts Institute of Technology, USA

CAMBRIDGE
UNIVERSITY PRESS

University Printing House, Cambridge CB2 8BS, United Kingdom

One Liberty Plaza, 20th Floor, New York, NY 10006, USA

477 Williamstown Road, Port Melbourne, VIC 3207, Australia

314-321, 3rd Floor, Plot 3, Splendor Forum, Jasola District Centre, New Delhi - 110025, India

79 Anson Road, #06-04/06, Singapore 079906

Cambridge University Press is part of the University of Cambridge.

It furthers the University's mission by disseminating knowledge in the pursuit of education, learning and research at the highest international levels of excellence.

www.cambridge.org
Information on this title: www.cambridge.org/9781107038608

First published 2014
3rd printing 2016

A catalogue record for this publication is available from the British Library

ISBN 978-1-107-03860-8 Hardback

Additional resources for this publication at www.cambridge.org/vetterli and www.fourierandwavelets.org

To Marie-Laure, for her ∞ patience and many other qualities,
Thomas and Noémie, whom I might still convince of the beauty of this material,
and my parents, who gave me all the opportunities one can wish for.

— *MV*

To Danica and Giovanni, who make life beautiful.
To my parents, who made me who I am.

— *JK*

To Allie, Sundeep, and my family, who encourage me unceasingly, and
to the educators who made me want to be one of them.

— *VKG*

The cover illustration captures an experiment first described by Isaac Newton in *Opticks* in 1730, showing that white light can be split into its color components and then synthesized back into white light. It is a physical implementation of a decomposition of white light into its Fourier components – the colors of the rainbow – followed by a synthesis to recover the original.

Contents

Quick reference

Abbreviations

AR	Autoregressive
ARMA	Autoregressive moving average
AWGN	Additive white Gaussian noise
BIBO	Bounded input, bounded output
CDF	Cumulative distribution function
DCT	Discrete cosine transform
DFT	Discrete Fourier transform
DTFT	Discrete-time Fourier transform
DWT	Discrete wavelet transform
FFT	Fast Fourier transform
FIR	Finite impulse response
i.i.d.	Independent and identically distributed
IIR	Infinite impulse response
KLT	Karhunen–Loève transform
LMMSE	Linear minimum mean-squared error
LPSV	Linear periodically shift-varying
LSI	Linear shift-invariant
MA	Moving average
MAP	Maximum a posteriori probability
ML	Maximum likelihood
MMSE	Minimum mean-squared error
MSE	Mean-squared error
PDF	Probability density function
PMF	Probability mass function
POCS	Projection onto convex sets
rad	Radians
ROC	Region of convergence
SNR	Signal-to-noise ratio
SVD	Singular value decomposition
WSCS	Wide-sense cyclostationary
WSS	Wide-sense stationary

Abbreviations used in tables and captions but not in the text

FT	Fourier transform
FS	Fourier series
LT	Laplace transform

Sets

natural numbers	$\mathbb{N}$	$0, 1, \ldots$
integers	$\mathbb{Z}$	$\ldots, -1, 0, 1, \ldots$
positive integers	$\mathbb{Z}^+$	$1, 2, \ldots$
rational numbers	$\mathbb{Q}$	$p/q, \quad p, q \in \mathbb{Z}, \quad q \neq 0$
real numbers	$\mathbb{R}$	$(-\infty, \infty)$
positive real numbers	$\mathbb{R}^+$	$(0, \infty)$
complex numbers	$\mathbb{C}$	$a + jb$ or $re^{j\theta}$ with $a, b, r, \theta \in \mathbb{R}$
a generic index set	$\mathcal{I}$	
a generic vector space	V	
a generic Hilbert space	H	
closure of set S	$\overline{S}$	

Real and complex analysis

sequence	x_n	argument n is an integer, $n \in \mathbb{Z}$
function	$x(t)$	argument t is continuous-valued, $t \in \mathbb{R}$
ordered sequence	$(x_n)_n$	
set containing x_n	$\{x_n\}_n$	
vector x with x_n as elements	$[x_n]$	
Kronecker delta sequence	δ_n	$\delta_n = 1$ for $n = 0$; $\delta_n = 0$ otherwise
Dirac delta function	$\delta(t)$	$\int_{-\infty}^{\infty} x(t)\,\delta(t)\,dt = x(0)$ for x continuous at 0
indicator function of interval I	1_I	$1_I(t) = 1$ for $t \in I$; $1_I(t) = 0$ otherwise
integration by parts		$\int u\,dv = uv - \int v\,du$
complex number	z	$a + jb$, $re^{j\theta}$, $a, b \in \mathbb{R}$, $r \in [0, \infty)$, $\theta \in [0, 2\pi)$
conjugation	z^*	$a - jb$, $re^{-j\theta}$
real part of	$\Re(\,\cdot\,)$	$\Re(a + jb) = a, \quad a, b \in \mathbb{R}$
imaginary part of	$\Im(\,\cdot\,)$	$\Im(a + jb) = b, \quad a, b \in \mathbb{R}$
conjugation of coefficients	$X_*(z)$	$X^*(z^*)$
principal root of unity	W_N	$e^{-j2\pi/N}$

Asymptotic notation

big O	$x \in O(y)$	$0 \leq x_n \leq \gamma y_n$ for all $n \geq n_0$; some n_0 and $\gamma > 0$
little o	$x \in o(y)$	$0 \leq x_n \leq \gamma y_n$ for all $n \geq n_0$; some n_0, any $\gamma > 0$
Omega	$x \in \Omega(y)$	$x_n \geq \gamma y_n$ for all $n \geq n_0$; some n_0 and $\gamma > 0$
Theta	$x \in \Theta(y)$	$x \in O(y)$ and $x \in \Omega(y)$
asymptotic equivalence	$x \asymp y$	$\lim_{n\to\infty} x_n/y_n = 1$

Standard vector spaces

Hilbert space of square-summable sequences	$\ell^2(\mathbb{Z})$	$\left\{x : \mathbb{Z} \to \mathbb{C} \,\middle|\, \sum_n \lvert x_n \rvert^2 < \infty\right\}$ with inner product $\langle x, y \rangle = \sum_n x_n y_n^*$
Hilbert space of square-integrable functions	$\mathcal{L}^2(\mathbb{R})$	$\left\{x : \mathbb{R} \to \mathbb{C} \,\middle|\, \int \lvert x(t) \rvert^2 \, dt < \infty\right\}$ with inner product $\langle x, y \rangle = \int x(t)\, y(t)^* \, dt$
normed vector space of sequences with finite ℓ^p norm, $1 \le p < \infty$	$\ell^p(\mathbb{Z})$	$\left\{x : \mathbb{Z} \to \mathbb{C} \,\middle|\, \sum_n \lvert x_n \rvert^p < \infty\right\}$ with norm $\lVert x \rVert_p = \left(\sum_n \lvert x_n \rvert^p\right)^{1/p}$
normed vector space of functions with finite $\mathcal{L}^p$ norm, $1 \le p < \infty$	$\mathcal{L}^p(\mathbb{R})$	$\left\{x : \mathbb{R} \to \mathbb{C} \,\middle|\, \int \lvert x(t) \rvert^p \, dt < \infty\right\}$ with norm $\lVert x \rVert_p = \left(\int \lvert x(t) \rvert^p \, dt\right)^{1/p}$
normed vector space of bounded sequences with supremum norm	$\ell^\infty(\mathbb{Z})$	$\left\{x : \mathbb{Z} \to \mathbb{C} \,\middle|\, \sup_n \lvert x_n \rvert < \infty\right\}$ with norm $\lVert x \rVert_\infty = \sup_n \lvert x_n \rvert$
normed vector space of bounded functions with supremum norm	$\mathcal{L}^\infty(\mathbb{R})$	$\left\{x : \mathbb{R} \to \mathbb{C} \,\middle|\, \operatorname{ess\,sup}_t \lvert x(t) \rvert < \infty\right\}$ with norm $\lVert x \rVert_\infty = \operatorname{ess\,sup}_t \lvert x(t) \rvert$

Bases and frames for sequences

standard basis	$\{e_k\}$	$e_{k,n} = 1$ for $n = k$; $e_{k,n} = 0$ otherwise
vector, element of basis or frame	φ	when applicable, a *column* vector
basis or frame	Φ	set of vectors $\{\varphi_k\}$
operator	Φ	concatenation of $\{\varphi_k\}$ in a linear operator: $[\varphi_0 \;\; \varphi_1 \;\; \ldots \;\; \varphi_{N-1}]$
vector, element of dual basis or frame	$\widetilde{\varphi}$	when applicable, a *column* vector
	$\widetilde{\Phi}$	set of vectors $\{\widetilde{\varphi}_k\}$
operator	$\widetilde{\Phi}$	concatenation of $\{\widetilde{\varphi}_k\}$ in a linear operator: $[\widetilde{\varphi}_0 \;\; \widetilde{\varphi}_1 \;\; \ldots \;\; \widetilde{\varphi}_{N-1}]$
expansion with a basis or frame	$x = \Phi\widetilde{\Phi}^* x$	

Discrete-time signal processing

Sequence	x_n	signal, vector
Convolution		
linear	$h * x$	$\sum_{k\in\mathbb{Z}} x_k h_{n-k}$
circular (N-periodic sequences)	$h \circledast x$	$\sum_{k=0}^{N-1} x_k h_{(n-k) \bmod N}$
	$(h * x)_n$	convolution result at n
Eigensequence	v_n	eigenvector
infinite time	$v_n = e^{j\omega n}$	$h * v = H(e^{j\omega})\, v$
finite time	$v_n = e^{j2\pi kn/N}$	$h \circledast v = H_k v$
Frequency response		eigenvalue corresponding to v_n
infinite time	$H(e^{j\omega})$	$\sum_{n\in\mathbb{Z}} h_n e^{-j\omega n}$
finite time	H_k	$\sum_{n=0}^{N-1} h_n e^{-j2\pi kn/N} = \sum_{n=0}^{N-1} h_n W_N^{kn}$

Continuous-time signal processing

Function	$x(t)$	signal
Convolution		
linear	$h * x$	$\int_{-\infty}^{\infty} x(\tau)\, h(t-\tau)\, d\tau$
circular (T-periodic functions)	$h \circledast x$	$\int_{-T/2}^{T/2} x(\tau)\, h(t-\tau)\, d\tau$
	$(h * x)(t)$	convolution result at t
Eigenfunction	$v(t)$	eigenvector
infinite time	$v(t) = e^{j\omega t}$	$h * v = H(\omega)\, v$
finite time	$v(t) = e^{j2\pi kt/T}$	$h \circledast v = H_k v$
Frequency response		eigenvalue corresponding to $v(t)$
infinite time	$H(\omega)$	$\int_{-\infty}^{\infty} h(t)\, e^{-j\omega t}\, dt$
finite time	H_k	$\int_{-T/2}^{T/2} h(\tau)\, e^{-j2\pi k\tau/T}\, d\tau$

Spectral analysis

Fourier transform	$x(t) \stackrel{\text{FT}}{\longleftrightarrow} X(\omega)$	$X(\omega) = \int_{-\infty}^{\infty} x(t) e^{-j\omega t}\, dt$
inverse		$x(t) = \frac{1}{2\pi} \int_{-\infty}^{\infty} X(\omega)\, e^{j\omega t}\, d\omega$
Fourier series coefficients	$x(t) \stackrel{\text{FS}}{\longleftrightarrow} X_k$	$X_k = \frac{1}{T} \int_{-T/2}^{T/2} x(t)\, e^{-j(2\pi/T)kt}\, dt$
reconstruction		$x(t) = \sum_{k\in\mathbb{Z}} X_k e^{j(2\pi/T)kt}$
discrete-time Fourier transform	$x_n \stackrel{\text{DTFT}}{\longleftrightarrow} X(e^{j\omega})$	$X(e^{j\omega}) = \sum_{n\in\mathbb{Z}} x_n e^{-j\omega n}$
inverse		$x_n = \frac{1}{2\pi} \int_{-\pi}^{\pi} X(e^{j\omega})\, e^{j\omega n}\, d\omega$
discrete Fourier transform	$x_n \stackrel{\text{DFT}}{\longleftrightarrow} X_k$	$X_k = \sum_{n=0}^{N-1} x_n W_N^{kn}$
inverse		$x_n = \frac{1}{N} \sum_{n=0}^{N-1} X_k W_N^{-kn}$
z-transform	$x_n \stackrel{\text{ZT}}{\longleftrightarrow} X(z)$	$X(z) = \sum_{n\in\mathbb{Z}} x_n z^{-n}$

Acknowledgments

This book would not exist without the help of many people, whom we attempt to list below. We apologize for any omissions and welcome corrections and suggestions.

We are grateful to Professor Libero Zuppiroli of EPFL and Christiane Grimm for the photograph that graces the cover; Professor Zuppiroli proposed an experiment from Newton's treatise, *Opticks* [70], as emblematic of the book, and Ms. Grimm beautifully photographed the apparatus that he designed. Françoise Behn, Jocelyne Plantefol, and Jacqueline Aeberhard typed parts of the manuscript, Eric Strattman assisted with some of the figures, Krista Van Guilder designed and implemented the initial book web site, and Jorge Albaladejo Pomares designed and implemented the book blog. We thank them for their diligence and patience. S. Grace Chang and Yann Barbotin helped organize and edit the problem companion, while Patrick Vandewalle designed and implemented a number of MATLAB® problems available on the book website.[1] We thank them for their expertise and insight. We are indebted to George Beck for his *Mathematica*® editorial comments on Wolfram Demonstrations inspired by this book. We are grateful to Giovanni Pacifici for the many useful scripts for automating various book-writing processes. We thank John Cozzens of the US National Science Foundation for his unwavering support over the last two decades. At Cambridge University Press, Steven Holt held us to high standards in our writing and typesetting through many corrections and suggestions; Elizabeth Horne and Jonathan Ratcliffe coordinated production expertly; and, last but not least, Phil Meyler repeatedly (and most gracefully) reminded us of our desire to finish the book.

Many instructors have gamely tested pre-alpha, alpha, and beta versions of this manuscript. Of these, Amina Chebira, Filipe Condessa, Mihailo Kolundžija, Yue M. Lu, Truong-Thao Nguyen, Reza Parhizkar, and Jayakrishnan Unnikrishnan have done far more than their share in providing invaluable comments and suggestions. We also thank Robert Gray for reviewing the manuscript and providing many important suggestions, in particular a better approach to covering the Dirac delta function; Thierry Blu for, among other things, providing a simple proof for a particular case of the Strang–Fix theorem; Matthew Fickus for consulting on some finer mathematical points; Michael Unser for his notes and teaching approach; and Zoran Cvetković, Minh N. Do, and Philip Schniter for teaching with the manuscript and providing many constructive comments. Useful comments have also been provided

[1] http://www.fourierandwavelets.org

by Pedro Aguilar, Ganesh Ajjanagadde, A. Avudainayagam, Jerry Bauch, Aniruddha Bhargava, Ozan Çağlayan, André Tomaz de Carvalho, Gustavo Corach, Todd Doucet, S. Esakkirajan, Germán González, Alexandre Haehlen, Mina Karzand, Ullrich Mönich, Juan Pablo Muszkats, Hossein Rouhani, Noah D. Stein, John Z. Sun, Christophe Tournery, and Lav R. Varshney.

Martin Vetterli thanks current and former EPFL graduate students and postdocs who helped develop the material, solve problems, catch typos, and suggest improvements, among other things. They include Florence Bénézit, Amina Chebira, Minh N. Do, Ivan Dokmanić, Pier Luigi Dragotti, Ali Hormati, Ivana Jovanović, Mihailo Kolundžija, Jérôme Lebrun, Yue M. Lu, Pina Marziliano, Fritz Menzer, Reza Parhizkar, Paolo Prandoni, Juri Ranieri, Olivier Roy, Rahul Shukla, Jayakrishnan Unnikrishnan, Patrick Vandewalle and Vladan Velisavljević. He gratefully acknowledges support from EPFL, the Swiss NSF through awards 2000-063664, 200020-103729, and 200021-121935, and the European Research Council through award SPARSAM 247006. In addition, the support from Qualcomm, in particular from Dr. Chong Lee, is gratefully acknowledged.

Jelena Kovačević thanks her present and past graduate students Ramamurthy Bhagavatula, Amina Chebira, Kuan-Chieh Jackie Chen, Siheng Chen, Filipe Condessa, Charles Jackson, Xindian Long, Michael T. McCann, Anupama Kuruvilla, Thomas E. Merryman, Vivek Oak, Aliaksei Sandryhaila, and Gowri Srinivasa, many of whom served as TAs for her classes at CMU, together with Pablo Hennings Yeomans. Thanks are due also to all the students in the following classes at CMU: 42-431/18-496, 42-540/42-698, 42-703/18-799, 42-731/18-795, and 42-732/18-790, taught from 2003 to 2013. She gratefully acknowledges support from the US NSF through awards 1017278, 1130616, 515152, 633775, and 331657; the NIH through awards EB008870, EB009875, and 1DC010283; the Carnegie Institute of Technology at CMU through an Infrastructure for Large-Scale Biomedical Computing award; the Pennsylvania State Tobacco Settlement, the Kamlet–Smith Bioinformatics Grant; and the Philip and Marsha Dowd Teaching Fellowship.

Vivek Goyal thanks TAs and students at MIT and BU for suggestions, in particular Giuseppe Bombara, Baris Erkmen, Ying-zong Huang, Zahi Karam, Ahmed Kirmani, Srilalitha Kumaresan, Ranko Sredojević, Ramesh Sridharan, Watcharapan Suwansantisuk, Vincent Tan, Archana Venkataraman, Adam Zelinski, and Serhii Zhak. He gratefully acknowledges support from the US NSF through awards 0643836, 0729069, 1101147, 1115159, and 1161413; Texas Instruments through its Leadership University Program; Hewlett–Packard; Qualcomm; Siemens; MIT through the Esther and Harold E. Edgerton Career Development Chair; and the Centre Bernoulli of EPFL.

Preface

Our main goals in this book and its companion volume, *Fourier and Wavelet Signal Processing (FWSP)* [57], are to enable an understanding of state-of-the-art signal processing methods and techniques, as well as to provide a solid foundation for those hoping to advance the theory and practice of signal processing. We believe that the best way to grasp and internalize the fundamental concepts in signal processing is through the geometry of Hilbert spaces, as this leverages the great innate human capacity for spatial reasoning. While using geometry should ultimately *simplify* the subject, the connection between signals and geometry is not innate. The reader will have to invest effort to see signals as vectors in Hilbert spaces before reaping the benefits of this view; we believe that effort to be well placed.

Many of the results and techniques presented in the two volumes, while rooted in classic Fourier techniques for signal representation, first appeared during a flurry of activity in the 1980s and 1990s. New constructions of local Fourier transforms and orthonormal wavelet bases during that period were motivated both by theoretical interest and by applications, multimedia communications in particular. New bases with specified time–frequency behavior were found, with impact well beyond the original fields of application. Areas as diverse as computer graphics and numerical analysis embraced some of the new constructions – no surprise given the pervasive role of Fourier analysis in science and engineering.

Many of these new tools for signal processing were developed in the applied harmonic analysis community. The resulting high level of mathematical sophistication was a barrier to entry for many signal processing practitioners. Now that the dust has settled, some of what was new and esoteric has become fundamental; we want to bring these new fundamentals to a broader audience. The Hilbert space formalism gives us a way to begin with the classical Fourier analysis of signals and systems and reach structured representations with time–frequency locality and their varied applications. Whenever possible, we use explanations rooted in elementary analysis over those that would require more advanced background (such as measure theory). We hope to have balanced the competing virtues of accessibility to the student, rigor, and adequate analytical power to reach important conclusions.

The book can be used as a self-contained text on the foundations of signal processing, where discrete and continuous time are treated on equal footing. All the necessary mathematical background is included, with examples illustrating the applicability of the results. In addition, the book serves as a precursor to *FWSP*, which relies on the framework built here; the two books are thus integrally related.

Foundations of Signal Processing This book covers the foundations for an in-depth understanding of modern signal processing. It contains material that many readers may have seen before scattered across multiple sources, but without the Hilbert space interpretations, which are essential in signal processing. Our aim is to teach signal processing *with geometry*, that is, to extend Euclidean geometric insights to abstract signals; we use Hilbert space geometry to accomplish that. With this approach, fundamental concepts – such as properties of bases, Fourier representations, sampling, interpolation, approximation, and compression – are often unified across finite dimensions, discrete time, and continuous time, thus making it easier to point out the few essential differences. Unifying results geometrically helps generalize beyond Fourier-domain insights, pushing the understanding farther, faster.

Chapter 2, *From Euclid to Hilbert*, is our main vehicle for drawing out unifying commonalities; it develops the basic geometric intuition central to Hilbert spaces, together with the necessary tools underlying the constructions of bases and frames.

The next two chapters cover signal processing on discrete-time and continuous-time signals, specializing general concepts from Chapter 2. Chapter 3, *Sequences and discrete-time systems*, is a crash course on processing signals in discrete time or discrete space together with spectral analysis with the discrete-time Fourier transform and discrete Fourier transform. Chapter 4, *Functions and continuous-time systems*, is its continuous-time counterpart, including spectral analysis with the Fourier transform and Fourier series.

Chapter 5, *Sampling and interpolation*, presents the critical link between discrete and continuous domains given by sampling and interpolation theorems. Chapter 6, *Approximation and compression*, veers from exact representations to approximate ones. The final chapter in the book, Chapter 7, *Localization and uncertainty*, considers time–frequency behavior of the abstract representation objects studied thus far. It also discusses issues arising in applications as well as ways of adapting the previously introduced tools for use in the real world.

Fourier and Wavelet Signal Processing The companion volume focuses on signal representations using local Fourier and wavelet bases and frames. It covers the two-channel filter bank in detail, and then uses it as the implementation vehicle for all sequence representations that follow. The local Fourier and wavelet methods are presented side-by-side, without favoring any one in particular; the truth is that each representation is a tool in the toolbox of the practitioner, and the problem or application at hand ultimately determines the appropriate one to use. We end with examples of state-of-the-art signal processing and communication problems, with sparsity as a guiding principle.

Teaching points Our aim is to present a synergistic view of signal representations and processing, starting from basic mathematical principles and going all the way to actual constructions of bases and frames, always with an eye on concrete applications. While the benefit is a self-contained presentation, the cost is a rather sizable manuscript. Referencing in the main text is sparse; pointers to the bibliography are given in *Further reading* at the end of each chapter.

The material grew out of teaching signal processing, wavelets, and applications in various settings. Two of us (MV and JK) authored a graduate textbook, *Wavelets and Subband Coding* (originally with Prentice Hall in 1995, now open access[2]), which we and others used to teach graduate courses at various US and European institutions. With the maturing of the field and the interest arising from and for these topics, the time was right for the three of us to write entirely new texts geared toward a broader audience. We and others have taught with these books, in their entirety or in parts, a number of times and to a number of different audiences: from senior undergraduate to graduate level, and from engineering to mixes that include life-science students.

The books and their website[3] provide a number of features for teaching and learning:

- *Exercises* are an integral part of the material and come in two forms: *solved exercises* with explicit solutions within the text, and *regular exercises* that allow students to test their knowledge. Regular exercises are marked with $^{(1)}$, $^{(2)}$, or $^{(3)}$ in increasing order of difficulty.
- Numerous *examples* illustrate concepts throughout the book.
- An *electronic version* of the text is provided with the printed copy. It includes PDF hyperlinks and an additional color to enhance interpretation of figures.
- A *free electronic version* of the text without PDF hyperlinks, exercises or solved exercises, and with figures in grayscale, is available at the book website.
- A *Mathematica®* *companion*, which contains the code to produce all numerical figures in the book, is provided with the printed version.
- Several interactive *Mathematica®* *demonstrations* using the free CDF player are available at the book website.
- Additional material, such as lecture slides, is available at the book website.
- To instructors, we provide a *Solutions Manual*, with solutions to all regular exercises in the book.

Notational points To traverse the book efficiently, it will help to know the various numbering conventions that we have employed. In each chapter, a single counter is used for definitions, theorems, and corollaries (which are all shaded) and another for examples (which are slightly indented). Equations, figures, and tables are also numbered within each chapter. A prefix E$n.m$– in the number of an equation, figure, or table indicates that it is part of Solved Exercise $n.m$ in Chapter n. The letter P is used similarly for statements of regular exercises.

Martin Vetterli, Jelena Kovačević, and Vivek K Goyal

[2] http://waveletsandsubbandcoding.org/
[3] http://www.fourierandwavelets.org/

Chapter 1

On rainbows and spectra

"One can enjoy a rainbow without necessarily forgetting the forces that made it."

— *Mark Twain*

In the late thirteenth century, Theodoric of Freiberg, a Dominican monk, theologian, and physicist, performed a simple experiment: with his back to the sun, he held a spherical bottle filled with water in the sunlight. By following the trajectory of the refracted and reflected light and having the bottle play the same role as a single water drop, he gave a scientific explanation of rainbows, including secondary rainbows with weaker, reversed colors. His geometric analysis, described in his famous treatise *De iride* (*On the Rainbow*, c. 1310), was "perhaps the most dramatic development of fourteenth- and fifteenth-century optics" [60].

Theodoric of Freiberg fell short of a complete understanding of the rainbow phenomenon because, like many of his contemporaries, he believed that colors were simply intensities between black and white. A full understanding emerged three hundred years later when René Descartes and Isaac Newton explained that dispersion decomposes white light into spectral components of different wavelengths – the colors of the rainbow. In 1730, Newton, in his landmark book on optics [70], describes what is often called the *experimentum crucis* (crucial experiment) to prove that white light can be decomposed into constituent colors and then recombined into white light. This experiment is a physical implementation of decomposing light into its Fourier components – pure frequencies or colors of the rainbow, followed by a synthesis to recover the original; the cover photograph of the book depicts this experiment.[4]

This bit of history evokes two central themes of this book: geometric thinking

[4]This is a realization of Figure 7 in Part II of Newton's *First Book of Opticks*.

is a great tool in deducing explanations of phenomena; and decomposing an entity into its constituent components can be a key step in understanding its essential character, as well as an enabling tool in modifying these components prior to recombination. The rainbow's appearance is explained by the fact that sunlight contains a combination of all wavelengths within the visible range; separation of white light by wavelength, as with a prism, enables modifications prior to recombination. The collection of wavelengths is, as we will see, the spectrum.

A French physicist and mathematician, Joseph Fourier, formalized the notion of the spectrum in the early nineteenth century. He was interested in the heat equation – the differential equation governing the diffusion of heat. Fourier's key insight was to decompose a periodic function $x(t) = x(t+T)$ into an infinite sum of sines and cosines of periods T/k, $k \in \mathbb{Z}^+$. Since these sine and cosine components are eigenfunctions of the heat equation, the solution of the problem is simplified: one can analyze the differential equation for each component separately and combine the intermediate results, thanks to the linearity of the system. Fourier's decomposition earned him a coveted prize from the French Academy of Sciences, but with a mention that his work lacked rigor. Indeed, the question of which functions admit a Fourier decomposition is a deep one, and it took many years to settle. Fourier's work is one of the foundational blocks of signal processing and at the heart of the present book as well as its companion volume [57]. Fourier techniques have been joined in the past two decades by new tools such as wavelets, the other pillar we cover.

Signal representations The idea of a decomposition and a possible modification in the decomposed state leads to signal representations, where signals can be sequences (discrete domain) or functions (continuous domain). Similarly to what Fourier did, where he used sines and cosines for decomposition, we can imagine using other functions with particular properties. Call these basis vectors and denote them by φ_k, $k \in \mathbb{Z}$. Then

$$x = \sum_{k\in\mathbb{Z}} X_k \varphi_k \tag{1.1}$$

is called an expansion of x with respect to $\{\varphi_k\}_{k\in\mathbb{Z}}$, with $\{X_k\}$ the expansion coefficients.

Orthonormal bases When the basis vectors form an orthonormal set, the coefficients X_k are obtained from the function x and the basis vectors φ_k through an inner product

$$X_k = \langle x, \varphi_k \rangle. \tag{1.2}$$

For example, Fourier's construction of a series representation for periodic functions with period $T = 1$ can be written as

$$x(t) = \sum_{k\in\mathbb{Z}} X_k e^{j2\pi kt}, \tag{1.3a}$$

where

$$X_k = \int_0^1 x(t) e^{-j2\pi kt}\, dt. \tag{1.3b}$$

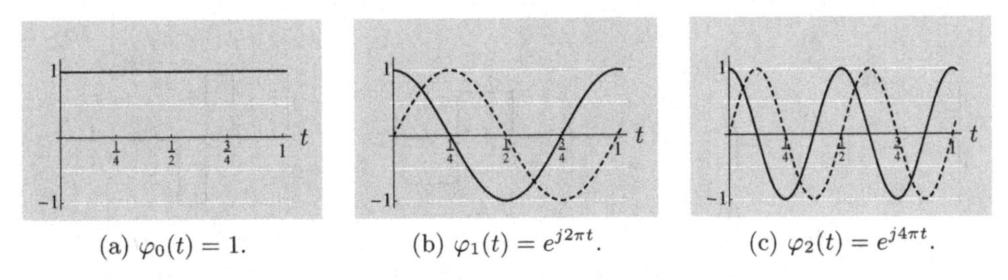

(a) $\varphi_0(t) = 1$. (b) $\varphi_1(t) = e^{j2\pi t}$. (c) $\varphi_2(t) = e^{j4\pi t}$.

Figure 1.1 Example Fourier series basis functions for the interval $[0, 1)$. Real parts are shown with solid lines and imaginary parts are shown with dashed lines.

We can define basis vectors φ_k, $k \in \mathbb{Z}$, on the interval $[0, 1)$, as

$$\varphi_k(t) \;=\; e^{j2\pi kt}, \qquad 0 \le t < 1, \tag{1.4}$$

and the Fourier series coefficients as

$$X_k \;=\; \langle x, \varphi_k \rangle \;=\; \int_0^1 x(t)\varphi_k^*(t)\,dt \;=\; \int_0^1 x(t)e^{-j2\pi kt}\,dt,$$

exactly the same as (1.3b). The basis vectors form an orthonormal set (the first few are shown in Figure 1.1):

$$\langle \varphi_k, \varphi_i \rangle \;=\; \int_0^1 e^{j2\pi kt}e^{-j2\pi it}\,dt \;=\; \begin{cases} 1, & \text{for } i = k; \\ 0, & \text{otherwise.} \end{cases} \tag{1.5}$$

While the Fourier series is certainly a key orthonormal basis with many outstanding properties, other bases exist, some of which have their own favorable properties. Early in the twentieth century, Alfred Haar proposed a basis which looks quite different from Fourier's. It is based on a function $\psi(t)$ defined as

$$\psi(t) \;=\; \begin{cases} 1, & \text{for } 0 \le t < \frac{1}{2}; \\ -1, & \text{for } \frac{1}{2} \le t < 1; \\ 0, & \text{otherwise.} \end{cases} \tag{1.6}$$

For the interval $[0, 1)$, we can build an orthonormal system by scaling $\psi(t)$ by powers of 2, and then shifting the scaled versions appropriately, yielding

$$\psi_{m,n}(t) \;=\; 2^{-m/2}\psi\!\left(\frac{t - n2^m}{2^m}\right), \tag{1.7}$$

with $m \in \{0, -1, -2, \ldots\}$ and $n \in \{0, 1, \ldots, 2^{-m} - 1\}$ (a few are shown in Figure 1.2). It is quite clear from the figure that the various basis functions are indeed orthogonal to each other, as they either do not overlap, or when they do, one changes sign over the constant span of the other. We will spend a considerable amount of time studying this system in the companion volume to this book, [57].

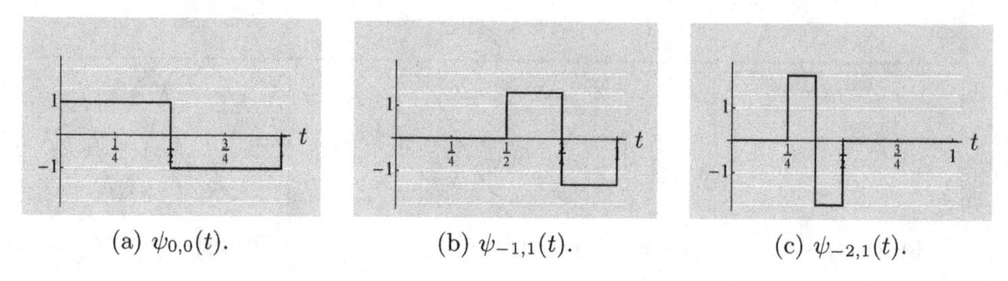

(a) $\psi_{0,0}(t)$. (b) $\psi_{-1,1}(t)$. (c) $\psi_{-2,1}(t)$.

Figure 1.2 Example Haar series basis functions for the interval $[0, 1)$. The prototype function is $\psi(t) = \psi_{0,0}(t)$.

While the system (1.7) is orthonormal, it cannot be a basis for all functions on $[0, 1)$; for example, there would be no way to reconstruct a constant 1. We remedy that by adding the function

$$\varphi_0(t) = \begin{cases} 1, & \text{for } 0 \leq t < 1; \\ 0, & \text{otherwise,} \end{cases} \tag{1.8}$$

into the mix, yielding an orthonormal basis for the interval $[0, 1)$. This is a very different basis from the Fourier one; for example, instead of being infinitely differentiable, no $\psi_{m,n}$ is even continuous. We can now define an expansion as in (1.3),

$$x(t) = \langle x, \varphi_0 \rangle \varphi_0(t) + \sum_{m=-\infty}^{0} \sum_{n=0}^{2^{-m}-1} X_{m,n} \psi_{m,n}(t), \tag{1.9a}$$

where

$$X_{m,n} = \int_0^1 x(t)\, \psi_{m,n}(t)\, dt. \tag{1.9b}$$

It is natural to ask which basis is better. Such a question does not have a simple answer, and the answer will depend on the class of functions or sequences we wish to represent, as well as our goals in the representation. Furthermore, we will have to be careful in describing what we mean by equality in an expansion such as (1.3a); otherwise we could be misled the same way Fourier was.

Approximation One way to assess the quality of a basis is to see how well it can approximate a given function with a finite number of terms. History is again enlightening. Fourier series became such a useful tool during the nineteenth century that researchers built elaborate mechanical devices to compute a function based on Fourier series coefficients. They built analog computers, based on harmonically related rotating wheels, where amplitudes of Fourier coefficients could be set and the sum computed. One such machine, the Harmonic Integrator, was designed by the physicists Albert Michelson and Samuel Stratton, and it could compute a series with 80 terms. To the designers' dismay, the synthesis of a square wave from its Fourier series led to oscillations around the discontinuity that would not go away

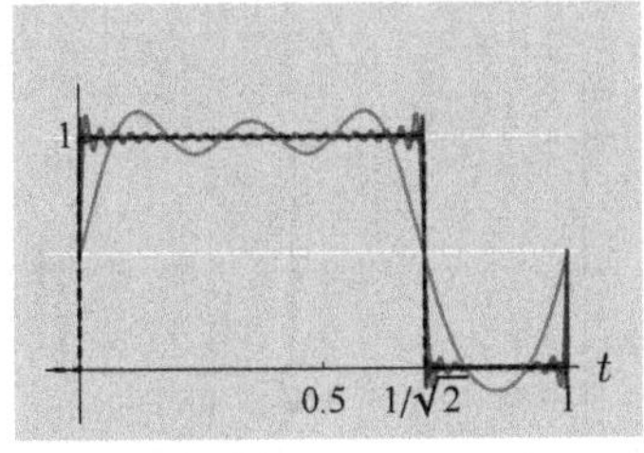

(a) Series with 9, 65, and 513 terms.

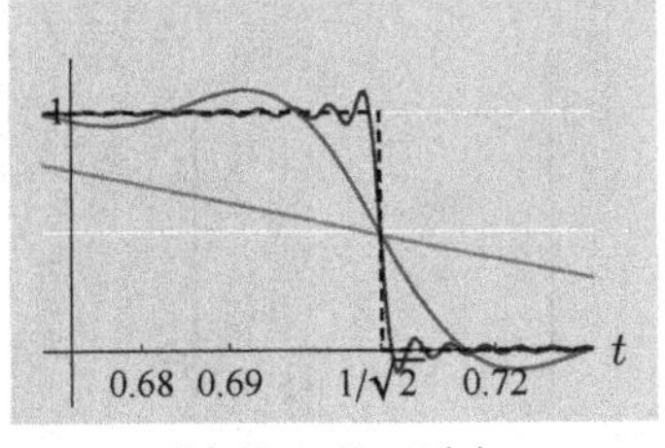

(b) Detail of (a).

Figure 1.3 Approximations of a box function (dashed lines) with a Fourier series basis using 9, 65, and 513 terms (solid lines, from lightest to darkest). The plots illustrate the Gibbs phenomenon – oscillations that do not diminish in amplitude when approximating a discontinuous function with truncated Fourier series.

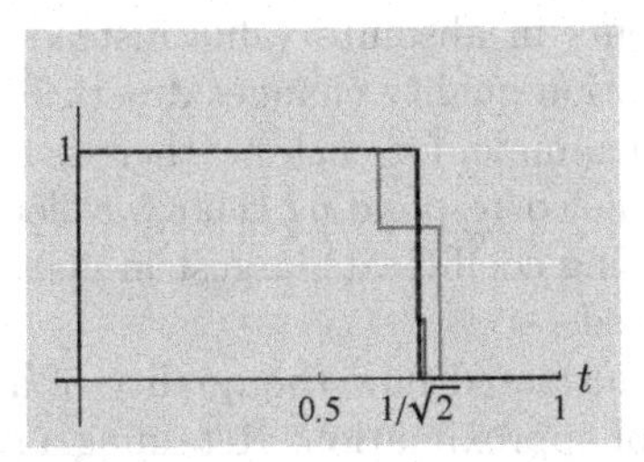

(a) Series with 8, 64, and 512 terms.

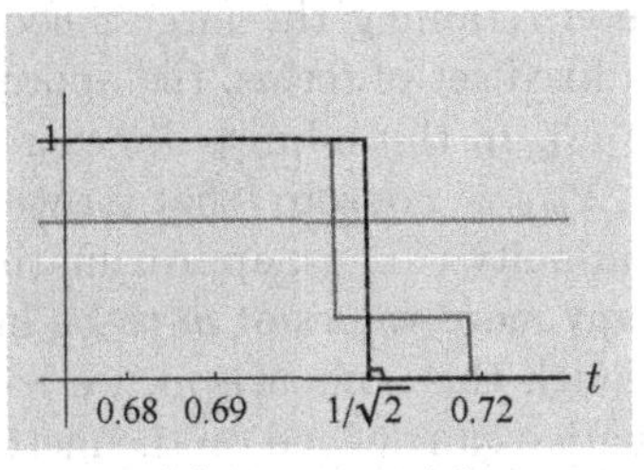

(b) Detail of (a).

Figure 1.4 Approximation of a box function (dashed lines) with a Haar basis using the first 8 ($m = 0, -1, -2$), 64 ($m = 0, -1, \ldots, -5$), and 512 ($m = 0, -1, \ldots, -8$), terms (solid lines, from lightest to darkest), with $n = 0, 1, \ldots, 2^{-m} - 1$. The discontinuity is at the irrational point $1/\sqrt{2}$.

even as they increased the number of terms; they concluded that a mechanical problem was at fault. Not until 1899, when Josiah Gibbs proved that Fourier series of discontinuous functions cannot converge uniformly, was this myth dispelled. The phenomenon was termed the *Gibbs phenomenon*, referring to the oscillations appearing around the discontinuity when using any finite number of terms. Figure 1.3 shows approximations of a box function with a Fourier series basis (1.3a) using X_k, $k = -K, -K+1, \ldots, K-1, K$.

So what would the Haar basis provide in this case? Surely, it seems more appropriate for a box function. Unfortunately, taking the first 2^{-m} terms in the natural ordering (the term corresponding to the function $\varphi_0(t)$ plus 2^{-m} terms corresponding to each scale $m = 0, -1, -2, \ldots$) leads to a similarly poor performance, shown in Figure 1.4. This poor performance is dependent on the position of the discontinuity; approximating a box function with a discontinuity at an integer multiple of 2^{-k} for some $k \in \mathbb{Z}$ would lead to a much better performance.

However, changing the approximation procedure slightly makes a big differ-

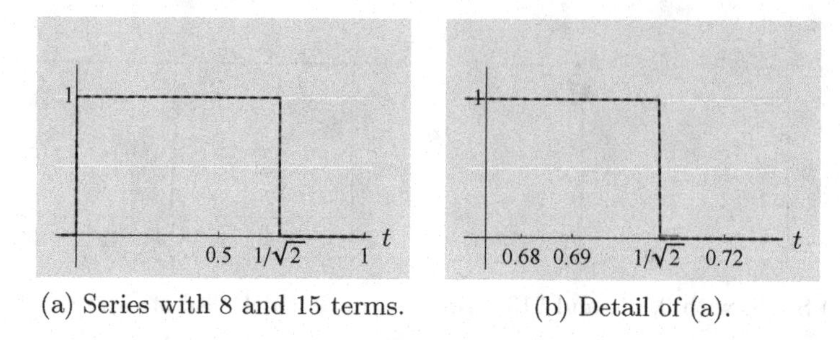

(a) Series with 8 and 15 terms.

(b) Detail of (a).

Figure 1.5 Approximation of a box function (dashed lines) with a Haar basis using the 8 (light) and 15 (dark) largest-magnitude terms. The 15-term approximation is visually indistinguishable from the target function.

ence. Upon retaining the largest coefficients in absolute value instead of simply keeping a fixed set of terms, the approximation quality changes drastically, as seen in Figure 1.5. In this admittedly extreme example, for each m, there is only one n such that $X_{m,n}$ is nonzero (that for which the corresponding Haar wavelet straddles the discontinuity). Thus, approximating using coefficients largest in absolute value allows many more values of m to be included.

Through this comparison, we have illustrated how the quality of a basis for approximation can depend on the method of approximation. Retaining a predefined set of terms, as in the Fourier example case (Figure 1.3) or the first Haar example (Figure 1.4) is called linear approximation. Retaining an adaptive set of terms instead, as in the second Haar example (Figure 1.5), is called nonlinear approximation and leads to a superior approximation quality.

Overview of the book The purpose of this book is to develop the framework for the methods just described, namely expansions and approximations, as well as to show practical examples where these methods are used in engineering and applied sciences. In particular, we will see that expansions and approximations are closely related to the essential signal processing tasks of sampling, filtering, estimation, and compression.

Chapter 2, From Euclid to Hilbert, introduces the basic machinery of Hilbert spaces. These are vector spaces endowed with operations that induce intuitive geometric properties. In this general setting, we develop the notion of signal representations, which are essentially coordinate systems for the vector space. When a representation is complete and not redundant, it provides a *basis* for the space; when it is complete and redundant, it provides a *frame* for the space. A key virtue for a basis is orthonormality; its counterpart for a frame is tightness.

Chapters 3 and 4 focus our attention on sequence and function spaces for which the domain can be associated with *time*, leading to an inherent ordering not necessarily present in a general Hilbert space. In **Chapter 3, Sequences and discrete-time systems**, a vector is a sequence that depends on *discrete time*, and an important class of linear operators on these vectors is those that are invariant to time shifts; these are convolution operators. These operators lead naturally to signal representations using the discrete-time Fourier transform and, for circularly extended finite-length sequences, the discrete Fourier transform.

Chapter 4, Functions and continuous-time systems, parallels Chapter 3; a vector is now a function that depends on *continuous time*, and an important class of linear operators on these vectors are again those that are invariant to time shifts; these are convolution operators. These operators lead naturally to signal representations using the Fourier transform and, for circularly extended finite-length functions, or periodic functions, the Fourier series. The four Fourier representations from these two chapters exemplify the diagonalization of linear, shift-invariant operators, or convolutions, in the various domains.

Chapter 5, Sampling and interpolation, makes fundamental connections between Chapters 3 and 4. Associating a discrete-time sequence with a given continuous-time function is *sampling*, and the converse is *interpolation*; these are central concepts in signal processing since digital computations on continuous-domain phenomena must be performed in a discrete domain.

Chapter 6, Approximation and compression, introduces many types of approximations that are central to making computationally practical tools. Approximation by polynomials and by truncations of series expansions are studied, along with the basic principles of compression.

Chapter 7, Localization and uncertainty, introduces time, frequency, scale, and resolution properties of individual vectors; these properties build our intuition for what might or might not be captured by a single representation coefficient. We then study these properties for sets of vectors used to represent signals. In particular, time and frequency localization lead to the concept of a time–frequency plane, where essential differences between Fourier techniques and wavelet techniques become evident: Fourier techniques use vectors with equal spacing in frequency while wavelet techniques use vectors with power-law spacing in frequency; furthermore, Fourier techniques use vectors at equal scale while wavelet techniques use geometrically spaced scales. We end with examples with real signals to develop intuition about various signal representations.

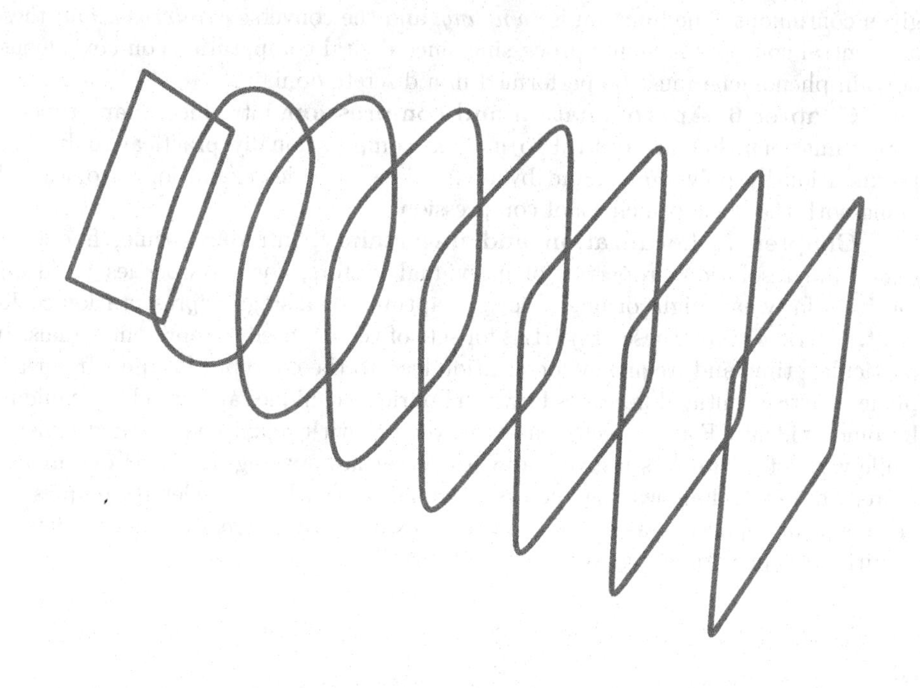

Chapter 2

From Euclid to Hilbert

"Mathematics is the art of giving the same name to different things."

— *Henri Poincaré*

Contents

We start our journey into signal processing with different backgrounds and perspectives. This chapter aims to establish a common language, develop the foundations for our study, and begin to draw out key themes.

There will be more formal definitions in this chapter than in any other, to approach the ideal of a self-contained treatment. However, we must assume some background in common: On the one hand, we expect the reader to be familiar with linear algebra at the level of [93, Ch. 1–5] (see also Appendix 2.B) and probability at the level of [6, Ch. 1–4] (see also Appendix 2.C). (The textbooks we have cited are just examples; nothing unique to those books is necessary.) On the other hand, we are not assuming prior knowledge of general vector space abstractions or mathematical analysis beyond basic calculus; we develop these topics here to extend geometric intuition from ordinary Euclidean space to spaces of sequences and functions. For more details on abstract vector spaces, we recommend books by Kreyszig [59], Luenberger [64], and Young [111].

2.1 Introduction

This section introduces many topics of the chapter through the familiar setting of the real plane. In the more general treatment of subsequent sections, the intuition we have developed through years of dealing with the Euclidean spaces around us ($\mathbb{R}^2$ and $\mathbb{R}^3$) will generalize to some not-so-familiar spaces. Readers comfortable with vector spaces, inner products, norms, projections, and bases may skip this section; otherwise, this will be a gentle introduction to Euclid's world.

Real plane as a vector space

Let us start with a look at the familiar setting of $\mathbb{R}^2$, that is, real vectors with two coordinates. We adopt the convention of vectors being columns and often write them compactly as transposes of rows, such as $x = \begin{bmatrix} x_0 & x_1 \end{bmatrix}^\top$. The first entry is the horizontal component and the second entry is the vertical component.

Adding two vectors in the plane produces a third one also in the plane; multiplying a vector by a real scalar produces a second vector also in the plane. These two ingrained facts make the real plane be a *vector space.*

Inner product and norm

The *inner product* of vectors $x = \begin{bmatrix} x_0 & x_1 \end{bmatrix}^\top$ and $y = \begin{bmatrix} y_0 & y_1 \end{bmatrix}^\top$ in the real plane is

$$\langle x, y \rangle \;=\; x_0 y_0 + x_1 y_1. \tag{2.1}$$

Other names for inner product are *scalar product* and *dot product.* The inner product of a vector with itself is simply

$$\langle x, x \rangle \;=\; x_0^2 + x_1^2,$$

a nonnegative quantity that is zero when $x_0 = x_1 = 0$. The *norm* of a vector x is

$$\|x\| \;=\; \sqrt{\langle x, x \rangle} \;=\; \sqrt{x_0^2 + x_1^2}. \tag{2.2}$$

While the norm is sometimes called the *length,* we avoid this usage because length can also refer to the number of components in a vector. A vector of norm 1 is called a *unit vector.*

In (2.1), the inner product computation depends on the choice of coordinate axes. Let us now derive an expression in which the coordinates disappear. Consider

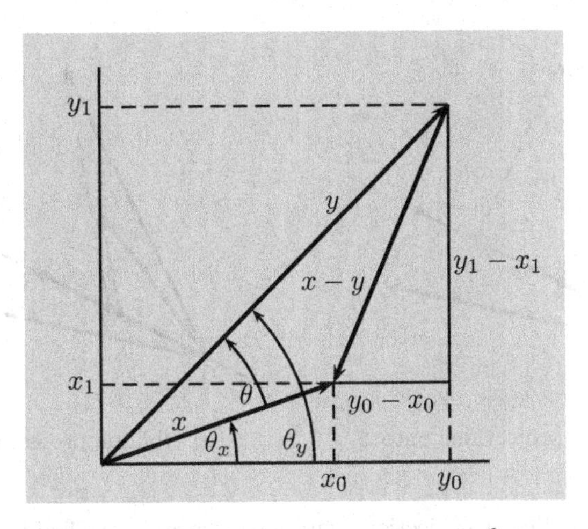

Figure 2.1 A pair of vectors in $\mathbb{R}^2$.

x and y as shown in Figure 2.1. Define the angle between x and the positive horizontal axis as θ_x (measured counterclockwise), and define θ_y similarly. Using a little algebra and trigonometry, we get

$$\begin{aligned}\langle x,\, y\rangle &= x_0 y_0 + x_1 y_1\\ &= (\|x\|\cos\theta_x)(\|y\|\cos\theta_y) + (\|x\|\sin\theta_x)(\|y\|\sin\theta_y)\\ &= \|x\|\,\|y\|(\cos\theta_x\cos\theta_y + \sin\theta_x\sin\theta_y)\\ &= \|x\|\,\|y\|\cos(\theta_x - \theta_y). \end{aligned} \tag{2.3}$$

Thus, the inner product of the two vectors is the product of their norms and the cosine of the angle $\theta = \theta_x - \theta_y$ between them.

The inner product measures both the norms of the vectors and the similarity of their orientations. For fixed vector norms, the greater the inner product, the closer the vectors are in orientation. The orientations are closest when the vectors are collinear and pointing in the same direction, that is, when $\theta = 0$; they are the farthest when the vectors are antiparallel, that is, when $\theta = \pi$. When $\langle x,\, y\rangle = 0$, the vectors are called *orthogonal* or *perpendicular*. From (2.3), we see that $\langle x,\, y\rangle$ is zero only when the norm of one vector is zero (meaning that one of the vectors is the vector $\begin{bmatrix}0 & 0\end{bmatrix}^\top$) or the cosine of the angle between them is zero ($\theta = \pm\frac{1}{2}\pi$). So, at least in the latter case, this is consistent with the conventional concept of perpendicularity.

The *distance* between two vectors is defined as the norm of their difference:

$$d(x, y) \;=\; \|x - y\| \;=\; \sqrt{\langle x - y,\, x - y\rangle} \;=\; \sqrt{(x_0 - y_0)^2 + (x_1 - y_1)^2}. \tag{2.4}$$

Subspaces and projections

A line through the origin is the simplest case of a *subspace*, and *projection* to a subspace is intimately related to inner products.

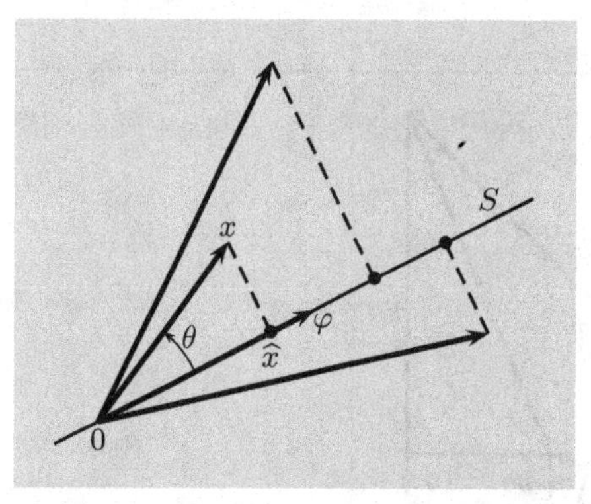

(a) Orthogonal projections onto S.

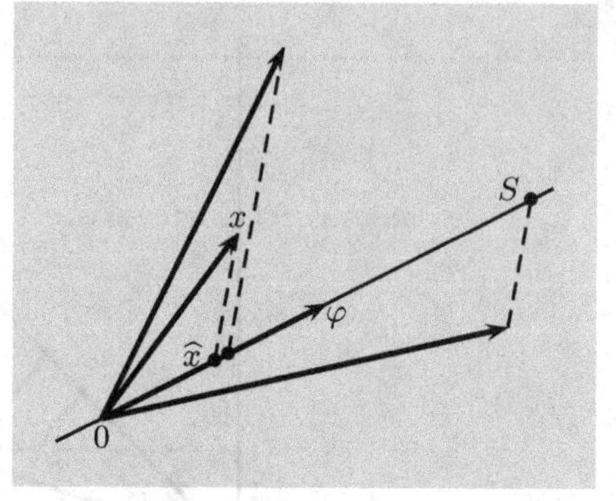

(b) Oblique projections onto S.

Figure 2.2 Examples of projections onto a subspace S specified by a unit vector φ.

Starting with a vector x and applying an *orthogonal projection operator* onto some subspace results in the vector $\widehat{x}$ closest (among all vectors in the subspace) to x. The connection to orthogonality is that the difference between the vector and its orthogonal projection $x - \widehat{x}$ is orthogonal to every vector in the subspace. Orthogonal projection is illustrated in Figure 2.2(a); the subspace S is formed by the scalar multiples of the vector φ, and three orthogonal projections onto S are shown. As depicted, the action of the operator is like looking at the shadow that the input vector casts on S when light rays are orthogonal to S. This operation is linear, meaning that the orthogonal projection of $x + y$ equals the sum of the orthogonal projections of x and y. Also, the orthogonal projection operator leaves vectors in S unchanged.

Given a unit vector φ, the *orthogonal projection* onto the subspace specified by φ is $\widehat{x} = \langle x, \varphi\rangle\varphi$. This can also be written as

$$\widehat{x} = \langle x, \varphi\rangle\varphi = (\|x\|\,\|\varphi\|\cos\theta)\varphi \stackrel{(a)}{=} (\|x\|\cos\theta)\varphi, \tag{2.5}$$

where (a) uses $\|\varphi\| = 1$, and θ is the angle measured counterclockwise from φ to x, as marked in Figure 2.2(a). When φ is not of unit norm, the orthogonal projection onto the subspace specified by φ is

$$\widehat{x} \stackrel{(a)}{=} (\|x\|\cos\theta)\frac{\varphi}{\|\varphi\|} = (\|x\|\,\|\varphi\|\cos\theta)\frac{\varphi}{\|\varphi\|^2} \stackrel{(b)}{=} \frac{1}{\|\varphi\|^2}\langle x, \varphi\rangle\varphi, \tag{2.6}$$

where (a) expresses the orthogonal projection using the unit vector $\varphi/\|\varphi\|$; and (b) uses (2.3).

Projection is more general than orthogonal projection; for example, Figure 2.2(b) illustrates *oblique projection.* The operator is still linear and vectors in the subspace are still left unchanged; however, the difference $(x - \widehat{x})$ is no longer orthogonal to S.

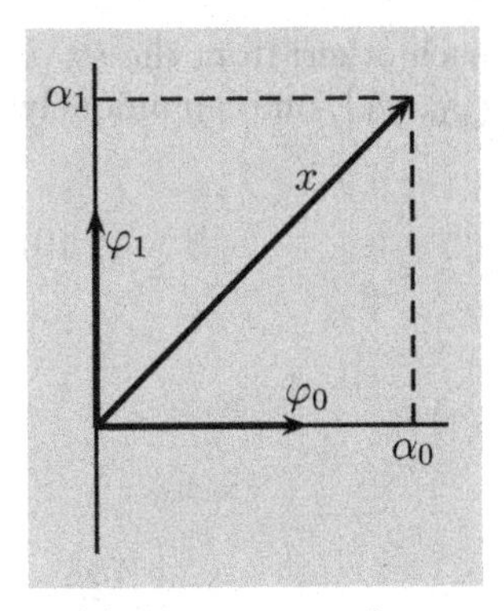

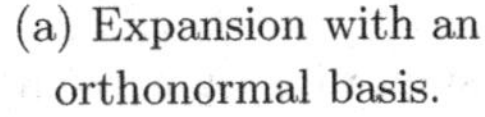
(a) Expansion with an orthonormal basis.

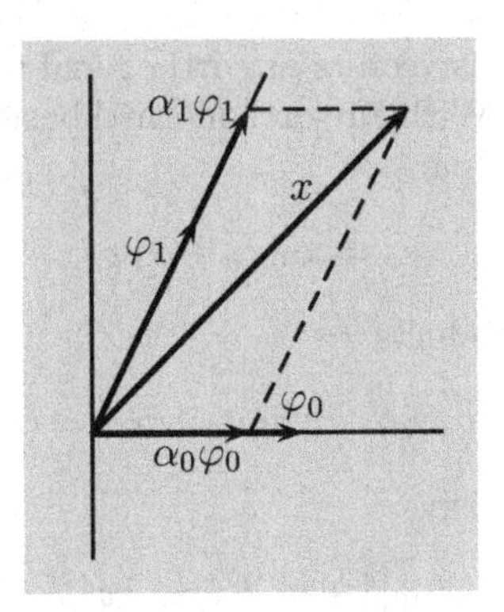

(b) Expansion with a nonorthogonal basis.

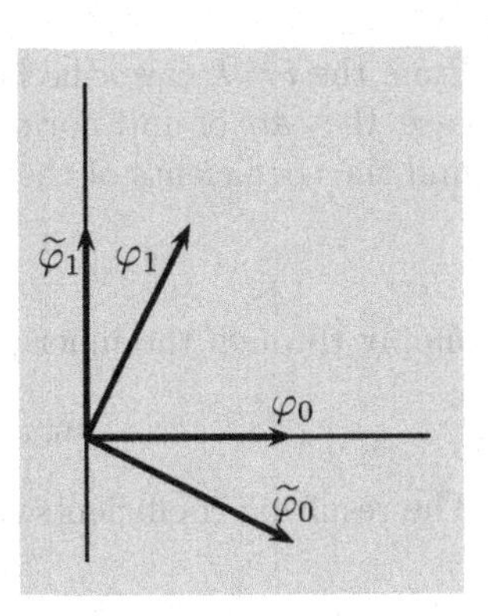

(c) Basis $\{\varphi_0, \varphi_1\}$ and its dual $\{\widetilde{\varphi}_0, \widetilde{\varphi}_1\}$.

Figure 2.3 Expansions in $\mathbb{R}^2$.

Bases and coordinates

We defined the real plane as a vector space using coordinates: the first coordinate is the signed distance as measured from left to right, and the second coordinate is the signed distance as measured from bottom to top. In doing so, we implicitly used the *standard basis* $e_0 = \begin{bmatrix} 1 & 0 \end{bmatrix}^\top$, $e_1 = \begin{bmatrix} 0 & 1 \end{bmatrix}^\top$, which is a particular orthonormal basis for $\mathbb{R}^2$. Expressing vectors in a variety of bases is central to our study, and vectors' coordinates will differ depending on the choice of basis.

Orthonormal bases Vectors $e_0 = \begin{bmatrix} 1 & 0 \end{bmatrix}^\top$ and $e_1 = \begin{bmatrix} 0 & 1 \end{bmatrix}^\top$ constitute the standard basis and are depicted in Figure 2.3(a). They are orthogonal and of unit norm, and are thus called *orthonormal*. We have been using this basis implicitly in that

$$x = \begin{bmatrix} x_0 \\ x_1 \end{bmatrix} = x_0 \begin{bmatrix} 1 \\ 0 \end{bmatrix} + x_1 \begin{bmatrix} 0 \\ 1 \end{bmatrix} = x_0 e_0 + x_1 e_1 \tag{2.7}$$

is an *expansion* of x with respect to the basis $\{e_0, e_1\}$. For this basis, it is obvious that an expansion exists for any x because the *coefficients* of the expansion x_0 and x_1 are simply the entries of x.

The general condition for $\{\varphi_0, \varphi_1\}$ to be an orthonormal basis for $\mathbb{R}^2$ is

$$\langle \varphi_i, \varphi_k \rangle = \delta_{i-k} \qquad \text{for } i, k \in \{0, 1\}, \tag{2.8}$$

where δ_{i-k} is a convenient shorthand defined as[5]

$$\delta_{i-k} = \begin{cases} 1, & \text{for } i = k; \\ 0, & \text{otherwise.} \end{cases} \tag{2.9}$$

[5] δ_n is called the *Kronecker delta* sequence and is formally defined in Chapter 3, (3.8).

From the $i \neq k$ case, the basis vectors are orthogonal to each other; from the $i = k$ case, they are of unit norm. With any orthonormal basis $\{\varphi_0, \varphi_1\}$, one can uniquely find the coefficients of the expansion

$$x = \alpha_0 \varphi_0 + \alpha_1 \varphi_1 \tag{2.10}$$

simply through the inner products

$$\alpha_0 = \langle x, \varphi_0 \rangle \qquad \text{and} \qquad \alpha_1 = \langle x, \varphi_1 \rangle.$$

The resulting coefficients satisfy

$$|\alpha_0|^2 + |\alpha_1|^2 = \|x\|^2 \tag{2.11}$$

by the Pythagorean theorem, because α_0 and α_1 form the sides of a right triangle with hypotenuse of length $\|x\|$ (see Figure 2.3(a)). The equality (2.11) is an example of a Parseval equality[6] and is related to Bessel's inequality; these will be formally introduced in Section 2.5.2.

An expansion like (2.10) is often termed a *change of basis*, since it expresses x with respect to $\{\varphi_0, \varphi_1\}$, rather than in the standard basis $\{e_0, e_1\}$. In other words, the coefficients (α_0, α_1) are the coordinates of x in this new basis $\{\varphi_0, \varphi_1\}$.

Biorthogonal pairs of bases Expansions like (2.10) do not need $\{\varphi_0, \varphi_1\}$ to be orthonormal. As an example, consider the problem of representing an arbitrary vector $x = \begin{bmatrix} x_0 & x_1 \end{bmatrix}^\top$ as an expansion $\alpha_0\varphi_0 + \alpha_1\varphi_1$ with respect to $\varphi_0 = \begin{bmatrix} 1 & 0 \end{bmatrix}^\top$ and $\varphi_1 = \begin{bmatrix} \frac{1}{2} & 1 \end{bmatrix}^\top$ (see Figure 2.3(b)). This is not a trivial exercise such as the one of expanding with the standard basis, but in this particular case we can still come up with an intuitive procedure.

Since φ_0 has no vertical component, we should use φ_1 to match the vertical component of x, yielding $\alpha_1 = x_1$. (This is illustrated with the diagonal dashed line in Figure 2.3(b).) Then, we need $\alpha_0 = x_0 - \frac{1}{2}x_1$ for the horizontal component to be correct. We can express what we have just done with inner products as

$$\alpha_0 = \langle x, \widetilde{\varphi}_0 \rangle \quad \text{and} \quad \alpha_1 = \langle x, \widetilde{\varphi}_1 \rangle,$$

where the vectors

$$\widetilde{\varphi}_0 = \begin{bmatrix} 1 & -\frac{1}{2} \end{bmatrix}^\top \quad \text{and} \quad \widetilde{\varphi}_1 = \begin{bmatrix} 0 & 1 \end{bmatrix}^\top$$

are shown in Figure 2.3(c). We have thus just derived an instance of the expansion formula

$$x = \alpha_0\varphi_0 + \alpha_1\varphi_1 = \langle x, \widetilde{\varphi}_0 \rangle \varphi_0 + \langle x, \widetilde{\varphi}_1 \rangle \varphi_1, \tag{2.12}$$

where $\{\widetilde{\varphi}_0, \widetilde{\varphi}_1\}$ is the basis *dual* to the basis $\{\varphi_0, \varphi_1\}$, and the two bases form a *biorthogonal pair of bases*. For any basis, the dual basis is unique. The defining characteristic for a biorthogonal pair is

$$\langle \widetilde{\varphi}_i, \varphi_k \rangle = \delta_{i-k} \qquad \text{for } i, k \in \{0, 1\}. \tag{2.13}$$

[6]What we call the Parseval equality in this book is sometimes called Plancherel's equality.

We can check that this is satisfied in our example and that any orthonormal basis is its own dual. Clearly, designing a biorthogonal basis pair has more degrees of freedom than designing an orthonormal basis. The disadvantage is that (2.11) does not hold, and, furthermore, computations can become numerically unstable if φ_0 and φ_1 are too close to collinear.

Frames The signal expansion (2.12) has the minimum possible number of terms to work for every $x \in \mathbb{R}^2$, namely two terms because the dimension of the space is two. It can also be useful to have an expansion of the form

$$x = \langle x, \widetilde{\varphi}_0 \rangle \varphi_0 + \langle x, \widetilde{\varphi}_1 \rangle \varphi_1 + \langle x, \widetilde{\varphi}_2 \rangle \varphi_2. \tag{2.14}$$

Here, an expansion will exist as long as $\{\varphi_0, \varphi_1, \varphi_2\}$ are not collinear. Then, even after the set $\{\varphi_0, \varphi_1, \varphi_2\}$ has been fixed, there are infinitely many dual sets $\{\widetilde{\varphi}_0, \widetilde{\varphi}_1, \widetilde{\varphi}_2\}$ such that (2.14) holds for all $x \in \mathbb{R}^2$. Such redundant sets are called *frames* and their (nonunique) dual sets are called *dual frames.* This flexibility can be used in various ways. For example, setting a component of $\widetilde{\varphi}_i$ to zero could save a multiplication and an addition in computing an expansion, or, the dual, which is not unique, could be chosen to make the coefficients as small as possible.

As an example, let us start with the standard basis $\{\varphi_0 = e_0, \varphi_1 = e_1\}$, add a vector $\varphi_2 = -e_0 - e_1$ to it,

$$\varphi_0 = \begin{bmatrix} 1 \\ 0 \end{bmatrix}, \qquad \varphi_1 = \begin{bmatrix} 0 \\ 1 \end{bmatrix}, \qquad \varphi_2 = \begin{bmatrix} -1 \\ -1 \end{bmatrix}, \tag{2.15}$$

and see what happens (see Figure 2.4(a)). As there are now three vectors in $\mathbb{R}^2$, they are linearly dependent; indeed, as defined, $\varphi_2 = -\varphi_0 - \varphi_1$. Moreover, these three vectors must be able to represent every vector in $\mathbb{R}^2$ since each two-element subset is able to do so. To show that, we use the expansion $x = \langle x, \varphi_0 \rangle \varphi_0 + \langle x, \varphi_1 \rangle \varphi_1$ and add a zero to it to give

$$x = \langle x, \varphi_0 \rangle \varphi_0 + \langle x, \varphi_1 \rangle \varphi_1 + \underbrace{(\langle x, \varphi_1 \rangle - \langle x, \varphi_1 \rangle)\varphi_0 + (\langle x, \varphi_1 \rangle - \langle x, \varphi_1 \rangle)\varphi_1}_{=0}.$$

We now rearrange it slightly:

$$x = \langle x, \varphi_0 + \varphi_1 \rangle \varphi_0 + \langle x, 2\varphi_1 \rangle \varphi_1 + \langle x, \varphi_1 \rangle (-\varphi_0 - \varphi_1) = \sum_{k=0}^{2} \langle x, \widetilde{\varphi}_k \rangle \varphi_k,$$

with $\widetilde{\varphi}_0 = \varphi_0 + \varphi_1$, $\widetilde{\varphi}_1 = 2\varphi_1$, $\widetilde{\varphi}_2 = \varphi_1$. This expansion is exactly of the form (2.14) and is reminiscent of the one for biorthogonal pairs of bases which we have seen earlier, except that the vectors involved in the expansion are now linearly dependent. This shows that we can indeed expand any $x \in \mathbb{R}^2$ in terms of the frame $\{\varphi_0, \varphi_1, \varphi_2\}$ and one of its possible dual frames $\{\widetilde{\varphi}_0, \widetilde{\varphi}_1, \widetilde{\varphi}_2\}$.

Can we now get a frame to somehow mimic an orthonormal basis? Consider

$$\varphi_0 = \begin{bmatrix} \sqrt{\frac{2}{3}} \\ 0 \end{bmatrix}, \qquad \varphi_1 = \begin{bmatrix} -\frac{1}{\sqrt{6}} \\ \frac{1}{\sqrt{2}} \end{bmatrix}, \qquad \varphi_2 = \begin{bmatrix} -\frac{1}{\sqrt{6}} \\ -\frac{1}{\sqrt{2}} \end{bmatrix}, \tag{2.16}$$

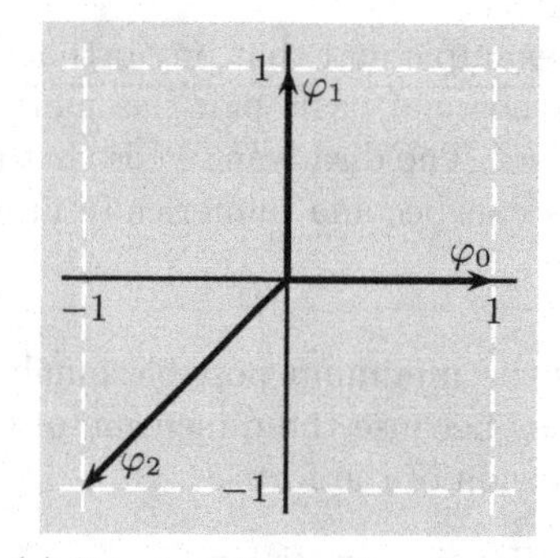

(a) Standard basis plus a vector.

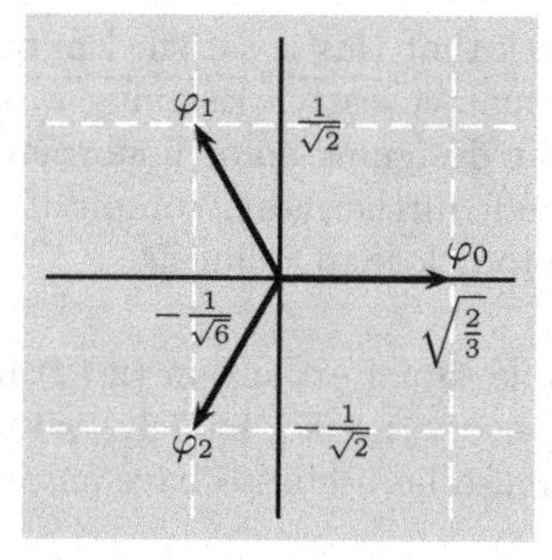

(b) Tight frame.

Figure 2.4 Illustrations of overcomplete sets of vectors (frames).

shown in Figure 2.4(b). By expanding an arbitrary $x = \begin{bmatrix} x_0 & x_1 \end{bmatrix}^\top$, we can verify that $x = \sum_{k=0}^{2} \langle x, \varphi_k \rangle \varphi_k$ holds for any x. The expansion looks like the orthonormal basis one, where the same set of vectors plays both roles (inside the inner product and outside). The norm is preserved similarly to what happens with orthonormal bases ($\sum_{k=0}^{2} |\langle x, \varphi_k \rangle|^2 = \|x\|^2$), except that the norms of the frame vectors are not 1, but rather $\sqrt{2/3}$. A frame with this property is called a *tight frame.* We could have renormalized the frame vectors by $\sqrt{3/2}$ to make them unit-norm vectors, in which case $\sum_{k=0}^{2} |\langle x, \varphi_k \rangle|^2 = \frac{3}{2}\|x\|^2$, where $\frac{3}{2}$ indicates the redundancy of the frame (we have $\frac{3}{2}$ times more vectors than needed for an expansion in $\mathbb{R}^2$).

Matrix view of bases and frames An expansion with a basis or frame involves operations that can be expressed conveniently with matrices.

Take the biorthogonal basis expansion formula (2.12). The coefficients in the expansion are the inner products

$$\begin{aligned} \alpha_0 &= \langle x, \widetilde{\varphi}_0 \rangle = \widetilde{\varphi}_{00} x_0 + \widetilde{\varphi}_{01} x_1, \\ \alpha_1 &= \langle x, \widetilde{\varphi}_1 \rangle = \widetilde{\varphi}_{10} x_0 + \widetilde{\varphi}_{11} x_1, \end{aligned}$$

where $\widetilde{\varphi}_0 = \begin{bmatrix} \widetilde{\varphi}_{00} & \widetilde{\varphi}_{01} \end{bmatrix}^\top$ and $\widetilde{\varphi}_1 = \begin{bmatrix} \widetilde{\varphi}_{10} & \widetilde{\varphi}_{11} \end{bmatrix}^\top$. Rewrite the above as a matrix–vector product,

$$\alpha = \begin{bmatrix} \alpha_0 \\ \alpha_1 \end{bmatrix} = \begin{bmatrix} \langle x, \widetilde{\varphi}_0 \rangle \\ \langle x, \widetilde{\varphi}_1 \rangle \end{bmatrix} = \underbrace{\begin{bmatrix} \widetilde{\varphi}_{00} & \widetilde{\varphi}_{01} \\ \widetilde{\varphi}_{10} & \widetilde{\varphi}_{11} \end{bmatrix}}_{\widetilde{\Phi}^\top} \begin{bmatrix} x_0 \\ x_1 \end{bmatrix} = \widetilde{\Phi}^\top x.$$

The matrix $\widetilde{\Phi}^\top$ with $\widetilde{\varphi}_0^\top$ and $\widetilde{\varphi}_1^\top$ as rows is called the *analysis operator*, and left multiplying a vector x by it computes the expansion coefficients (α_0, α_1) with respect to the basis $\{\varphi_0, \varphi_1\}$. The reconstruction of x from (α_0, α_1) is through

$$x = \alpha_0 \varphi_0 + \alpha_1 \varphi_1.$$

This can be written with a matrix–vector product as

$$x = \begin{bmatrix} x_0 \\ x_1 \end{bmatrix} = \alpha_0 \begin{bmatrix} \varphi_{00} \\ \varphi_{01} \end{bmatrix} + \alpha_1 \begin{bmatrix} \varphi_{10} \\ \varphi_{11} \end{bmatrix} = \underbrace{\begin{bmatrix} \varphi_{00} & \varphi_{10} \\ \varphi_{01} & \varphi_{11} \end{bmatrix}}_{\Phi} \begin{bmatrix} \alpha_0 \\ \alpha_1 \end{bmatrix} = \begin{bmatrix} \varphi_0 & \varphi_1 \end{bmatrix} \begin{bmatrix} \alpha_0 \\ \alpha_1 \end{bmatrix}$$

$$= \Phi\alpha = \Phi\widetilde{\Phi}^\top x,$$

where $\varphi_0 = \begin{bmatrix} \varphi_{00} & \varphi_{01} \end{bmatrix}^\top$ and $\varphi_1 = \begin{bmatrix} \varphi_{10} & \varphi_{11} \end{bmatrix}^\top$. The matrix Φ with φ_0 and φ_1 as columns is called the *synthesis operator*, and left multiplying an expansion coefficient vector α by it performs the reconstruction of x from (α_0, α_1).

The matrix view makes it obvious that the expansion formula (2.12) holds for any $x \in \mathbb{R}^2$ when $\Phi\widetilde{\Phi}^\top$ is the identity matrix. In other words, we must have $\Phi^{-1} = \widetilde{\Phi}^\top$, which is equivalent to (2.13). The inverse exists whenever $\{\varphi_0, \varphi_1\}$ is a basis, and inverting Φ determines the dual basis $\{\widetilde{\varphi}_0, \widetilde{\varphi}_1\}$.

In the case of an orthonormal basis, $\Phi^{-1} = \Phi^\top$, that is, the matrix–vector equations above hold with $\widetilde{\Phi} = \Phi$.

The case of a three-element frame is similar, with matrices Φ and $\widetilde{\Phi}$ each having two rows and three columns. The validity of the expansion (2.14) hinges on Φ being a left inverse of $\widetilde{\Phi}^\top$. In the example we saw earlier,

$$\Phi = \begin{bmatrix} 1 & 0 & -1 \\ 0 & 1 & -1 \end{bmatrix}, \tag{2.17a}$$

and its dual frame was

$$\widetilde{\Phi} = \begin{bmatrix} 1 & 0 & 0 \\ 1 & 2 & 1 \end{bmatrix}. \tag{2.17b}$$

Such a left inverse, $\widetilde{\Phi}^\top$, is never unique; thus dual frames are not unique. For example, the dual frame

$$\widetilde{\Phi} = \begin{bmatrix} 0 & -1 & -1 \\ -1 & 0 & -1 \end{bmatrix} \tag{2.17c}$$

would work as well.

Chapter outline

The next several sections follow the progression of topics in this brief introduction. In Section 2.2, we formally introduce vector spaces and equip them with inner products and norms. We also give several examples of common vector spaces. In Section 2.3, we discuss the concept of completeness that turns an inner product space into a Hilbert space. More importantly, we define the central concept of orthogonality and then introduce linear operators. We follow with approximations, projections, and decompositions in Section 2.4. In Section 2.5, we define bases and frames. This step gives us the tools to analyze signals and to create approximate representations. Section 2.5.5 develops the matrix view of basis and frame expansions. Section 2.6 discusses a few algorithms pertaining to the material covered. The first three appendices review some elements of analysis and topology, linear algebra, and probability. The final appendix discusses some finer mathematical points on the concept of a basis.

2.2 Vector spaces

Sets of mathematical objects can be highly abstract, and imposing the axioms of a normed vector space is amongst the simplest ways to induce useful structure. Furthermore, we will see that images, audio signals, and many other types of signals can be modeled and manipulated well using vector space models. This section introduces vector spaces formally, including inner products, norms, and metrics. We give pointers to reference texts in the *Further reading*.

2.2.1 Definition and properties

A vector space is any set with objects, called vectors, that can be added and scaled while staying within the set. For a formal definition, the field of scalars must be specified, and properties required of the vector addition and scalar multiplication operations must be stated.

DEFINITION 2.1 (VECTOR SPACE) A *vector space* over a field of scalars $\mathbb{C}$ (or $\mathbb{R}$) is a set of vectors, V, together with operations of vector addition and scalar multiplication. For any x, y, z in V and α, β in $\mathbb{C}$ (or $\mathbb{R}$), these operations must satisfy the following properties:

(i) *Commutativity:* $x + y = y + x$.

(ii) *Associativity:* $(x + y) + z = x + (y + z)$ and $(\alpha\beta)x = \alpha(\beta x)$.

(iii) *Distributivity:* $\alpha(x + y) = \alpha x + \alpha y$ and $(\alpha + \beta)x = \alpha x + \beta x$.

Furthermore, the following hold:

(iv) *Additive identity:* There exists an element $\mathbf{0}$ in V such that $x + \mathbf{0} = \mathbf{0} + x = x$ for every x in V.

(v) *Additive inverse:* For each x in V, there exists a unique element $-x$ in V such that $x + (-x) = (-x) + x = \mathbf{0}$.

(vi) *Multiplicative identity:* For every x in V, $1 \cdot x = x$.

We have used the bold $\mathbf{0}$ to emphasize that the zero vector is different than the zero scalar. In later chapters we will drop this distinction. The definition of a vector space requires the field of scalars to be specified; we opted to carry real and complex numbers in parallel. This will be true for a number of other definitions in this chapter as well. We now discuss some common vector spaces.

$\mathbb{C}^N$: Vector space of complex-valued finite-dimensional vectors

$$\mathbb{C}^N = \left\{ x = \begin{bmatrix} x_0 & x_1 & \ldots & x_{N-1} \end{bmatrix}^\top \;\middle|\; x_n \in \mathbb{C},\ n \in \{0, 1, \ldots, N-1\} \right\}, \quad (2.18a)$$

where the vector addition and scalar multiplication are defined componentwise,

$$\begin{aligned} x+y &= \begin{bmatrix} x_0 & x_1 & \dots & x_{N-1} \end{bmatrix}^\top + \begin{bmatrix} y_0 & y_1 & \dots & y_{N-1} \end{bmatrix}^\top \\ &= \begin{bmatrix} x_0+y_0 & x_1+y_1 & \dots & x_{N-1}+y_{N-1} \end{bmatrix}^\top, \\ \alpha x &= \alpha \begin{bmatrix} x_0 & x_1 & \dots & x_{N-1} \end{bmatrix}^\top = \begin{bmatrix} \alpha x_0 & \alpha x_1 & \dots & \alpha x_{N-1} \end{bmatrix}^\top. \end{aligned}$$

It is easy to verify that the six properties in Definition 2.1 hold; $\mathbb{C}^N$ is thus a vector space (see also Solved exercise 2.1). The definition of the standard Euclidean space, $\mathbb{R}^N$, follows similarly, except that it applies over $\mathbb{R}$.

$\mathbb{C}^\mathbb{Z}$: Vector space of complex-valued sequences over $\mathbb{Z}$

$$\mathbb{C}^\mathbb{Z} = \left\{ x = \begin{bmatrix} \dots & x_{-1} & \boxed{x_0} & x_1 & \dots \end{bmatrix}^\top \;\middle|\; x_n \in \mathbb{C},\ n \in \mathbb{Z} \right\}, \tag{2.18b}$$

where the vector addition and scalar multiplication are defined componentwise.[7]

$\mathbb{C}^\mathbb{R}$: Vector space of complex-valued functions over $\mathbb{R}$

$$\mathbb{C}^\mathbb{R} = \{ x \mid x(t) \in \mathbb{C},\ t \in \mathbb{R} \}, \tag{2.18c}$$

with the natural addition and scalar multiplication operations:

$$(x+y)(t) = x(t) + y(t), \tag{2.19a}$$

$$(\alpha x)(t) = \alpha x(t). \tag{2.19b}$$

Other vector spaces of sequences and functions can be denoted similarly, for example, $\mathbb{C}^\mathbb{N}$ for complex-valued sequences indexed from 0, $\mathbb{C}^{\mathbb{R}^+}$ for complex-valued functions on the positive real line, $\mathbb{C}^{[a,b]}$ for complex-valued functions on the interval $[a, b]$, etc.

The operations of vector addition and scalar multiplication seen above can be used to define many other vector spaces. For example, componentwise addition and multiplication can be used to define the vector space of matrices, while the natural operations of addition and scalar multiplication of functions can be used to define the vector space of polynomials:

Example 2.1 (Vector space of polynomials) Fix a positive integer N and consider the real-valued polynomials of degree at most $(N-1)$, $x(t) = \sum_{k=0}^{N-1} \alpha_k t^k$. These form a vector space over $\mathbb{R}$ under the natural addition and multiplication operations. Since each polynomial is specified by its coefficients, polynomials combine exactly like vectors in $\mathbb{R}^N$.

[7] When writing an infinite sequence as a column vector, the entry with index zero is boxed to serve as a reference point. Similarly, the row zero, column zero entry of an infinite matrix is boxed.

Definition 2.2 (Subspace) A nonempty subset S of a vector space V is a *subspace* when it is closed under the operations of vector addition and scalar multiplication:

(i) For all x and y in S, $x + y$ is in S.
(ii) For all x in S and α in $\mathbb{C}$ (or $\mathbb{R}$), αx is in S.

A subspace S is itself a vector space over the same field of scalars as V and with the same vector addition and scalar multiplication operations as V.

Example 2.2 (Subspaces)

(i) Let x be a vector in a vector space V. The set of vectors of the form αx with $\alpha \in \mathbb{C}$ is a subspace.
(ii) In the vector space of complex-valued sequences over $\mathbb{Z}$, the sequences that are zero outside of $\{2, 3, 4, 5\}$ form a subspace. The same can be said with $\{2, 3, 4, 5\}$ replaced by any finite or infinite subset of the domain $\mathbb{Z}$.
(iii) In the vector space of real-valued functions on $\mathbb{R}$, the functions that are constant on intervals $[k - \frac{1}{2}, k + \frac{1}{2})$, $k \in \mathbb{Z}$, form a subspace. This is because the sum of two functions each of which is constant on $[k - \frac{1}{2}, k + \frac{1}{2})$ is also a function constant on $[k - \frac{1}{2}, k + \frac{1}{2})$, while a function constant on $[k - \frac{1}{2}, k + \frac{1}{2})$ multiplied by a scalar is also a function constant on $[k - \frac{1}{2}, k + \frac{1}{2})$.
(iv) In the vector space of real-valued functions on the interval $[-\frac{1}{2}, \frac{1}{2}]$ under the natural operations of addition and scalar multiplication (2.19), the set of odd functions,

$$S_{\text{odd}} = \left\{x \,\middle|\, x(t) = -x(-t) \text{ for all } t \in [-\tfrac{1}{2}, \tfrac{1}{2}]\right\}, \tag{2.20a}$$

is a subspace. Similarly, the set of even functions,

$$S_{\text{even}} = \left\{x \,\middle|\, x(t) = x(-t) \text{ for all } t \in [-\tfrac{1}{2}, \tfrac{1}{2}]\right\}, \tag{2.20b}$$

is also a subspace. Either is easily checked because the sum of two odd (even) functions yields an odd (even) function; scalar multiplication of an odd (even) function yields again an odd (even) function.

Definition 2.3 (Affine subspace) A subset T of a vector space V is an *affine subspace* when there exist a vector $x \in V$ and a subspace $S \subset V$ such that any $t \in T$ can be written as $x + s$ for some $s \in S$.

Beware that an affine subspace is not necessarily a subspace; it is a subspace if and only if it includes $\mathbf{0}$. Affine subspaces generalize the concept of a plane in Euclidean geometry; subspaces correspond only to those planes that include the origin. Affine subspaces are *convex sets*, meaning that if vectors x and y are in the set, so is any vector $\lambda x + (1 - \lambda)y$ for $\lambda \in [0, 1]$.

Example 2.3 (Affine subspaces)

(i) Let x and y be vectors in a vector space V. The set of vectors of the form $x + \alpha y$ with $\alpha \in \mathbb{C}$ is an affine subspace.

(ii) In the vector space of complex-valued sequences over $\mathbb{Z}$, the sequences that equal 1 outside of $\{2, 3, 4, 5\}$ form an affine subspace.

The definition of a subspace is suggestive of one way in which subspaces arise – by combining a finite number of vectors in V.

Definition 2.4 (Span) The *span* of a set of vectors S is the set of all finite linear combinations of vectors in S:

$$\text{span}(S) = \left\{ \sum_{k=0}^{N-1} \alpha_k \varphi_k \;\middle|\; \alpha_k \in \mathbb{C} \text{ (or } \mathbb{R}), \varphi_k \in S, \text{ and } N \in \mathbb{N} \right\}.$$

Note that a span is always a subspace and that the sum has a finite number of terms even if the set S is infinite.

Example 2.4 (Proper subspace) Proper subspaces (those that do no equal the entire space) arise in linear algebra when one looks at matrix–vector products with tall matrices or rank-deficient square matrices. Let A be an $M \times N$ real-valued matrix with rank $K < M$, and let $S = \{y = Ax \mid x \in \mathbb{R}^N\}$. Applying the conditions in the definition of a subspace (Definition 2.2) to S and using the properties of matrix multiplication, we verify that S is indeed a K-dimensional subspace of the vector space $\mathbb{R}^M$. As per (2.222a) in Appendix 2.B, this subspace is the span of the columns of A.

Many different sets can have the same span, and it can be of fundamental interest to find the smallest set with a particular span. This leads to the dimension of a vector space, which depends on the concept of linear independence.

Definition 2.5 (Linearly independent set) The set of vectors $\{\varphi_0, \varphi_1, \ldots, \varphi_{N-1}\}$ is called *linearly independent* when $\sum_{k=0}^{N-1} \alpha_k \varphi_k = \mathbf{0}$ is true only if $\alpha_k = 0$ for all k. Otherwise, the set is linearly dependent. An infinite set of vectors is called linearly independent when every finite subset is linearly independent.

Example 2.5 (Linearly independent set) Consider the following set of vectors in $\mathbb{C}^N$:[8]

$$e_k = [\, \underbrace{0 \;\; 0 \;\; \ldots \;\; 0}_{k\ \text{0s}} \;\; 1 \;\; \underbrace{0 \;\; 0 \;\; \ldots \;\; 0}_{(N-k-1)\ \text{0s}} \,]^{\top}, \qquad k = 0, 1, \ldots, N-1.$$

[8] We will see in Example 2.29 that this set is called a *standard basis*.

We now show that this set is linearly independent. For any complex numbers $\{\alpha_k\}_{k=0}^{N-1}$,

$$\sum_{k=0}^{N-1} \alpha_k e_k = \alpha_0 \begin{bmatrix} 1 \\ 0 \\ \vdots \\ 0 \end{bmatrix} + \alpha_1 \begin{bmatrix} 0 \\ 1 \\ \vdots \\ 0 \end{bmatrix} + \cdots + \alpha_{N-1} \begin{bmatrix} 0 \\ 0 \\ \vdots \\ 1 \end{bmatrix} = \begin{bmatrix} \alpha_0 \\ \alpha_1 \\ \vdots \\ \alpha_{N-1} \end{bmatrix}.$$

The above vector is the $\mathbf{0}$ vector only if $\alpha_k = 0$ for all k.

DEFINITION 2.6 (DIMENSION) A vector space V is said to have *dimension* N when it contains a linearly independent set with N elements and every set with $N+1$ or more elements is linearly dependent. If no such finite N exists, the vector space is infinite-dimensional.

EXAMPLE 2.6 ($\mathbb{R}^N$ IS OF DIMENSION N) To show that $\mathbb{R}^N$ is of dimension N, we must first show that it contains a linearly independent set with N elements; indeed, we can choose the set $\{e_k\}_{k=0}^{N-1}$ from Example 2.5, which we showed to be linearly independent. The next step is to demonstrate that every set with $N+1$ or more elements is linearly dependent. With $M > N$, choose any set $\{\varphi_k\}_{k=0}^{M-1}$. If this set does not contain a subset of N linearly independent vectors, we are done. If it does, without loss of generality, assume that these are the first N vectors. Then, suppose that

$$\mathbf{0} = \sum_{k=0}^{M-1} \alpha_k \varphi_k = \sum_{k=0}^{N-1} \alpha_k \varphi_k + \sum_{k=N}^{M-1} \alpha_k \varphi_k. \tag{2.21}$$

By subtracting $\sum_{k=N}^{M-1} \alpha_k \varphi_k$ from both sides we obtain

$$\underbrace{\begin{bmatrix} \varphi_{0,0} & \cdots & \varphi_{N-1,0} \\ \varphi_{0,1} & \cdots & \varphi_{N-1,1} \\ \vdots & \ddots & \vdots \\ \varphi_{0,N-1} & \cdots & \varphi_{N-1,N-1} \end{bmatrix}}_{\Phi} \underbrace{\begin{bmatrix} \alpha_0 \\ \vdots \\ \alpha_{N-1} \end{bmatrix}}_{\alpha} = \underbrace{\begin{bmatrix} \varphi_{N,0} & \cdots & \varphi_{M-1,0} \\ \varphi_{N,1} & \cdots & \varphi_{M-1,1} \\ \vdots & \ddots & \vdots \\ \varphi_{N,N-1} & \cdots & \varphi_{M-1,N-1} \end{bmatrix}}_{\bar{\Phi}} \underbrace{\begin{bmatrix} -\alpha_N \\ \vdots \\ -\alpha_{M-1} \end{bmatrix}}_{-\bar{\alpha}},$$

where we expressed the sums using matrix–vector products. The matrix Φ is invertible because the first N vectors are linearly independent. Thus,

$$\alpha = -\Phi^{-1}\bar{\Phi}\bar{\alpha}.$$

Since we can choose $\bar{\alpha}$ to be a nonzero vector and solve for α, we see that (2.21) has solutions with some nonzero α_k; the set is thus linearly dependent.

Example 2.7 (Dimension of space of polynomials) The dimension of the space of polynomials of degree at most $N-1$ is N. To show this, we could proceed as we did in Examples 2.5 and 2.6 and show that this space contains a linearly independent set with N elements and every set with $N+1$ or more elements is linearly dependent. An easier way is to make the correspondence between the space of polynomials of degree at most $N-1$ and $\mathbb{R}^N$. We do that by forming a vector of polynomial coefficients $a = \begin{bmatrix} a_0 & a_1 & \ldots & a_{N-1} \end{bmatrix}^\top$ and observing that $a \in \mathbb{R}^N$. Since addition and scalar multiplication in the polynomial space correspond exactly to addition and scalar multiplication of vectors of coefficients and $\mathbb{R}^N$ has dimension N, so does the space of polynomials.

2.2.2 Inner product

Our intuition from Euclidean spaces goes farther than just adding and multiplying. It has geometric notions of orientation and orthogonality as well as metric notions of norm and distance. In this and the next subsection, we extend these to our abstract spaces.

As visualized in Figure 2.2, an inner product is like a signed norm of an orthogonal projection of one vector onto a subspace spanned by another. It thus measures norm along with relative orientation.

Definition 2.7 (Inner product) An *inner product* on a vector space V over $\mathbb{C}$ (or $\mathbb{R}$) is a complex-valued (or real-valued) function $\langle \cdot, \cdot \rangle$ defined on $V \times V$, with the following properties for any $x, y, z \in V$ and $\alpha \in \mathbb{C}$ (or $\mathbb{R}$):

(i) *Distributivity:* $\langle x+y, z\rangle = \langle x, z\rangle + \langle y, z\rangle$.

(ii) *Linearity in the first argument:* $\langle \alpha x, y\rangle = \alpha\langle x, y\rangle$.

(iii) *Hermitian symmetry:* $\langle x, y\rangle^* = \langle y, x\rangle$.

(iv) *Positive definiteness:* $\langle x, x\rangle \geq 0$, and $\langle x, x\rangle = 0$ if and only if $x = \mathbf{0}$.

Note that (ii) and (iii) imply that $\langle x, \alpha y\rangle = \alpha^*\langle x, y\rangle$. Thus, along with being linear in the first argument, the inner product is conjugate-linear in the second argument.[9] Also note that the inner product having $\mathbb{C}$ (or $\mathbb{R}$) as a codomain excludes the possibility of it being a nonconvergent expression or infinite quantity. Thus, an inner product on V must return a unique, finite number for every pair of vectors in V. This constrains both the functional form of an inner product and the set of vectors to which it can be applied.

Example 2.8 (Inner product) Consider the vector space $\mathbb{C}^2$.

(i) $\langle x, y\rangle = x_0 y_0^* + 5 x_1 y_1^*$ is a valid inner product; it satisfies all the conditions of Definition 2.7.

[9]We have used the convention that is dominant in mathematics; in physics, the inner product is defined to be linear in the second argument and conjugate-linear in the first.

(ii) $\langle x, y\rangle = x_0^* y_0 + x_1^* y_1$ is not a valid inner product; it violates Definition 2.7(ii). For example, if $x = y = \begin{bmatrix} 0 & 1 \end{bmatrix}^\top$ and $\alpha = j$, then $\langle \alpha x, y\rangle = -j$ and $\alpha\langle x, y\rangle = j$, and thus $\langle \alpha x, y\rangle \neq \alpha\langle x, y\rangle$.

(iii) $\langle x, y\rangle = x_0 y_0^*$ is not a valid inner product; it violates Definition 2.7(iv) because $x = \begin{bmatrix} 0 & 1 \end{bmatrix}^\top$ is nonzero yet yields $\langle x, x\rangle = 0$.

Standard inner product on $\mathbb{C}^N$ The standard inner product on $\mathbb{C}^N$ is

$$\langle x, y\rangle = \sum_{n=0}^{N-1} x_n y_n^* = y^* x, \tag{2.22a}$$

where the second equality uses matrix–vector multiplication to express the sum, with vectors implicitly column vectors and * denoting the Hermitian transpose operation. While we will use this inner product frequently and without special mention, this is not the only valid inner product for $\mathbb{C}^N$ (or $\mathbb{R}^N$) (see Exercise 2.7).

Standard inner product on $\mathbb{C}^\mathbb{Z}$ The standard inner product on the vector space of complex-valued sequences over $\mathbb{Z}$ is

$$\langle x, y\rangle = \sum_{n\in\mathbb{Z}} x_n y_n^* = y^* x, \tag{2.22b}$$

where we are taking the unusual step of using matrix product notation with an infinite row vector y^* and an infinite column vector x. As stated above, the sum must converge to a finite number for the inner product to be valid, restricting the set of sequences on which we can operate. For the sum over all $n \in \mathbb{Z}$ to be uniquely defined without specifying an order of summation, the series must be absolutely convergent (see Appendix 2.A.2 for a discussion of convergence of series); we will see in Section 2.2.4 that this required absolute convergence follows from x and y each having a finite norm.

Standard inner product on $\mathbb{C}^\mathbb{R}$ The standard inner product on the vector space of complex-valued functions over $\mathbb{R}$ is

$$\langle x, y\rangle = \int_{-\infty}^{\infty} x(t) y^*(t)\, dt. \tag{2.22c}$$

We must be careful that the integral exists and is finite for the inner product to be valid, restricting the set of functions on which we can operate. We restrict this set even further to those functions with a countable number of discontinuities, thus eliminating a number of subtle technical issues.[10]

[10] For the inner product to be positive definite, as per Definition 2.7(iv), we must identify any function satisfying $\int_{-\infty}^{\infty} |x(t)|^2\, dt = 0$ with 0. From this point on, we restrict our attention to

Orthogonality

An inner product endows a space with geometric properties that arise from angles, such as perpendicularity and relative orientation. In particular, an inner product being zero has special significance.

DEFINITION 2.8 (ORTHOGONALITY)

(i) Vectors x and y are said to be *orthogonal* when $\langle x, y\rangle = 0$, written as $x \perp y$.
(ii) A set of vectors S is called *orthogonal* when $x \perp y$ for every x and y in S such that $x \neq y$.
(iii) A set of vectors S is called *orthonormal* when it is orthogonal and $\langle x, x\rangle = 1$ for every x in S.
(iv) A vector x is said to be *orthogonal* to a set of vectors S when $x \perp s$ for all $s \in S$, written as $x \perp S$.
(v) Two sets S_0 and S_1 are said to be *orthogonal* when every vector s_0 in S_0 is orthogonal to the set S_1, written as $S_0 \perp S_1$.
(vi) Given a subspace S of a vector space V, the *orthogonal complement* of S, denoted $S^\perp$, is the set $\{x \in V \mid x \perp S\}$.

Note that the set $S^\perp$ is a subspace as well. Also, vectors in an orthonormal set $\{\varphi_k\}_{k\in\mathcal{K}}$ are linearly independent since $\mathbf{0} = \sum_{k\in\mathcal{K}} \alpha_k\varphi_k$ implies that

$$0 = \langle \mathbf{0}, \varphi_i\rangle = \left\langle \sum_{k\in\mathcal{K}} \alpha_k\varphi_k, \varphi_i\right\rangle \stackrel{(a)}{=} \sum_{k\in\mathcal{K}} \alpha_k\langle\varphi_k, \varphi_i\rangle \stackrel{(b)}{=} \sum_{k\in\mathcal{K}} \alpha_k\delta_{i-k} \stackrel{(c)}{=} \alpha_i \quad (2.23)$$

for any $i \in \mathcal{K}$, where (a) follows from the linearity in the first argument of the inner product; (b) from orthonormality of the set; and (c) from $\delta_{i-k} = 0$ for $k \neq i$.[11]

EXAMPLE 2.9 (ORTHOGONALITY) Consider the set of vectors $\Phi = \{\varphi_k\}_{k\in\mathbb{N}} \subset \mathbb{C}^{[-1/2,1/2]}$, where

$$\varphi_0(t) = 1, \quad (2.24a)$$
$$\varphi_k(t) = \sqrt{2}\cos(2\pi kt), \qquad k = 1, 2, \ldots. \quad (2.24b)$$

Lebesgue measurable functions, and all integrals should be seen as Lebesgue integrals. In other words, we exclude from consideration those functions that are not well behaved in the above sense. This restriction is not unduly stringent for any practical purpose. We follow the creed of R. W. Hamming [41]: "... if whether an airplane would fly or not depended on whether some function ... was Lebesgue but not Riemann integrable, then I would not fly in it."

[11] We will often use $\mathcal{K}$ for an arbitrary countable index set. Technically, we can write a sum over $\mathcal{K}$ without indicating the order of the terms only when reordering of terms does not affect the result. Here that is the case for the first sum in (2.23) because the terms are orthogonal and for the second and third sums because of absolute convergence. We will usually not comment on this technicality because of the sufficiency of the Riesz basis condition for unconditional convergence of series; see Section 2.5.1 and Appendix 2.D.

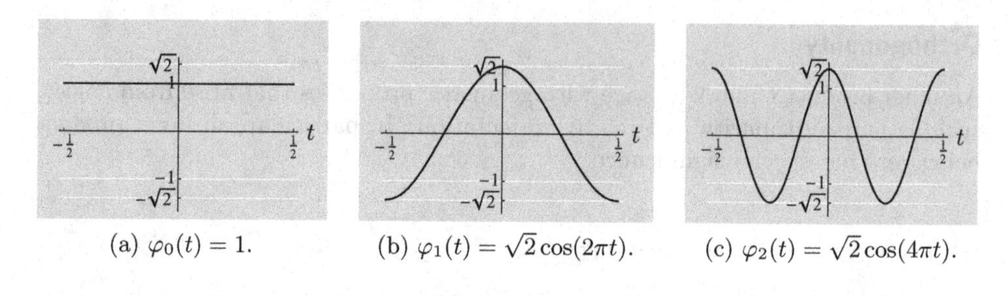

(a) $\varphi_0(t) = 1$. (b) $\varphi_1(t) = \sqrt{2}\cos(2\pi t)$. (c) $\varphi_2(t) = \sqrt{2}\cos(4\pi t)$.

Figure 2.5 Example functions from (2.24).

The functions φ_0, φ_1, and φ_2 are shown in Figure 2.5. Using the inner product

$$\langle x, y \rangle = \int_{-1/2}^{1/2} x(t) y^*(t)\, dt,$$

we have the following properties:

(i) For any $k, m \in \mathbb{Z}^+$ with $k \neq m$, vectors φ_k and φ_m are orthogonal because

$$\begin{aligned} \langle \varphi_k, \varphi_m \rangle &= 2 \int_{-1/2}^{1/2} \cos(2\pi k t) \cos(2\pi m t)\, dt \\ &\stackrel{(a)}{=} \int_{-1/2}^{1/2} [\cos(2\pi(k+m)t) + \cos(2\pi(k-m)t)]\, dt \\ &= \frac{1}{2\pi} \left[\frac{1}{k+m} \sin(2\pi(k+m)t) \Big|_{-1/2}^{1/2} + \frac{1}{k-m} \sin(2\pi(k-m)t) \Big|_{-1/2}^{1/2} \right] \\ &= 0, \end{aligned}$$

where (a) follows from the trigonometric identity for the product of cosines. It is easy to check that for any $k \in \mathbb{Z}^+$, the vectors φ_0 and φ_k are orthogonal.

(ii) The set of vectors Φ is orthogonal because, using (i), $\varphi_k \perp \varphi_m$ for every φ_k and φ_m in Φ such that $\varphi_k \neq \varphi_m$.

(iii) The set of vectors Φ is orthonormal because it is orthogonal, as we have just shown in (ii), and for any $k = 1, 2, \ldots$,

$$\begin{aligned} \langle \varphi_k, \varphi_k \rangle &= 2 \int_{-1/2}^{1/2} \cos^2(2\pi k t)\, dt \stackrel{(a)}{=} 2 \int_{-1/2}^{1/2} \frac{1 + \cos(4\pi k t)}{2}\, dt \\ &= \left[t \Big|_{-1/2}^{1/2} + \frac{1}{4\pi k} \sin(4\pi k t) \Big|_{-1/2}^{1/2} \right] = 1, \end{aligned}$$

where (a) follows from the double-angle formula for cosine. The inner product of φ_0 with itself is trivially 1.

(iv) For any $k \in \mathbb{Z}^+$, the vector φ_k is orthogonal to the set of odd functions

S_{odd} defined in (2.20a) because φ_k is orthogonal to every $s \in S_{\text{odd}}$:

$$\begin{aligned}
\langle \varphi_k, s\rangle &= \int_{-1/2}^{1/2} \sqrt{2}\cos(2\pi kt)s(t)\,dt \\
&= \sqrt{2}\left(\int_{-1/2}^{0} \cos(2\pi kt)s(t)\,dt + \int_{0}^{1/2} \cos(2\pi kt)s(t)\,dt\right) \\
&\stackrel{(a)}{=} \sqrt{2}\left(-\int_{-1/2}^{0} \cos(2\pi kt)s(-t)\,dt + \int_{0}^{1/2} \cos(2\pi kt)s(t)\,dt\right) \\
&\stackrel{(b)}{=} \sqrt{2}\left(-\int_{0}^{1/2} \cos(2\pi k\tau)s(\tau)\,d\tau + \int_{0}^{1/2} \cos(2\pi kt)s(t)\,dt\right) = 0,
\end{aligned}$$

where (a) follows from the definition of an odd function; and (b) from the change of variable $\tau = -t$ and the fact that cosine is an even function, that is, $\cos(-2\pi k\tau) = \cos(2\pi k\tau)$. It is also easy to check that φ_0 is orthogonal to S_{odd}.

(v) The set Φ is orthogonal to the set of odd functions S_{odd} because each vector in Φ is orthogonal to S_{odd}, using (iv).

Inner product spaces

A vector space equipped with an inner product from Definition 2.7 becomes an *inner product space* (sometimes also called a *pre-Hilbert space*). As we have mentioned, on $\mathbb{C}^{\mathbb{Z}}$ and $\mathbb{C}^{\mathbb{R}}$ we must exercise caution and choose the subspace for which the inner product is finite.

2.2.3 Norm

A norm is a function that assigns a length, or size, to a vector (analogously to the magnitude of a scalar).

Definition 2.9 (Norm) A *norm* on a vector space V over $\mathbb{C}$ (or $\mathbb{R}$) is a real-valued function $\|\cdot\|$ defined on V, with the following properties for any $x, y \in V$ and $\alpha \in \mathbb{C}$ (or $\mathbb{R}$):

(i) *Positive definiteness:* $\|x\| \geq 0$, and $\|x\| = 0$ if and only if $x = \mathbf{0}$.
(ii) *Positive scalability:* $\|\alpha x\| = |\alpha|\,\|x\|$.
(iii) *Triangle inequality:* $\|x+y\| \leq \|x\| + \|y\|$, with equality if and only if $y = \alpha x$.

Note that the comments we made about the finiteness and validity of an inner product apply to the norm as well.

In the above, the triangle inequality got its name because it has the following geometric interpretation: the length of any side of a triangle is smaller than or equal to the sum of the lengths of the other two sides; equality occurs only when two sides are collinear, that is, when the triangle degenerates into a line segment. For example, if $V = \mathbb{C}$ and the standard norm is used, the triangle inequality becomes

$$|x+y| \;\leq\; |x| + |y| \qquad \text{for any} \quad x,\, y \in \mathbb{C}. \tag{2.25}$$

An inner product can be used to define a norm, in which case we say that the norm is *induced* by the inner product. The three inner products we have seen in (2.22) induce corresponding standard norms on $\mathbb{C}^N$, $\mathbb{C}^{\mathbb{Z}}$, and $\mathbb{C}^{\mathbb{R}}$, respectively. Not all norms are induced by inner products. We will see examples of this both here and in Section 2.2.4.

EXAMPLE 2.10 (NORM) Consider the vector space $\mathbb{C}^2$.

(i) $\|x\| = |x_0|^2 + 5|x_1|^2$ is a valid norm; it satisfies all the conditions of Definition 2.9. It is induced by the inner product from Example 2.8(i).

(ii) $\|x\| = |x_0| + |x_1|$ is a valid norm. However, it is not induced by any inner product.

(iii) $\|x\| = |x_0|$ is not a valid norm; it violates Definition 2.9(i) because $x = \begin{bmatrix} 0 & 1 \end{bmatrix}^\top$ is nonzero yet yields $\|x\| = 0$.

Standard norm on $\mathbb{C}^N$ The standard norm on $\mathbb{C}^N$, induced by the inner product (2.22a), is

$$\|x\| \;=\; \sqrt{\langle x,\, x\rangle} \;=\; \left(\sum_{n=0}^{N-1} |x_n|^2\right)^{1/2}. \tag{2.26a}$$

This is also called the *Euclidean norm* and yields the conventional notion of length.

Standard norm on $\mathbb{C}^{\mathbb{Z}}$ The standard norm on $\mathbb{C}^{\mathbb{Z}}$, induced by the inner product (2.22b), is

$$\|x\| \;=\; \sqrt{\langle x,\, x\rangle} \;=\; \left(\sum_{n\in\mathbb{Z}} |x_n|^2\right)^{1/2}. \tag{2.26b}$$

Standard norm on $\mathbb{C}^{\mathbb{R}}$ The standard norm on $\mathbb{C}^{\mathbb{R}}$, induced by the inner product (2.22c), is

$$\|x\| \;=\; \sqrt{\langle x,\, x\rangle} \;=\; \left(\int_{-\infty}^{\infty} |x(t)|^2\, dt\right)^{1/2}. \tag{2.26c}$$

Properties of norms induced by an inner product

The following facts hold in any inner product space.

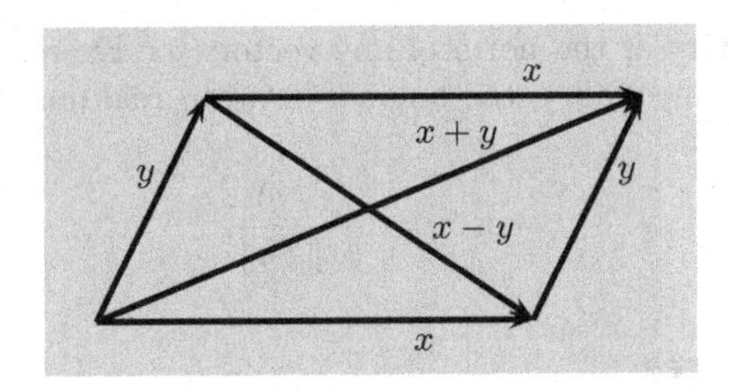

Figure 2.6 Illustration of the parallelogram law.

Pythagorean theorem This theorem generalizes a well-known fact from Euclidean geometry to any inner product space. The statement learned in elementary school involves the sides of a right triangle. In its more general form the theorem states that[12]

$$x \perp y \qquad \text{implies} \qquad \|x+y\|^2 \;=\; \|x\|^2 + \|y\|^2. \tag{2.27a}$$

Among many possible proofs of the theorem, one follows from expanding $\langle x+y,\, x+y\rangle$ into four terms and noting that $\langle x,\, y\rangle = \langle y,\, x\rangle = 0$ because of orthogonality. By induction, the Pythagorean theorem holds in a more general form for any countable set of orthogonal vectors:

$$\{x_k\}_{k\in\mathcal{K}} \text{ orthogonal} \qquad \text{implies} \qquad \Big\|\sum_{k\in\mathcal{K}} x_k\Big\|^2 \;=\; \sum_{k\in\mathcal{K}}\|x_k\|^2. \tag{2.27b}$$

Parallelogram law The parallelogram law of Euclidean geometry generalizes the Pythagorean theorem, and it too can be generalized to any inner product space. It states that

$$\|x+y\|^2 + \|x-y\|^2 \;=\; 2(\|x\|^2 + \|y\|^2). \tag{2.28}$$

From the illustration in Figure 2.6, it is clear that (2.28) holds due to the Pythagorean theorem when the parallelogram is a rectangle. Note that, even though no inner product or requirement of orthogonality appears in the parallelogram law, it necessarily holds only in an inner product space. In fact, (2.28) is a necessary and sufficient condition for a norm to be induced by an inner product (see Exercise 2.11).

Cauchy–Schwarz inequality This widely used inequality states that[13]

$$|\langle x,\, y\rangle| \;\le\; \|x\|\,\|y\|, \tag{2.29}$$

with equality if and only if $x = \alpha y$ for some scalar α. One way to prove the Cauchy–Schwarz inequality begins with expanding $\|\beta x + y\|^2$, where β is any scalar, and

[12]The theorem was found on a Babylonian tablet dating from c. 1900–1600 B.C., and it is not clear whether Pythagoras himself or one of his disciples stated and later proved the theorem. The first written proof and reference to the theorem are in Euclid's *Elements* [42].

[13]The result for sums is due to Cauchy, while the result for integrals is due to Schwarz. Bunyakovsky published the result for integrals six years earlier than Schwarz; thus, the integral version is sometimes referred to as the *Bunyakovsky inequality*.

using the nonnegativity of the norm of any vector (see Exercise 2.11). Because of the Cauchy–Schwarz inequality, the angle θ between real nonzero vectors x and y can be defined through

$$\cos\theta = \frac{\langle x, y\rangle}{\|x\|\,\|y\|}. \tag{2.30}$$

Normed vector spaces

A vector space equipped with a norm becomes a *normed vector space.* As with the inner product, we must exercise caution and choose the subspace for which the norm is finite.

Metric

Intuitively, the length of a vector can be thought of as the vector's distance from the origin. This extends naturally to a metric induced by a norm, or a distance.

DEFINITION 2.10 (METRIC, OR DISTANCE) In a normed vector space, the *metric*, or *distance*, between vectors x and y is the norm of their difference:

$$d(x, y) = \|x - y\|.$$

Much as norms induced by inner products are a small fraction of all possible norms, metrics induced by norms are a small fraction of all possible metrics. In this book, we will have no need for more general concepts of metric; for the interested reader, Exercise 2.13 gives the axioms that a metric must satisfy and explores metrics that are not induced by norms.

2.2.4 Standard spaces

We now discuss some standard vector spaces: first inner product spaces (which are also normed vector spaces, since their inner products induce the corresponding norms), followed by other normed vector spaces (which have norms not induced by inner products).

Standard inner product spaces

The first three spaces, $\mathbb{C}^N$, $\ell^2(\mathbb{Z})$, and $\mathcal{L}^2(\mathbb{R})$, are the spaces most often used in this book.[14] For each, the inner product and norm have been defined already in (2.22) and (2.26); we repeat them for each space for easy reference.

[14] The reasoning behind naming $\ell^2(\mathbb{Z})$ and $\mathcal{L}^2(\mathbb{R})$ will become clear shortly in the section on standard normed vector spaces.

$\mathbb{C}^N$: **Space of complex-valued finite-dimensional vectors** This is the normed vector space of complex-valued finite-dimensional vectors, and it uses the inner product (2.22a) and the norm (2.26a),

$$\langle x, y \rangle = \sum_{n=0}^{N-1} x_n y_n^*, \qquad \|x\| = \left(\sum_{n=0}^{N-1} |x_n|^2 \right)^{1/2}. \tag{2.31}$$

The above norm is not the only norm possible on $\mathbb{C}^N$; in the next subsection, we will introduce p norms as possible alternatives.

$\ell^2(\mathbb{Z})$: **Space of square-summable sequences** This is the normed vector space of square-summable complex-valued sequences, and it uses the inner product (2.22b) and the norm (2.26b),

$$\langle x, y \rangle = \sum_{n\in\mathbb{Z}} x_n y_n^*, \qquad \|x\| = \left(\sum_{n\in\mathbb{Z}} |x_n|^2 \right)^{1/2}. \tag{2.32}$$

This space is often referred to as the space of *finite-energy sequences.*

By the Cauchy–Schwarz inequality (2.29), the finiteness of $\|x\|$ and $\|y\|$ for any x and y in $\ell^2(\mathbb{Z})$ implies that the inner product $\langle x, y\rangle$ is finite, provided that the sum in the inner product is well defined. A somewhat technical point is that the square-summability condition that determines which sequences are in $\ell^2(\mathbb{Z})$ also ensures that the sum in the inner product is indeed well defined; see Exercise 2.14.

$\mathcal{L}^2(\mathbb{R})$: **Space of square-integrable functions** This is the normed vector space of square-integrable complex-valued functions, and it uses the inner product (2.22c) and the norm (2.26c),

$$\langle x, y \rangle = \int_{-\infty}^{\infty} x(t) y^*(t)\, dt, \qquad \|x\| = \left(\int_{-\infty}^{\infty} |x(t)|^2\, dt \right)^{1/2}. \tag{2.33}$$

This space is often referred to as the space of *finite-energy functions.* According to Definition 2.6, this space is infinite-dimensional; for example, $\{e^{-t^2}, te^{-t^2}, t^2e^{-t^2}, \ldots\}$ is a linearly independent set. As in the case of $\ell^2(\mathbb{Z})$, the inner product is always well defined.

We can restrict the domain $\mathbb{R}$ to just an interval $[a, b]$, in which case the space becomes $\mathcal{L}^2([a, b])$, that is, the space of complex-valued square-integrable functions on the interval $[a, b]$. The inner product and norm follow naturally from (2.33):

$$\langle x, y \rangle = \int_{a}^{b} x(t) y^*(t)\, dt, \qquad \|x\| = \left(\int_{a}^{b} |x(t)|^2\, dt \right)^{1/2}. \tag{2.34}$$

In $\mathcal{L}^2([a, b])$, the Cauchy–Schwarz inequality (2.29) becomes

$$\left| \int_a^b x(t) y^*(t)\, dt \right| \le \left(\int_a^b |x(t)|^2\, dt \right)^{1/2} \left(\int_a^b |y(t)|^2\, dt \right)^{1/2}, \tag{2.35}$$

with equality if and only if x and y are linearly dependent. By setting $y(t) = 1$ and squaring both sides, we get another useful fact:

$$\left| \int_a^b x(t)\, dt \right|^2 \leq (b - a) \int_a^b |x(t)|^2\, dt, \tag{2.36}$$

with equality if and only if x is constant on $[a, b]$.

$C^q([a, b])$: **Spaces of continuous functions with q continuous derivatives** For any finite a and b, the space $C([a, b])$ is defined as the space of complex-valued continuous functions over $[a, b]$ with inner product and norm given in (2.34). The space of complex-valued continuous functions over $[a, b]$ that are further restricted to have q continuous derivatives is denoted $C^q([a, b])$, so $C([a, b]) = C^0([a, b])$. Moreover, $C^q([a, b]) \subset C^p([a, b])$ when $q > p$.

Each $C^q([a, b])$ space is a subspace of $\mathbb{C}^{[a, b]}$. This is true because having q continuous derivatives is preserved by the vector space operations. Similarly, the set of polynomial functions forms a subspace of $C^q([a, b])$ for any $a, b \in \mathbb{R}$ and $q \in \mathbb{N}$ because the set of polynomials is closed under the vector space operations and polynomials are infinitely differentiable.

A $C^q([a, b])$ space is a subspace of $\mathcal{L}^2([a, b])$. As we discuss in Section 2.3.2, the requirement of q continuous derivatives makes a $C^q([a, b])$ space not complete and hence not a Hilbert space.

Spaces of random variables The set of complex random variables defined in some probabilistic model forms a vector space over the complex numbers with the standard addition and scalar multiplication operations; all the properties required in Definition 2.1 are inherited from the complex numbers, with the constant 0 as the additive identity.

A useful inner product to define on this vector space is

$$\langle \mathrm{x}, \mathrm{y} \rangle = \mathrm{E}[\, \mathrm{x}\mathrm{y}^* \,]. \tag{2.37}$$

It clearly satisfies the properties of Definition 2.7(i)–(iii), and also

$$\langle \mathrm{x}, \mathrm{x} \rangle = \mathrm{E}[\, \mathrm{x}\mathrm{x}^* \,] = \mathrm{E}[\, |\mathrm{x}|^2 \,] \geq 0$$

for the first part of Definition 2.7(iv). Only the second part of Definition 2.7(iv) is subtle. It is indeed true that $\mathrm{E}[\, \mathrm{x}\mathrm{x}^* \,] = 0$ implies that $\mathrm{x} = 0$. This is because of the sense of equality for random variables reviewed in Appendix 2.C. The norm induced by this inner product is

$$\|\mathrm{x}\| = \sqrt{\langle \mathrm{x}, \mathrm{x} \rangle} = \sqrt{\mathrm{E}[\, |\mathrm{x}|^2 \,]}. \tag{2.38}$$

This shows that if we restrict our attention to random variables with finite second moment, $\mathrm{E}[\, |\mathrm{x}|^2 \,] < \infty$, we have a normed vector space.

When $\mathrm{E}[\, \mathrm{x} \,] = 0$, the norm $\|\mathrm{x}\|$ is the standard deviation of x. When $\mathrm{E}[\, \mathrm{y} \,] = 0$ also, $\langle \mathrm{x}, \mathrm{y} \rangle$ is the covariance of x and y; the normalized inner product $\langle \mathrm{x}, \mathrm{y} \rangle / (\|\mathrm{x}\|\, \|\mathrm{y}\|)$ equals the correlation coefficient.

Standard normed vector spaces

$\mathbb{C}^N$ spaces As we said earlier, we can define other norms on $\mathbb{C}^N$. For example, the p *norm* is defined as

$$\|x\|_p = \left(\sum_{n=0}^{N-1} |x_n|^p \right)^{1/p}, \tag{2.39a}$$

for $p \in [1, \infty)$. Since the sum above has a finite number of terms, there is no doubt that it converges. Thus, we take as a vector space of interest the entire $\mathbb{C}^N$; note that this contrasts with some of the examples we will see shortly (such as $\ell^p(\mathbb{Z})$ spaces), where we must restrict the set of vectors to those with finite norm.

For $p = 1$, this norm is called the *taxicab norm* or *Manhattan norm* because $\|x\|_1$ represents the driving distance from the origin to x following a rectilinear street grid. For $p = 2$, we get our usual Euclidean square norm from (2.39a), and only in that case is a p norm induced by an inner product. The natural extension of (2.39a) to $p = \infty$ (see Exercise 2.15) defines the ∞ norm as

$$\|x\|_\infty = \max(|x_0|, |x_1|, \ldots, |x_{N-1}|). \tag{2.39b}$$

Using (2.39a) for $p \in (0, 1)$ does not give a norm but can still be a useful quantity. The failure to satisfy the requirements of a norm and an interpretation of (2.39a) with $p \to 0$ are explored in Exercise 2.16.

All norms on finite-dimensional spaces are equivalent in the sense that any two norms bound each other within constant factors (see Exercise 2.17). This is a crude equivalence that leaves significant differences in which vectors are considered larger than others, and it does not extend to infinite-dimensional spaces. Figure 2.7 shows this pictorially by showing the sets of unit-norm vectors for different p norms. All vectors ending on the curves have unit norm in the corresponding p norm. For example, with the usual Euclidean norm, unit-norm vectors fall on a circle; on the other hand, in 1 norm they fall on the diamond-shaped polygon. Note that only for $p = 2$ is the set of unit-norm vectors invariant to rotation of the coordinate system.

$\ell^p(\mathbb{Z})$ spaces We can define other norms on $\mathbb{C}^\mathbb{Z}$ as well (like we did for $\mathbb{C}^N$). However, because the space is infinite-dimensional, the choice of the norm and the requirement that it be finite restricts $\mathbb{C}^\mathbb{Z}$ to a smaller set. For example, for $p \in [1, \infty)$, the ℓ^p *norm* is

$$\|x\|_p = \left(\sum_{n \in \mathbb{Z}} |x_n|^p \right)^{1/p}. \tag{2.40a}$$

Analogously to (2.39b), we extend this to the ℓ^∞ *norm* as

$$\|x\|_\infty = \sup_{n \in \mathbb{Z}} |x_n|. \tag{2.40b}$$

We have already introduced the ℓ^p norm for $p = 2$ in (2.32); only in that case is an ℓ^p norm induced by an inner product. We can now define the spaces associated with the ℓ^p norms.

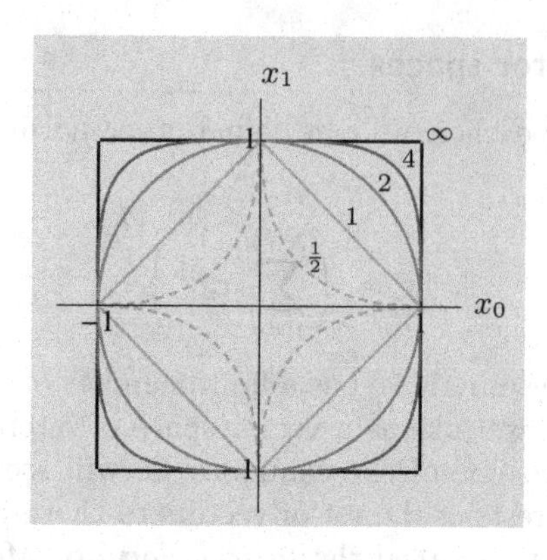

Figure 2.7 Sets of unit-norm vectors for different p norms: $p = \infty$, $p = 4$, $p = 2$, and $p = 1$ (solid lines, from darkest to lightest), as well as for $p = \frac{1}{2}$ (dashed line), which is not a norm. Vectors ending on the curves are of unit norm in the corresponding p norm.

DEFINITION 2.11 ($\ell^p(\mathbb{Z})$) For any $p \in [1, \infty]$, the normed vector space $\ell^p(\mathbb{Z})$ is the subspace of $\mathbb{C}^{\mathbb{Z}}$ consisting of vectors with finite ℓ^p norm.

Solved exercise 2.2 shows that the subset of $\mathbb{C}^{\mathbb{Z}}$ consisting of vectors with finite $\ell^p(\mathbb{Z})$ norm forms a subspace. Since $\ell^p(\mathbb{Z})$ is defined as a subspace of $\mathbb{C}^{\mathbb{Z}}$, it inherits the operations of vector addition and scalar multiplication from $\mathbb{C}^{\mathbb{Z}}$. The norm is the ℓ^p norm (2.40).

EXAMPLE 2.11 (NESTING OF $\ell^p(\mathbb{Z})$ SPACES) Consider the sequence x given by

$$x_n = \begin{cases} 0, & \text{for } n \leq 0; \\ 1/n^a, & \text{for } n > 0, \end{cases}$$

for some real number $a \geq 0$. Let us determine which of the $\ell^p(\mathbb{Z})$ spaces contain x. To check whether x is in $\ell^p(\mathbb{Z})$ for $p \in [1, \infty)$, we need to determine whether

$$\|x\|_p^p = \sum_{n=1}^{\infty} \left| \frac{1}{n^a} \right|^p = \sum_{n=1}^{\infty} \frac{1}{n^{pa}}$$

converges. The necessary and sufficient condition for convergence is $pa > 1$, so we conclude that $x \in \ell^p(\mathbb{Z})$ for $p > 1/a$ and $a > 0$. For $a = 0$, the above does not converge. For $x \in \ell^\infty(\mathbb{Z})$, x must be bounded, which occurs for all $a \geq 0$.

This example illustrates a simple inclusion property proven as Exercise 2.18:

$$p < q \qquad \text{implies} \qquad \ell^p(\mathbb{Z}) \subset \ell^q(\mathbb{Z}). \tag{2.41}$$

This can loosely be visualized with Figure 2.7: the larger the value of p, the larger the set of vectors with norm less than or equal to a given number. In particular,

$\ell^1(\mathbb{Z}) \subset \ell^2(\mathbb{Z})$. In other words, if a sequence has a finite ℓ^1 norm, then it has a finite ℓ^2 norm. Beware that the opposite is not true; if a sequence has a finite ℓ^2 norm, it does not follow that it has a finite ℓ^1 norm.

Example 2.12 (Sequence in $\ell^2(\mathbb{Z})$ but not in $\ell^1(\mathbb{Z})$) Consider the sequence $x_n = 1/n$, for $n \in \mathbb{Z}^+$. Since

$$\|x\|_2^2 = \sum_{n=1}^{\infty} \left|\frac{1}{n}\right|^2 = \frac{1}{6}\pi^2 \text{ converges, while } \|x\|_1 = \sum_{n=1}^{\infty} \left|\frac{1}{n}\right| \text{ diverges,}$$

we conclude that $x \in \ell^2(\mathbb{Z})$ and $x \notin \ell^1(\mathbb{Z})$.

$\mathcal{L}^p(\mathbb{R})$ **spaces** Like for sequences, we can define other norms on $\mathbb{C}^{\mathbb{R}}$ as well. Again, because the space is infinite-dimensional, the choice of the norm and the requirement that it be finite restricts $\mathbb{C}^{\mathbb{R}}$ to a smaller set. For example, for $p \in [1, \infty)$, the $\mathcal{L}^p$ *norm* is

$$\|x\|_p = \left(\int_{-\infty}^{\infty} |x(t)|^p \, dt\right)^{1/p}. \tag{2.42a}$$

The extension to $p = \infty$ leads to the $\mathcal{L}^\infty$ *norm* as

$$\|x\|_\infty = \operatorname*{ess\,sup}_{t\in\mathbb{R}} |x(t)|. \tag{2.42b}$$

We have already introduced the $\mathcal{L}^p$ norm for $p = 2$ in (2.33); only in that case is an $\mathcal{L}^p$ norm induced by an inner product. We can now define the spaces associated with the $\mathcal{L}^p$ norms.

Definition 2.12 ($\mathcal{L}^p(\mathbb{R})$) For any $p \in [1, \infty]$, the normed vector space $\mathcal{L}^p(\mathbb{R})$ is the subspace of $\mathbb{C}^{\mathbb{R}}$ consisting of vectors with finite $\mathcal{L}^p$ norm.

Since $\mathcal{L}^p(\mathbb{R})$ is defined as a subspace of $\mathbb{C}^{\mathbb{R}}$, it inherits the operations of vector addition and scalar multiplication from $\mathbb{C}^{\mathbb{R}}$. The norm is the $\mathcal{L}^p$ norm (2.42).

One can also use the same norms on different domains; for example, we can define the domain to be $[a, b]$ and use a finite $\mathcal{L}^p$ norm on it to yield the space $\mathcal{L}^p([a, b])$ (like we did for $\mathcal{L}^2(\mathbb{R})$ and $\mathcal{L}^2([a, b])$). We will use this in Chapters 3 and 4, where we will often be operating on $\mathcal{L}^p([-\pi, \pi])$.

2.3 Hilbert spaces

We will do most of our work in Hilbert spaces. These are inner product spaces seen in the previous section, with the additional requirement of *completeness*. Completeness is somewhat technical, and for a basic understanding it will suffice to trust that we work in vector spaces of sequences and functions in which convergence makes sense. We will furthermore be mostly concerned with *separable* Hilbert spaces because these spaces have countable bases.

2.3.1 Convergence

Convergence of a sequence of numbers should be a familiar concept; it is reviewed in Appendix 2.A.2. Convergence of a sequence of vectors requires a metric, and we limit our attention to metrics induced by norms.

DEFINITION 2.13 (CONVERGENT SEQUENCE OF VECTORS) A sequence of vectors $x_0, x_1, \ldots$ in a normed vector space V is said to *converge* to $v \in V$ when $\lim_{k\to\infty}\|v - x_k\| = 0$. In other words, given any $\varepsilon > 0$, there exists a K_ε such that

$$\|v - x_k\| < \varepsilon \qquad \text{for all} \quad k > K_\varepsilon.$$

The elements of a convergent sequence eventually stay arbitrarily close to v. Not only does the definition of convergence of sequences of vectors use a norm, but whether a sequence converges can depend on the choice of norm. This is illustrated in the following example.

EXAMPLE 2.13 (CONVERGENCE IN DIFFERENT NORMS)

(i) For each $k \in \mathbb{Z}^+$, let

$$x_k(t) = \begin{cases} 1, & \text{for } t \in [0, 1/k]; \\ 0, & \text{otherwise.} \end{cases}$$

Also, let $v(t) = 0$ for all t. For any $p \in [1, \infty)$, using the expression for the $\mathcal{L}^p$ norm, (2.42a),

$$\|v - x_k\|_p = \left(\int_{-\infty}^{\infty} |v(t) - x_k(t)|^p \, dt\right)^{1/p} = \left(\frac{1}{k}\right)^{1/p} \stackrel{k\to\infty}{\longrightarrow} 0,$$

so $x_1, x_2, \ldots$ converges to v. For $p = \infty$, using the expression for the $\mathcal{L}^\infty$ norm, (2.42b), $\|v - x_k\|_\infty = 1$ for all k, so the sequence does not converge to v under the $\mathcal{L}^\infty$ norm.

(ii) Let $\alpha \in (0, 1)$, and for each $k \in \mathbb{Z}^+$, let

$$x_{k,n} = \begin{cases} 1/k^\alpha, & \text{for } n \in \{1, 2, \ldots, k\}; \\ 0, & \text{otherwise.} \end{cases}$$

Also, let $v_n = 0$ for all n. Using the expression for the ℓ^p norm, (2.40a),

$$\|v - x_k\|_p = \left(\sum_{n=1}^{k} (1/k^\alpha)^p\right)^{1/p} = \left(\frac{1}{k}\right)^{(\alpha p - 1)/p},$$

so $x_1, x_2, \ldots$ converges to v when $p \in (1/\alpha, \infty)$. For $p = \infty$, using the expression for the ℓ^∞ norm, (2.40b), $\|v - x_k\|_\infty = 1/k^\alpha$ for all k, so the sequence converges to x under the ℓ^∞ norm as well.

As reviewed in Appendix 2.A.1, a set of real numbers is closed if and only if it contains the limits of all its convergent sequences; for example, $(0, 1]$ is not closed in $\mathbb{R}$ since the sequence of numbers $x_k = 1/k$, $k \in \mathbb{Z}^+$, lies in $(0, 1]$ and converges, but its limit point 0 is not in $(0, 1]$. Carrying this over to the Hilbert space setting yields the following definition.

DEFINITION 2.14 (CLOSED SUBSPACE) A subspace S of a normed vector space V is called *closed* when it contains all limits of sequences of vectors in S.

Subspaces of finite-dimensional normed vector spaces are always closed. Exercise 2.20 gives an example of a subspace of an infinite-dimensional normed vector space that is not closed.

Subspaces often arise as the span of a set of vectors. As the following example shows, the span of an infinite set of vectors is not necessarily closed. For this reason, we frequently work with the closure of a span.

EXAMPLE 2.14 (SPAN NEED NOT BE CLOSED) Consider the following infinite set of vectors from $\ell^2(\mathbb{N})$: For each $k \in \mathbb{N}$, let the sequence s_k be 0 except for a 1 in the kth position. Recall from Definition 2.4 that the span is all finite linear combinations, even when the set of vectors is infinite. Thus, $\text{span}(\{s_0, s_1, \ldots\})$ is the subspace of all vectors in $\ell^2(\mathbb{N})$ that have a finite number of nonzero entries. To prove that the span is not closed, we must find a sequence of vectors in the span (each having finitely many nonzero entries) that converges to a vector not in the span (having infinitely many nonzero entries). For example, let v be any sequence in $\ell^2(\mathbb{N})$ with infinite support, for example $v_n = 1/(n+1)$ for $n \in \mathbb{N}$. Then, for each $k \in \mathbb{N}$, define the vector $x_k \in \ell^2(\mathbb{N})$ by

$$x_{k,n} = \begin{cases} v_n, & \text{for } n = 0, 1, \ldots, k; \\ 0, & \text{otherwise.} \end{cases}$$

For each $k \in \mathbb{N}$, the vector x_k is a linear combination of $\{s_0, s_1, \ldots, s_k\}$. While the sequence $x_0, x_1, \ldots$ converges to v (under the $\ell^2(\mathbb{N})$ norm), its limit v is not in the span.

Since the closure of a set is the set of all limit points of convergent sequences in the set, the closure of the span of an infinite set of vectors is the set of all convergent infinite linear combinations:

$$\overline{\text{span}}(\{\varphi_k\}_{k\in\mathcal{K}}) = \left\{ \sum_{k\in\mathcal{K}} \alpha_k \varphi_k \;\middle|\; \alpha_k \in \mathbb{C} \text{ and the sum converges} \right\}.$$

The closure of the span of a set of vectors is always a closed subspace.

2.3.2 Completeness

It is awkward to do analysis in the set of rational numbers $\mathbb{Q}$ instead of in $\mathbb{R}$ because $\mathbb{Q}$ has infinitesimal gaps that can be limit points of sequences in $\mathbb{Q}$. A

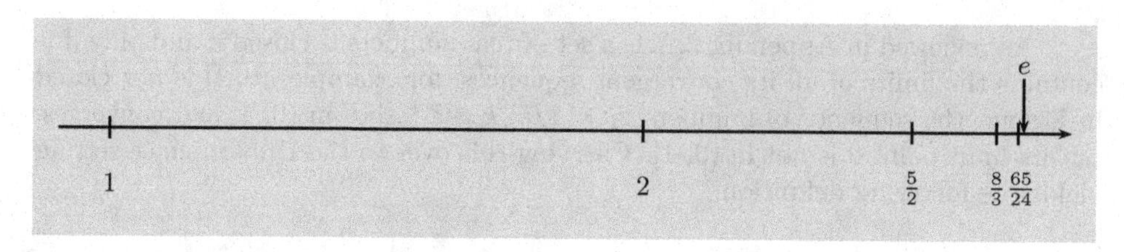

Figure 2.8 The first few partial sums of $\sum_{n=0}^{\infty} 1/n!$, each rational, converging to the irrational number e.

familiar example is that $x_k = \sum_{n=0}^{k} 1/n!$ is rational for every nonnegative integer k, but the limit of the sequence is the irrational number e (see Figure 2.8). If we want the limit of any sequence to make sense, we need to work in $\mathbb{R}$, which is the *completion* of $\mathbb{Q}$. Working only in $\mathbb{Q}$, it would be hard to distinguish between those sequences that converge to an irrational number and those that do not converge at all – neither would have a limit point in the space.

Completeness of a space is the property that ensures that any sequence that intuitively ought to converge indeed does converge to a limit in the same space. The intuition of what ought to converge is formalized by the concept of a Cauchy sequence.

DEFINITION 2.15 (CAUCHY SEQUENCE OF VECTORS) A sequence of vectors x_0, x_1, ... in a normed vector space is called a *Cauchy sequence* when, given any $\varepsilon > 0$, there exists a K_ε such that

$$\|x_k - x_m\| < \varepsilon \qquad \text{for all } k,\, m > K_\varepsilon.$$

The elements of a Cauchy sequence eventually stay arbitrarily close to each other. Thus it might be intuitive that a Cauchy sequence must converge; this is in fact true for real-valued sequences. This is not true in all vector spaces, however, and it gives us important terminology.

DEFINITION 2.16 (COMPLETENESS AND HILBERT SPACE) A normed vector space V is said to be *complete* when every Cauchy sequence in V converges to a vector in V. A complete inner product space is called a *Hilbert space.*

A complete normed vector space is called a *Banach space.*

EXAMPLE 2.15 ($\mathbb{Q}$ IS NOT COMPLETE) Ignoring for the moment that Definition 2.1 restricts the set of scalars to $\mathbb{R}$ or $\mathbb{C}$, consider $\mathbb{Q}$ as a normed vector space over the scalars $\mathbb{Q}$, with ordinary addition and multiplication and norm

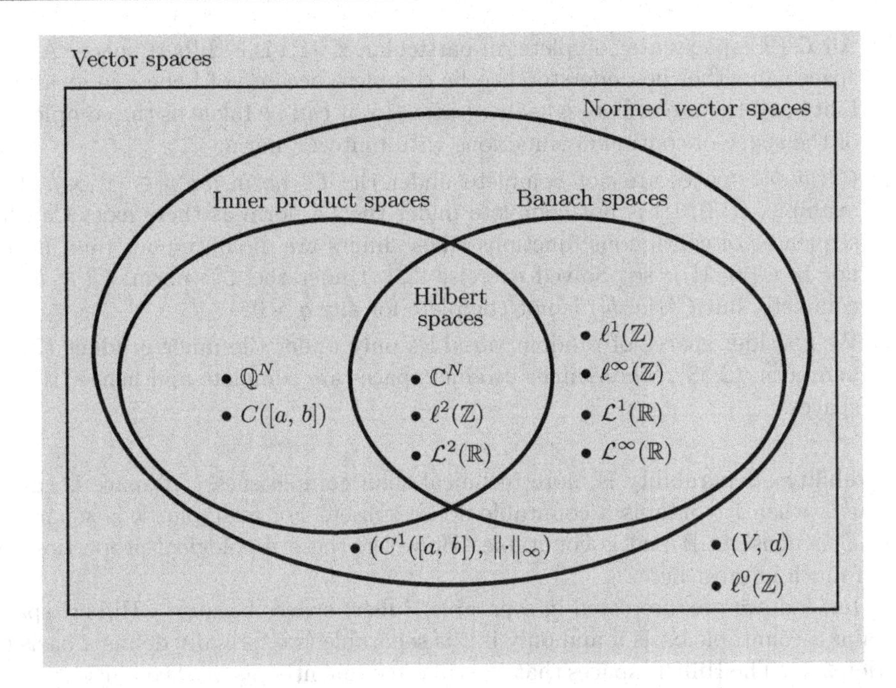

Figure 2.9 Relationships between types of vector spaces. Several examples of vector spaces are marked. For $\mathbb{Q}^N$, $\mathbb{C}^N$, and $C([a, b])$, we assume the standard inner product. (V, d) represents any vector space with the discrete metric as described in Exercise 2.13. $(C^1([a, b]), \|\cdot\|_\infty)$ represents $C^1([a, b])$ with the $\mathcal{L}^2$ norm replaced by the $\mathcal{L}^\infty$ norm. $\ell^0(\mathbb{Z})$ is described in Exercise 2.20.

$\|x\| = |x|$. This vector space is not complete because there exist rational sequences with irrational limits, such as the example of the number e we have just seen (see Figure 2.8).

Standard spaces

From Definition 2.16, completeness makes sense only in a normed vector space. We now comment on the completeness of the standard spaces we discussed in Section 2.2.4 (see Figure 2.9).

(i) All finite-dimensional spaces are complete.[15] For example, $\mathbb{C}$ as a normed vector space over $\mathbb{C}$ with ordinary addition and multiplication and with norm $\|x\| = |x|$ is complete. This can be used to show that $\mathbb{C}^N$ is complete under ordinary addition and multiplication and with any p norm; see Exercise 2.23. $\mathbb{C}^N$ under the 2 norm is a Hilbert space.

(ii) All $\ell^p(\mathbb{Z})$ spaces are complete; in particular, $\ell^2(\mathbb{Z})$ is a Hilbert space.

[15] Recall the restriction of the set of scalars to $\mathbb{R}$ or $\mathbb{C}$. Without this restriction, there are finite-dimensional vector spaces that are not complete, as in Example 2.15.

(iii) All $\mathcal{L}^p(\mathbb{R})$ spaces are complete; in particular, $\mathcal{L}^2(\mathbb{R})$ is a Hilbert space. An $\mathcal{L}^p$ space can either be understood to be complete because of Lebesgue measurability and the use of Lebesgue integration, or it can be taken as the completion of the space of continuous functions with finite $\mathcal{L}^p$ norm.

(iv) $C^q([a, b])$ spaces are not complete under the $\mathcal{L}^p$ norm for $p \in [1, \infty)$. For example, $C([0, 1])$ is not complete under the $\mathcal{L}^2$ norm as there exist Cauchy sequences of continuous functions whose limits are discontinuous (and hence not in $C([0, 1])$); see Solved exercise 2.3. Under the $\mathcal{L}^\infty$ norm, $C([a, b])$ is complete, but $C^q([a, b])$ is not complete for any $q > 0$.

(v) We consider spaces of random variables only under the inner product (2.37) and norm (2.38). These inner product spaces are complete and hence Hilbert spaces.

Separability Separability is more technical than completeness. A space is called *separable* when it contains a countable dense subset. For example, $\mathbb{R}$ is separable since $\mathbb{Q}$ is dense in $\mathbb{R}$ and is countable. However, these topological properties are not of much interest here.

Instead, we are interested in separable Hilbert spaces because a Hilbert space contains a countable basis if and only if it is separable (we formally define a basis in Section 2.5). The Hilbert spaces that we will use frequently (as marked in Figure 2.9) are all separable. Also, a closed subspace of a separable Hilbert space is separable, so it contains a countable basis as well.

2.3.3 Linear operators

Having dispensed with technicalities, we are now ready to develop operational Hilbert-space machinery. We start with linear operators, which generalize finite-dimensional matrices (see Appendix 2.B for more details).

DEFINITION 2.17 (LINEAR OPERATOR) A function $A : H_0 \to H_1$ is called a *linear operator* from H_0 to H_1 when, for all x, y in H_0 and α in $\mathbb{C}$ (or $\mathbb{R}$), the following hold:

(i) *Additivity:* $A(x + y) = Ax + Ay$.

(ii) *Scalability:* $A(\alpha x) = \alpha(Ax)$.

When the domain H_0 and the codomain H_1 are the same, A is also called a linear operator on H_0.

Note the convention of writing Ax instead of $A(x)$, just as is done for matrix multiplication. In fact, linear operators from $\mathbb{C}^N$ to $\mathbb{C}^M$ and matrices in $\mathbb{C}^{M\times N}$ are exactly the same thing.

Many concepts from finite-dimensional linear algebra extend to linear operators on Hilbert spaces in rather obvious ways. For example, the *null space* or *kernel* of a linear operator $A : H_0 \to H_1$ is the subspace of H_0 that A maps to $\mathbf{0}$:

$$\mathcal{N}(A) = \{x \in H_0 \mid Ax = \mathbf{0}\}. \tag{2.43}$$

The *range* of a linear operator $A : H_0 \to H_1$ is a subspace of H_1:

$$\mathcal{R}(A) = \{Ax \in H_1 \mid x \in H_0\}. \tag{2.44}$$

Some generalizations to Hilbert spaces are simplified by limiting attention to bounded linear operators.

DEFINITION 2.18 (OPERATOR NORM AND BOUNDED LINEAR OPERATOR) The *operator norm* of A, denoted by $\|A\|$, is defined as

$$\|A\| = \sup_{\|x\|=1} \|Ax\|. \tag{2.45}$$

A linear operator is called *bounded* when its operator norm is finite.

It is implicit in the definition that $\|x\|$ uses the norm of the domain of A and $\|Ax\|$ uses the norm of the codomain of A. This concept applies equally well when the domain and codomain are Hilbert spaces or, more generally, any normed vector spaces.

EXAMPLE 2.16 (UNBOUNDED OPERATOR) Consider $A : \mathbb{C}^{\mathbb{Z}} \to \mathbb{C}^{\mathbb{Z}}$ defined by

$$(Ax)_n = |n|\, x_n \qquad \text{for all} \quad n \in \mathbb{Z}.$$

While this operator is linear, it is not bounded. We provide a proof by contradiction. Suppose that the operator norm $\|A\|$ is finite. Then there is an integer M larger than $\|A\|$. For the input sequence x that is 0 except for a 1 in the Mth position, $Ax = Mx$, which implies $\|A\| \geq M$ – a contradiction. Hence, the operator norm $\|A\|$ is not finite.

Linear operators with finite-dimensional domains are always bounded. Conversely, by limiting attention to bounded linear operators we are able to extend concepts to Hilbert spaces while maintaining most intuitions from finite-dimensional linear algebra. For example, bounded linear operators are continuous:

$$\text{if} \quad x_k \xrightarrow{k\to\infty} v \quad \text{then} \quad Ax_k \xrightarrow{k\to\infty} Av.$$

DEFINITION 2.19 (INVERSE) A bounded linear operator $A : H_0 \to H_1$ is called *invertible* if there exists a bounded linear operator $B : H_1 \to H_0$ such that

$$BAx = x, \quad \text{for every } x \text{ in } H_0, \quad \text{and} \tag{2.46a}$$

$$ABy = y, \quad \text{for every } y \text{ in } H_1. \tag{2.46b}$$

When such a B exists, it is unique, is denoted by A^{-1}, and is called the *inverse* of A; B is called a *left inverse* of A when (2.46a) holds, and B is called a *right inverse* of A when (2.46b) holds.

For $A : \mathbb{C}^N \to \mathbb{C}^M$, basic linear algebra gives tests for the invertibility of A and methods to find left and right inverses when they exist; see Section 2.6.4 and Appendix 2.B. There are no general procedures for operators with other Hilbert spaces as the domain and codomain.

Example 2.17 (Linear operators)

(i) Ordinary matrix multiplication by the matrix

$$A = \begin{bmatrix} 3 & 1 \\ 1 & 3 \end{bmatrix}$$

defines a linear operator on $\mathbb{R}^2$, $A : \mathbb{R}^2 \to \mathbb{R}^2$. It is bounded, and its operator norm (assuming the 2 norm for both the domain and the codomain) is 4. We show here how to obtain the norm of A by direct computation (we could also use the relationship among eigenvalues, singular values, and the operator norm, which is explored in Exercises 2.25 and 2.56):

$$\begin{aligned} \|A\| &= \sup_{\|x\|=1} \|Ax\| = \sup_{\theta} \left\| \begin{bmatrix} 3 & 1 \\ 1 & 3 \end{bmatrix} \begin{bmatrix} \cos\theta \\ \sin\theta \end{bmatrix} \right\| = \sup_{\theta} \left\| \begin{bmatrix} 3\cos\theta + \sin\theta \\ \cos\theta + 3\sin\theta \end{bmatrix} \right\| \\ &= \sup_{\theta} \sqrt{(3\cos\theta + \sin\theta)^2 + (\cos\theta + 3\sin\theta)^2} \\ &= \sup_{\theta} \sqrt{10\cos^2\theta + 10\sin^2\theta + 12\sin\theta\cos\theta} \\ &= \sup_{\theta} \sqrt{10 + 6\sin 2\theta} = 4. \end{aligned}$$

The null space of A is only the vector $\mathbf{0}$, the range of A is all of $\mathbb{R}^2$, and

$$A^{-1} = \begin{bmatrix} \frac{3}{8} & -\frac{1}{8} \\ -\frac{1}{8} & \frac{3}{8} \end{bmatrix}.$$

(ii) Ordinary matrix multiplication by the matrix

$$A = \begin{bmatrix} 1 & j & 0 \\ 1 & 0 & j \end{bmatrix}$$

defines a linear operator $A : \mathbb{C}^3 \to \mathbb{C}^2$. It is bounded, and its operator norm (assuming the 2 norm for both the domain and the codomain) is $\sqrt{3}$. Its null space is $\left\{ \begin{bmatrix} x_0 & jx_0 & jx_0 \end{bmatrix}^{\top} \mid x_0 \in \mathbb{C} \right\}$, its range is all of $\mathbb{C}^2$, and it is not invertible. (There exists B satisfying (2.46b), but no B can satisfy (2.46a).)

(iii) For some fixed complex-valued sequence $(\alpha_k)_{k\in\mathbb{Z}}$, consider the componentwise multiplication

$$(Ax)_k = \alpha_k x_k, \qquad k \in \mathbb{Z}, \tag{2.47a}$$

as a linear operator on $\ell^2(\mathbb{Z})$. We can write this with infinite vectors and matrices as

$$Ax = \begin{bmatrix} \ddots & 0 & & & \\ 0 & \alpha_{-1} & 0 & & \\ & 0 & \boxed{\alpha_0} & 0 & \\ & & 0 & \alpha_1 & 0 \\ & & & 0 & \ddots \end{bmatrix} \begin{bmatrix} \vdots \\ x_{-1} \\ \boxed{x_0} \\ x_1 \\ \vdots \end{bmatrix}. \tag{2.47b}$$

It is easy to check that Definition 2.17(i) and (ii) are satisfied, but we must constrain α to ensure that the result is in $\ell^2(\mathbb{Z})$. For example,

$$\|\alpha\|_\infty = M < \infty$$

ensures that Ax is in $\ell^2(\mathbb{Z})$ for any x in $\ell^2(\mathbb{Z})$. Furthermore, the operator is bounded and $\|A\| = M$. The operator is invertible when $\inf_k |\alpha_k| > 0$. In this case, the inverse is given by

$$A^{-1}y = \begin{bmatrix} \ddots & 0 & & & \\ 0 & 1/\alpha_{-1} & 0 & & \\ & 0 & \boxed{1/\alpha_0} & 0 & \\ & & 0 & 1/\alpha_1 & 0 \\ & & & 0 & \ddots \end{bmatrix} \begin{bmatrix} \vdots \\ y_{-1} \\ \boxed{y_0} \\ y_1 \\ \vdots \end{bmatrix}.$$

Adjoint operator Finite-dimensional linear algebra has many uses for transposes and conjugate transposes. The conjugate transpose (or Hermitian transpose) is generalized by the adjoint of an operator.

DEFINITION 2.20 (ADJOINT AND SELF-ADJOINT OPERATOR) The linear operator $A^* : H_1 \to H_0$ is called the *adjoint* of the linear operator $A : H_0 \to H_1$ when

$$\langle Ax, y\rangle_{H_1} = \langle x, A^*y\rangle_{H_0}, \qquad \text{for every } x \text{ in } H_0 \text{ and } y \text{ in } H_1. \tag{2.48}$$

When $A = A^*$, the operator A is called *self-adjoint* or *Hermitian.*

Note that the adjoint gives a third meaning to *, the first two being the complex conjugate of a scalar and the Hermitian transpose of a matrix. These meanings are consistent, as we verify in the first two parts of the following example.

EXAMPLE 2.18 (ADJOINT OPERATORS)

(i) For any Hilbert space H, consider $A : H \to H$ given by $Ax = \alpha x$ for some scalar α. For any x and y in H,

$$\langle Ax, y\rangle = \langle \alpha x, y\rangle \stackrel{(a)}{=} \alpha\langle x, y\rangle \stackrel{(b)}{=} \langle x, \alpha^* y\rangle,$$

where (a) follows from the linearity in the first argument of the inner product; and (b) from the conjugate linearity in the second argument of the inner product. Upon comparison with (2.48), the adjoint of A is $A^*y = \alpha^* y$. Put simply, the adjoint of multiplication by a scalar is multiplication by the conjugate of the scalar, which is consistent with using * for conjugation of a scalar.

(ii) Consider a linear operator $A : \mathbb{C}^N \to \mathbb{C}^M$. The $\mathbb{C}^N$ and $\mathbb{C}^M$ inner products can both be written as $\langle x, y\rangle = y^* x$, where * represents the Hermitian transpose. Thus, for any $x \in \mathbb{C}^N$ and $y \in \mathbb{C}^M$,

$$\langle Ax, y\rangle_{\mathbb{C}^M} = y^*(Ax) = (y^*A)x \stackrel{(a)}{=} (A^*y)^*x = \langle x, A^*y\rangle_{\mathbb{C}^N},$$

where in (a) we use A^* to represent the Hermitian transpose of the matrix A. Upon comparison with (2.48), it seems we have reached a tautology, but this is because the uses of A^* as the adjoint of linear operator A and as the Hermitian transpose of matrix A are consistent. Put simply, the adjoint of multiplication by a matrix is multiplication by the Hermitian transpose of the matrix, which is consistent with using * for the Hermitian transpose of a matrix.

(iii) Consider the linear operator defined in (2.47). For any x and y in $\ell^2(\mathbb{Z})$,

$$\langle Ax, y\rangle_{\ell^2} \stackrel{(a)}{=} \sum_{n\in\mathbb{Z}} (\alpha_n x_n) y_n^* \stackrel{(b)}{=} \sum_{n\in\mathbb{Z}} x_n (\alpha_n^* y_n)^*,$$

where (a) follows from (2.47) and the definition of the $\ell^2(\mathbb{Z})$ inner product, (2.22b); and (b) from the commutativity and the associativity of scalar multiplication along with $(\alpha_n^*)^* = \alpha_n$. Our goal in expanding $\langle Ax, y\rangle$ above is to see the result as the inner product between x and some linear operator applied to y. Upon comparing the final expression with the definition of the $\ell^2(\mathbb{Z})$ inner product, we conclude that the componentwise multiplication $(A^*y)_n = \alpha_n^* y_n$ defines the adjoint.

The above examples are amongst the simplest, and they do not necessarily clearly reveal the role of an adjoint. In Hilbert spaces, the relationships between vectors are measured by inner products. The defining relation of the adjoint (2.48) shows that the action of A on H_0 is mimicked by the action of the adjoint A^* on H_1; this mimicry is visible only through the applicable inner products, so the adjoint itself depends on these inner products. Loosely, if A has some effect, A^* preserves the geometry of that effect while acting with a reversed domain and codomain.

In general, finding an adjoint requires some ingenuity, as we now show.

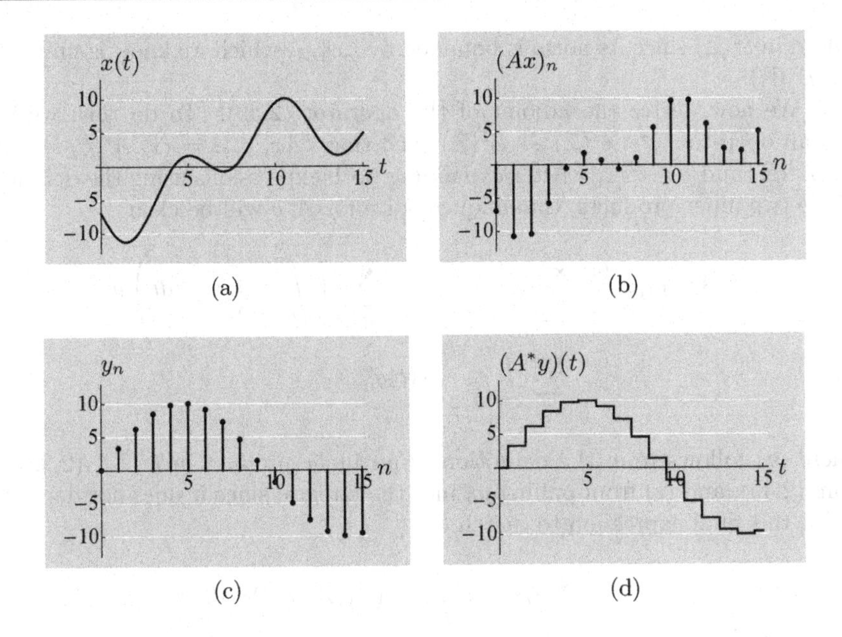

Figure 2.10 Illustration of the adjoint of an operator. (a) We start with a function x in $\mathcal{L}^2(\mathbb{R})$. (b) The local averaging operator A in (2.49) gives a sequence in $\ell^2(\mathbb{Z})$. (c) y is an arbitrary sequence in $\ell^2(\mathbb{Z})$. (d) The adjoint A^* is a linear operator from $\ell^2(\mathbb{Z})$ to $\mathcal{L}^2(\mathbb{R})$ that uniquely preserves geometry in that $\langle Ax, y\rangle_{\ell^2} = \langle x, A^*y\rangle_{\mathcal{L}^2}$. The adjoint of local averaging is to form a piecewise-constant function as in (2.52).

Example 2.19 (Local averaging and its adjoint) The operator A given by

$$(Ax)_n = \int_{n-1/2}^{n+1/2} x(t)\, dt, \qquad n \in \mathbb{Z}, \tag{2.49}$$

takes local averages of the function $x(t)$ to yield a sequence $(Ax)_n$. (This operation is depicted in Figure 2.10 and is a form of sampling, which will be covered in detail in Chapter 5.) We will first verify that A is a linear operator from $\mathcal{L}^2(\mathbb{R})$ to $\ell^2(\mathbb{Z})$ and then find its adjoint.

The operator A clearly satisfies Definition 2.17(i) and (ii); we just need to be sure that the result is in $\ell^2(\mathbb{Z})$. Given $x \in \mathcal{L}^2(\mathbb{R})$, let us compute the ℓ^2 norm of Ax:

$$\begin{aligned}\|Ax\|_{\ell^2}^2 &\overset{(a)}{=} \sum_{n\in\mathbb{Z}} |(Ax)_n|^2 \overset{(b)}{=} \sum_{n\in\mathbb{Z}} \left| \int_{n-1/2}^{n+1/2} x(t)\, dt \right|^2 \\ &\overset{(c)}{\le} \sum_{n\in\mathbb{Z}} \int_{n-1/2}^{n+1/2} |x(t)|^2\, dt = \int_{-\infty}^{\infty} |x(t)|^2\, dt \overset{(d)}{=} \|x\|_{\mathcal{L}^2}^2,\end{aligned}$$

where (a) follows from the definition of the ℓ^2 norm, (2.32); (b) from (2.49); (c) from (2.36); and (d) from the definition of the $\mathcal{L}^2$ norm, (2.33). Thus, Ax is indeed in $\ell^2(\mathbb{Z})$ since its norm is bounded by $\|x\|_{\mathcal{L}^2}$, which we know is finite since $x \in \mathcal{L}^2(\mathbb{R})$.

We now derive the adjoint of the operator (2.49). To do this, we must find an operator $A^* : \ell^2(\mathbb{Z}) \to \mathcal{L}^2(\mathbb{R})$ such that $\langle Ax, y\rangle_{\ell^2} = \langle x, A^*y\rangle_{\mathcal{L}^2}$ for any $x \in \mathcal{L}^2(\mathbb{R})$ and $y \in \ell^2(\mathbb{Z})$. After expanding both expressions using the definitions of the two inner products, the unique choice for A^*y will be clear:

$$\begin{aligned}\langle Ax, y\rangle_{\ell^2} &\overset{(a)}{=} \sum_{n\in\mathbb{Z}} (Ax)_n y_n^* \overset{(b)}{=} \sum_{n\in\mathbb{Z}} \left(\int_{n-1/2}^{n+1/2} x(t)\, dt\right) y_n^* \\ &\overset{(c)}{=} \sum_{n\in\mathbb{Z}} \int_{n-1/2}^{n+1/2} x(t) y_n^*\, dt, \end{aligned} \tag{2.50}$$

where (a) follows from the definition of an inner product in $\ell^2(\mathbb{Z})$, (2.32); (b) from (2.49); and (c) from pulling y_n into the integral since it does not depend on t. For this final expression to match

$$\langle x, A^*y\rangle_{\mathcal{L}^2} = \int_{-\infty}^{\infty} x(t)((A^*y)(t))^*\, dt \tag{2.51}$$

for arbitrary x and y, we must define A^*y as the piecewise-constant function

$$(A^*y)(t) = y_n \qquad \text{for} \quad t \in [n - \tfrac{1}{2}, n + \tfrac{1}{2}). \tag{2.52}$$

Then, the integral in (2.51) breaks into the sum of integrals in (2.50).

The following theorem summarizes several key properties of the adjoint.

THEOREM 2.21 (ADJOINT PROPERTIES) Let $A : H_0 \to H_1$ be a bounded linear operator.

(i) The adjoint A^*, defined through (2.48), exists.
(ii) The adjoint A^* is unique.
(iii) The adjoint of A^* equals the original operator, $(A^*)^* = A$.
(iv) The operators AA^* and A^*A are self-adjoint.
(v) The operator norms of A and A^* are equal, $\|A^*\| = \|A\|$.
(vi) If A is invertible, the adjoint of the inverse and the inverse of the adjoint are equal, $(A^{-1})^* = (A^*)^{-1}$.
(vii) Let $B : H_0 \to H_1$ be a bounded linear operator. Then, $(A+B)^* = A^* + B^*$.
(viii) Let $B : H_1 \to H_2$ be a bounded linear operator. Then, $(BA)^* = A^*B^*$.

Proof. Parts (i) and (v) are the most technically challenging and go beyond our scope; proofs of these based on the Riesz representation theorem can be found in texts such as [59]. Parts (ii) and (iii) are proven below, with the remaining parts left for Exercise 2.27.

(ii) Suppose that B and C are adjoints of A. Then, for any x in H_0 and y in H_1,

$$\begin{aligned} 0 &= \langle Ax, y\rangle - \langle Ax, y\rangle \stackrel{(a)}{=} \langle x, By\rangle - \langle x, Cy\rangle \\ &\stackrel{(b)}{=} \langle x, By - Cy\rangle \stackrel{(c)}{=} \langle x, (B-C)y\rangle, \end{aligned}$$

where (a) follows from (2.48) and B and C being adjoints of A; (b) from the distributivity of the inner product; and (c) from the additivity of the operators. Since this holds for every x in H_0, it holds in particular for $x = (B-C)y$. By the positive definiteness of the inner product, we must have $(B-C)y = 0$ for every y in H_1. This implies that $By = Cy$ for every y in H_1, so the adjoint is unique.

(iii) For any x in H_1 and y in H_0,

$$\langle A^*x, y\rangle \stackrel{(a)}{=} \langle y, A^*x\rangle^* \stackrel{(b)}{=} \langle Ay, x\rangle^* \stackrel{(c)}{=} \langle x, Ay\rangle,$$

where (a) follows from the Hermitian symmetry of the inner product; (b) from (2.48); and (c) from the Hermitian symmetry of the inner product.

The adjoint of a bounded linear operator provides key relationships between subspaces (Figure 2.36 in Appendix 2.B illustrates the case when the operator is a finite-dimensional matrix):

$$\mathcal{R}(A)^{\perp} = \mathcal{N}(A^*), \tag{2.53a}$$

$$\overline{\mathcal{R}(A)} = \mathcal{N}(A^*)^{\perp}. \tag{2.53b}$$

To see that $\mathcal{N}(A^*) \subseteq \mathcal{R}(A)^{\perp}$, first let $y \in \mathcal{N}(A^*)$ and $y' \in \mathcal{R}(A)$. Then, since $y' = Ax$ for some x in the domain of A, we can compute

$$\langle y', y\rangle = \langle Ax, y\rangle \stackrel{(a)}{=} \langle x, A^*y\rangle \stackrel{(b)}{=} \langle x, \mathbf{0}\rangle = 0,$$

where (a) follows from the definition of the adjoint; and (b) from $y \in \mathcal{N}(A^*)$. This shows that $y \perp \mathcal{R}(A)$, so $y \in \mathcal{R}(A)^{\perp}$. Conversely, to see that $\mathcal{R}(A)^{\perp} \subseteq \mathcal{N}(A^*)$, first let $y \in \mathcal{R}(A)^{\perp}$ and let x be any vector in the domain of A. Then,

$$0 \stackrel{(a)}{=} \langle Ax, y\rangle \stackrel{(b)}{=} \langle x, A^*y\rangle, \tag{2.54}$$

where (a) follows from $y \in \mathcal{R}(A)^{\perp}$; and (b) from the definition of the adjoint. Since (2.54) holds for all x, we can choose $x = A^*y$ so that (2.54) implies $A^*y = 0$ by the positive definiteness of the inner product. This shows that $y \in \mathcal{N}(A^*)$. These arguments prove (2.53a). The subtleties of infinite dimensions, for example that the range of a bounded linear operator need not be closed, make proving (2.53b) a bit more difficult. Note that linear operators that arise in later chapters have closed ranges.

Unitary operators Unitary operators are important because they preserve geometry (lengths and angles) when mapping one Hilbert space to another.

Definition 2.22 (Unitary operator) A bounded linear operator $A : H_0 \to H_1$ is called *unitary* when

(i) it is *invertible*; and
(ii) it *preserves inner products*,

$$\langle Ax, Ay\rangle_{H_1} = \langle x, y\rangle_{H_0} \quad \text{for every } x, y \text{ in } H_0. \tag{2.55}$$

Preservation of inner products leads to preservation of norms:

$$\|Ax\|^2 = \langle Ax, Ax\rangle = \langle x, x\rangle = \|x\|^2. \tag{2.56}$$

In Fourier theory, this is called the *Parseval equality*; it is used extensively in this book and its companion volume [57].

The following theorem provides conditions equivalent to the definition of a unitary operator above. These conditions are reminiscent of the standard definition of a unitary matrix.

Theorem 2.23 (Unitary linear operator) A bounded linear operator $A : H_0 \to H_1$ is unitary if and only if $A^{-1} = A^*$.

Proof. We first derive two intermediate results and then prove the theorem. Condition (2.55) is equivalent to A^* being a left inverse of A:

$$A^*A = I \quad \text{on} \quad H_0. \tag{2.57a}$$

To see that (2.55) implies (2.57a), note that for every x, y in H_0,

$$\langle A^*Ax, y\rangle \overset{(a)}{=} \langle Ax, Ay\rangle \overset{(b)}{=} \langle x, y\rangle,$$

where (a) follows from the definition of the adjoint; and (b) from (2.55). Conversely, to see that (2.57a) implies (2.55), note that

$$\langle Ax, Ay\rangle \overset{(a)}{=} \langle x, A^*Ay\rangle \overset{(b)}{=} \langle x, y\rangle,$$

where (a) follows from the definition of the adjoint; and (b) from (2.57a).

Combining the invertibility of A with condition (2.55) implies that A^* is a right inverse of A:

$$AA^* = I \quad \text{on} \quad H_1. \tag{2.57b}$$

To see this, note that for every x, y in H_1,

$$\langle AA^*x, y\rangle = \langle AA^*x, AA^{-1}y\rangle \overset{(a)}{=} \langle A^*x, A^{-1}y\rangle \overset{(b)}{=} \langle x, AA^{-1}y\rangle = \langle x, y\rangle,$$

where (a) follows from (2.55); and (b) from the definition of the adjoint.

The equivalence in the theorem now follows: If A is unitary, then by definition A is invertible and (2.55) holds, so both conditions (2.57) hold; thus, $A^{-1} = A^*$. Conversely, if $A^{-1} = A^*$, then A is invertible and (2.57a) holds, the latter implying (2.55) holds; thus, A is unitary.

Eigenvalues and eigenvectors The concept of an *eigenvector* generalizes from finite-dimensional linear algebra to our Hilbert space setting. Like other concepts that apply to matrices only when they are square, the generalization applies when the domain and codomain of a linear operator are the same Hilbert space. We call an eigenvector an *eigensequence* when the signal domain is $\mathbb{Z}$ or a subset of $\mathbb{Z}$ (for example, in Chapter 3); we call it an *eigenfunction* when the signal domain is $\mathbb{R}$ or an interval $[a, b]$ (for example, in Chapter 4).

DEFINITION 2.24 (EIGENVECTOR OF A LINEAR OPERATOR) An *eigenvector* of a linear operator $A : H \to H$ is a nonzero vector $v \in H$ such that

$$Av = \lambda v, \tag{2.58}$$

for some $\lambda \in \mathbb{C}$. The constant λ is called the corresponding *eigenvalue* and (λ, v) is called an *eigenpair.*

The eigenvalues and eigenvectors of a self-adjoint operator A have several useful properties:

(i) All eigenvalues are real: If (λ, v) is an eigenpair of A,

$$\begin{aligned}\lambda\langle v, v\rangle &\overset{(a)}{=} \langle \lambda v, v\rangle \overset{(b)}{=} \langle Av, v\rangle \overset{(c)}{=} \langle v, A^* v\rangle \\ &\overset{(d)}{=} \langle v, Av\rangle \overset{(e)}{=} \langle v, \lambda v\rangle \overset{(f)}{=} \lambda^* \langle v, v\rangle,\end{aligned}$$

where (a) follows from the linearity in the first argument of the inner product; (b) from (2.58); (c) from the definition of adjoint; (d) from the definition of self-adjoint; (e) from (2.58); and (f) from the conjugate linearity in the second argument of the inner product. Since $\lambda = \lambda^*$, λ is real.

(ii) Eigenvectors corresponding to distinct eigenvalues are orthogonal: If (λ_0, v_0) and (λ_1, v_1) are eigenpairs of A,

$$\begin{aligned}\lambda_0\langle v_0, v_1\rangle &\overset{(a)}{=} \langle \lambda_0 v_0, v_1\rangle \overset{(b)}{=} \langle Av_0, v_1\rangle \overset{(c)}{=} \langle v_0, A^* v_1\rangle \\ &\overset{(d)}{=} \langle v_0, Av_1\rangle \overset{(e)}{=} \langle v_0, \lambda_1 v_1\rangle \overset{(f)}{=} \lambda_1^* \langle v_0, v_1\rangle,\end{aligned}$$

where (a) follows from the linearity in the first argument of the inner product; (b) from (2.58); (c) from the definition of adjoint; (d) from the definition of self-adjoint; (e) from (2.58); and (f) from the conjugate linearity in the second argument of the inner product. Thus, $\lambda_0 \neq \lambda_1$ implies $\langle v_0, v_1\rangle = 0$.

Positive definite operators Positive definiteness can also be generalized from square Hermitian matrices to self-adjoint operators on a general Hilbert space.

DEFINITION 2.25 (DEFINITE LINEAR OPERATOR) A self-adjoint operator $A : H \to H$ is called

(i) *positive semidefinite* or *nonnegative definite*, written $A \geq 0$, when

$$\langle Ax, x\rangle \geq 0 \quad \text{for all} \quad x \in H; \tag{2.59a}$$

(ii) *positive definite*, written $A > 0$, when

$$\langle Ax, x\rangle > 0 \quad \text{for all nonzero} \quad x \in H; \tag{2.59b}$$

(iii) *negative semidefinite* or *nonpositive definite*, written $A \leq 0$, when $-A$ is positive semidefinite; and

(iv) *negative definite*, written $A < 0$, when $-A$ is positive definite.

As suggested by the notation, positive definiteness defines a partial order on self-adjoint operators. When A and B are self-adjoint operators defined on the same Hilbert space, $A \geq B$ means that $A - B \geq 0$; that is, $A - B$ is a positive semidefinite operator.

As noted above, all eigenvalues of a self-adjoint operator are real. Positive definiteness is equivalent to all eigenvalues being positive; positive semidefiniteness is equivalent to all eigenvalues being nonnegative. Exercise 2.28 develops a proof of these facts.

2.4 Approximations, projections, and decompositions

Many of the linear operators that we encounter in later chapters are projection operators, in particular orthogonal projection operators. As we will see in this section, orthogonal projection operators find best approximations from within a subspace, that is, approximations that minimize a Hilbert space norm of the error. An orthogonal projection generates a decomposition of a vector into components in two orthogonal subspaces. We will also see how the more general oblique projection operators generate decompositions of vectors into components in two subspaces that are not necessarily orthogonal.

Best approximation, orthogonal projection, and orthogonal decomposition

Let S be a closed subspace of a Hilbert space H and let x be a vector in H. The *best approximation problem* is to find the vector in S that is closest to x:

$$\widehat{x} = \arg\min_{s \in S} \|x - s\|. \tag{2.60}$$

Most commonly, the Hilbert space norm is the 2 norm, ℓ^2 norm, or $\mathcal{L}^2$ norm and thus involves squaring, in which case $\widehat{x}$ is called a *least-squares approximation*. Of

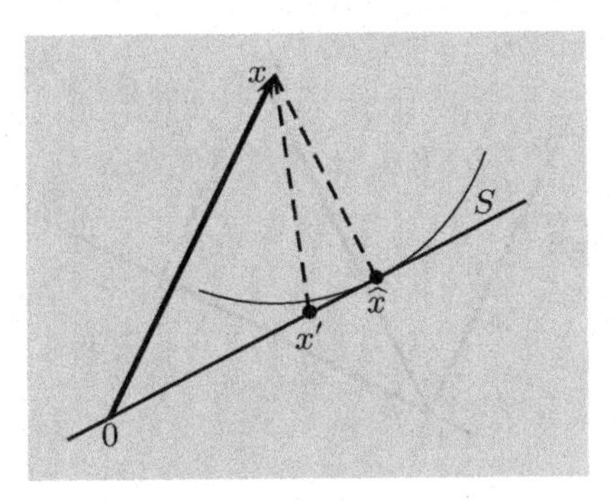

Figure 2.11 Illustration of best approximation. In Euclidean geometry, the best approximation of x on the line S is obtained with error $x - \widehat{x}$ orthogonal to S; any candidate x' such that $x - x'$ is not orthogonal to S is farther from x. This holds more generally in Hilbert spaces.

course, when x is in S then $\widehat{x} = x$ uniquely solves the problem – it makes the approximation error $\|x - \widehat{x}\|$ zero.[16] The interesting case is when x is not in S.

Figure 2.11 illustrates the problem and its solution in Euclidean geometry. The point on the line S that is closest to a point x not on the line is uniquely determined by finding the circle centered at x that is tangent to the line. Any other candidate x' on the line lies outside the circle and is thus farther from x. Since a line tangent to a circle is always perpendicular to a line segment from the tangent point to the center of the circle, the solution $\widehat{x}$ satisfies the following orthogonality property: $x - \widehat{x} \perp S$. The projection theorem extends this geometric result to arbitrary Hilbert spaces. The subspace S uniquely determines the orthogonal decomposition of x into the approximation $\widehat{x}$ and the residual $x - \widehat{x}$, and the orthogonality of $x - \widehat{x}$ and S uniquely determines $\widehat{x}$.

2.4.1 Projection theorem

The solution to the best approximation problem in a general Hilbert space is described by the following theorem.

THEOREM 2.26 (PROJECTION THEOREM) Let S be a closed subspace of a Hilbert space H, and let x be a vector in H.

(i) *Existence:* There exists $\widehat{x} \in S$ such that $\|x - \widehat{x}\| \leq \|x - s\|$ for all $s \in S$.
(ii) *Orthogonality:* $x - \widehat{x} \perp S$ is necessary and sufficient for determining $\widehat{x}$.
(iii) *Uniqueness:* The vector $\widehat{x}$ is unique.
(iv) *Linearity:* $\widehat{x} = Px$, where P is a linear operator that depends on S and not on x.

[16] Recall from Definition 2.9(i) that $\|x - \widehat{x}\|$ is nonnegative and zero if and only if $x - \widehat{x} = 0$.

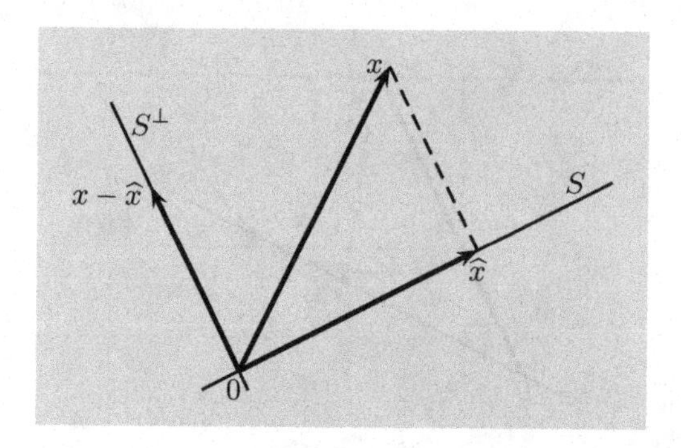

Figure 2.12 The best approximation of $x \in H$ within a closed subspace S is uniquely determined by $x - \widehat{x} \perp S$. The solution generates an orthogonal decomposition of x into $\widehat{x} \in S$ and $x - \widehat{x} \in S^{\perp}$.

(v) *Idempotency:* $P(Px) = Px$ for all $x \in H$.

(vi) *Self-adjointness:* $P = P^*$.

Proof. We prove existence last since it is the most technical and is the only part that requires completeness of the space. (Orthogonality and uniqueness hold with H replaced by any inner product space and S replaced by any subspace.)

(ii) *Orthogonality:* Suppose that $\widehat{x}$ minimizes $\|x - \widehat{x}\|$ but that $x - \widehat{x} \not\perp S$. Then, there exists a unit vector $\varphi \in S$ such that $\langle x - \widehat{x}, \varphi\rangle = \varepsilon \neq 0$. Let $s = \widehat{x} + \varepsilon\varphi$ and note that s is in S since x and φ are in S and S is a subspace. The calculation

$$\begin{aligned} \|x - s\|^2 &= \|x - \widehat{x} - \varepsilon\varphi\|^2 \\ &= \|x - \widehat{x}\|^2 - \underbrace{\langle x - \widehat{x}, \varepsilon\varphi\rangle}_{= |\varepsilon|^2} - \underbrace{\langle \varepsilon\varphi, x - \widehat{x}\rangle}_{= |\varepsilon|^2} + \underbrace{\|\varepsilon\varphi\|^2}_{= |\varepsilon|^2} \\ &= \|x - \widehat{x}\|^2 - |\varepsilon|^2 \; < \; \|x - \widehat{x}\|^2 \end{aligned}$$

then shows that $\widehat{x}$ is not the minimizing vector. This contradiction implies that $x - \widehat{x} \perp S$. This can also be written as $x - \widehat{x} \in S^{\perp}$; see Figure 2.12.

To prove sufficiency, we assume that $x - \widehat{x} \perp S$, or $\langle x - \widehat{x}, s\rangle = 0$ for all $s \in S$. Because $\widehat{x} \in S$ and S is a subspace, $\widehat{x} - s \in S$ as well, and thus $\langle x - \widehat{x}, \widehat{x} - s\rangle = 0$; that is, $x - \widehat{x}$ is orthogonal to $\widehat{x} - s$. Then,

$$\|x - s\|^2 \;=\; \|(x - \widehat{x}) + (\widehat{x} - s)\|^2 \overset{(a)}{=} \|x - \widehat{x}\|^2 + \|\widehat{x} - s\|^2 \overset{(b)}{\geq} \|x - \widehat{x}\|^2,$$

for all $s \in S$, where (a) follows from the Pythagorean theorem; and (b) from the positive definiteness of the norm.

(iii) *Uniqueness:* Suppose that $x - \widehat{x} \perp S$. For any $s \in S$,

$$\|x - s\|^2 \;=\; \|(x - \widehat{x}) + (\widehat{x} - s)\|^2 \overset{(a)}{=} \|x - \widehat{x}\|^2 + \|\widehat{x} - s\|^2,$$

where $\widehat{x} - s \in S$ implies that $x - \widehat{x} \perp \widehat{x} - s$, which allows an application of the Pythagorean theorem in (a). Since $\|\widehat{x} - s\| > 0$ for any $s \neq \widehat{x}$ by the positive definiteness of the norm, this shows that $\|x - s\| > \|x - \widehat{x}\|$ for any $s \neq \widehat{x}$.

(iv) *Linearity:* Let α be any scalar, and denote the best approximations in S of x_1 and x_2 by $\widehat{x}_1$ and $\widehat{x}_2$, respectively. The orthogonality property implies that $x_1 - \widehat{x}_1 \in S^\perp$ and $x_2 - \widehat{x}_2 \in S^\perp$. Since S is a subspace, $\widehat{x}_1 + \widehat{x}_2 \in S$; and since $S^\perp$ is a subspace, $(x_1 - \widehat{x}_1) + (x_2 - \widehat{x}_2) \in S^\perp$. With $(x_1 + x_2) - (\widehat{x}_1 + \widehat{x}_2) \in S^\perp$ and $\widehat{x}_1 + \widehat{x}_2 \in S$, the uniqueness property shows that $\widehat{x}_1 + \widehat{x}_2$ is the best approximation of $x_1 + x_2$. This shows additivity. Similarly, since S is a subspace, $\alpha\widehat{x}_1 \in S$; and since $S^\perp$ is a subspace, $\alpha(x_1 - \widehat{x}_1) \in S^\perp$. With $\alpha x_1 - \alpha\widehat{x}_1 \in S^\perp$ and $\alpha\widehat{x}_1 \in S$, the uniqueness property shows that $\alpha\widehat{x}_1$ is the best approximation of αx_1. This shows scalability.

(v) *Idempotency:* The property to check is that the operator P leaves Px unchanged. This follows from two facts: $Px \in S$ and $Pu = u$ for all $u \in S$. That Px is in S is part of the definition of $\widehat{x}$. For the second fact, let $u \in S$ and suppose that $\widehat{u}$ satisfies

$$\|u - \widehat{u}\| \le \|u - s\| \qquad \text{for all} \quad s \in S.$$

By the uniqueness property, there can be only one such $\widehat{u}$, and since $\widehat{u} = u$ gives $\|u - \widehat{u}\| = 0$ and the norm is nonnegative, we must have $\widehat{u} = u$.

(vi) *Self-adjointness:* We would like to show that $\langle Px, y\rangle = \langle x, Py\rangle$ for all $x, y \in H$:

$$\langle Px, y\rangle = \langle Px, Py + (y - Py)\rangle \overset{(a)}{=} \langle Px, Py\rangle + \langle Px, y - Py\rangle \overset{(b)}{=} \langle Px, Py\rangle,$$

where (a) follows from the distributivity of the inner product; and (b) from $Px \in S$ and $y - Py \in S^\perp$. Similarly,

$$\langle x, Py\rangle = \langle Px + (x - Px), Py\rangle = \langle Px, Py\rangle + \langle x - Px, Py\rangle = \langle Px, Py\rangle.$$

(i) *Existence:* We finally show the existence of a minimizing $\widehat{x}$. If x is in S, then $\widehat{x} = x$ achieves the minimum, so there is no question of existence. We thus restrict our attention to $x \notin S$. Let $\varepsilon = \inf_{s \in S} \|x - s\|$. Then, there exists a sequence of vectors $s_0, s_1, \ldots$ in S such that $\|x - s_k\| \to \varepsilon$; the challenge is to show that the infimum is achieved by some $\widehat{x} \in S$. We do this by showing that $\{s_k\}_{k\ge0}$ is a Cauchy sequence and thus converges, within the closed subspace S, to the desired $\widehat{x}$.

By applying the parallelogram law (2.28) to $x - s_j$ and $s_i - x$,

$$\|(x - s_j) + (s_i - x)\|^2 + \|(x - s_j) - (s_i - x)\|^2 = 2\|x - s_j\|^2 + 2\|s_i - x\|^2.$$

Canceling x in the first term and moving the second term to the right yields

$$\|s_i - s_j\|^2 = 2\|x - s_j\|^2 + 2\|s_i - x\|^2 - 4\left\|x - \frac{1}{2}(s_i + s_j)\right\|^2. \tag{2.61}$$

Now, since S is a subspace, $\frac{1}{2}(s_i + s_j)$ is in S. Thus by the definition of ε we have $\|x - \frac{1}{2}(s_i + s_j)\| \ge \varepsilon$. Substituting into (2.61) and using the nonnegativity of the norm gives

$$0 \le \|s_i - s_j\|^2 \le 2\|x - s_j\|^2 + 2\|s_i - x\|^2 - 4\varepsilon^2.$$

With the convergence of $\|x - s_j\|^2$ and $\|s_i - x\|^2$ to ε^2, we conclude that $\{s_k\}_{k\ge0}$ is a Cauchy sequence. Now, since S is a closed subspace of a complete space, $\{s_k\}_{k\ge0}$ converges to $\widehat{x} \in S$. Since a norm is continuous, convergence of $\{s_k\}_{k\ge0}$ to $\widehat{x}$ implies that $\|x - \widehat{x}\| = \varepsilon$.

The projection theorem leads to a simple and unified methodology for computing best approximations through the normal equations. We develop this in detail after the introduction of bases in Section 2.5. The following example provides a preview.

EXAMPLE 2.20 (PROJECTION THEOREM) Consider the function $x(t) = \cos(\frac{3}{2}\pi t)$ in the Hilbert space $\mathcal{L}^2([0,\, 1])$. To find the degree-1 polynomial closest to x directly (without using the projection theorem) would require solving

$$\min_{a_0,\, a_1} \int_0^1 \left|\cos\left(\frac{3}{2}\pi t\right) - (a_0 + a_1 t)\right|^2 dt.$$

Noting that the degree-1 polynomials form a closed subspace in this Hilbert space, the projection theorem shows that $(a_0,\, a_1)$ is determined uniquely by requiring

$$x(t) - \widehat{x}(t) \;=\; \cos\left(\frac{3}{2}\pi t\right) - (a_0 + a_1 t)$$

to be orthogonal to the entire subspace of degree-1 polynomials. Imposing orthogonality to 1 and to t gives two linearly independent equations to solve:

$$0 = \langle x(t) - \widehat{x}(t),\, 1\rangle = \int_0^1 \left(\cos\left(\frac{3}{2}\pi t\right) - (a_0 + a_1 t)\right) \cdot 1\, dt = -\frac{2}{3\pi} - a_0 - \frac{1}{2}a_1,$$

$$0 = \langle x(t) - \widehat{x}(t),\, t\rangle = \int_0^1 \left(\cos\left(\frac{3}{2}\pi t\right) - (a_0 + a_1 t)\right) \cdot t\, dt = \frac{4 + 6\pi}{9\pi^2} - \frac{1}{2}a_0 - \frac{1}{3}a_1.$$

Their solution is

$$a_0 \;=\; \frac{8 + 4\pi}{3\pi^2}, \qquad a_1 \;=\; -\frac{16 + 12\pi}{3\pi^2}.$$

Figure 2.13(a) shows the function and its degree-1 polynomial approximation.

Best approximations among degree-K polynomials for $K = 1,\, 2,\, 3,\, 4$ are shown in Figure 2.13(b). Increasing the degree increases the size of the subspace of polynomials, so the quality of approximation is naturally improved. We will see in Chapter 6 (Theorem 6.6) that, since $x(t)$ is continuous, the error is driven to zero by letting K grow without bound.

The effect of the operator P that arises in the projection theorem is to move the input vector x in a direction orthogonal to the subspace S until S is reached at $\widehat{x}$. In the following two sections, we show that P has the defining characteristics of what we will call an orthogonal projection operator, and we describe the mapping of x to an approximation $\widehat{x}$ and a residual $x - \widehat{x}$ as what we will call an orthogonal decomposition. As we will see shortly, projections and decompositions have nonorthogonal (oblique) versions as well.

2.4.2 Projection operators

The operator P that arises from solving the best approximation problem is an orthogonal projection operator, as per the following definition.[17]

[17]We use the term projection only with respect to Hilbert space geometry; many other mathematical and scientific meanings are inconsistent with this. Some authors use *projection* to mean

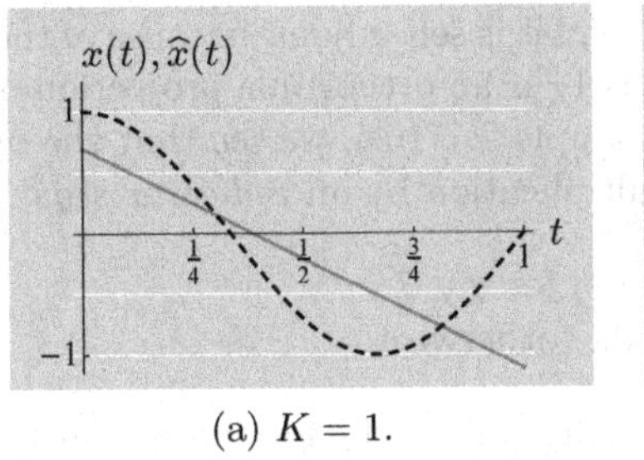

(a) $K = 1$.

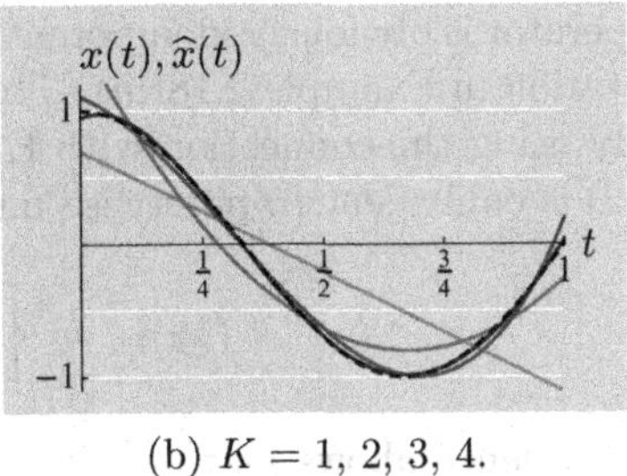

(b) $K = 1, 2, 3, 4$.

Figure 2.13 Best approximation $\widehat{x}_K(t)$ (solid lines, from lightest to darkest) of $x(t)$ (dashed lines) among degree-K polynomials; approximation quality is measured by the $\mathcal{L}^2$ norm on $([0, 1])$ as in Example 2.20. Allowing higher-degree polynomials increases the size of the subspace to which $x(t)$ is orthogonally projected and decreases the approximation error.

DEFINITION 2.27 (PROJECTION OPERATOR)

(i) An *idempotent* operator P is an operator such that $P^2 = P$.
(ii) A *projection* operator is a bounded linear operator that is idempotent.
(iii) An *orthogonal projection* operator is a projection operator that is self-adjoint.
(iv) An *oblique projection* operator is a projection operator that is not self-adjoint.

An operator is idempotent when applying it twice is no different than applying it once. For example, setting certain components of a vector to a constant value is an idempotent operation, and when that constant is zero this operation is linear. The following example introduces a notation for the basic class of orthogonal projection operators that set a portion of a vector to zero.

EXAMPLE 2.21 (PROJECTION VIA DOMAIN RESTRICTION) Let $\mathcal{I}$ be a subset of $\mathbb{Z}$, and define the linear operator $1_{\mathcal{I}} : \ell^2(\mathbb{Z}) \to \ell^2(\mathbb{Z})$ by

$$y = 1_{\mathcal{I}}\, x, \qquad \text{where} \qquad y_k = \begin{cases} x_k, & \text{for } k \in \mathcal{I}; \\ 0, & \text{otherwise.} \end{cases} \tag{2.62}$$

This is a special case of the linear operator in Example 2.17(iii), with

$$\alpha_k = \begin{cases} 1, & \text{for } k \in \mathcal{I}; \\ 0, & \text{otherwise.} \end{cases}$$

orthogonal projection. We will *not* adopt this potentially confusing shorthand because many important properties and uses of projection operators hold for all projections – both orthogonal and oblique.

This operator is obviously idempotent, and it is self-adjoint because of the adjoint computation in Example 2.18(iii). Thus $1_{\mathcal{I}}$ is an orthogonal projection operator.

By using the connection with Example 2.17(iii), we see that the operation in (2.62) is equivalent to pointwise multiplication by an *indicator sequence*

$$(1_{\mathcal{I}})_k = \begin{cases} 1, & \text{for } k \in \mathcal{I}; \\ 0, & \text{otherwise.} \end{cases} \tag{2.63}$$

The equivalence allows us to use the same symbol $1_{\mathcal{I}}$ for the operator and the sequence with little chance for ambiguity.

The same notation is used for vector spaces with domains other than $\mathbb{Z}$. For example, with $\mathcal{I}$ a subset of $\mathbb{R}$, we define the linear operator $1_{\mathcal{I}} : \mathcal{L}^2(\mathbb{R}) \to \mathcal{L}^2(\mathbb{R})$ by

$$y = 1_{\mathcal{I}}\, x, \qquad \text{where} \qquad y(t) = \begin{cases} x(t), & \text{for } t \in \mathcal{I}; \\ 0, & \text{otherwise.} \end{cases} \tag{2.64}$$

The operator results in pointwise multiplication of its input by the *indicator function*

$$1_{\mathcal{I}}(t) = \begin{cases} 1, & \text{for } t \in \mathcal{I}; \\ 0, & \text{otherwise.} \end{cases} \tag{2.65}$$

Exercise 2.30 establishes properties of this operator.

The following example gives an explicit expression for projection operators onto one-dimensional subspaces. This was discussed informally in Section 2.1.

Example 2.22 (Orthogonal projection onto 1-dimensional subspace) Given a vector $\varphi \in H$ of unit norm, let

$$Px = \langle x, \varphi \rangle \varphi. \tag{2.66}$$

This is a linear operator because of the distributivity and the linearity in the first argument of the inner product. To use Theorem 2.28 to show that P is the orthogonal projection operator onto the subspace of scalar multiples of φ, we verify the idempotency and self-adjointness of P. Idempotency is proven by the following computation:

$$P^2 x = \langle \langle x, \varphi \rangle \varphi, \varphi \rangle \varphi \overset{(a)}{=} \langle x, \varphi \rangle \langle \varphi, \varphi \rangle \varphi \overset{(b)}{=} \langle x, \varphi \rangle \varphi = Px,$$

where (a) follows from the linearity in the first argument of the inner product; and (b) from $\langle \varphi, \varphi \rangle = 1$. Self-adjointness is proven by the following computation:

$$\begin{aligned} \langle Px, y \rangle &= \langle \langle x, \varphi \rangle \varphi, y \rangle \overset{(a)}{=} \langle x, \varphi \rangle \langle \varphi, y \rangle = \langle \varphi, y \rangle \langle x, \varphi \rangle \\ &\overset{(b)}{=} \langle y, \varphi \rangle^* \langle x, \varphi \rangle \overset{(c)}{=} \langle x, \langle y, \varphi \rangle \varphi \rangle = \langle x, Py \rangle, \end{aligned}$$

where (a) follows from the linearity in the first argument of the inner product; (b) from the conjugate symmetry of the inner product; and (c) from the conjugate linearity in the second argument of the inner product.

Collections of one-dimensional projections are central to representations using bases, which are introduced in Section 2.5 and developed in several subsequent chapters. Solved exercise 2.4 extends the one-dimensional example to orthogonal projection onto subspaces of higher dimensions.

The following theorem uses the orthogonality of certain vectors to prove that an operator is an orthogonal projection operator. This complements the projection theorem, since here an operator is specified rather than a subspace. After the proof, we discuss the subspace that is implicit in the theorem.

THEOREM 2.28 (ORTHOGONAL PROJECTION OPERATOR) A bounded linear operator P on a Hilbert space H satisfies

$$\langle x - Px,\, Py\rangle = 0 \qquad \text{for all } x,\, y \in H \tag{2.67}$$

if and only if P is an orthogonal projection operator.

Proof. Condition (2.67) is equivalent to having

$$0 = \langle x - Px,\, Py\rangle \overset{(a)}{=} \langle P^*(x - Px),\, y\rangle = \langle P^*(I-P)x,\, y\rangle \qquad \text{for all } x,\, y \in H,$$

where (a) follows from the definition of the adjoint. This then implies that $P^*(I-P) = 0$, so

$$P^* = P^*P. \tag{2.68}$$

We will show that (2.68) is equivalent to P being an orthogonal projection operator.

First assume that (2.68) holds. Then,

$$P = (P^*)^* \overset{(a)}{=} (P^*P)^* = P^*P \overset{(b)}{=} P^*,$$

where (a) and (b) follow from (2.68). Thus, P is self-adjoint. Furthermore,

$$P^2 \overset{(a)}{=} P^*P \overset{(b)}{=} P^* \overset{(c)}{=} P,$$

where (a) and (c) follow from P being self-adjoint; and (b) from (2.68). Thus, P is idempotent. Therefore, (2.68) implies P is an orthogonal projection operator.

For the converse, assume P is an orthogonal projection operator. Then,

$$P^*P \overset{(a)}{=} P^2 \overset{(b)}{=} P \overset{(c)}{=} P^*,$$

where (a) and (c) follow from P being self-adjoint; and (b) from P being idempotent. Therefore, P being an orthogonal projection operator implies (2.68).

The range of any linear operator is a subspace. In the setting of the preceding theorem, we may associate with P the closed subspace $S = \mathcal{R}(P)$. Then, we have that P is the orthogonal projection operator onto S. The orthogonality equation (2.67) is a restatement of the projection residual $x - Px$ being orthogonal to S.

The final theorem of the section establishes important connections among inverses, adjoints and projections; the proof is simple and left for Exercise 2.31.

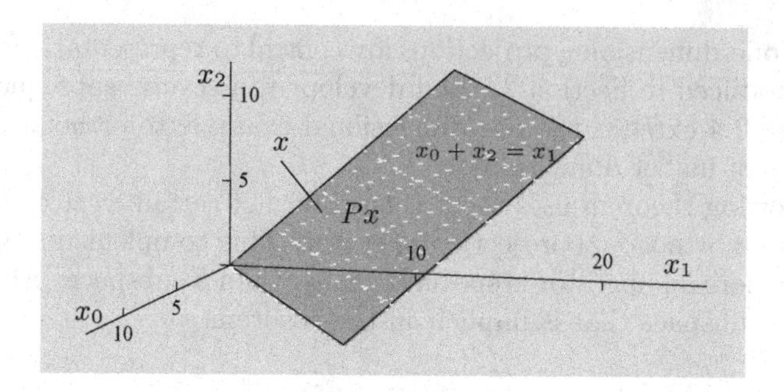

Figure 2.14 The two-dimensional range of the oblique projection operator P from Example 2.23 is the plane $x_0 + x_2 = x_1$. For example, the vector $x = \begin{bmatrix} 6 & 6 & 8 \end{bmatrix}^\top$ is projected via P onto $Px = \begin{bmatrix} 2 & 6 & 4 \end{bmatrix}^\top$, which is not an orthogonal projection.

THEOREM 2.29 (PROJECTION OPERATORS, ADJOINTS, AND INVERSES) Let $A : H_0 \to H_1$ and $B : H_1 \to H_0$ be bounded linear operators. If A is a left inverse of B, then BA is a projection operator. If additionally $B = A^*$, then $BA = A^*A$ is an orthogonal projection operator.

EXAMPLE 2.23 (PROJECTION ONTO A SUBSPACE) Let

$$A = \frac{1}{2}\begin{bmatrix} 1 & 1 & -1 \\ -1 & 1 & 1 \end{bmatrix} \quad \text{and} \quad B = \begin{bmatrix} 1 & 0 \\ 1 & 1 \\ 0 & 1 \end{bmatrix}.$$

Since A is a left inverse of B, we know from Theorem 2.29 that $P = BA$ is a projection operator. Explicitly,

$$P = BA = \frac{1}{2}\begin{bmatrix} 1 & 1 & -1 \\ 0 & 2 & 0 \\ -1 & 1 & 1 \end{bmatrix},$$

from which one can verify $P^2 = P$. A description of the two-dimensional range of this projection operator is most transparent from B: it is the set of three-tuples with middle component equal to the sum of the first and last (see Figure 2.14). Note that $P \neq P^*$, so the projection is oblique. Exercise 2.32 finds all matrices A such that BA is a projection operator onto the range of B.

The next example draws together some earlier results to give an orthogonal projection operator on $\mathcal{L}^2(\mathbb{R})$. It also illustrates the basics of sampling and interpolation, to which we will return in Chapter 5.

EXAMPLE 2.24 (ORTHOGONAL PROJECTION OPERATOR ON $\mathcal{L}^2(\mathbb{R})$) Let $A : \mathcal{L}^2(\mathbb{R}) \to \ell^2(\mathbb{Z})$ be the local averaging operator (2.49) and let $A^* : \ell^2(\mathbb{Z}) \to \mathcal{L}^2(\mathbb{R})$

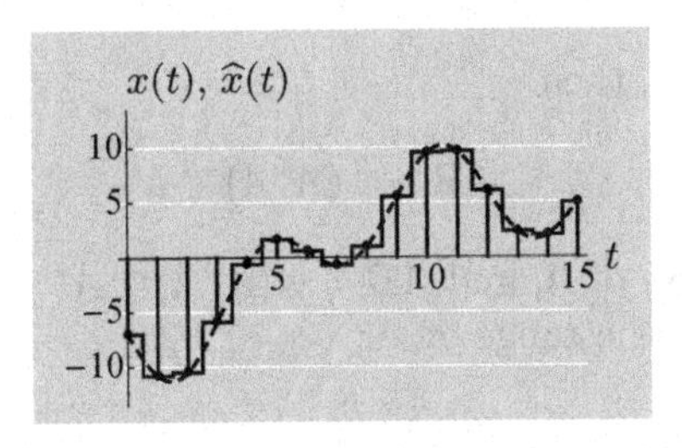

Figure 2.15 Illustration of an orthogonal projection operator on $\mathcal{L}^2(\mathbb{R})$. The linear operator A and its adjoint A^* illustrated in Figure 2.10 satisfy $AA^* = I$, so A^*A is an orthogonal projection operator. The range of A^* is the subspace of $\mathcal{L}^2(\mathbb{R})$ consisting of functions that are constant on all intervals $[n - \frac{1}{2}, n + \frac{1}{2})$, $n \in \mathbb{Z}$. Thus, $\widehat{x} = A^*Ax$ (solid line) is the best approximation of x (dashed line) in this subspace.

be its adjoint, as derived in Example 2.19. If we verify that A is a left inverse of A^*, we will have, as a consequence of Theorem 2.29, that A^*A is an orthogonal projection operator.

To check that A is a left inverse of A^*, consider the application of AA^* to an arbitrary sequence in $\ell^2(\mathbb{Z})$. (Recall that the separate effects of A and A^* are as illustrated in Figure 2.10.) Remembering to compose from right to left, AA^* starts with a sequence y, creates a function equal to y_n on each interval $[n - \frac{1}{2}, n + \frac{1}{2})$, and then recovers the original sequence by finding the average value of the function on each interval $[n - \frac{1}{2}, n + \frac{1}{2})$. So AA^* is indeed an identity operator. One conclusion to draw by combining the projection theorem with A^*A being an orthogonal projection operator is the following: Given a function $x \in \mathcal{L}^2(\mathbb{R})$, the function in the subspace of piecewise-constant functions $A^*\ell^2(\mathbb{Z})$ that is closest to x in $\mathcal{L}^2$ norm is the one obtained by replacing $x(t)$, $t \in [n-\frac{1}{2}, n+\frac{1}{2})$, by its local average $\int_{n-1/2}^{n+1/2} x(t)\, dt$. The result of applying A^*A is depicted in Figure 2.15.

Pseudoinverse operators By Definition 2.27, a projection operator will be an orthogonal projection operator if, in addition to being idempotent, it is orthogonal. We thus have the following as an extension of Theorem 2.29.

THEOREM 2.30 (ORTHOGONAL PROJECTION VIA PSEUDOINVERSE) Let $A : H_0 \to H_1$ be a bounded linear operator.

(i) If AA^* is invertible, then

$$B = A^*(AA^*)^{-1} \tag{2.69a}$$

is the *pseudoinverse* of A, and $BA = A^*(AA^*)^{-1}A$ is the orthogonal projection operator onto the range of A^*.

(ii) If A^*A is invertible, then

$$B = (A^*A)^{-1}A^* \tag{2.69b}$$

is the *pseudoinverse* of A, and $AB = A(A^*A)^{-1}A^*$ is the orthogonal projection operator onto the range of A.

Because of this connection to orthogonal projection, the pseudoinverse is useful for best approximation problems; we discuss this in more detail in Chapter 5. In addition, the pseudoinverse is useful for understanding dual bases and canonical dual frames; we will see that in Sections 2.5.3 and 2.5.4 and Chapters 5 and 6. When A has an infinite-dimensional domain and codomain, finding the pseudoinverse of A by inversion of AA^* or A^*A can be difficult. As we will see in Chapters 5 and 6, we can avoid explicit inversion by finding a system of equations satisfied by the desired dual.

2.4.3 Direct sums and subspace decompositions

In the projection theorem, the best approximation $\widehat{x}$ is uniquely determined by the orthogonality of $\widehat{x}$ and $x - \widehat{x}$. Thus, the projection theorem proves that x can be written uniquely as

$$x = x_S + x_{S^\perp}, \qquad \text{where} \quad x_S \in S \quad \text{and} \quad x_{S^\perp} \in S^\perp, \tag{2.70}$$

because we must have $x_S = \widehat{x}$ and $x_{S^\perp} = x - \widehat{x}$.

Being able to uniquely write any x as a sum defines a decomposition. Having an orthogonal pair of subspaces as in (2.70) is an important special case that yields an orthogonal decomposition.

DEFINITION 2.31 (DIRECT SUM AND DECOMPOSITION) A vector space V is a *direct sum* of subspaces S and T, denoted $V = S \oplus T$, when any nonzero vector $x \in V$ can be written uniquely as

$$x = x_S + x_T, \qquad \text{where} \quad x_S \in S \quad \text{and} \quad x_T \in T. \tag{2.71}$$

The subspaces S and T form a *decomposition* of V, and the vectors x_S and x_T form a *decomposition* of x. When S and T are orthogonal, each decomposition is called an *orthogonal decomposition.*

A general direct-sum decomposition $V = S \oplus T$ is illustrated in Figure 2.16(a).

When S is a closed subspace of a Hilbert space H, the projection theorem generates the unique decomposition (2.70); thus, $H = S \oplus S^\perp$. It is tempting to write $V = S \oplus S^\perp$ for any (not necessarily closed) subspace of any (not necessarily complete) vector space. However, this is not always possible. The following example

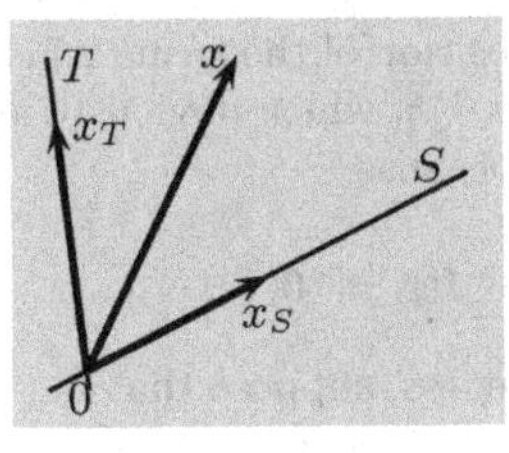

(a) Decomposition.

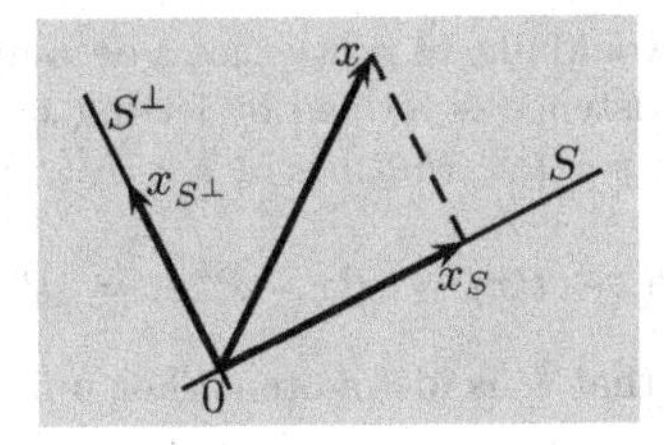

(b) Orthogonal projection.

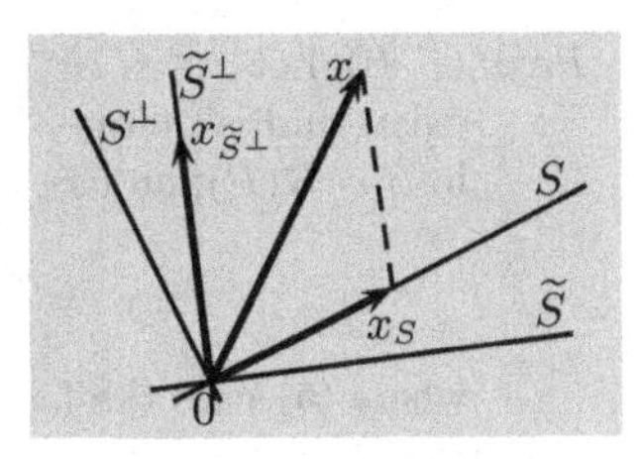

(c) Oblique projection.

Figure 2.16 Decompositions and projections. (a) A vector space V is decomposed as a direct sum $S \oplus T$ when any $x \in V$ can be written uniquely as a sum of components in S and T. (b) An orthogonal projection operator generates an orthogonal direct-sum decomposition of a Hilbert space. It decomposes the vector x into $x_S \in S$ and $x_{S^\perp} \in S^\perp$. (c) An oblique projection operator generates a nonorthogonal direct-sum decomposition of a Hilbert space. It decomposes the vector x into $x_S \in S$ and $x_{\widetilde{S}^\perp} \in \widetilde{S}^\perp$.

highlights the necessity of working with closed subspaces of a complete space. As noted before Example 2.14, we frequently work with the closure of a span to avoid the pitfalls of subspaces that are not closed.

EXAMPLE 2.25 (FAILURE OF DIRECT SUM $S \oplus S^\perp$) As in Example 2.14, consider Hilbert space $\ell^2(\mathbb{N})$ and, for each $k \in \mathbb{N}$, let the sequence s_k be 0 except for a 1 in the kth position. Then, $S = \text{span}(\{s_0, s_1, \ldots\})$ consists of all vectors in $\ell^2(\mathbb{N})$ that have a finite number of nonzero entries, and S is a subspace. Let $x \in \ell^2(\mathbb{N})$ be a nonzero vector. Then, $x_n \neq 0$ for some $n \in \mathbb{N}$, so $x \not\perp s_n$, which implies that $x \notin S^\perp$. Since no nonzero vector is in $S^\perp$, we have that $S^\perp = \{\mathbf{0}\}$. Since S itself is not all of $\ell^2(\mathbb{N})$, one cannot write every $x \in \ell^2(\mathbb{N})$ as in (2.70).

The main goal of this section is to extend the connection between decompositions and projections to the general (oblique) case. The following theorem establishes that a projection operator P generates a direct-sum decomposition as illustrated in Figure 2.16(c). The dashed line shows the effect of the operator, $x_S = Px$, and the residual $x_{\widetilde{S}^\perp} = x - x_S$ is in a subspace we denote $\widetilde{S}^\perp$ rather than T for reasons that will become clear shortly.

THEOREM 2.32 (DIRECT-SUM DECOMPOSITION FROM PROJECTION OPERATOR) Let H be a Hilbert space.

(i) Let P be a projection operator on H. It generates a direct-sum decomposition of H into its range $\mathcal{R}(P)$ and null space $\mathcal{N}(P)$: $H = S \oplus T$, where $S = \mathcal{R}(P)$ and $T = \mathcal{N}(P)$.

(ii) Conversely, let closed subspaces S and T satisfy $H = S \oplus T$. Then, there exists a projection operator on H such that $S = \mathcal{R}(P)$ and $T = \mathcal{N}(P)$.

Proof. (i) Let $x \in H$. We would like to prove that a decomposition of the form (2.71) exists and is unique. Existence is verified by letting $x_S = Px$, which obviously is in $S = \mathcal{R}(P)$; and $x_T = x - Px$, which is in $T = \mathcal{N}(P)$ because

$$Px_T = P(x - Px) = Px - P^2x \stackrel{(a)}{=} Px - Px = \mathbf{0},$$

where (a) uses the fact that P is idempotent. For uniqueness, suppose that

$$x = x'_S + x'_T, \qquad \text{where} \quad x'_S \in S \quad \text{and} \quad x'_T \in T.$$

Upon equating the two expansions of x and applying P, we have

$$\begin{aligned}\mathbf{0} &= P\left((x_S - x'_S) + (x_T - x'_T)\right) = P(x_S - x'_S) + P(x_T - x'_T) \\ &\stackrel{(a)}{=} P(x_S - x'_S) \stackrel{(b)}{=} x_S - x'_S,\end{aligned}$$

where (a) follows from $x_T - x'_T$ lying in T, the null space of P; and (b) from $x_S - x'_S$ lying in S and P equaling the identity on S. From this, $x'_S = x_S$ and $x'_T = x_T$ follow.

(ii) Define the desired projection operator P from the unique decomposition of any $x \in H$ of the form (2.71) through $Px = x_S$. The linearity of P follows easily from the assumed uniqueness of decompositions of vectors. By construction, the range of P is contained in S. It is actually all of S because any $x \in S$ can be uniquely decomposed as $x + \mathbf{0}$ with $x \in S$ and $\mathbf{0} \in T$. Similarly, the null space of P contains T because any $x \in T$ can be uniquely decomposed as $\mathbf{0} + x$ with $\mathbf{0} \in S$ and $x \in T$, showing that $Px = \mathbf{0}$. The null space of P is not larger than T because any vector $x \in H \setminus T$ can be written uniquely as in (2.71) with $x_S \neq 0$, so $Px \neq \mathbf{0}$. It remains only to verify that P is idempotent. This follows from $Px \in S$ and the fact that P equals the identity on S.

The following example makes explicit the form of a (possibly oblique) projection when S is a one-dimensional subspace. For consistency with later developments, the illustration of Theorem 2.32 in Figure 2.16(c) uses $\widetilde{S}^{\perp}$ in place of T. Since S and $T = \widetilde{S}^{\perp}$ have complementary dimension (adding to the whole Hilbert space), we have that S and $\widetilde{S}$ are of the same dimension. When $S = \widetilde{S}$, the projection and decomposition are orthogonal, and Figure 2.16(c) reduces to Figure 2.12. In the example, this corresponds to $\varphi = \widetilde{\varphi}$.

EXAMPLE 2.26 (OBLIQUE PROJECTION ONTO ONE-DIMENSIONAL SUBSPACE) Let S be the scalar multiples of unit vector $\varphi \in H$, and let $\widetilde{S}$ be the scalar multiples of an arbitrary vector $\widetilde{\varphi} \in H$. The operator

$$Px = \langle x, \widetilde{\varphi} \rangle \varphi \tag{2.72}$$

is linear and has range contained in S. We will find conditions under which it generates a decomposition $H = S \oplus \widetilde{S}^{\perp}$.

Since

$$P^2x = \langle \langle x, \widetilde{\varphi} \rangle \varphi, \widetilde{\varphi} \rangle \varphi \stackrel{(a)}{=} \langle x, \widetilde{\varphi} \rangle \langle \varphi, \widetilde{\varphi} \rangle \varphi,$$

where (a) uses the linearity in the first argument of the inner product, P is idempotent if and only if $\langle \varphi, \widetilde{\varphi} \rangle = 1$. Under this condition, we have $H =$

$\mathcal{R}(P) \oplus \mathcal{N}(P)$ by Theorem 2.32. This can also be written as $H = S \oplus \widetilde{S}^{\perp}$ because $\mathcal{N}(P)$ and $\widetilde{S}^{\perp}$ are both precisely the set $\{x \in H \mid \langle x, \widetilde{\varphi} \rangle = 0\}$.

If $\langle \varphi, \widetilde{\varphi} \rangle \neq 1$ but also $\langle \varphi, \widetilde{\varphi} \rangle \neq 0$, a simple adjustment of the length of $\widetilde{\varphi}$ will make P a projection operator. However, if $\langle \varphi, \widetilde{\varphi} \rangle = 0$, it is not possible to decompose H as desired. In fact, S and $\widetilde{S}$ are orthogonal, so S and $\widetilde{S}^{\perp}$ are the same subspace.

2.4.4 Minimum mean-squared error estimation

The set of complex random variables can be viewed as a vector space over the complex numbers (see Section 2.2.4). With the restriction of finite second moments – which is implicit throughout the remainder of this section – this set forms a Hilbert space under the inner product (2.37),

$$\langle \mathrm{x}, \mathrm{y} \rangle = \mathrm{E}[\, \mathrm{x}\mathrm{y}^* \,].$$

The square of the norm of the difference between vectors in this vector space is the *mean-squared error* (MSE) between the random variables,

$$\|\mathrm{x} - \widehat{\mathrm{x}}\|^2 = \mathrm{E}\big[\, |\mathrm{x} - \widehat{\mathrm{x}}|^2 \,\big]. \tag{2.73}$$

Since minimizing the MSE is equivalent to minimizing a Hilbert space norm, many minimum MSE (MMSE) estimation problems are solved easily with the projection theorem. Throughout this section, MMSE estimators are called *optimal*, irrespective of whether or not the estimator is constrained to a particular form.

Linear estimation Let x and $\mathrm{y}_1, \mathrm{y}_2, \ldots, \mathrm{y}_K$ be jointly distributed complex random variables. A linear estimator[18] of x from $\{\mathrm{y}_k\}_{k=1}^K$ is a random variable of the form

$$\widehat{\mathrm{x}} = \alpha_0 + \alpha_1 \mathrm{y}_1 + \alpha_2 \mathrm{y}_2 + \cdots + \alpha_K \mathrm{y}_K. \tag{2.74}$$

When the coefficients are chosen to minimize the MSE, the estimator is called the *linear minimum mean-squared error* (LMMSE) estimator. Since (2.74) places $\widehat{\mathrm{x}}$ in a closed subspace S of a Hilbert space of random variables, the projection theorem dictates that the optimal estimator $\widehat{\mathrm{x}}$ must be such that the error $\mathrm{x} - \widehat{\mathrm{x}}$ is orthogonal to the subspace: $\mathrm{x} - \widehat{\mathrm{x}} \perp S$.

Instead of trying to express that $\mathrm{x} - \widehat{\mathrm{x}}$ is orthogonal to every vector in S, it suffices to write enough linearly independent equations to be able to solve for $\{\alpha_k\}_{k=0}^K$ in (2.74). Since constant random variables are in S (by setting $\alpha_1 = \alpha_2 = \cdots = \alpha_K = 0$), we must have

$$\begin{aligned} 0 &\overset{(a)}{=} \langle \mathrm{x} - \widehat{\mathrm{x}}, 1 \rangle \overset{(b)}{=} \mathrm{E}[\, \mathrm{x} - \widehat{\mathrm{x}} \,] \overset{(c)}{=} \mathrm{E}[\, \mathrm{x} \,] - \mathrm{E}[\, \widehat{\mathrm{x}} \,] \\ &\overset{(d)}{=} \mathrm{E}[\, \mathrm{x} \,] - (\alpha_0 + \alpha_1 \mathrm{E}[\, \mathrm{y}_1 \,] + \alpha_2 \mathrm{E}[\, \mathrm{y}_2 \,] + \cdots + \alpha_K \mathrm{E}[\, \mathrm{y}_K \,]), \end{aligned} \tag{2.75a}$$

[18] A function of the form $f(x_1, x_2, \ldots, x_K) = \alpha_0 + \alpha_1 x_1 + \alpha_2 x_2 + \cdots + \alpha_K x_K$, where each α_k is a constant, is called *affine*. The estimator in (2.74) is called linear even though $\widehat{\mathrm{x}}$ is an affine function of $\mathrm{y}_1, \mathrm{y}_2, \ldots, \mathrm{y}_K$ because $\widehat{\mathrm{x}}$ is a linear function of $1, \mathrm{y}_1, \mathrm{y}_2, \ldots, \mathrm{y}_K$, and 1 can be viewed as a random variable.

where (a) follows from the desired orthogonality; (b) from (2.37); and (c) and (d) from the linearity of the expectation. We also have that each y_k is in S, so by analogous steps

$$\begin{aligned} 0 &= \langle x - \widehat{x}, y_k \rangle = E[(x-\widehat{x})y_k^*] = E[xy_k^*] - E[\widehat{x}y_k^*] \\ &= E[xy_k^*] - (\alpha_0 E[y_k^*] + \alpha_1 E[y_1 y_k^*] + \cdots + \alpha_K E[y_K y_k^*]), \end{aligned} \tag{2.75b}$$

for $k = 1, 2, \ldots, K$. Equations (2.75) can be rearranged using a matrix–vector product as

$$\begin{bmatrix} 1 & E[y_1] & E[y_2] & \cdots & E[y_K] \\ E[y_1^*] & E[|y_1|^2] & E[y_2 y_1^*] & \cdots & E[y_K y_1^*] \\ E[y_2^*] & E[y_1^* y_2^*] & E[|y_2|^2] & \cdots & E[y_K y_2^*] \\ \vdots & \vdots & \vdots & \ddots & \vdots \\ E[y_K^*] & E[y_1 y_K^*] & E[y_2 y_K^*] & \cdots & E[|y_K|^2] \end{bmatrix} \begin{bmatrix} \alpha_0 \\ \alpha_1 \\ \alpha_2 \\ \vdots \\ \alpha_K \end{bmatrix} = \begin{bmatrix} E[x] \\ E[xy_1^*] \\ E[xy_2^*] \\ \vdots \\ E[xy_K^*] \end{bmatrix}. \tag{2.76}$$

This system of equations will usually have a unique solution. The solution fails to be unique if and only if $\{1, y_1, \ldots, y_K\}$ is a linearly dependent set. In this case, (2.76) will have multiple solutions $\{\alpha_k\}_{k=0}^{K}$, but all the solutions yield the same estimator.

It is critical in this result that the set of estimators form a subspace, but this does not mean that the estimator must be an affine function of the observed data. For example, the estimator of x from a single scalar y,

$$\widehat{x} = \beta_0 + \beta_1 y + \beta_2 y^2 + \cdots + \beta_K y^K, \tag{2.77}$$

fits the form of (2.74) with y_k set to y^k. Thus, assuming that x has a finite second moment and y has finite moments up to order $2K$, (2.76) can be used to optimize the estimator. Assuming that y is real, (2.76) simplifies to

$$\begin{bmatrix} 1 & E[y] & E[y^2] & \cdots & E[y^K] \\ E[y] & E[y^2] & E[y^3] & \cdots & E[y^{K+1}] \\ E[y^2] & E[y^3] & E[y^4] & \cdots & E[y^{K+2}] \\ \vdots & \vdots & \vdots & \ddots & \vdots \\ E[y^K] & E[y^{K+1}] & E[y^{K+2}] & \cdots & E[y^{2K}] \end{bmatrix} \begin{bmatrix} \beta_0 \\ \beta_1 \\ \beta_2 \\ \vdots \\ \beta_K \end{bmatrix} = \begin{bmatrix} E[x] \\ E[xy^1] \\ E[xy^2] \\ \vdots \\ E[xy^K] \end{bmatrix}. \tag{2.78}$$

EXAMPLE 2.27 (LINEAR MMSE ESTIMATORS) Suppose that the joint distribution of x and y is uniform over the region shaded in Figure 2.17. Since the area of the shaded region is $\frac{1}{3}$, the joint probability density function of x and y is

$$f_{x,y}(s,t) = \begin{cases} 3, & \text{for } s \in [0, 1] \text{ and } t \in [0, s^2]; \\ 0, & \text{otherwise.} \end{cases}$$

We wish to find estimators of x from y, namely

$$\widehat{x}_1 = \alpha_0 + \alpha_1 y$$

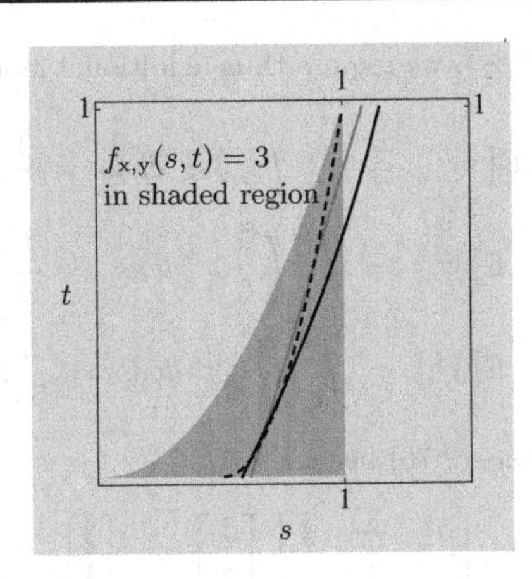

Figure 2.17 Minimum mean-squared error estimators of x from y from Examples 2.27 and 2.28. The joint distribution of x and y is uniform over the shaded region, the upper boundary of which is $t = s^2$. The optimal estimates of the form $\widehat{x}_1 = \alpha_0 + \alpha_1 y$ (solid gray line) and $\widehat{x}_2 = \beta_0 + \beta_1 y + \beta_2 y^2$ (solid line) are derived in Example 2.27. The optimal estimator is $\frac{1}{2}(1 + \sqrt{y})$ (dashed line) and is derived in Example 2.28.

and

$$\widehat{x}_2 = \beta_0 + \beta_1 y + \beta_2 y^2$$

that are optimal over the choices of coefficients $\{\alpha_0, \alpha_1\}$ and $\{\beta_0, \beta_1, \beta_2\}$.

To form the system of equations (2.76), we perform the following computations:

$$\begin{aligned}
E[x] &= \int_0^1 \int_0^{s^2} 3s\,dt\,ds = \frac{3}{4}, \\
E[y] &= \int_0^1 \int_0^{s^2} 3t\,dt\,ds = \frac{3}{10}, \\
E[xy] &= \int_0^1 \int_0^{s^2} 3st\,dt\,ds = \frac{1}{4}, \\
E[y^2] &= \int_0^1 \int_0^{s^2} 3t^2\,dt\,ds = \frac{1}{7}.
\end{aligned}$$

Then, $\{\alpha_0, \alpha_1\}$ is determined by solving

$$\begin{bmatrix} 1 & \frac{3}{10} \\ \frac{3}{10} & \frac{1}{7} \end{bmatrix} \begin{bmatrix} \alpha_0 \\ \alpha_1 \end{bmatrix} = \begin{bmatrix} \frac{3}{4} \\ \frac{1}{4} \end{bmatrix}$$

to obtain $\alpha_0 = 45/74$ and $\alpha_1 = 35/74$. The estimate $\widehat{x}_1$ is shown as a function of the observation $y = t$ in Figure 2.17 (solid gray line).

To find $\{\beta_0, \beta_1, \beta_2\}$, we require three additional moments:

$$\begin{aligned} \mathrm{E}[\,\mathrm{x}\mathrm{y}^2\,] &= \int_0^1 \int_0^{s^2} 3st^2\, dt\, ds = \frac{1}{8}, \\ \mathrm{E}[\,\mathrm{y}^3\,] &= \int_0^1 \int_0^{s^2} 3t^3\, dt\, ds = \frac{1}{12}, \\ \mathrm{E}[\,\mathrm{y}^4\,] &= \int_0^1 \int_0^{s^2} 3t^4\, dt\, ds = \frac{3}{55}. \end{aligned}$$

The system of equations (2.76) becomes

$$\begin{bmatrix} 1 & \frac{3}{10} & \frac{1}{7} \\ \frac{3}{10} & \frac{1}{7} & \frac{1}{12} \\ \frac{1}{7} & \frac{1}{12} & \frac{3}{55} \end{bmatrix} \begin{bmatrix} \beta_0 \\ \beta_1 \\ \beta_2 \end{bmatrix} = \begin{bmatrix} \frac{3}{4} \\ \frac{1}{4} \\ \frac{1}{8} \end{bmatrix},$$

which yields $\beta_0 = 12\,915/22\,558$, $\beta_1 = 17\,745/22\,558$, and $\beta_2 = -4\,620/11\,279$. The estimate $\widehat{\mathrm{x}}_2$ is shown as a function of the observation $\mathrm{y} = t$ in Figure 2.17 (solid line).

General optimal estimation The fact that estimators of the form (2.77) form a subspace hints at a more general fact: the set of *all* functions of a random variable forms a subspace in a vector space of random variables. While this might seem surprising or counterintuitive, verification of the properties required by Definition 2.2 is trivial. The subspace of functions of a random variable is furthermore closed, so several properties of general (not necessarily linear) MMSE estimation follow from the projection theorem.

As the simplest example, the constant c that minimizes $\mathrm{E}[\,(\mathrm{x}-c)^2\,]$ can be interpreted as the best estimator of x that does not depend on anything random. We must have

$$0 \overset{(a)}{=} \langle \mathrm{x} - c,\, 1\rangle \overset{(b)}{=} \mathrm{E}[\,\mathrm{x} - c\,] \overset{(c)}{=} \mathrm{E}[\,\mathrm{x}\,] - c,$$

where (a) follows from the orthogonality of the error $\mathrm{x} - c$ to the deterministic function 1; (b) from (2.37); and (c) from the linearity of the expectation. This derives the well-known fact that $c = \mathrm{E}[\,\mathrm{x}\,]$ is the constant that minimizes $\mathrm{E}[\,(\mathrm{x}-c)^2\,]$; see Appendix 2.C.3.

Now consider estimating x from observation $\mathrm{y} = t$. Conditioning on $\mathrm{y} = t$ yields a valid probability law, so

$$\langle \mathrm{x}, \mathrm{z}\rangle = \mathrm{E}[\,\mathrm{x}\mathrm{z}^* \,|\, \mathrm{y} = t\,] \tag{2.79}$$

is an inner product on the set of random variables with finite conditional second moment given $\mathrm{y} = t$. Orthogonality of the error $\mathrm{x} - \widehat{\mathrm{x}}_{\mathrm{MMSE}}(t)$ and any function of t under the inner product (2.79) yields

$$0 \overset{(a)}{=} \mathrm{E}[\,\mathrm{x} - \widehat{\mathrm{x}}_{\mathrm{MMSE}}(t) \,|\, \mathrm{y} = t\,] \overset{(b)}{=} \mathrm{E}[\,\mathrm{x} \,|\, \mathrm{y} = t\,] - \widehat{\mathrm{x}}_{\mathrm{MMSE}}(t),$$

where (a) follows from considering specifically the function 1; and (b) from $\widehat{x}_{\text{MMSE}}(t)$ being a (deterministic) function of t. Thus, the optimal estimate is the conditional mean:

$$\widehat{x}_{\text{MMSE}}(t) = \mathrm{E}[\,x \mid y = t\,], \tag{2.80a}$$

which is also written as

$$\widehat{x}_{\text{MMSE}} = \mathrm{E}[\,x \mid y\,]. \tag{2.80b}$$

EXAMPLE 2.28 (MMSE ESTIMATOR, EXAMPLE 2.27 CONTINUED) Consider x and y jointly distributed as in Example 2.27 (see Figure 2.17). Given an observation $y = t$, the conditional distribution of x is uniform on $[\sqrt{t}, 1]$. The mean of this conditional distribution gives the optimal estimator

$$\widehat{x}_{\text{MMSE}}(t) = \frac{1}{2}\left(1 + \sqrt{t}\right).$$

This optimal estimate is shown in Figure 2.17 (dashed line).

Orthogonality and optimal estimation of random vectors

Use of the inner product (2.37) has given us a geometric interpretation for *scalar* random variables with valuable ramifications for optimal estimation. One can define various inner products for random *vectors* as well. However, we will see that more useful estimation results come from generalizing the concept of orthogonality rather than from using a single inner product.

One valid inner product for complex random vectors of length N is obtained from the sum of inner products between components of the vectors, $\langle x_n, y_n \rangle$, using the inner product (2.37) between scalar random variables:

$$\langle x, y \rangle = \sum_{n=0}^{N-1} \mathrm{E}[\,x_n y_n^*\,].$$

This is identical to the expectation of the standard inner product on $\mathbb{C}^N$, (2.22a), or $\langle x, y \rangle = \mathrm{E}[\,y^* x\,]$.

With the projection theorem, one could optimize estimators of x from $y^{(1)}$, $y^{(2)}$, ..., $y^{(K)}$ of the form

$$\widehat{x} = \alpha_0 \mathbf{1} + \alpha_1 y^{(1)} + \alpha_2 y^{(2)} + \cdots + \alpha_K y^{(K)}, \tag{2.81}$$

where every component of $\mathbf{1} \in \mathbb{C}^N$ is 1. This is exactly as in (2.74), but now each element of $\{x, y^{(1)}, y^{(2)}, \ldots, y^{(K)}\}$ is a vector rather than a scalar.[19] Optimal coefficients are determined by solving a system of equations analogous to (2.76).

A weakness of the estimator $\widehat{x}$ in (2.81) is that any single component of $\widehat{x}$ depends only on the corresponding components of $\{y^{(1)}, y^{(2)}, \ldots, y^{(K)}\}$; other dependences are not exploited. To take a simple example, suppose that x has the

[19] Superscripts are introduced so we do not confuse indexing of the set with indexing of the components of a single vector.

uniform distribution over the unit square $[0, 1]^2$ and $\begin{bmatrix} y_1 & y_2 \end{bmatrix} = \begin{bmatrix} x_2 & x_1 \end{bmatrix}$. The vector x can be estimated perfectly from the vector y, but y_1 is useless in estimating x_1 and y_2 is useless in estimating x_2. Thus estimators more general than (2.81) are commonly used.

Linear estimation Let x be a $\mathbb{C}^N$-valued random vector, and let y be a $\mathbb{C}^M$-valued random vector. Suppose that all components of both vectors have finite second moments. Consider the estimator of x from y given by[20]

$$\widehat{x} = Ay, \tag{2.82}$$

where $A \in \mathbb{C}^{N \times M}$ is a constant matrix to be designed to minimize the MSE $E[\, \|x - \widehat{x}\|^2 \,]$. Note that, unlike in (2.81), every component of $\widehat{x}$ depends on every component of y.

Since each row of A determines a different component of $\widehat{x}$ and the MSE decouples across components as

$$E[\, \|x - \widehat{x}\|^2 \,] = \sum_{n=0}^{N-1} E[\, |x_n - \widehat{x}_n|^2 \,],$$

we can consider the design of each row of A separately. Then, for any fixed $n \in \{0, 1, \ldots, N-1\}$, the minimization of $E[\, |x_n - \widehat{x}_n|^2 \,]$ through the choice of the nth row of A is a problem we have already solved: it is the scalar linear MMSE estimation problem. The solution is characterized by orthogonality of the error and the data as in (2.75).

The orthogonality of the nth error component $x_n - \widehat{x}_n$ and the mth data component y_m can be expressed as

$$0 \overset{(a)}{=} E[\, (x_n - \widehat{x}_n) y_m^* \,] \overset{(b)}{=} E[\, (x_n - a_n^\top y) y_m^* \,],$$

where (a) follows from the inner product (2.37); and (b) introduces $a_n^\top$ as the nth row of A. Gathering these equations for $m = 0, 1, \ldots, M-1$ into one row gives

$$\mathbf{0}_{1 \times M} = E[\, (x_n - a_n^\top y) y^* \,], \qquad n = 0, 1, \ldots, N-1.$$

Now stacking these equations into a matrix gives

$$\mathbf{0}_{N \times M} = E[\, (x - Ay) y^* \,]. \tag{2.83}$$

Using linearity of expectation, a necessary and sufficient condition for optimality is thus

$$E[\, x y^* \,] = A\, E[\, y y^* \,]. \tag{2.84}$$

In most cases, $E[\, y y^* \,]$ is invertible; the optimal estimator is then

$$\widehat{x}_{\text{MMSE}} = E[\, x y^* \,] (E[\, y y^* \,])^{-1} y. \tag{2.85}$$

When $E[\, y y^* \,]$ is not invertible, solutions to (2.84) are not unique but yield the same estimator.

[20] In contrast to estimators (2.74) and (2.81), we have omitted a constant term from the estimator. There is no loss of generality because one can augment y with a constant random variable.

Orthogonality of random vectors Inspired by the usefulness of (2.83) in optimal estimation, we define a new orthogonality concept for random vectors.

DEFINITION 2.33 (ORTHOGONAL RANDOM VECTORS) Random vectors x and y are said to be *orthogonal* when $\mathrm{E}[\,\mathrm{x}\mathrm{y}^*\,] = \mathbf{0}$.

Note that $\mathrm{E}[\,\mathrm{x}\mathrm{y}^*\,]$ is not an inner product because it is not a scalar (except in the degenerate case where the random vectors have dimension 1 and are thus scalar random variables). Instead, random vectors x and y are orthogonal when every combination of components is orthogonal under the inner product (2.37):

$$\mathrm{E}[\,\mathrm{x}_n \mathrm{y}_m^*\,] \;=\; 0 \qquad \text{for every } m \text{ and } n. \tag{2.86}$$

In (2.83) we have an instance of a more general fact: Whenever an estimator $\widehat{\mathrm{x}}$ of random vector x is optimized over a closed subspace of possible estimators S, the optimal estimator will be determined by $\mathrm{x} - \widehat{\mathrm{x}} \perp S$ under the sense of orthogonality in Definition 2.33. We will apply this to optimal LMMSE estimation of discrete-time random processes in Section 3.8.5.

2.5 Bases and frames

The variety of bases and frames for sequences in $\ell^2(\mathbb{Z})$ and functions in $\mathcal{L}^2(\mathbb{R})$ is at the heart of this book and, to an even greater extent, its companion volume [57]. In this section, we develop general properties of bases and frames, with an emphasis on representing vectors in Hilbert spaces using bases. Bases will come in two flavors, orthonormal and biorthogonal; analogously, frames will come in two flavors, tight and general. In later chapters and in the companion volume [57], we will see that the choice of a basis or frame can have dramatic effects on the computational complexity, stability, and approximation accuracy of signal expansions.

Prominent in the developments of Section 2.4 were closed subspaces: We saw best approximation in a closed subspace, projection onto a closed subspace, and direct-sum decomposition into a pair of closed subspaces. In this section, we will see that a basis induces a direct-sum decomposition into possibly infinitely many one-dimensional subspaces; and that bases, especially orthonormal ones, facilitate the computations of projections and approximations. These developments reduce our level of abstraction and bring us closer to computational tools for signal processing. Specifically, Section 2.5.5 shows how representations with bases replace general vector space computations with matrix computations, albeit possibly infinite ones.

2.5.1 Bases and Riesz bases

In a finite-dimensional vector space, a basis is a linearly independent set of vectors that is used to uniquely represent any vector in that vector space as a linear combination of basis elements. For a definition to apply in infinite-dimensional spaces as well, the essential properties are existence and uniqueness of representations; linear independence is technically subtle, and we do not require it. In what follows, we consider any normed vector space, including any Hilbert space (where the norm is induced by an inner product).

DEFINITION 2.34 (BASIS) The set of vectors $\Phi = \{\varphi_k\}_{k\in\mathcal{K}} \subset V$, where $\mathcal{K}$ is finite or countably infinite, is called a *basis* for a normed vector space V when

(i) it is *complete* in V, meaning that, for any $x \in V$, there is a sequence $\alpha \in \mathbb{C}^{\mathcal{K}}$ such that

$$x = \sum_{k\in\mathcal{K}} \alpha_k \varphi_k; \tag{2.87}$$

and

(ii) for any $x \in V$, the sequence α that satisfies (2.87) is unique.

When $\mathcal{K}$ is infinite, the sum (2.87) lacks a specific meaning unless an order of summation is specified; for example, for $\mathcal{K} = \mathbb{N}$ taken in increasing order, (2.87) means that

$$\lim_{n\to\infty} \left\| x - \sum_{k=0}^{n} \alpha_k \varphi_k \right\| = 0.$$

Thus, $\mathcal{K}$ must have a fixed, singly infinite order, such as $0, -1, 1, -2, 2, \ldots$ for $\mathcal{K} = \mathbb{Z}$. The basis is called *unconditional* when the sum in (2.87) converges to x in one order of summation if and only if it converges to x in every order of summation. We dispense with the technicality of fixing an ordering of $\mathcal{K}$ because we almost exclusively limit our attention to Riesz bases (to be defined shortly), which are unconditional. Appendix 2.D explores other definitions of bases and their relationships to Definition 2.34.

When the index set $\mathcal{K}$ is finite, (2.87) shows that x is in the span of the basis. When $\mathcal{K}$ is infinite, the closure of the span is needed to allow infinitely many terms in the linear combination,[21]

$$V = \overline{\operatorname{span}}(\Phi), \tag{2.88}$$

which is equivalent to part (i) of Definition 2.34. The coefficients $(\alpha_k)_{k\in\mathcal{K}}$ are called the *expansion coefficients*[22] of x with respect to the basis Φ. We use a set notation (unordered) for bases and a sequence notation (ordered) for expansion coefficients, keeping in mind that the indexing of basis elements in (2.87) must maintain the correct correspondence between basis vectors and coefficients. The notation emphasizes that once an ordering of basis elements has been fixed, the ordering of the coefficients does matter.

[21]Recall that the span of a set of vectors is the set of all finite linear combinations of vectors from the set (Definition 2.4).

[22]Expansion coefficients are sometimes called *Fourier coefficients* or *generalized Fourier coefficients*, but we will avoid these terms except when the expansions are with respect to the specific bases that yield the various Fourier transforms. They are also called *transform coefficients* or *subband coefficients* in the source-coding literature.

Suppose that $\Phi = \{\varphi_k\}_{k=0}^{N-1}$ is a basis for V for some finite N. To show that Φ is linearly independent, we argue by contradiction. Suppose that Φ is not linearly independent. Then there exist $(\beta_k)_{k=0}^{N-1}$ not all equal to zero such that $\sum_{k=0}^{N-1} \beta_k \varphi_k = \mathbf{0}$. Since Φ is a basis, any $x \in V$ has an expansion (2.87), but this expansion is not unique because

$$\sum_{k=0}^{N-1} (\alpha_k + \beta_k)\varphi_k = \sum_{k=0}^{N-1} \alpha_k \varphi_k + \sum_{k=0}^{N-1} \beta_k \varphi_k = x + \mathbf{0} = x.$$

The lack of uniqueness contradicts Φ being a basis, so Φ must be linearly independent. Furthermore, the existence of an expansion (2.87) for any $x \in V$ implies that we cannot add another vector to Φ and have a linearly independent set. Therefore, V has dimension N. Putting these facts together, a finite set of linearly independent vectors is always a basis for its span.

EXAMPLE 2.29 (STANDARD BASIS FOR $\mathbb{C}^N$) The standard basis for $\mathbb{R}^2$ was introduced in Section 2.1. It is easily extended to $\mathbb{C}^N$ (or $\mathbb{R}^N$) with e_k as in Example 2.5. The set $\{e_k\}_{k=0}^{N-1}$ is both linearly independent (see Example 2.5) and complete in $\mathbb{C}^N$, and it is thus a basis for $\mathbb{C}^N$. For completeness, note that any vector $v = \begin{bmatrix} v_0 & v_1 & \cdots & v_{N-1} \end{bmatrix}^\top \in \mathbb{C}^N$ is precisely the finite linear combination $v = \sum_{k=0}^{N-1} v_k e_k$.

The infinite-dimensional Hilbert spaces that we consider have countably infinite bases because they are separable (see Section 2.3.2). For normed vector spaces that are not Hilbert spaces, the norm affects whether a particular set Φ is a basis.

EXAMPLE 2.30 (STANDARD BASIS FOR $\mathbb{C}^{\mathbb{Z}}$) The standard basis concept extends to some normed vector spaces of complex-valued sequences over $\mathbb{Z}$. Consider $E = \{e_k\}_{k \in \mathbb{Z}}$, where $e_k \in \mathbb{C}^{\mathbb{Z}}$ is the sequence that is 0 except for a 1 in the k-indexed position. If an expansion with respect to E exists, it is clearly unique, but whether such an expansion necessarily exists depends on the norm.

First consider a vector space $\ell^p(\mathbb{Z})$ with $p \in [1, \infty)$. The set E is complete for this vector space, meaning that $\ell^p(\mathbb{Z}) = \overline{\operatorname{span}}(E)$ when the closure of the span is defined with the ℓ^p norm. To show this, we establish $\overline{\operatorname{span}}(E) \subseteq \ell^p(\mathbb{Z})$ and $\ell^p(\mathbb{Z}) \subseteq \overline{\operatorname{span}}(E)$. The first inclusion, $\overline{\operatorname{span}}(E) \subseteq \ell^p(\mathbb{Z})$, holds because $\ell^p(\mathbb{Z})$ is a complete vector space; see Section 2.3.2. It remains to show that $\ell^p(\mathbb{Z}) \subseteq \overline{\operatorname{span}}(E)$ holds. We do this by showing that for an arbitrary $x \in \ell^p(\mathbb{Z})$, there exists a sequence of vectors in $\operatorname{span}(E)$ that converges (under the ℓ^p norm) to x. For each $M \in \mathbb{N}$, let $y_M = \sum_{n=-M}^{M} x_n e_n$. Then

$$\lim_{M\to\infty} \|x - y_M\|_p^p = \lim_{M\to\infty} \left(\sum_{n=-\infty}^{-M-1} |x_n|^p + \sum_{n=M+1}^{\infty} |x_n|^p \right) \stackrel{(a)}{=} 0,$$

where (a) holds because the meaning of $x \in \ell^p(\mathbb{Z})$ is that $\sum_{n\in\mathbb{Z}} |x_n|^p$ is convergent. Since each y_M is in $\operatorname{span}(E)$, the limit of the sequence x is in $\overline{\operatorname{span}}(E)$. Thus $\ell^p(\mathbb{Z}) \subseteq \overline{\operatorname{span}}(E)$.

Changing the norm changes the meaning of the closure of the span and thus can make E not be a basis. The set E is not complete under the ℓ^∞ norm since $\ell^\infty(\mathbb{Z}) \not\subseteq \overline{\operatorname{span}}(E)$. To show this, let x be the all-1s sequence, which is in $\ell^\infty(\mathbb{Z})$. Every sequence in $\operatorname{span}(E)$ has a finite number of nonzero entries and is therefore at least distance 1 from x under the ℓ^∞ norm. Therefore x is not in $\overline{\operatorname{span}}(E)$.

Riesz bases While the previous example provides an important note of caution on the meaning of a basis, dependence on the choice of norm is not our focus. In fact, we will primarily focus on the ℓ^2 and $\mathcal{L}^2$ Hilbert space norms. The more important complication with infinite-dimensional spaces is that a basis can be prohibitively ill suited to numerical computations. Specifically, it is not practical to allow coefficients in a linear combination to be unbounded or to require very small coefficients to be distinguished from zero; the concept of a Riesz basis restricts bases to avoid these pitfalls.

DEFINITION 2.35 (RIESZ BASIS) The set of vectors $\Phi = \{\varphi_k\}_{k\in\mathcal{K}} \subset H$, where $\mathcal{K}$ is finite or countably infinite, is called a *Riesz basis* for a Hilbert space H when

(i) it is a *basis* for H; and
(ii) there exist *stability constants* $\lambda_{\min}$ and $\lambda_{\max}$ satisfying $0 < \lambda_{\min} \leq \lambda_{\max} < \infty$ such that, for any x in H, the expansion of x with respect to the basis Φ, $x = \sum_{k\in\mathcal{K}} \alpha_k \varphi_k$, satisfies

$$\lambda_{\min}\|x\|^2 \leq \sum_{k\in\mathcal{K}} |\alpha_k|^2 \leq \lambda_{\max}\|x\|^2. \tag{2.89}$$

The largest such $\lambda_{\min}$ and smallest such $\lambda_{\max}$ are called *optimal stability constants* for Φ.

In $\mathbb{C}^N$ or $\ell^2(\mathbb{Z})$, the standard basis is a Riesz basis with $\lambda_{\min} = \lambda_{\max} = 1$ (see Exercise 2.33). Conversely, Riesz bases with $\lambda_{\min} = \lambda_{\max} = 1$ are orthonormal bases, as developed in Section 2.5.2. As we introduce a variety of bases for different purposes, it will be a virtue to have $\lambda_{\min} \approx \lambda_{\max}$, though we may relax this requirement to achieve other objectives.

EXAMPLE 2.31 (RIESZ BASES IN $\mathbb{R}^2$) Any two vectors φ_0 and φ_1 are a basis for $\mathbb{R}^2$ as long as there is no scalar α such that $\varphi_1 = \alpha\varphi_0$. We fix $\varphi_0 = e_0$ and vary φ_1 in two ways to illustrate deviations from the standard basis.

(i) Let $\varphi_1 = a e_1$ with $a \in (0, \infty)$, as illustrated in Figure 2.18(a). The unique expansion of $\begin{bmatrix} x_0 & x_1 \end{bmatrix}^\top$ is then

$$x = x_0\varphi_0 + (x_1/a)\varphi_1 = \alpha_0\varphi_0 + \alpha_1\varphi_1.$$

The largest $\lambda_{\min}$ such that (2.89) holds is

$$\lambda_{\min} = \inf_{x\in\mathbb{R}^2} \frac{x_0^2 + (x_1/a)^2}{x_0^2 + x_1^2} = \begin{cases} 1, & \text{for } a \in (0, 1]; \\ 1/a^2, & \text{for } a \in (1, \infty). \end{cases}$$

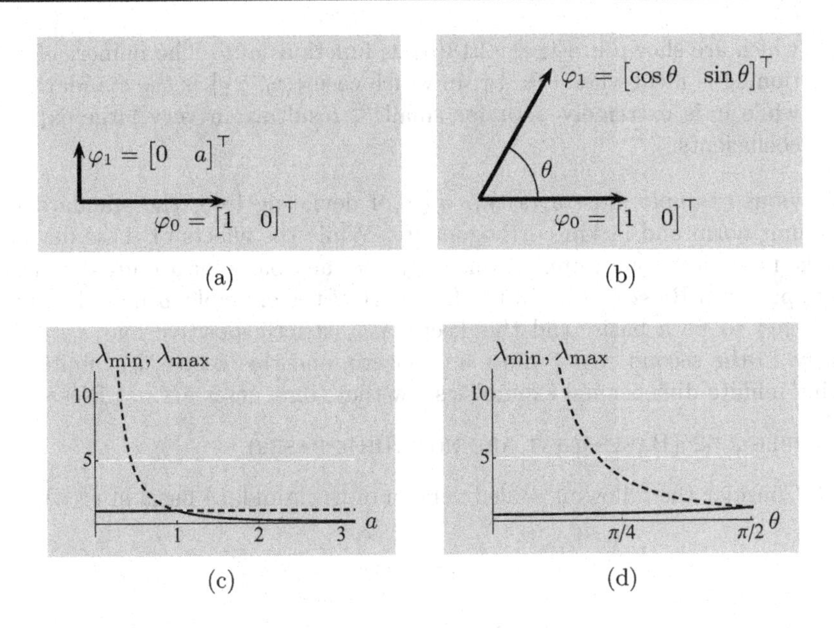

Figure 2.18 Two families of bases in $\mathbb{R}^2$ that deviate from the standard basis $\{e_0, e_1\}$ and their optimal stability constants λ_{min} (solid lines) and λ_{max} (dashed lines). (a) φ_1 is orthogonal to φ_0 but not necessarily of unit length. (b) φ_1 is of unit length but not necessarily orthogonal to φ_0. (c) λ_{min} and λ_{max} for the basis in (a) as a function of a. (d) λ_{min} and λ_{max} for the basis in (b) as a function of θ.

This means that, when a is very large, the basis becomes numerically ill conditioned in the sense that there are nonzero vectors x with very small expansion coefficients.

Similarly, the smallest λ_{max} such that (2.89) holds is

$$\lambda_{max} = \sup_{x \in \mathbb{R}^2} \frac{x_0^2 + (x_1/a)^2}{x_0^2 + x_1^2} = \begin{cases} 1/a^2, & \text{for } a \in (0,\, 1]; \\ 1, & \text{for } a \in (1,\, \infty). \end{cases}$$

This means that, when a is close to zero, the basis becomes numerically ill conditioned in the sense that there are vectors x with very large expansion coefficients. Figure 2.18(c) shows λ_{min} and λ_{max} as functions of a.

(ii) Let $\varphi_1 = \begin{bmatrix} \cos\theta & \sin\theta \end{bmatrix}^\top$ with $\theta \in (0,\, \frac{1}{2}\pi]$, as illustrated in Figure 2.18(b). The unique expansion of $\begin{bmatrix} x_0 & x_1 \end{bmatrix}^\top$ is then

$$x = (x_0 - \cot\theta\, x_1)\varphi_0 + (\csc\theta\, x_1)\varphi_1 = \alpha_0\varphi_0 + \alpha_1\varphi_1.$$

Using trigonometric identities, one can show that the largest λ_{min} and smallest λ_{max} such that (2.89) holds are

$$\lambda_{min} = \frac{1}{2}\sec^2\left(\frac{\theta}{2}\right) \quad \text{and} \quad \lambda_{max} = \frac{1}{2}\csc^2\left(\frac{\theta}{2}\right),$$

which are shown in Figure 2.18(d) as functions of θ. The numerical conditioning is ideal when $\theta = \frac{1}{2}\pi$, in which case $\{\varphi_0, \varphi_1\}$ is the standard basis, while it is extremely poor for small θ, resulting in very large expansion coefficients.

The previous example illustrates two ways of deviating from the standard basis: lacking unit norm and lacking orthogonality. While the effects of these deviations can make numerical conditioning arbitrarily bad, any basis for a finite-dimensional Hilbert space is a Riesz basis. (In the first part of the example, a must be nonzero for $\{\varphi_0, \varphi_1\}$ to be a basis, and this keeps $\lambda_{\min}$ strictly positive and $\lambda_{\max}$ finite. Similarly, in the second part θ must be nonzero, and this keeps $\lambda_{\max}$ finite.) The following infinite-dimensional examples show that some bases are not Riesz bases.

EXAMPLE 2.32 (BASES THAT ARE NOT RIESZ BASES)

(i) Consider the following scaled version of the standard basis in $\ell^2(\mathbb{Z})$:

$$\varphi_k = \frac{1}{|k|+1} e_k, \qquad k \in \mathbb{Z}.$$

As $|k| \to \infty$, the ratio of lengths of elements, $\|\varphi_0\|/\|\varphi_k\|$, is unbounded, similar to $\|\varphi_0\|/\|\varphi_1\|$ with $a \to 0$ in Example 2.31(i). The set $\Phi = \{\varphi_k\}_{k\in\mathbb{Z}}$ is a basis for $\ell^2(\mathbb{Z})$, but it is not a Riesz basis.

To prove that Φ is not a Riesz basis, we show that no finite $\lambda_{\max}$ satisfies (2.89). Suppose that there is a finite $\lambda_{\max}$ such that (2.89) holds for all $x \in \ell^2(\mathbb{Z})$. Then, we can choose an integer $M > \sqrt{\lambda_{\max}}$ and let $x \in \ell^2(\mathbb{Z})$ be the sequence that is 0 except for a 1 in the M-indexed position. The unique representation of this x using the basis Φ is $x = (|M|+1)\varphi_M$; that is, the coefficients in the expansion are

$$\alpha_k = \begin{cases} |M|+1, & \text{for } k = M; \\ 0, & \text{otherwise.} \end{cases}$$

The second inequality of (2.89) is contradicted, so the desired finite $\lambda_{\max}$ does not exist.

(ii) Consider the following vectors defined using the standard basis in $\ell^2(\mathbb{N})$:

$$\varphi_k = \sum_{i=0}^{k} \frac{1}{\sqrt{i+1}} e_i, \qquad k \in \mathbb{N}.$$

The angle between consecutive elements approaches zero as $k \to \infty$ because $\langle \varphi_k, \varphi_{k+1}\rangle / (\|\varphi_k\| \, \|\varphi_{k+1}\|) \to 1$, so this is intuitively similar to letting $\theta \to 0$ in Example 2.31(ii). Proving that the set $\Phi = \{\varphi_k\}_{k\in\mathbb{N}}$ is a basis for $\ell^2(\mathbb{N})$ but is not a Riesz basis is left for Exercise 2.34.

In the subsequent developments, it will be desirable for all bases to be Riesz bases. In particular, if $\{\varphi_k\}_{k\in\mathcal{K}}$ is a Riesz basis for a Hilbert space and the coefficient sequence α is in $\ell^2(\mathcal{K})$, then $\sum_{k\in\mathcal{K}} \alpha_k \varphi_k$ converges unconditionally, meaning that it converges for any ordering of the terms of the sum. We implicitly use this somewhat technical fact when we write sums over $\mathcal{K}$ without specifying the order of the terms. Where a sum is written without specifying the order of the terms and convergence would not otherwise be apparent, we assume a basis to be a Riesz basis.

Operators associated with bases Given a Riesz basis $\{\varphi_k\}_{k\in\mathcal{K}}$ satisfying (2.89), the expansion formula (2.87) can be viewed as mapping a coefficient sequence α to a vector x. This mapping is clearly linear. Let us suppose that the coefficient sequence has finite $\ell^2(\mathcal{K})$ norm. The first inequality of (2.89) implies that the vector x given by (2.87) has finite norm, at most $\|\alpha\|/\sqrt{\lambda_{\min}}$, and is thus legitimately in the Hilbert space H.

DEFINITION 2.36 (BASIS SYNTHESIS OPERATOR) Given a Riesz basis $\{\varphi_k\}_{k\in\mathcal{K}}$ for a Hilbert space H, the *synthesis operator* associated with it is

$$\Phi : \ell^2(\mathcal{K}) \to H, \qquad \text{with} \qquad \Phi\alpha = \sum_{k\in\mathcal{K}} \alpha_k \varphi_k. \tag{2.90}$$

The norm of the synthesis operator is the supremum of the ratio $\|\Phi\alpha\|/\|\alpha\|$ for nonzero $\alpha \in \ell^2(\mathcal{K})$. The first inequality of (2.89) implies that $\lambda_{\min}\|\Phi\alpha\|^2 \leq \|\alpha\|^2$, so, by rearranging the inequality and taking the square root, the norm of this linear operator is at most $1/\sqrt{\lambda_{\min}}$; the operator Φ is thus not only linear but bounded as well.

The adjoint of Φ maps from H to a sequence in $\ell^2(\mathcal{K})$. To derive the adjoint, consider the following computation for arbitrary $\alpha \in \ell^2(\mathcal{K})$ and $y \in H$:

$$\langle \Phi\alpha, y\rangle \overset{(a)}{=} \left\langle \sum_{k\in\mathcal{K}} \alpha_k\varphi_k, y\right\rangle \overset{(b)}{=} \sum_{k\in\mathcal{K}} \alpha_k \langle \varphi_k, y\rangle \overset{(c)}{=} \sum_{k\in\mathcal{K}} \alpha_k \langle y, \varphi_k\rangle^*,$$

where (a) follows from (2.90); (b) from the linearity in the first argument of the inner product; and (c) from the Hermitian symmetry of the inner product. The final expression is the $\ell^2(\mathcal{K})$ inner product between α and a sequence of inner products $(\langle y, \varphi_k\rangle)_{k\in\mathcal{K}}$. The adjoint is called the analysis operator.

DEFINITION 2.37 (BASIS ANALYSIS OPERATOR) Given a Riesz basis $\{\varphi_k\}_{k\in\mathcal{K}}$ for a Hilbert space H, the *analysis operator* associated with it is

$$\Phi^* : H \to \ell^2(\mathcal{K}), \qquad \text{with} \qquad (\Phi^* x)_k = \langle x, \varphi_k\rangle, \quad k \in \mathcal{K}. \tag{2.91}$$

Equation (2.91) holds since $\langle \Phi\alpha, y\rangle_H = \langle \alpha, \Phi^* y\rangle_{\ell^2}$. The norm of the analysis operator is also at most $1/\sqrt{\lambda_{\min}}$ because $\|A\| = \|A^*\|$ for all bounded linear operators A.

In Definitions 2.36 and 2.37, the order in which the elements of the index set $\mathcal{K}$ are used implicitly affects the operators. Usually, $\mathcal{K}$ will be the set of integers $\mathbb{Z}$ or a finite subset of $\mathbb{Z}$, implying a natural ordering of $\mathcal{K}$.

2.5.2 Orthonormal bases

An *orthonormal basis* is a basis of orthogonal, unit-norm vectors.

DEFINITION 2.38 (ORTHONORMAL BASIS) The set of vectors $\Phi = \{\varphi_k\}_{k\in\mathcal{K}} \subset H$, where $\mathcal{K}$ is finite or countably infinite, is called an *orthonormal basis* for a Hilbert space H when

(i) it is a *basis* for H; and
(ii) it is *orthonormal*,

$$\langle \varphi_i, \varphi_k\rangle = \delta_{i-k} \qquad \text{for every } i,\, k \in \mathcal{K}. \tag{2.92}$$

Since orthonormality implies uniqueness of expansions, we could alternatively say that a set $\Phi = \{\varphi_k\}_{k\in\mathcal{K}} \subset H$ satisfying (2.92) is an orthonormal basis whenever it is complete, that is, $\overline{\text{span}}(\Phi) = H$.

Standard bases are orthonormal bases. Two more examples follow, and we will see many more examples throughout the book.

EXAMPLE 2.33 (FINITE-DIMENSIONAL ORTHONORMAL BASIS) The vectors

$$\varphi_0 = \frac{1}{\sqrt{2}}\begin{bmatrix}1\\1\\0\end{bmatrix}, \quad \varphi_1 = \frac{1}{\sqrt{6}}\begin{bmatrix}-1\\1\\2\end{bmatrix}, \quad \text{and} \quad \varphi_2 = \frac{1}{\sqrt{3}}\begin{bmatrix}1\\-1\\1\end{bmatrix}$$

are orthonormal, as can be verified by direct computation. Since three linearly independent vectors in $\mathbb{C}^3$ always form a basis for $\mathbb{C}^3$, $\{\varphi_0, \varphi_1, \varphi_2\}$ is an orthonormal basis for $\mathbb{C}^3$.

EXAMPLE 2.34 (ORTHONORMAL BASIS OF COSINE FUNCTIONS) Consider $\Phi = \{\varphi_k\}_{k\in\mathbb{N}} \subset \mathcal{L}^2([-\frac{1}{2}, \frac{1}{2}])$ defined in (2.24). The first three functions in this set were shown in Figure 2.5. Example 2.9(iii) showed that Φ satisfies the orthonormality condition (2.92). Since orthonormality also implies uniqueness of expansions, Φ is an orthonormal basis for $S = \overline{\text{span}}(\Phi)$. (Remember that S is itself a Hilbert space.) The set Φ is not, however, an orthonormal basis for $\mathcal{L}^2([-\frac{1}{2}, \frac{1}{2}])$ because S is a proper subspace of $\mathcal{L}^2([-\frac{1}{2}, \frac{1}{2}])$; for example, no odd functions are in S.

Expansion and inner product computation Expansion coefficients with respect to an orthonormal basis are obtained by using the same basis for signal analysis.

THEOREM 2.39 (ORTHONORMAL BASIS EXPANSIONS) Let $\Phi = \{\varphi_k\}_{k\in\mathcal{K}}$ be an orthonormal basis for a Hilbert space H. The unique expansion with respect to Φ of any x in H has expansion coefficients

$$\alpha_k = \langle x, \varphi_k \rangle \qquad \text{for } k \in \mathcal{K}, \quad \text{or,} \tag{2.93a}$$

$$\alpha = \Phi^* x. \tag{2.93b}$$

Synthesis with these coefficients yields

$$x = \sum_{k\in\mathcal{K}} \langle x, \varphi_k \rangle \varphi_k \tag{2.94a}$$

$$= \Phi\alpha = \Phi\Phi^* x. \tag{2.94b}$$

Proof. The existence of a unique linear combination of the form (2.87) is guaranteed by Φ being a basis. The validity of (2.94a) with coefficients (2.93a) follows from the computation

$$\langle x, \varphi_k \rangle \overset{(a)}{=} \left\langle \sum_{i\in\mathcal{K}} \alpha_i \varphi_i, \varphi_k \right\rangle \overset{(b)}{=} \sum_{i\in\mathcal{K}} \alpha_i \langle \varphi_i, \varphi_k \rangle \overset{(c)}{=} \sum_{i\in\mathcal{K}} \alpha_i \delta_{i-k} \overset{(d)}{=} \alpha_k,$$

where (a) follows from (2.87); (b) from the linearity in the first argument of the inner product; (c) from the orthonormality of the set Φ, (2.92); and (d) from the definition of the Kronecker delta sequence, (2.9).

The expressions (2.93b) and (2.94b) are equivalent to (2.93a) and (2.94a) using the operators defined in (2.90) and (2.91).

Since (2.94b) holds for all x in H,

$$\Phi\Phi^* = I \quad \text{on} \quad H. \tag{2.95}$$

This leads to the frequently used properties in the following theorem.

THEOREM 2.40 (PARSEVAL EQUALITIES) Let $\Phi = \{\varphi_k\}_{k\in\mathcal{K}}$ be an orthonormal basis for a Hilbert space H. Expansion with coefficients (2.93) satisfies the *Parseval equality*,

$$\|x\|^2 = \sum_{k\in\mathcal{K}} |\langle x, \varphi_k \rangle|^2 \tag{2.96a}$$

$$= \|\Phi^* x\|^2 = \|\alpha\|^2, \tag{2.96b}$$

and the *generalized Parseval equality*,

$$\langle x, y \rangle = \sum_{k\in\mathcal{K}} \langle x, \varphi_k \rangle \langle y, \varphi_k \rangle^* \tag{2.97a}$$

$$= \langle \Phi^* x, \Phi^* y \rangle = \langle \alpha, \beta \rangle. \tag{2.97b}$$

Proof. Recall the equivalence of (2.55) and (2.57a). Thus (2.95) is equivalent to (2.97b). Setting $x = y$ in (2.97b) yields (2.96b). Equalities (2.96a) and (2.97a) are the same facts expanded with the definition of Φ^*.

Proving this using operator notation and properties in (2.95) is much less tedious than the direct proof. To see this for the example of (2.97), write

$$\langle x,\, y\rangle \overset{(a)}{=} \left\langle \sum_{k\in\mathcal{K}} \langle x,\, \varphi_k\rangle\varphi_k,\, y \right\rangle \overset{(b)}{=} \sum_{k\in\mathcal{K}} \langle x,\, \varphi_k\rangle\langle\varphi_k,\, y\rangle \overset{(c)}{=} \sum_{k\in\mathcal{K}} \langle x,\, \varphi_k\rangle\langle y,\, \varphi_k\rangle^*,$$

where (a) follows from expanding x with (2.94a); (b) from the linearity in the first argument of the inner product; and (c) from the Hermitian symmetry of the inner product.

The simple equality (2.97) captures an important role played by any orthonormal basis: it turns an abstract inner product computation into a computation with sequences. When $x = \sum_{k\in\mathcal{K}} \alpha_k\varphi_k$ and $y = \sum_{k\in\mathcal{K}} \beta_k\varphi_k$ as in the theorem,

$$\langle x,\, y\rangle_H \;=\; \langle \alpha,\, \beta\rangle_{\ell^2(\mathcal{K})} \;=\; \sum_{k\in\mathcal{K}} \alpha_k\beta_k^*, \tag{2.98}$$

where the final computation is an $\ell^2(\mathcal{K})$ inner product even though the first inner product is in an arbitrary Hilbert space. We might view this more concretely with matrix multiplication as

$$\langle x,\, y\rangle \;=\; \beta^*\alpha, \tag{2.99}$$

where α and β are column vectors.

Example 2.35 (Inner product computation by expansion sequences) Let α and β be sequences in $\ell^2(\mathbb{N})$. Then, the functions

$$x(t) \;=\; \alpha_0 + \sum_{k=1}^{\infty} \alpha_k\sqrt{2}\cos(2\pi kt), \tag{2.100a}$$

$$y(t) \;=\; \beta_0 + \sum_{k=1}^{\infty} \beta_k\sqrt{2}\cos(2\pi kt), \tag{2.100b}$$

are in $\mathcal{L}^2([-\frac{1}{2},\, \frac{1}{2}])$, and their inner product can be simplified as follows:

$$\langle x,\, y\rangle \;=\; \int_{-1/2}^{1/2} \left(\alpha_0 + \sum_{k=1}^{\infty} \alpha_k\sqrt{2}\cos(2\pi kt)\right)\left(\beta_0^* + \sum_{\ell=1}^{\infty} \beta_\ell^*\sqrt{2}\cos(2\pi\ell t)\right) dt$$

$$= \alpha_0\beta_0^* + \alpha_0 \sum_{\ell=1}^{\infty} \beta_\ell^* \underbrace{\int_{-1/2}^{1/2} \sqrt{2}\cos(2\pi\ell t)\, dt}_{=0}$$

$$+ \beta_0^* \sum_{k=1}^{\infty} \alpha_k \underbrace{\int_{-1/2}^{1/2} \sqrt{2}\cos(2\pi k t)\, dt}_{=0}$$

$$+ \sum_{k=1}^{\infty}\sum_{\ell=1}^{\infty} \alpha_k\beta_\ell^* \underbrace{\int_{-1/2}^{1/2} 2\cos(2\pi k t)\cos(2\pi\ell t)\, dt}_{=\delta_{k-\ell}}$$

$$= \alpha_0\beta_0^* + \sum_{k=1}^{\infty}\sum_{\ell=1}^{\infty} \alpha_k\beta_\ell^*\,\delta_{k-\ell} \stackrel{(a)}{=} \alpha_0\beta_0^* + \sum_{k=1}^{\infty}\alpha_k\beta_k^* = \sum_{k=0}^{\infty}\alpha_k\beta_k^*$$

$$\stackrel{(b)}{=} \langle \alpha, \beta \rangle,$$

where (a) follows from the definition of the Kronecker delta sequence, (2.9); and (b) from the definition of the ℓ^2 inner product between sequences.

Recalling the orthonormal basis from Example 2.34, the computation above is an explicit verification of (2.97b). We have shown that a more complicated (integral) inner product in $\mathcal{L}^2(\mathbb{R})$ can be computed via a simpler (series) inner product between expansion coefficients in $\ell^2(\mathbb{Z})$.

Unitary synthesis and analysis We will show that

$$\Phi^*\Phi = I \qquad \text{on} \quad \ell^2(\mathcal{K}). \tag{2.101}$$

Combined with (2.95), this establishes that the analysis and synthesis operators associated with an orthonormal basis are unitary.

To verify (2.101), do the following computation for any sequence α in $\ell^2(\mathcal{K})$:

$$\Phi^*\Phi\alpha \stackrel{(a)}{=} \Phi^*\sum_{i\in\mathcal{K}}\alpha_i\varphi_i \stackrel{(b)}{=} \left(\langle \textstyle\sum_{i\in\mathcal{K}}\alpha_i\varphi_i, \varphi_k\rangle\right)_{k\in\mathcal{K}} \stackrel{(c)}{=} \left(\textstyle\sum_{i\in\mathcal{K}}\alpha_i\langle\varphi_i, \varphi_k\rangle\right)_{k\in\mathcal{K}}$$

$$\stackrel{(d)}{=} \left(\textstyle\sum_{i\in\mathcal{K}}\alpha_i\delta_{i-k}\right)_{k\in\mathcal{K}} \stackrel{(e)}{=} (\alpha_k)_{k\in\mathcal{K}} = \alpha, \tag{2.102}$$

where (a) follows from (2.90); (b) from (2.91); (c) from the linearity in the first argument of the inner product; (d) from the orthonormality of the set $\{\varphi_k\}_{k\in\mathcal{K}}$, (2.92); and (e) from the definition of the Kronecker delta sequence, (2.9).

Isometry of separable Hilbert spaces and $\ell^2(\mathcal{K})$ The fact that the synthesis and analysis operators Φ and Φ^* associated with an orthonormal basis are unitary leads to key intuitions about separable Hilbert spaces. A unitary operator between Hilbert spaces puts Hilbert spaces in one-to-one correspondence while preserving the geometries (that is, inner products) in the spaces. Since Hilbert spaces that we consider are separable, they contain orthonormal bases. Therefore, these

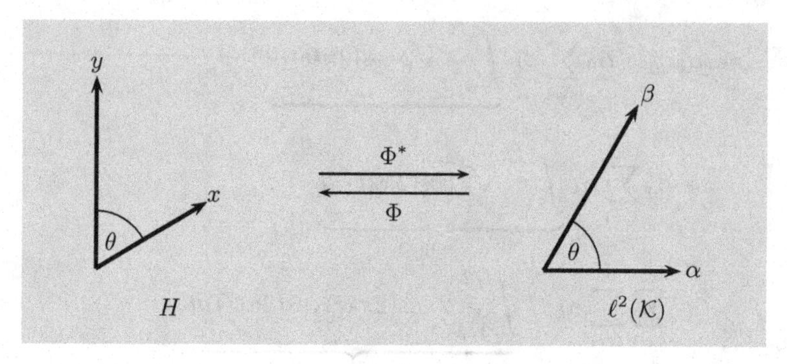

Figure 2.19 Conceptual illustration of the isometry between a separable Hilbert space H and sequence space $\ell^2(\mathcal{K})$ induced by an orthonormal basis Φ. It preserves geometry as $\langle x, y\rangle = \langle \alpha, \beta\rangle$.

Hilbert spaces can all be put in one-to-one correspondence with $\mathbb{C}^N$ if they are finite-dimensional or with $\ell^2(\mathbb{Z})$ if they are infinite-dimensional, as illustrated in Figure 2.19. (The notation $\ell^2(\mathcal{K})$, with $\mathcal{K}$ finite or countably infinite, unifies the two cases.)

Orthogonal projection Truncating the orthonormal expansion (2.94a) to $k \in \mathcal{I}$, where $\mathcal{I}$ is an index set that is a subset of the full index set $\mathcal{K}$, $\mathcal{I} \subset \mathcal{K}$, gives an orthogonal projection. As in the alternate proof of Theorem 2.40, this can be verified through somewhat tedious manipulations of sums and inner products; it is simpler to extend the definitions of the synthesis and analysis operators to apply for $\mathcal{I} \subset \mathcal{K}$ and then use these new operators.

Define the *synthesis operator* associated with $\{\varphi_k\}_{k\in\mathcal{I}}$ as

$$\Phi_{\mathcal{I}} : \ell^2(\mathcal{I}) \to H, \qquad \text{with} \qquad \Phi_{\mathcal{I}}\,\alpha \;=\; \sum_{k\in\mathcal{I}} \alpha_k \varphi_k. \tag{2.103}$$

This follows the form of (2.90) exactly, but the subscript $\mathcal{I}$ emphasizes that $\{\varphi_k\}_{k\in\mathcal{I}}$ need not be a basis. The adjoint of $\Phi_{\mathcal{I}}$, called the *analysis operator* associated with $\{\varphi_k\}_{k\in\mathcal{I}}$, is

$$\Phi_{\mathcal{I}}^* : H \to \ell^2(\mathcal{I}), \qquad \text{with} \qquad (\Phi_{\mathcal{I}}^*\, x)_k \;=\; \langle x, \varphi_k\rangle, \quad k \in \mathcal{I}. \tag{2.104}$$

Following the same steps as in (2.102), the orthonormality of the set $\{\varphi_k\}_{k\in\mathcal{I}}$ is equivalent to

$$\Phi_{\mathcal{I}}^*\, \Phi_{\mathcal{I}} \;=\; I \qquad \text{on} \quad \ell^2(\mathcal{I}). \tag{2.105}$$

However, we cannot conclude that $\Phi_{\mathcal{I}}$ and $\Phi_{\mathcal{I}}^*$ are unitary because the product $\Phi_{\mathcal{I}}\,\Phi_{\mathcal{I}}^*$ is not, in general, the identity operator on H; $\Phi_{\mathcal{I}}$ not being a basis, it cannot reconstruct every $x \in H$. Instead, $\Phi_{\mathcal{I}}\,\Phi_{\mathcal{I}}^*$ is an orthogonal projection operator that is an identity only when $\mathcal{I} = \mathcal{K}$, that is, when $\{\varphi_k\}_{k\in\mathcal{I}}$ is a basis. This is formalized in the following theorem.

THEOREM 2.41 (ORTHOGONAL PROJECTION ONTO A SUBSPACE) Let $\{\varphi_k\}_{k\in\mathcal{I}}$ be an orthonormal set in a Hilbert space H. Then for any $x \in H$,

$$P_{\mathcal{I}}\, x = \sum_{k\in\mathcal{I}} \langle x, \varphi_k\rangle \varphi_k \tag{2.106a}$$

$$= \Phi_{\mathcal{I}}\, \Phi_{\mathcal{I}}^*\, x \tag{2.106b}$$

is the orthogonal projection of x onto $S_{\mathcal{I}} = \overline{\operatorname{span}}(\{\varphi_k\}_{k\in\mathcal{I}})$.

Proof. From its definition (2.106a), $P_{\mathcal{I}}$ is clearly a linear operator on H with range contained in $S_{\mathcal{I}}$. To prove that $P_{\mathcal{I}}$ is an orthogonal projection operator, we show that it is idempotent and self-adjoint (see Definition 2.27).

The operator $P_{\mathcal{I}}$ is idempotent because, for any $x \in H$,

$$\begin{aligned} P_{\mathcal{I}}(P_{\mathcal{I}}\, x) &\overset{(a)}{=} \Phi_{\mathcal{I}}\, \Phi_{\mathcal{I}}^*\, (\Phi_{\mathcal{I}}\, \Phi_{\mathcal{I}}^*\, x) \overset{(b)}{=} \Phi_{\mathcal{I}}\, (\Phi_{\mathcal{I}}^*\, \Phi_{\mathcal{I}}) \Phi_{\mathcal{I}}^*\, x \\ &\overset{(c)}{=} \Phi_{\mathcal{I}}\, \Phi_{\mathcal{I}}^*\, x \overset{(d)}{=} P_{\mathcal{I}}\, x, \end{aligned}$$

where (a) follows from (2.106b); (b) from the associativity of linear operators; (c) from (2.105); and (d) from (2.106b). This shows that $P_{\mathcal{I}}$ is a projection operator. The operator $P_{\mathcal{I}}$ is self-adjoint because

$$P_{\mathcal{I}}^* = (\Phi_{\mathcal{I}}\, \Phi_{\mathcal{I}}^*)^* \overset{(a)}{=} (\Phi_{\mathcal{I}}^*)^* \Phi_{\mathcal{I}}^* \overset{(b)}{=} \Phi_{\mathcal{I}}\, \Phi_{\mathcal{I}}^* = P_{\mathcal{I}},$$

where (a) follows from Theorem 2.21(viii); and (b) from Theorem 2.21(iii). Combined with the previous computation, this shows that $P_{\mathcal{I}}$ is an orthogonal projection operator.

The previous theorem can be used to simplify the computation of an orthogonal projection, provided that $\{\varphi_k\}_{k\in\mathcal{I}}$ is an orthonormal basis for the subspace of interest.

EXAMPLE 2.36 (ORTHOGONAL PROJECTION WITH ORTHONORMAL BASIS) Consider the orthonormal basis for $\mathbb{C}^3$ from Example 2.33. The two-dimensional subspace

$$S = \left\{ \begin{bmatrix} x_0 & x_1 & x_2 \end{bmatrix}^\top \in \mathbb{C}^3 \,\middle|\, x_1 = x_0 + x_2 \right\}$$

is $\operatorname{span}(\{\varphi_0, \varphi_1\})$. Therefore, using (2.106a), the orthogonal projection onto S is given by

$$P_S\, x = \sum_{k=0}^{1} \langle x, \varphi_k\rangle \varphi_k.$$

To see explicitly that this is an orthogonal projection operator:

$$\begin{aligned} P_S\, x &= \langle x, \varphi_0\rangle \varphi_0 + \langle x, \varphi_1\rangle \varphi_1 \overset{(a)}{=} \varphi_0 \langle x, \varphi_0\rangle + \varphi_1 \langle x, \varphi_1\rangle \\ &\overset{(b)}{=} \varphi_0 \varphi_0^* x + \varphi_1 \varphi_1^* x \overset{(c)}{=} (\varphi_0 \varphi_0^* + \varphi_1 \varphi_1^*)\, x \\ &= \frac{1}{3} \begin{bmatrix} 2 & 1 & -1 \\ 1 & 2 & 1 \\ -1 & 1 & 2 \end{bmatrix} x, \end{aligned}$$

where (a) follows from the inner product being a scalar; (b) from writing the inner product as a product of a row vector and a column vector; and (c) from the distributivity of matrix–vector multiplication. The matrix representation of P_S is idempotent (as can be verified with a straightforward computation) and obviously Hermitian.

Orthogonal decomposition Given the orthogonal projection interpretation from Theorem 2.41, term k in the orthonormal expansion formula (2.94a) is the orthogonal projection of x to the subspace $S_{\{k\}} = \text{span}(\varphi_k)$ (recall also Example 2.22). So (2.94a) writes any x uniquely as a sum of its orthogonal projections onto orthogonal one-dimensional subspaces $\{S_{\{k\}}\}_{k\in\mathcal{K}}$. In other words, an orthonormal basis induces an orthogonal decomposition

$$H = \bigoplus_{k\in\mathcal{K}} S_{\{k\}} \tag{2.107}$$

while providing a simple way to compute the components of the decomposition of any $x \in H$. The expansion formula (2.94a) will be applied countless times in subsequent chapters, and orthogonal decompositions of Hilbert spaces $\ell^2(\mathbb{Z})$ and $\mathcal{L}^2(\mathbb{R})$ will be a recurring theme.

Best approximation The simple form of (2.106a) makes certain sequences of orthogonal projections extremely easy to compute. Let $\widehat{x}^{(k)}$ denote the best approximation of x in the subspace spanned by the orthonormal set $\{\varphi_0, \varphi_1, \ldots, \varphi_{k-1}\}$. Then, $\widehat{x}^{(0)} = \mathbf{0}$ and

$$\widehat{x}^{(k+1)} = \widehat{x}^{(k)} + \langle x, \varphi_k\rangle \varphi_k \qquad \text{for } k = 0, 1, \ldots, \tag{2.108}$$

that is, the new best approximation is the sum of the previous best approximation plus the orthogonal projection onto the span of the added vector φ_k. This follows from the projection theorem (Theorem 2.26) and comparing (2.106a) with index sets $\{0, 1, \ldots, k-1\}$ and $\{0, 1, \ldots, k\}$.

The recursive computation (2.108) is called *successive approximation*; it arises from the interest in nested subspaces and having orthonormal bases for those subspaces. Nested subspaces arise in practice quite frequently. For example, suppose that we wish to find an approximation of a function x by a polynomial of minimal degree that meets an approximation error criterion. Then, if $\{\varphi_k\}_{k\in\mathbb{N}}$ is an orthonormal set such that, for each M, $\{\varphi_0, \varphi_1, \ldots, \varphi_M\}$ is a basis for degree-M polynomials, we can apply the recursion (2.108) until the error criterion is met. Gram–Schmidt orthogonalization, discussed below, is a way to find the desired set $\{\varphi_k\}_{k\in\mathbb{N}}$, and approximation by polynomials is covered in detail in Section 6.2.

Bessel's inequality While Bessel's inequality is related to the Parseval equality (2.96), it holds for any orthonormal set – even if that set is not a basis. When it holds with equality (for *all* vectors in a Hilbert space, giving the Parseval equality), the orthonormal set must be a basis.

THEOREM 2.42 (BESSEL'S INEQUALITY) Let $\Phi = \{\varphi_k\}_{k\in\mathcal{I}}$ be an orthonormal set in a Hilbert space H. Then for any $x \in H$, *Bessel's inequality* holds:

$$\|x\|^2 \geq \sum_{k\in\mathcal{I}} |\langle x, \varphi_k\rangle|^2 \tag{2.109a}$$

$$= \|\Phi_{\mathcal{I}}^* x\|^2. \tag{2.109b}$$

Equality for every x in H implies that the set Φ is complete in H, so the orthonormal set is an orthonormal basis for H; (2.109) is then the Parseval equality (2.96).

Proof. Let $S = \overline{\text{span}}(\Phi)$ and $x_S = \Phi_{\mathcal{I}} \Phi_{\mathcal{I}}^* x$. By (2.106b), x_S is the orthogonal projection of x onto S. Thus, by the projection theorem (Theorem 2.26), $x - x_S \perp x_S$. From this we conclude

$$\|x\|^2 \overset{(a)}{=} \|x_S\|^2 + \|x - x_S\|^2 \overset{(b)}{\geq} \|x_S\|^2 = \|\Phi_{\mathcal{I}} \Phi_{\mathcal{I}}^* x\|^2 \overset{(c)}{=} \|\Phi_{\mathcal{I}}^* x\|^2 \overset{(d)}{=} \sum_{k\in\mathcal{I}} |\langle x, \varphi_k\rangle|^2,$$

where (a) follows from the Pythagorean theorem (2.27a); (b) from the nonnegativity of the norm of a vector; (c) from (2.105); and (d) from the definition of the analysis operator, (2.104).

Step (b) holds with equality for every x in H if and only if $x = x_S$ for every x in H. This occurs if and only if $S = H$, in which case we have that the set Φ is complete and thus an orthonormal basis for H.

For the case when the orthonormal set $\{\varphi_k\}_{k\in\mathcal{I}}$ is not complete, Bessel's inequality is especially easy to understand by extending the set to an orthonormal basis $\{\varphi_k\}_{k\in\mathcal{K}}$ with $\mathcal{K} \supset \mathcal{I}$. Then, Bessel's inequality follows from the Parseval equality because $\sum_{k\in\mathcal{I}} |\langle x, \varphi_k\rangle|^2$ simply omits some nonnegative terms from $\sum_{k\in\mathcal{K}} |\langle x, \varphi_k\rangle|^2$. The following example illustrates this in $\mathbb{R}^3$.

EXAMPLE 2.37 (BESSEL'S INEQUALITY) Let $\varphi_0 = \begin{bmatrix} 1 & 0 & 0 \end{bmatrix}^\top$ and $\varphi_1 = \begin{bmatrix} 0 & 1 & 0 \end{bmatrix}^\top$. These vectors are the first two elements of the standard basis in $\mathbb{R}^3$, and they are orthonormal. As illustrated in Figure 2.20, the norm of a vector $x \in \mathbb{R}^3$ is at least as large as the norm of its projection onto the (φ_0, φ_1)-plane, x_{01}:

$$\|x\|^2 \geq \|x_{01}\|^2 = |\langle x, \varphi_0\rangle|^2 + |\langle x, \varphi_1\rangle|^2.$$

Adding $\varphi_2 = \begin{bmatrix} 0 & 0 & 1 \end{bmatrix}^\top$ to the set gives an orthonormal basis (the standard basis), and adding the square of the length of the orthogonal projection of x onto the span of φ_2 yields the Parseval equality,

$$\|x\|^2 = \sum_{k=0}^{2} |\langle x, \varphi_k\rangle|^2.$$

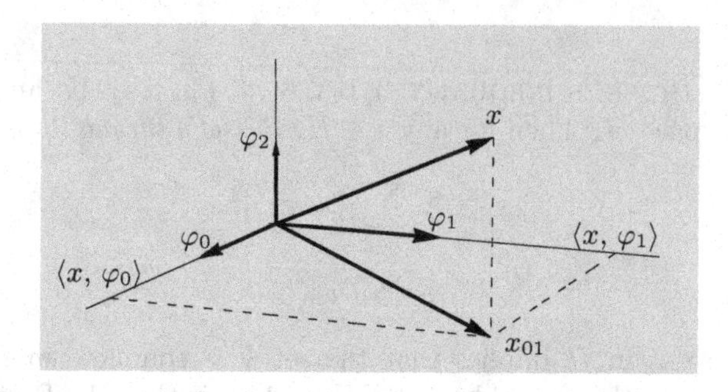

Figure 2.20 Illustration of Bessel's inequality in $\mathbb{R}^3$.

Gram–Schmidt orthogonalization We have thus far discussed properties of orthonormal bases and checked whether a set is an orthonormal basis. We now show how to construct an orthonormal basis for a space specified by a sequence of linearly independent vectors $(x_k)_{k\in\mathcal{K}}$. For notational convenience, assume that $\mathcal{K}$ is a set of consecutive integers starting at 0, so $\mathcal{K} = \{0, 1, \ldots, N-1\}$ or $\mathcal{K} = \mathbb{N}$.

The goal is to find an orthonormal set $\{\varphi_k\}_{k\in\mathcal{K}}$ with

$$\overline{\operatorname{span}}(\{\varphi_k\}_{k\in\mathcal{K}}) = \overline{\operatorname{span}}(\{x_k\}_{k\in\mathcal{K}}). \tag{2.110a}$$

Thus, when $\{x_k\}_{k\in\mathcal{K}}$ is a basis for H, the constructed set $\{\varphi_k\}_{k\in\mathcal{K}}$ is an orthonormal basis for H; otherwise, $\{\varphi_k\}_{k\in\mathcal{K}}$ is an orthonormal basis for the smaller space $\overline{\operatorname{span}}(\{x_k\}_{k\in\mathcal{K}})$, which is itself a Hilbert space.

There are many orthonormal bases for $\overline{\operatorname{span}}(\{x_k\}_{k\in\mathcal{K}})$. Upon requiring a stronger condition

$$\operatorname{span}(\{\varphi_k\}_{k=0}^{i}) = \operatorname{span}(\{x_k\}_{k=0}^{i}) \qquad \text{for every } i \in \mathbb{N}, \tag{2.110b}$$

the solution becomes essentially unique. Furthermore, enforcing (2.110b) for increasing values of i leads to a simple recursive procedure. Figure 2.21 illustrates the orthogonalization procedure for two vectors in a plane (initial, nonorthonormal basis). For example, for $i = 0$, (2.110b) holds when φ_0 is a scalar multiple of x_0. For φ_0 to have unit norm, it is natural to choose

$$\varphi_0 = x_0/\|x_0\|,$$

as illustrated in Figure 2.21(b), and the set of all possible solutions is obtained by including a unit-modulus scalar factor. Then, for (2.110b) to hold for $i = 1$, the vector φ_1 must be aligned with the component of x_1 orthogonal to φ_0, as illustrated in Figure 2.21(c). This is achieved when φ_1 is a scalar multiple of the residual from orthogonally projecting x_1 to the subspace spanned by φ_0,

$$\varphi_1 = \frac{x_1 - \langle x_1, \varphi_0\rangle\varphi_0}{\|x_1 - \langle x_1, \varphi_0\rangle\varphi_0\|},$$

as illustrated in Figure 2.21(d). In general, φ_k is determined by normalizing the residual of x_k orthogonally projected to $\operatorname{span}(\{\varphi_0, \varphi_1, \ldots, \varphi_{k-1}\})$. The residual is

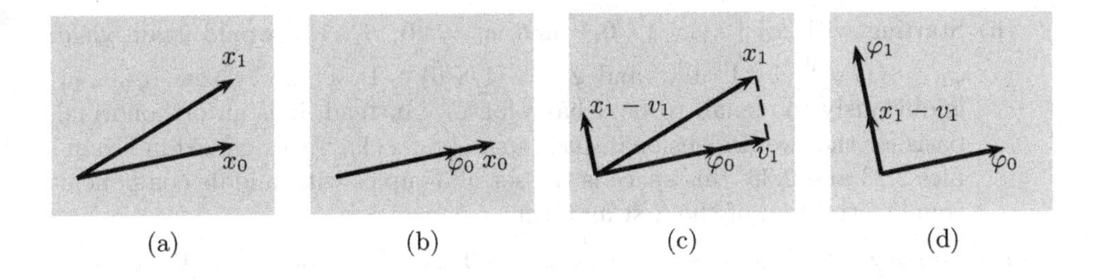

Figure 2.21 Illustration of Gram–Schmidt orthogonalization. (a) Input vectors (x_0, x_1). (b) The first output vector φ_0 is a normalized version of x_0. (c) The projection of x_1 onto the subspace spanned by φ_0 is subtracted from x_1 to obtain a residual $x_1 - v_1$. (d) The second output vector φ_1 is a normalized version of the residual.

Gram–Schmidt orthogonalization
Input: An ordered sequence of linearly independent vectors $(x_k)_{k \in \mathcal{K}}$
Output: Orthonormal vectors $\{\varphi_k\}_{k \in \mathcal{K}}$, with $\operatorname{span}(\{\varphi_k\}) = \operatorname{span}(\{x_k\})$

$\{\varphi_k\}$ = **GramSchmidt**($\{x_k\}$)
$\varphi_0 = x_0/\|x_0\|$
$k = 1$
while $k < |\mathcal{K}|$ **do**
  project $v_k = \sum_{i=0}^{k-1} \langle x_k, \varphi_i \rangle \varphi_i$
  normalize $\varphi_k = (x_k - v_k)/\|x_k - v_k\|$
  increment k
end while
return $\{\varphi_k\}$

Table 2.1 Gram–Schmidt orthogonalization algorithm.

nonzero because otherwise the linear independence of $\{x_0, x_1, \ldots, x_k\}$ is contradicted. The full recursive computation is summarized in Table 2.1.

Example 2.38 (Gram–Schmidt orthogonalization)

(i) Let $x_0 = \begin{bmatrix} 1 & 1 & 0 \end{bmatrix}^\top$, $x_1 = \begin{bmatrix} 0 & 1 & 1 \end{bmatrix}^\top$, and $x_2 = \begin{bmatrix} 1 & 1 & 1 \end{bmatrix}^\top$. These are linearly independent, and following the steps in Table 2.1 first yields $\varphi_0 = (1/\sqrt{2}) \begin{bmatrix} 1 & 1 & 0 \end{bmatrix}^\top$, then $v_1 = \frac{1}{2} \begin{bmatrix} 1 & 1 & 0 \end{bmatrix}^\top$, then $x_1 - v_1 = \frac{1}{2} \begin{bmatrix} -1 & 1 & 2 \end{bmatrix}^\top$, and $\varphi_1 = (1/\sqrt{6}) \begin{bmatrix} -1 & 1 & 2 \end{bmatrix}^\top$. For the final basis vector, $v_2 = \frac{1}{3} \begin{bmatrix} 2 & 4 & 2 \end{bmatrix}^\top$, $x_2 - v_2 = \frac{1}{3} \begin{bmatrix} 1 & -1 & 1 \end{bmatrix}^\top$, and $\varphi_2 = (1/\sqrt{3}) \begin{bmatrix} 1 & -1 & 1 \end{bmatrix}^\top$.

The set $\{\varphi_0, \varphi_1, \varphi_1\}$ is the orthonormal basis from Examples 2.33 and 2.36. Since $\operatorname{span}(\{\varphi_0, \varphi_1\}) = \operatorname{span}(\{x_0, x_1\})$ and the latter span is plainly the range of matrix B in Example 2.23, we can retrospectively see that the projection operators in Examples 2.23 and 2.36 project to the same subspace. (One projection operator is orthogonal and the other is oblique.)

(ii) Starting with $x_0 = \begin{bmatrix}1 & 1 & 0\end{bmatrix}^\top$ and $x_1 = \begin{bmatrix}0 & 1 & 1\end{bmatrix}^\top$ would again yield $\varphi_0 = (1/\sqrt{2})\begin{bmatrix}1 & 1 & 0\end{bmatrix}^\top$ and $\varphi_1 = (1/\sqrt{6})\begin{bmatrix}-1 & 1 & 2\end{bmatrix}^\top$. Now $\{\varphi_0, \varphi_1\}$ is obviously too small to be a basis for $\mathbb{C}^3$. Instead, it is an orthonormal basis for the two-dimensional space $\operatorname{span}(\{x_0, x_1\})$. As discussed in Examples 2.23 and 2.36, this space is the set of 3-tuples with middle component equal to the sum of the first and last.

(iii) Starting with $x_0 = \begin{bmatrix}1 & 1 & 1\end{bmatrix}^\top$, $x_1 = \begin{bmatrix}1 & 1 & 0\end{bmatrix}^\top$, and $x_2 = \begin{bmatrix}0 & 1 & 1\end{bmatrix}^\top$, the same set of vectors as in (i), but in a different order, yields

$$\varphi_0 = \frac{1}{\sqrt{3}}\begin{bmatrix}1\\1\\1\end{bmatrix}, \quad \varphi_1 = \frac{1}{\sqrt{6}}\begin{bmatrix}1\\1\\-2\end{bmatrix}, \quad \text{and} \quad \varphi_2 = \frac{1}{\sqrt{2}}\begin{bmatrix}-1\\1\\0\end{bmatrix}.$$

There is no obvious relationship between this orthonormal basis and the one found in part (i).

Solved exercise 2.5 applies Gram–Schmidt orthogonalization to derive normalized Legendre polynomials, which are polynomials orthogonal in $\mathcal{L}^2([-1, 1])$.

2.5.3 Biorthogonal pairs of bases

Orthonormal bases have several advantages over nonorthonormal bases, including the simple expressions for expansion in (2.94) and orthogonal projection in (2.106). While there are no general disadvantages caused directly by orthonormality, in some settings nonorthonormal bases have their advantages, too. For example, of the bases

$$\left\{\begin{bmatrix}1\\1\\0\end{bmatrix}, \begin{bmatrix}0\\1\\1\end{bmatrix}, \begin{bmatrix}1\\1\\1\end{bmatrix}\right\} \quad \text{and} \quad \left\{\frac{1}{\sqrt{2}}\begin{bmatrix}1\\1\\0\end{bmatrix}, \frac{1}{\sqrt{6}}\begin{bmatrix}-1\\1\\2\end{bmatrix}, \frac{1}{\sqrt{3}}\begin{bmatrix}1\\-1\\1\end{bmatrix}\right\}$$

from Example 2.38(i), the nonorthogonal basis is easier to store and compute with. Solved Exercise 2.5 provides a more dramatic example, since the set of functions $\{1, t, t^2, \ldots, t^N\}$ is certainly simpler for many purposes than the Legendre polynomials up to degree N.

A basis does not have to be orthonormal to provide unique expansions. The sacrifice we must make is that we cannot ask for a single set of vectors to serve the analysis role in $x \mapsto \alpha = (\langle x, \varphi_k\rangle)_{k\in\mathcal{K}}$ and the synthesis role in $\alpha \mapsto \sum_{k\in\mathcal{K}} \alpha_k \varphi_k$. This leads us to the concept of a *biorthogonal pair of bases*, or *dual bases.*

Definition 2.43 (Biorthogonal pair of bases) The sets of vectors $\Phi = \{\varphi_k\}_{k\in\mathcal{K}} \subset H$ and $\widetilde{\Phi} = \{\widetilde{\varphi}_k\}_{k\in\mathcal{K}} \subset H$, where $\mathcal{K}$ is finite or countably infinite, are called a *biorthogonal pair of bases* for a Hilbert space H when

(i) each is a *basis* for H; and

(ii) they are *biorthogonal*, meaning that

$$\langle \varphi_i, \widetilde{\varphi}_k\rangle = \delta_{i-k} \quad \text{for every } i, k \in \mathcal{K}. \tag{2.111}$$

Since the inner product has Hermitian symmetry and δ_{i-k} is real, the roles of the sets Φ and $\widetilde{\Phi}$ can be reversed with no change in whether (2.111) holds. We will generally maintain a convention of using the basis Φ in synthesis and the basis $\widetilde{\Phi}$ in analysis, with the understanding that the bases can be swapped in any of the results that follow. With each basis we associate synthesis and analysis operators defined through (2.90) and (2.91); a biorthogonal pair of bases thus yields four operators: Φ, Φ^*, $\widetilde{\Phi}$, and $\widetilde{\Phi}^*$.

EXAMPLE 2.39 (BIORTHOGONAL PAIR OF BASES IN FINITE DIMENSIONS) The sets

$$\varphi_0 = \begin{bmatrix} 1 \\ 1 \\ 0 \end{bmatrix}, \ \varphi_1 = \begin{bmatrix} 0 \\ 1 \\ 1 \end{bmatrix}, \ \varphi_2 = \begin{bmatrix} 1 \\ 1 \\ 1 \end{bmatrix} \quad \text{and} \quad \widetilde{\varphi}_0 = \begin{bmatrix} 0 \\ 1 \\ -1 \end{bmatrix}, \ \widetilde{\varphi}_1 = \begin{bmatrix} -1 \\ 1 \\ 0 \end{bmatrix}, \ \widetilde{\varphi}_2 = \begin{bmatrix} 1 \\ -1 \\ 1 \end{bmatrix}$$

are a biorthogonal pair of bases for $\mathbb{C}^3$, as can be verified by direct computation.

EXAMPLE 2.40 (BIORTHOGONAL PAIR OF BASES OF COSINE FUNCTIONS) Define $\Psi = \{\psi_k\}_{k\in\mathbb{N}} \subset \mathcal{L}^2([-\frac{1}{2}, \frac{1}{2}])$ and $\widetilde{\Psi} = \{\widetilde{\psi}_k\}_{k\in\mathbb{N}} \subset \mathcal{L}^2([-\frac{1}{2}, \frac{1}{2}])$ by

$$\begin{aligned} \psi_0(t) &= 1 \\ &= \varphi_0(t), && (2.112a) \\ \psi_k(t) &= \sqrt{2}\cos(2\pi k t) + \frac{1}{2}\sqrt{2}\cos(2\pi(k+1)t) \\ &= \varphi_k(t) + \frac{1}{2}\varphi_{k+1}(t), \qquad k = 1, 2, \ldots, && (2.112b) \\ \widetilde{\psi}_0(t) &= 1 \\ &= \varphi_0(t), && (2.112c) \\ \widetilde{\psi}_k(t) &= \sum_{m=1}^{k} \left(-\frac{1}{2}\right)^{k-m} \sqrt{2}\cos(2\pi m t) \\ &= \sum_{m=1}^{k} \left(-\frac{1}{2}\right)^{k-m} \varphi_m(t), \qquad k = 1, 2, \ldots, && (2.112d) \end{aligned}$$

where $\{\varphi_k\}_{k\in\mathbb{N}}$ are the orthonormal basis functions from (2.24). The first few functions in each of these sets are shown in Figure 2.22. Verifying that (2.111) holds is only part of proving that the sets Ψ (solid lines) and $\widetilde{\Psi}$ (dashed lines) form a biorthogonal pair of bases; this is left for Exercise 2.38. We must also verify that each set is a basis for the same subspace of $\mathcal{L}^2([-\frac{1}{2}, \frac{1}{2}])$.

By construction, $\{\varphi_k\}_{k\in\mathbb{N}}$ forms an orthonormal basis for the closure of its span S. We can use this to show that the sets Ψ and $\widetilde{\Psi}$ are also bases for S. The closure of the span of Ψ and S are equal: $\overline{\text{span}}(\Psi) \subset S$ because each ψ_k is a linear combination of one or two elements of Φ; and $S \subset \overline{\text{span}}(\Psi)$ because $\varphi_0 = \psi_0$ and, for each $k \in \mathbb{Z}^+$, φ_k can be written as an infinite linear combination of elements of Ψ. Furthermore, these expansions with respect to Ψ are unique; a detailed argument is left for Exercise 2.38. An analogous argument shows that the set $\widetilde{\Psi}$ is a basis for S.

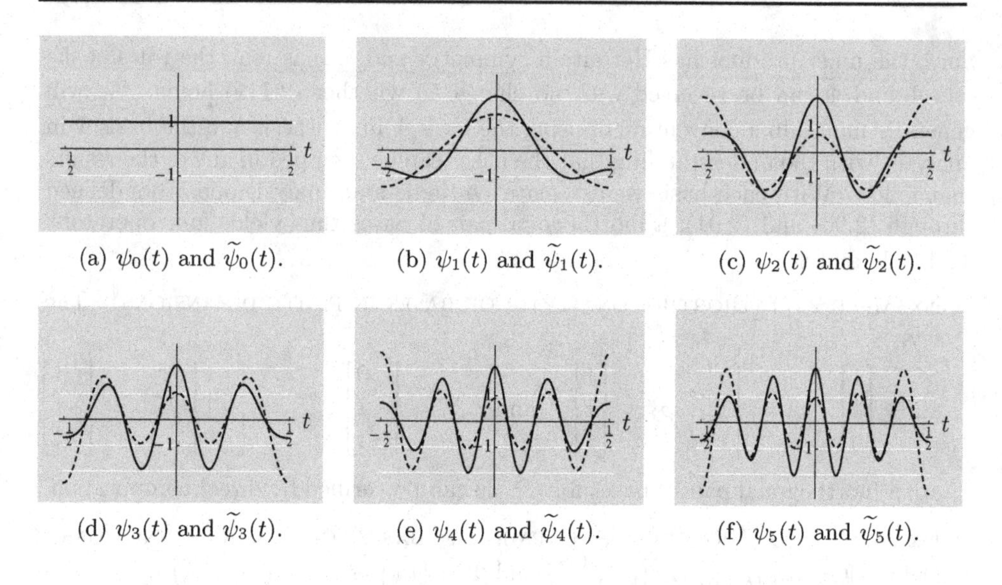

(a) $\psi_0(t)$ and $\widetilde{\psi}_0(t)$. (b) $\psi_1(t)$ and $\widetilde{\psi}_1(t)$. (c) $\psi_2(t)$ and $\widetilde{\psi}_2(t)$.

(d) $\psi_3(t)$ and $\widetilde{\psi}_3(t)$. (e) $\psi_4(t)$ and $\widetilde{\psi}_4(t)$. (f) $\psi_5(t)$ and $\widetilde{\psi}_5(t)$.

Figure 2.22 Elements of the biorthogonal pair of bases Ψ (solid lines) and $\widetilde{\Psi}$ (dashed lines) in Example 2.40.

Expansion and inner product computation With a biorthogonal pair of bases, expansion coefficients with respect to one basis are computed using the other basis.

THEOREM 2.44 (BIORTHOGONAL BASIS EXPANSIONS) Let $\Phi = \{\varphi_k\}_{k\in\mathcal{K}}$ and $\widetilde{\Phi} = \{\widetilde{\varphi}_k\}_{k\in\mathcal{K}}$ be a biorthogonal pair of bases for a Hilbert space H. The unique expansion with respect to the basis Φ of any x in H has expansion coefficients

$$\alpha_k = \langle x, \widetilde{\varphi}_k\rangle \quad \text{for } k \in \mathcal{K}, \quad \text{or}, \tag{2.113a}$$

$$\alpha = \widetilde{\Phi}^* x. \tag{2.113b}$$

Synthesis with these coefficients yields

$$x = \sum_{k\in\mathcal{K}} \langle x, \widetilde{\varphi}_k\rangle \varphi_k \tag{2.114a}$$

$$= \Phi\alpha = \Phi\widetilde{\Phi}^* x. \tag{2.114b}$$

Proof. The proof parallels the proof of Theorem 2.39 with minor modifications based on replacing the orthonormality condition (2.92) with the biorthogonality condition (2.111).

The existence of a unique linear combination of the form (2.87) is guaranteed by the set Φ being a basis. The validity of (2.114a) with coefficients (2.113a) follows from the following computation:

$$\langle x, \widetilde{\varphi}_k \rangle \overset{(a)}{=} \left\langle \sum_{i \in \mathcal{K}} \alpha_i \varphi_i, \widetilde{\varphi}_k \right\rangle \overset{(b)}{=} \sum_{i \in \mathcal{K}} \alpha_i \langle \varphi_i, \widetilde{\varphi}_k \rangle \overset{(c)}{=} \sum_{i \in \mathcal{K}} \alpha_i \delta_{i-k} \overset{(d)}{=} \alpha_k,$$

where (a) follows from (2.87); (b) from the linearity in the first argument of the inner product; (c) from the biorthogonality of the sets Φ and $\widetilde{\Phi}$, (2.111); and (d) from the definition of the Kronecker delta sequence, (2.9).

The expressions (2.113b) and (2.114b) are equivalent to (2.113a) and (2.114a) using the operators defined in (2.90) and (2.91).

Reversing the roles of the bases Φ and $\widetilde{\Phi}$ gives expansion coefficients with respect to the basis $\widetilde{\Phi}$:

$$\widetilde{\alpha}_k = \langle x, \varphi_k \rangle \quad \text{for } k \in \mathcal{K}, \quad \text{or} \quad \widetilde{\alpha} = \Phi^* x, \tag{2.115}$$

with the corresponding expansion

$$x = \sum_{k \in \mathcal{K}} \langle x, \varphi_k \rangle \widetilde{\varphi}_k = \widetilde{\Phi} \Phi^* x. \tag{2.116}$$

The theorem shows that a biorthogonal pair of bases can together do the job of an orthonormal basis in terms of signal expansion. The most interesting properties of the synthesis and analysis operators involve both bases of the pair. Since (2.114b) and (2.116) hold for all x in H,

$$\Phi \widetilde{\Phi}^* = I \quad \text{as well as} \quad \widetilde{\Phi} \Phi^* = I \quad \text{on} \quad H. \tag{2.117}$$

This leads to an analogue of Theorem 2.40.

Theorem 2.45 (Parseval equalities for biorthogonal pairs of bases) Let $\Phi = \{\varphi_k\}_{k \in \mathcal{K}}$ and $\widetilde{\Phi} = \{\widetilde{\varphi}_k\}_{k \in \mathcal{K}}$ be a biorthogonal pair of bases for a Hilbert space H. Expansion with respect to the bases Φ and $\widetilde{\Phi}$ with coefficients (2.113) and (2.115) satisfies

$$\|x\|^2 = \sum_{k \in \mathcal{K}} \langle x, \varphi_k \rangle \langle x, \widetilde{\varphi}_k \rangle^* \tag{2.118a}$$

$$= \langle \Phi^* x, \widetilde{\Phi}^* x \rangle = \langle \widetilde{\alpha}, \alpha \rangle. \tag{2.118b}$$

More generally,

$$\langle x, y \rangle = \sum_{k \in \mathcal{K}} \langle x, \varphi_k \rangle \langle y, \widetilde{\varphi}_k \rangle^* \tag{2.119a}$$

$$= \langle \Phi^* x, \widetilde{\Phi}^* y \rangle = \langle \widetilde{\alpha}, \beta \rangle. \tag{2.119b}$$

Proof. We will prove (2.119b); (2.119a) is the same fact expanded with the definitions of Φ^* and $\widetilde{\Phi}^*$, and equalities (2.118) follow upon setting $x = y$. For any x and y in H,

$$\langle \Phi^* x, \widetilde{\Phi}^* y\rangle \stackrel{(a)}{=} \langle x, \Phi\widetilde{\Phi}^* y\rangle \stackrel{(b)}{=} \langle x, y\rangle,$$

where (a) follows from the definition of the adjoint; and (b) from (2.117).

Gram matrix Theorem 2.45 is not nearly as useful as Theorem 2.40 because it involves expansions with respect to both bases of the pair. In (2.119), $x = \sum_{k\in\mathcal{K}} \widetilde{\alpha}_k \widetilde{\varphi}_k$ and $y = \sum_{k\in\mathcal{K}} \beta_k \varphi_k$ (note the use of different bases) so that

$$\langle x, y\rangle = \langle \widetilde{\alpha}, \beta\rangle = \sum_{k\in\mathcal{K}} \widetilde{\alpha}_k \beta_k^*.$$

More often, one wants all expansions to be with respect to one basis of the pair; the other basis of the pair serves as a helper in computing the expansion coefficients. If $x = \Phi\alpha$ and $y = \Phi\beta$ (both expansions with respect to the basis Φ), then

$$\langle x, y\rangle = \langle \Phi\alpha, \Phi\beta\rangle \stackrel{(a)}{=} \langle \Phi^*\Phi\alpha, \beta\rangle \stackrel{(b)}{=} \langle G\alpha, \beta\rangle, \tag{2.120}$$

where (a) follows from the meaning of the adjoint; and (b) from introducing the *Gram matrix* or *Gramian* G,

$$G = \Phi^*\Phi, \tag{2.121a}$$

$$G_{ik} = \langle \varphi_k, \varphi_i\rangle \quad \text{for every } i, k \in \mathcal{K}, \tag{2.121b}$$

$$G = \begin{bmatrix} & \vdots & \vdots & \vdots & \\ \cdots & \langle \varphi_{-1}, \varphi_{-1}\rangle & \langle \varphi_0, \varphi_{-1}\rangle & \langle \varphi_1, \varphi_{-1}\rangle & \cdots \\ \cdots & \langle \varphi_{-1}, \varphi_0\rangle & \boxed{\langle \varphi_0, \varphi_0\rangle} & \langle \varphi_1, \varphi_0\rangle & \cdots \\ \cdots & \langle \varphi_{-1}, \varphi_1\rangle & \langle \varphi_0, \varphi_1\rangle & \langle \varphi_1, \varphi_1\rangle & \cdots \\ & \vdots & \vdots & \vdots & \end{bmatrix}. \tag{2.121c}$$

The order of factors in (2.120) evokes a product of three terms:

$$\langle x, y\rangle = \beta^* G\alpha, \tag{2.122}$$

where as before α and β are column vectors. When the set Φ is an orthonormal basis, G simplifies to the identity operator on $\ell^2(\mathcal{K})$ and (2.122) simplifies to (2.99).

Example 2.41 ($\mathbb{C}^3$ inner product computation with bases) Consider the basis $\{\varphi_0, \varphi_1, \varphi_2\} \subset \mathbb{C}^3$ from Example 2.39. The Gram matrix of this basis is

$$G = \Phi^*\Phi = \begin{bmatrix} 1 & 1 & 0 \\ 0 & 1 & 1 \\ 1 & 1 & 1 \end{bmatrix} \begin{bmatrix} 1 & 0 & 1 \\ 1 & 1 & 1 \\ 0 & 1 & 1 \end{bmatrix} = \begin{bmatrix} 2 & 1 & 2 \\ 1 & 2 & 2 \\ 2 & 2 & 3 \end{bmatrix}.$$

For any x and y in $\mathbb{C}^3$, the expansions with respect to the basis Φ are $\alpha = \widetilde{\Phi}^* x$ and $\beta = \widetilde{\Phi}^* y$, where

$$\widetilde{\Phi}^* = \begin{bmatrix} 0 & 1 & -1 \\ -1 & 1 & 0 \\ 1 & -1 & 1 \end{bmatrix}$$

is the analysis operator associated with the basis $\{\widetilde{\varphi}_0, \widetilde{\varphi}_1, \widetilde{\varphi}_2\} \subset \mathbb{C}^3$ from Example 2.39. Then, $\langle x, y \rangle = \beta^* G \alpha$ by using (2.122).

In $\mathbb{C}^N$, it is often natural and easy to use the standard basis for inner product computations. Thus, the previous example might seem to be a complicated way to achieve a simple result. In fact, we have

$$\beta^* G \alpha = (\widetilde{\Phi}^* y)^* (\Phi^* \Phi) (\widetilde{\Phi}^* x) = y^* \widetilde{\Phi}\, \Phi^* \Phi\, \widetilde{\Phi}^* x = y^* (\Phi \widetilde{\Phi}^*)^* (\Phi \widetilde{\Phi}^*) x,$$

so, in light of (2.117), the inner product $y^* x$ has been altered only by the insertion of identity operators. The use of (2.122) is more valuable when expansion with respect to a biorthogonal basis is natural and precomputation of G avoids a laborious inner product computation, as we illustrate next.

EXAMPLE 2.42 (POLYNOMIAL INNER PRODUCT COMPUTATION WITH BASES) Consider the polynomials of degree at most 3 under the $\mathcal{L}^2([-1, 1])$ inner product. The basis $\{1, t, t^2, t^3\}$ is easy to use because the expansion coefficients of a polynomial $x(t) = \alpha_0 + \alpha_1 t + \alpha_2 t^2 + \alpha_3 t^3$ are read off directly as $\alpha = \begin{bmatrix} \alpha_0 & \alpha_1 & \alpha_2 & \alpha_3 \end{bmatrix}^\top$. However, since the basis is not orthonormal, we cannot compute inner products with (2.99). Instead, since

$$\langle t^k, t^i \rangle = \int_{-1}^{1} t^k t^i \, dt = \frac{1}{i+k+1} \left(1 - (-1)^{i+k+1}\right),$$

the Gram matrix of the basis is

$$G = \begin{bmatrix} 2 & 0 & \frac{2}{3} & 0 \\ 0 & \frac{2}{3} & 0 & \frac{2}{5} \\ \frac{2}{3} & 0 & \frac{2}{5} & 0 \\ 0 & \frac{2}{5} & 0 & \frac{2}{7} \end{bmatrix},$$

and inner products can be computed without any integration using (2.122).

Inverse synthesis and analysis Equation (2.117) shows that $\widetilde{\Phi}^*$ is a right inverse of Φ. It is also true that

$$\widetilde{\Phi}^* \Phi = I \quad \text{on } \ell^2(\mathcal{K}), \tag{2.123}$$

making $\widetilde{\Phi}^*$ a left inverse of Φ and furthermore showing that $\widetilde{\Phi}^*$ is the unique inverse of Φ. To verify (2.123), make the following computation for any sequence α in $\ell^2(\mathcal{K})$:

$$\begin{aligned} \widetilde{\Phi}^* \Phi \alpha &\overset{(a)}{=} \widetilde{\Phi}^* \sum_{i \in \mathcal{K}} \alpha_i \varphi_i \overset{(b)}{=} \left(\langle \textstyle\sum_{i \in \mathcal{K}} \alpha_i \varphi_i, \widetilde{\varphi}_k \rangle \right)_{k \in \mathcal{K}} \overset{(c)}{=} \left(\textstyle\sum_{i \in \mathcal{K}} \alpha_i \langle \varphi_i, \widetilde{\varphi}_k \rangle \right)_{k \in \mathcal{K}} \\ &\overset{(d)}{=} \left(\textstyle\sum_{i \in \mathcal{K}} \alpha_i \delta_{i-k} \right)_{k \in \mathcal{K}} \overset{(e)}{=} (\alpha_k)_{k \in \mathcal{K}} = \alpha, \end{aligned} \tag{2.124}$$

where (a) follows from (2.90); (b) from (2.91); (c) from the linearity in the first argument of the inner product; (d) from the biorthogonality of the sets $\{\varphi_k\}_{k\in\mathcal{K}}$ and $\{\widetilde{\varphi}_k\}_{k\in\mathcal{K}}$, (2.111); and (e) from the definition of the Kronecker delta sequence, (2.9).

Knowing that operators associated with a biorthogonal pair of bases satisfy

$$\widetilde{\Phi}^* = \Phi^{-1} \tag{2.125}$$

can be used to determine $\widetilde{\Phi}$ from Φ such that the sets Φ and $\widetilde{\Phi}$ form a biorthogonal pair of bases. A simple special case is when the Hilbert space H is $\mathbb{C}^N$ (or $\mathbb{R}^N$). Then, the synthesis operator Φ is an $N \times N$ matrix with the basis vectors as columns, and linear independence of the basis implies that Φ is invertible. Setting $\widetilde{\Phi} = (\Phi^{-1})^*$ means that the vectors of the dual basis are the conjugate transposes of the rows of Φ^{-1}. It is a valuable exercise to check that $\widetilde{\Phi}$ in Example 2.39 can be seen as derived from Φ in this manner.

Example 2.43 (Biorthogonal pair of bases, Example 2.31 continued) Take the basis Φ from Example 2.31(ii). Assuming that $\theta \neq 0$, and using (2.125), the synthesis operators associated with the basis Φ and its dual basis $\widetilde{\Phi}$ are

$$\Phi = \begin{bmatrix} 1 & \cos\theta \\ 0 & \sin\theta \end{bmatrix}, \qquad \widetilde{\Phi} = \begin{bmatrix} 1 & 0 \\ -\cot\theta & \csc\theta \end{bmatrix}; \tag{2.126}$$

these bases are shown in Figures 2.23(a) and (b). We can easily check that the biorthogonality condition (2.111) holds. Figure 2.23 also illustrates how a unit-norm basis does not necessarily lead to a unit-norm dual basis.

We have already computed the optimal stability constants $\lambda_{\min}$ and $\lambda_{\max}$ of the basis in Figure 2.23(a) in Example 2.31(ii); we can similarly find the corresponding optimal stability constants $\widetilde{\lambda}_{\min}$ and $\widetilde{\lambda}_{\max}$ of the dual basis in Figure 2.23(b). It turns out that these are reciprocals of $\lambda_{\min}$ and $\lambda_{\max}$:

$$\widetilde{\lambda}_{\min} = \frac{1}{\lambda_{\max}}, \qquad \widetilde{\lambda}_{\max} = \frac{1}{\lambda_{\min}}. \tag{2.127}$$

Clearly, the pair is best behaved for $\theta = \frac{1}{2}\pi$, when it reduces to an orthonormal basis. As θ approaches 0, the basis vectors in Φ become close to collinear, destroying the basis property.

When the Hilbert space is not $\mathbb{C}^N$ (or $\mathbb{R}^N$), the simplicity of the equation $\widetilde{\Phi}^* = \Phi^{-1}$ is deceptive. The operators $\widetilde{\Phi}^*$ and Φ^{-1} are mappings from H to $\ell^2(\mathcal{K})$. The analysis operator $\widetilde{\Phi}^*$ maps from H to $\ell^2(\mathcal{K})$ through the inner products with $\{\widetilde{\varphi}_k\}_{k\in\mathcal{K}}$. To determine $\{\widetilde{\varphi}_k\}_{k\in\mathcal{K}}$ from Φ^{-1} is to interpret the operation of Φ^{-1} as computing inner products with some set of vectors. We derive that set of vectors next.

Dual basis So far we have derived properties of a biorthogonal pair of bases. Now we show how to find the unique basis $\widetilde{\Phi}$ that completes a biorthogonal pair for

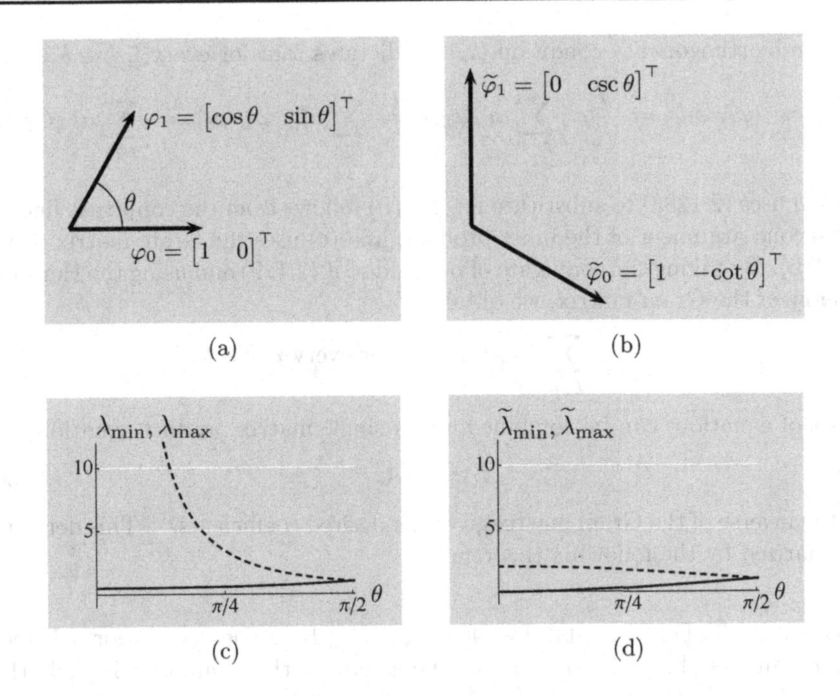

Figure 2.23 A biorthogonal pair of bases in $\mathbb{R}^2$ and their optimal stability constants. (a) The basis Φ from Figure 2.18(b). (b) Its corresponding dual basis $\widetilde{\Phi}$. (c) $\lambda_{\min}$ and $\lambda_{\max}$ for the basis in (a) as a function of θ (same as Figure 2.18(d)). (d) $\widetilde{\lambda}_{\min}$ and $\widetilde{\lambda}_{\max}$ for the dual basis in (b) as a function of θ. The Riesz basis constants here are the reciprocals of the ones in (c).

a given Riesz basis Φ. As noted above, when Φ is a basis[23] for $\mathbb{C}^N$, finding the appropriate $\widetilde{\Phi}$ is as simple as inverting the matrix Φ^*. In general, we find $\widetilde{\Phi}$ by imposing two key properties: Φ and $\widetilde{\Phi}$ span the same space H, and the sets are biorthogonal.

Let $\Phi = \{\varphi_k\}_{k\in\mathcal{K}} \subset H$ be a Riesz basis for a Hilbert space H. To ensure that $\overline{\operatorname{span}}(\widetilde{\Phi}) \subseteq \overline{\operatorname{span}}(\Phi)$, let

$$\widetilde{\varphi}_k \;=\; \sum_{\ell\in\mathcal{K}} a_{\ell,k}\varphi_\ell, \qquad \text{for each } k \in \mathcal{K}. \tag{2.128a}$$

This set of equations can be combined into a single matrix product equation to express the synthesis operator $\widetilde{\Phi}$ as

$$\widetilde{\Phi} \;=\; \Phi A, \tag{2.128b}$$

where the (ℓ, k) entry of $A : \ell^2(\mathcal{K}) \to \ell^2(\mathcal{K})$ is $a_{\ell,k}$. Determining the coefficients $a_{\ell,k}$ for $k,\ \ell \in \mathcal{K}$ specifies the dual basis $\widetilde{\Phi}$ through either of the forms of (2.128).

[23]Recall that, in any finite-dimensional space, any basis is a Riesz basis.

The biorthogonality condition (2.111) dictates that for every $i,\, k \in \mathcal{K}$,

$$\delta_{i-k} \;=\; \langle \varphi_i,\, \widetilde{\varphi}_k \rangle \;\overset{(a)}{=}\; \left\langle \varphi_i,\, \sum_{\ell\in\mathcal{K}} a_{\ell,k}\varphi_\ell \right\rangle \;\overset{(b)}{=}\; \sum_{\ell\in\mathcal{K}} a_{\ell,k}^* \langle \varphi_i,\, \varphi_\ell \rangle \;\overset{(c)}{=}\; \sum_{\ell\in\mathcal{K}} a_{\ell,k}^* G_{\ell,i}, \tag{2.129}$$

where (a) uses (2.128a) to substitute for $\widetilde{\varphi}_k$; (b) follows from the conjugate linearity in the second argument of the inner product; and (c) uses the Gram matrix defined in (2.121b). By taking the conjugate of both sides of (2.129) and using the Hermitian symmetry of the Gram matrix, we obtain

$$\delta_{i-k} \;=\; \sum_{\ell\in\mathcal{K}} G_{i,\ell}\, a_{\ell,k}, \qquad \text{for every } i,\, k \in \mathcal{K}. \tag{2.130a}$$

This set of equations can be combined into a single matrix product equation

$$I \;=\; GA. \tag{2.130b}$$

Thus the inverse of the Gram matrix gives the desired coefficients.[24] This derivation is summarized by the following theorem.

Theorem 2.46 (Dual basis) Let $\Phi = \{\varphi_k\}_{k\in\mathcal{K}}$ be a Riesz basis for a Hilbert space H, and let $A : \ell^2(\mathcal{K}) \to \ell^2(\mathcal{K})$ be the inverse of the Gram matrix of Φ, that is, $A = (\Phi^*\Phi)^{-1}$. Then, the set $\widetilde{\Phi} = \{\widetilde{\varphi}_k\}_{k\in\mathcal{K}}$ defined via

$$\widetilde{\varphi}_k \;=\; \sum_{\ell\in\mathcal{K}} a_{\ell,k}\varphi_\ell, \qquad \text{for each } k \in \mathcal{K}, \tag{2.131a}$$

together with Φ forms a biorthogonal pair of bases for H. The synthesis operator for this basis is given by

$$\widetilde{\Phi} \;=\; \Phi A \;=\; \Phi(\Phi^*\Phi)^{-1}, \tag{2.131b}$$

the pseudoinverse of Φ^*.

Recall from (2.125) that the synthesis operator associated with the dual basis can be written as $\widetilde{\Phi} = (\Phi^{-1})^*$. However, the inverse and adjoint in this expression are difficult to interpret. In contrast, the key virtue of (2.131) is that the inversion is of the Gram matrix, which is easier to interpret because it is an operator from $\ell^2(\mathcal{K})$ to $\ell^2(\mathcal{K})$. This is illustrated in the following finite-dimensional example. When $\mathcal{K}$ is infinite, we still prefer to avoid explicit inversion; examples of dual basis computations for infinite $\mathcal{K}$ arise in Chapter 5.

Example 2.44 (Dual to basis of periodic triangle functions) Let

$$\varphi_0(t) \;=\; \begin{cases} t, & \text{for } t \in [0,\, 1]; \\ 2-t, & \text{for } t \in (1,\, 2]; \\ 0, & \text{for } t \in (2,\, 3] \end{cases} \tag{2.132}$$

[24] The Riesz basis condition on Φ ensures that the inverse exists.

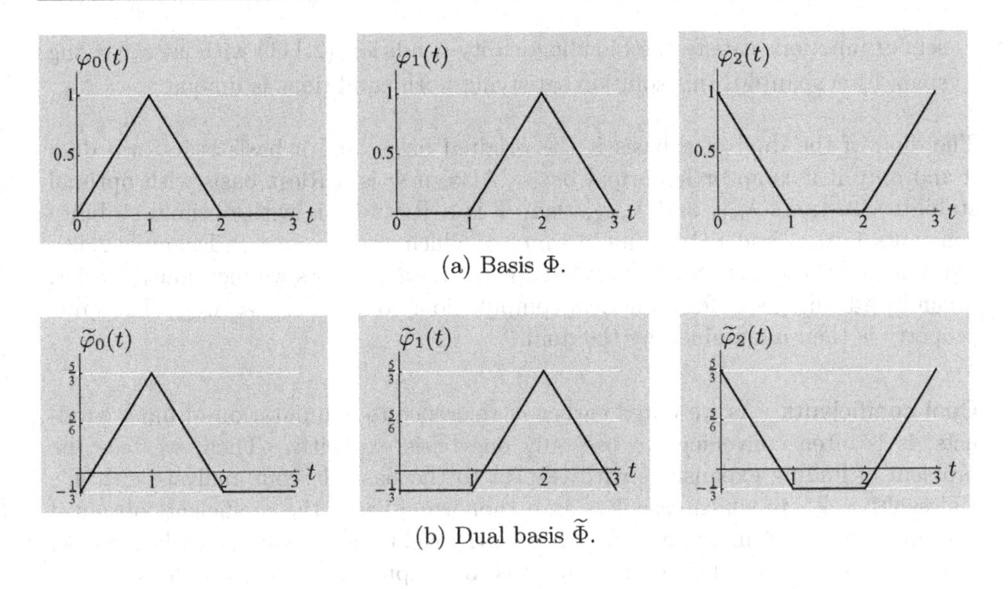

Figure 2.24 (a) The basis $\Phi = \{\varphi_0, \varphi_1, \varphi_2\}$ obtained by circularly shifting the triangle function φ_0 by 1 and 2. (b) The dual set $\widetilde{\Phi} = \{\widetilde{\varphi}_0, \widetilde{\varphi}_1, \widetilde{\varphi}_2\}$ is derived in Example 2.44.

in $\mathcal{L}^2([0, 3])$. The function and its circular shifts by 1 and 2 are shown in Figure 2.24(a). The set $\Phi = \{\varphi_0, \varphi_1, \varphi_2\}$ is a basis for the subspace $S = \overline{\text{span}}(\Phi) \subset \mathcal{L}^2([0, 3])$. This subspace is the set of functions x that satisfy $x(0) = x(3)$ and are piecewise-linear and continuous on $[0, 3]$ with breakpoints at 1 and 2.

We wish to find the basis $\widetilde{\Phi} = \{\widetilde{\varphi}_0, \widetilde{\varphi}_1, \widetilde{\varphi}_2\}$ that forms a biorthogonal pair with Φ. The Gram matrix of Φ is

$$G = \begin{bmatrix} \frac{2}{3} & \frac{1}{6} & \frac{1}{6} \\ \frac{1}{6} & \frac{2}{3} & \frac{1}{6} \\ \frac{1}{6} & \frac{1}{6} & \frac{2}{3} \end{bmatrix}. \tag{2.133}$$

Using its inverse in (2.131a) yields

$$\begin{aligned} \widetilde{\varphi}_0 &= \frac{5}{3}\varphi_0 - \frac{1}{3}\varphi_1 - \frac{1}{3}\varphi_2, \\ \widetilde{\varphi}_1 &= -\frac{1}{3}\varphi_0 + \frac{5}{3}\varphi_1 - \frac{1}{3}\varphi_2, \\ \widetilde{\varphi}_2 &= -\frac{1}{3}\varphi_0 - \frac{1}{3}\varphi_1 + \frac{5}{3}\varphi_2. \end{aligned}$$

These functions are depicted in Figure 2.24(b). Since each $\widetilde{\varphi}_k$ is a linear combination of $\{\varphi_0, \varphi_1, \varphi_2\}$, it is clear that $\text{span}(\widetilde{\Phi}) \subseteq \text{span}(\Phi)$. One can also show that $\text{span}(\Phi) \subseteq \text{span}(\widetilde{\Phi})$. For an intuitive understanding, note that, similarly to φ_k, each $\widetilde{\varphi}_k$ satisfies $\widetilde{\varphi}_k(0) = \widetilde{\varphi}_k(3)$ and is piecewise-linear on $[0, 3]$ with breakpoints at 1 and 2; thus, the sets span the same subspace. While many sets of functions satisfy the biorthogonality condition (2.111) without satisfying $\operatorname{span}(\widetilde{\Phi}) = \operatorname{span}(\Phi)$, this solution satisfying both conditions is unique.

The dual of the dual of a basis is the original basis, and a basis is its own dual if and only if it is an orthonormal basis. Also, if Φ is a Riesz basis with optimal stability constants $\lambda_{\min}$ and $\lambda_{\max}$, then $\widetilde{\Phi}$ is a Riesz basis with optimal stability constants $1/\lambda_{\max}$ and $1/\lambda_{\min}$ (an instance of which we have seen in Example 2.43). Establishing these facts formally is left for Exercise 2.39. As we mentioned earlier, it can be advantageous for numerical computations to have $\lambda_{\min} \approx \lambda_{\max}$. The same property is then maintained by the dual.

Dual coefficients As we noted earlier in reference to computation of inner products, it is often convenient to use only one basis explicitly. Then, we face the problem of finding expansions with respect to the basis Φ from analysis with Φ^*. Unless $\Phi^* = \widetilde{\Phi}^*$, in which case Φ is an orthonormal basis, the coefficients obtained with analysis by Φ^* must be adjusted to be the right ones to use in synthesis by Φ. The adjustment of coefficients is analogous to computation of the dual basis.

To have an expansion with respect to the basis Φ from analysis with Φ^*, we seek an operator $A : \ell^2(\mathcal{K}) \to \ell^2(\mathcal{K})$ such that

$$x = \Phi A \Phi^* x \qquad \text{for every } x \in H.$$

It is easy to verify that $A = (\Phi^*\Phi)^{-1}$, the inverse of the Gram matrix, is the desired operator. In terms of the coefficient sequences defined in (2.113b) and (2.115), A maps $\widetilde{\alpha}$ to α, while the Gram matrix maps α to $\widetilde{\alpha}$.

Oblique projection Similarly to the truncation of an orthonormal expansion giving an orthogonal projection (Theorem 2.41), truncation of (2.114a) or (2.115) gives an oblique projection. Proof of the following result is left for Exercise 2.40.

THEOREM 2.47 (OBLIQUE PROJECTION) Let $\Phi_{\mathcal{I}} = \{\varphi_k\}_{k\in\mathcal{I}}$ and $\widetilde{\Phi}_{\mathcal{I}} = \{\widetilde{\varphi}_k\}_{k\in\mathcal{I}}$ be sets in Hilbert space H that satisfy

$$\langle \varphi_i, \widetilde{\varphi}_k \rangle = \delta_{i-k} \qquad \text{for every } i,\, k \in \mathcal{I}.$$

Then for any $x \in H$,

$$P_{\mathcal{I}}\, x = \sum_{k\in\mathcal{I}} \langle x, \widetilde{\varphi}_k \rangle \varphi_k \tag{2.134a}$$

$$= \Phi_{\mathcal{I}}\, \widetilde{\Phi}_{\mathcal{I}}^*\, x \tag{2.134b}$$

is a projection of x onto $S_{\mathcal{I}} = \overline{\operatorname{span}}(\{\varphi_k\}_{k\in\mathcal{I}})$. The residual satisfies $x - P_{\mathcal{I}}\, x \perp \widetilde{S}_{\mathcal{I}}$, where $\widetilde{S}_{\mathcal{I}} = \overline{\operatorname{span}}(\{\widetilde{\varphi}_k\}_{k\in\mathcal{I}})$.

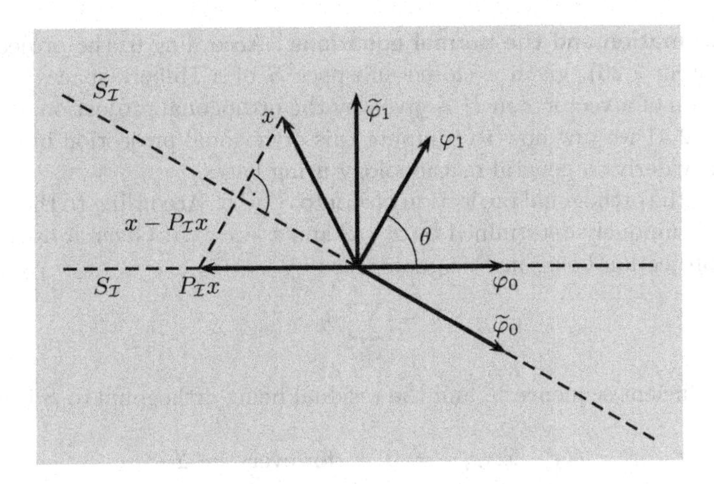

Figure 2.25 Example of an oblique projection. The projection is onto $S_{\mathcal{I}}$, the subspace spanned by φ_0. The projection is orthogonal to $\widetilde{S}_{\mathcal{I}}$, the subspace spanned by the biorthogonal vector $\widetilde{\varphi}_0$.

Example 2.45 (Oblique projection, Example 2.43 continued) We continue our discussion of Example 2.43 and illustrate oblique projection. Define $P_{\mathcal{I}}$ via (2.134) as

$$P_{\mathcal{I}}\, x \;=\; \langle x,\, \widetilde{\varphi}_0\rangle \varphi_0 \;=\; \Phi_{\mathcal{I}}\, \widetilde{\Phi}_{\mathcal{I}}^*\, x,$$

with

$$\Phi_{\mathcal{I}} \;=\; \begin{bmatrix}1 & 0\end{bmatrix}^{\top}, \qquad \widetilde{\Phi}_{\mathcal{I}} \;=\; \begin{bmatrix}1 & -\cot\theta\end{bmatrix}^{\top}.$$

Figure 2.25 illustrates the projection (not orthogonal anymore), the subspace $S_{\mathcal{I}}$, the residual $x - P_{\mathcal{I}}\, x$, and the subspace $\widetilde{S}_{\mathcal{I}}$.

While the above theorem gives an important property, it is not as useful as Theorem 2.41 because oblique projections do not solve best approximation problems.

Decomposition By applying Theorem 2.47 with any single-element set $\mathcal{I}$, we see that any one term of (2.114a) or (2.115) is an oblique projection onto a one-dimensional subspace. Thus, a biorthogonal pair of bases induces a pair of decompositions

$$H \;=\; \bigoplus_{k\in\mathcal{K}} S_{\{k\}} \qquad \text{and} \qquad H \;=\; \bigoplus_{k\in\mathcal{K}} \widetilde{S}_{\{k\}}, \tag{2.135}$$

where $S_{\{k\}} = \operatorname{span}(\varphi_k)$ and $\widetilde{S}_{\{k\}} = \operatorname{span}(\widetilde{\varphi}_k)$. Actually, because of linear independence and completeness, any basis gives a decomposition of the form above. A key merit of a decomposition that comes from a biorthogonal pair of bases is that the expansion coefficients are determined simply as in (2.113).

Best approximation and the normal equations According to the projection theorem (Theorem 2.26), given a closed subspace S of a Hilbert space H, the best approximation of a vector x in H is given by the orthogonal projection of x onto S. In Theorem 2.41 we saw how to compute this orthogonal projection in one special case. We now derive a general methodology using bases.

Denote the orthogonal projection of x onto S by $\widehat{x}$. According to the projection theorem, $\widehat{x}$ is uniquely determined by $\widehat{x} \in S$ and $x - \widehat{x} \perp S$. Given a basis $\{\varphi_k\}_{k\in\mathcal{I}}$ for S, the projection being in S is guaranteed by

$$\widehat{x} = \sum_{k\in\mathcal{I}} \beta_k \varphi_k \tag{2.136a}$$

for some coefficient sequence β, and the residual being orthogonal to S is expressed as

$$\langle x - \widehat{x}, \varphi_i\rangle = 0 \qquad \text{for every } i \in \mathcal{I}. \tag{2.136b}$$

Rearranging (2.136b) and substituting into (2.136a) gives

$$\langle x, \varphi_i\rangle = \langle \widehat{x}, \varphi_i\rangle = \langle \textstyle\sum_{k\in\mathcal{I}} \beta_k\varphi_k, \varphi_i\rangle = \displaystyle\sum_{k\in\mathcal{I}} \beta_k \langle \varphi_k, \varphi_i\rangle \qquad \text{for every } i \in \mathcal{I}.$$

Solving these equations gives the following result.

THEOREM 2.48 (NORMAL EQUATIONS) Given a vector x and a Riesz basis $\{\varphi_k\}_{k\in\mathcal{I}}$ for a closed subspace S in a separable Hilbert space H, the vector closest to x in S is

$$\widehat{x} = \sum_{k\in\mathcal{I}} \beta_k \varphi_k \tag{2.137a}$$

$$= \Phi\beta, \tag{2.137b}$$

where β is the unique solution to the system of equations

$$\sum_{k\in\mathcal{I}} \beta_k \langle \varphi_k, \varphi_i\rangle = \langle x, \varphi_i\rangle \qquad \text{for every } i \in \mathcal{I}, \quad \text{or,} \tag{2.138a}$$

$$\Phi^*\Phi\beta = \Phi^* x. \tag{2.138b}$$

Equations (2.138) are called *normal equations* because they express the normality (orthogonality) of the residual and the subspace (from (2.136b)). Invertibility of the Gram matrix $\Phi^*\Phi$ follows from $\{\varphi_k\}_{k\in\mathcal{I}}$ being a Riesz basis for S. In operator notation, using this invertibility in combining (2.137b) and (2.138b) leads to

$$\widehat{x} = \Phi(\Phi^*\Phi)^{-1}\Phi^* x = Px. \tag{2.139}$$

It is then easy to check that P is an orthogonal projection operator (see Theorem 2.29). If the set $\{\varphi_k\}_{k\in\mathcal{I}}$ is not a basis, the projection theorem still ensures that $\widehat{x}$ is unique, but naturally its expansion with respect to $\{\varphi_k\}_{k\in\mathcal{I}}$ is not unique. We illustrate these concepts with an example.

EXAMPLE 2.46 (NORMAL EQUATIONS IN $\mathbb{R}^3$) Let $\varphi_0 = \begin{bmatrix} 1 & 1 & 0 \end{bmatrix}^\top$ and $\varphi_1 = \begin{bmatrix} 0 & 1 & 1 \end{bmatrix}^\top$. Given a vector $x = \begin{bmatrix} 1 & 1 & 1 \end{bmatrix}^\top$, according to Theorem 2.48, the vector in $\operatorname{span}(\{\varphi_0, \varphi_1\})$ closest to x is

$$\widehat{x} = \beta_0\varphi_0 + \beta_1\varphi_1,$$

with β the unique solution to (2.138b), which simplifies to

$$\begin{bmatrix} 2 & 1 \\ 1 & 2 \end{bmatrix} \begin{bmatrix} \beta_0 \\ \beta_1 \end{bmatrix} = \begin{bmatrix} 2 \\ 2 \end{bmatrix}.$$

Solving the system yields $\beta_0 = \beta_1 = \frac{2}{3}$, leading to

$$\widehat{x} = \frac{2}{3}(\varphi_0 + \varphi_1) = \begin{bmatrix} \frac{2}{3} \\ \frac{4}{3} \\ \frac{2}{3} \end{bmatrix}.$$

We can easily check that the residual $x - \widehat{x}$ is orthogonal to $\operatorname{span}(\{\varphi_0, \varphi_1\})$:

$$x - \widehat{x} = \begin{bmatrix} \frac{2}{3} \\ -\frac{2}{3} \\ \frac{2}{3} \end{bmatrix} \perp \alpha_0\varphi_0 + \alpha_1\varphi_1 = \begin{bmatrix} \alpha_0 \\ \alpha_0 + \alpha_1 \\ \alpha_1 \end{bmatrix}.$$

Now let $\varphi_2 = \begin{bmatrix} 1 & 0 & -1 \end{bmatrix}^\top$. The vector in $\operatorname{span}(\{\varphi_0, \varphi_1, \varphi_2\})$ closest to x is

$$\widehat{x} = \beta_0\varphi_0 + \beta_1\varphi_1 + \beta_2\varphi_2,$$

where β satisfies

$$\begin{bmatrix} 2 & 1 & 1 \\ 1 & 2 & -1 \\ 1 & -1 & 2 \end{bmatrix} \begin{bmatrix} \beta_0 \\ \beta_1 \\ \beta_2 \end{bmatrix} = \begin{bmatrix} 2 \\ 2 \\ 0 \end{bmatrix}.$$

The solutions for β are not unique, but all solutions yield $\widehat{x} = \begin{bmatrix} \frac{2}{3} & \frac{4}{3} & \frac{2}{3} \end{bmatrix}^\top$ as before. This is as expected, since $\operatorname{span}(\{\varphi_0, \varphi_1, \varphi_2\}) = \operatorname{span}(\{\varphi_0, \varphi_1\})$.

In the special case when $\{\varphi_k\}_{k\in\mathcal{I}}$ is an orthonormal set, the normal equations (2.138) simplify greatly:

$$\langle x, \varphi_i \rangle \overset{(a)}{=} \sum_{k\in\mathcal{I}} \beta_k \langle \varphi_k, \varphi_i \rangle \overset{(b)}{=} \sum_{k\in\mathcal{I}} \beta_k \delta_{k-i} \overset{(c)}{=} \beta_i, \qquad \text{for every } i \in \mathcal{I},$$

where (a) is (2.138); (b) follows from orthonormality; and (c) from the definition of the Kronecker delta sequence, (2.9). Thus the coefficients of the expansion of $\widehat{x}$ with respect to $\{\varphi_k\}_{k\in\mathcal{I}}$ come from analysis with the same set of vectors, exactly as in Theorem 2.41.

Solving the normal equations is also simplified by having $\{\widetilde{\varphi}_k\}_{k\in\mathcal{I}}$ that together with $\{\varphi_k\}_{k\in\mathcal{I}}$ forms a biorthogonal pair of bases for the subspace of interest. Then by combination of (2.131b) and (2.139), the best approximation is $\widehat{x} = \widetilde{\Phi}\Phi^* x$.

In general, $\{\varphi_k\}_{k\in\mathcal{I}}$ is not an orthonormal set, and we might want to express both x and $\widehat{x}$ through expansions with some different basis $\{\psi_k\}_{k\in\mathcal{K}}$ that is not necessarily orthonormal itself,

$$x = \sum_{k\in\mathcal{K}} \alpha_k \psi_k \qquad \text{and} \qquad \widehat{x} = \sum_{k\in\mathcal{K}} \widehat{\alpha}_k \psi_k.$$

Properties of the mapping from α to $\widehat{\alpha}$ that make $\widehat{x}$ the orthogonal projection of x onto $\overline{\text{span}}(\{\varphi_k\}_{k\in\mathcal{I}})$ are established in Exercise 2.41. In particular, orthogonal projection in H corresponds to orthogonal projection in the coefficient space if and only if expansions are with respect to an orthonormal basis.

Successive approximation Continuing our discussion of best approximation, now consider the computation of a sequence of best approximations in subspaces of increasing dimension. Let $\{\varphi_i\}_{i\in\mathbb{N}}$ be a linearly independent set, and for each $k \in \mathbb{N}$, let $S_k = \text{span}(\{\varphi_0, \varphi_1, \ldots, \varphi_{k-1}\})$. Let $\widehat{x}^{(k)}$ denote the best approximation of x in S_k.[25] In Section 2.5.2, we found a simple recursive computation for the expansion of $\widehat{x}^{(k)}$ with respect to $\{\varphi_i\}_{i\in\mathbb{N}}$ for the case that $\{\varphi_i\}_{i\in\mathbb{N}}$ is an orthonormal set; see (2.108). Here, recursive computation is made more complicated by the lack of orthonormality of the basis.

The spaces S_k and S_{k+1} are nested with $S_k \subset S_{k+1}$, so approximating x with a vector from S_{k+1} instead of one from S_k cannot make the approximation quality worse; improvement is obtained by capturing the component of x that could not be captured before. The nesting of subspaces can be expressed as

$$S_{k+1} = S_k \oplus T_k, \tag{2.140}$$

where the one-dimensional subspace T_k is not uniquely specified. If we choose T_k to make this direct sum an orthogonal decomposition, then the increment $\widehat{x}^{(k+1)} - \widehat{x}^{(k)}$ will simply be the orthogonal projection of x onto T_k. The decomposition (2.140) is orthogonal when $T_k = \text{span}(\psi_k)$ with $\psi_k \perp S_k$, so we get the desired direct sum by choosing ψ_k parallel to the residual in orthogonally projecting φ_k to S_k. This approach simplifies the computation of the increment at the cost of requiring ψ_k. It can yield computation savings when the entire sequence of approximations is desired and the $\{\psi_k\}$ are computed recursively through Gram–Schmidt orthogonalization.

Let $\widehat{x}^{(0)} = \mathbf{0}$, and, for $k = 0, 1, \ldots,$ perform the following computations. First, compute ψ_k orthogonal to S_k and, for convenience in other computations, of unit norm,

$$v_k = \sum_{i=0}^{k-1} \langle \varphi_k, \psi_i \rangle \psi_i, \tag{2.141a}$$

$$\psi_k = \frac{\varphi_k - v_k}{\|\varphi_k - v_k\|}. \tag{2.141b}$$

In this computation, v_k is the orthogonal projection of φ_k onto S_k since $\{\psi_i\}_{i=0}^{k-1}$ is an orthonormal basis for S_k; see (2.106a). With this intermediate orthogonalization,

[25] By these definitions, $S_0 = \{\mathbf{0}\}$ and $\widehat{x}^{(0)} = \mathbf{0}$.

we have

$$\widehat{x}^{(k+1)} = \widehat{x}^{(k)} + \langle x, \psi_k \rangle \psi_k. \tag{2.141c}$$

Exercise 2.42 explores the connection between this algorithm and the normal equations.

2.5.4 Frames

Bases are sets of vectors that are complete and yield unique expansions (see Definition 2.34). In a finite-dimensional space, the existence of expansions lower bounds the number of vectors by the dimension of the space, while uniqueness upper bounds the number of vectors by the dimension of the space; thus, there are exactly as many vectors as the dimension of the space. Frames are more general than bases because they are complete, but the expansions they yield are not necessarily unique. In a finite-dimensional space, a frame must have at least as many vectors as the dimension of the space. In infinite-dimensional spaces, a frame must have infinitely many vectors, and imposing something analogous to the Riesz basis condition (2.89) prevents certain pathologies.

Why would we want more than the minimum number of vectors for completeness? There are several possible disadvantages: uniqueness of expansions is lost, and it would seem at first glance that having a larger set of vectors implies more computations both in analysis and in synthesis. The primary advantages come from flexibility in design: fixing analysis leaves flexibility in synthesis and vice versa.

DEFINITION 2.49 (FRAME) The set of vectors $\Phi = \{\varphi_k\}_{k \in \mathcal{J}} \subset H$, where $\mathcal{J}$ is finite or countably infinite, is called a *frame* for a Hilbert space H when there exist positive real numbers $\lambda_{\min}$ and $\lambda_{\max}$ such that

$$\lambda_{\min} \|x\|^2 \le \sum_{k \in \mathcal{J}} |\langle x, \varphi_k \rangle|^2 \le \lambda_{\max} \|x\|^2, \qquad \text{for every } x \text{ in } H. \tag{2.142}$$

The constants $\lambda_{\min}$ and $\lambda_{\max}$ are called *frame bounds*. The largest such $\lambda_{\min}$ and smallest such $\lambda_{\max}$ are called *optimal frame bounds* for Φ.

A frame is sometimes called a *Riesz sequence*. This highlights the similarity of (2.142) to condition (2.89) in Definition 2.35 for Riesz bases as well as the fact that a frame is not necessarily a basis.

Let us immediately compare and contrast Definitions 2.35 and 2.49.

(i) The notation for the index set has been changed from $\mathcal{K}$ to $\mathcal{J}$ to reflect the fact that these are not generally the same size when H has finite dimension.

(ii) The definition of a frame in Definition 2.49 uses the set Φ in the analysis of x; by contrast, the definition of a basis in Definition 2.35 uses the set Φ in the synthesis of x. Nevertheless, both bases and frames can be used for both analysis and synthesis. A frame generally lacks uniqueness of expansions of the form $x = \sum_{k \in \mathcal{J}} \alpha_k \varphi_k$; this prevents a closer parallel in the definitions.

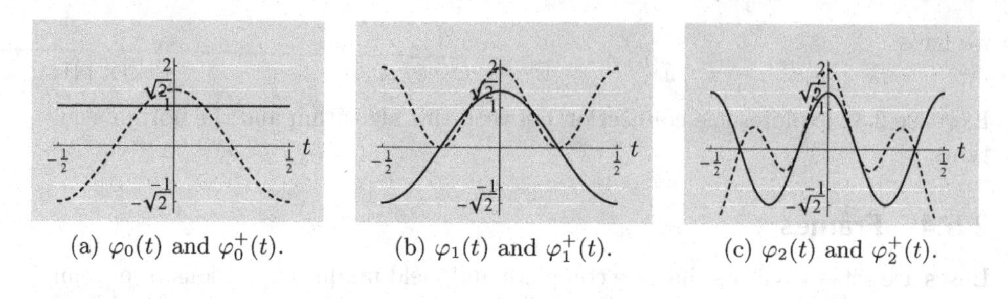

(a) $\varphi_0(t)$ and $\varphi_0^+(t)$.

(b) $\varphi_1(t)$ and $\varphi_1^+(t)$.

(c) $\varphi_2(t)$ and $\varphi_2^+(t)$.

Figure 2.26 Example frame functions from $\Phi \cup \Phi^+$ (solid lines for functions from Φ and dashed lines for functions from Φ^+).

(iii) If Φ and $\widetilde{\Phi}$ form a biorthogonal pair of bases, then the unique expansion with respect to $\widetilde{\Phi}$ is obtained through analysis with Φ; see (2.116). In this case, comparison of Definitions 2.35 and 2.49 shows that $\widetilde{\Phi}$ being a Riesz basis with stability constants $\lambda_{\min}$ and $\lambda_{\max}$ implies that Φ is a frame with frame bounds $\lambda_{\min}$ and $\lambda_{\max}$. Since the dual of a Riesz basis is a Riesz basis, we reach the simple conclusion that any Riesz basis is a frame.

Uses of frames in analysis and synthesis will be established shortly. Exercise 2.43 explores the differences between Definitions 2.35 and 2.49 further.

Example 2.47 (Frame of cosine functions) Starting with $\Phi = \{\varphi_k\}_{k\in\mathbb{N}} \subset \mathcal{L}^2([-\frac{1}{2}, \frac{1}{2}])$ from (2.24), define the set of functions $\Phi^+ = \{\varphi_k^+\}_{k\in\mathbb{N}}$ by multiplying each φ_k by $\sqrt{2}\cos(2\pi t)$,

$$\varphi_k^+(t) = \sqrt{2}\cos(2\pi t)\,\varphi_k(t), \qquad k \in \mathbb{N}. \tag{2.143}$$

A few functions from $\Phi \cup \Phi^+$ are shown in Figure 2.26.

We know already from Example 2.34 that Φ is an orthonormal basis for the closure of its span, $S = \overline{\text{span}}(\Phi)$. The union $\Phi \cup \Phi^+$ is a frame for S. To see that the closure of the span of $\Phi \cup \Phi^+$ is not larger than S, note that each φ_k^+ can be written as a linear combination of elements of Φ. The first element of Φ^+ is

$$\varphi_0^+(t) = \sqrt{2}\cos(2\pi t)\,\varphi_0(t) = \varphi_1(t). \tag{2.144a}$$

For $k \in \mathbb{Z}^+$,

$$\begin{aligned} \varphi_k^+(t) &= \sqrt{2}\cos(2\pi t)\varphi_k(t) = 2\cos(2\pi t)\cos(2\pi k t) \\ &= \cos(2\pi(k-1)t) + \cos(2\pi(k+1)t) \\ &= \begin{cases} \varphi_0(t) + (1/\sqrt{2})\varphi_2(t), & \text{for } k = 1; \\ (1/\sqrt{2})\varphi_{k-1}(t) + (1/\sqrt{2})\varphi_{k+1}(t), & \text{for } k = 2, 3, \ldots. \end{cases} \end{aligned} \tag{2.144b}$$

Computing the frame bounds of this frame is left for Exercise 2.44.

Operators associated with frames Analogously to bases, we can define the *synthesis operator* associated with $\{\varphi_k\}_{k\in\mathcal{J}}$ to be

$$\Phi : \ell^2(\mathcal{J}) \to H, \qquad \text{with} \qquad \Phi\alpha \;=\; \sum_{k\in\mathcal{J}} \alpha_k\varphi_k. \tag{2.145}$$

The second inequality of (2.142) implies that the norm of this linear operator is finite and the operator thus bounded.

Similarly, we define the *analysis operator* associated with $\{\varphi_k\}_{k\in\mathcal{J}}$ to be

$$\Phi^* : H \to \ell^2(\mathcal{J}), \qquad \text{with} \qquad (\Phi^*x)_k \;=\; \langle x,\, \varphi_k\rangle, \quad k \in \mathcal{J}. \tag{2.146}$$

The norm of the analysis operator is the same as that of the synthesis operator.

The power of the operator notation can be seen in rephrasing (2.142) as

$$\lambda_{\min} I \;\leq\; \Phi\Phi^* \;\leq\; \lambda_{\max} I. \tag{2.147}$$

The first inequality can be derived as follows:

$$\begin{aligned}\langle(\Phi\Phi^* - \lambda_{\min}I)x,\, x\rangle &\overset{(a)}{=} \langle\Phi\Phi^*x,\, x\rangle - \langle\lambda_{\min}Ix,\, x\rangle \overset{(b)}{=} \langle\Phi\Phi^*x,\, x\rangle - \lambda_{\min}\langle x,\, x\rangle \\ &\overset{(c)}{=} \langle\Phi^*x,\, \Phi^*x\rangle - \lambda_{\min}\langle x,\, x\rangle \;=\; \|\Phi^*x\|^2 - \lambda_{\min}\|x\|^2 \overset{(d)}{\geq} 0,\end{aligned}$$

where (a) follows from the distributivity of the inner product; (b) from the linearity in the first argument of the inner product and the meaning of the identity operator; (c) from the definition of the adjoint; and (d) from the first inequality of (2.142). The second inequality of (2.147) can be derived similarly.

Because $\Phi\Phi^*$ is a Hermitian operator, the operator analogue of (2.243) from Appendix 2.B.2 holds; thus, the optimal frame bounds are the smallest and largest eigenvalues of $\Phi\Phi^*$. This gives an easy way to find the optimal frame bounds, as we illustrate in the following example.

EXAMPLE 2.48 (FRAMES IN $\mathbb{R}^2$) In (2.15), we defined a frame, $\varphi_0 = \begin{bmatrix}1 & 0\end{bmatrix}^\top$, $\varphi_1 = \begin{bmatrix}0 & 1\end{bmatrix}^\top$, and $\varphi_2 = \begin{bmatrix}-1 & -1\end{bmatrix}^\top$. These vectors are clearly not linearly independent; however, they do satisfy (2.142). To compute the optimal frame bounds, we could follow the path from Example 2.31: Determine an expression for $\sum_{k\in\mathcal{J}}|\langle x,\, \varphi_k\rangle|^2$ and find the optimal frame bounds as the infimum and supremum of $\sum_{k\in\mathcal{J}}|\langle x,\, \varphi_k\rangle|^2/(x_0^2+x_1^2)$. An easier way is to use the operator notation, where Φ is given in (2.17a). This Φ is a short rectangular matrix, so Φ^* is a tall rectangular matrix, illustrating the fact that a frame expansion is overcomplete. Then, $\Phi\Phi^*$ is

$$\Phi\Phi^* \;=\; \begin{bmatrix}2 & 1\\ 1 & 2\end{bmatrix} \;=\; \underbrace{\frac{1}{\sqrt{2}}\begin{bmatrix}1 & -1\\ 1 & 1\end{bmatrix}}_{V}\,\underbrace{\begin{bmatrix}3 & 0\\ 0 & 1\end{bmatrix}}_{\Lambda}\,\underbrace{\frac{1}{\sqrt{2}}\begin{bmatrix}1 & 1\\ -1 & 1\end{bmatrix}}_{V^{-1}},$$

where we have performed an eigendecomposition of the Hermitian matrix $\Phi\Phi^*$ via (2.227a). We can immediately read the smallest and largest eigenvalues, $\lambda_{\min} = 1$ and $\lambda_{\max} = 3$, as the optimal frame bounds.

In many ways, including expansion and inner product computation, frames play the same roles as bases. When a frame lacks uniqueness of expansions, it cannot induce a subspace decomposition because of the uniqueness requirement in Definition 2.31. The connection between frames and projections is more subtle. We now develop these ideas further, covering the special case of tight frames first.

Tight frames

DEFINITION 2.50 (TIGHT FRAME) The frame $\Phi = \{\varphi_k\}_{k \in \mathcal{J}} \subset H$, where $\mathcal{J}$ is finite or countably infinite, is called a *tight frame*, or a *λ-tight frame*, for a Hilbert space H when its optimal frame bounds are equal, $\lambda_{\min} = \lambda_{\max} = \lambda$.

For a λ-tight frame, (2.147) simplifies to

$$\Phi\Phi^* = \lambda I. \tag{2.148}$$

A tight frame is a counterpart of an orthonormal basis, as we will see shortly.

EXAMPLE 2.49 (FINITE-DIMENSIONAL TIGHT FRAME) Take the following three vectors as a frame for $\mathbb{R}^2$:

$$\varphi_0 = \begin{bmatrix} 1 \\ 0 \end{bmatrix}, \quad \varphi_1 = \begin{bmatrix} -\frac{1}{2} \\ \frac{\sqrt{3}}{2} \end{bmatrix}, \quad \varphi_2 = \begin{bmatrix} -\frac{1}{2} \\ -\frac{\sqrt{3}}{2} \end{bmatrix}, \tag{2.149a}$$

$$\Phi = \begin{bmatrix} 1 & -\frac{1}{2} & -\frac{1}{2} \\ 0 & \frac{\sqrt{3}}{2} & -\frac{\sqrt{3}}{2} \end{bmatrix}. \tag{2.149b}$$

Computing its optimal frame bounds as we did in Example 2.48, we find that

$$\Phi\Phi^* = \frac{3}{2} I,$$

and thus the eigenvalues are $\lambda_{\min} = \lambda_{\max} = \frac{3}{2}$, and the frame is tight. Note that this frame is just a normalized version of the one in (2.16), which is a 1-tight frame.

We can normalize any λ-tight frame by pulling $1/\sqrt{\lambda}$ into the sum in (2.142) to yield a 1-tight frame,

$$\sum_k \left|\langle x, \lambda^{-1/2}\widetilde{\varphi}_k\rangle\right|^2 = \sum_k \left|\langle x, \widetilde{\varphi}'_k\rangle\right|^2 = \|x\|^2. \tag{2.150}$$

Because of this normalization, we can associate a 1-tight frame with any tight frame. Note that orthonormal bases are 1-tight frames with all unit-norm vectors. In general, the vectors in a 1-tight frame do not have unit norms or even equal norms.

Expansion and inner product computation Expansion coefficients with respect to a 1-tight frame can be obtained by using the same 1-tight frame for signal analysis.

THEOREM 2.51 (1-TIGHT FRAME EXPANSIONS) Let $\Phi = \{\varphi_k\}_{k\in\mathcal{J}}$ be a 1-tight frame for a Hilbert space H. Analysis of any x in H gives expansion coefficients in $\ell^2(\mathcal{J})$

$$\alpha_k = \langle x, \varphi_k\rangle \qquad \text{for } k \in \mathcal{J}, \text{ or,} \tag{2.151a}$$

$$\alpha = \Phi^* x. \tag{2.151b}$$

Synthesis with these coefficients yields

$$x = \sum_{k\in\mathcal{J}} \langle x, \varphi_k\rangle \varphi_k \tag{2.152a}$$

$$= \Phi\alpha = \Phi\Phi^* x. \tag{2.152b}$$

Note the apparent similarity of this theorem to Theorem 2.39. The equations in these theorems are identical, and each theorem shows that analysis and synthesis with the same set of vectors yields an identity on H. In the orthonormal basis case, the expansion is *unique*; in the 1-tight frame case it is generally not. The α given by (2.151b) can be replaced by any $\alpha' = \alpha + \alpha^{\perp}$, where $\alpha^{\perp}$ is in the null space of Φ, while maintaining $x = \Phi\alpha'$.

The theorem follows from two simple facts: $\alpha \in \ell^2(\mathcal{J})$ because Φ^* is a bounded operator; and

$$\Phi\Phi^* = I \quad \text{on} \quad H \tag{2.153}$$

by setting $\lambda = 1$ in (2.148). This leads to Parseval equalities for 1-tight frames.

THEOREM 2.52 (PARSEVAL EQUALITIES FOR 1-TIGHT FRAMES) Let $\Phi = \{\varphi_k\}_{k\in\mathcal{J}}$ be a 1-tight frame for a Hilbert space H. Expansion with coefficients (2.151) satisfies

$$\|x\|^2 = \sum_{k\in\mathcal{J}} |\langle x, \varphi_k\rangle|^2 \tag{2.154a}$$

$$= \|\Phi^* x\|^2 = \|\alpha\|^2. \tag{2.154b}$$

More generally,

$$\langle x, y\rangle = \sum_{k\in\mathcal{J}} \langle x, \varphi_k\rangle \langle y, \varphi_k\rangle^* \tag{2.155a}$$

$$= \langle \Phi^* x, \Phi^* y\rangle = \langle \alpha, \beta\rangle. \tag{2.155b}$$

This theorem looks formally the same as Theorem 2.40, but it applies to a different type of a set of vectors. Because of this theorem, frames that are 1-tight are called *Parseval tight frames.*

The norm-preservation property of Theorem 2.52 could be misleading. We often work with tight frames of unit-norm vectors rather than with unit frame bound; thus, norm preservation is replaced by a constant scaling, as illustrated in the following example.

EXAMPLE 2.50 (PARSEVAL EQUALITIES FOR TIGHT FRAMES) Let us continue with the frame from (2.16). Its vectors are all of norm $\frac{2}{3}$. Normalizing it so that all of its vectors are of unit norm yields the frame in (2.149). Computing the norm squared of the expansion coefficient vector $\|\alpha\|^2$ for this frame yields

$$\|\alpha\|^2 = \|\Phi^* x\|^2 = \left\| \begin{bmatrix} 1 & 0 \\ -\frac{1}{2} & \frac{\sqrt{3}}{2} \\ -\frac{1}{2} & -\frac{\sqrt{3}}{2} \end{bmatrix} \begin{bmatrix} x_0 \\ x_1 \end{bmatrix} \right\|^2 = \left\| \begin{bmatrix} x_0 \\ -\frac{x_0 - \sqrt{3}x_1}{2} \\ -\frac{x_0 + \sqrt{3}x_1}{2} \end{bmatrix} \right\|^2 = \frac{3}{2}\|x\|^2.$$

This tells us that, for this tight frame with all unit-norm vectors, the norm of the expansion coefficients is $\frac{3}{2}$ times larger than that of the vector itself. This is intuitive as we have $\frac{3}{2}$ times more vectors than needed for an expansion in $\mathbb{R}^2$.

This example generalizes to all finite-dimensional tight frames with unit-norm vectors. For such frames, the factor appearing in the analogue to the Parseval equality denotes the *redundancy* of the frame.

Inverse synthesis and analysis For a 1-tight frame, (2.153) shows that the synthesis operator is a left inverse of the analysis operator. Unlike with an orthonormal basis, the synthesis operator associated with a 1-tight frame is generally not a right inverse (hence, not an inverse) of the analysis operator because $\Phi^*\Phi \neq I$. In finite dimensions, this can be seen easily from the rank of $\Phi^*\Phi$; the rank of $\Phi^*\Phi$ is the dimension of H, but $\Phi^*\Phi$ is an operator on $\mathbb{C}^{|\mathcal{J}|}$, where $|\mathcal{J}|$ might be larger than the dimension of H.

EXAMPLE 2.51 (INVERSE RELATIONSHIP FOR FRAME OPERATORS) Let Φ be the 1-tight frame from (2.16). We have already seen that $\Phi\Phi^* = I_{2\times 2}$. We also have

$$\Phi^*\Phi = \begin{bmatrix} \frac{2}{3} & -\frac{1}{3} & -\frac{1}{3} \\ -\frac{1}{3} & \frac{2}{3} & -\frac{1}{3} \\ -\frac{1}{3} & -\frac{1}{3} & \frac{2}{3} \end{bmatrix} \neq I_{3\times 3}.$$

The rank of $\Phi^*\Phi$ is 2.

Orthogonal projection Since a frame for H generally has more than the minimum number of vectors needed to span H, omitting some terms from the synthesis sum (2.152) does not necessarily restrict the result to a proper subspace of H. Thus, a frame (even a 1-tight frame) does not yield a result analogous to Theorem 2.41 for computing orthogonal projections on H.

A different orthogonal projection property is easy to verify for any 1-tight frame: $\Phi^*\Phi : \ell^2(\mathcal{J}) \to \ell^2(\mathcal{J})$ is the orthogonal projection onto $\mathcal{R}(\Phi^*)$. This has important consequences for robustness to noise, such as in oversampled analog-to-digital conversion.

General frames

Tight frames are a small class of frames as defined in Definition 2.49. A frame might generally have optimal frame bounds that differ, the distance between which gives us information about the quality of the frame.

Dual frame pairs and expansion When a frame Φ is not 1-tight, to find expansion coefficients with respect to Φ with a linear operator requires a second frame, in analogy to biorthogonal pairs of bases.

DEFINITION 2.53 (DUAL PAIR OF FRAMES) The sets of vectors $\Phi = \{\varphi_k\}_{k\in\mathcal{J}} \subset H$ and $\widetilde{\Phi} = \{\widetilde{\varphi}_k\}_{k\in\mathcal{J}} \subset H$, where $\mathcal{J}$ is finite or countably infinite, are called a *dual pair of frames* for a Hilbert space H when

(i) each is a *frame* for H; and

(ii) for any x in H,

$$x = \sum_{k\in\mathcal{K}} \langle x, \widetilde{\varphi}_k\rangle \varphi_k \tag{2.156a}$$

$$= \Phi\widetilde{\Phi}^* x. \tag{2.156b}$$

Note that this definition combines the roles of Definition 2.43 and Theorem 2.44 for general frames. This is necessary because no simple pairwise condition between vectors like (2.111) will imply (2.156).

EXAMPLE 2.52 (DUAL PAIRS OF FRAMES IN $\mathbb{R}^2$) Let Φ be the frame for $\mathbb{R}^2$ defined in (2.15). Since synthesis operator Φ is a 2×3 matrix with rank 2, it has infinitely many right inverses; any right inverse specifies a frame that forms a dual pair with Φ. Examples include the following:

$$\left\{\begin{bmatrix}1\\1\end{bmatrix}, \begin{bmatrix}0\\2\end{bmatrix}, \begin{bmatrix}0\\1\end{bmatrix}\right\}, \quad \left\{\begin{bmatrix}0\\1\end{bmatrix}, \begin{bmatrix}-1\\2\end{bmatrix}, \begin{bmatrix}-1\\1\end{bmatrix}\right\}, \quad \left\{\begin{bmatrix}\frac{2}{3}\\-\frac{1}{3}\end{bmatrix}, \begin{bmatrix}-\frac{1}{3}\\\frac{2}{3}\end{bmatrix}, \begin{bmatrix}-\frac{1}{3}\\-\frac{1}{3}\end{bmatrix}\right\}.$$

The first of these examples demonstrates that a frame can have collinear elements; a frame can furthermore include the same vector multiple times.[26] The third of these examples is the canonical dual, which will be defined shortly.

[26] Allowing multiplicities generalizes the concept of a set to *multisets*, but we will continue to use the simpler term.

Inner product computation Suppose that Φ is a frame for H, and $x = \Phi\alpha$ and $y = \Phi\beta$. Then, just as in (2.122), we can write

$$\langle x, y \rangle = \langle \Phi\alpha, \Phi\beta \rangle = \langle G\alpha, \beta \rangle = \beta^* G\alpha, \tag{2.157}$$

where $G = \Phi^*\Phi$ is the Gram matrix defined in (2.121). This shows how to use frame expansion coefficients to convert an inner product in H into an inner product in $\ell^2(\mathcal{J})$.

The key difference between the Gram matrix of a frame and the Gram matrix of a basis is that G is now not necessarily invertible. In fact, it is an invertible bounded operator if and only if the frame is a Riesz basis.

Inverse analysis and synthesis Condition (2.156b) shows that if sets Φ and $\widetilde{\Phi}$ are a dual pair of frames, synthesis operator Φ is a left inverse of analysis operator $\widetilde{\Phi}^*$. As we saw before with 1-tight frames, Φ is generally not a right inverse (hence, not an inverse) of $\widetilde{\Phi}^*$ because $\widetilde{\Phi}^*\Phi \neq I$.

The roles of the two frames in a dual pair of frames can be reversed, so (2.156) becomes

$$x = \sum_{k \in \mathcal{K}} \langle x, \varphi_k \rangle \widetilde{\varphi}_k \tag{2.158a}$$

$$= \widetilde{\Phi}\Phi^* x. \tag{2.158b}$$

Thus, the synthesis operator $\widetilde{\Phi}$ is the left inverse of the analysis operator Φ^*. This and several other elementary properties of dual pairs of frames are established in Exercise 2.46.

Oblique projection If sets Φ and $\widetilde{\Phi}$ are a dual pair of frames, the operator $P = \widetilde{\Phi}^*\Phi$ is a projection operator. Checking the idempotency of P is straightforward:

$$P^2 = (\widetilde{\Phi}^*\Phi)(\widetilde{\Phi}^*\Phi) = \widetilde{\Phi}^*(\Phi\widetilde{\Phi}^*)\Phi \overset{(a)}{=} \widetilde{\Phi}^* I \Phi = \widetilde{\Phi}^*\Phi = P, \tag{2.159}$$

where (a) follows from synthesis operator Φ being a left inverse of analysis operator $\widetilde{\Phi}^*$.

Canonical dual frame So far we have derived properties of a dual pair of frames without regard for how to find such a pair. Given one frame Φ, there are infinitely many frames $\widetilde{\Phi}$ that complete a dual pair with Φ. There is a unique choice called the *canonical dual frame*[27] that is important because it leads to an orthogonal projection operator on $\ell^2(\mathcal{J})$.

For sets Φ and $\widetilde{\Phi}$ to form a dual pair of frames requires the associated operators to satisfy $\Phi\widetilde{\Phi}^* = I$ on H; see (2.156b). As established in (2.159), this makes

[27]Some authors use *dual* to mean *canonical dual.* We will *not* adopt this potentially confusing shorthand because it obscures the possible advantages that come from flexibility in the choice of a dual.

$P = \widetilde{\Phi}^*\Phi$ a projection operator. When, in addition, P is self-adjoint, it is an *orthogonal* projection operator. Setting

$$\widetilde{\Phi} = (\Phi\Phi^*)^{-1}\,\Phi \tag{2.160a}$$

satisfies (2.156b) and yields

$$P = \widetilde{\Phi}^*\Phi = \left((\Phi\Phi^*)^{-1}\,\Phi\right)^*\Phi = \Phi^*\,(\Phi\Phi^*)^{-1}\,\Phi,$$

which is self-adjoint. From (2.160a), the elements of the canonical dual are

$$\widetilde{\varphi}_k = (\Phi\Phi^*)^{-1}\,\varphi_k, \qquad k \in \mathcal{J}. \tag{2.160b}$$

When $H = \mathbb{C}^N$ (or $\mathbb{R}^N$), the computations in (2.160) are straightforward; for example, the third dual frame in Example 2.52 is a canonical dual. In general, it is difficult to make these computations without first expressing the linear operator $\Phi\Phi^*$ using a basis.

2.5.5 Matrix representations of vectors and linear operators

A basis for H creates a one-to-one correspondence between vectors in H and sequences in $\ell^2(\mathcal{K})$. As discussed in Section 2.5.2, an orthonormal basis preserves geometry (inner products) in this correspondence; see Figure 2.19. Even without orthonormality, using a basis is a key step toward computational feasibility because a basis allows us to do all computations with sequences. Here our intuition from finite dimensions might get in the way of appreciating what we have gained because we take the basis in finite dimensions for granted. Computations in the Hilbert spaces $\mathbb{C}^N$ are relatively straightforward in part because we use the standard basis automatically. Computations in other Hilbert spaces can be considerably more complicated; for example, integrating to compute an $\mathcal{L}^2(\mathbb{R})$ inner product can be difficult. With sequences, the greatest difficulty is that if the space is infinite-dimensional, the computation might require some truncation. Limiting our attention to sequences in $\ell^2(\mathcal{K})$ ensures that the truncation can be done with small relative error; details are deferred to Chapter 6.

We get the most benefit from our experience with finite-dimensional linear algebra by thinking of sequences in $\ell^2(\mathcal{K})$ as (possibly infinite) column vectors. A linear operator can then be represented with ordinary matrix–vector multiplication by a (possibly infinite) matrix. One goal in the choice of bases for the domain and codomain of the operator is to make this matrix simple. Like a basis, a frame also enables representations using sequences; however, lack of uniqueness of the representation creates some additional intricacies that are explored in Exercises 2.43–2.48.

Change of basis: Orthonormal bases Let $\Phi = \{\varphi_k\}_{k\in\mathcal{K}}$ and $\Psi = \{\psi_k\}_{k\in\mathcal{K}}$ be orthonormal bases for a Hilbert space H. Since bases provide unique representations, for any x in H we can use synthesis operators to write $x = \Phi\alpha$ and $x = \Psi\beta$ for unique α and β in $\ell^2(\mathcal{K})$. The operator $C_{\Phi,\Psi} : \ell^2(\mathcal{K}) \to \ell^2(\mathcal{K})$ that maps α to β is a *change of basis* from Φ to Ψ.

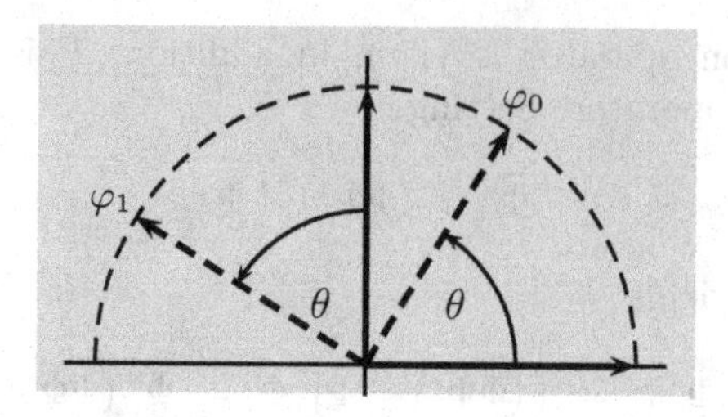

Figure 2.27 An orthonormal basis in $\mathbb{R}^2$ (dashed lines) generated by rotation of the standard basis (solid lines).

Since Ψ has inverse Ψ^*, we could simply write $C_{\Phi,\Psi} = \Psi^*\Phi$. This solves our problem because

$$C_{\Phi,\Psi}\alpha \;=\; (\Psi^*\Phi)\alpha \;=\; \Psi^*(\Phi\alpha) \;=\; \Psi^* x \;=\; \beta.$$

In a finite-dimensional setting, this is a perfectly adequate solution because we know how to interpret $\Psi^*\Phi$ as a product of matrices. We illustrate this in the following example.

EXAMPLE 2.53 (CHANGE OF BASIS BY ROTATION) Let $\{\varphi_0, \varphi_1\}$ be the basis for $\mathbb{R}^2$ shown in Figure 2.27,

$$\varphi_0 = \begin{bmatrix} \cos\theta \\ \sin\theta \end{bmatrix}, \qquad \varphi_1 = \begin{bmatrix} -\sin\theta \\ \cos\theta \end{bmatrix}.$$

Let $\{\psi_0, \psi_1\}$ be the standard basis for $\mathbb{R}^2$. The change of basis matrix from Φ to Ψ is

$$C_{\Phi,\Psi} \;=\; \Psi^*\Phi \;=\; \begin{bmatrix} 1 & 0 \\ 0 & 1 \end{bmatrix} \begin{bmatrix} \cos\theta & -\sin\theta \\ \sin\theta & \cos\theta \end{bmatrix} \;=\; \begin{bmatrix} \cos\theta & -\sin\theta \\ \sin\theta & \cos\theta \end{bmatrix}.$$

Consider the vector in $\mathbb{R}^2$ that has representation $\alpha = \begin{bmatrix} 1 & 0 \end{bmatrix}^\top$ *with respect to* Φ (not with respect to the standard basis). This means that the vector is

$$x \;=\; 1 \cdot \varphi_0 + 0 \cdot \varphi_1 \;=\; \varphi_0 \;=\; \begin{bmatrix} \cos\theta \\ \sin\theta \end{bmatrix},$$

where the final expression is with respect to the standard basis Ψ. This agrees with the result of the multiplication $C_{\Phi,\Psi}\alpha$.

Multiplying by $C_{\Phi,\Psi}$ is a counterclockwise rotation by angle θ. This agrees with the fact that the basis Φ is the standard basis Ψ rotated counterclockwise by θ.

In the previous example, since multiplication by a 2×2 matrix is simple, it makes little difference whether we interpret $\Psi^*\Phi$ as a composition of two operators or as a single operator. In general, $C_{\Phi,\Psi}$ should not be implemented as a composition of

Φ followed by Ψ^* because we do not want to return to computations in H, which might be more complicated than computations on coefficient sequences. Instead, we would like to think of $C_{\Phi,\Psi}$ as a $|\mathcal{K}| \times |\mathcal{K}|$ matrix, even if $|\mathcal{K}|$ is not finite.

Because of linearity, we can form the matrix $C_{\Phi,\Psi}$ by finding $C_{\Phi,\Psi}\alpha$ for particular values of α. Let $\alpha = e_k$, where e_k is the element of the standard basis for $\ell^2(\mathcal{K})$ with a 1 in position k. Then, $x = \Phi\alpha = \varphi_k$. Since Ψ is an orthonormal basis, the unique expansion of x with respect to Ψ is

$$x = \sum_{i\in\mathcal{K}} \langle x, \psi_i\rangle \psi_i = \sum_{i\in\mathcal{K}} \langle \varphi_k, \psi_i\rangle \psi_i,$$

from which we read off the ith coefficient of β in $x = \Psi\beta$ as $\beta_i = \langle \varphi_k, \psi_i\rangle$. This implies that column k of matrix $C_{\Phi,\Psi}$ is $(\langle \varphi_k, \psi_i\rangle)_{i\in\mathcal{K}}$. The full matrix, written for the case of $\mathcal{K} = \mathbb{Z}$, is

$$C_{\Phi,\Psi} = \begin{bmatrix} & \vdots & \vdots & \vdots & \\ \cdots & \langle\varphi_{-1}, \psi_{-1}\rangle & \langle\varphi_0, \psi_{-1}\rangle & \langle\varphi_1, \psi_{-1}\rangle & \cdots \\ \cdots & \langle\varphi_{-1}, \psi_0\rangle & \boxed{\langle\varphi_0, \psi_0\rangle} & \langle\varphi_1, \psi_0\rangle & \cdots \\ \cdots & \langle\varphi_{-1}, \psi_1\rangle & \langle\varphi_0, \psi_1\rangle & \langle\varphi_1, \psi_1\rangle & \cdots \\ & \vdots & \vdots & \vdots & \end{bmatrix}. \tag{2.161}$$

Example 2.54 (Change to standard basis) Let $\Phi = \{\varphi_k\}_{k\in\mathbb{Z}}$ be any orthonormal basis for $\ell^2(\mathbb{Z})$, and let Ψ be the standard basis for $\ell^2(\mathbb{Z})$. Then for any integers k and i,

$$\langle \varphi_k, \psi_i\rangle = \varphi_{k,i},$$

the ith-indexed entry of the kth vector of Φ. The change of basis operator (2.161) simplifies to

$$C_{\Phi,\Psi} = \begin{bmatrix} & \vdots & \vdots & \vdots & \\ \cdots & \varphi_{-1,-1} & \varphi_{0,-1} & \varphi_{1,-1} & \cdots \\ \cdots & \varphi_{-1,0} & \boxed{\varphi_{0,0}} & \varphi_{1,0} & \cdots \\ \cdots & \varphi_{-1,1} & \varphi_{0,1} & \varphi_{1,1} & \cdots \\ & \vdots & \vdots & \vdots & \end{bmatrix},$$

a matrix with the initial basis elements as columns.

Change of basis: Biorthogonal pairs of bases We now derive the change of basis operator without assuming that the bases are orthonormal. Let $\Phi = \{\varphi_k\}_{k\in\mathcal{K}}$ and $\Psi = \{\psi_k\}_{k\in\mathcal{K}}$ be bases for a Hilbert space H. For any x in H, we can again write $x = \Phi\alpha$ and $x = \Psi\beta$ for unique α and β in $\ell^2(\mathcal{K})$.

Since Ψ must be invertible, we could simply write $C_{\Phi,\Psi} = \Psi^{-1}\Phi$ because then

$$C_{\Phi,\Psi}\alpha = (\Psi^{-1}\Phi)\alpha = \Psi^{-1}(\Phi\alpha) = \Psi^{-1}x = \beta.$$

As before, we would not want to implement $C_{\Phi,\Psi}$ as a composition of two operators, where the first returns computations to H. Here we have the additional complication that Φ^{-1} might be difficult to interpret.

Because of linearity, we can again form the matrix $C_{\Phi,\Psi}$ by finding $C_{\Phi,\Psi}\alpha$ for particular values of α. Let $\alpha = e_k$, where e_k is the element of the standard basis for $\ell^2(\mathcal{K})$ with a 1 in position k. Then, $x = \Phi\alpha = \varphi_k$. If Ψ and $\widetilde{\Psi}$ form a biorthogonal pair of bases for H, the unique expansion of x with respect to Ψ is

$$x = \sum_{i\in\mathcal{K}} \langle x, \widetilde{\psi}_i\rangle \psi_i = \sum_{i\in\mathcal{K}} \langle \varphi_k, \widetilde{\psi}_i\rangle \psi_i,$$

from which we read off the ith coefficient of β as $\beta_i = \langle \varphi_k, \widetilde{\psi}_i\rangle$. This implies that column k of matrix $C_{\Phi,\Psi}$ is $(\langle \varphi_k, \widetilde{\psi}_i\rangle)_{i\in\mathcal{K}}$. The full matrix, written for the case of $\mathcal{K} = \mathbb{Z}$, is

$$C_{\Phi,\Psi} = \begin{bmatrix} & \vdots & \vdots & \vdots & \\ \cdots & \langle \varphi_{-1}, \widetilde{\psi}_{-1}\rangle & \langle \varphi_0, \widetilde{\psi}_{-1}\rangle & \langle \varphi_1, \widetilde{\psi}_{-1}\rangle & \cdots \\ \cdots & \langle \varphi_{-1}, \widetilde{\psi}_0\rangle & \boxed{\langle \varphi_0, \widetilde{\psi}_0\rangle} & \langle \varphi_1, \widetilde{\psi}_0\rangle & \cdots \\ \cdots & \langle \varphi_{-1}, \widetilde{\psi}_1\rangle & \langle \varphi_0, \widetilde{\psi}_1\rangle & \langle \varphi_1, \widetilde{\psi}_1\rangle & \cdots \\ & \vdots & \vdots & \vdots & \end{bmatrix}. \tag{2.162}$$

Note that $C_{\Phi,\Psi}$ depends on only one dual – the dual of the new representation basis Ψ. If the dual $\widetilde{\Psi}$ were not already available, computation of $C_{\Phi,\Psi}$ could be written in terms of the inner products in (2.161) and the Gram matrix of Ψ.

Matrix representation of linear operator with orthonormal bases Consider a Hilbert space H with orthonormal basis $\Phi = \{\varphi_k\}_{k\in\mathcal{K}}$, and let $A : H \to H$ be a linear operator. A matrix representation Γ allows A to be computed directly on coefficient sequences in the following sense: If

$$y = Ax, \tag{2.163a}$$

where

$$x = \sum_{i\in\mathcal{K}} \alpha_i \varphi_i \tag{2.163b}$$

and

$$y = \sum_{k\in\mathcal{K}} \beta_k \varphi_k, \tag{2.163c}$$

then

$$\beta = \Gamma\alpha. \tag{2.163d}$$

These relationships are depicted in Figure 2.28.

To find the matrix representation Γ, note that the kth coefficient of the expansion of y with respect to Φ is

$$\begin{aligned} \beta_k &\overset{(a)}{=} \langle y, \varphi_k\rangle \overset{(b)}{=} \langle Ax, \varphi_k\rangle \overset{(c)}{=} \langle A(\textstyle\sum_{i\in\mathcal{K}} \alpha_i\varphi_i), \varphi_k\rangle \\ &\overset{(d)}{=} \langle \textstyle\sum_{i\in\mathcal{K}} \alpha_i A\varphi_i, \varphi_k\rangle \overset{(e)}{=} \displaystyle\sum_{i\in\mathcal{K}} \alpha_i \langle A\varphi_i, \varphi_k\rangle, \end{aligned} \tag{2.164}$$

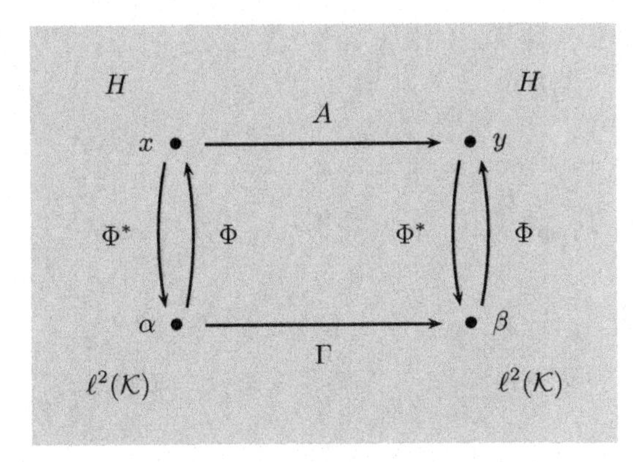

Figure 2.28 Conceptual illustration of the computation of a linear operator $A : H \to H$ using a matrix multiplication $\Gamma : \ell^2(\mathcal{K}) \to \ell^2(\mathcal{K})$. The abstract $y = Ax$ can be replaced by the more concrete $\beta = \Gamma\alpha$, where α and β are the representations of x and y with respect to orthonormal basis Φ of H.

where (a) follows from the expression for expansion coefficients with respect to an orthonormal basis, (2.93a); (b) from (2.163a); (c) from (2.163b); (d) from the linearity of A; and (e) from the linearity in the first argument of the inner product. This computation of one component of β as a linear combination of components of α determines one row of the matrix Γ. By gathering the equations (2.164) for all $k \in \mathcal{K}$ (and assuming that $\mathcal{K} = \mathbb{Z}$ for concreteness), we obtain

$$\Gamma = \begin{bmatrix} & \vdots & \vdots & \vdots & \\ \cdots & \langle A\varphi_{-1}, \varphi_{-1}\rangle & \langle A\varphi_0, \varphi_{-1}\rangle & \langle A\varphi_1, \varphi_{-1}\rangle & \cdots \\ \cdots & \langle A\varphi_{-1}, \varphi_0\rangle & \boxed{\langle A\varphi_0, \varphi_0\rangle} & \langle A\varphi_1, \varphi_0\rangle & \cdots \\ \cdots & \langle A\varphi_{-1}, \varphi_1\rangle & \langle A\varphi_0, \varphi_1\rangle & \langle A\varphi_1, \varphi_1\rangle & \cdots \\ & \vdots & \vdots & \vdots & \end{bmatrix}. \tag{2.165}$$

To check that (2.165) makes sense in a simple special case, let $H = \mathbb{C}^N$ and let Φ be the standard basis. For any k and i in $\{0, 1, \ldots, N-1\}$,

$$\Gamma_{i,k} = \langle A\varphi_k, \varphi_i\rangle = A_{i,k},$$

because $A\varphi_k$ is the kth column of A, and taking the inner product with φ_i picks out the ith entry. Thus the conventional use of matrices for linear operators on $\mathbb{C}^N$ is consistent with (2.165). This extends also to the use of the standard basis for $\ell^2(\mathbb{Z})$.

For a given operator A, a frequent goal in choosing the basis is to make Γ simple, for example, diagonal.

Example 2.55 (Diagonalizing basis) Let $H = \mathbb{R}^N$, and consider a linear operator $A : H \to H$ given by a symmetric matrix. Such a matrix can be decomposed as $A = \Phi\Lambda\Phi^\top$, where the columns of unitary matrix Φ are eigenvectors of A

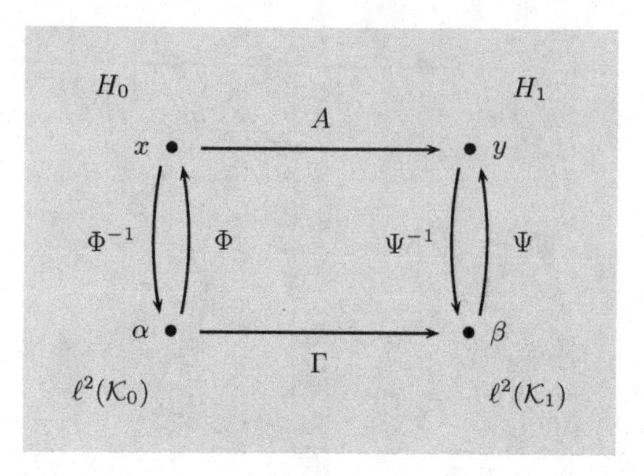

Figure 2.29 Conceptual illustration of the computation of a linear operator $A : H_0 \to H_1$ using a matrix multiplication $\Gamma : \ell^2(\mathcal{K}_0) \to \ell^2(\mathcal{K}_1)$. The abstract $y = Ax$ can be replaced by the more concrete $\beta = \Gamma\alpha$, where α is the representation of x with respect to basis Φ of H_0, and β is the representation of y with respect to basis Ψ of H_1.

and Λ is the diagonal matrix of corresponding eigenvalues, $\{\lambda_0, \lambda_1, \ldots, \lambda_{N-1}\}$; see (2.241b). Upon expressing the operator with respect to the orthonormal basis Φ we obtain

$$\Gamma_{i,k} \stackrel{(a)}{=} \langle A\varphi_k, \varphi_i\rangle \stackrel{(b)}{=} \langle \lambda_k\varphi_k, \varphi_i\rangle \stackrel{(c)}{=} \lambda_k\langle \varphi_k, \varphi_i\rangle \stackrel{(d)}{=} \lambda_k\delta_{i-k},$$

where (a) follows from (2.165); (b) from (λ_k, φ_k) being an eigenpair of A; (c) from the linearity in the first argument of the inner product; and (d) from the orthonormality of Φ. Thus, the representation is diagonal: multiplication of a vector by a matrix A is replaced by pointwise multiplication of expansion coefficients of x by the eigenvalues of A.

This simple example is fundamental since many basis changes aim to diagonalize operators. For example, we will see in Chapter 3 that the discrete Fourier transform diagonalizes the circular convolution operator because it is formed from the eigenvectors of the circular convolution operator. As in the example, multiplication by a dense matrix becomes a pointwise multiplication in a new basis.

Moving to cases where the domain and codomain of the linear operator are not necessarily the same Hilbert space, consider a linear operator $A : H_0 \to H_1$, and let $\Phi = \{\varphi_k\}_{k\in\mathcal{K}_0}$ be an orthonormal basis for H_0 and $\Psi = \{\psi_k\}_{k\in\mathcal{K}_1}$ be an orthonormal basis for H_1. We would like to implement A as an operation on sequence representations with respect to Φ and Ψ. The concept is depicted in Figure 2.29, where orthonormality of the bases gives $\Phi^{-1} = \Phi^*$ and $\Psi^{-1} = \Psi^*$. The computation $y = Ax$ is replaced by $\beta = \Gamma\alpha$, where α is the representation of x with respect to Φ and β is the representation of y with respect to Ψ.

Mimicking the derivation of (2.164) leads to a counterpart for (2.165) that uses both bases:

$$\Gamma = \begin{bmatrix} & \vdots & \vdots & \vdots & \\ \cdots & \langle A\varphi_{-1}, \psi_{-1}\rangle & \langle A\varphi_0, \psi_{-1}\rangle & \langle A\varphi_1, \psi_{-1}\rangle & \cdots \\ \cdots & \langle A\varphi_{-1}, \psi_0\rangle & \boxed{\langle A\varphi_0, \psi_0\rangle} & \langle A\varphi_1, \psi_0\rangle & \cdots \\ \cdots & \langle A\varphi_{-1}, \psi_1\rangle & \langle A\varphi_0, \psi_1\rangle & \langle A\varphi_1, \psi_1\rangle & \cdots \\ & \vdots & \vdots & \vdots & \end{bmatrix}. \tag{2.166}$$

Note the information upon which Γ is determined: by linearity of A and completeness of the basis Φ, the effect of A on any $x \in H_0$ can be computed from its effect on each basis element of the domain space, $\{A\varphi_k\}_{k\in\mathcal{K}_0}$; and the expansion coefficients of any one of these results are determined by inner products with the basis in the codomain space, $\{\langle A\varphi_k, \psi_i\rangle\}_{i\in\mathcal{K}_1}$.

EXAMPLE 2.56 (AVERAGING OPERATOR) Consider the operator $A : H_0 \to H_1$ that replaces a function by its average over intervals of length 2,

$$y(t) = Ax(t) = \frac{1}{2}\int_{2\ell}^{2(\ell+1)} x(\tau)\, d\tau, \qquad \text{for } 2\ell \le t < 2(\ell+1), \quad \ell \in \mathbb{Z}, \tag{2.167}$$

where H_0 is the space of piecewise-constant, finite-energy functions with breakpoints at integers and H_1 the space of piecewise-constant, finite-energy functions with breakpoints at even integers. As orthonormal bases for H_0 and H_1, we choose normalized indicator functions over unit and double-unit intervals, respectively:

$$\Phi = \{\varphi_k(t)\}_{k\in\mathbb{Z}} = \left\{1_{[k,k+1)}(t)\right\}_{k\in\mathbb{Z}},$$

$$\Psi = \{\psi_i(t)\}_{i\in\mathbb{Z}} = \left\{\frac{1}{\sqrt{2}}1_{[2i,2(i+1))}(t)\right\}_{i\in\mathbb{Z}}.$$

To evaluate Γ from (2.166) requires $\langle A\varphi_k, \psi_i\rangle$ for all integers k and i. Since $\varphi_0(t)$ is nonzero only for $t \in [0, 1)$,

$$A\varphi_0(t) = \begin{cases} \frac{1}{2}, & \text{for } 0 \le t < 2; \\ 0, & \text{otherwise,} \end{cases}$$

from which

$$\langle A\varphi_0, \psi_0\rangle = \int_0^2 \frac{1}{2}\frac{1}{\sqrt{2}}\, d\tau = \frac{1}{\sqrt{2}}$$

and

$$\langle A\varphi_0, \psi_i\rangle = 0 \qquad \text{for all } i \neq 0.$$

Since A integrates over intervals of the form $[2\ell, 2(\ell+1)]$, $A\varphi_1 = A\varphi_0$, so

$$\langle A\varphi_1, \psi_i\rangle = \begin{cases} 1/\sqrt{2}, & \text{for } i = 0; \\ 0, & \text{otherwise.} \end{cases}$$

Continuing the computation to cover every φ_k yields Γ:

$$\Gamma = \frac{1}{\sqrt{2}} \begin{bmatrix} & \vdots & \vdots & \vdots & \vdots & \vdots & \vdots & \\ \cdots & 1 & 1 & 0 & 0 & 0 & 0 & \cdots \\ \cdots & 0 & 0 & \boxed{1} & 1 & 0 & 0 & \cdots \\ \cdots & 0 & 0 & 0 & 0 & 1 & 1 & \cdots \\ & \vdots & \vdots & \vdots & \vdots & \vdots & \vdots & \end{bmatrix}. \tag{2.168}$$

Multiplying by Γ is thus a very simple operation.

Matrix representation of linear operator with biorthogonal pairs of bases As above, consider a linear operator $A : H_0 \to H_1$. Assume that Φ and $\widetilde{\Phi}$ form a biorthogonal pair of bases for H_0, and Ψ and $\widetilde{\Psi}$ form a biorthogonal pair of bases for H_1. We would like to implement A as an operation on sequence representations with respect to Φ and Ψ as in Figure 2.29, where biorthogonality of the bases gives $\Phi^{-1} = \widetilde{\Phi}^*$ and $\Psi^{-1} = \widetilde{\Psi}^*$.

Derivation of Γ, the matrix representation of the operator A, is almost unchanged from the orthonormal case, but we repeat the key computation to show the role of having biorthogonal bases. When

$$x = \sum_{i \in \mathcal{K}_0} \alpha_i \varphi_i, \tag{2.169}$$

the expansion of

$$y = Ax \tag{2.170}$$

with respect to Ψ has the kth coefficient

$$\beta_k \overset{(a)}{=} \langle y, \widetilde{\psi}_k \rangle \overset{(b)}{=} \langle Ax, \widetilde{\psi}_k \rangle \overset{(c)}{=} \left\langle A(\textstyle\sum_{i \in \mathcal{K}_0} \alpha_i \varphi_i), \widetilde{\psi}_k \right\rangle$$
$$\overset{(d)}{=} \left\langle \textstyle\sum_{i \in \mathcal{K}_0} \alpha_i A\varphi_i, \widetilde{\psi}_k \right\rangle \overset{(e)}{=} \sum_{i \in \mathcal{K}_0} \alpha_i \langle A\varphi_i, \widetilde{\psi}_k \rangle,$$

where (a) follows from the expression for expansion coefficients with a biorthogonal pair of bases, (2.113a); (b) from (2.170); (c) from (2.169); (d) from the linearity of A; and (e) from the linearity in the first argument of the inner product. Thus the matrix representation (assuming that $\mathcal{K}_0 = \mathcal{K}_1 = \mathbb{Z}$ for concreteness) is

$$\Gamma = \begin{bmatrix} & \vdots & \vdots & \vdots & \\ \cdots & \langle A\varphi_{-1}, \widetilde{\psi}_{-1} \rangle & \langle A\varphi_0, \widetilde{\psi}_{-1} \rangle & \langle A\varphi_1, \widetilde{\psi}_{-1} \rangle & \cdots \\ \cdots & \langle A\varphi_{-1}, \widetilde{\psi}_0 \rangle & \boxed{\langle A\varphi_0, \widetilde{\psi}_0 \rangle} & \langle A\varphi_1, \widetilde{\psi}_0 \rangle & \cdots \\ \cdots & \langle A\varphi_{-1}, \widetilde{\psi}_1 \rangle & \langle A\varphi_0, \widetilde{\psi}_1 \rangle & \langle A\varphi_1, \widetilde{\psi}_1 \rangle & \cdots \\ & \vdots & \vdots & \vdots & \end{bmatrix}. \tag{2.171}$$

Comparing (2.166) and (2.171), the only difference is the use of the dual basis for the codomain space; this is natural since we require expansions of $\{A\varphi_k\}_{k\in\mathcal{K}_0}$ with respect to Ψ. Also note the similarity between (2.162) and (2.171); a change of basis operator is a special case of a matrix representation of a linear operator, where $H_0 = H_1 = H$ and the operator is the identity on H.

The next example points to how differential operators can be implemented as matrix multiplications once bases for the domain and codomain spaces are available.

EXAMPLE 2.57 (DERIVATIVE OPERATOR) Consider the derivative operator $A : H_0 \to H_1$, with H_0 the space of piecewise-linear, continuous, finite-energy functions with breakpoints at integers and H_1 the space of piecewise-constant, finite-energy functions with breakpoints at integers. As a basis for H_0, choose the triangle function

$$\varphi(t) = \begin{cases} 1-|t|, & \text{for } |t|<1; \\ 0, & \text{otherwise} \end{cases} \tag{2.172}$$

and its integer shifts:

$$\Phi = \{\varphi_k(t)\}_{k\in\mathbb{Z}} = \{\varphi(t-k)\}_{k\in\mathbb{Z}}.$$

(This is an infinite-dimensional analogue to the basis in Example 2.44 and an example of a spline, discussed in detail in Section 6.3.) For H_1, we can choose the same orthonormal basis as in Example 2.56:

$$\Psi = \{\psi_i(t)\}_{i\in\mathbb{Z}} = \{1_{[i,i+1)}(t)\}_{i\in\mathbb{Z}}.$$

To evaluate Γ from (2.171) requires $\langle A\varphi_k, \widetilde{\psi}_i\rangle$ for all integers k and i. Since

$$A\varphi(t) = \varphi'(t) = \begin{cases} 1, & \text{for } -1<t<0; \\ -1, & \text{for } 0<t<1; \\ 0, & \text{for } |t|>1, \end{cases}$$

it follows that

$$\langle A\varphi_0, \widetilde{\psi}_i\rangle = \begin{cases} 1, & \text{for } i=-1; \\ -1, & \text{for } i=0; \\ 0, & \text{otherwise.} \end{cases}$$

Shifting $\varphi(t)$ by k simply shifts the derivative:

$$\langle A\varphi_k, \widetilde{\psi}_i\rangle = \begin{cases} 1, & \text{for } i=k-1; \\ -1, & \text{for } i=k; \\ 0, & \text{otherwise.} \end{cases}$$

Gathering these computations into a matrix yields

$$\Gamma = \begin{bmatrix} & \vdots & \vdots & \vdots & \vdots & \vdots & \vdots & \\ \cdots & 0 & -1 & 1 & 0 & 0 & 0 & \cdots \\ \cdots & 0 & 0 & \boxed{-1} & 1 & 0 & 0 & \cdots \\ \cdots & 0 & 0 & 0 & -1 & 1 & 0 & \cdots \\ & \vdots & \vdots & \vdots & \vdots & \vdots & \vdots & \end{bmatrix}. \tag{2.173}$$

Figure 2.30 gives an example of a derivative operator and its computation. The input function and expansion coefficients with respect to the basis Φ are

$$\begin{aligned} x(t) &= \varphi(t) - \varphi(t-1), \\ \alpha &= \begin{bmatrix} \cdots & 0 & \boxed{1} & -1 & 0 & 0 & \cdots \end{bmatrix}^\top, \end{aligned}$$

while its derivative and expansion coefficients with respect to the basis Ψ are

$$\begin{aligned} x'(t) &= \psi(t+1) - 2\psi(t) + \psi(t-1), \\ \beta &= \begin{bmatrix} \cdots & 0 & 1 & \boxed{-2} & 1 & 0 & \cdots \end{bmatrix}^\top. \end{aligned}$$

Then, indeed

$$\beta = \begin{bmatrix} \vdots \\ 0 \\ 1 \\ \boxed{-2} \\ 1 \\ 0 \\ \vdots \end{bmatrix} = \begin{bmatrix} & \vdots & \vdots & \vdots & \vdots & \vdots & \vdots & \\ \cdots & 0 & -1 & 1 & 0 & 0 & 0 & \cdots \\ \cdots & 0 & 0 & \boxed{-1} & 1 & 0 & 0 & \cdots \\ \cdots & 0 & 0 & 0 & -1 & 1 & 0 & \cdots \\ & \vdots & \vdots & \vdots & \vdots & \vdots & \vdots & \end{bmatrix} \begin{bmatrix} \vdots \\ 0 \\ 0 \\ \boxed{1} \\ -1 \\ 0 \\ 0 \\ \vdots \end{bmatrix} = \Gamma\alpha.$$

Matrix representation of the adjoint Example 2.18(ii) confirmed that the adjoint of a linear operator $A : \mathbb{C}^N \to \mathbb{C}^M$ given by a finite matrix is the Hermitian transpose of the matrix; implicit in this was the use of the standard bases for $\mathbb{C}^N$ and $\mathbb{C}^M$. The connection between the adjoint and the Hermitian transpose of a matrix extends to arbitrary Hilbert spaces and linear operators when orthonormal bases are used.

Consider a linear operator $A : H_0 \to H_1$, and let $\Phi = \{\varphi_k\}_{k\in\mathcal{K}_0}$ be an orthonormal basis for H_0 and $\Psi = \{\psi_k\}_{k\in\mathcal{K}_1}$ be an orthonormal basis for H_1. Let Γ be the matrix representation of A with respect to Φ and Ψ, as given by (2.166). The adjoint A^* is an operator $H_1 \to H_0$, and we would like to find its matrix representation with respect to Ψ and Φ. Applying (2.166) to A^*, the entry in row i, column k is

$$\langle A^*\psi_k, \varphi_i\rangle \stackrel{(a)}{=} \langle \psi_k, A\varphi_i\rangle \stackrel{(b)}{=} \langle A\varphi_i, \psi_k\rangle^* \stackrel{(c)}{=} \Gamma^*_{k,i}, \qquad (2.174)$$

where (a) follows from the definition of the adjoint; (b) from the Hermitian symmetry of the inner product; and (c) from (2.166). Thus, the matrix representation of A^* is indeed the Hermitian transpose of the matrix representation of A.

Now remove the assumption that Φ and Ψ are orthonormal bases and denote the respective dual bases by $\widetilde{\Phi}$ and $\widetilde{\Psi}$. Let Γ be the matrix representation of A with respect to Φ and Ψ, as given by (2.171). The matrix representation of A^* has a

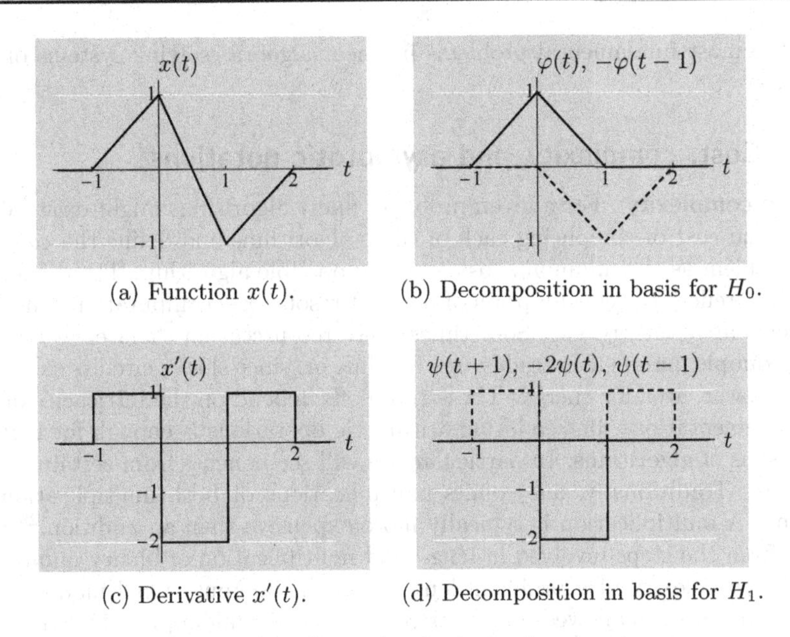

(a) Function $x(t)$. (b) Decomposition in basis for H_0.

(c) Derivative $x'(t)$. (d) Decomposition in basis for H_1.

Figure 2.30 Example of a derivative operator.

simple form with respect to the *duals* $\widetilde{\Psi}$ for Ψ in H_1 and $\widetilde{\Phi}$ for Φ in H_0. Applying (2.171) to A^*, the entry in row i, column k is

$$\langle A^* \widetilde{\psi}_k, \varphi_i \rangle \overset{(a)}{=} \langle \widetilde{\psi}_k, A\varphi_i \rangle \overset{(b)}{=} \langle A\varphi_i, \widetilde{\psi}_k \rangle^* \overset{(c)}{=} \Gamma^*_{k,i}, \tag{2.175}$$

where (a) follows from the definition of the adjoint; (b) from the Hermitian symmetry of the inner product; and (c) from (2.171). Thus, the matrix representation of A^* is the Hermitian transpose of the matrix representation of A *when the bases are switched to the duals.* To represent A^* with respect to Ψ and Φ rather than with respect to their duals is a bit more complicated.

2.6 Computational aspects

The cost of an algorithm is generally measured by the number of operations needed and the precision requirements both for the input data and for the intermediate results. These cost metrics are of primary interest and enable comparisons that are independent of the computation platform. Running time and hardware resources (such as chip area) can be traded off through parallelization and are also affected by more subtle algorithmic properties.

We start with the basics of using operation counts to express the complexity of a problem and the cost of an algorithm. We then discuss the precision of the computation in terms of the arithmetic representation of a number, followed by conditioning as the sensitivity of the solution to changes in the data. We close with one of the most fundamental problems in linear algebra: solving systems of linear equations.

2.6.1 Cost, complexity, and asymptotic notations

Cost and complexity For a given problem, many algorithms might exist. We can measure the *cost* of computing each of these algorithms and define the *complexity* of the problem as the minimum cost over any possible algorithm. The definition of cost should reflect the consumption of relevant resources – computation time, memory, circuit area, energy, etc. Sometimes these resources can themselves be traded off; for example, parallelism trades area for time or, since slower circuits can operate at lower power, area for energy. These trade-offs depend on the intricacies of hardware implementations, but counting arithmetic operations is enough for high-level comparisons of algorithms. In particular, we will see benefits from certain problem structures. Traditionally, one counts multiplications or both multiplications and additions. A multiplication is typically more expensive than an addition,[28] as can be seen from the steps involved in long-hand multiplication of binary numbers.

The complexity of a problem depends on the computational model (namely which operations are allowed and their costs), the possible inputs, and the format of the input. In the following example, the costs are the number of multiplications μ and the number of additions ν, and we see an impact from the format of the problem input.

EXAMPLE 2.58 (COMPLEXITY OF POLYNOMIAL EVALUATION) There are several algorithms to evaluate

$$x(t) = a_0 + a_1 t + a_2 t^2 + a_3 t^3. \tag{2.176}$$

The most obvious is through

$$\text{output: } a_0 + a_1 \cdot t + (a_2 \cdot t) \cdot t + ((a_3 \cdot t) \cdot t) \cdot t,$$

which has $\mu = 6$ multiplications and $\nu = 3$ additions. This is wasteful because powers of t could have been saved and reused. Specifically, the computations

$$t_2 = t \cdot t; \qquad t_3 = t_2 \cdot t; \qquad \text{output: } a_0 + a_1 \cdot t + a_2 \cdot t_2 + a_3 \cdot t_3$$

give the same final result with $\mu = 5$ and $\nu = 3$. An even cheaper algorithm is

$$\text{output: } a_0 + t \cdot (a_1 + t \cdot (a_2 + t \cdot a_3)),$$

with $\mu = 3$ and $\nu = 3$. In fact, $\mu = 3$ and $\nu = 3$ are the minimum possible multiplicative and additive costs (and hence the problem complexity) for *arbitrary* input (a_0, a_1, a_2, a_3, t). Restrictions on the input could reduce the complexity.

Other formats for the same polynomial lead to different algorithms with different costs. For example, if the polynomial is given in its factored form

$$x(t) = b_0(t + b_1)(t + b_2)(t + b_3), \tag{2.177}$$

[28]This is certainly true for fixed-point arithmetic; for floating-point arithmetic, the situation is more complicated.

it will have a natural implementation with $\mu = 3$ and $\nu = 3$, matching the complexity of the problem. However, a real polynomial could have complex roots. When using real operations to measure cost, one could assign the costs $(\mu, \nu) = (4, 2)$ to a complex multiplication and $(\mu, \nu) = (0, 2)$ to a complex addition.[29] The algorithm based on the factored form (2.177) then has higher cost than that based on the expanded form (2.176).

This example illustrates that mathematically equivalent expressions need not be equivalent for computation. We will revisit this, again in the context of polynomials, when we discuss precision.

The scaling of cost and complexity with the problem size is typically of interest. For example, the complexity determined in Example 2.58 generalizes to μ and ν equaling the degree of the polynomial. Finding exact complexities is usually very difficult, and we are satisfied with coarse descriptions expressed with asymptotic notations.

Asymptotic notation The most common asymptotic notation is the *big O*, which is rooted in the word *order*. While we define it and other asymptotic notations for sequences indexed by $n \in \mathbb{N}$ with $n \to \infty$, the same notation is used for functions with the argument approaching any finite or infinite limit. Informally, $x = O(y)$ means that x_n is eventually (for large enough n) bounded from above by a constant multiple of y_n.

DEFINITION 2.54 (ASYMPTOTIC NOTATION) Let x and y be sequences defined on $\mathbb{N}$. We say

(i) x is O of y and write $x \in O(y)$ or $x = O(y)$ when there exist constants $\gamma > 0$ and $n_0 \in \mathbb{N}$ such that

$$0 \leq x_n \leq \gamma y_n, \qquad \text{for all } n \geq n_0; \tag{2.178}$$

(ii) x is o of y and write $x \in o(y)$ or $x = o(y)$ when, for any $\gamma > 0$, there exists a constant $n_0 \in \mathbb{N}$ such that (2.178) holds;

(iii) x is Ω of y and write $x \in \Omega(y)$ or $x = \Omega(y)$ when there exist constants $\gamma > 0$ and $n_0 \in \mathbb{N}$ such that

$$\gamma y_n \leq x_n, \qquad \text{for all } n \geq n_0;$$

(iv) x is Θ of y and write $x \in \Theta(y)$ or $x = \Theta(y)$ when x is both O of y and Ω of y, that is, $x \in O(y) \cap \Omega(y)$.

These convenient asymptotic notations necessitate a few notes of caution: The use of an equal sign is an abuse of that symbol because asymptotic notations are not

[29] These costs are based on the obvious implementation of complex multiplication as $(a+jb)(c+jd) = (ac - bd) + j(ad + bc)$; a complex multiplication can also be computed with three real multiplications and five real additions, as in Exercise 2.49.

symmetric relations. Also, if the argument over which one is taking a limit is not clear from the context, it should be written explicitly; for example, $2^m n^2 = O_n(n^2)$ is correct when m does not depend on n because the added subscript specifies that we are interested only in the scaling with respect to n. Finally, all the asymptotic notations omit constant factors that can be critical in assessing and comparing algorithms.

Computing the cost of an algorithm Most often, we will compute the cost of an algorithm in terms of the number of operations, *multiplications* μ and *additions* ν, for a total cost of

$$C = \mu + \nu, \tag{2.179}$$

followed by an asymptotic estimation of its behavior in O notation.

EXAMPLE 2.59 (MATRIX MULTIPLICATION) We illustrate cost and complexity with one of the most basic operations in linear algebra, matrix multiplication. Using the definition for matrix multiplication, (2.219), directly, the product $Q = AB$, with $A \in \mathbb{C}^{M \times N}$ and $B \in \mathbb{C}^{N \times P}$, requires N multiplications and $N - 1$ additions for each Q_{ik}, for a total of $\mu = MNP$ multiplications and $\nu = M(N-1)P$ additions, and a total cost of $C = M(2N-1)P$. Setting $M = P = N$, the cost for multiplying two $N \times N$ matrices is

$$C_{\text{mat-mat}} = 2N^3 - N^2; \tag{2.180a}$$

and, setting $M = N$ and $P = 1$, the cost of multiplying an $N \times N$ matrix by an $N \times 1$ vector is

$$C_{\text{mat-vec}} = 2N^2 - N. \tag{2.180b}$$

By specifying that (2.219) is to be used, we have implicitly identified a particular algorithm for matrix multiplication. Other algorithms might be preferable. There are algorithms that reduce the number of multiplications at the expense of having more additions; for example, for the multiplication of 2×2 matrices, (2.219) gives a cost of eight multiplications, but we now show that the computation can be accomplished with seven multiplications. The product

$$\begin{bmatrix} Q_{00} & Q_{01} \\ Q_{10} & Q_{11} \end{bmatrix} = \begin{bmatrix} A_{00} & A_{01} \\ A_{10} & A_{11} \end{bmatrix} \begin{bmatrix} B_{00} & B_{01} \\ B_{10} & B_{11} \end{bmatrix}$$

can be computed from intermediate results, namely

$$\begin{aligned} h_0 &= (A_{01} - A_{11})(B_{10} + B_{11}), & h_4 &= A_{00}(B_{01} + B_{11}), \\ h_1 &= (A_{00} + A_{11})(B_{00} + B_{11}), & h_5 &= A_{11}(B_{10} + B_{00}), \\ h_2 &= (A_{00} - A_{10})(B_{00} + B_{01}), & h_6 &= (A_{10} + A_{11})B_{00}, \\ h_3 &= (A_{00} + A_{01})B_{11}, \end{aligned}$$

as

$$\begin{aligned} Q_{00} &= h_0 + h_1 - h_3 + h_5, & Q_{01} &= h_3 + h_4, \\ Q_{10} &= h_5 + h_6, & Q_{11} &= h_1 - h_2 + h_4 - h_6. \end{aligned}$$

The $\mu = 7$ multiplications is the minimum possible, but the number of additions is increased to $\nu = 18$. This procedure is known as *Strassen's algorithm* (see Solved Exercise 2.6).

2.6.2 Precision

Counting operations is important, but so is the trade-off between the cost of a specific operation and its precision. In digital computation, operations are over a (possibly huge) finite set of values. Not all operations are possible in the sense that many operations will have results outside the finite set. How these cases are handled – primarily through rounding and saturation – introduces inaccuracies into computations. The properties of this inaccuracy depend heavily on the specific arithmetic representation of a number. We assume binary arithmetic and discuss two dominant cases, namely fixed-point and floating-point arithmetic.

Fixed-point arithmetic Fixed-point arithmetic deals with operations over a finite set of evenly spaced values. Consider the values to be the integers between 0 and $2^B - 1$, where B is the number of bits used in the representation; then, a binary string $(b_0, b_1, \ldots, b_{B-1})$ defines an integer in that range, given by

$$x = b_0 + b_1 \cdot 2 + b_2 \cdot 2^2 + \cdots + b_{B-1} \cdot 2^{B-1}. \tag{2.181}$$

Negative numbers are handled with an extra bit and various formats; this does not change the basic issues.

The sum of two B-bit numbers is in $\{0, 1, \ldots, 2^B - 2\}$. The result can thus be represented exactly unless there is *overflow*, meaning that the result is too large for the number format. Overflow could result in an error, saturation (setting the result to the largest number $2^B - 1$), or wraparound (returning the remainder in dividing the correct result by 2^B). All of these make an algorithm difficult to analyze and are generally to be avoided. One could guarantee that there is no overflow in the computation $x + y$ by limiting x and y to the lower half of the valid numbers (requiring the most significant bit b_{B-1} to be 0). Reducing the range of possible inputs to ensure the accuracy of the computation has its limitations; for example, to avoid overflow in the product $x \cdot y$ one must restrict x and y to be less than $2^{B/2}$ (requiring half of the input bits to be zero).

Many other operations, such as taking a square root and division by a nonzero number, have results that cannot be represented exactly. Results of these operations will generally be rounded to the nearest valid representation point. Thus, assuming that there is no overload, the error from a fixed-point computation is bounded through

$$\frac{|x - \widehat{x}|}{x_{\max} - x_{\min}} \leq \frac{1/2}{2^B - 1} \approx 2^{-(B+1)}, \tag{2.182}$$

where x is the exact result, $\widehat{x}$ is the result of the fixed-point computation, and $x_{\max} - x_{\min}$ is the full range of valid representation points. The error bound is written with normalization by the range to highlight the role of the number of bits B. To contrast with what we will see for floating-point arithmetic, note that the error could be small relative to x (if $|x|$ is large) or large relative to x (if x is near 0).

Floating-point arithmetic Floating-point representations use numbers spread over a vastly larger range so that overload is mostly avoided. A binary floating-point representation has a *mantissa*, or *significand*, and an *exponent*. The exponent is a fixed-point binary number used to scale the significand by a power of 2 such that the significand lies in [1, 2). Written as a fixed-point binary number, the significand thus has a 1 to the left of the fraction point followed by some number of bits. Consider only strictly positive numbers and suppose that B bits are divided into B_S bits for the significand and $B_E = B - B_S$ bits for the exponent. Then, a number x can be represented in floating-point binary form as

$$x = \left(1 + \sum_{n=1}^{B_S} b_n 2^{-n}\right) 2^E, \tag{2.183}$$

where E is a fixed-point binary number having B_E bits chosen so that the leading bit of the significand is 1. Of course, this representation still covers a finite range of possible numbers, but, because of the significand/exponent decomposition, this range is larger than that in fixed-point arithmetic.

EXAMPLE 2.60 (32-BIT ARITHMETIC) In the IEEE 754-2008 standard for 32-bit floating-point arithmetic, 1 bit is reserved for the sign, 8 for the exponent, and 23 for the significand. Since the leading 1 to the left of the binary point is assumed, the significand effectively has 24 bits. The value of the number is given by

$$x = (-1)^{\text{sign}} \left(1 + \sum_{n=1}^{23} b_n 2^{-n}\right) 2^{E-127} \tag{2.184}$$

for $E \in \{1, 2, \ldots, 254\}$. The two remaining values of E are used differently: $E = 255$ is used for $\pm\infty$ and "not a number" (NaN); and $E = 0$ is used for 0 if the significand is zero and for *subnormal* numbers if the significand is nonzero. The subnormal numbers extend the range of representable positive numbers below 2^{-126} through

$$x = (-1)^{\text{sign}} \left(\sum_{n=1}^{23} b_n 2^{-n}\right) 2^{-126}.$$

All the positive numbers that can be represented through (2.184) lie in

$$[2^{-126}, 2 \cdot 2^{127}] \approx [1.18 \cdot 10^{-38}, 3.4 \cdot 10^{38}],$$

and similarly for negative numbers. To this we add subnormal numbers, the minimum of which is approximately $1.4 \cdot 10^{-45}$. Comparing this with $[0, 4.3 \cdot 10^9]$ (with integer spacing) for 32-bit fixed-point arithmetic shows an advantage of floating-point arithmetic.

In floating-point arithmetic following (2.183), the difference between a real number and the closest valid representation might be large – but only if the number itself is large. Suppose that x is positive and not too large to be represented. Then, its representation $\widehat{x}$ will satisfy

$$\widehat{x} = (1 + \varepsilon)x, \qquad \text{where } |\varepsilon| < 2^{-B_S}. \tag{2.185}$$

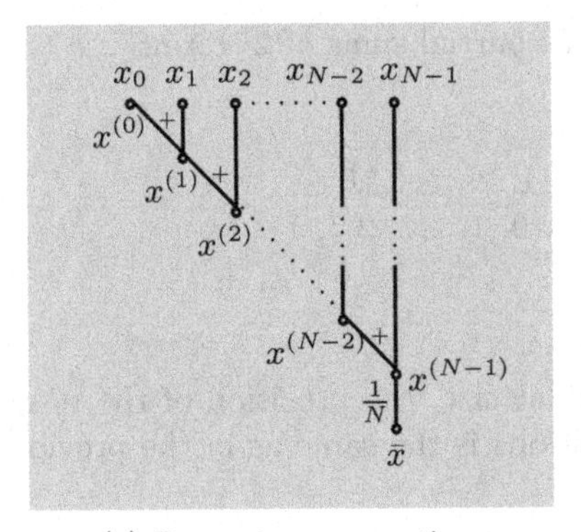

(a) Recursive summation.

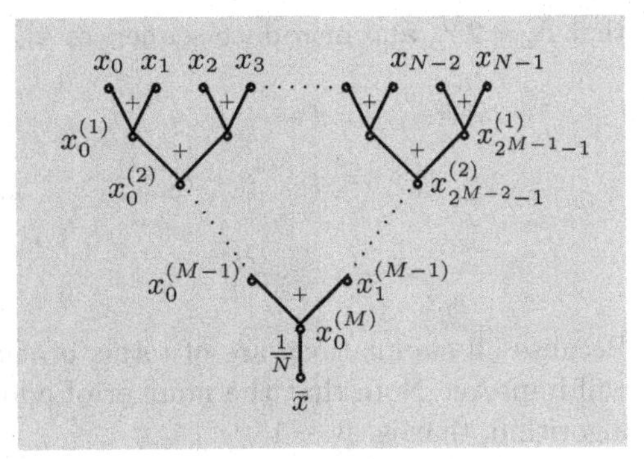

(b) Tree-based summation.

Figure 2.31 Two algorithms for computing an average.

This is very different than (2.182); the error $|x - \widehat{x}|$ might be large, but it is not large relative to x.

Multiplication of floating-point numbers amounts to taking the product of the significands and the sum of the exponents, followed possibly by rescaling. Addition is more involved, since, when the exponents are not equal, the significands have to adjust the fractional point so that they can be added. When the exponents are very different, a smaller term will lose precision, and could even be set to zero. Nevertheless, adding positive numbers or multiplying positive numbers, within the range of the number system, will have error satisfying (2.185). Much more troublesome is that subtracting numbers that are nearly equal can result in cancellation of many leading bits. This is called a *loss of significance* error.

Example 2.61 (Computing an average) We now highlight how the choice of an algorithm affects precision in one of the simplest operations, computation of the average of N numbers,

$$\bar{x} = \frac{1}{N} \sum_{n=0}^{N-1} x_n.$$

An obvious algorithm is the recursive procedure illustrated in Figure 2.31(a),

$$\begin{aligned} x^{(0)} &= x_0, \\ x^{(n)} &= x^{(n-1)} + x_n, \qquad n = 1, 2, \ldots, N, \\ \bar{x} = x^{(N)} &= \frac{1}{N} x^{(N-1)}. \end{aligned}$$

When all $\{x_k\}$ are close in value, the summands in step n above differ in size by a factor of n (since $x^{(n-1)}$ is the partial sum of the first n numbers). With N large, this becomes problematic as n grows.

A simple alternative is summing on a tree as in Figure 2.31(b). Assume that $N = 2^M$, and introduce sequences $x_n^{(i)}$ as partial sums of 2^i terms,

$$
\begin{aligned}
x_n^{(0)} &= x_n, \\
x_n^{(i)} &= x_{2n}^{(i-1)} + x_{2n+1}^{(i-1)}, \qquad i = 1, 2, \ldots, M, \quad n = 0, 1, \ldots, 2^{M-i} - 1, \\
\bar{x} &= \frac{1}{N} x_0^{(M)}.
\end{aligned}
$$

Because all summations are of terms of similar size, the precision of the result will improve. Note that the number of additions is the same as in the previous algorithm, that is, $N - 1$.

2.6.3 Conditioning

So far, we have discussed two issues: algorithmic efficiency, or the number of operations required to solve a given problem, and the precision of the computation, which is linked both to machine precision and to the algorithmic structure (as in Example 2.61). We now discuss the conditioning of a problem, which describes the sensitivity of the solution to changes in the data. In an ill-conditioned problem, the solution can vary widely with small changes in the input. Ill-conditioned problems also tend to be more sensitive to algorithmic choices that would be immaterial with exact arithmetic. We study conditioning by looking at the solution of systems of linear equations.

Given a system of linear equations $y = Ax$, where x is a column vector of length N and A is an $N \times N$ matrix of full rank, we know that a unique solution exists, $x = A^{-1}y$. The condition number we introduce shortly will roughly say how sensitive the solution x will be to small changes in y. In particular, if the condition number is large, a tiny change (error) in y can lead to a large change (error) in x. Conversely, a small condition number signifies that the error in x will be of the same order as the error in y.

For this discussion, we use the 2 norm as defined in (2.234d) in Appendix 2.B.1:

$$
\|A\|_2 = \|A\|_{2,2} = \sup_{\|x\|_2=1} \|Ax\|_2 = \sigma_{\max}(A) = \sqrt{\lambda_{\max}(A^*A)}, \tag{2.188a}
$$

where $\sigma_{\max}$ denotes the maximum singular value and $\lambda_{\max}$ denotes the maximum eigenvalue. Similarly,

$$
\|A^{-1}\|_2 = \frac{1}{\sigma_{\min}(A)} = \frac{1}{\sqrt{\lambda_{\min}(A^*A)}}, \tag{2.188b}
$$

where $\sigma_{\min}$ denotes the minimum singular value and $\lambda_{\min}$ denotes the minimum eigenvalue. For now, consider the matrix A to be exact, and let us see how changes in y, expressed as $\widehat{y} = y + \Delta y$, affect the changes in the solution x, expressed as $\widehat{x} = x + \Delta x$. Since $\widehat{x} = A^{-1}\widehat{y}$ and $x = A^{-1}x$,

$$
\Delta x = \widehat{x} - x = A^{-1}\widehat{y} - A^{-1}y = A^{-1}(\widehat{y} - y) = A^{-1}\Delta y.
$$

Using the fact that $\|Ax\| \leq \|A\|\|x\|$ for any norm,

$$\|\Delta x\|_2 = \|A^{-1}\Delta y\|_2 \leq \|A^{-1}\|_2\|\Delta y\|_2. \tag{2.189}$$

To find the relative error, we divide the norm of Δx by the norm of $\widehat{x}$. Using (2.189), the relative error is bounded as

$$\frac{\|\Delta x\|_2}{\|\widehat{x}\|_2} \leq \|A^{-1}\|_2\frac{\|\Delta y\|_2}{\|\widehat{x}\|_2} = \|A^{-1}\|_2\|A\|_2\frac{\|\Delta y\|_2}{\|A\|_2\|\widehat{x}\|_2} \leq \kappa(A)\frac{\|\Delta y\|_2}{\|\widehat{y}\|_2}, \tag{2.190}$$

where $\kappa(A)$ is called a *condition number of a matrix* A,

$$\kappa(A) = \|A\|_2\|A^{-1}\|_2 = \frac{\sigma_{\min}(A)}{\sigma_{\min}(A)} = \sqrt{\frac{\lambda_{\max}(A^*A)}{\lambda_{\min}(A^*A)}}. \tag{2.191a}$$

When A is a basis synthesis operator, $\lambda_{\min}(A^*A)$ and $\lambda_{\max}(A^*A)$ are the constants in Definition 2.35 (Riesz basis). For a Hermitian matrix (or more generally a normal matrix), the condition number is simply

$$\kappa(A) = \left|\frac{\lambda_{\max}(A)}{\lambda_{\min}(A)}\right|, \tag{2.191b}$$

where $\lambda_{\max}(A)$ and $\lambda_{\min}(A)$ are the eigenvalues of largest and smallest magnitude, respectively (since the eigenvalues can be negative or complex). From (2.190), we see that $\kappa(A)$ measures the sensitivity of the solution: a small amount of noise on a data vector y might grow by a factor $\kappa(A)$ in the solution vector x. In some ill-conditioned problems, the ratio between the largest and smallest eigenvalues in (2.191) can be several orders of magnitude, sometimes leading to a useless solution. At the other extreme, the best-conditioned problems appear when A is unitary, since then $\kappa(A) = 1$; the error in the solution is similar to the error in the input.

Poor conditioning can come from a large $|\lambda_{\max}|$, but it more often comes from a small $|\lambda_{\min}|$; that is, it occurs when the matrix A is almost singular. We would like to find out how close A is to being a singular matrix. In other words, can a perturbation ΔA lead to a singular matrix $(A + \Delta A)$? It can be shown that the *minimum relative perturbation* of A, or $\min(\|\Delta A\|_2/\|A\|_2)$, such that $A + \Delta A$ becomes singular equals $1/\kappa(A)$. We show this in a simple case, namely when both A and its perturbation are diagonalizable by the same unitary matrix U (this happens for certain structured matrices). Then,

$$U^*AU = \Lambda \quad \text{and} \quad U^*\Delta AU = \Delta\Lambda, \tag{2.192}$$

where Λ is the diagonal matrix of eigenvalues of A and $\Delta\Lambda$ is the diagonal matrix of eigenvalues of the perturbation matrix ΔA. The minimum to perturb A into a singular matrix is

$$\min\frac{\|\Delta A\|_2}{\|A\|_2} \stackrel{(a)}{=} \min\frac{\|U^*\Delta AU\|_2}{\|U^*AU\|_2} \stackrel{(b)}{=} \frac{\min\|\Delta\Lambda\|_2}{\|\Lambda\|_2} \stackrel{(c)}{=} \left|\frac{\lambda_{\min}}{\lambda_{\max}}\right| \stackrel{(d)}{=} \frac{1}{\kappa(A)}, \tag{2.193}$$

where (a) follows from U being unitary; (b) from (2.192) and the fact that the optimization is over the perturbation, not over A; (c) from (2.188a) and $\Lambda - \text{diag}(0, 0, \ldots, 0, \lambda_{\min})$ being the singular matrix closest to Λ and thus $\min \|\Delta\Lambda\|_2 = |\lambda_{\min}|$; and (d) from (2.191b).

EXAMPLE 2.62 (CONDITIONING OF MATRICES, EXAMPLE 2.31 CONTINUED) Take the matrix associated with the basis in Example 2.31(i),

$$A = \begin{bmatrix} \varphi_0 & \varphi_1 \end{bmatrix} = \begin{bmatrix} 1 & 0 \\ 0 & a \end{bmatrix}, \qquad a \in (0, \infty).$$

The eigenvalues of A are 1 and a, from which the condition number follows as

$$\kappa(A) = \begin{cases} a, & \text{for } a \geq 1; \\ 1/a, & \text{for } a < 1. \end{cases}$$

For Example 2.31(ii),

$$A = \begin{bmatrix} \varphi_0 & \varphi_1 \end{bmatrix} = \begin{bmatrix} 1 & \cos\theta \\ 0 & \sin\theta \end{bmatrix}, \qquad \theta \in (0, \tfrac{1}{2}\pi]. \tag{2.194}$$

The singular value decomposition (2.230) of this matrix A leads to the condition number

$$\kappa(A) = \sqrt{\frac{1+\cos\theta}{1-\cos\theta}}, \tag{2.195}$$

which is plotted in Figure 2.32(a) on a log scale; A is ill conditioned as $\theta \to 0$, as expected.

EXAMPLE 2.63 (AVERAGING, EXAMPLE 2.32 CONTINUED) We take another look at Example 2.32(ii) from a matrix conditioning point of view. Multiplication $y = Ax$ with the $N \times N$ matrix

$$A = \begin{bmatrix} 1 & & & & \\ & \frac{1}{2} & & & \\ & & \frac{1}{3} & & \\ & & & \ddots & \\ & & & & \frac{1}{N} \end{bmatrix} \begin{bmatrix} 1 & 0 & 0 & \cdots & 0 \\ 1 & 1 & 0 & \cdots & 0 \\ 1 & 1 & 1 & \cdots & 0 \\ \vdots & \vdots & \vdots & \ddots & \vdots \\ 1 & 1 & 1 & \cdots & 1 \end{bmatrix} \tag{2.196}$$

computes the successive averages

$$y_{k-1} = \frac{1}{k}\sum_{n=0}^{k-1} x_n, \qquad k = 1, 2, \ldots, N.$$

While the matrix A is nonsingular (being a product of a diagonal matrix and a lower-triangular matrix, both with positive diagonal entries), solving $y = Ax$ (finding the original values from the averages) is an ill-conditioned problem because the dependence of y on x_n diminishes with increasing n. Figure 2.32(b) shows the condition number $\kappa(A)$ for $N = 2, 3, \ldots, 50$ on a log–log scale.

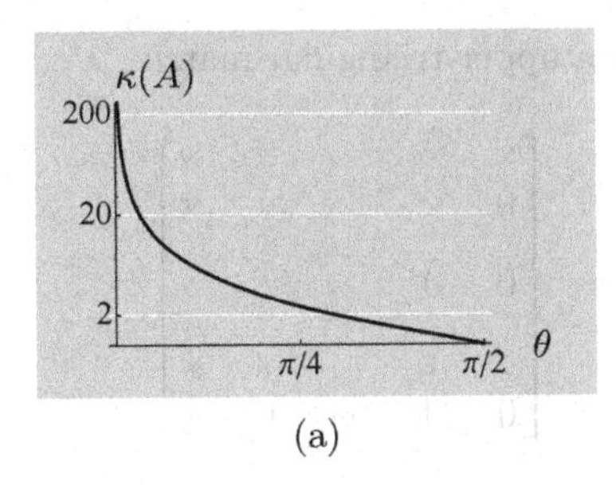

(a)

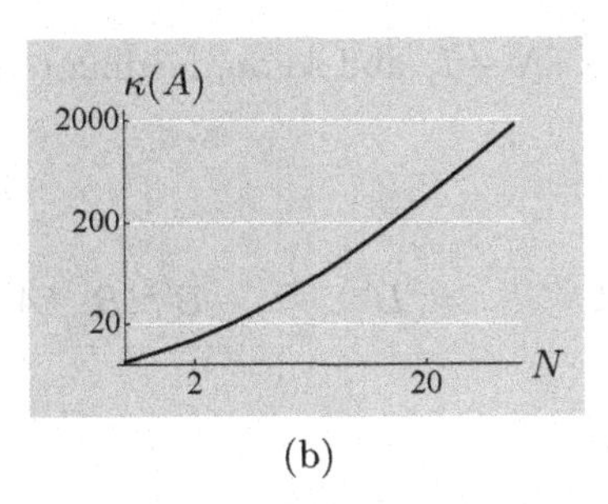

(b)

Figure 2.32 Behaviors of the condition numbers of matrices. (a) Condition numbers of the matrices from (2.194), expressed in (2.195), plotted on a log scale. (b) Condition numbers of the matrices from (2.196), plotted on a log–log scale.

2.6.4 Solving systems of linear equations

Having discussed conditioning of systems of linear equations, let us consider algorithms to compute the solution of $y = Ax$ where A is an $N \times N$ matrix. In general, as discussed in Appendix 2.B.1, the existence and uniqueness of a solution depend on whether the vector y belongs to the range (column space) of A. If it does not, there is no possible solution. If it does, and the columns are linearly independent, the solution is unique; otherwise there are infinitely many solutions.

Gaussian elimination The standard algorithm to solve a system of linear equations is Gaussian elimination. The algorithm uses elementary row operations to create a new system of equations $y' = A'x$, where A' is an upper-triangular matrix. Because of the upper-triangular form, the last entry of x is determined from the last entry of y'. Back substitution of this value reduces the size of the system of linear equations. Repeating the process, one finds the unknown vector x one entry at a time, from last to first.

We can easily obtain the upper-triangular A', by working on one column at a time and using the orthogonality of length-2 vectors. For example, given a vector $\begin{bmatrix} a_0 & a_1 \end{bmatrix}^\top$, then $\begin{bmatrix} a_1 & -a_0 \end{bmatrix}^\top$ is automatically orthogonal to it. To transform the first column, we premultiply A by the matrix $B^{(1)}$ with entries

$$B^{(1)} = \begin{bmatrix} 1 & 0 & 0 & \cdots & 0 \\ a_{1,0} & -a_{0,0} & 0 & \cdots & 0 \\ a_{2,0} & 0 & -a_{0,0} & \cdots & 0 \\ \vdots & \vdots & \vdots & \ddots & \vdots \\ a_{N-1,0} & 0 & 0 & \cdots & -a_{0,0} \end{bmatrix}, \tag{2.197}$$

leading to a new matrix $A^{(1)} = B^{(1)}A$ with first column $a_0^{(1)} = \begin{bmatrix} a_{0,0} & 0 & \ldots & 0 \end{bmatrix}^T$. We can continue the process by iterating on the lower-right submatrix of size

$(N-1) \times (N-1)$ and so on, leading to an upper-triangular matrix

$$A^{(N-1)} = B^{(N-1)} \cdots B^{(2)} B^{(1)} A = \begin{bmatrix} \times & \times & \cdots & \times & \times \\ 0 & \times & \cdots & \times & \times \\ 0 & 0 & \ddots & \vdots & \vdots \\ \vdots & \vdots & \ddots & \times & \times \\ 0 & 0 & \cdots & 0 & \times \end{bmatrix}. \tag{2.198}$$

The initial system of equations is thus transformed into a triangular one,

$$B^{(N-1)} \cdots B^{(2)} B^{(1)} y = A^{(N-1)} x,$$

which is solved easily by back substitution.

EXAMPLE 2.64 (TRIANGULARIZATION AND BACK SUBSTITUTION) Given a 3×3 system $y = Ax$ with a rank-3 matrix A, we can show that

$$B^{(1)} = \begin{bmatrix} 1 & 0 & 0 \\ a_{1,0} & -a_{0,0} & 0 \\ a_{2,0} & 0 & -a_{0,0} \end{bmatrix},$$

$$B^{(2)} = \begin{bmatrix} 1 & 0 & 0 \\ 0 & 1 & 0 \\ 0 & a_{2,0}a_{0,1} - a_{0,0}a_{2,1} & a_{0,0}a_{1,1} - a_{1,0}a_{0,1} \end{bmatrix}.$$

The new system is now of the form

$$y' = B^{(2)} B^{(1)} y = B^{(2)} B^{(1)} A x = A' x, \tag{2.199}$$

with the matrix A' upper-triangular and $a'_{i,i} \neq 0$ for $i \in \{0, 1, 2\}$ (because A is of full rank). We can solve for x using back substitution,

$$\begin{aligned} x_2 &= \frac{1}{a'_{2,2}} y'_2, \\ x_1 &= \frac{1}{a'_{1,1}} (y'_1 - a'_{1,2} x_2), \\ x_0 &= \frac{1}{a'_{0,0}} (y'_0 - a'_{0,1} x_1 - a'_{0,2} x_2). \end{aligned}$$

The form of the solution indicates that conditioning is a key issue, since, if some $a'_{i,i}$ is close to zero, the solution might be ill behaved.

In the above discussion, we did not discuss ordering of operations, or choosing a particular row to zero out the entries of a particular column. In practice, this choice, called choosing the *pivot*, is important for the numerical behavior of the algorithm.

Cost of Gaussian elimination The cost of Gaussian elimination is dominated by the cost of forming the product $B^{(N-1)} \cdots B^{(2)} B^{(1)} A$, resulting in the upper-triangular matrix $A^{(N-1)}$. Multiplying $B^{(1)}$ and A uses $\Theta(N^2)$ multiplications and additions. Similarly, multiplying $B^{(2)}$ and $B^{(1)} A$ also uses $\Theta(N^2)$ multiplications and additions. Having a total of $N - 1$ such multiplications, the algorithm forms $A^{(N-1)}$ with $\Theta(N^3)$ multiplications and additions. The other steps are cheaper: forming $B^{(N-1)} \cdots B^{(2)} B^{(1)} y$ has $\Theta(N^2)$ cost; and back substitution requires N divisions and $\Theta(N^2)$ multiplications and additions. A careful accounting gives a total multiplicative cost of about $\frac{1}{3}N^3$.

One can use Gaussian elimination to calculate the inverse of a matrix A. Solving $Ax = e_k$, where e_k is the kth vector of the standard basis, gives the kth column of A^{-1}. The cost of finding one column in this manner is $\Theta(N^3)$, so the overall cost of an inversion algorithm that finds each column independently is $\Theta(N^4)$. However, this algorithm is very inefficient because the matrix A is repeatedly transformed to the same upper-triangular form. An inversion algorithm that forms and saves the *LU decomposition* of A while solving $Ax = e_0$ with Gaussian elimination (and then uses this decomposition to efficiently solve $Ax = e_k$ for $k \in \{1, 2, \ldots, N-1\}$) has cost approximately $\frac{4}{3}N^3$. Note that we rarely calculate A^{-1} explicitly; solving $y = Ax$ is no cheaper with A^{-1} than with the LU decomposition of A.

Sparse matrices and iterative solutions of systems of linear equations If the matrix–vector product Ax is easy to compute, that is, with a cost substantially smaller than N^2, then iterative solvers can be considered. This is the case when A is sparse or banded as in (2.247) and has only $o(N^2)$ nonzero entries.

An iterative algorithm computes a new approximate solution from an old approximate solution with an update step; if properly designed, it will converge to the solution. The basic idea is to write $A = D - B$, transforming $y = Ax$ into

$$Dx = Bx + y. \tag{2.200}$$

The update step is

$$x^{(k+1)} = D^{-1}(Bx^{(k)} + y), \tag{2.201}$$

which has the desired solution as a fixed point. If D^{-1} is easy to compute (for example, it is diagonal), then this is a valid approach.

To study the error, let $e^{(k)} = x - x^{(k)}$. Subtracting

$$Dx^{(k+1)} = Bx^{(k)} + y$$

from (2.200) yields

$$e^{(k+1)} = D^{-1}Be^{(k)} = (D^{-1}B)^{k+1}e^{(0)},$$

with $e^{(0)}$ the initial error. The algorithm will converge to the true solution for any initial guess $x^{(0)}$ if and only if $(D^{-1}B)^{k+1} \to 0$ as $k \to \infty$, which happens when all the eigenvalues of $D^{-1}B$ are smaller than 1 in absolute value.

EXAMPLE 2.65 (ITERATIVE SOLUTION OF A TOEPLITZ SYSTEM) Take the Toeplitz matrix A from (2.246) and write it as $D - B = I - (I - A)$. Then, (2.201) reduces to

$$x^{(k+1)} = (I - A)x^{(k)} + y. \tag{2.202}$$

Note that $B = I - A$ is still Toeplitz, allowing a fast multiplication for evaluating $Bx^{(k)}$, as will be seen in Section 3.9. If the eigenvalues of $D^{-1}B = I - A$ are smaller than 1 in absolute value, the iterative algorithm will converge.

As an example, consider the matrix describing a two-point sum,

$$A = \begin{bmatrix} 1 & 0 & 0 & 0 \\ 1 & 1 & 0 & 0 \\ 0 & 1 & 1 & 0 \\ 0 & 0 & 1 & 1 \end{bmatrix},$$

in the system $Ax = y$ with $y = \begin{bmatrix} 1 & 3 & 5 & 7 \end{bmatrix}^{\top}$. The eigenvalues of $(I - A)$ are all 0, and thus the algorithm will converge. For example, start with an all-zero vector $x^{(0)}$. The iterative procedure (2.202) produces

$$x^{(1)} = \begin{bmatrix} 1 \\ 3 \\ 5 \\ 7 \end{bmatrix}, \quad x^{(2)} = \begin{bmatrix} 1 \\ 2 \\ 2 \\ 2 \end{bmatrix}, \quad x^{(3)} = \begin{bmatrix} 1 \\ 2 \\ 3 \\ 5 \end{bmatrix}, \quad x^{(4)} = \begin{bmatrix} 1 \\ 2 \\ 3 \\ 4 \end{bmatrix},$$

and converges in the fourth step ($x^{(n)} = x^{(4)}$ for $n \geq 5$).

Among iterative solvers of large systems of linear equations, Kaczmarz's algorithm has an intuitive geometric interpretation.

EXAMPLE 2.66 (KACZMARZ'S ALGORITHM) Consider a square system of linear equations $y = Ax$ with A real and of full rank. We can look for the solution $x = \begin{bmatrix} x_0 & x_1 & \ldots & x_{N-1} \end{bmatrix}^{\top}$ in two ways, concentrating on either the columns or the rows of A. When concentrating on the columns $\{v_0, v_1, \ldots, v_{N-1}\}$, we see the solution x as giving the coefficients to form y as a linear combination of columns:

$$\sum_{n=0}^{N-1} x_n v_n = y. \tag{2.203a}$$

When concentrating on the rows $\{r_0^{\top}, r_1^{\top}, \ldots, r_{N-1}^{\top}\}$, we see the solution x as the vector that has all the correct inner products:

$$\langle x, r_n \rangle = y_n, \qquad n = 0, 1, \ldots, N-1. \tag{2.203b}$$

Kaczmarz's algorithm uses the row-based view. Normalize r_n to be of unit norm, $\gamma_n = r_n / \|r_n\|$. Then, (2.203b) becomes

$$\langle x, \gamma_n \rangle = \frac{y_n}{\|r_n\|} = y'_n, \qquad n = 0, 1, \ldots, N-1. \tag{2.204}$$

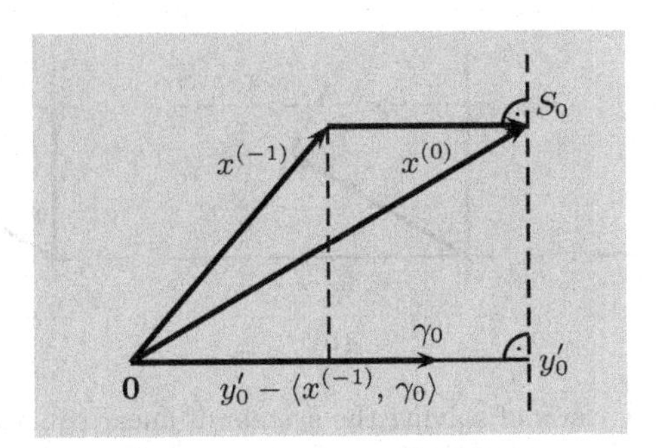

Figure 2.33 One step of Kaczmarz's algorithm. The update $x^{(0)}$ is the orthogonal projection of the initial guess $x^{(-1)}$ onto the affine subspace S_0 orthogonal to the subspace spanned by γ_0 and at the distance y_0' from the origin $\mathbf{0}$.

The idea of Kaczmarz's algorithm is to iteratively satisfy the constraints (2.204). Starting with an initial guess $x^{(-1)}$, the first update step is N computations

$$x^{(n)} = x^{(n-1)} + (y_n' - \langle x^{(n-1)}, \gamma_n\rangle)\gamma_n, \qquad n = 0, 1, \ldots, N-1, \tag{2.205}$$

called a sweep. With the update (2.205), $x^{(n)}$ satisfies

$$\langle x^{(n)}, \gamma_n\rangle = \langle x^{(n-1)}, \gamma_n\rangle + y_n' \underbrace{\langle \gamma_n, \gamma_n\rangle}_{=1} - \underbrace{\langle \gamma_n, \gamma_n\rangle}_{=1}\langle x^{(n-1)}, \gamma_n\rangle = y_n',$$

as desired. At the end of this sweep, it is most likely that $x^{(N-1)}$ will not satisfy $\langle x^{(N-1)}, \gamma_0\rangle = y_0'$, and thus, further sweeps are required.

To understand the algorithm geometrically, note that the update $x^{(0)}$ is the orthogonal projection of the initial guess $x^{(-1)}$ onto the affine subspace S_0 orthogonal to the subspace spanned by γ_0 and at distances y_0' from the origin as in Figure 2.33. The desired solution is $x = \cap_{i=1}^{n} S_i$. Convergence is geometric in the number of sweeps, with a constant depending on how close to orthogonal the vectors γ_n are. When the rows of A are orthogonal, convergence occurs in one sweep (see Exercise 2.51). The iterative algorithm presented in Section 6.6.3 is an extension of Kaczmarz's algorithm to a setting where entries of y are not known exactly.

In Figure 2.34, we show three different interpretations of solving the system of linear equations

$$\begin{bmatrix} \frac{\sqrt{3}}{2} & \frac{1}{2} \\ 0 & 1 \end{bmatrix} \begin{bmatrix} x_0 \\ x_1 \end{bmatrix} = \begin{bmatrix} \frac{\sqrt{3}+1}{2} \\ 1 \end{bmatrix}, \tag{2.206}$$

which has the solution $x = \begin{bmatrix} 1 & 1 \end{bmatrix}^\top$. The rows are already of norm 1, and thus $\gamma_n = r_n$ and $y_n' = y_n$. Figure 2.34(a) shows the solution as a linear combination (2.203a) of column vectors $\{v_0, v_1\}$. In this particular case, it turns out that $y = v_0 + v_1$ exactly since $x_0 = x_1 = 1$. Figure 2.34(b) shows the solution as the intersection of linear constraints (2.203b), that is, the intersection of the two

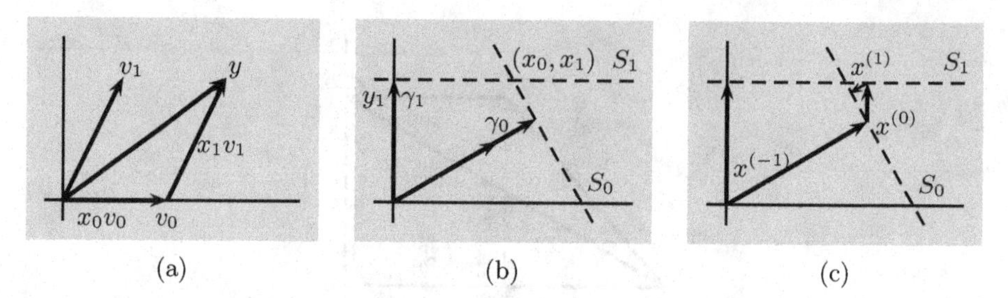

Figure 2.34 Different views of solving the system of linear equations in (2.206): (a) as a linear combination (2.203a) of column vectors $\{v_0, v_1\}$; (b) as the intersection of linear constraints (2.203b); and (c) as the solution to the iterative algorithm (2.205), starting with $x^{(-1)} = 0$.

subspaces S_0 and S_1, orthogonal to the spans of γ_0 and γ_1, and at the distances from the origin y_0 and y_1, respectively. The intersection of S_0 and S_1 is exactly the solution $x = \begin{bmatrix} 1 & 1 \end{bmatrix}^{\top}$. Finally, Figure 2.34(c) shows a few steps of the iterative algorithm (2.205), starting with $x^{(-1)} = 0$.

Complexity of solving a system of linear equations The cost of the Gaussian elimination algorithm (or any other algorithm) provides an upper bound to the complexity of solving a general system of linear equations. The precise multiplicative complexity is not known.

If the matrix A is structured, computational savings can be achieved. For example, we will see in the next chapter that when the matrix is circulant as in (2.245), the cost of a simple algorithm is $O(N \log_2 N)$ as in (3.271), because the discrete Fourier transform (DFT) diagonalizes the circulant convolution operator, and many fast algorithms for computing the DFT exist. Also, solvers with cost $O(N^2)$ exist for the cases where the matrix is Toeplitz as in (2.246) or Vandermonde as in (2.248). The *Further reading* gives pointers to literature on these algorithms.

Appendix 2.A Elements of analysis and topology

This appendix reviews some basic elements of real analysis (under Lebesgue measure as applicable) and the standard topology on the real line. Some material has been adapted from [66,85].

2.A.1 Basic definitions

Sets Let W be a subset of $\mathbb{R}$. An *upper bound* is a number M such that every w in W satisfies $w \leq M$. The smallest of all upper bounds is called the *supremum* of W and denoted $\sup W$; if no upper bound exists, $\sup W = \infty$. A *lower bound* is a number m such that every w in W satisfies $w \geq m$. The largest of all lower bounds is called the *infimum* of W and denoted $\inf W$; if no lower bound exists, $\inf W = -\infty$.

The essential supremum and essential infimum are defined similarly but are based on bounds that can be violated by a countable number of points. An *essential upper bound* is a number M such that at most a countable number of w in W violates $w \leq M$. The smallest of all essential upper bounds is called the *essential supremum* of W and denoted $\operatorname{ess\,sup} W$; if no essential upper bound exists, $\operatorname{ess\,sup} W = \infty$. An *essential lower bound* is a number m such that at most a countable number of w in W violates $w \geq m$. The largest of all essential lower bounds is called the *essential infimum* of W and denoted $\operatorname{ess\,inf} W$; if no essential lower bound exists, $\operatorname{ess\,inf} W = -\infty$.

Topology Let W be a subset of $\mathbb{R}$. An element $w \in W$ is an *interior point* if there is an $\varepsilon > 0$ such that $(w - \varepsilon,\ w + \varepsilon) \subset W$. A set is *open* if all its points are interior points. Facts about open sets include the following:

(i) $\mathbb{R}$ is open.
(ii) $\emptyset$ is open.
(iii) The union of *any* collection of open sets is open.
(iv) The intersection of *finitely many* open sets is open.

A set is *closed* when its complement is open. Upon complementing the sets in the list above, facts about closed sets include the following:

(i) $\emptyset$ is closed.
(ii) $\mathbb{R}$ is closed.
(iii) The intersection of *any* collection of closed sets is closed.
(iv) The union of *finitely many* closed sets is closed.

The *closure* of a set W, denoted by $\overline{W}$, is the intersection of all closed sets containing W. It is also the set of all limit points of convergent sequences in the set. A set is closed if and only if it is equal to its closure. Also, a set is closed if and only if it contains the limit of every convergent sequence in the set.

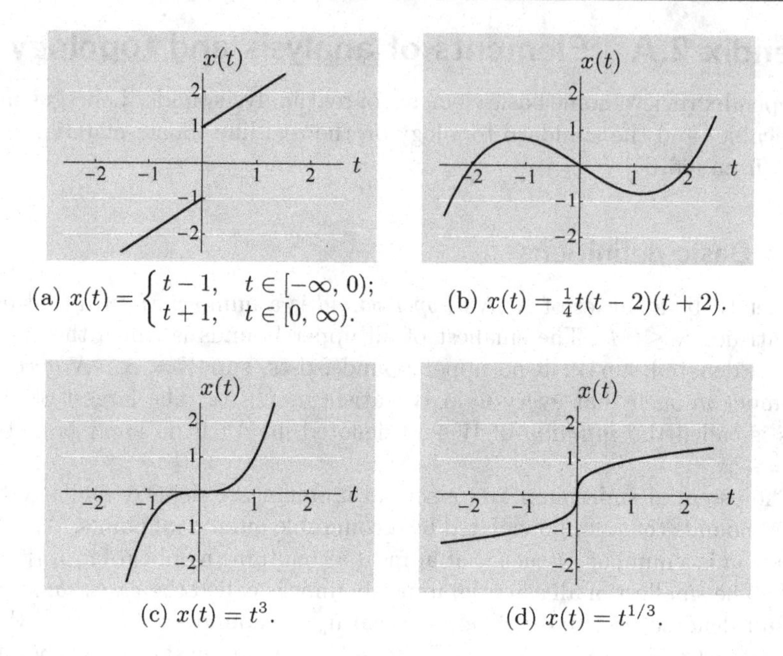

(a) $x(t) = \begin{cases} t-1, & t \in [-\infty, 0); \\ t+1, & t \in [0, \infty). \end{cases}$ (b) $x(t) = \frac{1}{4}t(t-2)(t+2)$.

(c) $x(t) = t^3$. (d) $x(t) = t^{1/3}$.

Figure 2.35 Examples of different types of functions $x : \mathbb{R} \to \mathbb{R}$. (a) Injective, but not surjective; the range is $\mathcal{R} = (\infty, -1] \cup [1, \infty)$. (b) Surjective, but not injective. (c) Bijective (both injective and surjective). (d) Inverse of the bijective function from (c).

Functions A *function* x takes an *argument* (input) t and produces a *value* (output) $x(t)$. The acceptable values of the argument form the *domain*, while the possible function values form the *range*, which is also called the *image*. If the range is a subset of a larger set, that set is termed the *codomain*. The notation $x : D \to C$ indicates that x is a function with the domain D and the codomain C. A *composition* of functions uses the output of one function as the input to another. A function that maps a vector space into a vector space is called an *operator*.

A function is *injective* if $x(t_1) = x(t_2)$ implies that $t_1 = t_2$. In other words, different values of the function must have been produced by different arguments. A function is *surjective* if the range equals the codomain, that is, if, for every $y \in C$, there exists a $t \in D$ such that $x(t) = y$. A function is *bijective* if it is both injective and surjective. A bijective function $x : D \to C$ has an *inverse* $x^{-1} : C \to D$ such that $x^{-1}(x(t)) = t$ for all $t \in D$ and $x(x^{-1}(y)) = y$ for all $y \in C$. These concepts are illustrated in Figure 2.35.

2.A.2 Convergence

Sequences A sequence of numbers $a_0, a_1, \ldots$ is said to *converge* to the number a (written $\lim_{k\to\infty} a_k = a$) when the following holds:

for any $\varepsilon > 0$ there exists a number K_ε such that $|a_k - a| < \varepsilon$ for every $k > K_\varepsilon$.

The sequence is said to *diverge* if it does not converge to any (finite) number. It *diverges to* ∞ (written $\lim_{k\to\infty} a_k = \infty$) when the following holds:

for any M there exists a number K_M such that $a_k > M$ for every $k > K_M$.

Similarly, it *diverges to* $-\infty$ (written $\lim_{k\to\infty} a_k = -\infty$) when the following holds:

for any M there exists a number K_M such that $a_k < M$ for every $k > K_M$.

A few properties of convergence of sequences are derived in Exercise 2.52.

Series Let $a_0, a_1, \ldots$ be numbers. The numbers $s_n = \sum_{k=0}^{n} a_k$, $n = 0, 1, \ldots$, are called *partial sums* of the *(infinite) series* $\sum_{k=0}^{\infty} a_k$. The series is said to *converge* when the sequence of partial sums converges. We write $\sum_{k=0}^{\infty} a_k = \infty$ when the partial sums diverge to ∞ and $\sum_{k=0}^{\infty} a_k = -\infty$ when the partial sums diverge to $-\infty$.

The series $\sum_{k=0}^{\infty} a_k$ is said to *converge absolutely* when $\sum_{k=0}^{\infty} |a_k|$ converges. A series that converges but does not converge absolutely is said to *converge conditionally*. The definition of convergence takes the terms of a series in a particular order. When a series is absolutely convergent, its terms can be reordered without altering its convergence or its value; otherwise not.[30] The *doubly infinite series* $\sum_{k=-\infty}^{\infty} a_k$ does not have a single natural choice of partial sums, so it is said to converge when it converges absolutely.

Tests for convergence of series are reviewed in Exercise 2.53, and a few useful series are explored in Exercise 2.54.

Functions A sequence of real-valued functions $x_0, x_1, \ldots$ *converges pointwise* when, for any fixed t, the sequence of numbers $x_0(t), x_1(t), \ldots$ converges. More explicitly, suppose that the functions have a common domain D. They converge pointwise to function x with the domain D when, for any $\varepsilon > 0$ and $t \in D$, there exists a number $K_{\varepsilon,t}$ (depending on ε and t) such that

$$|x_k(t) - x(t)| < \varepsilon \qquad \text{for all } k > K_{\varepsilon,t}.$$

A more restrictive form of convergence does not allow $K_{\varepsilon,t}$ to depend on t. A sequence $x_0, x_1, \ldots$ of real-valued functions on some domain D *converges uniformly* to $x : D \to \mathbb{R}$ when, for any $\varepsilon > 0$, there exists a number K_ε (depending on ε) such that

$$|x_k(t) - x(t)| < \varepsilon \qquad \text{for all } t \in D \text{ and all } k > K_\varepsilon.$$

Uniform convergence implies pointwise convergence. Furthermore, if a sequence of continuous functions is uniformly convergent, the limit function is necessarily continuous.

[30] A strange and wonderful fact known as the Riemann series theorem is that a conditionally convergent series can be rearranged to converge to any desired value or to diverge.

2.A.3 Interchange theorems

Many derivations in analysis involve interchanging the order of sums, integrals, and limits without changing the result. Without appropriate caution, this might be simply incorrect; refer again to Footnote 30 on page 137.

Two nested summations can be seen as a single sum over a two-dimensional index set. Since absolute convergence allows rearrangement of the terms in a sum, it allows changing the order of summations, yielding *Fubini's theorem for sequences*:

$$\sum_{n=0}^{\infty}\sum_{k=0}^{\infty}|x_{n,k}| < \infty$$

implies that

$$\sum_{n=0}^{\infty}\sum_{k=0}^{\infty} x_{n,k} = \sum_{k=0}^{\infty}\sum_{n=0}^{\infty} x_{n,k}. \tag{2.207}$$

When x takes nonnegative values, (2.207) holds without assuming absolute convergence as well; this result, *Tonelli's theorem for sequences*, implies that summing in one order diverges to ∞ if and only if summing in the other order diverges to ∞. These facts extend to doubly infinite summations and more than two summations.

The analogous result for absolutely integrable functions is called *Fubini's theorem for functions*:

$$\int_{-\infty}^{\infty}\int_{-\infty}^{\infty} |x(t_1, t_2)|\, dt_1\, dt_2 < \infty$$

implies that

$$\int_{-\infty}^{\infty}\int_{-\infty}^{\infty} x(t_1, t_2)\, dt_1\, dt_2 = \int_{-\infty}^{\infty}\int_{-\infty}^{\infty} x(t_1, t_2)\, dt_2\, dt_1. \tag{2.208}$$

When x takes nonnegative values, (2.208) holds without assuming absolute integrability; this result, *Tonelli's theorem for functions*, implies that integrating in one order diverges to ∞ if and only if integrating in the other order diverges to ∞. These facts extend to more than two integrals. Tonelli's theorem is often used to check absolute summability or integrability before applying Fubini's theorem.

Interchange of summation and integration can be justified by uniform convergence. Suppose that a sequence of partial sums $s_n(t) = \sum_{k=0}^{n} x_k(t)$, $n = 0, 1, \ldots$, is uniformly convergent to $s(t)$ on $[a, b]$. Then, the series may be integrated term by term:

$$\int_a^b \sum_{k=0}^{\infty} x_k(t)\, dt = \sum_{k=0}^{\infty} \int_a^b x_k(t)\, dt. \tag{2.209}$$

This result extends to infinite intervals as well.

Uniform convergence is rather restrictive as a justification for (2.209), since changing any of the x_k functions on a set of zero measure would not change either side of the equality. Another result on interchanging summation and integration is as follows. If $|x_k(t)| \le y_k(t)$ for all $k \in \mathbb{N}$ and almost all $t \in [a, b]$, and $\sum_{k=0}^{\infty} y_k(t)$ converges for almost all $t \in [a, b]$ and $\sum_{k=0}^{\infty} \int_a^b y_k(t)\, dt < \infty$, then (2.209) holds.

At the heart of this result is an application of a theorem, called the dominated convergence theorem, to the sequence of partial sums $s_n(t) = \sum_{k=0}^{n} x_k(t)$, $n = 0, 1, \ldots$.

The *dominated convergence theorem* enables interchange of a limit with an integral: Let $x_0, x_1, \ldots$ be real-valued functions such that $\lim_{k\to\infty} x_k(t) = x(t)$ for almost all $t \in \mathbb{R}$. If there exists a (nonnegative) real-valued function y such that. for all $k \in \mathbb{N}$,

$$|x_k(t)| \leq y(t) \quad \text{holds for almost all } t \in \mathbb{R} \qquad \text{and} \qquad \int_{-\infty}^{\infty} y(t)\, dt < \infty,$$

then x is integrable, and

$$\int_{-\infty}^{\infty} \left(\lim_{k\to\infty} x_k(t) \right) dt = \lim_{k\to\infty} \int_{-\infty}^{\infty} x_k(t)\, dt. \tag{2.210}$$

2.A.4 Inequalities

See [100] for elementary proofs and further details.

Minkowski's inequality For any $p \in [1, \infty)$,

$$\left(\sum_{k\in\mathbb{Z}} |x_k + y_k|^p \right)^{1/p} \leq \left(\sum_{k\in\mathbb{Z}} |x_k|^p \right)^{1/p} + \left(\sum_{k\in\mathbb{Z}} |y_k|^p \right)^{1/p}. \tag{2.211a}$$

This establishes that the ℓ^p norm (2.40a) satisfies the triangle inequality in Definition 2.9. Also, for any $p \in [1, \infty)$,

$$\left(\int_a^b |x(t) + y(t)|^p\, dt \right)^{1/p} \leq \left(\int_a^b |x(t)|^p\, dt \right)^{1/p} + \left(\int_a^b |y(t)|^p\, dt \right)^{1/p}, \tag{2.211b}$$

establishing that the $\mathcal{L}^p$ norm (2.42a) satisfies the triangle inequality.

Analogues of (2.211) hold for ℓ^∞ and $\mathcal{L}^\infty$ as well:

$$\sup_{k\in\mathbb{Z}} |x_k + y_k| \leq \sup_{k\in\mathbb{Z}} |x_k| + \sup_{k\in\mathbb{Z}} |y_k| \tag{2.212a}$$

and

$$\operatorname*{ess\,sup}_{t\in\mathbb{R}} |x(t) + y(t)| \leq \operatorname*{ess\,sup}_{t\in\mathbb{R}} |x(t)| + \operatorname*{ess\,sup}_{t\in\mathbb{R}} |y(t)|. \tag{2.212b}$$

Hölder's inequality Let p and q in $[1, \infty]$ satisfy $1/p + 1/q = 1$ with the convention that $1/\infty = 0$ is allowed. Then, p and q are called *Hölder conjugates*, and

$$\|xy\|_1 \leq \|x\|_p\, \|y\|_q \tag{2.213}$$

for sequences or functions x and y, with equality if and only if $|x|^p$ and $|y|^q$ are scalar multiples of each other. The case of $p = q = 2$ is the Cauchy–Schwarz inequality, (2.29).

Specializing (2.213) to sequences gives

$$\sum_{k\in\mathbb{Z}} |x_k y_k| \;\leq\; \left(\sum_{k\in\mathbb{Z}} |x_k|^p\right)^{1/p} \left(\sum_{k\in\mathbb{Z}} |y_k|^q\right)^{1/q} \tag{2.214a}$$

for finite p and q, and

$$\sum_{k\in\mathbb{Z}} |x_k y_k| \;\leq\; \left(\sup_{k\in\mathbb{Z}} |x_k|\right) \left(\sum_{k\in\mathbb{Z}} |y_k|\right) \tag{2.214b}$$

for $p = \infty$. Similarly, for functions

$$\int_{-\infty}^{\infty} |x(t)\, y(t)|\, dt \;\leq\; \left(\int_{-\infty}^{\infty} |x(t)|^p\, dt\right)^{1/p} \left(\int_{-\infty}^{\infty} |y(t)|^q\, dt\right)^{1/q} \tag{2.215a}$$

for finite p and q, and

$$\int_{-\infty}^{\infty} |x(t)\, y(t)|\, dt \;\leq\; \left(\operatorname*{ess\,sup}_{t\in\mathbb{R}} |x(t)|\right) \left(\int_{-\infty}^{\infty} |y(t)|\, dt\right) \tag{2.215b}$$

for $p = \infty$.

Other integral inequalities If $x(t) \leq y(t)$ for all $t \in [a, b]$, then

$$\int_a^b x(t)\, dt \;\leq\; \int_a^b y(t)\, dt. \tag{2.216}$$

The limits of integration may be $a = -\infty$ or $b = \infty$. If the inequality between $x(t)$ and $y(t)$ is strict for all $t \in [a, b]$, then (2.216) holds with strict inequality.

The inequality

$$\left|\int_a^b x(t)\, dt\right| \;\leq\; \int_a^b |x(t)|\, dt \tag{2.217}$$

holds, where the limits of integration may be $a = -\infty$ or $b = \infty$. For Riemann-integrable functions, this follows from the triangle inequality and taking limits. The generalization to Lebesgue-integrable functions holds as well.

2.A.5 Integration by parts

Integration by parts transforms an integral into another integral, which is then possibly easier to solve. It can be written very compactly as

$$\int u\, dv \;=\; uv - \int v\, du, \tag{2.218a}$$

or, more explicitly, as

$$\int_a^b u(t)\, v'(t)\, dt \;=\; u(t)\, v(t)\Big|_{t=a}^{t=b} - \int_a^b v(t)\, u'(t)\, dt. \tag{2.218b}$$

Appendix 2.B Elements of linear algebra

This appendix reviews basic concepts in linear algebra. Good sources for more details include [45, 93]. Contrary to the standard convention in finite-dimensional linear algebra, we start all indexing at 0 rather than 1; this facilitates consistency throughout the book.

2.B.1 Basic definitions and properties

We say a matrix A is $M \times N$ or in $\mathbb{C}^{M\times N}$ when it has M rows and N columns. It is a linear operator mapping $\mathbb{C}^N$ into $\mathbb{C}^M$. When $M = N$, the matrix is called *square*; otherwise it is *rectangular*.[31] An $M \times 1$ matrix is called a *column vector*, a $1 \times N$ matrix a *row vector*, and a 1×1 matrix a *scalar*. Unless stated explicitly otherwise, $A_{m,n}$ denotes the row-m, column-n entry of matrix A.

Basic operations Addition of matrices is element-by-element, so matrices can be added only if they have the same dimensions. The product of $A \in \mathbb{C}^{M\times P}$ and $B \in \mathbb{C}^{Q\times N}$ is defined only when $P = Q$, in which case it is given by

$$(AB)_{m,n} = \sum_{k=0}^{P-1} A_{m,k}B_{k,n}, \qquad \begin{array}{rcl} m & = & 0, 1, \ldots, M-1, \\ n & = & 0, 1, \ldots, N-1. \end{array} \tag{2.219}$$

Entries $A_{m,n}$ are on the *(main) diagonal* if $m = n$. A square matrix with unit diagonal entries and zero off-diagonal entries is called an *identity matrix* and denoted by I. It is the identity element under matrix multiplication. For square matrices A and B, if $AB = I$ and $BA = I$, B is called the *inverse* of A and is written as A^{-1}. If no such B exists, A is called *singular*. When the inverses exist, $(AB)^{-1} = B^{-1}A^{-1}$. Rectangular matrices do not have inverses. Instead, a short matrix can have a *right* inverse B, so $AB = I$; similarly, a tall matrix can have a *left* inverse B, so $BA = I$.

If $A_{m,n} = B_{n,m}$ for all m and n, we write $A = B^\top$; we call B the *transpose* of A. If $A_{m,n} = B^*_{n,m}$ for all m and n, we write $A = B^*$; we call B the *Hermitian transpose* of A. Here, * denotes both complex conjugation of a scalar and the combination of complex conjugation and transposition of a matrix. In general, $(AB)^\top = B^\top A^\top$ and $(AB)^* = B^*A^*$.

Determinant The *determinant* maps a square matrix into a scalar value. It is defined recursively, with $\det a = a$ for any scalar a and

$$\det A = \sum_{k=0}^{N-1} (-1)^{i+k}(\det M_{i,k})A_{i,k} = \sum_{k=0}^{N-1} C_{i,k}A_{i,k} \tag{2.220}$$

for $A \in \mathbb{C}^{N\times N}$, where the *minor* $M_{i,k}$ is the $(N-1)\times(N-1)$ matrix obtained by deleting the ith row and the kth column of A; the *cofactor* $C_{i,k} = (-1)^{i+k}\det M_{i,k}$ will be used later to define the adjugate. This definition is valid because the same

[31] We sometimes call a matrix with $M > N$ *tall*; similarly, we call a matrix with $M < N$ *short*.

result is obtained for any choice of $i \in \{0, 1, \ldots, N-1\}$; to simplify computations, one may choose i in such a way so as to minimize the number of nonzero terms in the sum.

The determinant of $A \in \mathbb{C}^{N \times N}$ has several useful properties, including the following:

(i) For any scalar α, $\det(\alpha A) = \alpha^N \det A$.
(ii) If B is obtained by interchanging two rows or two columns of A, then $\det B = -\det A$.
(iii) $\det A^\top = \det A$.
(iv) $\det(AB) = (\det A)(\det B)$.
(v) If A is triangular, that is, all of its elements above or below the main diagonal are 0, $\det A$ is the product of the diagonal elements of A.
(vi) A is singular if and only if $\det A = 0$.

The final property relating the determinant to invertibility has both a geometric interpretation and a connection to a formula for a matrix inverse.

(i) When the matrix is real, the determinant is the volume of the parallelepiped that has the column vectors of the matrix as edges. Thus, a zero determinant indicates linear dependence of the columns of the matrix, since the parallelepiped is not of full dimension. (The row vectors lead to a different parallelepiped with the same volume.)
(ii) The inverse of a nonsingular matrix A is given by *Cramer's formula*:

$$A^{-1} = \frac{\operatorname{adj} A}{\det A}, \tag{2.221}$$

where the *adjugate* of A is the transpose of the matrix of cofactors of A: $(\operatorname{adj} A)_{i,k} = C_{k,i}$. Cramer's formula is useful for finding inverses of small matrices by hand and as an analytical tool; it does not yield computationally efficient techniques for inversion.

Range, null space, and rank Associated with any matrix $A \in \mathbb{R}^{M \times N}$ are four fundamental subspaces. The *range* or *column space* of A is the span of the columns of A and thus a subspace of $\mathbb{R}^M$; it can be written as

$$\mathcal{R}(A) = \operatorname{span}(\{a_0, a_1, \ldots, a_{N-1}\}) = \{y \in \mathbb{R}^M \mid y = Ax \text{ for some } x \in \mathbb{R}^N\}, \tag{2.222a}$$

where $a_0, a_1, \ldots, a_{N-1}$ are the columns of A. Linear combinations of rows of A are all row vectors $y^\top A$, where $y \in \mathbb{R}^M$. Taking these as column vectors gives the *row space* of A, which is the range of $A^\top$ and a subspace of $\mathbb{R}^N$:

$$\mathcal{R}(A^\top) = \operatorname{span}(\{b_0^\top, b_1^\top, \ldots, b_{M-1}^\top\}) = \{x \in \mathbb{R}^N \mid x = A^\top y \text{ for some } y \in \mathbb{R}^M\}, \tag{2.222b}$$

where $b_0, b_1, \ldots, b_{M-1}$ are the rows of A. The *null space* or *kernel* of A is the set of vectors that A maps to $\mathbf{0}$ (a subspace of $\mathbb{R}^N$):

$$\mathcal{N}(A) = \{x \in \mathbb{R}^N \mid Ax = \mathbf{0}\}. \tag{2.222c}$$

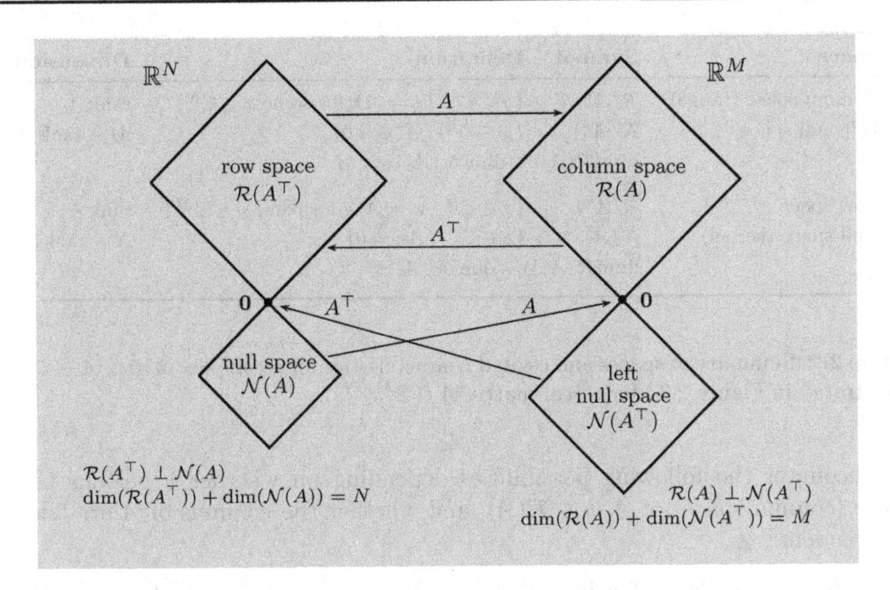

Figure 2.36 The four fundamental subspaces associated with a real matrix $A \in \mathbb{R}^{M \times N}$. The matrix determines an orthogonal decomposition of $\mathbb{R}^N$ into the row space of A and the null space of A, and an orthogonal decomposition of $\mathbb{R}^M$ into the column space (range) of A and the left null space of A. The column and row spaces of A have the same dimension, which equals the rank of A. (Figure inspired by the cover of [94].)

The *left null space* is the set of transposes of vectors mapped to zero when multiplied on the right by A. Since $y^\top A = \mathbf{0}$ is equivalent to $A^\top y = \mathbf{0}$, the left null space of A is the null space of $A^\top$ (a subspace of $\mathbb{R}^M$):

$$\mathcal{N}(A^\top) \;=\; \{y \in \mathbb{R}^M \mid A^\top y = \mathbf{0}\}. \tag{2.222d}$$

The four fundamental subspaces provide orthogonal decompositions of $\mathbb{R}^N$ and $\mathbb{R}^M$ as depicted in Figure 2.36. As shown, the null space is the orthogonal complement of the row space; the left null space is the orthogonal complement of the range (column space); A maps the null space to $\mathbf{0}$; $A^\top$ maps the left null space to $\mathbf{0}$; and A and $A^\top$ map between the row space and column space, which are of equal dimension. Properties of the subspaces are summarized for the complex case in Table 2.2.

The *rank* is defined by

$$\operatorname{rank} A \;=\; \dim(\mathcal{R}(A)). \tag{2.223}$$

It satisfies $\operatorname{rank} A = \operatorname{rank} A^*$ and $\operatorname{rank}(AB) \leq \min(\operatorname{rank} A, \operatorname{rank} B)$.

Systems of linear equations and least squares The product Ax describes a linear combination of the columns of A weighted by the entries of x. In solving a system of linear equations,

$$Ax \;=\; y, \qquad \text{where } A \in \mathbb{R}^{M \times N}, \tag{2.224}$$

Space	Symbol	Definition	Dimension
Column space (range)	$\mathcal{R}(A)$	$\{y \in \mathbb{C}^M \mid y = Ax \text{ for some } x \in \mathbb{C}^N\}$	$\operatorname{rank} A$
Left null space	$\mathcal{N}(A^*)$	$\{y \in \mathbb{C}^M \mid A^*y = 0\}$	$M - \operatorname{rank} A$
	$\dim(\mathcal{R}(A)) + \dim(\mathcal{N}(A^*)) = M$		
Row space	$\mathcal{R}(A^*)$	$\{x \in \mathbb{C}^N \mid x = A^*y \text{ for some } y \in \mathbb{C}^M\}$	$\operatorname{rank} A$
Null space (kernel)	$\mathcal{N}(A)$	$\{x \in \mathbb{C}^N \mid Ax = 0\}$	$N - \operatorname{rank} A$
	$\dim(\mathcal{R}(A^*)) + \dim(\mathcal{N}(A)) = N$		

Table 2.2 Summary of spaces and related characteristics for a complex matrix $A \in \mathbb{C}^{M\times N}$ (illustrated in Figure 2.36 for a real matrix $A \in \mathbb{R}^{M\times N}$).

we encounter the following possibilities depending on whether y belongs to the range (column space) of A, $y \in \mathcal{R}(A)$, and whether the columns of A are linearly independent:

(i) *Unique solution:* If y belongs to the range of A and the columns of A are linearly independent ($\operatorname{rank} A = N$), there is a unique solution.

(ii) *Infinitely many solutions:* If y belongs to the range of A and the columns of A are not linearly independent ($\operatorname{rank} A < N$), there are infinitely many solutions.

(iii) *No solution:* If y does not belong to the range of A, there is no solution. Only approximations are possible.

Cases with and without solutions are unified by looking for a *least-squares* solution $\widehat{x}$, meaning one that minimizes $\|y - \widehat{y}\|_2$, where $\widehat{y} = A\widehat{x}$. This is obtained from the orthogonality principle: the error $y - \widehat{y}$ is orthogonal to the range of A, leading to the *normal equations*,

$$A^\top A\widehat{x} = A^\top y. \tag{2.225a}$$

When $A^\top A$ is invertible ($\operatorname{rank} A = N$), the unique least-squares solution is

$$\widehat{x} = (A^\top A)^{-1}A^\top y. \tag{2.225b}$$

When A is square, the invertibility of $A^\top A$ implies that $y \in \mathcal{R}(A)$ and the least-squares solution simplifies to the exact solution $\widehat{x} = A^{-1}y$.

When $A^\top A$ is not invertible ($\operatorname{rank} A < N$), the minimization of $\|y - A\widehat{x}\|_2$ does not have a unique solution, so we additionally minimize $\|\widehat{x}\|_2$. When $AA^\top$ is invertible ($\operatorname{rank} A = M$), this solution is

$$\widehat{x} = A^\top(AA^\top)^{-1}y. \tag{2.225c}$$

The solutions (2.225b) and (2.225c) show the two forms of the *pseudoinverse* of A for $\operatorname{rank} A = \min(M, N)$. Multiplication by the pseudoinverse solves the least-squares problem for the case of $\operatorname{rank} A < \min(M, N)$ as well; the pseudoinverse is conveniently expressed using the singular value decomposition of A below. Figure 2.37 illustrates the discussion.

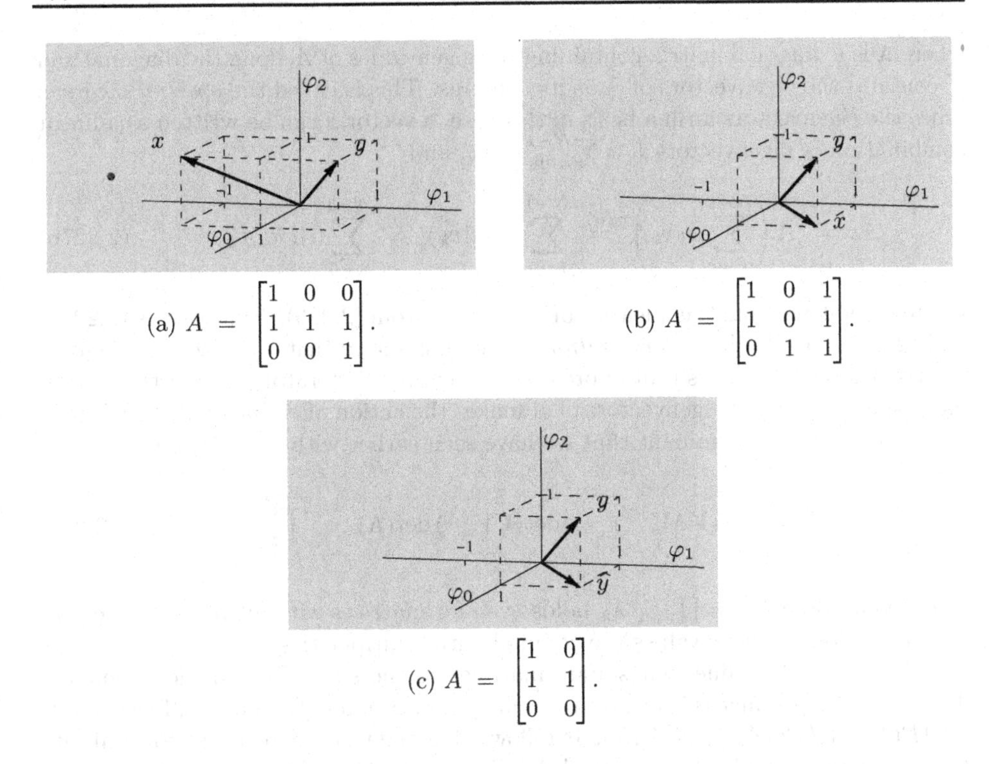

$$\text{(a) } A = \begin{bmatrix} 1 & 0 & 0 \\ 1 & 1 & 1 \\ 0 & 0 & 1 \end{bmatrix}. \qquad \text{(b) } A = \begin{bmatrix} 1 & 0 & 1 \\ 1 & 0 & 1 \\ 0 & 1 & 1 \end{bmatrix}. \qquad \text{(c) } A = \begin{bmatrix} 1 & 0 \\ 1 & 1 \\ 0 & 0 \end{bmatrix}.$$

Figure 2.37 Illustration of solutions to $Ax = y$ in $\mathbb{R}^3$, with $y = \begin{bmatrix} 1 & 1 & 1 \end{bmatrix}^\top$. (a) Unique solution: $y \in \mathcal{R}(A)$ and the columns of A are linearly independent. The unique solution is $x = \begin{bmatrix} 1 & -1 & 1 \end{bmatrix}^\top$. (b) Infinitely many solutions: $y \in \mathcal{R}(A)$ and the columns of A are not linearly independent. One of the possible solutions is $x = \begin{bmatrix} 1 & 1 & 0 \end{bmatrix}^\top$. (c) No solution: $y \notin \mathcal{R}(A)$ and the columns of A are linearly independent. The unique approximate solution of minimum 2 norm among the vectors that minimize the error is $\widehat{x} = \begin{bmatrix} 1 & 0 \end{bmatrix}^\top$, which yields $\widehat{y} = A\widehat{x} = \begin{bmatrix} 1 & 1 & 0 \end{bmatrix}^\top$.

Eigenvalues, eigenvectors, and spectral decomposition A number λ and a nonzero vector v are called an *eigenvalue* and an *eigenvector* of a square matrix A (they are also known as an *eigenpair*) when

$$Av = \lambda v, \tag{2.226}$$

as seen for general linear operators in (2.58). The eigenvalues are the roots of the *characteristic polynomial* $\det(xI - A)$. When all eigenvalues of A are real, $\lambda_{\max}(A)$ denotes the largest eigenvalue and $\lambda_{\min}(A)$ the smallest eigenvalue. When the eigenvalues are real, it is conventional to list them in nonincreasing order, $\lambda_0(A) \geq \lambda_1(A) \geq \cdots \geq \lambda_{N-1}(A)$.

If an $N \times N$ matrix A has N linearly independent eigenvectors, then it can be written as

$$A = V \Lambda V^{-1}, \tag{2.227a}$$

where Λ is a diagonal matrix containing the eigenvalues of A along the diagonal and V contains the eigenvectors of A as its columns. This is called the *spectral theorem.* Since the eigenvectors form a basis in this case, a vector x can be written as a linear combination of eigenvectors $x = \sum_{k=0}^{N-1} \alpha_k v_k$, and

$$Ax = A\left(\sum_{k=0}^{N-1} \alpha_k v_k\right) \stackrel{(a)}{=} \sum_{k=0}^{N-1} \alpha_k (Av_k) \stackrel{(b)}{=} \sum_{k=0}^{N-1} (\alpha_k \lambda_k) v_k, \tag{2.227b}$$

where (a) follows from the linearity of A; and (b) from (2.226). Expressions (2.227a) and (2.227b) are both *diagonalizations.* The first shows that $V^{-1}AV$ is a diagonal matrix; the second shows that expressing the input to operator A using the coordinates specified by the eigenvectors of A makes the action of A diagonal. Combining properties of the determinant that we have seen earlier with (2.227a) gives

$$\det A = \det(V\Lambda V^{-1}) = \det(VV^{-1})\det(\Lambda) = \prod_{k=0}^{N-1} \lambda_k. \tag{2.228}$$

The conclusion $\det A = \prod_{k=0}^{N-1} \lambda_k$ holds even for matrices without full sets of eigenvectors, as long as eigenvalues are counted with multiplicities.

The *trace* is defined for square matrices as the sum of the diagonal entries. The trace of a product is invariant to cyclic permutation of the factors, for example $\mathrm{tr}(ABC) = \mathrm{tr}(BCA) = \mathrm{tr}(CAB)$. It follows that the trace is invariant to similarity transformations: $\mathrm{tr}(BAB^{-1}) = \mathrm{tr}(AB^{-1}B) = \mathrm{tr}\, A$. The trace is given by the sum of eigenvalues (counted with multiplicities),

$$\mathrm{tr}\, A = \sum_{k=0}^{N-1} \lambda_k, \tag{2.229}$$

which is justified by (2.227a) for diagonalizable A.

Singular value decomposition *Singular value decomposition (SVD)* provides a diagonalization that applies to any rectangular or square matrix. An $M \times N$ real or complex matrix A can be factored as follows:

$$A = U\Sigma V^*, \tag{2.230}$$

where U is an $M \times M$ unitary matrix, V is an $N \times N$ unitary matrix, and Σ is an $M \times N$ matrix with nonnegative real values $\{\sigma_k\}_{k=0}^{\min(M,N)-1}$ called *singular values* on the main diagonal and zeros elsewhere. The columns of U are called *left singular vectors* and the columns of V are called *right singular vectors.* As for eigenvalues, $\sigma_{\max}(A)$ denotes the largest singular value and $\sigma_{\min}(A)$ the smallest singular value. Also as for eigenvalues, it is conventional to list singular values in nonincreasing order, $\sigma_{\max}(A) = \sigma_0(A) \geq \sigma_1(A) \geq \cdots \geq \sigma_{N-1}(A) = \sigma_{\min}(A)$. The number of nonzero singular values is the rank of A. The pseudoinverse of A is

$$A^\dagger = V\Sigma^\dagger U^*, \tag{2.231}$$

where $\Sigma^\dagger$ is the $N \times M$ matrix with $1/\sigma_k$ in the (k,k) position for each nonzero singular value and zeros elsewhere.

The following fact relates singular value decomposition and eigendecomposition (see also Exercise 2.56): Using the singular value decomposition (2.230),

$$\begin{aligned} AA^* &= (U\Sigma V^*)(V\Sigma^* U^*) = U\Sigma^2 U^*, \\ A^*A &= (V\Sigma^* U^*)(U\Sigma V^*) = V\Sigma^2 V^*, \end{aligned}$$

so the squares of the singular values of A are the nonzero eigenvalues of AA^* and A^*A; that is,

$$\sigma^2(A) = \lambda(AA^*) = \lambda(A^*A), \qquad \text{for } \lambda \neq 0. \tag{2.232}$$

Matrix norms Norms on matrices must satisfy the conditions in Definition 2.9. Many commonly used norms on $M \times N$ matrices are *operator norms* induced by norms on M- and N-dimensional vectors, as in Definition 2.18. Using the vector norms defined in (2.39a) and (2.39b),

$$\|A\|_{p,q} = \sup_{\|x\|_p=1} \|Ax\|_q. \tag{2.233}$$

The $p = q$ case $\|A\|_{p,p}$ is denoted $\|A\|_p$. A few of these norms simplify as follows:

$$\|A\|_1 = \|A\|_{1,1} = \max_{0\le j\le N-1} \sum_{i=0}^{M-1} |A_{i,j}|, \tag{2.234a}$$

$$\|A\|_{1,2} = \max_{0\le j\le N-1} \left(\sum_{i=0}^{M-1} |A_{i,j}|^2\right)^{1/2}, \tag{2.234b}$$

$$\|A\|_{1,\infty} = \max_{0\le i\le M-1,\, 0\le j\le N-1} |A_{i,j}|, \tag{2.234c}$$

$$\|A\|_2 = \|A\|_{2,2} = \sigma_{\max}(A) = \sqrt{\lambda_{\max}(A^*A)}, \tag{2.234d}$$

$$\|A\|_{2,\infty} = \max_{0\le i\le M-1} \left(\sum_{j=0}^{N-1} |A_{i,j}|^2\right)^{1/2}, \tag{2.234e}$$

$$\|A\|_\infty = \|A\|_{\infty,\infty} = \max_{0\le i\le M-1} \sum_{j=0}^{N-1} |A_{i,j}|. \tag{2.234f}$$

The most common matrix norm that is not an operator norm is the *Frobenius norm*:

$$\|A\|_F = \sqrt{\sum_{i=0}^{M-1}\sum_{j=0}^{N-1} |A_{i,j}|^2} = \sqrt{\operatorname{tr}(AA^*)}. \tag{2.235}$$

2.B.2 Special matrices

Unitary and orthogonal matrices A square matrix U is called *unitary* when it satisfies

$$U^*U = UU^* = I. \tag{2.236}$$

Its inverse U^{-1} equals its Hermitian transpose U^*. A real unitary matrix satisfies

$$U^\top U = UU^\top = I, \tag{2.237}$$

and is called *orthogonal*.[32]

Unitary matrices preserve norms for all complex vectors,

$$\|Ux\| = \|x\|,$$

and more generally preserve inner products,

$$\langle Ux, Uy\rangle = \langle x, y\rangle.$$

Each eigenvalue of a unitary matrix has unit modulus, and all its eigenvectors are orthogonal. Each eigenvalue of an orthogonal matrix is ± 1 or part of a complex conjugate pair $e^{\pm j\theta}$. From (2.191a), its condition number is $\kappa(U) = 1$.

Rotations and rotoinversions From (2.237), the determinant of an orthogonal matrix satisfies $(\det U)^2 = 1$. When $\det U = 1$, the orthogonal matrix is called a *rotation*; when $\det U = -1$, it is called an *improper rotation* or *rotoinversion*. In $\mathbb{R}^2$, a rotation is always of the form

$$\begin{bmatrix} \cos\theta & -\sin\theta \\ \sin\theta & \cos\theta \end{bmatrix}, \tag{2.238a}$$

and a rotoinversion is always of the form

$$\begin{bmatrix} \cos\theta & \sin\theta \\ \sin\theta & -\cos\theta \end{bmatrix}. \tag{2.238b}$$

The rotoinversion can be interpreted as a composition of a rotation and a reflection of one coordinate. In $\mathbb{R}^N$, a rotation can always be written as a product of $N(N-1)/2$ matrices that each performs a planar rotation in one pair of coordinates. For example, any rotation in $\mathbb{R}^3$ can be written as

$$\begin{bmatrix} \cos\theta_{01} & -\sin\theta_{01} & 0 \\ \sin\theta_{01} & \cos\theta_{01} & 0 \\ 0 & 0 & 1 \end{bmatrix} \begin{bmatrix} \cos\theta_{02} & 0 & -\sin\theta_{02} \\ 0 & 1 & 0 \\ \sin\theta_{02} & 0 & \cos\theta_{02} \end{bmatrix} \begin{bmatrix} 1 & 0 & 0 \\ 0 & \cos\theta_{12} & -\sin\theta_{12} \\ 0 & \sin\theta_{12} & \cos\theta_{12} \end{bmatrix}.$$

A general rotoinversion can be written similarly with one planar rotation replaced by a planar rotoinversion.

Hermitian, symmetric, and normal matrices A *Hermitian* matrix is equal to its adjoint,

$$A = A^*. \tag{2.239a}$$

Such a matrix must be square and is also called *self-adjoint*. A real Hermitian matrix is equal to its transpose,

$$A = A^\top, \tag{2.239b}$$

[32] It is sometimes a source of confusion that an orthogonal matrix has *orthonormal* (not merely orthogonal) columns (or rows).

and is called *symmetric.*

The 2 norm of a Hermitian matrix is

$$\|A\|_2 = |\lambda_{\max}|. \tag{2.240}$$

All the eigenvalues of a Hermitian matrix are real. All the eigenvectors corresponding to distinct eigenvalues are orthogonal. When an eigenvalue has multiplicity K, the corresponding eigenvectors form a K-dimensional subspace that is orthogonal to all other eigenvectors; one can find K eigenvectors that form an orthonormal basis for that K-dimensional subspace. A Hermitian matrix can be diagonalized as

$$A = U\Lambda U^*, \tag{2.241a}$$

where U is a unitary matrix with eigenvectors of A as columns and Λ is the diagonal matrix of corresponding eigenvalues; this is the spectral theorem for Hermitian matrices. For the case of A real (symmetric), U is real (orthogonal); thus,

$$A = U\Lambda U^\top. \tag{2.241b}$$

Equation (2.241a) further means that any Hermitian matrix can be factored as $A = QQ^*$, with $Q = U\sqrt{\Lambda}$. Its condition number is given in (2.191b).

A matrix A is called *normal* when it satisfies $A^*A = AA^*$; in words instead of symbols, it commutes with its Hermitian transpose. Hermitian matrices are obviously normal. A matrix is normal if and only if it can be unitarily diagonalized as in (2.241a).

For a normal (Hermitian) matrix A, for each eigenvalue λ_k there is a singular value $\sigma_k = |\lambda_k|$, where the singular values are listed in nonincreasing order, but the eigenvalues need not be.

Positive definite matrices A Hermitian matrix A is called *positive semidefinite* when, for all nonzero vectors x, the following is satisfied:

$$x^*Ax \geq 0. \tag{2.242}$$

This is also written as $A \geq 0$. If, furthermore, (2.242) holds with strict inequality, A is called *positive definite*, which is written as $A > 0$. When a Hermitian matrix A has smallest and largest eigenvalues of $\lambda_{\min}$ and $\lambda_{\max}$, the matrices $\lambda_{\max}I - A$ and $A - \lambda_{\min}I$ are positive semidefinite,

$$\lambda_{\min}I \leq A \leq \lambda_{\max}I. \tag{2.243}$$

All eigenvalues of a positive definite matrix are positive. For any positive definite matrix A, there exists a nonsingular matrix W such that $A = W^*W$, where W is a matrix generalization of the square root of A. One possible way to choose such a square root is to diagonalize A as

$$A = Q\Lambda Q^*, \tag{2.244}$$

and, since all the eigenvalues are positive, choose $W^* = Q\sqrt{\Lambda}$, where the square root is applied to each diagonal element of Λ.

Circulant matrices A (right) *circulant* matrix is a matrix where each row is obtained by a (right) circular shift of the previous row,

$$C = \begin{bmatrix} c_0 & c_{N-1} & \cdots & c_1 \\ c_1 & c_0 & \cdots & c_2 \\ \vdots & \vdots & \ddots & \vdots \\ c_{N-1} & c_{N-2} & \cdots & c_0 \end{bmatrix}. \qquad (2.245)$$

A circulant matrix is diagonalized by the DFT matrix (3.164), as we will see in (3.181b). This means that the columns of the DFT matrix are the eigenvectors of this circulant matrix, and, since the DFT matrix is unitary, these eigenvectors are orthonormal.

Toeplitz matrices A *Toeplitz* T matrix is a matrix whose entry T_{ki} depends only on the value of $k - i$. A Toeplitz matrix is thus constant along diagonals,

$$T = \begin{bmatrix} t_0 & t_1 & t_2 & \cdots & t_{N-1} \\ t_{-1} & t_0 & t_1 & \cdots & t_{N-2} \\ t_{-2} & t_{-1} & t_0 & \cdots & t_{N-3} \\ \vdots & \vdots & \vdots & \ddots & \vdots \\ t_{-N+1} & t_{-N+2} & t_{-N+3} & \cdots & t_0 \end{bmatrix}. \qquad (2.246)$$

A matrix in which blocks follow the form above is called a *block Toeplitz* matrix.

Band matrices A *band* or *banded* matrix is a square matrix with nonzero entries only in a band around the main diagonal. The band need not be symmetric; there might be N_r occupied diagonals on the right side and N_ℓ on the left side. For example, a 5×5 matrix with $N_r = 2$ and $N_\ell = 1$ is of the following form:

$$B = \begin{bmatrix} b_{00} & b_{01} & b_{02} & 0 & 0 \\ b_{10} & b_{11} & b_{12} & b_{13} & 0 \\ 0 & b_{21} & b_{22} & b_{23} & b_{24} \\ 0 & 0 & b_{32} & b_{33} & b_{34} \\ 0 & 0 & 0 & b_{43} & b_{44} \end{bmatrix}. \qquad (2.247)$$

Many sets of special matrices are subsets of the band matrices. For example, diagonal matrices have $N_r = N_\ell = 0$, tridiagonal matrices have $N_r = N_\ell = 1$, upper-triangular matrices have $N_\ell = 0$, and lower-triangular matrices have $N_r = 0$. Square matrices have a well-defined *main antidiagonal* running from the lower-left corner to the upper-right corner. An *antidiagonal matrix* has nonzero entries only in the main antidiagonal. A useful matrix is the *unit antidiagonal matrix*, which has ones on the main antidiagonal.

Vandermonde matrices A *Vandermonde matrix* is a matrix of the form

$$V = \begin{bmatrix} 1 & \alpha_0 & \alpha_0^2 & \cdots & \alpha_0^{N-1} \\ 1 & \alpha_1 & \alpha_1^2 & \cdots & \alpha_1^{N-1} \\ \vdots & \vdots & \vdots & \ddots & \vdots \\ 1 & \alpha_{M-1} & \alpha_{M-1}^2 & \cdots & \alpha_{M-1}^{N-1} \end{bmatrix}. \quad (2.248)$$

When $M = N$, the determinant of the matrix is

$$\det V = \prod_{0 \le i < j \le N-1} (\alpha_i - \alpha_j). \quad (2.249)$$

Many useful concepts in sequence processing use Vandermonde matrices, such as the DFT matrix in (3.164a).

Appendix 2.C Elements of probability

This appendix reviews basic concepts in the theory of probability, with an emphasis on continuous random variables. See [6] for a thorough but elementary introduction or [40] for an introduction with more mathematical sophistication.

2.C.1 Basic definitions

Probabilistic models A *probability law* $\mathrm{P}(\cdot)$ assigns probabilities to *events*, which are subsets of the *outcomes* of an experiment. The set of all outcomes is called the *sample space* and denoted Ω. A probability law satisfies the following axioms:

(i) *Nonnegativity:* $\mathrm{P}(A) \ge 0$ for every event A.
(ii) *Additivity:* If A and B are disjoint events, then $\mathrm{P}(A \cup B) = \mathrm{P}(A) + \mathrm{P}(B)$; this additivity extends to countable unions of disjoint events.
(iii) *Normalization:* $\mathrm{P}(\Omega) = 1$.

The *conditional probability* of event A, given event B with $\mathrm{P}(B) > 0$, is defined as

$$\mathrm{P}(A \,|\, B) = \frac{\mathrm{P}(A \cap B)}{\mathrm{P}(B)}. \quad (2.250)$$

Conditioning on B is a restriction of the sample space to B, with rescaling of probabilities such that $\mathrm{P}(\cdot \,|\, B)$ satisfies the normalization axiom, and thus is a probability law. If events A and B both have positive probability, then writing $\mathrm{P}(A \cap B) = \mathrm{P}(A \,|\, B)\mathrm{P}(B)$ and $\mathrm{P}(A \cap B) = \mathrm{P}(B \,|\, A)\mathrm{P}(A)$ yields *Bayes' rule*:

$$\mathrm{P}(A \,|\, B) = \frac{\mathrm{P}(B \,|\, A)\mathrm{P}(A)}{\mathrm{P}(B)}. \quad (2.251)$$

Events A and B are called *independent* when $\mathrm{P}(A \cap B) = \mathrm{P}(A)\,\mathrm{P}(B)$.

Continuous random variables A real, continuous random variable x has a *probability density function (PDF)* $f_{\rm x}$ defined on the real line such that

$$\mathrm{P}(\mathrm{x} \in A) = \int_A f_{\rm x}(t)\, dt \tag{2.252a}$$

is the probability that x falls in the set $A \subset \mathbb{R}$.[33] The *cumulative distribution function (CDF)* of x is

$$F_{\rm x}(t) = \mathrm{P}(\mathrm{x} \leq t) = \int_{-\infty}^{t} f_{\rm x}(s)\, ds. \tag{2.252b}$$

Since probabilities are nonnegative, we must have $f_{\rm x}(t) \geq 0$ for all $t \in \mathbb{R}$. Since x takes some real value, (2.252b) implies the following normalization of the PDF:

$$\int_{-\infty}^{\infty} f_{\rm x}(t)\, dt = 1. \tag{2.252c}$$

Elementary properties of the CDF include

$$\lim_{t\to-\infty} F_{\rm x}(t) = 0 \quad \text{and} \quad \lim_{t\to\infty} F_{\rm x}(t) = 1,$$

and that

$$\frac{d}{dt} F_{\rm x}(t) = f_{\rm x}(t), \qquad \text{where the derivative exists.}$$

By calling $f_{\rm x}$ a function, we are excluding Dirac delta components from $f_{\rm x}$; the CDF $F_{\rm x}$ is then continuous because it is the integral of the PDF.[34] Allowing Dirac delta components in $f_{\rm x}$ would introduce jumps in the CDF. This is necessary for describing discrete or mixed random variables.

Expectation, moments, and variance The *expectation* of a function $g(\mathrm{x})$ is defined as

$$\mathrm{E}[\,g(\mathrm{x})\,] = \int_{-\infty}^{\infty} g(t)\, f_{\rm x}(t)\, dt. \tag{2.253a}$$

In particular, for any $k \in \mathbb{N}$, $\mathrm{E}[\,\mathrm{x}^k\,]$ is called the *kth moment.* The zeroth moment must be 1, and other moments do not always exist. The first moment is called the *mean*, and the *variance* is obtained from the first and second moments as follows:

$$\mathrm{var}(\mathrm{x}) = \mathrm{E}[\,(\mathrm{x} - \mathrm{E}[\,\mathrm{x}\,])^2\,] = \mathrm{E}[\,\mathrm{x}^2\,] - (\mathrm{E}[\,\mathrm{x}\,])^2. \tag{2.253b}$$

The variance is nonnegative. The expectation is linear in that

$$\mathrm{E}[\,\alpha_0 g_0(\mathrm{x}) + \alpha_1 g_1(\mathrm{x})\,] = \alpha_0 \mathrm{E}[\,g_0(\mathrm{x})\,] + \alpha_1 \mathrm{E}[\,g_1(\mathrm{x})\,] \tag{2.253c}$$

[33] Formally, $\mathrm{x} : \Omega \to \mathbb{R}$, and $\{\omega \in \Omega \mid \mathrm{x}(\omega) \in A\}$ must be an event. There are technical subtleties in the functions $f_{\rm x}$ and sets A that should be allowed. It is adequate to assume that $f_{\rm x}$ has a countable number of discontinuities and that A is a countable union of intervals [40]. We refer the reader to Footnote 10 on page 25 for our philosophy on this type of mathematical technicality.

[34] The Dirac delta function and its properties are discussed in Appendix 3.A.4.

for any constants α_0 and α_1 and any functions g_0 and g_1. From this it follows that

$$\operatorname{var}(\alpha_0 \mathrm{x} + \alpha_1) = \alpha_0^2 \operatorname{var}(\mathrm{x}) \tag{2.253d}$$

for any constants α_0 and α_1.

Random variables x and y are said to have the same distribution when $\mathrm{E}[\, g(\mathrm{x})\,] = \mathrm{E}[\, g(\mathrm{y})\,]$ for any function g. This requires their CDFs to be equal, though their PDFs might differ at a countable number of points.[35]

Jointly distributed random variables Real, continuous random variables x and y have a *joint PDF* $f_{\mathrm{x,y}}$ defined such that

$$\mathrm{P}((\mathrm{x}, \mathrm{y}) \in A) = \iint_A f_{\mathrm{x,y}}(s, t)\, dt\, ds \tag{2.254a}$$

is the probability that (x, y) falls in $A \subset \mathbb{R}^2$ and a *joint CDF*

$$F_{\mathrm{x,y}}(s, t) = \mathrm{P}(\mathrm{x} \le s,\ \mathrm{y} \le t) = \int_{-\infty}^{s} \int_{-\infty}^{t} f_{\mathrm{x,y}}(u, v)\, dv\, du. \tag{2.254b}$$

The expectation of a function $g(\mathrm{x}, \mathrm{y})$ is now a double integral:

$$\mathrm{E}[\, g(\mathrm{x}, \mathrm{y})\,] = \int_{-\infty}^{\infty} \int_{-\infty}^{\infty} g(s, t) f_{\mathrm{x,y}}(s, t)\, dt\, ds. \tag{2.255a}$$

The *covariance* of x and y is defined as $\mathrm{E}[\,(\mathrm{x} - \mathrm{E}[\,\mathrm{x}\,])(\mathrm{y} - \mathrm{E}[\,\mathrm{y}\,])\,]$, and their *correlation coefficient* is defined as

$$\rho = \frac{\mathrm{E}[\,(\mathrm{x} - \mathrm{E}[\,\mathrm{x}\,])(\mathrm{y} - \mathrm{E}[\,\mathrm{y}\,])\,]}{\sqrt{\operatorname{var}(\mathrm{x})}\sqrt{\operatorname{var}(\mathrm{y})}}. \tag{2.255b}$$

The *marginal PDF* of x is

$$f_{\mathrm{x}}(s) = \int_{-\infty}^{\infty} f_{\mathrm{x,y}}(s, t)\, dt, \tag{2.256a}$$

and the *marginal CDF* of x is

$$F_{\mathrm{x}}(s) = \lim_{t \to \infty} F_{\mathrm{x,y}}(s, t). \tag{2.256b}$$

The *conditional PDF* of x given y is defined as

$$f_{\mathrm{x}|\mathrm{y}}(s \,|\, t) = \frac{f_{\mathrm{x,y}}(s, t)}{f_{\mathrm{y}}(t)} \quad \text{for } t \text{ such that } f_{\mathrm{y}}(t) \ne 0. \tag{2.257a}$$

The *conditional expectation* is defined with the conditional PDF:

$$\mathrm{E}[\, g(\mathrm{x}) \,|\, \mathrm{y} = t\,] = \int_{-\infty}^{\infty} g(s)\, f_{\mathrm{x}|\mathrm{y}}(s \,|\, t)\, ds. \tag{2.257b}$$

[35] This is analogous to equality in $\mathcal{L}^p$ for $1 \le p < \infty$: equality of CDFs F_X and F_Y implies that $\|f_X - f_Y\|_{\mathcal{L}^p} = 0$ for any $p \in [1, \infty)$.

When $f_{x,y}$ is separable as

$$f_{x,y}(s,t) = f_x(s) f_y(t) \tag{2.258}$$

for PDFs f_x and f_y, the random variables x and y are called *independent.* An immediate ramification of independence is that $f_{x|y}(s\,|\,t) = f_x(s)$ for every t such that $f_{x|y}(s\,|\,t)$ is defined. These definitions extend to any number of random variables, with some subtleties for infinite collections.

A complex random variable has real and imaginary parts that are jointly distributed real random variables. A random vector has components that are jointly distributed scalar random variables. The mean of an N-dimensional random vector x is a vector $\mu_x = E[\,x\,] \in \mathbb{C}^N$. The *covariance matrix* is defined as

$$\Sigma_x = E[\,(x - \mu_x)(x - \mu_x)^*\,]. \tag{2.259}$$

2.C.2 Standard distributions

Uniform random variables For any real numbers a and b with $a < b$, a random variable with PDF

$$f_x(t) = \begin{cases} 1/(b-a), & \text{for } t \in [a,\, b]; \\ 0, & \text{otherwise,} \end{cases} \tag{2.260a}$$

is called *uniform on* $[a,\, b]$. This is denoted $x \sim \mathcal{U}(a,b)$. Simple computations yield the CDF

$$F_x(t) = \begin{cases} 0, & \text{for } t < a; \\ (t-a)/(b-a), & \text{for } t \in [a,\, b]; \\ 1, & \text{for } t > b, \end{cases} \tag{2.260b}$$

the mean $E[\,x\,] = (a+b)/2$, and the variance $\mathrm{var}(x) = (b-a)^2/12$.

Gaussian random variables and vectors For any real μ and positive σ, a random variable with PDF

$$f_x(t) = \frac{1}{\sqrt{2\pi}\sigma} e^{-\frac{1}{2}(t-\mu)^2/\sigma^2} \tag{2.261}$$

is called *Gaussian* or *normal* with mean μ and variance σ^2. This is denoted $x \sim \mathcal{N}(\mu, \sigma^2)$. When $\mu = 0$ and $\sigma = 1$, the random variable is called *standard.* There is no elementary expression for the CDF of a Gaussian random variable.

For any $\mu \in \mathbb{R}^N$ and symmetric, positive definite $\Sigma \in \mathbb{R}^{N \times N}$, a random vector $x = \begin{bmatrix} x_0 & x_1 & \ldots & x_{N-1} \end{bmatrix}^\top$ with joint PDF

$$f_x(t) = \frac{1}{(2\pi)^{N/2} (\det(\Sigma))^{1/2}} e^{-\frac{1}{2}(t-\mu)^\top \Sigma^{-1}(t-\mu)} \tag{2.262}$$

is called *(jointly) Gaussian* or *multivariate normal* with mean μ and covariance Σ. This is denoted $x \sim \mathcal{N}(\mu, \Sigma)$.

Gaussianity is invariant to affine transformations: when x is jointly Gaussian, $Ax + b$ is also jointly Gaussian for any constant matrix $A \in \mathbb{R}^{M \times N}$ of rank M and constant vector $b \in \mathbb{R}^M$;[36] the new mean is $A\mu + b$ and the new covariance matrix is $A\Sigma A^\top$.

The marginal distributions and conditional distributions are jointly Gaussian also. Partition x, μ, and Σ (in a dimensionally compatible manner) as

$$\mathrm{x} = \begin{bmatrix} \mathrm{y} \\ \mathrm{z} \end{bmatrix}, \qquad \mu = \begin{bmatrix} \mu_{\mathrm{y}} \\ \mu_{\mathrm{z}} \end{bmatrix}, \qquad \text{and} \qquad \Sigma = \begin{bmatrix} \Sigma_{\mathrm{y}} & \Sigma_{\mathrm{y,z}} \\ \Sigma_{\mathrm{z,y}} & \Sigma_{\mathrm{z}} \end{bmatrix}.$$

The symmetry of Σ implies that $\Sigma_{\mathrm{y}} = \Sigma_{\mathrm{y}}^\top$, $\Sigma_{\mathrm{z}} = \Sigma_{\mathrm{z}}^\top$, and $\Sigma_{\mathrm{y,z}} = \Sigma_{\mathrm{z,y}}^\top$. The marginal distribution of y is jointly Gaussian with mean μ_{y} and covariance Σ_{y}, and the conditional distribution of y given $\mathrm{z} = t$ is jointly Gaussian with mean $\mu_{\mathrm{y}} + \Sigma_{\mathrm{y,z}}\Sigma_{\mathrm{z}}^{-1}(t - \mu_{\mathrm{z}})$ and covariance $\Sigma_{\mathrm{y}} - \Sigma_{\mathrm{y,z}}\Sigma_{\mathrm{z}}^{-1}\Sigma_{\mathrm{z,y}}$. These and other properties of jointly Gaussian vectors are developed in Exercise 2.59.

Gaussian random variables are very common in modeling physical phenomena because they arise from the accumulation of a large number of small, independent, random effects. This is made precise by the *central limit theorem*, a simple version of which is as follows:

> Let $\mathrm{x}_1, \mathrm{x}_2, \ldots$ be independent, identically distributed (i.i.d.) random variables with mean μ and variance σ^2. For each $n \in \mathbb{Z}^+$, define a shifted and scaled version of the sample mean of $\{\mathrm{x}_k\}_{k=1}^n$:
>
> $$\mathrm{z}_n = \frac{\sqrt{n}}{\sigma}\left(\left(\frac{1}{n}\sum_{k=1}^{n}\mathrm{x}_k\right) - \mu\right). \tag{2.263a}$$
>
> These random variables converge in distribution to a standard normal random variable z:
>
> $$\lim_{n\to\infty} F_{\mathrm{z}_n}(t) = F_{\mathrm{z}}(t) \qquad \text{for all } t \in \mathbb{R}. \tag{2.263b}$$

Similar results hold under conditions that allow weak dependence of the variables.

2.C.3 Estimation

Estimation is the process of forming estimates of parameters of interest from observations that are probabilistically related to the parameters. Bayesian and non-Bayesian (classical) techniques are distinguished by whether the parameters are considered to be random variables; observations are random in either case. For simplicity, generalities below are stated for a continuous scalar parameter and continuous scalar observations. Some examples demonstrate extensions to vectors.

---

[36] Some authors require the covariance matrix of a jointly Gaussian vector to be merely positive semidefinite rather than positive definite. In this case, the rank condition on A can be removed, and the PDF does not necessarily exist. A jointly Gaussian distribution with singular covariance matrix is called *degenerate*.

Bayesian estimation In *Bayesian estimation*, the parameter of interest is assumed to be a random variable x. Its distribution is called the *prior distribution* to emphasize that it describes x without use of observations. The conditional distribution of the observation y, given the parameter x, follows a distribution $f_{\mathrm{y}|\mathrm{x}}$ called the *likelihood.* After observing $\mathrm{y} = t$, Bayes' Rule, (2.251), specifies the *posterior distribution* of x to be

$$f_{\mathrm{x}|\mathrm{y}}(s \,|\, t) \;=\; \frac{f_{\mathrm{x}}(s) f_{\mathrm{y}|\mathrm{x}}(t \,|\, s)}{f_{\mathrm{y}}(t)} \;=\; \frac{f_{\mathrm{x}}(s) f_{\mathrm{y}|\mathrm{x}}(t \,|\, s)}{\int_{-\infty}^{\infty} f_{\mathrm{x}}(s) f_{\mathrm{y}|\mathrm{x}}(t \,|\, s)\, ds}.$$

Bayesian estimators are derived by using the posterior distribution to optimize a criterion of interest to find the best function $\widehat{\mathrm{x}} = g(\mathrm{y})$.

A common performance criterion is the MSE $\mathrm{E}[\,(\mathrm{x} - \widehat{\mathrm{x}})^2\,]$. In the trivial case of having no observation available, $\widehat{\mathrm{x}}$ is simply some constant c; the MSE is minimized by $c = \mathrm{E}[\,\mathrm{x}\,]$. This is verified through the following computation:

$$\mathrm{E}[\,(\mathrm{x} - c)^2\,] \overset{(a)}{=} \mathrm{var}(\mathrm{x} - c) + (\mathrm{E}[\,\mathrm{x} - c\,])^2 \overset{(b)}{=} \mathrm{var}(\mathrm{x}) + (\mathrm{E}[\,\mathrm{x}\,] - c)^2 \overset{(c)}{\geq} \mathrm{var}(\mathrm{x}),$$

where (a) follows from (2.253b); (b) from (2.253c) and (2.253d); and (c) holds with equality if and only if $c = \mathrm{E}[\,\mathrm{x}\,]$. When $\mathrm{y} = t$ has been observed, x is conditionally distributed as $f_{\mathrm{x}|\{\mathrm{y}=t\}}$. The *MMSE estimator* is thus

$$\widehat{\mathrm{x}}_{\mathrm{MMSE}}(t) \;=\; \arg\min_{g} \mathrm{E}[\,(\mathrm{x} - g(t))^2\,] \;=\; \mathrm{E}[\,\mathrm{x} \,|\, \mathrm{y} = t\,], \tag{2.264}$$

where the minimization is over any function g that depends only on t.

Another common approach is *maximum a posteriori probability* (MAP), meaning that one chooses the maximizer of the posterior distribution of x, so the *MAP estimator* is

$$\widehat{\mathrm{x}}_{\mathrm{MAP}}(t) \;=\; \arg\max_{s} f_{\mathrm{x}|\mathrm{y}}(s \,|\, t). \tag{2.265}$$

When x is a discrete random variable, the MAP estimate maximizes the probability that the estimate is exactly correct (and hence minimizes the probability that the estimate is in error). When x is a continuous random variable, a similar interpretation is problematic since any estimate is incorrect with probability 1; nevertheless, the MAP estimate is often useful.

MMSE and MAP estimates are each generally difficult to compute. The following example establishes an important and surprising special case: when the parameters and observations are jointly Gaussian vectors, the MMSE and MAP estimates coincide and are linear functions of the observations. The same optimal estimator arises for general (non-Gaussian) distributions when the estimator is restricted to being linear; see Exercise 2.60.

EXAMPLE 2.67 (BAYESIAN ESTIMATION: GAUSSIAN CASE) Let x and y be jointly Gaussian vectors, meaning that their concatenation $\begin{bmatrix}\mathrm{x}^\top & \mathrm{y}^\top\end{bmatrix}^\top$ has a PDF of the form (2.262). Assume that $\mathrm{E}[\,\mathrm{x}\,] = \mu_{\mathrm{x}}$ and $\mathrm{E}[\,\mathrm{y}\,] = \mu_{\mathrm{y}}$, and write the covariance matrix as

$$\mathrm{E}\left[\left(\begin{bmatrix}\mathrm{x}\\ \mathrm{y}\end{bmatrix} - \begin{bmatrix}\mu_{\mathrm{x}}\\ \mu_{\mathrm{y}}\end{bmatrix}\right)\left(\begin{bmatrix}\mathrm{x}\\ \mathrm{y}\end{bmatrix} - \begin{bmatrix}\mu_{\mathrm{x}}\\ \mu_{\mathrm{y}}\end{bmatrix}\right)^\top\right] = \begin{bmatrix}\Sigma_{\mathrm{x}} & \Sigma_{\mathrm{x},\mathrm{y}}\\ \Sigma_{\mathrm{x},\mathrm{y}}^\top & \Sigma_{\mathrm{y}}\end{bmatrix},$$

where Σ_x is the covariance of x, Σ_y is the covariance of y, and $\Sigma_{x,y}$ is the crosscovariance between x and y, $\Sigma_{x,y} = E[(x - \mu_x)(y - \mu_y)^T]$.

The conditional PDF of x given $y = t$ is jointly Gaussian with mean $\mu_x + \Sigma_{xy}\Sigma_y^{-1}(t - \mu_y)$ and covariance $\Sigma_x - \Sigma_{x,y}\Sigma_y^{-1}\Sigma_{y,x}^T$ (see Appendix 2.C.2). Since the conditional mean minimizes the MSE, we have

$$\widehat{x}_{MMSE}(t) = \mu_x + \Sigma_{x,y}\Sigma_y^{-1}(t - \mu_y). \tag{2.266a}$$

Its resulting MSE is

$$E[\|x - \widehat{x}_{MMSE}\|^2] = \operatorname{tr}(\Sigma_x - \Sigma_{x,y}\Sigma_y^{-1}\Sigma_{y,x}^T). \tag{2.266b}$$

Since the PDF of a jointly Gaussian random vector is maximum at the mean value (from the minimum of the quadratic form $(t - \mu)^T\Sigma^{-1}(t - \mu)$ in (2.262)), we have

$$\widehat{x}_{MAP}(t) = \mu_x + \Sigma_{x,y}\Sigma_y^{-1}(t - \mu_y),$$

which is exactly the same as the MMSE estimator in (2.266a).

As a special case, suppose that $y = x + z$, where $z \sim \mathcal{N}(0, \sigma_z^2 I)$ and x and z are independent. We say that y is an observation of x with additive white Gaussian noise (AWGN). Then, $\mu_y = \mu_x$, $\Sigma_{x,y} = E[(x - \mu_x)(x + z - \mu_x)^T] = \Sigma_x$, and $\Sigma_y = E[(x + z - \mu_x)(x + z - \mu_x)^T] = \Sigma_x + \sigma_z^2 I$. The optimal estimator from observation $y = y$ and its performance simplify to

$$\widehat{x}_{MMSE}(t) = \mu_x + \Sigma_x(\Sigma_x + \sigma_z^2 I)^{-1}(t - \mu_x), \tag{2.267a}$$

$$E[\|x - \widehat{x}_{MMSE}\|^2] = \operatorname{tr}(\Sigma_x - \Sigma_x(\Sigma_x + \sigma_z^2 I)^{-1}\Sigma_x). \tag{2.267b}$$

Specializing further, suppose that x is scalar with mean zero and variance σ_x^2. Then

$$\widehat{x}_{MMSE}(t) = \frac{\sigma_x^2}{\sigma_x^2 + \sigma_z^2} t \tag{2.268a}$$

and

$$E[\|x - \widehat{x}_{MMSE}\|^2] = \sigma_x^2\left(1 - \frac{\sigma_x^2}{\sigma_x^2 + \sigma_z^2}\right). \tag{2.268b}$$

Classical estimation In *classical estimation*, the parameter of interest is treated as an unknown nonrandom quantity. The observation y is random, with a *likelihood* (distribution) $f_{y;x}$ that depends on the parameter x.[37] An estimator $\widehat{x}(y)$ produces an estimate of x from the observation. Since it is a function of the random variable y, an estimate is also a random variable with a distribution that depends on x. The dependence on x is emphasized with a subscript in the following.

The *error* of an estimator is $\widehat{x}(y) - x$, and the *bias* is the expected error:

$$b_x(\widehat{x}(y)) = E_x[\widehat{x}(y) - x] = E_x[\widehat{x}(y)] - x.$$

[37] Some authors write this as $f_{y|x}$, with the potential to confuse random and nonrandom quantities.

An *unbiased* estimator has $b_x(\widehat{\mathrm{x}}(\mathrm{y})) = 0$ for all x. The *mean-squared error* of an estimator is

$$\mathrm{E}_x\left[\,|\widehat{\mathrm{x}}(\mathrm{y}) - x|^2\,\right].$$

It depends on x, and it can be expanded as the sum of the variance and the square of the bias:

$$\mathrm{E}_x\left[\,|\widehat{\mathrm{x}}(\mathrm{y}) - x|^2\,\right] \;=\; \mathrm{var}_x(\widehat{\mathrm{x}}(\mathrm{y})) + (b_x(\widehat{\mathrm{x}}(\mathrm{y})))^2.$$

Sometimes attention is limited to unbiased estimators, in which case the MSE is minimized by minimizing the variance of the estimator. This results in the *minimum-variance unbiased estimator.*

Example 2.68 (Classical MMSE estimation: Gaussian case) Let $x \in \mathbb{R}^N$ be a parameter of interest, and let $\mathrm{y} = Ax + \mathrm{z}$, where $A \in \mathbb{R}^{M\times N}$ is a known matrix and $\mathrm{z} \sim \mathcal{N}(0, \Sigma)$. Since z has zero mean, the estimator $\widehat{\mathrm{x}}(\mathrm{y}) = B\mathrm{y}$ is unbiased whenever $BA = I$. Assuming that $BA = I$, the MSE of the estimator is

$$\mathrm{E}_x\left[\,\|B(Ax + \mathrm{z}) - x\|^2\,\right] \;=\; \mathrm{E}_x\left[\,\|B\mathrm{z}\|^2\,\right] \;=\; \mathrm{tr}(B\Sigma B^\top).$$

Since this MSE does not depend on x, it can be minimized through the choice of B to yield a valid estimator. The MSE is minimized by $B = (A^\top\Sigma^{-1}A)^{-1}A^\top\Sigma^{-1}$. The resulting MSE is $\mathrm{tr}((A^\top\Sigma^{-1}A)^{-1})$.

Note that $A^\top\Sigma^{-1}A$ must be invertible for the estimator above to exist. If $\mathrm{rank}\, A < N$, it is hopeless to form an estimate of x without prior information; the component of x in the null space of A is unobserved.

As a special case, suppose that $\Sigma = \sigma_{\mathrm{z}}^2 I$. Then, the optimal estimator simplifies to $B = (A^\top A)^{-1}A^\top$, the pseudoinverse of A.

Another common approach is *maximum likelihood* (ML), meaning that one chooses the estimate to maximize the likelihood function,

$$\widehat{\mathrm{x}}_{\mathrm{ML}}(t) \;=\; \arg\max_s f_{\mathrm{y};x}(t\,;\,s). \tag{2.269}$$

Notice the reversal of the roles of x and y relative to the MAP estimator in (2.265). Using Bayes' rule, the ML estimator is equivalent to the MAP estimator that would arise if the prior distribution f_{x} were constant.[38]

Example 2.69 (ML estimation: Gaussian case) As in the previous example, let $x \in \mathbb{R}^N$ be a parameter of interest, and let $\mathrm{y} = Ax + \mathrm{z}$ where $A \in \mathbb{R}^{M\times N}$ is a known matrix and $\mathrm{z} \sim \mathcal{N}(0, \Sigma)$. Assume that A is a tall matrix $(M > N)$ with a left inverse, and let $S = \mathcal{R}(A)$. In a classical setting, no distribution for x is assumed, so, prior to any observation, we have no way of telling which vectors in the N-dimensional subspace S are more or less likely values for Ax. After observing $\mathrm{y} = t$, the likelihood of $x = s$ is the likelihood that $\mathrm{z} = t - As$,

$$f_{\mathrm{y};x}(t\,;\,s) \;=\; \frac{1}{(2\pi)^{M/2}(\det\Sigma)^{1/2}}e^{-\frac{1}{2}(t-As)^\top\Sigma^{-1}(t-As)},$$

[38]Note that a continuous random variable cannot have a constant density on an unbounded set while maintaining valid normalization, (2.252c).

where we have used the joint PDF in (2.262). Maximizing this likelihood over s is equivalent to minimizing $(t-As)^\top\Sigma^{-1}(t-As)$ over s. Since Σ^{-1} is a symmetric and positive definite matrix, $(t-As)^\top\Sigma^{-1}(t-As) = \|t-As\|^2$ for an appropriate Hilbert space norm (see Exercise 2.7). Thus, according to the projection theorem (Theorem 2.26), the ML estimate of Ax is the orthogonal projection of t onto the closed subspace S under this norm. Then, the ML estimate of x is obtained by applying the left inverse of A. When $\Sigma^{-1} = cI_M$, the Hilbert space norm used here is a scalar multiple of the standard norm, and thus the orthogonal projection follows standard Euclidean geometry.

Appendix 2.D Basis concepts

There are several ways to define the basis of a vector space. These coincide for finite-dimensional vector spaces but not necessarily for infinite-dimensional ones, so their distinctions are subtle.

Hamel basis In basic linear algebra, a set of vectors $\Phi = \{\varphi_k\}_{k\in\mathcal{K}} \subset V$ is called a basis for a vector space V when

(i) Φ is linearly independent; and
(ii) $V = \text{span}(\Phi)$.

To distinguish this from other definitions, a set satisfying these conditions is called a *Hamel basis* or an *algebraic basis*. The Hamel basis concept is not suitable for our purposes because infinite-dimensional Hilbert spaces do not have countable Hamel bases and an expansion with respect to a Hamel basis need not be unique when it has infinitely many terms.

To see why Hamel bases might be uncountable, consider $\mathbb{C}^\mathbb{Z}$. Recall that the span of a set of vectors is the set of *finite* linear combinations of those vectors (see Definition 2.4). The set $E = \{e_k\}_{k\in\mathbb{Z}}$ introduced in Example 2.30 is not a Hamel basis for $\mathbb{C}^\mathbb{Z}$ (under any vector space norm) because $\mathbb{C}^\mathbb{Z}$ contains sequences with infinitely many nonzero entries, which clearly cannot be formed by finite linear combinations of E. In fact, a Hamel basis for any infinite-dimensional Banach space must be uncountable.

An expansion with respect to a Hamel basis with infinitely many terms need not be unique because linear independence is too easily satisfied in infinite-dimensional spaces. Recall that linear independence of an infinite set of vectors is the linear independence of every *finite* subset of those vectors (see Definition 2.5). Consider $\Phi = E \cup \{x\}$, where E is the set introduced in Example 2.30 and $x \in \mathbb{C}^\mathbb{Z}$ has infinitely many nonzero entries. Then, Φ is a linearly independent set because x cannot be expressed as a finite linear combination of E. However, expansions with respect to Φ with infinitely many terms are not unique because

$$\gamma x + \sum_{k\in\mathcal{K}} \alpha_k e_k = 0x + \sum_{k\in\mathbb{Z}} (\gamma x_k + \alpha_k) e_k,$$

for any $\gamma \in \mathbb{C}$ and $\alpha \in \mathbb{C}^\mathbb{Z}$.

Schauder basis Assuming the normed vector space V to be complete, the definition of basis adopted in this book (Definition 2.34) is equivalent to what is called a *Schauder basis.* The existence of the expansion (2.87) for any $x \in V$ requires the span of Φ to be *dense* in V, so $\overline{\text{span}}(\Phi) = V$, as in (2.88). The uniqueness of these expansions is similar to requiring Φ to be linearly independent – but more restrictive.

We now compare concepts of independence.

(i) $\{\varphi_k\}_{k\in\mathcal{K}}$ is called *linearly independent* when every finite subset is linearly independent (see Definition 2.5).

(ii) $\{\varphi_k\}_{k\in\mathcal{K}}$ is called *ω-independent* when $\sum_{k\in\mathcal{K}} \alpha_k \varphi_k = \mathbf{0}$ implies that $\alpha_k = 0$ for every $k \in \mathcal{K}$. (The sum must be taken in some fixed order, so when $\mathcal{K}$ is infinite, some singly infinite ordering of $\mathcal{K}$ is assumed.)

(iii) $\{\varphi_k\}_{k\in\mathcal{K}}$ is called *minimal* when $\varphi_k \notin \overline{\text{span}}(\{\varphi_\ell\}_{\ell\neq k})$ for every $k \in \mathcal{K}$.

The set $\{\varphi_k\}_{k\in\mathcal{K}}$ being a (Schauder) basis implies that it is minimal; being minimal implies that it is ω-independent; and being ω-independent implies that it is linearly independent. None of the reverse implications hold in general.

A basis is called *unconditional* when, for every $x \in V$, the unique expansion (2.87) of x converges unconditionally. Equivalently, a basis is unconditional when every one of its permutations is also a basis. Since all Riesz bases are unconditional bases and we focus our attention on Riesz bases, we can usually drop the requirement of assuming a fixed singly infinite ordering of $\mathcal{K}$.

Orthonormal basis When attention is limited to Hilbert spaces, rather than using Definition 2.38, which indirectly uses the Schauder basis definition, an *orthonormal basis* $\{\varphi_k\}_{k\in\mathcal{K}}$ for H can be defined directly: this is an orthonormal set (see Definition 2.8) with span that is dense in H; that is, $\overline{\text{span}}(\{\varphi_k\}_{k\in\mathcal{K}}) = H$. An orthonormal basis defined in this way is a Riesz basis and hence an unconditional Schauder basis. When the H is infinite-dimensional, an orthonormal basis for H is not a Hamel basis.

Chapter at a glance

In this chapter, we found representations given by linear operators such that

$$x = \Phi\widetilde{\Phi}^* x.$$

After finding Φ and $\widetilde{\Phi}$ such that $\Phi\widetilde{\Phi}^* = I$, we call

$$\alpha = \widetilde{\Phi}^* x, \qquad x = \Phi\alpha = \Phi\widetilde{\Phi}^* x,$$

a *decomposition* and a *reconstruction*, respectively. Also, $\Phi\alpha$ is often called a *representation* of a signal. The elements of α are called *expansion coefficients* or *transform coefficients* and include Fourier, wavelet, and Gabor coefficients, as well as many others. We decompose signals to look into their properties in the *transform domain*. After analysis or manipulations such as compression, transmission, etc., we reconstruct the signal from its expansion coefficients. We studied cases distinguished by the properties of Φ; we review them in finite dimensions:

(i) If Φ is square and nonsingular, then the columns of Φ form a basis, and the columns of $\widetilde{\Phi}$ form its dual basis.

(ii) If Φ is unitary, then the columns of Φ form an orthonormal basis, and $\widetilde{\Phi} = \Phi$.

(iii) If Φ is rectangular and short, of full rank, then the columns of Φ form a frame, and the columns of $\widetilde{\Phi}$ form a dual frame.

(iv) If Φ is rectangular and short, with $\Phi\Phi^* = I$, then the columns of Φ form a 1-tight frame, and $\widetilde{\Phi} = \Phi$ is a valid choice for $\widetilde{\Phi}$.

Which of these options we will choose depends on the application and the criteria for designing such matrices (representations).

Property	Orthogonal basis	Biorthogonal bases	λ-tight frame	General frame
Expansion set	$\Phi = \{\varphi_k\}_{k=0}^{N-1}$	$\Phi = \{\varphi_k\}_{k=0}^{N-1}$ $\widetilde{\Phi} = \{\widetilde{\varphi}_k\}_{k=0}^{N-1}$	$\Phi = \{\varphi_k\}_{k=0}^{M-1}$	$\Phi = \{\varphi_k\}_{k=0}^{M-1}$ $\widetilde{\Phi} = \{\widetilde{\varphi}_k\}_{k=0}^{M-1}$
	$\varphi_k \in \mathbb{C}^N$	$\varphi_k, \widetilde{\varphi}_k \in \mathbb{C}^N$	$\varphi_k \in \mathbb{C}^N, M \geq N$	$\varphi_k, \widetilde{\varphi}_k \in \mathbb{C}^N, M \geq N$
Structure	$\langle \varphi_i, \varphi_k \rangle = \delta_{i-k}$	$\langle \varphi_i, \widetilde{\varphi}_k \rangle = \delta_{i-k}$	None	None
Expansion	$\sum_{k=0}^{N-1} \langle x, \varphi_k \rangle \varphi_k$	$\sum_{k=0}^{N-1} \langle x, \widetilde{\varphi}_k \rangle \varphi_k$	$\frac{1}{\lambda} \sum_{k=0}^{M-1} \langle x, \varphi_k \rangle \varphi_k$	$\sum_{k=0}^{M-1} \langle x, \widetilde{\varphi}_k \rangle \varphi_k$
Matrix view	Φ of size $N \times N$	Φ of size $N \times N$	Φ of size $N \times M$	Φ of size $N \times M$
	Φ unitary $\Phi\Phi^* = \Phi^*\Phi = I$	Φ full-rank N $\Phi\widetilde{\Phi}^* = I$ $\widetilde{\Phi} = (\Phi^*)^{-1}$	rows of Φ orthog. $\Phi\Phi^* = \lambda I$	Φ full-rank N $\Phi\widetilde{\Phi}^* = I$
Norm preservation	Yes, $\|x\|^2 =$ $\sum_{k=0}^{N-1} \lvert\langle x, \varphi_k \rangle\rvert^2$	No	Yes, $\|x\|^2 =$ $\frac{1}{\lambda} \sum_{k=0}^{M-1} \lvert\langle x, \varphi_k \rangle\rvert^2$	No
Redundant	No	No	Yes	Yes

Table 2.3 Signal representations in finite-dimensional spaces.

Historical remarks

The choice of the title for this chapter requires at least some acknowledgment of the two mathematical giants figuring in it: Euclid and Hilbert.

Little is known about **Euclid (*c.* 300 B.C.)** apart from his writings. He was a Greek mathematician who lived and worked in Alexandria, Egypt. His book *Elements* [42] (in fact, 13 books) "not only was the earliest major Greek mathematical work to come down to us, but also the most influential textbook of all times" [9]. In it, he introduces and discusses many topics, most of which have taken hold in our consciousness as immutable truths, such as the principles of Euclidean geometry. He also has numerous results in number theory, including a simple proof that there are infinitely many prime numbers and a procedure for finding the greatest common divisor of two numbers. Moreover, it was Euclid who introduced the axiomatic method upon which all mathematical knowledge today is based.

David Hilbert (1862–1943) was a German mathematician, known for an axiomatization of geometry supplanting Euclid's five original axioms. His contributions were extraordinarily broad, spanning functional analysis, number theory, mathematical physics, and many other branches of mathematics. At the turn of the twentieth century, he produced a list of 23 unsolved problems, which is generally thought to be the most thoughtful and comprehensive such list ever. He worked closely with another famous mathematician, Minkowski, and had as students or assistants such illustrious names as Weyl, von Neumann, and Courant, among many others. He taught all his life, first at the University of Königsberg and then at the University of Göttingen, where he died in 1943. On his tombstone, one of his famous sayings is inscribed: *Wir müssen wissen. Wir werden wissen.*[39]

Further reading

Linear algebra There are many good textbooks on linear algebra, for example, those by Strang [93,94]. Good reviews are also provided by Kailath [51] and by Vaidyanathan [105]. Parameterizations of unitary matrices in various forms, such as using Givens rotations or Householder building blocks, are given in [105].

Functional analysis and abstract vector spaces Books by Kreyszig [59], Luenberger [64], Gohberg and Goldberg [35], and Young [111] provide details on abstract vector spaces. In particular, parts of our proof of the projection theorem (Theorem 2.26) follow [64] closely. Many technical details on bases, in particular that Riesz bases are unconditional bases and justifications for statements in Appendix 2.D, are provided by Heil [43], while more on frames can be found in [14, 19, 55, 56].

Probability Textbooks by Bertsekas and Tsitsiklis [6] and by Papoulis [74] are recommended for elementary introductions to probability. A more mathematically rigorous introduction is provided by Grimmett and Stirzaker [40].

[39] We must know. We will know.

Exercises with solutions

2.1. *Vector space* $\mathbb{C}^N$

(i) Prove that $\mathbb{C}^N$ is a vector space.

(ii) Prove that the finite cross product of vector spaces, $V = V_0 \times V_1 \times \cdots \times V_{N-1}$, where each of $V_0, V_1, \ldots, V_{N-1}$ is a vector space, is a vector space as well. The finite cross product is defined as the set of sequences $x = \begin{bmatrix} x_0 & x_1 & \ldots & x_{N-1} \end{bmatrix}^\top$, where $x_0 \in V_0, x_1 \in V_1, \ldots, x_{N-1} \in V_{N-1}$.

Solution:

(i) To prove that $\mathbb{C}^N$ is a vector space, we need to check that the conditions stated in Definition 2.1 are satisfied. These properties follow from the way that operations are defined on $\mathbb{C}^N$ and the vector space properties of $\mathbb{C}$. We prove the following for any x, y, and z in $\mathbb{C}^N$.

1. *Commutativity:*

$$\begin{aligned}
x + y &= \begin{bmatrix} x_0 & x_1 & \ldots & x_{N-1} \end{bmatrix}^\top + \begin{bmatrix} y_0 & y_1 & \ldots & y_{N-1} \end{bmatrix}^\top \\
&= \begin{bmatrix} x_0 + y_0 & x_1 + y_1 & \ldots & x_{N-1} + y_{N-1} \end{bmatrix}^\top \\
&\stackrel{(a)}{=} \begin{bmatrix} y_0 + x_0 & y_1 + x_1 & \ldots & y_{N-1} + x_{N-1} \end{bmatrix}^\top \\
&= \begin{bmatrix} y_0 & y_1 & \ldots & y_{N-1} \end{bmatrix}^\top + \begin{bmatrix} x_0 & x_1 & \ldots & x_{N-1} \end{bmatrix}^\top = y + x,
\end{aligned}$$

where (a) follows from the commutative property of addition on $\mathbb{C}$.

2. *Associativity:*

$$\begin{aligned}
&(x + y) + z \\
&= \begin{bmatrix} x_0 + y_0 & x_1 + y_1 & \ldots & x_{N-1} + y_{N-1} \end{bmatrix}^\top + \begin{bmatrix} z_0 & z_1 & \ldots & z_{N-1} \end{bmatrix}^\top \\
&= \begin{bmatrix} (x_0 + y_0) + z_0 & (x_1 + y_1) + z_1 & \ldots & (x_{N-1} + y_{N-1}) + z_{N-1} \end{bmatrix}^\top \\
&\stackrel{(a)}{=} \begin{bmatrix} x_0 + (y_0 + z_0) & x_1 + (y_1 + z_1) & \ldots & x_{N-1} + (y_{N-1} + z_{N-1}) \end{bmatrix}^\top \\
&= \begin{bmatrix} x_0 & x_1 & \ldots & x_{N-1} \end{bmatrix}^\top + \begin{bmatrix} y_0 + z_0 & y_1 + z_1 & \ldots & y_{N-1} + z_{N-1} \end{bmatrix}^\top \\
&= x + (y + z),
\end{aligned}$$

where (a) follows from the associative property of addition on $\mathbb{C}$, and

$$\begin{aligned}
(\alpha\beta)x &= \begin{bmatrix} (\alpha\beta)x_0 & (\alpha\beta)x_1 & \ldots & (\alpha\beta)x_{N-1} \end{bmatrix}^\top \\
&\stackrel{(a)}{=} \begin{bmatrix} \alpha(\beta x_0) & \alpha(\beta x_1) & \ldots & \alpha(\beta x_{N-1}) \end{bmatrix}^\top = \alpha(\beta x),
\end{aligned}$$

where (a) follows from the associate property of multiplication on $\mathbb{C}$.

3. *Distributivity:* Follows similarly to the two properties above.

4. *Additive identity:* The element $\mathbf{0} = \begin{bmatrix} 0 & 0 & \ldots & 0 \end{bmatrix}^\top \in \mathbb{C}^N$ is the additive identity since

$$\begin{aligned}
x + \mathbf{0} &= \begin{bmatrix} x_0 + 0 & x_1 + 0 & \ldots & x_{N-1} + 0 \end{bmatrix}^\top \\
&= \begin{bmatrix} x_0 & x_1 & \ldots & x_{N-1} \end{bmatrix}^\top = x,
\end{aligned}$$

and similarly $\mathbf{0} + x = x$.

5. *Additive inverse:* For any $x \in \mathbb{C}^N$, the element

$$(-x) = \begin{bmatrix} -x_0 & -x_1 & \ldots & -x_{N-1} \end{bmatrix}^\top \in \mathbb{C}^N$$

is the unique additive inverse since

$$\begin{aligned}
x + (-x) &= \begin{bmatrix} x_0 + (-x_0) & x_1 + (-x_1) & \ldots & x_{N-1} + (-x_{N-1}) \end{bmatrix}^\top \\
&= \begin{bmatrix} 0 & 0 & \ldots & 0 \end{bmatrix}^\top = \mathbf{0},
\end{aligned}$$

and similarly $(-x) + x = \mathbf{0}$; uniqueness follows from the uniqueness of additive inverses in $\mathbb{C}$.

6. *Multiplicative identity:* Follows similarly to additive identity.

(ii) All the arguments in the previous part rely on the fact that $\mathbb{C}$ itself is a vector space, and thus addition and scalar multiplication satisfy all the necessary properties. This part is similar, relying on $V_0, V_1, \ldots, V_{N-1}$ being vector spaces. Denote the cross product of the vector spaces by $V = V_0 \times V_1 \times \cdots \times V_{N-1}$. Then, V is a vector space with the following operations:

$$x + y \;=\; \begin{bmatrix} x_0 \oplus_{V_0} y_0 & x_1 \oplus_{V_1} y_1 & \cdots & x_{N-1} \oplus_{V_{N-1}} y_{N-1} \end{bmatrix}^\top,$$

where $\oplus_{V_i}$ is the addition in V_i for $i = 0, 1, \ldots, N-1$, and

$$\alpha x \;=\; \begin{bmatrix} \alpha \odot_{V_0} x_0 & \alpha \odot_{V_1} x_1 & \cdots & \alpha \odot_{V_{N-1}} x_{N-1} \end{bmatrix}^\top,$$

where $\odot_{V_i}$ is the scalar multiplication in V_i for $i = 0, 1, \ldots, N-1$.

2.2. *Vector space $\ell^p(\mathbb{Z})$*
Show that, for any $p \in [1, \infty)$, the vectors in $\mathbb{C}^{\mathbb{Z}}$ with finite $\ell^p(\mathbb{Z})$ norm form a vector space.
(*Hint:* Use Minkowski's inequality (2.211a).)

Solution: Since $\ell^p(\mathbb{Z})$ is a subset of $\mathbb{C}^{\mathbb{Z}}$, we are using the vector addition and scalar multiplication operations of $\mathbb{C}^{\mathbb{Z}}$. Thus, we need not check commutativity, associativity, distributivity, and multiplicative identity properties. The additive identity $\mathbf{0} \in \mathbb{C}^{\mathbb{Z}}$ has ℓ^p norm 0 and thus is in the subset under consideration. For any $x \in \mathbb{C}^{\mathbb{Z}}$ with finite ℓ^p norm, $-x$ also has finite ℓ^p norm, so the subset under consideration has the additive inverse property.

What remains is to show that, if $x, y \in \ell^p(\mathbb{Z})$ and $\alpha \in \mathbb{C}$, then,

(i) $x + y \in \ell^p(\mathbb{Z})$, which follows immediately from Minkowski's inequality; and
(ii) $\alpha x \in \ell^p(\mathbb{Z})$, which is shown as follows:

$$\|\alpha x\|_p^p \;=\; \sum_{n \in \mathbb{Z}} |\alpha x_n|^p \;=\; |\alpha|^p \sum_{n \in \mathbb{Z}} |x_n|^p \;=\; |\alpha|^p \, \|x\|_p^p \;<\; \infty.$$

2.3. *Incompleteness of $C([0, 1])$*
Consider the sequence of functions $(x_k)_{k \geq 2}$ on $[0, 1]$ defined by

$$x_k(t) \;=\; \begin{cases} 0, & t \in [0, \tfrac{1}{2} - 1/k); \\ k(t - \tfrac{1}{2}) + 1, & t \in [\tfrac{1}{2} - 1/k, \tfrac{1}{2}); \\ 1, & t \in [\tfrac{1}{2}, 1]. \end{cases} \tag{E2.3-1}$$

(i) Sketch this sequence of functions.
(ii) Show that (x_k) is a Cauchy sequence under the $\mathcal{L}^2$ norm.
(iii) Show that $x_k \to x$ under the $\mathcal{L}^2$ norm for a discontinuous function x, which would show that $C([0, 1])$ is not complete since $x \notin C([0, 1])$.

Solution:

(i) Figure E2.3-1 plots three representative functions x_3, x_{10}, and x_{30}.
(ii) $(x_n)_{n \geq 2}$ is a Cauchy sequence under the $\mathcal{L}^2$ norm when, given any $\varepsilon > 0$, there exists a number K_ε such that

$$\|x_k - x_m\| \;<\; \varepsilon \qquad \text{for all } k, m > K_\varepsilon.$$

Assume that $k \geq m$ and consider the function $x_k - x_m$,

$$x_k(t) - x_m(t) \;=\; \begin{cases} 0, & t \in [0, \tfrac{1}{2} - 1/m) \cup [\tfrac{1}{2}, 1]; \\ -m(t - \tfrac{1}{2}) - 1, & t \in [\tfrac{1}{2} - 1/m, \tfrac{1}{2} - 1/k); \\ (k - m)(t - \tfrac{1}{2}), & t \in [\tfrac{1}{2} - 1/k, \tfrac{1}{2}]. \end{cases}$$

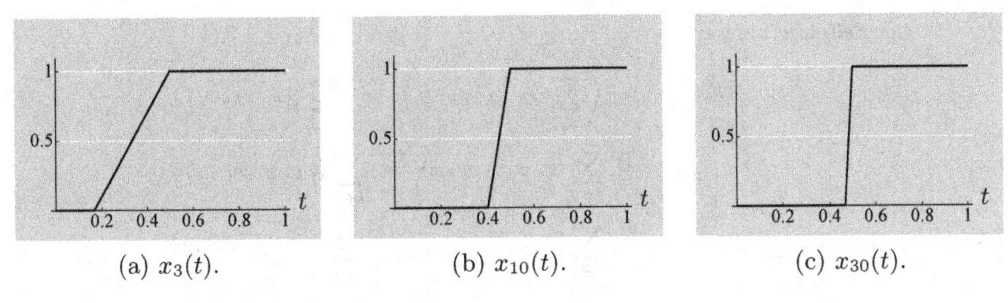

(a) $x_3(t)$. (b) $x_{10}(t)$. (c) $x_{30}(t)$.

Figure E2.3-1 Example functions $x_k(t)$ from (E2.3-1).

Since our sequence starts at $m = 2$, for any $\varepsilon > 0$ we define $K_\varepsilon = \max(2, \lceil 4/(3\varepsilon^2) \rceil)$. Then, for any $k \geq m \geq K_\varepsilon$ the following holds:

$$\begin{aligned}
\|x_k - x_m\|_2^2 &= \int_{1/2-1/m}^{1/2} (x_k - x_m)^2 \, dt \\
&= \int_{1/2-1/m}^{1/2-1/k} \left(m\left(t - \frac{1}{2}\right) + 1 \right)^2 dt + \int_{1/2-1/k}^{1/2} (k-m)^2 \left(t - \frac{1}{2}\right)^2 dt \\
&= \frac{(1-m/k)^3}{3m} + \frac{(k-m)^2}{3k^3} = \frac{(k-m)^2}{3k^2 m} \leq \frac{(k+m)^2}{3k^2 m} \\
&\leq \frac{(k+k)^2}{3k^2 m} = \frac{4}{3m} \leq \frac{4}{3K_\varepsilon} \leq \varepsilon^2,
\end{aligned}$$

and thus (x_m) is a Cauchy sequence under the $\mathcal{L}^2$ norm.

(iii)

$$\lim_{k\to\infty} x_k(t) = x(t) = \begin{cases} 0, & t \in [0, \frac{1}{2}); \\ 1, & t \in [\frac{1}{2}, 1]. \end{cases}$$

Since x has a discontinuity at $\frac{1}{2}$, $x \notin C([0, 1])$.

2.4. *Orthogonal projection to span of an orthonormal set*
Let $\{\varphi_k\}_{k\in\mathcal{I}}$ be an orthonormal set. Prove that

$$Px = \sum_{k\in\mathcal{I}} \langle x, \varphi_k \rangle \varphi_k$$

is an orthogonal projection operator onto $\overline{\text{span}}(\{\varphi_k\}_{k\in\mathcal{I}})$.
Solution: The linearity of P follows from the distributivity and the linearity in the first argument of the inner product. Verifying idempotency and self-adjointness establishes that P is an orthogonal projection operator, by application of Theorem 2.26.

(i) *Idempotency:*

$$\begin{aligned}
P^2 x &= \sum_{m\in\mathcal{I}} \left\langle \sum_{k\in\mathcal{I}} \langle x, \varphi_k\rangle \varphi_k, \varphi_m \right\rangle \varphi_m \overset{(a)}{=} \sum_{m\in\mathcal{I}} \sum_{k\in\mathcal{I}} \langle \langle x, \varphi_k\rangle \varphi_k, \varphi_m\rangle \varphi_m \\
&\overset{(b)}{=} \sum_{m\in\mathcal{I}} \sum_{k\in\mathcal{I}} \langle x, \varphi_k\rangle \langle \varphi_k, \varphi_m\rangle \varphi_m \overset{(c)}{=} \sum_{m\in\mathcal{I}} \sum_{k\in\mathcal{I}} \delta_{m-k} \langle x, \varphi_k\rangle \varphi_m \\
&= \sum_{k\in\mathcal{I}} \langle x, \varphi_k\rangle \varphi_k = Px,
\end{aligned}$$

where (a) follows from the distributivity of the inner product; (b) from the linearity in the first argument of the inner product; and (c) from the orthonormality of the set $\{\varphi_k\}_{k\in\mathcal{I}}$.

(ii) *Self-adjointness:*

$$\begin{aligned}\langle Px, y\rangle &= \left\langle \sum_{k\in\mathcal{I}} \langle x, \varphi_k\rangle\varphi_k, y\right\rangle \stackrel{(a)}{=} \sum_{k\in\mathcal{I}} \langle\langle x, \varphi_k\rangle\varphi_k, y\rangle \\ &\stackrel{(b)}{=} \sum_{k\in\mathcal{I}} \langle x, \varphi_k\rangle\langle\varphi_k, y\rangle \stackrel{(c)}{=} \sum_{k\in\mathcal{I}} \langle\varphi_k, y\rangle\langle x, \varphi_k\rangle \\ &\stackrel{(d)}{=} \sum_{k\in\mathcal{I}} \langle y, \varphi_k\rangle^*\langle x, \varphi_k\rangle \stackrel{(e)}{=} \sum_{k\in\mathcal{I}} \langle x, \langle y, \varphi_k\rangle\varphi_k\rangle \\ &\stackrel{(f)}{=} \left\langle x, \sum_{k\in\mathcal{I}} \langle y, \varphi_k\rangle\varphi_k\right\rangle = \langle x, Py\rangle,\end{aligned}$$

where (a) follows from the distributivity of the inner product; (b) from the linearity in the first argument of the inner product; (c) from the commutativity of the multiplication of scalars; (d) from the Hermitian symmetry of the inner product; (e) from the conjugate linearity in the second argument of the inner product; and (f) from the distributivity of the inner product.

2.5. *Legendre polynomials*
Consider the vectors $1, t, t^2, t^3, \ldots$ in the vector space $\mathcal{L}^2([-1, 1])$. Using Gram–Schmidt orthogonalization, find an orthonormal set with the same span.
Solution: Let $x_k(t) = t^k$ for $k \in \mathbb{N}$. We initiate the Gram–Schmidt procedure with

$$\varphi_0(t) = \frac{x_0(t)}{\|x_0\|} = \frac{1}{(\int_{-1}^{1} 1\, d\tau)^{1/2}} = \frac{1}{\sqrt{2}}.$$

Continuing the Gram–Schmidt orthogonalization,

$$\begin{aligned}v_1(t) &= \langle x_1, \varphi_0\rangle\,\varphi_0(t) = \left(\int_{-1}^{1} \frac{1}{\sqrt{2}}\tau\, d\tau\right)\frac{1}{\sqrt{2}} = 0, \\ \varphi_1(t) &= \frac{x_1(t) - v_1(t)}{\|x_1 - v_1\|} = \frac{t}{(\int_{-1}^{1} \tau^2\, d\tau)^{1/2}} = \sqrt{\frac{3}{2}}t, \\ v_2(t) &= \langle x_2, \varphi_0\rangle\,\varphi_0(t) + \langle x_2, \varphi_1\rangle\,\varphi_1(t) \\ &= \left(\int_{-1}^{1} \frac{1}{\sqrt{2}}\tau^2\, d\tau\right)\frac{1}{\sqrt{2}} + \left(\int_{-1}^{1} \sqrt{\frac{3}{2}}\tau^3\, d\tau\right)\sqrt{\frac{3}{2}}t = \frac{1}{3}, \\ \varphi_2(t) &= \frac{x_2(t) - v_2(t)}{\|x_2 - v_2\|} = \frac{t^2 - \frac{1}{3}}{(\int_{-1}^{1} (\tau^2 - \frac{1}{3})^2\, d\tau)^{1/2}} = \frac{3\sqrt{5}}{2\sqrt{2}}\left(t^2 - \frac{1}{3}\right).\end{aligned}$$

This process can be continued to find an orthonormal set of arbitrary size. It can be proven by induction that

$$\varphi_k(t) = \sqrt{\frac{2k+1}{2}}\,\frac{(-1)^k}{2^k k!}\,\frac{d^k}{dt^k}\left[(1-t^2)^k\right];$$

see Section 6.2.1 and Exercise 6.1.

2.6. *Complexity of matrix multiplication and Strassen's algorithm*
Example 2.59 presented an algorithm for multiplying two 2×2 matrices using seven rather than eight multiplications. To derive an algorithm for multiplication of $N \times N$ matrices with cost that grows slower than the $\Theta(N^3)$ cost of regular matrix multiplication, perform the following sequence of steps:

1. Take $N \times N$ matrices with $N = 2^k$.
2. Consider block-matrix multiplication:

$$Q = AB = \begin{bmatrix} A_{0,0} & A_{0,1} \\ A_{1,0} & A_{1,1}\end{bmatrix}\begin{bmatrix} B_{0,0} & B_{0,1} \\ B_{1,0} & B_{1,1}\end{bmatrix} = \begin{bmatrix} Q_{0,0} & Q_{0,1} \\ Q_{1,0} & Q_{1,1}\end{bmatrix},$$

where the blocks $A_{i,j}$, $B_{i,j}$, and $Q_{i,j}$ are all of size $N/2 \times N/2$. Apply Strassen's algorithm, which will use only seven products of $N/2 \times N/2$ submatrices.

3. Iterate the algorithm on the seven subproducts until they are all of size 1.

(i) Show that this algorithm requires $\mu = N^{\log_2 7}$ multiplications.

(ii) Compute the number of additions and compare it with the number of additions of regular matrix multiplication.

(iii) Write out the algorithm in pseudocode. Numerically compare running times for regular versus fast matrix multiplication for matrices of size $2^k \times 2^k$. Comment on the number of operations and running time.

Solution:

(i) The recursion for the number of multiplications for a matrix of size N is

$$\mu(N) = 7\mu\left(\frac{N}{2}\right). \tag{E2.6-1}$$

For $N = 2^k$, by repeated application of (E2.6-1),

$$\begin{aligned}\mu(2^k) &= 7\mu(2^{k-1}) = 7^2\mu(2^{k-2}) = \cdots = 7^k\mu(1)\\ &\overset{(a)}{=} 7^{\log_2 N} \overset{(b)}{=} 7^{(\log_7 N)(\log_2 7)} = N^{\log_2 7} \approx N^{2.8073},\end{aligned}$$

where (a) follows from $k = \log_2 N$ and $\mu(1) = 1$; and (b) from the change of base in the logarithm, $\log_b x = (\log_k x)/(\log_k b)$. For large values of N, this is a substantial improvement over N^3.

(ii) The recursion for the number of additions for a matrix of size N is

$$\nu(N) = 7\nu\left(\frac{N}{2}\right) + 18\left(\frac{N}{2}\right)^2. \tag{E2.6-2}$$

For $N = 2^k$, by repeated application of (E2.6-2),

$$\begin{aligned}\nu(2^k) &= 7\nu(2^{k-1}) + 18\cdot 2^{2(k-1)}\\ &= 7\left(7\nu(2^{k-2}) + 18\cdot 2^{2(k-2)}\right) + 18\cdot 2^{2(k-1)}\\ &\;\;\vdots\\ &= 7^k\nu(1) + 7^{k-1}18\cdot 2^0 + 7^{k-2}18\cdot 2^{2\cdot 2} + \cdots + 7^0 18\cdot 2^{2(k-1)}\\ &\overset{(a)}{=} 18\sum_{\ell=0}^{k-1} 7^{k-1-\ell}2^{2\ell} \overset{(b)}{=} 18\cdot 7^{k-1}\sum_{\ell=0}^{k-1} 7^{-\ell}2^{2\ell}\\ &= 18\cdot 7^{k-1}\sum_{\ell=0}^{k-1}\left(\frac{4}{7}\right)^\ell \overset{(c)}{=} 18\cdot 7^{k-1}\frac{1-(\frac{4}{7})^k}{1-\frac{4}{7}}\\ &= 6\cdot 7^k\left(1-\left(\frac{4}{7}\right)^k\right) \overset{(d)}{=} 6\cdot N^{\log_2 7}\left(1-\left(\frac{4}{7}\right)^{\log_2 N}\right),\end{aligned}$$

where (a) follows from $\nu(1) = 0$; (b) from pulling 7^{k-1} in front of the sum; (c) from the formula for the finite geometric series, (P2.54-1); and (d) from an identical sequence of manipulations as for the number of multiplications in (i).

Using Strassen's algorithm in place of ordinary matrix multiplication for $N = 2$, we are trading 1 multiplication for 14 additions. As derived above, as N increases the numbers of multiplications and additions both grow slower than N^3. Thus, for large enough N both the number of additions and the total number of scalar operations will be less than for ordinary matrix multiplication. Considering only powers of 2, Strassen's algorithm starts having fewer operations in total at $N = 1024$.

(iii) See Table E2.6-1 for pseudocode.

We implemented Strassen's algorithm and a regular multiplication procedure in Matlab. Compared with the regular implementation, the divide-and-conquer

Strassen's multiplication algorithm
Input: Square matrices A and B with power-of-two dimensions and a minimum block size b_{min}
Output: A matrix product $Q = AB$

$Q = \textbf{Strassen}(A, B, b_{\text{min}})$
if $\dim(A) > b_{\text{min}}$

partition $A = \begin{bmatrix} A_{11} & A_{12} \\ A_{21} & A_{22} \end{bmatrix}$ and $B = \begin{bmatrix} B_{11} & B_{12} \\ B_{21} & B_{22} \end{bmatrix}$ in four quadrants of equal dimensions.

$M_1 \leftarrow \text{Strassen}(A_{11} + A_{22}, B_{11} + B_{22}, b_{\text{min}})$
$M_2 \leftarrow \text{Strassen}(A_{21} + A_{22}, B_{11}, b_{\text{min}})$
$M_3 \leftarrow \text{Strassen}(A_{11}, B_{12} - B_{22}, b_{\text{min}})$
$M_4 \leftarrow \text{Strassen}(A_{22}, B_{21} - B_{11}, b_{\text{min}})$
$M_5 \leftarrow \text{Strassen}(A_{11} + A_{12}, B_{22}, b_{\text{min}})$
$M_6 \leftarrow \text{Strassen}(A_{21} - A_{11}, B_{11} + B_{12}, b_{\text{min}})$
$M_7 \leftarrow \text{Strassen}(A_{12} - A_{22}, B_{21} + B_{22}, b_{\text{min}})$

$Q \leftarrow \begin{bmatrix} M_1 + M_4 - M_5 + M_7 & M_3 + M_5 \\ M_2 + M_4 & M_1 - M_2 + M_3 + M_6 \end{bmatrix}$

else
$Q \leftarrow AB$
return Q

Table E2.6-1 Strassen's multiplication algorithm.

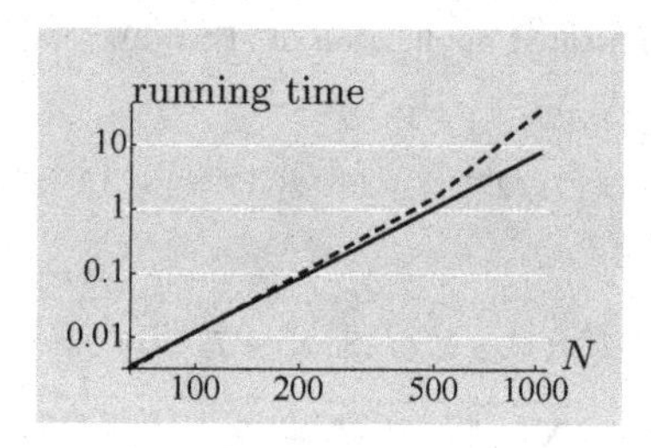

Figure E2.6-1 Execution times of Strassen's algorithm with a minimum block size of 64 (solid line) compared with a regular implementation (dashed line). Strassen's algorithm uses the regular implementation to multiply blocks of minimum size.

mechanism of Strassen's algorithm is used only on matrices with sizes larger than a given minimum block size. For smaller matrices, we use the regular algorithm. For a minimum block size of 64, the results are shown in Figure E2.6-1.

Strassen's algorithm is asymptotically more efficient than the regular approach. In modern computers, the bottleneck is in data access time and storage, and thus, Strassen's algorithm and the more advanced, but similar, Winograd algorithm are seldom used in practice.

Exercises

2.1. *Multiplication by an orthogonal matrix.*(1)
Consider the vector space $\mathbb{R}^N$ with the standard norm and standard inner product. Prove that

(i) multiplication by an orthogonal matrix U preserves lengths, that is,

$$\|Ux\| = \|x\|,$$

for any x;

(ii) multiplication by an orthogonal matrix U preserves angles, that is,

$$\langle Ux, Uy\rangle = \langle x, y\rangle,$$

for any x and y; and

(iii) the eigenvalues of U have unit absolute value.

2.2. *Bases and frames in* $\mathbb{R}^2$.(1)
Given are the following sets of vectors:

$$\Phi_1 = \{\varphi_{1,0}, \varphi_{1,1}\} = \left\{\begin{bmatrix}\frac{1}{2}\\ \frac{\sqrt{3}}{2}\end{bmatrix}, \begin{bmatrix}0\\1\end{bmatrix}\right\}, \tag{P2.2-1a}$$

$$\Phi_2 = \{\varphi_{2,0}, \varphi_{2,1}, \varphi_{2,2}, \varphi_{2,3}\} = \left\{\begin{bmatrix}\frac{1}{2\sqrt{2}}\\ \frac{\sqrt{3}}{2\sqrt{2}}\end{bmatrix}, \begin{bmatrix}-\frac{\sqrt{3}}{2\sqrt{2}}\\ \frac{1}{2\sqrt{2}}\end{bmatrix}, \begin{bmatrix}\frac{1}{\sqrt{2}}\\ 0\end{bmatrix}, \begin{bmatrix}0\\ \frac{1}{\sqrt{2}}\end{bmatrix}\right\}, \tag{P2.2-1b}$$

$$\Phi_3 = \{\varphi_{3,0}, \varphi_{3,1}\} = \left\{\begin{bmatrix}\frac{1}{2}\\ \frac{\sqrt{3}}{2}\end{bmatrix}, \begin{bmatrix}-\frac{\sqrt{3}}{2}\\ \frac{1}{2}\end{bmatrix}\right\}, \tag{P2.2-1c}$$

$$\Phi_4 = \{\varphi_{4,0}, \varphi_{4,1}, \varphi_{4,2}\} = \left\{\begin{bmatrix}1\\0\end{bmatrix}, \begin{bmatrix}\frac{1}{\sqrt{2}}\\ \frac{1}{\sqrt{2}}\end{bmatrix}, \begin{bmatrix}0\\1\end{bmatrix}\right\}. \tag{P2.2-1d}$$

For each of the sets of vectors above, do the following:

(i) Write the matrix representation for the set, that is, the synthesis operator associated with the set.

(ii) Find the dual basis or canonical dual frame. Sketch the original sets and their duals.

(iii) If the set is a basis, specify whether it is an orthonormal basis; otherwise, specify whether the frame is a tight frame.

(iv) For $x = \begin{bmatrix}2 & 0\end{bmatrix}^\top$, write down the projection coefficients, $\alpha_{i,k} = \langle x, \widetilde{\varphi}_{i,k}\rangle$.

(v) For the same x, verify the expansion formula $x = \sum_k \alpha_{i,k}\varphi_{i,k}$.

(vi) Verify the matrix version of the expansion formula $\Phi\widetilde{\Phi}^\top = I$.

(vii) Specify whether the expansion preserves the norm, that is, whether it is true that $\|x\|^2 = \sum_k |\alpha_{i,k}|^2$.

(viii) Specify whether the expansion is redundant. Justify your answer.

2.3. *Best approximation in* $\mathbb{R}^3$.(1)
Mimic what we did in $\mathbb{R}^2$ in Section 2.1, and find the best approximation in $\mathbb{R}^3$ of the vector x from Figure P2.3-1 in the (e_0, e_1)-plane. Find the difference between the vector and its approximation, and prove that it is the smallest possible.

2.4. *Matrices representing bases and frames.*(1)
Given is the following matrix:

$$\Phi = \begin{bmatrix}\varphi_0 & \varphi_1 & \varphi_2\end{bmatrix} = \begin{bmatrix}\sqrt{\frac{2}{3}} & -\frac{1}{\sqrt{6}} & -\frac{1}{\sqrt{6}}\\ 0 & \frac{1}{\sqrt{2}} & -\frac{1}{\sqrt{2}}\\ \frac{1}{\sqrt{3}} & \frac{1}{\sqrt{3}} & \frac{1}{\sqrt{3}}\end{bmatrix}.$$

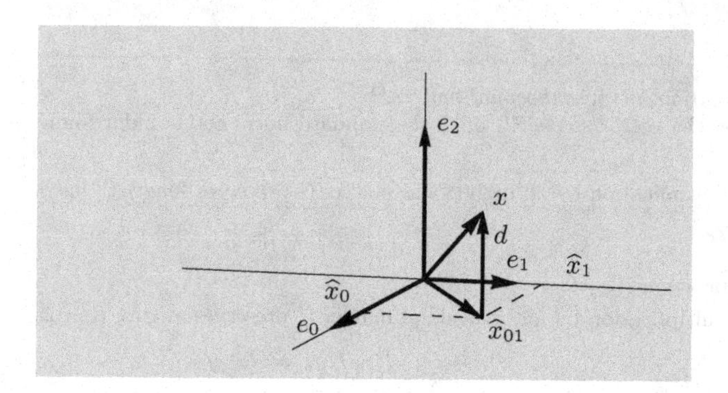

Figure P2.3-1 Orthogonal projection onto a subspace. Here, $x \in \mathbb{R}^3$, and $\widehat{x}_{01}$ is its projection onto the span of $\{e_0, e_1\}$. Note that $x - \widehat{x}_{01}$ is orthogonal to the span $\{e_0, e_1\}$, and that $\widehat{x}_{01}$ is the closest vector to x in the span of $\{e_0, e_1\}$.

(i) Is this matrix the synthesis operator associated with a basis? If yes, is the basis an orthonormal basis? If no, why not?

(ii) Each $\varphi_i = \begin{bmatrix} x_i & y_i & z_i \end{bmatrix}^\top$, $i = 0, 1, 2$, is a vector in a three-dimensional space $\mathbb{R}^3$. Project each φ_i onto the (x, y)-plane, that is, the two-dimensional space $\mathbb{R}^2$. Write the resulting matrix Φ', where each vector is now in $\mathbb{R}^2$. Is this matrix the synthesis operator associated with a frame? If yes, is it a tight frame? If not, why not?

2.5. *Linear independence.*(1)
Find the values of the parameter $a \in \mathbb{C}$ such that the following set is linearly independent:

$$U = \left\{ \begin{bmatrix} 0 & a^2 \\ 0 & j \end{bmatrix}, \begin{bmatrix} 0 & 1 \\ 1 & a-1 \end{bmatrix}, \begin{bmatrix} 0 & 0 \\ ja & 1 \end{bmatrix} \right\}.$$

For $a = j$, express the matrix

$$\begin{bmatrix} 0 & 5 \\ 2 & j-2 \end{bmatrix}$$

as a linear combination of elements of U.

2.6. *Continuity of the inner product.*(2)
Show that an inner product is continuous in both of its variables, that is, that

$$\lim_{\|h_1\|,\|h_2\| \to 0} |\langle x + h_1, y + h_2 \rangle - \langle x, y \rangle| = 0.$$

2.7. *Inner product on $\mathbb{C}^N$.*(1)
Prove that $\langle x, y \rangle = y^* A x$ is a valid inner product on $\mathbb{C}^N$ if and only if A is a Hermitian, positive definite matrix.

2.8. *Norms on $\mathbb{C}^N$.*(1)
Consider the vector space $\mathbb{C}^N$ of finite sequences $x = \begin{bmatrix} x_0 & x_1 & \ldots & x_{N-1} \end{bmatrix}^\top$. Prove that v_1 and v_2 are norms on $\mathbb{C}^N$, where

$$v_1(x) = \sum_{k=0}^{N-1} |x_k|, \qquad v_2(x) = \left(\sum_{k=0}^{N-1} |x_k|^2 \right)^{1/2}.$$

(*Hint:* For v_2, use Minkowski's inequality (2.211a).)

2.9. *Norms on $C([0, 1])$.*(1)
Consider the vector space $C([0, 1])$ of continuous functions on $[0, 1]$. Prove that v_1 and v_2 are norms on $C([0, 1])$, where

$$v_1(x) = \int_0^1 |x(t)|\, dt, \qquad v_2(x) = \left(\int_0^1 |x(t)|^2\, dt \right)^{1/2}.$$

(*Hint:* For v_2, use Minkowski's inequality (2.211b).)

2.10. *Orthogonal transforms and ∞ norm.*(3)
Orthogonal transforms conserve the 2 norm, but not others, in general. Consider the ∞ norm (2.39b).

(i) Consider the set of real orthogonal transforms T_2 on $\mathbb{R}^2$, that is, plane rotations and rotoinversions (2.238). Give the best lower and upper bounds a_2 and b_2 such that

$$a_2 \leq \|T_2 x\|_\infty \leq b_2 \tag{P2.10-1}$$

holds for all orthogonal T_2 and all vectors x of unit 2 norm.

(ii) Extend (P2.10-1) by giving the best lower and upper bounds a_N and b_N for the general case of real orthogonal transforms T_N on $\mathbb{R}^N$ with $N \geq 2$.

2.11. *Cauchy–Schwarz inequality, triangle inequality, and parallelogram law.*(3)
Prove the following:

(i) Cauchy–Schwarz inequality given in (2.29).

(ii) Triangle inequality given in Definition 2.9.

(iii) Parallelogram law given in (2.28).

(iv) In a normed vector space over the scalars $\mathbb{R}$, the inner product given by the *real polarization identity*

$$\langle x, y\rangle = \frac{1}{4}\left(\|x+y\|^2 - \|x-y\|^2\right) \tag{P2.11-1}$$

satisfies the distributivity, Hermitian symmetry, and positive definiteness properties. (More difficult to prove is the linearity in the first argument property, which also holds; together, these verify that (P2.11-1) is a valid inner product. Similarly, for a normed vector space over the scalars $\mathbb{C}$, the *complex polarization identity*

$$\langle x, y\rangle = \frac{1}{4}\left(\|x+y\|^2 - \|x-y\|^2 + j\|x+jy\|^2 - j\|x-jy\|^2\right) \tag{P2.11-2}$$

gives a valid inner product.)

2.12. *Norm induced by an inner product.*(1)
Let the mapping $\langle\cdot,\cdot\rangle : V \times V \to \mathbb{C}$ be an inner product defined on the vector space V. Show that the function

$$v(x) = \sqrt{\langle x, x\rangle}, \qquad x \in V,$$

defines a norm on V.

2.13. *Distances not necessarily induced by norms.*(2)
A *distance*, or *metric*, $d : V \times V \to \mathbb{R}$ is a function with the following properties:

(i) *Nonnegativity:* $d(x,y) \geq 0$ for every x, y in V.

(ii) *Symmetry:* $d(x,y) = d(y,x)$ for every x, y in V.

(iii) *Triangle inequality:* $d(x,y) + d(y,z) \geq d(x,z)$ for every x, y, z in V.

(iv) *Identity of indiscernibles:* $d(x,x) = 0$ and $d(x,y) = 0$ implies that $x = y$.

The *discrete metric* is given by

$$d(x,y) = \begin{cases} 0, & \text{if } x = y; \\ 1, & \text{if } x \neq y. \end{cases}$$

Show that the discrete metric is a valid distance and is not induced by any norm.

2.14. *Convergence of the inner product in $\ell^2(\mathbb{Z})$.*(3)
Let x and y be sequences in $\ell^2(\mathbb{Z})$. Show that the series (2.22b) defining $\langle x, y\rangle$ converges absolutely. Since convergence of doubly infinite series requires absolute convergence, this shows that the $\ell^2(\mathbb{Z})$ inner product is always well defined for vectors in $\ell^2(\mathbb{Z})$ (see Appendix 2.A.2).

2.15. *Definition of ∞ norm.*(2)
Show that the ∞ norm in (2.39b) is the natural extension of the p norm in (2.39a) by proving

$$\lim_{p\to\infty} \|x\|_p = \max_{i=0,1,\ldots,N-1} |x_i| \qquad \text{for any } x \in \mathbb{R}^N.$$

(*Hint:* Normalize x by dividing it by the entry of the largest magnitude. Compute the limit for the resulting vector.)

2.16. *Quasinorms with* $p < 1$.[(3)]
Equation (2.39a) does not yield a valid norm when $p < 1$.

(i) Show that Definition 2.9(iii) fails to hold for (2.39a) with $p = \frac{1}{2}$.

(ii) Show that, for $x \in \mathbb{R}^N$, $\lim_{p \to 0} \|x\|_p^p$ gives the number of nonzero components in x.

2.17. *Equivalence of norms on finite-dimensional spaces.*[(3)]
Two norms $\|\cdot\|_a$ and $\|\cdot\|_b$ on a vector space V are called *equivalent* when there exist finite constants c_a and c_b such that

$$\|v\|_a \leq c_a \|v\|_b \quad \text{and} \quad \|v\|_b \leq c_b \|v\|_a \quad \text{for all } v \in V.$$

(i) Show that the 1 norm, 2 norm, and ∞ norm are equivalent by proving

$$\|x\|_\infty \leq \|x\|_2 \leq \|x\|_1$$

and

$$\|x\|_1 \leq \sqrt{N}\|x\|_2 \leq N\|x\|_\infty,$$

for all $x \in \mathbb{R}^N$.

(ii) Show by counterexamples that the 1, 2, and ∞ norms are not equivalent for infinite-dimensional spaces.

2.18. *Nesting of* ℓ^p *spaces.*[(3)]
Prove that if $x \in \ell^p(\mathbb{Z})$ and $p < q$, then $x \in \ell^q(\mathbb{Z})$. This proves (2.41).

2.19. $\mathcal{L}^p([0, 1])$ *spaces.*[(2)]
Show that the parallelogram law holds in $\mathcal{L}^p([0, 1])$ only for $p = 2$. This shows that, among all $\mathcal{L}^p([0, 1])$ spaces, only $\mathcal{L}^2([0, 1])$ is a candidate for being a Hilbert space.
(*Hint:* To show that it does not hold for $p \neq 2$, use functions $x(t) = t$ and $y(t) = 1 - t$.)

2.20. *Closed subspaces and* $\ell^0(\mathbb{Z})$.[(3)]
Let $\ell^0(\mathbb{Z})$ denote the set of complex-valued sequences with a finite number of nonzero entries.

(i) Show that $\ell^0(\mathbb{Z})$ is a subspace of $\ell^2(\mathbb{Z})$.

(ii) Show that $\ell^0(\mathbb{Z})$ is not a closed subspace of $\ell^2(\mathbb{Z})$.

2.21. *Infinite sequences and completeness.*[(3)]
Let

$$\varphi_0 = \begin{bmatrix} \cdots & 0 & 0 & \boxed{\frac{1}{\sqrt{2}}} & \frac{1}{\sqrt{2}} & 0 & 0 & \cdots \end{bmatrix}^\top,$$

and for nonzero $k \in \mathbb{Z}$, let φ_k be φ_0 shifted by $2k$,

$$\varphi_{k,n} = \varphi_{0,n-2k}, \qquad n \in \mathbb{Z}.$$

Is the set of sequences $\{\varphi_k\}_{k \in \mathbb{Z}}$ a basis for $\ell^2(\mathbb{Z})$? Why?

2.22. *Completeness.*[(2)]
Let $\mathcal{P} \subset \mathcal{L}^2([0, 1])$ be the inner product space of polynomials with

$$\langle p, q \rangle = \int_0^1 p(t)\, q^*(t)\, dt,$$

and let (p_k) be a Cauchy sequence in $\mathcal{P}$,

$$p_k(t) = \sum_{i=0}^{k} \frac{1}{2^i} t^i.$$

Use $\lim_{k \to \infty} p_k$ to prove that $\mathcal{P}$ is not a Hilbert space.

2.23. *Completeness of* $\mathbb{C}^N$.[(2)]
Prove that $\mathbb{C}^N$, equipped with the p norm, is a complete vector space for any $p \in [1, \infty]$. You may assume that $\mathbb{C}$ itself is complete.
(*Hint:* Show that having a Cauchy sequence in $\mathbb{C}^N$ implies that each of the N components is a Cauchy sequence.)

2.24. *Cauchy sequences.*$^{(2)}$
Show that, in a normed vector space, every convergent sequence is a Cauchy sequence.

2.25. *Norms of operators.*$^{(2)}$

(i) Calculate $\|A\|$ and $\|A^{-1}\|$ for the following symmetric matrix:

$$A = \begin{bmatrix} 1 & 3 \\ 3 & 1 \end{bmatrix}.$$

(*Hint:* Calculate the eigenvalues of A.)

(ii) Find the norm of $A : \ell^2(\mathbb{Z}) \to \ell^2(\mathbb{Z})$ given by

$$(Ax)_n = e^{j\Theta_n} x_n, \qquad n \in \mathbb{Z},$$

where Θ_n is a real-valued sequence.

(iii) Find the norm of $A : \ell^2(\mathbb{Z}) \to \ell^2(\mathbb{Z})$ given by

$$(Ax)_{2n} = x_{2n} + x_{2n+1}, \qquad (Ax)_{2n+1} = x_{2n} - x_{2n+1}, \qquad n \in \mathbb{Z}.$$

2.26. *Relation between operator norm and eigenvalues.*$^{(2)}$
Show that, for any $A \in \mathbb{C}^{M \times N}$,

$$\|A\|_2 = \sqrt{\lambda_{\max}(A^* A)},$$

where $\lambda_{\max}$ denotes the largest eigenvalue.
(*Hint:* Apply the diagonalization (2.241a) to $A^* A$.)

2.27. *Adjoint operators.*$^{(2)}$
Prove parts (iv) and (vi)–(viii) of Theorem 2.21.

2.28. *Eigenvalues of definite operators.*$^{(2)}$
Let A be a self-adjoint operator on a Hilbert space H.

(i) Let λ be an eigenvalue of A. Show that A being positive semidefinite implies that $\lambda \geq 0$; furthermore, A being positive definite implies that $\lambda > 0$.

(ii) Show that the existence of a nonpositive eigenvalue implies that A is not positive definite; furthermore, the existence of a negative eigenvalue implies that A is not positive semidefinite.

2.29. *Operator expansion.*$^{(2)}$
Let $A : H \to H$ be a bounded linear operator with $\|A\| < 1$.

(i) Show that $I - A$ is invertible.

(ii) Show that, for every y in H,

$$(I - A)^{-1} y = \sum_{k=0}^{\infty} A^k y.$$

(iii) In practice one can compute only a finite number of terms in the series. For $\|y\| = 1$ and K terms in the expansion, find an upper bound on the error

$$\varepsilon_K = \left\| (I - A)^{-1} y - \sum_{k=0}^{K-1} A^k y \right\|.$$

2.30. *Projection via domain restriction.*$^{(2)}$
Recall the definition of $1_{\mathcal{I}} : \mathcal{L}^2(\mathbb{R}) \to \mathcal{L}^2(\mathbb{R})$ in (2.64).

(i) Show that $1_{\mathcal{I}}$ is an orthogonal projection operator.

(ii) Show that if $\mathcal{I}_1$ and $\mathcal{I}_2$ are disjoint subsets of $\mathbb{R}$, then the ranges of the associated operators, $\mathcal{R}(1_{\mathcal{I}_1})$ and $\mathcal{R}(1_{\mathcal{I}_2})$, are orthogonal.

(iii) Under what condition does the orthogonal decomposition $\mathcal{L}^2(\mathbb{R}) = \mathcal{R}(1_{\mathcal{I}_1}) \oplus \mathcal{R}(1_{\mathcal{I}_2})$ hold?

2.31. *Inverses, adjoints, and projection operators.*$^{(2)}$
Prove Theorem 2.29.

2.32. *Projection operators.*$^{(2)}$
Let

$$B = \begin{bmatrix} 1 & 0 \\ 1 & 1 \\ 0 & 1 \end{bmatrix}.$$

Find all projection operators onto the range of B; specify the orthogonal one.

2.33. *Riesz bases.*$^{(2)}$

(i) Prove that the standard basis in $\ell^2(\mathbb{Z})$ is a Riesz basis with optimal stability constants $\lambda_{\min} = \lambda_{\max} = 1$.

(ii) Let $\{e_k\}_{k\in\mathbb{Z}}$ denote the standard basis in $\ell^2(\mathbb{Z})$ and define the following scaled version:

$$\varphi_k = 2^k e_k, \qquad k \in \mathbb{Z}.$$

Prove that $\{\varphi_k\}_{k\in\mathbb{Z}}$ is a basis, but there is neither a positive $\lambda_{\min}$ nor a finite $\lambda_{\max}$ such that (2.89) in the definition of the Riesz basis holds.

(iii) Let

$$\psi_k = (\cos k) e_k, \qquad k \in \mathbb{Z}.$$

Prove or disprove that $\{\psi_k\}_{k\in\mathbb{Z}}$ is a basis for $\ell^2(\mathbb{Z})$, and prove or disprove that $\{\psi_k\}_{k\in\mathbb{Z}}$ is a Riesz basis for $\ell^2(\mathbb{Z})$.

2.34. *Basis that is not a Riesz basis.*$^{(2)}$
Complete Example 2.32(ii) by showing that

$$\varphi_k = \sum_{i=0}^{k} (i+1)^{-1/2} e_i, \qquad k \in \mathbb{N},$$

is a basis for $\ell^2(\mathbb{N})$ but not a Riesz basis.

2.35. *p norms in different bases.*$^{(2)}$
Let $x \in \mathbb{R}^2$ have expansion coefficient vector $\alpha = (\alpha_0, \alpha_1)$ with respect to basis $\{\varphi_0, \varphi_1\}$ for $\mathbb{R}^2$.

(i) Why is it true that $\|x\|_2 = \|\alpha\|_2$ for any orthonormal choice for the basis? This is called invariance of the 2 norm to the choice of orthonormal basis.

(ii) Let $p \in [1, \infty]$ with $p \neq 2$. Show that $\|x\|_p \neq \|\alpha\|_p$ for some choice of x and orthonormal basis. Thus, p norms with $p \neq 2$ are not invariant to the choice of orthonormal basis.

(iii) Suppose that the basis is not orthonormal. Show that $\|x\|_2 \neq \|\alpha\|_2$ for some choice of x. Thus, the 2 norm is not invariant to bases that are not orthonormal.

2.36. *Even and odd functions.*$^{(1)}$
Let

$$\Phi = \left\{ \frac{1}{\sqrt{2\pi}}, \frac{1}{\sqrt{\pi}} \cos kt, \frac{1}{\sqrt{\pi}} \sin kt \right\}_{k=1}^{\infty}$$

be an orthonormal basis for $\mathcal{L}^2([-\pi, \pi])$, and let S_{even} and S_{odd} be subspaces of even and odd functions, respectively, as in (2.20).

(i) Show that any $x \in \mathcal{L}^2([-\pi, \pi])$ can be written as $x = x_{\text{even}} + x_{\text{odd}}$, with $x_{\text{even}} \in S_{\text{even}}$ and $x_{\text{odd}} \in S_{\text{odd}}$.

(ii) Find orthonormal bases for S_{even} and S_{odd}.

(iii) Show that $\mathcal{L}^2([-\pi, \pi]) = S_{\text{even}} \oplus S_{\text{odd}}$.

2.37. *Least-squares approximation with an orthonormal basis.*$^{(1)}$
Let $\{\varphi_0, \varphi_1, \ldots, \varphi_{N-1}\}$ be an orthonormal basis for $\mathbb{R}^N$. For any $x \in \mathbb{R}^N$,

$$x = \sum_{i=0}^{N-1} \alpha_i \varphi_i, \qquad \text{with} \qquad \alpha_i = \langle x, \varphi_i \rangle.$$

Let $\widehat{x}$ be the best approximation of x onto $\operatorname{span}(\{\varphi_0, \varphi_1, \ldots, \varphi_{k-1}\})$,

$$\widehat{x} = \sum_{i=0}^{k-1} \beta_i \varphi_i, \qquad \text{with} \qquad \beta_i = \langle x, \varphi_i \rangle.$$

Prove that $\beta_i = \alpha_i$, for $i = 0, 1, \ldots, k$, by showing that these values minimize $\|x - \widehat{x}\|$.

2.38. *Biorthogonal pair of bases of cosine functions.*(2)
Let $\Phi = \{\varphi_k\}_{k\in\mathbb{N}}$ be the set defined in Example 2.34, and let $\Psi = \{\psi_k\}_{k\in\mathbb{N}}$ and $\widetilde{\Psi} = \{\widetilde{\psi}_k\}_{k\in\mathbb{N}}$ be the sets defined in Example 2.40.

(i) Show that Ψ and $\widetilde{\Psi}$ satisfy the biorthogonality condition (2.111).

(ii) Show that $\overline{\operatorname{span}}(\Psi) = \overline{\operatorname{span}}(\widetilde{\Psi}) = \overline{\operatorname{span}}(\Phi)$.

2.39. *Dual bases.*(2)
Let $\Phi = \{\varphi_k\}_{k\in\mathcal{K}}$ be a Riesz basis for Hilbert space H with optimal stability constants $\lambda_{\min}$ and $\lambda_{\max}$.

(i) Show that the dual of the dual of Φ is Φ.

(ii) Show that the dual of Φ is Φ if and only if Φ is an orthonormal basis.

(iii) Show that the dual of Φ is a Riesz basis with optimal stability constants $1/\lambda_{\max}$ and $1/\lambda_{\min}$.

2.40. *Oblique projection property.*(2)
Prove Theorem 2.47.

2.41. *Orthogonal projection in coefficient space.*(2)
Let $\{\psi_k\}_{k\in\mathcal{K}}$ be a basis for a Hilbert space H, and let $\{\varphi_k\}_{k\in\mathcal{I}}$ be another set in H. Denote the orthogonal projection of x onto $\overline{\operatorname{span}}(\{\varphi_k\}_{k\in\mathcal{I}})$ by $\widehat{x}$, and denote representations of x and $\widehat{x}$ with respect to $\{\psi_k\}_{k\in\mathcal{K}}$ by $x = \Psi\alpha$ and $\widehat{x} = \Psi\widehat{\alpha}$.

(i) Find an expression for $P : \ell^2(\mathcal{K}) \to \ell^2(\mathcal{K})$ that maps α to $\widehat{\alpha}$. You may use the operators Ψ, $\widetilde{\Psi}$, and Φ.

(ii) Show that P is a projection operator.

(iii) Show that P is an orthogonal projection operator if and only if $\{\psi_k\}_{k\in\mathcal{K}}$ is an orthonormal basis for H.

2.42. *Successive approximation with nonorthogonal basis.*(3)
Let $\{\varphi_i\}_{i\in\mathbb{N}}$ be a linearly independent set in the Hilbert space H. For each $k \in \mathbb{N}$, let $S_k = \operatorname{span}(\{\varphi_0, \varphi_1, \ldots, \varphi_{k-1}\})$, and let $\widehat{x}^{(k)}$ denote the best approximation of x in S_k. Prove that the recursive algorithm given in (2.141) provides a sequence of best approximations that all satisfy the normal equations (2.138a).
(*Hint:* Use induction over k.)

2.43. *Exploring the definition of a frame.*(3)
Let $\Phi = \{\varphi_k\}_{k\in\mathcal{J}}$ be a frame for a Hilbert space H.

(i) Show that, if φ_j is a linear combination of $\{\varphi_k\}_{k\in\mathcal{J}\setminus\{j\}}$ for some $j \in \mathcal{J}$, the following property of a Riesz basis is *not* satisfied by $\{\varphi_k\}_{k\in\mathcal{J}}$: For any expansion $x = \sum_{k\in\mathcal{J}} \alpha_k\varphi_k$, condition (2.89) holds.

(ii) Show that, for any $x \in H$, there exists an expansion $x = \sum_{k\in\mathcal{J}} \alpha_k\varphi_k$ such that condition (2.89) holds.

2.44. *Frame of cosine functions.*(3)
Find the optimal lower and upper frame bounds for the frame

$$\Phi \cup \Phi^+ = \{\varphi_k\}_{k\in\mathbb{N}} \cup \{\varphi_k^+\}_{k\in\mathbb{N}},$$

with φ_k from (2.24) and φ_k^+ from (2.143). If you cannot determine the optimal frame bounds analytically, find numerical estimates.

2.45. *Dual frame.*(2)
Let $E = \{e_k\}_{k=0}^{3}$ be the standard basis for $\mathbb{R}^4$ and let

$$\Psi = \{e_0 - e_1, e_1 - e_2, e_2 - e_3, e_3 - e_0\},$$
$$\Phi = \{e_0 + e_1, e_1 + e_2, e_2 + e_3, e_3 + e_0\}.$$

(i) Show that Ψ and Φ are not bases for $\mathbb{R}^4$. Which vectors from $\mathbb{R}^4$ are not in $\text{span}(\Psi)$?
(ii) Show that $F_1 = E \cup \Psi$ and $F_2 = \Phi \cup \Psi$ are frames. Compute their frame bounds.
(iii) Find the canonical dual frames to F_1 and F_2.

2.46. *Properties of dual pair of frames.*$^{(2)}$
This exercise develops results for dual pairs of frames that parallel the results for biorthogonal pairs of bases established in Section 2.5.3. Let $\Phi = \{\varphi_k\}_{k\in\mathcal{J}}$ and $\widetilde{\Phi} = \{\widetilde{\varphi}_k\}_{k\in\mathcal{J}}$ be a dual pair of frames for a Hilbert space H.
(i) Show that (2.156) holding for every x in H implies that (2.158) holds for every x in H. This establishes that the roles of the two frames in a dual pair can be reversed.
(ii) Let x and y be any vectors in H. Show that if $\widetilde{\alpha} = \Phi^* x$ and $\beta = \widetilde{\Phi}^* y$, then $\langle x, y\rangle = \langle \widetilde{\alpha}, \beta\rangle$. This shows how frame expansions enable computation of an inner product in H through an $\ell^2(\mathcal{J})$ inner product.
(iii) Let $\mathcal{I} \subseteq \mathcal{J}$ and recall the notations (2.103) and (2.104) for restricted synthesis and analysis operators. Determine a sufficient condition under which $\Phi_{\mathcal{I}}\widetilde{\Phi}_{\mathcal{I}}^*$ is a projection operator. (Forcing Φ and $\widetilde{\Phi}$ to form a biorthogonal pair of bases is not an adequate answer.)

2.47. *Tight frame with nonequal-norm vectors.*$^{(2)}$
Assume that $\alpha \in \mathbb{R}$, $\alpha \neq 0$, and let

$$\varphi_0 = \begin{bmatrix} 0 \\ \alpha \end{bmatrix}, \qquad \varphi_1 = \begin{bmatrix} \cos\theta \\ \sin\theta \end{bmatrix}, \qquad \varphi_2 = \begin{bmatrix} -\cos\theta \\ \sin\theta \end{bmatrix}.$$

For which values of θ and α is the above set a tight frame?

2.48. *Tight frame of affine functions.*$^{(2)}$
In $\mathcal{L}^2([0, 1])$, consider the subspace S of affine functions. Find a four-element tight frame Φ for S that includes $\varphi_0(t) = 1$ and $\varphi_1(t) = \sqrt{3}t$. Find the corresponding optimal frame bound λ and the canonical dual frame.
(*Hint:* A union of orthonormal bases is always a tight frame.)

2.49. *Complex multiplication.*$^{(3)}$
Multiplying two complex numbers

$$e + jf = (a + jb)(c + jd)$$

can be written as

$$\begin{bmatrix} e \\ f \end{bmatrix} = \begin{bmatrix} c & -d \\ d & c \end{bmatrix} \begin{bmatrix} a \\ b \end{bmatrix},$$

which takes four multiplications and two additions. Show that the same operation can be computed with three multiplications and five additions.

2.50. *Gaussian elimination.*$^{(1)}$
Our aim is to solve the system of linear equations $Ax = y$. (General conditions for the existence of a solution are given in Appendix 2.B.1.) Comment on whether a solution to each of the following systems of equations exists, and, if it does, find it.
(i)

$$A = \begin{bmatrix} 1 & 0 & 3 \\ 4 & 5 & 2 \\ -1 & -1 & 2 \end{bmatrix}, \qquad y = \begin{bmatrix} 10 \\ 20 \\ 3 \end{bmatrix}.$$

(ii)

$$A = \begin{bmatrix} 1 & 0 & 2 \\ 4 & 5 & 8 \\ -1 & -1 & -2 \end{bmatrix}, \qquad y = \begin{bmatrix} 7 \\ 38 \\ -9 \end{bmatrix}.$$

(iii)

$$A = \begin{bmatrix} 1 & 0 & 2 \\ 4 & 5 & 8 \\ -1 & -1 & -2 \end{bmatrix}, \qquad y = \begin{bmatrix} 1 \\ 2 \\ 3 \end{bmatrix}.$$

2.51. *Kaczmarz's algorithm.*(3)
Consider a square system of linear equations $y = Ax$ with a matrix $A \in \mathbb{R}^{N \times N}$ of full rank as in Example 2.66.

(i) Show that the intersection of the affine subspaces S_n, $n = 0, 1, \ldots, N-1$, is unique and specifies the solution of the system of linear equations $y = Ax$.

(ii) Prove that when the rows are orthogonal, Kaczmarz's algorithm converges in N steps or only one sweep.

2.52. *Convergence of sequences.*(3)
Let $(a_k)_{k=0}^{\infty}$ converge to a and $(b_k)_{k=0}^{\infty}$ converge to b. Prove the following:

(i) If $c \neq 0$ is some real number, $(ca_k)_{k=0}^{\infty}$ converges to ca.

(ii) $(a_k + b_k)_{k=0}^{\infty}$ converges to $a + b$.

(iii) $(a_k b_k)_{k=0}^{\infty}$ converges to ab.

(iv) If $a_k \neq 0$ for each $k \in \mathbb{N}$ and $a \neq 0$, $(b_k/a_k)_{k=0}^{\infty}$ converges to b/a.

2.53. *Convergence tests.*(1)
Let $(a_k)_{k=1}^{\infty}$ and $(b_k)_{k=1}^{\infty}$ be sequences that satisfy $0 \leq a_k \leq b_k$ for every $k \in \mathbb{Z}^+$. Then

- if $\sum_{k=1}^{\infty} b_k$ converges, $\sum_{k=1}^{\infty} a_k$ converges as well; and
- if $\sum_{k=1}^{\infty} a_k$ diverges, $\sum_{k=1}^{\infty} b_k$ diverges as well.

The *ratio test* for convergence of $\sum_{k=1}^{\infty} a_k$ states that, given

$$\lim_{k\to\infty} \left| \frac{a_{k+1}}{a_k} \right| = L, \tag{P2.53-1}$$

- if $0 \leq L < 1$, the series converges absolutely;
- if $1 < L$ or $L = \infty$, the series diverges; and
- if $L = 1$, the test is inconclusive, in that the series could converge or diverge. The convergence needs to be determined another way.

Based on the above, determine whether $\sum_{k=1}^{\infty} c_k$ converges for each of the following sequences:

(i) $c_k = k^2/(2k^4 - 3)$;
(ii) $c_k = (\log k)/k$;
(iii) $c_k = k^k/k!$; and
(iv) $c_k = a^k/k!$.

2.54. *Useful series.*(1)
In this exercise, we explore a few useful series.

(i) *Finite geometric series:* Prove the following formula:

$$\sum_{k=0}^{N-1} t^k = \frac{1 - t^N}{1 - t}. \tag{P2.54-1}$$

(ii) *Geometric series:* Determine conditions on when

$$\sum_{k=0}^{\infty} t^k \tag{P2.54-2}$$

converges. Prove that when it converges its sum is given by

$$\frac{1}{1-t}. \tag{P2.54-3}$$

(iii) *Power series:* Determine whether

$$\sum_{k=1}^{\infty} a_k t^k \tag{P2.54-4}$$

converges, as well as when and how.

(iv) *Taylor series:* If a function x has $(n+1)$ continuous derivatives, then it can be expanded into a *Taylor series* around a point t_0 as follows:

$$x(t) = \sum_{k=0}^{n} \frac{(t-t_0)^k}{k!} x^{(k)}(t_0) + R_n, \qquad R_n = \frac{(t-t_0)^{(n+1)}}{(n+1)!} x^{(n+1)}(\xi) \quad \text{(P2.54-5)}$$

for some ξ between t and t_0. Find the Taylor series expansion of $x(t) = 1/(1-t)$ around a point t_0. We discuss Taylor series expansion in more detail in Chapter 6.

(v) *MacLaurin series:* For $t_0 = 0$, the Taylor series is called the *MacLaurin series*:

$$x(t) = \sum_{k=0}^{n} \frac{t^k}{k!} x^{(k)}(0) + R_n. \qquad \text{(P2.54-6)}$$

Find the MacLaurin series expansion of $x(t) = 1/(1-t)$.

Function	**Expansion**	**Function**	**Expansion**
$\sin t$	$\sum_{k=0}^{\infty}(-1)^k \frac{t^{(2k+1)}}{(2k+1)!}$	$\cos t$	$\sum_{k=0}^{\infty}(-1)^k \frac{t^{(2k)}}{(2k)!}$
e^t	$\sum_{k=0}^{\infty} \frac{t^k}{(k)!}$	a^t	$\sum_{k=0}^{\infty} \frac{(t \ln a)^k}{(k)!}$
$\sinh t$	$\sum_{k=0}^{\infty} \frac{t^{(2k+1)}}{(2k+1)!}$	$\cosh t$	$\sum_{k=0}^{\infty} \frac{t^{(2k)}}{(2k)!}$
$\ln(1+t)$	$\sum_{k=1}^{\infty}(-1)^{(k+1)} \frac{t^k}{k}$	$\ln\left(\frac{1+t}{1-t}\right)$	$\sum_{k=1}^{\infty} 2\frac{t^{(2k-1)}}{2k-1}$

Table P2.54-1 Useful MacLaurin series expansions.

2.55. *Eigenvalues and eigenvectors.*[1]
Let

$$A = \begin{bmatrix} 1 & 2 \\ 2 & 1 \end{bmatrix} \quad \text{and} \quad B = \begin{bmatrix} \alpha & \beta \\ \beta & \alpha \end{bmatrix},$$

where α and β are not both equal to 0.

(i) Find the eigenvalues and eigenvectors for A and B (make sure that the eigenvectors are of norm 1).

(ii) Show that $A = V \Lambda V^\top$, where the columns of V correspond to the eigenvectors of A and Λ is a diagonal matrix whose main diagonal corresponds to the eigenvalues of A. Is V a unitary matrix? Why?

(iii) Compute the determinant of A. Is A invertible? If it is, give its inverse; if not, say why.

(iv) Compute the determinant of B. When is B invertible? Give its inverse when it is.

2.56. *Operator norm, singular values, and eigenvalues.*[2]
For matrix A (a bounded linear operator), show the following:

(i) If the matrix A is Hermitian, for each nonnegative eigenvalue λ_k, there is a corresponding identical singular value $\sigma_k = \lambda_k$. For each negative eigenvalue λ_k, there is a corresponding singular value $\sigma_k = |\lambda_k|$.

(ii) If the matrix A is Hermitian with eigenvalues $\{\lambda_k\}$, then

$$\|A\|_2 = \max\{|\lambda_k|\}.$$

(iii) Call μ_k the eigenvalues of A^*A; then

$$\|A\|_2 = \sqrt{\max\{\mu_k\}} = \max\{\sigma_k\}.$$

2.57. *Least-squares solution to a system of linear equations.*(1)
The general solution to this problem was given in (2.225).

(i) Show that if y belongs to the column space of A, then $\widehat{y} = y$.
(ii) Show that if y is orthogonal to the column space of A, then $\widehat{y} = 0$.
(iii) Show that for the least-squares solution, the partial derivatives $\partial(\|y - \widehat{y}\|^2)/\partial \widehat{x}_i$ are all zero.

2.58. *Power of a matrix.*(1)
Given a square, invertible matrix A, find an expression for A^k as a function of the eigenvectors and eigenvalues of A.

2.59. *Properties of jointly Gaussian vectors.*(2)
For establishing certain properties, it is convenient to use the following definition: A random vector $\mathrm{x} = \begin{bmatrix} \mathrm{x}_0 & \mathrm{x}_1 & \ldots & \mathrm{x}_{N-1} \end{bmatrix}^\top$ is jointly Gaussian when $\sum_{k=0}^{N-1} \alpha_k \mathrm{x}_k$ is a (univariate) Gaussian random variable for *any* choice of real coefficients $\{\alpha_k\}_{k=0}^{N-1}$. In this, we consider a random variable with zero variance to be Gaussian, and the definition is more general than (2.262) because it allows degenerate distributions where x does not have a joint PDF.

(i) Let x be a jointly Gaussian random variable with mean μ_{x} and covariance matrix Σ_{x}, and let A be a matrix with dimensions such that $A\mathrm{x}$ is defined. Show that $A\mathrm{x}$ is a jointly Gaussian vector with mean $A\mu_{\mathrm{x}}$ and covariance matrix $A\Sigma_{\mathrm{x}}A^\top$.

For the remaining parts, let $\mathrm{x} = \begin{bmatrix} \mathrm{y}^\top & \mathrm{z}^\top \end{bmatrix}^\top$ be a jointly Gaussian vector with

$$\mu_{\mathrm{x}} = E\left[\begin{bmatrix} \mathrm{y} \\ \mathrm{z} \end{bmatrix}\right] = \begin{bmatrix} \mu_{\mathrm{y}} \\ \mu_{\mathrm{z}} \end{bmatrix}, \quad \Sigma_{\mathrm{x}} = E\left[\left(\begin{bmatrix} \mathrm{y} \\ \mathrm{z} \end{bmatrix} - \begin{bmatrix} \mu_{\mathrm{y}} \\ \mu_{\mathrm{z}} \end{bmatrix}\right)\left(\begin{bmatrix} \mathrm{y} \\ \mathrm{z} \end{bmatrix} - \begin{bmatrix} \mu_{\mathrm{y}} \\ \mu_{\mathrm{z}} \end{bmatrix}\right)^\top\right] = \begin{bmatrix} \Sigma_{\mathrm{y}} & \Sigma_{\mathrm{y,z}} \\ \Sigma_{\mathrm{z,y}} & \Sigma_{\mathrm{z}} \end{bmatrix}.$$

(ii) Why must we have $\Sigma_{\mathrm{y}} = \Sigma_{\mathrm{y}}^\top$, $\Sigma_{\mathrm{z}} = \Sigma_{\mathrm{z}}^\top$, and $\Sigma_{\mathrm{y,z}} = \Sigma_{\mathrm{z,y}}^\top$?
(iii) Show that y is jointly Gaussian with mean μ_{y} and covariance matrix Σ_{y}.
(iv) Show that the conditional distribution of y given $\mathrm{z} = t$ is jointly Gaussian with mean $\mu_{\mathrm{y}} + \Sigma_{\mathrm{y,z}}\Sigma_{\mathrm{z}}^{-1}(t - \mu_{\mathrm{z}})$ and covariance matrix $\Sigma_{\mathrm{y}} - \Sigma_{\mathrm{y,z}}\Sigma_{\mathrm{z}}^{-1}\Sigma_{\mathrm{z,y}}$. You may assume that x has a joint PDF.

2.60. *Bayesian linear MMSE estimation via the orthogonality principle.*(1)
As in Example 2.67, let x and y be jointly distributed random vectors with $E[\mathrm{x}] = \mu_{\mathrm{x}}$, $E[\mathrm{y}] = \mu_{\mathrm{y}}$, and

$$E\left[\left(\begin{bmatrix} \mathrm{x} \\ \mathrm{y} \end{bmatrix} - \begin{bmatrix} \mu_{\mathrm{x}} \\ \mu_{\mathrm{y}} \end{bmatrix}\right)\left(\begin{bmatrix} \mathrm{x}^* & \mathrm{y}^* \end{bmatrix} - \begin{bmatrix} \mu_{\mathrm{x}}^* & \mu_{\mathrm{y}}^* \end{bmatrix}\right)\right] = \begin{bmatrix} \Sigma_{\mathrm{x}} & \Sigma_{\mathrm{x,y}} \\ \Sigma_{\mathrm{x,y}}^* & \Sigma_{\mathrm{y}} \end{bmatrix}.$$

Use the projection theorem to find the LMMSE estimator of x as a function of y, that is, the optimal estimator of the form $\widehat{\mathrm{x}} = A\mathrm{y} + b$.

2.61. *An inner product on random vectors.*(3)
Consider the set of complex random vectors of length N with the restriction that each component has finite variance. This set forms a vector space over $\mathbb{C}$.

(i) Show that $\langle \mathrm{x}, \mathrm{y} \rangle = \sum_{k=0}^{N-1} E[\mathrm{x}_k \mathrm{y}_k^*]$ is a valid inner product on this set. In doing so, explain the meaning of $\mathrm{x} = \mathbf{0}$.
(ii) What condition makes random vectors orthogonal under this inner product?
(iii) Zero-mean random vectors x and y are called uncorrelated when the matrix $E[\mathrm{x}\mathrm{y}^\top]$ is all 0s. Are orthogonal vectors uncorrelated? Are uncorrelated vectors orthogonal?
(iv) Let x and y be zero-mean Gaussian vectors. Does $\langle \mathrm{x}, \mathrm{y} \rangle = 0$ imply that x and y are independent?

Note that the inner product defined in this exercise is not often useful in deriving optimal estimators; see Exercise 2.60.

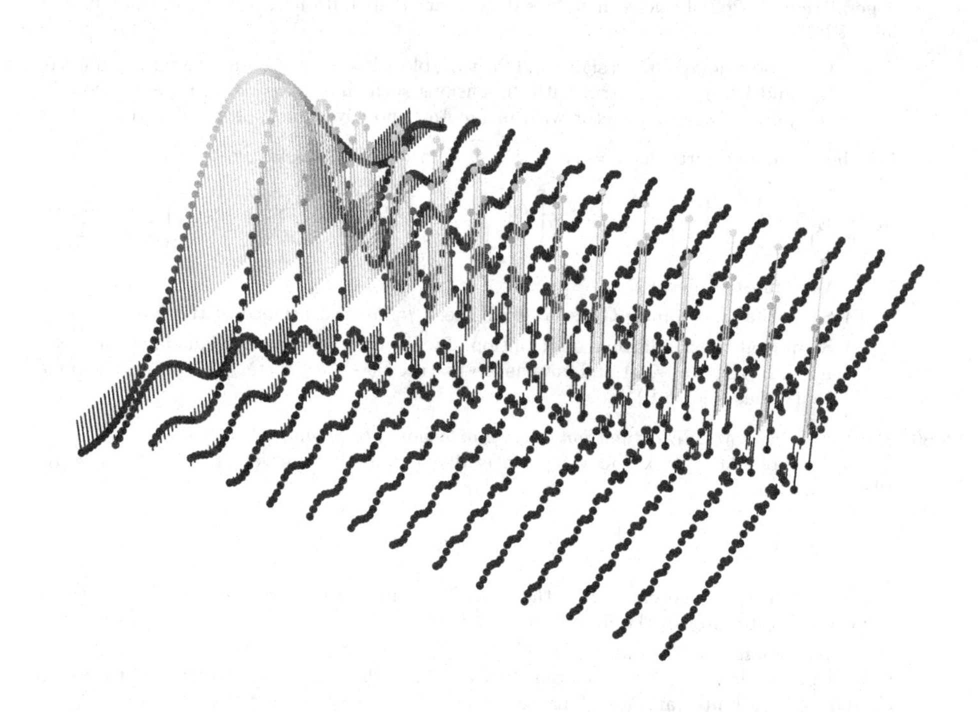

Chapter 3

Sequences and discrete-time systems

"Divide and conquer."

— *Julius Caesar*

Contents

Time is ordered – from past to future. Any countably infinite set of times can be indexed by the integers to maintain this order, and associating the integers with *discrete time* prompts us to refer to doubly infinite sequences as *discrete-time signals.* As we saw in the previous chapter, these sequences form the vector space $\mathbb{C}^{\mathbb{Z}}$ (assuming that they are complex-valued). Operators that map a sequence to a sequence are called *discrete-time systems.*

Some important classes of sequences and discrete-time systems have physical interpretations. For example, restrictions of sequences to the normed vector spaces $\ell^2(\mathbb{Z})$ and $\ell^\infty(\mathbb{Z})$ correspond to the physical properties of finite energy and boundedness. Also, many physical systems are described by time-invariant differential equations; with uniform discretization of time, this corresponds to a shift-invariance property for discrete-time systems. Linearity and shift invariance allow a system to be described uniquely by *convolution* with the system's *impulse response*. Once the convolution operation has been defined, spectral theory allows us to construct an appropriate Fourier transform. The Fourier transform is derived from the convolution operator, so the convolution property – which is central to signal processing – holds naturally. Shift-invariant systems, convolution, and the discrete-time Fourier transform also have many uses that need not have a physical underpinning.

The above discussion implicitly assumed that the underlying domain, time, is infinite. In practice we observe a finite portion of time, so we discuss handling a finite amount of data throughout the chapter.

3.1 Introduction

Suppose that some physical quantity is of interest, like the temperature in degrees Celsius in front of your house, at noon every day. For mathematical convenience, we look at this sequence as two-sided infinite,[40]

$$x = \begin{bmatrix} \cdots & x_{-2} & x_{-1} & \boxed{x_0} & x_1 & x_2 & \cdots \end{bmatrix}^\top, \tag{3.1}$$

with some arbitrary choice for time 0 (say January 14th). Implicit in the index is the fact that x_n corresponds to the temperature (at noon) on the nth day. A sequence is also known under the names discrete-time signal (in signal processing) and time series (in statistics).

In real life, we observe only a finite portion of an infinite-length sequence. Moreover, computations are always done on finite inputs. For example, consistent temperature recordings started in the eighteenth century and necessarily stop at the present time, producing a sequence of length N for some finite $N \in \mathbb{N}$,

$$x = \begin{bmatrix} \boxed{x_0} & x_1 & x_2 & \cdots & x_{N-1} \end{bmatrix}^\top. \tag{3.2}$$

Having only this data but methods that apply to all time, what do we do about days with no measurements? In effect, we are forced to assign some values; two techniques stand out to achieve this.

The first technique is to set $x_n = 0$ for all n outside of $\{0, 1, \ldots, N-1\}$. This is natural because, for any subsequent computation that uses x_n values linearly, this extension by zeros is equivalent to simply omitting measurements that are not available. However, the results are the same as if more data were available only when the signal is zero everywhere outside of $\{0, 1, \ldots, N-1\}$.

[40]The boxing of the time origin is intended to serve as a reference point, which is essential when dealing with infinite vectors/matrices.

A second, less obvious technique is to extend the signal circularly:[41] periodize the finite-length sequence, treating the observed values as one period of a periodic sequence of period $N \in \mathbb{N}$,

$$x = \begin{bmatrix} \cdots & x_{N-1} & \underbrace{\boxed{x_0} \; x_1 \; \cdots \; x_{N-1}}_{\text{one period}} & x_0 & x_1 & \cdots \end{bmatrix}^\top. \tag{3.3}$$

The consequences of this implicit periodization are central in digital signal processing. Other extensions are also possible, but they are much less common.

These considerations allow us to define two broad classes of sequences for which to develop our tools.

(i) *Infinite-length sequences* are the vector space $\mathbb{C}^{\mathbb{Z}}$ of sequences with the domain $\mathbb{Z}$, as defined in (2.18b). The support of a sequence might be a proper subset of $\mathbb{Z}$; for example, we will often consider infinite-length sequences that are nonzero only at nonnegative times.

(ii) *Finite-length sequences*, without loss of generality, have support in $\{0, 1, \ldots, N-1\}$. The tools we will develop do not treat the vector space of finite-length sequences as $\mathbb{C}^N$ generically, but rather as sequences defined on a circular domain.

Example 3.1 (Sequences)

(i) *Infinite-length sequences:* The sequence

$$x_n = \left(\frac{1}{2}\right)^n, \qquad n \in \mathbb{Z}, \qquad \text{or} \tag{3.4a}$$

$$x = \begin{bmatrix} \cdots & 4 & 2 & \boxed{1} & \frac{1}{2} & \frac{1}{4} & \cdots \end{bmatrix}^\top, \tag{3.4b}$$

is of infinite length and does not have finite ℓ^1, ℓ^2, or ℓ^∞ norm. If we made x_n nonzero only for $n \geq 0$, all these norms would be finite.

(ii) *Finite-length sequences:* A sequence obtained by observing N tosses of a fair coin, recording a 0 for heads and 1 for tails,

$$x = \begin{bmatrix} \boxed{0} & 0 & 1 & 1 & 0 & 1 & 0 & \cdots & 0 \end{bmatrix}^\top,$$

is of finite length. There is no extension of this sequence outside of the N observed values that is particularly natural.

A sinusoidal function sampled at N samples per period,

$$x_n = \sin\left(\frac{2\pi}{N}n + \theta\right), \qquad n \in \mathbb{Z}, \qquad \text{or}$$

$$x = \begin{bmatrix} \cdots & \underbrace{\boxed{\sin\theta} \; \sin\left(\frac{2\pi}{N} + \theta\right) \; \cdots \; \sin\left(\frac{2\pi}{N}(N-1) + \theta\right)}_{\text{one period}} & \sin\theta & \cdots \end{bmatrix}^\top,$$

[41] Another name for a circular extension is a periodic extension.

is an infinite-length periodic sequence. Taking N samples

$$x = \begin{bmatrix} \boxed{x_0} & x_1 & x_2 & \ldots & x_{N-1} \end{bmatrix}^\top$$

gives a finite-length sequence for which circular extension is quite natural.

Given a vector in a vector space, one can apply an operator to obtain another vector. When the domain and the codomain of the operator constitute a vector space of discrete-time signals (that is, a vector space of sequences), we call the operator a *discrete-time system*.

The basic building block of a discrete-time system is the *shift-by-1 operator* (also known as the unit delay, introduced formally in (3.39)),

$$y_n = x_{n-1}, \qquad n \in \mathbb{Z}. \tag{3.5}$$

Repeated applications of this operator or its inverse produce shifts of a sequence forward or backward in time while maintaining the ordering of the entries of the sequence; the ordering of the domain is essential in associating it with time.

Among discrete-time systems, we will focus almost exclusively on linear ones. Even more restricted is the class of *linear shift-invariant* systems (defined later in this chapter), an example of which is the moving-average filter.

EXAMPLE 3.2 (MOVING-AVERAGE FILTER) Consider our temperature example, and assume that we want to detect seasonal trends. The day-by-day variation might be too erratic, so we compute a local average

$$y_n = \frac{1}{N} \sum_{k=-(N-1)/2}^{(N-1)/2} x_{n-k}, \qquad n \in \mathbb{Z}, \tag{3.6}$$

where N is a small, odd positive integer. The local average reduces daily variations. This simple system is linear and shift-invariant since the same local averaging is performed at all n.

Chapter outline

The next several sections follow the progression of topics in this brief introduction. In Section 3.2, we start by formally defining the various types of sequences we discussed above. Section 3.3 considers linear discrete-time systems, especially of the shift-invariant kind, which correspond to difference equations, the discrete-time analogue of differential equations. Next, in Sections 3.4–3.6, we develop the tools to analyze discrete-time signals and systems, in particular the discrete-time Fourier transform, the z-transform, and the discrete Fourier transform. We discuss the fundamental result relating filtering to multiplication in the Fourier domain – the convolution property. Section 3.7 looks into discrete-time systems that operate with different rates – multirate systems, which are key for filter-bank developments in the companion volume [57]. This is followed by discrete-time stochastic processes and systems in Section 3.8, while important algorithms for discrete-time processing, such as the fast Fourier transform, are covered in Section 3.9. Appendix 3.A discusses topics from analysis, such as complex numbers, difference equations, convergence of certain sums, and the Dirac delta function, while Appendix 3.B discusses some elements of algebra, in particular, polynomial sequences.

Notation used in this chapter. We assume sequences to be complex in general, at the risk of having to use more cumbersome notation at times. Thus, Hermitian transposition is used often. We will be using $\|\cdot\|$ to denote the 2 norm; any other norm, such as the 1 norm, $\|\cdot\|_1$, will be explicitly specified. □

3.2 Sequences

3.2.1 Infinite-length sequences

The set of sequences in (3.1), where x_n is either real or complex, together with vector addition and scalar multiplication, forms a vector space (see Definition 2.1). The inner product between two infinite-length sequences is defined in (2.22b) and induces the standard ℓ^2 (or Euclidean) norm (2.26b). Other norms of interest are the ℓ^1 norm from (2.40a) with $p = 1$, and the ∞ norm from (2.40b).

As opposed to generic infinite-dimensional spaces, where ordering of indices does not matter in general, discrete-time signals belong to an infinite-dimensional space where ordering of indices is important since it represents time. Note that, in some instances later in the book, we will be dealing with vectors of sequences, for example, $x = \begin{bmatrix} x_0 & x_1 \end{bmatrix}^\top$, where x_0 and x_1 are sequences as well. We now look into a few spaces of interest.

Sequence spaces

Space of square-summable sequences $\ell^2(\mathbb{Z})$ The constraint of a finite square norm is necessary for turning the vector space $\mathbb{C}^{\mathbb{Z}}$ defined in (2.18b) into the Hilbert space of *finite-energy* sequences $\ell^2(\mathbb{Z})$. This space affords a geometric view; we now recall a few such geometric facts from Chapter 2.

(i) The angle between real nonzero sequences x and y is

$$\cos\theta = \frac{\langle x, y\rangle}{\|x\|\,\|y\|}.$$

(ii) As in Definition 2.8, if the inner product is zero,

$$\langle x, y\rangle = 0,$$

the sequences are said to be orthogonal to each other.

(iii) As in (2.66), given a unit-norm sequence y,

$$\widehat{x} = \langle x, y\rangle y$$

is the orthogonal projection of the sequence x onto the subspace of $\ell^2(\mathbb{Z})$ spanned by the sequence y.

Space of bounded sequences $\ell^\infty(\mathbb{Z})$ A looser constraint than finite energy is to bound the magnitude of the samples. The space of bounded sequences contains all sequences x such that, for some finite M, $|x_n| \le M$ for all $n \in \mathbb{Z}$. This space is denoted $\ell^\infty(\mathbb{Z})$ since it consists of sequences with finite ℓ^∞ norm.

Space of absolutely summable sequences $\ell^1(\mathbb{Z})$ A more restrictive constraint than finite energy is to require absolute summability (remember that $\ell^1(\mathbb{Z}) \subset \ell^2(\mathbb{Z})$ from (2.41)). By definition, sequences in $\ell^1(\mathbb{Z})$ have a finite ℓ^1 norm.

EXAMPLE 3.3 (SEQUENCE SPACES) For $\alpha \in \mathbb{R}$, the geometric sequence

$$x_n = \begin{cases} 0, & \text{for } n < 0; \\ \alpha^n, & \text{for } n \ge 0 \end{cases} \tag{3.7}$$

is in the following spaces:

$$x \in \begin{cases} \ell^2(\mathbb{Z}), \ell^1(\mathbb{Z}), \ell^\infty(\mathbb{Z}), & \text{for } |\alpha| < 1; \\ \ell^\infty(\mathbb{Z}), & \text{for } |\alpha| = 1; \\ \text{none of these}, & \text{for } |\alpha| > 1. \end{cases}$$

Special sequences

We now introduce certain sequences often used in the book.

Kronecker delta sequence The simplest nonzero sequence is the *Kronecker delta* sequence,

$$\delta_n = \begin{cases} 1, & \text{for } n = 0; \\ 0, & \text{otherwise}, \end{cases} \qquad n \in \mathbb{Z}, \qquad \text{or} \tag{3.8a}$$

$$\delta = \begin{bmatrix} \ldots & 0 & \boxed{1} & 0 & \ldots \end{bmatrix}^\top . \tag{3.8b}$$

Shifting the single 1 in the sequence to position k gives what is called the Kronecker delta sequence at location k, which is δ_{n-k}. The set of Kronecker delta sequences $\{\delta_{n-k}\}_{k \in \mathbb{Z}}$ forms an orthonormal basis for $\ell^2(\mathbb{Z})$; we called it the standard basis in Chapter 2. Table 3.1 lists some properties of the Kronecker delta sequence. (The shifting property uses convolution, which is defined in (3.61).)

Sinc function and sequences The *sinc function* appears frequently in signal processing and approximation. It is defined as

$$\operatorname{sinc} t = \begin{cases} (\sin t)/t, & \text{for } t \ne 0; \\ 1, & \text{for } t = 0. \end{cases} \tag{3.9a}$$

Evaluation of $\lim_{t \to 0} \operatorname{sinc} t$ using l'Hôpital's rule confirms the continuity of the function. Scaling the sinc function with $1/\sqrt{\pi}$ makes it of unit norm; that is,

$$\left\| \frac{1}{\sqrt{\pi}} \operatorname{sinc} t \right\| = 1. \tag{3.9b}$$

Kronecker delta sequence	
Normalization	$\sum_{n\in\mathbb{Z}} \delta_n = 1$
Sifting	$\sum_{n\in\mathbb{Z}} x_{n+n_0}\delta_n = \sum_{n\in\mathbb{Z}} x_n\delta_{n-n_0} = x_{n_0}$
Sampling	$x_n\delta_n = x_0\delta_n$
Restriction	$x_n\delta_n = 1_{\{0\}}\, x$
Shifting	$x_n *_n \delta_{n-n_0} = x_{n-n_0}$

Table 3.1 Properties of the Kronecker delta sequence.

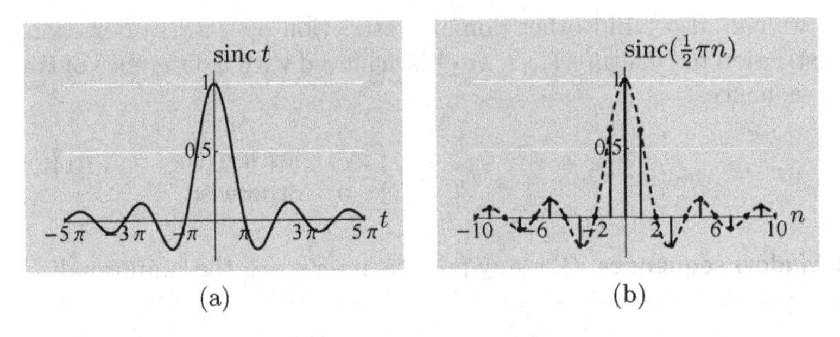

Figure 3.1 (a) The sinc function sinc t. (b) The sinc sequence $\operatorname{sinc}(\frac{1}{2}\pi n)$.

The sinc function is zero at $t = n\pi$ for nonzero integers n; together with the value at $t = 0$, this gives

$$\operatorname{sinc}(n\pi) \;=\; \delta_n, \qquad n \in \mathbb{Z}. \tag{3.9c}$$

The sinc function is illustrated in Figure 3.1(a).

For any positive T, we can obtain a sinc sequence

$$\frac{1}{\sqrt{T}}\operatorname{sinc}\left(\frac{\pi n}{T}\right) \;=\; \frac{1}{\sqrt{T}}\frac{\sin(\pi n/T)}{\pi n/T}. \tag{3.10}$$

This sequence is of unit norm and is in $\ell^\infty(\mathbb{Z})$ and in $\ell^2(\mathbb{Z})$. For general values of T, it is not in $\ell^1(\mathbb{Z})$, since it decays as $1/n$ (see Example 2.11, illustrating the inclusion property (2.41) of $\ell^p(\mathbb{Z})$ spaces). For $T = 1$ (or more generally for $T = 1/k$ for some $k \in \mathbb{Z}^+$) the sinc sequence in (3.10) reduces to the Kronecker delta sequence and thus is in $\ell^1(\mathbb{Z})$. The sinc sequence is zero for values of n such that n/T is a nonzero integer; this is illustrated in Figure 3.1(b) for $T = 2$.

Heaviside sequence The *Heaviside* or *unit-step* sequence is defined as

$$u_n \;=\; \begin{cases} 1, & \text{for } n \in \mathbb{N}; \\ 0, & \text{otherwise,} \end{cases} \qquad n \in \mathbb{Z}, \qquad \text{or} \tag{3.11a}$$

$$u \;=\; \begin{bmatrix} \cdots & 0 & \boxed{1} & 1 & \cdots \end{bmatrix}^\top. \tag{3.11b}$$

This sequence is bounded by 1, so it belongs to $\ell^\infty(\mathbb{Z})$. It belongs to neither $\ell^1(\mathbb{Z})$ nor $\ell^2(\mathbb{Z})$. The Kronecker delta and Heaviside sequences are related via

$$u_n = \sum_{k=-\infty}^{n} \delta_k.$$

Pointwise multiplication by the Heaviside sequence implements the domain restriction operator (2.62) for restriction from all the integers to just the nonnegative integers,

$$1_{\mathbb{N}} x = \begin{cases} x_n, & \text{for } n \in \mathbb{N}; \\ 0, & \text{otherwise} \end{cases} = u_n x_n, \qquad n \in \mathbb{Z}.$$

From this we can also build other domain restriction operators. For example, the domain restriction to $\{n_0, n_0+1, \ldots, n_1\}$ is achieved with a difference of two shifted Heaviside sequences,

$$1_{\{n_0,\ldots,n_1\}} x = (u_{n-n_0} - u_{n-n_1-1})x_n = \begin{cases} x_n, & \text{for } n \in \{n_0, \ldots, n_1\}; \\ 0, & \text{otherwise.} \end{cases} \tag{3.12}$$

Box and window sequences For any positive integer n_0, the (unnormalized) *right-sided box sequence* is defined as

$$w_n = \begin{cases} 1, & \text{for } 0 \le n \le n_0 - 1; \\ 0, & \text{otherwise,} \end{cases} \qquad n \in \mathbb{Z}, \qquad \text{or} \tag{3.13a}$$

$$w = \begin{bmatrix} \ldots & 0 & \boxed{1} & \underbrace{1 \;\; \ldots \;\; 1}_{n_0} & 0 & \ldots \end{bmatrix}^\top. \tag{3.13b}$$

For odd n_0, the *centered* and *normalized* box sequence is defined as

$$w_n = \begin{cases} 1/\sqrt{n_0}, & \text{for } |n| \le \frac{1}{2}(n_0 - 1); \\ 0, & \text{otherwise,} \end{cases} \qquad n \in \mathbb{Z}, \qquad \text{or} \tag{3.14a}$$

$$w = \begin{bmatrix} \ldots & 0 & \frac{1}{\sqrt{n_0}} & \ldots & \boxed{\frac{1}{\sqrt{n_0}}} & \ldots & \frac{1}{\sqrt{n_0}} & 0 & \ldots \end{bmatrix}^\top. \tag{3.14b}$$

Box sequences are also called *rectangular window* sequences.

Often, a finite-length sequence is treated as a glimpse of an infinite-length sequence. One way to state this is by using pointwise multiplication with a window sequence. Upon multiplying an arbitrary sequence x with the right-sided window w given in (3.13), we obtain a windowed version of x:

$$\widehat{x}_n = x_n w_n, \qquad n \in \mathbb{Z}, \qquad \text{or} \tag{3.15a}$$

$$\widehat{x} = \begin{bmatrix} \ldots & 0 & \boxed{x_0} & x_1 & \ldots & x_{n_0-1} & 0 & \ldots \end{bmatrix}^\top. \tag{3.15b}$$

With this use of an unnormalized rectangular window, $\widehat{x}_n$ equals x_n for $n \in \{0, 1, \ldots, n_0 - 1\}$ and is zero otherwise. We sometimes study x through the finite-length sequence $\widehat{x}$ that coincides with x over a window of interest.

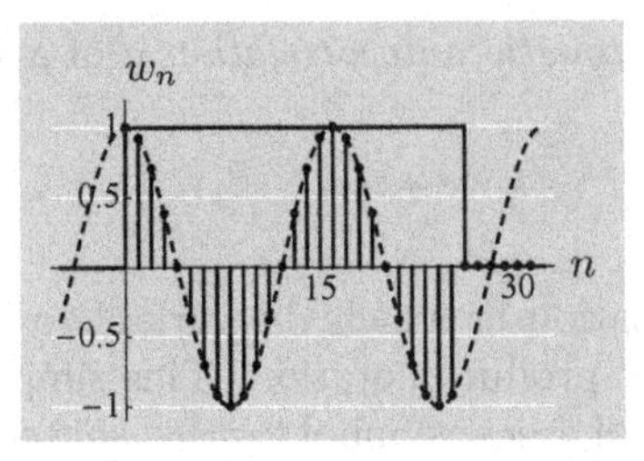

(a) Rectangular window (3.13).

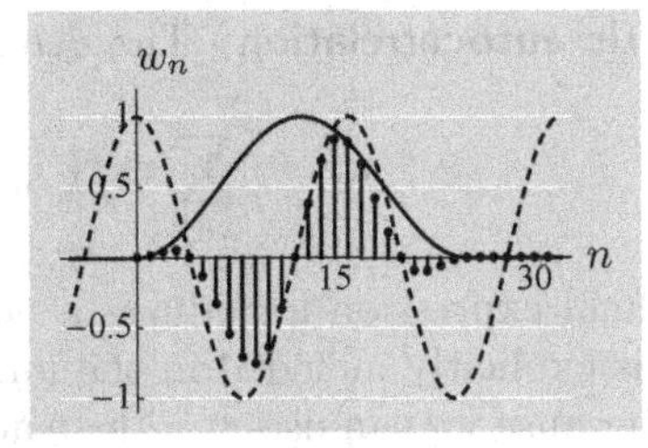

(b) Raised cosine window (3.16).

Figure 3.2 A sinusoidal sequence $x_n = \sin(\frac{1}{8}\pi n + \frac{1}{2}\pi)$ (dashed lines) and its windowed versions $w_n x_n$ (black stems) with two different windows of length $n_0 = 26$ (solid lines).

How good is the window we just used? For example, if x is smooth,[42] its windowed version $\widehat{x}$ is not because of the abrupt boundaries of the rectangular window. We might thus decide to use a different window to smooth the boundaries, an example of which we now discuss.

EXAMPLE 3.4 (WINDOWS) Consider an infinite-length sinusoidal sequence of frequency ω_0 and phase θ,

$$x_n = \sin(\omega_0 n + \theta),$$

and the following two windows:

(i) a rectangular, length-n_0 window, as in (3.13); and
(ii) a *raised cosine* window,[43] also of length n_0,

$$w_n = \begin{cases} \frac{1}{2}(1 - \cos(2\pi n/(n_0 - 1))), & \text{for } 0 \le n \le n_0 - 1; \\ 0, & \text{otherwise.} \end{cases} \quad (3.16)$$

The raised cosine window tapers off smoothly at the boundaries, while the rectangular one does not. The trade-off between the two windows is obvious from Figure 3.2: the rectangular window does not modify the sequence inside the window, but has abrupt transitions at the boundary, while the raised cosine window has smooth transitions at the boundary, but at the price of modifying the sequence inside the window.

Deterministic correlation

We now discuss two operations on sequences, both deterministic, that appear throughout the chapter. Stochastic versions of both operations will be given in Section 3.8.1.

[42]There is no formal definition of smoothness for a sequence. We use the word loosely, to mean that the envelope of the sequence is smooth.

[43]This is also known as a Hann or Hanning window, after Julius von Hann.

Deterministic autocorrelation The *deterministic autocorrelation* a of a sequence x is

$$a_n = \sum_{k\in\mathbb{Z}} x_k x_{k-n}^* = \langle x_k, x_{k-n}\rangle_k, \tag{3.17}$$

where the final expression introduces a notation in which the variable over which to sum, k, is explicitly included in the inner product notation. This simplifies our discussion because we can use x_{k-n} instead of a new symbol for this shifted version of x. The deterministic autocorrelation satisfies

$$a_n = a_{-n}^*, \tag{3.18a}$$

$$a_0 = \sum_{k\in\mathbb{Z}} |x_k|^2 = \|x\|^2, \tag{3.18b}$$

the proof of which is left for Exercise 3.2. The deterministic autocorrelation measures the similarity of a sequence with respect to shifts of itself, and it is Hermitian symmetric as in (3.18a). For a real x,

$$a_n = \sum_{k\in\mathbb{Z}} x_k x_{k-n} = a_{-n}. \tag{3.18c}$$

When we need to specify the sequence involved, we write $a_{x,n}$.

Example 3.5 (Deterministic autocorrelation) Assume that x is the box sequence from (3.14a) with $n_0 = 3$, that is, a constant sequence of length 3 and height $1/\sqrt{3}$. Using (3.17), we compute its deterministic autocorrelation to be

$$a_x = \begin{bmatrix} \ldots & 0 & \frac{1}{3} & \frac{2}{3} & \boxed{1} & \frac{2}{3} & \frac{1}{3} & 0 & \ldots \end{bmatrix}^\top. \tag{3.19}$$

This sequence is clearly symmetric, satisfying (3.18c).

Deterministic crosscorrelation The *deterministic crosscorrelation* c of two sequences x and y is

$$c_n = \sum_{k\in\mathbb{Z}} x_k y_{k-n}^* = \langle x_k, y_{k-n}\rangle_k \tag{3.20}$$

and is written as $c_{x,y,n}$ to specify the sequences involved. It satisfies

$$c_{x,y,n} = \left(\sum_{k\in\mathbb{Z}} y_{k-n} x_k^*\right)^* \stackrel{(a)}{=} \left(\sum_{m\in\mathbb{Z}} y_m x_{m+n}^*\right)^* = c_{y,x,-n}^*, \tag{3.21a}$$

where (a) follows from the change of variable $m = k - n$ (see also Exercise 3.2). For real x and y,

$$c_{x,y,n} = \sum_{k\in\mathbb{Z}} x_k y_{k-n} = c_{y,x,-n}. \tag{3.21b}$$

Example 3.6 (Deterministic crosscorrelation) Assume that x is the box sequence from (3.14a) with $n_0 = 3$, as in Example 3.5, and

$$y = \begin{bmatrix} \ldots & 0 & 0 & \boxed{\sqrt{\frac{2}{3}}} & \frac{1}{\sqrt{3}} & 0 & 0 & \ldots \end{bmatrix}^\top .$$

Using (3.20), we compute the deterministic crosscorrelations

$$c_{x,y} = \begin{bmatrix} \ldots & 0 & \frac{1}{3} & \frac{1+\sqrt{2}}{3} & \boxed{\frac{1+\sqrt{2}}{3}} & \frac{\sqrt{2}}{3} & 0 & 0 & \ldots \end{bmatrix}^\top , \tag{3.22a}$$

$$c_{y,x} = \begin{bmatrix} \ldots & 0 & 0 & \frac{\sqrt{2}}{3} & \boxed{\frac{1+\sqrt{2}}{3}} & \frac{1+\sqrt{2}}{3} & \frac{1}{3} & 0 & \ldots \end{bmatrix}^\top , \tag{3.22b}$$

satisfying (3.21b).

Deterministic autocorrelation of vector sequences Consider a vector of N sequences, $x = \begin{bmatrix} x_0 & x_1 & \ldots & x_{N-1} \end{bmatrix}^\top$, which is an infinite matrix whose $(k+1)$st row is the sequence

$$x_k = \begin{bmatrix} \ldots & x_{k,-1} & \boxed{x_{k,0}} & x_{k,1} & \ldots \end{bmatrix}.$$

Its deterministic autocorrelation is a sequence of matrices given by

$$A_n = \begin{bmatrix} a_{0,n} & c_{0,1,n} & \cdots & c_{0,N-1,n} \\ c_{1,0,n} & a_{1,n} & \cdots & c_{1,N-1,n} \\ \vdots & \vdots & \ddots & \vdots \\ c_{N-1,0,n} & c_{N-1,1,n} & \cdots & a_{N-1,n} \end{bmatrix}; \tag{3.23}$$

that is, a matrix with individual sequence deterministic autocorrelations $a_{i,n}$ on the diagonal and the pairwise deterministic crosscorrelations $c_{i,k,n}$ off the diagonal, for $i, k = 0, 1, \ldots, N-1$, $i \neq k$. Because of (3.18a) and (3.21a), A_n satisfies

$$A_n = \begin{bmatrix} a_{0,n} & c_{0,1,n} & \cdots & c_{0,N-1,n} \\ c^*_{0,1,-n} & a_{1,n} & \cdots & c_{1,N-1,n} \\ \vdots & \vdots & \ddots & \vdots \\ c^*_{0,N-1,-n} & c^*_{1,N-1,-n} & \cdots & a_{N-1,n} \end{bmatrix} = A^*_{-n}; \tag{3.24a}$$

that is, it is a Hermitian matrix (see (2.239a)). For a real x, it is a symmetric matrix,

$$A_n = A^\top_{-n}. \tag{3.24b}$$

Example 3.7 (Deterministic autocorrelation) Assume that we are given a vector of two sequences $x = \begin{bmatrix} x_0 & x_1 \end{bmatrix}^\top$ with $x_0 = x$ and $x_1 = y$ from Example 3.6. Its deterministic autocorrelation is then

$$A_n = \begin{bmatrix} a_{0,n} & c_{0,1,n} \\ c_{1,0,n} & a_{1,n} \end{bmatrix}.$$

We have already computed three out of four entries in the above matrix: the deterministic autocorrelation sequence $a_0 = a_x$ from (3.19) and the deterministic crosscorrelation sequences $c_{0,1} = c_{x,y}$ from (3.22a) and $c_{1,0} = c_{y,x}$ from (3.22b). The only entry left to compute is the deterministic autocorrelation sequence a_y,

$$a_y = \begin{bmatrix} \ldots & 0 & \frac{\sqrt{2}}{3} & \boxed{1} & \frac{\sqrt{2}}{3} & 0 & \ldots \end{bmatrix}^\top. \tag{3.25}$$

Because of the symmetries we already observed in Examples 3.5 and 3.6 and the symmetry of a_y, (3.24b) is satisfied. A few entries in the sequence of matrices A_n, including all the nonzero entries, are

$$\left\{ \ldots, \begin{bmatrix} \frac{1}{3} & \frac{1}{3} \\ 0 & 0 \end{bmatrix}, \begin{bmatrix} \frac{2}{3} & \frac{1+\sqrt{2}}{3} \\ \frac{\sqrt{2}}{3} & \frac{\sqrt{2}}{3} \end{bmatrix}, \boxed{\begin{bmatrix} 1 & \frac{1+\sqrt{2}}{3} \\ \frac{1+\sqrt{2}}{3} & 1 \end{bmatrix}}, \begin{bmatrix} \frac{2}{3} & \frac{\sqrt{2}}{3} \\ \frac{1+\sqrt{2}}{3} & \frac{\sqrt{2}}{3} \end{bmatrix}, \begin{bmatrix} \frac{1}{3} & 0 \\ \frac{1}{3} & 0 \end{bmatrix}, \ldots \right\}.$$

3.2.2 Finite-length sequences

Finite-length sequences as in (3.2) are those with the domain

$$n \in \{0, 1, \ldots, N-1\}$$

for some positive integer N. A finite-length sequence can be seen either as an infinite-length sequence that happens to take nonzero values only inside $\{0, 1, \ldots, N-1\}$ or as a period of a periodic sequence with

$$x_{n+kN} = x_n, \qquad k \in \mathbb{Z}. \tag{3.26}$$

Sequence spaces

In the case of a periodic sequence, it is useful to think of the domain itself as wrapping around into a circle, with $N-1$ next to 0. On this discrete circle domain, incrementing the time index is not ordinary addition but rather addition modulo N, so we could refer to the domain as $\mathbb{Z}_N$ and to the vector space of these sequences as $\mathbb{C}^{\mathbb{Z}_N}$. We do not actually adopt the notation $\mathbb{C}^{\mathbb{Z}_N}$ because the standard vector space operations (see Definition 2.1) are the same as for $\mathbb{C}^N$.

Differences between periodic sequences (those defined on a circular domain) and infinite sequences with finite support emerge with operations that we introduce later. For periodic sequences, there is a circular form of convolution. Applying spectral theory to this convolution leads to the discrete Fourier transform. As part of a more general theory, other convolution operators would lead to different Fourier transforms; more details on this topic can be found in the *Further reading*.

Special sequences

Periodic Kronecker delta sequences A periodic version of the Kronecker delta sequence is obtained by adding all shifts of δ by integer multiples of N,

$$\varphi_n = \sum_{\ell \in \mathbb{Z}} \delta_{n-\ell N}, \qquad n \in \mathbb{Z}.$$

The resulting sequence is

$$\varphi_n = \begin{cases} 1, & \text{for } n = \ell N,\ \ell \in \mathbb{Z}; \\ 0, & \text{otherwise}, \end{cases} \qquad n \in \mathbb{Z}, \qquad \text{or}$$

$$\varphi = \begin{bmatrix} \cdots & 0 & \underbrace{\boxed{1} \ 0 \ \cdots \ 0}_{N} & 1 & 0 & \cdots \end{bmatrix}^{\top}.$$

The set of N sequences generated from this φ by shifts in $\{0, 1, \ldots, N-1\}$ span the space of N-periodic sequences.

Complex exponential sequences As we will see in Section 3.6, the complex exponential sequences form a natural basis for N-periodic sequences. These are the N sequences φ_k, $k \in \{0, 1, \ldots, N-1\}$, given by

$$\varphi_{k,n} = \frac{1}{\sqrt{N}} e^{j(2\pi/N)kn}, \qquad k \in \{0, 1, \ldots, N-1\}, \quad n \in \mathbb{Z}. \tag{3.27}$$

Each of these sequences is periodic with period N. In Solved exercise 3.1, we explore a few properties of complex exponential sequences.

3.2.3 Two-dimensional sequences

Today, one of the most widespread devices is the digital camera. In our notation, a digital picture is a two-dimensional sequence, $x_{n,m}$. It can be seen either as an infinite-length sequence with a finite number of nonzero samples,

$$x_{n,m}, \qquad n, m \in \mathbb{Z}, \tag{3.28}$$

or as a sequence with the domain $n \in \{0, 1, \ldots, N-1\}$, $m \in \{0, 1, \ldots, M-1\}$, conveniently expressed as a matrix:

$$x = \begin{bmatrix} x_{0,0} & x_{0,1} & \cdots & x_{0,M-1} \\ x_{1,0} & x_{1,1} & \cdots & x_{1,M-1} \\ \vdots & \vdots & \ddots & \vdots \\ x_{N-1,0} & x_{N-1,1} & \cdots & x_{N-1,M-1} \end{bmatrix}. \tag{3.29}$$

While circularly extending the image at the borders is perhaps not natural (the top of the image appears next to the bottom), it is the extension that leads to the use of the discrete Fourier transform, as we will see later in this chapter. Each element $x_{n,m}$ is called a *pixel*, and the image has NM pixels. In reality, for $x_{n,m}$ to represent a color image, it must have more than one component; often, red, green and blue components are used (the RGB color space). Figure 3.3 gives examples of two-dimensional sequences.

Sequence spaces The spaces we introduced in one dimension generalize to multiple dimensions; for example, in two dimensions the standard inner product of two sequences x and y is

$$\langle x, y \rangle = \sum_{n\in\mathbb{Z}} \sum_{m\in\mathbb{Z}} x_{n,m} y_{n,m}^*, \tag{3.30}$$

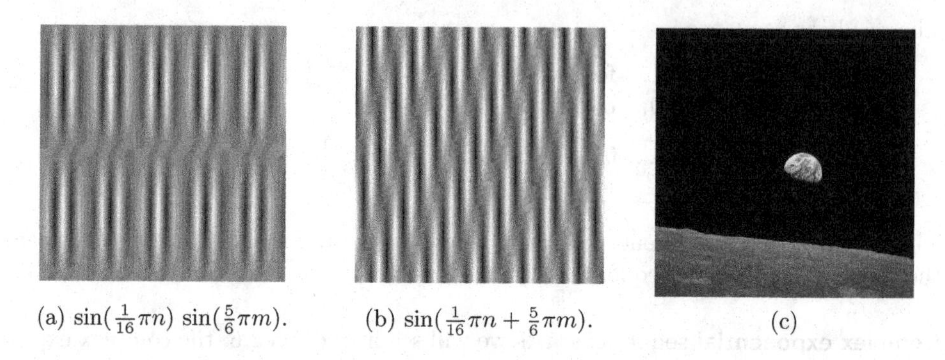

(a) $\sin(\frac{1}{16}\pi n)\sin(\frac{5}{6}\pi m)$. (b) $\sin(\frac{1}{16}\pi n + \frac{5}{6}\pi m)$. (c)

Figure 3.3 Two-dimensional sequences. (a) Separable sinusoidal sequence. (b) Nonseparable sinusoidal sequence. (c) Earth visible above the lunar surface, taken by Apollo 8 crew member Bill Anders on December 24, 1968.

Sequence space	Symbol	Finite norm
Absolutely summable	$\ell^1(\mathbb{Z}^2)$	$\|x\|_1 = \sum_{n,m\in\mathbb{Z}} \lvert x_{n,m}\rvert$
Square-summable/finite-energy	$\ell^2(\mathbb{Z}^2)$	$\|x\| = \left(\sum_{n,m\in\mathbb{Z}} \lvert x_{n,m}\rvert^2\right)^{1/2}$
Bounded	$\ell^\infty(\mathbb{Z}^2)$	$\|x\|_\infty = \sup_{n,m\in\mathbb{Z}} \lvert x_{n,m}\rvert$

Table 3.2 Norms and two-dimensional sequence spaces.

while the ℓ^2 norm and the appropriate space $\ell^2(\mathbb{Z}^2)$ are given in Table 3.2, together with other relevant norms and spaces. For example, a digital picture, having finite size and pixel values that are bounded, clearly belongs to all three spaces defined in Table 3.2. Infinite-length multidimensional sequences, on the other hand, can be harder to analyze.

EXAMPLE 3.8 (NORMS OF TWO-DIMENSIONAL SEQUENCES) Let

$$x_{n,m} = \frac{1}{2^n \cdot 3^m}, \qquad n, m \in \mathbb{N}.$$

Its squared ℓ^2 norm can be evaluated as[44]

$$\langle x, x\rangle = \sum_{n\in\mathbb{N}}\sum_{m\in\mathbb{N}} \frac{1}{4^n}\frac{1}{9^m} = \left(\sum_{n\in\mathbb{N}} \frac{1}{4^n}\right)\left(\sum_{m\in\mathbb{N}} \frac{1}{9^m}\right) = \frac{4}{3}\cdot\frac{9}{8} = \frac{3}{2},$$

yielding $\|x\|_2 = \sqrt{3/2}$. Similarly, $\|x\|_1 = 3$ and $\|x\|_\infty = 1$.

[44] We interchange summations freely, which can be done because each one-dimensional sequence involved is absolutely summable. When this is not the case, one has to be careful, as discussed in Appendix 2.A.3.

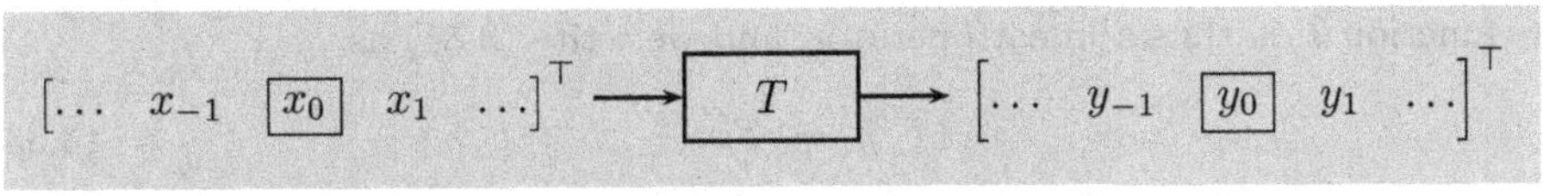

$$\begin{bmatrix} \cdots & x_{-1} & \boxed{x_0} & x_1 & \cdots \end{bmatrix}^{\top} \longrightarrow \boxed{T} \longrightarrow \begin{bmatrix} \cdots & y_{-1} & \boxed{y_0} & y_1 & \cdots \end{bmatrix}^{\top}$$

Figure 3.4 A discrete-time system.

3.3 Systems

Discrete-time systems are operators having discrete-time signals (sequences) as their inputs and outputs. Among all discrete-time systems, we will concentrate on those that are linear and shift-invariant. This subclass is both important in practice and amenable to analysis. The moving-average filter in (3.6) is such a linear, shift-invariant system. After an introduction to difference equations, which are natural descriptions of discrete-time systems, we study linear, shift-invariant systems in detail.

3.3.1 Discrete-time systems and their properties

A discrete-time system is an operator T that maps an input sequence $x \in V$ into an output sequence $y \in V$,

$$y = T(x), \tag{3.31}$$

as shown in Figure 3.4. As we have seen in the previous section, the sequence space V is typically $\ell^2(\mathbb{Z})$ or $\ell^\infty(\mathbb{Z})$. At times, the input or the output is in a subspace of such spaces.

Types of systems

Discrete-time systems can have a number of useful properties, which we will also encounter for continuous-time systems in Chapter 4. After defining key properties, we will illustrate them on certain basic systems.

Linear systems Similarly to Definition 2.17, linearity[45] combines two properties: additivity (the output of a sum of sequences is the sum of the outputs of the sequences) and scaling (the output of a scaled sequence is the scaled output of the sequence).

DEFINITION 3.1 (LINEAR SYSTEM) A discrete-time system T is called *linear* when, for any inputs x and y and any α, $\beta \in \mathbb{C}$,

$$T(\alpha x + \beta y) = \alpha T(x) + \beta T(y). \tag{3.32}$$

[45] In the engineering literature, linearity and the *superposition principle* are often used interchangeably.

The function T is thus a linear operator, and we write (3.31) as

$$y = Tx. \tag{3.33}$$

We will often use a matrix representation for a linear system, especially when the structure of the matrix reveals properties of the system.

As discussed in Section 2.5.5, a linear operator has a unique matrix representation once bases have been chosen for the domain and the codomain of the operator. Throughout this chapter, matrix representations of linear systems will be with respect to the standard basis (the Kronecker delta sequence and its shifts) both for the inputs and for the outputs. The general form of the matrix representation then follows from (2.165): column k holds the output that results from taking the shifted Kronecker delta sequence δ_{n-k} as the input. To be more explicit, for each $k \in \mathbb{Z}$, let input $x^{(k)}$ result in output $y^{(k)}$, where

$$x_n^{(k)} = \delta_{n-k}, \qquad n \in \mathbb{Z}.$$

Then, the matrix representation of the system is

$$\begin{bmatrix} & \vdots & \vdots & \vdots & \\ \cdots & y_{-1}^{(-1)} & y_{-1}^{(0)} & y_{-1}^{(1)} & \cdots \\ \cdots & y_{0}^{(-1)} & \boxed{y_{0}^{(0)}} & y_{0}^{(1)} & \cdots \\ \cdots & y_{1}^{(-1)} & y_{1}^{(0)} & y_{1}^{(1)} & \cdots \\ & \vdots & \vdots & \vdots & \end{bmatrix}. \tag{3.34}$$

Memoryless systems Certain simple systems are *instantaneous* in that they act based solely on the current input sample. It follows that if two inputs agree at a time index k, the corresponding outputs must also agree at time index k. For a mathematical representation of memorylessness, we use the domain-restriction operator defined in (2.62).

DEFINITION 3.2 (MEMORYLESS SYSTEM) A discrete-time system T is called *memoryless* when, for any integer k and inputs x and x',

$$1_{\{k\}}\, x = 1_{\{k\}}\, x' \quad \Rightarrow \quad 1_{\{k\}}\, T(x) = 1_{\{k\}}\, T(x'). \tag{3.35}$$

In a matrix representation of a linear and memoryless system, the matrix will be diagonal; we will illustrate this and other properties of matrix representations of linear systems in several examples shortly.

Causal systems The output of a causal system at time index k depends on the input only up to time index k. It follows that if two inputs agree up to time k, the corresponding outputs must agree up to time k.

Definition 3.3 (Causal system) A discrete-time system T is called *causal* when, for any integer k and inputs x and x',

$$1_{\{-\infty,\ldots,k\}}\, x \;=\; 1_{\{-\infty,\ldots,k\}}\, x' \quad\Rightarrow\quad 1_{\{-\infty,\ldots,k\}}\, T(x) \;=\; 1_{\{-\infty,\ldots,k\}}\, T(x'). \tag{3.36}$$

In a matrix representation of a linear and causal system, the matrix will be lower-triangular.

Since a computation cannot depend on inputs that will be provided only in the future, causality can seem to be a property that is required of any implemented system. However, this view takes the concept of the time index representing time too literally. First, the discrete time index might represent something else entirely, like a physical location along a line; the data can then be processed in any order. Second, when the time index does indeed represent time, the time origins of the input and output need not coincide; then, causality sometimes amounts to nothing more than a convenient convention for aligning time indices of the input and output.

Shift-invariant systems In a shift-invariant system, shifting the input has the effect of shifting the output by the same amount.

Definition 3.4 (Shift-invariant system) A discrete-time system T is called *shift-invariant* when, for any integer k and input x,

$$y \;=\; T(x) \quad\Rightarrow\quad y' \;=\; T(x'), \qquad \text{where } x'_n = x_{n-k} \text{ and } y'_n = y_{n-k}. \tag{3.37}$$

In a matrix representation of a linear and shift-invariant system, the matrix will be Toeplitz.

Shift invariance (or, when it corresponds to time, time invariance) is often a desirable property. For example, an MP3 player should produce the same music from the same file on Tuesday as on Monday. Moreover, *linear shift-invariant (LSI)* or *linear time-invariant (LTI)* systems have desirable mathematical properties. Much of the remainder of this section and Sections 3.4 and 3.5 are devoted to the powerful analysis techniques that apply to LSI systems. Sections 3.6 and 3.7 include variations on shift invariance and the corresponding techniques.

Stable systems A critical property for a discrete-time system is its stability. While various definitions exist, they all require that the system remain well behaved when presented with a certain class of inputs. We define *bounded-input, bounded-output (BIBO)* stability here, because it is both practical and easy to check in cases of interest.

DEFINITION 3.5 (BIBO-STABLE SYSTEM) A discrete-time system T is called *bounded-input, bounded-output stable* when a bounded input x produces a bounded output $y = T(x)$:

$$x \in \ell^\infty(\mathbb{Z}) \quad \Rightarrow \quad y \in \ell^\infty(\mathbb{Z}). \tag{3.38}$$

In a matrix representation of a linear and BIBO-stable system, every row of the matrix will be absolutely summable. The corresponding result for LSI systems is developed fully in Section 3.3.3.

The definition of BIBO stability involves the ℓ^∞ norm, so we can see immediately that a system that is linear and BIBO-stable is a bounded linear operator from $\ell^\infty(\mathbb{Z})$ to $\ell^\infty(\mathbb{Z})$. The absolute-summability condition on the system that ensures BIBO stability also ensures that the system is a bounded linear operator from $\ell^2(\mathbb{Z})$ to $\ell^2(\mathbb{Z})$. Thus, when we limit attention to BIBO stable systems, we are able to use the various results for bounded linear operators on a Hilbert space that were developed in Chapter 2.

Basic systems

We now discuss a few basic discrete-time systems. These include some basic building blocks that we will use frequently. Their properties are summarized in Table 3.3.

Shift The shift-by-1 operator, or *delay*, is defined as

$$y_n = x_{n-1}, \qquad n \in \mathbb{Z}, \qquad \text{or} \tag{3.39a}$$

$$y = \begin{bmatrix} \vdots \\ y_{-1} \\ \boxed{y_0} \\ y_1 \\ \vdots \end{bmatrix} = \begin{bmatrix} \vdots \\ x_{-2} \\ \boxed{x_{-1}} \\ x_0 \\ \vdots \end{bmatrix} = \begin{bmatrix} & \vdots & \vdots & \vdots & \\ \cdots & 0 & 0 & 0 & \cdots \\ \cdots & 1 & \boxed{0} & 0 & \cdots \\ \cdots & 0 & 1 & 0 & \cdots \\ & \vdots & \vdots & \vdots & \end{bmatrix} \begin{bmatrix} \vdots \\ x_{-1} \\ \boxed{x_0} \\ x_1 \\ \vdots \end{bmatrix}. \tag{3.39b}$$

It is an LSI operator, causal and BIBO-stable, but not memoryless; the matrix is Toeplitz, with a single nonzero off diagonal. A shift by k, $k > 0$, is obtained by applying the delay operator k times.

The *advance-by-1* operator, which maps x_n into x_{n+1}, is the inverse of the shift-by-1 operator (3.39),

$$y_n = x_{n+1}, \qquad n \in \mathbb{Z}, \qquad \text{or} \tag{3.40a}$$

$$y = \begin{bmatrix} \vdots \\ y_{-1} \\ \boxed{y_0} \\ y_1 \\ \vdots \end{bmatrix} = \begin{bmatrix} \vdots \\ x_0 \\ \boxed{x_1} \\ x_2 \\ \vdots \end{bmatrix} = \begin{bmatrix} & \vdots & \vdots & \vdots & \\ \cdots & 0 & 1 & 0 & \cdots \\ \cdots & 0 & \boxed{0} & 1 & \cdots \\ \cdots & 0 & 0 & 0 & \cdots \\ & \vdots & \vdots & \vdots & \end{bmatrix} \begin{bmatrix} \vdots \\ x_{-1} \\ \boxed{x_0} \\ x_1 \\ \vdots \end{bmatrix}. \tag{3.40b}$$

It is an LSI operator and BIBO-stable, but it is neither memoryless nor causal; the matrix is Toeplitz and upper-triangular with a single nonzero off diagonal. While it is obvious that the matrix in (3.40b) is the transpose of the one in (3.39b), it is also true that these matrices are inverses of each other. (Caution: Any finite-sized truncation of the matrix in (3.39b) or (3.40b), centered at the origin, is not invertible.)

Modulator Consider pointwise multiplication of a sequence x_n by $(-1)^n$,

$$y_n = (-1)^n x_n = \begin{cases} x_n, & \text{for even } n; \\ -x_n, & \text{for odd } n, \end{cases} \qquad n \in \mathbb{Z}, \qquad \text{or} \tag{3.41a}$$

$$\begin{bmatrix} \vdots \\ y_{-1} \\ \boxed{y_0} \\ y_1 \\ \vdots \end{bmatrix} = \begin{bmatrix} \vdots \\ -x_{-1} \\ \boxed{x_0} \\ -x_1 \\ \vdots \end{bmatrix} = \begin{bmatrix} & \vdots & \vdots & \vdots & \\ \cdots & -1 & 0 & 0 & \cdots \\ \cdots & 0 & \boxed{1} & 0 & \cdots \\ \cdots & 0 & 0 & -1 & \cdots \\ & \vdots & \vdots & \vdots & \end{bmatrix} \begin{bmatrix} \vdots \\ x_{-1} \\ \boxed{x_0} \\ x_1 \\ \vdots \end{bmatrix}. \tag{3.41b}$$

This is the simplest example of *modulation*, that is, a change of *frequency*[46] of a sequence. We use the term modulation to refer to multiplication of a sequence by a complex exponential or sinusoidal sequence; this is equivalent to *amplitude modulation* (AM) in communications. Here, modulation turns a constant sequence $x_n = 1$ into a fast-varying (high-frequency) sequence $y_n = (-1)^n$,

$$\begin{bmatrix} \cdots & 1 & 1 & \boxed{1} & 1 & 1 & \cdots \end{bmatrix}^\top \rightarrow \begin{bmatrix} \cdots & 1 & -1 & \boxed{1} & -1 & 1 & \cdots \end{bmatrix}^\top.$$

We will see in (3.90) that this operation is a shift of frequency from 0 to π. This operator is linear, causal, memoryless, and BIBO-stable, but not shift-invariant; the matrix is diagonal.

A more general version of (3.41a) would involve a sequence $\alpha \in \ell^\infty(\mathbb{Z})$ multiplying the input,

$$y_n = \alpha_n x_n, \qquad n \in \mathbb{Z}, \qquad \text{or} \tag{3.42a}$$

$$\begin{bmatrix} \vdots \\ y_{-1} \\ \boxed{y_0} \\ y_1 \\ \vdots \end{bmatrix} = \begin{bmatrix} \vdots \\ \alpha_{-1}x_{-1} \\ \boxed{\alpha_0 x_0} \\ \alpha_1 x_1 \\ \vdots \end{bmatrix} = \begin{bmatrix} & \vdots & \vdots & \vdots & \\ \cdots & \alpha_{-1} & 0 & 0 & \cdots \\ \cdots & 0 & \boxed{\alpha_0} & 0 & \cdots \\ \cdots & 0 & 0 & \alpha_1 & \cdots \\ & \vdots & \vdots & \vdots & \end{bmatrix} \begin{bmatrix} \vdots \\ x_{-1} \\ \boxed{x_0} \\ x_1 \\ \vdots \end{bmatrix}. \tag{3.42b}$$

Like (3.41), this operator is linear, causal, memoryless, and BIBO-stable, but not shift-invariant; the matrix is again diagonal.

[46] While we have not defined the notion of frequency yet, you may think of it as a rate of variation in a sequence; the more the sequence varies in a given interval, the higher the frequency.

Accumulator The output of the accumulator is akin to the integral of the input,

$$y_n = \sum_{k=-\infty}^{n} x_k, \qquad n \in \mathbb{Z}, \qquad \text{or} \tag{3.43a}$$

$$\begin{bmatrix} \vdots \\ y_{-1} \\ \boxed{y_0} \\ y_1 \\ \vdots \end{bmatrix} = \begin{bmatrix} & \vdots & \vdots & \vdots & \\ \cdots & 1 & 0 & 0 & \cdots \\ \cdots & 1 & \boxed{1} & 0 & \cdots \\ \cdots & 1 & 1 & 1 & \cdots \\ & \vdots & \vdots & \vdots & \end{bmatrix} \begin{bmatrix} \vdots \\ x_{-1} \\ \boxed{x_0} \\ x_1 \\ \vdots \end{bmatrix}. \tag{3.43b}$$

This is an LSI, causal operator, but it is neither memoryless nor BIBO-stable; the matrix is Toeplitz and lower-triangular.

If the input signal is restricted to be 0 for $n < 0$, (3.43) reduces to

$$y_n = \sum_{k=0}^{n} x_k, \qquad n \in \mathbb{N}, \qquad \text{or} \tag{3.44a}$$

$$\begin{bmatrix} \boxed{y_0} \\ y_1 \\ y_2 \\ \vdots \end{bmatrix} = \begin{bmatrix} \boxed{1} & 0 & 0 & \cdots \\ 1 & 1 & 0 & \cdots \\ 1 & 1 & 1 & \cdots \\ \vdots & \vdots & \vdots & \ddots \end{bmatrix} \begin{bmatrix} \boxed{x_0} \\ x_1 \\ x_2 \\ \vdots \end{bmatrix}. \tag{3.44b}$$

This is an LSI, causal operator, but it is neither memoryless nor BIBO-stable; the matrix is Toeplitz and lower-triangular.

Weighting by dividing (3.44a) by the number of terms involved turns the accumulator into a running average,

$$y_n = \frac{1}{n+1} \sum_{k=0}^{n} x_k, \qquad n \in \mathbb{N}, \qquad \text{or} \tag{3.45a}$$

$$\begin{bmatrix} \boxed{y_0} \\ y_1 \\ y_2 \\ \vdots \end{bmatrix} = \begin{bmatrix} \boxed{1} & 0 & 0 & \cdots \\ \frac{1}{2} & \frac{1}{2} & 0 & \cdots \\ \frac{1}{3} & \frac{1}{3} & \frac{1}{3} & \cdots \\ \vdots & \vdots & \vdots & \ddots \end{bmatrix} \begin{bmatrix} \boxed{x_0} \\ x_1 \\ x_2 \\ \vdots \end{bmatrix}. \tag{3.45b}$$

This is a linear operator that is also causal and BIBO-stable, but it is neither shift-invariant nor memoryless; the matrix is lower-triangular.

Other weight functions are possible, such as a decaying geometric weighting of the entries with factor $\alpha \in (0,1)$,

$$y_n = \sum_{k=0}^{n} \alpha^{n-k} x_k, \qquad n \in \mathbb{N}, \qquad \text{or} \tag{3.46a}$$

$$\begin{bmatrix} \boxed{y_0} \\ y_1 \\ y_2 \\ \vdots \end{bmatrix} = \begin{bmatrix} \boxed{1} & 0 & 0 & \cdots \\ \alpha & 1 & 0 & \cdots \\ \alpha^2 & \alpha & 1 & \cdots \\ \vdots & \vdots & \vdots & \ddots \end{bmatrix} \begin{bmatrix} \boxed{x_0} \\ x_1 \\ x_2 \\ \vdots \end{bmatrix}. \tag{3.46b}$$

This is an LSI, causal operator, but it is not memoryless; the matrix is Toeplitz and lower-triangular. It is BIBO-stable because $|\alpha| < 1$.

Averaging operators Consider a system that averages neighboring values, for example,

$$y_n = \frac{1}{3}(x_{n-1} + x_n + x_{n+1}), \qquad n \in \mathbb{Z}, \qquad \text{or} \tag{3.47a}$$

$$\begin{bmatrix} \vdots \\ y_{-1} \\ \boxed{y_0} \\ y_1 \\ \vdots \end{bmatrix} = \frac{1}{3} \begin{bmatrix} & \vdots & \vdots & \vdots & \vdots & \vdots & \\ \cdots & 1 & 1 & 1 & 0 & 0 & \cdots \\ \cdots & 0 & 1 & \boxed{1} & 1 & 0 & \cdots \\ \cdots & 0 & 0 & 1 & 1 & 1 & \cdots \\ & \vdots & \vdots & \vdots & \vdots & \vdots & \end{bmatrix} \begin{bmatrix} \vdots \\ x_{-2} \\ x_{-1} \\ \boxed{x_0} \\ x_1 \\ x_2 \\ \vdots \end{bmatrix}. \tag{3.47b}$$

As we have seen in Example 3.2, this is a moving-average filter with $N = 3$. It is called *moving-average* since we look at the sequence through a window of size 3, compute the average value, and then move the window to compute the next average. This operator is LSI and BIBO-stable, but it is neither memoryless nor causal; the matrix is Toeplitz.

For odd N, we obtain a causal version by simply delaying the moving-average in (3.6) by $(N-1)/2$ samples. For $N = 3$ as here, this results in

$$y_n = \frac{1}{3}(x_{n-2} + x_{n-1} + x_n), \tag{3.48a}$$

$$\begin{bmatrix} \vdots \\ y_{-1} \\ \boxed{y_0} \\ y_1 \\ \vdots \end{bmatrix} = \frac{1}{3} \begin{bmatrix} & \vdots & \vdots & \vdots & \vdots & \vdots & \\ \cdots & 1 & 1 & 0 & 0 & 0 & \cdots \\ \cdots & 1 & 1 & \boxed{1} & 0 & 0 & \cdots \\ \cdots & 0 & 1 & 1 & 1 & 0 & \cdots \\ & \vdots & \vdots & \vdots & \vdots & \vdots & \end{bmatrix} \begin{bmatrix} \vdots \\ x_{-2} \\ x_{-1} \\ \boxed{x_0} \\ x_1 \\ x_2 \\ \vdots \end{bmatrix}, \tag{3.48b}$$

which is a delayed-by-1 version of (3.47). This operator is again LSI and BIBO-stable but also causal, while still not being memoryless; the matrix is Toeplitz and lower-triangular.

An alternative is a *block average*,

$$y_n = \frac{1}{3}(x_{3n-1} + x_{3n} + x_{3n+1}), \tag{3.49a}$$

$$\begin{bmatrix} \vdots \\ y_{-1} \\ \boxed{y_0} \\ y_1 \\ \vdots \end{bmatrix} = \frac{1}{3} \begin{bmatrix} & \vdots & \vdots & \vdots & \vdots & \vdots & \vdots & \\ \cdots & 0 & 0 & 0 & 0 & 0 & 0 & \cdots \\ \cdots & 1 & \boxed{1} & 1 & 0 & 0 & 0 & \cdots \\ \cdots & 0 & 0 & 0 & 1 & 1 & 1 & \cdots \\ & \vdots & \vdots & \vdots & \vdots & \vdots & \vdots & \end{bmatrix} \begin{bmatrix} \vdots \\ x_{-1} \\ \boxed{x_0} \\ x_1 \\ x_2 \\ x_3 \\ x_4 \\ \vdots \end{bmatrix}. \tag{3.49b}$$

It is easy to see that (3.49a) is simply (3.47a) evaluated at multiples of 3. Similarly, the matrix in (3.49b) contains only every third row of the one in (3.47b). This is a linear and BIBO-stable operator; it is not shift-invariant, not memoryless, and not causal; the matrix is block diagonal.

A nonlinear version of the averaging operator could be

$$y_n = \text{median}\big(\big[x_{n-1} \quad x_n \quad x_{n+1}\big]\big). \tag{3.50}$$

Instead of the average of the three terms, this operator takes the median value. This operator is shift-invariant and BIBO-stable, but it is clearly not linear, not causal, and not memoryless.

Maximum operator This simple operator computes the maximum value of the input up to the current time,

$$y_n = \max\big(\big[\ldots \quad x_{n-2} \quad x_{n-1} \quad x_n\big]\big). \tag{3.51}$$

This operator is clearly neither linear nor memoryless, but it is causal, shift-invariant, and BIBO-stable.

3.3.2 Difference equations

An important class of discrete-time systems can be described by *linear difference equations* that relate the input sequence and past outputs to the current output,

$$y_n = \sum_{k \in \mathbb{Z}} b_k^{(n)} x_{n-k} - \sum_{k=1}^{\infty} a_k^{(n)} y_{n-k}. \tag{3.52}$$

If we require shift invariance, then the coefficients $a_k^{(n)}$ and $b_k^{(n)}$ are constant (do not depend on n), and we get a *linear, constant-coefficient difference equation*,

$$y_n = \sum_{k \in \mathbb{Z}} b_k x_{n-k} - \sum_{k=1}^{\infty} a_k y_{n-k}. \tag{3.53}$$

		Linear Def. 3.1	Shift inv. Def. 3.4	Causal Def. 3.3	Memoryless Def. 3.2	BIBO-stable Def. 3.5
Shift delay	(3.39)	✓	✓	✓	×	✓
advance	(3.40)	✓	✓	×	×	✓
Modulator	(3.41)	✓	×	✓	✓	✓
general	(3.42)	✓	×	✓	✓	✓
Accumulator	(3.43)	✓	✓	✓	×	×
restricted input	(3.44)	✓	✓	✓	×	×
weighted	(3.45)	✓	×	✓	×	✓
exp. weighted	(3.46)	✓	×	✓	×	✓ ($\|\alpha\| < 1$)
Averaging operator	(3.47)	✓	✓	×	×	✓
causal	(3.48)	✓	✓	✓	×	✓
block	(3.49)	✓	×	×	×	✓
median	(3.50)	×	✓	×	×	✓
Maximum operator	(3.51)	×	✓	✓	×	✓
Matrix representation		✓	Toeplitz	Lower-triangular	Diagonal	Rows absolutely summable

Table 3.3 Basic discrete-time systems and their properties. Matrix representation assumes linearity.

Such an equation does not determine whether a system is causal. However, (3.53) is suggestive of a recursive computation of the output, forward in time (increasing n); we will concentrate on such solutions. To make the system causal, we restrict the dependence on x to the current and past values, leading to

$$y_n = \sum_{k=0}^{\infty} b_k x_{n-k} - \sum_{k=1}^{\infty} a_k y_{n-k}. \tag{3.54}$$

Realizable systems will have only a finite number of nonzero coefficients a_k, $k \in \{1, 2, \ldots, N\}$, and b_k, $k \in \{0, 1, \ldots, M\}$, reducing (3.54) to

$$y_n = \sum_{k=0}^{M} b_k x_{n-k} - \sum_{k=1}^{N} a_k y_{n-k}, \tag{3.55}$$

as illustrated in Figure 3.5. We discuss finding solutions to such difference equations in Appendix 3.A.2.

Example 3.9 (Difference equation of the accumulator) As an example, consider the accumulator seen in (3.43a),

$$y_n = \sum_{k=-\infty}^{n} x_k = x_n + \sum_{k=-\infty}^{n-1} x_k = x_n + y_{n-1}, \tag{3.56}$$

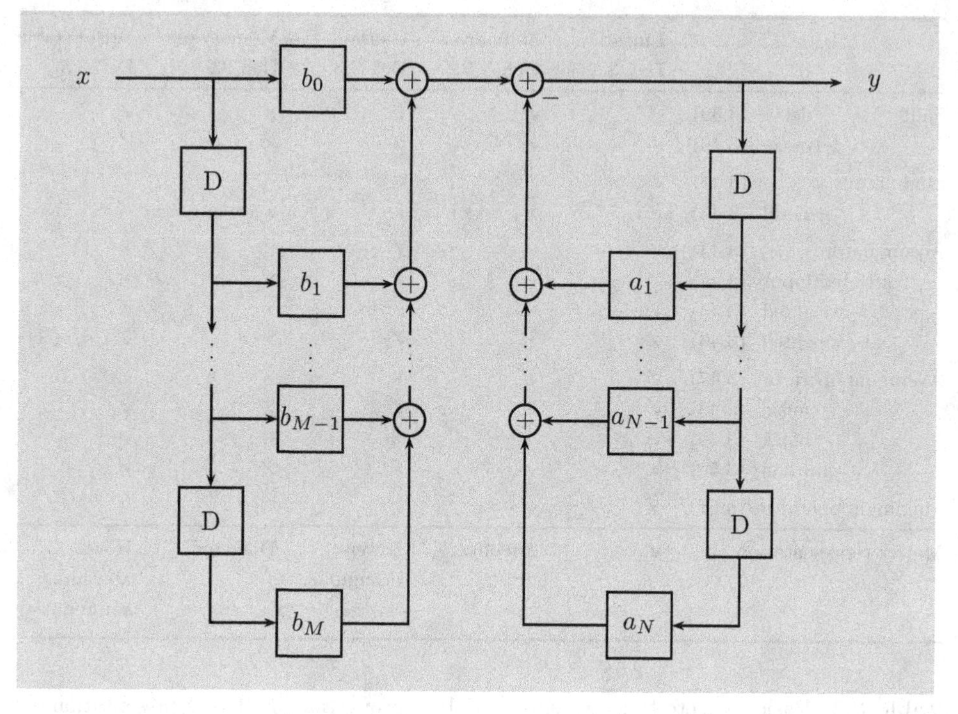

Figure 3.5 System representation of the difference equation (3.55), where D stands for *unit delay*, or shift-by-1 operator.

which is of the form (3.55), with $b_0 = 1$ and $a_1 = -1$. The infinite sum has been turned into a recursive formula (3.56), showing also how one could implement the accumulator: to obtain the current output y_n, add the current input x_n to the previously saved output y_{n-1}.

Let us take $x_n = \delta_n$, and see what the accumulator does. Assume that we are given $y_{-1} = \beta$. Then, for $n \geq 0$,

$$y_0 = x_0 + y_{-1} = 1 + \beta, \quad y_1 = x_1 + y_0 = 1 + \beta, \quad \ldots, \quad y_n = 1 + \beta, \quad \ldots.$$

Thus, the accumulator does exactly what it is supposed to do: at time $n = 0$, it adds the value of the input $x_0 = 1$ to the previously saved output $y_{-1} = \beta$, and then stays constant as the input for all $n > 0$ is zero. For $n < 0$, we can solve (3.56) by expressing $y_{n-1} = y_n - x_n$; it is easy to see that $y_n = \beta$, for all $n < 0$. Together, the expressions for $n \geq 0$ and $n < 0$, lead to

$$y_n = \beta + u_n, \tag{3.57}$$

that is, the initial value before the input is applied plus the input from the moment it is applied on.

From the above example, we see that, unless the initial conditions are zero, the system is not linear; for example, one could have a zero input producing a nonzero output. Similarly, the system is shift-invariant only if the initial conditions are zero. These properties are fundamental and hold beyond the case of the accumulator: difference equations as in (3.55) are linear and shift-invariant if and only if the initial conditions are zero. This also means that the homogeneous solution is necessarily zero (see Appendix 3.A.2 and Exercise 3.4).

3.3.3 Linear shift-invariant systems

Impulse response

A linear operator is specified by its outputs in response to each element of a basis for its domain space (see Section 2.5.5). As we saw in (3.34), this allows matrix representation of a linear discrete-time system. Such a matrix representation has columns that are the output sequences in response to the Kronecker delta sequence and its shifts as inputs. When, in addition, the system is shift-invariant, to satisfy (3.37) all these output sequences are themselves related by shifting. Thus, the system is specified completely by the output sequence resulting from the Kronecker delta sequence as the input.

DEFINITION 3.6 (IMPULSE RESPONSE) A sequence h is called the *impulse response* of LSI discrete-time system T when input δ produces output h.

The impulse response h of a causal linear system always satisfies $h_n = 0$ for all $n < 0$. This is required because, according to (3.36), the output in response to input δ must match on $\{-\infty, \ldots, -2, -1\}$ to the $\mathbf{0}$ output sequence that results from the $\mathbf{0}$ input sequence.

EXAMPLE 3.10 (IMPULSE RESPONSE FROM A DIFFERENCE EQUATION) The linear, constant-coefficient difference equation (3.54) with zero initial conditions represents an LSI system. An impulse response of the system is an output that results from Kronecker delta input, $x = \delta$. Thus, an impulse response satisfies

$$h_n \overset{(a)}{=} \sum_{k=0}^{\infty} b_k \delta_{n-k} - \sum_{k=1}^{\infty} a_k h_{n-k} \overset{(b)}{=} b_n - \sum_{k=1}^{\infty} a_k h_{n-k}, \tag{3.58}$$

where (a) follows from (3.54); and (b) from the sifting property of the Kronecker delta sequence (see Table 3.1).

If we restrict our attention to causal systems, then the difference equation uniquely specifies the system. The impulse response satisfies $h_n = 0$ for all $n < 0$, and h_n can be computed for all $n \geq 0$ by using (3.58) recursively for $n = 0, 1, \ldots$.

Convolution

The impulse response and its shifts form the columns of the matrix representation of an LSI system, as in (3.34). Expressing this as a summation is instructive and introduces the key concept of convolution.

Since an arbitrary input x to an LSI system T can be written as

$$x_n = \sum_{k\in\mathbb{Z}} x_k \delta_{n-k} \tag{3.59}$$

for all $n \in \mathbb{Z}$ (see the sifting property of the Kronecker delta sequence in Table 3.1), we can express the output as

$$y = Tx = T\sum_{k\in\mathbb{Z}} x_k\delta_{n-k} \overset{(a)}{=} \sum_{k\in\mathbb{Z}} x_k T\delta_{n-k} \overset{(b)}{=} \sum_{k\in\mathbb{Z}} x_k h_{n-k} = h * x, \tag{3.60}$$

where (a) follows from linearity; and (b) from shift invariance and the definition of the impulse response, defining the *convolution*.[47]

DEFINITION 3.7 (CONVOLUTION) The *convolution* between sequences h and x is defined as

$$(Hx)_n = (h * x)_n = \sum_{k\in\mathbb{Z}} x_k h_{n-k} = \sum_{k\in\mathbb{Z}} x_{n-k} h_k, \tag{3.61}$$

where H is called the *convolution operator* associated with h.

When it is not clear from the context, we will use a subscript on the convolution operator, such as $*_n$, to denote the argument over which we perform the convolution (for example, $x_{n-m} *_n h_{\ell-n} = \sum_k x_{k-m} h_{\ell-n+k}$).

EXAMPLE 3.11 (SOLUTION TO AN LSI DIFFERENCE EQUATION) Let us go back to the accumulator defined in (3.56). Either by using (3.56) directly or by using (3.58) with $b_0 = 1$ and $a_0 = -1$, we can determine the impulse response for this system by computing the output due to the input $x = \delta$,

$$h = \begin{bmatrix} \ldots & 0 & \boxed{1} & 1 & 1 & 1 & \ldots \end{bmatrix}^\top.$$

The convolution (3.61) expresses the output at time n due to an arbitrary input x as a linear combination of shifted versions of h. To illustrate this, suppose that x is supported on $\{0, 1 \ldots, L\}$. Then, (3.61) expresses the output y at all times

[47] Convolution is sometimes called linear convolution to distinguish it from the circular convolution of finite-length sequences as in Definition 3.9.

through a linear combination of h and its shifts by 1, 2, ..., L,

$$\begin{bmatrix} \vdots \\ y_{-1} \\ \boxed{y_0} \\ y_1 \\ y_2 \\ \vdots \\ y_L \\ y_{L+1} \\ \vdots \end{bmatrix} = x_0 \begin{bmatrix} \vdots \\ 0 \\ \boxed{1} \\ 1 \\ 1 \\ \vdots \\ 1 \\ 1 \\ \vdots \end{bmatrix} + x_1 \begin{bmatrix} \vdots \\ 0 \\ \boxed{0} \\ 1 \\ 1 \\ \vdots \\ 1 \\ 1 \\ \vdots \end{bmatrix} + x_2 \begin{bmatrix} \vdots \\ 0 \\ \boxed{0} \\ 0 \\ 1 \\ \vdots \\ 1 \\ 1 \\ \vdots \end{bmatrix} + \cdots + x_L \begin{bmatrix} \vdots \\ 0 \\ \boxed{0} \\ 0 \\ 0 \\ \vdots \\ 1 \\ 1 \\ \vdots \end{bmatrix} = \begin{bmatrix} \vdots \\ 0 \\ \boxed{x_0} \\ x_0 + x_1 \\ x_0 + x_1 + x_2 \\ \vdots \\ x_0 + \cdots + x_L \\ x_0 + \cdots + x_L \\ \vdots \end{bmatrix};$$

this is indeed the accumulator output for an input supported on $\{0, 1, \ldots, L\}$.

Properties The convolution (3.61) satisfies the following properties:

(i) *Connection to the inner product:*

$$(h * x)_n = \sum_{k \in \mathbb{Z}} x_k h_{n-k} = \langle x_k, h^*_{n-k} \rangle_k. \tag{3.62a}$$

(ii) *Commutativity:*

$$h * x = x * h. \tag{3.62b}$$

(iii) *Associativity:*

$$g * (h * x) = g * h * x = (g * h) * x. \tag{3.62c}$$

(iv) *Deterministic autocorrelation:*

$$a_n = \sum_{k \in \mathbb{Z}} x_k x^*_{k-n} = x_n *_n x^*_{-n}. \tag{3.62d}$$

(v) *Shifting:* For any $k \in \mathbb{Z}$,

$$x_n *_n \delta_{n-k} = x_{n-k}. \tag{3.62e}$$

Properties (i)–(iv) above depend on the sums – whether written explicitly or implicitly – converging. Convergence of the convolution is discussed in Appendix 3.A.3. The following example illustrates the apparent failure of the associative property when a convolution sum does not converge.

EXAMPLE 3.12 (WHEN CONVOLUTION IS NOT ASSOCIATIVE) Since a convolution might fail to converge, one needs to be careful about associativity. For g, choose the Heaviside sequence from (3.11), $g_n = u_n$, for h choose the first-order differencing sequence, $h_n = \delta_n - \delta_{n-1}$, and for x choose the constant sequence, $x_n = 1$. Now,

$$g * (h * x) \overset{(a)}{=} u * \mathbf{0} = \mathbf{0}, \quad \text{while} \quad (g * h) * x \overset{(b)}{=} \delta * 1 = 1,$$

where (a) follows because convolving a constant with the differencing operator yields a zero sequence; and (b) because convolving a Heaviside sequence with the differencing operator yields a Kronecker delta sequence. This failure of associativity occurs because $g * h * x$ is not well defined; for it to be well defined requires absolute convergence of

$$\sum_{m \in \mathbb{Z}} \sum_{k \in \mathbb{Z}} g_{n-m} h_{m-k} x_k$$

for every $n \in \mathbb{Z}$, which does not hold.

Filters The impulse response of a system is often called a *filter* and convolution with the impulse response is called *filtering*. Here are some basic classes of filters:

(i) *Causal filters* are such that $h_n = 0$ for all $n < 0$.
(ii) *Anticausal filters* are such that $h_n = 0$ for all $n > 0$.
(iii) *Two-sided filters* are neither causal nor anticausal.
(iv) *Finite impulse response (FIR) filters* have only a finite number of coefficients h_n different from zero.
(v) *Infinite impulse response (IIR) filters* have infinitely many nonzero terms.

For example, the impulse response of the accumulator in Example 3.11 is causal and IIR.

Stability We now discuss the stability of LSI systems.

THEOREM 3.8 (BIBO STABILITY) An LSI system is BIBO-stable if and only if its impulse response is absolutely summable.

Proof. To prove sufficiency (absolute summability implies BIBO stability), consider an absolutely summable impulse response $h \in \ell^1(\mathbb{Z})$, so $\|h\|_1 < \infty$, and a bounded input $x \in \ell^\infty(\mathbb{Z})$, so $\|x\|_\infty < \infty$. The absolute value of any one sample at the output can be bounded as follows:

$$|y_n| \overset{(a)}{=} \left| \sum_{k \in \mathbb{Z}} h_k x_{n-k} \right| \overset{(b)}{\le} \sum_{k \in \mathbb{Z}} |h_k| \, |x_{n-k}| \overset{(c)}{\le} \|x\|_\infty \sum_{k \in \mathbb{Z}} |h_k| \overset{(d)}{=} \|x\|_\infty \, \|h\|_1 < \infty,$$

where (a) follows from (3.61); (b) from the triangle inequality (Definition 2.9(iii)); (c) from bounding each $|x_{n-k}|$ by $\|x\|_\infty$; and (d) from the definition of the ℓ^1 norm. This proves that y is bounded.[48]

We prove necessity (BIBO stability implies absolute summability) by contradiction. For any h that is not absolutely summable we choose a particular input x (which depends on h) to create an unbounded output. Consider a real impulse response[49] h, and define the input sequence to be

$$x_n = \operatorname{sgn} h_{-n}, \qquad \text{where} \qquad \operatorname{sgn} t = \begin{cases} -1, & \text{for } t < 0; \\ 0, & \text{for } t = 0; \\ 1, & \text{for } t > 0 \end{cases}$$

is the sign function. Now, compute the convolution of x with h at $n = 0$,

$$y_0 = \sum_{k\in\mathbb{Z}} h_k x_{-k} = \sum_{k\in\mathbb{Z}} |h_k| = \|h\|_1, \tag{3.63}$$

which is unbounded when h is not in $\ell^1(\mathbb{Z})$.

The impulse response of the accumulator, for example, does not belong to $\ell^1(\mathbb{Z})$; a bounded input to the accumulator can lead to an unbounded output. Limiting attention to filters in $\ell^1(\mathbb{Z})$ avoids technical difficulties since it guarantees both the convergence of the convolution sum and that the resulting sequence is in a suitable sequence space. When $h \in \ell^1(\mathbb{Z})$ and $x \in \ell^p(\mathbb{Z})$ for any $p \in [1, \infty]$, the result of $h * x$ is in $\ell^p(\mathbb{Z})$ as well; see Solved exercise 3.2.

Matrix view As we have shown in Section 2.5.5, any linear operator can be expressed in matrix form. We may visualize (3.61) as

$$y = \begin{bmatrix} \vdots \\ y_{-2} \\ y_{-1} \\ \boxed{y_0} \\ y_1 \\ y_2 \\ \vdots \end{bmatrix} = \underbrace{\begin{bmatrix} & \vdots & \vdots & \vdots & \vdots & \vdots & \\ \cdots & h_0 & h_{-1} & h_{-2} & h_{-3} & h_{-4} & \cdots \\ \cdots & h_1 & h_0 & h_{-1} & h_{-2} & h_{-3} & \cdots \\ \cdots & h_2 & h_1 & \boxed{h_0} & h_{-1} & h_{-2} & \cdots \\ \cdots & h_3 & h_2 & h_1 & h_0 & h_{-1} & \cdots \\ \cdots & h_4 & h_3 & h_2 & h_1 & h_0 & \cdots \\ & \vdots & \vdots & \vdots & \vdots & \vdots & \end{bmatrix}}_{H} \begin{bmatrix} \vdots \\ x_{-2} \\ x_{-1} \\ \boxed{x_0} \\ x_1 \\ x_2 \\ \vdots \end{bmatrix} = Hx. \tag{3.64}$$

This again shows that the terms LSI discrete-time system, linear operator (on sequences), filter, and (doubly infinite) matrix are all synonyms. The key elements in (3.64) are the time reversal of the impulse response (in each row of the matrix, the impulse response goes from right to left), and the Toeplitz structure of the matrix (each row is a shifted version of the previous row, and the matrix is constant along diagonals; see (2.246)). In Figure 3.6, an example convolution is computed graphically, emphasizing time reversal.

[48] This boundedness is equivalent to the convergence of the convolution sum as discussed in Appendix 3.A.3.

[49] For a complex-valued impulse response, a slight modification, using $x_n = h^*_{-n}/|h_{-n}|$ for $|h_{-n}| \neq 0$, and $x_n = 0$ otherwise, leads to the same result.

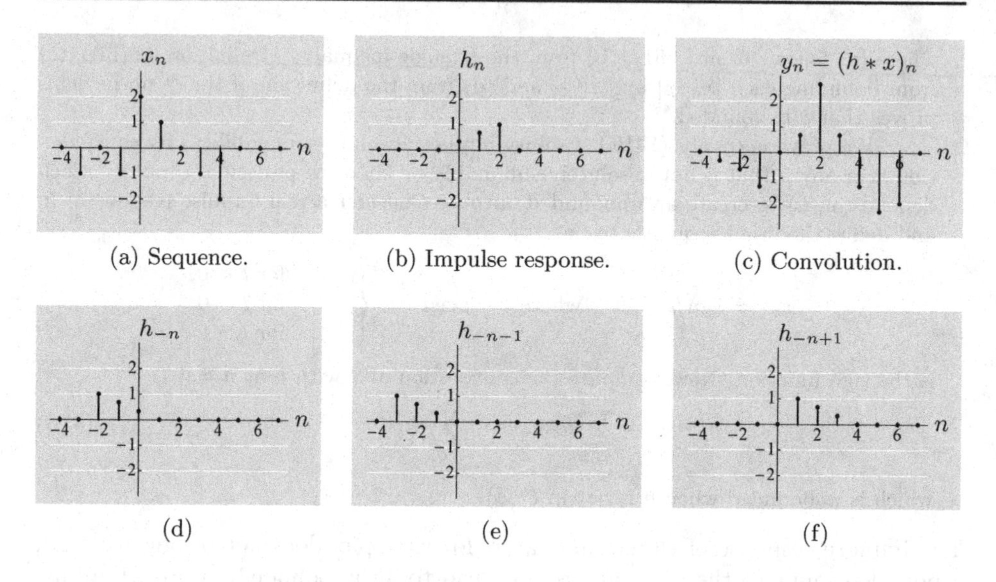

(a) Sequence. (b) Impulse response. (c) Convolution.

(d) (e) (f)

Figure 3.6 Example of the convolution of a sequence and a filter. (a) Sequence x. (b) Impulse response h. (c) Result of convolution $y = h * x$. (d) Time-reversed version of the impulse response, h_{-n}. (e)–(f) Two time-reversed and shifted versions of the impulse response involved in computing the convolution.

Adjoint The adjoint of the convolution operator H is the unique H^* satisfying (2.48):

$$\langle Hx, y\rangle = \langle x, H^*y\rangle. \tag{3.65}$$

From the matrix form of the convolution operator, (3.64), we can find the matrix form of the adjoint by Hermitian transposition. The result is

$$H^* = \begin{bmatrix} & \vdots & \vdots & \vdots & \vdots & \vdots & \\ \cdots & h_0^* & h_1^* & h_2^* & h_3^* & h_4^* & \cdots \\ \cdots & h_{-1}^* & h_0^* & h_1^* & h_2^* & h_3^* & \cdots \\ \cdots & h_{-2}^* & h_{-1}^* & \boxed{h_0^*} & h_1^* & h_2^* & \cdots \\ \cdots & h_{-3}^* & h_{-2}^* & h_{-1}^* & h_0^* & h_1^* & \cdots \\ \cdots & h_{-4}^* & h_{-3}^* & h_{-2}^* & h_{-1}^* & h_0^* & \cdots \\ & \vdots & \vdots & \vdots & \vdots & \vdots & \end{bmatrix}, \tag{3.66}$$

which is the convolution operator associated with the time-reversed and conjugated version of h. It is instructive to also verify this algebraically,

$$\begin{aligned} \langle x, H^*y\rangle &\overset{(a)}{=} \langle Hx, y\rangle = \sum_{n\in\mathbb{Z}} (h*x)_n y_n^* = \sum_{n\in\mathbb{Z}} \left(\sum_{k\in\mathbb{Z}} x_k h_{n-k}\right) y_n^* \\ &= \sum_{k\in\mathbb{Z}} x_k \sum_{n\in\mathbb{Z}} h_{n-k} y_n^* = \sum_{k\in\mathbb{Z}} x_k \left(\sum_{n\in\mathbb{Z}} h_{n-k}^* y_n\right)^*, \end{aligned} \tag{3.67}$$

where (a) follows from (2.48). This implies that

$$H^* y = \sum_{n\in\mathbb{Z}} h_{n-k}^* y_n \qquad \text{for all } y \in \ell^2(\mathbb{Z}),$$

so the adjoint operator is the convolution with the time-reversed and conjugated version of h.

Circular convolution

We now consider what happens with our second class of sequences, namely those that are of finite length and circularly extended.

Linear convolution with circularly extended signal Given a sequence x with circular extension as in (3.26) and a filter h in $\ell^1(\mathbb{Z})$, we can compute the convolution as usual,

$$y_n = (h * x)_n = \sum_{k\in\mathbb{Z}} x_k h_{n-k} = \sum_{k\in\mathbb{Z}} h_k x_{n-k}. \tag{3.68}$$

Since x is N-periodic, y is N-periodic as well,

$$y_{n+N} = \sum_{k\in\mathbb{Z}} h_k x_{n+N-k} \overset{(a)}{=} \sum_{k\in\mathbb{Z}} h_k x_{n-k} = y_n,$$

where (a) follows from the periodicity of x.

Let us now define a periodized version of h, with period N, as

$$h_{N,n} = \sum_{k\in\mathbb{Z}} h_{n-kN}, \tag{3.69}$$

where the sum converges for each n because $h \in \ell^1(\mathbb{Z})$. While this periodization is not equivalent to circular extension in general, it is equivalent when h is an FIR filter with support in $\{0, 1, \ldots, N-1\}$. We now want to show how we can express the convolution (3.68) in terms of what we will define as a *circular* convolution in Definition 3.9,

$$\begin{aligned}
(h * x)_n &= \sum_{k\in\mathbb{Z}} h_k x_{n-k} \overset{(a)}{=} \sum_{\ell\in\mathbb{Z}} \sum_{k=\ell N}^{(\ell+1)N-1} h_k x_{n-k} \\
&\overset{(b)}{=} \sum_{\ell\in\mathbb{Z}} \sum_{k'=0}^{N-1} h_{k'+\ell N} x_{n-k'-\ell N} \overset{(c)}{=} \sum_{\ell\in\mathbb{Z}} \sum_{k=0}^{N-1} h_{k+\ell N} x_{n-k} \\
&\overset{(d)}{=} \sum_{k=0}^{N-1} \underbrace{\sum_{\ell\in\mathbb{Z}} h_{k+\ell N}}_{=h_{N,k}} x_{n-k} = \sum_{k=0}^{N-1} h_{N,k} x_{n-k} \\
&\overset{(e)}{=} \sum_{k=0}^{N-1} h_{N,k} x_{(n-k) \bmod N} = (h_N \circledast x)_n,
\end{aligned} \tag{3.70}$$

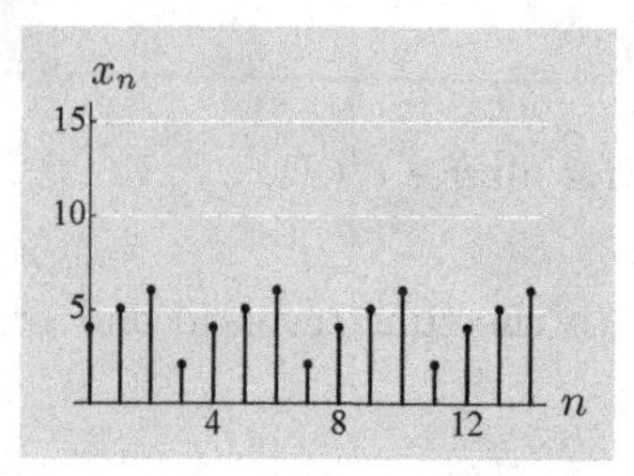

(a) Finite-length x, and circularly extended.

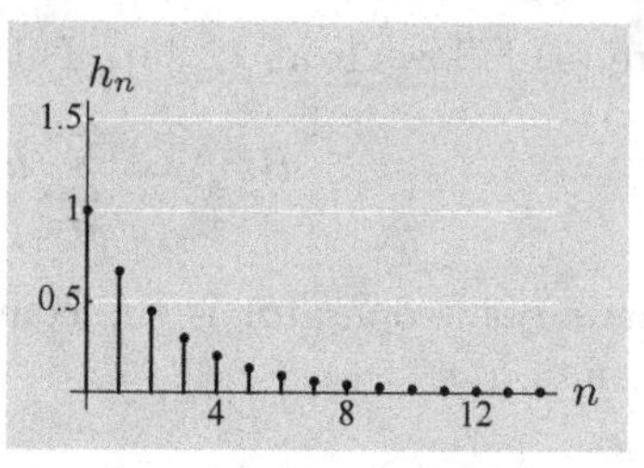

(b) Impulse response h.

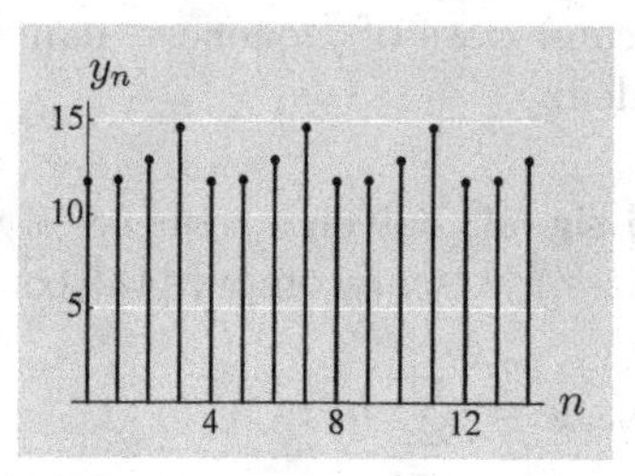

(c) Result of convolution $y = h * x$.

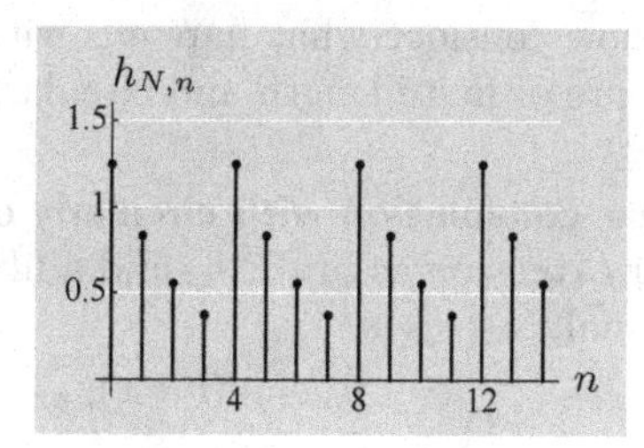

(d) Equivalent, periodized filter h_N.

Figure 3.7 Example of the convolution, $y = h * x$, of a periodic sequence x with period $N = 4$ and a filter h. Circular convolution of x and periodized h_N, $y = h_N \circledast x$, leads to the same output y as in (c).

where in (a) we split the set of integers into length-N segments; (b) follows from the change of variable $k' = k - \ell N$; (c) follows from the periodicity of x and the change of variable $k = k'$; in (d) we were allowed to exchange order of summation because $h \in \ell^1(\mathbb{Z})$ (see Appendix 2.A.3); and (e) follows from the periodicity of x. The expression above tends to be more convenient as it involves only one period both of x and of the periodized version h_N of the impulse response h. The equivalence between the convolution of h and a circularly extended finite-length sequence x, and the circular convolution of h_N and the sequence x is illustrated in Figure 3.7.

Example 3.13 (Impulse response and its periodized version) Let

$$h_n = \left(\frac{2}{3}\right)^n u_n, \qquad n \in \mathbb{Z},$$

as shown in Figure 3.7(b). We compute its periodized version as in (3.69),

$$h_{N,n} = \sum_{k\in\mathbb{Z}} h_{n-kN} = \sum_{k\in\mathbb{Z}} \left(\frac{2}{3}\right)^{n-kN} u_{n-kN}.$$

For $n \in \{0, 1, \ldots, N-1\}$,

$$h_{N,n} = \sum_{k=-\infty}^{0} \left(\frac{2}{3}\right)^{n-kN} \stackrel{(a)}{=} \sum_{m=0}^{\infty} \left(\frac{2}{3}\right)^{n+mN}$$
$$= \left(\frac{2}{3}\right)^{n} \sum_{m=0}^{\infty} \left(\left(\frac{2}{3}\right)^{N}\right)^{m} = \left(\frac{2}{3}\right)^{n} \frac{1}{1-(\frac{2}{3})^{N}},$$

where (a) follows from the change of variable $m = -k$. For $N = 4$, this becomes

$$h_{4,n} = \left(\frac{2}{3}\right)^{n} \frac{1}{1-(\frac{2}{3})^{4}} = \left(\frac{2}{3}\right)^{n} \frac{1}{1-16/81} = \left(\frac{2}{3}\right)^{n} \frac{81}{65},$$

as shown in Figure 3.7(d).

Definition of the circular convolution Above in (3.70), we implicitly defined a new form of convolution of a length-N input sequence x and a length-N impulse response h.

DEFINITION 3.9 (CIRCULAR CONVOLUTION) The *circular convolution* between length-N sequences h and x is defined as

$$(Hx)_n = (h \circledast x)_n = \sum_{k=0}^{N-1} x_k h_{(n-k) \bmod N} = \sum_{k=0}^{N-1} x_{(n-k) \bmod N} h_k, \quad (3.71)$$

where H is called the *circular convolution operator* associated with h.

The result of the circular convolution is a length-N sequence. While circular convolution is a separate concept from linear convolution, we have just seen that the two are related when one sequence in a linear convolution is periodic and the other is not. We made the connection by periodizing the aperiodic sequence.

Equivalence of circular and linear convolutions We have just seen that linear and circular convolutions are related; we now see that there are instances when the two are equivalent. Assume that we have a length-M input x and a length-L impulse response h,

$$x = \begin{bmatrix} \cdots & 0 & \boxed{x_0} & x_1 & \cdots & x_{M-1} & 0 & \cdots \end{bmatrix}^{\top}, \quad (3.72a)$$

$$h = \begin{bmatrix} \cdots & 0 & \boxed{h_0} & h_1 & \cdots & h_{L-1} & 0 & \cdots \end{bmatrix}^{\top}. \quad (3.72b)$$

The result of the linear convolution (3.61) has at most $L+M-1$ nonzero samples,

$$y = \begin{bmatrix} \cdots & 0 & \boxed{y_0} & y_1 & \cdots & y_{L+M-2} & 0 & \cdots \end{bmatrix}^{\top}.$$

as can be verified by using the length restrictions in (3.72). While we have chosen to write the sequences as infinite-length vectors, we could have chosen to write each as a finite-length vector with appropriate length; however, as these lengths are all different, we would have had to choose a common vector length N. Choosing this common length is exactly the crucial point in determining when the linear and circular convolutions are equivalent, as we show next.

THEOREM 3.10 (EQUIVALENCE OF CIRCULAR AND LINEAR CONVOLUTIONS) Linear and circular convolutions between a length-M sequence x and a length-L sequence h are equivalent when the period of the circular convolution N satisfies

$$N \geq L + M - 1. \tag{3.73}$$

Proof. Take x and h as in (3.72). The linear and circular convolutions, $y^{(\mathrm{lin})}$ and $y^{(\mathrm{circ})}$, are given by (3.61) and (3.71), respectively:

$$y_n^{(\mathrm{lin})} = (h_0 x_n + \cdots + h_n x_0) + (h_{n+1} x_{-1} + \cdots + h_{L-1} x_{-L+1+n}), \tag{3.74a}$$

$$y_n^{(\mathrm{circ})} = (h_0 x_n + \cdots + h_n x_0) + (h_{n+1} x_{N-1} + \cdots + h_{L-1} x_{N-L+1+n}), \tag{3.74b}$$

for $n \in \{0, 1, \ldots, N-1\}$. In the above, we broke each convolution sum into positive indices of x_n and the rest (negative ones for the linear convolution, and mod N for the circular convolution). Note that in (3.74b) the index goes from 0 to $(N-1)$, but stops at $(L-1)$ since h is zero after that.

Since x has no nonzero values for negative values of n, the second sum in (3.74a) is zero, and so must the second sum in (3.74b) be (and that for every $n = 0, 1, \ldots, N-1$), if (3.74a) and (3.74b) are to be equal. This, in turn, is possible only if $x_{N-L+1+n}$ (the last x term in the second sum of the circular convolution) has an index that is outside of the range of nonzero values of x, that is, if $N - L + 1 + n \geq M$, for every $n = 0, 1, \ldots, N-1$. As this is true for $n = 0$ by assumption (3.73), it will be true for all larger n as well.

Figure 3.8 depicts this equivalence and Example 3.14 examines it in matrix notation for $M = 4$, $L = 3$.

Matrix view As we have done for linear convolution in (3.64), we visualize circular convolution (3.71) using matrices,

$$y = \begin{bmatrix} y_0 \\ y_1 \\ y_2 \\ \vdots \\ y_{N-1} \end{bmatrix} = \underbrace{\begin{bmatrix} h_0 & h_{N-1} & h_{N-2} & \cdots & h_1 \\ h_1 & h_0 & h_{N-1} & \cdots & h_2 \\ h_2 & h_1 & h_0 & \cdots & h_3 \\ \vdots & \vdots & \vdots & \ddots & \vdots \\ h_{N-1} & h_{N-2} & h_{N-3} & \cdots & h_0 \end{bmatrix}}_{H} \begin{bmatrix} x_0 \\ x_1 \\ x_2 \\ \vdots \\ x_{N-1} \end{bmatrix} = Hx. \tag{3.75}$$

H is a circulant matrix as in (2.245) with h as its first column, and it represents the circular convolution operator when both the sequence x and the impulse response

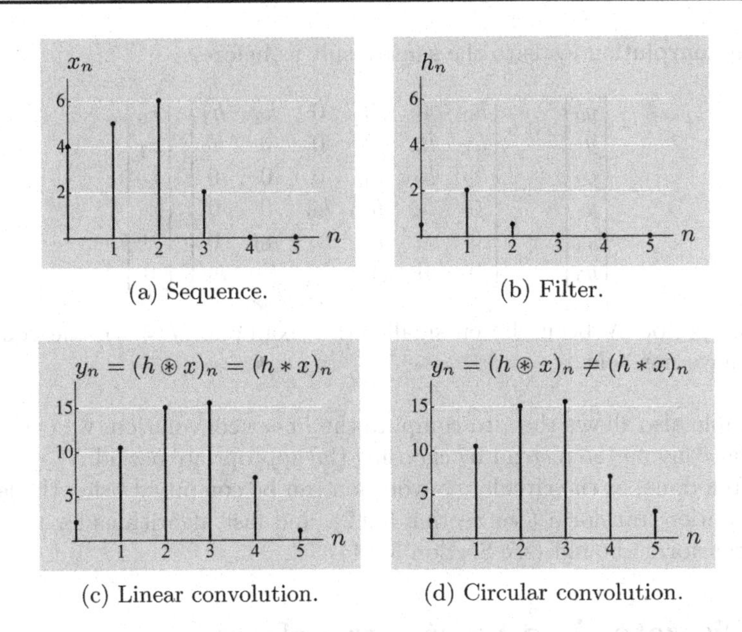

(a) Sequence. (b) Filter.

(c) Linear convolution. (d) Circular convolution.

Figure 3.8 Equivalence of circular and linear convolutions. (a) Sequence x of length $M = 4$. (b) Filter h of length $L = 3$. (c) Linear convolution results in a sequence of length $L + M - 1 = 6$, the same as a circular convolution with a period $N \geq L + M - 1$, $N = 6$ in this case. (d) Circular convolution with a smaller period, $N = 5$, does not lead to the same result.

h are finite; when the impulse response is not finite, the elements of H would be samples of the periodized impulse response h_N.

EXAMPLE 3.14 (EQUIVALENCE OF CIRCULAR AND LINEAR CONVOLUTIONS) We now look at a length-3 filter convolved with a length-4 sequence. The result of the linear convolution is of length 6,

$$\begin{bmatrix} \vdots \\ 0 \\ \boxed{y_0} \\ y_1 \\ y_2 \\ y_3 \\ y_4 \\ y_5 \\ 0 \\ \vdots \end{bmatrix} = \begin{bmatrix} & \vdots & \vdots & \vdots & \vdots & \vdots & \vdots & \vdots & \vdots & \\ \cdots & h_0 & 0 & 0 & 0 & 0 & 0 & 0 & 0 & \cdots \\ \cdots & h_1 & \boxed{h_0} & 0 & 0 & 0 & 0 & 0 & 0 & \cdots \\ \cdots & h_2 & h_1 & h_0 & 0 & 0 & 0 & 0 & 0 & \cdots \\ \cdots & 0 & h_2 & h_1 & h_0 & 0 & 0 & 0 & 0 & \cdots \\ \cdots & 0 & 0 & h_2 & h_1 & h_0 & 0 & 0 & 0 & \cdots \\ \cdots & 0 & 0 & 0 & h_2 & h_1 & h_0 & 0 & 0 & \cdots \\ \cdots & 0 & 0 & 0 & 0 & h_2 & h_1 & h_0 & 0 & \cdots \\ \cdots & 0 & 0 & 0 & 0 & 0 & h_2 & h_1 & h_0 & \cdots \\ & \vdots & \vdots & \vdots & \vdots & \vdots & \vdots & \vdots & \vdots & \end{bmatrix} \begin{bmatrix} \vdots \\ 0 \\ \boxed{x_0} \\ x_1 \\ x_2 \\ x_3 \\ 0 \\ 0 \\ 0 \\ \vdots \end{bmatrix}. \quad (3.76a)$$

To calculate circular convolution, we choose $N = M + L - 1 = 6$, and form a 6×6 circulant matrix H as in (3.75) by using h as its first column. Then, the circular convolution leads to the same result as before,

$$\begin{bmatrix} y_0 \\ y_1 \\ y_2 \\ y_3 \\ y_4 \\ y_5 \end{bmatrix} = \begin{bmatrix} h_0 & 0 & 0 & 0 & h_2 & h_1 \\ h_1 & h_0 & 0 & 0 & 0 & h_2 \\ h_2 & h_1 & h_0 & 0 & 0 & 0 \\ 0 & h_2 & h_1 & h_0 & 0 & 0 \\ 0 & 0 & h_2 & h_1 & h_0 & 0 \\ 0 & 0 & 0 & h_2 & h_1 & h_0 \end{bmatrix} \begin{bmatrix} x_0 \\ x_1 \\ x_2 \\ x_3 \\ 0 \\ 0 \end{bmatrix}. \tag{3.76b}$$

Had the period N been chosen smaller (for example, $N = 5$), the equivalence would have not held.

This example also shows that, to compute the linear convolution, we can compute the circular convolution instead by choosing the appropriate period $N \geq M + L - 1$. This is often done, as the circular convolution can be computed using the length-N discrete Fourier transform (see Section 3.9.2), and fast algorithms for the discrete Fourier transform abound (see Section 3.9.1).

3.4 Discrete-time Fourier transform

In the present section and the next two sections, we introduce various ways to analyze sequences and discrete-time systems. They range from the analytical to the computational and are all variations of the Fourier transform. Why do Fourier methods play such a prominent role? Simply because they are based on eigensequences of LSI systems (convolution operators). Thus far, we have seen two convolution operators (linear and circular). We will see that these have different sets of eigensequences, which lead to different Fourier transforms for sequences. The eigensequence property leads to the *convolution property* – an equivalence between convolving sequences and multiplying Fourier transforms of the sequences. This is also interpreted as *diagonalization of convolution operators* by the Fourier transform.

In this section, we introduce the *discrete-time Fourier transform (DTFT)* – the Fourier transform for infinite-length discrete-time signals. It is a 2π-periodic function of frequency $\omega \in \mathbb{R}$ that we write as $X(e^{j\omega})$, both to stress the periodicity and to create a unified notation for the DTFT and the *z-transform* $X(z)$, which we discuss in Section 3.5. The z-transform has argument $z \in \mathbb{C}$, where z can have any modulus; the DTFT is related to the z-transform by restricting the domain to the unit circle. In Section 3.6, we focus on the *discrete Fourier transform (DFT)* – the Fourier transform both for infinite-length periodic sequences and for circularly extended length-N sequences (both of these can be viewed as existing on a discrete circle of length N). The DFT is an N-dimensional vector we write as X_k.

3.4.1 Definition of the DTFT

Eigensequences of the convolution operator We start with a fundamental property of LSI systems: they have all unit-modulus complex exponential sequences as eigensequences. This follows from the convolution representation of LSI systems (3.61) and a simple computation.

Consider a complex exponential sequence

$$v_n = e^{j\omega n}, \qquad n \in \mathbb{Z}, \tag{3.77}$$

where ω is any real number. The quantity ω is called the *angular frequency*; it is measured in radians per second. With $\omega = 2\pi f$, the quantity f is called *frequency*; it is is measured in hertz, or the number of cycles per second. The sequence v is bounded since $|v_n| = 1$ for all $n \in \mathbb{Z}$. If the impulse response h is in $\ell^1(\mathbb{Z})$, then, according to Theorem 3.8, the output $h * v$ is bounded as well. Along with being bounded, $h * v$ takes a particular form:

$$\begin{aligned}(Hv)_n &= (h * v)_n = \sum_{k\in\mathbb{Z}} v_{n-k} h_k = \sum_{k\in\mathbb{Z}} e^{j\omega(n-k)} h_k \\ &= \underbrace{\sum_{k\in\mathbb{Z}} h_k e^{-j\omega k}}_{\lambda_\omega} \underbrace{e^{j\omega n}}_{v_n}.\end{aligned} \tag{3.78}$$

This shows that applying the convolution operator H to the complex exponential sequence v gives a scalar multiple of v; in other words, v is an eigensequence of H with the corresponding eigenvalue λ_ω. We denote this eigenvalue by $H(e^{j\omega})$ using the *frequency response* of the system, which is defined formally in (3.110a) in the discussion of filters. We can thus rewrite (3.78) as

$$Hv = h * v = H(e^{j\omega})\, v. \tag{3.79}$$

DTFT Finding the appropriate Fourier transform now amounts to projecting onto the subspaces generated by each of the eigensequences.

DEFINITION 3.11 (DISCRETE-TIME FOURIER TRANSFORM) The *discrete-time Fourier transform* of a sequence x is

$$X(e^{j\omega}) = \sum_{n\in\mathbb{Z}} x_n e^{-j\omega n}, \qquad \omega \in \mathbb{R}. \tag{3.80a}$$

It exists when (3.80a) converges for all $\omega \in \mathbb{R}$; we then call it the *spectrum* of x. The *inverse DTFT* of a 2π-periodic function $X(e^{j\omega})$ is

$$x_n = \frac{1}{2\pi}\int_{-\pi}^{\pi} X(e^{j\omega}) e^{j\omega n}\, d\omega, \qquad n \in \mathbb{Z}. \tag{3.80b}$$

When the DTFT exists, we denote the DTFT pair as

$$x_n \overset{\text{DTFT}}{\longleftrightarrow} X(e^{j\omega}).$$

Since $e^{-j\omega n}$ is a 2π-periodic function of ω for every $n \in \mathbb{Z}$, the DTFT is always a 2π-periodic function, which is emphasized by the notation $X(e^{j\omega})$. Note that the sum in (3.80a) is formally equivalent to an $\ell^2(\mathbb{Z})$ inner product, although the sequence $e^{j\omega n}$ has no decay and is thus not in $\ell^2(\mathbb{Z})$. We now discuss limitations on the inputs and the corresponding types of convergence.

3.4.2 Existence and convergence of the DTFT

The existence of the DTFT depends on the sequence x. When a doubly infinite series as in (3.80a) is given without a specification of how to interpret it as a limiting process, one must consider the series well defined only when it converges absolutely (see Appendix 2.A.2). This immediately implies the existence of the DTFT for all sequences in $\ell^1(\mathbb{Z})$. To extend the discussion beyond $\ell^1(\mathbb{Z})$, we must specify a limiting process and a sense of convergence. Consider the partial sums

$$X_N(e^{j\omega}) = \sum_{n=-N}^{N} x_n e^{-j\omega n}, \qquad N = 0, 1, \ldots. \tag{3.81}$$

If we consider the DTFT to exist whenever this sequence of partial sums converges under the $\mathcal{L}^2([-\pi, \pi))$ norm, then the DTFT exists for all sequences in $\ell^2(\mathbb{Z})$. Note that the DTFT can be a useful tool even when (3.80a) diverges to ∞ for some values of ω; this, however, requires more caution.

Sequences in $\ell^1(\mathbb{Z})$ If $x \in \ell^1(\mathbb{Z})$, then (3.80a) converges absolutely for every ω, since

$$\sum_{n\in\mathbb{Z}} \left|x_n e^{j\omega n}\right| = \sum_{n\in\mathbb{Z}} |x_n| \left|e^{j\omega n}\right| = \sum_{n\in\mathbb{Z}} |x_n| = \|x\|_1 < \infty.$$

This tells us that the DTFT of x exists. Moreover, as a consequence of absolute convergence for all ω, the limit $X(e^{j\omega})$ is a continuous function of ω.[50]

Since the DTFT itself is well defined, we can verify the inversion formula by substituting (3.80a) into (3.80b). First,

$$\frac{1}{2\pi}\int_{-\pi}^{\pi}\left(\sum_{k\in\mathbb{Z}} x_k e^{-j\omega k}\right) e^{j\omega n}\, d\omega \overset{(a)}{=} \sum_{k\in\mathbb{Z}} x_k \frac{1}{2\pi}\int_{-\pi}^{\pi} e^{j\omega(n-k)}\, d\omega, \tag{3.82a}$$

where in (a) we are allowed to exchange the order of summation and integration because $x \in \ell^1(\mathbb{Z})$ (see Section 2.A.3). The integral $\int_{-\pi}^{\pi} e^{j\omega(n-k)}\, d\omega$ must be treated separately for $n = k$ and $n \neq k$. Each case gives an elementary computation, and the result is

$$\int_{-\pi}^{\pi} e^{j\omega(n-k)}\, d\omega = 2\pi\delta_{n-k} \tag{3.82b}$$

[50] Absolute convergence of (3.80a) implies uniform convergence of the sequence of functions X_N in (3.81) to X. Looking at the DTFT as a function defined on a compact (closed and bounded) domain such as $[-\pi, \pi]$, the uniform convergence and the continuity of each X_N imply that X is continuous.

using the Kronecker delta sequence to combine the cases. We can then rewrite (3.82a) as

$$\frac{1}{2\pi}\int_{-\pi}^{\pi}\left(\sum_{k\in\mathbb{Z}} x_k e^{-j\omega k}\right) e^{j\omega n}\, d\omega \overset{(a)}{=} \sum_{k\in\mathbb{Z}} x_k \delta_{n-k} \overset{(b)}{=} x_n,$$

where (a) follows from (3.82b); and (b) from the definition of the Kronecker delta sequence (3.8), proving the inversion.

Sequences in $\ell^2(\mathbb{Z})$ For sequences not in $\ell^1(\mathbb{Z})$, the DTFT series (3.80a) might fail to converge for some values of ω. Nevertheless, convergence can be extended to the larger space of sequences $\ell^2(\mathbb{Z})$ by changing the sense of convergence.

If $x \in \ell^2(\mathbb{Z})$, the partial sum $X_N(e^{j\omega})$ in (3.81) converges to a function $X(e^{j\omega}) \in \mathcal{L}^2([-\pi, \pi))$ in the sense that

$$\lim_{N\to\infty} \|X(e^{j\omega}) - X_N(e^{j\omega})\| = 0. \tag{3.83}$$

This convergence in $\mathcal{L}^2([-\pi, \pi))$ norm[51] implies convergence of (3.80a) for almost all values of ω, but there is no guarantee of the convergence being uniform or the limit function $X(e^{j\omega})$ being continuous.

The sense in which the inversion formula holds changes subtly as well. We return to this in Section 4.5.2.

Example 3.15 (Mean-square convergence of DTFT) Take the sinc sequence from Figure 3.1(b),

$$x_n = \frac{1}{\sqrt{2}}\operatorname{sinc}\left(\frac{1}{2}\pi n\right) = \frac{1}{\sqrt{2}}\frac{\sin(\frac{1}{2}\pi n)}{\frac{1}{2}\pi n}. \tag{3.84}$$

It decays too slowly to be absolutely summable but fast enough to be square summable; that is, $x \in \ell^2(\mathbb{Z})$ but $x \notin \ell^1(\mathbb{Z})$. Thus, we cannot guarantee that (3.80a) converges for every ω, but the DTFT still converges in mean square. To see this, in Figure 3.9 we plot the DTFT partial sum (3.81),

$$X_N(e^{j\omega}) = \frac{1}{\sqrt{2}}\sum_{n=-N}^{N}\frac{\sin(\frac{1}{2}\pi n)}{\frac{1}{2}\pi n}e^{-j\omega n}, \tag{3.85}$$

for various values of N. As Figure 3.9 suggests, convergence in mean square is to

$$X(e^{j\omega}) = \begin{cases} \sqrt{2}, & \text{for } |\omega| \in [0, \frac{1}{2}\pi); \\ 0, & \text{for } |\omega| \in (\frac{1}{2}\pi, \pi], \end{cases}$$

and the convergence as $N \to \infty$ is nonuniform: it is very slow near $\omega = \frac{1}{2}\pi$ and faster farther away. In fact, while there is no convergence at $\omega = \frac{1}{2}\pi$, lack of convergence at this isolated point does not prevent convergence in mean square.

[51] This is also called *convergence in the mean-square sense* or *convergence in mean square*.

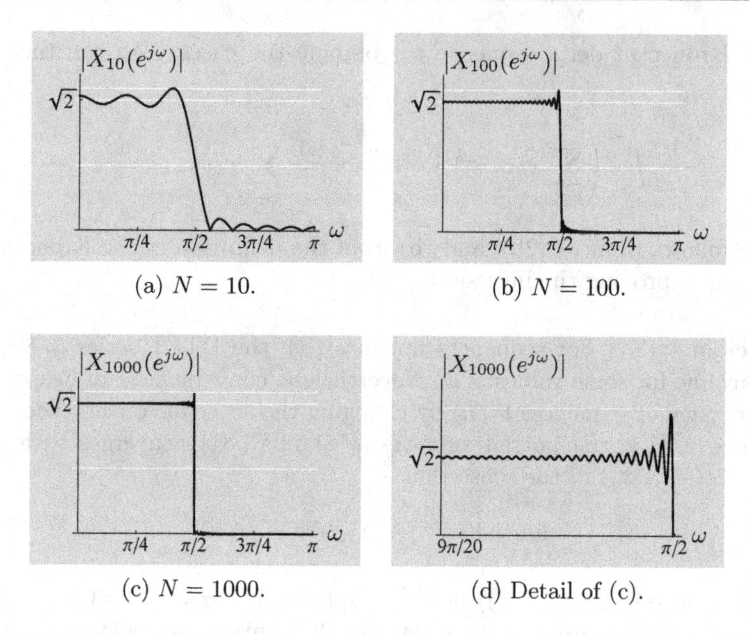

(a) $N = 10$. (b) $N = 100$.

(c) $N = 1000$. (d) Detail of (c).

Figure 3.9 Truncated DTFT of the sinc sequence, illustrating the Gibbs phenomenon. Shown are $|X_N(e^{j\omega})|$ from (3.85) with different N. Observe how oscillations narrow from (a) to (c), but their amplitude remains constant (the topmost grid line in every plot), at $1.089\sqrt{2}$.

The partial sum $X_N(e^{j\omega})$ oscillates near the points of discontinuity, with oscillations becoming narrower as N increases but not decreasing in size. This overshoot and undershoot is of the order of 9%, and is called the *Gibbs phenomenon* (see also Figure 1.3).[52]

Using the DTFT without convergence The DTFT is still a useful tool even in some cases where it converges neither pointwise over ω nor in mean square. These are cases where an expression for the DTFT involving a Dirac delta function makes sense because evaluating the inverse DTFT gives the desired result. As with other uses of the Dirac delta function, we must be cautious; for more details and properties of the Dirac delta function, see Appendix 3.A.4.

Example 3.16 (DTFT of constant sequence) Let $x_n = 1$ for all $n \in \mathbb{Z}$. This sequence belongs to neither $\ell^1(\mathbb{Z})$ nor $\ell^2(\mathbb{Z})$, so neither of our previous discussions of convergence applies. In fact, there is no value of ω for which the DTFT series (3.80a) converges. However, the lack of convergence is not the same for all values of ω. When ω is an integer multiple of 2π, (3.80a) diverges to ∞ because every term in the sum is 1. For other values of ω, it is tempting (but *not* mathematically correct; see Appendix 2.A.2) to assign the value of zero to

[52]For any piecewise continuously differentiable function with a discontinuity of height α, the overshoot is 0.089α, roughly 9% higher than the original height.

the sum because the terms in (3.80a) all lie on the unit circle, with no direction preferred.

Despite the lack of convergence of the DTFT, the expression

$$X(e^{j\omega}) \;=\; 2\pi\delta(\omega), \qquad \omega \in [-\pi,\, \pi], \tag{3.86a}$$

with δ the Dirac delta function, proves useful. It is valid in the sense that substituting it into the inverse DTFT, (3.80b), recovers the constant sequence with which we started, $x_n = 1$, for all $n \in \mathbb{Z}$. The same DTFT can also be written as

$$X(e^{j\omega}) \;=\; 2\pi \sum_{k=-\infty}^{\infty} \delta(\omega - k2\pi) \tag{3.86b}$$

to capture the 2π-periodicity of $X(e^{j\omega})$.

Example 3.17 (DTFT of cosine sequence) Let

$$X(e^{j\omega}) \;=\; \pi\left(\delta(\omega+\omega_0) + \delta(\omega-\omega_0)\right), \qquad \omega \in [-\pi,\, \pi], \tag{3.87a}$$

for some $\omega_0 \in (-\pi,\, \pi)$. Then, by evaluating the inverse DTFT, for any $n \in \mathbb{Z}$,

$$\begin{aligned}
x_n &\overset{(a)}{=} \frac{1}{2\pi}\int_{-\pi}^{\pi} X(e^{j\omega})e^{j\omega n}\, d\omega \overset{(b)}{=} \frac{1}{2\pi}\int_{-\pi}^{\pi} \pi\left(\delta(\omega+\omega_0) + \delta(\omega-\omega_0)\right) e^{j\omega n} d\omega \\
&= \frac{1}{2}\left(\int_{-\pi}^{\pi} \delta(\omega+\omega_0)e^{j\omega n} d\omega + \int_{-\pi}^{\pi} \delta(\omega-\omega_0)e^{j\omega n} d\omega\right) \\
&\overset{(c)}{=} \frac{1}{2}\left(e^{-j\omega_0 n} + e^{j\omega_0 n}\right) \overset{(d)}{=} \cos\omega_0 n,
\end{aligned} \tag{3.87b}$$

where (a) follows from (3.80b); (b) from (3.87a); (c) from the sifting property of the Dirac delta function, (3.293); and (d) from (3.286). We may also derive this DTFT pair from (3.86) and the linearity and shift in frequency properties in the following section.

3.4.3 Properties of the DTFT

We list here basic properties of the DTFT; Table 3.4 summarizes these, together with symmetries as well as a few standard transform pairs. Of course, all the expressions must be well defined for these properties to hold.

Linearity The DTFT operator F is a linear operator, or

$$\alpha x_n + \beta y_n \quad \overset{\text{DTFT}}{\longleftrightarrow} \quad \alpha X(e^{j\omega}) + \beta Y(e^{j\omega}). \tag{3.88}$$

Shift in time The DTFT pair corresponding to a shift in time by n_0 is

$$x_{n-n_0} \quad \overset{\text{DTFT}}{\longleftrightarrow} \quad e^{-j\omega n_0} X(e^{j\omega}). \tag{3.89}$$

DTFT properties	Time domain	DTFT domain
Basic properties		
Linearity	$\alpha x_n + \beta y_n$	$\alpha X(e^{j\omega}) + \beta Y(e^{j\omega})$
Shift in time	x_{n-n_0}	$e^{-j\omega n_0} X(e^{j\omega})$
Shift in frequency	$e^{j\omega_0 n} x_n$	$X(e^{j(\omega-\omega_0)})$
Scaling in time		
Downsampling	x_{Nn}	$\frac{1}{N}\sum_{k=0}^{N-1} X(e^{j(\omega-2\pi k)/N})$
Upsampling	$x_{n/N},\ n/N \in \mathbb{Z}$	$X(e^{jN\omega})$
Time reversal	x_{-n}	$X(e^{-j\omega})$
Differentiation in frequency	$(-jn)^k x_n$	$\frac{\partial^k X(e^{j\omega})}{\partial\omega^k}$
Moments	$m_k = \sum_{n\in\mathbb{Z}} n^k x_n = (-j)^k \frac{\partial X(e^{j\omega})}{\partial\omega}\Big\vert_{\omega=0}$	
Convolution in time	$(h * x)_n$	$H(e^{j\omega})\, X(e^{j\omega})$
Circular convolution in frequency	$h_n x_n$	$\frac{1}{2\pi}(H \circledast X)(e^{j\omega})$
Deterministic autocorrelation	$a_n = \sum_{k\in\mathbb{Z}} x_k x^*_{k-n}$	$A(e^{j\omega}) = \vert X(e^{j\omega})\vert^2$
Deterministic crosscorrelation	$c_n = \sum_{k\in\mathbb{Z}} x_k y^*_{k-n}$	$C(e^{j\omega}) = X(e^{j\omega})\, Y^*(e^{j\omega})$
Parseval equality	$\Vert x\Vert^2 = \sum_{n\in\mathbb{Z}} \vert x_n\vert^2 = \frac{1}{2\pi}\int_{-\pi}^{\pi} \vert X(e^{j\omega})\vert^2\, d\omega = \frac{1}{2\pi}\Vert X\Vert^2$	
Related sequences		
Conjugate	x^*_n	$X^*(e^{-j\omega})$
Conjugate, time-reversed	x^*_{-n}	$X^*(e^{j\omega})$
Real part	$\Re(x_n)$	$(X(e^{j\omega}) + X^*(e^{-j\omega}))/2$
Imaginary part	$\Im(x_n)$	$(X(e^{j\omega}) - X^*(e^{-j\omega}))/(2j)$
Conjugate-symmetric part	$(x_n + x^*_{-n})/2$	$\Re(X(e^{j\omega}))$
Conjugate-antisymmetric part	$(x_n - x^*_{-n})/(2j)$	$\Im(X(e^{j\omega}))$
Symmetries for real x		
X conjugate symmetric		$X(e^{j\omega}) = X^*(e^{-j\omega})$
Real part of X even		$\Re(X(e^{j\omega})) = \Re(X(e^{-j\omega}))$
Imaginary part of X odd		$\Im(X(e^{j\omega})) = -\Im(X(e^{-j\omega}))$
Magnitude of X even		$\vert X(e^{j\omega})\vert = \vert X(e^{-j\omega})\vert$
Phase of X odd		$\arg X(e^{j\omega}) = -\arg X(e^{-j\omega})$
Common transform pairs		
Kronecker delta sequence	δ_n	1
Shifted Kronecker delta sequence	δ_{n-n_0}	$e^{-j\omega n_0}$
Constant sequence	1	$2\pi \sum_{k=-\infty}^{\infty} \delta(\omega - 2\pi k)$
Geometric sequence	$\alpha^n u_n, \quad \vert\alpha\vert < 1$	$1/(1-\alpha e^{-j\omega})$
Arithmetic–geometric sequence	$n\alpha^n u_n, \quad \vert\alpha\vert < 1$	$\alpha e^{-j\omega}/(1-\alpha e^{-j\omega})^2$
Sinc sequence (ideal lowpass filter)	$\sqrt{\frac{\omega_0}{2\pi}}\,\mathrm{sinc}(\tfrac{1}{2}\omega_0 n)$	$\begin{cases} \sqrt{2\pi/\omega_0}, & \vert\omega\vert \le \tfrac{1}{2}\omega_0; \\ 0, & \text{otherwise.} \end{cases}$
Box sequence	$\begin{cases} 1/\sqrt{n_0}, & \vert n\vert \le \tfrac{1}{2}(n_0-1); \\ 0, & \text{otherwise.} \end{cases}$	$\sqrt{n_0}\frac{\mathrm{sinc}(\tfrac{1}{2}n_0\omega)}{\mathrm{sinc}(\tfrac{1}{2}\omega)}$

Table 3.4 Properties of the discrete-time Fourier transform.

Shift in frequency The DTFT pair corresponding to a shift in frequency by ω_0 is

$$e^{j\omega_0 n} x_n \quad \overset{\text{DTFT}}{\longleftrightarrow} \quad X(e^{j(\omega-\omega_0)}). \tag{3.90}$$

A shift in frequency is often referred to as *modulation*. The shift in time and shift in frequency are the first of several Fourier transform properties that are *duals* in that swapping the roles of time and frequency results in a pair of similar statements.[53]

Scaling in time Scaling in time appears in two flavors.

(i) The DTFT pair corresponding to scaling in time by N is

$$x_{Nn} \quad \overset{\text{DTFT}}{\longleftrightarrow} \quad \frac{1}{N} \sum_{k=0}^{N-1} X(e^{j(\omega-2\pi k)/N}). \tag{3.91}$$

This type of scaling is referred to as *downsampling*; we will discuss it in more detail in Section 3.7.

(ii) The DTFT pair corresponding to scaling in time by $1/N$ is

$$\begin{cases} x_{n/N}, & \text{for } n/N \in \mathbb{Z}; \\ 0, & \text{otherwise} \end{cases} \quad \overset{\text{DTFT}}{\longleftrightarrow} \quad X(e^{jN\omega}). \tag{3.92}$$

This type of scaling is referred to as *upsampling*; we will discuss it in more detail in Section 3.7.

Time reversal The DTFT pair corresponding to time reversal x_{-n} is

$$x_{-n} \quad \overset{\text{DTFT}}{\longleftrightarrow} \quad X(e^{-j\omega}). \tag{3.93}$$

For a real x_n, the DTFT of the time-reversed version x_{-n} is $X^*(e^{j\omega})$.

Differentiation The DTFT pair corresponding to differentiation in frequency is

$$(-jn)^k x_n \quad \overset{\text{DTFT}}{\longleftrightarrow} \quad \frac{\partial^k X(e^{j\omega})}{\partial \omega^k}. \tag{3.94}$$

Moments Computing the kth moment using the DTFT results in

$$m_k = \sum_{n\in\mathbb{Z}} n^k x_n = \left.\left(\sum_{n\in\mathbb{Z}} n^k x_n e^{-j\omega n}\right)\right|_{\omega=0} = (-j)^k \left.\frac{\partial X(e^{j\omega})}{\partial \omega}\right|_{\omega=0}, \qquad k \in \mathbb{N}, \tag{3.95a}$$

as a direct application of (3.94). The first two moments are

$$m_0 = \sum_{n\in\mathbb{Z}} x_n = \left.\left(\sum_{n\in\mathbb{Z}} x_n e^{-j\omega n}\right)\right|_{\omega=0} = X(0), \tag{3.95b}$$

$$m_1 = \sum_{n\in\mathbb{Z}} n x_n = \left.\left(\sum_{n\in\mathbb{Z}} n x_n e^{-j\omega n}\right)\right|_{\omega=0} = -j \left.\frac{\partial X(e^{j\omega})}{\partial \omega}\right|_{\omega=0}. \tag{3.95c}$$

[53] In this section, time is discrete and frequency is continuous; dualities are more transparent when both are discrete (see Section 3.6) or both are continuous (see Section 4.4).

Convolution in time The DTFT pair corresponding to convolution in time is

$$(h * x)_n \quad \overset{\text{DTFT}}{\longleftrightarrow} \quad H(e^{j\omega})\, X(e^{j\omega}). \tag{3.96}$$

We first present a direct algebraic proof: The spectrum $Y(e^{j\omega})$ of the sequence $y = h * x$ can be written as

$$\begin{aligned} Y(e^{j\omega}) &\overset{(a)}{=} \sum_{n\in\mathbb{Z}} y_n e^{-j\omega n} \overset{(b)}{=} \sum_{n\in\mathbb{Z}} \left(\sum_{k\in\mathbb{Z}} x_k h_{n-k} \right) e^{-j\omega n} \\ &= \sum_{n\in\mathbb{Z}} \sum_{k\in\mathbb{Z}} x_k e^{-j\omega k} h_{n-k} e^{-j\omega(n-k)} \\ &\overset{(c)}{=} \sum_{k\in\mathbb{Z}} x_k e^{-j\omega k} \sum_{n\in\mathbb{Z}} h_{n-k} e^{-j\omega(n-k)} \overset{(d)}{=} X(e^{j\omega})\, H(e^{j\omega}), \end{aligned} \tag{3.97}$$

where (a) follows from the definition of the DTFT; (b) from the definition of convolution; (c) from interchanging the order of summation, which is an allowed operation since absolute summability is implied by $h * x$ being well defined; and (d) from the definition of the DTFT.

This key result is also a direct consequence of the eigensequence property of complex exponential sequences v from (3.77): when x is written as a combination of spectral components, the effect of the convolution operator is to simply scale each spectral component by the corresponding eigenvalue of the convolution operator. The DTFT thus diagonalizes the convolution operator, and, furthermore, this is the motivation for the definition of the DTFT.

Circular convolution in frequency The DTFT pair corresponding to circular convolution in frequency is

$$h_n x_n \quad \overset{\text{DTFT}}{\longleftrightarrow} \quad \frac{1}{2\pi} (H \circledast X)(e^{j\omega}), \tag{3.98}$$

where we have introduced the circular convolution between 2π-periodic functions

$$(H \circledast X)(e^{j\omega}) \;=\; \int_{-\pi}^{\pi} X(e^{j\theta})\, H(e^{j(\omega-\theta)})\, d\theta. \tag{3.99}$$

The circular convolution in frequency property (3.98) is dual to the convolution in time property (3.96) (see Exercise 3.6).

Deterministic autocorrelation The DTFT pair corresponding to the deterministic autocorrelation of a sequence x is

$$a_n \;=\; \sum_{k\in\mathbb{Z}} x_k x^*_{k-n} \quad \overset{\text{DTFT}}{\longleftrightarrow} \quad A(e^{j\omega}) \;=\; |X(e^{j\omega})|^2 \tag{3.100}$$

and satisfies

$$A(e^{j\omega}) \;=\; A^*(e^{j\omega}), \tag{3.101a}$$

$$A(e^{j\omega}) \;\geq\; 0. \tag{3.101b}$$

Thus, $A(e^{j\omega})$ is not only real, (3.101a), but also positive semidefinite, (3.101b). To verify (3.100), express the deterministic autocorrelation as a convolution of x and its time-reversed version as in (3.62d), $x_n * x^*_{-n}$. We know from Table 3.4 that the DTFT of x^*_{-n} is $X^*(e^{j\omega})$. Then, using the convolution property (3.96), we obtain (3.100). For a real x,

$$A(e^{j\omega}) = |X(e^{j\omega})|^2 = A(e^{-j\omega}), \tag{3.101c}$$

since $X(e^{-j\omega}) = X^*(e^{j\omega})$.

The quantity $A(e^{j\omega})$ is called the *energy spectral density* (it is the deterministic counterpart of the *power spectral density* for WSS sequences[54] in (3.242)). The *energy* is the average of the energy spectral density over the frequency range,

$$E = \frac{1}{2\pi}\int_{-\pi}^{\pi} A(e^{j\omega})\, d\omega = \frac{1}{2\pi}\int_{-\pi}^{\pi} |X(e^{j\omega})|^2\, d\omega = \sum_{n\in\mathbb{Z}} |x_n|^2 = a_0. \tag{3.102}$$

Thus, the energy spectral density measures the distribution of energy over the frequency range. Mimicking the relationship between the energy spectral density for deterministic sequences and the power spectral density for WSS sequences, (3.102) is the deterministic counterpart of the *power* for WSS sequences (3.243).

Deterministic crosscorrelation The DTFT pair corresponding to the deterministic crosscorrelation of sequences x and y is

$$c_n = \sum_{k\in\mathbb{Z}} x_k y^*_{k-n} \quad \overset{\text{DTFT}}{\longleftrightarrow} \quad C_{x,y}(e^{j\omega}) = X(e^{j\omega})\, Y^*(e^{j\omega}), \tag{3.103}$$

the proof of which is left for Exercise 3.2. The deterministic crosscorrelation satisfies

$$C_{x,y}(e^{j\omega}) = C^*_{y,x}(e^{j\omega}). \tag{3.104a}$$

For real x and y,

$$C_{x,y}(e^{j\omega}) = X(e^{j\omega})\, Y(e^{-j\omega}) = C_{y,x}(e^{-j\omega}). \tag{3.104b}$$

Deterministic autocorrelation of vector sequences The DTFT pair corresponding to the deterministic autocorrelation of a length-N vector sequence x is

$$A_n \quad \overset{\text{DTFT}}{\longleftrightarrow} \quad A(e^{j\omega}) = \begin{bmatrix} A_0(e^{j\omega}) & C_{0,1}(e^{j\omega}) & \cdots & C_{0,N-1}(e^{j\omega}) \\ C_{1,0}(e^{j\omega}) & A_1(e^{j\omega}) & \cdots & C_{1,N-1}(e^{j\omega}) \\ \vdots & \vdots & \ddots & \vdots \\ C_{N-1,0}(e^{j\omega}) & C_{N-1,1}(e^{j\omega}) & \cdots & A_{N-1}(e^{j\omega}) \end{bmatrix}, \tag{3.105}$$

[54] WSS stands for *wide-sense stationary*, defined in Section 3.8.3.

where A_n is given in (3.23). Because of (3.101a) and (3.104a), this energy spectral density matrix is Hermitian, that is,

$$A(e^{j\omega}) = \begin{bmatrix} A_0(e^{j\omega}) & C_{0,1}(e^{j\omega}) & \cdots & C_{0,N-1}(e^{j\omega}) \\ C_{0,1}^*(e^{j\omega}) & A_1(e^{j\omega}) & \cdots & C_{1,N-1}(e^{j\omega}) \\ \vdots & \vdots & \ddots & \vdots \\ C_{0,N-1}^*(e^{j\omega}) & C_{1,N-1}^*(e^{j\omega}) & \cdots & A_{N-1}(e^{j\omega}) \end{bmatrix} = A^*(e^{j\omega}). \tag{3.106a}$$

For a real x,

$$A(e^{j\omega}) = A^\top(e^{-j\omega}). \tag{3.106b}$$

Parseval equality As noted earlier, from the form of (3.80a), the DTFT is a linear operator from the space of sequences to the space of 2π-periodic functions. Let us denote this through $X = Fx$. We have $F : \ell^2(\mathbb{Z}) \to \mathcal{L}^2([-\pi, \pi))$ because $x \in \ell^2(\mathbb{Z})$ implies that $X(e^{j\omega})$ has finite $\mathcal{L}^2([-\pi, \pi))$ norm. Specifically,

$$\begin{aligned} \|X\|^2 &\overset{(a)}{=} \int_{-\pi}^{\pi} |X(e^{j\omega})|^2 \, d\omega = \int_{-\pi}^{\pi} X(e^{j\omega}) X^*(e^{j\omega}) \, d\omega \\ &\overset{(b)}{=} \int_{-\pi}^{\pi} \left(\sum_{n\in\mathbb{Z}} x_n e^{j\omega n} \right) \left(\sum_{k\in\mathbb{Z}} x_k e^{j\omega k} \right)^* d\omega \\ &\overset{(c)}{=} \sum_{n\in\mathbb{Z}} \sum_{k\in\mathbb{Z}} \int_{-\pi}^{\pi} x_n x_k^* e^{j\omega(n-k)} \, d\omega = \sum_{n\in\mathbb{Z}} \sum_{k\in\mathbb{Z}} x_n x_k^* \int_{-\pi}^{\pi} e^{j\omega(n-k)} \, d\omega \\ &\overset{(d)}{=} \sum_{n\in\mathbb{Z}} \sum_{k\in\mathbb{Z}} x_n x_k^* 2\pi \delta_{n-k} \overset{(e)}{=} 2\pi \sum_{n\in\mathbb{Z}} x_n x_n^* \\ &= 2\pi \sum_{n\in\mathbb{Z}} |x_n|^2 \overset{(f)}{=} 2\pi \, \|x\|^2, \end{aligned} \tag{3.107}$$

where (a) follows from the definition of the $\mathcal{L}^2([-\pi, \pi))$ norm; (b) from the definition of the DTFT; (c) from an interchange that is allowed because $x \in \ell^2(\mathbb{Z})$ implies absolute convergence of the sums in the integrand; (d) from (3.82b); (e) from the definition of the Kronecker delta sequence, (2.9); and (f) from the definition of the $\ell^2(\mathbb{Z})$ norm.

If it were not for the 2π factor, the equality (3.107) would be like the equality (2.56) for a unitary operator; (3.107) is the version of the *Parseval equality* for the DTFT. The Parseval equality is often termed the *energy conservation property*, as the energy (3.102) is the integral of the energy spectral density over the frequency range.

A computation similar to (3.107) shows that $F/\sqrt{2\pi}$ is a unitary operator (see (2.55)):

$$\left\langle \frac{1}{\sqrt{2\pi}} Fx, \frac{1}{\sqrt{2\pi}} Fy \right\rangle = \langle x, y \rangle \qquad \text{for every } x \text{ and } y \text{ in } \ell^2(\mathbb{Z}),$$

or,

$$\langle x, y \rangle = \frac{1}{2\pi} \langle X, Y \rangle \qquad \text{for every } x \text{ and } y \text{ in } \ell^2(\mathbb{Z}), \tag{3.108}$$

where X and Y are the DTFTs of x and y. This is the version of the generalized Parseval equality for the DTFT, which follows immediately from the convolution in frequency property (3.98) and the fact that the DTFT of x_n^* is $X^*(e^{j\omega})$ (see Table 3.4).

Adjoint The adjoint of the DTFT, $F^* : \mathcal{L}^2([-\pi, \pi)) \to \ell^2(\mathbb{Z})$, is determined uniquely by

$$\langle Fx, y \rangle = \langle x, F^* y \rangle \qquad \text{for every } x \in \ell^2(\mathbb{Z}) \text{ and } y \text{ in } \mathcal{L}^2([-\pi, \pi)).$$

Since we have already concluded that $F/\sqrt{2\pi}$ is a unitary operator, by Theorem 2.23,

$$\left(\frac{1}{\sqrt{2\pi}}F\right)^* = \left(\frac{1}{\sqrt{2\pi}}F\right)^{-1} = \sqrt{2\pi}F^{-1}.$$

Thus,

$$F^* = 2\pi F^{-1}, \tag{3.109}$$

with F^{-1} given by (3.80b).

3.4.4 Frequency response of filters

The DTFT of a filter (the impulse response of an LSI system) h is called the *frequency response*:

$$H(e^{j\omega}) = \sum_{n\in\mathbb{Z}} h_n e^{-j\omega n}, \qquad \omega \in \mathbb{R}. \tag{3.110a}$$

The inverse DTFT of the frequency response recovers the impulse response,

$$h_n = \frac{1}{2\pi}\int_{-\pi}^{\pi} H(e^{j\omega})e^{j\omega n}\, d\omega, \qquad n \in \mathbb{Z}. \tag{3.110b}$$

We often write the magnitude and phase separately:

$$H(e^{j\omega}) = |H(e^{j\omega})|e^{j\arg(H(e^{j\omega}))},$$

where the *magnitude response* $|H(e^{j\omega})|$ is a 2π-periodic real-valued, nonnegative function, and the *phase response* $\arg(H(e^{j\omega}))$ is a 2π-periodic real-valued function between $-\pi$ and π.[55] A filter is said to have *zero phase* when its frequency response is real; this is equivalent to the phase response taking only values that are integer multiples of π. A filter is said to have *generalized linear phase* when its frequency response can be written in the form

$$H(e^{j\omega}) = r(\omega)e^{j(\alpha\omega+\beta)}, \tag{3.111}$$

where $r(\omega)$ is real-valued and α and β are real numbers; this corresponds to a phase response that is affine in ω (straight lines with slope α) except where there are jumps by 2π. When furthermore $\beta = 0$, the filter is said to have *linear phase*. Solved exercise 3.3 explores filters as projections through their frequency response.

[55] The argument of the complex number $H(e^{j\omega})$ can be equally well defined to be on $[0, 2\pi)$.

Ideal lowpass filter

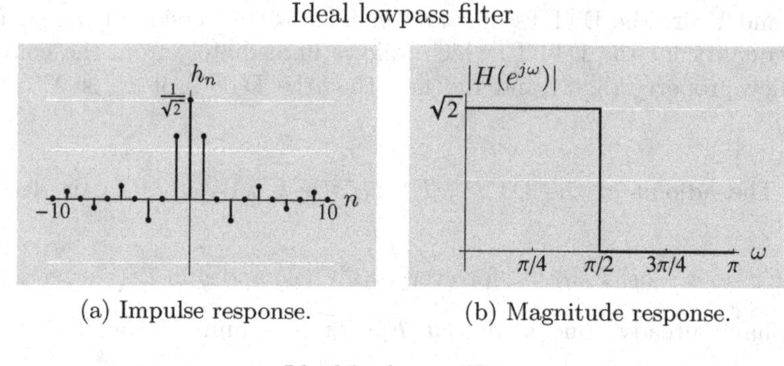

(a) Impulse response. (b) Magnitude response.

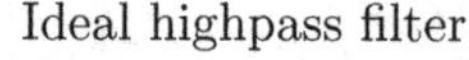

Ideal highpass filter

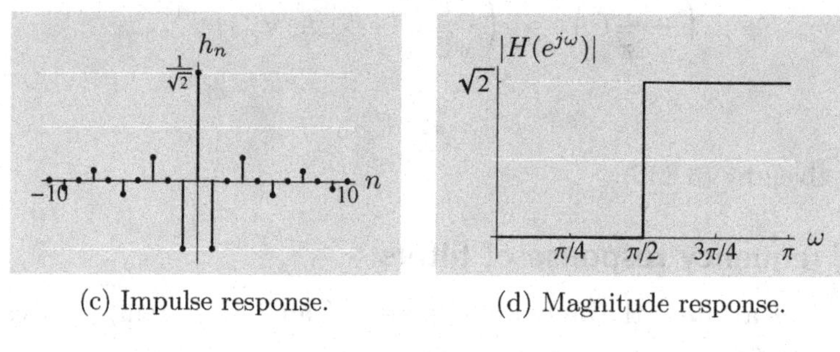

(c) Impulse response. (d) Magnitude response.

Ideal bandpass filter

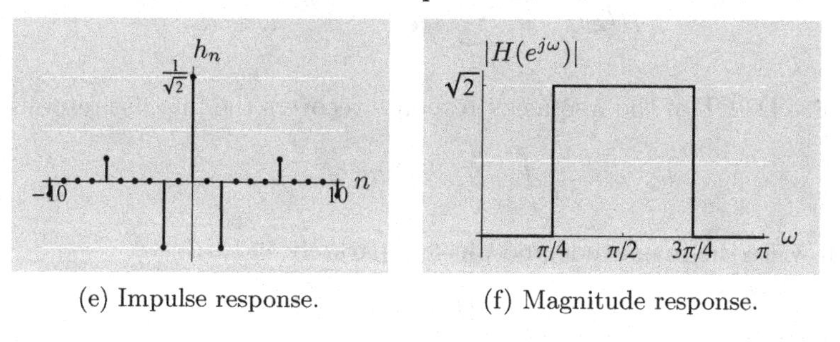

(e) Impulse response. (f) Magnitude response.

Figure 3.10 Impulse responses and magnitude responses of ideal filters.

Ideal filters The frequency response of a filter is typically used to design a filter with specific properties, where we want to let certain frequencies pass – the *passband*, while blocking others – the *stopband*. The magnitude response of an *ideal filter* is constant in its passband and zero outside of its passband. For example, an ideal lowpass filter passes frequencies below some cutoff frequency $\frac{1}{2}\omega_0$ and blocks the others; its passband is thus the interval $[-\frac{1}{2}\omega_0, \frac{1}{2}\omega_0]$. Figure 3.10(b) gives an example for $\omega_0 = \pi$. Figures 3.10(d) and (f) show magnitude responses for ideal highpass and bandpass filters.

To find the impulse response of an ideal lowpass filter, we start with the desired

Ideal filters	Time domain	DTFT domain
Ideal lowpass filter	$\sqrt{\frac{\omega_0}{2\pi}}\,\mathrm{sinc}\left(\frac{1}{2}\omega_0 n\right)$	$\begin{cases}\sqrt{2\pi/\omega_0}, & \lvert\omega\rvert \le \frac{1}{2}\omega_0;\\ 0, & \text{otherwise}\end{cases}$
Ideal Nth-band filter	$\frac{1}{\sqrt{N}}\,\mathrm{sinc}\left(\frac{\pi n}{N}\right)$	$\begin{cases}\sqrt{N}, & \lvert\omega\rvert \le \pi/N;\\ 0, & \text{otherwise}\end{cases}$
Ideal half-band lowpass filter	$\frac{1}{\sqrt{2}}\,\mathrm{sinc}\left(\frac{1}{2}\pi n\right)$	$\begin{cases}\sqrt{2}, & \lvert\omega\rvert \le \frac{1}{2}\pi;\\ 0, & \text{otherwise}\end{cases}$

Table 3.5 Ideal filters with unit-norm impulse responses.

frequency response:

$$H(e^{j\omega}) = \begin{cases}\sqrt{2\pi/\omega_0}, & \text{for } |\omega| \le \frac{1}{2}\omega_0;\\ 0, & \text{otherwise.}\end{cases} \tag{3.112a}$$

This is a zero-phase filter and a box function in frequency.[56] Upon applying the inverse DTFT, we obtain the impulse response as

$$h_n = \frac{1}{\sqrt{2\pi\omega_0}}\int_{-\omega_0/2}^{\omega_0/2} e^{j\omega n}\,d\omega = \sqrt{\frac{\omega_0}{2\pi}}\,\mathrm{sinc}\left(\frac{1}{2}\omega_0 n\right) \tag{3.112b}$$

by elementary integrations, with the $n = 0$ and $n \neq 0$ cases separated. This impulse response is of unit norm. A case of particular interest is the *half-band* filter that arises from $\omega_0 = \pi$; it has a passband of $[-\frac{1}{2}\pi, \frac{1}{2}\pi]$, half of the full band $[-\pi, \pi]$. The impulse response

$$h_n = \frac{1}{\sqrt{2}}\,\mathrm{sinc}\left(\frac{1}{2}\pi n\right) = \frac{1}{\sqrt{2}}\frac{\sin(\frac{1}{2}\pi n)}{\frac{1}{2}\pi n} \tag{3.113}$$

and magnitude response are shown in Figures 3.10(a) and (b). More generally, an Nth-band filter has $\omega_0 = 2\pi/N$. These ideal filters are summarized in Table 3.5. Their impulse responses decay slowly as $O(1/n)$ and are thus not absolutely summable. This lack of absolute summability of the impulse response h is unavoidable when the desired frequency response H is discontinuous; see Section 3.4.2.

FIR filters Ideal filters are not realizable; thus, we now explore a few examples of filters with realizable frequency responses. We start with an FIR filter we have already seen in Example 3.2.

Example 3.18 (Moving-average filter, Example 3.2 continued) The impulse response of the moving-average filter in (3.6) is (we assumed that N is odd)

$$h_n = \begin{cases}1/N, & \text{for } |n| \le \frac{1}{2}(N-1);\\ 0, & \text{otherwise,}\end{cases}$$

[56] Table 4.5 in Chapter 4 summarizes box and sinc functions in time and frequency.

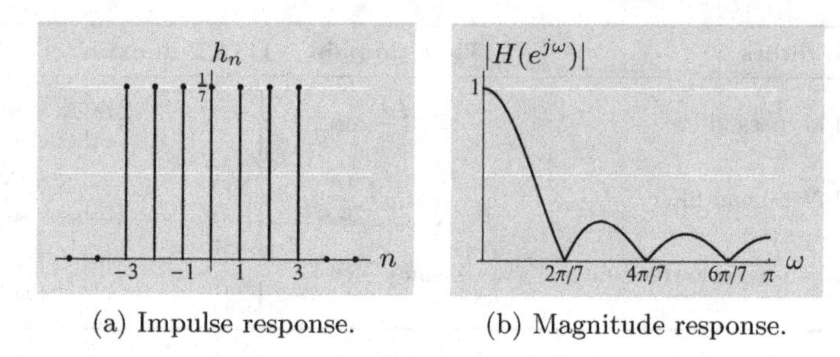

(a) Impulse response. (b) Magnitude response.

Figure 3.11 Moving-average filter (3.6) with $N = 7$.

which is the same, within scaling, as the box sequence from (3.14a). Its frequency response is

$$\begin{aligned} H(e^{j\omega}) &= \frac{1}{N} \sum_{n=-(N-1)/2}^{(N-1)/2} e^{-j\omega n} \stackrel{(a)}{=} \frac{1}{N} e^{j\omega(N-1)/2} \sum_{k=0}^{N-1} e^{-j\omega k} \\ &\stackrel{(b)}{=} \frac{1}{N} e^{j\omega(N-1)/2} \frac{1 - e^{-j\omega N}}{1 - e^{-j\omega}} = \frac{1}{N} \frac{e^{j\omega N/2} - e^{-j\omega N/2}}{e^{j\omega/2} - e^{-j\omega/2}} = \frac{1}{N} \frac{\sin(\frac{1}{2}N\omega)}{\sin(\frac{1}{2}\omega)}, \end{aligned}$$

where (a) follows from the change of variable $k = n + \frac{1}{2}(N-1)$; and (b) from the formula for the finite geometric series, (P2.54-1). Figure 3.11 shows the impulse response and magnitude response of this filter for $N = 7$.

Linear-phase filters Real-valued FIR filters have linear phase when they are symmetric or antisymmetric. Consider causal filters with length L, so the support is $\{0, 1, \ldots, L-1\}$. These filters then satisfy

$$\begin{array}{ll} \text{symmetric} & \text{antisymmetric} \\ h_n = h_{L-1-n} & h_n = -h_{L-1-n} \end{array} \tag{3.114}$$

These symmetries are illustrated in Figure 3.12 for L even and odd. Let us now show that an even-length, symmetric filter as in part (a) of Figure 3.12 does indeed lead to linear phase; other cases follow similarly. We compute the frequency response of h_n,

$$\begin{aligned} H(e^{j\omega}) &= \sum_{n=0}^{L-1} h_n e^{-j\omega n} \stackrel{(a)}{=} \sum_{n=0}^{L/2-1} h_n \left(e^{-j\omega n} + e^{-j\omega(L-1-n)}\right) \\ &= \sum_{n=0}^{L/2-1} h_n e^{-j\omega(L-1)/2} \left(e^{j\omega(n-(L-1)/2)} + e^{-j\omega(n-(L-1)/2)}\right) \\ &\stackrel{(b)}{=} 2 \sum_{n=0}^{L/2-1} h_n \cos\left(\omega\left(n - \frac{L-1}{2}\right)\right) e^{-j\omega((L-1)/2)} \\ &= r(\omega) e^{j\alpha\omega}, \end{aligned} \tag{3.115a}$$

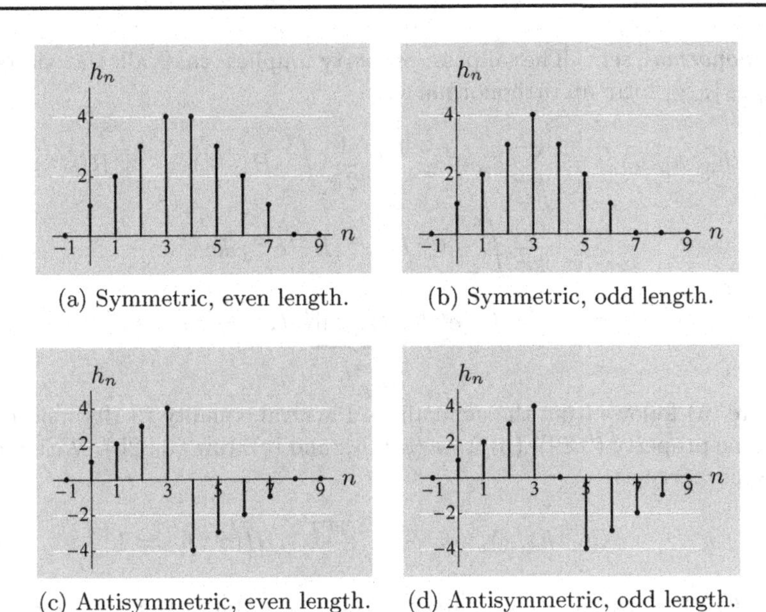

(a) Symmetric, even length. (b) Symmetric, odd length.

(c) Antisymmetric, even length. (d) Antisymmetric, odd length.

Figure 3.12 Filters with symmetries.

with

$$r(\omega) = 2 \sum_{n=0}^{L/2-1} h_n \cos\left(\omega\left(n - \frac{L-1}{2}\right)\right) \quad \text{and} \quad \alpha = -\frac{L-1}{2}, \tag{3.115b}$$

where (a) follows from gathering factors with the same h_n because of symmetry in (3.114); and (b) from using (3.286). This frequency response fits the form of (3.111), so the filter does indeed have linear phase.

Allpass filters Another important class is filters with unit magnitude response, that is,

$$|H(e^{j\omega})| = 1. \tag{3.116}$$

Since all frequencies go through without change of magnitude, a filter satisfying (3.116) is called an *allpass filter.* Allpass filters have some interesting properties.

(i) *Energy conservation:* The allpass property corresponds to energy conservation, since we have

$$\begin{aligned}\|y\|^2 &\overset{(a)}{=} \frac{1}{2\pi}\|Y\|^2 = \frac{1}{2\pi}\|HX\|^2 = \frac{1}{2\pi}\int_{-\pi}^{\pi} |H(e^{j\omega})X(e^{j\omega})|^2\, d\omega \\ &\overset{(b)}{=} \frac{1}{2\pi}\int_{-\pi}^{\pi} |X(e^{j\omega})|^2\, d\omega \overset{(c)}{=} \|x\|^2,\end{aligned}$$

where (a) follows from the Parseval equality (3.107); (b) from (3.116); and (c) from the Parseval equality again.

(ii) *Orthonormal set:* The allpass property implies that all the shifts of h, $\{h_{n-k}\}_{k\in\mathbb{Z}}$, form an orthonormal set:

$$\begin{aligned}\langle h_n, h_{n-k}\rangle_n &= \sum_{n\in\mathbb{Z}} h_n h_{n-k}^* \overset{(a)}{=} \frac{1}{2\pi}\int_{-\pi}^{\pi} H(e^{j\omega})\left(e^{-j\omega k}H(e^{j\omega})\right)^* d\omega \\ &= \frac{1}{2\pi}\int_{-\pi}^{\pi} e^{j\omega k} H(e^{j\omega})H^*(e^{j\omega})\, d\omega \\ &\overset{(b)}{=} \frac{1}{2\pi}\int_{-\pi}^{\pi} e^{j\omega k} \underbrace{|H(e^{j\omega})|^2}_{=1} d\omega \overset{(c)}{=} \delta_k, \end{aligned} \tag{3.117}$$

where (a) follows from the generalized Parseval equality (3.108) and the shift in time property (3.89); (b) from (3.116); and (c) from (3.82b). We summarize this property as

$$\langle h_n, h_{n-k}\rangle_n = \delta_k \quad \overset{\text{DTFT}}{\longleftrightarrow} \quad |H(e^{j\omega})| = 1. \tag{3.118}$$

(iii) *Orthonormal basis:* The allpass property implies that $\{\varphi_k\}_{k\in\mathbb{Z}}$, where $\varphi_{k,n} = h_{n-k}$, $n \in \mathbb{Z}$, is an orthonormal basis for $\ell^2(\mathbb{Z})$. Having already shown in (3.117) that the set is orthonormal, we can use Theorem 2.42 to show that the set is a basis for $\ell^2(\mathbb{Z})$. For $x \in \ell^2(\mathbb{Z})$, let β denote the coefficient sequence obtained by analysis with $\{\varphi_k\}_{k\in\mathbb{Z}}$. Then

$$\beta_k = \langle x, \varphi_k\rangle = \langle x_n, h_{n-k}\rangle_n \overset{(a)}{=} x_n *_n h_{k-n}^*, \qquad k \in \mathbb{Z},$$

where (a) follows from (3.62a). To apply Theorem 2.42, we would like to show that $\|\beta\|^2 = \|x\|^2$. This equality does indeed hold

$$\|\beta\|^2 \overset{(a)}{=} \frac{1}{2\pi}\|X(e^{j\omega})H^*(e^{j\omega})\|^2 \overset{(b)}{=} \frac{1}{2\pi}\|X(e^{j\omega})\|^2 \overset{(c)}{=} \|x\|^2,$$

where (a) follows from the Parseval equality (3.107), the convolution property (3.96), and using (3.93) for the time reversal of h; (b) from (3.116); and (c) from the Parseval equality again.

This discussion contains a piece of good news – there exist shift-invariant orthonormal bases for $\ell^2(\mathbb{Z})$, as well as a piece of bad news – these bases have no frequency selectivity (they are allpass sequences). This is one of the main reasons to search for more general orthonormal bases for $\ell^2(\mathbb{Z})$, as we do in the companion volume to this book, [57].

EXAMPLE 3.19 (ALLPASS FILTERS) Consider the simple shift-by-k filter given in (3.39a) with the impulse response $h_n = \delta_{n-k}$. By evaluating (3.110a), the frequency response is $H(e^{j\omega}) = e^{-j\omega k}$. Thus, h is an allpass filter,

$$|H(e^{j\omega})| = 1, \qquad \arg(H(e^{j\omega})) = -\omega k \bmod 2\pi.$$

This filter has linear phase with a slope $-k$ given by the delay.

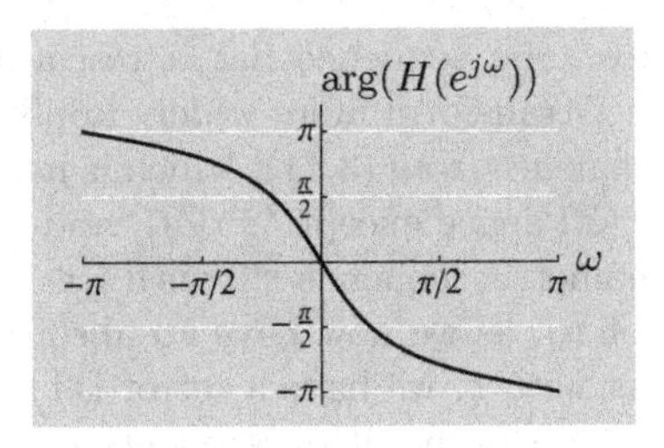

Figure 3.13 Phase of a first-order allpass filter as in (3.119) with $\alpha = \frac{1}{2}$.

We now look at a more sophisticated allpass filter. It provides an example where key properties not plainly visible in the time domain become obvious in the frequency domain. Start with a sequence

$$g_n = \alpha^n u_n, \qquad g = \begin{bmatrix} \dots & 0 & \boxed{1} & \alpha & \alpha^2 & \alpha^3 & \dots \end{bmatrix}^\top,$$

with $\alpha \in \mathbb{C}$ and $|\alpha| < 1$, and u_n is the Heaviside sequence from (3.11). Suppose that h satisfies

$$h_n = -\alpha^* g_n + g_{n-1}, \qquad n \in \mathbb{Z}.$$

We now show that h is an allpass filter, so filtering a sequence x with h will not change its magnitude; moreover, h is of norm 1 and orthogonal to all its shifts as in (3.118). To start, find the frequency response of g_n,

$$G(e^{j\omega}) = \sum_{n\in\mathbb{Z}} \alpha^n e^{-j\omega n} \stackrel{(a)}{=} \frac{1}{1-\alpha e^{-j\omega}},$$

where (a) follows from the formula for the infinite geometric series (P2.54-3). Then,

$$H(e^{j\omega}) = -\alpha^* G(e^{j\omega}) + e^{-j\omega} G(e^{j\omega}) = \frac{e^{-j\omega} - \alpha^*}{1 - \alpha e^{-j\omega}}. \tag{3.119}$$

The magnitude squared of $H(e^{j\omega})$ is

$$|H(e^{j\omega})|^2 = H(e^{j\omega})H^*(e^{j\omega}) = \frac{(e^{-j\omega} - \alpha^*)(e^{j\omega} - \alpha)}{(1-\alpha e^{-j\omega})(1-\alpha^* e^{j\omega})} = 1,$$

and thus $|H(e^{j\omega})| = 1$ for all ω. The phase response is shown in Figure 3.13.

3.5 z-transform

While the DTFT has many nice properties, its use is limited by the convergence issues discussed in Section 3.4.2. The z-transform introduces a set of rescalings of sequences so that almost any sequence has a rescaling such that the DTFT converges. This makes the z-transform more widely applicable than the DTFT.

Take the Heaviside sequence from (3.11), which is neither in $\ell^1(\mathbb{Z})$ nor in $\ell^2(\mathbb{Z})$ (and is not in any other $\ell^p(\mathbb{Z})$ space except $\ell^\infty(\mathbb{Z})$), and thus has no DTFT. If we were to multiply it by a geometric sequence r^n, with $r \in [0, 1)$, yielding $x_n = r^n u_n$, we could take the DTFT of x_n, as we now have an absolutely summable sequence. By controlling the rescaling with r, we have a set of DTFTs indexed by r, and we can think of this as a new transform with arguments r and ω. Upon combining r and ω through $z = re^{j\omega}$, we obtain a transform with argument $z \in \mathbb{C}$, where z need not have unit modulus.

Because of the close connection to the DTFT, we will see that a convolution property holds as well as many other properties similar to those in Section 3.4.3, but now for more general sequences. This is the essential motivation behind extending the analysis that uses the unit-norm complex exponential sequences in (3.77) to more general complex exponential sequences $v_n = z^n = (re^{j\omega})^n$. We will also see that convolution of causal, finite-length sequences becomes polynomial multiplication in the z-transform domain.

3.5.1 Definition of the z-transform

Eigensequences of the convolution operator The eigensequence property (3.78) extends from complex exponentials with unit modulus to those with any modulus. Consider the sequence

$$v_n = z^n = (re^{j\omega})^n, \qquad n \in \mathbb{Z}, \tag{3.120}$$

where $r \in [0, \infty)$ and $\omega \in \mathbb{R}$, so z is any complex number. Like a complex exponential sequence with unit modulus, this is also an eigensequence of the convolution operator H associated with the LSI system with impulse response h since

$$\begin{aligned}(Hv)_n &= (h * v)_n = \sum_{k\in\mathbb{Z}} v_{n-k} h_k = \sum_{k\in\mathbb{Z}} z^{n-k} h_k \\ &= \underbrace{\sum_{k\in\mathbb{Z}} h_k z^{-k}}_{\lambda_z} \, \underbrace{z^n}_{v_n}.\end{aligned} \tag{3.121}$$

This shows that applying the convolution operator H to the sequence v gives a scalar multiple of v; in other words, v is an eigensequence of H with the corresponding eigenvalue λ_z. We denote this eigenvalue by $H(z)$ using the transfer function defined formally in (3.154). We can thus rewrite (3.121) as

$$Hv = h * v = H(z)\, v. \tag{3.122}$$

The key distinction from (3.78) is that the set of impulse responses h for which the sum (3.121) converges now depends on $|z|$.

z-transform The z-transform is defined similarly to the DTFT in Definition 3.11.

DEFINITION 3.12 (z-TRANSFORM) The *z-transform* of a sequence x is

$$X(z) = \sum_{n\in\mathbb{Z}} x_n z^{-n}, \qquad z \in \mathbb{C}. \tag{3.123}$$

It exists when (3.123) converges absolutely for some values of z; these values of z are called the *region of convergence (ROC)*,

$$\text{ROC} = \{z \mid |X(z)| < \infty\}. \tag{3.124}$$

When the z-transform exists, we denote the z-transform pair as

$$x_n \stackrel{\text{ZT}}{\longleftrightarrow} X(z),$$

where the ROC is part of the specification of $X(z)$.

Relation of the z-transform to the DTFT Given a sequence x and its z-transform $X(z)$ with an ROC that includes the unit circle $|z| = 1$, the z-transform evaluated on the unit circle is equal to the DTFT of the same sequence,

$$X(z)\big|_{z=e^{j\omega}} = X(e^{j\omega}). \tag{3.125}$$

Conversely, suppose that $y_n = r^{-n}x_n$ has DTFT $Y(e^{j\omega})$. Then

$$Y(e^{j\omega}) = \sum_{n\in\mathbb{Z}} r^{-n}x_n e^{-j\omega n} = \sum_{n\in\mathbb{Z}} x_n \left(re^{j\omega}\right)^{-n} = X(re^{j\omega}),$$

so

$$X(z)\big|_{z=re^{j\omega}} = Y(e^{j\omega}), \tag{3.126}$$

and the circle $|z| = r$ is in the ROC of $X(z)$.

3.5.2 Existence and convergence of the z-transform

Convergence For the z-transform to exist and have $z = re^{j\omega}$ in its ROC, (3.123) must converge absolutely. Since

$$\sum_{n\in\mathbb{Z}} \left|x_n z^{-n}\right| = \sum_{n\in\mathbb{Z}} \left|x_n r^{-n}\right| \left|e^{-j\omega n}\right| = \sum_{n\in\mathbb{Z}} \left|x_n r^{-n}\right|,$$

absolute summability of $x_n r^{-n}$ is necessary and sufficient for the circle $|z| = r$ to be in the ROC of $X(z)$. Thus, the ROC is a ring of the form (see also Table 3.6)

$$\text{ROC} = \{z \mid 0 \le r_1 < |z| < r_2 \le \infty\}. \tag{3.127}$$

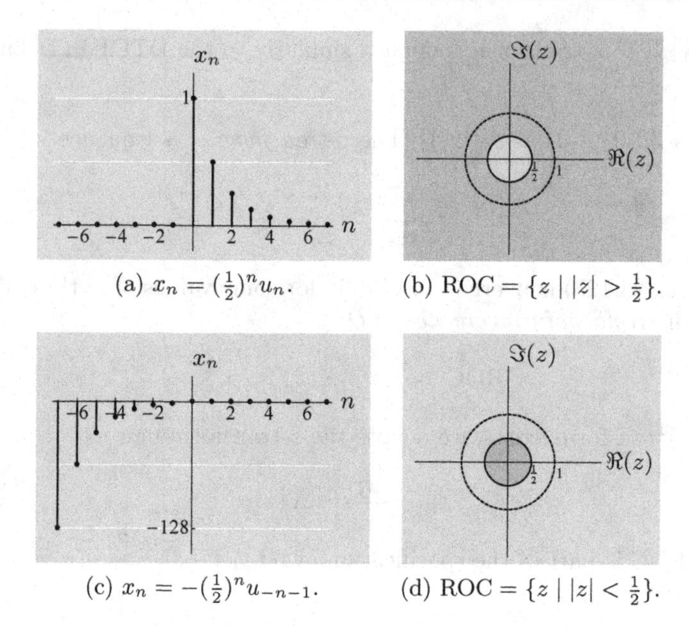

(a) $x_n = (\frac{1}{2})^n u_n$. (b) ROC $= \{z \mid |z| > \frac{1}{2}\}$.

(c) $x_n = -(\frac{1}{2})^n u_{-n-1}$. (d) ROC $= \{z \mid |z| < \frac{1}{2}\}$.

Figure 3.14 Illustration of Example 3.20. (a) The right-sided geometric series sequence and (b) the associated ROC of its z-transform. (c) The left-sided geometric series sequence sequence and (d) the associated ROC of its z-transform. The unit circle is marked in both (b) and (d) for reference.

By convention, the ROC concept is extended to $|z| = \infty$ by including $|z| = \infty$ in the ROC when $x_n = 0$ for all $n < 0$ and excluding it otherwise. Similarly, $z = 0$ is in the ROC when $x_n = 0$ for all $n > 0$ and not in the ROC otherwise. Exercise 3.8 explores a number of properties of the ROC.

EXAMPLE 3.20 (ROCS) To develop intuition, we look at a few examples.

(i) *Shift-by-n_0 sequence:*

$$x_n = \delta_{n-n_0} \quad \stackrel{\text{ZT}}{\longleftrightarrow} \quad X(z) = z^{-n_0}; \qquad \text{ROC} = \begin{cases} |z| > 0, & \text{for } n_0 > 0; \\ \text{all } z, & \text{for } n_0 = 0; \\ |z| < \infty, & \text{for } n_0 < 0. \end{cases}$$

The shift-by-one maps to z^{-1}, which is why z^{-1} is often called a delay operator. It also follows that

$$x_{n-n_0} \quad \stackrel{\text{ZT}}{\longleftrightarrow} \quad z^{-n_0} X(z); \qquad \text{ROC} = \text{ROC}_x,$$

with the only possible changes to the ROC at 0 or ∞.

(ii) *Right-sided geometric sequence:*

$$x_n = \alpha^n u_n \quad \stackrel{\text{ZT}}{\longleftrightarrow} \quad X(z) = \frac{1}{1 - \alpha z^{-1}}; \qquad \text{ROC} = \{z \mid |z| > |\alpha|\}. \tag{3.128a}$$

Now, $z = \alpha$ is a zero of the denominator of the complex function $X(z)$, and we see that the ROC is bounded from inside by a circle containing the $z = \alpha$. This is a general property, since the ROC cannot contain a singularity (a z such that $X(z)$ does not exist).

(iii) *Left-sided geometric sequence:*

$$x_n = -\alpha^n u_{-n-1} \quad \overset{\text{ZT}}{\longleftrightarrow} \quad X(z) = \frac{1}{1-\alpha z^{-1}}; \qquad \text{ROC} = \{z \mid |z| < |\alpha|\}. \tag{3.128b}$$

The expression for $X(z)$ is exactly as in the previous case; the only difference is in the ROC. Had we been given only this $X(z)$ without the associated ROC, we would not have been able to tell whether it originated from x in (3.128a) or in (3.128b). This shows why the z-transform and its ROC form a pair that should not be broken.

A standard way of showing the ROC is a plot of the complex plane, as in Figure 3.14. Marking the unit circle establishes the scale of the plot, and the DTFT converges for all ω when the unit circle is in the ROC.

Rational z-transforms An important class of z-transforms consists of those that are rational functions, since transfer functions of most realizable systems (systems that can be built and used in practice) are rational. We will see in Section 3.5.4 that these are directly related to difference equations with a finite number of coefficients, as in (3.55). Such transfer functions are of the form

$$H(z) = \frac{B(z)}{A(z)}, \tag{3.129}$$

where $A(z)$ and $B(z)$ are polynomials in z^{-1} with no common roots, of degree N and M, respectively. The degrees satisfy $M \leq N$, otherwise, polynomial division would lead to a sum of a polynomial and a rational function satisfying this constraint. The zeros of the numerator $B(z)$ and denominator $A(z)$ are called the *zeros* and *poles* of the rational transfer function $H(z)$; they are typically shown on a *pole–zero plot* with "o" for zeros and "×" for poles (see Figure 3.15 on Page 248 for an example). Many properties of LSI systems depend on the zeros and poles and their multiplicities.

Consider a finite-length sequence $h = \begin{bmatrix} h_0 & h_1 & \ldots & h_M \end{bmatrix}^\top$. Then, $H(z) = \sum_{k=0}^{M} h_k z^{-k}$, which has M poles at $z = 0$ and M zeros at the roots $\{z_k\}_{k=1}^{M}$ of the polynomial $H(z)$.[57] Therefore, $H(z)$ can be written as

$$H(z) = h_0 \prod_{k=1}^{M} (1 - z_k z^{-1}), \qquad |z| > 0, \tag{3.130}$$

[57] The fundamental theorem of algebra (Theorem 3.23) states that a degree-M polynomial has M complex roots.

where the form of the factorization shows explicitly both the roots and the multiplicative factor h_0. The rational z-transform in (3.129) can thus be written as

$$H(z) = \frac{b_0 \prod_{k=1}^{M}(1 - z_k z^{-1})}{a_0 \prod_{k=1}^{N}(1 - p_k z^{-1})}, \tag{3.131}$$

where $\{z_k\}_{k=1}^{M}$ are zeros and $\{p_k\}_{k=1}^{N}$ poles (remember that $M \leq N$). The ROC cannot contain any poles and is thus, assuming a right-sided sequence, all z outside of the pole largest in magnitude. If M is smaller than N, then $H(z)$ has $N - M$ additional zeros at 0. This can be seen in our previous example (3.128a), which can be rewritten as $1/(1 - \alpha z^{-1}) = z/(z - \alpha)$ and has thus a pole at $z = \alpha$ and a zero at $z = 0$.

Inversion

Given a z-transform and its ROC, how do we invert the z-transform? The general inversion formula for the z-transform involves contour integration, which is a standard topic of complex analysis. However, most z-transforms encountered in practice can be inverted using simpler methods, which we now discuss; the *Further reading* gives pointers for a more detailed treatment of the inverse z-transform.

Inversion by inspection This method is just a way of recognizing certain z-transform pairs. For example, from Table 3.6 in Section 3.5.3, we see that the z-transform

$$X(z) = \frac{1}{1 - \frac{1}{4}z^{-1}}$$

has the form of $1/(1 - az^{-1})$, with $a = \frac{1}{4}$. From Table 3.6, we can then read the sequence that generated it as one of the following two:

$$\left(\frac{1}{4}\right)^n u_n, \qquad \text{if ROC} = \left\{z \,\middle|\, |z| > \tfrac{1}{4}\right\}$$

or

$$-\left(\frac{1}{4}\right)^n u_{-n-1}, \qquad \text{if ROC} = \left\{z \,\middle|\, |z| < \tfrac{1}{4}\right\};$$

no other ROC is possible.

Inversion using partial fraction expansion When the z-transform is given as a rational function, partial fraction expansion results in a sum of terms, each of which can be inverted by inspection. Here we consider cases in which the numerator and denominator are polynomials in z^{-1}, as in (3.131).

(i) $M < N$, simple poles: If all the N poles are of first order, we can express $X(z)$ as

$$X(z) = \sum_{k=1}^{N} \frac{A_k}{1 - p_k z^{-1}}, \qquad A_k = (1 - p_k z^{-1})X(z)\big|_{z=p_k}. \tag{3.133a}$$

Each term has a simple inverse z-transform, which depends on the ROC of $X(z)$. The ROC takes one of the following forms:

$$\text{ROC} = \begin{cases} \{z \mid |z| < |p_1|\}, & \\ \{z \mid |p_k| < |z| < |p_{k+1}|\} & \text{for some } k, \text{ or,} \\ \{z \mid |z| > |p_N|\}, & \end{cases}$$

where we have assumed that $|p_1| \leq |p_2| \leq \cdots \leq |p_N|$ for simplicity. Each distinct ROC corresponds to a different sequence. Often the ROC is $\{z \mid |z| > |p_N|\}$, resulting in

$$x_n = \sum_{k=1}^{N} A_k (p_k)^n u_n. \tag{3.133b}$$

(ii) $M < N$, poles with multiplicity: Suppose that $X(z)$ has pole p_i of order $s > 1$. Then, in general, the ith term in (3.133a) is replaced by s terms

$$\sum_{k=1}^{s} \frac{C_k}{(1 - p_i z^{-1})^k}.$$

The $k = 1$ term is inverted as before, and the terms for $k > 1$ are inverted using the differentiation rule from Table 3.6.

(iii) $M \geq N$: Assume that all poles are of first order; multiplicities can be treated as above. Using polynomial division, we can write $X(z)$ as

$$X(z) = \sum_{k=0}^{M-N} B_k z^{-k} + \sum_{k=1}^{N} \frac{A_k}{1 - p_k z^{-1}}. \tag{3.134a}$$

The first summation of (3.134a) is clearly the z-transform of the sequence

$$\begin{bmatrix} \cdots & 0 & \boxed{B_0} & B_1 & B_2 & \cdots & B_{M-N} & 0 & \cdots \end{bmatrix}^{\top}.$$

There are many possible ROCs, each determining a distinct sequence corresponding to the second summation in (3.134a). When the ROC is outside of the largest pole, putting together both summations of (3.134a) yields

$$x_n = \sum_{k=0}^{M-N} B_k \delta_{n-k} + \sum_{k=1}^{N} A_k (p_k)^n u_n. \tag{3.134b}$$

We illustrate the method with an example:

Example 3.21 (Inversion using partial fraction expansion) Given

$$X(z) = \frac{1 - z^{-1}}{1 - 5z^{-1} + 6z^{-2}} = \frac{1 - z^{-1}}{(1 - 2z^{-1})(1 - 3z^{-1})},$$

with poles at $z = 2$ and $z = 3$, we compute the coefficients as in (3.133a) to be

$$A_1 = \left.\frac{1 - z^{-1}}{1 - 3z^{-1}}\right|_{z=2} = -1, \qquad A_2 = \left.\frac{1 - z^{-1}}{1 - 2z^{-1}}\right|_{z=3} = 2,$$

yielding

$$X(z) = \frac{-1}{1 - 2z^{-1}} + \frac{2}{1 - 3z^{-1}}.$$

The original sequence is then

$$x_n = \begin{cases} (2^n - 2 \cdot 3^n)u_{-n-1}, & \text{if ROC} = \{z \mid |z| < 2\}; \\ -2^n u_n - 2 \cdot 3^n u_{-n-1}, & \text{if ROC} = \{z \mid 2 < |z| < 3\}; \\ (-2^n + 2 \cdot 3^n)u_n, & \text{if ROC} = \{z \mid |z| > 3\}. \end{cases}$$

Inversion using power-series expansion This method is most useful for finite-length sequences. For example, given $X(z) = (1 - z^{-1})(1 - 2z^{-1})$, we can expand it in its power-series form as

$$X(z) = 1 - 3z^{-1} + 2z^{-2}.$$

Knowing that each of the elements in this power series corresponds to a delayed Kronecker delta sequence, we can read off the sequence directly,

$$x_n = \delta_n - 3\delta_{n-1} + 2\delta_{n-2}.$$

EXAMPLE 3.22 (INVERSION USING POWER-SERIES EXPANSION) Suppose that

$$X(z) = \log(1 + 2z^{-1}); \qquad \text{ROC} = \{z \mid |z| > 2\}. \tag{3.135}$$

To invert this z-transform, we use its power-series expansion from Table P2.54-1. By substituting $x = 2z^{-1}$, we confirm that $|x| = |2z^{-1}| < 2 \cdot \frac{1}{2} = 1$, and thus the series expansion

$$\log(1 + 2z^{-1}) = \sum_{n=1}^{\infty} (-1)^{n+1} \frac{2^n}{n} z^{-n}$$

holds for the z values of interest. Thus, the desired inverse z-transform is

$$x_n = \begin{cases} (-1)^{n+1} 2^n n^{-1}, & \text{for } n \geq 1; \\ 0, & \text{otherwise.} \end{cases}$$

3.5.3 Properties of the z-transform

The z-transform has the same properties as the DTFT, but for a larger class of sequences. The main new twist is the need to properly account for ROCs. As an example, the convolution of two sequences can be computed as a product in the transform domain even when the sequences do not have proper DTFTs, provided that the sequences have some part of their ROCs in common. A summary of z-transform properties can be found in Table 3.6. As both convolution in frequency and the Parseval equality involve contour integration, we opt not to state them here; a number of standard texts cover those.

Linearity The z-transform is a linear operator, or

$$\alpha x_n + \beta y_n \quad \stackrel{\text{ZT}}{\longleftrightarrow} \quad \alpha X(z) + \beta Y(z); \qquad \text{ROC}_{\alpha x+\beta y} \supset \text{ROC}_x \cap \text{ROC}_y. \tag{3.136}$$

Shift in time The z-transform pair corresponding to a shift in time by n_0 is

$$x_{n-n_0} \quad \stackrel{\text{ZT}}{\longleftrightarrow} \quad z^{-n_0} X(z), \tag{3.137}$$

with no changes to the ROC except possibly at $z = 0$ or $|z| = \infty$.

Scaling in time Scaling in time appears in two flavors.

(i) The z-transform pair corresponding to scaling in time by N is

$$x_{Nn} \quad \stackrel{\text{ZT}}{\longleftrightarrow} \quad \frac{1}{N} \sum_{k=0}^{N-1} X(W_N^k z^{1/N}); \qquad (\text{ROC}_x)^{1/N}. \tag{3.138}$$

We have already seen this operation of downsampling in (3.91), and will discuss it in more detail in Section 3.7.

(ii) The z-transform pair corresponding to scaling in time by $1/N$ is

$$\begin{cases} x_{n/N}, & \text{for } n/N \in \mathbb{Z}; \\ 0, & \text{otherwise} \end{cases} \quad \stackrel{\text{ZT}}{\longleftrightarrow} \quad X(z^N); \qquad (\text{ROC}_x)^N. \tag{3.139}$$

We have already seen this operation of upsampling in (3.92), and will discuss it in more detail in Section 3.7.

Scaling in z The z-transform pair corresponding to scaling in z by α^{-1} is

$$\alpha^n x_n \quad \stackrel{\text{ZT}}{\longleftrightarrow} \quad X(\alpha^{-1} z); \qquad |\alpha| \, \text{ROC}_x. \tag{3.140}$$

Time reversal The z-transform pair corresponding to time reversal x_{-n} is

$$x_{-n} \quad \stackrel{\text{ZT}}{\longleftrightarrow} \quad X(z^{-1}); \qquad \frac{1}{\text{ROC}_x}. \tag{3.141}$$

Differentiation The z-transform pair corresponding to differentiation in z is

$$n^k x_n \quad \stackrel{\text{ZT}}{\longleftrightarrow} \quad (-1)^k z^k \frac{\partial^k X(z)}{\partial z^k}; \qquad \text{ROC}_x. \tag{3.142}$$

Moments Computing the kth moment using the z-transform results in

$$m_k = \sum_{n\in\mathbb{Z}} n^k x_n = \left(\sum_{n\in\mathbb{Z}} n^k x_n z^{-n} \right) \Bigg|_{z=1} = (-1)^k \frac{\partial^k X(z)}{\partial z^k} \Bigg|_{z=1}, \qquad k \in \mathbb{N}, \tag{3.143a}$$

as a direct application of (3.142). The first two moments are

$$m_0 = \sum_{n\in\mathbb{Z}} x_n = \left.\left(\sum_{n\in\mathbb{Z}} x_n z^{-n}\right)\right|_{z=1} = X(0), \tag{3.143b}$$

$$m_1 = \sum_{n\in\mathbb{Z}} n x_n = \left.\left(\sum_{n\in\mathbb{Z}} n x_n z^{-n}\right)\right|_{z=1} = -\left.\frac{\partial X(z)}{\partial z}\right|_{z=1}. \tag{3.143c}$$

Convolution in time The z-transform pair corresponding to convolution in time is

$$(h * x)_n \quad \stackrel{\text{ZT}}{\longleftrightarrow} \quad H(z)\,X(z); \qquad \text{ROC}_{h*x} \supset \text{ROC}_h \cap \text{ROC}_x. \tag{3.144}$$

This key result is the z-transform analogue of the DTFT property (3.96). The z-transform of $y = h * x$ can be obtained with slight modifications of (3.97):

$$\begin{aligned}
Y(z) &\stackrel{(a)}{=} \sum_{n\in\mathbb{Z}} y_n z^{-n} \stackrel{(b)}{=} \sum_{n\in\mathbb{Z}} \left(\sum_{k\in\mathbb{Z}} x_k h_{n-k}\right) z^{-n} \\
&= \sum_{n\in\mathbb{Z}} \sum_{k\in\mathbb{Z}} x_k z^{-k} h_{n-k} z^{-(n-k)} \\
&\stackrel{(c)}{=} \sum_{k\in\mathbb{Z}} x_k z^{-k} \sum_{n\in\mathbb{Z}} h_{n-k} z^{-(n-k)} \stackrel{(d)}{=} X(z)H(z),
\end{aligned} \tag{3.145}$$

where (a) follows from the definition of the z-transform; (b) from the definition of convolution; (c) from interchanging the order of summation; and (d) from the definition of the z-transform. The distinction from (3.97) is that (c) might hold when DTFTs of x and h do not exist; when $z \in \text{ROC}_h \cap \text{ROC}_x$, each series following (c) is absolutely convergent, enabling the interchange. The wider applicability of (3.144) than (3.96) is a key feature of the z-transform.

Example 3.23 (z-transform convolution property) For some $\alpha \in \mathbb{R}^+$, consider

$$x_n = u_n, \qquad h_n = \alpha^n u_n.$$

We cannot use the DTFT to compute the convolution $y = h * x$ because x does not have a DTFT, and, for $\alpha \geq 1$, neither does h. The z-transform exists for both x and h, and it can be used to compute the convolution, provided that the ROCs overlap. The z-transforms are

$$X(z) = \frac{1}{1 - z^{-1}}; \qquad \text{ROC}_x = \{z \mid |z| > 1\},$$

and

$$H(z) = \frac{1}{1 - \alpha z^{-1}}; \qquad \text{ROC}_h = \{z \mid |z| > \alpha\},$$

and thus

$$Y(z) = \frac{1}{(1 - \alpha z^{-1})(1 - z^{-1})}; \qquad \text{ROC}_y \supset \{z \mid |z| > \max\{\alpha, 1\}\}.$$

z-Transform properties	Time domain	z-Transform domain	ROC
ROC properties	General sequence		An open annulus
	Finite-length sequence		All z, except possibly 0, ∞
	Right-sided sequence		$\|z\| >$ largest pole
	Left-sided sequence		$\|z\| <$ smallest pole
	BIBO-stable		$\supset \|z\| = 1$
Basic properties			
Linearity	$\alpha x_n + \beta y_n$	$\alpha X(z) + \beta Y(z)$	$\supset \mathrm{ROC}_x \cap \mathrm{ROC}_y$
Shift in time	x_{n-n_0}	$z^{-n_0} X(z)$	ROC_x
Scaling in time			
Downsampling	x_{Nn}	$\frac{1}{N}\sum_{k=0}^{N-1} X(W_N^k z^{1/N})$	$(\mathrm{ROC}_x)^{1/N}$
Upsampling	$\begin{cases} x_{n/N}, & n/N \in \mathbb{Z}; \\ 0, & \text{otherwise.}\end{cases}$	$X(z^N)$	$(\mathrm{ROC}_x)^N$
Scaling in z	$\alpha^n x_n$	$X(\alpha^{-1} z)$	$\|\alpha\| \, \mathrm{ROC}_x$
Time reversal	x_{-n}	$X(z^{-1})$	$1/\mathrm{ROC}_x$
Differentiation	$n^k x_n$	$(-1)^k z^k \partial^k X(z)/\partial z^k$	ROC_x
	$m_k = \sum_{n\in\mathbb{Z}} n^k x_n =$	$(-1)^k \partial^k X(z)/\partial z^k \big\vert_{z=1}$	
Convolution in time	$(h * x)_n$	$H(z)\, X(z)$	$\supset \mathrm{ROC}_h \cap \mathrm{ROC}_x$
Deterministic autocorrelation	$a_n = \sum_{k\in\mathbb{Z}} x_k x^*_{k-n}$	$A(z) = X(z)\, X_*(z^{-1})$	$\mathrm{ROC}_x \cap 1/\mathrm{ROC}_x$
Deterministic crosscorrelation	$c_n = \sum_{k\in\mathbb{Z}} x_k y^*_{k-n}$	$C(z) = X(z)\, Y_*(z^{-1})$	$1/\mathrm{ROC}_x \cap \mathrm{ROC}_y$
Related sequences			
Conjugate	x^*_n	$X^*(z^*)$	ROC_x
Conjugate, time-reversed	x^*_{-n}	$X_*(z^{-1})$	$1/\mathrm{ROC}_x$
Real part	$\Re(x_n)$	$(X(z) + X^*(z^*))/2$	ROC_x
Imaginary part	$\Im(x_n)$	$(X(z) - X^*(z^*))/(2j)$	ROC_x
Symmetries for real x			
X conjugate symmetric		$X(z) = X^*(z^*)$	
Common transform pairs			
Kronecker delta sequence	δ_n	1	All z
Shifted Kronecker delta sequence	δ_{n-n_0}	z^{-n_0}	All z, except possibly 0, ∞
Geometric sequence	$\alpha^n u_n$	$1/(1-\alpha z^{-1})$	$\|z\| > \|\alpha\|$
	$-\alpha^n u_{-n-1}$		$\|z\| < \|\alpha\|$
Arithmetic–geometric sequence	$n\alpha^n u_n$	$\alpha z^{-1}/(1-\alpha z^{-1})^2$	$\|z\| > \|\alpha\|$
	$-n\alpha^n u_{-n-1}$		$\|z\| < \|\alpha\|$

Table 3.6 Properties of the z-transform ($X_*(z)$ denotes $X^*(z^*)$).

By partial fraction expansion, we can rewrite $Y(z)$ as

$$Y(z) = \frac{-\alpha/(1-\alpha)}{1-\alpha z^{-1}} + \frac{1/(1-\alpha)}{1-z^{-1}},$$

leading to

$$y_n = -\frac{\alpha}{1-\alpha}\alpha^n u_n + \frac{1}{1-\alpha}u_n = \frac{1-\alpha^{n+1}}{1-\alpha}u_n.$$

As a check, we can compute the time-domain convolution directly,

$$y_n = \sum_{k\in\mathbb{Z}} x_k h_{n-k} = \sum_{k=0}^{\infty} h_{n-k} = \sum_{k=0}^{n} \alpha^{n-k} = \frac{1-\alpha^{n+1}}{1-\alpha}u_n.$$

When $\alpha \in [0,1)$, the DTFT of y exists, but we nevertheless needed the z-transform to compute the convolution because the DTFT of x does not exist. When $\alpha > 1$, the DTFT of y does not exist, while the z-transform $Y(z)$ exists with ROC $\{z \mid |z| > \alpha\}$.

Example 3.24 (Failure of z-transform convolution property) Here is an example where the convolution sum converges, but even the z-transform does not help in computing it:

$$x_n = 1, \quad n \in \mathbb{Z}, \qquad h_n = \alpha^n u_n, \quad 0 < \alpha < 1.$$

We can compute the convolution directly,

$$y_n = h * x = \sum_{n\in\mathbb{Z}} h_n x_{k-n} = \sum_{n\in\mathbb{N}} \alpha^n = \frac{1}{1-\alpha}.$$

However, there are no values of z such that the z-transform of x converges; that is, the ROC is empty. This prohibits the use of the z-transform for the computation of this convolution.

For finite-length, right-sided sequences, (3.144) connects convolution with polynomial multiplication. Given a length-N sequence x and a length-M impulse response h, the z-transforms of x and h are

$$H(z) = \sum_{n=0}^{M-1} h_n z^{-n}, \qquad X(z) = \sum_{n=0}^{N-1} x_n z^{-n}.$$

Each is a polynomial in z^{-1}. The product polynomial $H(z)X(z)$ has powers of z^{-1} going from 0 to $M+N-2$, and its nth coefficient is obtained from the coefficients in $H(z)$ and $X(z)$ that have powers summing to n, that is, the convolution $h * x$ given in (3.61).

Deterministic autocorrelation The z-transform pair corresponding to the deterministic autocorrelation of a sequence x is

$$a_n = \sum_{k\in\mathbb{Z}} x_k x_{k-n}^* \quad \overset{\text{ZT}}{\longleftrightarrow} \quad A(z) = X(z)\,X_*(z^{-1}); \qquad \text{ROC}_x \cap \frac{1}{\text{ROC}_x}, \tag{3.146}$$

where $X_*(z)$ denotes $X^*(z^*)$, which amounts to conjugating coefficients but not z. This z-transform satisfies

$$A(z) = A_*(z^{-1}). \tag{3.147a}$$

For a real x,

$$A(z) = X(z)\,X(z^{-1}) = A(z^{-1}). \tag{3.147b}$$

Proof of (3.147b) is left for Exercise 3.11. We know that, on the unit circle, the deterministic autocorrelation is the squared magnitude of the spectrum $|X(e^{j\omega})|^2$ as in (3.100). This quadratic form, when extended to the z-plane, leads to a particular symmetry of poles and zeros when $A(z)$ is a rational function.

THEOREM 3.13 (RATIONAL AUTOCORRELATION) A rational function $A(z)$ is the z-transform of the deterministic autocorrelation of a stable real sequence x, if and only if

(i) its complex poles and zeros appear in quadruples:

$$\{z_i, z_i^*, z_i^{-1}, (z_i^{-1})^*\}, \qquad \{p_i, p_i^*, p_i^{-1}, (p_i^{-1})^*\}; \tag{3.148a}$$

(ii) its real poles and zeros appear in pairs:

$$\{z_i, z_i^{-1}\}, \qquad \{p_i, p_i^{-1}\}; \tag{3.148b}$$

and

(iii) its zeros on the unit circle are double zeros:

$$\{z_i, z_i^*, z_i^{-1}, (z_i^{-1})^*\} = \{e^{j\omega_i}, e^{-j\omega_i}, e^{-j\omega_i}, e^{j\omega_i}\}, \tag{3.148c}$$

with possibly double zeros at $z = \pm 1$. There are no poles on the unit circle.

Proof. The proof is based on the following two facts:

1. Since the sequence x is real, its deterministic autocorrelation a is real. From Table 3.6, this implies that

$$A^*(z) = A(z^*) \quad \Rightarrow \quad \begin{array}{l} p_i \text{ pole} \;\Rightarrow\; p_i^* \text{ pole}, \\ z_i \text{ zero} \;\Rightarrow\; z_i^* \text{ zero}. \end{array} \tag{3.149a}$$

2. Since any deterministic autocorrelation satisfies $a_{-n} = a_n^*$ and a is real, $a_{-n} = a_n$. From Table 3.6, this implies that

$$A(z^{-1}) = A(z) \quad \Rightarrow \quad \begin{array}{l} p_i \text{ pole} \;\Rightarrow\; p_i^{-1} \text{ pole}, \\ z_i \text{ zero} \;\Rightarrow\; z_i^{-1} \text{ zero}. \end{array} \tag{3.149b}$$

We now proceed to prove that $A(z)$ being the z-transform of the autocorrelation of a stable and real sequence x implies (i)–(iii). The converse follows similarly.

(i) From (3.149a) and (3.149b), we have that

$$p_i \text{ pole} \quad \Rightarrow \quad \begin{matrix} p_i^* \text{ pole} \\ p_i^{-1} \text{ pole} \end{matrix} \quad \Rightarrow \quad (p_i^*)^{-1} \text{ pole},$$

and similarly for zeros, and we obtain the pole/zero quadruples in (3.148a).

(ii) If a zero/pole is real, it is its own conjugate, and thus quadruples in (3.148a) become pairs in (3.148b).

(iii) Since x is stable, there are no poles on the unit circle. Since x is real, $X^*(z) = X(z^*)$. Thus, a rational $A(z)$ has only zeros on the unit circle from $X(z)$ and $X(z^{-1})$.

$$z_i \text{ zero of } X(z) \Rightarrow \begin{matrix} z_i^{-1} & \text{zero of} & X(z^{-1}) \\ z_i^* & \text{zero of} & X(z) \end{matrix}$$
$$\Rightarrow (z_i^*)^{-1} \text{ zero of } X(z^{-1}) \Rightarrow z_i = (z_i^*)^{-1} \text{ zero of } X(z).$$

Thus, both $X(z)$ and $X(z^{-1})$ have z_i as a zero, leading to double zeros on the unit circle.

Deterministic crosscorrelation The z-transform pair corresponding to the deterministic crosscorrelation of sequences x and y is

$$c_n = \sum_{k\in\mathbb{Z}} x_k y_{k-n}^* \quad \stackrel{\text{ZT}}{\longleftrightarrow} \quad C_{x,y}(z) = X(z)Y_*(z^{-1}); \qquad \text{ROC}_x \cap \frac{1}{\text{ROC}_y}, \tag{3.150}$$

and satisfies

$$C_{x,y}(z) = C_{y,x*}(z^{-1}). \tag{3.151a}$$

For real x and y,

$$C_{x,y}(z) = X(z)Y(z^{-1}) = C_{y,x}(z^{-1}). \tag{3.151b}$$

Deterministic autocorrelation of vector sequences The z-transform pair corresponding to the deterministic autocorrelation of a length-N vector sequence x is

$$A_n \quad \stackrel{\text{ZT}}{\longleftrightarrow} \quad A(z) = \begin{bmatrix} A_0(z) & C_{0,1}(z) & \cdots & C_{0,N-1}(z) \\ C_{1,0}(z) & A_1(z) & \cdots & C_{1,N-1}(z) \\ \vdots & \vdots & \ddots & \vdots \\ C_{N-1,0}(z) & C_{N-1,1}(z) & \cdots & A_{N-1}(z) \end{bmatrix}, \tag{3.152}$$

where A_n is given in (3.23). Because of (3.21a), $A(z)$ satisfies

$$A(z) = \begin{bmatrix} A_0(z) & C_{0,1}(z) & \cdots & C_{0,N-1}(z) \\ C_{0,1*}(z^{-1}) & A_1(z) & \cdots & C_{1,N-1}(z) \\ \vdots & \vdots & \ddots & \vdots \\ C_{0,N-1*}(z^{-1}) & C_{1,N-1*}(z^{-1}) & \cdots & A_{N-1}(z) \end{bmatrix} = A_*(z^{-1}). \tag{3.153a}$$

Here, $A_*(z) = A^*(z^*)$ extends the previous notation to mean transposition of A and conjugation of coefficients, but not of z.[58] For a real x,

$$A(z) \;=\; A^{\top}(z^{-1}). \tag{3.153b}$$

Spectral factorization The particular pattern of poles and zeros which characterizes a rational autocorrelation in Theorem 3.13 leads to a key procedure called *spectral factorization.* This amounts to taking the square root of $A(e^{j\omega})$, and, by extension, of $A(z)$, factoring it into rational factors $X(z)$ and $X(z^{-1})$,[59] as a direct corollary of Theorem 3.13.

COROLLARY 3.14 (SPECTRAL FACTORIZATION) A rational z-transform $A(z)$ is the deterministic autocorrelation of a stable real sequence x if and only if it can be factored as $A(z) = X(z)X(z^{-1})$.

Spectral factorization amounts to assigning poles and zeros from quadruples and pairs (3.148a)–(3.148c) to $X(z)$ and $X(z^{-1})$. For the poles, there is a unique rule: take all poles inside the unit circle and assign them to $X(z)$. This is because stability of x requires $X(z)$ to have only poles inside the unit circle (see Theorem 3.15), while x being real requires that conjugate pairs be kept together. For the zeros, there is a choice, since we are not forced to assign only zeros inside the unit circle to $X(z)$. Doing so, however, creates a unique solution called the *minimum-phase solution.*[60] It is now clear why it is important that the zeros on the unit circle appear in pairs: it allows the assignment of one of each to $X(z)$ and $X(z^{-1})$.

EXAMPLE 3.25 (SPECTRAL FACTORIZATION) We now illustrate both the procedure and how we can recognize a deterministic autocorrelation of a real and stable sequence (see Figure 3.15).

(i) The first sequence we examine is a finite-length, symmetric sequence a_n with its associated z-transform,

$$\begin{aligned} a_n &= 2\delta_{n+1} + 5\delta_n + 2\delta_{n-1}, \\ A(z) &= 5 + 2(z + z^{-1}) \;=\; (1 + 2z^{-1})(1 + 2z), \end{aligned}$$

which is depicted in Figures 3.15(a) and (b). This sequence is a deterministic autocorrelation since it has two zeros, $z = -\frac{1}{2}$ and $z = -2$, which appear in a pair as per Theorem 3.13. As we said above, we have a choice of whether to assign $-\frac{1}{2}$ or -2 to $X(z)$; the minimum-phase solution assigns $-\frac{1}{2}$ to $X(z)$ and $-2 = (-\frac{1}{2})^{-1}$ to $X(z^{-1})$.

[58]Note that in (3.106) we could have written the elements below the diagonal, for example, $C^*_{0,1}(e^{j\omega})$, as $C_{0,1*}(e^{-j\omega})$ to parallel the z-transform. Here, the subscript $*$ would just mean conjugation of coefficients, as conjugation of $e^{j\omega}$ is taken care of by negation.

[59]Note that, since $A(e^{j\omega})$ is real and nonnegative, one could write $X(e^{j\omega}) = \sqrt{A(e^{j\omega})}$. However, such a spectral root will in general not be rational.

[60]The name stems from the fact that, among the various solutions, this one will create a minimal delay, or that the sequence is most concentrated toward the origin of time.

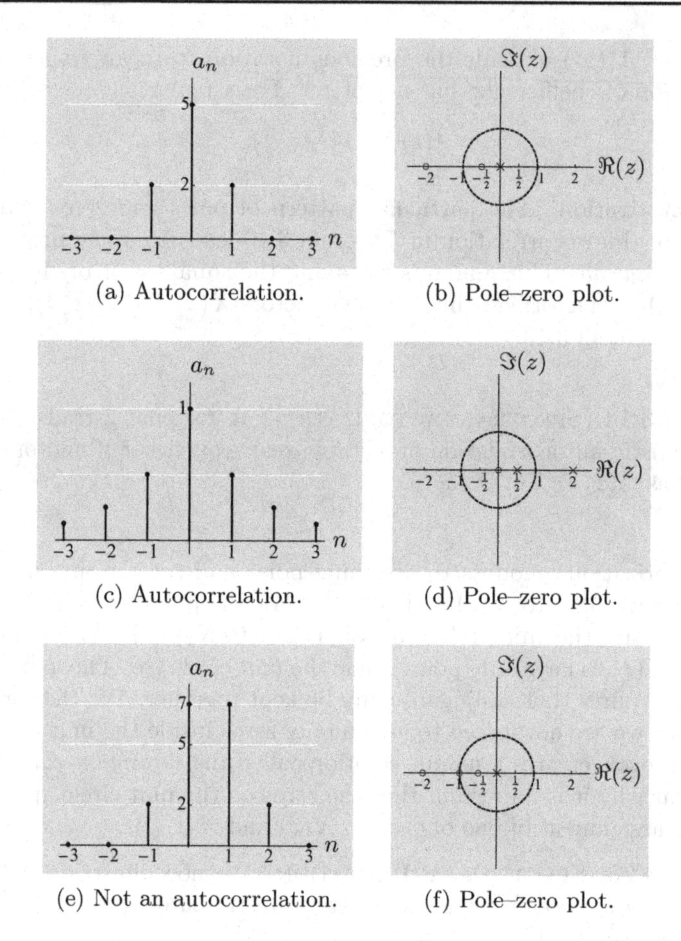

(a) Autocorrelation. (b) Pole–zero plot.

(c) Autocorrelation. (d) Pole–zero plot.

(e) Not an autocorrelation. (f) Pole–zero plot.

Figure 3.15 Pole–zero plots of sequences that are or are not deterministic autocorrelations. (a)–(b) Finite-length, symmetric sequence that is a deterministic autocorrelation. (c)–(d) Infinite-length, symmetric sequence that is a deterministic autocorrelation. (e)–(f) Finite-length, symmetric sequence that is not a deterministic autocorrelation.

(ii) The second sequence is an infinite-length, symmetric sequence a_n with its associated z-transform that we find from Table 3.6,

$$a_n = \left(\frac{1}{2}\right)^n u_n + 2^n u_{-n-1},$$

$$A(z) = \frac{1}{1-\frac{1}{2}z^{-1}} - \frac{1}{1-2z^{-1}} = -\frac{\frac{3}{2}z^{-1}}{(1-\frac{1}{2}z^{-1})(1-2z^{-1})},$$

which is depicted in Figures 3.15(c) and (d). This sequence is a deterministic autocorrelation since it has two poles, $z = \frac{1}{2}$ and $z = 2$, which appear in a pair as per Theorem 3.13. We now have no choice but to assign $\frac{1}{2}$ to

$X(z)$, since for a stable sequence all its poles must be inside the unit circle; the other pole, $2 = (\frac{1}{2})^{-1}$, goes to $X(z^{-1})$.

(iii) Finally, we examine the following finite-length, symmetric sequence a_n with its associated z-transform,

$$\begin{aligned} a_n &= 2\delta_{n+1} + 7\delta_n + 7\delta_{n-1} + 2\delta_{n-2}, \\ A(z) &= 7(1+z^{-1}) + 2(z+z^{-2}) = \left(1+\frac{1}{2}z^{-1}\right)(1+2z^{-1})(1+z^{-1}), \end{aligned}$$

which is depicted in Figures 3.15(e) and (f). This sequence is not a deterministic autocorrelation since it has three zeros, two appearing in a pair as in part (i) and the third, a single zero on the unit circle, violating Theorem 3.13. The DTFT of a_n is not real; for example, $A(e^{j\pi/2}) = -\frac{5}{2}(1+j)$.

3.5.4 z-transform of filters

For filters,

$$H(z) = \sum_{n\in\mathbb{Z}} h_n z^{-n} \tag{3.154}$$

is the counterpart of the frequency response in (3.110a); it is well defined for values of z for which $h_n z^{-n}$ is absolutely summable. As mentioned previously, there is a one-to-one relationship between a rational z-transform and a realizable difference equation (one with a finite number of coefficients). After revisiting this relationship, we establish a necessary and sufficient condition for stability of causal systems with rational transfer functions.

Difference equations with finite number of coefficients Consider a causal solution of a difference equation with a finite number of terms as in (3.55) with zero initial conditions. Assuming that x and y have well-defined z-transforms $X(z)$ and $Y(z)$, and using the fact that x_{n-k} and $z^{-k}X(z)$ are a z-transform pair, we can rewrite (3.55) as

$$Y(z) = \left(\sum_{k=0}^{M} b_k z^{-k}\right) X(z) - \left(\sum_{k=1}^{N} a_k z^{-k}\right) Y(z).$$

The *transfer function* is given by

$$H(z) = \frac{Y(z)}{X(z)} = \frac{\sum_{k=0}^{M} b_k z^{-k}}{1+\sum_{k=1}^{N} a_k z^{-k}}. \tag{3.155}$$

In other words, a linear discrete-time system satisfying difference equation (3.55) has a rational transfer function $H(z)$ in the z-transform domain; that is, the z-transform of the impulse response of the system is a rational function.

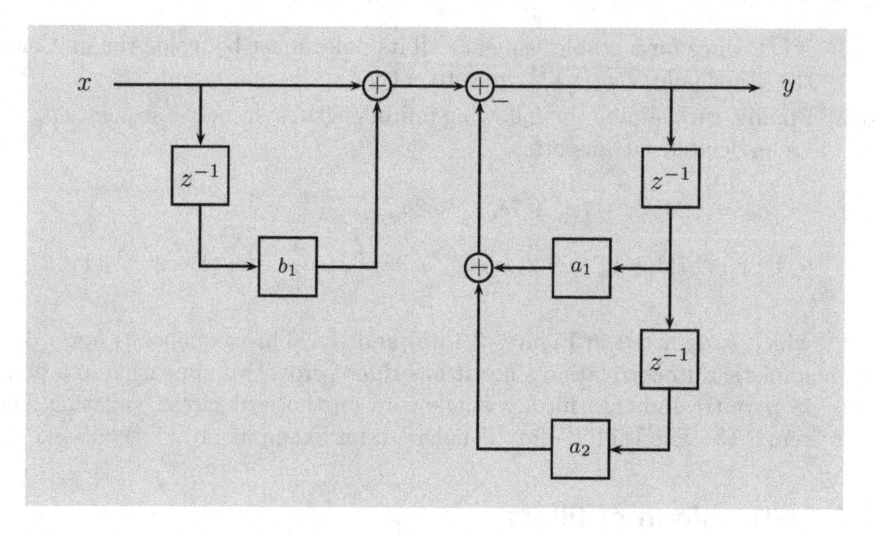

Figure 3.16 A simple discrete-time system, where z^{-1} stands for *unit delay.*

EXAMPLE 3.26 (RATIONAL TRANSFER FUNCTION) Consider the simple system in Figure 3.16. The output is given by

$$y_n = x_n + b_1 x_{n-1} - a_1 y_{n-1} - a_2 y_{n-2},$$

or, in the z-transform domain,

$$Y(z) = X(z) + b_1 z^{-1} X(z) - a_1 z^{-1} Y(z) - a_2 z^{-2} Y(z),$$

yielding the system transfer function

$$H(z) = \frac{1 + b_1 z^{-1}}{1 + a_1 z^{-1} + a_2 z^{-2}}.$$

We now discuss stability for systems with rational transfer functions.

THEOREM 3.15 (BIBO STABILITY WITH RATIONAL TRANSFER FUNCTIONS) A causal, LSI discrete-time system with a rational transfer function is BIBO-stable if and only if the poles of its transfer function are inside the unit circle.

Proof. Using the partial fraction inversion method described in Section 3.5.2, the impulse response of a causal, LSI discrete-time system with rational transfer function is a linear combination of right-sided geometric sequences as in (3.133b) – possibly with multiplication by n^k factors (stemming from multiplicities of poles) and with additional terms that are shifted Kronecker delta sequences (from the numerator having higher degree than the denominator as in (3.134b)). When each pole is inside the unit circle, each term in this linear combination is absolutely summable, so the impulse response is absolutely summable as well; thus, according to Theorem 3.8, the system is BIBO-stable. Conversely, if any pole is outside the unit circle, the impulse response is not absolutely summable; thus, according to Theorem 3.8, the system is not BIBO stable.

Filters

A major application of the z-transform is in the analysis and design of filters. With the restriction to rational functions for realizability, designing a desirable filter is essentially the problem of strategically placing poles and zeros in the z-plane. Although this might sound simple, filter design is a rather sophisticated problem, and it has led to a vast literature and numerous numerical procedures. Here we briefly discuss the z-transform domain properties of certain useful classes of filters. The design of linear-phase FIR filters will be discussed in Section 6.2.6.

FIR filters The z-transform of a length-L FIR filter is a polynomial in z^{-1},

$$H(z) = \sum_{n=0}^{L-1} h_n z^{-n},$$

and is given in its factored form as (3.130).

Linear-phase filters In the z-transform domain, the symmetries from (3.114) become

$$\begin{array}{lcl} \text{symmetric} \\ h_n = h_{L-1-n} & \stackrel{\text{ZT}}{\longleftrightarrow} & H(z) = z^{-L+1} H(z^{-1}), \end{array} \tag{3.156a}$$

$$\begin{array}{lcl} \text{antisymmetric} \\ h_n = -h_{L-1-n} & \stackrel{\text{ZT}}{\longleftrightarrow} & H(z) = -z^{-L+1} H(z^{-1}), \end{array} \tag{3.156b}$$

In the z-transform domain, $H(z^{-1})$ reverses the filter, z^{-L+1} makes it causal again, and $\pm$ determines the type of symmetry.

Allpass filters The basic single-zero, single-pole allpass building block given in (3.119) has the z-transform

$$H(z) = \frac{z^{-1} - \alpha^*}{1 - \alpha z^{-1}}, \tag{3.157}$$

with the zero $1/\alpha^*$ and pole α. For stability in the causal case, $|\alpha| < 1$ is required. A more general allpass filter is formed by cascading these elementary building blocks as

$$H(z) = \prod_{i=1}^{N} \frac{z^{-1} - \alpha_i^*}{1 - \alpha_i z^{-1}} = z^{-N} \frac{B_*(z^{-1})}{B(z)}, \tag{3.158}$$

where $B(z) = \prod_{i=1}^{N} (1 - \alpha_i z^{-1})$. The z-transform of the deterministic autocorrelation of such an allpass filter is given by

$$A(z) = H(z) H_*(z^{-1}) = \prod_{i=1}^{N} \frac{z^{-1} - \alpha_i^*}{1 - \alpha_i z^{-1}} \prod_{i=1}^{N} \frac{z - \alpha_i}{1 - \alpha_i^* z} = 1,$$

and thus an allpass filter has the deterministic autocorrelation sequence $a_m = \delta_m$. Poles and zeros appear in pairs as $\{\alpha, 1/\alpha^*\} = \{r_0 e^{j\omega_0}, (1/r_0)e^{j\omega_0}\}$ for some real $r_0 \in (0,1)$ and angle ω_0. They appear across the unit circle at the same angle and at reciprocal magnitudes, and thus the magnitude $|H(e^{j\omega})|$ is not influenced while the phase is, as was shown in Figure 3.13.

3.6 Discrete Fourier transform

As was already mentioned, one way in which a finite-length sequence arises is from one period of an infinite-length periodic sequence. The version of the Fourier transform designed for finite-length sequences treats all finite-length sequences this way, so effectively we are circularly extending any finite-length sequence. As we have seen in Section 3.3.3, the circular convolution operator (3.71) is the appropriate description of LSI systems operating on circularly extended finite-length inputs.

The version of the Fourier transform for this combination of a sequence space and convolution is the discrete Fourier transform. We will introduce the DFT using eigensequences of the circular convolution operator; this is analogous to our discussion on eigensequences of the linear convolution operator leading to the definition of the DTFT. As expected from this construction, the DFT diagonalizes the circular convolution operator.

The use of the DFT extends far beyond the analysis of periodic sequences. One important use arises from the connection between circular and linear convolution; specifically, we will see that the DFT is a tool for fast computation of linear convolution. The DFT computes the DTFT of a finite segment of an infinite-length signal at a finite number of frequencies; it is thus the operational tool for computing the DTFT, which cannot be computed in full.

3.6.1 Definition of the DFT

Eigensequences of the circular convolution operator Upon mimicking what we did for the DTFT, the DFT arises from identifying the unit-modulus eigensequences of the circular convolution operator defined in (3.71). We can guess that any unit-modulus eigensequence is a complex exponential of the form $v_n = e^{j\omega n}$ like in Section 3.4.1. In addition, since we will be representing sequences of period N, we can guess that any eigensequence should be periodic with period N as well, and thus,

$$v_{n+N} = e^{j\omega(n+N)} = v_n \quad \Leftrightarrow \quad e^{j\omega N} = 1, \tag{3.159}$$

so $\omega = 2\pi k/N$ for some $k \in \mathbb{Z}$. Let us check that

$$v_n = e^{j(2\pi/N)kn} = W_N^{-kn}, \qquad v = \begin{bmatrix} 1 & W_N^{-k} & \ldots & W_N^{-(N-1)k} \end{bmatrix}^\top, \tag{3.160}$$

where $W_N = e^{-j2\pi/N}$ (for details, see (3.287) in Appendix 3.A.1) is indeed an eigensequence of the circular convolution operator H from (3.71),

$$\begin{aligned}(Hv)_n &= (h \circledast v)_n = \sum_{i=0}^{N-1} v_{(n-i) \bmod N} h_i = \sum_{i=0}^{N-1} W_N^{-k[(n-i) \bmod N]} h_i \\ &\overset{(a)}{=} \sum_{i=0}^{N-1} W_N^{-k(n-i)} h_i = \underbrace{\sum_{i=0}^{N-1} h_i W_N^{ki}}_{\lambda_k} \underbrace{W_N^{-kn}}_{v_n},\end{aligned} \tag{3.161}$$

where (a) follows from the fact that $W_N^N = 1$. Thus, applying the convolution operator H to the complex exponential sequence v does indeed result in the same sequence, albeit scaled by the corresponding eigenvalue λ_k. We denote this eigenvalue by H_k using the *frequency response* of the system, which is defined formally in (3.180a). We can thus rewrite (3.161) as

$$Hv = h \circledast v = H_k \, v. \tag{3.162}$$

The quantity k is called the *discrete frequency*. Since k and $k + \ell N$, $\ell \in \mathbb{Z}$, lead to the same complex exponential sequence, we have exactly N distinct complex exponential sequences of period N, indexed by $k \in \{0, 1, \ldots, N-1\}$.

DFT Finding the appropriate Fourier transform now amounts to projecting onto the subspaces generated by each of the eigensequences.

DEFINITION 3.16 (DISCRETE FOURIER TRANSFORM) The *discrete Fourier transform* of a length-N sequence x is

$$X_k = (Fx)_k = \sum_{n=0}^{N-1} x_n W_N^{kn}, \qquad k \in \{0, 1, \ldots, N-1\}; \tag{3.163a}$$

we call it the *spectrum* of x. The *inverse DFT* of a length-N sequence X is

$$x_n = \frac{1}{N}(F^* X)_n = \frac{1}{N} \sum_{k=0}^{N-1} X_k W_N^{-kn}, \qquad n \in \{0, 1, \ldots, N-1\}. \tag{3.163b}$$

We denote the DFT pair as

$$x_n \overset{\text{DFT}}{\longleftrightarrow} X_k.$$

Within the definition, we have introduced $F : \mathbb{C}^N \to \mathbb{C}^N$ to represent the linear DFT operator. The relationship between the inverse and the adjoint in (3.163b) is verified shortly.

Matrix view The DFT expression (3.163a) is the vector–vector product

$$X_k = \begin{bmatrix} 1 & W_N^k & \dots & W_N^{(N-1)k} \end{bmatrix} x,$$

where as usual $x \in \mathbb{C}^N$ is a column vector. By stacking the results for $k \in \{0, 1, \dots, N-1\}$ to produce the column vector $X \in \mathbb{C}^N$, we have

$$X = \begin{bmatrix} 1 & 1 & 1 & \cdots & 1 \\ 1 & W_N & W_N^2 & \cdots & W_N^{N-1} \\ 1 & W_N^2 & W_N^4 & \cdots & W_N^{2(N-1)} \\ \vdots & \vdots & \vdots & \ddots & \vdots \\ 1 & W_N^{N-1} & W_N^{2(N-1)} & \cdots & W_N^{(N-1)^2} \end{bmatrix} x.$$

Thus, the matrix $F \in \mathbb{C}^{N \times N}$ introduced in (3.163a) is

$$F = \begin{bmatrix} 1 & 1 & 1 & \cdots & 1 \\ 1 & W_N & W_N^2 & \cdots & W_N^{N-1} \\ 1 & W_N^2 & W_N^4 & \cdots & W_N^{2(N-1)} \\ \vdots & \vdots & \vdots & \ddots & \vdots \\ 1 & W_N^{N-1} & W_N^{2(N-1)} & \cdots & W_N^{(N-1)^2} \end{bmatrix}. \tag{3.164a}$$

Using orthogonality of the the roots of unity (see (3.288c) in Appendix 3.A.1), we can easily verify that

$$F^{-1} = \frac{1}{N} \begin{bmatrix} 1 & 1 & 1 & \cdots & 1 \\ 1 & W_N^{-1} & W_N^{-2} & \cdots & W_N^{-(N-1)} \\ 1 & W_N^{-2} & W_N^{-4} & \cdots & W_N^{-2(N-1)} \\ \vdots & \vdots & \vdots & \ddots & \vdots \\ 1 & W_N^{-(N-1)} & W_N^{-2(N-1)} & \cdots & W_N^{-(N-1)^2} \end{bmatrix}. \tag{3.164b}$$

Thus,

$$F^{-1} = \frac{1}{N} F^*, \tag{3.164c}$$

as asserted in (3.163b). This shows that the DFT is a unitary operator (up to a scaling factor).[61] Note also that F is a Vandermonde matrix (see (2.248)).

DFT as analysis with an orthogonal basis The expression (3.163a) is an inner product $X_k = \langle x, \varphi_k \rangle$, where

$$\varphi_k = \begin{bmatrix} 1 & W_N^{-k} & \dots & W_N^{-(N-1)k} \end{bmatrix}^\top = \begin{bmatrix} 1 & e^{j(2\pi/N)k} & \dots & e^{j(2\pi/N)(N-1)k} \end{bmatrix}^\top. \tag{3.165}$$

[61] A normalized version uses a factor of $1/\sqrt{N}$ on both F and its inverse.

Thus, the DFT of a length-N sequence is a set of N inner products obtained by applying the analysis operator associated with the basis $\{\varphi_k\}_{k=0}^{N-1}$ of $\mathbb{C}^N$ (see (2.91)).

The basis $\{\varphi_k\}_{k=0}^{N-1}$ is orthogonal (see (3.288c) in Appendix 3.A.1), and each element has norm $\sqrt{N}$. The associated dual basis is

$$\widetilde{\varphi}_k = \frac{1}{N}\varphi_k, \qquad k \in \{0, 1, \ldots, N-1\},$$

and the inverse DFT (3.163b) is obtained by applying the synthesis operator associated with this basis (see (2.90)).

Relation of the DFT to the DTFT Given a length-N sequence x to analyze, we might first turn to what we already have for infinite sequences – the DTFT. However, since x is only N-dimensional, we should not need a function of a continuous variable $X(e^{j\omega})$ to characterize it. Choosing any N distinct samples within one 2π period of $X(e^{j\omega})$ allows recovery of x and thus contains all the information present. Choosing those sampling points to be

$$\omega_k = \frac{2\pi}{N}k, \qquad k \in \{0, 1, \ldots, N-1\} \tag{3.166a}$$

gives the DFT:

$$\begin{aligned} X(e^{j\omega})\big|_{\omega=\omega_k} &\overset{(a)}{=} X(e^{j(2\pi/N)k}) \overset{(b)}{=} \sum_{n\in\mathbb{Z}} x_n e^{-j(2\pi/N)kn} \\ &\overset{(c)}{=} \sum_{n=0}^{N-1} x_n e^{-j(2\pi/N)kn} \overset{(d)}{=} X_k, \end{aligned} \tag{3.166b}$$

where (a) follows from the choice of sampling points; (b) from the definition of the DTFT, (3.80a); (c) from x being finite of length N; and (d) from the definition of the DFT (3.163a). Thus, sampling the DTFT uniformly results in the DFT.

3.6.2 Properties of the DFT

We list here basic properties of the DFT; Table 3.7 summarizes these, together with symmetries as well as standard transform pairs, while Exercise 3.15 explores proofs for some of the properties.

Linearity The DFT operator F is a linear operator, or

$$\alpha x_n + \beta y_n \quad \overset{\text{DFT}}{\longleftrightarrow} \quad \alpha X_k + \beta Y_k. \tag{3.167}$$

Circular shift in time The DFT pair corresponding to a circular shift in time by n_0 is

$$x_{(n-n_0) \bmod N} \quad \overset{\text{DFT}}{\longleftrightarrow} \quad W_N^{kn_0} X_k. \tag{3.168}$$

DFT properties	Time domain	DFT domain
Basic properties		
Linearity	$\alpha x_n + \beta y_n$	$\alpha X_k + \beta Y_k$
Circular shift in time	$x_{(n-n_0) \bmod N}$	$W_N^{k n_0} X_k$
Circular shift in frequency	$W_N^{-k_0 n} x_n$	$X_{(k-k_0) \bmod N}$
Circular time reversal	$x_{-n \bmod N}$	$X_{-k \bmod N}$
Circular convolution in time	$(h \circledast x)_n$	$H_k X_k$
Circular convolution in frequency	$h_n x_n$	$\frac{1}{N}(H \circledast X)_k$
Circular deterministic autocorrelation	$a_n = \sum_{k=0}^{N-1} x_k x^*_{(k-n) \bmod N}$	$A_k = \|X_k\|^2$
Circular deterministic crosscorrelation	$c_n = \sum_{k=0}^{N-1} x_k y^*_{(k-n) \bmod N}$	$C_k = X_k Y_k^*$
Parseval equality	$\\|x\\|^2 = \sum_{n=0}^{N-1} \|x_n\|^2 = \frac{1}{N}\sum_{k=0}^{N-1} \|X_k\|^2 = \frac{1}{N}\\|X\\|^2$	
Related sequences		
Conjugate	x_n^*	$X^*_{-k \bmod N}$
Conjugate, time-reversed	$x^*_{-n \bmod N}$	X_k^*
Real part	$\Re(x_n)$	$(X_k + X^*_{-k \bmod N})/2$
Imaginary part	$\Im(x_n)$	$(X_k - X^*_{-k \bmod N})/(2j)$
Conjugate-symmetric part	$(x_n + x^*_{-n \bmod N})/2$	$\Re(X_k)$
Conjugate-antisymmetric part	$(x_n - x^*_{-n \bmod N})/(2j)$	$\Im(X_k)$
Symmetries for real x		
X conjugate symmetric		$X_k = X^*_{-k \bmod N}$
Real part of X even		$\Re(X_k) = \Re(X_{-k \bmod N})$
Imaginary part of X odd		$\Im(X_k) = -\Im(X_{-k \bmod N})$
Magnitude of X even		$\|X_k\| = \|X_{-k \bmod N}\|$
Phase of X odd		$\arg X_k = -\arg X_{-k \bmod N}$
Common transform pairs		
Kronecker delta sequence	δ_n	1
Shifted Kronecker delta sequence	$\delta_{(n-n_0) \bmod N}$	$W_N^{k n_0}$
Constant sequence	1	$N\delta_k$
Geometric sequence	α^n	$(1-\alpha W_N^{kN})/(1-\alpha W_N^k)$
Periodic sinc sequence (ideal lowpass filter)	$\sqrt{\frac{k_0}{N}}\frac{\operatorname{sinc}(\pi n k_0/N)}{\operatorname{sinc}(\pi n/N)}$	$\begin{cases} \sqrt{\frac{N}{k_0}}, & \left\|k-\frac{N}{2}\right\| \geq \frac{k_0-1}{2}; \\ 0, & \text{otherwise.} \end{cases}$
Box sequence	$\begin{cases} \frac{1}{\sqrt{n_0}}, & \left\|n-\frac{N}{2}\right\| \geq \frac{n_0-1}{2}; \\ 0, & \text{otherwise.} \end{cases}$	$\sqrt{n_0}\,\frac{\operatorname{sinc}(\pi n_0 k/N)}{\operatorname{sinc}(\pi k/N)}$

Table 3.7 Properties of the discrete Fourier transform. (For all sequences in common transform pairs, $n = 0, 1, \ldots, N-1$, and $k = 0, 1, \ldots, N-1$.)

Circular shift in frequency The DFT pair corresponding to a circular shift in frequency by k_0 is

$$W_N^{-k_0 n} x_n \quad \overset{\text{DFT}}{\longleftrightarrow} \quad X_{(k-k_0) \bmod N}. \tag{3.169}$$

As for the DTFT, a shift in frequency is often referred to as *modulation.*

Circular time reversal The DFT pair corresponding to circular time reversal $x_{-n \bmod N}$ is

$$x_{-n \bmod N} \quad \overset{\text{DFT}}{\longleftrightarrow} \quad X_{-k \bmod N}. \tag{3.170}$$

For a real x_n, the DFT of the time-reversed version $x_{-n \bmod N}$ is X_k^*.

Circular convolution in time The DFT pair corresponding to circular convolution in time is

$$(h \circledast x)_n \quad \overset{\text{DFT}}{\longleftrightarrow} \quad H_k X_k. \tag{3.171}$$

We have seen analogous properties for linear convolution: (3.96) for the DTFT and (3.144) for the z-transform. We can derive (3.171) similarly; the proof is left for Exercise 3.15.

At an abstract level, the properties (3.96) (for linear convolution and the DTFT) and (3.171) (for circular convolution and the DFT) have identical justifications: when x is written as a combination of spectral components, the effect of the convolution operator is to simply scale each spectral component by the corresponding eigenvalue of the convolution operator; thus, using the appropriate Fourier transform has diagonalized the convolution operator. The diagonalization represented by (3.171) gains extra significance from two facts: linear convolution of a finite-length sequence and a finite-length filter can be computed using a circular convolution of appropriate length as in Theorem 3.10; and there are fast algorithms for computing the DFT, as discussed in Section 3.9.1. These together yield fast algorithms for computing linear convolution, as discussed in Section 3.9.2.

Circular convolution in frequency The DFT pair corresponding to circular convolution in frequency is

$$h_n x_n \quad \overset{\text{DFT}}{\longleftrightarrow} \quad \frac{1}{N}(H \circledast X)_k. \tag{3.172}$$

The circular convolution in frequency property (3.172) is dual to the convolution in time property (3.171).

Circular deterministic autocorrelation The DFT pair corresponding to the *circular deterministic autocorrelation* of a sequence x is

$$a_n = \sum_{k=0}^{N-1} x_k x^*_{(k-n) \bmod N} \quad \overset{\text{DFT}}{\longleftrightarrow} \quad A_k = |X_k|^2 \tag{3.173}$$

and satisfies

$$A_k = A_k^*, \tag{3.174a}$$

$$A_k \geq 0. \tag{3.174b}$$

For a real x,

$$A_k = |X_k|^2 = A_{-k \bmod N}. \tag{3.174c}$$

Circular deterministic crosscorrelation The DFT pair corresponding to the *circular deterministic crosscorrelation* of sequences x and y is

$$c_n = \sum_{k=0}^{N-1} x_k y^*_{(k-n) \bmod N} \quad \overset{\text{DFT}}{\longleftrightarrow} \quad C_k = X_k Y_k^* \tag{3.175}$$

and satisfies

$$C_{x,y,k} = C^*_{y,x,k}. \tag{3.176a}$$

For real x and y,

$$C_{x,y,k} = X_k Y_{-k \bmod N} = C_{y,x,-k \bmod N}. \tag{3.176b}$$

Circular deterministic autocorrelation of vector sequences The DFT pair corresponding to the *circular deterministic autocorrelation* of a length-N vector sequence x is

$$A_n \quad \overset{\text{DFT}}{\longleftrightarrow} \quad A_k = \begin{bmatrix} A_{0,k} & C_{0,1,k} & \cdots & C_{0,N-1,k} \\ C_{1,0,k} & A_{1,k} & \cdots & C_{1,N-1,k} \\ \vdots & \vdots & \ddots & \vdots \\ C_{N-1,0,k} & C_{N-1,1,k} & \cdots & A_{N-1,k} \end{bmatrix}, \tag{3.177}$$

and satisfies

$$A_k = \begin{bmatrix} A_{0,k} & C_{0,1,k} & \cdots & C_{0,N-1,k} \\ C^*_{0,1,k} & A_{1,k} & \cdots & C_{1,N-1,k} \\ \vdots & \vdots & \ddots & \vdots \\ C^*_{0,N-1,k} & C^*_{1,N-1,k} & \cdots & A_{N-1,k} \end{bmatrix} = A_k^*. \tag{3.178a}$$

For a real x,

$$A_k = A^\top_{-k \bmod N}. \tag{3.178b}$$

Parseval equality The DFT operator F is a unitary operator (up to scaling) and thus preserves the Euclidean norm (up to scaling); see (2.56):

$$\|x\|^2 = \sum_{n=0}^{N-1} |x_n|^2 = \frac{1}{N} \sum_{k=0}^{N-1} |X_k|^2 = \frac{1}{N}\|X\|^2 = \frac{1}{N}\|Fx\|^2. \tag{3.179}$$

This follows from $F/\sqrt{N}$ being a unitary matrix since $F^*F = NI$.

3.6.3 Frequency response of filters

The DFT of a length-N filter (impulse response of an LSI system) h is called the *frequency response*:

$$H_k = \sum_{n=0}^{N-1} h_n W_N^{kn}, \qquad k \in \{0, 1, \ldots, N-1\}. \tag{3.180a}$$

The inverse DFT of the frequency response recovers the impulse response,

$$h_n = \frac{1}{N} \sum_{k=0}^{N-1} H_k W_N^{-kn}, \qquad n \in \{0, 1, \ldots, N-1\}. \tag{3.180b}$$

We can again denote the magnitude and phase as

$$H_k = |H_k| e^{j \arg(H_k)}, \qquad k \in \{0, 1, \ldots, N-1\},$$

where the *magnitude response* $|H_k|$ is an N-periodic real-valued, nonnegative sequence, and the *phase response* $\arg(H_k)$ is an N-periodic, real-valued sequence between $-\pi$ and π.

Diagonalization of the circular convolution operator Let H be the circular convolution operator associated with the length-N filter h (see (3.71) and the matrix representation in (3.75)). The frequency response of h gives a diagonal form for the operator H. Specifically, let

$$\Lambda = \operatorname{diag}(H_0, H_1, \ldots, H_{N-1}), \qquad \text{where } H_k = \sum_{n=0}^{N-1} h_n W_N^{kn}.$$

Also, let X_k denote the spectrum of length-N sequence x_n. Then, since the circular convolution property (3.171) shows that the DFT of $h \circledast x$ is $H_k X_k$, we have

$$F(Hx) = \Lambda F x,$$

where F is the DFT operator as in (3.164a). Since this is true for any x,

$$H = F^{-1} \Lambda F, \tag{3.181a}$$

so the DFT operator F diagonalizes the circular convolution operator H,

$$F H F^{-1} = \Lambda. \tag{3.181b}$$

We illustrate this diagonalization in Figure 3.17, and explore it further in Solved exercise 3.5.

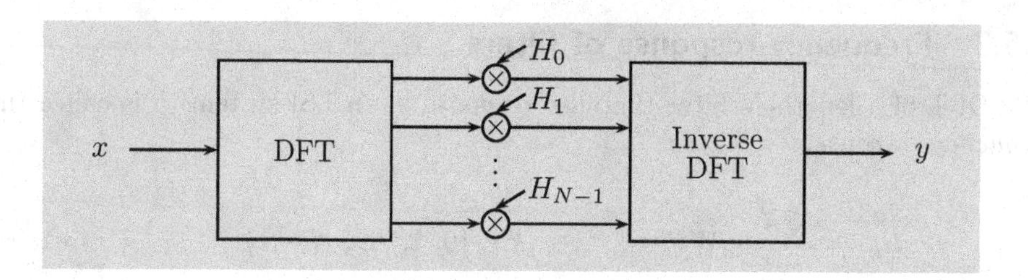

Figure 3.17 Diagonalization property of the DFT. The circular convolution operation $y = Hx = h \circledast x$ is implemented by a pointwise multiplication of the spectrum of x by the frequency response of h.

DFT analysis of infinite sequences and LSI systems While the DFT is intimately related to circular convolution, we should keep in mind that LSI systems described by linear convolution are of primary interest. These systems are characterized by the DTFT of the system impulse response. Since the DFT computes the DTFT at only N frequencies, analysis with the DFT can be incomplete or misleading. In particular, the DFT involves an implicit periodic extension, which might introduce features that require careful interpretation. Furthermore, windowing a longer sequence to length N so that the DFT can be applied also has important consequences.

The following example shows that looking at the frequency response of a filter at only N frequencies can give incorrect impressions about properties of the filter.

Example 3.27 (Inferring DTFT properties from the DFT)

(i) The DFT could indicate allpass behavior despite a nonconstant magnitude response of the DTFT. The length-N DFT H_k of a filter h can satisfy $|H_k| = 1$ for all $k \in \{0, 1, \ldots, N-1\}$ while $\arg(H_k)$ takes arbitrary values. For almost every choice of $\arg(H_k)$ values, the filter h is not allpass. Figures 3.18(a)–(c) show an example of a filter of length 8, along with the magnitude and phase of its DTFT and DFT. Between any two samples of the DTFT that are calculated by the DFT, the magnitude response $|H(e^{j\omega})|$ is not constant. The example was generated with antisymmetric phase $\arg(H_k) = -\arg(H_{-k \bmod 8})$ so that H_k has the conjugate symmetry $H_k = H^*_{-k \bmod 8}$ associated with h being real.

Actually, the only allpass filters having finite length are pure delays or advances, $h_n = \delta_{n-n_0}$ for some $n_0 \in \mathbb{Z}$. This follows from the discussion of the poles and zeros of allpass filters in Section 3.5.4.

(ii) Similarly, the DFT could indicate linear phase despite a nonlinear phase response of the DTFT. Figures 3.18(d)–(f) show an example of a filter g of length 8 generated by setting the phase to be

$$\arg(G_{k \bmod 8}) = \begin{cases} -\frac{1}{4}\pi k, & \text{for } k \in \{-3, -2, \ldots, 3\}; \\ 0, & \text{for } k = 4, \end{cases}$$

and setting the magnitude to satisfy $|G_k| = |G_{-k \bmod 8}|$. The full DTFT phase response $\arg(G(e^{j\omega}))$ is not linear.

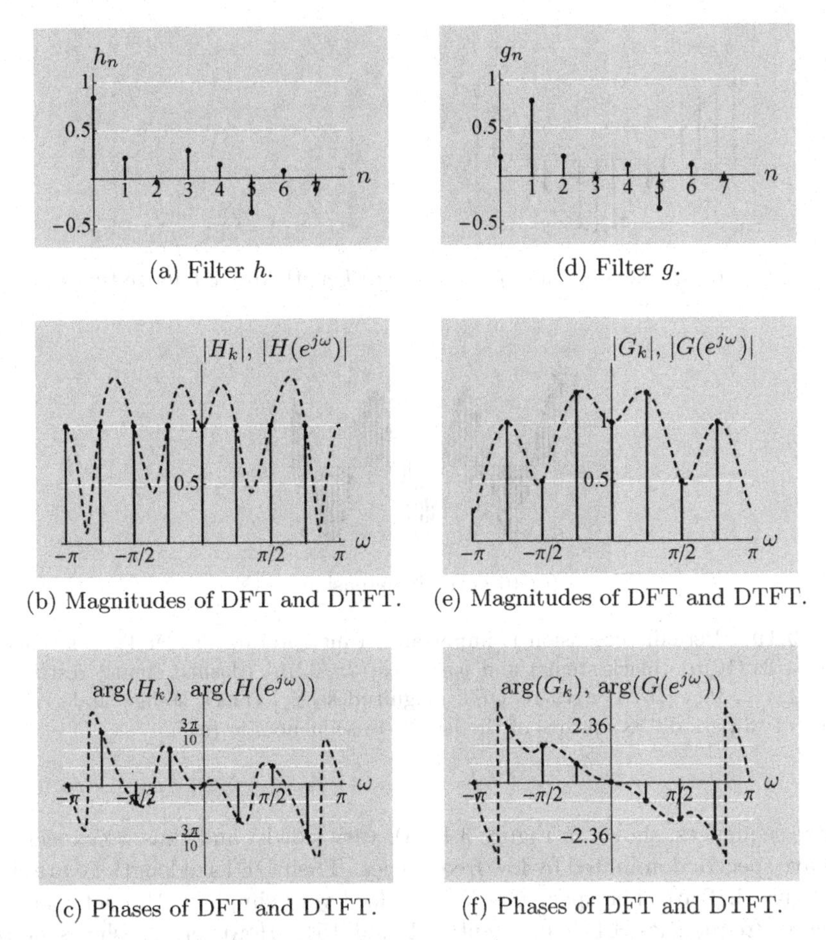

(a) Filter h.

(b) Magnitudes of DFT and DTFT.

(c) Phases of DFT and DTFT.

(d) Filter g.

(e) Magnitudes of DFT and DTFT.

(f) Phases of DFT and DTFT.

Figure 3.18 DFT analysis of a filter can be misleading. (a)–(c) Example of a length-8 filter h that is not an allpass filter although its length-8 DFT satisfies $|H_k| = 1$ for all $k \in \{0, 1, \ldots, 7\}$. (d)–(f) Example of a length-8 filter g that does not have linear phase although its length-8 DFT satisfies $\arg(G_{k \bmod 8}) = -\frac{1}{4}\pi k$ for all $k \in \{-3, -2, \ldots, 3\}$. In (b)–(f), the discrete frequency k of the DFT (solid stems) is converted to the continuous frequency scale of the DTFT (dashed lines) using (3.166a).

The next example illustrates the importance of periodic extension in interpreting the DFT.

Example 3.28 (Frequencies present in the DFT) Let

$$x_n = \cos\left(\frac{2\pi}{16}n\right) = \frac{1}{2}\left(e^{-j(2\pi/16)n} + e^{j(2\pi/16)n}\right),$$

$$y_n = \cos\left(\frac{2\pi}{32}n\right) = \frac{1}{2}\left(e^{-j(2\pi/32)n} + e^{j(2\pi/32)n}\right).$$

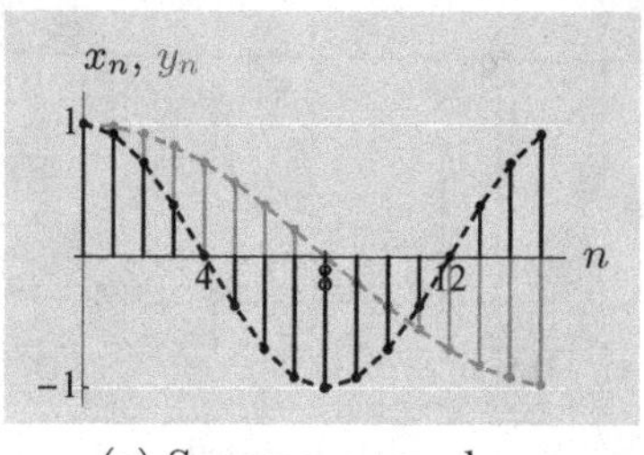

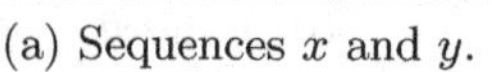

(a) Sequences x and y.

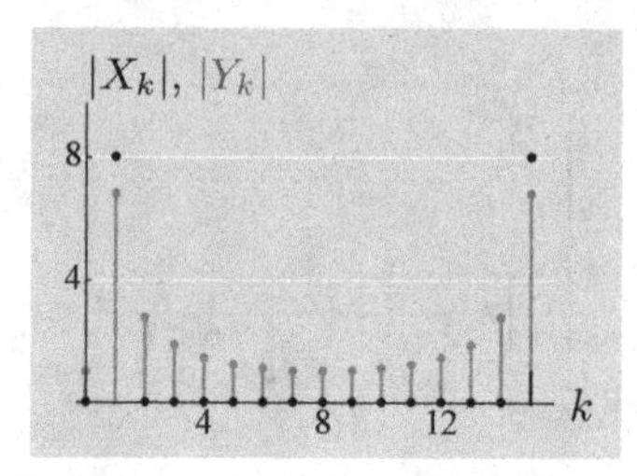

(b) Length-16 DFT magnitudes.

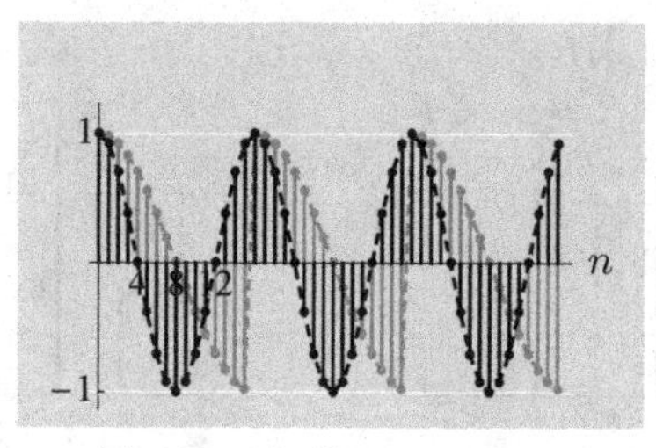

(c) 16-periodic extensions.

Figure 3.19 Periodic extension is important in interpreting the DFT. (a) Sequences $x_n = \cos((2\pi/16)n)$ (black stems) and $y_n = \cos((2\pi/32)n)$ (dashed stems) restricted to $n \in \{0, 1, \ldots, 15\}$. (b) Length-16 DFT magnitudes $|X_k|$ (black stems) and $|Y_k|$ (gray stems). (c) 16-periodic extensions of the length-16 sequences in (a).

These sequences, shown in Figure 3.19(a), vary slowly, and thus we expect them to have spectra dominated by low frequencies. Their DFTs of length 16 are shown in Figure 3.19(b). The spectrum X_k is indeed very simple, with content only at discrete frequencies ± 1 (equivalently, 1 and 15). However, Y_k shows nonzero content at every discrete frequency, even though y_n is even more slowly varying than x_n. The periodic extension of y_n with $N = 16$ is not a single sinusoid; as shown in Figure 3.19(c), it has a large jump at every multiple of 16, and thus significant high-frequency content. The extreme simplicity of X_k is because the period of the single sinusoid in x_n is a divisor of the length of the DFT.

Since the DFT is applied to a finite number of data samples, it implicitly involves the windowing of a sequence. This is important when interpreting the DFT as giving samples of the DTFT, as we illustrate in the following example.

Example 3.29 (Length of DFT, spectral resolution, and windowing) Let

$$x_n = \cos\left(\frac{2\pi}{10}n\right) + \frac{1}{2}\cos\left(\frac{2\pi}{3}n\right), \qquad n \in \mathbb{Z}, \tag{3.182}$$

and suppose that we wish to estimate the frequencies of the sinusoidal components of x from the DFT applied to a block of samples of x. We will see that the number of data samples used and the length of the DFT influence the result differently.

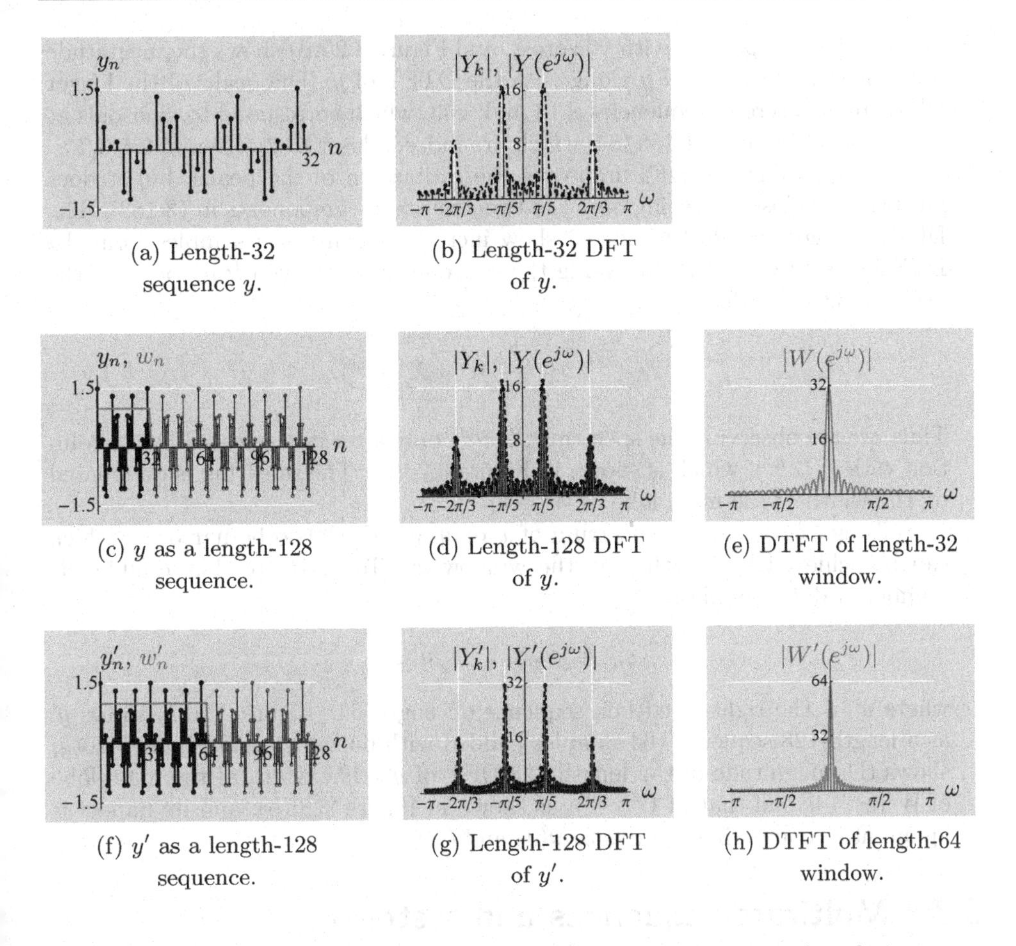

(a) Length-32 sequence y.

(b) Length-32 DFT of y.

(c) y as a length-128 sequence.

(d) Length-128 DFT of y.

(e) DTFT of length-32 window.

(f) y' as a length-128 sequence.

(g) Length-128 DFT of y'.

(h) DTFT of length-64 window.

Figure 3.20 Effect of windowing and length of DFT on spectral analysis.

When we apply the DFT of length N, we use the values of the sequence at $n \in \{0, 1, \ldots, N-1\}$. We can express this as windowing. Specifically, let

$$y_n = w_n x_n, \qquad n \in \mathbb{Z},$$

where w is the right-sided box sequence of length 32 (see (3.13)). The sequence y is now of length 32, as shown in Figure 3.20(a), and the magnitude of its length-32 DFT is shown in Figure 3.20(b). Since y is real, its spectral magnitude exhibits the symmetry $|Y_k| = |Y_{-k \bmod 32}|$. Two pairs of peaks are apparent in the DFT, at discrete frequencies ± 3 and ± 11. These correspond to sinusoidal components at frequencies $3\pi/16$ and $11\pi/16$, which do not quite match the frequencies $2\pi/10$ and $2\pi/3$ in (3.182). Figure 3.20(b) also shows the DTFT magnitude $|Y(e^{j\omega})|$, and we can see that the DFT samples miss the peaks of $|Y(e^{j\omega})|$.

To have finer resolution of frequencies in $Y(e^{j\omega})$, we can compute a longer DFT of the same data segment. Figure 3.20(c) shows y as a length-128 sequence

(32 samples of x padded with 96 zeros), and Figure 3.20(d) shows the magnitude of the length-128 DFT of y along with the DTFT of y. The peaks of the longer DFT are at discrete frequencies ± 13 and ± 43, which correspond to sinusoids at frequencies $13\pi/64$ and $43\pi/64$, which are indeed closer to $2\pi/10$ and $2\pi/3$.

Lengthening the DFT improves the estimation of the peaks, but it does not bring us closer to seeing only the two sinusoidal components in (3.182). The DFTs in Figures 3.20(b) and (d) show increasing density of samples from the DTFT $Y(e^{j\omega})$ (see (3.166)). Using the convolution in frequency property of the DTFT, (3.98), we have

$$Y(e^{j\omega}) = \frac{1}{2\pi}(W \circledast X)(e^{j\omega}).$$

Thus, we are observing the spectrum $X(e^{j\omega})$ only after it is smeared by convolution with $W(e^{j\omega})$, which is shown in Figure 3.20(e). The smearing is dominated by the width of the main lobe of $W(e^{j\omega})$.

To get closer to the spectrum of x requires $W(e^{j\omega})$ to be narrower, which can be achieved by lengthening the window w. To illustrate this, double the number of data samples: let

$$y'_n = w'_n x_n, \qquad n \in \mathbb{Z},$$

where w' is the right-sided box sequence of length 64. Figure 3.20(f) shows y' as a length-128 sequence (64 samples padded with 64 zeros), and Figure 3.20(g) shows the magnitude of the length-128 DFT of y'. The width of the main lobe of $W'(e^{j\omega})$ is half that of $W(e^{j\omega})$, as shown in Figure 3.20(h), and its impact is apparent in comparing Figures 3.20(d) and (g).

3.7 Multirate sequences and systems

In our study of sequences and discrete-time systems with time indexed by integers thus far, the time index has implicitly had the same physical meaning for all sequences. This is as if every physical process had been converted to a sequence by taking samples at the same regular intervals (for example, every second).

In multirate sequence processing, different sequences might have different time scales. Thus, the index n of the sequence might refer to different physical times for different sequences. We might ask both why one should do that and how one can go between these different scales. Let us look at a simple example. Start with a sequence x_n and derive a *downsampled* sequence y_n by dropping every other sample,

$$y_n = x_{2n}, \qquad n \in \mathbb{Z}, \qquad \text{or} \tag{3.183a}$$

$$y = \begin{bmatrix} \cdots & x_{-2} & \boxed{x_0} & x_2 & \cdots \end{bmatrix}^\top. \tag{3.183b}$$

If x_n is the sample of a physical process taken at time $t = n$, then y_n is a sample taken at time $t = 2n$. In other words, y has a timeline with intervals of 2 seconds when x has a timeline with intervals of 1 second; the clock of y is twice as slow. The

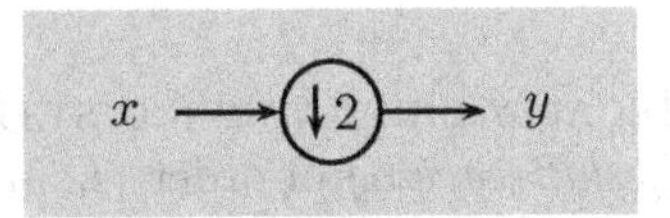

Figure 3.21 Block diagram representation of downsampling by 2.

operation above is called *downsampling* by 2, and it has a number of properties that we will study in detail. For example, it is irreversible: once we remove samples, we cannot get them back. It is also shift-varying, requiring more complicated analysis.

The dual operation to downsampling is *upsampling*. For example, upsampling a sequence x by 2 results in a new sequence y that is obtained by inserting a zero between every neighboring two samples,

$$y_n = \begin{cases} x_{n/2}, & \text{for } n \text{ even}; \\ 0, & \text{for } n \text{ odd}, \end{cases} \quad \text{or} \tag{3.184a}$$

$$y = \begin{bmatrix} \ldots & x_{-1} & 0 & \boxed{x_0} & 0 & x_1 & 0 & \ldots \end{bmatrix}^\top. \tag{3.184b}$$

The index of y_n corresponds to a time that is half of that for x_n. For example, if x has intervals of 1 second between samples, then y has intervals of 0.5 seconds; the clock of y is twice as fast.

What we just saw for rate changes by 2 can be done for any positive integer. By combining upsampling by N and downsampling by M, we can achieve any rational rate change. To smooth a sequence before dropping samples, downsampling is preceded by lowpass filtering, while to fill in the zeros, upsampling is followed by filtering; thus combinations of filtering with changes in sampling rate are important.

The purpose of this section is to study multirate operations and their consequences on sequences and their spectra. These operations create *linear periodically shift-varying (LPSV)* systems, which are represented by block-Toeplitz instead of Toeplitz matrices. For example, in Section 3.3.1, we encountered a block-averaging operator, (3.49), which is linear but not shift-invariant; it is, however, LPSV. While the LPSV nature of multirate systems does complicate analysis, we use a relatively simple and powerful tool called *polyphase analysis* to mitigate the problem.

3.7.1 Downsampling

Downsampling by 2 Downsampling[62] by 2, as introduced in (3.183), is represented by the block diagram in Figure 3.21. It is not shift-invariant because the output in response to the input $x_n = \delta_n$ is $y_n = \delta_n$, whereas the output in response to the shifted input $x_n = \delta_{n-1}$ is zero. However, this simple system does satisfy a less strict invariance property.

[62]Downsampling is often referred to as *subsampling* or *decimation*.

Definition 3.17 (Periodically shift-varying system) A discrete-time system T is called *periodically shift-varying* of order (L, M) when, for any integer k and input x,

$$y = T(x) \quad \Rightarrow \quad y' = T(x'), \qquad \text{where } x'_n = x_{n-Lk} \text{ and } y'_n = y_{n-Mk}. \tag{3.185}$$

Downsampling by 2 is periodically shift-varying of order $(2, 1)$ because shifting the input by $2k$ shifts the output by k.

Downsampling by 2 is a linear operator, so it has a matrix representation:

$$\begin{bmatrix} \vdots \\ y_{-1} \\ \boxed{y_0} \\ y_1 \\ y_2 \\ \vdots \end{bmatrix} = \underbrace{\begin{bmatrix} & \vdots & \vdots & \vdots & \vdots & \vdots & \vdots & \vdots & \\ \cdots & 1 & 0 & 0 & 0 & 0 & 0 & 0 & \cdots \\ \cdots & 0 & 0 & \boxed{1} & 0 & 0 & 0 & 0 & \cdots \\ \cdots & 0 & 0 & 0 & 0 & 1 & 0 & 0 & \cdots \\ \cdots & 0 & 0 & 0 & 0 & 0 & 0 & 1 & \cdots \\ & \vdots & \vdots & \vdots & \vdots & \vdots & \vdots & \vdots & \end{bmatrix}}_{D_2} \begin{bmatrix} \vdots \\ x_{-2} \\ x_{-1} \\ \boxed{x_0} \\ x_1 \\ x_2 \\ x_3 \\ x_4 \\ \vdots \end{bmatrix} = \begin{bmatrix} \vdots \\ x_{-2} \\ \boxed{x_0} \\ x_2 \\ x_4 \\ \vdots \end{bmatrix}, \tag{3.186a}$$

$$y = D_2 x, \tag{3.186b}$$

where D_2 stands for the downsampling-by-2 operator. Inspection of D_2 shows that it is similar to an identity matrix, but with the odd rows taken out. Intuitively, it is a rectangular operator, with the output space being a subspace of the input space (one would like to say that it is of half the size, but both are infinite-dimensional).

To find the z-transform $Y(z)$, we break downsampling into two steps: first x_n is transformed to a sequence $x_{0,n}$ having the even samples of x_n, with the odd ones set to zero; then, $x_{0,n}$ is contracted by removing those zeros to obtain y_n. Thus, let

$$x_{0,n} = \begin{cases} x_n, & \text{for } n \text{ even}; \\ 0, & \text{for } n \text{ odd}. \end{cases}$$

A clever way to express $x_{0,n}$ in terms of x_n without having two cases is

$$x_{0,n} = \frac{1}{2}(1 + (-1)^n)\, x_n, \qquad n \in \mathbb{Z}. \tag{3.187}$$

From (3.187), we can use linearity and the z-transform property (3.140) to obtain

$$\begin{aligned} X_0(z) &= \frac{1}{2}[X(z) + X(-z)] \\ &= \frac{1}{2}\left[(\cdots + x_0 + x_1 z^{-1} + x_2 z^{-2} + \cdots) + (\cdots + x_0 - x_1 z^{-1} + x_2 z^{-2} + \cdots)\right] \\ &= (\cdots + x_{-2} z^2 + x_0 + x_2 z^{-2} + \cdots) = \sum_{n \in \mathbb{Z}} x_{2n} z^{-2n}, \end{aligned}$$

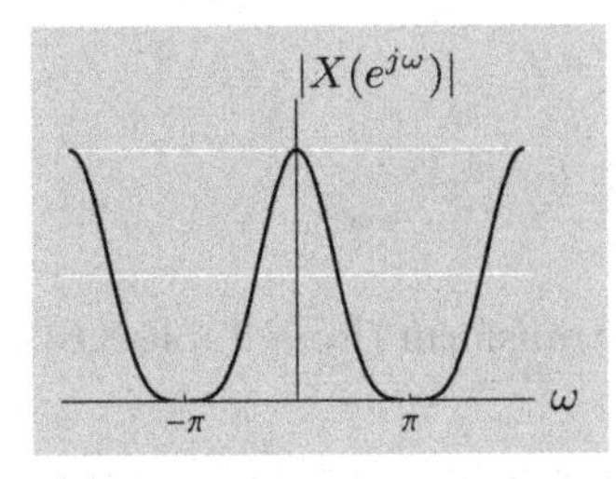

(a) Spectrum of a sequence.

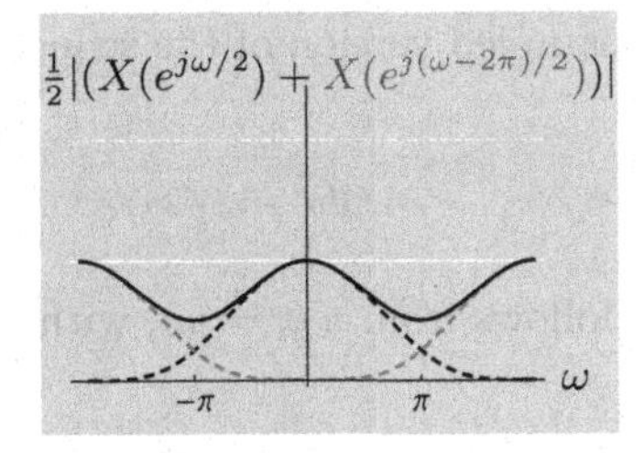

(b) Spectrum of downsampled version.

Figure 3.22 Effect of downsampling by 2 on the DTFT.

canceling out the odd powers of z and keeping the even ones. We now get $Y(z)$ by contracting $X_0(z)$ as

$$Y(z) = \sum_{n\in\mathbb{Z}} x_{2n}z^{-n} = X_0(z^{1/2}) = \frac{1}{2}\left[X(z^{1/2}) + X(-z^{1/2})\right]. \tag{3.188}$$

To find the DTFT $Y(e^{j\omega})$, we simply evaluate $Y(z)$ at $z = e^{j\omega}$,

$$Y(e^{j\omega}) = \frac{1}{2}\left[X(e^{j\omega/2}) + X(e^{j(\omega-2\pi)/2})\right], \tag{3.189}$$

where $-e^{j\omega/2}$ can be written as $e^{j(\omega-2\pi)/2}$ since $e^{-j\pi} = -1$. With the help of Figure 3.22, we now interpret this formula. First, $X(e^{j\omega/2})$ (black dashed line) is a stretched version of $X(e^{j\omega})$, by a factor of 2, and is thus 4π-periodic; since downsampling contracts time, it is natural that it expands frequency accordingly. Then, $X(e^{j(\omega-2\pi)/2})$ (gray dashed line) is not only a stretched version of $X(e^{j\omega})$, but also shifted by 2π. The sum is again 2π-periodic, since $Y(e^{j\omega}) = Y(e^{j(\omega-2k\pi)})$. Both stretching and shifting can create content at frequencies not present in the original sequence.[63] The shifted version $X(e^{j(\omega-2\pi)/2})$ is called the *aliased* version of the original (a ghost image).

EXAMPLE 3.30 (DOWNSAMPLING) Consider first a standard example illustrating the effect of downsampling: $x_n = (-1)^n = \cos(\pi n)$, the highest-frequency discrete sequence. Its downsampled version is $y_n = x_{2n} = \cos(2\pi n) = 1$, the lowest-frequency discrete sequence (a constant),

$$\begin{bmatrix}\cdots & 1 & -1 & \boxed{1} & -1 & 1 & \cdots\end{bmatrix}^\top \xrightarrow{2\downarrow} \begin{bmatrix}\cdots & 1 & 1 & \boxed{1} & 1 & 1 & \cdots\end{bmatrix}^\top,$$

completely changing the nature of the sequence.

Consider now the right-sided geometric series sequence, $x_n = \alpha^n u_n$ with $|\alpha| < 1$, from (3.128a). Its z-transform is (from Table 3.6)

$$X(z) = \frac{1}{1-\alpha z^{-1}}.$$

[63]This cannot happen with LSI processing. Because of the eigensequence property of complex exponentials, frequency components are only scaled by the system and thus cannot emerge from nothing.

The downsampled version of the sequence is

$$y_n = x_{2n} = \alpha^{2n} u_{2n} \stackrel{(a)}{=} \alpha^{2n} u_n,$$

where (a) follows from $u_{2n} = u_n$, with the z-transform (from Table 3.6)

$$Y(z) = \frac{1}{1 - \alpha^2 z^{-1}}.$$

We could have also obtained this z-transform using the expression for downsampling, (3.188), yielding

$$\frac{1}{2}\left(\frac{1}{1-\alpha z^{1/2}} + \frac{1}{1+\alpha z^{-1/2}}\right) = \frac{1}{1-\alpha^2 z^{-1}}.$$

The downsampled sequence is again exponential, but decays faster.

Downsampling by N We can generalize the discussion of downsampling by 2 to any positive integer N. A sequence downsampled by N and its z-transform are

$$y_n = x_{Nn} \quad \stackrel{\text{ZT}}{\longleftrightarrow} \quad Y(z) = \frac{1}{N} \sum_{k=0}^{N-1} X(W_N^k z^{1/N}). \tag{3.190}$$

The corresponding DTFT pair is

$$y_n = x_{Nn} \quad \stackrel{\text{DTFT}}{\longleftrightarrow} \quad Y(e^{j\omega}) = \frac{1}{N} \sum_{k=0}^{N-1} X(e^{j(\omega - 2\pi k)/N}), \tag{3.191}$$

using

$$W_N^k z^{1/N}\Big|_{z=e^{j\omega}} = e^{-j(2\pi/N)k} e^{j\omega/N} = e^{j(\omega-2\pi k)/N}.$$

We have already seen these expressions in Section 3.4.3 and Table 3.4 as scaling in time. The proof is an extension of the $N = 2$ case, and we leave it as Exercise 3.17. We denote the downsampling-by-N operator by D_N.

3.7.2 Upsampling

Upsampling by 2 Upsampling by 2, as introduced in (3.184), is represented by the block diagram in Figure 3.23. It is not shift-invariant; it is instead periodically shift-varying of order (1, 2) because shifting the input by k shifts the output by $2k$.

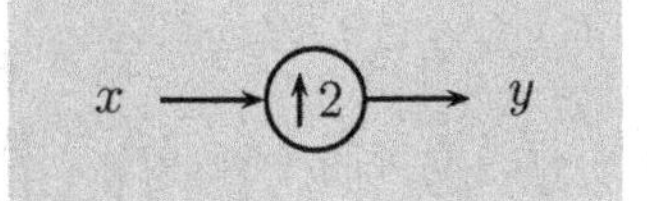

Figure 3.23 Block diagram representation of upsampling by 2.

Upsampling by 2 is a linear operator, so it has a matrix representation

$$\begin{bmatrix} \vdots \\ y_{-2} \\ y_{-1} \\ \boxed{y_0} \\ y_1 \\ y_2 \\ y_3 \\ y_4 \\ \vdots \end{bmatrix} = \underbrace{\begin{bmatrix} & \vdots & \vdots & \vdots & \vdots & \\ \cdots & 1 & 0 & 0 & 0 & \cdots \\ \cdots & 0 & 0 & 0 & 0 & \cdots \\ \cdots & 0 & \boxed{1} & 0 & 0 & \cdots \\ \cdots & 0 & 0 & 0 & 0 & \cdots \\ \cdots & 0 & 0 & 1 & 0 & \cdots \\ \cdots & 0 & 0 & 0 & 0 & \cdots \\ \cdots & 0 & 0 & 0 & 1 & \cdots \\ & \vdots & \vdots & \vdots & \vdots & \end{bmatrix}}_{U_2} \begin{bmatrix} \vdots \\ x_{-1} \\ \boxed{x_0} \\ x_1 \\ x_2 \\ \vdots \end{bmatrix} = \begin{bmatrix} \vdots \\ x_{-1} \\ 0 \\ \boxed{x_0} \\ 0 \\ x_1 \\ 0 \\ x_2 \\ \vdots \end{bmatrix}, \tag{3.192a}$$

$$y = U_2 x, \tag{3.192b}$$

where U_2 stands for the upsampling-by-2 operator. The matrix U_2 is an identity matrix with a row of zeros in between every two rows.

In the z-transform domain, the expression for upsampling by 2 is

$$Y(z) = \sum_{n\in\mathbb{Z}} y_n z^{-n} \stackrel{(a)}{=} \sum_{n\in\mathbb{Z}} y_{2n} z^{-2n} \stackrel{(b)}{=} \sum_{n\in\mathbb{Z}} x_n z^{-2n} = X(z^2), \tag{3.193}$$

where (a) follows from $y_n = 0$ for all odd n; and (b) from $y_{2n} = x_n$ following (3.184). To find the DTFT $Y(e^{j\omega})$, we simply evaluate $Y(z)$ at $z = e^{j\omega}$,

$$Y(e^{j\omega}) = X(e^{j2\omega}), \tag{3.194}$$

a contraction by a factor of 2 as shown in Figure 3.24.

Example 3.31 (Upsampling) Take the constant sequence $x_n = 1$. Its upsampled version

$$y = \begin{bmatrix} \ldots & 1 & 0 & \boxed{1} & 0 & 1 & \ldots \end{bmatrix}^\top$$

can be written as

$$y_n = \frac{1}{2}(1 + (-1)^n) = \frac{1}{2}(1 + \cos(\pi n)),$$

indicating that it contains both the original frequency (constant, zero frequency) and a new high frequency (at $\omega = \pi$, since $(-1)^n = \cos(\pi n)$).

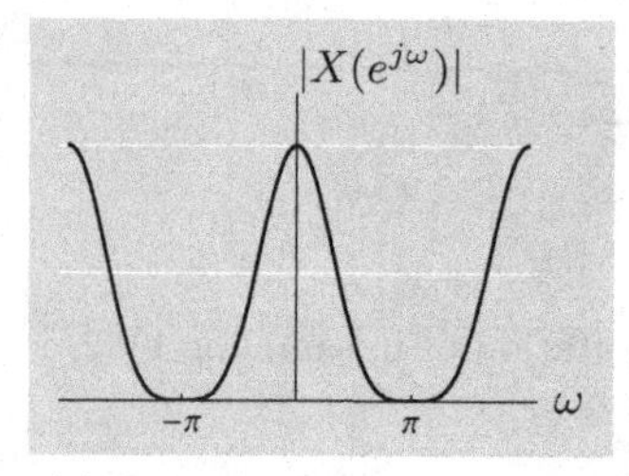

(a) Spectrum of a sequence.

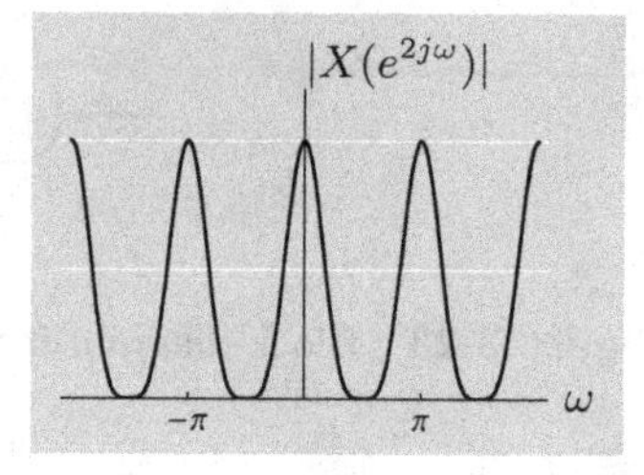

(b) Spectrum of upsampled version.

Figure 3.24 Effect of upsampling by 2 on the DTFT.

Upsampling by N We can generalize the discussion of upsampling by 2 to any positive integer N. A sequence upsampled by N and its z-transform are given by

$$y_n = \begin{cases} x_{n/N}, & \text{for } n/N \in \mathbb{Z}; \\ 0, & \text{otherwise} \end{cases} \quad \overset{\text{ZT}}{\longleftrightarrow} \quad Y(z) = X(z^N). \tag{3.195}$$

The corresponding DTFT pair is

$$y_n = \begin{cases} x_{n/N}, & \text{for } n/N \in \mathbb{Z}; \\ 0, & \text{otherwise} \end{cases} \quad \overset{\text{DTFT}}{\longleftrightarrow} \quad Y(e^{j\omega}) = X(e^{jN\omega}). \tag{3.196}$$

We denote the upsampling-by-N operator by U_N.

3.7.3 Combinations of downsampling and upsampling

By comparing (3.186a) and (3.192a), we see that the downsampling and upsampling operators are transposes of each other; since they have real entries, they are also adjoints of each other,

$$U_2 = D_2^\top = D_2^*. \tag{3.197}$$

Upsampling followed by downsampling Upsampling by 2 followed by downsampling by 2 results in the identity

$$D_2 U_2 = I, \tag{3.198}$$

since the zeros added by upsampling are at odd-indexed locations and are subsequently eliminated by downsampling.

Similarly, for any positive integer N,

$$D_N U_N = I.$$

Downsampling followed by upsampling Carrying out the operations in the reverse order is more interesting; downsampling by 2 followed by upsampling by 2 results in a sequence where all odd-indexed samples have been replaced by zeros, or

$$\begin{bmatrix} \ldots & x_{-1} & \boxed{x_0} & x_1 & x_2 & \ldots \end{bmatrix}^\top \xrightarrow{\downarrow 2} \xrightarrow{\uparrow 2} \begin{bmatrix} \ldots & 0 & \boxed{x_0} & 0 & x_2 & \ldots \end{bmatrix}^\top.$$

This operator,

$$P = U_2 D_2, \tag{3.199}$$

is an orthogonal projection operator onto the subspace of sequences with all odd-indexed samples equal to zero. To verify this, we check idempotency (see Definition 2.27),

$$P^2 = (U_2D_2)(U_2D_2) = U_2(D_2U_2)D_2 = U_2D_2 = P,$$

using (3.198), as well as self-adjointness,

$$P^* = (U_2D_2)^* = D_2^*U_2^* = U_2D_2 = P,$$

using (3.197). Applying this projection operator to a sequence x, we get the expressions in the DTFT and the z-transform domains as

$$Px \quad \begin{array}{cl} \overset{\text{DTFT}}{\longleftrightarrow} & \frac{1}{2}(X(e^{j\omega}) + X(e^{j(\omega+\pi)})), \\ \overset{\text{ZT}}{\longleftrightarrow} & \frac{1}{2}(X(z) + X(-z)). \end{array} \tag{3.200}$$

Similarly, for any positive integer N,

$$P_N = U_N D_N$$

is an orthogonal projection operator. Applying this projection operator to a sequence x, we get the expressions in the DTFT and the z-transform domains as

$$P_N x \quad \begin{array}{cl} \overset{\text{DTFT}}{\longleftrightarrow} & \frac{1}{N}\sum_{k=0}^{N-1} X(e^{j(\omega-2\pi k/N)}), \\ \overset{\text{ZT}}{\longleftrightarrow} & \frac{1}{N}\sum_{k=0}^{N-1} X(W_N^k z). \end{array} \tag{3.201}$$

Commutativity of upsampling and downsampling We have just seen that D_2U_2 and U_2D_2 are quite different – one is the identity and the other sets all odd-indexed samples to zero. Thus, upsampling and downsampling by the same factor do not commute. However, upsampling by N and downsampling by M commute if and only if N and M are relatively prime (that is, they have no common factors). The proof is the topic of Exercise 3.22, and a couple of illustrative examples are given in Exercise 3.23.

Example 3.32 (Commutativity of upsampling and downsampling) We look at upsampling by 2 and downsampling by 3. If we apply U_2 to x we get

$$U_2x = \begin{bmatrix} \cdots & \boxed{x_0} & 0 & x_1 & 0 & x_2 & 0 & x_3 & 0 & \cdots \end{bmatrix}^\top;$$

then, upon applying D_3, we get

$$D_3U_2x = \begin{bmatrix} \cdots & x_{-3} & 0 & \boxed{x_0} & 0 & x_3 & 0 & x_6 & \cdots \end{bmatrix}^\top.$$

Upon applying D_3 first we get

$$D_3x = \begin{bmatrix} \cdots & x_{-3} & \boxed{x_0} & x_3 & x_6 & \cdots \end{bmatrix}^\top,$$

which, when followed by U_2,a leads to the same result, $U_2D_3x = D_3U_2x$.

3.7.4 Combinations of downsampling, upsampling, and filtering

In multirate processing, the scaling of frequencies (up with downsampling and down with upsampling) is a consequence of time-scale change. However, in many applications, the introduction of new high-frequency components, as shown in Figures 3.22(b) and 3.24(b), is not desired. This is one of the reasons why downsampling is often preceded by filtering and upsampling is often followed by filtering. We now consider these two cases in more detail.

Downsampling preceded by filtering Consider downsampling by 2, preceded by filtering by $\widetilde{g}$,[64] with operator $\widetilde{G}$ as in (3.64), as illustrated in Figure 3.25(a). This can be written as (with a causal, length-4 FIR filter for illustration)

$$y = \begin{bmatrix} \vdots \\ y_{-1} \\ \boxed{y_0} \\ y_1 \\ y_2 \\ \vdots \end{bmatrix} = \underbrace{\begin{bmatrix} & \vdots & \vdots & \vdots & \vdots & \vdots & \vdots & \\ \cdots & \widetilde{g}_1 & \widetilde{g}_0 & 0 & 0 & 0 & 0 & \cdots \\ \cdots & \widetilde{g}_3 & \widetilde{g}_2 & \widetilde{g}_1 & \boxed{\widetilde{g}_0} & 0 & 0 & \cdots \\ \cdots & 0 & 0 & \widetilde{g}_3 & \widetilde{g}_2 & \widetilde{g}_1 & \widetilde{g}_0 & \cdots \\ \cdots & 0 & 0 & 0 & 0 & \widetilde{g}_3 & \widetilde{g}_2 & \cdots \\ & \vdots & \vdots & \vdots & \vdots & \vdots & \vdots & \end{bmatrix}}_{D_2\widetilde{G}} \begin{bmatrix} \vdots \\ x_{-3} \\ x_{-2} \\ x_{-1} \\ \boxed{x_0} \\ x_1 \\ x_2 \\ \vdots \end{bmatrix} = D_2\widetilde{G}x. \qquad (3.202)$$

The operator $D_2\widetilde{G}$ is the convolution operator $\widetilde{G}$ with the odd rows removed; it is block Toeplitz with blocks of size 1×2. From (3.202), and using (3.62a), we can also express y as

$$y_n = (\widetilde{g} * x)_{2n} = \sum_{k\in\mathbb{Z}} x_k \widetilde{g}_{2n-k} = \langle x_k, \widetilde{g}^*_{2n-k}\rangle_k, \qquad n \in \mathbb{Z}. \qquad (3.203)$$

In the z-transform and the DTFT domains, the output of downsampling by 2 preceded by filtering is

$$Y(z) = \frac{1}{2}\left[\widetilde{G}(z^{1/2})X(z^{1/2}) + \widetilde{G}(-z^{1/2})X(-z^{1/2})\right], \qquad (3.204a)$$

$$Y(e^{j\omega}) = \frac{1}{2}\left[\widetilde{G}(e^{j\omega/2})X(e^{j\omega/2}) + \widetilde{G}(e^{j(\omega-2\pi)/2})X(e^{j(\omega-2\pi)/2})\right]. \qquad (3.204b)$$

Figure 3.25 shows the effect of downsampling by 2 preceded by filtering in the DTFT domain. The input spectrum is as in Figure 3.24(a), the spectrum after filtering is shown in Figure 3.25(b), and the final output spectrum after downsampling is shown in Figure 3.25(c). In this example, the filter is an ideal lowpass filter with cutoff frequency $\frac{1}{2}\pi$. The spectrum of the input from $-\frac{1}{2}\pi$ to $\frac{1}{2}\pi$ is conserved, the rest is put to zero so that no aliasing occurs, and the central lowpass part of the spectrum is conserved in the downsampled version.

[64] From this point on, we use $\widetilde{g}$ to denote a filter when followed by a downsampler; similarly, we use g to denote a filter when preceded by an upsampler. In Chapter 5 and the companion volume, [57], this will be standard notation.

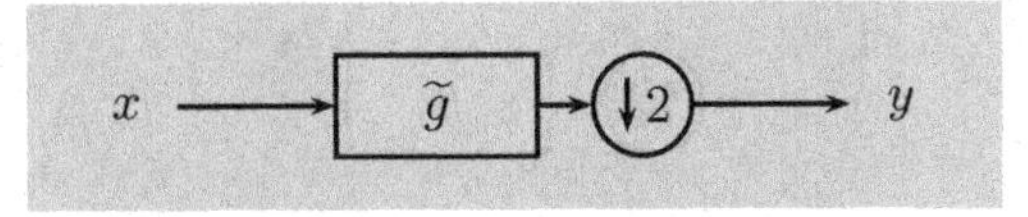

(a) Block diagram.

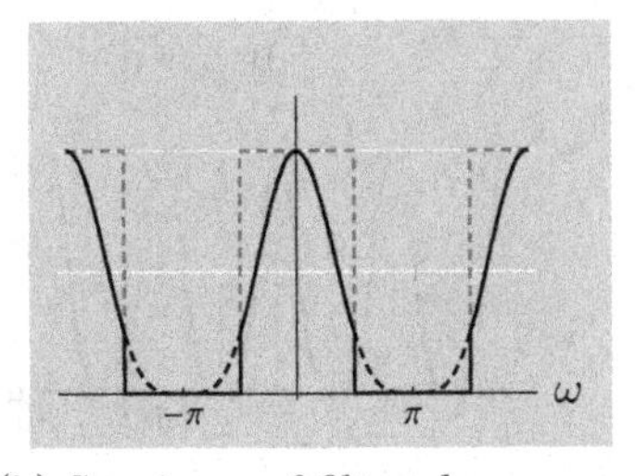

(b) Spectrum of filtered sequence.

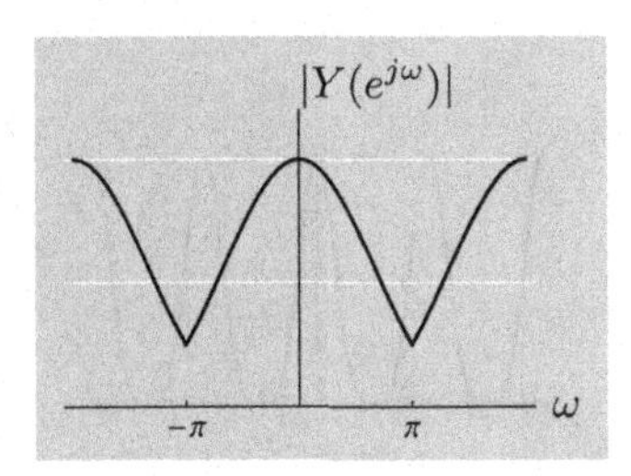

(c) Spectrum of output.

Figure 3.25 Downsampling preceded by filtering, and the effect on the DTFT.

EXAMPLE 3.33 (DOWNSAMPLING PRECEDED BY FILTERING) Consider the two-point averaging filter $\widetilde{g}_n = \frac{1}{2}(\delta_n + \delta_{n-1})$, whose output, downsampled by 2, $y = D_2\widetilde{G}x$, is

$$y_n = \frac{1}{2}(x_{2n} + x_{2n-1}), \qquad n \in \mathbb{Z}.$$

Because of filtering, all input samples influence the output, as opposed to downsampling without filtering, where the odd-indexed samples had no impact.

Upsampling followed by filtering Consider upsampling by 2 followed by filtering by g, with operator G as in (3.64), as illustrated in Figure 3.26(a). This can be written as (with a causal, length-4 FIR filter for illustration)

$$y = \begin{bmatrix} \vdots \\ y_{-2} \\ y_{-1} \\ \boxed{y_0} \\ y_1 \\ y_2 \\ y_3 \\ \vdots \end{bmatrix} = \underbrace{\begin{bmatrix} & \vdots & \vdots & \vdots & \vdots & \\ \cdots & g_2 & g_0 & 0 & 0 & \cdots \\ \cdots & g_3 & g_1 & 0 & 0 & \cdots \\ \cdots & 0 & g_2 & \boxed{g_0} & 0 & \cdots \\ \cdots & 0 & g_3 & g_1 & 0 & \cdots \\ \cdots & 0 & 0 & g_2 & g_0 & \cdots \\ \cdots & 0 & 0 & g_3 & g_1 & \cdots \\ & \vdots & \vdots & \vdots & \vdots & \end{bmatrix}}_{GU_2} \begin{bmatrix} \vdots \\ x_{-2} \\ x_{-1} \\ \boxed{x_0} \\ x_1 \\ \vdots \end{bmatrix} = GU_2 x. \qquad (3.205)$$

The operator GU_2 is the convolution operator G with the odd columns removed; it is block Toeplitz with blocks of size 2×1. From (3.205), we can also express y as

$$y_n = \sum_{k \in \mathbb{Z}} g_{n-2k} x_k, \qquad n \in \mathbb{Z}. \qquad (3.206)$$

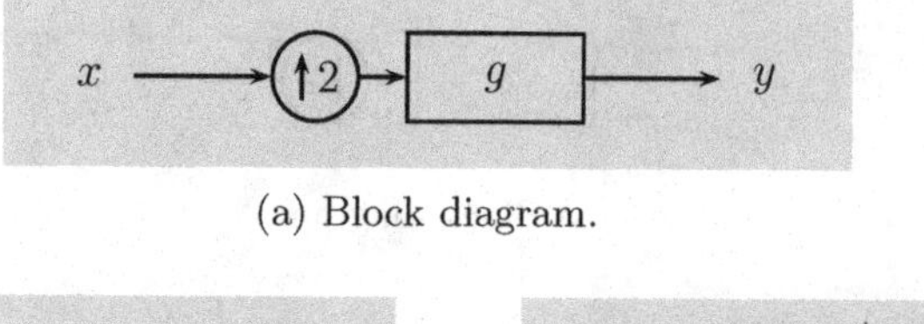

(a) Block diagram.

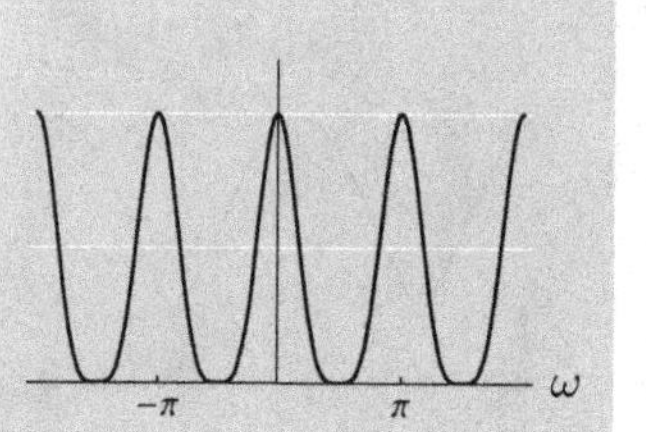

(b) Spectrum of upsampled sequence.

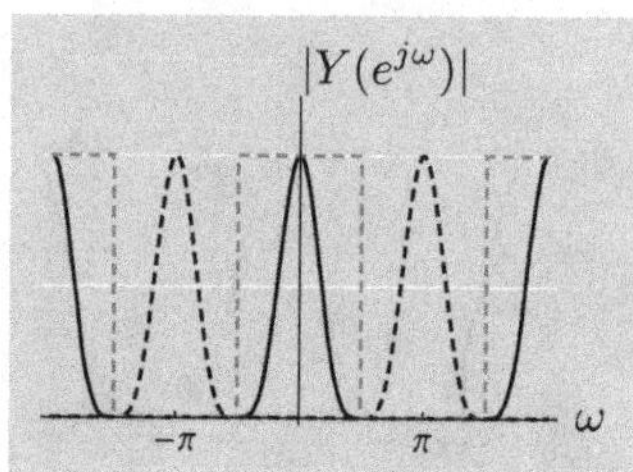

(c) Spectrum of output.

Figure 3.26 Upsampling followed by filtering, and the effect on the DTFT.

Another way to look at (3.205) and (3.206) is to see that each input sample x_k generates a response g_n delayed by $2k$ samples and weighted by x_k.

In the z-transform and the DTFT domains, the output of upsampling by 2 followed by filtering is

$$Y(z) = G(z)X(z^2), \tag{3.207a}$$
$$Y(e^{j\omega}) = G(e^{j\omega})X(e^{j2\omega}). \tag{3.207b}$$

Figure 3.26 shows the effect of upsampling by 2 followed by filtering in the DTFT domain. The input spectrum is again as in Figure 3.24(a), the spectrum after upsampling is shown in Figure 3.26(b), and the final output spectrum after ideal lowpass filtering is shown in Figure 3.26(c). In this example, the filter is an ideal lowpass filter with cutoff frequency $\frac{1}{2}\pi$. We see that the ghost spectrum at 2π produced by upsampling is removed; only the base spectrum around the origin remains.

EXAMPLE 3.34 (UPSAMPLING FOLLOWED BY FILTERING) Let $g_n = \delta_n + \delta_{n-1}$. The sequence x, upsampled by 2 and filtered with g, leads to

$$y = \begin{bmatrix} \cdots & x_{-1} & x_{-1} & \boxed{x_0} & x_0 & x_1 & x_1 & \cdots \end{bmatrix}^{\top}, \tag{3.208}$$

which is a staircase sequence, with stairs of height x_n and length 2. The filter thus performs piecewise-constant interpolation. A smoother interpolation is obtained with a linear interpolator: $g_n = \frac{1}{2}\delta_{n-1} + \delta_n + \frac{1}{2}\delta_{n+1}$. From (3.205) or (3.206), the even-indexed outputs are equal to input samples (at half the index), while

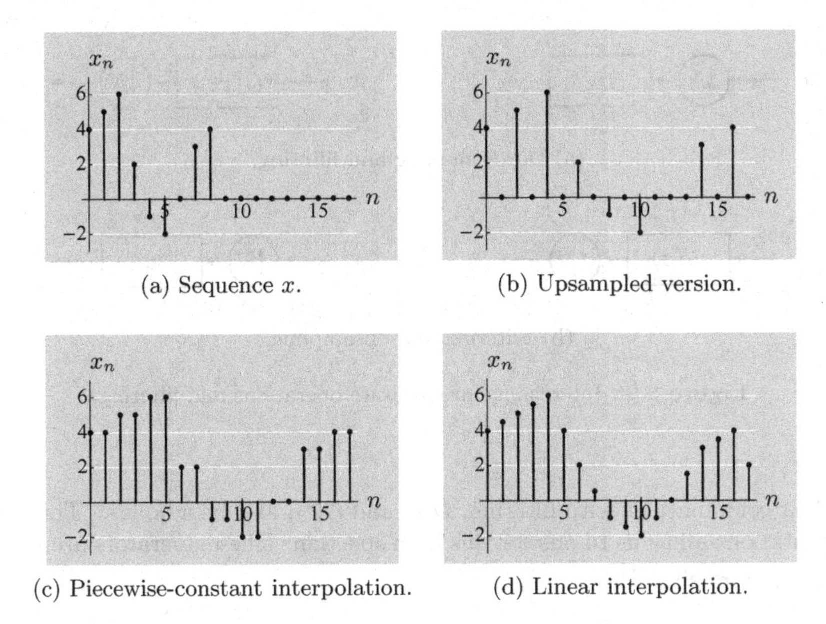

(a) Sequence x. (b) Upsampled version.

(c) Piecewise-constant interpolation. (d) Linear interpolation.

Figure 3.27 Examples of upsampling followed by filtering.

the odd-indexed outputs are averages of two input samples,

$$y_n = \begin{cases} x_{n/2}, & \text{for } n \text{ even;} \\ \frac{1}{2}\left(x_{(n+1)/2} + x_{(n-1)/2}\right), & \text{for } n \text{ odd,} \end{cases} \tag{3.209}$$

$$y = \begin{bmatrix} \cdots & x_{-1} & \frac{1}{2}(x_{-1} + x_0) & \boxed{x_0} & \frac{1}{2}(x_0 + x_1) & x_1 & \cdots \end{bmatrix}^{\top}.$$

Compare (3.209) with (3.208) to see why (3.209) is a smoother interpolation, and see Figure 3.27 for an example.

Interchange of multirate operations and filtering Filtering preceding downsampling and filtering following upsampling are the most frequent combinations of multirate operations. These combinations in reversed order satisfy simple, useful identities.

(i) Downsampling by N followed by filtering with $\widetilde{G}(z)$ is equivalent to filtering with $\widetilde{G}(z^N)$ followed by downsampling by N, as shown in Figure 3.28(a).

(ii) Filtering with $G(z)$ followed by upsampling by N is equivalent to upsampling by N followed by filtering with $G(z^N)$, as shown in Figure 3.28(b).

Proofs of these identities are left as Exercise 3.24.

Downsampling, upsampling and filtering Earlier, we noted the duality of downsampling and upsampling, which was made explicit in the adjoint relation (3.197).

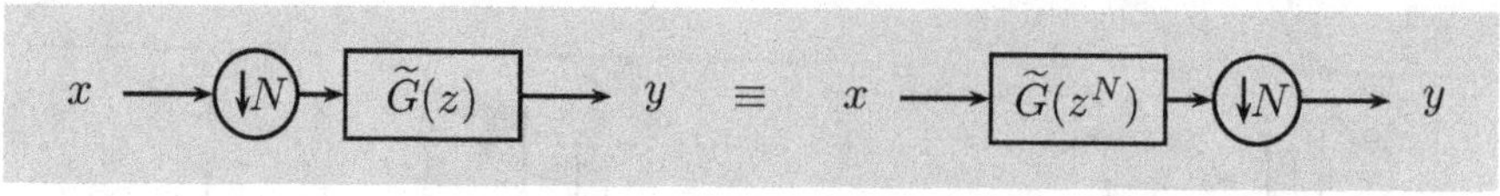

$$x \rightarrow \boxed{\downarrow N} \rightarrow \boxed{\widetilde{G}(z)} \rightarrow y \quad \equiv \quad x \rightarrow \boxed{\widetilde{G}(z^N)} \rightarrow \boxed{\downarrow N} \rightarrow y$$

(a) Downsampling and filtering.

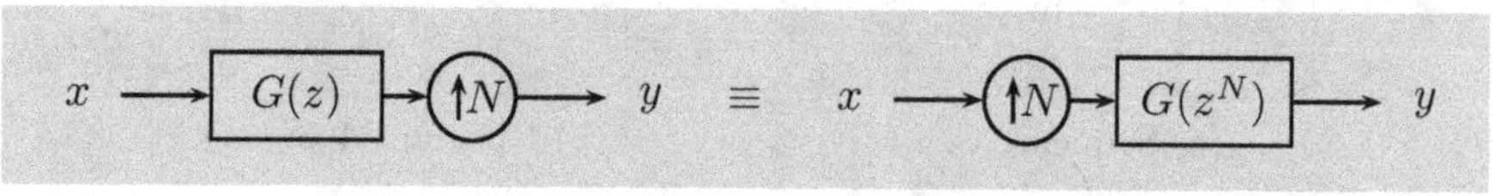

$$x \rightarrow \boxed{G(z)} \rightarrow \boxed{\uparrow N} \rightarrow y \quad \equiv \quad x \rightarrow \boxed{\uparrow N} \rightarrow \boxed{G(z^N)} \rightarrow y$$

(b) Filtering and upsampling.

Figure 3.28 Interchange of multirate operations and filtering.

Can their combinations with filtering, $D_2\widetilde{G}$ and GU_2, also be adjoints? The matrix representations allow us to answer this by inspection. These operators are adjoints when $\widetilde{g}_n^* = g_{-n}$, since then[65]

$$(D_2\widetilde{G})^* = \begin{bmatrix} & \vdots & \vdots & \vdots & \vdots & \vdots & \\ \cdots & \widetilde{g}_{-2}^* & \widetilde{g}_0^* & \widetilde{g}_2^* & \widetilde{g}_4^* & \widetilde{g}_6^* & \cdots \\ \cdots & \widetilde{g}_{-3}^* & \widetilde{g}_{-1}^* & \widetilde{g}_1^* & \widetilde{g}_3^* & \widetilde{g}_5^* & \cdots \\ \cdots & \widetilde{g}_{-4}^* & \widetilde{g}_{-2}^* & \boxed{\widetilde{g}_0^*} & \widetilde{g}_2^* & \widetilde{g}_4^* & \cdots \\ \cdots & \widetilde{g}_{-5}^* & \widetilde{g}_{-3}^* & \widetilde{g}_{-1}^* & \widetilde{g}_1^* & \widetilde{g}_3^* & \cdots \\ \cdots & \widetilde{g}_{-6}^* & \widetilde{g}_{-4}^* & \widetilde{g}_{-2}^* & \widetilde{g}_0^* & \widetilde{g}_2^* & \cdots \\ & \vdots & \vdots & \vdots & \vdots & \vdots & \end{bmatrix}$$

$$= \begin{bmatrix} & \vdots & \vdots & \vdots & \vdots & \vdots & \\ \cdots & g_2 & g_0 & g_{-2} & g_{-4} & g_{-6} & \cdots \\ \cdots & g_3 & g_1 & g_{-1} & g_{-3} & g_{-5} & \cdots \\ \cdots & g_4 & g_2 & \boxed{g_0} & g_{-2} & g_{-4} & \cdots \\ \cdots & g_5 & g_3 & g_1 & g_{-1} & g_{-3} & \cdots \\ \cdots & g_6 & g_4 & g_2 & g_0 & g_{-2} & \cdots \\ & \vdots & \vdots & \vdots & \vdots & \vdots & \end{bmatrix} = GU_2. \tag{3.210}$$

We could also show the above by using the definition of the adjoint operator in (2.48). That is, we want to find a $\widetilde{G}$ so that $D_2\widetilde{G}$ and GU_2 are adjoints of each other,

$$\langle D_2\widetilde{G}x,\, y\rangle = \langle x,\, GU_2y\rangle. \tag{3.211}$$

We thus write

$$\langle D_2\widetilde{G}x,\, y\rangle \overset{(a)}{=} \langle \widetilde{G}x,\, D_2^*y\rangle \overset{(b)}{=} \langle \widetilde{G}x,\, U_2y\rangle \overset{(c)}{=} \langle x,\, \widetilde{G}^*U_2y\rangle, \tag{3.212}$$

[65]Note that, unlike in (3.202) and (3.205), here we do not assume causal filters.

where (a) and (c) follow from the definition of the adjoint; and (b) from (3.197). Then, (3.211) will hold only if $\widetilde{G}^* = G$, that is, $\widetilde{g}_n^* = g_{-n}$.

The above proof that *Hermitian transposition equals time-reversal* will appear prominently in the analysis of sampling in Section 5.3. In particular, we will prove that when the impulse response of the filter g is orthogonal to its even shifts as in (3.213) below and $\widetilde{g}_n^* = g_{-n}$, then the operation of filtering with $\widetilde{g}$, downsampling by 2, upsampling by 2, and filtering with g is an orthogonal projection onto the subspace spanned by g and its even shifts.

Orthogonality of a filter's impulse response to its even shifts Filters that have impulse responses orthogonal to their even shifts,

$$\langle g_n, g_{n-2k}\rangle_n \;=\; \delta_k, \tag{3.213}$$

play an important role in the analysis of filter banks. Geometrically, (3.213) means that the columns of GU_2 in (3.205) are orthonormal to each other (similarly for the rows of $D_2\widetilde{G}$); that is,

$$I \;=\; (GU_2)^*(GU_2) \;=\; U_2^*G^*GU_2 \;=\; D_2G^*GU_2. \tag{3.214}$$

While we have written the above in its most general form using Hermitian transposition to allow for complex filters, most of the time, we will be dealing with real filters, and thus simple transposition.

We can see (3.213) as the deterministic autocorrelation of g downsampled by 2. Write the deterministic autocorrelation of g as in (3.17),

$$a_k \;=\; \langle g_n, g_{n-k}\rangle_n,$$

and note that it has a single nonzero even term, $g_0 = 1$,

$$a_{2k} \;=\; \delta_k. \tag{3.215}$$

Assuming now a real g, in the z-transform domain, $A(z) = G(z)G(z^{-1})$ using (3.146). Keeping only the even terms can be accomplished by adding $A(z)$ and $A(-z)$ and dividing by 2. Therefore, (3.215) can be expressed as

$$A(z) + A(-z) \;=\; G(z)G(z^{-1}) + G(-z)G(-z^{-1}) \;=\; 2, \tag{3.216}$$

which on the unit circle leads to

$$|G(e^{j\omega})|^2 + |G(e^{j(\omega+\pi)})|^2 = 2. \tag{3.217}$$

This *quadrature mirror formula*, also called *power complementarity*, is central in the design of orthonormal filter banks. In reaching (3.217) we have assumed that g is real, and used both

$$G(z)G(z^{-1})\big|_{z=e^{j\omega}} \;=\; G(e^{j\omega})G(e^{-j\omega}) \;=\; G(e^{j\omega})G^*(e^{j\omega}) \;=\; |G(e^{j\omega})|^2,$$

and

$$G(-z)G(-z^{-1})\big|_{z=e^{j\omega}} = |G(e^{j(\omega+\pi)})|^2.$$

In summary, a real filter satisfying any of the conditions below is called *orthogonal*:

$$\langle g_n, g_{n-2k}\rangle_n = \delta_k \quad \begin{array}{c}\xleftrightarrow{\text{Matrix View}}\\ \xleftrightarrow{\text{ZT}}\\ \xleftrightarrow{\text{DTFT}}\end{array} \quad \begin{array}{rcl} D_2G^\top GU_2 &=& I, \\ G(z)G(z^{-1})+G(-z)G(-z^{-1}) &=& 2, \\ |G(e^{j\omega})|^2+|G(e^{j(\omega+\pi)})|^2 = 2. && \end{array} \tag{3.218}$$

Compare this with the expression for the allpass filter in (3.118). An allpass impulse response h is orthogonal to all its shifts, at the expense of having no frequency selectivity; here we have a basis containing only even shifts, and some frequency selectivity exists (g is a half-band lowpass filter).

3.7.5 Polyphase representation

Multirate processing brings a major twist to signal processing: shift invariance is replaced by periodic shift variance, which replaces Toeplitz matrices with block-Toeplitz matrices. This section examines *polyphase representation*, a method to transform single-input, single-output linear periodically shift-varying systems into multiple-input, multiple-output linear shift-invariant systems. For simplicity, we introduce all of the concepts for period 2 and generalize to period N only at the end of the section.

Polyphase representation of sequences The *polyphase decomposition* with period 2 splits a sequence into its even- and odd-indexed subsequences, called even and odd *polyphase components*. We will see later why it is convenient to use both of two different conventions: for the odd subsequence to be either *advanced* or *delayed* relative to the even subsequence.

Using the convention of advancing the odd subsequence, a sequence x is decomposed into

$$\begin{bmatrix}\cdots & x_{-2} & \boxed{x_0} & x_2 & x_4 & \cdots\end{bmatrix}^\top \quad \text{and} \quad \begin{bmatrix}\cdots & x_{-1} & \boxed{x_1} & x_3 & x_5 & \cdots\end{bmatrix}^\top.$$

This splitting operation, using operations we have studied thus far, is illustrated in Figure 3.29: to get the even polyphase component, we simply downsample by 2 to remove the odd samples from x; to get the odd polyphase component, we shift x to the left by one (advance by one represented by z) and then downsample by 2 to remove the odd samples. The polyphase components and their z-transforms are

$$x_{0,n} = x_{2n} \quad \xleftrightarrow{\text{ZT}} \quad X_0(z) = \sum_{n\in\mathbb{Z}} x_{2n}z^{-n}, \tag{3.219a}$$

$$x_{1,n} = x_{2n+1} \quad \xleftrightarrow{\text{ZT}} \quad X_1(z) = \sum_{n\in\mathbb{Z}} x_{2n+1}z^{-n}. \tag{3.219b}$$

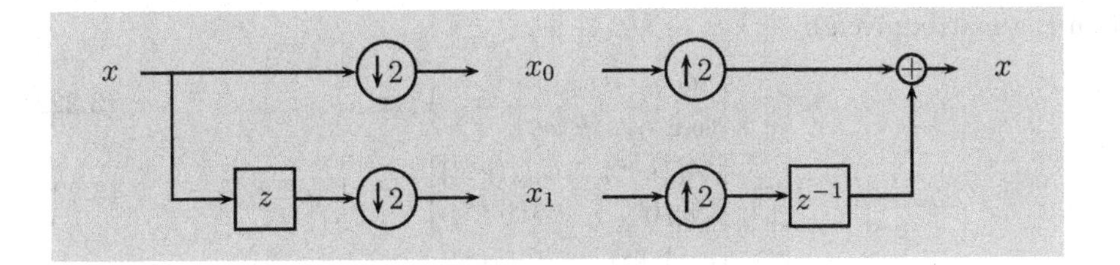

Figure 3.29 Forward and inverse polyphase transform.

To get the original sequence back, we interleave the two polyphase components. Figure 3.29 illustrates this: upsampling the even polyphase component by 2 gives the even-indexed samples of x; upsampling the odd polyphase component and shifting to the right by one (delay by one represented by z^{-1}) gives the odd-indexed samples of x,

$$x_n = \begin{cases} x_{0,n/2}, & \text{for } n \text{ even;} \\ x_{1,(n-1)/2}, & \text{for } n \text{ odd} \end{cases} \quad \overset{\text{ZT}}{\longleftrightarrow} \quad X(z) = X_0(z^2) + z^{-1}X_1(z^2). \tag{3.219c}$$

Deterministic autocorrelation Let us denote by $a_{0,n}$ the deterministic autocorrelation sequence of the polyphase component $x_{0,n}$, by $a_{1,n}$ the deterministic autocorrelation sequence of the polyphase component $x_{1,n}$, and by $c_{0,1,n}$ their deterministic crosscorrelation. Then, we can represent the polyphase components of the deterministic autocorrelation (3.146) via the deterministic autocorrelation and crosscorrelation of the polyphase components (for simplicity, we show it for a real sequence x and in the z-transform domain),

$$\begin{aligned} A(z) &= X(z)\,X(z^{-1}) \\ &= (X_0(z^2) + z^{-1}X_1(z^2))(X_0(z^{-2}) + zX_1(z^{-2})) \\ &= (A_0(z^2) + A_1(z^2)) + z^{-1}(C_{1,0}(z^2) + z^2C_{0,1}(z^2)), \end{aligned} \tag{3.220}$$

where the first and second terms are the first and second polyphase components of the deterministic autocorrelation, respectively, and satisfy

$$A_0(z^2) + A_1(z^2) = A_0(z^{-2}) + A_1(z^{-2}), \tag{3.221a}$$

$$C_{1,0}(z^2) + z^2C_{0,1}(z^2) = z^2(C_{1,0}(z^{-2}) + z^{-2}C_{0,1}(z^{-2})). \tag{3.221b}$$

The polyphase transform maps a sequence to a length-2 vector sequence with its polyphase components as elements. Using (3.24), its deterministic autocorrelation is a matrix given by

$$A_n = \begin{bmatrix} a_{0,n} & c_{0,1,n} \\ c^*_{0,1,-n} & a_{1,n} \end{bmatrix} = A^*_{-n}, \tag{3.222a}$$

$$A(e^{j\omega}) = \begin{bmatrix} A_0(e^{j\omega}) & C_{0,1}(e^{j\omega}) \\ C^*_{0,1}(e^{j\omega}) & A_1(e^{j\omega}) \end{bmatrix} = A^*(e^{j\omega}), \tag{3.222b}$$

$$A(z) = \begin{bmatrix} A_0(z) & C_{0,1}(z) \\ C_{0,1*}(z^{-1}) & A_1(z) \end{bmatrix} = A_*(z^{-1}). \tag{3.222c}$$

Polyphase representation of filtering Applying the polyphase decomposition to an impulse response g gives a pair of filters,

$$\begin{bmatrix} \cdots & g_{-2} & \boxed{g_0} & g_2 & g_4 & \cdots \end{bmatrix}^\top \quad \text{and} \quad \begin{bmatrix} \cdots & g_{-1} & \boxed{g_1} & g_3 & g_5 & \cdots \end{bmatrix}^\top,$$

leading to

$$g_{0,n} = g_{2n} \quad \overset{\text{ZT}}{\longleftrightarrow} \quad G_0(z) = \sum_{n\in\mathbb{Z}} g_{2n} z^{-n}, \tag{3.223a}$$

$$g_{1,n} = g_{2n+1} \quad \overset{\text{ZT}}{\longleftrightarrow} \quad G_1(z) = \sum_{n\in\mathbb{Z}} g_{2n+1} z^{-n}, \tag{3.223b}$$

$$G(z) = G_0(z^2) + z^{-1}G_1(z^2). \tag{3.223c}$$

When g is an FIR filter, its polyphase components are also FIR filters, each of about half the length of the original filter.

The polyphase decomposition can be used to break filtering of x with g into two steps: first, each polyphase component of x is filtered with each polyphase component of g; then, the results are appropriately combined. As we will verify, the appropriate way to combine the results is as depicted in Figure 3.30(b) using z-transform expressions. The filters in Figure 3.30(b) can be gathered into the *polyphase matrix*

$$G_\text{p}(z) = \begin{bmatrix} G_0(z) & z^{-1}G_1(z) \\ G_1(z) & G_0(z) \end{bmatrix} \tag{3.224}$$

to give the equivalent block diagram shown in Figure 3.30(c). The matrix $G_\text{p}(z)$ is pseudocirculant (see Appendix 3.B.2).

To check that the equivalence is correct, we perform the following calculation:

$$\begin{aligned}
Y(z) &\overset{(a)}{=} \begin{bmatrix} 1 & z^{-1} \end{bmatrix} G_\text{p}(z^2) \begin{bmatrix} X_0(z^2) \\ X_1(z^2) \end{bmatrix} \\
&= \begin{bmatrix} 1 & z^{-1} \end{bmatrix} \begin{bmatrix} G_0(z^2) & z^{-2}G_1(z^2) \\ G_1(z^2) & G_0(z^2) \end{bmatrix} \begin{bmatrix} X_0(z^2) \\ X_1(z^2) \end{bmatrix} \\
&= \begin{bmatrix} 1 & z^{-1} \end{bmatrix} \begin{bmatrix} G_0(z^2)X_0(z^2) + z^{-2}G_1(z^2)X_1(z^2) \\ G_1(z^2)X_0(z^2) + G_0(z^2)X_1(z^2) \end{bmatrix} \\
&= G_0(z^2)X_0(z^2) + z^{-2}G_1(z^2)X_1(z^2) + z^{-1}(G_1(z^2)X_0(z^2) + G_0(z^2)X_1(z^2)) \\
&= \left(G_0(z^2) + z^{-1}G_1(z^2)\right)\left(X_0(z^2) + z^{-1}X_1(z^2)\right) \\
&\overset{(b)}{=} G(z)\,X(z),
\end{aligned}$$

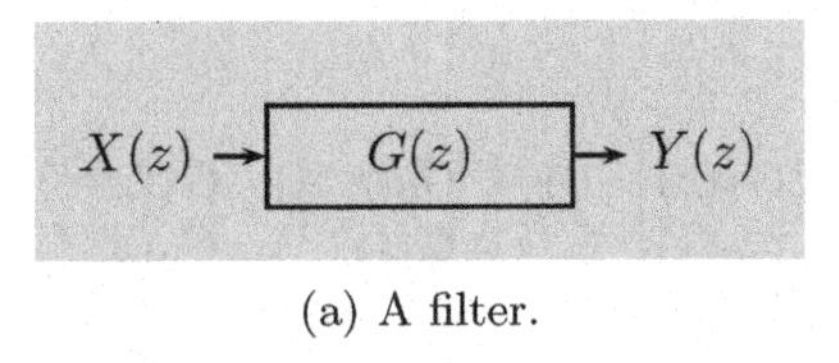

(a) A filter.

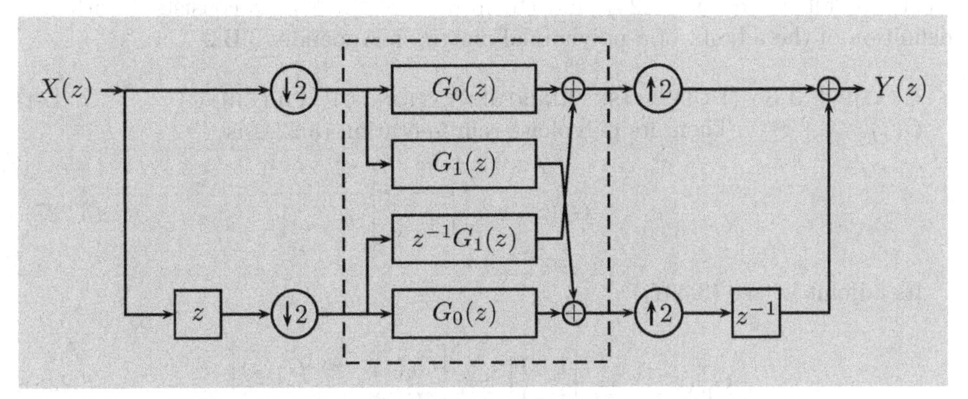

(b) An equivalent system using polyphase decompositions.

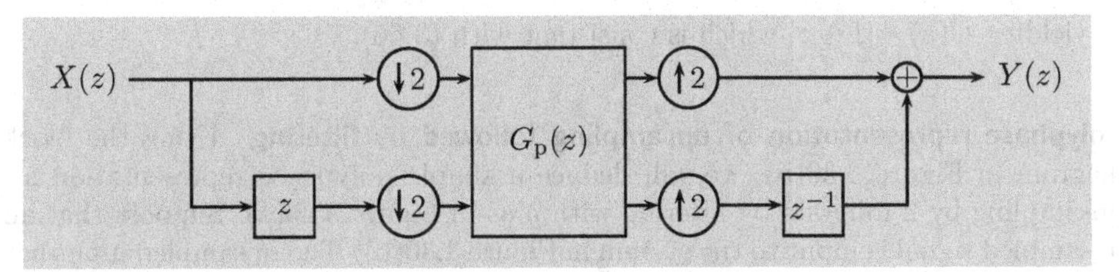

(c) An equivalent system using the polyphase matrix.

Figure 3.30 Polyphase representation and implementation of filtering.

where (a) follows from the inspection of Figure 3.30(c) and the effect of upsampling by 2 being the replacement of z with z^2; and (b) from (3.219c) and (3.223c).

We have seen in (3.66) that the adjoint of the filter operator G is $G^* = \widetilde{G}$ when $\widetilde{g}_n = g^*_{-n}$. Applying the polyphase decomposition to the time-reversed and conjugated version of g gives the pair of filters

$$\begin{bmatrix} \cdots & g_2^* & \boxed{g_0^*} & g_{-2}^* & g_{-4}^* & \cdots \end{bmatrix}^\top \quad \text{and} \quad \begin{bmatrix} \cdots & g_1^* & \boxed{g_{-1}^*} & g_{-3}^* & g_{-5}^* & \cdots \end{bmatrix}^\top.$$

Assuming real filter coefficients, this gives

$$\widetilde{G}_0(z) = G_0(z^{-1}) \quad \text{and} \quad \widetilde{G}_1(z) = zG_1(z^{-1}) \tag{3.225}$$

in the z-transform domain. We thus find the polyphase representation of the adjoint as

$$\widetilde{G}_{\rm p}(z) \overset{(a)}{=} \begin{bmatrix} \widetilde{G}_0(z) & z^{-1}\widetilde{G}_1(z) \\ \widetilde{G}_1(z) & \widetilde{G}_0(z) \end{bmatrix} \overset{(b)}{=} \begin{bmatrix} G_0(z^{-1}) & G_1(z^{-1}) \\ zG_1(z^{-1}) & G_0(z^{-1}) \end{bmatrix} = G_{\rm p}^{\top}(z^{-1}), \quad (3.226)$$

where (a) follows from (3.224); and (b) from (3.225). This is consistent with the definition of the adjoint of a polynomial matrix in Appendix 3.B.2.

EXAMPLE 3.35 (POLYPHASE REPRESENTATION OF FILTERING) Let $G(z) = 1 + z^{-1}$. Then, its polyphase representation (3.224) is

$$G_{\rm p}(z) = \begin{bmatrix} 1 & z^{-1} \\ 1 & 1 \end{bmatrix}. \quad (3.227)$$

Its adjoint is (see (3.307))

$$G_{\rm p}^{\top}(z^{-1}) = \begin{bmatrix} 1 & 1 \\ z & 1 \end{bmatrix} = \begin{bmatrix} \widetilde{G}_0(z) & z^{-1}\widetilde{G}_1(z) \\ \widetilde{G}_1(z) & \widetilde{G}_0(z) \end{bmatrix}, \quad (3.228)$$

yielding $\widetilde{G}(z) = 1 + z$, which is consistent with (3.66).

Polyphase representation of upsampling followed by filtering Using the block diagram in Figure 3.30(b), we will deduce a simple polyphase representation for upsampling by 2 followed by filtering with g as in Figure 3.26(a). Suppose that an upsampled signal is input to the system in Figure 3.30(b). The upsampled input has only an even polyphase component (that is, the odd polyphase component equals zero), so the output of the lower downsampler is zero. We may thus omit the third and fourth filters in the dashed box in Figure 3.30(b). In the top input branch, the downsampling by 2 cancels the upsampling by 2, so Figure 3.31(a) provides an equivalent block diagram.

For a polyphase expression of the output, it follows from Figure 3.31(a) that

$$Y(z) = G_0(z^2)X(z^2) + z^{-1}G_1(z^2)X(z^2) = Y_0(z^2) + z^{-1}Y_1(z^2),$$

with $Y_0(z) = G_0(z)X(z)$ and $Y_1(z) = G_1(z)X(z)$ the polyphase components of $Y(z)$.

The polyphase representation can lead to a reduction in computational complexity. In the direct implementation shown in Figure 3.26(a), the filtering with g is done after upsampling and thus at twice the rate of the filtering in Figure 3.31(a). Suppose that g is an FIR filter. Then, relative to the direct implementation, each of the two filters in the polyphase implementation operates at half the rate and with filters of half the length. This leads to computational savings by about a factor of two. Similar savings arise from polyphase representation of downsampling preceded by filtering. These are discussed in detail in Section 3.9.3.

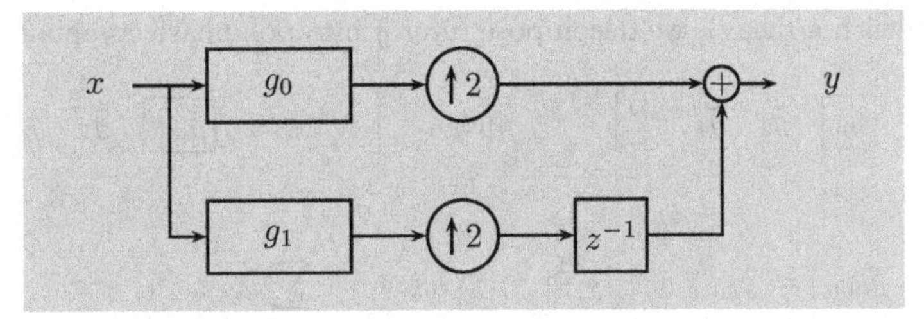

(a) Polyphase implementation of upsampling and filtering from Figure 3.26(a).

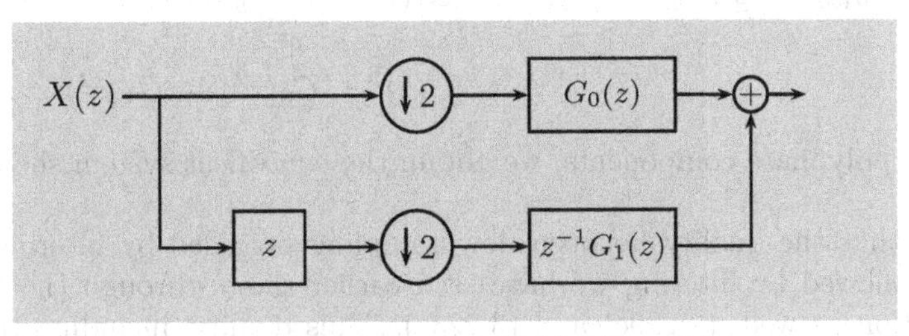

(b) System equivalent to system of Figure 3.30(b) followed by downsampling by 2.

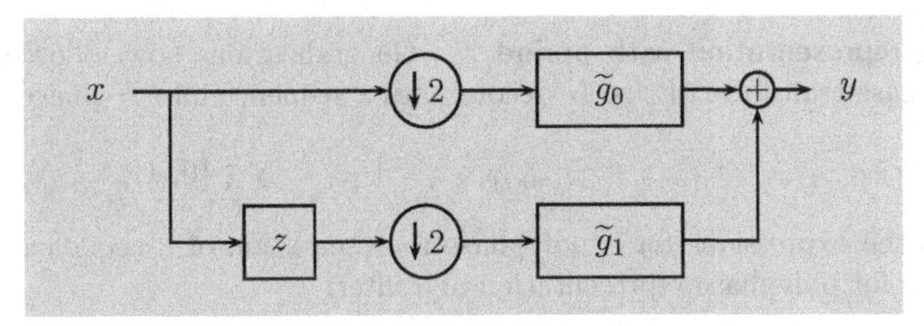

(c) Polyphase implementation of filtering and downsampling from Figure 3.25(a).

Figure 3.31 Polyphase representations of multirate operations. Note that the definitions of polyphase components in (a) and (c) are different; see (3.223) and (3.229).

Polyphase representation of downsampling preceded by filtering Similar reasoning with block diagrams leads us to a simple equivalent for filtering with $\widetilde{g}$ followed by downsampling by 2 as in Figure 3.25(a). Suppose that the output of the system in Figure 3.30(b) is downsampled by 2. Since the second and fourth filters inside the dashed box lead only to odd-indexed samples, which are discarded by downsampling by 2, we may omit the second and fourth filters. In the top output branch, the upsampling by 2 is canceled by the subsequent downsampling by 2, so we get the equivalent system shown in Figure 3.31(b).

The presence of the delay factor z^{-1} in the second filter $z^{-1}G_1(z)$ is inelegant and can be eliminated by a change of convention in the definition of polyphase components. Changing the numbering of the odd polyphase component by 1 (replacing an advance with a delay), we decompose filter $\widetilde{g}$ into polyphase components

$$\begin{bmatrix} \cdots & \widetilde{g}_{-2} & \boxed{\widetilde{g}_0} & \widetilde{g}_2 & \widetilde{g}_4 & \cdots \end{bmatrix}^\top \quad \text{and} \quad \begin{bmatrix} \cdots & \widetilde{g}_{-3} & \boxed{\widetilde{g}_{-1}} & \widetilde{g}_1 & \widetilde{g}_3 & \cdots \end{bmatrix}^\top,$$

leading to

$$\widetilde{g}_{0,n} = \widetilde{g}_{2n} \quad \stackrel{\text{ZT}}{\longleftrightarrow} \quad \widetilde{G}_0(z) = \sum_{n \in \mathbb{Z}} \widetilde{g}_{2n} z^{-n}, \tag{3.229a}$$

$$\widetilde{g}_{1,n} = \widetilde{g}_{2n-1} \quad \stackrel{\text{ZT}}{\longleftrightarrow} \quad \widetilde{G}_1(z) = \sum_{n \in \mathbb{Z}} \widetilde{g}_{2n-1} z^{-n}, \tag{3.229b}$$

$$\widetilde{G}(z) = \widetilde{G}_0(z^2) + z\widetilde{G}_1(z^2). \tag{3.229c}$$

With these polyphase components, we obtain the equivalent system shown in Figure 3.31(c).

Note that the duality between downsampling preceded by filtering and upsampling followed by filtering we have seen earlier shows through the polyphase decomposition as well, (3.229c) and (3.223c). This duality, including the change from z^{-1} to z, is related to the transposition and time reversal seen in (3.210).

Polyphase representation with period N Generalizations now follow naturally. The polyphase transform of size N decomposes a sequence into N phases,

$$\begin{bmatrix} \cdots & x_{N(n-1)+j} & x_{Nn+j} & x_{N(n+1)+j} & \cdots \end{bmatrix}^\top, \qquad j \in \{0, 1, \ldots, N-1\},$$

leading to the expressions for a polyphase representation of a sequence, and two conventions for polyphase representation of a filter:

$$x_{j,n} = x_{Nn+j} \quad \stackrel{\text{ZT}}{\longleftrightarrow} \quad X_j(z) = \sum_{n \in \mathbb{Z}} x_{Nn+j} z^{-n}, \tag{3.230a}$$

$$X(z) = \sum_{j=0}^{N-1} z^{-j} X_j(z^N), \tag{3.230b}$$

$$g_{j,n} = g_{Nn+j} \quad \stackrel{\text{ZT}}{\longleftrightarrow} \quad G_j(z) = \sum_{n \in \mathbb{Z}} g_{Nn+j} z^{-n}, \tag{3.230c}$$

$$G(z) = \sum_{j=0}^{N-1} z^{-j} G_j(z^N), \tag{3.230d}$$

$$\widetilde{g}_{j,n} = \widetilde{g}_{Nn-j} \quad \stackrel{\text{ZT}}{\longleftrightarrow} \quad \widetilde{G}_j(z) = \sum_{n \in \mathbb{Z}} \widetilde{g}_{Nn-j} z^{-n}, \tag{3.230e}$$

$$\widetilde{G}(z) = \sum_{j=0}^{N-1} z^{j} \widetilde{G}_j(z^N). \tag{3.230f}$$

Note the difference between how polyphase components of $\widetilde{g}$ are defined compared with the polyphase components of g. Those for g are numbered forward modulo N, that is, the zeroth polyphase component is the one at nN, the first is the one at $nN+1$, the second that at $nN+2$, and so on (the same as for sequences). Those for $\widetilde{g}$, on the other hand, are numbered in reverse modulo N, that is, the zeroth polyphase component is the one at nN, but the first is the one at $(Nn-1)$, the second is that at $(Nn-2)$, and so on, in reverse order from those for G. As an illustration, we give below both conventions for $N=3$:

$$
\begin{aligned}
g_{0,n} &= \begin{bmatrix} \cdots & g_{-3} & \boxed{g_0} & g_3 & g_6 & \cdots \end{bmatrix}^\top, \\
g_{1,n} &= \begin{bmatrix} \cdots & g_{-2} & \boxed{g_1} & g_4 & g_7 & \cdots \end{bmatrix}^\top, \\
g_{2,n} &= \begin{bmatrix} \cdots & g_{-1} & \boxed{g_2} & g_5 & g_8 & \cdots \end{bmatrix}^\top, \\
\widetilde{g}_{0,n} &= \begin{bmatrix} \cdots & \widetilde{g}_{-3} & \boxed{\widetilde{g}_0} & \widetilde{g}_3 & \widetilde{g}_6 & \cdots \end{bmatrix}^\top, \\
\widetilde{g}_{1,n} &= \begin{bmatrix} \cdots & \widetilde{g}_{-4} & \boxed{\widetilde{g}_{-1}} & \widetilde{g}_2 & \widetilde{g}_5 & \cdots \end{bmatrix}^\top, \\
\widetilde{g}_{2,n} &= \begin{bmatrix} \cdots & \widetilde{g}_{-5} & \boxed{\widetilde{g}_{-2}} & \widetilde{g}_1 & \widetilde{g}_4 & \cdots \end{bmatrix}^\top.
\end{aligned}
$$

3.8 Stochastic processes and systems

Many applications of signal processing involve resolving, reducing, or exploiting uncertainty. Resolving uncertainty includes identifying which out of a set of sequences was transmitted over a noisy channel; reducing uncertainty includes estimating parameters from noisy observations; and exploiting uncertainty includes cryptographic encoding in which the meanings of symbols are hidden from anyone lacking the key. Careful modeling of uncertainty is also exploited in compression when short descriptions are assigned to the most likely inputs.

One of the tools for modeling uncertainty is probability theory (see Appendix 2.C). In what follows, we discuss the use of probabilistic models for sequences within the context of discrete-time signal processing. Following the structure of the chapter in its entirety, we progress from discrete-time stochastic processes (random sequences) to the effect of systems (almost exclusively LSI systems) on stochastic processes in the time domain,[66] then the application of Fourier-domain analysis, and the analysis of multirate systems. Finally, we apply Hilbert space tools to minimum mean-squared error estimation of stochastic processes.

3.8.1 Stochastic processes

A discrete-time stochastic process x is a countably infinite collection of jointly distributed random variables $\{\ldots, \mathrm{x}_0, \mathrm{x}_1, \mathrm{x}_2, \ldots\}$. For example, our temperature example from the opening of the chapter – the temperature at noon in front of your house measured every day – could be modeled as a stochastic process. Especially

[66] We study systems that act deterministically on random signals. Some of this is analogous to systems acting randomly on deterministic signals, but we do not study the latter explicitly.

when the index represents time, a stochastic process is often called a *time series.* Other commonly studied time series include stock market closing prices.

We use the following notations for moments and related quantities defined on stochastic processes:

$$\begin{array}{lll} \text{mean} & \mu_{\mathrm{x},n} & \mathrm{E}[\,\mathrm{x}_n\,] \\ \text{variance} & \mathrm{var}(\mathrm{x}_n) & \mathrm{E}\big[\,|\mathrm{x}_n - \mu_{\mathrm{x},n}|^2\,\big] \\ \text{standard deviation} & \sigma_{\mathrm{x},n} & \sqrt{\mathrm{var}(\mathrm{x}_n)} \\ \text{autocorrelation} & a_{\mathrm{x},n,k} & \mathrm{E}\big[\,\mathrm{x}_n \mathrm{x}_{n-k}^*\,\big] \\ \text{crosscorrelation} & c_{\mathrm{x},\mathrm{y},n,k} & \mathrm{E}\big[\,\mathrm{x}_n \mathrm{y}_{n-k}^*\,\big] \end{array} \tag{3.231}$$

Since these are moments up to second order, they are often referred to as *second-order statistics.* The mean, variance, and standard deviation are (deterministic) sequences. The autocorrelation and crosscorrelation are (deterministic) two-dimensional sequences. The variance can be computed from the autocorrelation and mean through

$$\sigma_{\mathrm{x},n}^2 = a_{\mathrm{x},n,0} - |\mu_{\mathrm{x},n}|^2, \qquad n \in \mathbb{Z}. \tag{3.232}$$

The autocorrelation and crosscorrelation satisfy the symmetries

$$a_{\mathrm{x},n,k} = a_{\mathrm{x},n-k,-k}^*, \qquad k,\, n \in \mathbb{Z}, \tag{3.233a}$$

and

$$c_{\mathrm{y},\mathrm{x},n,k} = c_{\mathrm{x},\mathrm{y},n-k,-k}^*, \qquad k,\, n \in \mathbb{Z}. \tag{3.233b}$$

There is a common abbreviation to express that all the random variables in a stochastic process are *independent and identically distributed*: i.i.d. For an i.i.d. process, the mean, variance, and standard deviation are constant sequences,

$$\begin{aligned} \mu_{\mathrm{x},n} &= \mu_{\mathrm{x}}, & n \in \mathbb{Z}, \\ \mathrm{var}(\mathrm{x}_n) &= \sigma_{\mathrm{x}}^2, & n \in \mathbb{Z}, \\ \sigma_{\mathrm{x},n} &= \sigma_{\mathrm{x}}, & n \in \mathbb{Z}. \end{aligned}$$

The autocorrelation also has a restricted form,

$$\begin{aligned} a_{\mathrm{x},n,k} &= \begin{cases} \mathrm{E}\big[\,|\mathrm{x}_n|^2\,\big], & \text{for } k = 0; \\ \mathrm{E}[\,\mathrm{x}_n\,]\,\mathrm{E}\big[\,\mathrm{x}_{n-k}^*\,\big], & \text{for } k \neq 0 \end{cases} \\ &= \begin{cases} \sigma_{\mathrm{x}}^2 + \mu_{\mathrm{x}}\mu_{\mathrm{x}}^*, & \text{for } k = 0; \\ \mu_{\mathrm{x}}\mu_{\mathrm{x}}^*, & \text{for } k \neq 0 \end{cases} \\ &= |\mu_{\mathrm{x}}|^2 + \sigma_{\mathrm{x}}^2 \delta_k. \end{aligned}$$

Stationarity Stationarity generalizes the i.i.d. property by allowing dependence between the random variables in a stochastic process, but only in a manner that preserves invariance under a shift operator.

DEFINITION 3.18 (STATIONARY PROCESS) A discrete-time stochastic process x is called *stationary* when, for any finite set of time indices $\{n_0, n_1, \ldots, n_L\} \subset \mathbb{Z}$ and any time shift $k \in \mathbb{Z}$, the joint distributions of

$$(\mathrm{x}_{n_0}, \mathrm{x}_{n_1}, \ldots, \mathrm{x}_{n_L}) \quad \text{and} \quad (\mathrm{x}_{n_0+k}, \mathrm{x}_{n_1+k}, \ldots, \mathrm{x}_{n_L+k})$$

are identical.

Stationarity is a highly restrictive condition. Most of the time, we will assume the weaker condition of wide-sense stationarity, which depends only on second-order statistics.

DEFINITION 3.19 (WIDE-SENSE STATIONARY PROCESS) A discrete-time stochastic process x is called *wide-sense stationary (WSS)* when its mean sequence $\mu_{\mathrm{x},n}$ is a constant,

$$\mu_{\mathrm{x},n} = \mathrm{E}[\mathrm{x}_n] = \mu_{\mathrm{x}}, \qquad n \in \mathbb{Z}, \tag{3.234a}$$

and its autocorrelation depends only on the time difference k,

$$a_{\mathrm{x},n,k} = \mathrm{E}\big[\mathrm{x}_n \mathrm{x}^*_{n-k}\big] = a_{\mathrm{x},k}, \qquad k, n \in \mathbb{Z}. \tag{3.234b}$$

Stochastic processes x and y are called jointly WSS when each is WSS and their crosscorrelation depends only on the time difference k,

$$c_{\mathrm{x},\mathrm{y},n,k} = \mathrm{E}\big[\mathrm{x}_n \mathrm{y}^*_{n-k}\big] = c_{\mathrm{x},\mathrm{y},k}, \qquad k, n \in \mathbb{Z}. \tag{3.234c}$$

With wide-sense stationarity, (3.232) and (3.233a) simplify to

$$\sigma^2_{\mathrm{x},n} = a_{\mathrm{x},0} - |\mu_{\mathrm{x}}|^2 = \sigma^2_{\mathrm{x}}, \qquad n \in \mathbb{Z},$$

and

$$a_{\mathrm{x},k} = a^*_{\mathrm{x},-k}, \qquad k \in \mathbb{Z}.$$

With joint wide-sense stationarity, (3.233b) simplifies to the conjugate symmetry

$$c_{\mathrm{y},\mathrm{x},k} = c^*_{\mathrm{x},\mathrm{y},-k}, \qquad k \in \mathbb{Z}.$$

As with any other modeling assumption, wide-sense stationarity should be used with caution. It is often a useful approximation over a short enough time period; for example, many biological processes are approximately stationary over a period of milliseconds, while the noise in a communications channel might be approximately stationary over a much longer period of time.[67]

[67] Consideration of an appropriate time scale is an essential part of modeling. As John Maynard Keynes wrote, "The long run is a misleading guide to current affairs. In the long run we are all dead."

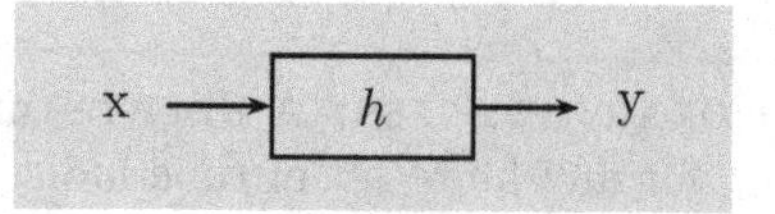

Figure 3.32 An LSI system with WSS input.

White noise A *white noise*[68] process x is a WSS stochastic process whose mean is zero and whose elements are uncorrelated,

$$\mu_{x,n} = 0; \qquad \text{var}(x_n) = \sigma_x^2; \qquad \sigma_{x,n} = \sigma_x; \qquad a_{x,k} = \sigma_x^2 \delta_k. \tag{3.235}$$

The random variables in a white noise process are not always independent. The term *whitening*, or *decorrelation*, is used to mean processing that results in a white noise process. It is basically a diagonalization of the covariance matrix.

Gaussian processes The distribution of a Gaussian process – a stochastic process consisting of jointly Gaussian random variables – is completely specified by its second-order statistics. Since jointly Gaussian random variables are uncorrelated if and only if they are independent, any white Gaussian process is i.i.d. White Gaussian processes often arise in physical models as additive noise; thus, the term *additive white Gaussian noise* (AWGN) is common.

3.8.2 Systems

Consider a BIBO-stable LSI system described by its impulse response h with WSS input sequence x, as depicted in Figure 3.32. What can we say about the output y? It is given by the convolution (3.61), so each y_n is a linear combination of random variables. We will demonstrate that y is a WSS process by deriving formulas for its second-order statistics.

We start with the mean,

$$\begin{aligned} \mu_{y,n} &= E[\, y_n \,] \overset{(a)}{=} E\left[\sum_{k\in\mathbb{Z}} x_k h_{n-k} \right] \overset{(b)}{=} \sum_{k\in\mathbb{Z}} E[\, x_k \,]\, h_{n-k} \overset{(c)}{=} \sum_{k\in\mathbb{Z}} \mu_{x,k} h_{n-k} \\ &\overset{(d)}{=} \sum_{k\in\mathbb{Z}} \mu_x h_{n-k} = \mu_x \sum_{k\in\mathbb{Z}} h_{n-k} \overset{(e)}{=} \mu_x H(e^{j0}) = \mu_y, \end{aligned} \tag{3.236a}$$

where (a) follows from (3.61); (b) from the linearity of the expectation; (c) from the definition of the mean sequence; (d) from x being WSS, (3.234a); and (e) from the frequency response of the LSI system (which exists because the system is BIBO-stable). The final equality emphasizes that the mean of the output is a constant,

[68] We will shortly see that the DTFT of the autocorrelation of a white noise is a constant, mimicking the spectrum of white light; thus the term white noise.

independent of n. The autocorrelation is

$$\begin{aligned}
a_{y,n,k} &= E\left[y_n y_{n-k}^*\right] \stackrel{(a)}{=} E\left[\sum_{m\in\mathbb{Z}} x_{n-m} h_m \sum_{\ell\in\mathbb{Z}} x_{n-k-\ell}^* h_\ell^*\right] \\
&\stackrel{(b)}{=} \sum_{m\in\mathbb{Z}}\sum_{\ell\in\mathbb{Z}} h_m h_\ell^* E\left[x_{n-m} x_{n-k-\ell}^*\right] \stackrel{(c)}{=} \sum_{m\in\mathbb{Z}}\sum_{\ell\in\mathbb{Z}} h_m h_\ell^* a_{x,n-m,k-(m-\ell)} \\
&\stackrel{(d)}{=} \sum_{m\in\mathbb{Z}}\sum_{\ell\in\mathbb{Z}} h_m h_\ell^* a_{x,k-(m-\ell)} \stackrel{(e)}{=} \sum_{p\in\mathbb{Z}}\left(\sum_{m\in\mathbb{Z}} h_m h_{m-p}^*\right) a_{x,k-p} \\
&\stackrel{(f)}{=} \sum_{p\in\mathbb{Z}} a_{h,p} a_{x,k-p} = a_{y,k},
\end{aligned} \tag{3.236b}$$

where (a) follows from (3.61); (b) from the linearity of the expectation; (c) from the definition of the autocorrelation; (d) from x being WSS, (3.234b); (e) from the change of variable $p = m - \ell$; and (f) from the definition of deterministic autocorrelation (3.17). The final step emphasizes the lack of dependence of $a_{y,n,k}$ on n. Combined with the lack of dependence of $\mu_{y,n}$ on n, we see that, when the input x is WSS, the output y is WSS as well. We also see that the autocorrelation of the output is the convolution of the autocorrelation of the input and the deterministic autocorrelation of the impulse response of the system,

$$a_{y,k} = a_{h,k} *_k a_{x,k} \tag{3.236c}$$

$$\stackrel{(a)}{=} h_k *_k h_{-k}^* *_k a_{x,k}, \tag{3.236d}$$

where (a) follows from (3.62d). The z-transform equivalents are

$$A_y(z) = A_h(z)\, A_x(z) \tag{3.236e}$$

$$= H(z)\, H_*(z^{-1})\, A_x(z), \tag{3.236f}$$

assuming that the regions of convergence of $A_h(z)$ and $A_x(z)$ have a nonempty intersection.

Computing the crosscorrelation between the input and the output shows that they are jointly WSS:

$$\begin{aligned}
c_{x,y,n,k} &= E\left[x_n y_{n-k}^*\right] \stackrel{(a)}{=} E\left[x_n \sum_{\ell\in\mathbb{Z}} h_\ell^* x_{n-k-\ell}^*\right] = E\left[\sum_{\ell\in\mathbb{Z}} h_\ell^* x_n x_{n-(k+\ell)}^*\right] \\
&\stackrel{(b)}{=} \sum_{\ell\in\mathbb{Z}} h_\ell^* E\left[x_n x_{n-(k+\ell)}^*\right] \stackrel{(c)}{=} \sum_{\ell\in\mathbb{Z}} h_\ell^* a_{x,n,k+\ell} \\
&\stackrel{(d)}{=} \sum_{\ell\in\mathbb{Z}} h_\ell^* a_{x,k+\ell} = c_{x,y,k},
\end{aligned} \tag{3.237a}$$

where (a) follows from (3.61); (b) from the linearity of the expectation; (c) from the definition of the autocorrelation; and (d) from x being WSS, (3.234b). The final

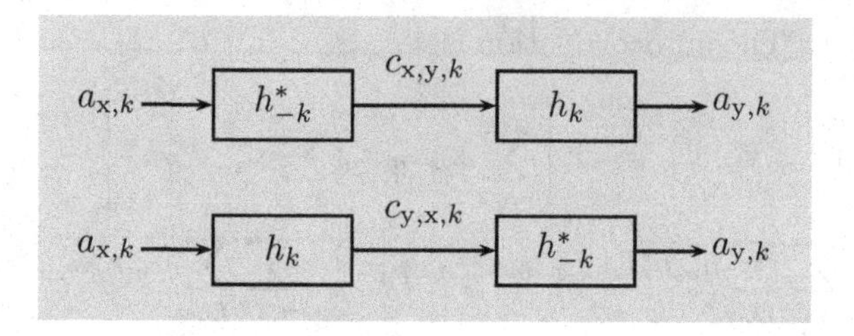

Figure 3.33 Block diagram representations for the autocorrelation and crosscorrelations resulting from filtering a WSS process with an LSI system as in Figure 3.32.

step emphasizes the lack of dependence of $c_{x,y,n,k}$ on n. This crosscorrelation can also be written as a convolution,

$$c_{x,y,k} = h^*_{-k} *_k a_{x,k} \quad \overset{\text{ZT}}{\longleftrightarrow} \quad C_{x,y}(z) = H_*(z^{-1})\, A_x(z). \tag{3.237b}$$

Similarly,

$$c_{y,x,k} = h_k *_k a_{x,k} \quad \overset{\text{ZT}}{\longleftrightarrow} \quad C_{y,x}(z) = H(z)\, A_x(z). \tag{3.237c}$$

Expressions (3.236d), (3.237b), and (3.237c) are represented by the block diagrams in Figure 3.33. We will use these expressions shortly to make some important observations in the Fourier domain.

Autoregressive moving-average process When a white noise process is input to a BIBO-stable, causal LSI system with a rational transfer function, the output is called an *autoregressive moving-average (ARMA) process.* LSI systems with rational transfer functions are those described by linear, constant-coefficient difference equations. A *generative model* for an ARMA process is thus

$$y_n = \sum_{k=0}^{M} b_k x_{n-k} + \sum_{k=1}^{N} a_k y_{n-k}, \tag{3.238}$$

where x is a white noise process with $\sigma_x^2 = 1$. This expression is the same as the linear, constant-coefficient difference equation (3.55) except for the sign in front of the second sum.

When every a_k is zero, the generative model simplifies to

$$y_n = \sum_{k=0}^{M} b_k x_{n-k}.$$

This sequence is called a *moving-average (MA)* process since it is the average over a moving or sliding window of input samples (see also Example 3.2). The number M of past input samples in the generative model is the *order* of the MA process.

EXAMPLE 3.36 (FIRST-ORDER MA PROCESS) An MA-1 process is generated by

$$y_n = b_0 x_n + b_1 x_{n-1}, \tag{3.239}$$

where x is a white noise process with $\sigma_x^2 = 1$. Let $h_n = b_0\delta_n + b_1\delta_{n-1}$. Then, y is generated by filtering white noise x by h. The mean of y is zero; this follows from (3.236a) and $\mu_x = 0$. The most convenient expression for finding the autocorrelation of y is (3.236f)

$$A_y(z) = (b_0 + b_1 z^{-1})(b_0^* + b_1^* z)\sigma_x^2 = b_0 b_1^* z + (|b_0|^2 + |b_1|^2) + b_0^* b_1 z^{-1}.$$

By taking the inverse z-transform, we have

$$a_{y,k} = b_0 b_1^* \delta_{k+1} + (|b_0|^2 + |b_1|^2)\delta_k + b_0^* b_1 z^{-1}\delta_{k-1}.$$

The variance of y is

$$\sigma_y^2 = a_{y,0} = |b_0|^2 + |b_1|^2.$$

When, instead, b_0 is nonzero and $b_k = 0$ for $k > 0$, the generative model simplifies to

$$y_n = b_0 x_n + \sum_{k=1}^{N} a_k y_{n-k}.$$

The sequence is called an *autoregressive (AR)* process. The number N of past output samples in the generative model is the *order* of the AR process.

EXAMPLE 3.37 (FIRST-ORDER AR PROCESS) An AR-1 process is generated by

$$y_n = b x_n + a y_{n-1}, \tag{3.240}$$

where x is a white noise process with $\sigma_x^2 = 1$. For this generative model to be BIBO-stable, we must have $|a| < 1$. As in the previous example, the mean of the process is zero. We may follow the same steps to compute the autocorrelation and variance. Instead, we illustrate a recursive computation.

Starting with (3.240),

$$\sigma_y^2 = \mathrm{E}\left[\,|b x_n + a y_{n-1}|^2\,\right] \stackrel{(a)}{=} |b|^2\sigma_x^2 + |a|^2 \mathrm{E}\left[\,|y_{n-1}|^2\,\right] \stackrel{(b)}{=} |b|^2 + |a|^2\sigma_y^2,$$

where (a) follows from the linearity of the expectation and x_n being uncorrelated with y_{n-1} (since x is white and the generative model is causal); and (b) from the wide-sense stationarity of y and $\sigma_x^2 = 1$. Thus,

$$a_{y,0} = \sigma_y^2 = \frac{|b|^2}{1 - |a|^2}.$$

For $k \in \mathbb{Z}^+$,

$$\begin{aligned} a_{y,k} &= \mathrm{E}\left[\,y_n y_{n-k}^*\,\right] \stackrel{(a)}{=} \mathrm{E}\left[\,(b x_n + a y_{n-1}) y_{n-k}^*\,\right] \\ &\stackrel{(b)}{=} b\mathrm{E}\left[\,x_n y_{n-k}^*\,\right] + a\mathrm{E}\left[\,y_{n-1} y_{n-k}^*\,\right] \stackrel{(c)}{=} a a_{y,k-1}, \end{aligned}$$

where (a) follows from (3.240); (b) from the linearity of the expectation; and (c) from x_n being uncorrelated with past values of y and the definition of the autocorrelation. Similarly, for $k \in \mathbb{Z}^-$,

$$\begin{aligned} a_{y,k} &= \mathrm{E}\big[y_n y_{n-k}^* \big] \overset{(a)}{=} \mathrm{E}[y_n (b x_{n-k} + a y_{n-k-1})^*] \\ &\overset{(b)}{=} b^* \mathrm{E}\big[y_n x_{n-k}^* \big] + a^* \mathrm{E}\big[y_n y_{n-k-1}^* \big] \overset{(c)}{=} a^* a_{y,k+1}, \end{aligned}$$

where (a) follows from (3.240); (b) from the linearity of the expectation; and (c) from x_{n-k} being uncorrelated with past values of y and the definition of the autocorrelation. By solving these recursions, we get

$$a_{y,k} = \begin{cases} a^k \sigma_y^2, & \text{for } k = 0, 1, \ldots; \\ (a^*)^{-k} \sigma_y^2, & \text{for } k = -1, -2, \ldots. \end{cases}$$

When a is real, the autocorrelation has the simpler form

$$a_{y,k} = a^{|k|} \sigma_y^2, \qquad k \in \mathbb{Z}.$$

To normalize to $\sigma_y^2 = 1$, set $b = \sqrt{1 - |a|^2}$, leading to the generating filter

$$h_n = \sqrt{1 - |a|^2} a^n u_n, \qquad n \in \mathbb{Z}. \tag{3.241}$$

3.8.3 Discrete-time Fourier transform

Just like for deterministic sequences, we can use Fourier techniques to gain insight into the behavior of discrete-time stochastic processes and systems. While we cannot take a DTFT of a stochastic process,[69] we can make assessments based on averages (moments), such as taking the DTFT of the autocorrelation.

Power spectral density Let x be a WSS stochastic process. The DTFT of its autocorrelation (3.234b) (which we assume to have sufficient decay so as to be absolutely summable) is

$$A_x(e^{j\omega}) = \sum_{k \in \mathbb{Z}} a_{x,k} e^{-j\omega k} = \sum_{k \in \mathbb{Z}} \mathrm{E}\big[x_n x_{n-k}^* \big] e^{-j\omega k}. \tag{3.242}$$

This is called the *power spectral density*, the counterpart of the *energy spectral density* for deterministic sequences in (3.100). The power spectral density exists if and only if x is WSS, which is a consequence of the Wiener–Khinchin theorem. When x is real, the power spectral density is nonnegative and thus admits a spectral factorization

$$A_x(e^{j\omega}) = U(e^{j\omega}) \, U^*(e^{j\omega}),$$

[69]The technical difficulties can be subtle, but the lack of decay of stationary processes puts them outside the classes of signals for which we considered convergence of the DTFT in Section 3.4.2.

Deterministic sequences	WSS discrete-time stochastic processes
Energy spectral density	Power spectral density
$A(e^{j\omega}) = \lvert X(e^{j\omega})\rvert^2$	$A(e^{j\omega}) = \sum_{k\in\mathbb{Z}} \mathrm{E}[\mathrm{x}_n \mathrm{x}_{n-k}^*] e^{-j\omega k}$
Energy	Power
$E = \frac{1}{2\pi}\int_{-\pi}^{\pi} A(e^{j\omega})\, d\omega$	$P = \frac{1}{2\pi}\int_{-\pi}^{\pi} A(e^{j\omega})\, d\omega$
$E = a_0 = \sum_{n\in\mathbb{Z}} \lvert x_n\rvert^2$	$P = a_0 = \mathrm{E}[\,\lvert \mathrm{x}_n\rvert^2\,]$

Table 3.8 Energy concepts for deterministic sequences and their counterpart power concepts for WSS discrete-time stochastic processes.

where $U(e^{j\omega})$ is its (nonunique) spectral root. The average of the power spectral density over the frequency range,

$$P_{\mathrm{x}} = \frac{1}{2\pi}\int_{-\pi}^{\pi} A_{\mathrm{x}}(e^{j\omega})\, d\omega = a_{\mathrm{x},0} = \mathrm{E}\big[\,\lvert \mathrm{x}_n\rvert^2\,\big], \tag{3.243}$$

is the *power*, the counterpart of the *energy* for deterministic sequences in (3.102). The power spectral density measures the distribution of power over the frequency range. These four concepts – energy and energy spectral density for deterministic sequences, and power and power spectral density for WSS processes – are summarized in Table 3.8.

Example 3.38 (First-order MA process, Example 3.36 continued) Consider a unit-power version of an MA-1 process as in (3.239) by setting $b_0 = b_1 = 1/\sqrt{2}$. Its power spectral density is

$$A_{\mathrm{y}}(e^{j\omega}) = \frac{1}{2}(1+e^{j\omega})(1+e^{-j\omega}) = 1 + \frac{1}{2}(e^{j\omega}+e^{-j\omega}) = 1+\cos\omega. \tag{3.244}$$

It is positive semidefinite; that is, $A_{\mathrm{y}}(e^{j\omega}) \geq 0$.

Example 3.39 (First-order AR process, Example 3.37 continued) Consider a unit-power version of an AR-1 process as generated by the filter in (3.241) with real-valued a. Its power spectral density is

$$A_{\mathrm{y}}(e^{j\omega}) = \frac{1-a^2}{(1-ae^{j\omega})(1-ae^{-j\omega})} = \frac{1-a^2}{\lvert 1-ae^{-j\omega}\rvert^2}, \qquad \lvert a\rvert < 1. \tag{3.245}$$

This function is positive definite; that is, $A_{\mathrm{y}}(e^{j\omega}) > 0$.

Power spectral density estimation is typically done by estimating local behavior, requiring some form of local Fourier transform.

White noise Using (3.235) and Table 3.4, we see that the power spectral density of white noise is a constant,

$$A(e^{j\omega}) = \sigma_x^2. \tag{3.246}$$

Its variance, or power, is

$$a_0 = \frac{1}{2\pi}\int_{-\pi}^{\pi} \sigma_x^2 \, d\omega = \sigma_x^2.$$

Effect of filtering Consider an LSI system with impulse response h, WSS input x, and WSS output y, as depicted in Figure 3.32. Using (3.236c) for the autocorrelation of y, the power spectral density of the output is given by

$$A_y(e^{j\omega}) = A_h(e^{j\omega})\, A_x(e^{j\omega}) = |H(e^{j\omega})|^2 A_x(e^{j\omega}), \tag{3.247}$$

where $A_h(e^{j\omega}) = |H(e^{j\omega})|^2$ is the DTFT of the deterministic autocorrelation of h, according to Table 3.4. The quantity

$$P_y = \mathrm{E}\big[y_n^2\big] = \frac{1}{2\pi}\int_{-\pi}^{\pi} A_y(e^{j\omega})\, d\omega = \frac{1}{2\pi}\int_{-\pi}^{\pi} |H(e^{j\omega})|^2 A_x(e^{j\omega})\, d\omega = a_y(0)$$

is the *output power*. Similarly to (3.247), using (3.237b) and (3.237c), we can express the *cross power spectral density* between the input and the output as

$$C_{x,y}(e^{j\omega}) = H^*(e^{j\omega})\, A_x(e^{j\omega}), \tag{3.248a}$$
$$C_{y,x}(e^{j\omega}) = H(e^{j\omega})\, A_x(e^{j\omega}). \tag{3.248b}$$

3.8.4 Multirate sequences and systems

When a discrete-time stochastic process makes its way through a multirate system, stationarity or wide-sense stationarity is in general not preserved. We will see that *periodic shift variance* for deterministic systems has its counterpart in *wide-sense cyclostationarity*[70] for stochastic systems.

DEFINITION 3.20 (WIDE-SENSE CYCLOSTATIONARY PROCESS) A stochastic process x is called *wide-sense cyclostationary* of period N (WSCS_N) when the vector of its polyphase components is WSS.

Our temperature example comes in handy again. Take the temperature sequence x and decompose it into its polyphase components modulo 365. Then, each calendar day can follow its own statistical behavior. For example, the temperature at noon on January 14th in New York City will likely be low, while the measurement taken on July 14th will likely be high. The notion of cyclostationarity for LPSV systems is intuitive, given their cyclic nature. We now discuss a few basic operations for illustration only; see the *Further reading* for pointers to the literature.

[70] Cyclostationarity is also called block-stationarity or N-stationarity.

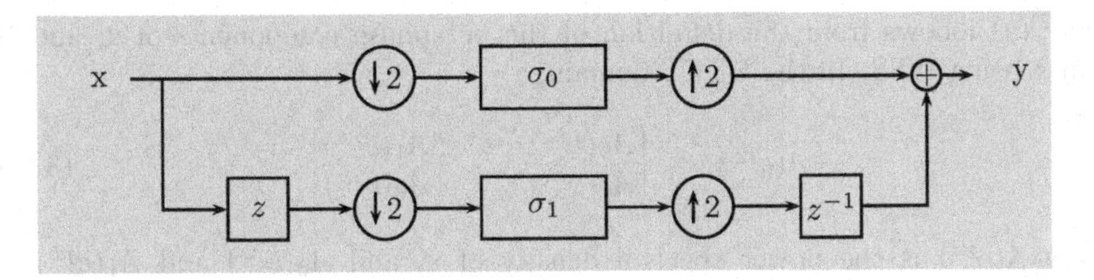

Figure 3.34 Generative model for the WSCS_2 sequence from Example 3.40. The input x is WSS, while the output y is WSCS_2.

The mean and autocorrelation of a WSCS_N sequence x satisfy

$$\mu_{\mathrm{x},n+N} = \mathrm{E}[\, \mathrm{x}_{n+N} \,] = \mathrm{E}[\, \mathrm{x}_n \,] = \mu_{\mathrm{x},n}, \tag{3.249a}$$

$$a_{\mathrm{x},n+N,k} = \mathrm{E}\big[\, \mathrm{x}_{n+N}\mathrm{x}_{n+N-k}^* \,\big] = \mathrm{E}\big[\, \mathrm{x}_n\mathrm{x}_{n-k}^* \,\big] = a_{\mathrm{x},n,k}, \tag{3.249b}$$

exhibiting a periodicity with respect to time index n, in contrast to the lack of dependence on n in (3.234a) and (3.234b). Note also that

$$\mathrm{x} \text{ is } \text{WSCS}_N \;\Rightarrow\; \mathrm{x} \text{ is } \text{WSCS}_{\ell N}, \quad \ell \in \mathbb{Z}^+.$$

Beware that (3.249b) does not imply that $a_{\mathrm{x},n,k}$ is periodic in the time lag k, as we now illustrate.

Example 3.40 (Generative model for a WSCS_2 sequence) Consider the system in Figure 3.34, where x is a white noise process with $\sigma_\mathrm{x}^2 = 1$. The system scales x_n by σ_0 for even n and by σ_1 for odd n, with σ_0 and σ_1 real numbers. Since the components of x are uncorrelated,

$$a_{\mathrm{y},2n,k} = \mathrm{E}\big[\, \mathrm{y}_{2n}\mathrm{y}_{2n-k}^* \,\big] = \sigma_0^2\delta_k, \qquad k,\, n \in \mathbb{Z},$$

$$a_{\mathrm{y},2n+1,k} = \mathrm{E}\big[\, \mathrm{y}_{2n+1}\mathrm{y}_{2n+1-k}^* \,\big] = \sigma_1^2\delta_k, \qquad k,\, n \in \mathbb{Z}.$$

It is easy to check that (3.249) holds, so the sequence y is WSCS with period 2. The autocorrelation is not, however, a periodic function of the time lag k,

$$a_{\mathrm{y},2n,k+2} = \mathrm{E}\big[\, \mathrm{y}_{2n}\mathrm{y}_{2n-k-2}^* \,\big] = \sigma_0^2\delta_{k+2} \neq \sigma_0^2\delta_k = a_{\mathrm{y},2n,k}.$$

As we have done earlier for deterministic sequences, we can characterize vector processes using autocorrelation matrices. For the sake of simplicity, let us consider a WSS sequence x and the vector of its polyphase components of period 2. Its matrix autocorrelation will be

$$A_k = \begin{bmatrix} a_{0,k} & c_{01,k} \\ c_{10,k} & a_{1,k} \end{bmatrix} = \begin{bmatrix} \mathrm{E}[\, \mathrm{x}_{0,n}\mathrm{x}_{0,n-k}^* \,] & \mathrm{E}[\, \mathrm{x}_{0,n}\mathrm{x}_{1,n-k}^* \,] \\ \mathrm{E}[\, \mathrm{x}_{1,n}\mathrm{x}_{0,n-k}^* \,] & \mathrm{E}[\, \mathrm{x}_{1,n}\mathrm{x}_{1,n-k}^* \,] \end{bmatrix}$$

$$\overset{(a)}{=} \begin{bmatrix} \mathrm{E}[\, \mathrm{x}_{2n}\mathrm{x}_{2n-2k}^* \,] & \mathrm{E}[\, \mathrm{x}_{2n}\mathrm{x}_{2n-2k+1}^* \,] \\ \mathrm{E}[\, \mathrm{x}_{2n+1}\mathrm{x}_{2n-2k}^* \,] & \mathrm{E}[\, \mathrm{x}_{2n+1}\mathrm{x}_{2n-2k+1}^* \,] \end{bmatrix} \overset{(b)}{=} \begin{bmatrix} a_{2k} & a_{2k-1} \\ a_{2k+1} & a_{2k} \end{bmatrix},$$

where (a) follows from the definition of the polyphase components of x; and (b) from x being WSS. In the DTFT domain,

$$A(e^{j\omega}) = \begin{bmatrix} A_0(e^{j\omega}) & e^{-j\omega}A_1(e^{j\omega}) \\ A_1(e^{j\omega}) & A_0(e^{j\omega}) \end{bmatrix}, \tag{3.250}$$

where $A(e^{j\omega})$ is the power spectral density of x, and $A_0(e^{j\omega})$ and $A_1(e^{j\omega})$ are the DTFTs of the polyphase components of a_k. The matrix $A(e^{j\omega})$ is positive semidefinite, which we now prove. For simplicity, we assume real entries. First, we know that $A(e^{j\omega})$ is an even function of ω, (3.101c), and nonnegative, (3.101b). Furthermore,

$$\begin{aligned} A_0(e^{2j\omega}) &= \tfrac{1}{2}[A(e^{j\omega}) + A(e^{j(\omega+\pi)})] &= A_0(e^{-2j\omega}), \\ A_1(e^{2j\omega}) &= e^{j\omega}\tfrac{1}{2}[A(e^{j\omega}) - A(e^{j(\omega+\pi)})] &= e^{-2j\omega}A_1(e^{-2j\omega}). \end{aligned}$$

We thus get that

$$\begin{aligned} A_0(e^{2j\omega}) + e^{-j\omega}A_1(e^{2j\omega}) &= A(e^{j\omega}) \overset{(a)}{\geq} 0, \\ A_0(e^{2j\omega}) - e^{-j\omega}A_1(e^{2j\omega}) &= A(e^{j(\omega+\pi)}) \overset{(b)}{\geq} 0, \end{aligned}$$

where (a) follows from (3.101b); and (b) from (3.101a) and (3.101b). Then,

$$\begin{aligned} &(A_0(e^{2j\omega}) + e^{-j\omega}A_1(e^{2j\omega}))(A_0(e^{2j\omega}) - e^{-j\omega}A_1(e^{2j\omega})) \\ &\quad = A_0^2(e^{2j\omega}) - e^{-2j\omega}A_1^2(e^{2j\omega}) \geq 0. \end{aligned} \tag{3.251}$$

To prove that $A(e^{j\omega})$ is positive semidefinite, we must prove that all of its principal minors have nonnegative determinants (see Section 2.B.2). In this case, that means that $A_0(e^{j\omega}) \geq 0$, which we know from (3.101b), as well as

$$\det(A(e^{j\omega})) = A_0^2(e^{j\omega}) - e^{-j\omega}A_1^2(e^{j\omega}) \overset{(a)}{\geq} 0,$$

where (a) follows from (3.251), proving that $A(e^{j\omega})$ is positive semidefinite.

Example 3.41 (First-order AR process, Example 3.39 continued) Consider the vector of polyphase components of period 2 of the AR-1 process y with power spectral density as in (3.245). The matrix $A(e^{j\omega})$ is then

$$A(e^{j\omega}) = \frac{1-a^4}{(1-a^2e^{-j\omega})(1-a^2e^{j\omega})} \begin{bmatrix} 1 & \frac{a}{1+a^2}(1+e^{j\omega}) \\ \frac{a}{1+a^2}(1+e^{-j\omega}) & 1 \end{bmatrix}. \tag{3.252}$$

To check that the matrix is positive definite, we compute $v^\top A(e^{j\omega})\, v$, for an arbitrary $v = \begin{bmatrix}\cos\theta & \sin\theta\end{bmatrix}^\top$:

$$\begin{bmatrix}\cos\theta & \sin\theta\end{bmatrix} A(e^{j\omega}) \begin{bmatrix}\cos\theta \\ \sin\theta\end{bmatrix} = (1-a^2)\frac{(1+a^2) + a(1+\cos\omega)\sin 2\theta}{1+a^4-2a^2\cos\omega}.$$

The denominator of the above expression is always positive since

$$1 + a^4 - 2a^2 \cos\omega \overset{(a)}{\geq} 1 + a^4 - 2a^2 = (1 - a^2)^2 > 0,$$

where (a) follows from $a^2 \geq 0$ and $|\cos\omega| \leq 1$. Then, we just have to show that the following expression is nonnegative:

$$(1 + a^2) + a(1 + \cos\omega)\sin 2\theta \overset{(a)}{\geq} 1 + a^2 - 2|a| = (1 - |a|)^2 \overset{(b)}{>} 0,$$

where (a) follows from $|(1 + \cos\omega)\sin 2\theta| \leq 2$; and (b) from $|a| < 1$.

We could have saved ourselves all this computation had we observed that

$$A(e^{j\omega}) = U(e^{j\omega})\, U^\top(e^{-j\omega}), \tag{3.253a}$$

with

$$U(e^{j\omega}) = \frac{\sqrt{1-a^2}}{1 - a^2 e^{-j\omega}} \begin{bmatrix} 1 & a \\ a e^{-j\omega} & 1 \end{bmatrix}, \tag{3.253b}$$

making it obvious that $A(e^{j\omega})$ is positive semidefinite.

Example 3.42 (First-order MA process, Example 3.38 continued) Consider the vector of polyphase components of period 2 of the MA-1 process y with power spectral density as in (3.244). The autocorrelation matrix is

$$\begin{aligned} A(e^{j\omega}) &= \begin{bmatrix} A_0(e^{j\omega}) & A_1(e^{j\omega}) \\ A_1^*(e^{-j\omega}) & A_0(e^{j\omega}) \end{bmatrix} = \begin{bmatrix} 1 & (1+e^{j\omega})/2 \\ (1+e^{-j\omega})/2 & 1 \end{bmatrix} \\ &= \frac{1}{\sqrt{2}} \begin{bmatrix} 1 & e^{j\omega} \\ 1 & 1 \end{bmatrix} \frac{1}{\sqrt{2}} \begin{bmatrix} 1 & 1 \\ e^{-j\omega} & 1 \end{bmatrix} = U(e^{j\omega})\, U^\top(e^{-j\omega}), \end{aligned}$$

and is thus clearly positive semidefinite since it admits a spectral factorization.

We now establish some basic results involving multirate operations with WSS or WSCS inputs; a summary is given in Table 3.9.

Downsampling Consider downsampling by N as in (3.190). If the input x is WSS, the output y is WSS as well,

$$a_{\mathrm{y},n,k} = \mathrm{E}\big[\mathrm{y}_n \mathrm{y}_{n-k}^*\big] \overset{(a)}{=} \mathrm{E}\Big[\mathrm{x}_{Nn} \mathrm{x}_{N(n-k)}^*\Big] \overset{(b)}{=} a_{\mathrm{x},Nk} = a_{\mathrm{y},k}, \tag{3.254}$$

where (a) follows from (3.190); and (b) from x being WSS.

Under the weaker condition that x is WSCS$_N$, y is again WSS,

$$a_{\mathrm{y},n,k} = \mathrm{E}\big[\mathrm{y}_n \mathrm{y}_{n-k}^*\big] \overset{(a)}{=} \mathrm{E}\Big[\mathrm{x}_{Nn} \mathrm{x}_{N(n-k)}^*\Big] \overset{(b)}{=} a_{\mathrm{x},Nk} = a_{\mathrm{y},k},$$

where again, (a) follows from (3.190); and (b) from x being WSCS$_N$.

The above two cases are special cases of the more general fact that if x is WSCS$_M$, then y is WSCS$_L$, with $L = M/\gcd(M, N)$. For this, and other special cases, see the *Further reading*.

System	Input x	Output y	
Downsampling by N	WSS	WSS	
	WSCS_N	WSS	
	WSCS_M	WSCS_L	$L = M/\text{gcd}(M, N)$
Upsampling by N	WSS	WSCS_N	
Filtering with h	WSS	WSS	
	WSCS_N	WSCS_N	
Downsampling by N preceded by filtering	WSS	WSS	
Upsampling by N followed by filtering	WSS	WSCS_N	
	WSS	WSS	Filter has alias-free support
Rational change (M up N down)	WSS	WSCS_L	$L = M/\text{gcd}(M, N)$

Table 3.9 Summary of results for multirate systems with stochastic inputs.

The power spectral density of the output is given by

$$A_{\text{y}}(e^{j\omega}) \overset{(a)}{=} \sum_{k\in\mathbb{Z}} a_{\text{y},k} e^{-j\omega k} \overset{(b)}{=} \sum_{k\in\mathbb{Z}} a_{\text{x},Nk} e^{-j\omega k} \overset{(c)}{=} \frac{1}{N} \sum_{n=0}^{N-1} A_{\text{x}}(e^{j(\omega - 2\pi n)/N}), \tag{3.255}$$

where (a) follows from the definition of the power spectral density (3.242); (b) from (3.254); and (c) from the expression for downsampling by N, (3.191).

Upsampling Consider upsampling by N as in (3.195). If the input x is WSS, the output y is WSCS_N. This is easily seen if we remember Definition 3.20: y will be WSCS_N if all of its polyphase components are WSS. All polyphase components of y, except for the first one, are zero, and are thus WSS. The first polyphase component is just the input sequence x, which is WSS by assumption.

The power spectral density of the output is given by

$$A_{\text{y}}(e^{j\omega}) = A_{\text{x}}(e^{jN\omega}), \tag{3.256}$$

from the expression for upsampling by N, (3.196).

Filtering We need one more element, a filter, to be able to build basic multirate systems. Given a WSCS_N input sequence x and an LPSV system with period N, the output y will also be WSCS_N,

$$\begin{aligned}
a_{\mathrm{y},n+N,k} &= \mathrm{E}\big[\mathrm{y}_{n+N}\mathrm{y}^*_{n+N-k}\big] \stackrel{(a)}{=} \mathrm{E}\Bigg[\sum_{m\in\mathbb{Z}} h_m \mathrm{x}_{n+N-m} \sum_{\ell\in\mathbb{Z}} h^*_\ell \mathrm{x}^*_{n+N-k-\ell}\Bigg] \\
&\stackrel{(b)}{=} \sum_{m\in\mathbb{Z}}\sum_{\ell\in\mathbb{Z}} h_m h^*_\ell \mathrm{E}\big[\mathrm{x}_{n+N-m}\mathrm{x}^*_{n+N-k-\ell}\big] \\
&\stackrel{(c)}{=} \sum_{m\in\mathbb{Z}}\sum_{\ell\in\mathbb{Z}} h_m h^*_\ell a_{\mathrm{x},n+N-m,k+\ell-m} \\
&\stackrel{(d)}{=} \sum_{m\in\mathbb{Z}}\sum_{\ell\in\mathbb{Z}} h_m h^*_\ell a_{\mathrm{x},n-m,k+\ell-m} \stackrel{(e)}{=} \sum_{m\in\mathbb{Z}}\sum_{\ell\in\mathbb{Z}} h_m h^*_\ell \mathrm{E}\big[\mathrm{x}_{n-m}\mathrm{x}^*_{n-k-\ell}\big] \\
&\stackrel{(f)}{=} \mathrm{E}\Bigg[\sum_{m\in\mathbb{Z}} h_m \mathrm{x}_{n-m} \sum_{\ell\in\mathbb{Z}} h^*_\ell \mathrm{x}^*_{n-k-\ell}\Bigg] \stackrel{(g)}{=} \mathrm{E}\big[\mathrm{y}_n\mathrm{y}^*_{n-k}\big] = a_{\mathrm{y},k},
\end{aligned}$$

where (a) and (g) follow from the convolution expression (3.61); (b) and (f) from linearity of the expectation and h being deterministic; (c) and (e) from the definition of the autocorrelation of x, (3.231); and (d) from x being WSCS_N, (3.249b).

Downsampling preceded by filtering Consider downsampling preceded by filtering as in Figure 3.25(a). We have already seen that downsampling does not change the nature of the sequence, and neither does filtering. Thus, if x is WSS, y is WSS as well. In the DTFT domain, using (3.247) and (3.255), we get

$$A_{\mathrm{y}}(e^{j\omega}) = \frac{1}{N}\sum_{n=0}^{N-1} A_{\widetilde{g}}(e^{j(\omega-2\pi n)/N})\, A_{\mathrm{x}}(e^{j(\omega-2\pi n)/N}), \tag{3.257}$$

where $A_{\widetilde{g}}(e^{j\omega}) = |\widetilde{G}(e^{j\omega})|^2$ is the DTFT of the deterministic autocorrelation of $\widetilde{g}$.

Upsampling followed by filtering We finally look at upsampling followed by filtering, as shown in Figure 3.26(a). If x is WSS, y is WSCS_N. To see that, we use what we have shown so far. We know that if x is WSS, the output of the upsampler will be WSCS_N. As this is the input to an LSI (and consequently LPSV) system, its output will be WSCS_N. We illustrate this with an example.

Example 3.43 (Upsampling and filtering, Example 3.34 continued) Consider the system in Figure 3.26(a), with $g_n = \delta_n + \delta_{n-1}$ as in Example 3.34 and x a white noise process with $\sigma_{\mathrm{x}}^2 = 1$. After upsampling, the first polyphase component is just x itself, so it is WSS, while the second polyphase component is all zero, and is thus WSS as well. The output of the upsampler is thus WSCS_2, but it is not WSS since (3.234b) is not satisfied. After filtering, the output is the staircase sequence from (3.208). This stochastic process is not WSS, but it is WSCS_2 as each of its two polyphase components is now equal to the WSS process x.

Rational sampling rate change Suppose now that we have a combination of upsampling by M, followed by filtering, followed by downsampling by N. We can say that if x is WSS, then y is WSCS_L, with $L = M/\gcd(M, N)$. This follows directly from the fact we just proved on upsampling followed by filtering and applying the result on downsampling.

3.8.5 Minimum mean-squared error estimation

The projection theorem establishes that orthogonality is central to optimal approximation in a Hilbert space (see Section 2.4.1). By applying this to Hilbert spaces of random variables under the standard inner product $\langle \mathrm{x}, \mathrm{y} \rangle = \mathrm{E}[\, \mathrm{x}\mathrm{y}^* \,]$, we saw that orthogonality is also central to minimum mean-squared error estimation of random variables and finite-dimensional random vectors (see Definition 2.33 and Section 2.4.4). We now consider MMSE estimation of discrete-time stochastic processes, with an emphasis on linear estimation of one process from another when the pair is jointly WSS.

Orthogonality of stochastic processes Orthogonality of vectors is defined by their inner product being zero. Just like for finite-dimensional random vectors (see Section 2.4.4), we need an extension of this concept to handle stochastic processes.

DEFINITION 3.21 (ORTHOGONAL STOCHASTIC PROCESSES) Discrete-time stochastic processes x and y are said to be *orthogonal* when

$$c_{\mathrm{x},\mathrm{y},k,n} = \mathrm{E}\big[\mathrm{x}_k \mathrm{y}_{k-n}^*\big] = 0, \qquad \text{for all } k,\, n \in \mathbb{Z}. \tag{3.258a}$$

For jointly WSS processes, the condition on crosscorrelation has no dependence on the time lag k,

$$c_{\mathrm{x},\mathrm{y},n} = \mathrm{E}\big[\mathrm{x}_k \mathrm{y}_{k-n}^*\big] = 0, \qquad \text{for all } n \in \mathbb{Z}, \tag{3.258b}$$

and can be written equivalently as a condition on the cross power spectral density,

$$C_{\mathrm{x},\mathrm{y}}(e^{j\omega}) = 0, \qquad \text{for all } \omega \in \mathbb{R}. \tag{3.258c}$$

Condition (3.258a) requires orthogonality of every pair of scalar random variables $(\mathrm{x}_k, \mathrm{y}_\ell)$, as in (2.86). Joint wide-sense stationarity reduces the number of independent conditions to (3.258b) since orthogonality of the pair $(\mathrm{x}_k, \mathrm{y}_\ell)$ implies the orthogonality of $(\mathrm{x}_{k+m}, \mathrm{y}_{\ell+m})$ for every $m \in \mathbb{Z}$.

Orthogonality properties both for deterministic sequences and for WSS discrete-time stochastic processes are summarized in Table 3.10.

Wiener filtering In Section 2.4.4, we derived the optimal linear estimator of $\mathbb{C}^N$-valued random vector x from $\mathbb{C}^M$-valued random vector y (see (2.85)). In principle, this derivation holds unchanged for discrete-time stochastic processes. However,

	Deterministic sequences	WSS discrete-time stochastic processes
Time	$c_{x,y,k} = \langle x_n, y_{n-k}\rangle_n = 0$	$c_{\mathrm{x},\mathrm{y},k} = \mathrm{E}[\,\mathrm{x}_n \mathrm{y}^*_{n-k}\,] = 0$
Frequency	$C_{x,y}(e^{j\omega}) = X(e^{j\omega})Y^*(e^{j\omega}) = 0$	$C_{\mathrm{x},\mathrm{y}}(e^{j\omega}) = 0$

Table 3.10 Orthogonality for deterministic sequences and WSS discrete-time stochastic processes.

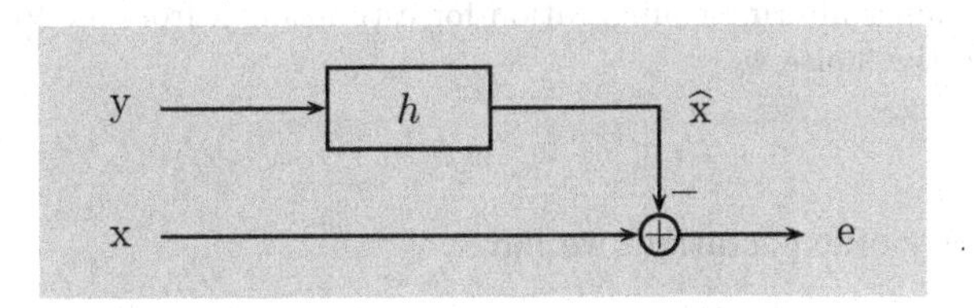

Figure 3.35 Wiener filtering is the optimal linear estimation of a WSS process from another WSS process, where the pair of processes is jointly WSS. Wiener filter h produces an MMSE estimate $\widehat{\mathrm{x}}$ of the WSS process x from y by minimizing the MSE $\mathrm{E}[\,\mathrm{e}_n^2\,]$.

when working with infinite-length sequences, we are motivated to have a structured linear estimator (one implemented with an LSI system) and a restricted model for the autocorrelations and crosscorrelation (wide-sense stationarity). Wiener filtering (or Wiener–Kolmogorov filtering) is the name for MMSE estimation in this setting. Here we derive the Wiener filter without concern for implementability; see the *Further reading* for pointers to results that require the filter to be causal or FIR.

Suppose that stochastic processes x and y are jointly WSS. We observe y and wish to estimate x from it. We want to find a filter h such that the filter output in response to input y,

$$\widehat{\mathrm{x}} = h * \mathrm{y},$$

is a linear MMSE estimate of x; that is, we wish to minimize the power of the estimation error $\mathrm{e} = \mathrm{x} - \widehat{\mathrm{x}}$ as shown in Figure 3.35:

$$\min_h \mathrm{E}\big[\,|\mathrm{e}_n|^2\,\big] = \min_h \mathrm{E}\big[\,|\mathrm{x}_n - \widehat{\mathrm{x}}_n|^2\big]. \tag{3.259}$$

It is convenient to assume that both x and y have mean zero; if not, subtract μ_{y} from y before filtering by h and add μ_{x} after filtering by h to reduce to the zero-mean case.

The minimization (3.259) can be solved by writing $\mathrm{E}[\,|\mathrm{e}_n|^2\,]$ as a function of the impulse response h and setting the derivatives with respect to each entry of h to zero (see Solved exercise 3.7). Instead, we use a geometric approach.

Since $\widehat{\mathrm{x}} = h * \mathrm{y}$, the estimate lies in the subspace $S = \overline{\mathrm{span}}(\{\mathrm{y}_{n-k}\}_{k\in\mathbb{Z}})$. The orthogonality principle states that, with the best estimator, the error e is orthogonal to S and, in particular, orthogonal to the estimate $\widehat{\mathrm{x}}$. From Definition 3.21, orthogonality of e and $\widehat{\mathrm{x}}$ is expressed as

$$\mathrm{E}\big[\,(\mathrm{x}_n - \widehat{\mathrm{x}}_n)\widehat{\mathrm{x}}^*_{n-k}\,\big] = 0, \qquad k \in \mathbb{Z}.$$

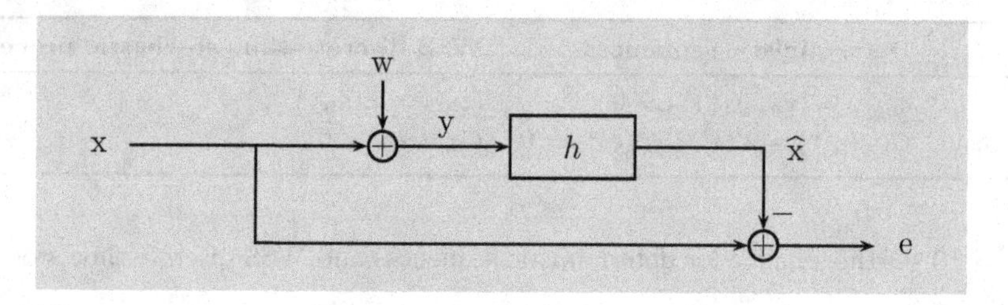

Figure 3.36 Wiener filtering configuration for estimating x from an observation that is corrupted by additive noise w.

Using the linearity of expectation, we have

$$\mathrm{E}\big[\mathrm{x}_n \widehat{\mathrm{x}}^*_{n-k}\big] = \mathrm{E}\big[\widehat{\mathrm{x}}_n \widehat{\mathrm{x}}^*_{n-k}\big], \qquad k \in \mathbb{Z},$$

or

$$c_{\mathrm{x},\widehat{\mathrm{x}},k} = a_{\widehat{\mathrm{x}},k}, \qquad k \in \mathbb{Z}. \tag{3.260a}$$

In the DTFT domain, this is

$$C_{\mathrm{x},\widehat{\mathrm{x}}}(e^{j\omega}) = A_{\widehat{\mathrm{x}}}(e^{j\omega}), \qquad \omega \in \mathbb{R}. \tag{3.260b}$$

Using the fact that $\widehat{\mathrm{x}}$ is the filtered version of x and (3.247)–(3.248a), (3.260b) is equivalent to

$$H^*(e^{j\omega}) C_{\mathrm{x},\mathrm{y}}(e^{j\omega}) = |H(e^{j\omega})|^2 A_{\mathrm{y}}(e^{j\omega}), \qquad \omega \in \mathbb{R},$$

so the Wiener filter is determined by

$$H(e^{j\omega}) = \frac{C_{\mathrm{x},\mathrm{y}}(e^{j\omega})}{A_{\mathrm{y}}(e^{j\omega})}, \qquad \omega \in \mathbb{R}. \tag{3.261}$$

EXAMPLE 3.44 (FILTERING TO REMOVE UNCORRELATED ADDITIVE NOISE) Consider the setting depicted in Figure 3.36, where x and w are uncorrelated WSS processes. Assume that x and w have mean zero so that y does as well. Finding filter h to minimize the power of the estimation error e is a Wiener filtering problem because $\mathrm{y} = \mathrm{x} + \mathrm{w}$ and x are jointly WSS. We can find the Wiener filter by specializing (3.261) to our setting.

Since x and w are uncorrelated and $\mathrm{y} = \mathrm{x} + \mathrm{w}$,

$$C_{\mathrm{x},\mathrm{y}}(e^{j\omega}) = A_{\mathrm{x}}(e^{j\omega}), \tag{3.262a}$$
$$A_{\mathrm{y}}(e^{j\omega}) = A_{\mathrm{x}}(e^{j\omega}) + A_{\mathrm{w}}(e^{j\omega}), \tag{3.262b}$$

detailed verification of which is left for Exercise 3.28. Substituting (3.262) into (3.261) gives

$$H(e^{j\omega}) = \frac{A_{\mathrm{x}}(e^{j\omega})}{A_{\mathrm{x}}(e^{j\omega}) + A_{\mathrm{w}}(e^{j\omega})}. \tag{3.263}$$

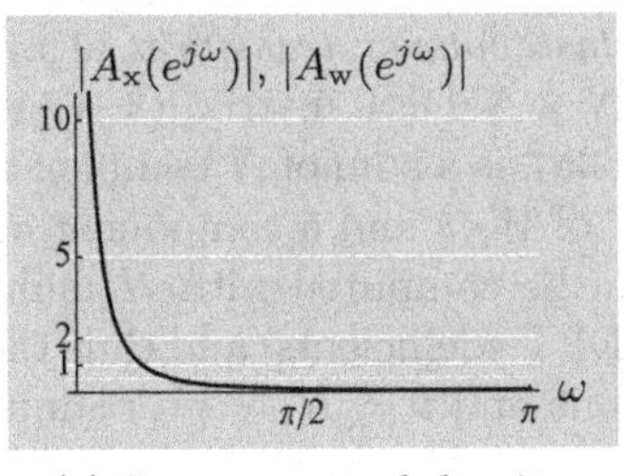

(a) Power spectral density.

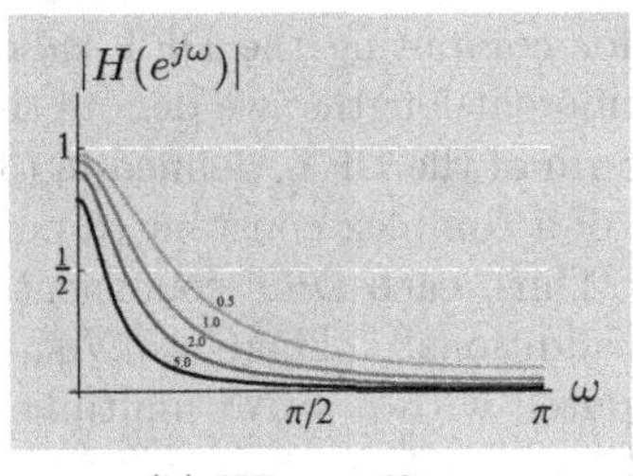

(b) Wiener filters.

Figure 3.37 Wiener filtering. (a) Power spectral density $A_x(e^{j\omega})$ of an AR-1 process with $a = 0.9$ in (3.245) and noise levels of $A_w(e^{j\omega}) = \sigma_w^2 \in \{0.5, 1, 2, 5\}$ (white grid lines). (b) Magnitude responses of the corresponding Wiener filters $H(e^{j\omega})$ for noise levels of $A_w(e^{j\omega}) = \sigma_w^2 \in \{0.5, 1, 2, 5\}$.

This has intuitive limiting behavior: if there is no noise ($A_w(e^{j\omega}) = 0$), the Wiener filter is the identity; if the signal x is weak relative to the noise w ($|A_x(e^{j\omega})/A_w(e^{j\omega})| \to 0$), the Wiener estimate is zero (the mean of x).

As a numerical example, suppose that x is an AR-1 process as in Example 3.39 with power spectral density as in (3.245). Assume that w is a white noise process with variance σ_w^2, so $A_w(e^{j\omega}) = \sigma_w^2$. Then, the Wiener filter is

$$H(e^{j\omega}) = \frac{1 - a^2}{1 - a^2 + \sigma_w^2 |1 - ae^{-j\omega}|^2}, \tag{3.264}$$

which is illustrated in Figure 3.37 for $a = 0.9$ and various noise levels. The magnitude response shows that the Wiener filter preserves the frequency components at which the signal is strong relative to the noise while it suppresses the frequency components at which the noise is strong relative to the signal. As the noise variance σ_w^2 increases, less of y is preserved.

When, in addition to being jointly WSS, x and y are jointly Gaussian, the linear MMSE estimator we have derived here also minimizes the MSE over all estimators, including those that are not linear and not shift-invariant. This is analogous to the finite-dimensional case discussed in Example 2.67 of Appendix 2.C.3.

3.9 Computational aspects

In this section we give an overview of fast Fourier transform (FFT) algorithms and their application to the computation of circular and linear convolutions. We then discuss the complexity of some multirate operations.

3.9.1 Fast Fourier transforms

We have thus far studied the DFT as the natural analysis tool for either periodic sequences or circularly extended finite-length sequences. We now study efficient algorithms for computing the DFT called *fast Fourier transform* algorithms. To work with different lengths, we denote the $N \times N$ DFT matrix, (3.164a), by F_N.

Each term of the DFT, defined in (3.163a), is a sum of N complex terms, each the product of a complex constant (a power of W_N) and a component of the input sequence x. Thus, each DFT coefficient can be computed with N multiplications and $(N-1)$ additions.[71] There are N such DFT coefficients, and thus the full DFT can be computed with $\mu = N^2$ multiplications and $\nu = N(N-1)$ additions, for a total cost of

$$C_{\text{DFT,direct}} = N(N-1) + N^2 = 2N^2 - N, \qquad (3.265)$$

which is exactly the cost of a direct matrix–vector multiplication of F_N by an $N \times 1$ vector x, as in (2.180b).

The special structure of the DFT allows its computation with far fewer than $2N^2$ operations. Savings are based on a *divide and conquer* approach, where a DFT is decomposed into smaller DFTs, and only a few simple operations are needed to combine the results of the smaller DFTs into the original DFT. For illustration purposes, we consider in detail the case of $N = 2^k$, $k \in \mathbb{Z}^+$, and only briefly comment on fast algorithms for other values of N.

Radix-2 FFT Starting with the definition of the DFT (3.163a), write

$$\begin{aligned} X_k &= \sum_{n=0}^{N-1} x_n W_N^{kn} \overset{(a)}{=} \sum_{n=0}^{N/2-1} x_{2n} W_N^{k(2n)} + \sum_{n=0}^{N/2-1} x_{2n+1} W_N^{k(2n+1)} \\ &\overset{(b)}{=} \sum_{n=0}^{N/2-1} x_{2n} W_{N/2}^{kn} + W_N^k \sum_{n=0}^{N/2-1} x_{2n+1} W_{N/2}^{kn}, \end{aligned} \qquad (3.266)$$

where (a) separates the summation over odd- and even-numbered terms; and (b) follows from $W_N^2 = W_{N/2}$. Recognize the first sum as the length-$N/2$ DFT of the sequence $\begin{bmatrix} x_0 & x_2 & \ldots & x_{N-2} \end{bmatrix}^\top$ and the second sum as the length-$N/2$ DFT of the sequence $\begin{bmatrix} x_1 & x_3 & \ldots & x_{N-1} \end{bmatrix}^\top$. It is now apparent that the length-N DFT computation can make use of $F_{N/2} D_2 x$ and $F_{N/2} C_2 x$, where D_2 is the downsampling-by-2 operator defined in (3.186b), and C_2 is a similar operator, except that it keeps the odd-indexed values. Since the length-$N/2$ DFT is $(N/2)$-periodic in k, (3.266) can be used both for $k \in \{0, 1, \ldots, sN/2 - 1\}$ and for $k \in \{N/2, N/2+1, \ldots, N-1\}$.

To get a compact matrix representation, we introduce the diagonal matrix

$$A_{N/2} = \operatorname{diag}(1, W_N, W_N^2, \ldots, W_N^{(N/2)-1})$$

[71]We are counting *complex* multiplications and *complex* additions. It is customary to not count multiplications by (-1) and thus lump together additions and subtractions.

and rewrite (3.266) as

$$\begin{bmatrix} X_0 \\ \vdots \\ X_{N/2-1} \end{bmatrix} = F_{N/2} D_2 x + A_{N/2} F_{N/2} C_2 x, \tag{3.267a}$$

$$\begin{bmatrix} X_{N/2} \\ \vdots \\ X_{N-1} \end{bmatrix} = F_{N/2} D_2 x - A_{N/2} F_{N/2} C_2 x, \tag{3.267b}$$

where the final twist was to realize that $W_N^k = -W_N^{k-N/2}$, leading to

$$\begin{aligned} X = F_N x &= \begin{bmatrix} I_{N/2} & A_{N/2} \\ I_{N/2} & -A_{N/2} \end{bmatrix} \begin{bmatrix} F_{N/2} & 0 \\ 0 & F_{N/2} \end{bmatrix} \begin{bmatrix} D_2 \\ C_2 \end{bmatrix} x \\ &= \begin{bmatrix} I_{N/2} & I_{N/2} \\ I_{N/2} & -I_{N/2} \end{bmatrix} \begin{bmatrix} I_{N/2} & 0 \\ 0 & A_{N/2} \end{bmatrix} \begin{bmatrix} F_{N/2} & 0 \\ 0 & F_{N/2} \end{bmatrix} \begin{bmatrix} D_2 \\ C_2 \end{bmatrix} x. \end{aligned} \tag{3.268}$$

If this turns out to be a useful factorization, then we can repeat it to represent $F_{N/2}$ using $F_{N/4}$, etc., until we reach F_2, which requires no multiplications, $\mu_2 = 0$, and only $\nu_2 = 2$ additions. Let us count computations in the factored form to see whether the factorization leads to savings.

With μ_N and ν_N the number of multiplications and additions in computing a length-N DFT, the factorization (3.268) shows that a length-N DFT can be computed using two length-$N/2$ DFTs, $N/2$ multiplications, and N additions. Iterating on the length-$N/2$ DFTs, then length-$N/4$ DFTs, and so on, leads to the following recursions:

$$\begin{aligned} \nu_N &= 2\nu_{N/2} + N \\ &= 2\left(2\nu_{N/2^2} + \frac{N}{2}\right) + N = 2^2 \nu_{N/2^2} + 2N \\ &= 2^2\left(2\nu_{N/2^3} + \frac{N}{4}\right) + 2N = 2^3 \nu_{N/2^3} + 3N \\ &\overset{(a)}{=} \frac{N}{2}\nu_2 + (\log_2 N - 1)N \overset{(b)}{=} N \log_2 N, \end{aligned} \tag{3.269a}$$

$$\begin{aligned} \mu_N &= 2\mu_{N/2} + \frac{N}{2} \\ &= 2\left(2\mu_{N/2^2} + \frac{N}{4}\right) + \frac{N}{2} = 2^2 \mu_{N/2^2} + 2\frac{N}{2} \\ &= 2^2\left(2\mu_{N/2^3} + \frac{N}{8}\right) + 2\frac{N}{2} = 2^3 \mu_{N/2^3} + 3\frac{N}{2} \\ &\overset{(c)}{=} \frac{N}{2}\mu_2 + (\log_2 N - 1)\frac{N}{2} \overset{(d)}{=} \frac{N}{2}\log_2 N - \frac{N}{2}, \end{aligned} \tag{3.269b}$$

where (a) and (c) follow from continuing the recursion; (b) from $\nu_2 = 2$; and (d) from $\mu_2 = 0$. Combining these into a total number of operations gives

$$C_{\text{DFT,radix-2}} = \frac{3}{2} N \log_2 N - \frac{N}{2}. \tag{3.270}$$

Thus, using the asymptotic notation $\Theta(\cdot)$ to express the dominating term in the dependence on N without regard for constant factors (see Definition 2.54), recursive application of (3.268) reduces the cost from $\Theta(N^2)$ in (3.265) to $\Theta(N \log_2 N)$ in (3.270). We illustrate the above procedure with a simple example.

EXAMPLE 3.45 (COMPUTATION OF THE LENGTH-4 FFT) We check that the factorization (3.268) does indeed equal the length-4 DFT:

$$
\begin{aligned}
F_4 &= \begin{bmatrix} I_2 & A_2 \\ I_2 & -A_2 \end{bmatrix} \begin{bmatrix} F_2 & 0 \\ 0 & F_2 \end{bmatrix} \begin{bmatrix} D_2 \\ C_2 \end{bmatrix} \\
&= \begin{bmatrix} 1 & 0 & 1 & 0 \\ 0 & 1 & 0 & j \\ 1 & 0 & -1 & 0 \\ 0 & 1 & 0 & -j \end{bmatrix} \begin{bmatrix} 1 & 1 & 0 & 0 \\ 1 & -1 & 0 & 0 \\ 0 & 0 & 1 & 1 \\ 0 & 0 & 1 & -1 \end{bmatrix} \begin{bmatrix} 1 & 0 & 0 & 0 \\ 0 & 0 & 1 & 0 \\ 0 & 1 & 0 & 0 \\ 0 & 0 & 0 & 1 \end{bmatrix} \\
&= \begin{bmatrix} 1 & 1 & 1 & 1 \\ 1 & j & -1 & -j \\ 1 & -1 & 1 & -1 \\ 1 & -j & -1 & j \end{bmatrix} = \begin{bmatrix} 1 & 1 & 1 & 1 \\ 1 & W_4 & W_4^2 & W_4^3 \\ 1 & W_4^2 & W_4^4 & W_4^6 \\ 1 & W_4^3 & W_4^6 & W_4^9 \end{bmatrix},
\end{aligned}
$$

which is exactly (3.164a). We can also write out (3.266),

$$
\begin{aligned}
X_k &= \sum_{n=0}^{3} x_n W_4^{kn} = \sum_{n=0}^{3} x_{2n} W_4^{k(2n)} + \sum_{n=0}^{3} x_{2n+1} W_4^{k(2n+1)} \\
&= \sum_{n=0}^{3} x_{2n} W_2^{kn} + W_4^k \sum_{n=0}^{3} x_{2n+1} W_2^{kn} \\
&= (x_0 W_2^0 + x_2 W_2^k) + W_4^k (x_1 W_2^0 + x_3 W_2^k) \\
&= (x_0 + (-1)^k x_2) + W_4^k (x_1 + (-1)^k x_3),
\end{aligned}
$$

which is equivalent to computing one DFT of length 2 on even samples, then one DFT of length 2 on odd samples, and finally multiplying the latter by a constant W_4^k.

Other FFT algorithms A famous class of FFT algorithms, the *Cooley–Tukey FFT*, works for any composite length $N = N_1 N_2$. The algorithm breaks down a length-N DFT into N_2 length-N_1 DFTs, N complex multiplications, and N_1 length-N_2 DFTs. Often, either N_1 or N_2 is a small factor called a *radix*.

The *Good–Thomas FFT* works for $N = N_1 N_2$, where N_1 and N_2 are coprime. It is based on the Chinese remainder theorem and avoids the complex factors of the Cooley–Tukey FFT. Thus, the algorithm breaks down a length-N DFT into N_2 length-N_1 DFTs and N_1 length-N_2 DFTs, equivalent to a two-dimensional length-$(N_1 \times N_2)$ DFT.

Rader's FFT works for prime-length N. It is based on mapping the computation of the DFT into a computation of a circular convolution of length $N-1$ (recall that (3.181b) shows that the DFT diagonalizes the circular convolution operator).

Winograd extended Rader's FFT to include powers of prime lengths, so Rader's FFT is sometimes considered a subclass of the Winograd FFT.

The *Winograd FFT* is often used for small lengths. It is based on considering N_1 length-N_2 DFTs as a two-dimensional DFT of length $(N_1 \times N_2)$, as we have seen for the Good–Thomas algorithm. If N_1 and N_2 are prime, we can use Rader's FFT on each N_1 and N_2. While it is less costly in terms of required additions and multiplications, it is also complicated and thus not often used.

The *split-radix FFT* is used for values of N that are multiples of 4. It recursively splits length-N DFTs into terms of one length-$N/2$ DFT and two length-$N/4$ DFTs and boasts the lowest operations count for $N = 2^k$, $k > 1$.

Remember that the cost in terms of additions and multiplications is just one measure of how fast an algorithm can be computed; many other factors come into play, including the specific computing platform. As a result, the Cooley–Tukey FFT is still the prevalent one, despite some of the other algorithms having lower multiplication, addition, or total operation counts. Most FFT algorithms have a cost of

$$C_{\text{DFT,FFT}} = \alpha N \log_2 N + \beta N = \Theta(N \log_2 N), \tag{3.271}$$

where α and β are small constants that depend on the choice of algorithm.

3.9.2 Convolution

In discussing convolution, for the first time we will encounter the distinction between computing on a finite-length input and on an infinite-length one. In the first case, we will be computing the cost per block of input samples (as we have done in Section 3.9.1), while in the second, we will be computing the cost per input sample.

Computing circular convolution Circular convolution of length-N sequences can be written as a matrix–vector product with an $N \times N$ matrix and an $N \times 1$ vector, as in (3.75). Treating it as a generic matrix multiplication, performed in the standard way, gives an algorithm with $\Theta(N^2)$ cost as in (2.180b). Since the DFT operator diagonalizes circular convolution, algorithms with lower cost are obtained from FFT algorithms. Using (3.181a), applying operator H requires a length-N DFT, N scalar multiplications for multiplication by the diagonal matrix, and a length-N inverse DFT. Using (3.271) for the cost of the DFT and inverse DFT, computing the circular convolution through this algorithm has a cost of

$$C_{\text{cconv,freq}} = 2\,(\alpha N \log_2 N + \beta N) + N = \Theta(N \log_2 N), \tag{3.272}$$

per N input samples, or $\Theta(\log_2 N)$ per input sample.

Computing linear convolution We start with a straight implementation of linear convolution in the time domain, (3.61), for a finite-length input x. Without loss of generality, assume that the input is of length M and the filter h is of length $L < M$. We need one multiplication and no additions to compute $y_0 = h_0 x_0$, two multiplications and one addition for $y_1 = h_1 x_0 + h_0 x_1$, all the way to L multiplications and

$(L-1)$ additions for $y_{L-1} = \sum_{k=0}^{L-1} x_k h_{L-1-k}$, $y_L, \ldots, y_{M-1}$, and then back down to one multiplication and no additions for $y_{M+L-1} = h_{L-1}x_{M-1}$, leading to

$$\begin{aligned}
C_{\text{lconv,time}} &= \nu + \mu \\
&= \left[2\sum_{k=1}^{L-2} k + (M-L+1)(L-1)\right] + \left[2\sum_{k=1}^{L-1} k + (M-L+1)L\right] \\
&= 2\frac{(L-2)(L-1)}{2} + 2\frac{(L-1)L}{2} + (M-L+1)(2L-1) \\
&= 2ML - M - L + 1
\end{aligned} \tag{3.273}$$

per M input samples, or $\Theta(L)$ per input sample.

In (3.272), we saw the cost of an efficient implementation of the circular convolution using FFTs. We will now show how to compute the linear convolution of an infinite-length input with a sequence of circular convolutions. Using the results we just developed for efficient circular convolution, we will have an efficient algorithm for linear convolution. To that end, we build upon Example 3.14 on the equivalence of circular and linear convolutions.

Example 3.46 (Linear convolution from circular convolution) In Example 3.14, we wanted to compute the linear convolution of a length-3 filter with a length-4 sequence, and we found that computing a circular convolution instead was equivalent – when the length of the circular convolution was at least as large as the length of the result of the linear convolution. We rewrite (3.76b) as follows:

$$\begin{aligned}
\begin{bmatrix} y_0 \\ y_1 \\ y_2 \\ y_3 \\ y_4 \\ y_5 \end{bmatrix} &= \begin{bmatrix} h_0 & 0 & 0 & 0 & h_2 & h_1 \\ h_1 & h_0 & 0 & 0 & 0 & h_2 \\ h_2 & h_1 & h_0 & 0 & 0 & 0 \\ 0 & h_2 & h_1 & h_0 & 0 & 0 \\ 0 & 0 & h_2 & h_1 & h_0 & 0 \\ 0 & 0 & 0 & h_2 & h_1 & h_0 \end{bmatrix} \begin{bmatrix} x_0 \\ x_1 \\ x_2 \\ x_3 \\ 0 \\ 0 \end{bmatrix} \\
&= \begin{bmatrix} h_0 & 0 & 0 & 0 & h_2 & h_1 \\ h_1 & h_0 & 0 & 0 & 0 & h_2 \\ h_2 & h_1 & h_0 & 0 & 0 & 0 \\ 0 & h_2 & h_1 & h_0 & 0 & 0 \\ 0 & 0 & h_2 & h_1 & h_0 & 0 \\ 0 & 0 & 0 & h_2 & h_1 & h_0 \end{bmatrix} \begin{bmatrix} 1 & 0 & 0 & 0 \\ 0 & 1 & 0 & 0 \\ 0 & 0 & 1 & 0 \\ 0 & 0 & 0 & 1 \\ 0 & 0 & 0 & 0 \\ 0 & 0 & 0 & 0 \end{bmatrix} \begin{bmatrix} x_0 \\ x_1 \\ x_2 \\ x_3 \end{bmatrix} \\
&= H_6 \begin{bmatrix} I_4 \\ 0_{2\times 4} \end{bmatrix} \begin{bmatrix} x_0 \\ x_1 \\ x_2 \\ x_3 \end{bmatrix},
\end{aligned} \tag{3.274}$$

where the input vector is now stated explicitly without the trailing zeros, H_6 is the 6×6 circulant matrix as in (3.75), I_4 is a 4×4 identity matrix, and $0_{2\times 4}$ is a 2×4 all-zero matrix.

Now, with filter h still of length 3, suppose that the input sequence x is of infinite length. The result y of the linear convolution $h * x$ is

$$
\begin{array}{ccccccc}
 & & & \vdots & & & \\
y_0 & = & h_2 x_{-2} & + & h_1 x_{-1} & + & \enclose{box}{h_0 x_0} \\
y_1 & = & h_2 x_{-1} & + & \enclose{box}{h_1 x_0} & + & \enclose{box}{h_0 x_1} \\
\enclose{circle}{y_2} & = & \enclose{box}{h_2 x_0} & + & \enclose{box}{h_1 x_1} & + & \enclose{box}{h_0 x_2} \\
\enclose{circle}{y_3} & = & \enclose{box}{h_2 x_1} & + & \enclose{box}{h_1 x_2} & + & \enclose{box}{h_0 x_3} \\
y_4 & = & \enclose{box}{h_2 x_2} & + & \enclose{box}{h_1 x_3} & + & \enclose{roundedbox}{h_0 x_4} \\
y_5 & = & \enclose{box}{h_2 x_3} & + & \enclose{roundedbox}{h_1 x_4} & + & \enclose{roundedbox}{h_0 x_5} \\
\enclose{circle}{y_6} & = & \enclose{roundedbox}{h_2 x_4} & + & \enclose{roundedbox}{h_1 x_5} & + & \enclose{roundedbox}{h_0 x_6} \\
\enclose{circle}{y_7} & = & \enclose{roundedbox}{h_2 x_5} & + & \enclose{roundedbox}{h_1 x_6} & + & \enclose{roundedbox}{h_0 x_7} \\
y_8 & = & \enclose{roundedbox}{h_2 x_6} & + & \enclose{roundedbox}{h_1 x_7} & + & h_0 x_8 \\
y_9 & = & \enclose{roundedbox}{h_2 x_7} & + & h_1 x_8 & + & h_0 x_9 \\
 & & & \vdots & & &
\end{array}
$$

The quantities in boxes depend on the input block (x_0, x_1, x_2, x_3), and the quantities in ovals depend on the next input block (x_4, x_5, x_6, x_7). The circled samples of y can be computed using just one input block; y_2 and y_3 can be computed from the first input block alone, while y_6 and y_7 can be computed from the second input block alone. The samples in between (y_4 and y_5) require the combination of contributions from the two input blocks. For an arbitrary number of input samples, the contribution from each input block of length 4 can be computed as in (3.274); these overlap and should be added to give the linear convolution.

We can write this procedure more formally using the linearity and shift invariance of the (linear) convolution operator. The output y is the sum of the outputs resulting from each input block

$$
\bar{x}_{k,n} = \begin{cases} x_n, & \text{for } n = 4k,\ 4k+1,\ 4k+2,\ 4k+3; \\ 0, & \text{otherwise,} \end{cases} \qquad k \in \mathbb{Z}.
$$

Specifically, since $x = \sum_{k\in\mathbb{Z}} \bar{x}_k$, we have

$$
y = \sum_{k\in\mathbb{Z}} h * \bar{x}_k. \tag{3.275}
$$

To implement the convolutions in (3.275) with (3.274), the input block should be supported on $\{0, 1, 2, 3\}$, so shift each block $\bar{x}_k$ by $4k$ to define

$$
\widetilde{x}_k = \begin{bmatrix} \enclose{box}{x_{4k}} & x_{4k+1} & x_{4k+2} & x_{4k+3} \end{bmatrix}^\top, \qquad k \in \mathbb{Z}.
$$

By shift invariance of the (linear) convolution operator, the computation (3.275) can be replaced by

$$
y_n = \sum_{k\in\mathbb{Z}} \widetilde{y}_{k,n-4k},
$$

where

$$\widetilde{y}_k = H_6 \begin{bmatrix} I_4 \\ 0_{2\times 4} \end{bmatrix} \widetilde{x}_k, \qquad k \in \mathbb{Z}.$$

In this computation, each input block of length 4 generates an intermediate block of length 6. These intermediate blocks are shifted by multiples of 4 in generating the output. We can write the full linear convolution as

$$y = AHEx, \qquad \text{with } A = \begin{bmatrix} \ddots & & & & & & & \\ & I_2 & 0 & 0 & & & & \\ & 0 & I_2 & 0 & & & & \\ & 0 & 0 & I_2 & I_2 & 0 & 0 & \\ & & & & 0 & I_2 & 0 & \\ & & & & 0 & 0 & I_2 & \\ & & & & & & & \ddots \end{bmatrix},$$

$$H = \begin{bmatrix} \ddots & & & \\ & H_6 & & \\ & & H_6 & \\ & & & \ddots \end{bmatrix}, \qquad \text{and} \qquad E = \begin{bmatrix} \ddots & & & \\ & I_4 & & \\ & 0_{2\times 4} & & \\ & & I_4 & \\ & & 0_{2\times 4} & \\ & & & \ddots \end{bmatrix}.$$

We now see why the computation of the convolution as above would be efficient, since multiplication by E is merely the insertion of zeros (*extension*), multiplication by H is circular convolution as we have just seen, and multiplication by A (*addition*) requires only two additions for each block of four input samples.

The *overlap–add algorithm* generalizes this simple example to any filter length and input block size. Given a length-L FIR filter h and input sequence x, we choose an input block length of M samples. For computation of the linear convolution on one block to coincide with the circular convolution, the length of the circular convolution should be $N = L + M - 1$ according to Theorem 3.10. (Using any larger value of N provides no benefit in principle, but N is often chosen to be the smallest integer power of 2 that is at least $L + M - 1$.) Then, the (linear) convolution $y = h * x$ can be computed with the following factorization:

$$y = AHEx, \tag{3.276}$$

where the extension matrix E is block-diagonal with

$$\begin{bmatrix} I_M \\ 0_{L-1\times M} \end{bmatrix}$$

on the diagonal, so it extends blocks of M input samples to N samples; H is a block-diagonal matrix with circular convolution operator H_N on the diagonal; and addition matrix A has blocks of I_N on the diagonal, offset by M rows.

We now compute the cost of the algorithm. As we said, E merely inserts zeros and thus has no cost, H costs $(2\alpha N \log_2 N + (2\beta+1)N)$ operations per block according to (3.272), and A requires $(N-M)$ additions per block, for a total cost of

$$C_{\text{lconv,overlap-add}} = \frac{2\alpha}{M} N \log_2 N + \frac{2\beta+2}{M} N - 1 = \Theta(\log_2 N) \qquad (3.277)$$

per input sample, assuming that N is proportional to M. This is a significant saving compared with (3.273) when L is large.

The *overlap–save algorithm* can be written with a similar factorization. Its cost is similar, with a small advantage of there being no additions in the final stage but a disadvantage that the DFTs are calculated on denser input vectors, which might make the FFTs more costly. This algorithm is developed in Exercise 3.30.

3.9.3 Multirate operations

The key to improving the computational efficiency in multirate signal processing is simple: always operate at the lowest possible sample rate. We now show this idea in action, bearing in mind that downsampling and upsampling cost nothing in terms of additions and multiplications (although they might require memory access).

Downsampling preceded by filtering We start with the time-domain computation. Assume that we directly compute the convolution $(h * x)$ of a length-M input x and a length-L filter h, and discard every other sample. Using (3.273), the cost is

$$C_{\text{time,direct}} = 2ML - M - L + 1 \qquad (3.278\text{a})$$

per M input samples. However, it is wasteful to compute samples that are subsequently discarded. A polyphase implementation, as shown in Figure 3.31(c), avoids this wastefulness and follows the principle of applying filtering at the lowest possible rate. The polyphase components of the input, x_0 and x_1, each of length $M/2$, are convolved with the polyphase components of the filter, $\widetilde{h}_0$ and $\widetilde{h}_1$, each of length $L/2$.[72] We thus have to compute two convolutions at half the length and add the results of these convolutions. Using (3.273), the cost is $ML - M - L + 2$ operations for the two convolutions plus $M/2 + L/2 - 1$ additions (since the result of each convolution is now of length $M/2 + L/2 - 1$), for a total cost of

$$C_{\text{time,poly}} = ML - \frac{M}{2} - \frac{L}{2} + 1 \qquad (3.278\text{b})$$

per M input samples, which, although it amounts to a saving of roughly 50%, is still $\Theta(ML)$.

Whether or not we use polyphase representations, we can replace the time-domain computation of convolution with a frequency-domain computation as in

[72]The lengths are actually $\lfloor M/2 \rfloor$, $\lceil M/2 \rceil$, $\lfloor L/2 \rfloor$, and $\lceil L/2 \rceil$. Accounting for this yields the same costs for the time-domain computations and slightly different costs for the frequency-domain computations.

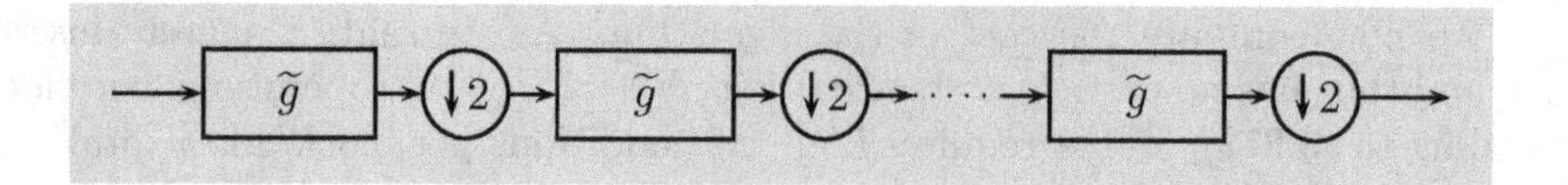

Figure 3.38 Cascade of K filters followed by downsamplers.

Section 3.9.2. Using circular convolution of the minimum length, $M+L-1$, implemented with a length-$(M+L-1)$ DFT, the cost without polyphase implementation is

$$C_{\text{freq,direct}} = 2\alpha(M+L-1)\log_2(M+L-1) + (2\beta+1)(M+L-1) \quad (3.278c)$$

per M input samples, by substitution into (3.272). With polyphase implementation, we need two convolutions at half the length and the addition of the results of these convolutions, for a total cost of

$$C_{\text{freq,poly}} = 2\alpha(M+L-2)\log_2(M+L-2) + \left(2\beta - 2\alpha + \frac{3}{2}\right)(M+L-2) \quad (3.278d)$$

per M input samples, giving a small additional saving that is dependent on the sizes M and L.

This discussion generalizes straightforwardly to downsampling factors larger than 2. The cost saving factor from using the polyphase implementation in the time-domain computations is equal to the downsampling factor.

EXAMPLE 3.47 (ITERATION OF DOWNSAMPLING PRECEDED BY FILTERING) Consider a cascade of K filters followed by downsamplers as in Figure 3.38. Using any of the expressions we just derived for the cost of one stage C, we can calculate the cost for K stages. For the second stage, it is $C/2$ (because it runs at half the rate of the input), for the third stage it is $C/4$, etc., leading to the total cost of

$$C + \frac{C}{2} + \frac{C}{4} + \cdots + \frac{C}{2^{K-1}} \stackrel{(a)}{=} \left(2 - \frac{1}{2^{K-1}}\right)C \; < \; 2C, \quad (3.279)$$

where (a) follows from (P2.54-1), the formula for a finite geometric series.

Upsampling followed by filtering Analysis of the cost of upsampling followed by filtering is similar to the analysis of the cost of downsampling preceded by filtering. In fact, these operations are dual, as exhibited by the transpose relationship between (3.202) and (3.205), so they have the same cost.

As before, the savings in using the polyphase implementation shown in Figure 3.31(a) comes from filtering at the lowest possible rate. The direct implementation is wasteful because it performs arithmetic operations on samples that equal zero.

Appendix 3.A Elements of analysis

3.A.1 Complex numbers

The imaginary unit j is defined as a number that satisfies $j^2 = -1$; since $j^2 = -1$ implies that $(-j)^2 = -1$, we are simply fixing one of the two solutions to have the name j.[73] A *complex number* $z \in \mathbb{C}$ is then a number of the form

$$z = a + jb, \qquad a, b \in \mathbb{R}. \tag{3.280}$$

In (3.280), a is called the *real part* while b is called the *imaginary part*. The *complex conjugate* of z is denoted by z^* and is by definition

$$z^* = a - jb. \tag{3.281}$$

The basic operations on complex numbers are as follows:

$$\begin{aligned} z_1 + z_2 &= (a_1 + a_2) + j(b_1 + b_2), \\ z_1 - z_2 &= (a_1 - a_2) + j(b_1 - b_2), \\ z_1 z_2 &= (a_1 a_2 - b_1 b_2) + j(b_1 a_2 + a_1 b_2), \\ \frac{z_1}{z_2} &= \frac{a_1 a_2 + b_1 b_2}{a_2^2 + b_2^2} + j\frac{a_2 b_1 - a_1 b_2}{a_2^2 + b_2^2}. \end{aligned}$$

Any complex number can be represented in *polar form*:

$$z = re^{j\theta}, \tag{3.282}$$

where r is called the *modulus* or *magnitude* and θ is the *argument* or *phase*. Using *Euler's formula*,

$$e^{j\theta} = \cos\theta + j\sin\theta, \tag{3.283}$$

we can express a complex number further as

$$z = re^{j\theta} = r(\cos\theta + j\sin\theta). \tag{3.284}$$

It allows us to easily find a power of a complex number as

$$(\cos\theta + j\sin\theta)^n = (e^{j\theta})^n = e^{jn\theta} = \cos n\theta + j\sin n\theta. \tag{3.285}$$

Euler's formula highlights that the argument of a complex number is not unique; adding any integer multiple of 2π to the argument does not change the number,

$$e^{j(\theta + k2\pi)} = e^{j\theta}e^{j2k\pi} = e^{j\theta}(e^{j2\pi})^k = e^{j\theta}, \qquad \text{for any } k \in \mathbb{Z},$$

since $e^{j2\pi} = 1$. Two other useful relations that can be derived using Euler's formula are

$$\cos\theta = \frac{e^{j\theta} + e^{-j\theta}}{2} \quad \text{and} \quad \sin\theta = \frac{e^{j\theta} - e^{-j\theta}}{2j}. \tag{3.286}$$

[73] Mathematicians and physicists typically use i for the imaginary unit, while j is more common in engineering.

Complex numbers are typically shown in the complex plane. The complex plane has a one-to-one correspondence with $\mathbb{R}^2$, with the real part shown horizontally and the imaginary part vertically. Conversion from polar form $e^{j\theta}$ to standard (or rectangular) form $a + jb$ is by

$$a = r\cos\theta, \qquad b = r\sin\theta.$$

Conversion from standard form to polar form is simple by just looking at the complex plane, but more complicated to write. One solution is as follows:

$$r = \sqrt{a^2+b^2}, \qquad \theta = \begin{cases} \arctan(b/a), & \text{for } a > 0; \\ \arctan(b/a) + \pi, & \text{for } a < 0,\, b \geq 0; \\ \arctan(b/a) - \pi, & \text{for } a < 0,\, b < 0; \\ \pi/2, & \text{for } a = 0,\, b > 0; \\ -\pi/2, & \text{for } a = 0,\, b < 0; \\ \text{undefined}, & \text{for } a = 0,\, b = 0, \end{cases}$$

where arctan returns a value in $(-\frac{1}{2}\pi,\, \frac{1}{2}\pi)$.

Roots of unity In the same way as j was defined as the square root of unity, we can define the principal Nth root of unity as

$$W_N = e^{-j2\pi/N}. \tag{3.287}$$

It is easy to check that W_N^k, for $k \in \{2, 3, \ldots, N\}$, are also Nth roots of unity, meaning that $(W_N^k)^N = 1$. If we drew all N roots of unity in the complex plane, we would see that they slice up the unit circle by equal angles; the choice of W_N as the principal root makes $W_N^0, W_N^1, \ldots, W_N^{N-1}$ consecutive in clockwise order. Figure 3.39 shows an example with $N = 8$.

Here are some useful identities involving the roots of unity:

$$W_N^N = 1, \tag{3.288a}$$

$$W_N^{kN+n} = W_N^n, \qquad \text{with } k, n \in \mathbb{Z}, \tag{3.288b}$$

$$\sum_{k=0}^{N-1} W_N^{nk} = \begin{cases} N, & \text{for } n = \ell N,\ \ell \in \mathbb{Z}; \\ 0, & \text{otherwise.} \end{cases} \tag{3.288c}$$

The last relation is often referred to as *orthogonality of the roots of unity.* To prove it, for any n not an integer multiple of N, use the finite sum formula from (P2.54-1),

$$\sum_{k=0}^{N-1} (W_N^n)^k = \frac{1 - W_N^{Nn}}{1 - W_N^n} = 0, \tag{3.289}$$

since the numerator is 0 and the denominator is nonzero. For $n = \ell N$ with $\ell \in \mathbb{Z}$, $W_N^{kn} = W_N^{k\ell N} = 1$, and thus, by direct substitution into (3.288c), we get N.

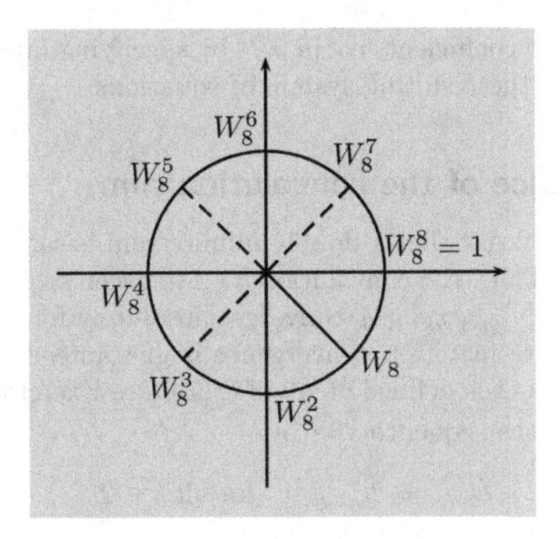

Figure 3.39 Roots of unity for $N = 8$. The principal one is denoted by a solid line.

3.A.2 Difference equations

Finding solutions to the linear difference equations introduced in Section 3.3.2 involves the following steps:

(i) *Homogeneous solution:* First, we find a solution to the *homogeneous equation*,

$$y_n^{(h)} = -\sum_{k=1}^{N} a_k y_{n-k}, \tag{3.290a}$$

which is obtained by setting the input x in (3.55) to zero. The solution is of the form

$$y_n^{(h)} = \sum_{k=1}^{N} \alpha_k \lambda_k^n, \tag{3.290b}$$

where λ_k, $k = 1, 2, \ldots, N$, are obtained by solving the characteristic equation of the system,

$$\sum_{k=0}^{N} a_k \lambda^{N-k} = 0. \tag{3.290c}$$

(ii) *Particular solution:* Then, any particular solution to (3.55), $y_n^{(p)}$, is found (independently of $y_n^{(h)}$). This is typically done by assuming that $y_n^{(p)}$ is of the same form as x_n, possibly scaled.

(iii) *Complete solution:* By superposition, the complete solution y_n is the sum of the homogeneous solution $y_n^{(h)}$ and the particular solution $y_n^{(p)}$:

$$y_n = y_n^{(h)} + y_n^{(p)} = \sum_{k=1}^{N} \alpha_k \lambda_k^n + y_n^{(p)}. \tag{3.290d}$$

We determine the coefficients α_k in $y_n^{(h)}$ by specifying initial conditions for y_n and then solving the resulting system of equations.

3.A.3 Convergence of the convolution sum

Recall from Appendix 2.A.2 that a doubly infinite sum is said to converge when it converges absolutely. Thus, the convolution $h * x$ between sequences h and x is well defined when the sum $\sum_{k\in\mathbb{Z}} x_k h_{n-k}$ converges absolutely for every value of n.

When h and x are in $\ell^2(\mathbb{Z})$, convergence is guaranteed by the fact that the standard ℓ^2 inner product is defined on all of $\ell^2(\mathbb{Z})$ (see Exercise 2.14). Specifically, for each $n \in \mathbb{Z}$, define the sequence $\widetilde{h}^{(n)}$ by

$$\widetilde{h}_k^{(n)} = h_{n-k}^* \qquad \text{for all } k \in \mathbb{Z}.$$

Then, each $\widetilde{h}^{(n)}$ is in $\ell^2(\mathbb{Z})$ because time reversal, shifting, and conjugation do not change the ℓ^2 norm. Thus, for every $n \in \mathbb{Z}$, the inner product $\langle x, \widetilde{h}^{(n)}\rangle = (h * x)_n$ is well defined, so the convolution is well defined.

In signal processing, we are not quite satisfied with restricting attention to sequences in $\ell^2(\mathbb{Z})$; for example, simple sequences like constants and sinusoids are not in $\ell^2(\mathbb{Z})$. To ensure convergence of the convolution sum while loosening the constraints on x requires tightening the constraints on h, or vice versa. Hölder's inequality for sequences (2.214) gives a simple condition: The convolution sum is guaranteed to converge absolutely when $h \in \ell^p(\mathbb{Z})$ and $x \in \ell^q(\mathbb{Z})$ for some p and q in $[1, \infty]$ satisfying $1/p + 1/q \geq 1$.[74] This use of Hölder's inequality gives a bound

$$|(h * x)(n)| \leq \|h\|_p \, \|x\|_q, \qquad \text{for any } n \in \mathbb{Z},$$

that is uniform over n and hence shows that, beyond being merely well defined, $h * x$ is in $\ell^\infty(\mathbb{Z})$.

We employ the $p = 1$, $q = \infty$ case often: By restricting an LSI system impulse response h to $\ell^1(\mathbb{Z})$, we can allow the input x to be any sequence in $\ell^\infty(\mathbb{Z})$ (that is, merely bounded) while ensuring that the output sequence $h * x$ is well defined and furthermore in $\ell^\infty(\mathbb{Z})$. Note that the condition $h \in \ell^1(\mathbb{Z})$ was already used in BIBO stability of the LSI system with impulse response h and is a common assumption.

3.A.4 Dirac delta function

DEFINITION 3.22 (DIRAC DELTA FUNCTION) The *Dirac delta function* δ satisfies

$$\int_{-\infty}^{\infty} x(t)\delta(t)\, dt = x(0), \qquad (3.291)$$

for any function x that is continuous at 0.

[74] Hölder's inequality is given with p and q as Hölder conjugates, $1/p + 1/q = 1$, and making p or q smaller makes the corresponding sequence space smaller; see (2.41).

The definition is based on what the Dirac delta function does as part of an integrand rather than an expression that is valid pointwise. This is necessary because no function δ can actually satisfy the defining property. In fact, the use of the word "function" is merely to distinguish the Dirac delta from a sequence, that is, to indicate that we integrate δ as if it were a function with domain $\mathbb{R}$.[75]

One immediate consequence of the definition is that

$$\int_{-\infty}^{\infty} \delta(t)\, dt = 1, \tag{3.292}$$

which follows from the continuity at 0 of $x(t) = 1$. Also, for any x that is continuous at t_0,

$$\int_{-\infty}^{\infty} x(t+t_0)\delta(t)\, dt = x(t_0) \tag{3.293a}$$

follows from the continuity at 0 of the shifted function $x(t+t_0)$; furthermore,

$$\int_{-\infty}^{\infty} x(t)\delta(t-t_0)\, dt = x(t_0) \tag{3.293b}$$

follows from changing the variable of integration to $\tau = t + t_0$.

Since the dependence of the integral in (3.291) on x is only through $x(0)$, the range of integration need not be $(-\infty, \infty)$. The limits of integration can be any a and b satisfying $a < 0 < b$.

An equation with a Dirac delta function that is not part of an integrand cannot be checked pointwise. Instead, it is a shorthand for having equality after multiplying both sides by a continuous function and integrating.[76] For example, for any function y that is continuous at 0, we say

$$y(t)\delta(t) = y(0)\delta(t). \tag{3.294}$$

To verify this, let x be any function that is continuous at 0. Then,

$$\int_{-\infty}^{\infty} x(t)y(t)\delta(t)\, dt \stackrel{(a)}{=} x(0)y(0) \stackrel{(b)}{=} \int_{-\infty}^{\infty} x(t)y(0)\delta(t)\, dt,$$

where (a) follows from the continuity of $x(t)y(t)$ at $t = 0$; and (b) from the continuity of $x(t)y(0)$ at $t = 0$. Similarly, for any $a \in \mathbb{R}^+$,

$$\delta(t/a) = a\delta(t).$$

To verify this, let x be any function that is continuous at 0. Then,

$$\int_{-\infty}^{\infty} x(t)\delta(t/a)\, dt \stackrel{(a)}{=} \int_{-\infty}^{\infty} x(a\tau)\delta(\tau)a\, d\tau = a\int_{-\infty}^{\infty} x(a\tau)\delta(\tau)\, d\tau \stackrel{(b)}{=} ax(0),$$

[75] Dirac called this δ an *improper function*, others have used the terms *generalized function* and *delta distribution*, and it is called a *measure* in measure theory.

[76] As Dirac wrote in introducing the δ function [21, p. 59]: "The use of improper functions ... does not involve any lack of rigour in the theory, but is merely a convenient notation, enabling us to express in a concise form certain relations which we could, if necessary, rewrite in a form not involving improper functions, but only in a cumbersome way which would tend to obscure the argument."

Dirac delta function	
Normalization	$\int_{-\infty}^{\infty} \delta(t)\, dt = 1$
Sifting	$\int_{-\infty}^{\infty} x(t+t_0)\,\delta(t)\, dt = \int_{-\infty}^{\infty} x(t)\,\delta(t-t_0)\, dt = x(t_0)$
Sampling	$x(t)\,\delta(t) = x(0)\,\delta(t)$
Scaling	$\delta(t/a) = \lvert a\rvert\,\delta(t)$ for any nonzero $a \in \mathbb{R}$
Shifting	$x(t) *_t \delta(t-t_0) = x(t-t_0)$

Table 3.11 Properties of the Dirac delta function. Sifting requires continuity of x at t_0 and sampling requires continuity of x at 0.

where (a) follows from the change of variable $\tau = t/a$; and (b) from the continuity at 0 of the scaled function $x(at)$. Combined with a similar argument for $a \in \mathbb{R}^-$, we have

$$\delta(t/a) = \lvert a\rvert\,\delta(t) \tag{3.295}$$

for any nonzero $a \in \mathbb{R}$.

Properties of the Dirac delta function are summarized in Table 3.11. Note that the shifting property uses the convolution operation defined in (4.31) in Section 4.3.3.

Appendix 3.B Elements of algebra

3.B.1 Polynomials

A *polynomial* is a function of the following form:

$$p(z) = \sum_{n=0}^{N} a_n z^n. \tag{3.296}$$

Assuming that $a_N \neq 0$, the *degree of the polynomial* is N. The set of polynomials with coefficients a_n from a given ring itself forms a ring.[77]

The *roots* of a polynomial are obtained by equating a *polynomial function* $p(z)$, a function obtained by evaluating a polynomial $p(z)$ over a given domain of z, to zero. The following theorem, formulated by Gauss, is a useful tool in algebra:

THEOREM 3.23 (FUNDAMENTAL THEOREM OF ALGEBRA) Every polynomial with complex coefficients of degree N possesses exactly N complex roots.

[77] A ring is a set together with two binary operations, addition and multiplication. The addition must be commutative and associative, while the multiplication must be associative, and distributive over addition. There exists an additive identity element and each element must have an additive inverse. A standard example of a ring is the set of integers, $\mathbb{Z}$, with ordinary addition and multiplication.

Thus, the degree of the polynomial is also the number of complex roots of that polynomial. For example, $p(z) = a_2z^2 + a_1z + a_0$ is a *quadratic* polynomial and has two roots. An *irreducible* quadratic polynomial is a quadratic polynomial with no real roots. For example, $p(z) = z^2 + 2$ has no real roots; rather, its roots are complex, $\pm j\sqrt{2}$.

We can factor any polynomial with real coefficients into a product of linear factors, $(z - b_n)$, and irreducible quadratic factors, $(z^2 + c_nz + d_n)$, where the coefficients b_n, c_n, and d_n are all real; for a polynomial with complex coefficients, the factors are all linear, $(z - z_n)$, where the coefficients z_n are complex,

$$p(z) = \sum_{n=0}^{N} a_n z^n = \begin{cases} a_N \prod_{n=0}^{N-2k-1}(z-b_n)\prod_{n=0}^{k-1}(z^2+c_nz+d_n), \\ a_n, b_n, c_n, d_n \in \mathbb{R}, \\ \quad \text{or} \\ a_N \prod_{n=0}^{N-1}(z-z_n), \\ a_n, z_n \in \mathbb{C}. \end{cases} \tag{3.297}$$

Two polynomials $p(z)$ and $q(z)$ are called *coprime*, written as $(p(z), q(z)) = 1$, when they have no common factors. The *Bézout identity* states that if $p(z)$ and $q(z)$ are coprime, there exist two other polynomials $a(z)$ and $b(z)$ such that

$$a(z)\,p(z) + b(z)\,q(z) = 1, \qquad \text{for all } z. \tag{3.298}$$

Euclid's algorithm is a constructive way of finding $a(z)$ and $b(z)$ in (3.298).

Laurent polynomials A *Laurent polynomial* is like a polynomial except that negative powers are allowed in (3.296),

$$p(z) = \sum_{n=-M}^{N} a_n z^n. \tag{3.299}$$

This can be written as

$$p(z) = z^{-M} q(z) \qquad \text{with} \qquad q(z) = \sum_{n=0}^{N+M} a_n z^n, \tag{3.300}$$

where $q(z)$ is now just an ordinary polynomial.

Ratios of polynomials A *rational function* $r(z)$ is a ratio of two polynomials,

$$r(z) = \frac{p(z)}{q(z)} = \frac{\sum_{n=0}^{N} a_n z^n}{\sum_{n=0}^{M} b_n z^n}. \tag{3.301}$$

In general, we assume that $M \geq N$, as otherwise, we could use polynomial division to write (3.301) as a sum of a polynomial and a ratio of polynomials with the numerator now of degree smaller or equal to M.

Assume that $p(z)$ and $q(z)$ are coprime; otherwise, we can cancel common factors and proceed. When $M = N$, by Theorem 3.23, (3.301) has N zeros and M

poles (zeros of the denominator $q(z)$) in the complex plane. When $M < N$, there are $N - M$ additional zeros at $z = \infty$. When $M > N$, there are $M - N$ additional poles at $z = \infty$, indicating that a rational function has $\max(M, N)$ poles and zeros, including ones at 0 and ∞.

Discrete polynomials A *polynomial sequence* is a sequence whose nth element is a finite sum of the following form:

$$p_n = \sum_{k=0}^{N} a_k n^k, \qquad n \in \mathbb{Z}. \tag{3.302}$$

For example, a *constant* polynomial sequence is of the form $p_n = a$, a *linear* polynomial sequence is of the form $p_n = a_0 + a_1 n$, and a *quadratic* polynomial sequence is of the form $p_n = a_0 + a_1 n + a_2 n^2$. The z-transform of such a sequence is

$$P(z) = \sum_{n\in\mathbb{Z}} p_n z^{-n} = \sum_{n\in\mathbb{Z}} \left(\sum_{k=0}^{N} a_k n^k \right) z^{-n}.$$

When we study wavelets and filter banks, we will be concerned with the moment-annihilating/preserving properties of such systems. The following fact will then be of use: Convolution of the polynomial sequence with a differencing filter $d_n = (\delta_n - \delta_{n-1})$, or multiplication of $P(z)$ by $D(z) = (1 - z^{-1})$, reduces the degree of the polynomial by 1, as in

$$\begin{aligned}
D(z)P(z) &= (1 - z^{-1}) \sum_{n\in\mathbb{Z}} p_n z^{-n} = (1 - z^{-1}) \sum_{n\in\mathbb{Z}} \left(\sum_{k=0}^{N} a_k n^k \right) z^{-n} \\
&= \sum_{n\in\mathbb{Z}} \sum_{k=0}^{N} a_k n^k z^{-n} - \sum_{n\in\mathbb{Z}} \sum_{k=0}^{N} a_k n^k z^{-(n+1)} \\
&= \sum_{n\in\mathbb{Z}} \sum_{k=0}^{N} a_k n^k z^{-n} - \sum_{n\in\mathbb{Z}} \sum_{k=0}^{N} a_k (n-1)^k z^{-n} \\
&= \sum_{n\in\mathbb{Z}} \sum_{k=0}^{N} a_k (n^k - (n-1)^k) z^{-n} \\
&\overset{(a)}{=} \sum_{n\in\mathbb{Z}} \left(\sum_{k=0}^{N-1} b_k n^k \right) z^{-n} = \sum_{n\in\mathbb{Z}} r_n z^{-n},
\end{aligned}$$

where r_n is a polynomial of degree $(N - 1)$, and (a) follows from $(n^N - (n-1)^N)$ being a polynomial of degree $(N - 1)$. The above process can be seen as applying a differencing filter with a zero at $z = 1$. Extending the above argument, we see that by repeatedly applying the differencing filter we will eventually reach a degree-0 polynomial sequence (a constant) and then the all-zero sequence.

3.B.2 Vectors and matrices of polynomials

Notions of vectors and matrices can be combined with polynomials and rational functions. For simplicity, we introduce all concepts on 2×1 vectors and 2×2 matrices.

A vector of polynomials, or *polynomial vector*, is given by

$$v(z) = \begin{bmatrix} \sum_{n=0}^{N} a_n z^n \\ \sum_{n=0}^{N} b_n z^n \end{bmatrix} = \begin{bmatrix} p(z) \\ q(z) \end{bmatrix} = \sum_{n=0}^{N} v_n z^n, \tag{3.303}$$

where each v_n is a 2×1 vector of scalars.

Similarly, a matrix of polynomials, or *polynomial matrix*, is given by

$$H(z) = \begin{bmatrix} \sum_{n=0}^{N} a_n z^n & \sum_{n=0}^{N} b_n z^n \\ \sum_{n=0}^{N} c_n z^n & \sum_{n=0}^{N} d_n z^n \end{bmatrix} = \begin{bmatrix} p(z) & q(z) \\ r(z) & s(z) \end{bmatrix} = \sum_{n=0}^{N} H_n z^n, \tag{3.304}$$

where each H_n is a 2×2 matrix of scalars. In both of the above expressions, N is the maximum degree of any of the entries.

Rank is more subtle for polynomial matrices than for ordinary ones. For example, if $\lambda = 3$,

$$H(z) = \begin{bmatrix} a + bz & 3(a + bz) \\ c + dz & \lambda(c + dz) \end{bmatrix}$$

is rank-deficient for every value of z. On the other hand, if $\lambda \neq 3$, then it is rank-deficient only if $z = -a/b$ or $z = -c/d$, leading to the notion of *normal rank*. The normal rank of $H(z)$ is the size of the largest minor that has a determinant that is not identically zero. In the above example, for $\lambda = 3$, the normal rank is 1, while for $\lambda \neq 3$, the normal rank is 2.

A square polynomial matrix of full normal rank has an inverse, which can be computed identically to a scalar matrix as in (2.221),

$$H^{-1}(z) = \frac{\operatorname{adj}(H(z))}{\det(H(z))}. \tag{3.305}$$

A polynomial matrix $H(z)$ is called *unimodular* if $|\det H(z)| = 1$ for all z. The product of two unimodular matrices is unimodular, and the inverse of a unimodular matrix is unimodular as well. A polynomial matrix is unimodular if and only if its inverse is a polynomial matrix. All these facts can be proven using properties of determinants.

EXAMPLE 3.48 (UNIMODULAR POLYNOMIAL MATRIX) The determinant of the polynomial matrix

$$H(z) = \begin{bmatrix} 1 + z & 2 + z \\ z & 1 + z \end{bmatrix}$$

is given by $\det H(z) = (1 + z)2 - z(2 + z) = 1$; it is thus unimodular. Its inverse is

$$H^{-1}(z) = \begin{bmatrix} 1 + z & -(2 + z) \\ -z & 1 + z \end{bmatrix},$$

which is also a unimodular polynomial matrix.

Vectors and matrices of Laurent polynomials Just as polynomials can be extended to Laurent polynomials, matrices of polynomials can be extended to matrices of Laurent polynomials, or *Laurent polynomial matrices*,

$$H(z) = \begin{bmatrix} \sum_{n=-N}^{N} a_n z^n & \sum_{n=-N}^{N} b_n z^n \\ \sum_{n=-N}^{N} c_n z^n & \sum_{n=-N}^{N} d_n z^n \end{bmatrix} = \sum_{n=-N}^{N} H_n z^n,$$

and similarly for vectors. The normal rank is defined as for polynomial matrices. A Laurent polynomial matrix $H(z)$ is called *Laurent unimodular* if $|\det H(z)| = cz^k$ for all $z \in \mathbb{C}$, for some $c \in \mathbb{C}$ and $k \in \mathbb{Z}$. The inverse of a Laurent polynomial matrix is again a Laurent polynomial matrix only if it is Laurent unimodular, since the adjugate in (3.305) is again a Laurent polynomial matrix, while the determinant is a monomial.

EXAMPLE 3.49 (LAURENT UNIMODULAR POLYNOMIAL MATRIX) The Laurent polynomial matrix

$$H(z) = \frac{1}{4z} \begin{bmatrix} 1+3z & 1-3z \\ 3+z & 3-z \end{bmatrix}$$

has determinant z^{-1}; it is thus unimodular. Its inverse is

$$H^{-1}(z) = \frac{1}{4} \begin{bmatrix} 3-z & -1+3z \\ -3-z & 1+3z \end{bmatrix},$$

which is also a Laurent unimodular polynomial matrix.

Vectors and matrices of ratios of polynomials A matrix of rational functions, or *rational matrix*, has entries that are ratios of polynomials,

$$H(z) = \begin{bmatrix} p_{00}(z)/q_{00}(z) & p_{01}(z)/q_{01}(z) \\ p_{10}(z)/q_{10}(z) & p_{11}(z)/q_{11}(z) \end{bmatrix},$$

where each $p_{ij}(z)$ and $q_{ij}(z)$ is a polynomial in z. The normal rank is defined as for polynomial matrices. The inverse of a rational matrix is again a rational matrix.

Adjoint of a polynomial vector or matrix We now discuss the adjoint of a vector or matrix of polynomials; extensions to vectors and matrices of Laurent polynomials and rational functions follow similarly. The adjoints of a vector of polynomials $v(z)$ and matrix of polynomials $H(z)$ are defined to be used with z-transform representations of LSI systems; the adjoint of $v(z)$ or $H(z)$ gives the z-transform representation of the adjoint of the corresponding system. The definitions also ensure that the products $v^*(z)v(z)$ and $H^*(z)H(z)$ are positive semidefinite on the unit circle, $|z| = 1$.[78] This is because we are extending the idea of autocorrelation (3.146) to vectors and matrices.

[78] A polynomial representing a z-transform turns into the DTFT on the unit circle.

For simplicity, we consider polynomials with real coefficients first. The adjoint of a 2×1 vector of polynomials (3.303) is

$$v^*(z) = v^\top(z^{-1}) = \begin{bmatrix} p(z^{-1}) & q(z^{-1}) \end{bmatrix}. \tag{3.306}$$

The product

$$v^*(z)\, v(z) = \begin{bmatrix} p(z^{-1}) & q(z^{-1}) \end{bmatrix} \begin{bmatrix} p(z) \\ q(z) \end{bmatrix} = p(z^{-1})p(z) + q(z^{-1})q(z)$$

is positive semidefinite on the unit circle since it is the sum of two positive semidefinite functions on the unit circle. The same holds for matrices of polynomials: the adjoint of a 2×2 matrix of polynomials (3.304) is

$$H^*(z) = H^\top(z^{-1}) = \begin{bmatrix} p(z^{-1}) & r(z^{-1}) \\ q(z^{-1}) & s(z^{-1}) \end{bmatrix}; \tag{3.307}$$

the product $H^*(z)\, H(z)$ is a Laurent polynomial matrix and is positive semidefinite on the unit circle.

An extension of the spectral factorization Corollary 3.14 states the following:

Theorem 3.24 (Extension of spectral factorization) Let $A(z)$ be a Laurent polynomial matrix. Then, it is positive semidefinite on the unit circle, that is, $A(e^{j\omega}) \geq 0$, if and only if

$$A(z) = H(z)\, H_*(z^{-1}), \tag{3.308}$$

where $H(z)$ is a polynomial matrix and $H_*(z) = H^*(z^*)$ amounts to conjugating the coefficients of H but not its argument.

Such a matrix is given in Example 3.41 and its factorization in (3.253).

When polynomial coefficients are complex, the adjoint of the matrix in (3.304) is

$$H^*(z) = H_*(z^{-1}) = \begin{bmatrix} p_*(z^{-1}) & r_*(z^{-1}) \\ q_*(z^{-1}) & s_*(z^{-1}) \end{bmatrix} = \sum_{n=0}^{N} H_n^* z^{-n}. \tag{3.309}$$

Paraunitary matrices A square matrix of polynomials, Laurent polynomials, or rational functions, $U(z)$, is called *unitary* when every value of the argument z gives a unitary matrix,

$$U^*(z)\, U(z) = I, \qquad \text{for all } z \in \mathbb{C}. \tag{3.310a}$$

When evaluated for $z = e^{j\omega}$, (3.310a) reduces to

$$U^*(e^{j\omega})\, U(e^{j\omega}) = I, \qquad \text{for all } \omega \in \mathbb{R}. \tag{3.310b}$$

The unitary condition (3.310a) is very restrictive; for example, it makes unitary matrices a subset of unimodular matrices. Extending the condition (3.310b) away from the unit circle in a less restrictive manner gives the paraunitary condition. A square matrix of polynomials, Laurent polynomials, or rational functions, $U(z)$, is called *paraunitary* when it satisfies

$$U_*(z^{-1})\, U(z) \;=\; I, \qquad \text{for all } z \in \mathbb{C}. \tag{3.311a}$$

Restricting (3.311a) to the unit circle is equivalent to (3.310b), giving the connection to unitary matrices. Since for a matrix with real coefficients $U_*(z^{-1}) = U^*(z^{-1}) = U^\top(z^{-1})$, a paraunitary matrix with real coefficients satisfies

$$U^\top(z^{-1})\, U(z) \;=\; I, \qquad \text{for all } z \in \mathbb{C}. \tag{3.311b}$$

When a paraunitary matrix has polynomial entries (not Laurent polynomials), it is called *lossless.*

EXAMPLE 3.50 (PARAUNITARY MATRIX) The matrix

$$U(z) \;=\; \frac{1}{\sqrt{2}} \begin{bmatrix} 1 & 1 \\ 1 & -1 \end{bmatrix} \begin{bmatrix} 1 & 0 \\ 0 & z \end{bmatrix} \begin{bmatrix} \frac{1}{2} & -\frac{\sqrt{3}}{2} \\ \frac{\sqrt{3}}{2} & \frac{1}{2} \end{bmatrix} \;=\; \frac{1}{\sqrt{2}} \begin{bmatrix} \frac{1+\sqrt{3}z}{2} & \frac{-\sqrt{3}+z}{2} \\ \frac{1-\sqrt{3}z}{2} & \frac{-\sqrt{3}-z}{2} \end{bmatrix}$$

is paraunitary since (3.311b) is satisfied. Since its entries are polynomials, it is lossless as well.

Pseudocirculant polynomial matrices The extension of circulant matrices (2.245) to polynomial matrices are *pseudocirculant* matrices, an example of which is (3.224). Such a matrix has polynomial entries and is circulant with entries above the diagonal multiplied by z^{-1} (thus pseudocirculant), for example,

$$H(z) \;=\; \begin{bmatrix} h_0(z) & z^{-1}h_2(z) & z^{-1}h_1(z) \\ h_1(z) & h_0(z) & z^{-1}h_2(z) \\ h_2(z) & h_1(z) & h_0(z) \end{bmatrix}. \tag{3.312}$$

3.B.3 Kronecker product

The Kronecker product of two matrices is defined as

$$\begin{bmatrix} a_{0,0} & \cdots & a_{0,N-1} \\ \vdots & \ddots & \vdots \\ a_{M-1,0} & \cdots & a_{M-1,N-1} \end{bmatrix} \otimes B \;=\; \begin{bmatrix} a_{0,0}B & \cdots & a_{0,N-1}B \\ \vdots & \ddots & \vdots \\ a_{M-1,0}B & \cdots & a_{M-1,N-1}B \end{bmatrix}, \tag{3.313}$$

where each a_{ij} is a scalar, B is a matrix, and neither matrix need be square. The Kronecker product has the following useful property with respect to the usual matrix product:

$$(A \otimes B)(C \otimes D) = (AC) \otimes (BD), \tag{3.314}$$

where all the matrix products have to be well defined. See Solved exercise 3.8 for an application of Kronecker products.

Chapter at a glance

We now summarize the main concepts and results seen in this chapter, some in a tabular form. One of the key elements was finding the appropriate Fourier transform for a given space of sequences (such as $\ell^2(\mathbb{Z})$ or circularly extended finite-length ones). That procedure can be summarized as follows:

(i) Start with a given time shift δ_{n-1}.
(ii) Determine the induced convolution operator $Tx = Hx = h * x$.
(iii) Find the eigensequences x_n of H ($e^{j\omega n}$ for infinite-length sequences and $e^{j(2\pi/N)kn}$ for finite-length sequences).
(iv) Identify the frequency response as the eigenvalues corresponding to the above eigensequences ($H(e^{j\omega n})$ for infinite-length sequences, H_k for finite-length sequences).
(v) The appropriate Fourier transform projects sequences on the spaces spanned by eigensequences identified in (iii) (discrete-time Fourier transform for infinite-length sequences, discrete Fourier transform for finite-length sequences).

Concept	Notation	Infinite-length sequences	Finite-length sequences
Shift	δ_{n-1}	linear	circular
Sequence vector	x_n	$n \in \mathbb{Z}$	$n \in \{0, 1, \ldots, N-1\}$
LSI system filter impulse response operator	h_n	$n \in \mathbb{Z}$	$n \in \{0, 1, \ldots, N-1\}$
Convolution	$h * x$	$\sum_{k\in\mathbb{Z}} x_k h_{n-k}$	$\sum_{k=0}^{N-1} x_k h_{N,(n-k) \bmod N}$
Eigensequence satisfies invariant space	v $h * v_\lambda = \lambda v_\lambda$ S_λ	$e^{j\omega n}$ $h * v_\omega = H(e^{j\omega})\, v_\omega$ $S_\omega = \{\alpha e^{j\omega n}\}$ $\alpha \in \mathbb{C},\ \omega \in \mathbb{R}$	$e^{j(2\pi/N)kn}$ $h * v_k = H_k v_k$ $S_k = \{\alpha e^{j(2\pi/N)kn}\}$ $\alpha \in \mathbb{C},\ k \in \mathbb{Z}$
Frequency response eigenvalue	λ	$\lambda_\omega = H(e^{j\omega})$ $\sum_{n\in\mathbb{Z}} h_n e^{-j\omega n}$	$\lambda_k = H_k$ $\sum_{n=0}^{N-1} h_n e^{-j(2\pi/N)kn}$
Fourier transform spectrum	X	DTFT $X(e^{j\omega}) = \sum_{n\in\mathbb{Z}} x_n e^{-j\omega n}$	DFT $X_k = \sum_{n=0}^{N-1} x_n e^{-j(2\pi/N)kn}$

Table 3.12 Concepts in discrete-time processing.

Concept	Expression
Sampling factor 2	
Input	$x_n,\ X(z),\ X(e^{j\omega})$
Downsampling	$y_n = x_{2n}$ $y = D_2 x$ $Y(z) = \frac{1}{2}\left[X(z^{1/2}) + X(-z^{1/2})\right]$ $Y(e^{j\omega}) = \frac{1}{2}\left[X(e^{j\omega/2}) + X(e^{j(\omega-2\pi)/2})\right]$
Upsampling	$y_n = \begin{cases} x_{n/2}, & n \text{ even;} \\ 0, & \text{otherwise.} \end{cases}$ $y = U_2 x$ $Y(z) = X(z^2)$ $Y(e^{j\omega}) = X(e^{j2\omega})$
Downsampling preceded by filtering	$y = D_2 H x$ $Y(z) = \frac{1}{2}\left[H(z^{1/2})X(z^{1/2}) + H(-z^{1/2})X(-z^{1/2})\right]$ $Y(e^{j\omega}) = \frac{1}{2}\left[H(e^{j\omega/2})X(e^{j\omega/2}) + H(e^{j(\omega-2\pi)/2})X(e^{j(\omega-2\pi)/2})\right]$
Upsampling followed by filtering	$y = G U_2 x$ $Y(z) = G(z)X(z^2)$ $Y(e^{j\omega}) = G(e^{j\omega})X(e^{j2\omega})$
Sampling factor N	
Input	$x_n,\ X(z),\ X(e^{j\omega})$
Downsampling	$y_n = x_{Nn}$ $y = D_N x$ $Y(z) = \frac{1}{N}\sum_{k=0}^{N-1} X(W_N^k z^{1/N})$ $Y(e^{j\omega}) = \frac{1}{N}\sum_{k=0}^{N-1} X(e^{j(\omega-2\pi k/N)})$
Upsampling	$y_n = \begin{cases} x_{n/N}, & n/N \in \mathbb{Z}; \\ 0, & \text{otherwise.} \end{cases}$ $y = U_N x$ $Y(z) = X(z^N)$ $Y(e^{j\omega}) = X(e^{jN\omega})$

Table 3.13 Concepts in multirate discrete-time processing.

Domain		Autocorrelation/crosscorrelation	Properties
Sequences		$x_n,\ y_n$	
Time	a_n	$\sum_{k\in\mathbb{Z}} x_k x^*_{k-n}$	$a_n = a^*_{-n}$
	c_n	$\sum_{k\in\mathbb{Z}} x_k y^*_{k-n}$	$c_{x,y,n} = c^*_{y,x,-n}$
DTFT	$A(e^{j\omega})$	$\lvert X(e^{j\omega})\rvert^2$	$A(e^{j\omega}) = A^*(e^{j\omega})$
	$C(e^{j\omega})$	$X(e^{j\omega})\, Y^*(e^{j\omega})$	$C_{x,y}(e^{j\omega}) = C^*_{y,x}(e^{j\omega})$
z-transform	$A(z)$	$X(z)\, X_*(z^{-1})$	$A(z) = A_*(z^{-1})$
	$C(z)$	$X(z)\, Y_*(z^{-1})$	$C_{x,y}(z) = C_{y,x*}(z^{-1})$
DFT	A_k	$\lvert X_k\rvert^2$	$A_k = A^*_{-k \bmod N}$
	C_k	$X_k Y^*_k$	$C_k = C^*_{y,x,-k \bmod N}$
Real sequences		$x_n,\ y_n$	
Time	a_n	$\sum_{k\in\mathbb{Z}} x_k x_{k-n}$	$a_n = a_{-n}$
	c_n	$\sum_{k\in\mathbb{Z}} x_k y_{k-n}$	$c_{x,y,n} = c_{y,x,-n}$
DTFT	$A(e^{j\omega})$	$\lvert X(e^{j\omega})\rvert^2$	$A(e^{j\omega}) = A(e^{-j\omega})$
	$C(e^{j\omega})$	$X(e^{j\omega})\, Y(e^{-j\omega})$	$C_{x,y}(e^{j\omega}) = C_{y,x}(e^{-j\omega})$
z-transform	$A(z)$	$X(z)\, X(z^{-1})$	$A(z) = A(z^{-1})$
	$C(z)$	$X(z)\, Y(z^{-1})$	$C_{x,y}(z) = C_{y,x}(z^{-1})$
DFT	A_k	$\lvert X_k\rvert^2$	$A_k = A_{-k \bmod N}$
	C_k	$X_k Y_k$	$C_k = C_{y,x,-k \bmod N}$
Vector of sequences		$\begin{bmatrix} x_{0,n} & x_{1,n} \end{bmatrix}^\top$	
Time	A_n	$\begin{bmatrix} a_{0,n} & c_{0,1,n} \\ c_{1,0,n} & a_{1,n} \end{bmatrix}$	$A_n = A^*_{-n}$ $A_n = A^\top_{-n}$
DTFT	$A(e^{j\omega})$	$\begin{bmatrix} A_0(e^{j\omega}) & C_{0,1}(e^{j\omega}) \\ C_{1,0}(e^{j\omega}) & A_1(e^{j\omega}) \end{bmatrix}$	$A(e^{j\omega}) = A^*(e^{j\omega})$ $A(e^{j\omega}) = A^\top(e^{-j\omega})$
z-transform	$A(z)$	$\begin{bmatrix} A_0(z) & C_{0,1}(z) \\ C_{1,0}(z) & A_1(z) \end{bmatrix}$	$A(z) = A_*(z^{-1})$ $A(z) = A^\top(z^{-1})$
DFT	A_k	$\begin{bmatrix} A_{0,k} & C_{0,1,k} \\ C_{1,0,k} & A_{1,k} \end{bmatrix}$	$A_k = A^*_{-k \bmod N}$ $A_k = A^\top_{-k \bmod N}$

Table 3.14 Summary of concepts related to deterministic autocorrelation and crosscorrelation of a sequence (upper half) and a vector of sequences (lower half). For vectors of sequences, an example for a vector of two sequences is given. Also for vectors of sequences, under properties, two are given each time: the first is for complex sequences and the second for real sequences. The DFT entries are for circular deterministic autocorrelation and crosscorrelation. Note that we overload the notations $A(e^{j\omega})$, $A(z)$, and A_k to mean either a scalar or a matrix depending on the sequence.

Historical remarks

The impact signal processing has had in practical terms is perhaps due in large part to the advent of **fast Fourier transform** algorithms, spurred on by the paper of Cooley and Tukey in 1965 [16]. It breaks the computation of the discrete Fourier transform of length $N = N_1 N_2$ into a recursive computation of smaller DFTs of lengths N_1 and N_2, respectively (see Section 3.9.1). Unbeknownst to them, a similar algorithm had been published by Gauss some 150 years earlier, in his attempt to track asteroid trajectories.

Further reading

Signal processing books The standard textbook on discrete-time signal processing is the one by Oppenheim and Schafer [72]. For multidimensional signal processing, see the book by Dudgeon and Mersereau [27]; for multirate signal processing, see books by Crochiere and Rabiner [18] and by Vaidyanathan [105]; for statistical signal processing, the book by Porat [80]; for fast algorithms for discrete-time signal processing, see the book by Blahut [7]; and for signal processing for communications, see the book by Prandoni and Vetterli [82].

Inverse z-transform via contour integration The formal inversion process for the z-transform is given by contour integration using Cauchy's integral formula when $X(z)$ is a rational function of z. When $X(z)$ is not rational, inversion can be quite difficult; a short account is given in [54], and more details can be found in [72].

Algebraic theory of signal processing A framework for signal processing in algebraic terms is a recent development whose foundations can be found in [83]; some of the observations in this chapter were inspired by this framework. It provides a recipe for starting with a sequence space, shift, and method of extension, and deriving from these an appropriate convolution operator and Fourier transform. For example, using a symmetric extension rather than the circular one in Section 3.6 can be more natural for signals whose domain represents space rather than time; this leads to the discrete cosine transform (DCT), which is well known in image processing. The existence and forms of fast algorithms for computing transforms, including the well-known trigonometric ones, all follow from this theory.

Pseudocirculant matrices We have seen the importance of these matrices in multirate systems, in particular in representing convolution in the polyphase domain. A thorough presentation of such matrices can be found in [106].

Stochastic multirate systems In Section 3.8.4, we examined only a few basic operations in stochastic multirate systems. A thorough discussion of other cases, including different periods for cyclostationarity of the input and the output, can be found in [86].

Exercises with solutions

3.1. *Properties of complex exponential sequences*
Let x be a complex exponential sequence of the form

$$x_n = e^{j\omega_0 n}, \qquad n \in \mathbb{Z}.$$

(i) Show that if $\alpha = \omega_0/(2\pi)$ is a rational number, $\alpha = p/q$, with p and q coprime integers, then x is periodic with period q.
(ii) Show that if $\alpha = \omega_0/(2\pi)$ is irrational, then x is not periodic.
(iii) Show that if x and y are two periodic sequences with periods M and N, respectively, then $x + y$ is periodic with period $\mathrm{lcm}(M, N)$.

Solution:

(i) If $\alpha = p/q = \omega_0/(2\pi)$ with $p, q \in \mathbb{Z}$, and $(p, q) = 1$, then $\omega_0 = 2\pi\alpha = 2\pi p/q$, and

$$x_{n+q} = e^{j(n+q)(2\pi p/q)} = e^{jn(2\pi p/q)}\, e^{j2\pi p} = e^{jn(2\pi p/q)} = x_n;$$

x is thus periodic with period q.
(ii) Suppose that there exists $N \in \mathbb{Z}$ such that $x_n = x_{n+N}$ for all $n \in \mathbb{Z}$. Then

$$e^{jn2\pi\alpha} = e^{j(n+N)2\pi\alpha} \Rightarrow 1 = e^{jN2\pi\alpha} \Rightarrow N\alpha = M \in \mathbb{Z} \Rightarrow \alpha = \frac{M}{N} \in \mathbb{Q}.$$

Thus, periodicity of x implies α is rational. By contraposition, α being irrational implies x is not periodic.
(iii) Let $z = x + y$ and $P = \mathrm{lcm}(M, N)$. Clearly

$$z_{n+P} = x_{n+P} + y_{n+P} = x_n + y_n,$$

since P is a multiple of both M and N. This holds for any pair of periodic sequences, not just those in the specified form.

3.2. *LSI system acting on signals in $\ell^p(\mathbb{Z})$*
Prove that if $x \in \ell^p(\mathbb{Z})$ and $h \in \ell^1(\mathbb{Z})$, the result of $h * x$ is in $\ell^p(\mathbb{Z})$ as well.
Solution: Given $x \in \ell^p(\mathbb{Z})$ and $h \in \ell^1(\mathbb{Z})$, we will prove that, for $y = h * x$,

$$\|y\|_p \leq \|x\|_p \|h\|_1,$$

implying that $h * x \in \ell^p(\mathbb{Z})$.

(i) $1 < p < \infty$. Choose q such that $1/p + 1/q = 1$. Then,

$$|y_n| = |(h * x)_n| \overset{(a)}{\leq} \sum_{k\in\mathbb{Z}} |x_k|\,|h_{n-k}| \overset{(b)}{\leq} \left(\sum_{k\in\mathbb{Z}} |x_k|^p |h_{n-k}|\right)^{1/p} \left(\sum_{k\in\mathbb{Z}} |h_{n-k}|\right)^{1/q},$$

where (a) follows from the triangle inequality, (2.25); and (b) from Hölder's inequality, (2.213), and $|h_{n-k}| = |h_{n-k}|^{1/p}|h_{n-k}|^{1/q}$. Upon raising both sides of the inequality to the pth power, summing over n, and changing the order of summations (which is allowed because the sums converge absolutely), we obtain

$$\sum_{n\in\mathbb{Z}} |y_n|^p \leq \|h\|_1^{p/q} \sum_{n\in\mathbb{Z}} \sum_{k\in\mathbb{Z}} |x_k|^p |h_{n-k}| \leq \|h\|_1^{p/q+1} \|x\|_p^p.$$

Taking the pth root then gives

$$\|y\|_p \leq \|x\|_p \,\|h\|_1^{1/p+1/q} = \|x\|_p \,\|h\|_1.$$

(ii) $p = 1$. We now have

$$\|y\|_1 \leq \sum_{n\in\mathbb{Z}} \sum_{k\in\mathbb{Z}} |x_k|\,|h_{n-k}| \leq \|x\|_1 \|h\|_1,$$

implying that $y \in \ell^1(\mathbb{Z})$.

(iii) $p = \infty$. We now have

$$|y_n| \leq \sum_{n\in\mathbb{Z}}\sum_{k\in\mathbb{Z}}|x_k|\,|h_{n-k}| \leq \|x\|_\infty \|h\|_1,$$

implying that $y \in \ell^\infty(\mathbb{Z})$.

3.3. *Filtering as a projection*
Using the frequency response, determine whether the convolution operator associated with each of the filters below is an orthogonal projection operator:

(i) $g_n = \delta_{n-k}$ for some $k \in \mathbb{Z}$;
(ii) $g_n = \frac{1}{2}\delta_{n+1} + \delta_n + \frac{1}{2}\delta_{n-1}$;
(iii) $g_n = (1/\sqrt{2})\,\mathrm{sinc}(\frac{1}{2}\pi n)$.

Also, give a precise characterization of the real-valued filters that perform an orthogonal projection. (Be more specific than just repeating the conditions for an operator to be an orthogonal projection.)

Solution: Denote by G the convolution operator associated with g, as in (3.64). To check that an operator is an orthogonal projection, we must check that it is idempotent and self-adjoint as in Definition 2.27. Checking idempotency is easy in the Fourier domain since $G^2 = G$ corresponds to $G^2(e^{j\omega}) = G(e^{j\omega})$. Checking self-adjointness is equivalent to checking that the matrix representation of G is Hermitian.

(i) We have $G(e^{j\omega}) = e^{-j\omega k}$ and $G^2(e^{j\omega}) = e^{-j\omega 2k}$. Thus, $G^2(e^{j\omega}) \neq G(e^{j\omega})$, unless $k = 0$; this filter is not a projection operator except in the trivial case when it is the identity operator.

(ii) We have

$$G(e^{j\omega}) = \frac{1}{2}e^{j\omega} + 1 + \frac{1}{2}e^{-j\omega},$$

$$G^2(e^{j\omega}) = \left(\frac{1}{2}e^{j\omega} + 1 + \frac{1}{2}e^{-j\omega}\right)^2 = \frac{1}{4}e^{j2\omega} + e^{j\omega} + \frac{3}{2} + e^{-j\omega} + \frac{1}{4}e^{-j2\omega}.$$

Thus, $G^2(e^{j\omega}) \neq G(e^{j\omega})$; this filter is not a projection operator either.

(iii) From Table 3.4, we can rewrite g_n as

$$g_n = \frac{1}{\sqrt{2}}\,\mathrm{sinc}\left(\frac{1}{2}\pi n\right) = \frac{1}{2\pi}\int_{-\pi/2}^{\pi/2} \sqrt{2}\,e^{j\omega n}\,d\omega,$$

yielding

$$G(e^{j\omega}) = \begin{cases} \sqrt{2}, & |\omega| \leq \frac{1}{2}\pi; \\ 0, & \text{otherwise}, \end{cases} \qquad G^2(e^{j\omega}) = \begin{cases} 2, & |\omega| \leq \frac{1}{2}\pi; \\ 0, & \text{otherwise}. \end{cases}$$

Again, $G^2(e^{j\omega}) \neq G(e^{j\omega})$; this filter is not a projection operator either. Note that for $g_n = \mathrm{sinc}(\frac{1}{2}\pi n)$ we would have had a projection.

Trying to satisfy idempotency in the most general case, we see that, for $G^2(e^{j\omega})$ to be equal to $G(e^{j\omega})$, $G(e^{j\omega})$ can only be 1 or 0,

$$G(e^{j\omega}) = \begin{cases} 1, & \text{for } \omega \in R_1 \cup R_2 \cup \cdots \cup R_n; \\ 0, & \text{otherwise}, \end{cases}$$

where $R_1, R_2, \ldots, R_n$ are disjoint intervals in $[-\pi, \pi)$.

Self-adjointness is satisfied if $g_n = g^*_{-n}$, or in the Fourier domain,

$$G(e^{j\omega}) = \sum_{n\in\mathbb{Z}} g_n e^{-j\omega n} = \sum_{n\in\mathbb{Z}} g^*_{-n} e^{-j\omega n} = \sum_{n\in\mathbb{Z}} g^*_n (e^{-j\omega n})^* = G^*(e^{j\omega}).$$

Since the requirement for idempotency makes $G(e^{j\omega})$ real, self-adjointness is automatically satisfied.

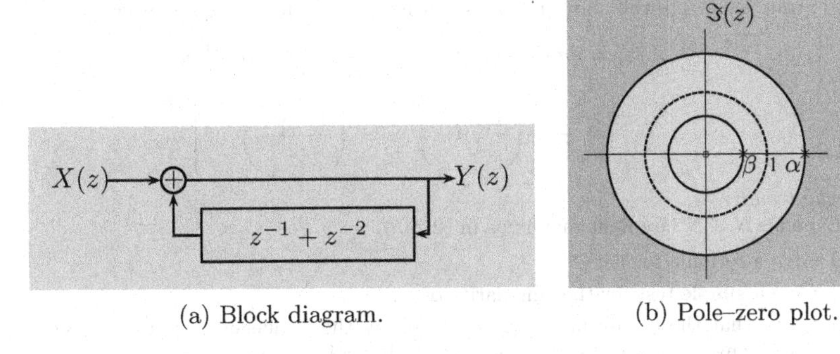

(a) Block diagram.

(b) Pole–zero plot.

Figure E3.4-1 Fibonacci filter.

3.4. *Fibonacci filter*

Let h be a filter given by a Fibonacci sequence,

$$h = \begin{bmatrix} \dots & 0 & \boxed{1} & 1 & 2 & 3 & 5 & 8 & 13 & \dots \end{bmatrix}^{\top}, \tag{E3.4-1}$$

obtained via the following recursion:

$$h_n = h_{n-1} + h_{n-2}, \qquad \text{for } n = 2, 3, \dots, \tag{E3.4-2}$$

with initial conditions $h_0 = h_1 = 1$.

(i) Is this filter FIR or IIR? Is it BIBO-stable?

(ii) Find the transfer function of the filter $H(z)$, draw a block diagram of an implementation, and show its pole–zero plot.

(iii) Show that h is a sum of two geometric series,

$$h_n = a\alpha^n + b\beta^n.$$

Solution: The *Fibonacci filter* is one of the oldest known digital filters.

(i) The filter is IIR because it is supported on all nonnegative integers. It is not BIBO-stable because it is not absolutely summable.

(ii) We first draw the block diagram of the system as in Figure E3.4-1(a) based on the recursion (E3.4-2). Then, we can easily find the transfer function from the block diagram as

$$Y(z) = \frac{1}{1 - z^{-1} - z^{-2}} X(z),$$

and thus

$$H(z) = \frac{1}{1 - z^{-1} - z^{-2}} = \frac{1}{(1 - \alpha z^{-1})(1 - \beta z^{-1})}$$

with

$$\alpha = \frac{1 + \sqrt{5}}{2}, \qquad \beta = \frac{1 - \sqrt{5}}{2}.$$

The constant α is called the *golden ratio*. The pole–zero plot of $H(z)$ is given in Figure E3.4-1(b).

(iii) We use partial fraction expansion to get

$$H(z) = \frac{1}{1 - z^{-1} - z^{-2}} = \frac{a}{1 - \alpha z^{-1}} + \frac{b}{1 - \beta z^{-1}}; \qquad \text{ROC} = \{z \mid |z| > \alpha\},$$

with

$$a = \frac{\alpha}{\sqrt{5}}, \qquad b = -\frac{\beta}{\sqrt{5}}.$$

From Table 3.6, the above is the z-transform of the following sequence:

$$h_n = (a\alpha^n + b\beta^n)u_n = \frac{1}{\sqrt{5}}(\alpha^{n+1} - \beta^{n+1})u_n$$
$$= \frac{1}{\sqrt{5}}\left[\left(\frac{1+\sqrt{5}}{2}\right)^{n+1} - \left(\frac{1-\sqrt{5}}{2}\right)^{n+1}\right]u_n.$$

3.5. *Circulant matrices*

Let C be an $N \times N$ circulant matrix as in (2.245).

(i) Give a formula for $\det C$.

(ii) Give a simple test for the singularity of C.

(iii) Prove that the eigenvalues of C are given by the frequency response (3.180a), the left eigenvectors are the rows of F, and the right eigenvectors are the columns of F^*/N.

(iv) Prove that C^{-1} is circulant.

(v) Given two circulant matrices C_1 and C_2, show that they commute, $C_1C_2 = C_2C_1$, and that the result is circulant as well.

Solution: The solution to this problem is based on the fact that the DFT diagonalizes the circulant convolution operator as in (3.181). Denote the DFT coefficients of $(c_0, c_1, \ldots, c_{N-1})$ by $(C_0, C_1, \ldots, C_{N-1})$, and let $\Lambda = \operatorname{diag}(C_0, C_1, \ldots, C_{N-1})$. The results now follow easily.

(i) As the determinant of a product is equal to the product of determinants, from (3.181a),

$$\det C = \det(F^{-1}\Lambda F) = (\det F^{-1})(\det \Lambda)(\det F)$$
$$= \underbrace{(\det F^{-1})(\det F)}_{=1}(\det \Lambda) = \prod_{k=0}^{N-1} C_k.$$

(ii) From (i), C is nonsingular if and only if every C_k is nonzero.

(iii) To show the eigenvector properties, write (3.181) as

$$CF^{-1} = F^{-1}\Lambda,$$
$$C\begin{bmatrix} r_0 & r_1 & \ldots & r_{N-1}\end{bmatrix} = \begin{bmatrix} r_0 & r_1 & \ldots & r_{N-1}\end{bmatrix}\operatorname{diag}(C_0, C_1, \ldots, C_{N-1})$$
$$= \begin{bmatrix} C_0r_0 & C_1r_1 & \ldots & C_{N-1}r_{N-1}\end{bmatrix},$$
$$Cr_k = C_kr_k, \qquad k = 0, 1, \ldots, N-1,$$

where we call r_k the columns of F^*/N. The above then implies that the columns of F^*/N are the right eigenvectors of C.

The argument follows similarly for left eigenvectors,

$$FC = \Lambda F,$$
$$\begin{bmatrix} r_0^* \\ r_1^* \\ \vdots \\ r_{N-1}^*\end{bmatrix} C = \operatorname{diag}(C_0, C_1, \ldots, C_{N-1})\begin{bmatrix} r_0^* \\ r_1^* \\ \vdots \\ r_{N-1}^*\end{bmatrix} = \begin{bmatrix} C_0r_0^* \\ C_1r_1^* \\ \vdots \\ C_{N-1}r_{N-1}^*\end{bmatrix},$$
$$r_k^*C = C_kr_k^*, \qquad k = 0, 1, \ldots, N-1,$$

where we call r_k^* the rows of F. The above then implies that the rows of F are the left eigenvectors of C.

(iv) Here we want to show that, if a matrix satisfies an equation of the same form as (3.181a), it is circulant as well. To do that, we start with (3.181a) and express the (i, j)th element of H as

$$h_{i,j} = \frac{1}{N}\sum_{k=0}^{N-1} H_k W_N^{k(j-i)}. \tag{E3.5-1}$$

For a matrix to be circulant, the following must hold:

$$h_{i+\ell,j+\ell} = h_{i,j},$$

with $\ell = 0, 1, \ldots, N-1$ and the addition performed mod$N$. It is easy to see that this is indeed the case for (E3.5-1).

Now write C^{-1} as

$$C^{-1} = (F\Lambda F^{-1})^{-1} = F\Lambda^{-1}F^{-1}.$$

Using the above argument, since C^{-1} satisfies an equation of the same form as (3.181a), it is circulant as well.

(v) This part follows similarly to the previous one,

$$C_1C_2 = F\Lambda_1 \underbrace{F^{-1}F}_{=I} \Lambda_2 F^{-1} = F(\Lambda_1\Lambda_2)F^{-1} = F\Lambda F^{-1},$$

again satisfying an equation of the same form as (3.181a); C_1C_2 is thus circulant as well. Because Λ_1 and Λ_2 are diagonal matrices, they commute, allowing us to reverse the process and show that C_1 and C_2 commute.

3.6. *Smoothing operators*

Let

$$g_n = -\frac{1}{16}\delta_{n+1} + \frac{1}{4}\delta_n - \frac{1}{16}\delta_{n-1}.$$

(i) Is g orthogonal to its even shifts?

(ii) Let $\widetilde{g}_n$ be such that $\{g_{n-2k}\}_{k\in\mathbb{Z}}$ and $\{\widetilde{g}_{n-2k}\}_{k\in\mathbb{Z}}$ are a pair of biorthogonal bases, where

$$\widetilde{g}_n = \delta_{n+2} + a\delta_{n+1} + b\delta_n + a\delta_{n-1} + \delta_{n-2}.$$

Let G and $\widetilde{G}$ be operators associated with g and $\widetilde{g}$, respectively. How is the biorthogonality condition (2.111) expressed in terms of G and $\widetilde{G}$? Find constants a and b for it to hold. Is $P = GU_2D_2\widetilde{G}$ an orthogonal projection operator?

Solution:

(i) Take g_n and g_{n-2}. They overlap at $n = 1$, with $\langle g_n, g_{n-2}\rangle_n = (1/16)^2$; thus, g_n is not orthogonal to its even shifts.

(ii) The biorthogonality condition can be expressed as

$$D_2\widetilde{G}GU_2 = I.$$

This is the counterpart of the orthogonality condition (3.214).

Writing the inner products for $k = 0$ and $k = 1$ gives the constraints

$$-\frac{1}{8}a + \frac{1}{4}b = 1,$$
$$-\frac{1}{16}a + \frac{1}{4} = 0,$$

yielding $a = 4$ and $b = 6$.

To check whether P is an orthogonal projection operator, write

$$P^2 = (GU_2D_2\widetilde{G})(GU_2D_2\widetilde{G}) = GU_2ID_2\widetilde{G} = P,$$
$$P^* = (GU_2D_2\widetilde{G})^* = \widetilde{G}^*(U_2D_2)^*G^* \neq P.$$

Thus P is an oblique projection operator – a projection operator that is not an orthogonal projection operator.

3.7. *Wiener filtering*

Let x and y be jointly WSS stochastic processes, and define estimation error process e as in Figure 3.35, where h is a BIBO-stable filter. By differentiating with respect to each filter coefficient h_n, show that the mean-squared error

$$\mathcal{E} = \mathrm{E}\big[\,|\mathrm{x}_n - \widehat{\mathrm{x}}_n|^2\,\big]$$

is minimized when $h * a_{\mathrm{y}} = c_{\mathrm{x,y}}$ is satisfied. This verifies the Wiener filter solution (3.261) without using the orthogonality principle.
Solution: Since x and $\widehat{\mathrm{x}}$ are jointly WSS, the MSE $\mathcal{E}$ is independent of n. Setting $n = 0$, we can write the MSE as

$$\begin{aligned}
\mathcal{E} &= \mathrm{E}[\,|\mathrm{x}_0 - \widehat{\mathrm{x}}_0|^2\,] \\
&= \mathrm{E}[\,(\mathrm{x}_0 - \widehat{\mathrm{x}}_0)(\mathrm{x}_0 - \widehat{\mathrm{x}}_0)^*\,] = \mathrm{E}[\,|\mathrm{x}_0|^2\,] - \mathrm{E}[\,\mathrm{x}_0\widehat{\mathrm{x}}_0^*\,] - \mathrm{E}[\,\widehat{\mathrm{x}}_0\mathrm{x}_0^*\,] + \mathrm{E}[\,\widehat{\mathrm{x}}_0\widehat{\mathrm{x}}_0^*\,] \\
&\overset{(a)}{=} \mathrm{E}[\,|\mathrm{x}_0|^2\,] - \mathrm{E}\left[\mathrm{x}_0 \sum_{k\in\mathbb{Z}} h_k^* \mathrm{y}_{-k}^*\right] - \mathrm{E}\left[\mathrm{x}_0^* \sum_{k\in\mathbb{Z}} h_k \mathrm{y}_{-k}\right] + \mathrm{E}\left[\sum_{k\in\mathbb{Z}} h_k \mathrm{y}_{-k} \sum_{m\in\mathbb{Z}} h_m^* \mathrm{y}_{-m}^*\right] \\
&\overset{(b)}{=} \mathrm{E}[\,|\mathrm{x}_0|^2\,] - \sum_{k\in\mathbb{Z}} h_k^* \mathrm{E}[\,\mathrm{x}_0\mathrm{y}_{-k}^*\,] - \sum_{k\in\mathbb{Z}} h_k \mathrm{E}[\mathrm{x}_0^*\mathrm{y}_{-k}\,] + \sum_{k\in\mathbb{Z}}\sum_{m\in\mathbb{Z}} h_k h_m^* \mathrm{E}[\,\mathrm{y}_{-k}\mathrm{y}_{-m}^*\,] \\
&\overset{(c)}{=} a_{\mathrm{x},0} - \sum_{k\in\mathbb{Z}} h_k^* c_{\mathrm{x,y},k} - \sum_{k\in\mathbb{Z}} h_k c_{\mathrm{x,y},k}^* + \sum_{k\in\mathbb{Z}}\sum_{m\in\mathbb{Z}} h_k h_m^* a_{\mathrm{y},m-k},
\end{aligned} \tag{E3.7-1}$$

where (a) follows from the definition of $\widehat{\mathrm{x}}$; (b) from the linearity of the expectation; and (c) from (3.234b) and (3.234c).

For simplicity, let us first look at the case where x and y are real-valued processes. Then, without loss of optimality, we may restrict h to have real coefficients, and the MSE (E3.7-1) simplifies to

$$\mathcal{E} = a_{\mathrm{x},0} - 2\sum_{k\in\mathbb{Z}} h_k c_{\mathrm{x,y},k} + \sum_{k\in\mathbb{Z}}\sum_{m\in\mathbb{Z}} h_k h_m a_{\mathrm{y},m-k}.$$

For any $n \in \mathbb{Z}$, differentiating gives

$$\frac{\partial \mathcal{E}}{\partial h_n} = -2c_{\mathrm{x,y},n} + 2\sum_{k\in\mathbb{Z}} h_k a_{\mathrm{y},n-k}, \tag{E3.7-2}$$

where the second term follows from

$$\frac{\partial}{\partial h_n} h_k h_m a_{\mathrm{y},m-k} = \begin{cases} 2h_n a_{\mathrm{y},0}, & \text{for } k = n \text{ and } m = n; \\ h_m a_{\mathrm{y},m-n}, & \text{for } k = n \text{ and } m \neq n; \\ h_k a_{\mathrm{y},n-k}, & \text{for } k \neq n \text{ and } m = n; \\ 0, & \text{for } k \neq n \text{ and } m \neq n, \end{cases}$$

and the symmetry of $a_{\mathrm{y},n}$. Setting the derivative in (E3.7-2) to zero for every $n \in \mathbb{Z}$ and recognizing the convolution in the second term of (E3.7-2) gives $h * a_{\mathrm{y}} = c_{\mathrm{x,y}}$ as desired.

The general (complex) case is similar. Let $h_n = a_n + jb_n$, where a_n and b_n are real sequences. Then, by applying

$$\frac{\partial \mathcal{E}}{\partial h_n} = \frac{\partial \mathcal{E}}{\partial a_n} + j\frac{\partial \mathcal{E}}{\partial b_n}$$

to (E3.7-1), one can arrive at (E3.7-2).

3.8. *Walsh–Hadamard transform*
Let $N = 2^n$. The Walsh matrix W_N of size $N \times N$ can be defined via the following recursion:

$$W_N = \begin{bmatrix} 1 & 1 \\ 1 & -1 \end{bmatrix} \otimes W_{N/2}, \qquad W_1 = \begin{bmatrix} 1 \end{bmatrix},$$

where $\otimes$ is the Kronecker product (3.313).

(i) Find W_2, W_4, and W_8.

(ii) Show that W_N is unitary up to a scale factor and find the scale factor.

(iii) Show that the infinite block matrix

$$T = \begin{bmatrix} W_1 & & & & \\ & 2^{-1/2}W_2 & & & \\ & & 2^{-1}W_4 & & \\ & & & 2^{-3/2}W_8 & \\ & & & & \ddots \end{bmatrix}$$

is unitary. Write out the upper-left corner of T.

(iv) The length-N Walsh–Hadamard transform of a vector is obtained by left-multiplication with W_N. Derive an algorithm that uses $N \log_2 N$ additions for a length-N transform.

Solution:

(i) The first three Walsh matrices are

$$W_2 = \begin{bmatrix} 1 & 1 \\ 1 & -1 \end{bmatrix},$$

$$W_4 = \begin{bmatrix} W_2 & W_2 \\ W_2 & -W_2 \end{bmatrix} = \begin{bmatrix} 1 & 1 & 1 & 1 \\ 1 & -1 & 1 & -1 \\ 1 & 1 & -1 & -1 \\ 1 & -1 & -1 & 1 \end{bmatrix},$$

$$W_8 = \begin{bmatrix} W_4 & W_4 \\ W_4 & -W_4 \end{bmatrix} = \begin{bmatrix} 1 & 1 & 1 & 1 & 1 & 1 & 1 & 1 \\ 1 & -1 & 1 & -1 & 1 & -1 & 1 & -1 \\ 1 & 1 & -1 & -1 & 1 & 1 & -1 & -1 \\ 1 & -1 & -1 & 1 & 1 & -1 & -1 & 1 \\ 1 & 1 & 1 & 1 & -1 & -1 & -1 & -1 \\ 1 & -1 & 1 & -1 & -1 & 1 & -1 & 1 \\ 1 & 1 & -1 & -1 & -1 & -1 & 1 & 1 \\ 1 & -1 & -1 & 1 & -1 & 1 & 1 & -1 \end{bmatrix}.$$

(ii) Because $W_1 = W_1^\top$ and $W_2 = W_2^\top$, one can recursively show that $W_N = W_N^\top$, so

$$\begin{aligned}
W_1^\top W_1 &= 1 = I_1, \\
W_2^\top W_2 &= \begin{bmatrix} 1 & 1 \\ 1 & -1 \end{bmatrix} \begin{bmatrix} 1 & 1 \\ 1 & -1 \end{bmatrix} = 2I_2, \\
W_4^\top W_4 &= \begin{bmatrix} W_2 & W_2 \\ W_2 & -W_2 \end{bmatrix} \begin{bmatrix} W_2 & W_2 \\ W_2 & -W_2 \end{bmatrix} = 4I_4, \\
&\ \vdots \\
W_N^\top W_N &= \begin{bmatrix} W_{N/2} & W_{N/2} \\ W_{N/2} & -W_{N/2} \end{bmatrix} \begin{bmatrix} W_{N/2} & W_{N/2} \\ W_{N/2} & -W_{N/2} \end{bmatrix} = NI_N.
\end{aligned}$$

Hence, $N^{-1/2}W_N$ is an $N \times N$ unitary matrix, for any $N = 2^n$, with $n \in \mathbb{N}$.

(iii) To show that T is unitary, write

$$T^\top T = \begin{bmatrix} W_1^\top W_1 & & & \\ & 2^{-1}W_2^\top W_2 & & \\ & & 2^{-2}W_4^\top W_4 & \\ & & & \ddots \end{bmatrix} = \begin{bmatrix} I_1 & & & \\ & I_2 & & \\ & & I_4 & \\ & & & \ddots \end{bmatrix} = I.$$

(iv) A length-N Walsh–Hadamard transform W_N is computed by evaluating

$$\begin{bmatrix} W_{N/2} & W_{N/2} \\ W_{N/2} & -W_{N/2} \end{bmatrix} \begin{bmatrix} X_N^{(0)} \\ X_N^{(1)} \end{bmatrix}.$$

From this, the addition cost ν_N of multiplying by W_N satisfies the recursion formula

$$\begin{aligned}
\nu_N &= 2\nu_{N/2} + 2\frac{N}{2} \\
&= 2\left(2\nu_{N/2^2} + 2\frac{N}{2^2}\right) + 2\frac{N}{2^1} \\
&\ \vdots \\
&= 2^n \nu_1 + 2^n \frac{N}{2^n} + \cdots + 2^1 \frac{N}{2^1} \\
&\overset{(a)}{=} N + (\log_2 N - 1)N = N \log_2 N,
\end{aligned}$$

where (a) follows from $\nu_1 = 1$.

Exercises

3.1. *Sinusoidal sequence.*(1)
Let x and y be the following sequences:

$$x_n = \sin\left(\frac{\pi}{8}n\right), \qquad y_n = x_n(u_n - u_{n-N}), \tag{P3.1-1}$$

where u_n is the Heaviside sequence.

(i) For $N = 8$, 12, and 16, sketch y_n.
(ii) For which of the above three values of N is

$$\sum_{k\in\mathbb{Z}} y_{n-Nk} = x_n$$

true?

3.2. *Deterministic autocorrelation and crosscorrelation.*(2)
Consider deterministic autocorrelation and crosscorrelation sequences and their DTFTs as in (3.100) and (3.103). Show the following:

(i) $a_n = a^*_{-n}$ as in (3.18a);
(ii) $|a_n| \le a_0$ for all $n \in \mathbb{Z}$;
(iii) though Hermitian symmetry $c_{x,y,n} = c^*_{y,x,-n}$ holds as in (3.21a), symmetry $c_{x,y,n} = c_{x,y,-n}$ does not hold in general; and
(iv) $C(e^{j\omega}) = X(e^{j\omega})\,Y^*(e^{j\omega})$.

3.3. *Discrete Laplacian operator.*(1)
Given is the following system:

$$y_n = x_{n-1} - 2x_n + x_{n+1}.$$

(i) Is this system linear? Is it shift-invariant? Is it causal? Is it memoryless? Is it BIBO-stable?
(ii) Does it have a matrix representation? If yes, write it down.
(iii) For each of the inputs,

$$x_{0,n} = c, \qquad x_{1,n} = \delta_n, \qquad x_{2,n} = u_n,$$

where δ_n is the Kronecker delta sequence and u_n is the Heaviside sequence, write and sketch the corresponding output of the system. Explain the effect of this system.

3.4. *Linear and shift-invariant difference equations.*(2)
Consider the difference equation (3.55).

(i) Show, possibly using a simple example, that, if the initial conditions are nonzero, the system is neither linear nor shift-invariant.
(ii) Show that, if the initial conditions are zero, then (a) the homogeneous solution is zero; (b) the system is linear; and (c) the system is shift-invariant.

3.5. *Geometric sequences and their properties.*(2)
Let x be a geometric sequence as in (3.7) with $\alpha \in \mathbb{R}$, $|\alpha| < 1$,

$$x_n = \begin{cases} 0, & \text{for } n < 0; \\ \gamma\alpha^n, & \text{for } n \ge 0. \end{cases}$$

(i) Show that $\gamma = \sqrt{1-\alpha^2}$ leads to a unit-norm sequence, that is, $\|x\| = 1$.
(ii) Compute the autocorrelation of this unit-norm geometric sequence.
(iii) Compute the convolution of this unit-norm geometric sequence with itself.
(iv) Let y be a different unit-norm geometric sequence. Compute the crosscorrelation between x and y, as well as their convolution.

3.6. *Circular convolution in frequency property of the DTFT.*(1)
Let x and h be sequences in $\ell^1(\mathbb{Z})$. Verify that the circular convolution in frequency property (3.98),

$$hx \quad \overset{\text{DTFT}}{\longleftrightarrow} \quad \frac{1}{2\pi} H \circledast X,$$

holds.

3.7. *Third-band filter.*(1)
Given is the following filter:

$$h_n \;=\; \sqrt{3}\,\frac{\sin(\frac{1}{3}\pi n)}{\pi n}.$$

(i) Find the DTFT of h. What kind of filter is it?
(ii) Given $x_n = \frac{1}{2}(\delta_n + \delta_{n-1})$ and $y = h * x$, sketch $|Y(e^{j\omega})|$.

3.8. *ROC of z-transform.*(1)
Compute the z-transforms and associated ROCs for the following sequences:

(i) δ_n;
(ii) δ_{n-k};
(iii) $\alpha^n u_n$;
(iv) $-\alpha^n u_{-n-1}$;
(v) $n\alpha^n u_n$;
(vi) $-n\alpha^n u_{-n-1}$;
(vii) $(\cos\omega_0 n)u_n$;
(viii) $(\sin\omega_0 n)u_n$; and
(ix) α^n for $0 \le n \le N$, and 0 otherwise.

3.9. *Orthogonality.*(2)
Let p be a sequence with z-transform $P(z)$. The goal is to find sequences satisfying the orthogonality constraint with respect to all shifts,

$$\langle p_n, p_{n-k}\rangle_n \;=\; \delta_k \quad \Leftrightarrow \quad P(z)P(z^{-1}) \;=\; 1. \tag{P3.9-1}$$

(i) If $P(z)$ is a polynomial, show that all solutions are of the form

$$P(z) \;=\; \pm z^{-\ell}, \qquad \ell \in \mathbb{Z}.$$

(ii) If $P(z)$ is a ratio of two polynomials, show that

$$P(z) \;=\; \frac{A(z)}{\widetilde{A}(z)},$$

where $A(z)$ is a polynomial of degree $(L-1)$ and $\widetilde{A}(z) = z^{-L+1}A(z^{-1})$ is a solution.

3.10. *Linear and circular convolution as polynomial products.*(3)
Given two polynomials of degree $N-1$,

$$A(z) \;=\; \sum_{n=0}^{N-1} a_n z^n, \qquad B(z) \;=\; \sum_{n=0}^{N-1} b_n z^n,$$

show that

$$D(z) \;=\; A(z)B(z) \bmod (z^N - 1) \;=\; \sum_{n=0}^{N-1} d_n z^n$$

is the z-transform of the circular convolution $d = a \circledast b$.
(*Hint:* The operation $A(z)B(z) \bmod (z^N-1)$ is the remainder of the division of $A(z)B(z)$ by (z^N-1).)

3.11. *Deterministic autocorrelation.*(2)
Let x be a real-valued sequence in $\ell^1(\mathbb{Z})$, and let a be its deterministic autocorrelation sequence.

(i) Show that the z-transform of a is $A(z) = X(z)X(z^{-1})$. Determine the region of convergence for $A(z)$.

(ii) Let $x_n = \alpha^n u_n$, with u the Heaviside sequence (3.11). Determine the region of convergence for $A(z)$. Find a by evaluating the inverse z-transform of $A(z)$.

(iii) Specify another sequence, y, that is not equal to x from (ii), but has the same deterministic autocorrelation sequence as that of x.

(iv) Specify a third sequence, v, that is not equal to either x or y but has the same deterministic autocorrelation sequence as that of x.

3.12. *Block circulant matrices.*(3)
A block-circulant matrix of size $NM \times NM$ is a circulant matrix of size $N \times N$ with elements that are blocks of size $M \times M$. Show that block-circulant matrices are block-diagonalized by block Fourier transforms of size $NM \times NM$ defined as

$$F = F_N \otimes I_M,$$

where F_N is an $N \times N$ Fourier matrix, I_M is an $M \times M$ identity matrix, and $\otimes$ is the Kronecker product (3.313).

3.13. *Pattern recognition.*(3)
In pattern recognition, it is sometimes useful to expand a sequence using the desired pattern and its shifts as basis sequences. Let x be a sequence of length N, and let p be a pattern of length N. As basis sequences choose p and its circular shifts,

$$\varphi_{k,n} = p_{(n-k) \bmod N}, \qquad k = 0, 1, \ldots, N-1.$$

(i) Derive a condition on p so that any x can be written as a linear combination of $\{\varphi_k\}_{K=0}^{N-1}$.

(ii) Assuming that the previous condition is met, find the coefficients α_k of the expansion

$$x_n = \sum_{k=0}^{N-1} \alpha_k \varphi_{k,n}.$$

3.14. *Computing linear convolution with the DFT.*(2)
Prove that the linear convolution of two sequences of length M and L can be computed using DFTs of length $N \geq M + L - 1$, and show how to do it.

3.15. *DFT properties.*(2)

(i) Prove the DFT pair property for time reversal (3.170).

(ii) Prove the DFT pair property for circular convolution in time (3.171).

(iii) Prove the DFT pair property for circular convolution in frequency (3.172).

(iv) Prove that the DFT of a real symmetric sequence (satisfying $x_n = x_{-n \bmod N}$) is real.

(v) Prove that the DFT of a real antisymmetric sequence (satisfying $x_n = -x_{-n \bmod N}$) is imaginary.

3.16. *Tight frames as projections from orthonormal bases.*(1)
We look at finite-dimensional frames.

(i) Let Φ be a frame in $\mathbb{R}^2$ formed from the following vectors:

$$\varphi_k = \begin{bmatrix} \Re\{W_N^k\} & \Im\{W_N^k\} \end{bmatrix}^\top, \qquad k = 0, 1, \ldots, N-1.$$

Plot the vectors of the frame on the unit circle for $N = 3$, 4, and 5.

(ii) Let Φ be a frame in $\mathbb{R}^M$ whose M rows are the first M columns of the DFT matrix (3.164a) normalized by $1/\sqrt{M}$. Show that Φ is a tight frame with redundancy factor N/M, that is,

$$\Phi\Phi^* = \frac{N}{M} I_M.$$

3.17. *Downsampling by N.*(1)
Prove the z-transform and the DTFT transform pairs for downsampling by N given by (3.190) and (3.191), respectively.

3.18. *Downsampling.*(2)
Given is a length-N sequence x. Let $N = m_0 M$, where m_0 and M are both positive integers, and let y be a sequence obtained by downsampling x by M, that is,

$$y_n = x_{Mn}, \qquad n = 0, 1, \ldots, m_0 - 1.$$

Let Y be the length-(N/M) DFT of y, and let X_k be the length-N DFT of x. Prove the following:

(i) For $M = 2$, the DFT of the downsampled sequence y is

$$Y_k = \frac{1}{2}\left(X_k + X_{k+N/2}\right), \qquad k = 0, 1, \ldots, \frac{N}{2} - 1.$$

(ii) For arbitrary M, the DFT of the downsampled sequence y is

$$Y_k = \frac{1}{M}\sum_{i=0}^{M-1} X_{k+iN/M}, \qquad k = 0, 1, \ldots, \frac{N}{M} - 1.$$

3.19. *Multirate system with different sampling rates.*(1)
Given is the following system:

$$y = HD_4GD_2FD_3x,$$

where the D_n are downsampling operators, and H, G, and F are filters.

(i) Draw a block diagram.

(ii) Derive an equivalent system consisting of a single filter block and a single downsampling operator. Write the z-transform of the equivalent system as a function of $F(z)$, $G(z)$, and $H(z)$.

3.20. *Multirate identities.*(2)

(i) Find the overall transfer function $Y(z)/X(z)$ of the system in Figure P3.20-1.

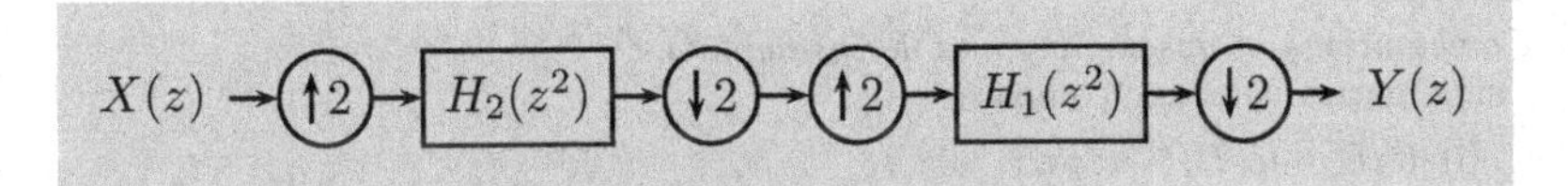

Figure P3.20-1 Multirate system.

(ii) In the system in Figure P3.20-2, if $H(z) = H_0(z^2) + z^{-1}H_1(z^2)$, prove that $Y(z) = X(z)H_0(z)$.

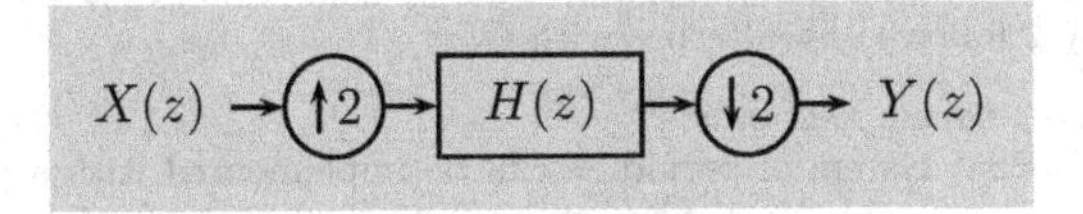

Figure P3.20-2 Multirate system.

(iii) Let $H(z)$, $F(z)$, and $G(z)$ be filters satisfying

$$H(z)G(z) + H(-z)G(-z) = 2, \qquad \text{(P3.20-1a)}$$

$$H(z)F(z) + H(-z)F(-z) = 0. \qquad \text{(P3.20-1b)}$$

Prove that $Y(z)/X(z) = 1$ for one of the systems in Figure P3.20-3, while $Y(z)/X(z) = 0$ for the other.

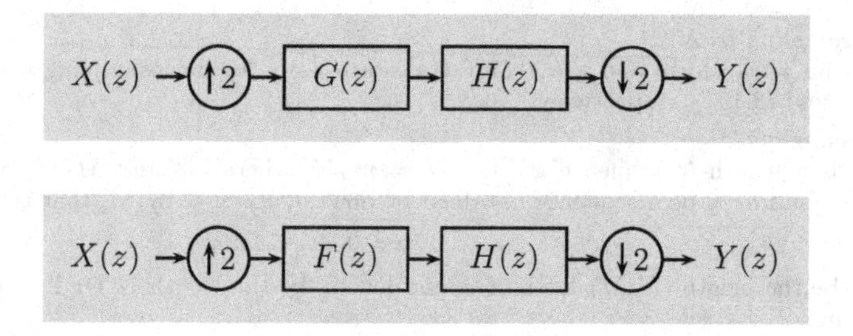

Figure P3.20-3 Multirate system.

3.21. *Interchange of multirate operations and filtering.*(2)
Consider the system given by the input–output relation

$$y = D_2AD_2AD_2Ax,$$

where A is a matrix representing a filter.

(i) With the use of identities for the interchange of multirate operations and filtering, find the simplest equivalent system, $y = D_N Hx$. Specify the downsampling factor N, and write H in the z-transform and Fourier domains.

(ii) If A is an ideal half-band lowpass filter, draw $|H(e^{j\omega})|$, clearly specifying the cutoff frequencies.

(iii) If A is an ideal half-band highpass filter, draw $|H(e^{j\omega})|$, clearly specifying the cutoff frequencies. Is this transfer function capturing the highest frequency content in the sequence x? Explain.

3.22. *Commutativity of upsampling and downsampling.*(3)
Prove that downsampling by M and upsampling by N commute if and only if M and N are coprime.

3.23. *Combinations of upsampling and downsampling.*(2)
Using matrix notation, compare

(i) U_3D_2x to D_2U_3x; and

(ii) U_4D_2x to D_2U_4x.

Explain the outcomes of these comparisons.

3.24. *Interchange of filtering and sampling rate change.*(2)

(i) Prove that downsampling by 2 followed by filtering with $\widetilde{G}(z)$ is equivalent to filtering with $\widetilde{G}(z^2)$ followed by downsampling by 2.

(ii) Prove that filtering with $G(z)$ followed by upsampling by 2 is equivalent to upsampling by 2 followed by filtering with $G(z^2)$.

3.25. *Periodically shift-varying systems.*(3)
Show that an LPSV system of period N can be implemented with a polyphase transform followed by upsampling by N, N filter operations, and a summation.

3.26. *Sequence with a zero-polyphase component.*(3)
Let $x \in \ell^2(\mathbb{Z})$ have all odd-indexed samples equal to 0. Prove that if its DTFT $X(e^{j\omega})$ is nonzero at $\omega = 0$, then it is nonzero at $\omega = \pi$ as well.

3.27. *Convolution and sum of discrete random variables.*(2)
A random variable is *discrete* when it takes values in a countable set. A discrete random variable x has a *probability mass function* (PMF) p_{x} defined by $p_{\mathrm{x}}(k) = \mathrm{P}(\mathrm{x} = k)$. The PMF of an integer-valued random variable is a nonnegative sequence in $\ell^1(\mathbb{Z})$. Let x and y be independent, integer-valued random variables with PMFs p_{x} and p_{y}. Show that $\mathrm{z} = \mathrm{x} + \mathrm{y}$ has PMF $p_{\mathrm{z}} = p_{\mathrm{x}} * p_{\mathrm{y}}$.

3.28. *Autocorrelation and crosscorrelation.*$^{(2)}$
Let x and w be jointly WSS stochastic processes that are uncorrelated ($c_{\mathrm{x,w},k} = 0$ for all $k \in \mathbb{Z}$). Let $\mathrm{y} = \mathrm{x} + \mathrm{w}$. Show that equations (3.262) hold.

3.29. *Toeplitz matrix–vector products.*$^{(3)}$
Given a size-$(N \times N)$ Toeplitz matrix T and a length-N vector x, show that the product Tx can be computed with $O(N \log_2 N)$ operations. The method includes extending T into a circulant matrix C. What is the minimum size of C, and how does it change if T is symmetric?

3.30. *Overlap–save convolution algorithm.*$^{(2)}$
Let h be a length-L filter, and let M and N be positive integers satisfying $N = L + M - 1$. Show that the factored form

$$y = E^{\top} H A^{\top} x$$

is a way to implement the convolution $y = h * x$, where E, H, and A are defined in Section 3.9.2. Multiplication by $A^{\top}$ forms blocks of length N from the input sequence x that *overlap* since the window advances by M samples per block. Multiplication by $E^{\top}$ *saves* the last M samples out of each block of $HA^{\top}x$ of length N.

3.31. *Sums and products.*$^{(1)}$
Compute the following:

(i) $\prod_{k=1}^{\infty} \exp\left(\frac{j2\pi}{k(k+1)}\right)$; and

(ii) $\sum_{k=0}^{1023} W_{16}^{k}$.

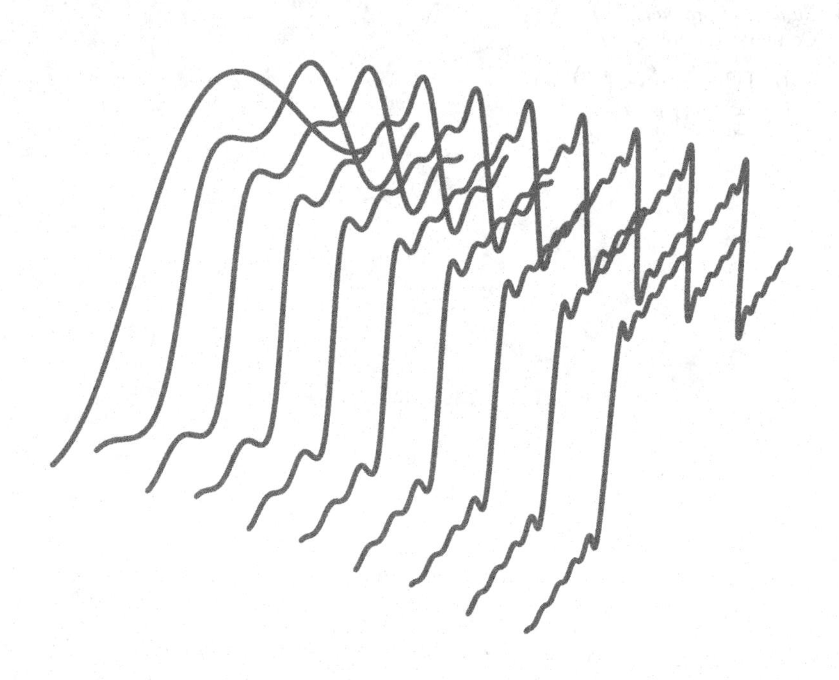

Chapter 4

Functions and continuous-time systems

"The profound study of nature is the most fertile source of mathematical discoveries."

— *Joseph Fourier*

Contents

In contrast to Chapter 3, *discrete time* is now replaced with *continuous time*. *Sequences* are replaced by *functions* defined on the domain of real numbers, which we associate with continuous time. As we saw in Chapter 2, these functions form the vector space $\mathbb{C}^{\mathbb{R}}$ (assuming that they are complex-valued). Operators that map a function to a function are called *continuous-time systems*.

As in Chapter 3, some important classes of functions and continuous-time systems have physical interpretations. Restricting functions to the normed vector spaces $\mathcal{L}^2(\mathbb{R})$ and $\mathcal{L}^\infty(\mathbb{R})$ corresponds to the physical properties of finite energy and boundedness. Shift-invariance of a continuous-time system sometimes follows directly from our assumption that physical laws do not change over time. A linear continuous-time system with the shift-invariance property is described by *convolution* with the system's *impulse response*. An impulse response is a function, and we will see that it is appropriate to require that it belong to $\mathcal{L}^1(\mathbb{R})$ or $\mathcal{L}^2(\mathbb{R})$. Once the convolution operation has been defined, spectral theory allows us to construct the *Fourier transform.*

The above discussion implicitly assumed that the underlying domain, time, is infinite. In practice we observe a finite portion of time. In Chapter 3, we dealt with this issue by assuming that a finite-length sequence was circularly extended, leading to the notion of *circular convolution* and the *discrete Fourier transform*; in this chapter, circularly extended finitely supported functions will also have an appropriate circular convolution as well as an appropriate Fourier transform, the mapping to *Fourier series coefficients.*

4.1 Introduction

In most of Chapter 3, we considered sequences; here, we look at functions defined for all times $t \in \mathbb{R}$. Such a function,

$$x(t), \qquad t \in \mathbb{R}, \tag{4.1}$$

could be the sound pressure sensed by a microphone, or the temperature at a sensor, etc.; the key is that a value exists at every time instant.

In real life, we observe only a finite portion of a function on the real line,

$$x(t), \qquad t \in [0, T). \tag{4.2}$$

Moreover, computations are always done on finite-length inputs, requiring a decision on how to treat the function at times that are not observed. As with sequences, there are two principal techniques: either set $x(t) = 0$ for all t outside of $[0, T)$ or extend x circularly (creating a periodic function in the process):

$$x(t+T) \;=\; x(t), \qquad t \in \mathbb{R}. \tag{4.3}$$

While finite-length functions in (4.2) and infinite-length periodic ones in (4.3) have a fundamentally different character, we will use the same types of tools to analyze them. Techniques designed explicitly for finite-length functions are mathematically rooted in treating the function as one period of an infinite-length periodic function. The consequences of this implicit periodization are central to signal processing.

As in Chapter 3, we thus define two broad classes of functions for which to develop our tools.

(i) *Functions on the real line* form the vector space $\mathbb{C}^{\mathbb{R}}$ of functions with the domain $\mathbb{R}$, as defined in (2.18c). The support of a function might be a proper subset of $\mathbb{R}$; for example, we will often consider functions on the real line that are nonzero only at nonnegative times.

(ii) *Functions on a finite interval,* without loss of generality, have support in $[0, T)$. The tools we will develop do not treat the vector space of functions with support in $[0, T)$ generically, but rather as functions defined on a circular domain.

The functions we consider are typically bounded, often smooth, and sometimes periodic.

An operator with a domain and codomain of continuous-time signals (that is, a vector space of functions) is called a *continuous-time system*. Many continuous-time systems, such as a microphone responding to pressure variations, are physical systems governed by differential equations; for example, the sound waves reaching a microphone obey the wave equation. Often, these differential equations have a smoothing effect, and functions with singularities (such as a point of discontinuity) are smoothed by the time they are observed. In the microphone example, a gunshot is first smoothed by the wave equation, and further smoothed by the microphone itself. We mention these effects to emphasize the difference between mathematical abstractions and observed phenomena. These differential equations are often linear, or even linear and shift-invariant; in Chapter 3, the same was true of difference equations.

Chapter outline

Section 4.2 discusses continuous-time signals, where we introduce function spaces of interest and comment on local and global smoothness. We follow with a short overview of continuous-time systems in Section 4.3, and in particular, linear, shift-invariant systems stemming from linear constant-coefficient differential equations. This discussion leads to the convolution operator and its properties, such as stability. Section 4.4 develops the Fourier transform and its properties. We emphasize the eigenfunction property of complex exponentials and give key relations of the Fourier transform, together with properties for certain function spaces. We briefly discuss the Laplace transform, an extension of the Fourier transform akin to the z-transform seen in the previous chapter, allowing us to deal with larger classes of functions. In Section 4.5, we discuss the natural orthonormal basis for periodic functions given by the Fourier series. We study circular convolution and the eigenfunction property of complex exponentials as well as properties of the Fourier series. Then, we explore the duality with the DTFT for sequences seen in Chapter 3. In Section 4.6, we study continuous-time stochastic processes and systems.

4.2 Functions

4.2.1 Functions on the real line

The set of functions in (4.1), where $x(t)$ is either real or complex, together with vector addition and scalar multiplication, forms a vector space (see Definition 2.1). The standard inner product between two functions on the real line was defined in (2.22c) and induces the standard $\mathcal{L}^2$ (or Euclidean) norm (2.26c). Other norms of interest are the $\mathcal{L}^1$ norm from (2.42a) with $p = 1$, and the $\mathcal{L}^\infty$ norm from (2.42b). We now look into a few spaces of interest.

Function spaces

Space of square-integrable functions $\mathcal{L}^2(\mathbb{R})$ The constraint of a finite square norm is necessary for turning the vector space $\mathbb{C}^{\mathbb{R}}$ defined in (2.18c) into the Hilbert space of *finite-energy* functions $\mathcal{L}^2(\mathbb{R})$. As for sequences, this space affords a geometric view, for example, the projection theorem (Theorem 2.26).

Space of bounded functions $\mathcal{L}^\infty(\mathbb{R})$ The space of bounded functions contains all functions $x(t)$ such that, for some finite M, $|x(t)| \leq M$ for all $t \in \mathbb{R}$. This space is denoted $\mathcal{L}^\infty(\mathbb{R})$ since it consists of functions with finite $\mathcal{L}^\infty$ norm.

Space of absolutely integrable functions $\mathcal{L}^1(\mathbb{R})$ The space of absolutely integrable functions consists of those with finite $\mathcal{L}^1$ norm.

The $\mathcal{L}^p$ spaces do not satisfy a nesting property as the ℓ^p sequence spaces do in (2.41). Therefore, when inclusion in different $\mathcal{L}^p$ spaces is needed to avoid technical difficulties, we restrict the discussion to the intersection of the spaces. For example, certain theorems apply only to functions that are both absolutely integrable and square integrable, that is, those belonging to $\mathcal{L}^1(\mathbb{R}) \cap \mathcal{L}^2(\mathbb{R})$.

Spaces of smooth functions To describe the global smoothness of a function, we use its continuity and the continuity of its derivative(s); these characterize the C^q spaces we saw in Section 2.2.4. Even a single point where the qth derivative does not exist or is not continuous prevents membership in C^q. Thus, global smoothness can fail to capture distinctions between important, frequently encountered types of functions and those that are quite esoteric. For example, the simple function $u(t) = 1$ for $t \geq 0$ and $u(t) = 0$ for $t < 0$ is infinitely differentiable at every nonzero t, but it fails to be even in C^0; in terms of global smoothness, it is no different than a function that is discontinuous everywhere. Therefore, to distinguish functions in terms of smoothness, we consider local smoothness as well.

In calculus, it is natural to look at differentiability at various points in the domain of the function. In signal processing, on the other hand, it is often preferable to define local smoothness using the global smoothness of a windowed version of a function. We illustrate this with an example.

Example 4.1 (Continuous and piecewise-linear function) Let $(x_0, x_1, \ldots, x_L)$ be a sequence of real numbers with $x_0 = x_L = 0$. Construct the function

$$x(t) = \begin{cases} x_n + (t-n)(x_{n+1} - x_n), & \text{for } n \leq t < n+1,\ n \in \{0, 1, \ldots, L-1\}; \\ 0, & \text{for } t \notin [0, L), \end{cases} \tag{4.4a}$$

a linear interpolation between the integer points as in Figure 4.1(a) such that $x(n) = x_n$. This function is in $\mathcal{L}^1(\mathbb{R})$, $\mathcal{L}^2(\mathbb{R})$, and $\mathcal{L}^\infty(\mathbb{R})$, since the sequence $\{x_n\}$ is finite and bounded. In terms of smoothness, looking only at a single linear piece, the function seems to be in C^∞, but, since the function is not differentiable at the integers, it is only in C^0.

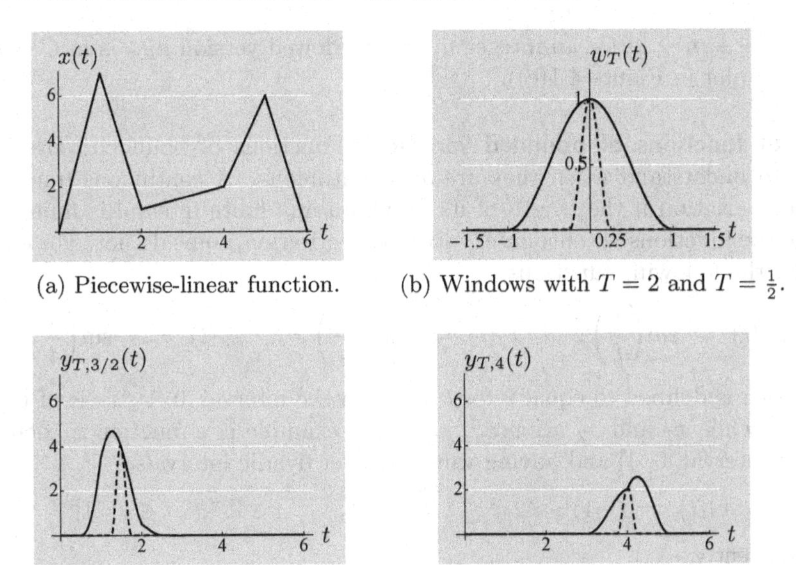

(a) Piecewise-linear function. (b) Windows with $T = 2$ and $T = \frac{1}{2}$.

(c) Windowed functions for $\tau = \frac{3}{2}$. (d) Windowed functions for $\tau = 4$.

Figure 4.1 (a) A piecewise-linear function $x(t)$ obtained by linearly interpolating a finite sequence of real values x_n via (4.4a). (b) Two different windows $w_T(t)$ from (4.4b) for widths $T = 2$ (solid line) and $T = \frac{1}{2}$ (dashed line). (c) Windowed versions $y_{2,3/2}$ (solid line) and $y_{1/2,3/2}$ (dashed line) obtained with the two different window widths and the same shift. Both are in C^0, but only $y_{1/2,3/2}$ is in C^1. (d) Windowed versions $y_{2,4}$ (solid line) and $y_{1/2,4}$ (dashed line), again obtained with the two window widths and the same shift. Both are in C^0, and neither is in C^1.

We investigate local smoothness using a window that is in C^1, for example,

$$w_T(t) = \begin{cases} \frac{1}{2}(1 + \cos(2\pi t/T)), & \text{for } |t| \le \frac{1}{2}T; \\ 0, & \text{otherwise,} \end{cases} \tag{4.4b}$$

where $T > 0$. The window support is of size T and centered around the origin. The window has one continuous derivative, that is, $w_T \in C^1$. Figure 4.1(b) shows w_T for two values of T.

For any fixed width parameter T and shift parameter τ, define the windowed version of $x(t)$ as

$$y_{T,\tau}(t) = x(t)\, w_T(t - \tau). \tag{4.4c}$$

The global smoothness of $y_{T,\tau}$ varies depending on the parameters T and τ and gives us information about the local smoothness of x. As a product of continuous functions, $y_{T,\tau}$ is always continuous (that is, it is in C^0). When $T > 1$, the support of the shifted window will always include at least one integer point, no matter what τ is; thus, $y_{T,\tau}$ will not be in C^1 (see the solid plots in Figures 4.1(c) and (d) for examples with $T = 2$). When $T < 1$, depending on T and τ, some of the windowed versions will be in C^1; for example, for $T = \frac{1}{2}$ and

$\tau \in [n + \frac{1}{4}, n + \frac{3}{4}]$ for an integer n, the windowed version $y_{T,\tau}$ is in C^1 (see the dashed plot in Figure 4.1(c)).

Space of functions of bounded variation Functions of bounded variation are easiest to understand when they are also continuous. A continuous function has bounded variation if the length of its graph on any finite interval is finite. While most of the functions we encounter satisfy this criterion, some do not. For example, consider the following functions:

$$x_1(t) = \sin\left(\frac{1}{t}\right), \qquad x_2(t) = t\sin\left(\frac{1}{t}\right), \qquad x_3(t) = t^2\sin\left(\frac{1}{t}\right),$$

where each is defined to equal 0 for $t = 0$. On the interval $[0, 1]$, x_3 is of bounded variation while x_1 and x_2 are not.[79] Another example is a function x_4 defined on the unit interval $[0, 1]$ and having value ± 1 over dyadic intervals,

$$x_4(t) = (-1)^i, \quad 2^{-i} \leq t < 2^{-i+1}, \qquad i \in \mathbb{Z}^+, \quad t \in [0, 1],$$

or, equivalently,

$$x_4(t) = (-1)^{\lceil \log_2(1/t) \rceil}.$$

This function is not of bounded variation either. All four functions are shown in Figure 4.2.

Formally, the total variation of a function x over $[a, b]$ is defined as

$$V_a^b(x) = \sup_{N \in \mathbb{N}} \sup_{t_0, t_1, \ldots, t_N} \sum_{k=0}^{N-1} |x(t_{k+1}) - x(t_k)|,$$

where the inner supremum is taken over all increasing sequences $(t_0, t_1, \ldots, t_N)$ in $[a, b]$. Then, a real-valued $x(t)$ is said to be of bounded variation over $[a, b]$ when $V_a^b(x)$ is finite.

Special functions

We now introduce certain functions often used in the book.

Heaviside function The *Heaviside* or *unit-step* function is defined as

$$u(t) = \begin{cases} 1, & \text{for } t \geq 0; \\ 0, & \text{otherwise,} \end{cases} \qquad t \in \mathbb{R}. \tag{4.5}$$

This function is bounded by 1, so it belongs to $\mathcal{L}^\infty(\mathbb{R})$. It belongs to neither $\mathcal{L}^1(\mathbb{R})$ nor $\mathcal{L}^2(\mathbb{R})$. The Dirac delta function defined in Appendix 3.A.4 and the Heaviside function are related via

$$u(t) = \int_{-\infty}^{t} \delta(\tau)\, d\tau, \qquad t \neq 0. \tag{4.6}$$

[79]The lack of bounded variation of x_1 on $[0, 1]$ follows immediately from $x_1(t)$ having infinitely many local extrema (of nondiminishing magnitude) as $t \to 0$. It is more subtle to show that $x_2(t)$ remains of unbounded variation despite the diminishing magnitudes of the extrema as $t \to 0$ and that $x_3(t)$ has bounded variation.

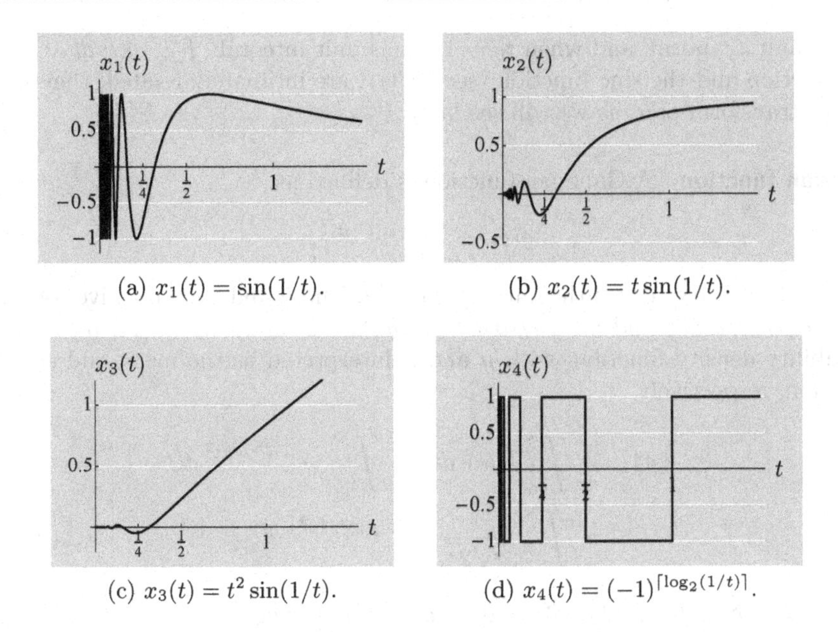

(a) $x_1(t) = \sin(1/t)$. (b) $x_2(t) = t\sin(1/t)$.

(c) $x_3(t) = t^2\sin(1/t)$. (d) $x_4(t) = (-1)^{\lceil \log_2(1/t) \rceil}$.

Figure 4.2 Functions illustrating the concept of bounded/unbounded variations. On the interval $[0, 1]$, only $x_3(t)$ is of bounded variation.

Pointwise multiplication by the Heaviside function implements the domain restriction operator (2.64) for restriction from all real numbers to just the nonnegative real numbers:

$$1_{\mathbb{R}^+} x \;=\; \begin{cases} x(t), & \text{for } t \geq 0; \\ 0, & \text{otherwise} \end{cases} \;=\; u(t)\,x(t), \qquad t \in \mathbb{R}.$$

From this we can also build other domain restriction operators. For example, the domain restriction to $[t_0, t_1)$ is achieved with a difference of two shifted Heaviside functions:

$$1_{[t_0,t_1)}\, x \;=\; (u(t-t_0) - u(t-t_1))\,x(t) \;=\; \begin{cases} x(t), & \text{for } t \in [t_0, t_1); \\ 0, & \text{otherwise.} \end{cases} \tag{4.7}$$

Box function For any positive real number t_0, the *centered* and *normalized box* function is given by

$$w(t) \;=\; \begin{cases} 1/\sqrt{t_0}, & \text{for } |t| \leq \frac{1}{2}t_0; \\ 0, & \text{otherwise.} \end{cases} \tag{4.8a}$$

This is a normalized indicator function of the interval $[-\frac{1}{2}t_0, \frac{1}{2}t_0]$,

$$w(t) \;=\; \frac{1}{\sqrt{t_0}} 1_{[-t_0/2,t_0/2]}(t). \tag{4.8b}$$

It is of unit $\mathcal{L}^2$ norm, and when $t_0 = 1$ it has unit integral: $\int_{-\infty}^{\infty} w(t)\, dt = 1$. The box function and the sinc function (see (3.9a)) are intimately related; they form a Fourier transform pair, as we will see later.

Gaussian function A Gaussian function is defined as

$$g(t) \;=\; \gamma e^{-\alpha(t-\mu)^2}, \tag{4.9a}$$

where μ shifts the center of the function to $t = \mu$, and α and γ are positive constants. When $\alpha = 1/(2\sigma^2)$ and $\gamma = 1/(\sigma\sqrt{2\pi})$, $\|g\|_1 = 1$, and thus g can be seen as a probability density function, with μ and σ interpreted as the mean and standard deviation, respectively,

$$\begin{aligned}\|g\|_1 \;&=\; \int_{-\infty}^{\infty} |g(t)|\, dt \;=\; \int_{-\infty}^{\infty} \gamma e^{-\alpha(t-\mu)^2}\, dt \\ &=\; \int_{-\infty}^{\infty} \frac{1}{\sigma\sqrt{2\pi}} e^{-(t-\mu)^2/(2\sigma^2)}\, dt \;=\; 1.\end{aligned} \tag{4.9b}$$

When $\gamma = (2\alpha/\pi)^{1/4}$, $\|g\|_2 = 1$; that is, g is of unit energy,

$$\begin{aligned}\|g\|_2 \;&=\; \left(\int_{-\infty}^{\infty} |g(t)|^2\, dt\right)^{1/2} \;=\; \left(\int_{-\infty}^{\infty} \gamma^2 e^{-2\alpha(t-\mu)^2}\, dt\right)^{1/2} \\ &=\; \left(\int_{-\infty}^{\infty} \sqrt{\frac{2\alpha}{\pi}} e^{-2\alpha(t-\mu)^2}\, dt\right)^{1/2} \;=\; 1.\end{aligned} \tag{4.9c}$$

Deterministic correlation

We now discuss two operations on functions, both deterministic, that appear throughout the chapter. These are analogous to the notions of deterministic autocorrelation and crosscorrelation for sequences defined in Section 3.2.1. Stochastic versions of both operations will be given in Section 4.6.1.

Deterministic autocorrelation The *deterministic autocorrelation* a of a function x is

$$a(\tau) \;=\; \int_{-\infty}^{\infty} x(t)\, x^*(t-\tau)\, dt \;=\; \langle x(t),\, x(t-\tau)\rangle_t, \qquad \tau \in \mathbb{R}. \tag{4.10}$$

The deterministic autocorrelation satisfies

$$a(\tau) \;=\; a^*(-\tau), \tag{4.11a}$$

$$a(0) \;=\; \int_{-\infty}^{\infty} |x(t)|^2\, dt \;=\; \|x\|^2, \tag{4.11b}$$

analogously to (3.18). The deterministic autocorrelation measures the similarity of a function with respect to shifts of itself, and it is Hermitian symmetric as in

(4.11a). For a real x,

$$a(\tau) \;=\; \int_{-\infty}^{\infty} x(t)\, x(t-\tau)\, dt \;=\; a(-\tau). \tag{4.11c}$$

When we need to specify the function involved, we write a_x.

Deterministic crosscorrelation The *deterministic crosscorrelation* c of two functions x and y is

$$c(\tau) \;=\; \int_{-\infty}^{\infty} x(t)\, y^*(t-\tau)\, dt \;=\; \langle x(t),\, y(t-\tau)\rangle_t, \qquad \tau \in \mathbb{R}, \tag{4.12}$$

and is written as $c_{x,y}$ when we want to specify the functions involved. It satisfies

$$c_{x,y}(\tau) \;=\; \left(\int_{-\infty}^{\infty} y(t-\tau)\, x^*(t)\, dt\right)^* \stackrel{(a)}{=} \left(\int_{-\infty}^{\infty} y(t')\, x^*(t'+\tau)\, dt'\right)^* \;=\; c_{y,x}^*(-\tau), \tag{4.13a}$$

where (a) follows from the change of variable $t' = t - \tau$. For real x and y,

$$c_{x,y}(\tau) \;=\; \int_{-\infty}^{\infty} x(t)\, y(t-\tau)\, dt \;=\; c_{y,x}(-\tau). \tag{4.13b}$$

4.2.2 Periodic functions

Periodic functions with period T satisfy

$$x(t+T) \;=\; x(t), \qquad t \in \mathbb{R}. \tag{4.14}$$

Such functions appear in many physical problems, most notably in the original work of Fourier on heat conduction in a circular wire. Such functions cannot have finite $\mathcal{L}^1$ or $\mathcal{L}^2$ norms unless they are zero almost everywhere. Instead, we consider functions with *one period* in $\mathcal{L}^1(\mathbb{R})$ or $\mathcal{L}^2(\mathbb{R})$; equivalently, the functions are in $\mathcal{L}^1([-\frac{1}{2}T,\, \frac{1}{2}T))$ or $\mathcal{L}^2([-\frac{1}{2}T,\, \frac{1}{2}T))$:[80]

$$\int_{-T/2}^{T/2} |x(t)|\, dt \;<\; \infty \qquad \text{or} \qquad \int_{-T/2}^{T/2} |x(t)|^2\, dt \;<\; \infty. \tag{4.15}$$

As we said earlier, another way to look at these functions is as functions on an interval, circularly extended, similarly to what we have seen in Chapter 3.

4.3 Systems

Continuous-time systems are operators having continuous-time signals (functions) as their inputs and outputs. Among all continuous-time systems, we will concentrate on those that are linear and shift-invariant. This subclass is both important in practice and amenable to analysis. After an introduction to differential equations, which are natural descriptions of continuous-time systems, we study linear, shift-invariant systems in detail.

[80] We write a half-closed interval $[-\frac{1}{2}T,\, \frac{1}{2}T)$, excluding $\frac{1}{2}T$, to emphasize that the function at $\frac{1}{2}T$ is determined by its value at $-\frac{1}{2}T$ by periodicity.

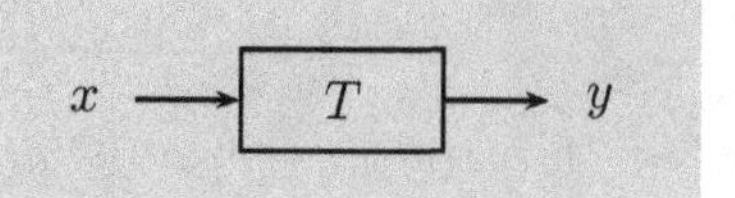

Figure 4.3 A continuous-time system.

4.3.1 Continuous-time systems and their properties

A continuous-time system is an operator T that maps an input function $x \in V$ into an output function $y \in V$,

$$y = T(x), \tag{4.16}$$

as shown in Figure 4.3. As we have seen in the previous section, the function space V is typically $\mathcal{L}^2(\mathbb{R})$ or $\mathcal{L}^\infty(\mathbb{R})$. At times, the input or the output is in a subspace of such spaces.

Types of systems

For each of the types of systems seen in the previous chapter, we have a continuous-time counterpart. As the concepts are identical, we list them with little elaboration. It is instructive to compare the discrete-time and continuous-time definitions.

Linear systems The definition of linearity of a continuous-time system is similar to Definition 2.17 of a linear operator and Definition 3.1 of a linear discrete-time system.

DEFINITION 4.1 (LINEAR SYSTEM) A continuous-time system T is called *linear* when, for any inputs x and y and any $\alpha, \beta \in \mathbb{C}$,

$$T(\alpha x + \beta y) = \alpha T(x) + \beta T(y). \tag{4.17}$$

The function T is thus a linear operator, and we write (4.16) as

$$y = Tx. \tag{4.18}$$

Memoryless systems The definition of a memoryless system closely follows Definition 3.2, with the system output at any instant depending only on the input at the same instant.

DEFINITION 4.2 (MEMORYLESS SYSTEM) A continuous-time system T is called *memoryless* when, for any real τ and inputs x and x',

$$1_{\{\tau\}} x = 1_{\{\tau\}} x' \quad \Rightarrow \quad 1_{\{\tau\}} T(x) = 1_{\{\tau\}} T(x'). \tag{4.19}$$

Causal systems The output of a causal system at time t depends on the input only up to time t. It follows that if two inputs agree up to time t, the corresponding outputs must agree up to time t.

DEFINITION 4.3 (CAUSAL SYSTEM) A continuous-time system T is called *causal* when, for any real τ and inputs x and x',

$$1_{(-\infty,\tau]}\, x \;=\; 1_{(-\infty,\tau]}\, x' \quad\Rightarrow\quad 1_{(-\infty,\tau]}\, T(x) \;=\; 1_{(-\infty,\tau]}\, T(x'). \tag{4.20}$$

As was discussed in Section 3.3.1, causality can seem to be a property that is required of any real system. When the time variable literally represents time and the same time origin is used for the input and output functions, causality is indeed necessary for accurate models of physical systems.

Shift-invariant systems In a shift-invariant system, shifting the input has the effect of shifting the output by the same amount.

DEFINITION 4.4 (SHIFT-INVARIANT SYSTEM) A continuous-time system T is called *shift-invariant* when, for any real τ and input x,

$$y \;=\; T(x) \Rightarrow y' \;=\; T(x'), \quad \text{where } x'(t) = x(t-\tau) \text{ and } y'(t) = y(t-\tau). \tag{4.21}$$

As in Chapter 3, *linear shift-invariant (LSI)* or *linear time-invariant (LTI)* systems have desirable mathematical properties.

Stable systems As for discrete-time systems, we consider *bounded-input, bounded-output (BIBO)* stability exclusively.

DEFINITION 4.5 (BIBO-STABLE SYSTEM) A continuous-time system T is called *bounded-input, bounded-output stable* when a bounded input x produces a bounded output $y = T(x)$:

$$x \in \mathcal{L}^\infty(\mathbb{R}) \quad\Rightarrow\quad y \in \mathcal{L}^\infty(\mathbb{R}). \tag{4.22}$$

Basic systems

We now discuss a few basic continuous-time systems.

Shift The shift-by-t_0 operator is defined as

$$y(t) \;=\; x(t-t_0), \qquad t \in \mathbb{R}, \tag{4.23}$$

and simply delays $x(t)$ by t_0. It is an LSI operator, causal (for $t_0 \geq 0$) and BIBO-stable, but not memoryless. While this is one of the simplest continuous-time systems, it is also the most important, as the entire concept of time processing is based on this simple operator. Compare this continuous-time shift operator with the discrete-time one defined in (3.39).

Modulator While the shift we just saw is the shift in time, modulation is shift in frequency (as we will see later in this chapter). A modulation by a complex exponential of frequency ω_0 is given by

$$y(t) = e^{j\omega_0 t} x(t), \qquad t \in \mathbb{R}. \tag{4.24}$$

This operator is linear, causal, memoryless, and BIBO-stable, but not shift-invariant. For those already familiar with Fourier analysis, (4.24) shifts the spectrum of x to the position ω_0 in frequency. Compare this continuous-time modulation operator with the discrete-time one defined in (3.41).

Integrator Similarly to the discrete-time accumulator (3.43), an integrator sums up the inputs up to the present time,

$$y(t) = \int_{-\infty}^{t} x(\tau)\, d\tau, \qquad t \in \mathbb{R}. \tag{4.25}$$

This is an LSI, causal operator, but it is neither memoryless nor BIBO-stable.

Averaging operators As in (3.47), for any fixed $T > 0$, we could consider a system that takes an average of the input,

$$y(t) = \frac{1}{T} \int_{t-T/2}^{t+T/2} x(\tau)\, d\tau, \qquad t \in \mathbb{R}. \tag{4.26}$$

This is a *moving-average* filter since we look at the function through a window of length T. This operator is LSI and BIBO-stable, but it is neither memoryless nor causal.

We could obtain a causal version by simply delaying the moving average in (4.26) by $\frac{1}{2}T$, resulting in

$$y(t) = \frac{1}{T} \int_{t-T}^{t} x(\tau)\, d\tau, \qquad t \in \mathbb{R}. \tag{4.27}$$

This operator is again LSI and BIBO-stable but it is also causal, while still not being memoryless.

Maximum operator This simple operator computes the maximum value of the input up to the current time,

$$y(t) = \max(1_{\{-\infty,\ldots,t\}}\, x). \tag{4.28}$$

This operator is clearly neither linear nor memoryless, but it is causal, shift-invariant, and BIBO-stable.

4.3.2 Differential equations

In Chapter 3, we have examined the basic principles behind difference equations. Just as in discrete time, where a linear difference equation relates an input sequence and an output sequence, in continuous time, a linear differential equation relates an input function and an output function. In particular, a linear constant-coefficient differential equation (compare with a linear constant-coefficient difference equation in (3.55)) describes an LSI system and is of the form

$$y(t) = \sum_{k=0}^{M} b_k \frac{d^k x(t)}{dt^k} - \sum_{k=1}^{N} a_k \frac{d^k y(t)}{dt^k}. \tag{4.29}$$

To find the solution, we follow the procedure outlined in Section 3.A.2: we find a solution $y^{(\mathrm{h})}(t)$ to the homogeneous equation obtained by setting the input $x(t)$ in (4.29) to zero; we then find any particular solution $y^{(\mathrm{p})}(t)$ to (4.29), typically by assuming the output to be of the same form as the input; and finally, the complete solution is formed by superposition of the solution to the homogeneous equation and the particular solution. The coefficients in the homogeneous solution are found by specifying initial conditions for $y(t)$ and then solving the system. A standard way of finding solutions to differential equations is to use Fourier and Laplace transforms.

4.3.3 Linear shift-invariant systems

Impulse response

The impulse response of an LSI continuous-time system is defined with the Dirac delta function playing the role that the Kronecker delta sequence plays in discrete time. The impulse response is sufficient for specifying the discrete-time system completely; this is clear from the fact that the Kronecker delta sequence and its shifts form a basis for $\ell^2(\mathbb{Z})$. While we cannot create a basis for $\mathcal{L}^2(\mathbb{R})$ with the Dirac delta function and its shifts, we will see shortly that an expression analogous to (3.59) leads again to specifying an LSI system through convolution with the impulse response.

DEFINITION 4.6 (IMPULSE RESPONSE) A function h is called the *impulse response* of LSI continuous-time system T when input δ produces output h.

The impulse response h of a causal linear system always satisfies $h(t) = 0$ for all $t < 0$. This is required because, according to (4.20), the output in response to input δ must match on $(-\infty, 0)$ to the $\mathbf{0}$ output function that results from the $\mathbf{0}$ input function.

Convolution

To parallel (3.59), we can express an arbitrary input x to an LSI system T as

$$x(t) = \int_{-\infty}^{\infty} x(\tau)\delta(t-\tau)\,d\tau, \tag{4.30}$$

where the equality holds for all $t \in \mathbb{R}$ at which x is continuous (see the sifting property of the Dirac delta function in (3.293)).[81] Then, the output resulting from this input is

$$y = Tx = T\int_{-\infty}^{\infty} x(\tau)\,\delta(t-\tau)\,d\tau \stackrel{(a)}{=} \int_{-\infty}^{\infty} x(\tau)\,T\delta(t-\tau)\,d\tau$$
$$\stackrel{(b)}{=} \int_{-\infty}^{\infty} x(\tau)\,h(t-\tau)\,d\tau = h * x,$$

where (a) follows from linearity; and (b) from shift invariance and the definition of the impulse response, defining the *convolution.*

DEFINITION 4.7 (CONVOLUTION) The *convolution* between functions h and x is defined as

$$(Hx)(t) = (h*x)(t) = \int_{-\infty}^{\infty} x(\tau)h(t-\tau)\,d\tau = \int_{-\infty}^{\infty} x(t-\tau)h(\tau)\,d\tau, \tag{4.31}$$

where H is called the *convolution operator* associated with h.

Properties The convolution (4.31) satisfies the following properties:

(i) *Connection to the inner product:*

$$(h*x)(t) = \int_{-\infty}^{\infty} x(\tau)h(t-\tau)\,d\tau = \langle x(\tau),\, h^*(t-\tau)\rangle_\tau. \tag{4.32a}$$

(ii) *Commutativity:*

$$h*x = x*h. \tag{4.32b}$$

(iii) *Associativity:*

$$g*(h*x) = g*h*x = (g*h)*x. \tag{4.32c}$$

---

[81] Since we are mostly interested in equalities in the $\mathcal{L}^2$ sense, we neglect what happens exactly at points of discontinuity of x. When x has a countable number of points of discontinuity, we can assign an arbitrary value (such as zero) to x at those points of discontinuity without changing the integrals in (4.31).

(iv) *Deterministic autocorrelation:*

$$a(\tau) = \int_{-\infty}^{\infty} x(t)\, x^*(t-\tau)\, dt = x(t) *_t x^*(-t). \tag{4.32d}$$

(v) *Shifting:* For any $\tau \in \mathbb{R}$,

$$x(t) *_t \delta(t-\tau) = x(t-\tau). \tag{4.32e}$$

These properties have discrete-time counterparts in (3.62) and are explored further in Solved exercise 4.2. Note that properties (i)–(iv) above depend on the integrals – whether written explicitly or implicitly – being well defined. We will not dwell on these technicalities; Appendix 3.A.3, though focused only on discrete time, has a related discussion.

Filters As in Chapter 3, the impulse response of a system is often called a *filter* and convolution with the impulse response is called *filtering.*

Stability Similarly to the absolute summability encountered in Chapter 3, BIBO stability of a continuous-time system is equivalent to the absolute integrability of its impulse response. The proof is similar to that for the discrete-time case (see Theorem 3.8) and is left for Exercise 4.2.

THEOREM 4.8 (BIBO STABILITY) An LSI system is BIBO-stable if and only if its impulse response is absolutely integrable.

If the input to a BIBO-stable system is in $\mathcal{L}^p(\mathbb{R})$, the output is in $\mathcal{L}^p(\mathbb{R})$ as well. This is a continuous-time analogue to Solved exercise 3.2, and can also be proven using Hölder's inequality.

Smoothing One key feature of many convolution operators is their smoothing effect. For example, when the impulse response has a nonzero mean (zeroth moment, see (4.60a)), the convolution will compute a local average, as we now show.

EXAMPLE 4.2 (LOCAL SMOOTHING BY CONVOLUTION) Choose the box function of width t_0, (4.8a), as the impulse response h, and a piecewise-constant function over integer intervals,

$$x(t) = x_n, \qquad \text{for } t \in [n,\, n+1), \quad n \in \mathbb{Z},$$

as the input (for some sequence x_n). The convolution $y = h * x$ is continuous for any $t_0 > 0$. For example, the output with $t_0 = 1$ is

$$y(t) = x_n + \left(t - n - \frac{1}{2}\right)(x_{n+1} - x_n), \qquad \text{for } t \in [n+\tfrac{1}{2},\, n+\tfrac{3}{2}], \quad n \in \mathbb{Z},$$

which is piecewise-linear and continuous. Thus, thanks to a smoothing impulse response, a discontinuous function x is transformed into a C^0 function y.

Adjoint Like in discrete time, the adjoint of a convolution operator is another convolution operator, where the filter is time-reversed and conjugated: If H is the convolution operator associated with the filter h, then its adjoint G is the convolution operator associated with the filter $g(t) = h^*(-t)$. Formal verification of this fact is left for Exercise 4.3.

Circular convolution

We now consider what happens with our second class of functions, those that are of finite length and circularly extended.

Linear convolution with circularly extended signal Given a bounded periodic function x as in (4.14) and a filter with impulse response h in $\mathcal{L}^1(\mathbb{R})$, we can compute the convolution as usual,

$$y(t) = (h * x)(t) = \int_{-\infty}^{\infty} x(\tau)h(t-\tau)\,d\tau = \int_{-\infty}^{\infty} h(\tau)x(t-\tau)\,d\tau. \tag{4.33}$$

Since x is T-periodic, y is T-periodic as well,

$$y(t+T) = \int_{-\infty}^{\infty} h(\tau)x(t+T-\tau)\,d\tau \overset{(a)}{=} \int_{-\infty}^{\infty} h(\tau)x(t-\tau)\,d\tau = y(t),$$

where (a) follows from the periodicity of x.

Let us now define a periodized version of h, with period T, as

$$h_T(t) = \sum_{n\in\mathbb{Z}} h(t-nT), \tag{4.34}$$

where the sum converges for each t because $h \in \mathcal{L}^1(\mathbb{R})$. We now want to show how we can express the convolution (4.33) in terms of what we will define as a *circular* convolution:

$$\begin{aligned}
(h * x)(t) &= \int_{-\infty}^{\infty} h(\tau)x(t-\tau)\,d\tau \overset{(a)}{=} \sum_{n\in\mathbb{Z}} \int_{nT-T/2}^{nT+T/2} h(\tau)x(t-\tau)\,d\tau \\
&\overset{(b)}{=} \sum_{n\in\mathbb{Z}} \int_{-T/2}^{T/2} h(\tau'+nT)x(t-\tau'-nT)\,d\tau' \\
&\overset{(c)}{=} \sum_{n\in\mathbb{Z}} \int_{-T/2}^{T/2} h(\tau+nT)x(t-\tau)\,d\tau \\
&\overset{(d)}{=} \int_{-T/2}^{T/2} \underbrace{\sum_{n\in\mathbb{Z}} h(\tau+nT)}_{h_T(\tau)}\, x(t-\tau)\,d\tau \\
&= \int_{-T/2}^{T/2} h_T(\tau)x(t-\tau)\,d\tau = (h_T \circledast x)(t),
\end{aligned} \tag{4.35}$$

where (a) follows from splitting the real line into length-T segments; (b) from the change of variable $\tau' = \tau - nT$; (c) from the periodicity of x and the change of variable $\tau = \tau'$; and (d) from (2.209). The expression above tends to be more convenient as it involves only one period of both x and the periodized version h_T of the impulse response h.

Definition of the circular convolution In computing the convolution of a periodic function x with an impulse response $h \in \mathcal{L}^1(\mathbb{R})$, we implicitly defined the *circular convolution* of a T-periodic function x and a T-periodic impulse response h.

DEFINITION 4.9 (CIRCULAR CONVOLUTION) The *circular convolution* between T-periodic functions h and x is defined as

$$(Hx)(t) = (h \circledast x)(t) = \int_{-T/2}^{T/2} x(\tau)h(t-\tau)\,d\tau = \int_{-T/2}^{T/2} x(t-\tau)h(\tau)\,d\tau, \quad (4.36)$$

where H is called the *circular convolution operator* associated with h.

The result of the circular convolution is again T-periodic. While circular convolution is a separate concept from linear convolution, we have just seen that the two are related when one function in a linear convolution is periodic and the other is not. We made the connection by periodizing the aperiodic function.

4.4 Fourier transform

As discussed in Section 3.4, the ubiquity of the Fourier transform is mostly due to the fact that complex exponentials are eigenfunctions of LSI systems (convolution operators). As in discrete time, this leads to the *convolution property* – an equivalence between convolving functions and multiplying Fourier transforms of the functions. This is also interpreted as diagonalization of convolution operators by the Fourier transform.

4.4.1 Definition of the Fourier transform

Eigenfunctions of the convolution operator LSI systems have all unit-modulus complex exponential functions as eigenfunctions. As in discrete time, this follows from the convolution representation of LSI systems (4.31) and a simple computation.

Consider a complex exponential function

$$v(t) = e^{j\omega t}, \qquad t \in \mathbb{R}, \quad (4.37)$$

where ω is any real number. The quantity ω is called the *angular frequency*; it is measured in radians per second. With $\omega = 2\pi f$, the quantity f is called *frequency*; it is is measured in hertz, or the number of cycles per second. The function v is bounded since $|v(t)| = 1$ for all $t \in \mathbb{R}$. If the impulse response h is in $\mathcal{L}^1(\mathbb{R})$, according to Theorem 4.8, the output $h * v$ is bounded as well. Along with being bounded, $h * v$ takes a particular form:

$$(Hv)(t) = (h * v)(t) = \int_{-\infty}^{\infty} v(t-\tau)h(\tau)\,d\tau = \int_{-\infty}^{\infty} e^{j\omega(t-\tau)}h(\tau)\,d\tau$$
$$= \underbrace{\int_{-\infty}^{\infty} h(\tau)e^{-j\omega\tau}\,d\tau}_{\lambda_\omega} \underbrace{e^{j\omega t}}_{v(t)}. \tag{4.38}$$

This shows that applying the convolution operator H to the complex exponential function v gives a scalar multiple of v; in other words, v is an eigenfunction of H with the corresponding eigenvalue λ_ω. We denote this eigenvalue by $H(\omega)$ using the *frequency response* of the system, which is defined formally in (4.79a). We can thus rewrite (4.38) as

$$Hv = h * v = H(\omega)\,v. \tag{4.39}$$

Fourier transform We are now ready to define the Fourier transform, which amounts to projecting onto the subspaces generated by each of the eigenfunctions.

DEFINITION 4.10 (FOURIER TRANSFORM) The *Fourier transform* of a function x is

$$X(\omega) = \int_{-\infty}^{\infty} x(t)e^{-j\omega t}\,dt, \qquad \omega \in \mathbb{R}. \tag{4.40a}$$

It exists when (4.40a) is defined and is finite for all $\omega \in \mathbb{R}$; we then call it the *spectrum* of x. The *inverse Fourier transform* of X is

$$x(t) = \frac{1}{2\pi}\int_{-\infty}^{\infty} X(\omega)e^{j\omega t}\,d\omega, \qquad t \in \mathbb{R}. \tag{4.40b}$$

It exists when (4.40b) is defined and is finite for all $t \in \mathbb{R}$. When the Fourier transform and inverse Fourier transform exist, we denote the Fourier transform pair as

$$x(t) \stackrel{\mathrm{FT}}{\longleftrightarrow} X(\omega).$$

Note that the integral in (4.40a) is formally equivalent to an $\mathcal{L}^2(\mathbb{R})$ inner product, although the function $e^{j\omega t}$ has no decay and thus is not in $\mathcal{L}^2(\mathbb{R})$. We now discuss limitations on the inputs and corresponding interpretations of the Fourier transform. Since the transform (4.40a) and inverse transform (4.40b) differ only by a constant factor and reversal of time, the technical conditions on their existence are identical.

4.4.2 Existence and inversion of the Fourier transform

We will use the Fourier transform as an analysis tool without getting bogged down in technicalities. First, we work toward a basic understanding of how certain restrictions are needed for statements to be mathematically rigorous. Integration has

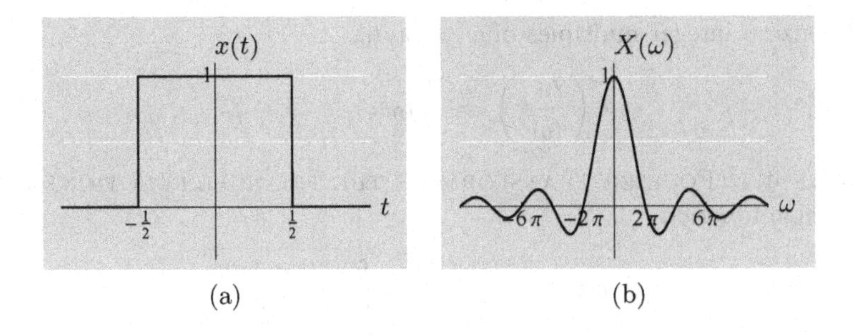

Figure 4.4 (a) The box function (4.8a) for $t_0 = 1$. (b) Its Fourier transform.

more technical subtleties than summation; accordingly there are more subtleties in the interpretation of the Fourier transform than there are in that of the DTFT, and we leave more of the details to the advanced texts listed in the *Further reading*.

Similarly to the development in Section 3.4.2, the existence of the Fourier transform is immediate for functions in $\mathcal{L}^1(\mathbb{R})$, but it is $\mathcal{L}^2(\mathbb{R})$ that allows the use of all the tools of Hilbert space theory; thus, we could conservatively restrict attention exclusively to functions in $\mathcal{L}^1(\mathbb{R}) \cap \mathcal{L}^2(\mathbb{R})$.[82] At the other extreme, certain Fourier transform expressions arise even when no rigorous sense of existence of the Fourier transform applies, and we treat these as useful to develop intuition rather than as mathematical statements.

Functions in $\mathcal{L}^1(\mathbb{R})$ If x is in $\mathcal{L}^1(\mathbb{R})$, then the integral in (4.40a) is defined and finite for all $\omega \in \mathbb{R}$.[83] Moreover, the Fourier transform X is bounded and continuous (see Solved exercise 4.3). We illustrate these facts through two examples.

EXAMPLE 4.3 (FOURIER TRANSFORM OF THE BOX FUNCTION) Let x be the box function from (4.8a). It is in $\mathcal{L}^1(\mathbb{R})$, so there is no question of whether its Fourier transform exists. The Fourier transform of x is

$$\begin{aligned} X(\omega) &= \frac{1}{\sqrt{t_0}} \int_{-t_0/2}^{t_0/2} e^{-j\omega t}\, dt = -\frac{1}{j\sqrt{t_0}\omega} e^{-j\omega t}\Big|_{-t_0/2}^{t_0/2} \\ &= \frac{e^{jt_0\omega/2} - e^{-jt_0\omega/2}}{j\omega\sqrt{t_0}} = \sqrt{t_0}\,\mathrm{sinc}\left(\frac{1}{2}t_0\omega\right). \end{aligned} \tag{4.41}$$

The function and its Fourier transform for $t_0 = 1$ are shown in Figure 4.4.

As is guaranteed by x being in $\mathcal{L}^1(\mathbb{R})$, the Fourier transform X is bounded and continuous. In this case, X is in $\mathcal{L}^2(\mathbb{R})$ but not in $\mathcal{L}^1(\mathbb{R})$.

Using (3.9c), we see that the Fourier transform of the box function is zero

[82] Recall from (2.41) that $\ell^1(\mathbb{Z}) \subset \ell^2(\mathbb{Z})$, so $\ell^1(\mathbb{Z}) \cap \ell^2(\mathbb{Z}) = \ell^1(\mathbb{Z})$, but no analogous inclusion holds for $\mathcal{L}^1(\mathbb{R})$ and $\mathcal{L}^2(\mathbb{R})$.

[83] Formally, this is because Lebesgue integrability of $|x(t)|$ implies the Lebesgue integrability of $x(t)e^{j\omega t}$ for every $\omega \in \mathbb{R}$.

at all nonzero integer multiples of $\omega = 2\pi/t_0$,

$$X\left(\frac{2\pi}{t_0}k\right) = \sqrt{t_0}\delta_k, \qquad k \in \mathbb{Z}. \tag{4.42}$$

EXAMPLE 4.4 (FOURIER TRANSFORM OF THE TRIANGLE FUNCTION) Let x be the triangle function,

$$x(t) = \begin{cases} 1-|t|, & \text{for } |t|<1; \\ 0, & \text{otherwise.} \end{cases} \tag{4.43}$$

Its Fourier transform is

$$X(\omega) = \int_{-1}^{0} (1+t)e^{-j\omega t}\,dt + \int_{0}^{1} (1-t)e^{-j\omega t}\,dt. \tag{4.44a}$$

Pulling the constant terms together gives

$$\int_{-1}^{1} e^{-j\omega t}\,dt = -\frac{1}{j\omega}\,e^{-j\omega t}\Big|_{-1}^{1} = \frac{e^{j\omega}-e^{-j\omega}}{j\omega}. \tag{4.44b}$$

The remaining part of the second integral in (4.44a) is

$$\begin{aligned}\int_{0}^{1} -te^{-j\omega t}\,dt &\stackrel{(a)}{=} \frac{1}{\omega^2}\int_{0}^{-j\omega} ue^{u}\,du \stackrel{(b)}{=} \frac{1}{\omega^2}(u-1)e^{u}\Big|_{0}^{-j\omega} \\ &= \frac{e^{-j\omega}}{j\omega} - \frac{e^{-j\omega}}{\omega^2} + \frac{1}{\omega^2},\end{aligned} \tag{4.44c}$$

where (a) follows from the change of variable $u = -j\omega t$; and (b) from the fact that a primitive of ue^u is $(u-1)e^u$.[84] By similar arguments, the remaining part of the first integral in (4.44a) is

$$\int_{-1}^{0} te^{-j\omega t}\,dt = -\frac{e^{j\omega}}{j\omega} - \frac{e^{j\omega}}{\omega^2} + \frac{1}{\omega^2}. \tag{4.44d}$$

Summing (4.44b)–(4.44d) completes the computation of $X(\omega)$ in (4.44a),

$$X(\omega) = \frac{1}{\omega^2}(2-e^{j\omega}-e^{-j\omega}) = \left(\frac{e^{j\omega/2}-e^{-j\omega/2}}{j\omega}\right)^2 = \operatorname{sinc}^2\left(\frac{1}{2}\omega\right). \tag{4.45}$$

The triangle function and its Fourier transform are shown in Figure 4.5.

As is guaranteed by x being in $\mathcal{L}^1(\mathbb{R})$, the Fourier transform X is bounded and continuous. In this case, X is in both $\mathcal{L}^1(\mathbb{R})$ and $\mathcal{L}^2(\mathbb{R})$.

The triangle function (4.43) is the convolution of two box functions from (4.8a) with $t_0 = 1$. Its Fourier transform (4.45) is the square of the Fourier transform of such a box function from (4.41). This is a preview of the convolution property (4.62): the Fourier transform of a convolution is the product of the Fourier transforms.

[84] A function $x(t)$ is called a *primitive* of $y(t)$ when $x'(t) = y(t)$. In this context, a synonym for primitive is *antiderivative*. We will often denote a primitive function with a superscript $^{(1)}$, as in $y^{(1)}$ for the primitive of y.

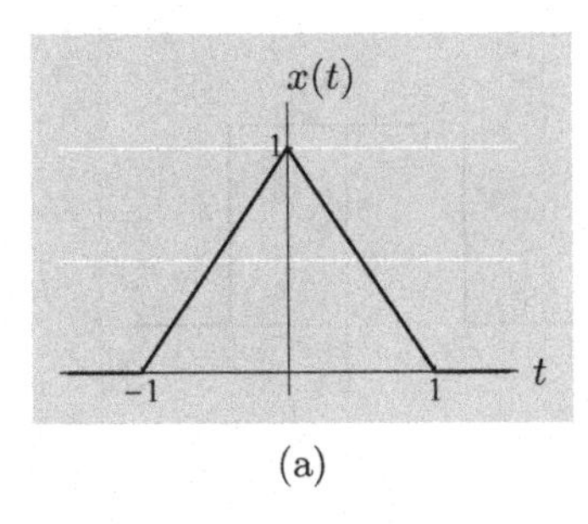

(a)

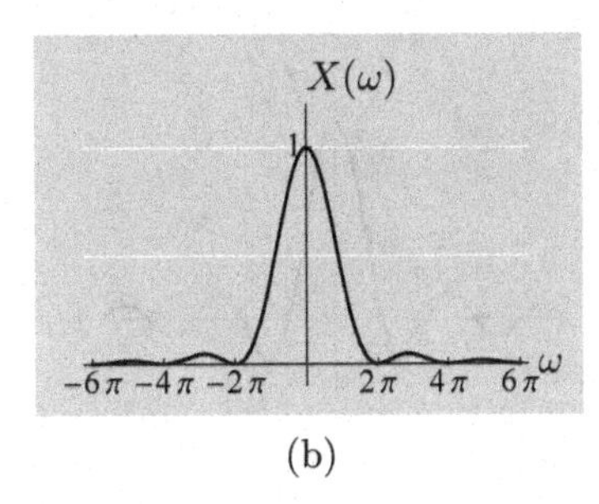

(b)

Figure 4.5 (a) The triangle function (4.43). (b) Its Fourier transform.

The box function (Example 4.3) illustrates an issue that does not arise with the triangle function (Example 4.4): a function x in $\mathcal{L}^1(\mathbb{R})$ can have a Fourier transform X that is not in $\mathcal{L}^1(\mathbb{R})$; thus, inversion by evaluation of (4.40b) is not guaranteed to work. This is part of what is alleviated with the extension of (4.40) to apply to any function in $\mathcal{L}^2(\mathbb{R})$.

Functions in $\mathcal{L}^2(\mathbb{R})$ If x is in $\mathcal{L}^2(\mathbb{R})$, the Fourier transform integral (4.40a) might or might not be defined for every $\omega \in \mathbb{R}$. The extension of the Fourier transform and its inverse from $\mathcal{L}^1(\mathbb{R})$ to $\mathcal{L}^2(\mathbb{R})$ is technically nontrivial; see the *Further reading* for texts with thorough discussions of this topic.[85]

Without understanding the technical details, one can still appreciate the extremely useful end result: The Fourier transform gives, for any x in $\mathcal{L}^2(\mathbb{R})$, a function X that is itself in $\mathcal{L}^2(\mathbb{R})$. The inverse Fourier transform can be defined similarly for any X in $\mathcal{L}^2(\mathbb{R})$, and it does indeed provide an inversion, so x can be recovered from X.

Knowing that the Fourier transform can be extended rigorously to all of $\mathcal{L}^2(\mathbb{R})$, we can put the technicalities aside. We can determine Fourier transform pairs by evaluating either the transform integral (4.40a) or the inverse transform integral (4.40b) – if either is tractable. This strategy will not cover every $x \in \mathcal{L}^2(\mathbb{R})$, but closed-form integration is often a matter of luck. We illustrate this through the following example.

EXAMPLE 4.5 (FOURIER TRANSFORM OF THE SINC FUNCTION) Let

$$x(t) = \sqrt{\frac{\omega_0}{2\pi}} \operatorname{sinc}\left(\frac{1}{2}\omega_0 t\right). \tag{4.46}$$

As in its discrete-time counterpart in Table 3.5, the factor $\sqrt{\omega_0/(2\pi)}$ is present to make $x(t)$ have unit $\mathcal{L}^2$ norm. Since x is in $\mathcal{L}^2(\mathbb{R})$, it has a Fourier transform, but, since x is not in $\mathcal{L}^1(\mathbb{R})$, we cannot necessarily find the Fourier transform by

[85] A typical route is to use the existence of the Fourier transform on $\mathcal{L}^1(\mathbb{R}) \cap \mathcal{L}^2(\mathbb{R})$ along with the density of $\mathcal{L}^1(\mathbb{R}) \cap \mathcal{L}^2(\mathbb{R})$ in $\mathcal{L}^2(\mathbb{R})$ to define the Fourier transform as the limit of a sequence of Fourier transforms of functions that converge to x.

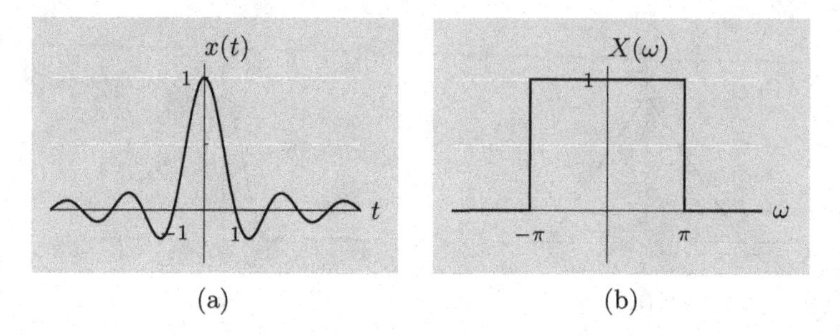

Figure 4.6 (a) The sinc function (4.46) for $\omega_0 = 2\pi$. (b) Its Fourier transform.

evaluating the integral (4.40a). We can avoid having to deal with a potentially difficult integral by simply recognizing from Example 4.3 that the sinc function and the box function are associated by the Fourier transform. Specifically, we can easily apply the inverse Fourier transform integral (4.40b) to

$$X(\omega) = \begin{cases} \sqrt{2\pi/\omega_0}, & \text{for } |\omega| \le \frac{1}{2}\omega_0; \\ 0, & \text{otherwise} \end{cases} \tag{4.47}$$

to confirm that (4.46) and (4.47) form a Fourier transform pair. The function and its Fourier transform for $\omega_0 = 2\pi$ are shown in Figure 4.6.

Inversion When x in $\mathcal{L}^1(\mathbb{R})$ has a Fourier transform X that is itself in $\mathcal{L}^1(\mathbb{R})$, the transform is inverted by (4.40b); see the *Further reading*. Specifically, this means that if applying the inverse Fourier transform to X yields $\widehat{x}$, then

$$\widehat{x}(t) = x(t) \qquad \text{for all } t \in \mathbb{R}. \tag{4.48}$$

Since $X \in \mathcal{L}^1(\mathbb{R})$ implies that x and $\widehat{x}$ are both continuous, it is perhaps not surprising that a pointwise match can be achieved.

The extension of the Fourier transform to $\mathcal{L}^2(\mathbb{R})$ uses sequences of functions converging under the $\mathcal{L}^2$ norm. The meaning of the inversion is thus changed accordingly. Let x be in $\mathcal{L}^2(\mathbb{R})$. It has a Fourier transform X (which is automatically in $\mathcal{L}^2(\mathbb{R})$), which in turn has an inverse Fourier transform $\widehat{x}$ (which is also automatically in $\mathcal{L}^2(\mathbb{R})$). The inverse relationship guarantees

$$\|x - \widehat{x}\| = 0, \tag{4.49}$$

but (4.48) does not necessarily hold.

Using the Fourier transform without existence The Fourier transform is still a useful tool even in some cases where neither its $\mathcal{L}^1(\mathbb{R})$ nor its $\mathcal{L}^2(\mathbb{R})$ definition applies. These are cases where an expression for the Fourier transform involving a Dirac delta function makes sense because evaluating the inverse Fourier transform integral gives the desired result, or vice versa. As with other uses of the Dirac delta function, we must be cautious.

EXAMPLE 4.6 (FOURIER TRANSFORM OF A CONSTANT FUNCTION) Let $x(t) = 1$ for all $t \in \mathbb{R}$. This function belongs to neither $\mathcal{L}^1(\mathbb{R})$ nor $\mathcal{L}^2(\mathbb{R})$, so our previous discussions of the existence of the Fourier transform do not apply. In fact, there is no value of ω for which the Fourier transform integral (4.40a) is defined and finite. However, there is a dependence on ω: for $\omega = 0$, (4.40a) diverges to ∞; for other values of ω, it is tempting (but *not* mathematically correct) to assign the value of zero to the integral because the real and imaginary parts of the integrand oscillate evenly between positive and negative values.

Despite the lack of existence of the Fourier transform, the expression

$$X(\omega) = 2\pi\delta(\omega) \tag{4.50}$$

proves useful. Substituting this into the inverse Fourier transform (4.40b) recovers the function with which we started, $x(t) = 1$, for all $t \in \mathbb{R}$.

Similar reasoning gives the Fourier transform pairs

$$e^{j\omega_0 t} \quad \stackrel{\text{FT}}{\longleftrightarrow} \quad 2\pi\delta(\omega - \omega_0), \tag{4.51}$$

for any $\omega_0 \in \mathbb{R}$, and

$$\delta(t) \quad \stackrel{\text{FT}}{\longleftrightarrow} \quad 1. \tag{4.52}$$

4.4.3 Properties of the Fourier transform

Basic properties

We list here basic properties of the Fourier transform; Table 4.1 summarizes these, together with symmetries as well as a few standard transform pairs. Of course, all the expressions must be well defined for these properties to hold.

Linearity The Fourier transform operator F is a linear operator, or

$$\alpha x(t) + \beta y(t) \quad \stackrel{\text{FT}}{\longleftrightarrow} \quad \alpha X(\omega) + \beta Y(\omega). \tag{4.53}$$

Shift in time The Fourier transform pair corresponding to a shift in time by t_0 is

$$x(t - t_0) \quad \stackrel{\text{FT}}{\longleftrightarrow} \quad e^{-j\omega t_0} X(\omega). \tag{4.54}$$

Shift in frequency The Fourier transform pair corresponding to a shift in frequency by ω_0 is

$$e^{j\omega_0 t} x(t) \quad \stackrel{\text{FT}}{\longleftrightarrow} \quad X(\omega - \omega_0). \tag{4.55}$$

A shift in frequency is often referred to as *modulation*.

FT properties	Time domain	FT domain
Basic properties		
Linearity	$\alpha x(t) + \beta y(t)$	$\alpha X(\omega) + \beta Y(\omega)$
Shift in time	$x(t - t_0)$	$e^{-j\omega t_0} X(\omega)$
Shift in frequency	$e^{j\omega_0 t} x(t)$	$X(\omega - \omega_0)$
Scaling in time and frequency	$x(\alpha t)$	$(1/\alpha) X(\omega/\alpha)$
Time reversal	$x(-t)$	$X(-\omega)$
Differentiation in time	$d^n x(t)/dt^n$	$(j\omega)^n X(\omega)$
Differentiation in frequency	$(-jt)^n x(t)$	$d^n X(\omega)/d\omega^n$
Integration in time	$\int_{-\infty}^{t} x(\tau)\, d\tau$	$X(\omega)/j\omega,\ X(0) = 0$
Moments	$m_k = \int_{-\infty}^{\infty} t^k x(t)\, dt = (j)^k \left. \frac{d^k X(\omega)}{d\omega^k} \right\|_{\omega=0}$	
Convolution in time	$(h * x)(t)$	$H(\omega)\, X(\omega)$
Convolution in frequency	$h(t)\, x(t)$	$\frac{1}{2\pi}(H * X)(\omega)$
Deterministic autocorrelation	$a(t) = \int_{-\infty}^{\infty} x(\tau)\, x^*(\tau - t)\, d\tau$	$A(\omega) = \|X(\omega)\|^2$
Deterministic crosscorrelation	$c(t) = \int_{-\infty}^{\infty} x(\tau)\, y^*(\tau - t)\, d\tau$	$C(\omega) = X(\omega) Y^*(\omega)$
Parseval equality	$\\|x\\|^2 = \int_{-\infty}^{\infty} \|x(t)\|^2\, dt = \frac{1}{2\pi} \int_{-\infty}^{\infty} \|X(\omega)\|^2\, d\omega = \frac{1}{2\pi} \\|X\\|^2$	
Related functions		
Conjugate	$x^*(t)$	$X^*(-\omega)$
Conjugate, time-reversed	$x^*(-t)$	$X^*(\omega)$
Real part	$\Re(x(t))$	$(X(\omega) + X^*(-\omega))/2$
Imaginary part	$\Im(x(t))$	$(X(\omega) - X^*(-\omega))/(2j)$
Conjugate-symmetric part	$(x(t) + x^*(-t))/2$	$\Re(X(\omega))$
Conjugate-antisymmetric part	$(x(t) - x^*(-t))/(2j)$	$\Im(X(\omega))$
Symmetries for real x		
X conjugate symmetric		$X(\omega) = X^*(-\omega)$
Real part of X even		$\Re(X(\omega)) = \Re(X(-\omega))$
Imaginary part of X odd		$\Im(X(\omega)) = -\Im(X(-\omega))$
Magnitude of X even		$\|X(\omega)\| = \|X(-\omega)\|$
Phase of X odd		$\arg X(\omega) = -\arg X(-\omega)$
Common transform pairs		
Dirac delta function	$\delta(t)$	1
Shifted Dirac delta function	$\delta(t - t_0)$	$e^{-j\omega_0 t}$
Dirac comb	$\sum_{n\in\mathbb{Z}} \delta(t - nT)$	$(2\pi/T) \sum_{k\in\mathbb{Z}} \delta(\omega - (2\pi/T)k)$
Constant function	1	$2\pi\delta(\omega)$
Exponential function	$e^{-\alpha\|t\|}$	$(2\alpha)/(\omega^2 + \alpha^2)$
Gaussian function	$e^{-\alpha t^2}$	$\sqrt{\pi/\alpha}\, e^{-\omega^2/\alpha}$
Sinc function (ideal lowpass filter)	$\sqrt{\frac{\omega_0}{2\pi}}\, \mathrm{sinc}(\frac{1}{2}\omega_0 t)$	$\begin{cases} \sqrt{2\pi/\omega_0}, & \|\omega\| \le \frac{1}{2}\omega_0; \\ 0, & \text{otherwise.} \end{cases}$
Box function	$\begin{cases} 1/\sqrt{t_0}, & \|t\| \le \frac{1}{2}t_0; \\ 0, & \text{otherwise.} \end{cases}$	$\sqrt{t_0}\, \mathrm{sinc}(\frac{1}{2}t_0\omega)$
Triangle function	$\begin{cases} 1 - \|t\|, & \|t\| < 1; \\ 0, & \text{otherwise} \end{cases}$	$\mathrm{sinc}^2(\frac{1}{2}\omega)$

Table 4.1 Properties of the Fourier transform.

Scaling in time and frequency The Fourier transform pair corresponding to scaling in time by α is scaling in frequency by $1/\alpha$:

$$x(\alpha t) \quad \overset{\text{FT}}{\longleftrightarrow} \quad \frac{1}{\alpha} X\Big(\frac{\omega}{\alpha}\Big). \tag{4.56a}$$

This is another of the key properties of the Fourier transform, where a stretch in time corresponds to a compaction in frequency, and vice versa. We often use normalized rescaling, namely, for $\alpha > 0$,

$$\sqrt{\alpha} x(\alpha t) \quad \overset{\text{FT}}{\longleftrightarrow} \quad \frac{1}{\sqrt{\alpha}} X\Big(\frac{\omega}{\alpha}\Big), \tag{4.56b}$$

which conserves the $\mathcal{L}^2$ norm of x (and thus of X), since

$$\|\sqrt{\alpha} x(\alpha t)\|^2 = \int_{-\infty}^{\infty} \alpha |x(\alpha t)|^2 \, dt \overset{(a)}{=} \alpha \int_{-\infty}^{\infty} |x(\tau)|^2 \, \frac{d\tau}{\alpha} = \|x\|^2, \tag{4.56c}$$

where (a) follows from the change of variable $\tau = \alpha t$.

Time reversal The Fourier transform pair corresponding to time reversal $x(-t)$ is

$$x(-t) \quad \overset{\text{FT}}{\longleftrightarrow} \quad X(-\omega). \tag{4.57}$$

For a real $x(t)$, the Fourier transform of the time-reversed version $x(-t)$ is $X^*(\omega)$.

Differentiation in time The Fourier transform pair corresponding to differentiation in time is

$$\frac{d^n x(t)}{dt^n} \quad \overset{\text{FT}}{\longleftrightarrow} \quad (j\omega)^n X(\omega), \tag{4.58}$$

assuming that the derivatives exist and are bounded, or equivalently, assuming that $\omega^n X(\omega)$ is absolutely integrable.

Differentiation in frequency The Fourier transform pair corresponding to differentiation in frequency is

$$(-jt)^n x(t) \quad \overset{\text{FT}}{\longleftrightarrow} \quad \frac{d^n X(\omega)}{d\omega^n}. \tag{4.59a}$$

This is obtained by multiple applications of

$$\int_{-\infty}^{\infty} t x(t) e^{-j\omega t} \, dt = j \frac{d}{d\omega} \left[\int_{-\infty}^{\infty} x(t) e^{-j\omega t} \, dt \right] = j \frac{dX(\omega)}{d\omega}, \tag{4.59b}$$

assuming that $t^k x(t)$ is integrable for $k = 0, 1, \ldots, n$ and using

$$j \frac{de^{-j\omega t}}{d\omega} = t e^{-j\omega t}.$$

This result is dual to differentiation in time above.

Moments Computing the kth moment using the Fourier transform results in

$$m_k = \int_{-\infty}^{\infty} t^k x(t)\, dt = (j)^k \left.\frac{d^k X(\omega)}{d\omega^k}\right|_{\omega=0}, \qquad k \in \mathbb{N}, \tag{4.60a}$$

as a direct application of (4.59a). The first two moments are

$$m_0 = \int_{-\infty}^{\infty} x(t)\, dt = \left.\int_{-\infty}^{\infty} x(t)e^{-j\omega t}\, dt\right|_{\omega=0} = X(0), \tag{4.60b}$$

$$m_1 = \int_{-\infty}^{\infty} tx(t)\, dt = \left.\int_{-\infty}^{\infty} tx(t)e^{-j\omega t}\, dt\right|_{\omega=0} = j\left.\frac{dX(\omega)}{d\omega}\right|_{\omega=0}. \tag{4.60c}$$

Integration The Fourier transform pair corresponding to the integral of a function with zero mean ($m_0 = X(0) = 0$) is

$$\int_{-\infty}^{t} x(\tau)\, d\tau \quad \stackrel{\text{FT}}{\longleftrightarrow} \quad \frac{1}{j\omega} X(\omega). \tag{4.61}$$

Convolution in time The Fourier transform pair corresponding to convolution in time is

$$(h * x)(t) \quad \stackrel{\text{FT}}{\longleftrightarrow} \quad H(\omega)X(\omega). \tag{4.62}$$

Thus, as with the DTFT and DFT in Chapter 3, convolution in the time domain corresponds to multiplication in the Fourier domain.

For a direct algebraic proof, assume that both h and x are in $\mathcal{L}^1(\mathbb{R})$. Then, $h * x$ is also in $\mathcal{L}^1(\mathbb{R})$, as noted following Theorem 4.8. The spectrum $Y(\omega)$ of the output function $y = h * x$ can be written as

$$\begin{aligned} Y(\omega) &\stackrel{(a)}{=} \int_{-\infty}^{\infty} y(t)e^{-j\omega t}\, dt \stackrel{(b)}{=} \int_{-\infty}^{\infty} \left(\int_{-\infty}^{\infty} x(\tau)h(t-\tau)\, d\tau\right) e^{-j\omega t}\, dt \\ &= \int_{-\infty}^{\infty}\int_{-\infty}^{\infty} x(\tau)e^{-j\omega\tau} h(t-\tau)e^{-j\omega(t-\tau)}\, d\tau\, dt \\ &\stackrel{(c)}{=} \int_{-\infty}^{\infty} x(\tau)e^{-j\omega\tau}\, d\tau \int_{-\infty}^{\infty} h(\tau')e^{-j\omega\tau'}\, d\tau' \stackrel{(d)}{=} X(\omega)H(\omega), \end{aligned}$$

where (a) follows from the definition of the Fourier transform; (b) from the definition of convolution; (c) from interchanging the order of integration, which is an allowed operation since absolute integrability is implied by $h * x$ being well defined, and change of variable $\tau' = t - \tau$; and (d) from the definition of the Fourier transform.

This key result is a direct consequence of the eigenfunction property of complex exponential functions v from (4.38): when x is written as a combination of spectral components, each spectral component is simply scaled by the corresponding eigenvalue of the convolution operator; thus, using the Fourier transform has diagonalized the convolution operator. While this is slightly more subtle than the case in discrete time because we have not been expressing convolution operators using matrices, the concept is unchanged.

EXAMPLE 4.7 (DIFFERENTIATION AND CONVOLUTION) Suppose that $h * x$ is differentiable. Using (4.58) with $n = 1$ and (4.62),

$$\frac{d}{dt}(h * x)(t) \quad \stackrel{\text{FT}}{\longleftrightarrow} \quad j\omega H(\omega)X(\omega).$$

Since the $j\omega$ factor can be associated either with $X(\omega)$ or with $H(\omega)$, upon using (4.58) with $n = 1$ again, we see that $(h * x)'$ can be written in either of the following two ways:

$$(h * x)' \;=\; h * x' \;=\; h' * x. \tag{4.63}$$

Similarly, by writing

$$H(\omega)X(\omega) \;=\; \left(\frac{1}{j\omega}H(\omega)\right)(j\omega X(\omega)) \;=\; (j\omega H(\omega))\left(\frac{1}{j\omega}X(\omega)\right)$$

and using (4.58) with $n = 1$, (4.61), and (4.62), we find

$$h * x \;=\; h^{(1)} * x' \;=\; h' * x^{(1)}, \tag{4.64}$$

where

$$h^{(1)}(t) \;=\; \int_{-\infty}^{t} h(\tau)\,d\tau \qquad \text{and} \qquad x^{(1)}(t) \;=\; \int_{-\infty}^{t} x(\tau)\,d\tau$$

are primitives of h and x. Justifying (4.64) with (4.61) requires $H(0) = 0$ or $X(0) = 0$, but it can be justified under looser conditions using integration by parts (see Exercise 4.4). A pictorial example is shown in Figure 4.7.

Convolution in frequency The Fourier transform pair corresponding to convolution in frequency is

$$h(t)\,x(t) \quad \stackrel{\text{FT}}{\longleftrightarrow} \quad \frac{1}{2\pi}(H * X)(\omega). \tag{4.65}$$

The convolution in frequency property (4.65) is dual to the convolution in time property (4.62).

Deterministic autocorrelation The Fourier transform pair corresponding to the deterministic autocorrelation of a function x is

$$a(t) \;=\; \int_{-\infty}^{\infty} x(\tau)\,x^*(\tau - t)\,d\tau \quad \stackrel{\text{FT}}{\longleftrightarrow} \quad A(\omega) \;=\; |X(\omega)|^2 \tag{4.66}$$

and satisfies

$$A(\omega) \;=\; A^*(\omega), \tag{4.67a}$$

$$A(\omega) \;\geq\; 0. \tag{4.67b}$$

To verify (4.66), express the deterministic autocorrelation a as the convolution of x and its conjugated and time-reversed version as in (4.32d), $x(t) * x^*(-t)$. We

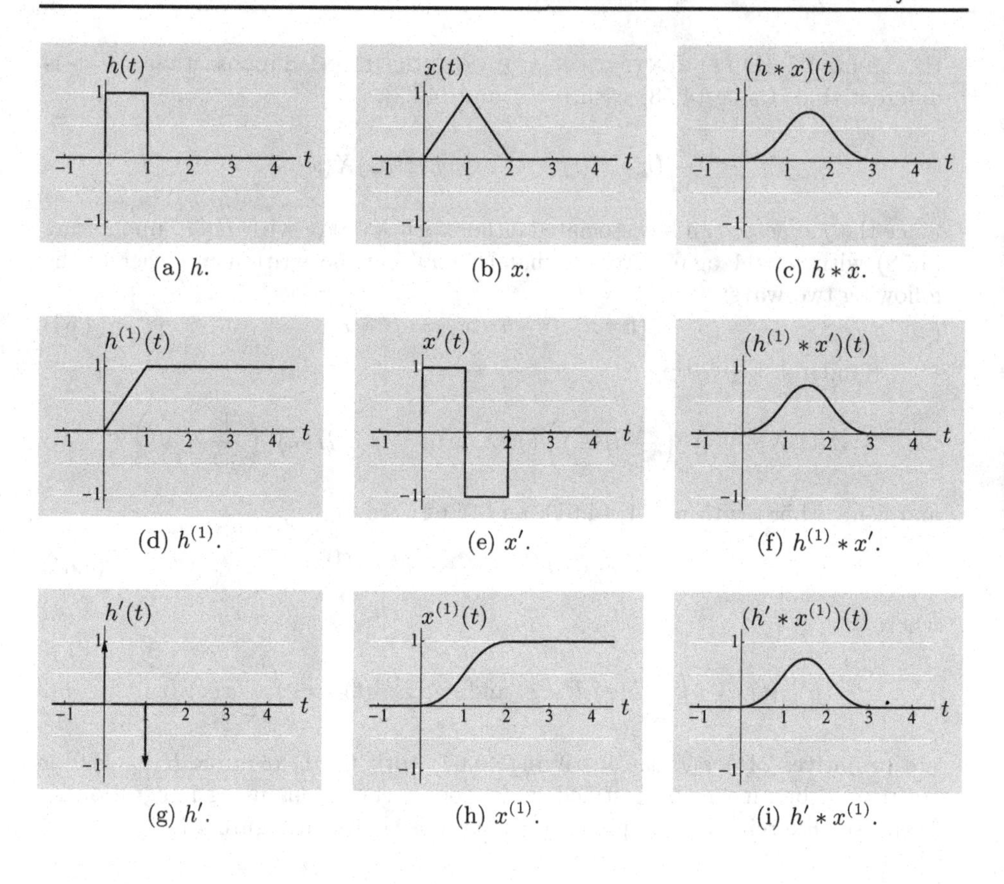

Figure 4.7 Computation of convolution through the equivalent of convolving primitive and derivative instead.

know from Table 4.1 that the Fourier transform of $x^*(-t)$ is $X^*(\omega)$. Then, using the convolution property (4.62), we obtain (4.66). For a real x,

$$A(\omega) = |X(\omega)|^2 = A(-\omega), \tag{4.67c}$$

since $X(-\omega) = X^*(\omega)$.

The quantity $A(\omega)$ is called the *energy spectral density* (this is the deterministic counterpart of the *power spectral density* for WSS continuous-time stochastic processes defined by (4.140) in Section 4.6.3). The *energy* is the normalized integral of the energy spectral density,

$$E = \frac{1}{2\pi}\int_{-\infty}^{\infty} A(\omega)\, d\omega = \frac{1}{2\pi}\int_{-\infty}^{\infty} |X(\omega)|^2\, d\omega = \int_{-\infty}^{\infty} |x(t)|^2\, dt = a_0. \tag{4.68}$$

Mimicking the relationship between the energy spectral density for deterministic functions and the power spectral density for WSS stochastic processes, (4.68) is the deterministic counterpart of the *power* for WSS continuous-time stochastic processes (4.141) that will be introduced in Section 4.6.3.

Deterministic crosscorrelation The Fourier transform pair corresponding to the deterministic crosscorrelation of functions x and y is

$$c(t) = \int_{-\infty}^{\infty} x(\tau)\, y^*(\tau - t)\, d\tau \quad \stackrel{\text{FT}}{\longleftrightarrow} \quad C_{x,y}(\omega) = X(\omega)Y^*(\omega) \tag{4.69}$$

and satisfies

$$C_{x,y}(\omega) = C^*_{y,x}(\omega). \tag{4.70a}$$

For real x and y,

$$C_{x,y}(\omega) = X(\omega)\, Y(-\omega) = C_{y,x}(-\omega). \tag{4.70b}$$

Parseval equality The Fourier transform is a unitary linear transformation on $\mathcal{L}^2(\mathbb{R})$, up to a 2π scaling factor. It satisfies Parseval equalities

$$\|x\|^2 = \frac{1}{2\pi}\|X\|^2 \tag{4.71a}$$

and

$$\langle x,\, y\rangle = \frac{1}{2\pi}\langle X,\, Y\rangle, \tag{4.71b}$$

where X and Y are the Fourier transforms of x and y. A proof for x and y in $\mathcal{L}^1(\mathbb{R}) \cap \mathcal{L}^2(\mathbb{R})$ is elementary (see Exercise 4.5), and the extension to the rest of $\mathcal{L}^2(\mathbb{R})$ also holds.

Adjoint The adjoint of the Fourier transform, $F^* : \mathcal{L}^2(\mathbb{R}) \to \mathcal{L}^2(\mathbb{R})$, is determined uniquely by

$$\langle Fx,\, y\rangle = \langle x,\, F^*y\rangle \qquad \text{for every } x, y \in \mathcal{L}^2(\mathbb{R}).$$

Since (4.71b) shows that $F/\sqrt{2\pi}$ is a unitary operator, by Theorem 2.23,

$$\left(\frac{1}{\sqrt{2\pi}}F\right)^* = \left(\frac{1}{\sqrt{2\pi}}F\right)^{-1} = \sqrt{2\pi}\, F^{-1}.$$

Thus,

$$F^* = 2\pi F^{-1}, \tag{4.72}$$

with F^{-1} given by (4.40b).

Transform pairs

Rather than develop an exhaustive list of Fourier transform pairs, we highlight a few that are routinely used. Table 4.1 summarizes these pairs (along with some others).

Dirac delta function We saw earlier that the Fourier transform pair

$$1 \quad \stackrel{\text{FT}}{\longleftrightarrow} \quad 2\pi\delta(\omega) \tag{4.73}$$

is justified by evaluating the inverse Fourier transform integral, and similarly

$$\delta(t) \quad \stackrel{\text{FT}}{\longleftrightarrow} \quad 1 \tag{4.74}$$

is justified by applying the Fourier transform integral. These, along with the various properties derived previously, enable the computation of many Fourier transforms, including the transforms of certain trigonometric functions.

Box function From Example 4.3, the Fourier transform of the box function (4.8a) is a sinc function,

$$x(t) \;=\; \begin{cases} 1/\sqrt{t_0}, & \text{for } |t| \leq \frac{1}{2}t_0; \\ 0, & \text{otherwise} \end{cases} \quad \stackrel{\text{FT}}{\longleftrightarrow} \quad X(\omega) \;=\; \sqrt{t_0}\,\operatorname{sinc}\left(\tfrac{1}{2}t_0\omega\right). \tag{4.75}$$

Sinc function From Example 4.5, the Fourier transform of a sinc function is a box function,

$$x(t) \;=\; \sqrt{\frac{\omega_0}{2\pi}}\,\operatorname{sinc}\left(\tfrac{1}{2}\omega_0 t\right) \quad \stackrel{\text{FT}}{\longleftrightarrow} \quad X(\omega) \;=\; \begin{cases} \sqrt{2\pi/\omega_0}, & \text{for } |\omega| \leq \frac{1}{2}\omega_0; \\ 0, & \text{otherwise.} \end{cases} \tag{4.76}$$

The scaling of the sinc was chosen so that the time-domain function has unit norm. Because of the prominent role of box and sinc functions, the transform pairs (4.75) and (4.76), together with their counterparts for sequences as well as for periodic functions, are included in Table 4.5 at the end of this chapter.

Heaviside function The Heaviside function was defined in (4.5). Its Fourier transform pair is

$$u(t) \;=\; \begin{cases} 1, & \text{for } t \geq 0; \\ 0, & \text{otherwise} \end{cases} \quad \stackrel{\text{FT}}{\longleftrightarrow} \quad U(\omega) \;=\; \pi\delta(\omega) + \frac{1}{j\omega}, \tag{4.77}$$

the derivation of which is beyond the scope of this book.

Gaussian function The Fourier transform of a Gaussian function in time is a Gaussian function in frequency (see Figure 4.8),

$$g(t) \;=\; \gamma e^{-\alpha t^2} \quad \stackrel{\text{FT}}{\longleftrightarrow} \quad G(\omega) \;=\; \gamma\sqrt{\frac{\pi}{\alpha}}e^{-\omega^2/(4\alpha)}, \tag{4.78}$$

where α and γ are positive real constants. That this is a Fourier transform pair can be proven in various ways. One way is to first observe that $e^{-\alpha t^2}$ is the solution of the differential equation

$$\frac{dx(t)}{dt} + 2\alpha t x(t) \;=\; 0.$$

Then, taking the Fourier transform and using (4.58) and (4.59b) leads to an equivalent differential equation in the Fourier domain which, when solved, yields (4.78) (see Exercise 4.6).

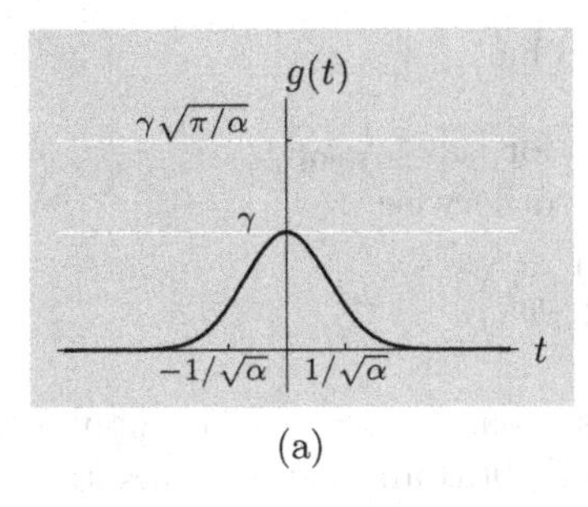

(a)

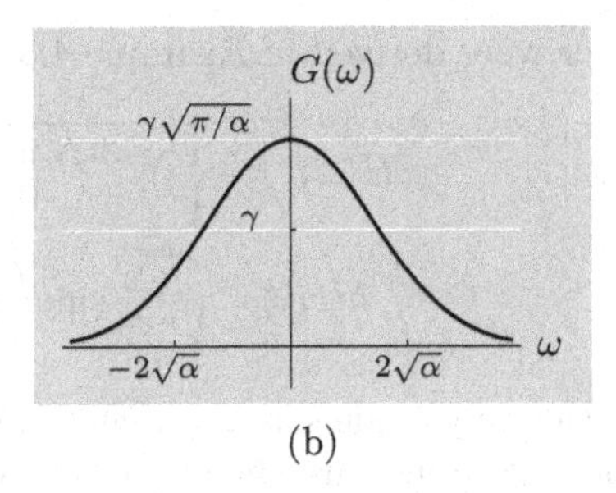

(b)

Figure 4.8 (a) The Gaussian function (4.9a) with $\mu = 0$. (b) Its Fourier transform. (Illustrated for $\alpha = 1$.)

4.4.4 Frequency response of filters

Like in discrete time, the Fourier transform of a function is called its spectrum, and the Fourier transform of a filter (impulse response of an LSI system) h is called the *frequency response*:

$$H(\omega) = \int_{-\infty}^{\infty} h(t)e^{-j\omega t}\, dt, \qquad \omega \in \mathbb{R}. \tag{4.79a}$$

The inverse Fourier transform of the frequency response recovers the impulse response,

$$h(t) = \frac{1}{2\pi}\int_{-\infty}^{\infty} H(\omega)e^{j\omega t}\, d\omega, \qquad t \in \mathbb{R}. \tag{4.79b}$$

We often write the magnitude and phase of the frequency response separately:

$$H(\omega) = |H(\omega)|e^{j\arg(H(\omega))},$$

where the *magnitude response* $|H(\omega)|$ is a real, nonnegative function, and the *phase response* $\arg(H(\omega))$ is a real function between $-\pi$ and π. The terms *zero phase*, *linear phase*, and *allpass* have the same meanings as in Section 3.4.4.

Ideal filters The frequency response of a filter is typically used to design a filter with specific properties, where we want to let certain frequencies pass – the *passband*, while blocking others – the *stopband*. An *ideal filter* is a filter whose magnitude response takes a single positive value in its passband. We limit attention to real-valued impulse responses (so the magnitude response is an even function) and real-valued frequency responses (so the impulse response is an even function, and hence noncausal).

An *ideal lowpass filter* passes frequencies below some cutoff frequency $\frac{1}{2}\omega_0$ and blocks the others; its passband is thus the interval $[-\frac{1}{2}\omega_0, \frac{1}{2}\omega_0]$, and the frequency response is a box function. The frequency response and impulse response of an ideal lowpass filter were derived in Example 4.5 to be

$$H(\omega) = \begin{cases} \sqrt{2\pi/\omega_0}, & \text{for } |\omega| \le \frac{1}{2}\omega_0; \\ 0, & \text{otherwise,} \end{cases} \tag{4.80a}$$

$$h(t) = \sqrt{\frac{\omega_0}{2\pi}}\, \operatorname{sinc}\left(\frac{1}{2}\omega_0 t\right). \tag{4.80b}$$

As earlier, the amplitude has been chosen so that $\|h\| = 1$. The impulse response and frequency response are shown in Figures 4.9(a) and (b) for $\omega_0 = \pi$.

An *ideal bandpass filter* passes frequencies with absolute value between $\frac{1}{2}\omega_0$ and $\frac{1}{2}\omega_1$ and blocks the others; its passband is thus the union of intervals $[-\frac{1}{2}\omega_1, -\frac{1}{2}\omega_0] \cup [\frac{1}{2}\omega_0, \frac{1}{2}\omega_1]$. By the Parseval equality, to have $\|h\| = 1$ we require $\|H\| = 2\pi$, so we set

$$H(\omega) = \begin{cases} \sqrt{2\pi/(\omega_1 - \omega_0)}, & \text{for } |\omega| \in [\frac{1}{2}\omega_0, \frac{1}{2}\omega_1]; \\ 0, & \text{otherwise.} \end{cases} \tag{4.81a}$$

An easy way to derive the impulse response is to recognize the frequency response as a difference between two scaled box functions. Using linearity of the inverse Fourier transform, this yields the impulse response

$$h(t) = \frac{1}{\sqrt{2\pi(\omega_1 - \omega_0)}} \left(\omega_1 \operatorname{sinc}\left(\frac{1}{2}\omega_1 t\right) - \omega_0 \operatorname{sinc}\left(\frac{1}{2}\omega_0 t\right)\right). \tag{4.81b}$$

The impulse response and frequency response are shown in Figures 4.9(c) and (d) for $\omega_0 = \pi$ and $\omega_1 = 2\pi$.

An *ideal highpass filter* passes frequencies above some cutoff frequency $\frac{1}{2}\omega_1$ and blocks the others; its stopband is thus the interval $[-\frac{1}{2}\omega_1, \frac{1}{2}\omega_1]$. Since the frequency response takes some nonzero value on the unbounded support of $(-\infty, -\frac{1}{2}\omega_1] \cup [\frac{1}{2}\omega_1, \infty)$, the $\mathcal{L}^2$ norm of H diverges to infinity. There is no way to choose an amplitude that gives an impulse response with unit norm, so let us choose a unit passband gain:

$$H(\omega) = \begin{cases} 1, & \text{for } |\omega| \ge \frac{1}{2}\omega_1; \\ 0, & \text{otherwise.} \end{cases} \tag{4.82a}$$

The inverse Fourier transform integral is not defined for this H, but by recognizing it as a constant minus a box function, we can associate with it the following impulse response:

$$h(t) = \delta(t) - \frac{\omega_1}{2\pi} \operatorname{sinc}\left(\frac{1}{2}\omega_1 t\right). \tag{4.82b}$$

The impulse response and frequency response are shown in Figures 4.9(e) and (f) for $\omega_1 = 2\pi$.

4.4.5 Regularity and spectral decay

We discussed earlier a distinction between the box function from Example 4.3 and the triangle function from Example 4.4: the triangle function has a spectrum $X(\omega)$

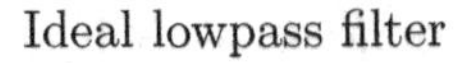

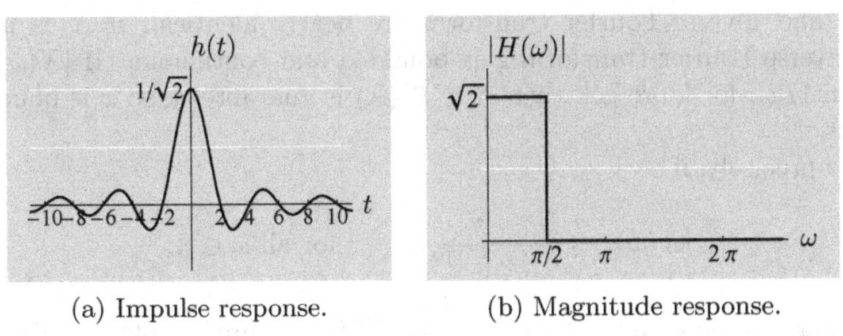

(a) Impulse response. (b) Magnitude response.

Ideal bandpass filter

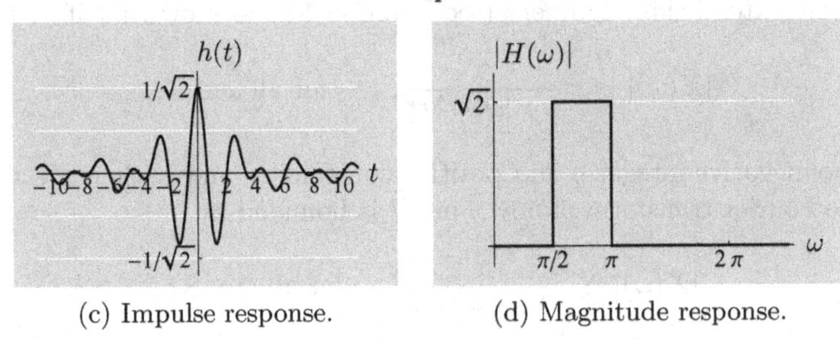

(c) Impulse response. (d) Magnitude response.

Ideal highpass filter

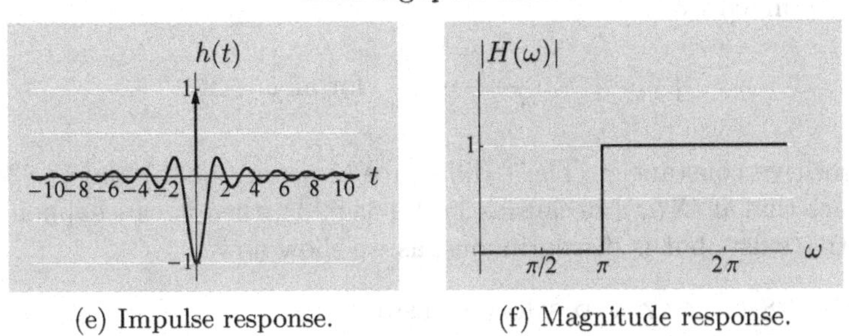

(e) Impulse response. (f) Magnitude response.

Figure 4.9 Impulse responses and magnitude responses of ideal filters. The magnitude responses are even functions; only positive frequencies are shown.

that decays fast enough as $|\omega| \to \infty$ for it to be in $\mathcal{L}^1(\mathbb{R})$, while the box function does not. This reflects the time-domain distinction that the triangle function is continuous, while the box function is not. Spectral decay is related to smoothness or regularity more generally, and characterizing regularity is an important analytical use of the Fourier transform.

C^q **regularity** The Fourier transform facilitates the characterization of functions with q continuous derivatives, that is, those functions belonging to C^q spaces (see Section 2.2.4). As mentioned in Section 4.4.2, if x is in $\mathcal{L}^1(\mathbb{R})$, then its Fourier transform X is bounded and continuous (see Solved exercise 4.3). Since the Fourier transform and inverse Fourier transform are nearly identical, if X is in $\mathcal{L}^1(\mathbb{R})$, then its inverse Fourier transform x is bounded and continuous. If $|X(\omega)|$ decays faster than $1/|\omega|$ for large $|\omega|$, then $X \in \mathcal{L}^1(\mathbb{R})$ is guaranteed, so x is bounded and continuous.

More precisely, if

$$|X(\omega)| \leq \frac{\gamma}{1+|\omega|^{1+\varepsilon}}, \qquad \text{for all } \omega \in \mathbb{R}, \tag{4.83a}$$

for some positive constants γ and ε, then $|X(\omega)|$ is integrable, or $X \in \mathcal{L}^1(\mathbb{R})$. Therefore, the inverse Fourier transform x is bounded and continuous, or $x \in C^0$. We can easily extend this argument (see Exercise 4.8) to show that if

$$|X(\omega)| \leq \frac{\gamma}{1+|\omega|^{q+1+\varepsilon}}, \qquad \text{for all } \omega \in \mathbb{R}, \tag{4.83b}$$

for some nonnegative integer q and positive constants γ and ε, then $x \in C^q$. Conversely, the Fourier transform of any x in C^q is bounded by

$$|X(\omega)| \leq \frac{\gamma}{1+|\omega|^{q+1}}, \qquad \text{for all } \omega \in \mathbb{R}, \tag{4.83c}$$

for some positive constant γ. Taking $q = 0$, the Fourier transform of any continuous function is bounded by

$$|X(\omega)| \leq \frac{\gamma}{1+|\omega|}, \qquad \text{for all } \omega \in \mathbb{R}, \tag{4.83d}$$

for some positive constant γ. The ε difference between (4.83b) and (4.83c) comes from the fact that if $|X(\omega)|$ decays as $1/(1+|\omega|^{q+1})$, then it can happen that the qth derivative exists but is discontinuous, as we show now.

Example 4.8 (Decay and smoothness)

(i) *Unit-width box function:* Let x be the box function in (4.8a) with $t_0 = 1$. From (4.41), its spectrum is $X(\omega) = \operatorname{sinc}(\frac{1}{2}\omega)$. To bound the magnitude of the spectrum from above in the form of (4.83), we need to use the boundedness of the sinc function for small $|\omega|$ and the decay of the sinc function for large $|\omega|$. For $|\omega| \geq 2$, the spectrum satisfies

$$|X(\omega)| = \left|\operatorname{sinc}\left(\tfrac{1}{2}\omega\right)\right| = \left|\frac{\sin\frac{1}{2}\omega}{\frac{1}{2}\omega}\right| \overset{(a)}{\leq} \frac{1}{|\frac{1}{2}\omega|} \overset{(b)}{\leq} \frac{3}{1+|\omega|}, \tag{4.84a}$$

where (a) follows from the fact that $|\sin\frac{1}{2}\omega| \leq 1$ for all ω; and (b) is an elementary consequence of $|\omega| > 2$. For $|\omega| < 2$,

$$\frac{3}{1+|\omega|} > \frac{3}{1+2} = 1 \overset{(a)}{\geq} \left|\operatorname{sinc}\left(\tfrac{1}{2}\omega\right)\right| = |X(\omega)|, \tag{4.84b}$$

where (a) follows from the fact that $|\text{sinc}(\frac{1}{2}\omega)| \leq 1$ for all ω. Thus, combining (4.84a) and (4.84b),

$$|X(\omega)| \leq \frac{\gamma}{1+|\omega|}, \qquad \text{for all } \omega \in \mathbb{R},$$

holds for $\gamma = 3$. While it is not important to find the smallest γ in this bound, we do want to know whether we could increase the exponent of $|\omega|$ in the denominator. We cannot: There are no positive constants γ and ε such that

$$|X(\omega)| \leq \frac{\gamma}{1+|\omega|^{1+\varepsilon}}, \qquad \text{for all } \omega \in \mathbb{R}. \tag{4.85}$$

To show this, fix any positive γ and ε, and we will find ω such that (4.85) is violated. The upper envelope of $|X(\omega)|$ connects the relative maxima, which occur when $\frac{1}{2}\omega$ is an odd multiple of $\frac{1}{2}\pi$. This upper envelope is $|2/\omega|$ since we are picking the values of ω such that $|\sin \frac{1}{2}\omega| = 1$. For the upper envelope to violate (4.85), we want

$$\frac{2}{|\omega|} > \frac{\gamma}{1+|\omega|^{1+\varepsilon}},$$

which does indeed happen when $|\omega|$ is large. Specifically, when $\frac{1}{2}\omega$ is an odd multiple of $\frac{1}{2}\pi$ and $|\omega| > (\frac{1}{2}\gamma)^{1/\varepsilon}$,

$$\frac{\gamma}{1+|\omega|^{1+\varepsilon}} < \frac{\gamma}{|\omega| \cdot |\omega|^{\varepsilon}} < \frac{\gamma}{|\omega| \cdot (\frac{1}{2}\gamma)} = \frac{2}{|\omega|} = |X(\omega)|.$$

Since we cannot apply (4.83b), we cannot conclude from the spectrum whether x is continuous. Of course, we know it is not.

(ii) *Triangle function:* By arguments similar to those given above for the box function, the spectrum of the triangle function given in (4.45) satisfies

$$|X(\omega)| \leq \frac{\gamma}{1+|\omega|^2}, \qquad \text{for all } \omega \in \mathbb{R},$$

for some positive constant γ, but there are no positive constants γ and ε such that

$$|X(\omega)| \leq \frac{\gamma}{1+|\omega|^{2+\varepsilon}}, \qquad \text{for all } \omega \in \mathbb{R}.$$

Therefore, we can apply (4.83b) with $q = 0$ (but not with any larger q) and conclude from the spectrum that x is continuous (and not more). Actually, x is almost continuously differentiable because there are only three points where the derivative does not exist.

(iii) Let

$$x(t) = e^{-|t|}. \tag{4.86}$$

This function is very smooth (infinitely differentiable) except at a single point; due to the lack of differentiability at $t = 0$, it is thus merely a C^0

function. Its Fourier transform is

$$\begin{aligned} X(\omega) &= \int_{-\infty}^{\infty} e^{-|t|} e^{-j\omega t}\, dt = \int_{-\infty}^{0} e^{(1-j\omega)t}\, dt + \int_{0}^{\infty} e^{-(1+j\omega)t}\, dt \\ &= \frac{1}{1-j\omega} + \frac{1}{1+j\omega} = \frac{2}{1+\omega^2}. \end{aligned}$$

The spectral decay is precisely as in (4.83c) with $q = 1$ but unable to satisfy (4.83b) for $q = 1$ and any positive ε. Thus, like for the triangle function, the spectrum allows us to conclude that x is a C^0 function.

These examples illustrate the power of the Fourier transform in characterizing regularity of functions; however, they all suffer from local properties (isolated points where the function is not continuous or not differentiable) determining the spectral decay. Wavelet tools allow one to perform local characterizations; see the *Further reading.*

Lipschitz regularity The C^q classes are defined for nonnegative integer values of q. Lipschitz regularity provides a generalization akin to differentiability of fractional order. It can be defined pointwise, over a subset of the domain, or globally (over the entire domain). The global property has a Fourier-domain characterization. Like for differentiability, wavelet tools again allow one to perform local characterizations; see the *Further reading.*

DEFINITION 4.11 (LIPSCHITZ REGULARITY) Let α be in $[0, 1)$. A function x is said to be *pointwise Lipschitz* of order α at t_0 when

$$|x(t) - x(t_0)| \leq c\,|t - t_0|^{\alpha}, \qquad \text{for all } t \in \mathbb{R}, \tag{4.87}$$

for some constant c. A function x is said to be *uniformly Lipschitz* of order α over $I \subseteq \mathbb{R}$ when it satisfies (4.87) for all $t_0 \in I$ with a constant c that does not depend on t_0.

The Lipschitz order is also called the *Lipschitz exponent* or *Hölder exponent.* Setting $\alpha = 0$ in (4.87) shows that being uniformly Lipschitz of order 0 over $\mathbb{R}$ is equivalent to being bounded. Setting $\alpha = 1$ in (4.87) gives a condition that implies differentiability at t_0. Higher-order characterization for $r = n + \alpha$, where $n \in \mathbb{N}$ and $\alpha \in [0, 1)$, can be obtained from this differentiability: x is Lipschitz of order r when the nth derivative of x is Lipschitz of order α.

Uniform Lipschitz regularity of order $r > n$ over $\mathbb{R}$, with n a positive integer, implies that a function is n times continuously differentiable. Differentiability almost everywhere is captured by the Lipschitz order being just shy of the corresponding integer; for example, the triangle function (4.43) and function (4.86) are uniformly Lipschitz of order $1 - \varepsilon$ over $\mathbb{R}$ for any positive ε.

Lipschitz regularity is manifested in the Fourier domain similarly to (4.83b). A function x is bounded and uniformly Lipschitz α over $\mathbb{R}$ when

$$\int_{-\infty}^{\infty} |X(\omega)|(1+|\omega|^{\alpha})\, d\omega \;<\; \infty, \tag{4.88}$$

providing a global characterization of regularity (see Exercise 4.9).

4.4.6 Laplace transform

In the previous chapter, we introduced the z-transform, which extends the DTFT to sequences for which the DTFT might not exist. Similarly, we now introduce the Laplace transform, which extends the Fourier transform to functions for which the Fourier transform might not exist. The z-transform and Laplace transform both replace unit-modulus complex exponentials with general complex exponentials.

Definition 4.12 (Laplace transform) The *Laplace transform* of a function x is

$$X(s) \;=\; \int_{-\infty}^{\infty} x(t)e^{-st}\, dt, \qquad s \in \mathbb{C}. \tag{4.89}$$

It exists when the integral in (4.89) is defined and finite for some values of s; these values of s are called the *region of convergence (ROC)*. When the Laplace transform exists, we denote the Laplace transform pair as

$$x(t) \;\stackrel{\text{LT}}{\longleftrightarrow}\; X(s),$$

where the ROC is part of the specification of $X(s)$.

When $s = j\omega$ with ω real, the Laplace transform is simply the Fourier transform; when $s = \sigma + j\omega$ with σ and ω real, the Laplace transform is the Fourier transform of $x_\sigma(t) = x(t)e^{-\sigma t}$, or

$$X(\sigma + j\omega) = \int_{-\infty}^{\infty} \underbrace{x(t)e^{-\sigma t}}_{x_\sigma(t)}\, e^{-j\omega t}\, dt. \tag{4.90}$$

Therefore, convergence of the Laplace transform depends solely on the exponent σ, and the ROC of the integral (4.89) consists of vertical strips. In particular, for σ such that $x(t)e^{-\sigma t}$ is absolutely integrable, the Laplace transform is a bounded and continuous function of s. Just as for the z-transform, the Laplace transform and its associated ROC define the time-domain function. The Laplace transform satisfies a number of properties which follow directly from their Fourier counterparts; a few are summarized in Table 4.2. We now illustrate the necessity of associating an ROC to a Laplace transform of a function.

LT properties	Time domain	LT domain	ROC
Linearity	$\alpha x(t) + \beta y(t)$	$\alpha X(s) + \beta Y(s)$	$\supset \mathrm{ROC}_x \cap \mathrm{ROC}_y$
Shift in time	$x(t - t_0)$	$e^{-st_0} X(s)$	ROC_x
Shift in s	$e^{s_0 t} x(t)$	$X(s - s_0)$	$\mathrm{ROC}_x + s_0$
Scaling in time	$x(\alpha t)$	$(1/\alpha) X(s/\alpha)$	$\alpha \mathrm{ROC}_x$
Convolution in time	$(h * x)(t)$	$H(s)\, X(s)$	$\supset \mathrm{ROC}_h \cap \mathrm{ROC}_x$

Table 4.2 Selected properties of the Laplace transform.

EXAMPLE 4.9 (LAPLACE TRANSFORM OF THE HEAVISIDE FUNCTION) Let x_1 be the Heaviside function defined in (4.5). Its Fourier transform does not exist in a strict sense but can be expressed using a Dirac delta function as in (4.77). Its Laplace transform

$$X_1(s) = \int_{-\infty}^{\infty} x_1(t) e^{-st}\, dt = \int_0^{\infty} e^{-st}\, dt$$

is well defined for $\sigma > 0$, yielding

$$x_1(t) \overset{\mathrm{LT}}{\longleftrightarrow} X_1(s) = \frac{1}{s}; \qquad \mathrm{ROC} = \{s \mid \Re(s) > 0\}.$$

Let x_2 be the negated and time-reversed Heaviside function, $x_2(t) = -x_1(-t)$. Its Laplace transform also exists and is also $1/s$, but with a different ROC,

$$x_2(t) \overset{\mathrm{LT}}{\longleftrightarrow} X_2(s) = \frac{1}{s}; \qquad \mathrm{ROC} = \{s \mid \Re(s) < 0\}.$$

This example shows two important points: a function for which the Fourier transform does not exist can have a well-defined Laplace transform; and a Laplace transform must have an associated ROC to uniquely specify a time-domain function.

4.5 Fourier series

Periodic functions arise in many settings, including through circular extension of a function defined on a finite interval. Series expansions of periodic functions in terms of Fourier coefficients are of great mathematical and scientific interest; questions of convergence occupied mathematicians for a long time after Fourier's original work, inspiring some of the major advances in functional analysis. Moreover, the Fourier series is the dual of the discrete-time Fourier transform seen in the previous chapter, and thus, its study is central to discrete-time signal processing as well.

Our treatment of the Fourier series will be brief; most of its properties are similar to those of the Fourier transform and the DFT. We follow a similar path to that used before: we start by defining an appropriate convolution for periodic functions, and the Fourier series emerges naturally from its eigenfunctions. We then define the Fourier series formally and proceed to discuss a few of its properties, including the duality with the DTFT. A key commonality with the DFT – not shared with the DTFT and Fourier transform – is that the Fourier series gives a countable basis for a Hilbert space. We develop this, specifically the completeness of the Fourier series basis.

4.5.1 Definition of the Fourier series

Eigenfunctions of the circular convolution operator The Fourier series arises from identifying the eigenfunctions of the circular convolution operator defined in (4.36). Like with all the other convolution operators we have studied, the eigenfunctions are of unit-modulus complex exponential form, and now we should expect the eigenfunctions to be periodic with period T. Thus, the eigenfunctions will be of the following form (compare with (4.37) for the Fourier transform and to (3.160) for the DFT):

$$v(t) = e^{j(2\pi/T)kt}, \qquad t \in \mathbb{R}, \tag{4.91}$$

where $k \in \mathbb{Z}$ is called the *discrete frequency*. It represents the multiple of the *fundamental frequency* $\omega_0 = 2\pi/T$ (expressed in radians per second). Unlike for the Fourier transform, however, there are countably many eigenfunctions.

To verify that v is an eigenfunction of the circular convolution operator H from (4.36) is a simple computation:

$$\begin{aligned}(Hv)(t) &= (h \circledast v)(t) = \int_{-T/2}^{T/2} v(t-\tau)h(\tau)\,d\tau = \int_{-T/2}^{T/2} e^{j(2\pi/T)k(t-\tau)}h(\tau)\,d\tau \\ &= \underbrace{\int_{-T/2}^{T/2} h(\tau)e^{-j(2\pi/T)k\tau}\,d\tau}_{\lambda_k}\ \underbrace{e^{j(2\pi/T)kt}}_{v(t)}.\end{aligned} \tag{4.92}$$

This shows that v is an eigenfunction of H with the corresponding eigenvalue λ_k. We denote this eigenvalue by H_k using the *frequency response* of the system, which is defined formally in (4.132a). We can thus rewrite (4.92) as

$$He^{j(2\pi/T)kt} = h \circledast e^{j(2\pi/T)kt} = H_k e^{j(2\pi/T)kt}. \tag{4.93}$$

Fourier series We are now ready to define the Fourier series, which amounts to projecting onto the subspaces generated by each of the eigenfunctions.

DEFINITION 4.13 (FOURIER SERIES) The *Fourier series coefficient sequence* of a periodic function x with period T is

$$X_k = \frac{1}{T}\int_{-T/2}^{T/2} x(t)e^{-j(2\pi/T)kt}\,dt, \qquad k \in \mathbb{Z}. \tag{4.94a}$$

It exists when (4.94a) is defined and finite for all $k \in \mathbb{Z}$; we then call it the *spectrum* of x. The *Fourier series* reconstruction from X is

$$x(t) = \sum_{k\in\mathbb{Z}} X_k e^{j(2\pi/T)kt}, \qquad t \in [-\tfrac{1}{2}T, \tfrac{1}{2}T). \tag{4.94b}$$

When the Fourier series coefficient sequence exists and the reconstruction converges, we denote the Fourier series pair as

$$x(t) \quad \overset{\text{FS}}{\longleftrightarrow} \quad X_k.$$

The integral in (4.94a) is the $\mathcal{L}^2([-\frac{1}{2}T,\, \frac{1}{2}T))$ inner product $\langle x,\, v\rangle$, and we will see shortly that by varying k over the integers we have an orthonormal basis for $\mathcal{L}^2([-\frac{1}{2}T,\, \frac{1}{2}T))$. Before considering the existence of the Fourier series coefficients and convergence of the reconstruction, we present two formal equivalences.

Relation of the Fourier series coefficients to the Fourier transform For a function that is identically zero outside of $[-\frac{1}{2}T,\, \frac{1}{2}T)$, the integral in (4.94a) is the same as the Fourier transform integral (4.40a) for frequency $\omega = (2\pi/T)k$. In particular, an arbitrary T-periodic function x can be restricted to $[-\frac{1}{2}T,\, \frac{1}{2}T)$ by defining $\widetilde{x} = 1_{[-T/2,T/2)}\, x$. It is then easy to verify (see Solved exercise 4.4) that

$$X_k \;=\; \frac{1}{T}\widetilde{X}\left(\frac{2\pi}{T}k\right), \qquad k \in \mathbb{Z}, \tag{4.95}$$

where $\widetilde{X}$ is the Fourier transform of $\widetilde{x}$; in other words, the Fourier series coefficients of a periodic function are scaled samples of the Fourier transform of the same function restricted to one period.

For a nonzero T-periodic function, the Fourier transform integral (4.40a) does not converge absolutely, so the Fourier transform does not exist in the strict sense developed in Section 4.4.2. We will see in Section 4.5.3 how the Fourier series coefficients lead to a Fourier transform expression that is a sum of weighted Dirac delta functions.

Duality of the Fourier series and the DTFT Consider the Fourier series coefficient expression (4.94a) for period $T = 2\pi$:

$$X_k \;=\; \frac{1}{2\pi}\int_{-\pi}^{\pi} x(t)e^{-jkt}\, dt, \qquad k \in \mathbb{Z}. \tag{4.96}$$

We can recognize this as the inverse DTFT in (3.80b). In other words, the inverse DTFT expresses the sequence x_n as the Fourier series coefficients of a 2π-periodic function $X(e^{j\omega})$. Conversely, a periodic function $x(t)$ can be seen as the DTFT of the Fourier series coefficient sequence X_k. Table 4.4 on page 401 summarizes this duality.

Real Fourier series For a real-valued periodic function, the real Fourier series gives an expansion with respect to sine and cosine functions with real coefficients. Exercise 4.10 explores the connection between the real Fourier series and (4.94b).

4.5.2 Existence and convergence of the Fourier series

Since the integral (4.94a) defining the Fourier series is over a finite interval, the existence of the Fourier series coefficients is usually perfectly clear. For example, when x is bounded, the integral is defined and finite for every $k \in \mathbb{Z}$. Furthermore, continuity implies boundedness, so any continuous periodic function will have Fourier series coefficients.

While there is little subtlety in the existence of the Fourier series coefficients of x, understanding the sense in which (4.94b) recovers x requires some care. Sometimes, exact (pointwise) recovery can be guaranteed, but we are generally satisfied with convergence in the $\mathcal{L}^2$ norm. This convergence follows from establishing the complex exponential sequences in (4.94b) as an orthogonal basis for $\mathcal{L}^2([-\frac{1}{2}T, \frac{1}{2}T))$. Some of this discussion should be reminiscent of Section 3.4.2 because of the duality between the DTFT and Fourier series.

Spectrum in $\ell^1(\mathbb{Z})$ Let the T-periodic function x have spectrum $X \in \ell^1(\mathbb{Z})$. Then, (4.94b) converges absolutely for every t since

$$\sum_{k\in\mathbb{Z}} \left|X_k e^{j(2\pi/T)kt}\right| = \sum_{k\in\mathbb{Z}} |X_k|\left|e^{j(2\pi/T)kt}\right| = \sum_{k\in\mathbb{Z}} |X_k| = \|X\|_1 < \infty.$$

Moreover, the series (4.94b) converges to a continuous function $\widehat{x}(t)$.[86] When x is continuous and $X \in \ell^1(\mathbb{Z})$,

$$x(t) = \widehat{x}(t), \qquad \text{for all } t \in \mathbb{R}; \tag{4.97}$$

this follows from $\ell^1(\mathbb{Z}) \subset \ell^2(\mathbb{Z})$ and the convergence in $\mathcal{L}^2([-\frac{1}{2}T, \frac{1}{2}T))$ norm that we will show shortly for $X \in \ell^2(\mathbb{Z})$.[87]

Fourier series as an orthonormal basis expansion The expression (4.94a) is an inner product $X_k = \langle x, \varphi_k \rangle$, where $\varphi_k(t) = (1/T)e^{j(2\pi/T)kt}$. The set $\{\varphi_k\}_{k\in\mathbb{Z}}$ is a basis for $\mathcal{L}^2([-\frac{1}{2}T, \frac{1}{2}T))$ with dual basis $\{\widetilde{\varphi}_k\}_{k\in\mathbb{Z}}$, where $\widetilde{\varphi} = e^{j(2\pi/T)kt}$. Thus, the Fourier series is a biorthogonal basis expansion as in Theorem 2.44. However, since φ_k and $\widetilde{\varphi}_k$ differ solely by a constant factor, a simple rescaling allows us to see the Fourier series as an orthonormal basis expansion. The benefits of orthonormality include successive approximation and simpler and more useful Parseval equalities (see Section 2.5.2).

THEOREM 4.14 (ORTHONORMAL BASIS FROM FOURIER SERIES) The set $\{\varphi_k\}_{k\in\mathbb{Z}}$ with

$$\varphi_k(t) = \frac{1}{\sqrt{T}} e^{j(2\pi/T)kt}, \qquad t \in [-\tfrac{1}{2}T, \tfrac{1}{2}T), \tag{4.98}$$

forms an orthonormal basis for $\mathcal{L}^2([-\frac{1}{2}T, \frac{1}{2}T))$.

---

[86] See Footnote 50 on page 218.

[87] Knowing x is continuous is not enough to imply that $X \in \ell^1(\mathbb{Z})$ and thus (4.97). However, x being continuously differentiable does imply (4.97).

Proof. It is easy to see that $\Phi = \{\varphi_k\}_{k\in\mathbb{Z}}$ is an orthonormal set: by separate elementary computations for $k = \ell$ and $k \neq \ell$,

$$\langle \varphi_k, \varphi_\ell \rangle \;=\; \frac{1}{T}\int_{-T/2}^{T/2} e^{j(2\pi/T)(k-\ell)t}\, dt \;=\; \delta_{k-\ell}. \tag{4.99}$$

It remains to show that Φ is a basis for $\mathcal{L}^2([-\frac{1}{2}T,\, \frac{1}{2}T))$, which requires that any $x \in \mathcal{L}^2([-\frac{1}{2}T,\, \frac{1}{2}T))$ is in $\overline{\operatorname{span}}(\Phi)$. Let

$$\alpha_k \;=\; \langle x, \varphi_k\rangle \;=\; \frac{1}{\sqrt{T}}\int_{-T/2}^{T/2} x(t) e^{-j(2\pi/T)kt}\, dt, \qquad k \in \mathbb{Z}, \tag{4.100}$$

and let $\widehat{x}_N$ be a $(2N+1)$-term approximation of x:

$$\widehat{x}_N(t) \;=\; \frac{1}{\sqrt{T}} \sum_{k=-N}^{N} \alpha_k e^{j(2\pi/T)kt}. \tag{4.101}$$

By showing

$$\lim_{N\to\infty} \|x - \widehat{x}_N\|^2 \;=\; \lim_{N\to\infty} \int_{-T/2}^{T/2} |x(t) - \widehat{x}_N(t)|^2\, dt \;=\; 0 \tag{4.102}$$

we establish $x \in \overline{\operatorname{span}}(\Phi)$ and hence the completeness of Φ as a basis for $\mathcal{L}^2([-\frac{1}{2}T,\, \frac{1}{2}T))$. Proving (4.102) for continuous functions is left for Exercise 4.11. The final desired result is then obtained by extending to general $\mathcal{L}^2(\mathbb{R})$ functions by the argument that continuous functions are dense in $\mathcal{L}^2(\mathbb{R})$.

Spectrum in $\ell^2(\mathbb{Z})$ Up to scaling by the constant $\sqrt{T}$, the Fourier series is an orthonormal basis expansion. The orthonormality leads to geometrically intuitive properties using Hilbert space tools. We summarize a few key results that follow from orthonormality in the following theorem.

THEOREM 4.15 (FOURIER SERIES ON $\mathcal{L}^2([-\frac{1}{2}T,\, \frac{1}{2}T))$) Let $x \in \mathcal{L}^2([-\frac{1}{2}T,\, \frac{1}{2}T))$ have Fourier series coefficients $X = Fx$. Then, the following hold:

(i) *$\mathcal{L}^2$ inversion:* Let $\widehat{x}$ be the Fourier series reconstruction from X. Then

$$\|x - \widehat{x}\| \;=\; 0 \tag{4.103}$$

using the $\mathcal{L}^2([-\frac{1}{2}T,\, \frac{1}{2}T))$ norm.

(ii) *Norm conservation:* The linear operator $\sqrt{T}F : \mathcal{L}^2([-\frac{1}{2}T,\, \frac{1}{2}T)) \to \ell^2(\mathbb{Z})$ is unitary. This yields the Parseval equalities

$$\|x\|^2 \;=\; \int_{-T/2}^{T/2} |x(t)|^2\, dt \;=\; T\sum_{k\in\mathbb{Z}} |X_k|^2, \tag{4.104a}$$

$$\langle x, y\rangle \;=\; \int_{-T/2}^{T/2} x(t) y^*(t)\, dt \;=\; T\sum_{k\in\mathbb{Z}} X_k Y_k^*, \tag{4.104b}$$

where the T-periodic function y has Fourier series coefficients Y.

(iii) *Least-squares approximation:* The function

$$\widehat{x}_N(t) = \sum_{k=-N}^{N} X_k e^{j(2\pi/T)kt} \tag{4.105}$$

is the least-squares approximation of x on the subspace spanned by $\{\varphi_k(t)\}_{k=-N}^{N}$.

These results follow from results on orthonormal basis expansion, Theorem 2.39; Parseval equalities, Theorem 2.40; and orthogonal projection onto a subspace, Theorem 2.41.

4.5.3 Properties of the Fourier series

Basic properties

We list here basic properties of the Fourier series; Table 4.3 summarizes these, together with symmetries as well as a few standard Fourier series pairs. Of course, all the expressions must be well defined for these properties to hold.

Linearity The Fourier series expansion operator F is a linear operator, or

$$\alpha x(t) + \beta y(t) \quad \overset{\text{FS}}{\longleftrightarrow} \quad \alpha X_k + \beta Y_k. \tag{4.106}$$

Shift in time The Fourier series pair corresponding to a shift in time by t_0 is

$$x(t - t_0) \quad \overset{\text{FS}}{\longleftrightarrow} \quad e^{-j(2\pi/T)kt_0} X_k. \tag{4.107}$$

Shift in frequency The Fourier series pair corresponding to a shift in frequency by k_0 is

$$e^{j(2\pi/T)k_0 t} x(t) \quad \overset{\text{FS}}{\longleftrightarrow} \quad X_{k-k_0}. \tag{4.108}$$

As in Chapter 3, a shift in frequency is called *modulation*.

Time reversal The Fourier series pair corresponding to time reversal $x(-t)$ is

$$x(-t) \quad \overset{\text{FS}}{\longleftrightarrow} \quad X_{-k}. \tag{4.109}$$

Differentiation The Fourier series pair corresponding to differentiation in time is

$$\frac{d^n x(t)}{dt^n} \quad \overset{\text{FS}}{\longleftrightarrow} \quad \left(j\frac{2\pi}{T}k\right)^n X_k, \tag{4.110}$$

FS properties	Time domain	FS domain
Basic properties		
Linearity	$\alpha x(t) + \beta y(t)$	$\alpha X_k + \beta Y_k$
Shift in time	$x(t - t_0)$	$e^{-j(2\pi/T)kt_0} X_k$
Shift in frequency	$e^{j(2\pi/T)k_0 t} x(t)$	$X(k - k_0)$
Time reversal	$x(-t)$	X_{-k}
Differentiation	$d^n x(t)/dt^n$	$(j2\pi k/T)^n X_k$
Integration	$\int_{-T/2}^{t} x(\tau)\, d\tau$	$(T/(j2\pi k))X_k,\ X_0 = 0$
Circular convolution in time	$(h \circledast x)(t)$	$T H_k X_k$
Convolution in frequency	$h(t)\, x(t)$	$(H * X)_k$
Circular deterministic autocorrelation	$a(t) = \int_{-T/2}^{T/2} x(\tau)\, x^*(\tau - t)\, d\tau$	$A_k = T\,\lvert X_k\rvert^2$
Circular deterministic crosscorrelation	$c(t) = \int_{-T/2}^{T/2} x(\tau)\, y^*(\tau - t)\, d\tau$	$C_k = T X_k Y_k^*$
Parseval equality	$\lVert x\rVert^2 = \int_{-T/2}^{T/2} \lvert x(t)\rvert^2\, dt = T \sum_{k\in\mathbb{Z}} \lvert X_k\rvert^2 = T\,\lVert X\rVert^2$	
Related functions		
Conjugate	$x^*(t)$	X_{-k}^*
Conjugate, time-reversed	$x^*(-t)$	X_k^*
Real part	$\Re(x(t))$	$(X_k + X_{-k}^*)/2$
Imaginary part	$\Im(x(t))$	$(X_k - X_{-k}^*)/(2j)$
Conjugate-symmetric part	$(x(t) + x^*(-t))/2$	$\Re(X_k)$
Conjugate-antisymmetric part	$(x(t) - x^*(-t))/(2j)$	$\Im(X_k)$
Symmetries for real x		
X conjugate symmetric		$X_k = X_{-k}^*$
Real part of X even		$\Re(X_k) = \Re(X_{-k})$
Imaginary part of X odd		$\Im(X_k) = -\Im(X_{-k})$
Magnitude of X even		$\lvert X_k\rvert = \lvert X_{-k}\rvert$
Phase of X odd		$\arg X_k = -\arg X_{-k}$
Common transform pairs		
Dirac comb	$\sum_{n\in\mathbb{Z}} \delta(t - nT)$	$1/T$
Periodic sinc function (ideal lowpass filter)	$\sqrt{\frac{k_0}{T}} \frac{\operatorname{sinc}(\pi k_0 t/T)}{\operatorname{sinc}(\pi t/T)}$	$\begin{cases} 1/\sqrt{k_0 T}, & \lvert k\rvert \le \frac{1}{2}(k_0 - 1); \\ 0, & \text{otherwise} \end{cases}$
Box function (one period)	$\begin{cases} 1/\sqrt{t_0}, & \lvert t\rvert \le \frac{1}{2}t_0; \\ 0, & \frac{1}{2}t_0 < \lvert t\rvert \le \frac{1}{2}T \end{cases}$	$\frac{\sqrt{t_0}}{T} \operatorname{sinc}(\pi k t_0/T)$
Square wave (one period with $T = 1$)	$\begin{cases} -1, & t \in [-\frac{1}{2}, 0); \\ 1, & t \in [0, \frac{1}{2}) \end{cases}$	$\begin{cases} -2j/(\pi k), & k \text{ odd}; \\ 0, & k \text{ even} \end{cases}$
Triangle wave (one period with $T = 1$)	$\frac{1}{2} - \lvert t\rvert,\quad \lvert t\rvert \le \frac{1}{2}$	$\begin{cases} 1/4, & k = 0; \\ 1/(\pi k)^2, & k \text{ odd}; \\ 0, & k \neq 0 \text{ even} \end{cases}$
Sawtooth wave (one period with $T = 1$)	$2t,\quad \lvert t\rvert \le \frac{1}{2}$	$\begin{cases} 0, & k = 0; \\ j(-1)^k/(\pi k), & k \neq 0 \end{cases}$

Table 4.3 Properties of the Fourier series.

assuming that the indicated nth derivative exists. This can be derived for $n = 1$ by computing the Fourier series coefficient of x' at discrete frequency k as follows:

$$\begin{aligned}
&\int_{-T/2}^{T/2} x'(t) e^{-j(2\pi/T)kt}\, dt \\
&\stackrel{(a)}{=} x(t) e^{-j(2\pi/T)kt} \Big|_{-T/2}^{T/2} - \int_{-T/2}^{T/2} x(t) \left(-j\frac{2\pi}{T}k\right) e^{-j(2\pi/T)kt}\, dt \\
&\stackrel{(b)}{=} j\frac{2\pi}{T}k \int_{-T/2}^{T/2} x(t) e^{-j(2\pi/T)kt}\, dt \stackrel{(c)}{=} \left(j\frac{2\pi}{T}k\right) X_k,
\end{aligned}$$

where (a) follows from integration by parts; (b) from $x(-\frac{1}{2}T) = x(\frac{1}{2}T)$; and (c) from the Fourier series integral, (4.94a). Repeating for higher-order derivatives gives (4.110).

An alternative derivation is to start with the Fourier series reconstruction

$$x(t) = \sum_{k\in\mathbb{Z}} X_k e^{j(2\pi/T)kt}$$

and differentiate term-by-term,

$$x'(t) = \sum_{k\in\mathbb{Z}} \left(j\frac{2\pi}{T}k\right) X_k e^{j(2\pi/T)kt}. \tag{4.111}$$

The coefficient of $e^{j(2\pi/T)kt}$ is the Fourier series coefficient of the derivative at discrete frequency k, as we wanted to show.

Integration The Fourier series pair corresponding to the integral of a periodic function with zero mean ($X_0 = 0$) is

$$\int_{-T/2}^{t} x(\tau)\, d\tau \quad \stackrel{\text{FS}}{\longleftrightarrow} \quad \frac{T}{j2\pi k} X_k, \quad \text{for } k \neq 0, \tag{4.112}$$

where the Fourier series coefficient for discrete frequency $k = 0$ must be computed separately. This can be derived similarly to (4.110). Exercise 4.12 develops this for real Fourier series.

Circular convolution in time The Fourier series pair corresponding to the circular convolution of T-periodic functions is

$$(h \circledast x)(t) \quad \stackrel{\text{FS}}{\longleftrightarrow} \quad T H_k X_k. \tag{4.113}$$

Thus, as we have seen many times, convolution in the time domain corresponds to multiplication in the Fourier domain. Proof of (4.113) is left for Exercise 4.13, which also establishes an analogous property for linear convolution of a periodic function x and an aperiodic filter h.

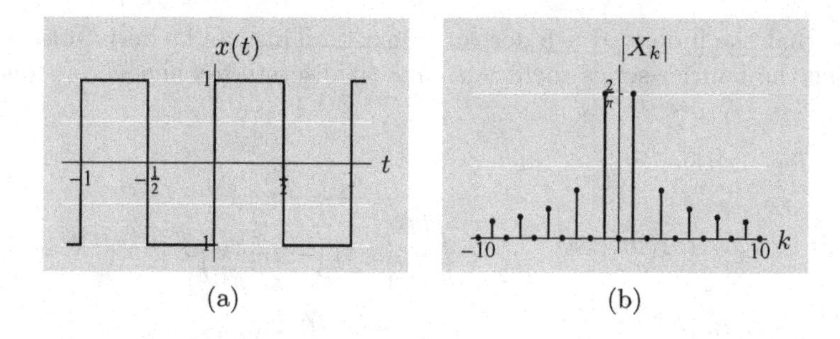

Figure 4.10 (a) The square wave (4.117). (b) Magnitude response of its Fourier series coefficients.

Convolution in frequency The Fourier series pair corresponding to convolution in frequency is

$$h(t)\, x(t) \quad \stackrel{\text{FS}}{\longleftrightarrow} \quad (H * X)_k. \tag{4.114}$$

The convolution in frequency property (4.114) is dual to the convolution in time property (4.113).

Circular deterministic autocorrelation The Fourier series pair corresponding to the *circular deterministic autocorrelation* of a T-periodic function x is

$$a(t) \;=\; \int_{-T/2}^{T/2} x(\tau)x^*(\tau - t)\, d\tau \quad \stackrel{\text{FS}}{\longleftrightarrow} \quad A_k \;=\; T|X_k|^2. \tag{4.115}$$

Circular deterministic crosscorrelation The Fourier series pair corresponding to the *circular deterministic crosscorrelation* of T-periodic functions x and y is

$$c(t) \;=\; \int_{-T/2}^{T/2} x(\tau)y^*(\tau - t)\, d\tau \quad \stackrel{\text{FS}}{\longleftrightarrow} \quad C_k \;=\; TX_kY_k^*. \tag{4.116}$$

Transform pairs

We derive Fourier series pairs of a few simple periodic functions to illustrate some of the properties seen above. Table 4.3 summarizes these pairs (along with some additional ones).

Square wave Let x be a square wave of period $T = 1$, with one period given by

$$x(t) \;=\; \begin{cases} -1, & \text{for } t \in [-\frac{1}{2},\, 0); \\ 1, & \text{for } t \in [0,\, \frac{1}{2}); \end{cases} \tag{4.117}$$

it is shown in Figure 4.10(a). This function is real and odd; thus, we expect to find

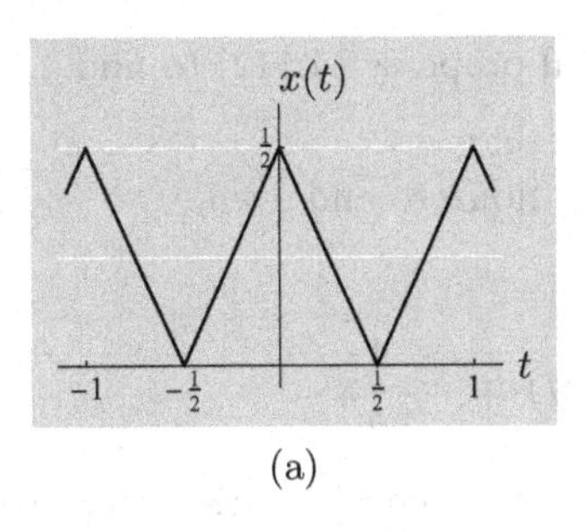

(a)

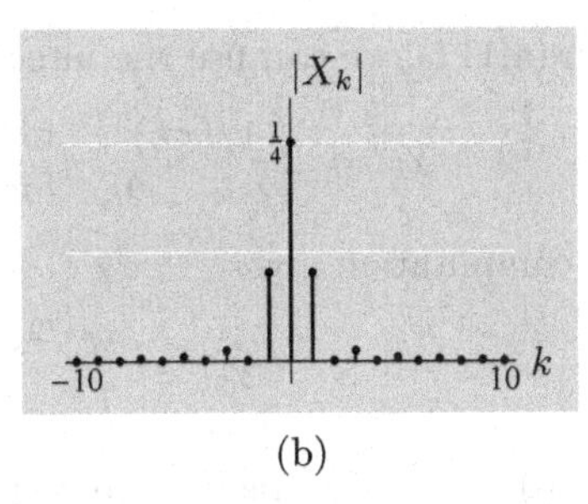

(b)

Figure 4.11 (a) The triangle wave (4.119). (b) Magnitude response of its Fourier series coefficients.

that its Fourier series coefficients are purely imaginary (see Table 4.3). Indeed,

$$X_0 = \int_{-1/2}^{1/2} x(t)\, dt = 0, \tag{4.118a}$$

and, for nonzero k,

$$\begin{aligned} X_k &= -\int_{-1/2}^{0} e^{-j2\pi kt}\, dt + \int_{0}^{1/2} e^{-j2\pi kt}\, dt \\ &= \frac{1}{j2\pi k}(1 - e^{j\pi k}) - \frac{1}{j2\pi k}(e^{j\pi k} - 1) \\ &= \frac{1}{j\pi k}\left(1 - (-1)^k\right) = \begin{cases} -2j/(\pi k), & \text{for } k \text{ odd;} \\ 0, & \text{for } k \text{ even.} \end{cases} \end{aligned} \tag{4.118b}$$

Figure 4.10(b) shows the magnitude response of Fourier series coefficients. Thus,

$$\begin{aligned} x(t) &= \frac{2}{\pi} \sum_{\ell \in \mathbb{Z}} \frac{-j}{2\ell + 1} e^{j2\pi(2\ell+1)t} \\ &= \frac{2}{\pi} \sum_{\ell=0}^{\infty} \frac{-j}{2\ell + 1} \left(e^{j2\pi(2\ell+1)t} - e^{-j2\pi(2\ell+1)t} \right) \\ &= \frac{4}{\pi} \sum_{\ell=0}^{\infty} \frac{1}{2\ell + 1} \sin(2\pi(2\ell+1)t). \end{aligned}$$

Triangle wave Let y be a triangle wave of period $T = 1$, with one period given by

$$y(t) = \frac{1}{2} - |t|, \qquad \text{for } t \in [-\tfrac{1}{2}, \tfrac{1}{2}); \tag{4.119}$$

it is shown in Figure 4.11(a). Since

$$y(t) = \int_{-1/2}^{t} x(\tau)\, d\tau, \qquad \text{for } t \in [-\tfrac{1}{2}, \tfrac{1}{2}),$$

with x from (4.117), we can use the integral property (4.112) to find

$$Y_k = \begin{cases} 1/(\pi k)^2, & \text{for } k \text{ odd;} \\ 0, & \text{for } k \text{ nonzero and even.} \end{cases} \tag{4.120a}$$

A separate computation gives

$$Y_0 = \int_{-1/2}^{1/2} y(t)\, dt = \frac{1}{4}. \tag{4.120b}$$

Figure 4.11(b) shows the magnitude response of Fourier series coefficients. Exercise 4.14 explores three alternatives for computing this Fourier series.

Dirac comb The *Dirac comb* or *picket-fence function* (with spacing T) is a sum of Dirac delta functions at uniformly spaced shifts nT for all $n \in \mathbb{Z}$,

$$s_T(t) = \sum_{n \in \mathbb{Z}} \delta(t - nT). \tag{4.121}$$

We can find its Fourier series coefficients by directly evaluating (4.94a):

$$\begin{aligned} S_{T,k} &= \frac{1}{T} \int_{-T/2}^{T/2} \sum_{n \in \mathbb{Z}} \delta(t - nT)\, e^{-j(2\pi/T)kt}\, dt \\ &\overset{(a)}{=} \frac{1}{T} \int_{-T/2}^{T/2} \delta(t)\, e^{-j(2\pi/T)kt}\, dt \overset{(b)}{=} \frac{1}{T}, \qquad k \in \mathbb{Z}, \end{aligned} \tag{4.122}$$

where (a) follows from only the $n = 0$ term of the sum affecting the integrand on $(-\frac{1}{2}T, \frac{1}{2}T)$; and (b) from the sifting property of the Dirac delta function, (3.293). Using these coefficients in the Fourier series reconstruction (4.94b) gives

$$s_T(t) = \frac{1}{T} \sum_{k \in \mathbb{Z}} e^{j(2\pi/T)kt}. \tag{4.123}$$

The series in (4.123) is problematic because it does not converge absolutely for any value of t.[88] Putting aside this concern and combining (4.121) and (4.123) gives

$$\frac{1}{T} \sum_{k \in \mathbb{Z}} e^{j(2\pi/T)kt} = \sum_{n \in \mathbb{Z}} \delta(t - nT). \tag{4.124}$$

While (4.124) cannot be justified using only elementary notions of convergence, it can guide us to other useful expressions. For example, if x is continuous at all points $\{nT\}_{n \in \mathbb{Z}}$, then

$$\frac{1}{T} \sum_{k \in \mathbb{Z}} \int_{-\infty}^{\infty} x(t) e^{j(2\pi/T)kt}\, dt \overset{(a)}{=} \sum_{n \in \mathbb{Z}} \int_{-\infty}^{\infty} x(t) \delta(t - nT)\, dt \overset{(b)}{=} \sum_{n \in \mathbb{Z}} x(nT),$$

assuming the sums converge absolutely, where (a) follows from (4.124); and (b) from the sifting property of the Dirac delta function, (3.293). We return to (4.124) shortly in deriving the Poisson sum formula.

[88]Recall from Appendix 2.A.2 that a doubly infinite series is said to converge when it converges absolutely.

Fourier transform of periodic functions

As noted earlier, the Fourier transform of a nonzero periodic function does not exist in the elementary sense since the integral (4.40a) does not exist for any value of ω. However, through the use of Dirac combs and Fourier series, we will arrive at a Fourier transform expression involving Dirac delta functions as well as the elegant and useful Poisson sum formula.

Let x be a T-periodic function, and let $\widetilde{x}$ be the restriction of x to $[-\frac{1}{2}T, \frac{1}{2}T)$,

$$\widetilde{x}(t) = 1_{[-T/2,T/2)}\, x(t), \qquad t \in \mathbb{R}.$$

When the Fourier transform of $\widetilde{x}$ exists, the Fourier series of x exists as well; they are related through (4.95). While the Fourier transform of x does not exist (except in the trivial case of $x = \mathbf{0}$), we can find a useful expression for it using Dirac delta functions (with all the usual cautions and caveats).

Using the Dirac comb, we can write

$$x(t) = (s_T * \widetilde{x})(t), \tag{4.125a}$$

where the equality holds for all $t \in \mathbb{R}$ at which x is continuous. (This can be checked by noting that – through linearity and the shifting property of convolution, (4.32e) – the Dirac delta component $\delta(t - nT)$ in s_T is responsible for the period of x on $[(n - \frac{1}{2})T, (n + \frac{1}{2})T)$.) Assuming that we may apply the convolution theorem (4.62) to (4.125a), we have

$$X(\omega) = S_T(\omega)\, \widetilde{X}(\omega), \qquad \omega \in \mathbb{R}. \tag{4.125b}$$

We thus require an expression for the Fourier transform of the Dirac comb.

Using the shift in time property from Table 4.1 and linearity, we can write the Fourier transform of the Dirac comb as

$$S_T(\omega) = \sum_{n\in\mathbb{Z}} e^{-j\omega n T}. \tag{4.126}$$

The Fourier transform S_T in (4.126) does not strictly make sense because it does not converge absolutely for any value of ω; this should come as no surprise since s_T in (4.121) is not really a function. Nevertheless, this leads to a useful expression for S_T. By changing variables in (4.124) and using properties of the Dirac delta function (see Exercise 4.16), S_T can be expressed as a scaled Dirac comb with spacing $2\pi/T$,

$$S_T(\omega) = \sum_{n\in\mathbb{Z}} e^{-j\omega n T} = \frac{2\pi}{T} \sum_{k\in\mathbb{Z}} \delta\left(\omega - \frac{2\pi}{T}k\right) = \frac{2\pi}{T} s_{2\pi/T}(\omega). \tag{4.127}$$

Substituting (4.127) into (4.125b) yields

$$X(\omega) = \frac{2\pi}{T} \sum_{k\in\mathbb{Z}} \widetilde{X}\left(\frac{2\pi}{T}k\right) \delta\left(\omega - \frac{2\pi}{T}k\right) \tag{4.128a}$$

by using the sampling property of the Dirac delta function, (3.294).[89] By comparison with (4.95), we may write this Fourier transform using the Fourier series coefficients of x,

$$X(\omega) = 2\pi \sum_{k\in\mathbb{Z}} X_k \delta\Big(\omega - \frac{2\pi}{T}k\Big). \tag{4.128b}$$

In other words, the Fourier transform of a T-periodic function is a weighted Dirac comb with spacing $2\pi/T$ and weights given by the Fourier series coefficients of the function scaled by 2π.

Recall that in (4.125a) we used convolution with a Dirac comb with spacing T to periodically extend a function with support limited to $[-\frac{1}{2}T, \frac{1}{2}T)$ to obtain a T-periodic function. Actually, even if we had not started with a function with support limited to $[-\frac{1}{2}T, \frac{1}{2}T)$, we still would have obtained a T-periodic function, provided that the convolution converges. This gives the Poisson sum formula, summarized in the following theorem.

THEOREM 4.16 (POISSON SUM FORMULA) Let s_T be the Dirac comb defined in (4.121), and let x be a function with decay sufficient for the periodization

$$(s_T * x)(t) = \sum_{n\in\mathbb{Z}} x(t - nT) \tag{4.129a}$$

to converge absolutely for all t. Then

$$\sum_{n\in\mathbb{Z}} x(t - nT) = \frac{1}{T} \sum_{k\in\mathbb{Z}} X\Big(\frac{2\pi}{T}k\Big) e^{j(2\pi/T)kt}, \tag{4.129b}$$

where X is the Fourier transform of x, which is assumed to be continuous at $\{2\pi k/T\}_{k\in\mathbb{Z}}$. Upon specializing to $T = 1$ and $t = 0$,

$$\sum_{n\in\mathbb{Z}} x(n) = \sum_{k\in\mathbb{Z}} X(2\pi k). \tag{4.129c}$$

Proof. Let $x_T = s_T * x$. Following the derivation of (4.128a),

$$X_T(\omega) = \frac{2\pi}{T} \sum_{k\in\mathbb{Z}} X\Big(\frac{2\pi}{T}k\Big)\, \delta\Big(\omega - \frac{2\pi}{T}k\Big). \tag{4.130}$$

[89] Use of the sampling property requires continuity of $\widetilde{X}$ at $\{2\pi k/T\}_{k\in\mathbb{Z}}$. This continuity follows from the finite support of $\widetilde{x}$.

Taking the inverse Fourier transform of (4.130), we get

$$\begin{aligned} x_T(t) &= \frac{1}{2\pi}\int_{-\infty}^{\infty} X_T(\omega)e^{j\omega t}\,d\omega = \frac{1}{T}\int_{-\infty}^{\infty}\sum_{k\in\mathbb{Z}} X\left(\frac{2\pi}{T}k\right)\delta\left(\omega-\frac{2\pi}{T}k\right)e^{j\omega t}\,d\omega \\ &= \frac{1}{T}\sum_{k\in\mathbb{Z}} X\left(\frac{2\pi}{T}k\right)\int_{-\infty}^{\infty}\delta\left(\omega-\frac{2\pi}{T}k\right)e^{j\omega t}\,d\omega \\ &\stackrel{(a)}{=} \frac{1}{T}\sum_{k\in\mathbb{Z}} X\left(\frac{2\pi}{T}k\right)e^{j(2\pi/T)kt}, \end{aligned}$$

where (a) follows from the sampling property of the Dirac delta function, (3.294). We could arrive at the same result by performing the inverse Fourier transform term-by-term in (4.130).

The Poisson sum formula has many applications; in signal processing, it is used in the proof of the sampling theorem (see Chapter 5), since the process of sampling can be described using multiplication of an input signal by a Dirac comb.

Regularity and spectral decay

Smoothness of a periodic function and decay of its Fourier series coefficients are related. A discontinuous periodic function such as the square wave has Fourier series coefficients decaying only as $O(1/k)$ (see (4.118)), while a continuous periodic function such as the triangle wave leads to an $O(1/k^2)$ decay (see (4.120)). This is analogous to the difference in decays of Fourier transforms for the corresponding aperiodic functions in Examples 4.8(i) and (ii).

More generally, one can relate the Fourier series coefficients of a periodic function with the Fourier transform of one period of the function using (4.95) and thus obtain regularity characterizations from those in Section 4.4.5. For example, (4.83) gives a characterization of C^q regularity. Suppose that the restricted version $\widetilde{x} = 1_{[-T/2,T/2)}\,x$ of $x \in \mathcal{L}^2([-\frac{1}{2}T, \frac{1}{2}T))$ has the Fourier transform satisfying

$$|\widetilde{X}(\omega)| \le \frac{\gamma}{1+|\omega|^{q+1+\varepsilon}} \qquad \text{for all } \omega \in \mathbb{R}, \tag{4.131a}$$

for some nonnegative integer q and positive constants γ and ε. Then, $\widetilde{x} \in C^q$ by (4.84b), and this can be extended to show that x has q continuous derivatives everywhere (see Exercise 4.18). A major difficulty with this approach, however, is that the restriction to $[-\frac{1}{2}T, \frac{1}{2}T)$ will usually create discontinuities.

A condition similar to (4.131a) but directly on the Fourier series (rather than on the Fourier transform of the restriction),

$$|X_k| \le \frac{\gamma}{1+|k|^{q+1+\varepsilon}} \qquad \text{for all } k \in \mathbb{Z}, \tag{4.131b}$$

for some nonnegative integer q and positive constants γ and ε, also implies that x has q continuous derivatives (see Exercise 4.19). Conversely, if the periodic function x has q continuous derivatives, then its Fourier series coefficients satisfy

$$|X_k| \le \frac{\gamma}{1+|k|^{q+1}} \qquad \text{for all } k \in \mathbb{Z}, \tag{4.131c}$$

for some positive constant γ.

4.5.4 Frequency response of filters

The Fourier series coefficient sequence of a T-periodic filter h specifying a circular convolution operator is called its *frequency response*:

$$H_k = \frac{1}{T}\int_{-T/2}^{T/2} h(t)e^{-j(2\pi/T)kt}\,dt, \qquad k \in \mathbb{Z}. \tag{4.132a}$$

The Fourier series reconstruction recovers the filter,

$$h(t) = \sum_{k\in\mathbb{Z}} H_k e^{j(2\pi/T)kt}, \qquad t \in \mathbb{R}. \tag{4.132b}$$

We can again denote the magnitude and phase as

$$H_k = |H_k|e^{j\arg(H_k)},$$

where the *magnitude response* $|H_k|$ is a real, nonnegative sequence and the *phase response* $\arg(H_k)$ is a real sequence between $-\pi$ and π.

Recall from Section 4.3.3 that circular convolution with T-periodic h arises from LSI processing when the system input is T-periodic. In this case, h represents the periodized version of the system impulse response, as in (4.34).

Diagonalization of the circular convolution operator Let H be the circular convolution operator associated with a T-periodic filter h. The frequency response of h gives a diagonal form for the operator H. This is a restatement of the Fourier series pair corresponding to circular convolution in time, (4.113), and we have made similar statements for the DTFT, DFT, and Fourier transform based on transform pairs (3.96), (3.171), and (4.62). Like for the DFT but unlike for the DTFT and Fourier transform, the diagonalization is literal because the transform is associated with a countable basis for our space.

Let x and $y = Hx$ be in $\mathcal{L}^2([-\frac{1}{2}T, \frac{1}{2}T))$, and let X and Y be their Fourier series coefficient sequences. Then

$$Y_k = H_k X_k, \qquad k \in \mathbb{Z},$$

by the circular convolution in time property, (4.113). Thus, the matrix representation Γ of the linear operator H using the Fourier series basis from (4.98) both for the domain and for the codomain is the infinite diagonal matrix

$$\Gamma = \operatorname{diag}(\ldots, H_{-1}, H_0, H_1, \ldots).$$

We illustrate this diagonalization in Figure 4.12.

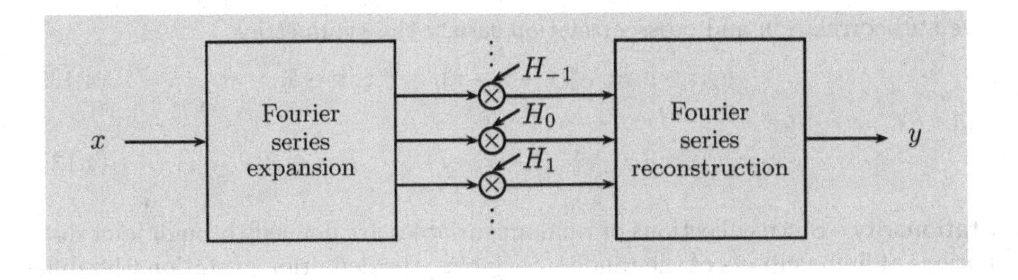

Figure 4.12 Diagonalization property of the Fourier series. The circular convolution operation $y = Hx = h \circledast x$ is implemented by pointwise multiplication of the spectrum of x by the frequency response of h.

4.6 Stochastic processes and systems

We now consider functions and continuous-time systems in the presence of uncertainty. Like in Chapter 3, we use probability theory to model uncertainty through random functions while considering only systems that act deterministically on these random functions.

This section follows the structure of the chapter in its entirety: We start with continuous-time stochastic processes (random functions), followed by the effect of systems (almost exclusively LSI systems) on stochastic processes in the time domain, and finally the application of Fourier-domain analysis. This section is brief in comparison with Section 3.8; in particular, we do not develop optimal estimation because modern systems do not implement estimation in continuous time.

4.6.1 Stochastic processes

A continuous-time stochastic process is an uncountably infinite collection of jointly distributed random variables $\{\mathrm{x}(t)\}_{t \in \mathbb{R}}$. For example, the sound pressure sensed by a microphone could be modeled as a stochastic process.

We use the following notations for moments and related quantities defined on stochastic processes:

$$\begin{array}{lll} \text{mean} & \mu_{\mathrm{x}}(t) & \mathrm{E}[\,\mathrm{x}(t)\,] \\ \text{variance} & \mathrm{var}(\mathrm{x}(t)) & \mathrm{E}[\,|\mathrm{x}(t) - \mu_{\mathrm{x}}(t)|^2\,] \\ \text{standard deviation} & \sigma_{\mathrm{x}}(t) & \sqrt{\mathrm{var}(\mathrm{x}(t))} \\ \text{autocorrelation} & a_{\mathrm{x}}(t, \tau) & \mathrm{E}[\,\mathrm{x}(t)\,\mathrm{x}^*(t-\tau)\,] \\ \text{crosscorrelation} & c_{\mathrm{x},\mathrm{y}}(t, \tau) & \mathrm{E}[\,\mathrm{x}(t)\,\mathrm{y}^*(t-\tau)\,] \end{array} \tag{4.133}$$

Like in discrete time, these are referred to as *second-order statistics.* The variance can be computed from the autocorrelation and mean through

$$\sigma_{\mathrm{x}}^2(t) = a_{\mathrm{x}}(t, 0) - |\mu_{\mathrm{x}}(t)|^2, \qquad t \in \mathbb{R}. \tag{4.134}$$

The autocorrelation and crosscorrelation satisfy the symmetries

$$a_{\mathrm{x}}(t,\tau) = a_{\mathrm{x}}^{*}(t-\tau,-\tau), \qquad t,\tau \in \mathbb{R}, \tag{4.135a}$$

and

$$c_{\mathrm{y,x}}(t,\tau) = c_{\mathrm{x,y}}^{*}(t-\tau,-\tau), \qquad t,\tau \in \mathbb{R}. \tag{4.135b}$$

Stationarity Since collections of random variables are defined through joint distributions of finite subsets of the random variables, the definition of stationarity differs between discrete time and continuous time only in the notation for the process.

DEFINITION 4.17 (STATIONARY PROCESS) A continuous-time stochastic process x is called *stationary* when, for any finite set of time indices $\{t_0, t_1, \ldots, t_L\} \subset \mathbb{R}$ and any time shift $\tau \in \mathbb{R}$, the joint distributions of

$$(\mathrm{x}(t_0), \mathrm{x}(t_1), \ldots, \mathrm{x}(t_L)) \quad \text{and} \quad (\mathrm{x}(t_0+\tau), \mathrm{x}(t_1+\tau), \ldots, \mathrm{x}(t_L+\tau))$$

are identical.

Stationarity is a highly restrictive condition. Most of the time, we will assume the weaker condition of wide-sense stationarity, which depends only on second-order statistics.

DEFINITION 4.18 (WIDE-SENSE STATIONARY PROCESS) A continuous-time stochastic process x is called *wide-sense stationary (WSS)* when its mean function $\mu_{\mathrm{x}}(t)$ is a constant,

$$\mu_{\mathrm{x}}(t) = \mathrm{E}[\,\mathrm{x}(t)\,] = \mu_{\mathrm{x}}, \qquad t \in \mathbb{R}, \tag{4.136a}$$

and its autocorrelation depends only on the time difference τ,

$$a_{\mathrm{x}}(t,\tau) = \mathrm{E}[\,\mathrm{x}(t)\,\mathrm{x}^{*}(t-\tau)\,] = a_{\mathrm{x}}(\tau), \qquad t,\tau \in \mathbb{R}. \tag{4.136b}$$

Stochastic processes x and y are called jointly WSS when each is WSS and their crosscorrelation depends only on the time difference τ,

$$c_{\mathrm{x,y}}(t,\tau) = \mathrm{E}[\,\mathrm{x}(t)\,\mathrm{y}^{*}(t-\tau)\,] = c_{\mathrm{x,y}}(\tau), \qquad t,\tau \in \mathbb{R}. \tag{4.136c}$$

With wide-sense stationarity, (4.134) and (4.135a) simplify to

$$\sigma_{\mathrm{x}}^{2}(t) = a_{\mathrm{x}}(0) - |\mu_{\mathrm{x}}|^{2} = \sigma_{\mathrm{x}}^{2}, \qquad t \in \mathbb{R},$$

and

$$a_{\mathrm{x}}(\tau) = a_{\mathrm{x}}^{*}(-\tau), \qquad \tau \in \mathbb{Z}.$$

With joint wide-sense stationarity, (4.135b) simplifies to the conjugate symmetry

$$c_{\mathrm{y,x}}(\tau) = c_{\mathrm{x,y}}^{*}(-\tau), \qquad \tau \in \mathbb{R}.$$

White noise A *white noise*[90] process x is a WSS stochastic process whose mean is zero and whose elements are uncorrelated,

$$\mu_{\mathrm{x}}(t) = 0; \qquad a_{\mathrm{x}}(t) = \sigma_{\mathrm{x}}^2 \delta(t). \tag{4.137}$$

Like in discrete time, the random variables in a white noise process are not always independent.

While comparing (3.235) and (4.137) reveals a close analogy between discrete- and continuous-time white noise, it also suggests that continuous-time white noise is a more subtle concept. The meaning of the Kronecker delta sequence is straightforward, while, as we have said before, the Dirac delta function is a convenient abstraction that makes sense when part of an integrand. Analogously, the concept of discrete-time white noise is straightforward, while continuous-time white noise makes sense after suitable filtering; see the *Further reading*.

To illustrate the distinction, note that a discrete-time white noise model is perfectly plausible; for example, using fair coin flips to generate a sequence of -1 and 1 values gives a white noise process. However, it is not plausible for any real physical process to generate uncorrelated $\mathrm{x}(t_0)$ and $\mathrm{x}(t_1)$ even when $|t_0 - t_1|$ is arbitrarily small. This mismatch from physical reality appears in its mathematical representation as well: one cannot define a continuous-time process that is white and has nonzero, finite variance. The presence of the Dirac delta function in $a_{\mathrm{x}}(t)$ in (4.137) reflects the fact that the variance of a continuous-time white noise process is undefined. As we will see below – and analogously to the discrete-time case – LSI filtering of a WSS process has the effect of altering the autocorrelation and crosscorrelations by convolutions, and the presence of the Dirac delta function in (4.137) does not pose a difficulty in such convolutions.

Gaussian processes The distribution of a Gaussian process – a stochastic process consisting of jointly Gaussian random variables – is completely specified by its second-order statistics. Since jointly Gaussian random variables are uncorrelated if and only if they are independent, a white Gaussian process would be i.i.d. if it were to exist. As described above, while a white Gaussian process cannot exist physically or mathematically, it can be a convenient mathematical abstraction. For example, noise with a flat spectrum over a bandwidth much greater than the effective bandwidth of a system can be treated as white without causing mathematical difficulties.

4.6.2 Systems

Consider a BIBO-stable LSI system described by its impulse response h with WSS input process x, as depicted in Figure 4.13. What can we say about the output y? It is given by the convolution (4.31), and we will demonstrate that it is a WSS process by deriving formulas for its second-order statistics.

---

[90] As in Chapter 3, the Fourier transform of the autocorrelation of white noise is a constant, mimicking the spectrum of white light; thus the term white noise.

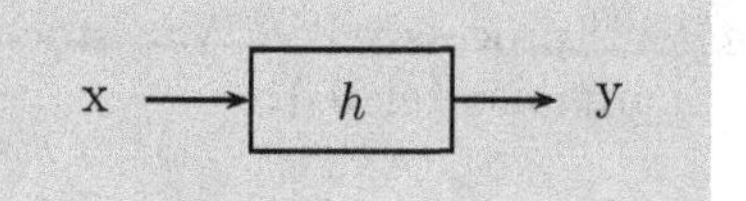

Figure 4.13 An LSI system with WSS input.

We start with the mean,

$$\begin{aligned}
\mu_{\mathrm{y}}(t) &= \mathrm{E}[\mathrm{y}(t)] \overset{(a)}{=} \mathrm{E}\left[\int_{-\infty}^{\infty} \mathrm{x}(\tau)\, h(t-\tau)\, d\tau\right] \overset{(b)}{=} \int_{-\infty}^{\infty} \mathrm{E}[\mathrm{x}(\tau)]\, h(t-\tau)\, d\tau \\
&\overset{(c)}{=} \int_{-\infty}^{\infty} \mu_{\mathrm{x}}(\tau) h(t-\tau)\, d\tau \overset{(d)}{=} \int_{-\infty}^{\infty} \mu_{\mathrm{x}} h(t-\tau)\, d\tau \\
&= \mu_{\mathrm{x}} \int_{-\infty}^{\infty} h(t-\tau)\, d\tau \overset{(e)}{=} \mu_{\mathrm{x}} H(0) = \mu_{\mathrm{y}},
\end{aligned} \tag{4.138a}$$

where (a) follows from (4.31); (b) from the linearity of the expectation; (c) from the definition of the mean function; (d) from x being WSS, (4.136a); and (e) from the frequency response of the LSI system (which exists because the system is BIBO-stable). The final equality emphasizes that the mean of the output is a constant, which is independent of t. The autocorrelation is

$$\begin{aligned}
a_{\mathrm{y}}(t,\tau) &= \mathrm{E}[\mathrm{y}(t)\, \mathrm{y}^*(t-\tau)] \\
&\overset{(a)}{=} \mathrm{E}\left[\int_{-\infty}^{\infty} \mathrm{x}(t-q) h(q)\, dq \int_{-\infty}^{\infty} \mathrm{x}^*(t-\tau-r) h^*(r)\, dr\right] \\
&\overset{(b)}{=} \int_{-\infty}^{\infty}\int_{-\infty}^{\infty} h(q) h^*(r)\, \mathrm{E}[\mathrm{x}(t-q)\, \mathrm{x}^*(t-\tau-r)]\, dr\, dq \\
&\overset{(c)}{=} \int_{-\infty}^{\infty}\int_{-\infty}^{\infty} h(q) h^*(r)\, a_{\mathrm{x}}(t-q, \tau-(q-r))\, dr\, dq \\
&\overset{(d)}{=} \int_{-\infty}^{\infty}\int_{-\infty}^{\infty} h(q) h^*(r)\, a_{\mathrm{x}}(\tau-(q-r))\, dr\, dq \\
&\overset{(e)}{=} \int_{-\infty}^{\infty}\left(\int_{-\infty}^{\infty} h(q) h^*(q-p)\, dq\right) a_{\mathrm{x}}(\tau-p)\, dp \\
&\overset{(f)}{=} \int_{-\infty}^{\infty} a_h(p)\, a_{\mathrm{x}}(\tau-p)\, dp = a_{\mathrm{y}}(\tau),
\end{aligned} \tag{4.138b}$$

where (a) follows from (4.31); (b) from the linearity of the expectation; (c) from the definition of the autocorrelation; (d) from x being WSS, (4.136b); (e) from the change of variable $p = q - r$; and (f) from the definition of deterministic autocorrelation (4.10). The final step emphasizes the lack of dependence of $a_{\mathrm{y}}(t,\tau)$ on t. Combined with the lack of dependence of $\mu_{\mathrm{y}}(t)$ on t, we see that, when the input x is WSS, the output y is WSS as well. We also see that the autocorrelation of the output is the convolution of the autocorrelation of the input and the deterministic autocorrelation of the impulse response of the system,

$$a_{\mathrm{y}} = a_h * a_{\mathrm{x}}. \tag{4.138c}$$

Computing the crosscorrelation between the input and the output shows that they are jointly WSS:

$$\begin{aligned} c_{\mathrm{x,y}}(t,\tau) &= \mathrm{E}[\,\mathrm{x}(t)\,\mathrm{y}^*(t-\tau)\,] \\ &\stackrel{(a)}{=} \mathrm{E}\left[\mathrm{x}(t)\int_{-\infty}^{\infty} h^*(r)\,\mathrm{x}^*(t-\tau-r)\,dr\right] \\ &= \mathrm{E}\left[\int_{-\infty}^{\infty} h^*(r)\,\mathrm{x}(t)\,\mathrm{x}^*(t-(\tau+r))\,dr\right] \\ &\stackrel{(b)}{=} \int_{-\infty}^{\infty} h^*(r)\,\mathrm{E}[\,\mathrm{x}(t)\,\mathrm{x}^*(t-(\tau+r))\,]\,dr \\ &\stackrel{(c)}{=} \int_{-\infty}^{\infty} h^*(r)\,a_{\mathrm{x}}(t,\tau+r)\,dr \\ &\stackrel{(d)}{=} \int_{-\infty}^{\infty} h^*(r)\,a_{\mathrm{x}}(\tau+r)\,dr \;=\; c_{\mathrm{x,y}}(\tau), \end{aligned} \tag{4.139a}$$

where (a) follows from (4.31); (b) from the linearity of the expectation; (c) from the definition of the autocorrelation; and (d) from x being WSS, (4.136b). The final step emphasizes the lack of dependence of $c_{\mathrm{x,y}}(t,\tau)$ on t. This crosscorrelation can also be written as the convolution between a_{x} and the time-reversed and conjugated impulse response,

$$c_{\mathrm{x,y}} = h^*(-\tau) *_\tau a_{\mathrm{x}}(\tau). \tag{4.139b}$$

Similarly,

$$c_{\mathrm{y,x}} = h * a_{\mathrm{x}}. \tag{4.139c}$$

We will use these expressions shortly to make some important observations in the Fourier domain.

4.6.3 Fourier transform

Like for deterministic functions, we can use Fourier techniques to gain insight into the behavior of continuous-time stochastic processes and systems. While we cannot take a Fourier transform of a stochastic process, we make assessments based on averages (moments), such as taking the Fourier transform of the autocorrelation.

Power spectral density Let x be a WSS stochastic process. The Fourier transform of its autocorrelation (4.136b) (which we assume to have sufficient decay for it to be absolutely or square integrable) is

$$A_{\mathrm{x}}(\omega) = \int_{-\infty}^{\infty} a_{\mathrm{x}}(t)\,e^{-j\omega t}\,dt = \int_{-\infty}^{\infty} \mathrm{E}[\,\mathrm{x}(t)\,\mathrm{x}^*(t-\tau)\,]\,e^{-j\omega t}\,dt. \tag{4.140}$$

This is called the *power spectral density*, the counterpart of the *energy spectral density* for deterministic functions in (4.10). The power spectral density exists if and only if x is WSS, which is a consequence of the Wiener–Khinchin theorem. When x is real, the power spectral density is nonnegative and thus admits a spectral factorization

$$A_{\mathrm{x}}(\omega) = U(\omega)\, U^*(\omega),$$

where $U(\omega)$ is its (nonunique) spectral root. The normalized integral of the power spectral density,

$$P_{\mathrm{x}} = \frac{1}{2\pi}\int_{-\infty}^{\infty} A_{\mathrm{x}}(\omega)\, d\omega = a_{\mathrm{x}}(0) = \mathrm{E}\left[\, |\mathrm{x}(t)|^2 \,\right], \tag{4.141}$$

is the *power*, the counterpart of the *energy* for deterministic functions in (4.68).

White noise Using (4.137) and Table 4.1, we see that the power spectral density of white noise is a constant:

$$A(\omega) = \sigma_{\mathrm{x}}^2. \tag{4.142}$$

As noted earlier, its variance, or power, is infinite.

Effect of filtering Consider an LSI system with impulse response h, WSS input x, and WSS output y, as depicted in Figure 4.13. Using (4.138c) for the autocorrelation of y, the power spectral density of the output is given by

$$A_{\mathrm{y}}(\omega) = A_h(\omega)\, A_{\mathrm{x}}(\omega) = |H(\omega)|^2 A_{\mathrm{x}}(\omega), \tag{4.143}$$

where $A_h(\omega) = |H(\omega)|^2$ is the Fourier transform of the deterministic autocorrelation of h, according to Table 4.1. The quantity

$$P_{\mathrm{y}} = \mathrm{E}\left[\, \mathrm{y}^2(t) \,\right] = \frac{1}{2\pi}\int_{-\infty}^{\infty} A_{\mathrm{y}}(\omega)\, d\omega, = \frac{1}{2\pi}\int_{-\infty}^{\infty} |H(\omega)|^2 A_{\mathrm{x}}(\omega)\, d\omega = a_{\mathrm{y}}(0)$$

is the *output power*. Similarly to (4.143), using (4.139b) and (4.139c), we can express the *cross spectral density* between the input and the output as

$$C_{\mathrm{x,y}}(\omega) = H^*(\omega)\, A_{\mathrm{x}}(\omega), \tag{4.144a}$$
$$C_{\mathrm{y,x}}(\omega) = H(\omega)\, A_{\mathrm{x}}(\omega). \tag{4.144b}$$

Chapter at a glance

We have now seen all the versions of the Fourier transform and series that will be used in what follows; they are summarized in Table 4.4 and Figure 4.14. These variants of the Fourier transform differ depending on the underlying space of sequences or functions.

Transform	Forward/inverse	Duality/periodicity
Fourier transform	$X(\omega) = \int_{-\infty}^{\infty} x(t)\, e^{-j\omega t}\, dt$ $x(t) = \frac{1}{2\pi}\int_{-\infty}^{\infty} X(\omega)\, e^{j\omega t}\, d\omega$	
Fourier series	$X_k = \frac{1}{T}\int_{-T/2}^{T/2} x(t)\, e^{-j(2\pi/T)kt}\, dt$ $x(t) = \sum_{k\in\mathbb{Z}} X_k e^{j(2\pi/T)kt}$	Dual with DTFT $x(t+T) = x(t)$
Discrete-time Fourier transform	$X(e^{j\omega}) = \sum_{n\in\mathbb{Z}} x_n e^{-j\omega n}$ $x_n = \frac{1}{2\pi}\int_{-\pi}^{\pi} X(e^{j\omega})\, e^{j\omega n}\, d\omega$	Dual with Fourier series $X(e^{j(\omega+2\pi)}) = X(e^{j\omega})$
Discrete Fourier transform	$X_k = \sum_{n=0}^{N-1} x_n e^{-j(2\pi/N)kn}$ $x_n = \frac{1}{N}\sum_{k=0}^{N-1} X_k e^{j(2\pi/N)kn}$	

Table 4.4 Various forms of Fourier transforms seen in this chapter and Chapter 3.

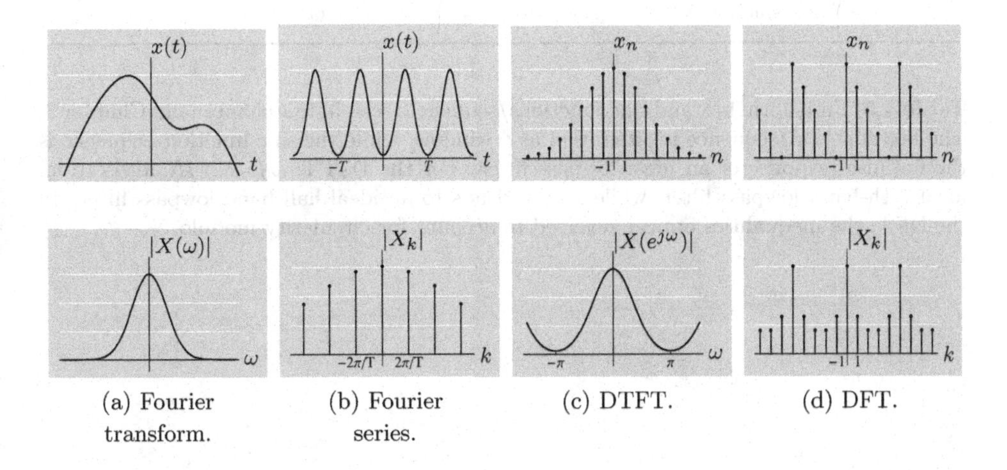

(a) Fourier transform. (b) Fourier series. (c) DTFT. (d) DFT.

Figure 4.14 Various forms of Fourier transforms seen in this chapter and Chapter 3.

In both this chapter and the previous one, box and sinc functions played a prominent role; for easy reference, they are summarized in Table 4.5.

Functions on the real line	$x(t),\quad t \in \mathbb{R},\quad \|x\| = 1$	**FT** $X(\omega),\quad \omega \in \mathbb{R},\quad \|X\| = \sqrt{2\pi}$
Box	$\begin{cases} 1/\sqrt{t_0}, & \lvert t\rvert \le \frac{1}{2}t_0; \\ 0, & \text{otherwise} \end{cases}$	$\sqrt{t_0}\,\mathrm{sinc}\big(\frac{1}{2}t_0\omega\big)$
Sinc	$\sqrt{\frac{\omega_0}{2\pi}}\,\mathrm{sinc}\big(\frac{1}{2}\omega_0 t\big)$	$\begin{cases} \sqrt{2\pi/\omega_0}, & \lvert\omega\rvert \le \frac{1}{2}\omega_0; \\ 0, & \text{otherwise} \end{cases}$
Periodic functions	$x(t),\quad t \in [-\frac{1}{2}T, \frac{1}{2}T),\quad \|x\| = 1$	**FS** $X_k,\quad k \in \mathbb{Z},\quad \|X\| = 1/\sqrt{T}$
Box	$\begin{cases} 1/\sqrt{t_0}, & \lvert t\rvert \le \frac{1}{2}t_0; \\ 0, & \text{otherwise} \end{cases}$	$\frac{\sqrt{t_0}}{T}\,\mathrm{sinc}(\pi k t_0/T)$
Sinc	$\sqrt{\frac{k_0}{T}}\,\frac{\mathrm{sinc}(\pi k_0 t/T)}{\mathrm{sinc}(\pi t/T)}$	$\begin{cases} 1/\sqrt{k_0 T}, & \lvert k\rvert \le \frac{1}{2}(k_0-1); \\ 0, & \text{otherwise} \end{cases}$
Infinite-length sequences	$x_n,\quad n \in \mathbb{Z},\quad \|x\| = 1$	**DTFT** $X(e^{j\omega}),\quad \omega \in [-\pi, \pi),\quad \|X\| = \sqrt{2\pi}$
Box (n_0 odd)	$\begin{cases} 1/\sqrt{n_0}, & \lvert n\rvert \le \frac{1}{2}(n_0-1); \\ 0, & \text{otherwise} \end{cases}$	$\sqrt{n_0}\,\frac{\mathrm{sinc}\big(\frac{1}{2}n_0\omega\big)}{\mathrm{sinc}\big(\frac{1}{2}\omega\big)}$
Sinc	$\sqrt{\frac{\omega_0}{2\pi}}\,\mathrm{sinc}\big(\frac{1}{2}\omega_0 n\big)$	$\begin{cases} \sqrt{2\pi/\omega_0}, & \lvert\omega\rvert \le \frac{1}{2}\omega_0; \\ 0, & \text{otherwise} \end{cases}$
Finite-length sequences	$x_n,\quad n \in \{0, 1, \ldots, N-1\},\quad \|x\| = 1$	**DFT** $X_k,\quad k \in \{0, 1, \ldots, N-1\},\quad \|X\| = \sqrt{N}$
Box (n_0 odd)	$\begin{cases} 1/\sqrt{n_0}, & \lvert n - N/2\rvert \ge \frac{1}{2}(n_0-1); \\ 0, & \text{otherwise} \end{cases}$	$\sqrt{n_0}\,\frac{\mathrm{sinc}(\pi n_0 k/N)}{\mathrm{sinc}(\pi k/N)}$
Sinc	$\sqrt{\frac{k_0}{N}}\,\frac{\mathrm{sinc}(\pi n k_0/N)}{\mathrm{sinc}(\pi n/N)}$	$\begin{cases} \sqrt{N/k_0}, & \lvert k - N/2\rvert \ge \frac{1}{2}(k_0-1); \\ 0, & \text{otherwise} \end{cases}$

Table 4.5 Unit-norm box and sinc functions/sequences seen in this chapter and Chapter 3. The box function/sequence is often used as a window, while the sinc function/sequence is the impulse response of an ideal lowpass filter. For the DTFT, $\omega_0 = 2\pi/N$ leads to an ideal Nth-band lowpass filter, while $\omega_0 = \pi$ leads to an ideal half-band lowpass filter. In the DFT, the inequalities appear reversed to account for circularity modulo N.

Historical remarks

Jean Baptiste Joseph Fourier (1768–1830), was a French mathematician and physicist, who proposed his famous Fourier series while working on the equations for heat flow. His interests were varied, his biography unusual. He followed Napoleon to Egypt and spent a few years in Cairo, even contributing a few papers to the Egyptian Institute that Napoleon founded. He served as a permanent secretary of the French Academy of Science. In 1822, he published *Théorie analytique de la chaleur*, in which he claimed that any function can be decomposed into a sum of sines and cosines; while we know that this is only partially true, it took mathematicians a long time to tighten the result. Lagrange and Dirichlet both worked on it, with Dirichlet first formulating conditions under which a Fourier series exists.

Josiah Willard Gibbs (1839–1903) was an American mathematician, physicist, and chemist known for many significant contributions (among others as the inventor of vector analysis, independently of Heaviside). He was also the one to remark upon the unusual way the Fourier series behaves at a discontinuity; the Fourier series overshoots significantly, though in a controlled manner. Moreover, we can get into trouble by trying to differentiate the Fourier series. In 1926, **Andrey Nikolaevich Kolmogorov (1903–1987)** proved that there exists a function, which is periodic and locally Lebesgue integrable, with a Fourier series that is divergent at all points. While this seemed to be another strike against Fourier series, **Lennart Carleson (1928–)**, a Swedish mathematician, showed in 1966 that every periodic, locally square-integrable function has a Fourier series that converges almost everywhere.

Further reading

Fourier analysis of signals and systems Mallat's book [66] on wavelets and signal processing is similar in outlook and scope to this text (together with the companion volume [57]), but more mathematical in style. It includes details on the inversion of the Fourier transform on $\mathcal{L}^1(\mathbb{R})$ and extension of the Fourier transform to $\mathcal{L}^2(\mathbb{R})$. The book by Brémaud [11] is a clean, self-contained text aimed at signal processing researchers. The text by Bracewell [10] is a classic; the material is written with engineering in mind, with plenty of intuition. The book by Papoulis [77] is another classic engineering text. More on the Dirac delta function can be found in another text by Papoulis [73] and in signals and systems book by Siebert [91]. Finally, the text by Folland [30] has been written from a physicist's point of view and offers an excellent treatment of partial differential equations.

Stochastic processes and systems Additional material on stochastics can be found in the book by Porat [80]. The subtlety of continuous-time white noise is discussed in the stochastic process book of Gallager [32].

Exercises with solutions

4.1. *Sifting property*
Let

$$d_n(t) = \begin{cases} n/2, & \text{for } |t| \le 1/n; \\ 0, & \text{otherwise,} \end{cases} \qquad \text{for } n = 1, 2, \ldots.$$

(i) Prove that $d_1, d_2, \ldots$ does not converge in $\mathcal{L}^2$ norm.
(*Hint:* Recall from Definition 2.13 that convergence in a vector space depends on the norm. It is adequate to show that $d_1, d_2, \ldots$ is not a Cauchy sequence under the $\mathcal{L}^2$ norm.)

(ii) Prove that if x is continuous at t_0, then

$$\lim_{n\to\infty} \int_{-\infty}^{\infty} x(t) d_n(t - t_0)\, dt = x(t_0).$$

Solution:

(i) If d were a Cauchy sequence, $\|d_n - d_m\|$ would converge to 0 as $\min(n, m)$ grows without bound. To show that this is not the case, let us start by computing $\|d_n - d_m\|$ explicitly for $n > m$:

$$\|d_n - d_m\|^2 = \int_{-\infty}^{\infty} (d_n(t) - d_m(t))^2\, dt = 2\left(\int_0^{1/n} \frac{(n-m)^2}{4}\, dt + \int_{1/n}^{1/m} \frac{m^2}{4}\, dt\right)$$
$$= 2\left(\frac{1}{n}\frac{(n-m)^2}{4} + \left(\frac{1}{m} - \frac{1}{n}\right)\frac{m^2}{4}\right) = \frac{n-m}{2}.$$

This does not shrink as $m \to \infty$ with $n > m$, so d is not a Cauchy sequence and hence does not converge.

(ii) Intuitively, the result holds because the integrand depends on x only in the interval $(t_0 - 1/n, t_0 + 1/n)$, and by continuity x cannot vary much from $x(t_0)$ on this interval when n is large. Specifically, we use the Weierstrass definition of continuity at t_0: for any $\varepsilon > 0$ there exists $\eta > 0$ such that

$$x(t_0) - \varepsilon < x(t) < x(t_0) + \varepsilon \qquad \text{for all } t \in (t_0 - \eta, t_0 + \eta). \tag{E4.1-1}$$

Thus, for any $\varepsilon > 0$, there exists an N_ε such that $\eta = 1/N_\varepsilon$ is small enough for (E4.1-1) to hold. For any $n > N_\varepsilon$,

$$\int_{-\infty}^{\infty} x(t) d_n(t - t_0)\, dt = \frac{n}{2} \int_{t_0 - 1/n}^{t_0 + 1/n} x(t)\, dt \tag{E4.1-2}$$

by substitution of d_n. Using the upper and lower bounds on $x(t)$ from (E4.1-1) upper and lower bounds the integral in (E4.1-2) by $x(t_0) - \varepsilon$ and $x(t_0) + \varepsilon$ since the $2/n$ length of the interval cancels the $n/2$ factor. Since ε can be arbitrarily small, the limit of the integral must be $x(t_0)$.

4.2. *Associative property of convolution*

(i) Prove the associativity property (4.32c) for functions in $\mathcal{L}^1(\mathbb{R})$.

(ii) Derive a counterexample to associativity for some functions not in $\mathcal{L}^1(\mathbb{R})$.

Solution:

(i) As stated in the text and analogous to Solved exercise 3.2, the result of convolving two functions in $\mathcal{L}^1(\mathbb{R})$ is a function in $\mathcal{L}^1(\mathbb{R})$. So when g, h, and x are in $\mathcal{L}^1(\mathbb{R})$, both $g * (h * x)$ and $(g * h) * x$ are well defined and finite. We have that

$$(g * (h * x))(t) = \int_{-\infty}^{\infty} g(t - \tau) \int_{-\infty}^{\infty} h(\tau - s)\, x(s)\, ds\, d\tau$$
$$\overset{(a)}{=} \int_{-\infty}^{\infty} \int_{-\infty}^{\infty} g(t - \tau) h(\tau - s)\, x(s)\, d\tau\, ds$$

$$= \int_{-\infty}^{\infty} x(s) \int_{-\infty}^{\infty} g(t-\tau)h(\tau - s)\, d\tau\, ds$$

$$\overset{(b)}{=} \int_{-\infty}^{\infty} x(s) \int_{-\infty}^{\infty} g(\tau')h(t-s-\tau')\, d\tau'\, ds$$

$$= \int_{-\infty}^{\infty} x(s)\,(g*h)(t-s)\, ds \;=\; ((g*h)*x)(t),$$

where (a) follows from Fubini's theorem (2.208); and (b) from the change of variable $\tau' = t - \tau$.

(ii) We are looking for a continuous-time analogue to Example 3.12. Let

$$g(t) \;=\; u(t), \qquad h(t) \;=\; u(t) - 2u(t-1) + u(t-2), \qquad x(t) \;=\; 1,$$

where u is the Heaviside function defined in (4.5). Then

$$g*(h*x) \;=\; u*\mathbf{0} \;=\; \mathbf{0}$$

because $\int_{-\infty}^{\infty} h(t)\, dt = 0$. However,

$$(g*h)*x \;=\; f*x \;=\; 1,$$

where $f = g*h$ is given by

$$f(t) \;=\; \begin{cases} t, & \text{for } t \in [0,\,1); \\ 2-t, & \text{for } t \in [1,\,2); \\ 0, & \text{otherwise.} \end{cases}$$

4.3. *Fourier transform of functions in $\mathcal{L}^1(\mathbb{R})$*
Let x be a function in $\mathcal{L}^1(\mathbb{R})$. Show the following properties of its Fourier transform X:

(i) X is bounded; and
(ii) X is continuous.

(*Hint:* For continuity, use the dominated convergence theorem, (2.210).)
Solution:

(i) The boundedness of X follows directly from $x \in \mathcal{L}^1(\mathbb{R})$:

$$|X(\omega)| \;=\; \left|\int_{-\infty}^{\infty} x(t)\, e^{-j\omega t}\, dt\right| \;\overset{(a)}{\leq}\; \int_{-\infty}^{\infty} |x(t)\, e^{-j\omega t}|\, dt$$

$$= \int_{-\infty}^{\infty} |x(t)|\, dt \;=\; \|x\|_1 \;\overset{(b)}{<}\; \infty,$$

where (a) follows from (2.217); and (b) from $x \in \mathcal{L}^1(\mathbb{R})$.

(ii) To show that X is continuous at ω, we will show that

$$\lim_{k\to\infty} |X(\omega) - X(\omega + \varepsilon_k)| \;=\; 0 \tag{E4.3-1}$$

for any sequence of real numbers $\{\varepsilon_k\}_{k\in\mathbb{N}}$ with $\varepsilon_k \to 0$. For any ε_k,

$$X(\omega) - X(\omega+\varepsilon_k) \;=\; \int_{-\infty}^{\infty} x(t)e^{-j\omega t}\, dt - \int_{-\infty}^{\infty} x(t)e^{-j(\omega+\varepsilon_k)t}\, dt$$

$$= \int_{-\infty}^{\infty} x(t)e^{-j\omega t}\left(1 - e^{-j\varepsilon_k t}\right) dt. \tag{E4.3-2}$$

Now, if $\varepsilon_k \to 0$ as $k \to \infty$, the integrand of (E4.3-2) also goes to zero as $k \to \infty$. Thus,

$$\int_{-\infty}^{\infty} \lim_{k\to\infty} x(t)e^{-j\omega t}\left(1 - e^{-j\varepsilon_k t}\right) dt \;=\; \int_{-\infty}^{\infty} 0\, dt \;=\; 0. \tag{E4.3-3}$$

For the desired expression (E4.3-1) to follow from (E4.3-3) requires the interchange of limit and integral. The interchange is valid by the dominated convergence theorem because

$$|x(t)e^{-j\omega t}\left(1 - e^{-j\varepsilon_k t}\right)| \;\leq\; 2\,|x(t)|$$

and

$$\int_{-\infty}^{\infty} 2\,|x(t)|\, dt \;=\; 2\,\|x\|_1 \;<\; \infty.$$

4.4. *Relation of the Fourier series coefficients to the Fourier transform*
Verify (4.95), that is, that the Fourier series coefficients of a T-periodic function are scaled samples of the Fourier transform of the same function restricted to the interval $[-\frac{1}{2}T, \frac{1}{2}T)$.

Solution: The Fourier series coefficients of a T-periodic x are given by (4.94a):

$$X_k = \frac{1}{T} \int_{-T/2}^{T/2} x(t)\, e^{-j(2\pi/T)kt}\, dt. \tag{E4.4-1}$$

On the other hand, when $\widetilde{x}$ is the restriction of x to the interval $[-\frac{1}{2}T, \frac{1}{2}T)$ (that is, $\widetilde{x}$ is equal to x on $[-\frac{1}{2}T, \frac{1}{2}T)$ and is 0 otherwise), its Fourier transform is

$$\widetilde{X}(\omega) = \int_{-\infty}^{\infty} \widetilde{x}(t)\, e^{-j\omega t}\, dt = \int_{-T/2}^{T/2} x(t)\, e^{-j\omega t}\, dt. \tag{E4.4-2}$$

Setting $\omega = (2\pi/T)k$ in the right-hand side of (E4.4-2) and multiplying the result by $1/T$ gives the right-hand side of (E4.4-1), so (4.95) is verified.

4.5. *Convolution and sum of continuous random variables*
Let x and y be independent continuous random variables with PDFs f_{x} and f_{y}. Show that $\mathrm{z} = \mathrm{x} + \mathrm{y}$ has PDF $f_{\mathrm{x}} * f_{\mathrm{y}}$.
Solution: For any $t \in \mathbb{R}$,

$$\begin{aligned} F_{\mathrm{z}}(t) &= \mathrm{P}(\mathrm{z} \le t) = \mathrm{P}(\mathrm{x}+\mathrm{y} \le t) \stackrel{(a)}{=} \int_{-\infty}^{\infty}\int_{-\infty}^{t-u} f_{\mathrm{x,y}}(s,u)\, ds\, du \\ &\stackrel{(b)}{=} \int_{-\infty}^{\infty}\int_{-\infty}^{t-u} f_{\mathrm{x}}(s)\, f_{\mathrm{y}}(u)\, ds\, du \stackrel{(c)}{=} \int_{-\infty}^{\infty} f_{\mathrm{y}}(u) \left(\int_{-\infty}^{t-u} f_{\mathrm{x}}(s)\, ds\right) du \\ &\stackrel{(d)}{=} \int_{-\infty}^{\infty} f_{\mathrm{y}}(u) F_{\mathrm{x}}(t-u)\, du, \end{aligned}$$

where (a) follows from expressing the region $\{(s,u) \mid s+u \le t\}$; (b) from the independence of x and y; (c) from $f_{\mathrm{y}}(y)$ not depending on s; and (d) from the definition of F_{x}. Differentiating with respect to t gives

$$f_{\mathrm{z}}(t) = \int_{-\infty}^{\infty} f_{\mathrm{y}}(u) f_{\mathrm{x}}(t-u)\, du,$$

showing that $f_{\mathrm{z}} = f_{\mathrm{x}} * f_{\mathrm{y}}$.

Exercises

4.1. *Derivative of the Dirac delta function.*(2)
Let x be a continuously differentiable function. Show that

$$\int_{-\infty}^{\infty} x(t)\, \delta'(t)\, dt = -x'(0)$$

is a property of the derivative of the Dirac delta function, assuming that integration by parts is valid for expressions involving δ and its derivative. Find similar expressions for higher-order derivatives of δ.

4.2. *BIBO stability.*(2)
Prove Theorem 4.8, that is, an LSI system is BIBO-stable if and only if its impulse response is absolutely integrable.

4.3. *Adjoint of the convolution operator.*(2)
Let $h \in \mathcal{L}^2(\mathbb{R})$ and $g(t) = h^*(-t)$. Also let G and H be the convolution operators associated with g and h. Show that $H^* = G$.

4.4. *Convolution of derivative and primitive.*(2)
Let h and x be differentiable functions, and let

$$h^{(1)}(t) = \int_{-\infty}^{t} h(\tau)\, d\tau \qquad \text{and} \qquad x^{(1)}(t) = \int_{-\infty}^{t} x(\tau)\, d\tau$$

be their primitives. Give a sufficient condition for (4.64),

$$h * x = h^{(1)} * x',$$

based on integration by parts.

4.5. *Generalized Parseval equality for the Fourier transform.*(2)
Prove that if x and y in $\mathcal{L}^1(\mathbb{R}) \cap \mathcal{L}^2(\mathbb{R})$ have Fourier transforms X and Y, then

$$\langle x, y \rangle = \frac{1}{2\pi} \langle X, Y \rangle.$$

(*Hint:* Express the inner product as a convolution of $x(t)$ and $y^*(-t)$ evaluated at the origin, and use the convolution theorem.)

4.6. *Fourier transform of a Gaussian function.*(3)
Prove that (4.78) forms a Fourier transform pair.
(*Hint:* See the discussion following (4.78). You can use the fact that

$$\int_{-\infty}^{\infty} e^{-t^2}\, dt = \sqrt{\pi} \tag{P4.6-1}$$

to specify the constant.)

4.7. *Function decay.*(2)
Let $a(t)$ be a function with the Fourier transform $A(\omega)$ given in Figure P4.7-1.

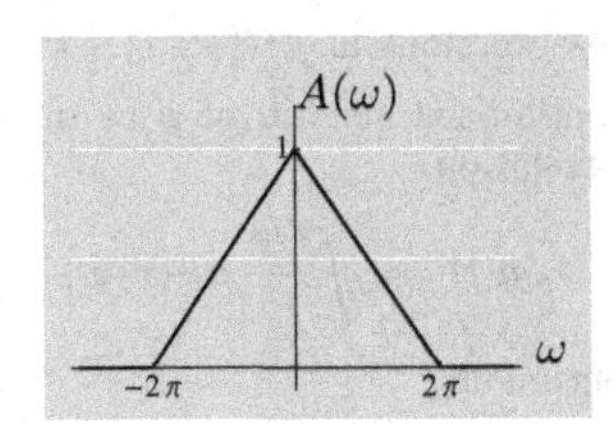

Figure P4.7-1 Fourier transform $A(\omega)$.

(i) Show that, for any $\Phi(\omega)$ that satisfies $|\Phi(\omega)|^2 = A(\omega)$, its inverse Fourier transform $\varphi(t)$ satisfies

$$\langle \varphi(t), \varphi(t-n) \rangle_t = \delta_n. \tag{P4.7-1}$$

(*Hint:* Perform the analysis in the frequency domain.)

(ii) Let $B(\omega)$ be the following box function in the frequency domain:

$$B(\omega) = \begin{cases} 1, & \text{for } |\omega| \in [0, \pi]; \\ 0, & \text{for } |\omega| \in (\pi, \infty). \end{cases}$$

Show that $A(\omega) = (1/(2\pi))B(\omega) * B(\omega)$. What can be said about the rate of decay of $a(t)$?

4.8. *Decay of Fourier transform and smoothness.*(3)
Prove that if $X(\omega)$ decays faster than $1/|\omega|^{q+1}$ for large $|\omega|$ as in (4.83b), then $x \in C^q$.
(*Hint:* Use (4.83a) and the differentiation property of the Fourier transform.)

4.9. *Lipschitz regularity.*(3)
Assume that $X(\omega)$ satisfies (4.88),

$$\int_{-\infty}^{\infty} |X(\omega)|(1 + |\omega|^{\alpha})\, d\omega < \infty.$$

(i) Show that $x(t)$ is bounded.

(ii) Show that $x(t)$ is Lipschitz α; that is, show (4.87):

$$\frac{|x(t) - x(t_0)|}{|t - t_0|^{\alpha}} \leq c,$$

for any t and t_0 and $0 \leq \alpha < 1$.
(*Hint:* Express the above ratio in terms of the inverse Fourier transform, and divide the integral into two parts: $|t - t_0|^{-1} \leq |\omega|$ and $|t - t_0|^{-1} > |\omega|$.)

(iii) Show how to extend the above characterization for $r = n + \alpha$, $n \in \mathbb{Z}^+$.

4.10. *Real Fourier series.*(1)
Let x be a 2π-periodic, real-valued function. Show that x can be expressed as a real Fourier series

$$x(t) = a_0 + \sum_{k=1}^{\infty} (a_k \cos kt + b_k \sin kt), \qquad \text{(P4.10-1)}$$

where

$$a_0 = \frac{1}{2\pi} \int_{-\pi}^{\pi} x(t)\, dt,$$

$$a_k = \frac{1}{\pi} \int_{-\pi}^{\pi} x(t) \cos kt\, dt, \qquad k \in \mathbb{Z}^+,$$

$$b_k = \frac{1}{\pi} \int_{-\pi}^{\pi} x(t) \sin kt\, dt, \qquad k \in \mathbb{Z}^+,$$

and indicate the relationship between $\{a_k, b_k\}$ and the Fourier series coefficients $\{X_k\}$.

4.11. *Completeness of Fourier series.*(2)
Prove (4.102) for continuous functions in $\mathcal{L}^2([-\frac{1}{2}T, \frac{1}{2}T))$ as follows:

(i) Introduce x_T as a periodized version of x as in (4.129a). Show that its circular deterministic autocorrelation,

$$a(t) = \int_{-T/2}^{T/2} x_T(\tau)\, x_T^*(\tau - t)\, d\tau,$$

is T-periodic and continuous.

(ii) Prove that its Fourier series coefficients are $A_k = |\alpha_k|^2$, with α_k as in (4.100).

With $\sum_{k\in\mathbb{Z}} |X_k|^2 < \infty$, (4.94b) and continuity of a, this leads to

$$a(t) = \sum_{k\in\mathbb{Z}} |\alpha_k|^2 e^{j(2\pi/T)kt}, \qquad \text{(P4.11-1)}$$

for all t. Setting $t = 0$ in (P4.11-1),

$$\int_{-T/2}^{T/2} |x(t)|^2\, dt = \sum_{k\in\mathbb{Z}} |\alpha_k|^2,$$

which, from the projection theorem, shows that the squared norm of the error in

$$\int_{-T/2}^{T/2} |x(t)|^2\, dt = \sum_{k=-N}^{N} |X_k|^2 + \int_{-T/2}^{T/2} |x(t) - \widehat{x}_N(t)|^2\, dt,$$

with $\widehat{x}_N$ as in (4.101), goes to 0 as $N \to \infty$, proving completeness.

4.12. *Fourier series of an integral.*(1)
Let x be a 2π-periodic, zero-mean, real-valued function with real Fourier series as in Exercise 4.10. Prove that the Fourier series pair corresponding to the integral of x is

$$\int_{-\pi}^{t} x(\tau)\, d\tau \quad \overset{\text{FS}}{\longleftrightarrow} \quad \sum_{k=1}^{\infty} (-1)^k \frac{b_k}{k} + \sum_{k=1}^{\infty} \left(\frac{a_k}{k} \sin kt - \frac{b_k}{k} \cos kt \right).$$

4.13. *Convolution for Fourier series.*(1)
Let x be a T-periodic function.

(i) Prove that the Fourier series pair corresponding to the circular convolution of a T-periodic x with a T-periodic h is given by (4.113).

(ii) Prove that the Fourier series pair corresponding to the linear convolution of a T-periodic x with an aperiodic, stable h is given by

$$(h * x)(t) \quad \overset{\text{FS}}{\longleftrightarrow} \quad H\left(\frac{2\pi}{T}k\right) X_k.$$

4.14. *Fourier series of a triangle wave.*(1)
In (4.120a), the Fourier series of the triangle wave was computed using the integral property. Obtain the same result using

(i) the definition of the Fourier series (4.94a);

(ii) the Fourier transform of one period of the triangle function, (4.45), and the relation between the Fourier series coefficients to the Fourier transform, (4.95); and

(iii) the convolution property.
(*Hint:* Use a square wave and a filter that is the indicator function of $[0, \frac{1}{2}]$.)

4.15. *Sawtooth wave and Gibbs phenomenon.*(1)
Let x be a sawtooth wave with period $T = 1$ as in Figure P4.15-1, with one period

$$x(t) \;=\; 2t, \qquad \text{for } t \in [-\tfrac{1}{2}, \tfrac{1}{2}). \tag{P4.15-1}$$

Compute the Fourier series coefficients of x and comment on a possible Gibbs phenomenon.

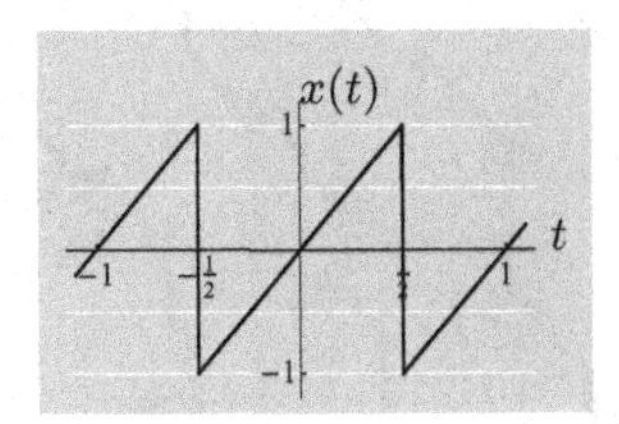

Figure P4.15-1 The sawtooth wave (P4.15-1).

4.16. *Fourier transform of Dirac comb.*(2)
Derive (4.127) from (4.124).

4.17. *Application of Poisson sum formula.*(3)
Let x be a function with spectrum given by $X(\omega) = 1/(1+|\omega|)^{\alpha}$ for some positive real number α. Determine a condition on α such that $\sum_{n\in\mathbb{Z}} x(n)$ converges. (You may assume that $\sum_{n\in\mathbb{Z}} x(t-n)$ converges absolutely for all t when it converges for $t = 0$.)

4.18. *Regularity of periodic functions and decay of Fourier transform.*(3)
Let $x \in \mathcal{L}^2([-\frac{1}{2}T, \frac{1}{2}T))$ and let $\widetilde{x} = 1_{[-T/2,T/2)}\, x$ be its restricted version. Show that if the Fourier transform of $\widetilde{x}$ satisfies (4.131a), then x has q continuous derivatives.

4.19. *Regularity of periodic functions and decay of Fourier series.*(3)
Let $x \in \mathcal{L}^2([-\frac{1}{2}T, \frac{1}{2}T))$ have Fourier series coefficient sequence X satisfying (4.131b). Show that x has q continuous derivatives.
(*Hint:* Rather than attempting to use (4.83b) and (4.95), instead parallel the arguments in Section 4.4.5.)

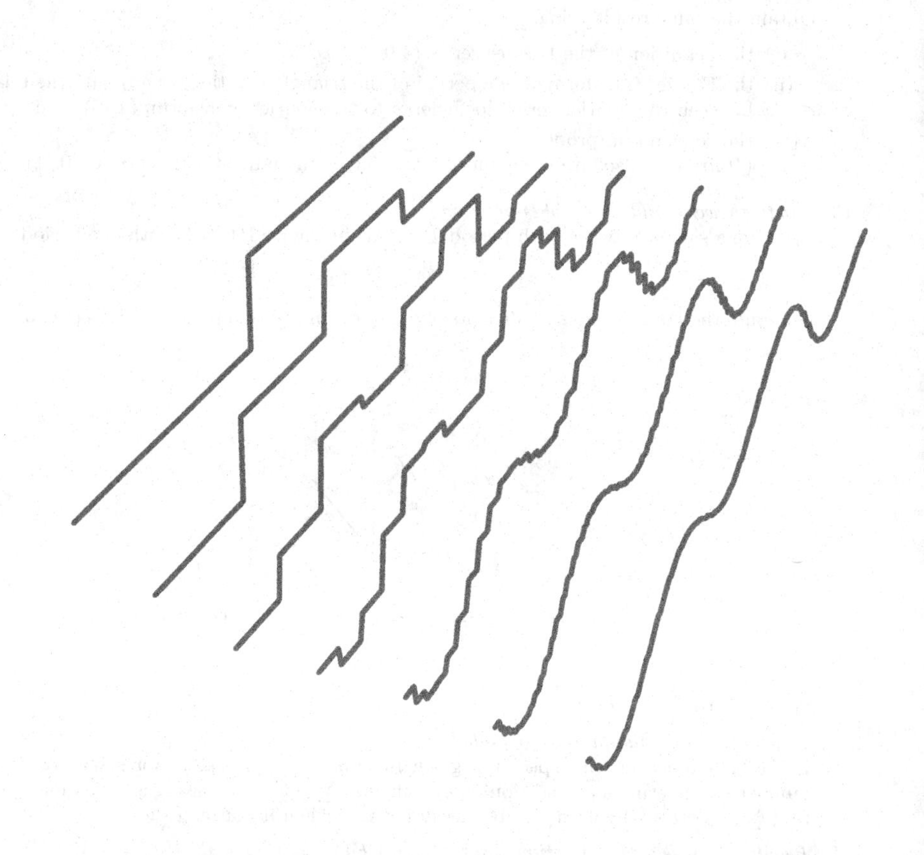

Chapter 5

Sampling and interpolation

"An experiment is a question which science poses to nature, and a measurement is the recording of nature's answer."

— *Max Planck*

Contents

The previous two chapters dealt with discrete-time signals (sequences indexed by integers) and continuous-time signals (functions of a real variable). The primary purpose of the present chapter is to link these two worlds. This is done through *sampling*, which produces a sequence from a function, and *interpolation*, which produces a function from a sequence. The ability to sample a function, manipulate the resulting sequence with a discrete-time system, and then interpolate to produce a function is the foundation of digital signal processing. Conversely, the ability to interpolate a sequence to create a function, manipulate the resulting function with a continuous-time system, and then sample to produce a sequence is the foundation of digital communications. These processes, illustrated in Figure 5.1, conceptually position this chapter as a bridge between Chapters 3 and 4.

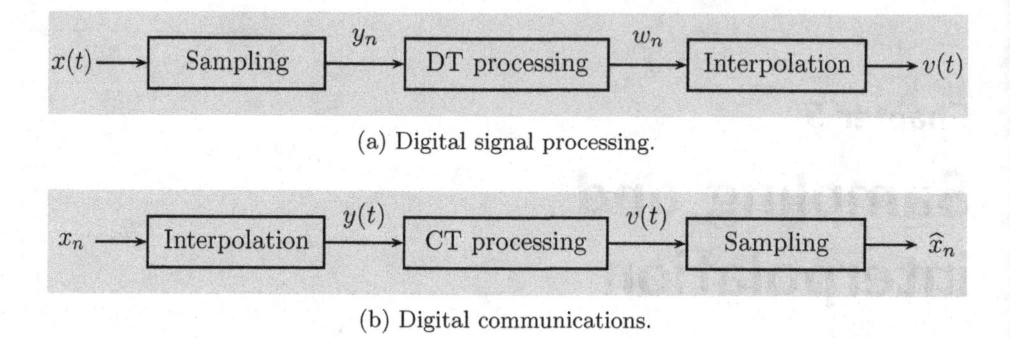

(a) Digital signal processing.

(b) Digital communications.

Figure 5.1 Sampling and interpolation in signal processing and communications. (a) Digital signal processing: sampling produces a discrete-time signal (sequence) from a continuous-time signal (function), which is then processed with a discrete-time system and interpolated. (b) Digital communications: interpolation produces a continuous-time signal (function) from a discrete-time signal (sequence), which is then processed with a continuous-time system and finally sampled.

Given a function, one can associate a sequence with it by simply taking samples (evaluating or measuring the function) uniformly in time. Classical sampling theory places a bandwidth restriction on the function so that the samples are a faithful representation of the function. This leads to the concept of the Nyquist rate, a minimum rate of sampling at which changes in the function are captured by these samples. We will develop this result in detail, but we will also see it as a special case of a more general theory involving shift-invariant subspaces.

Our approach treats any decrease in dimension via a linear operator as sampling and, conversely, any increase in dimension via a linear operator as interpolation. These are thus intimately tied to basis expansions and subspaces: sampling followed by interpolation projects to a subspace, while interpolation alone embeds information within a subspace of a higher-dimensional space. These concepts are important for sequences and finite-dimensional vectors as well as for functions, so our development follows a progression from finite-dimensional spaces to $\ell^2(\mathbb{Z})$ and to $\mathcal{L}^2(\mathbb{R})$.

5.1 Introduction

In Chapters 3 and 4, we saw a pair of bijections between sequences and functions:

$$\begin{aligned} \text{discrete-time signal} &\overset{\text{DTFT}}{\longleftrightarrow} \text{periodic Fourier-domain function}, \\ \text{periodic continuous-time signal} &\overset{\text{FS}}{\longleftrightarrow} \text{discrete Fourier-domain sequence}. \end{aligned}$$

The first associates a periodic function, the discrete-time Fourier transform $X(e^{j\omega})$, $\omega \in \mathbb{R}$, with a discrete-time signal x_n, $n \in \mathbb{Z}$, and the second associates a sequence, the Fourier series coefficients X_k, $k \in \mathbb{Z}$, with a periodic continuous-time signal $x(t)$, $t \in \mathbb{R}$. While these are important connections, they are different in spirit

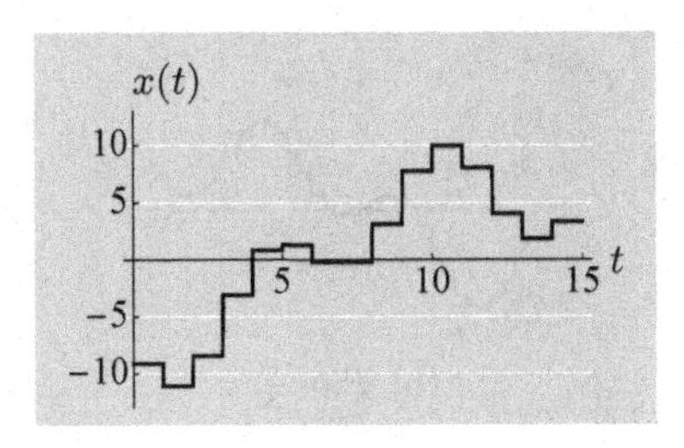

Figure 5.2 A piecewise-constant function that is constant over unit-length intervals $[n, n+1)$, $n \in \mathbb{Z}$.

from sampling and interpolation, both of which operate within the time domain. Typically, sampling and interpolation mean the following:

$$\text{discrete-time signal} \quad \underset{\text{sampling}}{\overset{\text{interpolation}}{\rightleftarrows}} \quad \text{continuous-time signal.}$$

They can equally well mean

$$\text{lower-rate discrete-time signal} \quad \underset{\text{sampling}}{\overset{\text{interpolation}}{\rightleftarrows}} \quad \text{higher-rate discrete-time signal,}$$

and, since the mathematical analogies are so strong, we expand the terms somewhat to include also

$$\text{shorter finite-length vector} \quad \underset{\text{sampling}}{\overset{\text{interpolation}}{\rightleftarrows}} \quad \text{longer finite-length vector.}$$

In this section, we first capture many main themes of the chapter for cases with orthonormal vectors through a representative example that expands upon Examples 2.17(i) and 2.18(iii). We then use examples in $\mathbb{R}^2$ to introduce the additional issues that arise from using nonorthogonal vectors in sampling and interpolation.

Subspace of functions Consider the set S of piecewise-constant functions that are constant over unit-length intervals such that

$$x(t) = x(n) \qquad \text{for all } t \in [n,\, n+1), \qquad n \in \mathbb{Z}. \tag{5.1}$$

One such function is shown in Figure 5.2. The set S is a closed subspace, and it is called shift-invariant with respect to integer shifts because, for any x in S and any integer k, the function $x(t-k)$ also belongs to S. Because of (5.1), functions in S are in one-to-one correspondence with sequences. If $g = 1_{[0,1)}$ – the indicator function of the unit interval – the set $\{g(t-k)\}_{k\in\mathbb{Z}}$ is an orthonormal basis for S.

Instead of unit-length intervals, we could consider intervals of any positive length T. Then, S becomes the space of piecewise-constant functions that are constant over intervals $[kT,\, (k+1)T)$, and $\{(1/\sqrt{T})g(t/T-k)\}_{k\in\mathbb{Z}}$ is an orthonormal

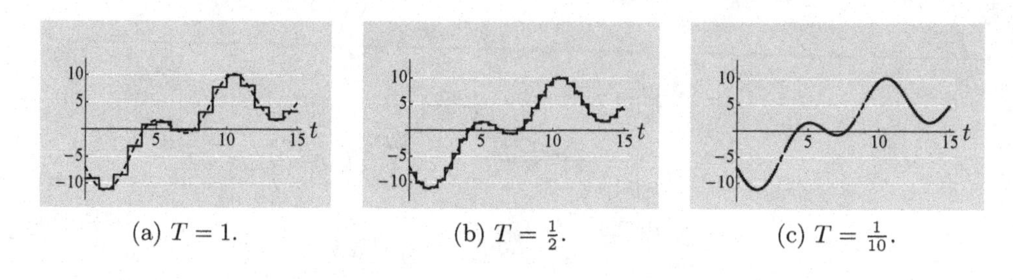

(a) $T = 1$. (b) $T = \frac{1}{2}$. (c) $T = \frac{1}{10}$.

Figure 5.3 Approximation by nearest piecewise-constant functions for different T.

basis for this new S. Adjusting T changes how closely an arbitrary function is approximated by the nearest function in S (see Figure 5.3); with smaller T, the error in approximating by the nearest function in S becomes smaller. Finding such nearest approximations through sampling and interpolation is a focus of this chapter.

Sampling Suppose that we want to measure a continuous-time signal x to obtain a sequence of real numbers y describing x. While the device we employ might be able to take a measurement of x at a single point in time t, this measurement would be sensitive to noise. Instead, it is both more technologically feasible and more robust to measure an integral. One way to do this is

$$\begin{aligned} y_n &= \int_n^{n+1} x(t)\,dt \stackrel{(a)}{=} \int_{-\infty}^{\infty} x(t)\,g^*(t-n)\,dt \\ &= \langle x(t),\, g(t-n)\rangle_t \stackrel{(b)}{=} (\Phi^* x)_n, \end{aligned} \tag{5.2}$$

for each $n \in \mathbb{Z}$, where (a) follows from $g = 1_{[0,1)}$;[91] and (b) introduces Φ^* to denote the basis analysis operator associated with $\{g(t-n)\}_{n\in\mathbb{Z}}$. From the connection between inner products and convolutions, (4.32a), the sample in (5.2) can be seen as evaluating the convolution of $x(t)$ and $g^*(-t)$ at $t = n$:

$$y_n = \int_{-\infty}^{\infty} x(t)\,g^*(t-n)\,dt = \left(g^*(-t) *_t x(t)\right)\Big|_{t=n}. \tag{5.3}$$

In other words, this sampling operator is implemented using filtering by $g^*(-t)$ and recording the result at integer time instants. The sampling operator Φ^* can thus be represented as in Figure 5.4(a), with a switch that closes at times that are integer multiples of T. Since square integrability of x implies the square summability of y (see Example 2.17(i)), Φ^* is a mapping from $\mathcal{L}^2(\mathbb{R})$ to $\ell^2(\mathbb{Z})$.

Interpolation Interpolation generates a function $\widehat{x}$ from a sequence y. One way to do this is to form

$$\widehat{x}(t) = \sum_{n\in\mathbb{Z}} y_n g(t-n) = (\Phi y)(t), \tag{5.4}$$

[91] Since the indicator function $1_{[0,1)}$ is real, the conjugation has no effect. It is included for consistency with the general case developed later.

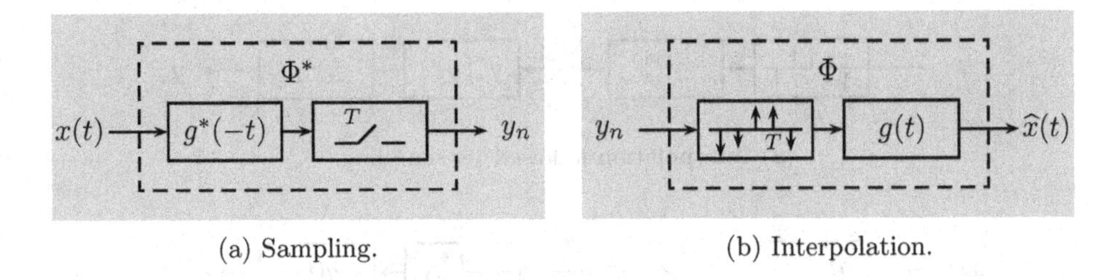

(a) Sampling.

(b) Interpolation.

Figure 5.4 Representations of sampling and interpolation operators. (a) Sampling as in (5.2) is equivalent to filtering by $g^*(-t)$ and recording the result at time instants that are integer multiples of T, as in (5.3) when $T = 1$. (b) Interpolation as in (5.4) is equivalent to forming a weighted Dirac comb followed by filtering by $g(t)$, as in (5.5) when $T = 1$.

where Φ denotes the basis synthesis operator associated with $\{g(t-n)\}_{n\in\mathbb{Z}}$. The function $\widehat{x}$ in (5.4) could be produced by filtering a *weighted Dirac comb*

$$\sum_{n\in\mathbb{Z}} y_n \delta(t-n)$$

with g,

$$\widehat{x}(t) = \left(\sum_{n\in\mathbb{Z}} y_n \delta(t-n)\right) *_t g(t). \tag{5.5}$$

The interpolation operator Φ is illustrated in Figure 5.4(b). Since $\widehat{x}$ is constant on intervals $[n, n+1)$, $n \in \mathbb{Z}$, a calculation similar to Example 2.17(i) shows that square summability of y implies the square integrability of $\widehat{x}$; thus, Φ is a mapping from $\ell^2(\mathbb{Z})$ to $\mathcal{L}^2(\mathbb{R})$. From Section 2.5.1 and Examples 2.17(i) and 2.18(iii), we know that the interpolation operator is the adjoint of the sampling operator; thus our choice to call it Φ.

Interpolation followed by sampling: Sequence recovery Figure 5.5(a) depicts interpolation followed by sampling. Because of the specific choice of sampling and interpolation operators, we have that $\Phi^*\Phi = I$; in other words, any sequence $y \in \ell^2(\mathbb{Z})$ is recovered perfectly when the interpolated function computed through (5.4) is used in the sampling formula (5.2),

$$\Phi^*\Phi y = y, \qquad \text{for } y \in \ell^2(\mathbb{Z}). \tag{5.6}$$

Another choice for sampling or interpolation would not necessarily have led to perfect recovery, as we discuss later in this chapter.

Sampling followed by interpolation: Function recovery Figure 5.5(b) depicts sampling followed by interpolation. Because of the specific choice of sampling and interpolation operators, when a function x belongs to S, it is recovered perfectly

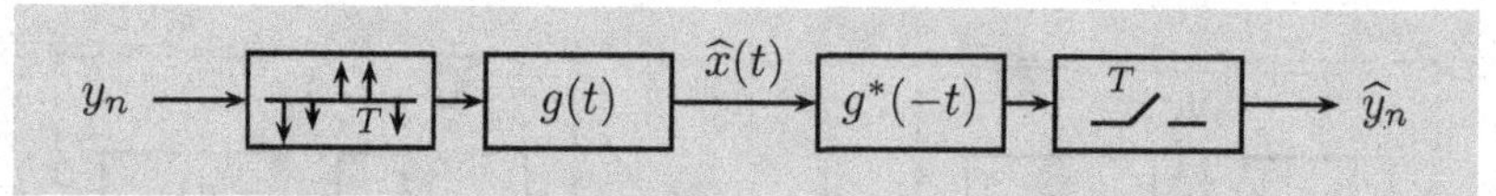

(a) Interpolation followed by sampling.

(b) Sampling followed by interpolation.

Figure 5.5 Combinations of sampling and interpolation. (a) With interpolation as in (5.4) followed by sampling as in (5.2), $\Phi^*\Phi = I$ recovers the input sequence perfectly for any $y \in \ell^2(\mathbb{Z})$; that is, $\widehat{y} = y$. (b) Sampling as in (5.2) followed by interpolation as in (5.4), $\Phi\Phi^*$, recovers the input function perfectly when $x \in S$; that is, $\widehat{x} = x$.

when the samples computed through (5.2) are used in the interpolation formula (5.4),

$$\Phi\Phi^* x = x, \qquad \text{for } x \in S \subset \mathcal{L}^2(\mathbb{R}). \tag{5.7}$$

Unlike for interpolation followed by sampling, here we needed to impose a restriction on the input function to guarantee perfect recovery. Both the sequence and the function recovery properties depend on having a proper match between sampling and interpolation operators, as we will discuss later in this chapter.

If the function were not in S, sampling followed by interpolation, $P = \Phi\Phi^*$, would not act as the identity on it, and we would obtain merely an approximation. Finding the closest function in S (where distance is measured with the $\mathcal{L}^2$ norm) is simple because of Hilbert-space geometry. We find that

$$P^2 = \Phi\Phi^*\Phi\Phi^* \stackrel{(a)}{=} \Phi\Phi^* = P, \tag{5.8a}$$

$$P^* = (\Phi\Phi^*)^* = \Phi\Phi^* = P, \tag{5.8b}$$

where (a) follows from $\Phi^*\Phi = I$. In other words, P is idempotent and self-adjoint; that is, P is an orthogonal projection operator. The projection theorem (Theorem 2.26) then states that, given an arbitrary $x \in \mathcal{L}^2(\mathbb{R})$, sampling followed by interpolation results in $\widehat{x} = \Phi\Phi^* x$, the least-squares approximation of x in S, that is, the best approximation under the $\mathcal{L}^2$ norm (see Figure 5.6).

Another way to verify the least-squares approximation property is to note that the set $\{g(t-k)\}_{k\in\mathbb{Z}}$ is an orthonormal basis for S and that (5.4) is an orthonormal basis expansion formula. Thus, the approximation property follows from Theorem 2.41. Finally, one can explicitly verify that computing the average of a function over an interval minimizes the $\mathcal{L}^2$ norm of the error of a piecewise-constant approximation to the function (see Solved exercise 5.1).

For any fixed $T > 0$, there are functions $x \in \mathcal{L}^2(\mathbb{R})$ that differ appreciably from the closest function $\widehat{x}_T$ that is piecewise-constant over intervals $[kT, (k+1)T)$.

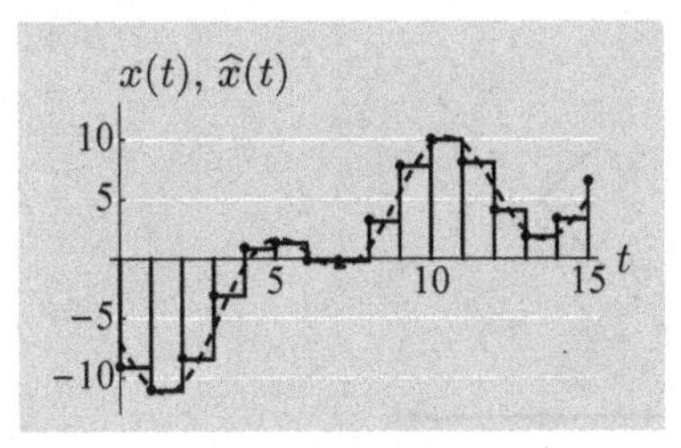

(a) Approximation of x by $\widehat{x} \in S$.

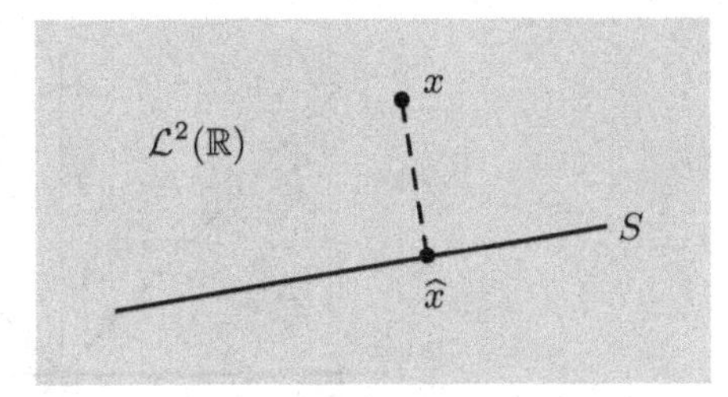

(b) Abstract view.

Figure 5.6 Least-squares approximation of an arbitrary function x (dashed line) by a piecewise-constant approximation $\widehat{x} \in S$ (solid line). (a) Example function and its piecewise-constant approximation. (b) Conceptual depiction of orthogonal projection.

However, considering all $T > 0$, these piecewise-constant functions are dense in $\mathcal{L}^2(\mathbb{R})$, and the approximation error between $\widehat{x}_T$ and x goes to zero as $T \to 0$. The rate at which this error goes to zero is an important parameter; it indicates the *approximation power* of the sampling and interpolation scheme (see Exercise 5.1).

Sampling and interpolation with nonorthogonal vectors in $\mathbb{R}^2$ In the discussion so far, we have had sampling and interpolation operators that are adjoints of each other, making their composition (in either order) a self-adjoint operator. This does not have to be the case, and throughout this chapter we will see instances of general sampling and interpolation operators $\widetilde{\Phi}^*$ and Φ, where we need not have $\widetilde{\Phi} = \Phi$.

The significance of the sampling and interpolation operators being an adjoint pair is closely related to whether the associated bases are orthonormal. To get a feel for this and a preview of the subspaces that arise in understanding sampling, we now look into the simple setting of vectors in $\mathbb{R}^2$. Here we call any linear operator $\mathbb{R}^2 \to \mathbb{R}$ a sampling operator and any linear operator $\mathbb{R} \to \mathbb{R}^2$ an interpolation operator. While in this setting there is no filtering as defined in Chapters 3 and 4 – contrasting with what we have discussed thus far in this section – it maintains the essential features of sampling and interpolation: sampling involves a reduction of dimension, and interpolation embeds in a higher-dimensional space. We start with an orthogonal case and follow it with a nonorthogonal one, allowing for a comparison of the two.

(i) Consider first the sampling and interpolation operators to be

$$\Phi^* = \frac{1}{\sqrt{2}} \begin{bmatrix} 1 & 1 \end{bmatrix}, \qquad \Phi = \frac{1}{\sqrt{2}} \begin{bmatrix} 1 \\ 1 \end{bmatrix}. \tag{5.9a}$$

The sampling operator Φ^* returns one sample that is the average of the two components of the input vector, scaled by $\sqrt{2}$. The interpolation operator creates a vector out of a single scalar sample by duplicating the sample, now with the scaling factor of $1/\sqrt{2}$. The null space of the sampling operator is

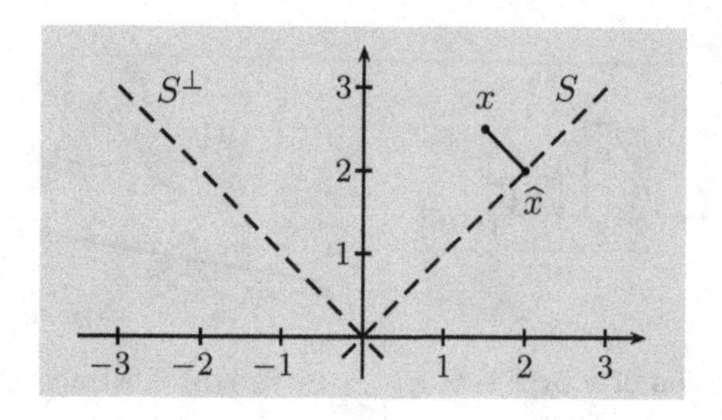

Figure 5.7 Least-squares approximation of $x = \begin{bmatrix} \frac{3}{2} & \frac{5}{2} \end{bmatrix}^\top$ via the orthogonal projection operator (5.10).

the orthogonal complement of the range of the interpolation operator,

$$S \;=\; \mathcal{R}(\Phi) \;=\; \left\{ \alpha \begin{bmatrix} 1 \\ 1 \end{bmatrix} \,\middle|\, \alpha \in \mathbb{C} \right\}, \tag{5.9b}$$

$$S^\perp \;=\; \mathcal{N}(\Phi^*) \;=\; \left\{ \alpha \begin{bmatrix} -1 \\ 1 \end{bmatrix} \,\middle|\, \alpha \in \mathbb{C} \right\}. \tag{5.9c}$$

Clearly, interpolation followed by sampling leads to identity,

$$\Phi^*\Phi \;=\; 1.$$

What is more interesting is that sampling followed by interpolation, $P = \Phi\Phi^*$, is the orthogonal projection operator onto S,

$$P \;=\; \frac{1}{2}\begin{bmatrix} 1 & 1 \\ 1 & 1 \end{bmatrix}, \qquad P^2 \;=\; P, \qquad P^* \;=\; P. \tag{5.10}$$

For any $x \in \mathbb{R}^2$, $\widehat{x} = Px$ is the least-squares approximation of x in S. In particular, $\widehat{x} = x$ when $x \in S$; otherwise, $\widehat{x}$ is the approximation in S that satisfies $x - \widehat{x} \in S^\perp$. Figure 5.7 illustrates these spaces as well as the approximation for $x = \begin{bmatrix} \frac{3}{2} & \frac{5}{2} \end{bmatrix}^\top$.

(ii) We keep the interpolation operator from (5.9a) and choose the sampling operator to be

$$\widetilde{\Phi}^* = \frac{1}{2\sqrt{2}} \begin{bmatrix} 1 & 3 \end{bmatrix}. \tag{5.11a}$$

The null space of the sampling operator and its orthogonal complement are

$$\widetilde{S}^\perp \;=\; \mathcal{N}(\widetilde{\Phi}^*) \;=\; \left\{ \alpha \begin{bmatrix} -3 \\ 1 \end{bmatrix} \,\middle|\, \alpha \in \mathbb{C} \right\}, \tag{5.11b}$$

$$\widetilde{S} \;=\; \left\{ \alpha \begin{bmatrix} 1 \\ 3 \end{bmatrix} \,\middle|\, \alpha \in \mathbb{C} \right\}. \tag{5.11c}$$

Note that the latter, $\widetilde{S}$, is no longer the same as the range of the interpolation operator, S. Again, interpolation followed by sampling leads to identity,

$$\widetilde{\Phi}^*\Phi \;=\; 1.$$

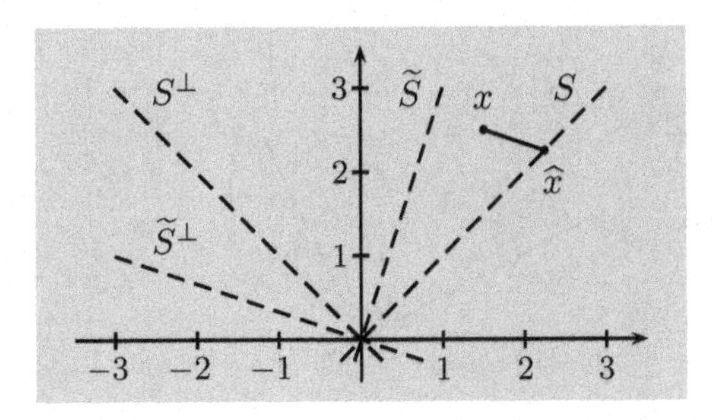

Figure 5.8 Approximation of $x = \begin{bmatrix} \frac{3}{2} & \frac{5}{2} \end{bmatrix}^\top$ via the projection operator (5.12).

However, sampling followed by interpolation, $P = \Phi\widetilde{\Phi}^*$, is no longer an orthogonal projection operator, though it is still a projection operator with range S,

$$P = \frac{1}{4}\begin{bmatrix} 1 & 3 \\ 1 & 3 \end{bmatrix}, \qquad P^2 = P, \qquad P^* \neq P. \tag{5.12}$$

In particular, $\widehat{x} = x$ when $x \in S$; otherwise, $\widehat{x}$ is the approximation in S that satisfies $x - \widehat{x} \in \widetilde{S}^\perp$ (but $x - \widehat{x} \in S^\perp$ is not satisfied). Figure 5.8 illustrates these spaces as well as the approximation for $x = \begin{bmatrix} \frac{3}{2} & \frac{5}{2} \end{bmatrix}^\top$.

(iii) We still keep the interpolation operator from (5.9a) the same and choose the sampling operator to be

$$\widetilde{\Phi}^* = \frac{1}{2\sqrt{2}}\begin{bmatrix} 1 & 2 \end{bmatrix}. \tag{5.13a}$$

The null space of the sampling operator and its orthogonal complement are

$$\widetilde{S}^\perp = \mathcal{N}(\widetilde{\Phi}^*) = \left\{ \alpha \begin{bmatrix} -2 \\ 1 \end{bmatrix} \,\middle|\, \alpha \in \mathbb{C} \right\},$$
$$\widetilde{S} = \left\{ \alpha \begin{bmatrix} 1 \\ 2 \end{bmatrix} \,\middle|\, \alpha \in \mathbb{C} \right\}.$$

This time, interpolation followed by sampling is not an identity on $\mathbb{C}$,

$$\widetilde{\Phi}^*\Phi \neq 1.$$

Sampling followed by interpolation, $P = \Phi\widetilde{\Phi}^*$, is no longer even a projection operator,

$$P = \frac{1}{4}\begin{bmatrix} 1 & 2 \\ 1 & 2 \end{bmatrix}, \qquad P^2 = \frac{3}{16}\begin{bmatrix} 1 & 2 \\ 1 & 2 \end{bmatrix} \neq P. \tag{5.14}$$

The range of P is S, and since P is not a projection operator, applying it a second time gives some P^2x in S that is different from Px. In this example,

$$P^k = \frac{1}{4}\left(\frac{3}{4}\right)^{k-1}\begin{bmatrix} 1 & 2 \\ 1 & 2 \end{bmatrix},$$

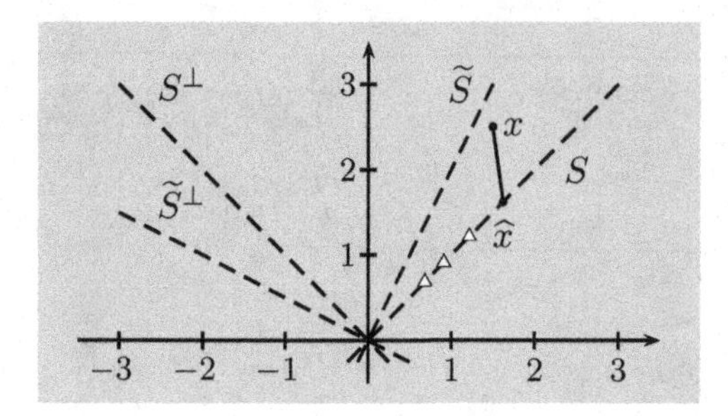

Figure 5.9 Operator (5.14) that is not a projection applied to $x = \begin{bmatrix} \frac{3}{2} & \frac{5}{2} \end{bmatrix}^\top$. Triangles denote P^2x, P^3x, and P^4x in descending order.

so $P^k x \to \mathbf{0}$ as k increases. Figure 5.9 illustrates this effect[92] both for $x = \begin{bmatrix} \frac{3}{2} & \frac{5}{2} \end{bmatrix}^\top$ and for the spaces involved.

Chapter outline

The bulk of this chapter, Sections 5.2–5.5, follows a progression through the Hilbert spaces developed in Chapters 2–4. In each case, we consider analysis and synthesis first with orthonormal vectors and then with nonorthogonal ones. Section 5.2 develops sampling and interpolation from the perspective of linear operators (matrices) in finite-dimensional spaces. Section 5.3 does the same for infinite-length sequences, with a restriction to LSI and LPSV operators. This restriction is important for practical implementations, and it enables the use of techniques from Chapter 3, including the DTFT. Section 5.4 progresses to functions on the real line, making use of the Fourier transform and other concepts from Chapter 4. The use of LSI filtering naturally leads to the sampling theory for shift-invariant subspaces, and the celebrated sampling theorem for bandlimited functions is presented both with a classical justification and as an instance of the general theory for shift-invariant subspaces. Section 5.5 develops sampling for periodic functions, making use of the Fourier series expansion, and finally, Section 5.6 concludes with a discussion of computational aspects.

5.2 Finite-dimensional vectors

As illustrated in Section 5.1, sampling and interpolation normally refer to operations applied to functions and sequences. In this chapter, we emphasize a geometric view of sampling and interpolation together with an investigation of the subspaces (range, null space) associated with sampling and interpolation operators. To set

[92]Here, the points move along S closer to the origin because the operator has eigenvalues smaller than 1 in absolute value. Choosing a different scaling in (5.13a), for example, $\frac{1}{2}$ instead of $1/(2\sqrt{2})$, would make the points move along S to infinity. Note that a scaling of $1/\sqrt{5}$ would make P idempotent, and thus a projection operator again, albeit still not an orthogonal one.

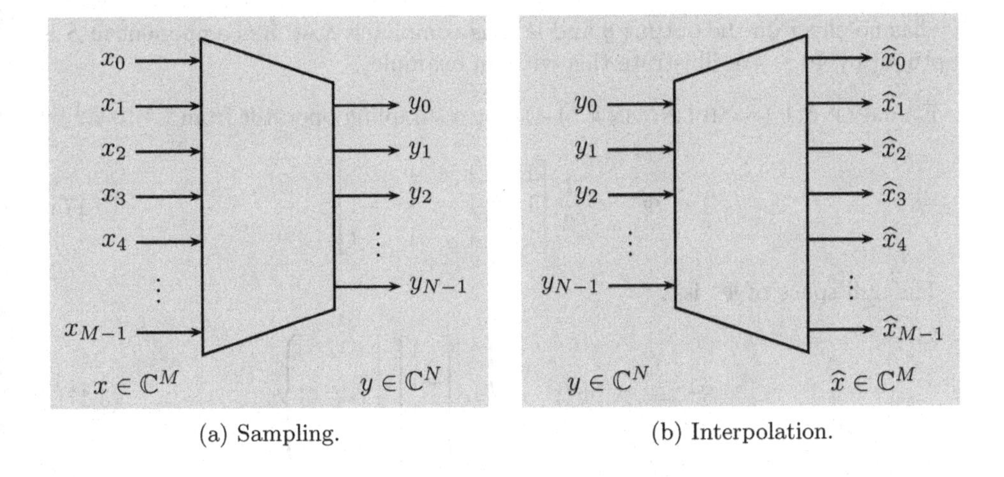

(a) Sampling. (b) Interpolation.

Figure 5.10 Sampling ($\mathbb{C}^M \to \mathbb{C}^N$) and interpolation ($\mathbb{C}^N \to \mathbb{C}^M$); $M > N$.

the stage for this view, we first look at sampling and interpolation operators in finite dimensions. Here we refer to any decrease in dimension as sampling and any increase in dimension as interpolation; sampling will take M values and produce $N < M$ values, while interpolation will take N values and produce $M > N$ values. These are depicted in Figure 5.10.

5.2.1 Sampling and interpolation with orthonormal vectors

Sampling Sampling is a linear operator from $\mathbb{C}^M$ to $\mathbb{C}^N$, $M > N$, so it can be represented by an $N \times M$ matrix Φ^*. For a given input vector $x \in \mathbb{C}^M$, the sampling output is a vector $y \in \mathbb{C}^N$,

$$y = \underbrace{\begin{bmatrix} \langle x, \varphi_0 \rangle \\ \langle x, \varphi_1 \rangle \\ \vdots \\ \langle x, \varphi_{N-1} \rangle \end{bmatrix}}_{N \times 1} = \underbrace{\begin{bmatrix} \varphi_0^* \\ \varphi_1^* \\ \vdots \\ \varphi_{N-1}^* \end{bmatrix}}_{N \times M} \underbrace{\begin{bmatrix} x_0 \\ x_1 \\ \vdots \\ x_{M-1} \end{bmatrix}}_{M \times 1} = \Phi^* x. \tag{5.15}$$

In the above matrix, φ_k^* is the kth row of Φ^*.

We first assume that $\{\varphi_k\}_{k=0}^{N-1}$ is an orthonormal set,

$$\langle \varphi_n, \varphi_k \rangle = \delta_{n-k} \quad \Leftrightarrow \quad \Phi^* \Phi = I, \tag{5.16}$$

and thus Φ^* has rank N, the largest possible for an $N \times M$ matrix with $N < M$ (that is, of full rank). Then, the sampling operator has an $(M-N)$-dimensional null space, $\mathcal{N}(\Phi^*)$; the set $\{\varphi_k\}_{k=0}^{N-1}$ spans its orthogonal complement, $S = \mathcal{N}(\Phi^*)^\perp = \operatorname{span}(\{\varphi_k\}_{k=0}^{N-1})$. When a vector $x \in \mathbb{C}^M$ is sampled, its component in the null space

$S^\perp$ has no effect on the output y and is thus completely lost; its component in S is captured by $\Phi^* x$. We illustrate this with an example.

EXAMPLE 5.1 (SAMPLING IN $\mathbb{C}^4$) Define a sampling operator from $\mathbb{C}^4$ to $\mathbb{C}^3$ by

$$\Phi^* = \frac{1}{2}\begin{bmatrix} 1 & 1 & 1 & 1 \\ 1 & -1 & 1 & -1 \\ 1 & 1 & -1 & -1 \end{bmatrix}. \tag{5.17a}$$

The null space of Φ^* is

$$S^\perp = \mathcal{N}(\Phi^*) = \left\{ \alpha \begin{bmatrix} 1 \\ -1 \\ -1 \\ 1 \end{bmatrix} \,\middle|\, \alpha \in \mathbb{C} \right\}, \tag{5.17b}$$

and its orthogonal complement is the span of the transposes of the rows of Φ^*,

$$S = \mathcal{N}(\Phi^*)^\perp = \left\{ \alpha_0 \begin{bmatrix} 1 \\ 1 \\ 1 \\ 1 \end{bmatrix} + \alpha_1 \begin{bmatrix} 1 \\ -1 \\ 1 \\ -1 \end{bmatrix} + \alpha_2 \begin{bmatrix} 1 \\ 1 \\ -1 \\ -1 \end{bmatrix} \,\middle|\, \alpha_0, \alpha_1, \alpha_2 \in \mathbb{C} \right\}. \tag{5.17c}$$

An arbitrary vector $x \in \mathbb{C}^4$ has an orthogonal decomposition $x = x_S + x_{S^\perp}$ with $x_S \in S$ and $x_{S^\perp} \in S^\perp$. The component $x_{S^\perp}$ cannot be recovered from $\Phi^* x$. For example,

$$\Phi^* \begin{bmatrix} 1+\alpha \\ 1-\alpha \\ 1-\alpha \\ 1+\alpha \end{bmatrix} = \Phi^* \Bigg(\underbrace{\begin{bmatrix} 1 \\ 1 \\ 1 \\ 1 \end{bmatrix}}_{\in S} + \alpha \underbrace{\begin{bmatrix} 1 \\ -1 \\ -1 \\ 1 \end{bmatrix}}_{\in S^\perp} \Bigg) = \Phi^* \begin{bmatrix} 1 \\ 1 \\ 1 \\ 1 \end{bmatrix} + \underbrace{\alpha \Phi^* \begin{bmatrix} 1 \\ -1 \\ -1 \\ 1 \end{bmatrix}}_{=0} = \begin{bmatrix} 2 \\ 0 \\ 0 \end{bmatrix},$$

showing that infinitely many choices of x in $\mathbb{C}^4$ result in the same value of $\Phi^* x$.

Interpolation Interpolation is a linear operator from $\mathbb{C}^N$ to $\mathbb{C}^M$, $N < M$, so it can be represented by an $M \times N$ matrix; we first choose that matrix to be the adjoint of the sampling operator, Φ, as we have done in the introductory example. For a given input vector $y \in \mathbb{C}^N$, the interpolation output is a vector $\widehat{x} \in \mathbb{C}^M$,

$$\widehat{x} = \underbrace{\begin{bmatrix} \varphi_0 & \varphi_1 & \cdots & \varphi_{N-1} \end{bmatrix}}_{M \times N} \underbrace{\begin{bmatrix} y_0 \\ y_1 \\ \vdots \\ y_{N-1} \end{bmatrix}}_{N \times 1} = \Phi y = \sum_{k=0}^{N-1} y_k \varphi_k. \tag{5.18}$$

In the above matrix, φ_k is the kth column of Φ.

We continue to assume that $\{\varphi_k\}_{k=0}^{N-1}$ is an orthonormal set. As was true for Φ^*, Φ has full rank, N. Thus, the interpolation operator has an N-dimensional range S, which is a proper subspace of $\mathbb{C}^M$ and is given by $S = \operatorname{span}(\{\varphi_k\}_{k=0}^{N-1})$. This subspace is, of course, the same as the orthogonal complement of the null space of the sampling operator, as we have seen earlier.

EXAMPLE 5.2 (INTERPOLATION TO $\mathbb{C}^4$) Continuing the previous example, the interpolation operator associated with the sampling operator in (5.17a) is

$$\Phi = \frac{1}{2}\begin{bmatrix} 1 & 1 & 1 \\ 1 & -1 & 1 \\ 1 & 1 & -1 \\ 1 & -1 & -1 \end{bmatrix}. \tag{5.19}$$

The range of this operator is S (the same as the orthogonal complement of the null space of the sampling operator in (5.17c)).

Interpolation followed by sampling Interpolation followed by sampling is described by $\Phi^*\Phi$, which maps from the smaller space, $\mathbb{C}^N$, to itself. Since by assumption (5.16) holds, $\Phi^*\Phi y = y$ for all $y \in \mathbb{C}^N$, so any input is perfectly recovered. Equation (5.16) also shows that the condition for perfect recovery is the same as the set of vectors $\{\varphi_k\}_{k=0}^{N-1}$ being orthonormal, as in (2.92). This set of vectors is not a basis for $\mathbb{C}^M$ (it has too few vectors); instead, it is an orthonormal basis for the N-dimensional subspace S that it spans.

Sampling followed by interpolation Sampling followed by interpolation is described by $P = \Phi\Phi^*$. While this operator has $\mathbb{C}^M$ both as its domain and as its codomain, it cannot act as identity on all inputs because there is an intermediate representation with only N numbers, where $N < M$. It can be an identity operator on at most an N-dimensional subspace of $\mathbb{C}^M$, depending on whether (5.16) holds.

When (5.16) does hold, P is idempotent by (5.8a); since the sampling and interpolation operators are adjoints, P is self-adjoint by (5.8b). Thus, P is an orthogonal projection operator. Then, by Theorem 2.26, Px is the best approximation of x in S, where S is the range of P; if $x \in S$, sampling followed by interpolation will perfectly recover x.

THEOREM 5.1 (RECOVERY FOR VECTORS, ORTHOGONAL) Let $N, M \in \mathbb{Z}^+$ with $N < M$, let the sampling operator $\Phi^* : \mathbb{C}^M \to \mathbb{C}^N$ be given in (5.15), and let the interpolation operator $\Phi : \mathbb{C}^N \to \mathbb{C}^M$ be given in (5.18). Denote the result of sampling followed by interpolation applied to $x \in \mathbb{C}^M$ by $\widehat{x} = Px = \Phi\Phi^* x$.

If orthonormality (5.16) is satisfied, then P is the orthogonal projection operator with range $S = \mathcal{R}(\Phi)$ and $\widehat{x}$ is the best approximation of x in S,

$$\widehat{x} = \arg\min_{x_S \in S} \|x - x_S\|, \qquad x - \widehat{x} \perp S;$$

in particular, $\widehat{x} = x$ when $x \in S$.

We now illustrate the above result by building upon Examples 5.1 and 5.2.

EXAMPLE 5.3 (SAMPLING FOLLOWED BY INTERPOLATION IN $\mathbb{C}^4$) Let Φ be given by (5.19), which has range S in (5.17c). Consider first any $x \in S$. It can be written as

$$x = \alpha_0\varphi_0 + \alpha_1\varphi_1 + \alpha_2\varphi_2$$

for some α_0, α_1, $\alpha_2 \in \mathbb{C}$. Then, the output from sampling followed by interpolation is

$$\begin{aligned}
\widehat{x} &= Px = \Phi\Phi^* x \overset{(a)}{=} \Phi\Phi^*(\alpha_0\varphi_0 + \alpha_1\varphi_1 + \alpha_2\varphi_2) \\
&\overset{(b)}{=} \left(\frac{1}{4}\begin{bmatrix} 3 & 1 & 1 & -1 \\ 1 & 3 & -1 & 1 \\ 1 & -1 & 3 & 1 \\ -1 & 1 & 1 & 3 \end{bmatrix}\right) \frac{1}{2}\left(\alpha_0 \begin{bmatrix} 1 \\ 1 \\ 1 \\ 1 \end{bmatrix} + \alpha_1 \begin{bmatrix} 1 \\ -1 \\ 1 \\ -1 \end{bmatrix} + \alpha_2 \begin{bmatrix} 1 \\ 1 \\ -1 \\ -1 \end{bmatrix}\right) \\
&= \frac{1}{2}\left(\alpha_0 \begin{bmatrix} 1 \\ 1 \\ 1 \\ 1 \end{bmatrix} + \alpha_1 \begin{bmatrix} 1 \\ -1 \\ 1 \\ -1 \end{bmatrix} + \alpha_2 \begin{bmatrix} 1 \\ 1 \\ -1 \\ -1 \end{bmatrix}\right) = x,
\end{aligned}$$

where (a) follows from substituting for x; and (b) from computing $\Phi\Phi^*$ from (5.19) and substituting for φ_0, φ_1, and φ_2. Note that the matrix–vector product is simple because φ_0, φ_1, and φ_2 are eigenvectors of P associated with eigenvalue 1.

Consider next a particular $x \notin S$, $x = \begin{bmatrix} 2 & 0 & 0 & 2 \end{bmatrix}^\top$ from Example 5.1. Applying $P = \Phi\Phi^*$ to x leads to

$$\widehat{x} = Px = \Phi\Phi^* x = \frac{1}{4}\begin{bmatrix} 3 & 1 & 1 & -1 \\ 1 & 3 & -1 & 1 \\ 1 & -1 & 3 & 1 \\ -1 & 1 & 1 & 3 \end{bmatrix} \begin{bmatrix} 2 \\ 0 \\ 0 \\ 2 \end{bmatrix} = \begin{bmatrix} 1 \\ 1 \\ 1 \\ 1 \end{bmatrix},$$

which clearly belongs to S since it is twice the first basis vector φ_0. It is also the vector in S that is closest to x, since the difference between x and $\widehat{x}$ is orthogonal to S,

$$x - \widehat{x} = \begin{bmatrix} 2 \\ 0 \\ 0 \\ 2 \end{bmatrix} - \begin{bmatrix} 1 \\ 1 \\ 1 \\ 1 \end{bmatrix} = \begin{bmatrix} 1 \\ -1 \\ -1 \\ 1 \end{bmatrix} \perp S.$$

As we saw in Chapter 3, finite-dimensional vectors can be seen as finite-length sequences. In that case, we know that the DFT will be an appropriate Fourier transform when these sequences are circularly extended (or viewed as periodic sequences). We could then define bandlimited subspaces of $\mathbb{C}^M$ similarly to what we will do in the following sections for sequences and functions. Exercise 5.2 explores sampling and interpolation in such subspaces.

5.2.2 Sampling and interpolation with nonorthogonal vectors

We now expand our sets of sampling and interpolation operators to be built from nonorthogonal vectors. As in our discussion of orthonormal and biorthogonal pairs of bases, nonorthogonal vectors make the geometry more complicated; the sampling and interpolation operators are no longer adjoints of each other, and the relevant spaces we discussed earlier – the range of the interpolation operator and the orthogonal complement of the null space of the sampling operator – are no longer the same.

Sampling Again, sampling is represented by an $N \times M$ matrix as in (5.15), but this time with rows that are not necessarily orthogonal. We call the kth row $\widetilde{\varphi}_k^*$ and the corresponding sampling matrix $\widetilde{\Phi}^*$. Thus, for a given input vector $x \in \mathbb{C}^M$, the sampling output is a vector $y \in \mathbb{C}^N$,

$$y = \underbrace{\begin{bmatrix} \langle x, \widetilde{\varphi}_0 \rangle \\ \langle x, \widetilde{\varphi}_1 \rangle \\ \vdots \\ \langle x, \widetilde{\varphi}_{N-1} \rangle \end{bmatrix}}_{N \times 1} = \underbrace{\begin{bmatrix} \widetilde{\varphi}_0^* \\ \widetilde{\varphi}_1^* \\ \vdots \\ \widetilde{\varphi}_{N-1}^* \end{bmatrix}}_{N \times M} \underbrace{\begin{bmatrix} x_0 \\ x_1 \\ \vdots \\ x_{M-1} \end{bmatrix}}_{M \times 1} = \widetilde{\Phi}^* x. \tag{5.20}$$

We again assume $\widetilde{\Phi}^*$ to have full rank, N. Thus, the sampling operator has an $(M - N)$-dimensional null space, $\mathcal{N}(\widetilde{\Phi}^*)$; the set $\{\widetilde{\varphi}_k\}_{k=0}^{N-1}$ spans its orthogonal complement, $\widetilde{S} = \mathcal{N}(\widetilde{\Phi}^*)^\perp = \operatorname{span}(\{\widetilde{\varphi}_k\}_{k=0}^{N-1})$. When a vector $x \in \mathbb{C}^M$ is sampled, its component in the null space $\widetilde{S}^\perp$ is completely lost; its component in $\widetilde{S}$ is captured by $\widetilde{\Phi}^* x$.

Example 5.4 (Sampling in $\mathbb{C}^4$) Define sampling of $x \in \mathbb{C}^4$ to obtain three samples $y \in \mathbb{C}^3$ as giving the midpoints of neighboring pairs of samples. For $k \in \{0, 1, 2\}$, the sample y_k is the average of x_k and x_{k+1}, so the sampling operator can be written as

$$\widetilde{\Phi}^* = \frac{1}{2} \begin{bmatrix} 1 & 1 & 0 & 0 \\ 0 & 1 & 1 & 0 \\ 0 & 0 & 1 & 1 \end{bmatrix}. \tag{5.21a}$$

The null space of $\widetilde{\Phi}^*$ is

$$\widetilde{S}^\perp = \mathcal{N}(\widetilde{\Phi}^*) = \left\{ \beta \begin{bmatrix} -1 \\ 1 \\ -1 \\ 1 \end{bmatrix} \,\middle|\, \beta \in \mathbb{C} \right\}, \tag{5.21b}$$

and its orthogonal complement is

$$\widetilde{S} = \mathcal{N}(\widetilde{\Phi}^*)^{\perp} = \left\{ \beta_0 \begin{bmatrix} 1 \\ 1 \\ 0 \\ 0 \end{bmatrix} + \beta_1 \begin{bmatrix} 0 \\ 1 \\ 1 \\ 0 \end{bmatrix} + \beta_2 \begin{bmatrix} 0 \\ 0 \\ 1 \\ 1 \end{bmatrix} \;\middle|\; \beta_0, \beta_1, \beta_2 \in \mathbb{C} \right\}. \tag{5.21c}$$

Interpolation Again, interpolation is represented by an $M \times N$ matrix Φ, as in (5.18), but this time it is not the adjoint of the sampling operator $\widetilde{\Phi}^*$. For a given input vector $y \in \mathbb{C}^N$, the interpolation output is a vector $\widehat{x} \in \mathbb{C}^M$; it will lie in the subspace

$$S = \mathcal{R}(\Phi) = \left\{ \sum_{k=0}^{N-1} \alpha_k \varphi_k \;\middle|\; \alpha \in \mathbb{C}^N \right\}. \tag{5.22}$$

When the interpolation operator is specially chosen so that

$$\Phi = \widetilde{\Phi}(\widetilde{\Phi}^*\widetilde{\Phi})^{-1}, \tag{5.23}$$

that is, it is the pseudoinverse of $\widetilde{\Phi}^*$, then $S = \widetilde{S}$, because

$$S = \mathcal{R}(\Phi) \overset{(a)}{=} \mathcal{R}(\widetilde{\Phi}(\widetilde{\Phi}^*\widetilde{\Phi})^{-1}) \overset{(b)}{=} \mathcal{R}(\widetilde{\Phi}) \overset{(c)}{=} \mathcal{N}(\widetilde{\Phi}^*)^{\perp} \overset{(d)}{=} \widetilde{S}, \tag{5.24}$$

where (a) follows from (5.23); (b) from $\mathcal{R}(AB) = \mathcal{R}(A)$ when B is invertible; (c) from (2.53a); and (d) from the definition of $\widetilde{S}$. Though the interpolation operator in (5.23) has other distinguishing properties that we will develop shortly, we do not assume the use of this particular interpolation operator in all cases.

Example 5.5 (Interpolation to $\mathbb{C}^4$) One possible interpolation operator from $\mathbb{C}^3$ to $\mathbb{C}^4$ is the pseudoinverse of $\widetilde{\Phi}^*$ in (5.21a),

$$\Phi_1 = \frac{1}{2} \begin{bmatrix} 3 & -2 & 1 \\ 1 & 2 & -1 \\ -1 & 2 & 1 \\ 1 & -2 & 3 \end{bmatrix}. \tag{5.25a}$$

The range of Φ_1 is

$$S = \left\{ \alpha_0 \begin{bmatrix} 3 \\ 1 \\ -1 \\ 1 \end{bmatrix} + \alpha_1 \begin{bmatrix} -2 \\ 2 \\ 2 \\ -2 \end{bmatrix} + \alpha_2 \begin{bmatrix} 1 \\ -1 \\ 1 \\ 3 \end{bmatrix} \;\middle|\; \alpha_0, \alpha_1, \alpha_2 \in \mathbb{C} \right\}. \tag{5.25b}$$

To demonstrate that $\widetilde{S}$ in (5.21c) and S in (5.25b) are the same, we can write each vector generating S as a linear combination of the vectors generating $\widetilde{S}$. For example,

$$\begin{bmatrix} 3 \\ 1 \\ -1 \\ 1 \end{bmatrix} = 3 \begin{bmatrix} 1 \\ 1 \\ 0 \\ 0 \end{bmatrix} - 2 \begin{bmatrix} 0 \\ 1 \\ 1 \\ 0 \end{bmatrix} + \begin{bmatrix} 0 \\ 0 \\ 1 \\ 1 \end{bmatrix},$$

and similarly for the others.

Interpolation followed by sampling Interpolation followed by sampling is described by $\widetilde{\Phi}^* \Phi$, which maps from the smaller space, $\mathbb{C}^N$, to itself. From the dimensions of the operators, it is possible to have $\widetilde{\Phi}^* \Phi y = y$ for all $y \in \mathbb{C}^N$; this happens when Φ is a right inverse of $\widetilde{\Phi}^*$,

$$\widetilde{\Phi}^* \Phi = I \quad \Leftrightarrow \quad \langle \varphi_n, \widetilde{\varphi}_k \rangle = \delta_{n-k}. \tag{5.26}$$

The sampling and interpolation operators are then called *consistent.* Choosing the pseudoinverse in (5.23) for Φ satisfies (5.26); since $M > N$, there exist infinitely many other right inverses that lead to consistency. Equation (5.26) also shows that the condition for perfect recovery is the same as the sets of vectors $\{\varphi_k\}_{k=0}^{N-1}$ and $\{\widetilde{\varphi}_k\}_{k=0}^{N-1}$ being biorthogonal, as in (2.111). These sets of vectors are not bases for $\mathbb{C}^M$ (there are too few vectors); instead, they are bases for the N-dimensional subspaces S and $\widetilde{S}$ that they span, respectively. Note that, while they are biorthogonal sets, they are not a biorthogonal pair of bases unless $S = \widetilde{S}$.

EXAMPLE 5.6 (INTERPOLATION FOLLOWED BY SAMPLING IN $\mathbb{C}^4$) The interpolation operator

$$\Phi_2 = \begin{bmatrix} 2 & -1 & 0 \\ 0 & 1 & 0 \\ 0 & 1 & 0 \\ 0 & -1 & 2 \end{bmatrix} \tag{5.27}$$

is a right inverse of the sampling operator in (5.21a),

$$\widetilde{\Phi}^* \Phi_2 = \left(\frac{1}{2} \begin{bmatrix} 1 & 1 & 0 & 0 \\ 0 & 1 & 1 & 0 \\ 0 & 0 & 1 & 1 \end{bmatrix} \right) \begin{bmatrix} 2 & -1 & 0 \\ 0 & 1 & 0 \\ 0 & 1 & 0 \\ 0 & -1 & 2 \end{bmatrix} = I.$$

Thus, Φ_2 and $\widetilde{\Phi}^*$ are consistent. Infinitely many other interpolation operators are consistent with $\widetilde{\Phi}^*$; one of these is Φ_1 in (5.25a), which is a right inverse of $\widetilde{\Phi}^*$ by construction. Interpolation with Φ_1 or Φ_2 followed by sampling with $\widetilde{\Phi}^*$ results in perfect recovery.

Sampling followed by interpolation Sampling followed by interpolation is described by $P = \Phi\widetilde{\Phi}^*$. Like in the orthogonal case, this cannot be an identity operator on $\mathbb{C}^M$ because there is an intermediate representation with only N numbers, with $N < M$. When the sampling and interpolation operators are consistent as in (5.26), then

$$P^2 = (\Phi\widetilde{\Phi}^*)(\Phi\widetilde{\Phi}^*) = \Phi(\widetilde{\Phi}^*\Phi)\widetilde{\Phi}^* \stackrel{(a)}{=} \Phi I \widetilde{\Phi}^* = \Phi\widetilde{\Phi}^* = P, \tag{5.28}$$

where (a) follows from consistency. In other words, the idempotency of P is guaranteed by consistency; consistency thus implies that P is a projection operator, albeit not necessarily an orthogonal one.[93] Figure 5.11 shows what happens in that case:

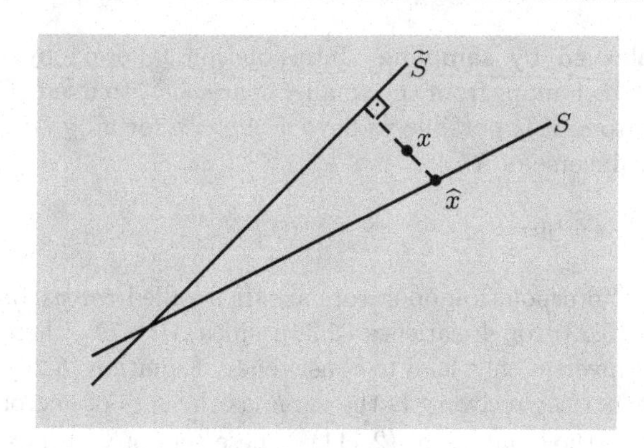

Figure 5.11 Subspaces defined in sampling and interpolation. $\widetilde{S}$ represents what can be measured; it is the orthogonal complement of the null space of the sampling operator $\widetilde{\Phi}^*$. S represents what can be produced; it is the range of the interpolation operator Φ. When sampling and interpolation are *consistent*, $\Phi\widetilde{\Phi}^*$ is a projection and $x - \widehat{x}$ is orthogonal to $\widetilde{S}$. When, furthermore, $S = \widetilde{S}$, the projection becomes an orthogonal projection and the sampling and interpolation are *ideally matched.*

P projects onto S, but the projection is not orthogonal. The approximation error $x - \widehat{x}$ is orthogonal to $\widetilde{S}$ but not to S.

It turns out that for P to be self-adjoint as well, Φ must be chosen to be the pseudoinverse of $\widetilde{\Phi}^*$, (5.23); in this case,

$$\begin{aligned} P^* &= (\Phi\widetilde{\Phi}^*)^* \stackrel{(a)}{=} (\widetilde{\Phi}(\widetilde{\Phi}^*\widetilde{\Phi})^{-1}\widetilde{\Phi}^*)^* = \widetilde{\Phi}((\widetilde{\Phi}^*\widetilde{\Phi})^{-1})^*\widetilde{\Phi}^* \\ &= \widetilde{\Phi}(\widetilde{\Phi}^*\widetilde{\Phi})^{-1}\widetilde{\Phi}^* \stackrel{(b)}{=} \Phi\widetilde{\Phi}^* = P, \end{aligned}$$

where (a) and (b) follow from (5.23). When Φ is the pseudoinverse of $\widetilde{\Phi}^*$, the subspaces S and $\widetilde{S}$ are the same, as shown in (5.24). When we have consistency combined with $S = \widetilde{S}$, the sampling and interpolation operators are called *ideally matched.* In other words, when sampling and interpolation operators are ideally matched, $P = \Phi\widetilde{\Phi}^*$ is an orthogonal projection operator, and, by Theorem 2.26, $\widehat{x} = Px$ is the best approximation of x in S.

The previous discussion can be summarized as follows (see also Figure 5.11):

THEOREM 5.2 (RECOVERY FOR VECTORS, NONORTHOGONAL) Let $N, M \in \mathbb{Z}^+$ with $N < M$, let the sampling operator $\widetilde{\Phi}^* : \mathbb{C}^M \to \mathbb{C}^N$ be given in (5.20), and let the interpolation operator $\Phi : \mathbb{C}^N \to \mathbb{C}^M$ be given in (5.18). Denote the result of sampling followed by interpolation applied to $x \in \mathbb{C}^M$ by $\widehat{x} = Px = \Phi\widetilde{\Phi}^*x$.

If consistency (5.26) is satisfied, then P is a projection operator with range $S = \mathcal{R}(\Phi)$ and $x - \widehat{x} \perp \widetilde{S} = \mathcal{N}(\widetilde{\Phi}^*)^\perp$; in particular, $\widehat{x} = x$ when $x \in S$.

[93] Recall the distinction between a projection and an orthogonal projection from Definition 2.27.

If, additionally, ideal matching (5.23) is satisfied, then $S = \widetilde{S}$, P is the orthogonal projection operator with range S, and $\widehat{x}$ is the best approximation of x in S,

$$\widehat{x} = \arg\min_{x_S \in S} \|x - x_S\|, \qquad x - \widehat{x} \perp S.$$

Example 5.7 (Sampling followed by interpolation in $\mathbb{C}^4$) Recall $\widetilde{\Phi}^*$ from (5.21a), Φ_1 from (5.25a), and Φ_2 from (5.27). As shown in Example 5.6, pairs $(\widetilde{\Phi}^*, \Phi_1)$ and $(\widetilde{\Phi}^*, \Phi_2)$ both satisfy consistency condition (5.26). However, only $(\widetilde{\Phi}^*, \Phi_1)$ is an ideally matched pair, since Φ_1 is the pseudoinverse of $\widetilde{\Phi}^*$. Thus, using Theorem 5.2, both $P_1 = \Phi_1\widetilde{\Phi}^*$ and $P_2 = \Phi_2\widetilde{\Phi}^*$ are projection operators, but only P_1 is an orthogonal projection operator.

To demonstrate these facts, compute

$$P_1 = \Phi_1\widetilde{\Phi}^* = \left(\frac{1}{2}\begin{bmatrix} 3 & -2 & 1 \\ 1 & 2 & -1 \\ -1 & 2 & 1 \\ 1 & -2 & 3 \end{bmatrix}\right)\left(\frac{1}{2}\begin{bmatrix} 1 & 1 & 0 & 0 \\ 0 & 1 & 1 & 0 \\ 0 & 0 & 1 & 1 \end{bmatrix}\right)$$
$$= \frac{1}{4}\begin{bmatrix} 3 & 1 & -1 & 1 \\ 1 & 3 & 1 & -1 \\ -1 & 1 & 3 & 1 \\ 1 & -1 & 1 & 3 \end{bmatrix},$$
$$P_2 = \Phi_2\widetilde{\Phi}^* = \begin{bmatrix} 2 & -1 & 0 \\ 0 & 1 & 0 \\ 0 & 1 & 0 \\ 0 & -1 & 2 \end{bmatrix}\left(\frac{1}{2}\begin{bmatrix} 1 & 1 & 0 & 0 \\ 0 & 1 & 1 & 0 \\ 0 & 0 & 1 & 1 \end{bmatrix}\right)$$
$$= \frac{1}{2}\begin{bmatrix} 2 & 1 & -1 & 0 \\ 0 & 1 & 1 & 0 \\ 0 & 1 & 1 & 0 \\ 0 & -1 & 1 & 2 \end{bmatrix}.$$

While both P_1 and P_2 are idempotent, only P_1 is self-adjoint as well. Projection operators P_1 and P_2 have the same null space, which was identified in Example 5.4 as being the one in (5.21b).

5.3 Sequences

The previous section gave us a firm grasp of the operator view of sampling and interpolation through the finite-dimensional (matrix) case. In the present section and in Sections 5.4 and 5.5, we move to other spaces developed in Chapters 3 and 4, with domains associated with discrete and continuous time. In these developments, we restrict the sampling and interpolation operators to those implemented with LSI filtering and emphasize the cases in which they are ideally matched.

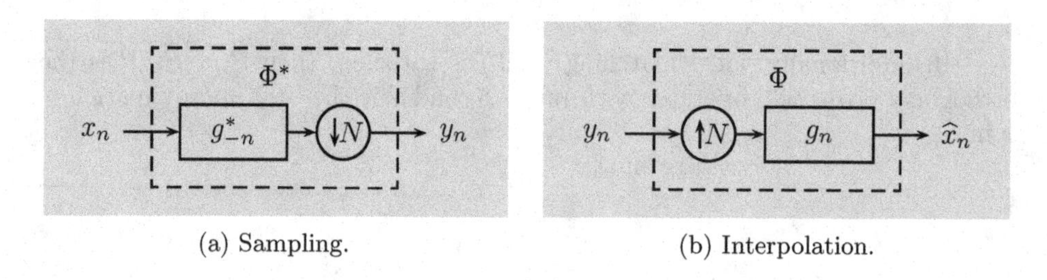

(a) Sampling.

(b) Interpolation.

Figure 5.12 Sampling and interpolation in $\ell^2(\mathbb{Z})$ with orthonormal sequences.

5.3.1 Sampling and interpolation with orthonormal sequences

Paralleling the development for finite-dimensional vectors, we start with the case when the sampling and interpolation operators are defined using orthonormal sequences. We also impose a relationship between filters through time reversal and conjugation to create an adjoint pair of operators. Together, these make sampling and interpolation equivalent to analysis and synthesis with the orthonormal basis $\{\varphi_k\}_{k\in\mathbb{Z}}$ for a subspace $S \subset \ell^2(\mathbb{Z})$.

Shift-invariant subspaces of sequences We start by introducing a class of subspaces of $\ell^2(\mathbb{Z})$ that will play a prominent role in the material that follows.

DEFINITION 5.3 (SHIFT-INVARIANT SUBSPACE OF $\ell^2(\mathbb{Z})$) A subspace $S \subset \ell^2(\mathbb{Z})$ is a *shift-invariant subspace* with respect to a shift $L \in \mathbb{Z}^+$ when $x_n \in S$ implies that $x_{n-kL} \in S$ for every integer k. In addition, $s \in \ell^2(\mathbb{Z})$ is called a *generator* of S when $S = \overline{\text{span}}(\{s_{n-kL}\}_{k\in\mathbb{Z}})$.

For example, for any particular filter, the outputs of upsampling by 2 followed by filtering, (3.206), form a shift-invariant subspace with respect to any positive integer multiple of 2. The same will be true for the interpolation operators we define shortly.

Sampling We refer to the operation depicted in Figure 5.12(a), involving filtering with g^*_{-n} and downsampling by integer $N > 1$, as *sampling of the sequence x_n with rate $1/N$ and prefilter g^*_{-n}*, and we denote it by $y = \Phi^* x$. Even though this operation results in infinitely many samples, it involves a dimensionality reduction because there is only one output sample per N input samples. We have seen this combination for $N = 2$ in Section 3.7.4 and an expression for its output in (3.203).

Generalizing (3.203) to an arbitrary N, for any time index $k \in \mathbb{Z}$, the output of sampling is

$$y_k \overset{(a)}{=} (\Phi^* x)_k \overset{(b)}{=} (g^*_{-n} *_n x_n)\big|_{n=kN} = \left(\sum_{m\in\mathbb{Z}} x_m g^*_{m-n}\right)\Bigg|_{n=kN}$$
$$= \sum_{m\in\mathbb{Z}} x_m g^*_{m-kN} = \langle x_m, g_{m-kN}\rangle_m \overset{(c)}{=} \langle x, \varphi_k\rangle, \tag{5.29}$$

where (a) follows from denoting the operator in Figure 5.12(a) by Φ^*; (b) from composing filtering by g^*_{-n} with downsampling by N; and (c) from defining the sequence φ_k to be g shifted by kN,

$$\varphi_{k,n} = g_{n-kN}, \qquad n \in \mathbb{Z}. \tag{5.30}$$

The final expression in (5.29) shows that calling the sampling operator Φ^* is consistent with the previous use of Φ^* as the analysis operator associated with $\{\varphi_k\}_{k\in\mathbb{Z}}$; see (2.91). The operation in (5.29) is a discrete-time counterpart to (5.2) and (5.3). As before, we first assume $\{\varphi_k\}_{k\in\mathbb{Z}}$ to be an orthonormal set,

$$\langle \varphi_k, \varphi_\ell\rangle = \delta_{k-\ell} \qquad \Leftrightarrow \qquad \langle g_{n-kN}, g_{n-\ell N}\rangle_n = \delta_{k-\ell}. \tag{5.31}$$

Since the output at time k is $\langle x, \varphi_k\rangle$, the sampling operator Φ^* produces the inner products with all the sequences in $\{\varphi_k\}_{k\in\mathbb{Z}}$. This operator can be written as an infinite matrix with rows equal to g^*_n and its shifts by integer multiples of N (see (3.202) for $N = 2$). Again, the sampling operator has a nontrivial null space, $\mathcal{N}(\Phi^*)$; the set $\{\varphi_k\}_{k\in\mathbb{Z}}$ spans its orthogonal complement, $S = \mathcal{N}(\Phi^*)^\perp = \overline{\operatorname{span}}(\{\varphi_k\}_{k\in\mathbb{Z}})$. The subspaces S and $S^\perp = \mathcal{N}(\Phi^*)$ are closed and shift-invariant with respect to shift N. We establish the shift invariance explicitly for $S^\perp$; for S, it is more natural to establish the shift invariance after introducing the interpolation operator. A sequence x is in $S^\perp$ when

$$\langle x, \varphi_k\rangle = 0, \qquad \text{for all } k \in \mathbb{Z}. \tag{5.32a}$$

Let x' be a shifted version of x,

$$x'_n = x_{n-\ell N}, \qquad n \in \mathbb{Z}, \tag{5.32b}$$

for some fixed $\ell \in \mathbb{Z}$. Then, for any $k \in \mathbb{Z}$,

$$\langle x', \varphi_k\rangle \overset{(a)}{=} \langle x_{n-\ell N}, \varphi_{k,n}\rangle_n \overset{(b)}{=} \langle x_{n-\ell N}, g_{n-kN}\rangle_n \overset{(c)}{=} \sum_{n\in\mathbb{Z}} x_{n-\ell N} g^*_{n-kN}$$
$$\overset{(d)}{=} \sum_{m\in\mathbb{Z}} x_m g^*_{m-(k-\ell)N} \overset{(e)}{=} \langle x_m, g_{m-(k-\ell)N}\rangle_m \overset{(f)}{=} \langle x, \varphi_{k-\ell}\rangle \overset{(g)}{=} 0,$$

where (a) follows from (5.32b); (b) and (f) from (5.30); (c) and (e) from the definition of the $\ell^2(\mathbb{Z})$ inner product; (d) from the change of variable $m = n - \ell N$; and (g) from (5.32a). This shows that x' is also in $S^\perp$, so $S^\perp$ is a shift-invariant subspace with respect to shift N.

When a sequence $x \in \ell^2(\mathbb{Z})$ is sampled, its component in the null space $S^\perp$ has no effect on the output y and is thus completely lost; its component in S is captured by $\Phi^* x$. We illustrate this with an example.

EXAMPLE 5.8 (SAMPLING IN $\ell^2(\mathbb{Z})$) Define a sampling operator on $\ell^2(\mathbb{Z})$ with $N = 2$ and the prefilter g^*_{-n},

$$\frac{1}{\sqrt{2}} \begin{bmatrix} \cdots & 0 & 1 & \boxed{1} & 0 & 0 & \cdots \end{bmatrix}^\top. \tag{5.33}$$

Then, from (3.202), the output is

$$\begin{bmatrix} \vdots \\ \boxed{y_0} \\ y_1 \\ \vdots \end{bmatrix} = \frac{1}{\sqrt{2}} \begin{bmatrix} & \vdots & \vdots & \vdots & \vdots & \\ \cdots & \boxed{1} & 1 & 0 & 0 & \cdots \\ \cdots & 0 & 0 & 1 & 1 & \cdots \\ & \vdots & \vdots & \vdots & \vdots & \end{bmatrix} \begin{bmatrix} \vdots \\ \boxed{x_0} \\ x_1 \\ x_2 \\ x_3 \\ \vdots \end{bmatrix} = \Phi^* x. \tag{5.34a}$$

For every two input samples x_{2k} and x_{2k+1}, we get one output sample $y_k = (x_{2k} + x_{2k+1})/\sqrt{2}$. The rows of matrix Φ^* clearly form an orthonormal set, since the nonzero entries do not overlap, and the rows have unit norm. The null space of Φ^* is

$$\begin{aligned} S^\perp &= \mathcal{N}(\Phi^*) = \{x \in \ell^2(\mathbb{Z}) \mid x_{2k} = -x_{2k+1} \text{ for all } k \in \mathbb{Z}\} \\ &= \left\{ \cdots + \alpha_{-1} \begin{bmatrix} \vdots \\ 0 \\ 1 \\ -1 \\ \boxed{0} \\ 0 \\ 0 \\ 0 \\ 0 \\ \vdots \end{bmatrix} + \alpha_0 \begin{bmatrix} \vdots \\ 0 \\ 0 \\ 0 \\ \boxed{1} \\ -1 \\ 0 \\ 0 \\ 0 \\ \vdots \end{bmatrix} + \alpha_1 \begin{bmatrix} \vdots \\ 0 \\ 0 \\ 0 \\ \boxed{0} \\ 0 \\ 1 \\ -1 \\ 0 \\ \vdots \end{bmatrix} + \cdots \;\middle|\; \alpha \in \ell^2(\mathbb{Z}) \right\}, \end{aligned}$$

and its orthogonal complement is the span of the transposes of the rows of Φ^*,

$$\begin{aligned} S &= \mathcal{N}(\Phi^*)^\perp = \{x \in \ell^2(\mathbb{Z}) \mid x_{2k} = x_{2k+1} \text{ for all } k \in \mathbb{Z}\} \\ &= \left\{ \cdots + \alpha_{-1} \begin{bmatrix} \vdots \\ 0 \\ 1 \\ 1 \\ \boxed{0} \\ 0 \\ 0 \\ 0 \\ 0 \\ \vdots \end{bmatrix} + \alpha_0 \begin{bmatrix} \vdots \\ 0 \\ 0 \\ 0 \\ \boxed{1} \\ 1 \\ 0 \\ 0 \\ 0 \\ \vdots \end{bmatrix} + \alpha_1 \begin{bmatrix} \vdots \\ 0 \\ 0 \\ 0 \\ \boxed{0} \\ 0 \\ 1 \\ 1 \\ 0 \\ \vdots \end{bmatrix} + \cdots \;\middle|\; \alpha \in \ell^2(\mathbb{Z}) \right\}. \end{aligned} \tag{5.34b}$$

Any $x \in \ell^2(\mathbb{Z})$ has an orthogonal decomposition $x = x_S + x_{S^\perp}$ with $x_S \in S$ and $x_{S^\perp} \in S^\perp$. The component $x_{S^\perp}$ cannot be recovered from $\Phi^* x$.

Interpolation We refer to the operation depicted in Figure 5.12(b), involving upsampling by integer $N > 1$ and filtering with g, as *interpolation of the sequence y with spacing N and postfilter g*, and we denote it by $\widehat{x} = \Phi y$. We have seen this combination for $N = 2$ in Section 3.7.4 and an expression for its output in (3.206). As developed in (3.212) for $N = 2$, but true also for any $N > 1$, choosing a prefilter and a postfilter that are related through time-reversed conjugation makes the sampling and interpolation operators adjoints of each other. This justifies our convention of calling one Φ^* and the other Φ in this case.

Upon generalizing (3.206) to an arbitrary N, for any time $n \in \mathbb{Z}$, the output of interpolation is

$$\widehat{x}_n \overset{(a)}{=} (\Phi y)_n \overset{(b)}{=} \sum_{k \in \mathbb{Z}} y_k g_{n-kN} \overset{(c)}{=} \left(\sum_{k \in \mathbb{Z}} y_k \varphi_k \right)_n, \tag{5.35}$$

where (a) follows from denoting the operator in Figure 5.12(b) by Φ; (b) from composing upsampling by N with filtering by g; and (c) from (5.30). This shows that calling the interpolation operator Φ is consistent with the previous use of Φ as the synthesis operator associated with $\{\varphi_k\}_{k \in \mathbb{Z}}$; see (2.90). The operation in (5.35) is a discrete-time counterpart to (5.4); a counterpart to (5.5) is

$$\widehat{x}_n = \left(\sum_{k \in \mathbb{Z}} y_k \delta_{n-kN} \right) *_n g_n.$$

The interpolation operator Φ can be written as an infinite matrix with columns equal to g and its shifts by integer multiples of N (see (3.205) for $N = 2$). Denoting the range of Φ by S as before, this subspace is the same as the orthogonal complement of the null space of the sampling operator, as we saw earlier.

Subspace S is a shift-invariant subspace with respect to shift N with generator g. Specifically, $\widehat{x} \in S$ means that

$$\widehat{x} = \sum_{k \in \mathbb{Z}} \alpha_k \varphi_k \tag{5.36a}$$

for some coefficient sequence $\alpha \in \ell^2(\mathbb{Z})$. If $\widehat{x}'$ is a shifted version of $\widehat{x}$,

$$\widehat{x}'_n = \widehat{x}_{n-\ell N}, \qquad n \in \mathbb{Z}, \tag{5.36b}$$

for some fixed $\ell \in \mathbb{Z}$, then

$$\begin{aligned} \widehat{x}'_n &\overset{(a)}{=} \sum_{k \in \mathbb{Z}} \alpha_k \varphi_{k,n-\ell N} \overset{(b)}{=} \sum_{k \in \mathbb{Z}} \alpha_k g_{n-\ell N-kN} \\ &\overset{(c)}{=} \sum_{k \in \mathbb{Z}} \alpha_k \varphi_{k+\ell,n} \overset{(d)}{=} \sum_{m \in \mathbb{Z}} \alpha_{m-\ell} \varphi_{m,n}, \end{aligned} \tag{5.36c}$$

where (a) follows from (5.36a) and (5.36b); (b) and (c) from (5.30); and (d) from the change of variable $m = k + \ell$. Thus, $\widehat{x}' \in S$; shifting $\widehat{x}$ by ℓN has shifted the coefficient sequence α by ℓ.

EXAMPLE 5.9 (INTERPOLATION TO $\ell^2(\mathbb{Z})$) Continuing Example 5.8, time reversal and conjugation of (5.33) give

$$g = \frac{1}{\sqrt{2}} \begin{bmatrix} \cdots & 0 & 0 & \boxed{1} & 1 & 0 & \cdots \end{bmatrix}^\top.$$

The output of interpolation with $N = 2$ and postfilter g is

$$\begin{bmatrix} \vdots \\ \boxed{\widehat{x}_0} \\ \widehat{x}_1 \\ \widehat{x}_2 \\ \widehat{x}_3 \\ \vdots \end{bmatrix} = \frac{1}{\sqrt{2}} \begin{bmatrix} & \vdots & \vdots & \\ \cdots & \boxed{1} & 0 & \cdots \\ \cdots & 1 & 0 & \cdots \\ \cdots & 0 & 1 & \cdots \\ \cdots & 0 & 1 & \cdots \\ & \vdots & \vdots & \end{bmatrix} \begin{bmatrix} \vdots \\ \boxed{y_0} \\ y_1 \\ \vdots \end{bmatrix} = \Phi y. \tag{5.37}$$

For every input sample y_k, we get two output samples $x_{2k} = x_{2k+1} = y_k/\sqrt{2}$. The range of this operator is S (the same as the orthogonal complement of the null space of the sampling operator in (5.34b)).

Interpolation followed by sampling Interpolation followed by sampling is described by $\Phi^*\Phi$ as in Figure 5.13(a). Using explicit expressions involving the sampling prefilter and interpolation postfilter, for any time $k \in \mathbb{Z}$, we have

$$\begin{aligned} \widehat{y}_k &\overset{(a)}{=} \sum_{n\in\mathbb{Z}} g^*_{n-kN} \widehat{x}_n \overset{(b)}{=} \sum_{n\in\mathbb{Z}} g^*_{n-kN} \sum_{\ell\in\mathbb{Z}} y_\ell g_{n-\ell N} \overset{(c)}{=} \sum_{\ell\in\mathbb{Z}} y_\ell \sum_{n\in\mathbb{Z}} g^*_{n-kN} g_{n-\ell N} \\ &\overset{(d)}{=} \sum_{\ell\in\mathbb{Z}} y_\ell \delta_{k-\ell} = y_k, \end{aligned} \tag{5.38}$$

where (a) follows from using (5.29) for sampling; (b) from using (5.35) for interpolation; (c) from interchanging the summations; and (d) from the orthonormality of $\{g_{n-kN}\}_{k\in\mathbb{Z}}$ in (5.31). Thus,

$$\Phi^*\Phi = I. \tag{5.39}$$

This shows that the condition for perfect recovery in interpolation followed by sampling is the same as the set of sequences $\{g_{n-kN}\}_{k\in\mathbb{Z}}$ being orthonormal, as in (5.31).

Using $\{\varphi_k\}_{k\in\mathbb{Z}}$ in place of $\{g_{n-Nk}\}_{k\in\mathbb{Z}}$, the identity expressed in (5.39) is precisely the property (2.101) established for analysis and synthesis operators associated with orthonormal bases. Here, $\{\varphi_k\}_{k\in\mathbb{Z}}$ is an orthonormal basis for the subspace S of $\ell^2(\mathbb{Z})$. We will thus exploit the more abstract view to avoid tedious computations.

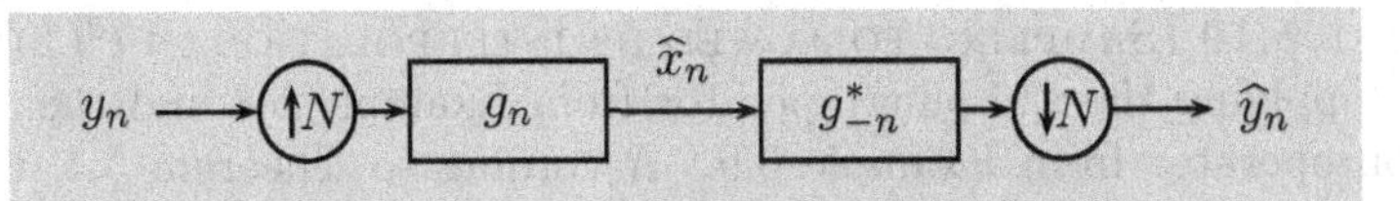

(a) Interpolation followed by sampling.

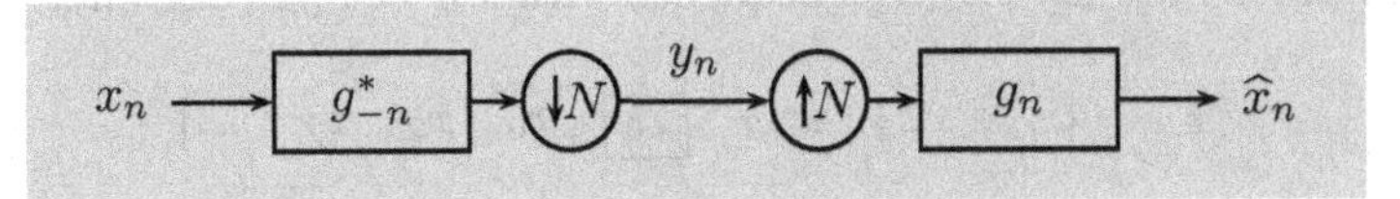

(b) Sampling followed by interpolation.

Figure 5.13 Sampling and interpolation in $\ell^2(\mathbb{Z})$ with orthonormal sequences.

Sampling followed by interpolation Sampling followed by interpolation is described by $P = \Phi\Phi^*$ as in Figure 5.13(b). We know this does not perfectly recover x in general, since sampling loses the component of x in the null space of the sampling operator, $S^\perp$.

As we saw for finite-dimensional vectors, given our choice of sampling and interpolation operators to satisfy (5.39), P is an orthogonal projection operator; see (5.8). This immediately gives the following analogue to Theorems 2.41 and 5.1:

THEOREM 5.4 (RECOVERY FOR SEQUENCES, ORTHOGONAL) Let $N \in \mathbb{Z}$ with $N > 1$, let the sampling operator $\Phi^* : \ell^2(\mathbb{Z}) \to \ell^2(\mathbb{Z})$ be given by

$$(\Phi^* x)_k = \sum_{m\in\mathbb{Z}} x_m g^*_{m-kN}, \qquad k \in \mathbb{Z}, \tag{5.40a}$$

and let the interpolation operator $\Phi : \ell^2(\mathbb{Z}) \to \ell^2(\mathbb{Z})$ be given by

$$(\Phi y)_n = \sum_{k\in\mathbb{Z}} y_k g_{n-kN}, \qquad n \in \mathbb{Z}. \tag{5.40b}$$

As depicted in Figure 5.13(b), denote the result of sampling followed by interpolation applied to $x \in \ell^2(\mathbb{Z})$ by $\widehat{x} = Px = \Phi\Phi^* x$.

If the filter g satisfies orthogonality with shifts by multiples of N, (5.31), then P is the orthogonal projection operator with range $S = \mathcal{R}(\Phi)$ and $\widehat{x}$ is the best approximation of x in S,

$$\widehat{x} = \arg\min_{x_S\in S} \|x - x_S\|, \qquad x - \widehat{x} \perp S;$$

in particular, $\widehat{x} = x$ when $x \in S$.

Example 5.10 (Sampling followed by interpolation in $\ell^2(\mathbb{Z})$) Consider first applying the sampling operator from Example 5.8 and then the interpolation operator from Example 5.9. According to Theorem 5.4, this should implement an orthogonal projection onto subspace S specified in (5.34b). Let us find the resulting sequences for an $x \in S$ and an $x \notin S$.

Consider first any $x \in S$, a sequence with $x_{2k} = x_{2k+1}$ for all $k \in \mathbb{Z}$,

$$x = \begin{bmatrix} \dots & x_{-2} & x_{-2} & \boxed{x_0} & x_0 & x_2 & x_2 & \dots \end{bmatrix}^\top.$$

Using (5.34a), the result of sampling is

$$y = \begin{bmatrix} \dots & \sqrt{2}\,x_{-2} & \boxed{\sqrt{2}\,x_0} & \sqrt{2}\,x_2 & \sqrt{2}\,x_4 & \dots \end{bmatrix}^\top.$$

Then, using (5.37), the result of interpolation is $\widehat{x} = x$, that is, perfect recovery of x.

Consider next a particular $x \notin S$,

$$x = \begin{bmatrix} \dots & 0 & \boxed{2} & 0 & 0 & 2 & 0 & \dots \end{bmatrix}^\top.$$

Using (5.34a), the result of sampling is

$$y = \begin{bmatrix} \dots & 0 & \boxed{\sqrt{2}} & \sqrt{2} & 0 & \dots \end{bmatrix}^\top.$$

Then, using (5.37), the result of interpolation is

$$\widehat{x} = \begin{bmatrix} \dots & 0 & \boxed{1} & 1 & 1 & 1 & 0 & \dots \end{bmatrix}^\top,$$

which clearly belongs to S. It is also the sequence in S closest to x, since the difference between x and $\widehat{x}$ is orthogonal to S,

$$x - \widehat{x} = \begin{bmatrix} \dots & 0 & \boxed{1} & -1 & -1 & 1 & 0 & \dots \end{bmatrix}^\top \perp S.$$

Example 5.11 (Best approximation by ramps) Let g be the filter shown in Figure 5.14(a). Since g has support $\{0, 1, \dots, 7\}$, it is orthogonal to its shifts by integer multiples of 8. The filter has also been chosen to be of unit norm, so it satisfies (5.31) with $N = 8$. According to Theorem 5.4, the system in Figure 5.13(b) with $N = 8$ performs the orthogonal projection to a shift-invariant subspace S.

The limited support of g makes the shifted versions $\{g_{n-8k}\}_{k\in\mathbb{Z}}$ have disjoint supports. Thus, the subspace S is easy to visualize: every sequence in S looks like a ramp starting at zero on every block of the form $\{8k, 8k+1, \dots, 8k+7\}$. Figures 5.14(b)–(d) show an arbitrary input x, the resulting samples y, and the interpolated samples $\widehat{x}$. Notice that $\widehat{x}$ fits the description of the sequences in S and it is the closest such sequence to x.

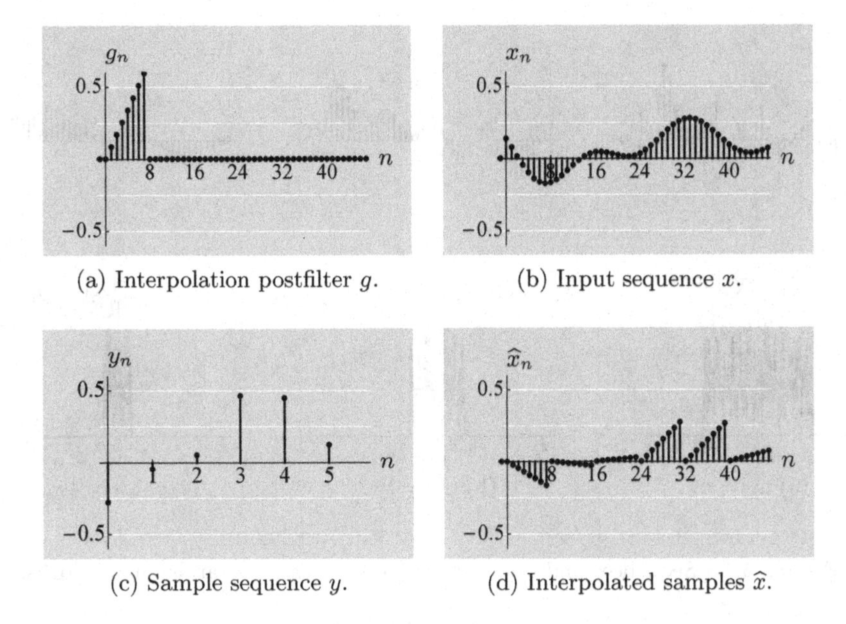

(a) Interpolation postfilter g.

(b) Input sequence x.

(c) Sample sequence y.

(d) Interpolated samples $\widehat{x}$.

Figure 5.14 Projection to a shift-invariant subspace by sampling followed by interpolation ($N = 8$).

5.3.2 Sampling and interpolation for bandlimited sequences

An important special case of sampling and interpolation of sequences arises from sequences with limited support in the DTFT domain. We study this both as an instance of the theory developed in Section 5.3.1 and more directly using DTFT tools developed in Section 3.4.

Subspaces of bandlimited sequences Shift-invariant subspaces of particular importance in signal processing are the subspaces of bandlimited sequences. To define such subspaces, we first need to define the bandwidth of a sequence.

DEFINITION 5.5 (BANDWIDTH OF SEQUENCE) A sequence x is called *bandlimited* when there exists $\omega_0 \in [0, 2\pi)$ such that its discrete-time Fourier transform X satisfies

$$X(e^{j\omega}) = 0 \qquad \text{for all } \omega \text{ with } |\omega| \in (\tfrac{1}{2}\omega_0, \pi].$$

The smallest such ω_0 is called the *bandwidth* of x. A sequence that is not bandlimited is called a *full-band* sequence.

The bandwidth of x is the width of the smallest zero-centered interval that contains the support of $X(e^{j\omega})$; the factor of 2 between the bandwidth ω_0 and the highest frequency $\frac{1}{2}\omega_0$ is introduced by counting positive and negative frequencies. This

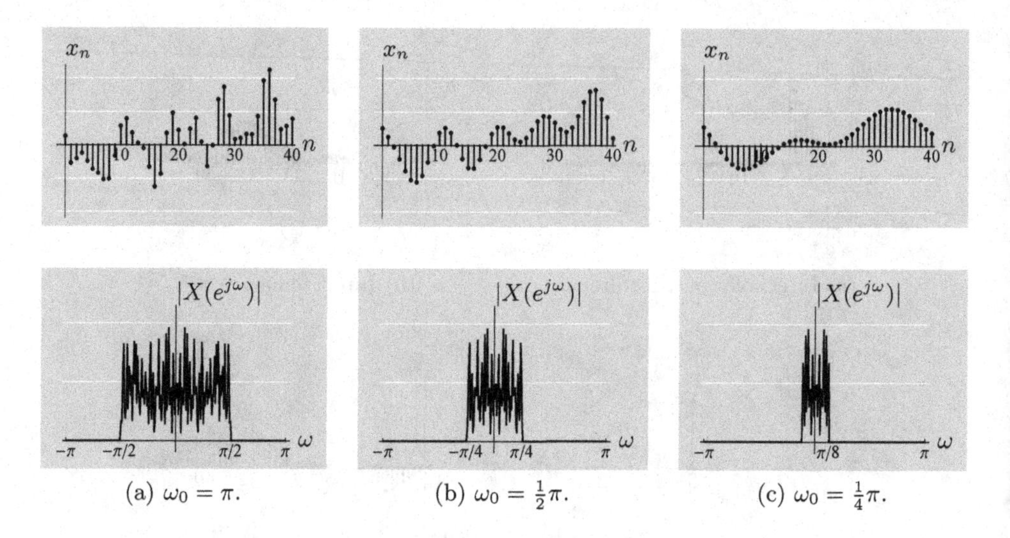

(a) $\omega_0 = \pi$. (b) $\omega_0 = \frac{1}{2}\pi$. (c) $\omega_0 = \frac{1}{4}\pi$.

Figure 5.15 Sequences and their respective DTFTs with various bandwidths.

definition is inspired by the even symmetry of the magnitude of the DTFT of a real sequence; alternate definitions might not count positive and negative frequencies or might not require the interval to be centered at zero. Figure 5.15 illustrates that decreasing bandwidth corresponds to slower variation, or increasing smoothness, of a sequence.

If sequences x and y both have bandwidth ω_0, then by the linearity of the DTFT we are assured that $x + y$ has bandwidth of at most ω_0. Thus, bandlimited sequences form subspaces. These subspaces are closed.

Definition 5.6 (Subspace of bandlimited sequences) The set of sequences in $\ell^2(\mathbb{Z})$ with bandwidth of at most ω_0 is a closed subspace denoted $\mathrm{BL}[-\frac{1}{2}\omega_0, \frac{1}{2}\omega_0]$.

A subspace of bandlimited sequences is shift-invariant for any shift $L \in \mathbb{Z}^+$. To see the shift invariance, take $x \in \mathrm{BL}[-\frac{1}{2}\omega_0, \frac{1}{2}\omega_0]$. Then, (3.89) states that

$$x_{n-kL} \quad \stackrel{\text{DTFT}}{\longleftrightarrow} \quad e^{-j\omega kL} X(e^{j\omega}). \tag{5.41}$$

The DTFT is multiplied by a complex exponential, not changing where its magnitude is nonzero and hence not changing the bandwidth of the shifted sequence.

Projection to bandlimited subspaces Since a subspace of bandlimited sequences is also a shift-invariant subspace, the techniques developed in Section 5.3.1 suggest a way to recover a bandlimited sequence from samples or compute the best approximation of a full-band sequence in a bandlimited subspace.

Fix integer $N > 1$ and let

$$g_n = \frac{1}{\sqrt{N}} \operatorname{sinc}\Big(\frac{\pi n}{N}\Big), \quad n \in \mathbb{Z} \qquad \overset{\text{DTFT}}{\longleftrightarrow} \qquad G(e^{j\omega}) = \begin{cases} \sqrt{N}, & |\omega| \le \pi/N; \\ 0, & \text{otherwise,} \end{cases} \tag{5.42}$$

where the DTFT pair can be found in Table 3.5. Using (5.41), it is clear that g and its shifts are in $\mathrm{BL}[-\pi/N, \pi/N]$, and the stronger statement that g is a generator with shift N of $\mathrm{BL}[-\pi/N, \pi/N]$ is also true. We can easily check that g satisfies (5.31),

$$\begin{aligned} \langle g_{n-kN}, g_{n-\ell N}\rangle_n &\overset{(a)}{=} \frac{1}{2\pi}\langle e^{-j\omega kN} G(e^{j\omega}), e^{-j\omega \ell N} G(e^{j\omega})\rangle_\omega \\ &\overset{(b)}{=} \frac{1}{2\pi}\int_{-\pi}^{\pi} e^{-j\omega(k-\ell)N} |G(e^{j\omega})|^2 \, d\omega \\ &\overset{(c)}{=} \frac{N}{2\pi}\int_{-\pi/N}^{\pi/N} e^{-j\omega(k-\ell)N} \, d\omega \overset{(d)}{=} \delta_{k-\ell}, \end{aligned}$$

where (a) follows from the generalized Parseval equality for the DTFT, (3.108); (b) from the definition of the $\mathcal{L}^2([-\pi, \pi))$ inner product; (c) from substituting the DTFT of g; and (d) from evaluating the integral (separately for $k = \ell$ and $k \neq \ell$). Thus, we get the following as a corollary to Theorem 5.4:

THEOREM 5.7 (PROJECTION TO BANDLIMITED SUBSPACE) Let $N \in \mathbb{Z}^+$, and let the system in Figure 5.13(b) have filter g from (5.42). Then

$$\widehat{x}_n = \frac{1}{\sqrt{N}} \sum_{k\in\mathbb{Z}} y_k \operatorname{sinc}\Big(\frac{\pi}{N}(n - kN)\Big), \qquad n \in \mathbb{Z}, \tag{5.43a}$$

with

$$y_k = \frac{1}{\sqrt{N}} \sum_{n\in\mathbb{Z}} x_n \operatorname{sinc}\Big(\frac{\pi}{N}(n - kN)\Big), \qquad k \in \mathbb{Z}, \tag{5.43b}$$

is the best approximation of x in $\mathrm{BL}[-\pi/N, \pi/N]$,

$$\widehat{x} = \operatorname*{arg\,min}_{x_{\mathrm{BL}} \in \mathrm{BL}[-\pi/N, \pi/N]} \|x - x_{\mathrm{BL}}\|, \qquad x - \widehat{x} \perp \mathrm{BL}[-\pi/N, \pi/N]. \tag{5.43c}$$

In particular, $\widehat{x} = x$ when $x \in \mathrm{BL}[-\pi/N, \pi/N]$.

In general, the effect of orthogonal projection to $\mathrm{BL}[-\pi/N, \pi/N]$ is a simple truncation of the spectrum of x to $[-\pi/N, \pi/N]$,

$$\widehat{X}(e^{j\omega}) = \begin{cases} X(e^{j\omega}), & \text{for } |\omega| \le \pi/N; \\ 0, & \text{otherwise.} \end{cases}$$

Exercise 5.4 explores bandlimited spaces with rational sampling rate changes.

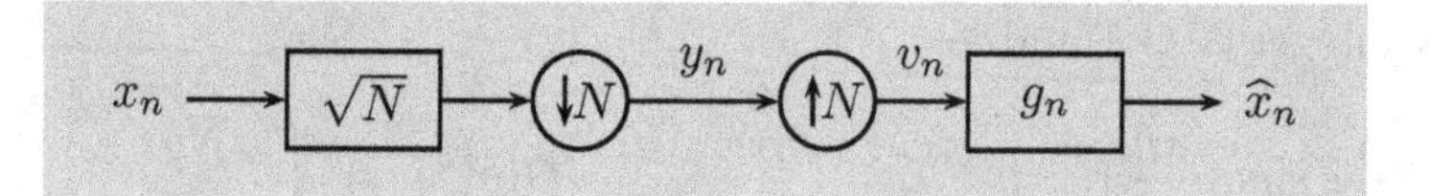

Figure 5.16 Sampling and interpolation in $\ell^2(\mathbb{Z})$ with no sampling prefilter. (The scalar multiplication by $\sqrt{N}$ could be incorporated into the interpolation postfilter g.)

Sampling without a prefilter followed by interpolation When the input to the system in Figure 5.13(b) is in $\mathrm{BL}[-\pi/N, \pi/N]$ and the filter g is given by (5.42), the sampling prefilter simply scales the input by $\sqrt{N}$. For these inputs, the system is equivalent to the one in Figure 5.16. This simpler system is worth further study – for both bandlimited and full-band inputs – because it is essentially a system without a sampling prefilter, and in practice there is often no opportunity to include a sampling prefilter.

The effect of downsampling followed by upsampling was studied in Section 3.7.3. Using (3.201) and the scaling by $\sqrt{N}$, we can write

$$V(e^{j\omega}) = \frac{1}{\sqrt{N}} \sum_{k=0}^{N-1} X(e^{j(\omega-2\pi k/N)}), \tag{5.44}$$

so

$$\widehat{X}(e^{j\omega}) = \frac{1}{\sqrt{N}} \sum_{k=0}^{N-1} G(e^{j\omega}) X(e^{j(\omega-2\pi k/N)}). \tag{5.45}$$

Using $G(e^{j\omega})$ from (5.42), it is clear that $\widehat{X} = X$ (equivalently, $\widehat{x} = x$) when x has bandwidth of at most $2\pi/N$. This discussion yields the sampling theorem for sequences.

THEOREM 5.8 (SAMPLING THEOREM FOR SEQUENCES) Let $N \in \mathbb{Z}^+$. If the sequence x is in $\mathrm{BL}[-\pi/N, \pi/N]$,

$$x_n = \sum_{k \in \mathbb{Z}} x_{kN} \operatorname{sinc}\Big(\frac{\pi}{N}(n-kN)\Big), \qquad n \in \mathbb{Z}. \tag{5.46}$$

Equation (5.45) also allows us to understand more: exactly what happens when x has bandwidth greater than $2\pi/N$, and the additional flexibility in g that comes from x having bandwidth less than $2\pi/N$.

Aliasing A component in x at a frequency $\omega \in [-\pi, \pi)$ appears in the sum (5.44) at frequencies $\{\omega - 2\pi k/N\}_{k=0}^{N-1}$. Modulo 2π, exactly one of these frequencies lies in $[-\pi/N, \pi/N)$; we denote that frequency by $\widetilde{\omega}$, as illustrated in Figure 5.17. The pointwise multiplication by $G(e^{j\omega})$ in (5.45) causes only components at frequencies

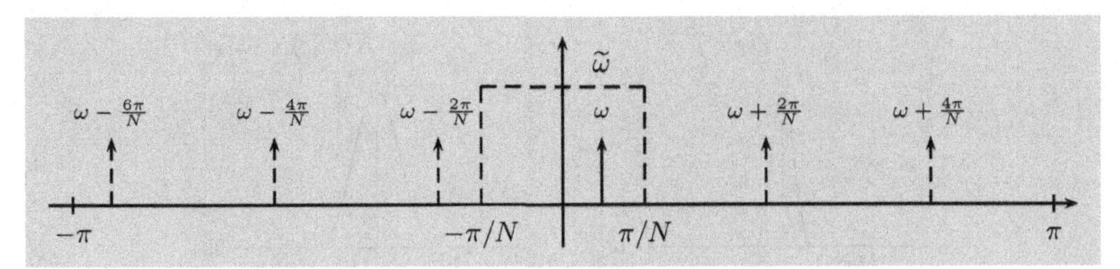

(a) No aliasing.

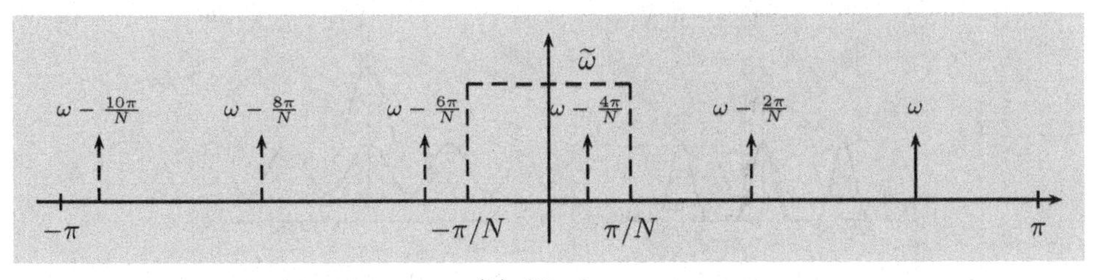

(b) Aliasing.

Figure 5.17 Replication of frequency ω and determination of the aliasing frequency $\widetilde{\omega}$. A frequency component in x at ω yields components in the sum (5.44) at N frequencies $\{\omega - 2\pi k/N\}_{k=0}^{N-1}$, reduced modulo 2π. The frequency in $[-\pi/N, \pi/N)$ is $\widetilde{\omega}$. Aliasing is present when $\widetilde{\omega} \neq \omega$, which occurs when $\omega \notin [-\pi/N, \pi/N)$. (Illustrated for $N = 6$.)

in $[-\pi/N, \pi/N)$ to pass, so the component of x at frequency ω affects $\widehat{x}$ at frequency $\widetilde{\omega}$ (and only at this frequency). A component of x at frequency ω affecting the output at frequency $\widetilde{\omega} \neq \omega$ is called *aliasing.*[94] When x has bandwidth ω_0 with $\omega_0 \geq 2\pi/N$, aliasing is present, as illustrated in Figure 5.18. The sampling rate $1/N$ must exceed $\omega_0/(2\pi)$ to avoid aliasing.

The $k \neq 0$ terms in (5.44) are called *spectral replicas*, and aliasing is the overlap of a spectral replica with the contribution from the $k = 0$ *base spectrum* term. When there is such overlap, no LSI filtering applied to v can recover every $x \in \mathrm{BL}[-\frac{1}{2}\omega_0, \frac{1}{2}\omega_0]$. Avoiding aliasing is the motivation for the sampling prefilter; with that, we remove the components outside $\mathrm{BL}[-\pi/N, \pi/N]$ so that they cannot create aliasing. We will discuss aliasing in more detail in Section 5.4.2.

Oversampling Thus far, there has been no flexibility in the choice of the interpolation postfilter g. Suppose that x has bandwidth ω_0 with $\omega_0 < 2\pi/N$. Then, spectral replicas do not overlap with the base spectrum in (5.44). The gap between $\frac{1}{2}\omega_0$ and π/N allows perfect recovery of any $x \in \mathrm{BL}[-\frac{1}{2}\omega_0, \frac{1}{2}\omega_0]$ without necessarily choosing g as in (5.42). From (5.45), the requirement on $G(e^{j\omega})$ for perfect recovery is

$$G(e^{j\omega}) = \begin{cases} \sqrt{N}, & \text{for } |\omega| \leq \frac{1}{2}\omega_0; \\ 0, & \text{for } |\omega| \in [2\pi/N - \frac{1}{2}\omega_0, \pi]. \end{cases}$$

[94]An *alias* is an assumed name or role, and here the frequency $\widetilde{\omega}$ is assuming the role of ω.

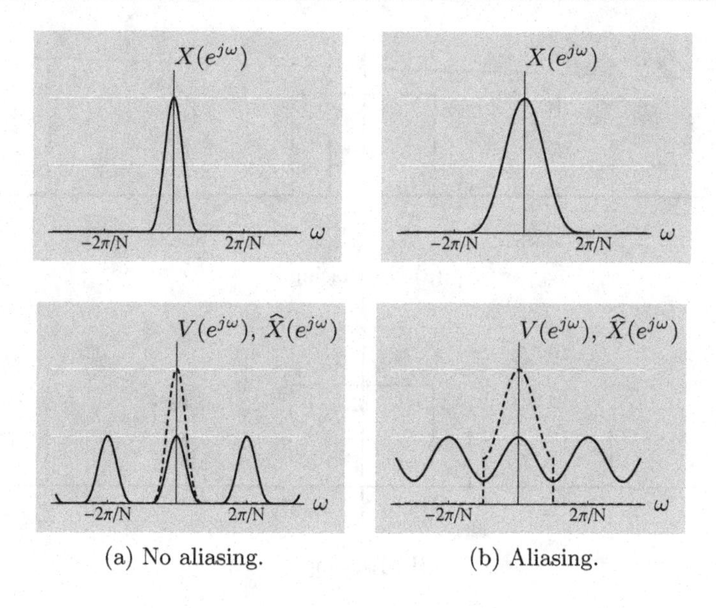

(a) No aliasing. (b) Aliasing.

Figure 5.18 Downsampling by N followed by upsampling by N of $x \in \mathrm{BL}[-\frac{1}{2}\omega_0, \frac{1}{2}\omega_0]$ as in Figure 5.16. The top panels show the spectrum of the input x; the bottom panels show the spectra of the upsampled signal v (solid line) and of the subsequently lowpass-filtered signal $\widehat{x}$ (dashed line). (a) When $\omega_0 \leq 2\pi/N$, spectral replicas do not overlap with the base spectrum; x can be recovered by lowpass filtering by g. (b) When $\omega_0 > 2\pi/N$, spectral replicas overlap with the base spectrum; x cannot be recovered without additional prior knowledge beyond its bandwidth. (Illustrated for $N = 4$.)

The flexibility in the *transition band* of $|\omega| \in (\frac{1}{2}\omega_0, 2\pi/N - \frac{1}{2}\omega_0)$ is essential in being able to approximate the desired interpolation postfilter response with a realizable filter.

5.3.3 Sampling and interpolation with nonorthogonal sequences

We now expand the sets of filters that we use in sampling and interpolation. We no longer require the interpolation postfilter and its shifts to be an orthonormal set, nor do we require the sampling prefilter to be the time-reversed and conjugated version of the interpolation postfilter. These make the geometry more complicated: the sampling and interpolation operators are no longer adjoints of each other, and the shift-invariant subspaces we discussed earlier – the range of the interpolation operator and the orthogonal complement of the null space of the sampling operator – are no longer the same. The sampling and interpolation operators are equivalent to analysis with basis $\{\widetilde{\varphi}_k\}_{k\in\mathbb{Z}}$ and synthesis with basis $\{\varphi_k\}_{k\in\mathbb{Z}}$. Under the consistency condition, these bases are biorthogonal sets; under the additional condition of the operators being ideally matched, these bases are a dual basis pair for the same subspace of $\ell^2(\mathbb{Z})$.

Sampling We now refer to the operation depicted in Figure 5.19(a), involving filtering with $\widetilde{g}$ and downsampling by integer $N > 1$, as *sampling of the sequence x with rate $1/N$ and prefilter $\widetilde{g}$*, and we denote it by $y = \widetilde{\Phi}^* x$. In contrast to Section 5.3.1, we do not make an assumption of orthonormality.

Similarly to (5.29), for any time index $k \in \mathbb{Z}$, the output of sampling is

$$y_k = (\widetilde{\Phi}^* x)_k = (\widetilde{g} * x)_{kN} = \sum_{m\in\mathbb{Z}} x_m \widetilde{g}_{kN-m} = \langle x_m, \widetilde{g}^*_{kN-m}\rangle_m \stackrel{(a)}{=} \langle x, \widetilde{\varphi}_k\rangle, \tag{5.47}$$

where in (a) we have defined the sequence $\widetilde{\varphi}_k$ to be the time-reversed and conjugated version of $\widetilde{g}$, shifted by kN,

$$\widetilde{\varphi}_{k,n} = \widetilde{g}^*_{kN-n}, \qquad n \in \mathbb{Z}. \tag{5.48}$$

Since the output at time k is $\langle x, \widetilde{\varphi}_k\rangle$, the sampling operator $\widetilde{\Phi}^*$ produces the inner products with all the sequences in $\{\widetilde{\varphi}_k\}_{k\in\mathbb{Z}}$. This operator is again an infinite matrix, now with rows equal to $\widetilde{g}_{-n}$ and its shifts by integer multiples of N (see (3.202) for $N = 2$). The null space of $\widetilde{\Phi}^*$ is a closed, shift-invariant subspace, as shown in Section 5.3.1; its orthogonal complement is denoted $\widetilde{S}$.

Example 5.12 (Sampling in $\ell^2(\mathbb{Z})$) Define a sampling operator on $\ell^2(\mathbb{Z})$ with $N = 2$ and the prefilter

$$\widetilde{g} = \frac{1}{8}\begin{bmatrix} \cdots & 0 & -1 & 2 & \boxed{6} & 2 & -1 & 0 & \cdots \end{bmatrix}^\top. \tag{5.49}$$

Then, from (3.202), the output is

$$\begin{bmatrix} \vdots \\ \boxed{y_0} \\ y_1 \\ \vdots \end{bmatrix} = \frac{1}{8}\begin{bmatrix} & \vdots & \vdots & \vdots & \vdots & \vdots & \vdots & \vdots & \\ \cdots & -1 & 2 & \boxed{6} & 2 & -1 & 0 & 0 & \cdots \\ \cdots & 0 & 0 & -1 & 2 & 6 & 2 & -1 & \cdots \\ & \vdots & \vdots & \vdots & \vdots & \vdots & \vdots & \vdots & \end{bmatrix} \begin{bmatrix} \vdots \\ x_{-2} \\ x_{-1} \\ \boxed{x_0} \\ x_1 \\ x_2 \\ x_3 \\ x_4 \\ \vdots \end{bmatrix} = \widetilde{\Phi}^* x. \tag{5.50a}$$

In contrast to Example 5.8, the matrix $\widetilde{\Phi}^*$ does not have orthogonal rows, and $\mathcal{N}(\widetilde{\Phi}^*)$ does not have an obvious form. From the horizontal symmetry of $\widetilde{\Phi}^*$, we can guess that $\mathcal{N}(\widetilde{\Phi}^*)$ contains some vector of the form

$$\begin{bmatrix} \cdots & 0 & 0 & a & \boxed{b} & c & b & a & \cdots \end{bmatrix}^\top.$$

Note that the center of symmetry of this sequence is at $n = 1$; if the symmetry were about $n = 0$, the shifts of the sequence by ± 4 could not be orthogonal to

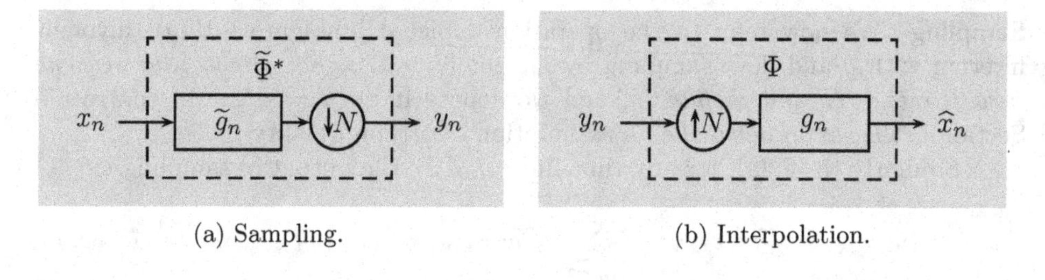

(a) Sampling. (b) Interpolation.

Figure 5.19 Sampling and interpolation in $\ell^2(\mathbb{Z})$ with nonorthogonal sequences.

$\widetilde{g}$. Solving a simple system of linear equations yields

$$s = \begin{bmatrix} \ldots & 0 & 0 & -1 & \boxed{-2} & 6 & -2 & -1 & \ldots \end{bmatrix}^{\top} \in \mathcal{N}(\widetilde{\Phi}^*). \tag{5.50b}$$

Thus $\mathcal{N}(\widetilde{\Phi}^*)$ is the shift-invariant subspace with respect to shift 2 with generator s. Its orthogonal complement $\widetilde{S} = \mathcal{N}(\widetilde{\Phi}^*)^{\perp}$ is the shift-invariant subspace with respect to shift 2 with generator $\widetilde{g}^*_{-n}$ (which in this case happens to be the same as $\widetilde{g}_n$ because it is real and symmetric).

Interpolation Again, we refer to the operation depicted in Figure 5.19(b), involving upsampling by integer $N > 1$ and filtering with g, as *interpolation of the sequence y with spacing N and postfilter g*, and we denote it by $\widehat{x} = \Phi y$. This is unchanged from Figure 5.12(b) in Section 5.3.1, and (5.35) again holds. We denote the range of Φ by S as before, and it is a closed, shift-invariant subspace with respect to shift N with generator g.

In contrast to Section 5.3.1, the interpolation operator Φ is not necessarily the adjoint of the sampling operator $\widetilde{\Phi}^*$, and S does not necessarily equal $\widetilde{S}$. When the interpolation operator is specially chosen so that it satisfies (5.23), that is, it is the pseudoinverse of $\widetilde{\Phi}^*$, then $S = \widetilde{S}$, by the same arguments as in (5.24). The pseudoinverse in (5.23) requires $\widetilde{\Phi}^*\widetilde{\Phi}$ to be invertible; this is guaranteed if $\{\widetilde{\varphi}_k\}_{k\in\mathbb{Z}}$ is a Riesz basis for $\widetilde{S}$ since $\widetilde{\Phi}^*\widetilde{\Phi}$ is the Gram matrix of $\{\widetilde{\varphi}_k\}_{k\in\mathbb{Z}}$ (recall that the Riesz basis condition implies the invertibility of the Gram matrix, see the footnote on page 94). For the rest of this section, we will assume that $\{\widetilde{\varphi}_k\}_{k\in\mathbb{Z}}$ is a Riesz basis. Finding a pseudoinverse is not an easy task when dealing with infinite-dimensional spaces; Example 5.13 provides an illustration.

Interpolation followed by sampling Interpolation followed by sampling is described by $\widetilde{\Phi}^*\Phi$ as in Figure 5.20(a). Analogously to (5.38),

$$\widehat{y}_k = \sum_{n\in\mathbb{Z}} \widetilde{g}_{kN-n}\widehat{x}_n = \sum_{n\in\mathbb{Z}} \widetilde{g}_{kN-n} \sum_{\ell\in\mathbb{Z}} y_\ell g_{n-\ell N} = \sum_{\ell\in\mathbb{Z}} y_\ell \sum_{n\in\mathbb{Z}} \widetilde{g}_{kN-n} g_{n-\ell N},$$

so $\widetilde{\Phi}^*\Phi y = y$ for all $y \in \ell^2(\mathbb{Z})$ if and only if

$$\langle g_{n-\ell N}, \widetilde{g}^*_{kN-n}\rangle_n = \delta_{\ell-k}. \tag{5.51}$$

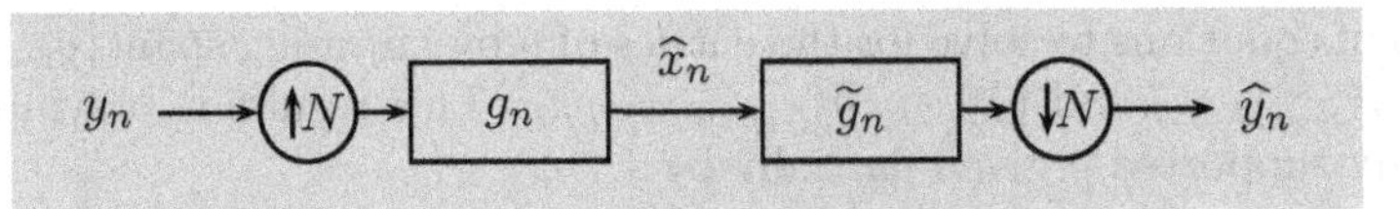

(a) Interpolation followed by sampling.

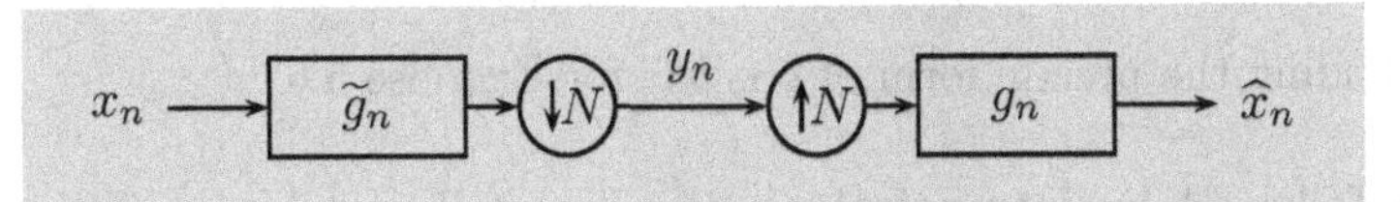

(b) Sampling followed by interpolation.

Figure 5.20 Sampling and interpolation in $\ell^2(\mathbb{Z})$ with nonorthogonal sequences.

Because of (5.30) and (5.48), this is equivalent to a counterpart for (5.31) and (5.39),

$$\widetilde{\Phi}^*\Phi = I \qquad \Leftrightarrow \qquad \langle \varphi_\ell, \widetilde{\varphi}_k \rangle = \delta_{\ell-k}. \tag{5.52}$$

The sampling and interpolation operators are then called *consistent.* Choosing the pseudoinverse in (5.23) for Φ would satisfy (5.52); there exist infinitely many other right inverses of $\widetilde{\Phi}^*$ that one could also use. Equation (5.52) also shows that the condition for perfect recovery is the same as the sets of sequences $\{\varphi_k\}_{k\in\mathbb{Z}}$ and $\{\widetilde{\varphi}_k\}_{k\in\mathbb{Z}}$ being biorthogonal, as in (2.111). These sets of sequences are not bases for $\ell^2(\mathbb{Z})$; instead, they are bases for the subspaces S and $\widetilde{S}$ that they span, respectively. They are a biorthogonal pair of bases when $S = \widetilde{S}$.

Example 5.13 (Interpolation followed by sampling in $\ell^2(\mathbb{Z})$) We would like to find an interpolation operator that is consistent with the sampling operator from Example 5.12. Inspired by the symmetry of $\widetilde{g}$ from (5.49) and attempting to find the shortest suitable interpolation postfilter, assume that g is of the following form:

$$g = \begin{bmatrix} \dots & 0 & a & \boxed{b} & a & 0 & \dots \end{bmatrix}^\top.$$

To satisfy (5.51), we get the following system of equations:

$$2a + 3b = 4, \qquad 2a - b = 0,$$

which has the solution $a = \frac{1}{2}$ and $b = 1$, so

$$g = \begin{bmatrix} \dots & 0 & \frac{1}{2} & \boxed{1} & \frac{1}{2} & 0 & \dots \end{bmatrix}^\top. \tag{5.53}$$

Finding sampling and interpolation operators that are pseudoinverses of each other can be somewhat difficult. Consider the interpolation operator Φ associated with spacing $N = 2$ and the postfilter (5.53) above. Evaluating $\Phi(\Phi^*\Phi)^{-1}$ is not straightforward because the matrices involved are infinite. One can find a system of equations to solve for the entries of $\widetilde{g}$ by imposing $\overline{\operatorname{span}}(\{g_{n-2k}\}_{k\in\mathbb{Z}}) = \overline{\operatorname{span}}(\{\widetilde{g}^*_{2k-n}\}_{k\in\mathbb{Z}})$ and $\langle g_{n-2k}, \widetilde{g}^*_{2\ell-n}\rangle_n = \delta_{k-\ell}$. This results in a symmetric and infinitely supported $\widetilde{g}$ given partially by

$$\widetilde{g} = \frac{1}{\sqrt{2}}\begin{bmatrix}\ldots & \boxed{1} & \sqrt{2}-1 & 2\sqrt{2}-3 & 7-5\sqrt{2} & \ldots\end{bmatrix}^{\top}, \tag{5.54}$$

where finding the precise form of $\widetilde{g}$ is left for Exercise 5.6.

Sampling followed by interpolation Sampling followed by interpolation is described by $P = \Phi\widetilde{\Phi}^*$ as in Figure 5.20(b). When the sampling and interpolation operators are consistent as in (5.52), P is idempotent by the computation in (5.28). It projects onto S, but the projection is not necessarily orthogonal. The approximation error $x - \widehat{x}$ is orthogonal to $\widetilde{S}$ but not to S (recall Figure 5.11 for a conceptual picture).

Again, for P to be self-adjoint too, Φ must be chosen to be the pseudoinverse of $\widetilde{\Phi}^*$, (5.23); we then have that the sampling and interpolation operators are *ideally matched*, and the subspaces S and $\widetilde{S}$ are identical. In other words, when the sampling and interpolation operators are ideally matched, $P = \Phi\widetilde{\Phi}^*$ is an orthogonal projection operator, and, by Theorem 2.26, $\widehat{x} = Px$ is the best approximation of x in S.

The previous discussion can be summarized by the following analogue to Theorem 5.2.

THEOREM 5.9 (RECOVERY FOR SEQUENCES, NONORTHOGONAL) Let $N \in \mathbb{Z}$ with $N > 1$, let the sampling operator $\widetilde{\Phi}^* : \ell^2(\mathbb{Z}) \to \ell^2(\mathbb{Z})$ be given by

$$(\widetilde{\Phi}^* x)_k = \sum_{m\in\mathbb{Z}} x_m \widetilde{g}_{kN-m}, \qquad k \in \mathbb{Z}, \tag{5.55a}$$

and let the interpolation operation $\Phi : \ell^2(\mathbb{Z}) \to \ell^2(\mathbb{Z})$ be given by

$$(\Phi y)_n = \sum_{k\in\mathbb{Z}} y_k g_{n-kN}, \qquad n \in \mathbb{Z}. \tag{5.55b}$$

As depicted in Figure 5.20(b), denote the result of sampling followed by interpolation applied to $x \in \ell^2(\mathbb{Z})$ by $\widehat{x} = Px = \Phi\widetilde{\Phi}^*$.

If the sampling prefilter $\widetilde{g}$ and interpolation postfilter g satisfy consistency condition (5.51), then P is a projection operator with range $S = \mathcal{R}(\Phi)$ and $x - \widehat{x} \perp \widetilde{S} = \mathcal{N}(\widetilde{\Phi}^*)^{\perp}$; in particular, $\widehat{x} = x$ when $x \in S$.

If, additionally, ideal matching (5.23) is satisfied, then $S = \widetilde{S}$, P is the orthogonal projection operator with range S, and $\widehat{x}$ is the best approximation of x in S,

$$\widehat{x} = \arg\min_{x_S\in S} \|x - x_S\|, \qquad x - \widehat{x} \perp S.$$

Example 5.14 (Sampling followed by interpolation in $\ell^2(\mathbb{Z})$) As shown in Example 5.13, the interpolation operator with $N = 2$ and postfilter g in (5.53) is consistent with sampling operators with $N = 2$ and prefilters $\widetilde{g}$ given in (5.49) and (5.54). Thus, with either of these sampling operators, sampling followed by interpolation implements a projection onto $S = \mathcal{R}(\Phi)$. In one case (with $\widetilde{g}$ from (5.54)), the projection is orthogonal; in the other (with $\widetilde{g}$ from (5.49)), the projection is oblique.

The shift-invariant subspace S in this example has a useful form. Any $\widehat{x} \in S$ can be written as $\widehat{x}_n = \sum_{k\in\mathbb{Z}} \alpha_k g_{n-2k}$ for some $\alpha \in \ell^2(\mathbb{Z})$. For any even time index, we have

$$\widehat{x}_{2n} = \sum_{k\in\mathbb{Z}} \alpha_k g_{2n-2k} \stackrel{(a)}{=} \sum_{k\in\mathbb{Z}} \alpha_k \delta_{n-k} = \alpha_n, \tag{5.56a}$$

where (a) follows from (5.53). Then, for any odd time index, we have

$$\begin{aligned}\widehat{x}_{2n+1} &= \sum_{k\in\mathbb{Z}} \alpha_k g_{2(n-k)+1} \stackrel{(a)}{=} \sum_{k\in\mathbb{Z}} \alpha_k \frac{1}{2}\left(\delta_{n-k} + \delta_{n-k+1}\right) \\ &= \frac{1}{2}\left(\alpha_n + \alpha_{n+1}\right) \stackrel{(b)}{=} \frac{1}{2}\left(\widehat{x}_{2n} + \widehat{x}_{2n+2}\right), \end{aligned} \tag{5.56b}$$

where (a) follows from (5.53); and (b) from (5.56a). So, for any $\widehat{x} \in S$, each odd-indexed entry is the average of the neighboring even-indexed entries.

For the input sequence in Figure 5.21(a), Figures 5.21(b) and (d) illustrate the orthogonal and oblique projections to S obtained with the two combinations of sampling and interpolation described above. The interpolated sequences are nearly equal; this is reinforced by comparing the difference sequences in Figures 5.21(c) and (d) and explained by the small difference between the sampling prefilters in (5.49) and (5.54). Thus, in this case, using the shortest symmetric $\widetilde{g}$ that is consistent with g results in little increase in approximation error as compared with the ideal orthogonal projection.

There are also interpolation prefilters very different from (5.54) that are consistent with g. For example,

$$\widetilde{g} = \begin{bmatrix} \cdots & 0 & \boxed{\tfrac{3}{2}} & -1 & \tfrac{1}{2} & 0 & \cdots \end{bmatrix}^{\top} \tag{5.57}$$

is consistent with g. For the same input sequence as before, Figure 5.21(f) shows the result of sampling followed by interpolation using this $\widetilde{g}$ and g. As shown in Figure 5.21(g), the approximation error is now much larger than before.

5.4 Functions

In this section, we study sampling and interpolation operators that map between continuous time, $\mathcal{L}^2(\mathbb{R})$, and discrete time, $\ell^2(\mathbb{Z})$. As before, we restrict the sampling and interpolation operators to those implemented with LSI filtering and emphasize the cases in which they are ideally matched. Our development closely parallels the discrete-time case in the previous section.

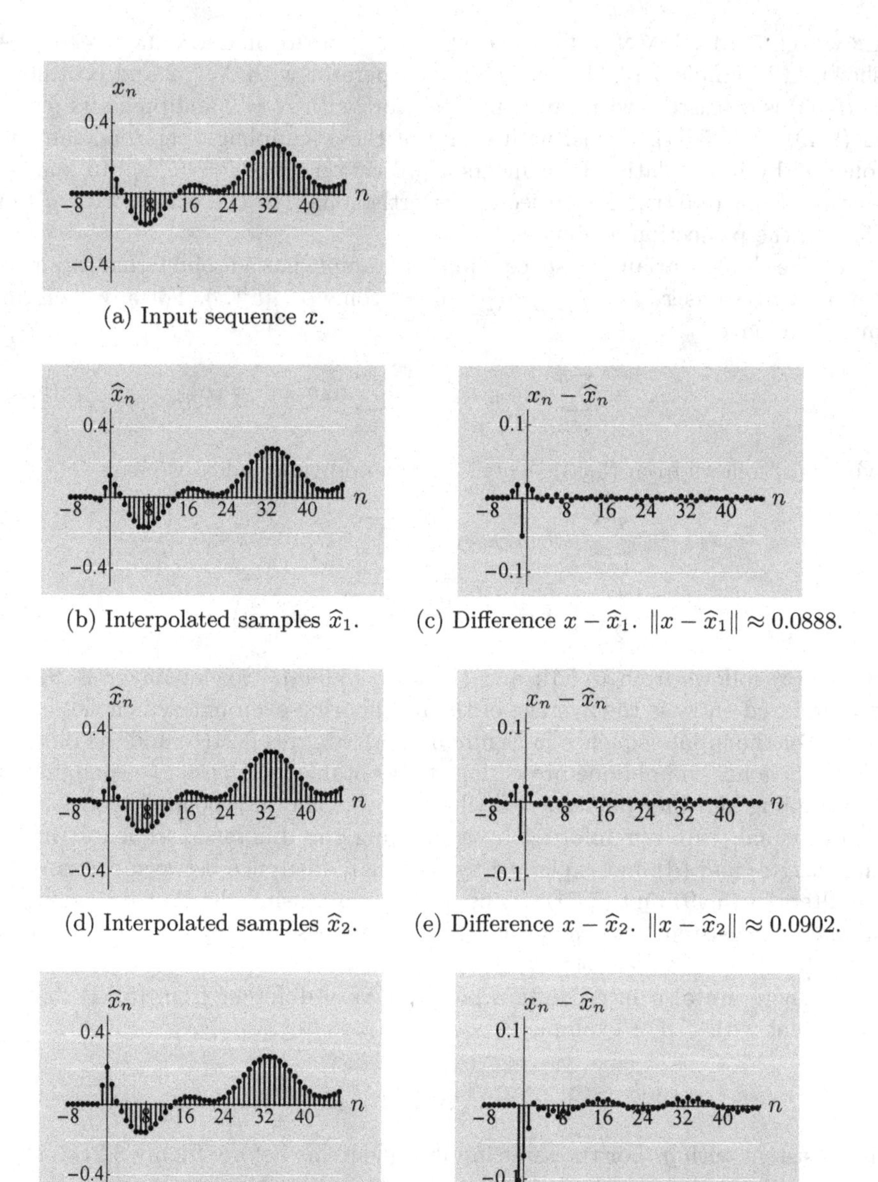

Figure 5.21 Comparison of orthogonal and oblique projections to the same shift-invariant space S. For any sequence in S, each odd-indexed entry is the average of the neighboring even-indexed entries. Orthogonal projection to S gives $\widehat{x}_1$. Oblique projection with sampling prefilter (5.49) and interpolation postfilter (5.53) gives $\widehat{x}_2$. Oblique projection with sampling prefilter (5.57) and interpolation postfilter (5.53) gives $\widehat{x}_3$.

5.4.1 Sampling and interpolation with orthonormal functions

In Section 5.1, we introduced sampling and interpolation operators and their combinations, operating on the shift-invariant space of piecewise-constant functions that are constant on unit-length intervals $[n, n+1)$, $n \in \mathbb{Z}$. This was an example where perfect recovery after sampling and interpolation was guaranteed by the specific choice of operators as well as the function subspace. We now generalize this discussion to any sampling interval T and other shift-invariant subspaces.

Like in Section 5.3.1 for sequences, we impose orthonormality of a filter and its shifts to have an orthonormal basis $\{\varphi_k\}_{k\in\mathbb{Z}}$ for a shift-invariant subspace S. We also impose a relationship between filters through time reversal and conjugation to create an adjoint pair of operators. Together, these make sampling and interpolation equivalent to analysis and synthesis with the orthonormal basis $\{\varphi_k\}_{k\in\mathbb{Z}}$ for $S \subset \mathcal{L}^2(\mathbb{R})$.

Shift-invariant subspaces of functions We start by introducing a class of subspaces of $\mathcal{L}^2(\mathbb{R})$ that will play a prominent role in the material that follows.

Definition 5.10 (Shift-invariant subspace of $\mathcal{L}^2(\mathbb{R})$) A subspace $S \subset \mathcal{L}^2(\mathbb{R})$ is a *shift-invariant subspace* with respect to a shift $T \in \mathbb{R}^+$ when $x(t) \in S$ implies that $x(t-kT) \in S$ for every integer k. In addition, $s \in \mathcal{L}^2(\mathbb{R})$ is called a *generator* of S when $S = \overline{\operatorname{span}}(\{s(t-kT)\}_{k\in\mathbb{Z}})$.

For example, the outputs of the interpolation operator in (5.4) form a shift-invariant subspace with respect to integer shifts. The same will be true for the interpolation operators we define shortly.

Sampling We refer to the operation depicted in Figure 5.4(a), involving filtering with $g^*(-t)$ and recording the result at time instants $t = nT$, $n \in \mathbb{Z}$, as *sampling of the function $x(t)$ with rate $1/T$ and prefilter $g^*(-t)$*, and we denote it by $y = \Phi^* x$. Through this operation, we move from the larger space $\mathcal{L}^2(\mathbb{R})$ into the smaller one $\ell^2(\mathbb{Z})$.

Similarly to the development in (5.2) and (5.3), for any time index $k \in \mathbb{Z}$, the output of sampling is

$$\begin{aligned} y_k &\overset{(a)}{=} (\Phi^* x)_k \overset{(b)}{=} \left(g^*(-t) *_t x(t)\right)\big|_{t=kT} = \left(\int_{-\infty}^{\infty} x(\tau)\, g^*(\tau - t)\, d\tau\right)\bigg|_{t=kT} \\ &= \int_{-\infty}^{\infty} x(\tau)\, g^*(\tau - kT)\, d\tau = \langle x(\tau),\, g(\tau - kT)\rangle_\tau \overset{(c)}{=} \langle x,\, \varphi_k\rangle, \end{aligned} \tag{5.58}$$

where (a) follows from denoting the operator in Figure 5.4(a) by Φ^*; (b) from composing filtering by $g^*(-t)$ with recording the result at time kT; and (c) from defining the function φ_k to be g shifted by kT,

$$\varphi_k(t) = g(t - kT), \qquad t \in \mathbb{R}. \tag{5.59}$$

The final expression in (5.58) shows that calling the sampling operator Φ^* is consistent with the previous use of Φ^* as the analysis operator associated with $\{\varphi_k\}_{k\in\mathbb{Z}}$; see (2.91). As before, we first assume $\{\varphi_k\}_{k\in\mathbb{Z}}$ to be an orthonormal set,

$$\langle \varphi_n, \varphi_k\rangle = \delta_{n-k} \quad\Leftrightarrow\quad \langle g(t-nT), g(t-kT)\rangle_t = \delta_{n-k}. \tag{5.60}$$

Since the output at discrete time k is $\langle x, \varphi_k\rangle$, the sampling operator Φ^* gives the inner products with all the functions in $\{\varphi_k\}_{k\in\mathbb{Z}}$. We cannot write the sampling operator Φ^* as an infinite matrix because the domain of Φ^* is functions (instead of sequences). However, we will continue to see a strong similarity to the development in Section 5.3.1 for sequences. As before, the sampling operator has a nontrivial null space, $\mathcal{N}(\Phi^*)$; the set $\{\varphi_k\}_{k\in\mathbb{Z}}$ spans its orthogonal complement, $S = \mathcal{N}(\Phi^*)^\perp = \overline{\operatorname{span}}(\{\varphi_k\}_{k\in\mathbb{Z}})$. The subspaces S and $S^\perp$ are closed and shift-invariant with respect to a shift T (see Solved exercise 5.3).

When a function $x \in \mathcal{L}^2(\mathbb{R})$ is sampled, its component in the null space $S^\perp$ has no effect on the output y and is thus completely lost; its component in S is captured by $\Phi^* x$. Section 5.1 illustrated this with $T = 1$ and $g = 1_{[0,1)}$, the indicator function of the unit interval.

Interpolation We refer to the operation depicted in Figure 5.4(b), involving weighting a Dirac comb s_T, (4.121), and filtering with g, as *interpolation of the sequence y with spacing T and postfilter g*, and we denote it by $\widehat{x} = \Phi y$.

Generalizing (5.5) for an arbitrary T, for any continuous time $t \in \mathbb{R}$, the output of interpolation is

$$\begin{aligned}\widehat{x}(t) &\overset{(a)}{=} (\Phi y)(t) \overset{(b)}{=} g(t) * \sum_{k\in\mathbb{Z}} y_k \delta(t-kT) \overset{(c)}{=} \sum_{k\in\mathbb{Z}} y_k \left(g(t) * \delta(t-kT)\right) \\ &\overset{(d)}{=} \sum_{k\in\mathbb{Z}} y_k g(t-kT) \overset{(e)}{=} \left(\sum_{k\in\mathbb{Z}} y_k \varphi_k\right)(t),\end{aligned} \tag{5.61}$$

where (a) follows from denoting the operator in Figure 5.4(b) by Φ; (b) from composing the generation of a weighted Dirac comb with filtering by g; (c) from the linearity of convolution; (d) from the shifting property of the Dirac delta function, (4.32e); and (e) from (5.59). This shows that calling the interpolation operator Φ is consistent with the previous use of Φ as the synthesis operator associated with $\{\varphi_k\}_{k\in\mathbb{Z}}$; see (2.90).

We cannot write the interpolation operator Φ as an infinite matrix because the codomain of Φ is functions (instead of sequences). Denoting the range of Φ by S as before, this subspace is the same as the orthogonal complement of the null space of the sampling operator, as we have seen earlier. Section 5.1 illustrated interpolation with $T = 1$ and $g = 1_{[0,1)}$.

Choosing a prefilter and a postfilter that are related through time-reversed conjugation makes the sampling and interpolation operators adjoints of each other:

for any $x \in \mathcal{L}^2(\mathbb{R})$ and $y \in \ell^2(\mathbb{Z})$,

$$\begin{aligned}
\langle \Phi^* x, y \rangle_{\ell^2} &\overset{(a)}{=} \left\langle \int_{-\infty}^{\infty} x(\tau)\, g^*(\tau - nT)\, d\tau,\, y_n \right\rangle_n \overset{(b)}{=} \sum_{n\in\mathbb{Z}} y_n^* \int_{-\infty}^{\infty} x(\tau)\, g^*(\tau - nT)\, d\tau \\
&\overset{(c)}{=} \int_{-\infty}^{\infty} x(\tau) \sum_{n\in\mathbb{Z}} y_n^* g^*(\tau - nT)\, d\tau \overset{(d)}{=} \langle x, \Phi y \rangle_{\mathcal{L}^2}, \qquad (5.62)
\end{aligned}$$

where (a) follows from (5.58); (b) from the definition of the ℓ^2 inner product; (c) from interchanging the order of summation and integration; and (d) from the definition of the $\mathcal{L}^2$ inner product and (5.61). This again justifies our convention of calling one Φ^* and the other Φ.

Interpolation followed by sampling Interpolation followed by sampling is described by $\Phi^*\Phi$ as in Figure 5.5(a). Using explicit expressions involving the sampling prefilter and interpolation postfilter, for any discrete time $n \in \mathbb{Z}$, we have

$$\begin{aligned}
\widehat{y}_n &\overset{(a)}{=} \int_{-\infty}^{\infty} \widehat{x}(\tau)\, g^*(\tau - nT)\, d\tau \overset{(b)}{=} \int_{-\infty}^{\infty} \left(\sum_{k\in\mathbb{Z}} y_k g(\tau - kT) \right) g^*(\tau - nT)\, d\tau \\
&\overset{(c)}{=} \sum_{k\in\mathbb{Z}} y_k \int_{-\infty}^{\infty} g(\tau - kT)\, g^*(\tau - nT)\, d\tau \overset{(d)}{=} \sum_{k\in\mathbb{Z}} y_k \delta_{n-k} = y_n, \qquad (5.63)
\end{aligned}$$

where (a) follows from using (5.58) for sampling; (b) from using (5.61) for interpolation; (c) from the interchange of summation and integration; and (d) from our assumption, (5.60), that $\{g(t - kT)\}_{k\in\mathbb{Z}}$ is an orthonormal set. Thus, sequence y is perfectly recovered from samples of its interpolated version, and

$$\Phi^*\Phi = I. \qquad (5.64)$$

This shows that the condition for perfect recovery is the same as the set of functions $\{g(t - kT)\}_{k\in\mathbb{Z}}$ being orthonormal, as in (2.92). The close parallel between (5.38) and (5.63) is because both are examples of the property (2.101) for analysis and synthesis operators associated with orthonormal bases. Section 5.1 illustrated interpolation followed by sampling with $T = 1$ and $g = 1_{[0,1)}$.

Sampling followed by interpolation Sampling followed by interpolation is described by $P = \Phi\Phi^*$ as in Figure 5.5(b). We know this does not perfectly recover x in general, since sampling loses the component of x in the null space of the sampling operator, $S^\perp$.

As we have now seen twice – for finite-dimensional vectors and for sequences – our choice of sampling and interpolation operators to satisfy (5.64) implies that P is an orthogonal projection operator; see (5.8). Thus, we have another analogue to Theorems 2.41, 5.1, and 5.4:

THEOREM 5.11 (RECOVERY FOR FUNCTIONS, ORTHOGONAL) Let $T \in \mathbb{R}^+$, the sampling operation $\Phi^* : \mathcal{L}^2(\mathbb{R}) \to \ell^2(\mathbb{Z})$ be given by

$$(\Phi^* x)_k = \int_{-\infty}^{\infty} x(\tau)\, g^*(\tau - kT)\, d\tau, \qquad k \in \mathbb{Z}, \tag{5.65a}$$

and let the interpolation operation $\Phi : \ell^2(\mathbb{Z}) \to \mathcal{L}^2(\mathbb{R})$ be given by

$$(\Phi y)(t) = \sum_{k\in\mathbb{Z}} y_k g(t - kT), \qquad t \in \mathbb{R}. \tag{5.65b}$$

As depicted in Figure 5.5(b), denote the result of sampling followed by interpolation applied to $x \in \mathcal{L}^2(\mathbb{R})$ by $\widehat{x} = Px = \Phi\Phi^* x$.

If the filter g satisfies orthogonality with shifts by multiples of T, (5.60), then P is the orthogonal projection operator with range $S = \mathcal{R}(\Phi)$ and $\widehat{x}$ is the best approximation of x in S:

$$\widehat{x} = \arg\min_{x_S \in S} \|x - x_S\|, \qquad x - \widehat{x} \perp S;$$

in particular, $\widehat{x} = x$ when $x \in S$.

Section 5.1 illustrated sampling followed by interpolation with $T = 1$, $g = 1_{[0,1)}$, and $x \in S$.

5.4.2 Sampling and interpolation for bandlimited functions

We now consider the most important special case of sampling theory: the sampling of functions with finitely supported Fourier transforms. We study this as a special case of the theory developed in Section 5.4.1 and more directly using the Fourier transform tools developed in Section 4.4.

Subspaces of bandlimited functions Shift-invariant subspaces of particular importance in signal processing are the subspaces of bandlimited functions. To define such subspaces, we first need to define the bandwidth of a function.

DEFINITION 5.12 (BANDWIDTH OF FUNCTION) A function x is called *bandlimited* when there exists $\omega_0 \in [0, \infty)$ such that its Fourier transform X satisfies

$$X(\omega) = 0 \qquad \text{for all } \omega \text{ with } |\omega| \in (\tfrac{1}{2}\omega_0, \infty). \tag{5.66}$$

The smallest such ω_0 is called the *bandwidth* of x. A function that is not bandlimited is called a *full-band* function or a function of *infinite bandwidth.*

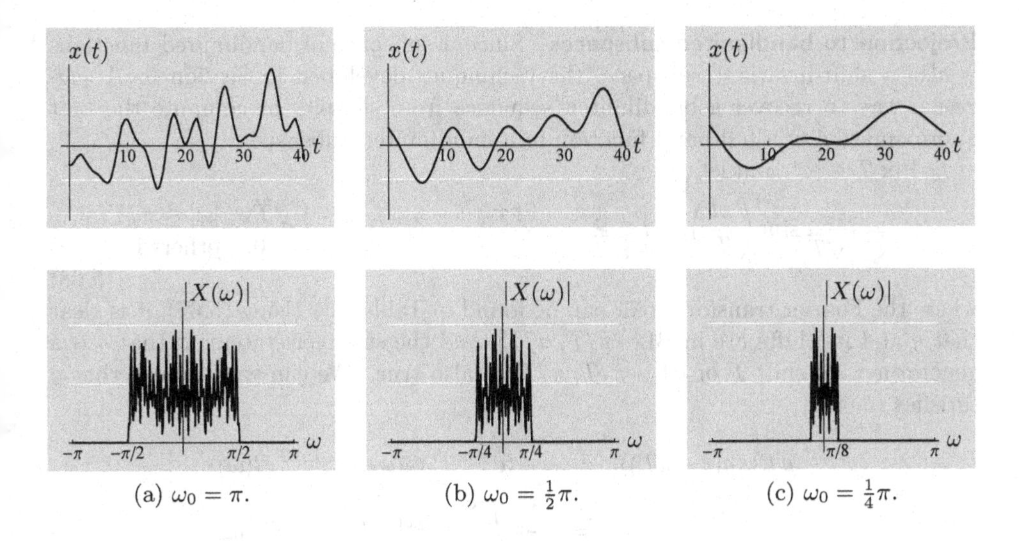

(a) $\omega_0 = \pi$.

(b) $\omega_0 = \frac{1}{2}\pi$.

(c) $\omega_0 = \frac{1}{4}\pi$.

Figure 5.22 Functions and their respective Fourier transforms with various bandwidths.

Like for sequences, the bandwidth of x is the width of the smallest zero-centered interval that contains the support of $X(\omega)$. This definition is inspired by the even symmetry of the magnitude of the Fourier transform of a real function; alternate definitions might not count positive and negative frequencies or might not require the interval to be centered at zero. Figure 5.22 illustrates that bandlimited functions are all infinitely differentiable, and decreasing bandwidth corresponds to slower variation.

If functions x and y both have bandwidth ω_0, then by linearity of the Fourier transform we are assured that $x+y$ has bandwidth of at most ω_0. Thus, bandlimited functions form subspaces. These subspaces are closed.

DEFINITION 5.13 (SUBSPACE OF BANDLIMITED FUNCTIONS) The set of functions in $\mathcal{L}^2(\mathbb{R})$ with bandwidth of at most ω_0 is a closed subspace denoted $\mathrm{BL}[-\frac{1}{2}\omega_0, \frac{1}{2}\omega_0]$.

A subspace of bandlimited functions is shift-invariant for any shift $T \in \mathbb{R}^+$. To see the shift invariance, take $x \in \mathrm{BL}[-\frac{1}{2}\omega_0, \frac{1}{2}\omega_0]$. Then, (4.54) states that

$$x(t - kT) \quad \stackrel{\mathrm{FT}}{\longleftrightarrow} \quad e^{-j\omega kT} X(\omega). \tag{5.67}$$

The Fourier transform is multiplied by a complex exponential, not changing where its magnitude is nonzero and hence not changing the bandwidth of the shifted function.

Projection to bandlimited subspaces Since a subspace of bandlimited functions is also a shift-invariant subspace, the techniques developed in Section 5.4.1 suggest a way to recover a bandlimited sequence from samples or compute the best approximation of a full-band function in a bandlimited subspace.

Fix $T \in \mathbb{R}^+$ and let

$$g(t) \;=\; \frac{1}{\sqrt{T}}\,\mathrm{sinc}\Big(\frac{\pi t}{T}\Big), \quad t \in \mathbb{R} \qquad \overset{\mathrm{FT}}{\longleftrightarrow} \qquad G(\omega) \;=\; \begin{cases} \sqrt{T}, & |\omega| \le \pi/T; \\ 0, & \text{otherwise}, \end{cases} \tag{5.68}$$

where the Fourier transform pair can be found in Table 4.1. Using (5.67), it is clear that g and its shifts are in $\mathrm{BL}[-\pi/T,\, \pi/T]$, and the stronger statement that g is a generator with shift T of $\mathrm{BL}[-\pi/T,\, \pi/T]$ is also true. We can easily check that g satisfies (5.60):

$$\begin{aligned}
\langle g(t-nT),\, g(t-kT)\rangle_t &\overset{(a)}{=} \frac{1}{2\pi}\langle e^{-j\omega nT}G(\omega),\, e^{-j\omega kT}G(\omega)\rangle_\omega \\
&\overset{(b)}{=} \frac{1}{2\pi}\int_{-\infty}^{\infty} e^{-j\omega(n-k)T}\,|G(\omega)|^2\, d\omega \\
&\overset{(c)}{=} \frac{T}{2\pi}\int_{-\pi/T}^{\pi/T} e^{-j\omega(n-k)T}\, d\omega \;\overset{(d)}{=}\; \delta_{n-k},
\end{aligned}$$

where (a) follows from the generalized Parseval equality for the Fourier transform, (4.71b); (b) from the definition of the $\mathcal{L}^2(\mathbb{R})$ inner product; (c) from substituting the Fourier transform of g; and (d) from evaluating the integral (separately for $n = k$ and $n \ne k$). Thus, we get the following as a corollary to Theorem 5.11:

THEOREM 5.14 (PROJECTION TO BANDLIMITED SUBSPACE) Let $T \in \mathbb{R}^+$, and let the system in Figure 5.5(b) have filter g from (5.68). Then

$$\widehat{x}(t) \;=\; \frac{1}{\sqrt{T}}\sum_{k\in\mathbb{Z}} y_k\,\mathrm{sinc}\Big(\frac{\pi}{T}(t-kT)\Big), \qquad t \in \mathbb{R}, \tag{5.69a}$$

with

$$y_k \;=\; \frac{1}{\sqrt{T}}\int_{-\infty}^{\infty} x(\tau)\,\mathrm{sinc}\Big(\frac{\pi}{T}(\tau-kT)\Big)\, d\tau, \qquad k \in \mathbb{Z}, \tag{5.69b}$$

is the best approximation of x in $\mathrm{BL}[-\pi/T,\, \pi/T]$,

$$\widehat{x} \;=\; \underset{x_{\mathrm{BL}}\in \mathrm{BL}[-\pi/T,\, \pi/T]}{\arg\min} \|x - x_{\mathrm{BL}}\|, \qquad x - \widehat{x} \perp \mathrm{BL}[-\pi/T,\, \pi/T]. \tag{5.69c}$$

In particular, $\widehat{x} = x$ when $x \in \mathrm{BL}[-\pi/T,\, \pi/T]$.

The effect of orthogonal projection to $\mathrm{BL}[-\pi/T,\, \pi/T]$ is a simple truncation of the spectrum of x to $[-\pi/T,\, \pi/T]$,

$$\widehat{X}(\omega) \;=\; \begin{cases} X(\omega), & \text{for } |\omega| \le \pi/T; \\ 0, & \text{otherwise}. \end{cases}$$

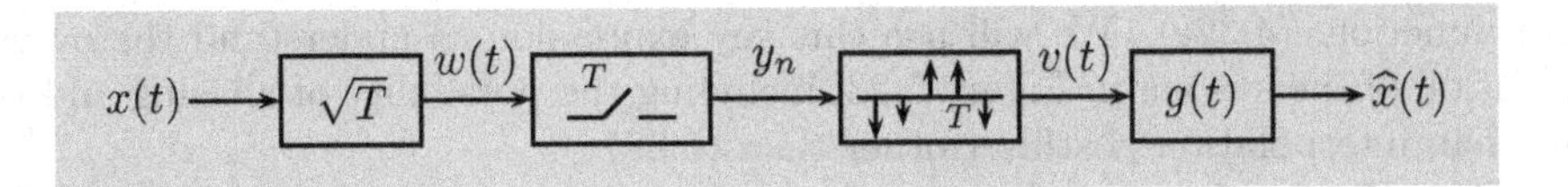

Figure 5.23 Sampling and interpolation in $\mathcal{L}^2(\mathbb{R})$ with no sampling prefilter. (The scalar multiplication by $\sqrt{T}$ could be incorporated into the interpolation postfilter.)

Sampling without a prefilter followed by interpolation When the input to the system in Figure 5.5(b) is in $\mathrm{BL}[-\pi/T, \pi/T]$ and the filter g is given in (5.68), the sampling prefilter simply scales the input by $\sqrt{T}$. For these inputs, the system is equivalent to the one in Figure 5.23. Analogously to Figure 5.16 for sequences, this simpler system is worth further study – for both bandlimited and full-band inputs – because it is essentially a system without a sampling prefilter, and often sampling is done without a prefilter. Before including any filtering (the scaling by $\sqrt{T}$ and the interpolation postfilter g), we relate v to w in the Fourier domain; we also relate v to y in the Fourier domain and thus obtain a relationship between y and w in the Fourier domain as a byproduct.

To relate v to w in the Fourier domain, first note that the weights of the Dirac delta components of v are the values of the sample sequence y, and these values are obtained by recording w at times that are integer multiples of T; that is,

$$v(t) = \sum_{n\in\mathbb{Z}} y_n \delta(t - nT) = \sum_{n\in\mathbb{Z}} w(nT)\delta(t - nT). \tag{5.70}$$

By the sampling property of the Dirac delta function, (3.294), we have

$$v(t) = s_T(t)\, w(t), \qquad t \in \mathbb{R},$$

where we have used the Dirac comb[95]

$$s_T(t) = \sum_{n\in\mathbb{Z}} \delta(t - nT) \quad \overset{\mathrm{FT}}{\longleftrightarrow} \quad S_T(\omega) = \frac{2\pi}{T}\sum_{k\in\mathbb{Z}} \delta\left(\omega - \frac{2\pi}{T}k\right) \tag{5.71}$$

and continuity of w at $\{nT\}_{n\in\mathbb{Z}}$ (which follows from w being bandlimited). Now the Fourier transform of v is given by

$$\begin{aligned} V(\omega) &\overset{(a)}{=} \frac{1}{2\pi}(S_T * W)(\omega) \overset{(b)}{=} \frac{1}{2\pi}\frac{2\pi}{T}\left(\sum_{k\in\mathbb{Z}} \delta\left(\omega - \frac{2\pi}{T}k\right)\right) * W(\omega) \\ &\overset{(c)}{=} \frac{1}{T}\sum_{k\in\mathbb{Z}} W\left(\omega - \frac{2\pi}{T}k\right), \end{aligned} \tag{5.72}$$

where (a) follows from expressing multiplication in time as convolution in frequency, (4.65); (b) from using (5.71) for S_T; and (c) from the shifting property of the Dirac

[95]The Dirac comb is defined in (4.121), and its Fourier transform is given in (4.127).

delta function, (4.32e). We will use this key expression to understand the overall operation of the system in Figure 5.23, including the possibility of aliasing and the use of an interpolation postfilter other than (5.68).

The Fourier-domain relations of v and w to y are also useful. The Fourier transform of the weighted Dirac comb v has a simple relationship with the DTFT of the sample sequence y,

$$\begin{aligned} V(\omega) &\overset{(a)}{=} \int_{-\infty}^{\infty} v(t)\, e^{-j\omega t}\, dt \overset{(b)}{=} \int_{-\infty}^{\infty} \sum_{n\in\mathbb{Z}} y_n \delta(t - nT)\, e^{-j\omega t}\, dt \\ &\overset{(c)}{=} \sum_{n\in\mathbb{Z}} y_n \int_{-\infty}^{\infty} \delta(t - nT)\, e^{-j\omega t}\, dt \overset{(d)}{=} \sum_{n\in\mathbb{Z}} y_n e^{-j\omega nT} \\ &\overset{(e)}{=} Y(e^{j\omega T}), \end{aligned} \tag{5.73a}$$

where (a) follows from the definition of the Fourier transform; (b) from the definition of v in (5.70); (c) from the interchange of integration and summation; (d) from the sifting property of the Dirac delta function, (3.293); and (e) from the definition of the DTFT. We may equivalently write

$$Y(e^{j\omega}) = V\left(\frac{\omega}{T}\right), \qquad \omega \in \mathbb{R}. \tag{5.73b}$$

These equations show that $V(\omega)$ and $Y(e^{j\omega})$ are related by dilation or contraction. In particular, $V(\omega)$ must be $2\pi/T$-periodic. By combining (5.73b) and (5.72), we can also relate the DTFT of the sample sequence y to the Fourier transform of the scaled input function w through

$$Y(e^{j\omega}) = \frac{1}{T} \sum_{k\in\mathbb{Z}} W\left(\frac{\omega}{T} - \frac{2\pi}{T}k\right). \tag{5.74}$$

The Fourier-domain relationships above are summarized in Table 5.1 in Section 5.5.2.

Sampling theorem We now consider the full system in Figure 5.23, including the scaling prior to sampling and the interpolation postfilter. Suppose that x has bandwidth of at most $2\pi/T$. The Fourier-domain implications of this case are illustrated in Figure 5.24(a). Using (5.72), the spectrum of v is the combination of $X(\omega)$ and spectral replicas spaced by $2\pi/T$, all scaled by $1/\sqrt{T}$. Since the bandwidth is less than $2\pi/T$, the base spectrum does not overlap with any of the spectral replicas, so x can be recovered by ideal lowpass filtering of v. Specifically,

$$\widehat{X}(\omega) = \frac{1}{\sqrt{T}} \sum_{k\in\mathbb{Z}} G(\omega) X\left(\omega - \frac{2\pi}{T}k\right) \tag{5.75}$$

follows from (5.72) and the fact that scaling by $\sqrt{T}$ and interpolation postfiltering both yield pointwise multiplications in the Fourier domain. Using $G(\omega)$ from (5.68), it is clear that $\widehat{X} = X$ (equivalently, $\widehat{x} = x$) when x has bandwidth of at most $2\pi/T$.

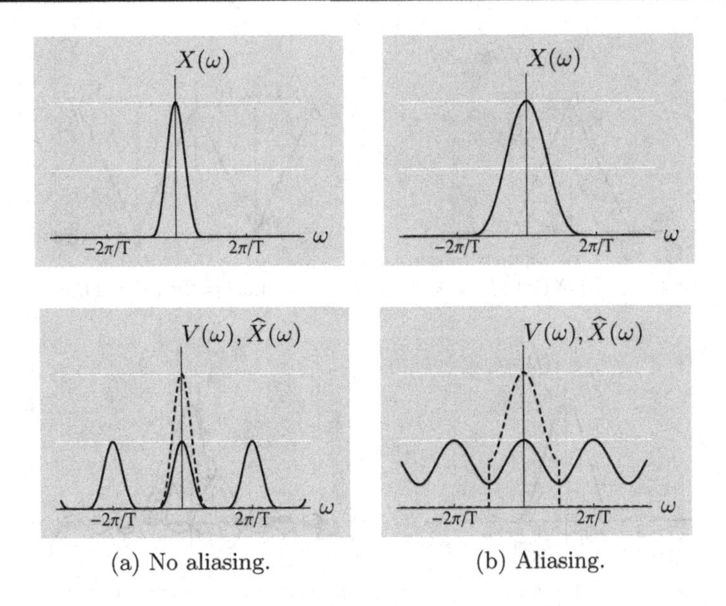

(a) No aliasing. (b) Aliasing.

Figure 5.24 Spectra corresponding to sampling with no prefilter as shown in Figure 5.23. The top panels show the spectrum of the input $x \in \mathrm{BL}[-\frac{1}{2}\omega_0, \frac{1}{2}\omega_0]$; the bottom panels show the spectra of the weighted Dirac comb v (solid line) and of the subsequently lowpass-filtered signal $\widehat{x}$ (dashed line). (a) When $\omega_0 \le 2\pi/T$, spectral replicas do not overlap with the base spectrum; x can be recovered by lowpass filtering by g. (b) When $\omega_0 > 2\pi/T$, spectral replicas overlap with the base spectrum; x cannot be recovered without additional prior knowledge beyond its bandwidth. (Illustrated for $T = 4$.)

This discussion yields one of the cornerstone results in signal processing, the sampling theorem for functions, which is analogous to Theorem 5.8.[96]

THEOREM 5.15 (SAMPLING THEOREM) Let $T \in \mathbb{R}^+$. If the function x is in $\mathrm{BL}[-\pi/T, \pi/T]$,

$$x(t) = \sum_{n\in\mathbb{Z}} x(nT)\,\mathrm{sinc}\Big(\frac{\pi}{T}(t - nT)\Big), \qquad t \in \mathbb{R}. \tag{5.76}$$

The sampling theorem gives a sufficient condition for reconstruction of a function from its samples by constructively demonstrating recovery of any function in $\mathcal{L}^2(\mathbb{R})$ with bandwidth of at most $2\pi/T$ from samples at rate $1/T$ Hz or *sampling frequency* $\omega_s = 2\pi/T$ rad/s. The bandwidth of a function is called its *Nyquist rate* because of this sufficient condition.[97]

[96]The sampling theorem was formulated and proved by a number of scientists and could bear all their names: Kotelnikov, Raabe, Shannon, Someya, Whittaker, and others; see the *Historical remarks* for more details.

[97]Not to be confused with the Nyquist rate, π/T is called the *Nyquist frequency* or *folding*

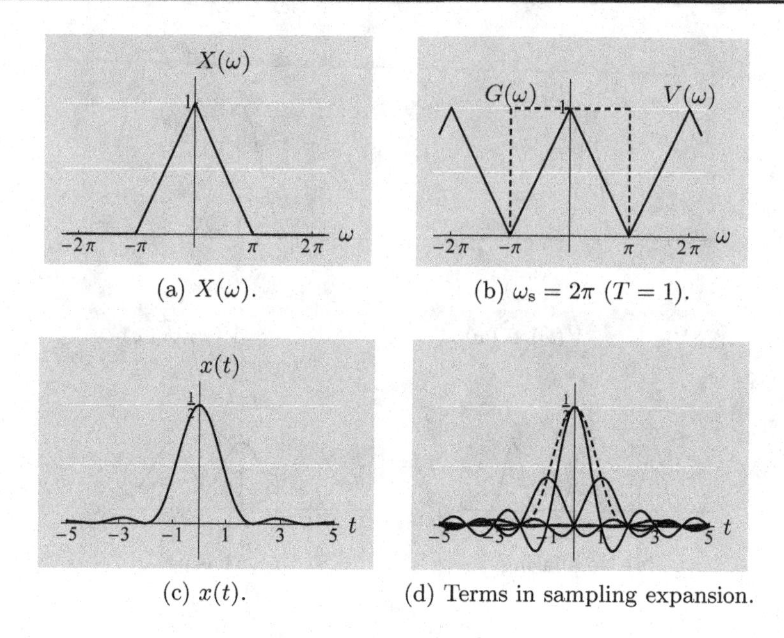

(a) $X(\omega)$.

(b) $\omega_s = 2\pi$ $(T = 1)$.

(c) $x(t)$.

(d) Terms in sampling expansion.

Figure 5.25 Illustration of the sampling theorem. (a) The Fourier-domain function X, supported on $[-\frac{1}{2}\omega_0, \frac{1}{2}\omega_0] = [-\pi, \pi]$. (b) Spectrum of the weighted Dirac comb v (solid line) and the ideal lowpass filter (dashed line) used to extract the base spectrum $[-\frac{1}{2}\omega_s, \frac{1}{2}\omega_s] = [-\pi, \pi]$. (c) The time-domain function x. (d) The original function x (dashed line) reconstructed using sinc interpolators (solid lines).

EXAMPLE 5.15 (NYQUIST SAMPLING THE SINC-SQUARED FUNCTION) Let X be the triangle function shown in Figure 5.25(a). Since $X(\omega) = 0$ for all ω with $|\omega| > \pi$, Theorem 5.15 applies with $T = 1$ to show that its inverse Fourier transform x can be recovered by interpolation using (5.76). Figure 5.25(b) illustrates why the recovery succeeds: the spectral replicas in $V(\omega)$ do not overlap with the base spectrum, and the ideal lowpass filter g has the correct cutoff frequency to preserve the base spectrum and remove the replicas. Since the sampling is at the Nyquist rate, any increase in T would create overlap between spectral replicas and the base spectrum, as in the next example.

By Example 4.4 and duality, the time-domain function x is some scaled and dilated sinc function squared. Specifically, because of the scaling by 2π in (4.65), we would like to find the box function that convolved with itself gives $2\pi X$ for X given in Figure 5.25(a). That box function is centered, with width π and amplitude $\sqrt{2}$ – precisely the box function in (4.47) with $\omega_0 = \pi$. Squaring

frequency. It is half the sampling frequency – a property of the sampling system rather than of its input. It is the frequency at which contributions from the $k = 0$ and $k = 1$ terms in (5.72) or (5.75) meet symmetrically. As shown in Figure 5.24, symmetry of $V(\omega)$ around π/T evokes folding of the spectrum.

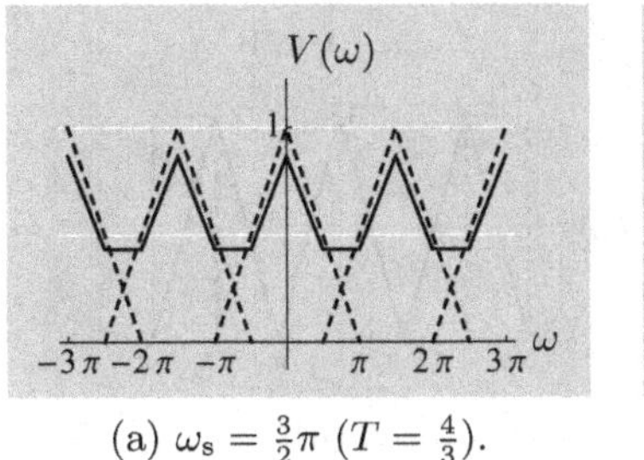

(a) $\omega_s = \frac{3}{2}\pi$ ($T = \frac{4}{3}$).

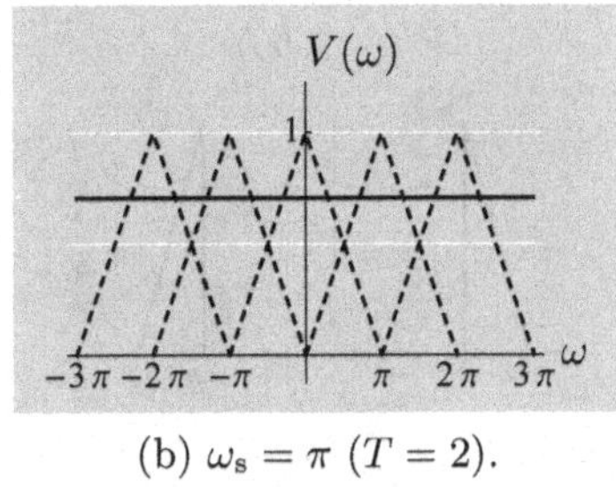

(b) $\omega_s = \pi$ ($T = 2$).

Figure 5.26 Undersampled versions of the triangle spectrum from Figure 5.25(a). Dashed lines represent shifted versions $X(\omega - (2\pi/T)k)$, $k \in \mathbb{Z}$, and solid lines their sum scaled by $1/\sqrt{T}$ to yield $V(\omega)$. (a) Spectrum of the undersampled function with sampling frequency $\omega_s = \frac{3}{2}\pi$. (b) Spectrum of the undersampled function with sampling frequency $\omega_s = \pi$.

the corresponding time-domain function in (4.46) gives

$$x(t) \;=\; \frac{1}{2}\,\mathrm{sinc}^2\left(\frac{1}{2}\pi t\right), \tag{5.77}$$

as shown in Figure 5.25(c). The sampling expansion (5.76) shows that $x(t)$ can be written as a linear combination of $\{\mathrm{sinc}(\pi(t-n))\}_{n\in\mathbb{Z}}$,

$$x(t) \;=\; \sum_{n\in\mathbb{Z}} x(n)\,\mathrm{sinc}(\pi(t-n)) \;=\; \sum_{n\in\mathbb{Z}} \frac{1}{2}\,\mathrm{sinc}^2\left(\frac{1}{2}\pi n\right)\,\mathrm{sinc}(\pi(t-n)).$$

Individual terms are shown with solid lines in Figure 5.25(d), combining to give the dashed line for x. A curious fact about this example is that every term takes positive and negative values, but the sum is nonnegative for all t.

Example 5.16 (Undersampling the sinc-squared function) Let $x(t) = \frac{1}{2}\,\mathrm{sinc}^2(\frac{1}{2}\pi t)$ as in the previous example. With sampling frequency equal to the Nyquist rate of x, $\omega_s = 2\pi$, the triangular base spectrum and spectral replicas meet each other with no overlap as in Figure 5.25(b).

If we undersample (use smaller ω_s or, equivalently, larger T), the spectral replicas will start to overlap with the base spectrum. The dashed lines in Figure 5.26(a) show a small amount of overlap from sampling at three quarters of the Nyquist rate, and Figure 5.26(b) shows more overlap from slower sampling (half of the Nyquist rate). Here, we can easily interpret the sum of the base spectrum and replicas because they are real. Take $\omega_s = \pi$ (that is, $T = 2$) as in Figure 5.26(b). In the Fourier domain, the triangles sum up to a constant, yielding $V(\omega) = 1/\sqrt{2}$ for all $\omega \in \mathbb{R}$. From (5.73), we conclude that $Y(e^{j\omega}) = 1/\sqrt{2}$ for all $\omega \in [-\pi, \pi]$ as well. Thus, the sample sequence is $y_n = (1/\sqrt{2})\delta_n$, which is consistent with evaluating $\sqrt{T}x(t)$ at even integers.

The Nyquist rate is necessary for recovery of a function in $\mathcal{L}^2(\mathbb{R})$ from samples when the bandwidth is the only thing known about the function a priori. One

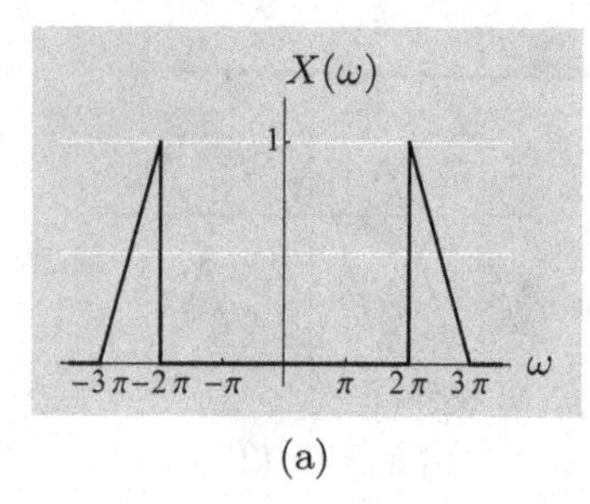

(a)

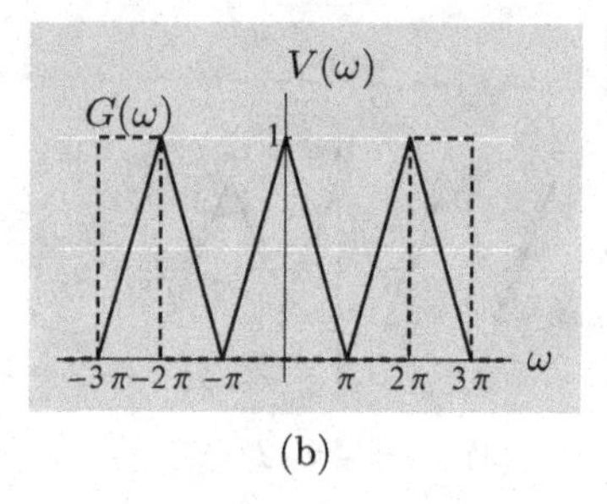

(b)

Figure 5.27 Bandpass sampling. (a) Original spectrum $X(\omega)$ in passbands $2\pi \leq |\omega| \leq 3\pi$. (b) Spectrum $X_s(\omega)$ of the sampled function with sampling frequency $\omega_s = 2\pi$ filtered with an ideal bandpass filter $G(\omega)$.

common way to know more than just the bandwidth is for the support of the Fourier transform to be given and not centered at zero. The Nyquist rate is then not necessary for recovery, as shown in the following example.

EXAMPLE 5.17 (BANDPASS SAMPLING) Let $x \in \mathcal{L}^2(\mathbb{R})$ be a function with Fourier transform supported on $[-3\pi, -2\pi] \cup [2\pi, 3\pi]$. One such Fourier transform is

$$X(\omega) = \begin{cases} 3 - |\omega|/\pi, & \text{for } |\omega| \in [2\pi, 3\pi]; \\ 0, & \text{otherwise,} \end{cases} \tag{5.78}$$

as shown in Figure 5.27(a). Since the maximum frequency is 3π, the bandwidth and the Nyquist rate are 6π, and one might think that a sampling frequency of $\omega_s = 6\pi$ is required for recovery of x from samples. A lower sampling frequency is sufficient when we choose the interpolation postfilter differently than in (5.68). Using the notation from Figure 5.23, sampling with $\omega_s = 2\pi$ (so $T = 1$) gives the spectrum $V(\omega)$ in (5.72). The spectral replicas ($k \neq 0$) do not overlap with the base spectrum ($k = 0$), but unlike what we have seen thus far, the base spectrum is not at the lowest frequencies. This is depicted in Figure 5.27(b) for the example input in Figure 5.27(a). By choosing the interpolation postfilter to be a bandpass filter with support $|\omega| \in [2\pi, 3\pi]$, we recover any x with the specified spectral support.

Exercise 5.9 explores bandpass sampling further, showing an orthonormal basis interpretation, a modulation-based solution, and generalizations. The main idea is that, for bandpass functions with the Fourier-domain support on two symmetric intervals of length $\frac{1}{2}\omega_0$ each, a sampling frequency of $\omega_s = \omega_0$ is sufficient. When only the Fourier-domain support is known and it comprises more intervals, having a sampling rate at least as large as the measure of the support is necessary, but it might be sufficient only when nonuniform sampling is allowed; for more details, see the *Further reading*.

Aliasing Figure 5.24(b) shows the effect of a spectral replica overlapping with the base spectrum, which makes it impossible to recover the input function x by any LSI filtering. As discussed for sequences in Section 5.3.2, confusion of frequencies caused by spectral replicas contributing to the reconstructed function is called aliasing.

EXAMPLE 5.18 (ALIASING OF COMPLEX EXPONENTIALS) Let $x(t) = e^{j(\pi/T)t}$ and $\widetilde{x}(t) = e^{-j(\pi/T)t}$. Then, x and $\widetilde{x}$ are indistinguishable from samples taken with period T: For any integer n,

$$x(nT) = e^{j(\pi/T)nT} = e^{j\pi n} = (-1)^n = e^{-j\pi n} = e^{-j(\pi/T)nT} = \widetilde{x}(nT).$$

Thus, frequencies of π/T and $-\pi/T$ are inevitably confused with each other in the absence of more prior knowledge. At first glance this is a boundary case of Theorem 5.15, but actually x and $\widetilde{x}$ are not in $\mathcal{L}^2(\mathbb{R})$ and hence not in $\mathrm{BL}[-\pi/T, \pi/T] \subset \mathcal{L}^2(\mathbb{R})$.

EXAMPLE 5.19 (ALIASING OF SINUSOIDS) Let $x(t) = \cos(\frac{1}{2}\omega_0 t)$, where ω_0 is the frequency in rad/s, and let $\omega_s = 2\pi/T$ be the sampling frequency. In the Fourier domain,

$$\cos(\tfrac{1}{2}\omega_0 t) = \frac{e^{j(\omega_0/2)t} + e^{-j(\omega_0/2)t}}{2} \quad \overset{\mathrm{FT}}{\longleftrightarrow} \quad \pi\left(\delta(\omega - \tfrac{1}{2}\omega_0) + \delta(\omega + \tfrac{1}{2}\omega_0)\right). \tag{5.79}$$

The weighted Dirac comb obtained from samples taken with period T is

$$\begin{aligned}
&\sum_{n\in\mathbb{Z}} \cos(\tfrac{1}{2}\omega_0 nT)\,\delta(t - nT) \\
&\quad \overset{\mathrm{FT}}{\longleftrightarrow} \quad \pi\sum_{k\in\mathbb{Z}} \left(\delta(\omega - k\omega_s - \tfrac{1}{2}\omega_0) + \delta(\omega - k\omega_s + \tfrac{1}{2}\omega_0)\right).
\end{aligned}$$

For any $\ell \in \mathbb{Z}$, the Fourier-domain expression is unchanged when $\frac{1}{2}\omega_0$ is replaced with $\ell\omega_s \pm \frac{1}{2}\omega_0$. Thus, the sampled functions are also left unchanged. We can easily verify this in the time domain:

$$\begin{aligned}
\cos\left((\ell\omega_s + \tfrac{1}{2}\omega_0)\, nT\right) &\overset{(a)}{=} \cos(2\pi\ell n + \tfrac{1}{2}\omega_0 nT) \overset{(b)}{=} \cos(\tfrac{1}{2}\omega_0 nT), \\
\cos\left((\ell\omega_s - \tfrac{1}{2}\omega_0)\, nT\right) &\overset{(c)}{=} \cos(2\pi\ell n - \tfrac{1}{2}\omega_0 nT) \overset{(d)}{=} \cos(-\tfrac{1}{2}\omega_0 nT) \\
&\overset{(e)}{=} \cos(\tfrac{1}{2}\omega_0 nT),
\end{aligned}$$

where (a) and (c) follow from $\omega_s T = 2\pi$; (b) and (d) from cosine being a 2π-periodic function; and (e) from cosine being an even function. See Figure 5.28 for an example.

The aliases that come from shifts by integer multiples of the sampling frequency ω_s are clear from the spectral replication in (5.72) or (5.75). The aliases that involve negation of the signal frequency $\frac{1}{2}\omega_0$ are due to the even symmetry of the magnitude of the Fourier transform of a real function. If $x(t)$ were replaced by $x(t) = \cos(\frac{1}{2}\omega_0 t + \theta)$ with nonzero phase θ, the aliases at frequencies $\ell\omega_s - \frac{1}{2}\omega_0$ would have phase $-\theta$. These are often the most important aliases, as in the following scenario. Suppose that a real function x has a component at frequency

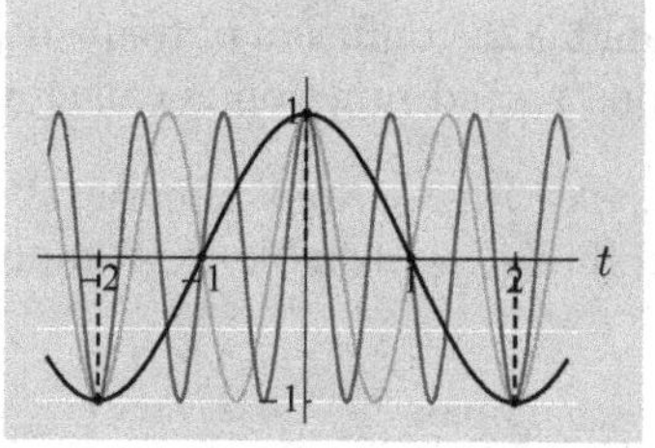

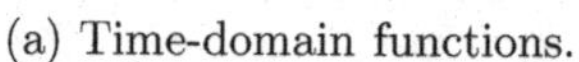
(a) Time-domain functions.

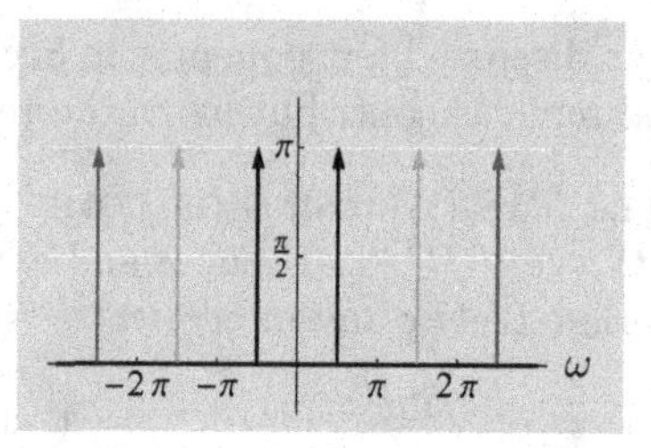

(b) Frequency spectra.

Figure 5.28 Illustration of aliasing. Sampling of three cosine functions produces the same samples. The three functions are $\cos(\frac{1}{2}\omega_0 t)$ (black) and $\cos((\omega_s \pm \frac{1}{2}\omega_0)t)$ (darker and lighter gray), with $\omega_0 = \pi$ and sampling frequency $\omega_s = 2\pi$, or $T = 1$.

$\frac{1}{2}\omega_s + \varepsilon$, slightly larger than $\frac{1}{2}\omega_s$, and any higher-frequency components are much weaker. All frequency components outside of the baseband interval $[-\frac{1}{2}\omega_s, \frac{1}{2}\omega_s]$ will have aliases in the baseband, but the strongest are at $\pm(\frac{1}{2}\omega_s - \varepsilon)$. This emphasizes the importance of *folding* around the folding frequency of $\frac{1}{2}\omega_s$, as shown in Figure 5.24.

Aliasing effects create some optical illusions. For example, in old Western movies, shot at 24 frames/s, the spokes of wagon wheels sometimes seem to be turning backwards (see Exercise 5.10).

Continuous-time processing using discrete-time operators The broad technological impact of sampling arises primarily from our ability to perform cheap and accurate digital processing. We present one example of how continuous-time processing of a function is emulated through computations performed on samples of the function.

THEOREM 5.16 (CT CONVOLUTION IMPLEMENTED USING DT PROCESSING) Let $T \in \mathbb{R}^+$ and $x \in \mathrm{BL}[-\pi/T, \pi/T]$. The continuous-time convolution $y = h * x$ can be computed using the system in Figure 5.29, where the interpolation postfilter g is the lowpass filter (5.68) and the discrete-time LSI filter $\widetilde{h}$ is given by

$$\widetilde{h}_n = \left\langle h(t), \operatorname{sinc}\left(\frac{\pi}{T}(t - nT)\right)\right\rangle_t, \qquad n \in \mathbb{Z}. \tag{5.80a}$$

The discrete-time filter input is

$$\widetilde{x}_n = \sqrt{T}x(nT), \qquad n \in \mathbb{Z}, \tag{5.80b}$$

and the system output in terms of the discrete-time filter output $\widetilde{y} = \widetilde{h} * \widetilde{x}$ is

$$y(t) = \sqrt{T}\sum_{n\in\mathbb{Z}} \widetilde{y}_n \operatorname{sinc}\left(\frac{\pi}{T}(t - nT)\right), \qquad t \in \mathbb{R}. \tag{5.80c}$$

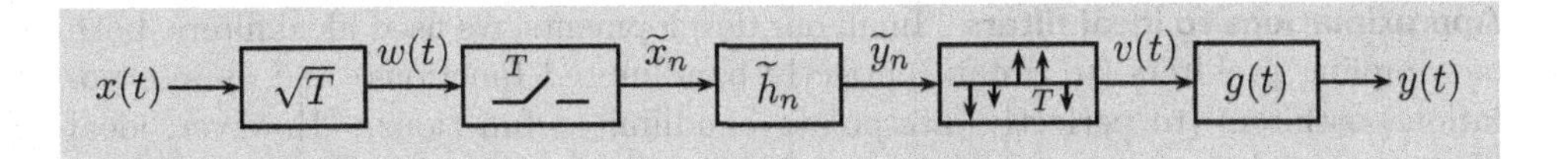

Figure 5.29 Continuous-time convolution implemented using discrete-time convolution of sequences of samples.

Proof. It is easiest to prove this in the Fourier domain. We want to show that

$$Y(\omega) = H(\omega)X(\omega)$$

can be obtained by interpolating the result of a discrete-time convolution. Since x belongs to $\mathrm{BL}[-\pi/T,\, \pi/T]$, the frequency response $H(\omega)$ matters only on this support.

Let us first incorporate the effect of filtering by an arbitrary $\widetilde{h}$ on $Y(\omega)$. Using (5.74) to relate $\widetilde{X}(e^{j\omega})$ to $W(\omega)$,

$$\widetilde{Y}(e^{j\omega}) = \widetilde{H}(e^{j\omega})\widetilde{X}(e^{j\omega}) = \widetilde{H}(e^{j\omega})\frac{1}{\sqrt{T}}\sum_{k\in\mathbb{Z}} X\left(\frac{\omega}{T} - \frac{2\pi}{T}k\right). \tag{5.81}$$

Now using (5.73a) to relate $V(\omega)$ to $\widetilde{Y}(e^{j\omega})$,

$$Y(\omega) = G(\omega)V(\omega) = G(\omega)\widetilde{H}(e^{j\omega T})\frac{1}{\sqrt{T}}\sum_{k\in\mathbb{Z}} X\left(\omega - \frac{2\pi}{T}k\right). \tag{5.82}$$

Since $x \in \mathrm{BL}[-\pi/T,\, \pi/T]$ and the support of $G(\omega)$ is $[-\pi/T,\, \pi/T]$, the $k \neq 0$ terms in (5.82) make no contribution, so

$$Y(\omega) = \frac{1}{\sqrt{T}}G(\omega)\widetilde{H}(e^{j\omega T})X(\omega). \tag{5.83}$$

This shows that the overall system is LSI; the only assumption for this conclusion is $x \in \mathrm{BL}[-\pi/T,\, \pi/T]$.

The equation defining $\widetilde{h}$, (5.80a), shows that $\widetilde{h}$ is obtained by sampling h with rate $1/T$ and prefilter $\sqrt{T}g(t)$. Thus, we can apply (5.74) again to obtain

$$\widetilde{H}(e^{j\omega}) = \frac{1}{T}\sum_{k\in\mathbb{Z}} \sqrt{T}G\left(\frac{\omega}{T} - \frac{2\pi}{T}k\right) H\left(\frac{\omega}{T} - \frac{2\pi}{T}k\right). \tag{5.84}$$

Substituting (5.84) into (5.83) gives

$$\begin{aligned} Y(\omega) &= \frac{1}{\sqrt{T}}G(\omega)\left[\frac{1}{T}\sum_{k\in\mathbb{Z}} \sqrt{T}G\left(\omega - \frac{2\pi}{T}k\right) H\left(\omega - \frac{2\pi}{T}k\right)\right] X(\omega) \\ &= \frac{1}{T}\sum_{k\in\mathbb{Z}} G(\omega)G\left(\omega - \frac{2\pi}{T}k\right) H\left(\omega - \frac{2\pi}{T}k\right) X(\omega) \overset{(a)}{=} \frac{1}{T}G^2(\omega)H(\omega)X(\omega), \end{aligned}$$

where in (a) all $k \neq 0$ terms are discarded because the support of G is $[-\pi/T,\, \pi/T]$. This is precisely the desired result because $(1/T)G^2(\omega) = 1$ on the support of $X(\omega)$.

While we showed the result for convolution, other continuous-time signal processing algorithms having a bandlimited result can also be implemented in discrete time; see Exercise 5.11.

Approximations to ideal filters In all our developments, we used ideal filters, both as sampling prefilters (to obtain perfectly bandlimited functions) and as interpolation postfilters (to perfectly interpolate bandlimited functions). However, ideal filters cannot be implemented because of their doubly infinite support; moreover, they have slow decay in the time domain, of the order $1/t$, so they are difficult to approximate by truncation. The solution is to use filters that are smoother in the frequency domain than the ideal filters. While such filters are more realistic, for bandlimited subspaces either they will lead to approximate reconstruction of the input function after sampling and interpolation or they will require oversampling.

EXAMPLE 5.20 (APPROXIMATIONS TO IDEAL FILTERS) Let g be a lowpass filter with the following frequency response:

$$G(\omega) = \begin{cases} 1, & \text{for } |\omega| \in [0, \frac{1}{2}\omega_0); \\ 1 - (|\omega| - \frac{1}{2}\omega_0)/\alpha, & \text{for } |\omega| \in [\frac{1}{2}\omega_0, \frac{1}{2}\omega_0 + \alpha); \\ 0, & \text{for } |\omega| \in [\frac{1}{2}\omega_0 + \alpha, \infty), \end{cases} \tag{5.85a}$$

for some $\alpha \in \mathbb{R}^+$, as in Figure 5.30(a). One way to obtain such a filter is to convolve two ideal filters in frequency, one with a cutoff frequency of $\frac{1}{2}\omega_0$ and gain 1, and the other with a cutoff frequency of α and gain $1/(2\alpha)$, where $\alpha \ll \omega_0$. Thus, in time, the corresponding impulse response is the product of two sinc functions,

$$g(t) = \frac{\omega_0}{4\pi^2} \operatorname{sinc}\left(\tfrac{1}{2}\omega_0 t\right) \operatorname{sinc}(\alpha t). \tag{5.85b}$$

As opposed to the impulse response of an ideal filter, this impulse response decays faster, as $1/t^2$, but uses more bandwidth, since the length of the support of G is $\omega_0 + 2\alpha$.

If X is bandlimited to $[-\frac{1}{2}\omega_0, \frac{1}{2}\omega_0]$, then, according to (5.72) and Figure 5.23, sampling and reconstruction using G as a postfilter yields

$$\widehat{X}(\omega) = G(\omega) \frac{1}{\sqrt{T}} \sum_{k \in \mathbb{Z}} X(\omega - k\omega_s). \tag{5.86}$$

If we sample with $\omega_s = \omega_0$, then, according to (5.86), the replicas of X will repeat at integer multiples of ω_0, and the larger support of G will catch the tails of the replicas centered at $\pm\omega_0$ (see Figure 5.30(b)). We solve the problem by oversampling with $\omega_s = \omega_0 + \alpha$, leading to perfect reconstruction. This oversampling by a factor of $(\omega_0 + \alpha)/\omega_0 = 1 + \alpha/\omega_0$ is the minimum needed to ensure perfect reconstruction for any bandlimited x.

As the example shows, having an interpolation filter that is smooth in frequency has a cost, since only an ideal filter allows critical sampling (sampling at the Nyquist rate) together with perfect reconstruction. In practice, most functions of interest have spectra that decay toward their band limits; thus, the effect of an imperfect reconstruction near the boundary (Figure 5.30(c)) is usually not severe. Solved exercise 5.4 explores this topic further.

We now describe a practical application of the concepts discussed so far: speech processing in mobile phones.

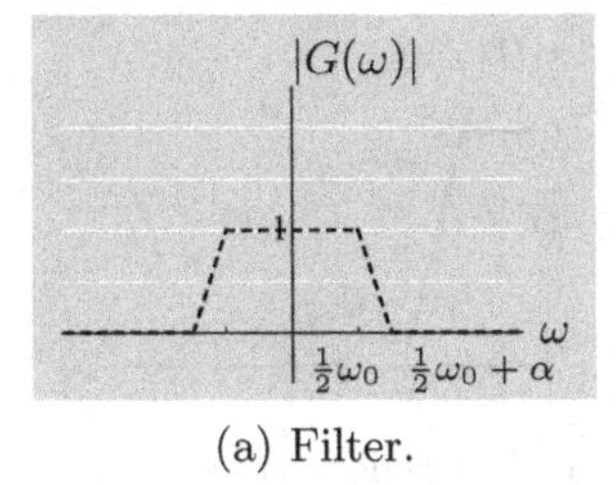

(a) Filter.

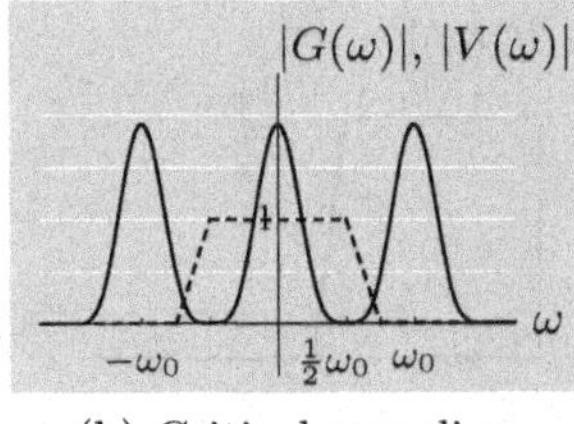

(b) Critical sampling.

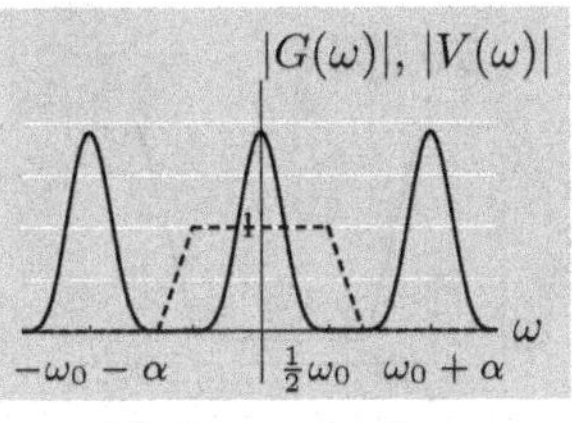

(c) Oversampling.

Figure 5.30 Sampling and reconstruction of x bandlimited to $[-\frac{1}{2}\omega_0, \frac{1}{2}\omega_0]$ with nonideal filters. (a) Filter G with continuous spectrum obtained from convolving two ideal filters of bandwidths ω_0 and 2α. With $\omega_s = \omega_0$, replicas of X appear at integer multiples of ω_0, causing G to catch the tails of the replicas centered at $\pm\omega_0$. (c) With $\omega_s = \omega_0 + \alpha$, replicas of X appear at integer multiples of $\omega_0 + \alpha$, allowing G to extract X perfectly.

Example 5.21 (Speech processing in mobile phones) The bandlimited assumption used in speech and audio processing is based on the fact that humans cannot hear frequencies above 20 kHz. Thus, music for compact disks is sampled at 44 kHz, with a lowpass filter having a passband from −20 kHz to 20 kHz and a transition band of 2 kHz. For speech, in telephone applications where bandwidth has always been at a premium, a passband from 0.3 to 3.4 kHz is sufficient for good quality. A sampling frequency of $f_s = 8$ kHz is used, or $T = 2\pi/\omega_s = 0.125$ ms as a sampling period. The continuous-time prefilters and postfilters have a passband of up to 3.4 kHz, followed by a smooth transition to very high attenuation for frequencies above 4 kHz. For the sake of this example, we assume that these continuous-time filters have frequency responses like G in (5.85a), with $\frac{1}{2}\omega_0 = 2\pi\cdot 3.4 = 6.8\pi$ krad/s and $\alpha = 1.2\pi$ krad/s. In the remainder of this example (including the figures), all angular frequencies of continuous-time filters are expressed in krad/s (that is, 10^3 rad/s).

We want to illustrate the process of implementing continuous-time filtering of speech in the discrete domain as an application of Theorem 5.16. Assume that we want to filter speech from its full spectrum of 4 kHz to half of that, 2 kHz, using a continuous-time filter $H(\omega)$ implemented in discrete time as in Figure 5.29. Normalizing the frequency axis so that $\omega_s = 16\pi$ krad/s in the Fourier transform corresponds to 2π in the DTFT is equivalent to designing a lowpass filter $\widetilde{H}(e^{j\omega})$ with cutoff at $\frac{1}{2}\pi$. The concessions that must be made to have an implementable discrete-time filter include the following:

(i) Instead of perfect stopband attenuation, we allow a stopband gain of 10^{-1}.
(ii) Instead of a sharp (discontinuous) drop off at the cutoff frequency, we allow a transition band from $\frac{3}{8}\pi$ to $\frac{5}{8}\pi$.

This leads to a discrete-time filter $\widetilde{h}$ with specifications in the frequency domain of

$$\widetilde{H}(e^{j\omega}) = \begin{cases} 1.0, & \text{for } |\omega| \le \frac{3}{8}\pi; \\ 0.1, & \text{for } \frac{5}{8}\pi \le |\omega| < \pi, \end{cases} \tag{5.87}$$

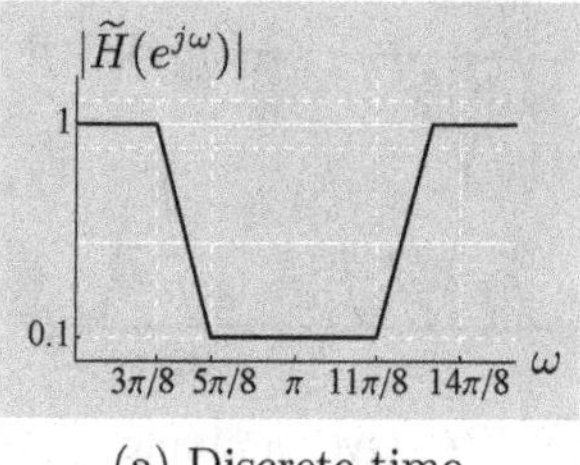

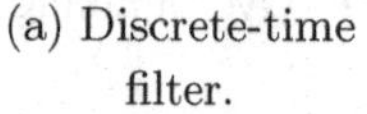

(a) Discrete-time filter.

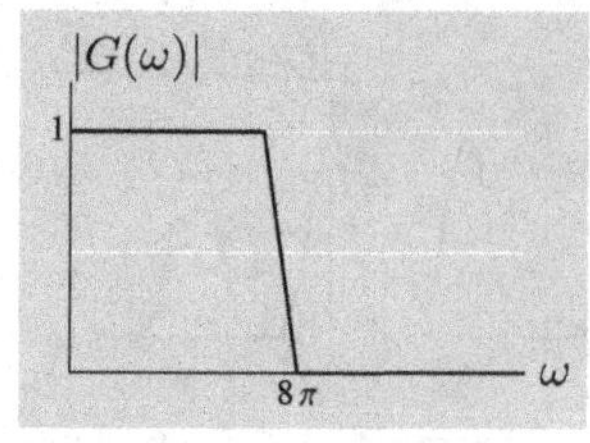

(b) Continuous-time pre/postfilter.

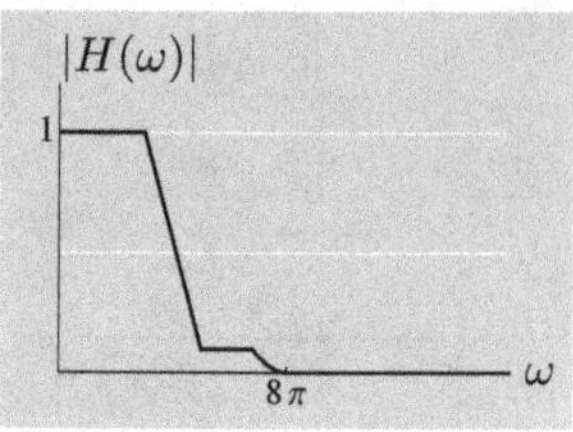

(c) Equivalent continuous-time filter.

Figure 5.31 Filters involved in continuous- and discrete-time implementation of filtering in Figure 5.29. The transition band in the discrete-time filter is one of many possible ones. The continuous-time pre/postfilter is shown as an idealized conceptual plot. The sampling period is $T = 0.125$ ms, or the sampling frequency is $\omega_s = 16\pi$ krad/s; the frequency axes in (b) and (c) are in krad/s.

the band $\frac{3}{8}\pi \leq |\omega| \leq \frac{5}{8}\pi$ being a transition band where the filter is unspecified; this filter, with one possible transition band, is depicted in Figure 5.31(a).

Before this discrete-time filter is applied, we need to sample the signal. This is done using a continuous-time lowpass filter G with cutoff at 4 kHz, or 8π krad/s, an example of which is shown in Figure 5.31(b). This filter has a transition from its passband to a perfect, infinite attenuation beyond the cut off frequency, which is an idealization.

Finally, the interpolation postfilter G removes the spectral replicas. Therefore, the overall effect in the continuous-time frequency domain is to multiply by $\widetilde{H}(e^{j\omega})$, rescaled as $\widetilde{H}(e^{j\omega T})$, and the square of G, since g is applied both as a prefilter and as a postfilter,

$$H(\omega) = \begin{cases} \widetilde{H}(e^{j\omega T})\, G^2(\omega), & \text{for } |\omega| \leq 8\pi \cdot 10^3; \\ 0, & \text{for } |\omega| > 8\pi \cdot 10^3. \end{cases}$$

The key in this process is the rescaling of the frequency axis, which maps the $[-\pi, \pi]$ interval of the DTFT to the $[-\frac{1}{2}\omega_0, \frac{1}{2}\omega_0]$ interval of the Fourier transform. Note that the scale factor from input to output, while mathematically specified by factors such as $1/T$ in (5.72) and gains in continuous-time and discrete-time filters, depends on implementation issues such as continuous-time amplifiers and continuous-to-discrete and discrete-to-continuous converter scaling factors.

Let us now consider what happens to an actual continuous-time signal x with a spectrum depicted in Figure 5.32(a) and given by

$$X(\omega) = e^{-2|\omega|/\omega_0}; \tag{5.88}$$

thus $X(0) = 1$ and $X(\pm\frac{1}{2}\omega_0) = 1/e$. This spectrum is chosen for illustration only and does not correspond to actual speech. Before sampling, we prefilter with G from (5.85a), depicted in Figure 5.31(b); the result of this operation is shown in Figure 5.32(b). We can now sample the resulting signal yielding a discrete-time

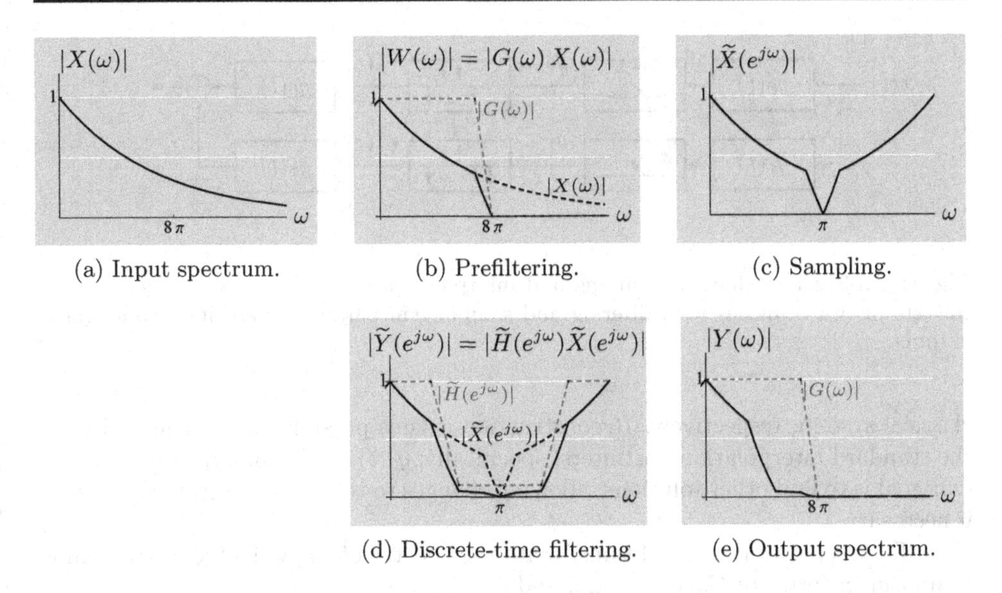

(a) Input spectrum. (b) Prefiltering. (c) Sampling.

(d) Discrete-time filtering. (e) Output spectrum.

Figure 5.32 Discrete-time implementation of continuous-time filtering, shown in the spectral domain. Filtering the input spectrum X in the continuous-time domain with a continuous-time filter H from Figure 5.31(c) yields the output Y. The same can be achieved by prefiltering with a continuous-time filter G from Figure 5.31(b), sampling, filtering with a discrete-time filter $\widetilde{H}$ from Figure 5.31(a), and interpolating using the postfilter G. The solid black lines are signals at each point in the system from Figure 5.29, the dashed black lines indicate the spectrum to be filtered, and the dashed gray lines indicate the corresponding filters; the frequency axes in (a), (b), and (e) are in krad/s.

signal with DTFT $\widetilde{X}$ depicted in Figure 5.32(c). The effect of filtering in the discrete-time domain is shown in Figure 5.32(d). Finally, the output spectrum is as in Figure 5.32(e). What should be clear is that, because of filters not being ideal, we are only roughly approximating an ideal, lowpass filter with cutoff at 2 kHz (4π krad/s).

Multichannel sampling The classical sampling result in Theorem 5.15 has many extensions and generalizations. An important example for practical data acquisition is to reconstruct a function from samples measured in multiple parallel channels. We develop this for the two-channel system in Figure 5.33.

Assume that the input to the system x is in $\mathrm{BL}[-\pi/T,\, \pi/T]$; this means that it can be reconstructed perfectly from samples with period T, using the interpolation postfilter g in (5.68). Since we have access to two sampled filter outputs, it is intuitive that, under certain conditions on the filters $\widetilde{g}_0$ and $\widetilde{g}_1$, sampling the two channels with sampling period $2T$ would be a sufficient representation of x. Indeed, one way to do this is to choose $\widetilde{g}_0$ to be the sinc filter used in the standard single-channel system, $\widetilde{g}_0(t) = g(t)$, and $\widetilde{g}_1$ to be the same with an advance of T, $\widetilde{g}_1(t) = g(t+T)$. Then, the sequences y_0 and y_1 are the even- and odd-indexed samples of the

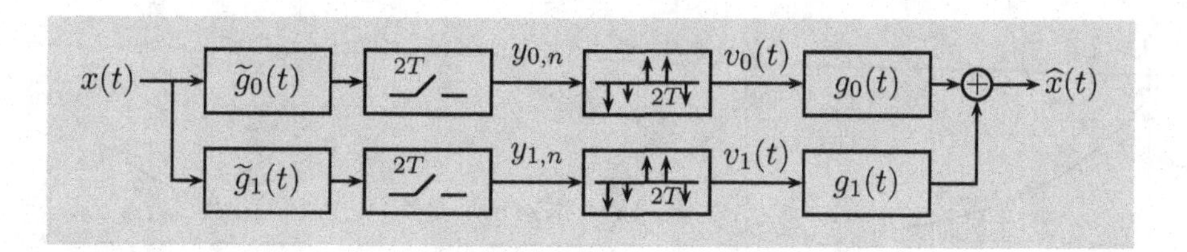

Figure 5.33 Two-channel sampling and interpolation system. The sampling operator consists of two branches with filtering and sampling in parallel, producing two sampled outputs.

classical system, respectively. Reconstruction is thus possible and is achieved with the standard interpolation postfilters; specifically, $g_0(t) = g(t)$ and $g_1(t) = g(t-T)$. Our goal is to find other solutions, allowing changes to the reconstruction processing if necessary.

To develop general conditions, write the spectra of the weighted Dirac combs v_0 and v_1 in terms of $\widetilde{G}_0(\omega)$, $\widetilde{G}_1(\omega)$, and $X(\omega)$,

$$V_i(\omega) \overset{(a)}{=} \frac{1}{2T}\sum_{k\in\mathbb{Z}} \widetilde{G}_i\Big(\omega + \frac{\pi}{T}k\Big)\, X\Big(\omega + \frac{\pi}{T}k\Big), \qquad i = 0,\, 1,$$

where (a) follows from (5.72), with sampling period $2T$. Since $V_i(\omega)$ is periodic with period π/T, we can consider only one interval $[0,\, \pi/T]$. Moreover, since $X(\omega)$ is bandlimited to $[-\pi/T,\, \pi/T]$, only two spectral components overlap on $[0,\, \pi/T]$,

$$\begin{aligned} V_0(\omega) &= \frac{1}{2T}\Big(\widetilde{G}_0(\omega)X(\omega) + \widetilde{G}_0(\omega - \pi/T)X(\omega - \pi/T)\Big), \\ V_1(\omega) &= \frac{1}{2T}\Big(\widetilde{G}_1(\omega)X(\omega) + \widetilde{G}_1(\omega - \pi/T)X(\omega - \pi/T)\Big). \end{aligned}$$

In matrix notation, for $\omega \in [0,\, \pi/T]$,

$$\begin{bmatrix} V_0(\omega) \\ V_1(\omega) \end{bmatrix} = \frac{1}{2T}\begin{bmatrix} \widetilde{G}_0(\omega) & \widetilde{G}_0(\omega - \pi/T) \\ \widetilde{G}_1(\omega) & \widetilde{G}_1(\omega - \pi/T) \end{bmatrix} \begin{bmatrix} X(\omega) \\ X(\omega - \pi/T) \end{bmatrix} = \widetilde{G}(\omega) \begin{bmatrix} X(\omega) \\ X(\omega - \pi/T) \end{bmatrix}, \tag{5.89}$$

where we have introduced the matrix $\widetilde{G}(\omega)$. As long as $\widetilde{G}(\omega)$ is nonsingular on the interval $[0,\, \pi/T]$, we can recover $X(\omega)$ on $[-\pi/T,\, \pi/T]$. The key is that, since x is bandlimited, undersampling it by a factor of 2 results in only one spectral replica overlapping with the base spectrum, and having two versions of the spectrum in (5.89) allows us to separate these spectra when the matrix $\widetilde{G}(\omega)$ is invertible. We illustrate this in the next two examples.

Example 5.22 (Periodic nonuniform sampling) Consider the two-channel system in Figure 5.33 with input $x \in \mathrm{BL}[-\pi,\, \pi]$ and $T = 1$. Let the sampling prefilters be the identity filter and a delay of $\tau \in (0,\, 2)$,

$$\widetilde{G}_0(\omega) = 1, \qquad \widetilde{G}_1(\omega) = e^{-j\omega\tau}.$$

By substitution into (5.89), the spectra of the weighted Dirac combs produced from the samples are given by

$$\begin{bmatrix} V_0(\omega) \\ V_1(\omega) \end{bmatrix} = \frac{1}{2} \begin{bmatrix} 1 & 1 \\ e^{-j\omega\tau} & e^{-j(\omega-\pi)\tau} \end{bmatrix} \begin{bmatrix} X(\omega) \\ X(\omega-\pi) \end{bmatrix}.$$

Since $\tau \in (0,2)$ implies that $\det(\widetilde{G}(\omega)) = \frac{1}{4}e^{-j\omega\tau}(e^{j\pi\tau}-1) \neq 0$, $\widetilde{G}(\omega)$ is invertible. Thus, x can be recovered from the sample sequences y_0 and y_1. However, inversion of $\widetilde{G}(\omega)$ becomes arbitrarily ill conditioned as τ approaches 0 or 2. This comes as no surprise, since, for either no delay ($\tau = 0$) or delay equal to the sampling period ($\tau = 2$), the samples in the two channels are the same, and a single channel is not adequate for recovering the input.

As a sanity check, choose $\tau = 1$, which leads to the usual sampling of $x(t)$ at $t \in \mathbb{Z}$, with even-indexed samples in channel 0 and odd-indexed ones in channel 1. Then,

$$\begin{bmatrix} V_0(\omega) \\ V_1(\omega) \end{bmatrix} = \frac{1}{2} \begin{bmatrix} 1 & 1 \\ e^{-j\omega} & -e^{-j\omega} \end{bmatrix} \begin{bmatrix} X(\omega) \\ X(\omega-\pi) \end{bmatrix}.$$

The matrix $\widetilde{G}(\omega)$ is not only invertible but also well conditioned for inversion since its columns are orthogonal.

Example 5.23 (Sampling a function and its derivative) Again consider the two-channel system in Figure 5.33 with input $x \in \text{BL}[-\pi, \pi]$ and $T = 1$. This time let the sampling prefilters be the identity filter $\widetilde{G}_0(\omega) = 1$ and the derivative filter $\widetilde{G}_1(\omega) = j\omega$. The spectra of the weighted Dirac combs produced from the samples are given by

$$\begin{bmatrix} V_0(\omega) \\ V_1(\omega) \end{bmatrix} = \frac{1}{2} \begin{bmatrix} 1 & 1 \\ j\omega & j(\omega-\pi) \end{bmatrix} \begin{bmatrix} X(\omega) \\ X(\omega-\pi) \end{bmatrix}.$$

The determinant of the above matrix, $\det(\widetilde{G}(\omega)) = -\frac{1}{4}j\pi$, is a nonzero constant, making the system invertible and showing that one can reconstruct a bandlimited function from twice undersampled versions of the function and its derivative.

The two-channel case in Figure 5.33 can be readily extended to N channels.

Theorem 5.17 (Multichannel sampling) Let $\omega_0 \in \mathbb{R}^+$ and $x \in \text{BL}[-\frac{1}{2}\omega_0, \frac{1}{2}\omega_0]$, and let $T \in (0, 2\pi/\omega_0]$. Consider an N-channel system with filters $\widetilde{g}_i$, $i = 0, 1, \ldots, N-1$, followed by uniform sampling with period NT. A necessary and sufficient condition for recovery of x is that the matrix

$$\widetilde{G}(\omega) = \begin{bmatrix} \widetilde{G}_0(\omega) & \widetilde{G}_0(\omega + \frac{2\pi}{NT}) & \cdots & \widetilde{G}_0\left(\omega + \frac{2\pi(N-1)}{NT}\right) \\ \widetilde{G}_1(\omega) & \widetilde{G}_1(\omega + \frac{2\pi}{NT}) & \cdots & \widetilde{G}_1\left(\omega + \frac{2\pi(N-1)}{NT}\right) \\ \vdots & \vdots & \ddots & \vdots \\ \widetilde{G}_{N-1}(\omega) & \widetilde{G}_{N-1}(\omega + \frac{2\pi}{NT}) & \cdots & \widetilde{G}_{N-1}\left(\omega + \frac{2\pi(N-1)}{NT}\right) \end{bmatrix} \tag{5.90}$$

be nonsingular for $\omega \in [0, 2\pi/(NT)]$.

The proof is a direct extension of what we saw for two channels; see the *Further reading*. Exercise 5.12 explores some ramifications of this result, in particular for periodic nonuniform sampling and derivatives.

Bandlimited continuous-time stochastic processes The sampling theorem extends from individual bandlimited functions to bandlimited stochastic processes. The concept of bandwidth does not apply directly to a WSS stochastic process because, with probability 1, the Fourier transform of a realization does not converge. Instead, the bandwidth of the stochastic process is defined to be the bandwidth of the autocorrelation function of the process. The main conclusion remains the same: a sampling frequency ω_s exceeding the process bandwidth is sufficient for recovering the process.

Sampling results for stochastic processes are mathematically more sophisticated than the results we have developed so far. We thus state results but omit technical justifications; see the *Further reading* for details and pointers.

THEOREM 5.18 (SAMPLING FOR CONTINUOUS-TIME STOCHASTIC PROCESSES) Let $\omega_0 \in \mathbb{R}^+$, and let x be a WSS continuous-time stochastic process with autocorrelation function $a_x \in \mathrm{BL}[-\frac{1}{2}\omega_0, \frac{1}{2}\omega_0]$. For any $T \in (0, 2\pi/\omega_0]$,

$$x(t) = \sum_{k\in\mathbb{Z}} x(nT)\,\mathrm{sinc}\Big(\frac{\pi}{T}(t-nT)\Big) \qquad \text{for all } t \in \mathbb{R}, \tag{5.91}$$

in the mean-square sense, meaning that

$$\lim_{N\to\infty} \mathrm{E}\left[\left|x(t) - \sum_{k=-N}^{N} x(nT)\,\mathrm{sinc}\Big(\frac{\pi}{T}(t-nT)\Big)\right|^2\right] = 0 \qquad \text{for all } t \in \mathbb{R}.$$

Convergence in the mean-square sense implies also convergence in probability. For any $T \in (0, 2\pi/\omega_0)$, (5.91) holds almost surely.

5.4.3 Sampling and interpolation with nonorthogonal functions

We now expand the set of filters that we use in sampling and interpolation. We no longer require the interpolation postfilter and its shifts to be an orthonormal set, nor do we require the sampling prefilter to be the time-reversed and conjugated version of the interpolation postfilter. These changes are precisely as in Section 5.3.3 for sequences, and they have the same effect on the geometry: the sampling and interpolation operators are no longer adjoints of each other, and the range of the interpolation operator and the orthogonal complement of the null space of the sampling operator are no longer the same. We will again find that the sampling and interpolation operators are equivalent to analysis with basis $\{\widetilde{\varphi}_k\}_{k\in\mathbb{Z}}$ and synthesis with basis $\{\varphi_k\}_{k\in\mathbb{Z}}$. Under the consistency condition, these bases are biorthogonal

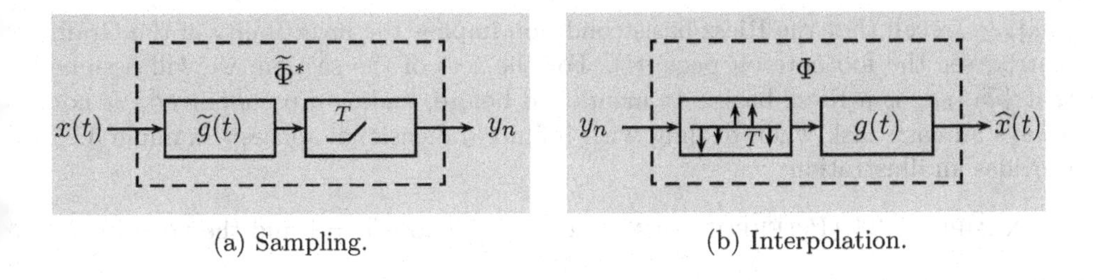

(a) Sampling. (b) Interpolation.

Figure 5.34 Sampling and interpolation in $\mathcal{L}^2(\mathbb{R})$ with nonorthogonal functions.

sets; under the additional condition of the operators being ideally matched, these bases are a dual basis pair for the same subspace of $\mathcal{L}^2(\mathbb{R})$.

Sampling We now refer to the operation depicted in Figure 5.34(a), involving filtering with $\widetilde{g}$ and recording the result at time instants $t = nT$, $n \in \mathbb{Z}$, as *sampling of the function x with rate $1/T$ and prefilter $\widetilde{g}$*, and we denote it by $y = \widetilde{\Phi}^* x$. In contrast to Section 5.4.1, we do not make an assumption of orthonormality.

Similarly to (5.58), for any time index $k \in \mathbb{Z}$, the output of sampling is

$$y_k = (\widetilde{\Phi}^* x)_k = \int_{-\infty}^{\infty} x(\tau)\, \widetilde{g}(kT-\tau)\, d\tau = \langle x(\tau),\, \widetilde{g}^*(kT-\tau)\rangle_\tau \stackrel{(a)}{=} \langle x,\, \widetilde{\varphi}_k\rangle, \quad (5.92)$$

where in (a) we have defined the function $\widetilde{\varphi}_k$ to be the time-reversed and conjugated version of $\widetilde{g}$, shifted by kT,

$$\widetilde{\varphi}_k(t) = \widetilde{g}^*(kT - t), \qquad t \in \mathbb{R}. \quad (5.93)$$

Since the output at time k is $\langle x, \widetilde{\varphi}_k\rangle$, the sampling operator $\widetilde{\Phi}^*$ gives the inner products with all the functions in $\{\widetilde{\varphi}_k\}_{k\in\mathbb{Z}}$. As before, its null space, $\widetilde{S}^\perp = \mathcal{N}(\widetilde{\Phi}^*)$, and the orthogonal complement of the null space, $\widetilde{S} = \mathcal{N}(\widetilde{\Phi}^*)^\perp = \overline{\text{span}}(\{\widetilde{\varphi}_k\}_{k\in\mathbb{Z}})$, are closed, shift-invariant subspaces.

Interpolation Again, we refer to the operation depicted in Figure 5.34(b), involving weighting a Dirac comb s_T, (4.121), and filtering with g, as *interpolation of the sequence y with spacing T and postfilter g*, and we denote it by $\widehat{x} = \Phi y$. This is unchanged from Figure 5.4(b), and (5.61) again holds. We denote the range of Φ by S as before, and it is a closed, shift-invariant subspace with respect to shift T with generator g.

In contrast to Section 5.4.1, the interpolation operator Φ is not necessarily the adjoint of the sampling operator $\widetilde{\Phi}^*$, and S does not necessarily equal $\widetilde{S}$. When the interpolation operator is specially chosen so that it satisfies (5.23), that is, it is the pseudoinverse of $\widetilde{\Phi}^*$, then $S = \widetilde{S}$, by the same arguments as in (5.24). The pseudoinverse in (5.23) requires $\widetilde{\Phi}^*\Phi$ to be invertible; as for sequences, this is guaranteed if $\{\widetilde{\varphi}_k\}_{k\in\mathbb{Z}}$ is a Riesz basis for $\widetilde{S}$ since $\widetilde{\Phi}^*\widetilde{\Phi}$ is the Gram matrix of

$\{\widetilde{\varphi}_k\}_{k\in\mathbb{Z}}$ (recall that the Riesz basis condition implies the invertibility of the Gram matrix, see the footnote on page 94). For the rest of the section, we will assume that $\{\widetilde{\varphi}_k\}_{k\in\mathbb{Z}}$ is a Riesz basis. As mentioned before, finding a pseudoinverse is not always an easy task when dealing with infinite-dimensional spaces; Example 5.26 provides an illustration.

EXAMPLE 5.24 (INTERPOLATION IN $\mathcal{L}^2(\mathbb{R})$) Choose $T = 1$ and the postfilter

$$g(t) \;=\; \begin{cases} 1-|t|, & \text{for } |t|<1; \\ 0, & \text{otherwise.} \end{cases} \tag{5.94}$$

The range of the interpolation operator, $S = \overline{\operatorname{span}}(\{g(t-k)\}_{k\in\mathbb{Z}})$, is a shift-invariant subspace with respect to integer shifts. Each element of S is a continuous, piecewise-linear function with changes of derivative at a subset of the integers.

Interpolation followed by sampling Interpolation followed by sampling is described by $\widetilde{\Phi}^*\Phi$, as in Figure 5.35(a). Analogously to (5.63),

$$\begin{aligned} \widehat{y}_n &= \int_{-\infty}^{\infty} \widehat{x}(t)\,\widetilde{g}(nT-t)\,dt \;=\; \int_{-\infty}^{\infty}\Bigl(\sum_{k\in\mathbb{Z}} y_k g(t-kT)\Bigr)\widetilde{g}(nT-t)\,dt \\ &= \sum_{k\in\mathbb{Z}} y_k \int_{-\infty}^{\infty} g(t-kT)\,\widetilde{g}(nT-t)\,dt, \end{aligned}$$

so $\widetilde{\Phi}^*\Phi y = y$ for all $y \in \ell^2(\mathbb{Z})$ if and only if

$$\langle g(t-kT),\, \widetilde{g}^*(nT-t)\rangle_t \;=\; \delta_{k-n}. \tag{5.95}$$

Because of (5.59) and (5.93), this is equivalent to a counterpart for (5.60) and (5.64),

$$\widetilde{\Phi}^*\Phi \;=\; I \qquad \Leftrightarrow \qquad \langle \varphi_k, \widetilde{\varphi}_n\rangle \;=\; \delta_{k-n}. \tag{5.96}$$

The sampling and interpolation operators are then called *consistent.* Choosing the pseudoinverse in (5.23) for Φ would satisfy (5.96); there exist infinitely many other right inverses of $\widetilde{\Phi}^*$ one could also use. Equation (5.96) also shows that the condition for perfect recovery is the same as the sets of functions $\{\varphi_k\}_{k\in\mathbb{Z}}$ and $\{\widetilde{\varphi}_k\}_{k\in\mathbb{Z}}$ being biorthogonal, as in (2.111). These sets of functions are not bases for $\mathcal{L}^2(\mathbb{R})$; instead, they are bases for the subspaces S and $\widetilde{S}$ that they span, respectively. They are a biorthogonal pair of bases when $S = \widetilde{S}$.

EXAMPLE 5.25 (INTERPOLATION FOLLOWED BY SAMPLING IN $\mathcal{L}^2(\mathbb{R})$) We would like to find a sampling operator that is consistent with the interpolation operator from Example 5.24. There exist many choices of $\widetilde{g}$ to satisfy (5.95) with $T = 1$. It simplifies the problem greatly to assume a short support for $\widetilde{g}$ because many of the inner products in (5.95) become zero, as desired, by virtue of the lack of overlap of $\widetilde{g}$ with shifted versions of g. For example, fix the support

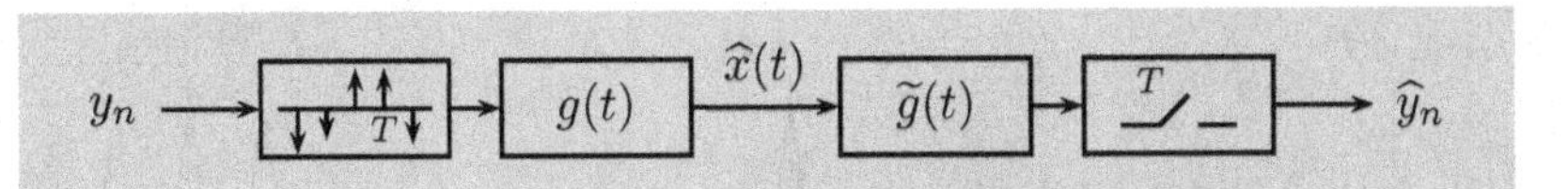

(a) Interpolation followed by sampling.

(b) Sampling followed by interpolation.

Figure 5.35 Sampling and interpolation in $\mathcal{L}^2(\mathbb{R})$ with nonorthogonal functions.

of $\widetilde{g}$ to be $[-\frac{1}{2}, \frac{1}{2}]$ so that (5.95) is satisfied for $|n-k| > 1$ solely from lack of overlap. With the restricted support of $\widetilde{g}$, we get three constraints from (5.95), corresponding to $n-k \in \{-1, 0, 1\}$. Upon further constraining $\widetilde{g}$ to be an even function, the constraints for $n-k = \pm 1$ will be the same, and many possible parameterizations of $\widetilde{g}$ with two parameters will yield a solution.

Assume $\widetilde{g}$ to be of the following form:

$$\widetilde{g}(t) = \begin{cases} a(b-|t|), & \text{for } |t| < \frac{1}{2}; \\ 0, & \text{otherwise.} \end{cases} \tag{5.97}$$

We thus get the following system of equations from (5.95):

$$1 = \langle g(t), \widetilde{g}^*(-t)\rangle_t = \int_{-1/2}^{1/2} (1-|t|)a(b-|t|)\,dt = \frac{1}{12}a(9b-2),$$

$$0 = \langle g(t), \widetilde{g}^*(1-t)\rangle_t = \int_{1/2}^{1} (1-t)a(b-(1-t))\,dt = \frac{1}{24}a(3b-1),$$

which has the solution $a = 12$, $b = \frac{1}{3}$, or

$$\widetilde{g}(t) = \begin{cases} 4-12|t|, & \text{for } |t| < \frac{1}{2}; \\ 0, & \text{otherwise.} \end{cases} \tag{5.98}$$

The arbitrariness of (5.97) suggests that many other choices of $\widetilde{g}$ will also yield consistent sampling and interpolation, even with the given arbitrary limitation of the support of $\widetilde{g}$.

The functions g from (5.94) and $\widetilde{g}$ from (5.98) are plotted in Figures 5.36(a) and (b). With their integer shifts, these functions span different spaces; that is, $S \neq \widetilde{S}$. The functions in S are continuous and might have a change of derivative at each integer. The functions in $\widetilde{S}$ are generally discontinuous at the integers and will have changes of derivative at all odd multiples of $\frac{1}{2}$. In fact, functions in $\widetilde{S}$ always look like saw blades; Figure 5.36(c) shows the example of $x(t) = \widetilde{g}^*(1-t) + \widetilde{g}^*(-t) + \widetilde{g}^*(-1-t)$.

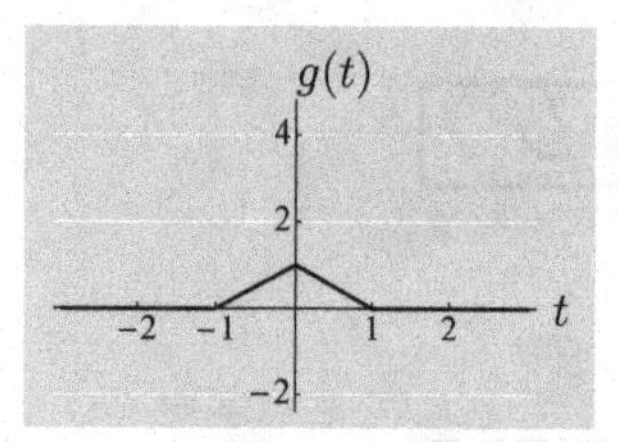

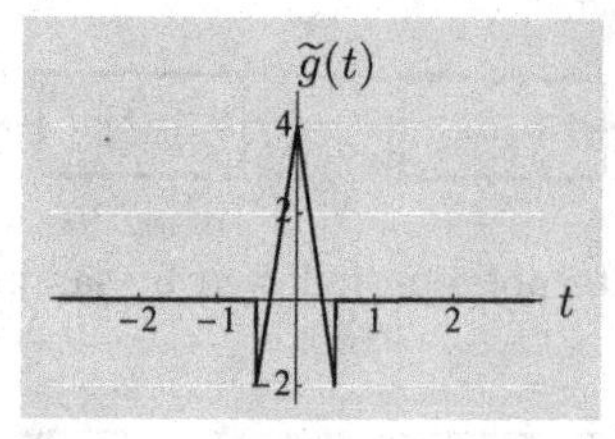

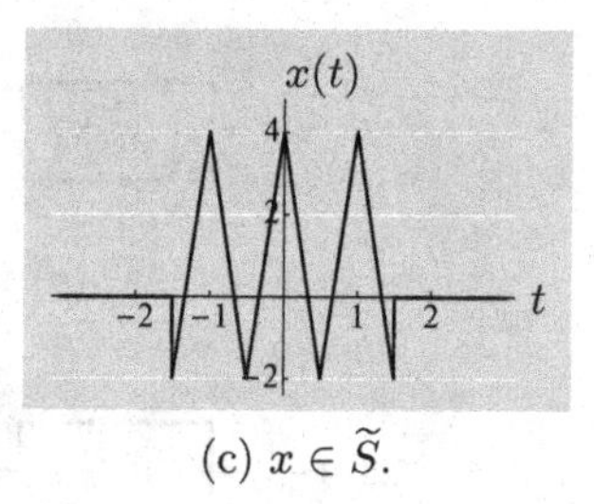

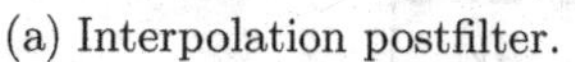

(a) Interpolation postfilter. (b) Sampling prefilter. (c) $x \in \widetilde{S}$.

Figure 5.36 Consistent sampling and interpolation for functions. (a) Interpolation postfilter g from (5.94). (b) Sampling prefilter $\widetilde{g}$ from (5.98), resulting in consistent sampling and interpolation operators. (c) A function $x(t) = \widetilde{g}^*(1-t) + \widetilde{g}^*(-t) + \widetilde{g}^*(-1-t)$ in $\widetilde{S}$ using $\widetilde{g}$ from (b).

Sampling followed by interpolation Sampling followed by interpolation is described by $P = \Phi\widetilde{\Phi}^*$, as in Figure 5.35(b). When the sampling and interpolation operators are consistent as in (5.96), P is idempotent by the computation in (5.28). It projects onto S, but the projection is not necessarily orthogonal. The approximation error $x - \widehat{x}$ is orthogonal to $\widetilde{S}$ but not to S (recall Figure 5.11 for a conceptual picture).

Again, for P to be self-adjoint as well, Φ must be chosen to be the pseudoinverse of $\widetilde{\Phi}^*$, (5.23); the sampling and interpolation operators are then *ideally matched*, and subspaces S and $\widetilde{S}$ are identical. In other words, when the sampling and interpolation operators are ideally matched, $P = \Phi\widetilde{\Phi}^*$ is an orthogonal projection operator, and, by Theorem 2.26, $\widehat{x} = Px$ is the best approximation of x in S.

The previous discussion can be summarized by the following analogue to Theorems 5.2 and 5.9.

THEOREM 5.19 (RECOVERY FOR FUNCTIONS, NONORTHOGONAL) Let $T \in \mathbb{R}^+$, let the sampling operation $\widetilde{\Phi}^* : \mathcal{L}^2(\mathbb{R}) \to \ell^2(\mathbb{Z})$ be given by

$$(\widetilde{\Phi}^* x)_k = \int_{-\infty}^{\infty} x(\tau)\, \widetilde{g}(kT - \tau)\, d\tau, \qquad k \in \mathbb{Z}, \tag{5.99a}$$

and let the interpolation operation $\Phi : \ell^2(\mathbb{Z}) \to \mathcal{L}^2(\mathbb{R})$ be given by

$$(\Phi y)(t) = \sum_{k\in\mathbb{Z}} y_k g(t - kT), \qquad t \in \mathbb{R}. \tag{5.99b}$$

As depicted in Figure 5.35(b), denote the result of sampling followed by interpolation applied to $x \in \mathcal{L}^2(\mathbb{R})$ by $\widehat{x} = Px = \Phi\widetilde{\Phi}^* x$.

If the sampling prefilter $\widetilde{g}$ and interpolation postfilter g satisfy consistency condition (5.95), then P is a projection operator with range $S = \mathcal{R}(\Phi)$ and $x - \widehat{x} \perp \widetilde{S} = \mathcal{N}(\widetilde{\Phi}^*)^{\perp}$; in particular, $\widehat{x} = x$ when $x \in S$.

If, additionally, ideal matching (5.23) is satisfied, then $S = \widetilde{S}$, P is the orthogonal projection operator with range S, and $\widehat{x}$ is the best approximation of x in S,

$$\widehat{x} = \arg\min_{x_S \in S} \|x - x_S\|, \qquad x - \widehat{x} \perp S.$$

EXAMPLE 5.26 (SAMPLING FOLLOWED BY INTERPOLATION IN $\mathcal{L}^2(\mathbb{R})$) As shown in Example 5.25, sampling with rate 1 and prefilter $\widetilde{g}$ given in (5.98) is consistent with interpolation with spacing 1 and postfilter g given in (5.94). Thus, $P = \Phi\widetilde{\Phi}^*$ for the associated sampling and interpolation operators is a projection operator. It is an oblique projection operator because $\widetilde{S} \neq S$.

For P to be an orthogonal projection operator, the operators Φ and $\widetilde{\Phi}^*$ must be ideally matched, that is, $\widetilde{S} = S$. For this, we cannot make arbitrary choices as in Example 5.25; in fact, $\widetilde{g}$ is then uniquely determined.

The key to finding $\widetilde{g}$ such that $\widetilde{S} = S$ is to choose $\widetilde{g}$ so that $\widetilde{\varphi}_0$ is in S. Then, since S is a shift-invariant subspace, every $\widetilde{\varphi}_k$, $k \in \mathbb{Z}$, will be in S. Thus, let

$$\widetilde{g}(t) = \sum_{\ell \in \mathbb{Z}} \alpha_\ell g^*(-t - \ell T) \tag{5.100}$$

for some sequence $\alpha \in \ell^2(\mathbb{Z})$ to be determined, since $\widetilde{\varphi}_0$ is the time-reversed and conjugated version of $\widetilde{g}$. Now the requirement of consistency can be written as

$$\begin{aligned}
\delta_k &\overset{(a)}{=} \langle g(t - kT), \widetilde{g}^*(-t) \rangle_t \overset{(b)}{=} \left\langle g(t - kT), \sum_{\ell \in \mathbb{Z}} \alpha_\ell^* g(t - \ell T) \right\rangle_t \\
&\overset{(c)}{=} \sum_{\ell \in \mathbb{Z}} \alpha_\ell \langle g(t - kT), g(t - \ell T) \rangle_t \overset{(d)}{=} \sum_{\ell \in \mathbb{Z}} \alpha_\ell a_{\ell - k},
\end{aligned} \tag{5.101}$$

where (a) follows from the consistency condition (5.95) with $n = 0$; (b) from (5.100); (c) from the conjugate linearity in the second argument of the inner product; and (d) from defining an autocorrelation sequence

$$a_m = \langle g(t), g(t - m) \rangle, \qquad m \in \mathbb{Z}, \tag{5.102}$$

which is the deterministic autocorrelation of g evaluated at the integers. To find the sequence α, we recognize the final expression of (5.101) as the convolution between sequence α and the time-reversed version of sequence a. In the z-transform domain, we can rephrase (5.101) as

$$\alpha(z) A(z^{-1}) = 1. \tag{5.103}$$

For g from (5.94), $A(z) = \frac{1}{6}(z+4+z^{-1})$, and thus

$$\begin{aligned}\alpha(z) &= \frac{1}{A(z^{-1})} = \frac{6}{z^{-1}+4+z}\\ &= \frac{6z^{-1}}{(1-(\sqrt{3}-2)z^{-1})(1-(\sqrt{3}-2)^{-1}z^{-1})}\\ &= \sqrt{3}\left(\frac{1}{1-(\sqrt{3}-2)z^{-1}} - \frac{1}{1-(\sqrt{3}-2)^{-1}z^{-1}}\right). \end{aligned} \tag{5.104}$$

Three different ROCs are possible: inside the pole of the smaller magnitude, between the two poles, and outside the pole of the larger magnitude. To have a stable sequence, we choose the ROC to contain the unit circle, that is, ROC $= \{z \mid |\sqrt{3}-2| < |z| < |\sqrt{3}-2|^{-1}\}$. From Table 3.6, the first summand then corresponds to a right-sided geometric sequence $(\sqrt{3}-2)^n u_n$ and the second to a left-sided geometric sequence $-(\sqrt{3}-2)^{-n}u_{-n-1}$. By combining the two, we get that the rational z-transform from (5.104) corresponds to the two-sided sequence

$$\begin{aligned}\alpha_n &= \sqrt{3}\left((\sqrt{3}-2)^n u_n + (\sqrt{3}-2)^{-n} u_{-n-1}\right)\\ &= \sqrt{3}(\sqrt{3}-2)^{|n|}, \qquad n \in \mathbb{Z},\end{aligned} \tag{5.105}$$

from which $\widetilde{g}$ follows according to (5.100); see Figure 5.37(b). We have thus proven that $\widetilde{S} \subseteq S$ (since $\widetilde{g}$ is a linear combination of time-reversed and shifted versions of g^*). The opposite is true as well, namely $S \subseteq \widetilde{S}$, since g is a linear combination of three time-reversed and shifted versions of $\widetilde{g}^*$. To see that, take the Fourier transform of (5.100):

$$\begin{aligned}\widetilde{G}(\omega) &= \int_{-\infty}^{\infty} \widetilde{g}(t)\, e^{-j\omega t}\, dt \overset{(a)}{=} \int_{-\infty}^{\infty} \sum_{\ell\in\mathbb{Z}} \alpha_\ell g^*(-t-\ell T)\, e^{-j\omega t}\, dt\\ &\overset{(b)}{=} \int_{-\infty}^{\infty} \sum_{\ell\in\mathbb{Z}} \alpha_\ell g^*(\tau)\, e^{j\omega(\tau+\ell T)}\, d\tau = \left(\sum_{\ell\in\mathbb{Z}} \alpha_\ell e^{j\omega\ell T}\right)\left(\int_{-\infty}^{\infty} g^*(\tau)\, e^{j\omega\tau}\, d\tau\right)\\ &= \left(\sum_{\ell\in\mathbb{Z}} \alpha_\ell e^{j\omega\ell T}\right)\left(\int_{-\infty}^{\infty} g(\tau)\, e^{-j\omega\tau}\, d\tau\right)^* \overset{(c)}{=} \alpha(e^{-j\omega T})G^*(\omega),\end{aligned}$$

where (a) follows from (5.100); (b) from the change of variable $\tau = -t - \ell T$; and (c) from recognizing the first factor as the DTFT of the sequence α evaluated at $-\omega T$ and the second factor as the conjugate of the Fourier transform of g. We can thus express $G(\omega)$ as

$$G(\omega) = \frac{\widetilde{G}^*(\omega)}{\alpha^*(e^{-j\omega T})} \overset{(a)}{=} \widetilde{G}^*(\omega)A^*(e^{j\omega T}) = \widetilde{G}^*(\omega)\,\frac{1}{6}\left(e^{-j\omega T}+4+e^{j\omega T}\right),$$

where (a) follows from (5.103). Then, from Table 4.1, we get the time-domain expression as

$$g(t) = \frac{1}{6}\left(\widetilde{g}^*(-t+T) + 4\widetilde{g}^*(-t) + \widetilde{g}^*(-t-T)\right).$$

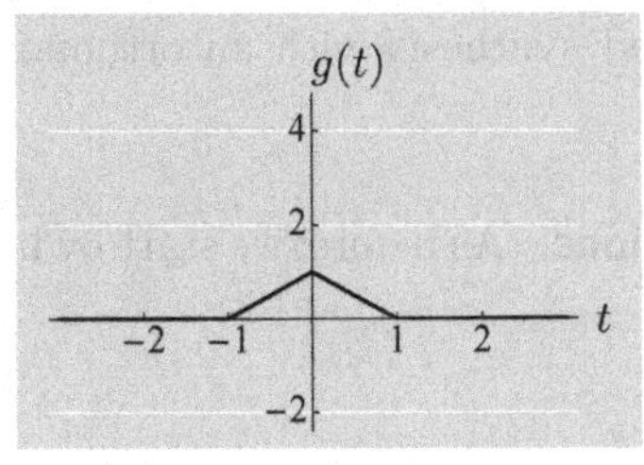

(a) Interpolation postfilter g.

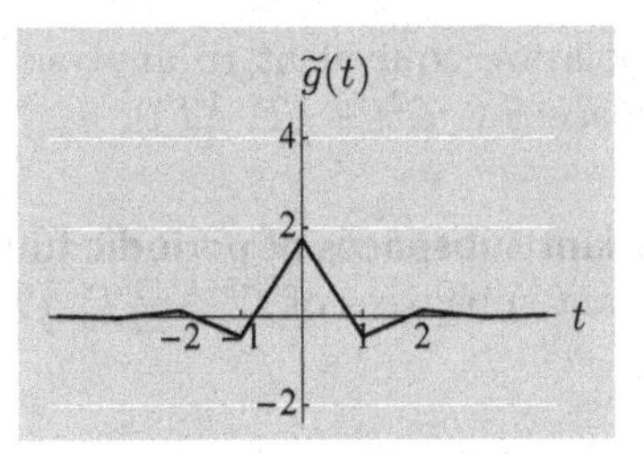

(b) Sampling prefilter $\widetilde{g}$.

Figure 5.37 Ideally matched sampling and interpolation for functions. (a) Interpolation postfilter $g(t)$ from (5.94). (b) Sampling prefilter $\widetilde{g}(t)$ from (5.100) with coefficients as in (5.105), resulting in orthogonal projection.

The interpolation postfilter g from Figure 5.37(a) produces a piecewise-linear interpolation of the sampled version of x. Having found $\widetilde{g}$ in Figure 5.37(b) such that the sampling and interpolation operators are ideally matched means that $P = \Phi\widetilde{\Phi}^*$ is an orthogonal projection operator, which further means that $\widehat{x} = Px$ is the best piecewise-linear approximation of x. This example is generalized by the elementary B-splines and their canonical duals, as developed in Section 6.3.

5.5 Periodic functions

Sections 5.2–5.4 introduced analogous theories of sampling and interpolation, first for finite-dimensional vectors, then for sequences in $\ell^2(\mathbb{Z})$, and finally for functions in $\mathcal{L}^2(\mathbb{R})$. The final vectors of interest are the T-periodic functions with a square-integrable period. We call the space of such functions $\mathcal{L}^2([-\frac{1}{2}T, \frac{1}{2}T))$, with the T-periodicity implicit.

As we have seen in Chapter 4, the (linear) convolution between a bounded, periodic input function and an absolutely integrable impulse response converges and equals the circular convolution between the input function and the periodized version of the impulse response (see (4.35)). We will express filtering using circular convolution exclusively, but parallel expressions could be written with linear convolution. Fewer results are developed here than in the previous sections since most key variations (sequences vs. functions, orthonormal, bandlimited) have been explored.

5.5.1 Sampling and interpolation with orthonormal periodic functions

Paralleling the previous developments, we start with the case when sampling and interpolation operators are defined using functions that are orthonormal with their shifts. We also again impose a relationship between filters through time reversal and conjugation to create an adjoint pair of operators. Together these make sampling and interpolation equivalent to analysis and synthesis with an orthonormal basis for a subspace $S \subset \mathcal{L}^2([-\frac{1}{2}T, \frac{1}{2}T))$.

Shift-invariant subspaces of periodic functions As before, we start by introducing shift-invariant subspaces of $\mathcal{L}^2([-\frac{1}{2}T, \frac{1}{2}T))$.

DEFINITION 5.20 (SHIFT-INVARIANT SUBSPACE OF $\mathcal{L}^2([-\frac{1}{2}T, \frac{1}{2}T))$) A subspace $S \subset \mathcal{L}^2([-\frac{1}{2}T, \frac{1}{2}T))$ is a *shift-invariant subspace* with respect to a shift $\tau \in (0, T]$ when $x(t) \in S$ implies that $x(t-k\tau) \in S$ for every integer k. In addition, $s \in \mathcal{L}^2([-\frac{1}{2}T, \frac{1}{2}T))$ is called a *generator* of S when $S = \overline{\text{span}}(\{s(t-k\tau)\}_{k\in\mathbb{Z}})$.

We will largely restrict our attention to shifts such that T/τ is an integer. Suppose that $T/\tau = N \in \mathbb{Z}^+$. Then, for any T-periodic function s and any integer m, we have

$$s(t-(k+Nm)\tau) \overset{(a)}{=} s(t-k\tau-mT) \overset{(b)}{=} s(t-k\tau),$$

where (a) follows from $T = N\tau$; and (b) from the T-periodicity of s. Therefore

$$\overline{\text{span}}(\{s(t-k\tau)\}_{k\in\mathbb{Z}}) = \overline{\text{span}}(\{s(t-k\tau)\}_{k=0}^{N-1}),$$

so the shift-invariant subspace generated by s has finite dimension N. By restricting attention to shifts τ such that T/τ is an integer, we are effectively only omitting the cases where T/τ is irrational; see Exercise 5.14.

Sampling We refer to the operation depicted in Figure 5.38(a), involving circular convolution with T-periodic function $g^*(-t)$ and recording the result at $t = nT_s$, $n = 0, 1, \ldots, N-1$, as *sampling of the T-periodic function $x(t)$ with rate $1/T_s$ and prefilter $g^*(-t)$*, and we denote it by $y = \Phi^* x$. We will restrict our attention to cases where $N = T/T_s$ is an integer. Since the result of the circular convolution is T-periodic, sampling results in an N-periodic sequence. Thus, through this sampling operation we move from the larger space $\mathcal{L}^2([-\frac{1}{2}T, \frac{1}{2}T))$ into the smaller one $\mathbb{C}^N$.

The output of sampling is

$$\begin{aligned} y_k &\overset{(a)}{=} (\Phi^* x)_k \overset{(b)}{=} \left(g^*(-t) \circledast_t x(t)\right)\Big|_{t=kT_s} = \left(\int_{-T/2}^{T/2} x(\tau)\, g^*(\tau - t)\, d\tau\right)\Bigg|_{t=kT_s} \\ &= \int_{-T/2}^{T/2} x(\tau)\, g^*(\tau - kT_s)\, d\tau = \langle x(\tau), g(\tau - kT_s)\rangle_\tau \overset{(c)}{=} \langle x, \varphi_k\rangle, \end{aligned} \tag{5.106}$$

where (a) follows from denoting the operator in Figure 5.38(a) by Φ^*; (b) from composing circular convolution by $g^*(-t)$ with recording the result at time kT_s; and (c) from defining the T-periodic function φ_k to be g shifted by kT_s,

$$\varphi_k(t) = g(t - kT_s), \qquad t \in \mathbb{R}, \tag{5.107}$$

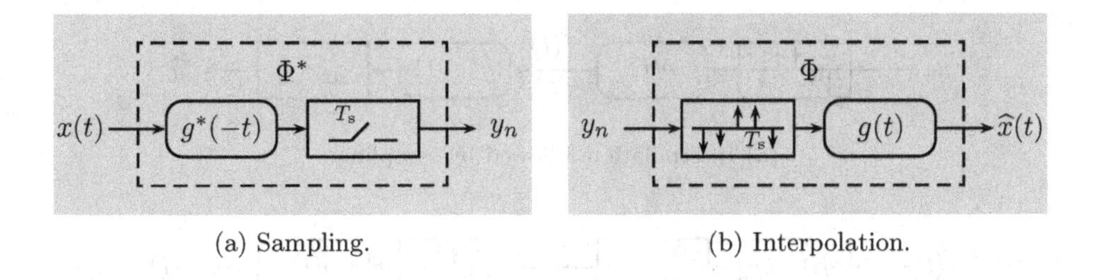

(a) Sampling. (b) Interpolation.

Figure 5.38 Sampling and interpolation in $\mathcal{L}^2([-\frac{1}{2}T, \frac{1}{2}T))$ with orthonormal periodic functions. This figure introduces the rounded box symbol for circular convolution to distinguish it from ordinary (linear) convolution.

and we are using the $\mathcal{L}^2([-\frac{1}{2}T, \frac{1}{2}T))$ inner product. The final expression in (5.106) shows that calling the sampling operator Φ^* is consistent with the previous use of Φ^* as the analysis operator associated with $\{\varphi_k\}_{k\in\mathbb{Z}}$; see (2.91). We develop only the case where $\{\varphi_k\}_{k=0}^{N-1}$ is an orthonormal set,

$$\langle \varphi_n, \varphi_k \rangle = \delta_{n-k} \quad \Leftrightarrow \quad \langle g(t-nT_s), g(t-kT_s)\rangle_t = \delta_{n-k}. \tag{5.108}$$

As before, the sampling operator Φ^* gives the inner products with all the functions in $\{\varphi_k\}_{k=0}^{N-1}$. It has a nontrivial null space, $S^\perp = \mathcal{N}(\Phi^*)$; the set $\{\varphi_k\}_{k=0}^{N-1}$ spans its orthogonal complement, $S = \mathcal{N}(\Phi^*)^\perp = \overline{\operatorname{span}}(\{\varphi_k\}_{k=0}^{N-1})$. The subspaces S and $S^\perp$ are closed and shift-invariant with respect to shift T_s. When a function $x \in \mathcal{L}^2([-\frac{1}{2}T, \frac{1}{2}T))$ is sampled, its component in the null space $S^\perp$ has no effect on the output y and is thus completely lost; its component in S is captured by $\Phi^* x$.

Interpolation Let $T \in \mathbb{R}^+$ and $T_s = T/N$ for some $N \in \mathbb{Z}^+$, and let y be an N-periodic sequence. Then, we refer to the operation depicted in Figure 5.38(b), involving weighting a Dirac comb s_{T_s}, (4.121), and circular convolution with a T-periodic function g, as *interpolation of the N-periodic sequence y with spacing T_s and postfilter g*, and we denote it by $\widehat{x} = \Phi y$. Since y is N-periodic, we can associate it with a vector in $\mathbb{C}^N$ by extracting one period, so this interpolation operation maps from $\mathbb{C}^N$ to $\mathcal{L}^2([-\frac{1}{2}T, \frac{1}{2}T))$.

The output of interpolation is

$$\begin{aligned}
\widehat{x}(t) &\overset{(a)}{=} (\Phi y)(t) \overset{(b)}{=} g(t) \circledast \sum_{k\in\mathbb{Z}} y_k \delta(t-kT_s) \\
&\overset{(c)}{=} g(t) \circledast \sum_{k=0}^{N-1} y_k \sum_{m\in\mathbb{Z}} \delta(t-kT_s-mT) \\
&\overset{(d)}{=} \sum_{k=0}^{N-1} y_k \left(g(t) \circledast \sum_{m\in\mathbb{Z}} \delta(t-kT_s-mT) \right) \\
&\overset{(e)}{=} \sum_{k=0}^{N-1} y_k g(t-kT_s) \overset{(f)}{=} \left(\sum_{k=0}^{N-1} y_k \varphi_k \right)(t),
\end{aligned} \tag{5.109}$$

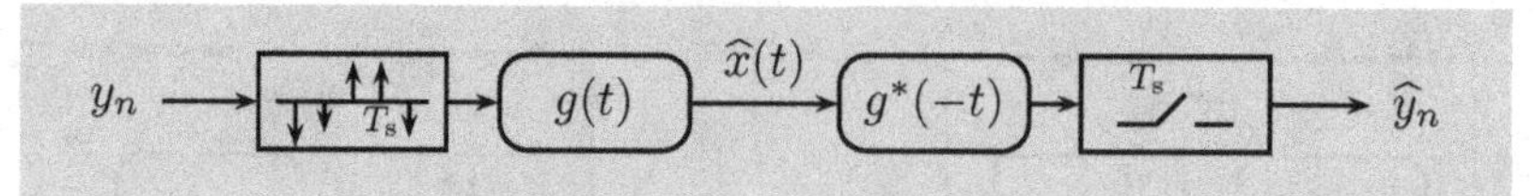

(a) Interpolation followed by sampling.

(b) Sampling followed by interpolation.

Figure 5.39 Sampling and interpolation in $\mathcal{L}^2([-\frac{1}{2}T, \frac{1}{2}T))$ with orthonormal periodic functions.

where (a) follows from denoting the operator in Figure 5.38(b) by Φ; (b) from composing the generation of a weighted Dirac comb with circular convolution by g; (c) from the N-periodicity of y and $NT_s = T$; (d) from the linearity of circular convolution; (e) from the shifting property of the Dirac delta function, (4.32e), extended to periodic convolution; and (f) from (5.107). This shows that calling the interpolation operator Φ is consistent with the previous use of Φ as the synthesis operator associated with $\{\varphi_k\}_{k\in\mathbb{Z}}$; see (2.90).

Denoting the range of Φ by S as before, this subspace is the same as the orthogonal complement of the null space of the sampling operator, as we have seen earlier. Because of how we chose a prefilter and a postfilter that are related through time-reversed conjugation, the sampling and interpolation operators are adjoints of each other; the proof mimics (5.62) and is left for Exercise 5.15.

Interpolation followed by sampling Interpolation followed by sampling is described by $\Phi^*\Phi$ as in Figure 5.39(a). By a computation analogous to (5.63), $\Phi^*\Phi = I$ on N-periodic input sequences, provided that $T_s = T/N$. The set $\{\varphi_k\}_{k=0}^{N-1}$ is an orthonormal basis for its span, and perfect recovery is again an example of the property (2.101) for analysis and synthesis operators associated with orthonormal bases.

Sampling followed by interpolation Sampling followed by interpolation is described by $P = \Phi\Phi^*$ as in Figure 5.39(b). We know that this does not perfectly recover x in general, since sampling loses the component of x in the null space of the sampling operator, $S^\perp$. As for finite-dimensional vectors, sequences, and aperiodic functions in previous sections, our choice of sampling and interpolation operators to satisfy $\Phi^*\Phi = I$ implies that P is an orthogonal projection operator; see (5.8). Thus, we have another analogue to Theorems 2.41, 5.1, 5.4, and 5.11.

THEOREM 5.21 (RECOVERY FOR PERIODIC FUNCTIONS, ORTHOGONAL) Let $T \in \mathbb{R}^+$, $N \in \mathbb{Z}^+$, and $T_s = T/N$, let the sampling operation $\Phi^* : \mathcal{L}^2([-\frac{1}{2}T, \frac{1}{2}T)) \to \mathbb{C}^N$ be given by

$$(\Phi^* x)_k = \int_{-T/2}^{T/2} x(\tau)\, g^*(\tau - kT_s)\, d\tau, \qquad k \in \{0, 1, \ldots, N-1\}, \tag{5.110a}$$

and let the interpolation operation $\Phi : \mathbb{C}^N \to \mathcal{L}^2([-\frac{1}{2}T, \frac{1}{2}T))$ be given by

$$(\Phi y)(t) = \sum_{k=0}^{N-1} y_k g(t - kT_s), \qquad t \in \mathbb{R}. \tag{5.110b}$$

As depicted in Figure 5.39(b), denote the result of sampling followed by interpolation applied to $x \in \mathcal{L}^2([-\frac{1}{2}T, \frac{1}{2}T))$ by $\widehat{x} = Px = \Phi\Phi^* x$. If the T-periodic filter g satisfies orthogonality with shifts by multiples of T_s, (5.108), then P is the orthogonal projection operator with range $S = \mathcal{R}(\Phi)$ and $\widehat{x}$ is the best approximation of x in S,

$$\widehat{x} = \arg\min_{x_S \in S} \|x - x_S\|, \qquad x - \widehat{x} \perp S;$$

in particular, $\widehat{x} = x$ when $x \in S$.

5.5.2 Sampling and interpolation for bandlimited periodic functions

We now consider sampling and interpolation of bandlimited periodic functions, both as an instance of the theory developed in Section 5.5.1 and more directly using Fourier series tools developed in Section 4.5.

Subspaces of bandlimited periodic functions As we have seen for sequences and aperiodic functions, shift-invariant subspaces of particular importance in signal processing are the subspaces of bandlimited periodic functions. To define such subspaces, we first need to define the bandwidth of a periodic function.

DEFINITION 5.22 (BANDWIDTH OF PERIODIC FUNCTION) A periodic function x is called *bandlimited* when there exists $k_0 \in \mathbb{N}$ such that its Fourier series coefficient sequence X satisfies

$$X_k = 0 \qquad \text{for all } k \text{ with } |k| > \tfrac{1}{2}(k_0 - 1). \tag{5.111}$$

The smallest such k_0 is called the *bandwidth* of x. A periodic function that is not bandlimited is called a *full-band* periodic function.

Except for $x = \mathbf{0}$ having bandwidth $k_0 = 0$, the bandwidth is always odd. Like the previous definitions of bandwidth, this definition is inspired by the even symmetry of the magnitude of the Fourier series coefficient sequence of a real periodic function (see Table 4.3).

If periodic functions x and y both have bandwidth k_0, then by linearity of the Fourier series we are assured that $x + y$ has bandwidth of at most k_0. Thus, bandlimited periodic functions form subspaces. These subspaces are closed.

DEFINITION 5.23 (SUBSPACE OF BANDLIMITED PERIODIC FUNCTIONS) The set of functions in $\mathcal{L}^2([-\frac{1}{2}T, \frac{1}{2}T))$ with bandwidth of at most k_0 is a closed subspace denoted $\mathrm{BL}\{-\frac{1}{2}(k_0 - 1), \ldots, \frac{1}{2}(k_0 - 1)\}$.

A subspace of bandlimited periodic functions is shift-invariant for any shift $\tau \in [0, T)$. To see the shift invariance, take $x \in \mathrm{BL}\{-\frac{1}{2}(k_0 - 1), \ldots, \frac{1}{2}(k_0 - 1)\}$. Then, from (4.107),

$$x(t - n\tau) \quad \stackrel{\mathrm{FS}}{\longleftrightarrow} \quad e^{-j(2\pi/T)kn\tau} X_k. \tag{5.112}$$

The Fourier series coefficient sequence is multiplied by a complex exponential, not changing where its magnitude is nonzero and hence not changing the bandwidth of the shifted function.

Dirichlet kernel The most important feature of the sinc function is that its Fourier transform is a centered box function; this leads to its use in Section 5.4.2. The Dirichlet kernel[98] gives a counterpart for periodic functions.

THEOREM 5.24 (DIRICHLET KERNEL) Let $K \in \mathbb{N}$ and $T \in \mathbb{R}^+$. The *Dirichlet kernel of order K and period T*, defined as

$$d(t) = \sum_{k=-K}^{K} e^{j(2\pi/T)kt}, \qquad t \in \mathbb{R}, \tag{5.113}$$

satisfies the following properties:

[98] *Kernel* has many meanings in mathematics. Here, it is used in the sense of an *integral kernel* that defines an integral transform as in (5.116b).

(i) It can be expressed as

$$\begin{aligned} d(t) &= 1 + 2\sum_{\ell=1}^{K} \cos\left(\frac{2\pi\ell}{T}t\right), \qquad \text{for } t \in \mathbb{R}, && (5.114a)\\ &= \begin{cases} \dfrac{\sin((2K+1)\pi t/T)}{\sin(\pi t/T)}, & \text{for } t/T \notin \mathbb{Z};\\ 2K+1, & \text{for } t/T \in \mathbb{Z}, \end{cases} && (5.114b)\\ &= \begin{cases} (2K+1)\dfrac{\operatorname{sinc}((2K+1)\pi t/T)}{\operatorname{sinc}(\pi t/T)}, & \text{for } t/T \notin \mathbb{Z};\\ 2K+1, & \text{for } t/T \in \mathbb{Z}. \end{cases} && (5.114c) \end{aligned}$$

(ii) Its Fourier series coefficient sequence is

$$D_k = 1_{\{-K,\dots,K\}} = \begin{cases} 1, & \text{for } k = -K, -K+1, \dots, K;\\ 0, & \text{otherwise.} \end{cases} \tag{5.115}$$

(iii) Let y be the T-periodic function obtained by truncating the Fourier series of the T-periodic function x through

$$Y_k = 1_{\{-K,\dots,K\}} X_k = \begin{cases} X_k, & \text{for } k = -K, -K+1, \dots, K;\\ 0, & \text{otherwise.} \end{cases} \tag{5.116a}$$

Then

$$y = \frac{1}{T} d \circledast x. \tag{5.116b}$$

(iv) Let $\{\tau_\ell\}_{\ell=0}^{2K} \subset [0, T)$ be a set of $2K+1$ distinct numbers. Then, $\{d(t-\tau_\ell)\}_{\ell=0}^{2K}$ is a basis for $\mathrm{BL}\{-K, \dots, K\} \subset \mathcal{L}^2([-\frac{1}{2}T, \frac{1}{2}T))$.

(v) An orthonormal basis for $\mathrm{BL}\{-K, \dots, K\} \subset \mathcal{L}^2([-\frac{1}{2}T, \frac{1}{2}T))$ is

$$\left\{ \frac{1}{\sqrt{T(2K+1)}} d\left(t - \frac{T}{2K+1}\ell\right) \right\}_{\ell=0}^{2K}. \tag{5.117}$$

The theorem is proven in Solved exercise 5.5.

Projection to bandlimited subspaces Since a subspace of periodic bandlimited functions is also a shift-invariant subspace, the techniques developed in Section 5.5.1 suggest a way to recover a bandlimited periodic function from samples or compute the best approximation of a full-band periodic function in a bandlimited subspace. To use Theorem 5.21, we need a T-periodic function g that satisfies (5.108) and, with its shifts by T_s with $T/T_s = N \in \mathbb{Z}^+$, generates a subspace of bandlimited functions. An appropriately scaled Dirichlet kernel does the trick.

Fix the period $T \in \mathbb{R}^+$ and odd bandwidth $k_0 \in \mathbb{N}$. To obtain an orthonormal basis for $\mathrm{BL}\{-\frac{1}{2}(k_0-1), \dots, \frac{1}{2}(k_0-1)\}$ with (5.117), we must use the Dirichlet

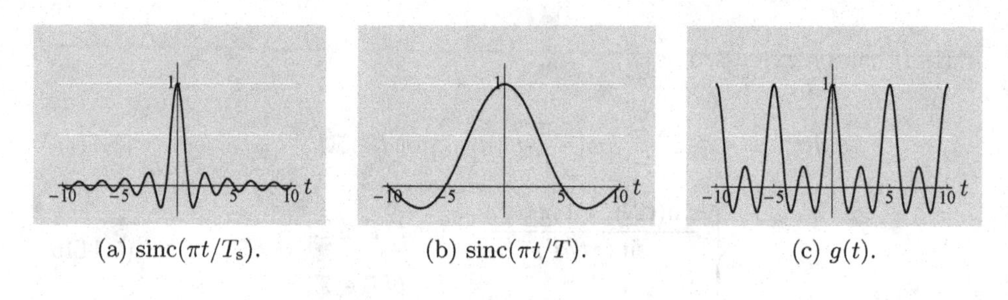

(a) $\operatorname{sinc}(\pi t/T_s)$. (b) $\operatorname{sinc}(\pi t/T)$. (c) $g(t)$.

Figure 5.40 The interpolation postfilter g from (5.118a) is the scaled ratio of sinc functions. The numerator, denominator, and resulting g are shown for $T = 5$ and $T_s = 1$.

kernel of order $\frac{1}{2}(k_0 - 1)$ and period T. Thus, let the interpolation postfilter be

$$g(t) \overset{(a)}{=} \frac{1}{\sqrt{Tk_0}} d(t) \overset{(b)}{=} \frac{1}{\sqrt{Tk_0}} \left(k_0 \frac{\operatorname{sinc}(k_0 \pi t/T)}{\operatorname{sinc}(\pi t/T)} \right) \overset{(c)}{=} \frac{1}{\sqrt{T_s}} \frac{\operatorname{sinc}(\pi t/T_s)}{\operatorname{sinc}(\pi t/T)}, \tag{5.118a}$$

where (a) follows from substitution in (5.117); (b) from (5.114c); and (c) from simplifying using sampling period $T_s = T/k_0$. An example of a function of this form is shown in Figure 5.40. Applying the same scaling to (5.115) gives

$$G_k = \begin{cases} \sqrt{T_s}/T, & \text{for } k = -\frac{1}{2}(T/T_s - 1), \ldots, \frac{1}{2}(T/T_s - 1); \\ 0, & \text{otherwise.} \end{cases} \tag{5.118b}$$

Since Theorem 5.24(v) verifies that g satisfies (5.108) for $T_s = T/k_0$, we get the following corollary to Theorem 5.21.

THEOREM 5.25 (PROJECTION TO BANDLIMITED SUBSPACE) Let $T, T_s \in \mathbb{R}^+$, and let the system in Figure 5.39(b) have T-periodic input x, filter g from (5.118a), and $k_0 = T/T_s = 2K + 1$ an odd positive integer. Then,

$$\widehat{x}(t) = \frac{1}{\sqrt{T_s}} \sum_{k=-K}^{K} y_k \frac{\operatorname{sinc}((\pi/T_s)(t - kT_s))}{\operatorname{sinc}((\pi/T)(t - kT_s))}, \qquad t \in \mathbb{R}, \tag{5.119a}$$

with

$$y_k = \frac{1}{\sqrt{T_s}} \int_{-T/2}^{T/2} x(\tau) \frac{\operatorname{sinc}\left(\frac{\pi}{T_s}(\tau - kT_s)\right)}{\operatorname{sinc}\left(\frac{\pi}{T}(\tau - kT_s)\right)} \, d\tau, \qquad k \in \{-K, \ldots, K\}, \tag{5.119b}$$

is the best approximation of x in $\mathrm{BL}\{-K, \ldots, K\}$,

$$\widehat{x} = \underset{x_{\mathrm{BL}} \in \mathrm{BL}\{-K, \ldots, K\}}{\arg\min} \|x - x_{\mathrm{BL}}\|, \tag{5.119c}$$

$$x - \widehat{x} \perp \mathrm{BL}\{-K, \ldots, K\}. \tag{5.119d}$$

In particular, $\widehat{x} = x$ when $x \in \mathrm{BL}\{-K, \ldots, K\}$.

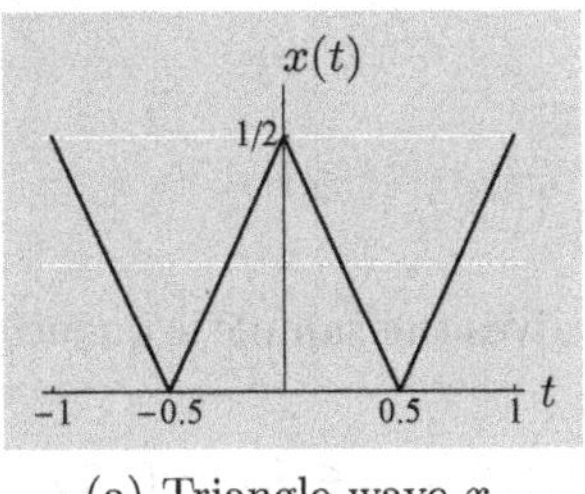

(a) Triangle wave x.

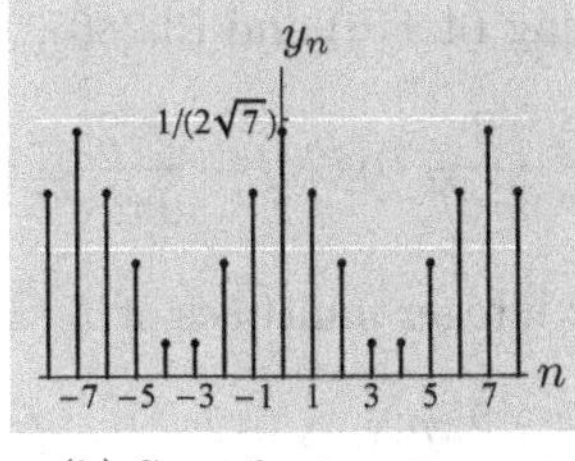

(b) Sample sequence y.

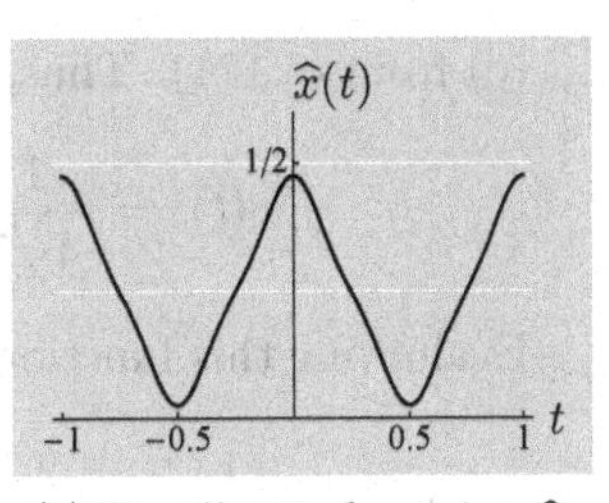

(c) Bandlimited version $\widehat{x}$.

Figure 5.41 Sampling the 1-periodic triangle wave with $T_s = 1/7$ results in a 7-periodic sample sequence. With Dirichlet kernel sampling prefilter and interpolation postfilter, the reconstructed function is the best approximation in BL$\{-3, \ldots, 3\}$.

The effect of orthogonal projection to BL$\{-K, \ldots, K\}$ is a simple truncation of the spectrum of x to $\{-K, \ldots, K\}$,

$$\widehat{X}_k = \begin{cases} X_k, & \text{for } k \in \{-K, \ldots, K\}; \\ 0, & \text{otherwise.} \end{cases} \tag{5.120}$$

Recall that this was obtained using a different basis for BL$\{-\frac{1}{2}(k_0 - 1), \ldots, \frac{1}{2}(k_0 - 1)\}$ in Theorem 4.15(iii).

Example 5.27 (Sampling the triangle wave) Consider 1-periodic x shown in Figure 5.41(a). It has one period given in (4.119) and Fourier series coefficients given in (4.120); we repeat both here for easy reference:

$$x(t) = \tfrac{1}{2} - |t|, \quad \text{for } |t| \le \tfrac{1}{2} \quad \overset{\text{FS}}{\longleftrightarrow} \quad X_k = \begin{cases} 1/4, & \text{for } k = 0; \\ 0, & \text{for } k \neq 0,\ k \text{ even}; \\ 1/(\pi k)^2, & \text{for } k \text{ odd}. \end{cases} \tag{5.121}$$

This function is not bandlimited because $X_k \neq 0$ for arbitrarily large odd k. By Theorem 5.25, the best approximation of x in BL$\{-3, \ldots, 3\}$ can be computed by a sampling and interpolation system that measures seven samples per period and uses a Dirichlet kernel as sampling prefilter and interpolation postfilter.

It is difficult to evaluate the circular convolution between $x(t)$ and $g^*(-t)$ directly in the time domain using the expression (5.118a) for g. Instead, we determine the sample sequence y and reconstruction $\widehat{x}$ using Fourier series. Let $w(t) = g^*(-t) \circledast_t x(t)$. Then the Fourier series coefficients of w are

$$W_k \overset{(a)}{=} G_k X_k \overset{(b)}{=} \frac{1}{\sqrt{7}} 1_{\{-3,\ldots,3\}} X_k \overset{(c)}{=} \begin{cases} 1/(4\sqrt{7}), & \text{for } k = 0; \\ 1/(\pi^2\sqrt{7}), & \text{for } k = \pm 1; \\ 1/(9\pi^2\sqrt{7}), & \text{for } k = \pm 3; \\ 0, & \text{otherwise,} \end{cases}$$

where (a) follows from the circular convolution in time property of the Fourier series, (4.113), along with the symmetry of g and $T = 1$; (b) from (5.118b); and

(c) from (5.121). Thus, using (4.94b) and (3.286),

$$w(t) = \frac{1}{4\sqrt{7}} + \frac{2}{\pi^2\sqrt{7}}\cos 2\pi t + \frac{2}{9\pi^2\sqrt{7}}\cos 6\pi t, \qquad t \in \mathbb{R}.$$

Evaluating this function at integer multiples of $T_s = \frac{1}{7}$ gives the sample sequence

$$y_n = \frac{1}{4\sqrt{7}} + \frac{2}{\pi^2\sqrt{7}}\cos\left(\frac{2\pi n}{7}\right) + \frac{2}{9\pi^2\sqrt{7}}\cos\left(\frac{6\pi n}{7}\right), \qquad n \in \mathbb{Z}, \tag{5.122}$$

which is shown in Figure 5.41(b).

We also use the Fourier series to evaluate $\widehat{x}$ in (5.119a),

$$\begin{aligned}
\widehat{X}_k &\overset{(a)}{=} \sum_{n=0}^{6} y_n e^{-j2\pi kn/7} G_k \overset{(b)}{=} \frac{1}{\sqrt{7}} \sum_{n=0}^{6} y_n e^{-j2\pi kn/7} \\
&\overset{(c)}{=} \frac{1}{\sqrt{7}} \sum_{n=0}^{6} \left(\frac{1}{4\sqrt{7}} + \frac{1}{\pi^2\sqrt{7}} \left(e^{-j2\pi n/7} + e^{j2\pi n/7}\right) \right. \\
&\qquad \left. + \frac{1}{9\pi^2\sqrt{7}} \left(e^{-j6\pi n/7} + e^{j2\pi n/7}\right) \right) e^{-j2\pi kn/7} \\
&\overset{(d)}{=} \begin{cases} 1/4, & \text{for } k = 0; \\ 1/(\pi^2), & \text{for } k = \pm 1; \\ 1/(9\pi^2), & \text{for } k = \pm 3; \\ 0, & \text{otherwise,} \end{cases}
\end{aligned} \tag{5.123}$$

where (a) follows from the shift in time property of the Fourier series, (4.107); (b) from (5.118b); (c) from (5.122) and (3.286); and (d) from evaluating the sum separately for each k. Thus, the reconstructed function is

$$\widehat{x}(t) = \frac{1}{4} + \frac{2}{\pi^2}\cos 2\pi t + \frac{2}{9\pi^2}\cos 6\pi t, \qquad t \in \mathbb{R},$$

which is shown in Figure 5.41(c).

Note that we could arrive directly at (5.123) by truncation of the Fourier series in (5.121), so this serves to show that Theorems 4.15(iii) and 5.25 are consistent. Also, although it is somewhat difficult to see in Figure 5.41(b), the sample sequence y is not within a scalar multiple of evaluating x at integer multiples of $\frac{1}{7}$; the sampling prefilter is needed to get the best bandlimited approximation from a linear interpolation of samples.

Sampling without a prefilter followed by interpolation When the input to the system in Figure 5.39(b) is in $\mathrm{BL}\{-\frac{1}{2}(k_0 - 1), \ldots, \frac{1}{2}(k_0 - 1)\}$ with $k_0 = T/T_s$ and the filter g is given in (5.118a), the sampling prefilter simply scales the input by $\sqrt{T_s}$; this follows from (5.118b) and the factor of T in the circular convolution in time property of the Fourier series, (4.113). For these inputs, the system is equivalent to the one in Figure 5.42. Analogously to Figure 5.16 for sequences and Figure 5.23 for aperiodic functions, this simpler system is worth further study – for

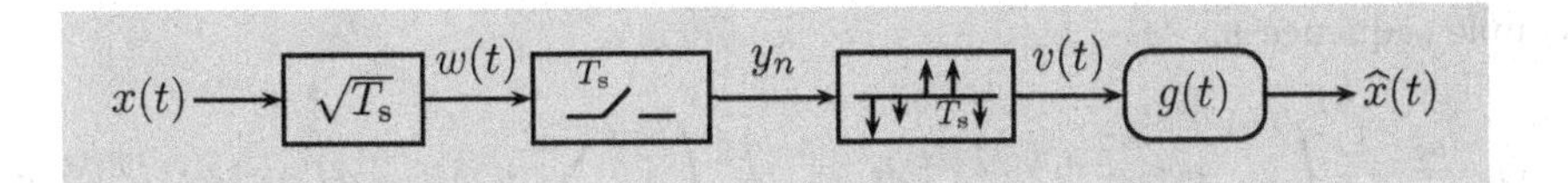

Figure 5.42 Sampling and interpolation in $\mathcal{L}^2([-\frac{1}{2}T, \frac{1}{2}T))$ with no sampling prefilter. (The scalar multiplication by $\sqrt{T_s}$ could be incorporated into the interpolation postfilter.)

both bandlimited and full-band inputs – because it is essentially a system without a sampling prefilter, and often sampling is done without a prefilter. Before including any filtering (the scaling by $\sqrt{T_s}$ and the interpolation postfilter g), we relate v to w in the Fourier domain; we also relate v to y and thus obtain a relationship between y and w as a byproduct.

We assume that $N = T/T_s$ is an integer. To relate v to w in the Fourier domain, first note that the weights of the Dirac delta components of v are the values of w recorded at times that are integer multiples of T_s; that is,

$$v(t) = \sum_{n\in\mathbb{Z}} w(nT_s)\delta(t - nT_s).$$

By the sampling property of the Dirac delta function, (3.294), we have

$$v(t) = s_{T_s}(t)\, w(t), \qquad t \in \mathbb{R},$$

where we have used the Dirac comb with spacing T_s

$$s_{T_s}(t) = \sum_{n\in\mathbb{Z}} \delta(t - nT_s) \quad \overset{\text{FS}}{\longleftrightarrow} \quad S_{T_s,k} = \frac{1}{T_s}\sum_{n\in\mathbb{Z}} \delta_{k-Nn}. \tag{5.124}$$

This Fourier series pair treats s_{T_s} as T-periodic, and its proof is left for Exercise 5.16.

Now the Fourier series of v is given by

$$V_k \overset{(a)}{=} (S_{T_s} * W)_k \overset{(b)}{=} \left(\frac{1}{T_s}\sum_{n\in\mathbb{Z}} \delta_{k-Nn}\right) *_k W_k \overset{(c)}{=} \frac{1}{T_s}\sum_{n\in\mathbb{Z}} W_{k-Nn}, \tag{5.125}$$

where (a) follows from expressing multiplication in time as convolution in frequency, (4.114); (b) from using (5.124) for S_{T_s}; and (c) from the shifting property of the Kronecker delta sequence, (3.62e). We will use this key expression to understand the overall operation of the system in Figure 5.42.

The Fourier-domain relations of v and w to y are also useful. The Fourier series of the weighted Dirac comb v has a simple relationship with the DFT of the sample sequence y,

$$\begin{aligned} V_k &\stackrel{(a)}{=} \frac{1}{T}\int_{-\epsilon}^{T-\epsilon} v(t)\, e^{-j(2\pi/T)kt}\, dt \stackrel{(b)}{=} \frac{1}{T}\int_{-\epsilon}^{T-\epsilon} \sum_{n\in\mathbb{Z}} y_n \delta(t-nT_{\rm s})\, e^{-j(2\pi/T)kt}\, dt \\ &\stackrel{(c)}{=} \frac{1}{T}\sum_{n=0}^{N-1} y_n \int_{-\epsilon}^{T-\epsilon} \delta(t-nT_{\rm s})\, e^{-j(2\pi/T)kt}\, dt \stackrel{(d)}{=} \frac{1}{T}\sum_{n=0}^{N-1} y_n e^{-j(2\pi/T)knT_{\rm s}} \\ &\stackrel{(e)}{=} \frac{1}{T} Y_k, \end{aligned} \tag{5.126}$$

where (a) follows from the definition of the Fourier series; (b) from the definition of v; (c) from the interchange of integration and summation, with only Dirac delta functions centered within the interval of integration retained; (d) from the sifting property of the Dirac delta function, (3.293); and (e) from the definition of the length-N DFT and $T/T_{\rm s} = N$.

By combining (5.125) and (5.126), we can also relate the DFT of the sample sequence y to the Fourier series of the scaled input function w,

$$Y_k = \frac{T}{T_{\rm s}} \sum_{n\in\mathbb{Z}} W_{k-Nn}. \tag{5.127}$$

This is illustrated in Figure 5.43. Table 5.1 summarizes these Fourier-domain relationships along with those derived for sampling aperiodic functions in Section 5.4.2.

We now consider the full system in Figure 5.42, including the scaling prior to sampling and the interpolation postfilter, for a T-periodic input x that is not necessarily bandlimited. Using (5.125), the spectrum of v is the combination of X_k and spectral replicas spaced by N; combining the $1/T_{\rm s}$ factor in (5.125) with the $\sqrt{T_{\rm s}}$ factor relating X and W gives

$$V_k = \frac{1}{\sqrt{T_{\rm s}}} \sum_{n\in\mathbb{Z}} X_{k-Nn}.$$

Incorporating the filtering by g – and not forgetting the factor of T in (4.113) – gives

$$\widehat{X}_k = \frac{T}{\sqrt{T_{\rm s}}} G_k \sum_{n\in\mathbb{Z}} X_{k-Nn}.$$

The G_k factor cancels the $T/\sqrt{T_{\rm s}}$ factor and restricts $\sum_{n\in\mathbb{Z}} X_{k-Nn}$ to $\{\frac{1}{2}(T/T_{\rm s}-1), \ldots, \frac{1}{2}(T/T_{\rm s}-1)\}$. Thus, the input x is recovered if and only if its spectral support is contained in $\{\frac{1}{2}(T/T_{\rm s}-1), \ldots, \frac{1}{2}(T/T_{\rm s}-1)\}$, since this condition ensures that the restriction does not discard any of the base spectrum and no spectral replicas interfere with the base spectrum.

This discussion yields the following sampling theorem for periodic functions, which is analogous to Theorems 5.8 and 5.15.

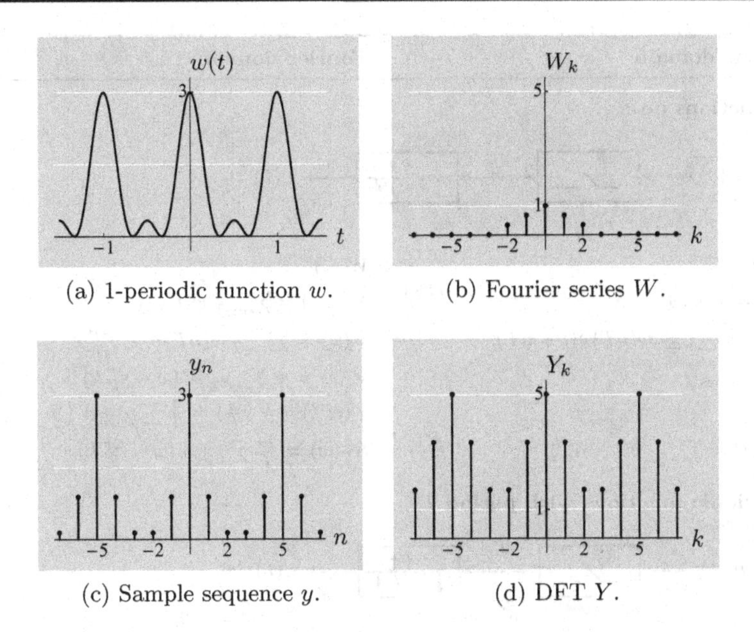

(a) 1-periodic function w. (b) Fourier series W.

(c) Sample sequence y. (d) DFT Y.

Figure 5.43 Illustration of the Fourier-domain relationships between periodic function w and periodic sequence y in Figure 5.42. (a) A bandlimited periodic function $w \in \mathrm{BL}\{-2, \ldots, 2\}$ with period $T = 1$. (b) Its Fourier series coefficients W. (c) The sampled version of w with $T_s = T/5 = \frac{1}{5}$. (d) Its length-5 DFT Y.

THEOREM 5.26 (SAMPLING THEOREM FOR PERIODIC FUNCTIONS) Let $T, T_s \in \mathbb{R}^+$ such that $T/T_s = 2K+1$ is an odd integer. If the T-periodic function x is in $\mathrm{BL}\{-K, \ldots, K\}$,

$$x(t) = \sum_{n=-K}^{K} x(nT_s) \frac{\mathrm{sinc}\left(\frac{\pi}{T_s}(t-nT_s)\right)}{\mathrm{sinc}\left(\frac{\pi}{T}(t-nT_s)\right)}, \qquad t \in \mathbb{R}. \tag{5.128}$$

The Dirichlet kernel plays exactly the interpolating role of the sinc filter in Theorem 5.15: looking only at times that are integer multiples of the sampling period, $t = kT_s$, the ratio of sincs gives δ_{n-k}; in between these times, the interpolation is smooth.

5.6 Computational aspects

This chapter includes some methods that are computational and others that are merely conceptual (or implementable only with an analog computer). Sampling and interpolation of finite-dimensional vectors is straightforward, involving only matrix multiplications. Sampling and interpolation of sequences involves filtering

Time domain		Fourier domain
Functions on $\mathbb{R}$		
$w(t)$	$\stackrel{\text{FT}}{\longleftrightarrow}$	$W(\omega)$
$y_n = w(nT)$	$\stackrel{\text{DTFT}}{\longleftrightarrow}$	$Y(e^{j\omega}) = \sum_{n\in\mathbb{Z}} w(nT)\, e^{-j\omega n}$
$v(t) = \sum_{n\in\mathbb{Z}} w(nT)\,\delta(t-nT)$	$\stackrel{\text{FT}}{\longleftrightarrow}$	$V(\omega) = \sum_{n\in\mathbb{Z}} w(nT)\, e^{-j\omega nT}$
		$V(\omega) = \frac{1}{T}\sum_{k\in\mathbb{Z}} W(\omega - \frac{2\pi}{T}k)$
		$Y(e^{j\omega}) = V(\frac{\omega}{T}) = \frac{1}{T}\sum_{k\in\mathbb{Z}} W(\frac{\omega}{T} - \frac{2\pi}{T}k)$
$s_T(t) = \sum_{n\in\mathbb{Z}} \delta(t-nT)$	$\stackrel{\text{FT}}{\longleftrightarrow}$	$S_T(\omega) = \frac{2\pi}{T}\sum_{k\in\mathbb{Z}} \delta(\omega - \frac{2\pi}{T}k)$
Periodic functions with period T		
$w(t)$	$\stackrel{\text{FS}}{\longleftrightarrow}$	W_k
$y_n = w(nT_s)$	$\stackrel{\text{DFT}}{\longleftrightarrow}$	$Y_k = \sum_{n=0}^{N-1} w(nT_s)\, e^{-j(2\pi/N)kn}$
$v(t) = \sum_{n\in\mathbb{Z}} w(nT_s)\,\delta(t-nT_s)$	$\stackrel{\text{FS}}{\longleftrightarrow}$	$V_k = \frac{1}{T}\sum_{n=0}^{N-1} w(nT_s)\, e^{-j(2\pi/N)kn}$
		$V_k = \frac{1}{T_s}\sum_{n\in\mathbb{Z}} W_{k-Nn}$
		$Y_k = TV_k = \frac{T}{T_s}\sum_{n\in\mathbb{Z}} W_{k-Nn}$
$s_{T_s}(t) = \sum_{n\in\mathbb{Z}} \delta(t-nT_s)$	$\stackrel{\text{FS}}{\longleftrightarrow}$	$S_{T_s,k} = \frac{1}{T_s}\sum_{n\in\mathbb{Z}} \delta_{k-Nn}$

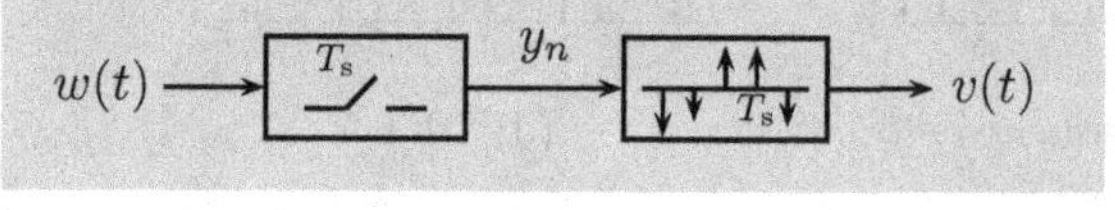

Table 5.1 Summary of the Fourier-domain sampling relationships. For the periodic case, the period T and sampling period T_s must be such that $N = T/T_s$ is an odd integer.

and multirate operations; the polyphase methods of Section 3.9.3 can be critical to achieving computationally efficient implementations. Sampling and interpolation of functions also involves filtering, but this filtering is inherently analog – not implementable with a digital computer.

For sequences, we have made no distinction between implementable and unimplementable filters. Infinite support necessitates recursive computation or truncation. Recursive computation is usually done in only one direction (forward in time) and for systems with rational transfer functions. Importantly, this excludes the ideal lowpass filters at the heart of Section 5.3.2. The approximation of these filters is discussed in Chapter 6.

This section presents a set of methods for reconstruction that apply when, in addition to samples, one is given constraints expressed as membership in one or more convex sets.

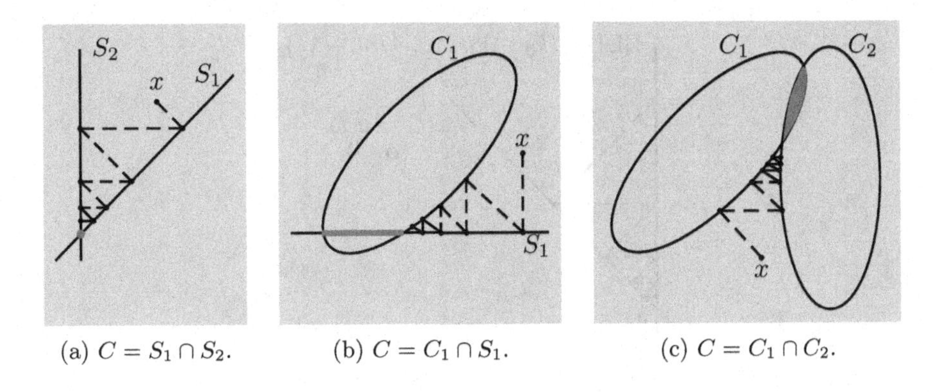

(a) $C = S_1 \cap S_2$. (b) $C = C_1 \cap S_1$. (c) $C = C_1 \cap C_2$.

Figure 5.44 Iterative solution of finding the nearest vector in a convex set using POCS. By iteratively projecting to the closest vector, a solution belonging to the intersection is found. (a) The convex sets are affine subspaces; the intersection is a single vector. (b) Intersection of a general convex set and an affine subspace; there are many vectors in the intersection. (c) Intersection of two general convex sets; there are many vectors in the intersection.

5.6.1 Projection onto convex sets

Given a vector in a Hilbert space, the closest vector to it in a closed subspace S is unique and given by the orthogonal projection onto S (see Theorem 2.26). More generally, given a vector in a Hilbert space, the closest vector to it in a closed convex set C is unique. Finding that vector is not always straightforward. A method called *projection onto convex sets (POCS)* or alternating projections applies to cases where C is the intersection of two convex sets upon which one can easily project.

Instead of trying to find the nearest vector that satisfies both set membership constraints directly, one satisfies the constraints alternately; because of convexity of the sets, the procedure is guaranteed to converge to a vector that belongs to the intersection of the convex sets, irrespective of whether C contains one vector or many (see Figure 5.44).

POCS-type algorithms are often used in problems involving bandlimited signals that have to satisfy some other convex constraint. They are used because they are simple and can be easily applied to large problems. POCS can be extended to the intersection of more than two convex sets, yielding sequential or cyclic projection algorithms. Actually, any closed convex set can be characterized by a set of half-space constraints, but infinitely many constraints might be needed.

Papoulis–Gerchberg algorithm Consider the reconstruction of a partially observed periodic function that is known to be bandlimited. Let $x \in \mathrm{BL}\{-\frac{1}{2}(k_0-1), \ldots, \frac{1}{2}(k_0-1)\}$ be a 1-periodic function, and suppose it is observed over only a part of one period. For example,

$$y(t) = \begin{cases} x(t), & \text{for } t \in [0, \alpha] \cup [\beta, 1); \\ 0, & \text{for } t \in (\alpha, \beta), \end{cases}$$

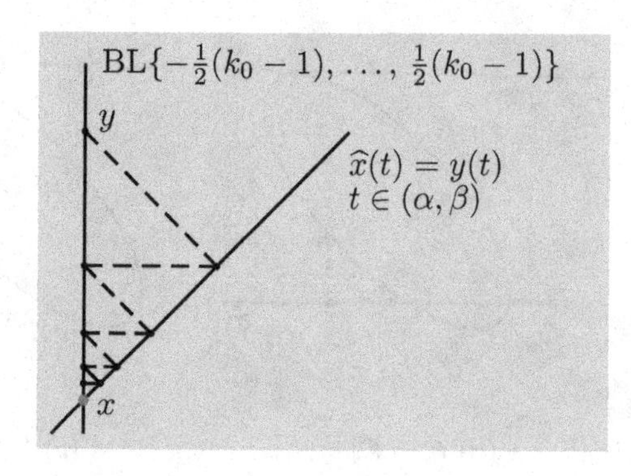

Figure 5.45 Conceptual illustration of the Papoulis–Gerchberg algorithm to reconstruct a bandlimited periodic function from observation of only one partial period.

with $0 < \alpha < \beta < 1$, is observed. Then, the information known about x is membership in two closed convex sets:

(i) $S_1 = \mathrm{BL}\{-\frac{1}{2}(k_0-1), \ldots, \frac{1}{2}(k_0-1)\}$ is a closed subspace (see Definition 5.23); and

(ii) $S_2 = \{x \in \mathcal{L}^2([0, 1)) \mid x(t) = y(t) \text{ for } t \in [0, \alpha] \cup [\beta, 1)\}$ is a closed affine subspace.

To reconstruct x, we seek the function in $C = S_1 \cap S_2$ that is closest to y.

The Papoulis–Gerchberg algorithm alternates between finding the nearest function in S_1 and the nearest function in S_2, each through a simple enforcement of equality constraints:

(i) $\widehat{x} \in S_1$ is achieved by setting $\widehat{X}_k = 0$ for $k \notin \{-\frac{1}{2}(k_0 - 1), \ldots, \frac{1}{2}(k_0 - 1)\}$, leaving $\widehat{X}_k$ unchanged otherwise, as in (5.120); and

(ii) $\widehat{x} \in S_2$ is achieved by setting $\widehat{x}(t) = y(t)$ for $t \in [0, \alpha] \cup [\beta, 1)$ and leaving $\widehat{x}(t)$ unchanged otherwise.

Figure 5.45 illustrates the procedure conceptually, and Figure 5.46 shows computed results. In the example, the Fourier series of x is supported only on $\{-10, \ldots, 10\}$, and observations on the interval $(\alpha, \beta) = (0.19, 0.39)$ are missing. Reconstructions and errors for 20 and 10 000 iterations are shown. Note the slow convergence at the boundary of the observation interval in Figure 5.46(b).

Inpainting The same algorithm can be used for images, to recover a part of an image that is missing. This is called *inpainting*.

Let x be an $N \times N$-size image, and assume that its two-dimensional DFT X is bandlimited to $k_{0,1} \times k_{0,2}$ lowpass coefficients. A region of the image is missing, so the observation y has $M < N^2$ pixels matching x, with the rest equal to zero. As long as $M \geq k_{0,1}k_{0,2}$ holds, the image x is uniquely determined (though finding it can be an ill-conditioned problem). This is because the bandlimitedness is

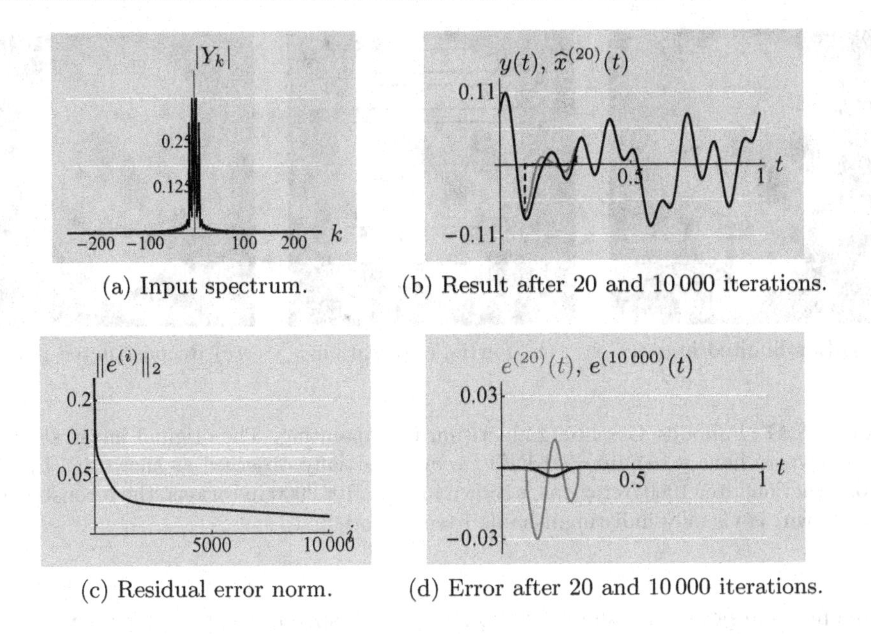

(a) Input spectrum.

(b) Result after 20 and 10 000 iterations.

(c) Residual error norm.

(d) Error after 20 and 10 000 iterations.

Figure 5.46 Papoulis–Gerchberg algorithm results for reconstructing a bandlimited 1-periodic function despite observations missing on the interval $(0.19, 0.39)$. The dashed line shows where observations are missing and the gray/black lines show the (b) reconstruction $\widehat{x}^{(i)}$ and (d) error $e^{(i)} = x - \widehat{x}^{(i)}$ after 20 and 10 000 iterations.

$N^2 - k_{0,1}k_{0,2}$ constraints and the observed pixel values are M additional linearly independent constraints, so in total we have at least N^2 constraints.

The Papoulis–Gerchberg algorithm alternates between enforcing two convex constraints as before:

(i) bandlimit $\widehat{x}$ by truncating the DFT; and
(ii) set $\widehat{x}_{n_1,n_2} = y_{n_1,n_2}$ for those (n_1, n_2) for which $y_{n_1,n_2} = x_{n_1,n_2}$; for the missing values, leave $\widehat{x}_{n_1,n_2}$ as it is.

Conceptually, the situation is again as in Figure 5.45. We show an example in Figure 5.47. The original image shown in Figure 5.47(a) has half-band DFT. It is observed with $N^2/10$ pixels missing in stripes, as shown in Figure 5.47(b). The Papoulis–Gerchberg algorithm was used on every column of the image, with the result after 1000 iterations shown in Figure 5.47(c). Ultimately, the reconstruction will be perfect. In this case, the inpainting looks like magic, since it completely recovers the missing stripes.

Analysis The Papoulis–Gerchberg algorithm will converge to a unique solution when specific conditions are satisfied. For example, for a unique solution that is independent of the starting point, the number of unknowns (the number of missing time-domain observations) should be at most the number of equations (known

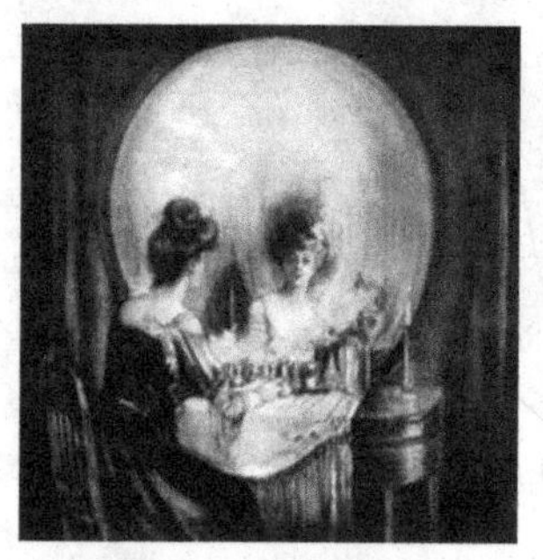

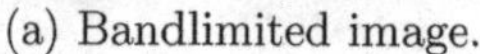

(a) Bandlimited image. (b) Partial observation. (c) Reconstructed image.

Figure 5.47 Papoulis–Gerchberg algorithm for inpainting. The original image shown in (a), known to have a bandlimited DFT, is only partially observed as shown in (b). The reconstruction after 1000 iterations is shown in (c); after 2000 iterations, the reconstruction (not shown) is visually indistinguishable from the original.

frequency components, usually set to zero). The speed of convergence depends on more than these counts.

Let $x \in \mathbb{C}^N$ be a discrete-time signal with DFT $X \in \mathbb{C}^N$. Partition x and X as

$$x = \begin{bmatrix} x_0 \\ x_1 \\ x_2 \end{bmatrix} \begin{matrix} \}\, s \\ \}\, q \\ \}\, N-s-q \end{matrix} \qquad \text{and} \qquad X = \begin{bmatrix} X_0 \\ X_1 \\ X_2 \end{bmatrix} \begin{matrix} \}\, \frac{1}{2}N-k \\ \}\, 2k+1 \\ \}\, \frac{1}{2}N-k-1 \end{matrix},$$

with dimensions as marked. With the above partitioning, X can be written as

$$\begin{bmatrix} X_0 \\ X_1 \\ X_2 \end{bmatrix} = \begin{bmatrix} F_{00} & F_{01} & F_{02} \\ F_{10} & F_{11} & F_{12} \\ F_{20} & F_{21} & F_{22} \end{bmatrix} \begin{bmatrix} x_0 \\ x_1 \\ x_2 \end{bmatrix}.$$

The discrete version of the Papoulis–Gerchberg algorithm solves the following problem: assuming that x_0 and x_2 are given (these are time-domain observations) and X_0 and X_2 are known (these are often zero), find x_1 and X_1. Note that for the algorithm to converge to a unique solution we need $q \leq N - 2k - 1$.

Denote the estimates of the missing parts at iteration n by $\widehat{x}_1^{(n)}$ and $\widehat{X}_1^{(n)}$. Then, the next iteration will produce

$$\widehat{X}^{(n+1)} = \begin{bmatrix} F_{10} & F_{11} & F_{11} \end{bmatrix} \begin{bmatrix} x_0 \\ \widehat{x}_1^{(n)} \\ x_2 \end{bmatrix},$$

$$\widehat{x}^{(n+1)} = \frac{1}{N} \begin{bmatrix} F_{01}^* & F_{11}^* & F_{21}^* \end{bmatrix} \begin{bmatrix} X_0 \\ \widehat{X}_1^{(n+1)} \\ X_2 \end{bmatrix}.$$

Define $Q = F_{11}F_{11}^*/N$ and $P = F_{11}^*F_{11}/N$; then, the updates are

$$\begin{aligned}\widehat{X}_1^{(n+1)} &= (I - Q)X_1 + Q\widehat{X}_1^{(n)},\\ \widehat{x}_1^{(n+1)} &= (I - P)x_1 + P\widehat{x}_1^{(n)},\end{aligned}$$

where we have used that $F_{10}x_0 + F_{12}x_2 = X_1 - F_{11}x_1$ and $(1/N)(F_{01}^*X_0 + F_{21}^*X_2) = x_1 - (1/N)F_{11}^*X_1$. With the initial condition $\widehat{X}_1^{(0)} = \mathbf{0}$,

$$\widehat{X}_1^{(n)} = (I - Q^n)X_1. \tag{5.129}$$

From (5.129), we see that the convergence depends on the operator norm of the matrix Q (its largest eigenvalue). For example, if one chooses $N = 32$, $s = 7$, $q = 15$ and $k = 6$, the largest eigenvalue of Q is 0.999 999 998, causing the algorithm to converge very slowly.

Chapter at a glance

We now summarize the main concepts we have seen in this chapter. They all revolve around sampling, interpolation, and their combinations, between a larger space and a smaller space. In Section 5.2, the larger space is the space of vectors $\mathbb{C}^M$, while the smaller space is $\mathbb{C}^N$, with $N < M$; in Section 5.3, the larger space is the space of sequences $\ell^2(\mathbb{Z})$ and the smaller space is its subspace; in Section 5.4, the larger space is the space of functions $\mathcal{L}^2(\mathbb{R})$ and the smaller space is the space of sequences $\ell^2(\mathbb{Z})$; and in Section 5.5, the larger space is the space of T-periodic functions $\mathcal{L}^2([-\frac{1}{2}T, \frac{1}{2}T))$ and the smaller space is the space of N-periodic sequences, which is equivalent to $\mathbb{C}^N$. By cascading interpolation followed by sampling, or sampling followed by interpolation, we are able to move from one space to the other and back. It is the match between sampling and interpolation that will determine the type of recovery possible in each case. These concepts were illustrated in a simple example in Figures 5.7–5.9.

Sampling and interpolation with orthonormal vectors/sequences/functions

• The *sampling* operator Φ^* takes an input x from a larger space and maps it into an output y in a smaller space. We assume orthonormality, that is, $\Phi^*\Phi = I$. We denote by S the orthogonal complement of the null space of Φ^*,

$$S = \mathcal{N}(\Phi^*)^{\perp}.$$

Inputs $x \in \mathcal{N}(\Phi^*)$ are mapped to $\mathbf{0}$, while inputs $x \in S$ can be recovered exactly. For inputs $x \notin S$, the component in $\mathcal{N}(\Phi^*)$ is lost due to sampling, while the rest is preserved through $\Phi^* x$.

Input space	Input	Output	Output space
$\mathbb{C}^M$	x	y	$\mathbb{C}^N$
$\ell^2(\mathbb{Z})$	x_n	y_n	Subset of $\ell^2(\mathbb{Z})$
$\mathcal{L}^2(\mathbb{R})$	$x(t)$	y_n	$\ell^2(\mathbb{Z})$
$\mathcal{L}^2([-\frac{1}{2}T, \frac{1}{2}T))$	$x(t)$	y_n	$\mathbb{C}^N$

Table 5.2 Sampling, $y = \Phi^* x$.

• The *interpolation* operator Φ takes an input y from a smaller space and maps it into an output $\widehat{x}$ in a larger space. We denote by S the range of the interpolation operator Φ.

Input space	Input	Output	Output space $S = \mathcal{R}(\Phi)$
$\mathbb{C}^N$	y	$\widehat{x}$	$S \subset \mathbb{C}^M$
$\ell^2(\mathbb{Z})$	y_n	$\widehat{x}_n$	$S \subset \ell^2(\mathbb{Z})$
$\ell^2(\mathbb{Z})$	y_n	$\widehat{x}(t)$	$S \subset \mathcal{L}^2(\mathbb{R})$
$\mathbb{C}^N$	y_n	$\widehat{x}(t)$	$S \subset \mathcal{L}^2([-\frac{1}{2}T, \frac{1}{2}T))$

Table 5.3 Interpolation, $\widehat{x} = \Phi y$.

• *Interpolation followed by sampling* leads to perfect recovery because of the assumption of orthonormality.

• *Sampling followed by interpolation* will recover the input perfectly only when $x \in S$. Because of the choice of sampling and interpolation operators, $P = \Phi\Phi^*$ is an orthogonal projection operator.

Input	Output	Reconstruction	Property
Interpolation followed by sampling ($y = \Phi^*\Phi y$)			
y	y	Perfect	$\Phi^*\Phi = I$
Sampling followed by interpolation ($\widehat{x} = \Phi\Phi^* x$)			
$x \in S$	x	Perfect	$\Phi\Phi^* = I$ for $x \in S$
$x \notin S$	$\widehat{x}$	Best approximation in S	$\Phi\Phi^*$ is the orthogonal projection onto S

Table 5.4 Sampling and interpolation for the orthonormal case.

Sampling and interpolation with nonorthogonal vectors/sequences/functions

• The *sampling* operator $\widetilde{\Phi}^*$ takes an input x from a larger space and maps it into an output y in a smaller space. We denote by $\widetilde{S}$ the orthogonal complement of the null space of $\widetilde{\Phi}^*$,

$$\widetilde{S} = \mathcal{N}(\widetilde{\Phi}^*)^{\perp}.$$

Inputs $x \in \mathcal{N}(\widetilde{\Phi}^*)$ are mapped to $\mathbf{0}$, while inputs $x \in \widetilde{S}$ can be recovered exactly. For inputs $x \notin \widetilde{S}$, the component in $\mathcal{N}(\widetilde{\Phi}^*)$ is lost due to sampling, while the rest is preserved through $\widetilde{\Phi}^* x$.

• The *interpolation* operator Φ takes an input y from a smaller space and maps it into an output $\widehat{x}$ in a larger space, as in the orthonormal case. We denote by S the range of the interpolation operator Φ,

$$S = \mathcal{R}(\Phi).$$

• *Interpolation followed by sampling*, $\widetilde{\Phi}^*\Phi$, will not always be identity; when it is, interpolation and sampling are called *consistent*, and the input is perfectly recovered.

• *Sampling followed by interpolation* will recover the input perfectly only with consistency and $x \in S$. When the interpolation operator is the pseudoinverse of the sampling one, $\Phi = \widetilde{\Phi}(\widetilde{\Phi}^*\widetilde{\Phi})^{-1}$, sampling and interpolation are consistent and furthermore *ideally matched.* When sampling and interpolation are ideally matched, $P = \Phi\widetilde{\Phi}^*$ is an orthogonal projection operator. When they are consistent but not ideally matched, P is an oblique projection operator.

Input	Output	Reconstruction	Property	
Interpolation followed by sampling ($\widehat{y} = \widetilde{\Phi}^*\Phi y$)				
y	y	Perfect	Consistent	$\widetilde{\Phi}^*\Phi = I$
y	$\widehat{y}$	Not perfect		$\widetilde{\Phi}^*\Phi \neq I$
Sampling followed by interpolation ($\widehat{x} = \Phi\widetilde{\Phi}^* x$)				
$x \in S$	x	Perfect	Consistent	$\widetilde{\Phi}^*\Phi = I$
$x \notin S$	$\widehat{x}$	Projection to S	Consistent	$\widetilde{\Phi}^*\Phi = I$
$x \notin S$	$\widehat{x}$	Orthogonal projection to S	Ideally matched	$\Phi = \widetilde{\Phi}(\widetilde{\Phi}^*\widetilde{\Phi})^{-1}$

Table 5.5 Sampling and interpolation for the general (nonorthonormal) case.

Historical remarks

The sampling theorem for bandlimited functions has an interesting history. Long before computers, functions were tabulated for sets of points in the domain, raising the question of interpolation between these points. Exact recovery of bandlimited functions by sinc interpolation was first shown by **Edmund T. Whittaker (1873–1956)**, a British mathematician, in 1915. **Harry Nyquist (1889–1976)**, a Swedish electrical engineer working at Bell Labs, was interested in signaling over bandlimited channels and formulated the celebrated criterion bearing his name in 1928: sampling at twice the maximum frequency uniquely specifies a bandlimited function. In the Russian literature, **Vladimir A. Kotelnikov (1908–2005)**, an information theorist working in the Soviet Union, proved the sampling theorem independently in 1933. **John M. Whittaker (1905–1984)**, son of the initial contributor, added further results to the interpolation theory on which his father had worked. Meanwhile, **Herbert P. Raabe (1909–2004)**, a German electrical engineer, wrote a dissertation in 1939 stating and proving sampling results for bandlimited functions. In 1949, **Isao Someya (1915–2007)** in Japan also proved the sampling theorem. In signal processing and communications, **Claude E. Shannon (1916–2001)** (pictured), an American mathematician and engineer, is most often connected to the sampling theorem, which frequently bears his name. In 1948, in his landmark treatise *A Mathematical Theory of Communication*, Shannon formulated the sampling theorem as the first step in digital communications, stating "this is a fact which is common knowledge in the communication art" [89]. Shannon is considered the father of information theory, which laid the foundation for digital communications and source compression, and ultimately, the information society as we know it.

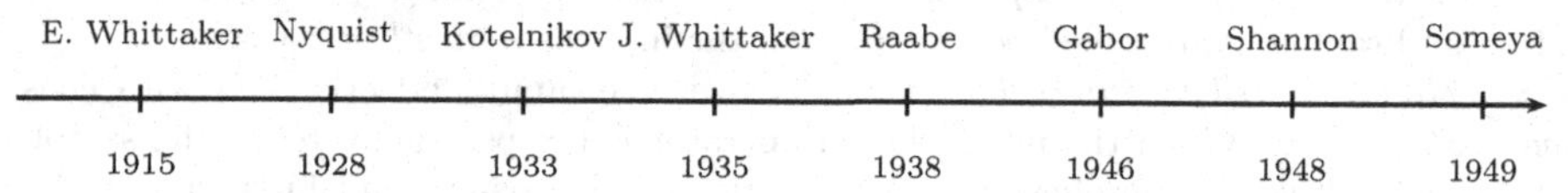

Further reading

Deterministic sampling results Many books cover sampling theory and its applications in signal processing and communications, for example, [67]. We also recommend the review papers by Jerri [49] and by Unser [103]; the latter develops sampling in shift-invariant subspaces. In [76], Papoulis introduced multichannel sampling and provided a result equivalent to Theorem 5.17. Nonuniform sampling in its general form is more difficult; briefly, the sampling density has to be at least the Nyquist rate, and the set of sample times cannot be too unevenly spread (for example, it should have no accumulation points). Vetterli, Marziliano, and Blu [108] introduced an alternate sampling theory that does not require finite bandwidth or shift invariance with a fixed shift T as in Definition 5.10. Rather, the sampling rate for exact reconstruction of a function is related to its *rate of innovation*, which is defined as the function's number of degrees of freedom per unit time; an overview can be found in [8].

Sampling stochastic processes For stationary bandlimited processes, Theorem 5.18 combines results of Balakrishnan [2] and Belyaev [4]; Lloyd [61] produced the earliest results with almost sure convergence. Note that $a_{\mathrm{x}} \in \mathrm{BL}[-\frac{1}{2}\omega_0, \frac{1}{2}\omega_0]$ implies that $a_{\mathrm{x}} \in \mathcal{L}^2(\mathbb{R})$, which excludes the existence of Dirac delta components from the power spectrum. Convergence in the mean-square sense can be proven even with Dirac delta components in the power spectrum, provided that the power spectrum is continuous at $\pm\frac{1}{2}\omega_0$. We have been guided by the historical review of Draščić [26] and also recommend [53, 80, 109].

Iterative reconstruction algorithms The alternating projection algorithm was introduced and analyzed by Cheney and Goldstein [13]; Papoulis [75] and Gerchberg [33] popularized it for reconstruction of bandlimited signals. For its convergence behavior when applied to discrete signals, see [50].

Exercises with solutions

5.1. *Best piecewise-constant approximation by sampling followed by interpolation*
Let $x \in \mathcal{L}^2(\mathbb{R})$ and let $S \subset \mathcal{L}^2(\mathbb{R})$ be the set of piecewise-constant functions that are constant over unit-length intervals $[n, n+1)$, $n \in \mathbb{Z}$. Show that $\min_{\widehat{x}\in S}\|x - \widehat{x}\|$ is achieved by

$$\widehat{x}(t) = \sum_{n\in\mathbb{Z}} \left(\int_n^{n+1} x(t)\, dt\right) 1_{[0,1)}(t-n). \tag{E5.1-1}$$

Use this to show that $P = \Phi\Phi^*$, with Φ^* the sampling operator from (5.2) and Φ the interpolation operator from (5.4), results in this least-squares approximation.
Solution: Any function $\widehat{x} \in S$ is of the form

$$\widehat{x}(t) = \alpha_n \qquad \text{for } t \in [n, n+1),\ n \in \mathbb{Z}, \tag{E5.1-2}$$

for some sequence $\alpha \in \ell^2(\mathbb{Z})$. Let x be an arbitrary function in $\mathcal{L}^2(\mathbb{R})$. Then

$$\begin{aligned} \|x - \widehat{x}\|^2 &\overset{(a)}{=} \int_{-\infty}^{\infty} |x(t) - \widehat{x}(t)|^2\, dt \overset{(b)}{=} \sum_{n\in\mathbb{Z}} \int_n^{n+1} |x(t) - \widehat{x}(t)|^2\, dt \\ &\overset{(c)}{=} \sum_{n\in\mathbb{Z}} \int_n^{n+1} |x(t) - \alpha_n|^2\, dt \overset{(d)}{=} \sum_{n\in\mathbb{Z}} \mathcal{E}_n, \end{aligned}$$

where (a) follows from the definition of the $\mathcal{L}^2(\mathbb{R})$ norm; (b) from breaking the interval of integration into subintervals of unit length; (c) from the definition of $\widehat{x}$, (E5.1-2); and (d) from introducing

$$\mathcal{E}_n = \int_n^{n+1} |x(t) - \alpha_n|^2\, dt,$$

for each $n \in \mathbb{Z}$.

Since $\mathcal{E}_n$ depends on α only through α_n, finding the best $\widehat{x}$ amounts to separately minimizing each $\mathcal{E}_n$ by choosing α_n depending on $x(t)$ for $t \in (n, n+1)$. To do that, compute the first and second derivatives of $\mathcal{E}_n$ with respect to α_n,

$$\begin{aligned} \frac{d\mathcal{E}_n}{d\alpha_n} &= -\int_n^{n+1} 2(x(t) - \alpha_n)\, dt = 2\alpha_n - 2\int_n^{n+1} x(t)\, dt, \\ \frac{d^2\mathcal{E}_n}{d\alpha_n^2} &= 2. \end{aligned}$$

In these computations, we have assumed that x and α are real; extension to the complex case is as in Solved exercise 3.7. Since the second derivative is positive, any point where the first derivative is zero gives a local minimum. Setting the first derivative to zero and solving for α_n gives

$$\alpha_n = \int_n^{n+1} x(t)\, dt, \qquad n \in \mathbb{Z}. \tag{E5.1-3}$$

Substituting (E5.1-3) into (E5.1-2) yields the desired expression, (E5.1-1).

All that remains is to associate (E5.1-1) with the sampling and interpolation operators in Section 5.1. The sequence α computed in (E5.1-3) is identical to y in (5.2). Then, since $g = 1_{[0,1)}$ in (5.4), (5.4) is identical to (E5.1-1).

5.2. *Correcting for inconsistent sampling and interpolation*
Suppose that the sampling operator $\widetilde{\Phi}^* \in \mathbb{C}^{N\times M}$ and the interpolation operator $\Phi \in \mathbb{C}^{M\times N}$, with $N < M$, are not consistent, meaning that $D = \widetilde{\Phi}^*\Phi \neq I$. We call C a *correction operator* when $P = \Phi C\widetilde{\Phi}^*$ is a projection operator with an N-dimensional range. When will such a C exist? Find C in terms of D and note any restrictions on D.
Solution: Our goal is to choose $C \in \mathbb{C}^{N\times N}$ so that $P = \Phi C\widetilde{\Phi}^*$ is idempotent and has an N-dimensional range. Since

$$P^2 = (\Phi C\widetilde{\Phi}^*)(\Phi C\widetilde{\Phi}^*) = \Phi CDC\widetilde{\Phi}^*,$$

idempotency is achieved when $CDC = C$. In addition, for P to have an N-dimensional range, C must have rank N and hence be invertible, allowing us to eliminate C from each side of $CDC = C$. Then, the correction operator exists if and only if D is invertible, and $C = D^{-1}$.

5.3. *Null space of sampling operator and range of interpolation operator in $\mathcal{L}^2(\mathbb{R})$*

(i) Let Φ^* denote sampling with rate $1/T$ and prefilter $g^*(-t)$ as depicted in Figure 5.4(a) and expressed by (5.58). Show that $S^\perp = \mathcal{N}(\Phi^*)$ is a shift-invariant subspace with respect to shift T.

(ii) Let Φ denote interpolation with spacing T and postfilter g as depicted in Figure 5.4(b) and expressed by (5.61). Show that $S = \mathcal{R}(\Phi)$ is a shift-invariant subspace with respect to shift T.

Solution:

(i) A function x is in $S^\perp$ when

$$\langle x, \varphi_k \rangle = 0, \qquad \text{for all } k \in \mathbb{Z}. \tag{E5.3-1}$$

Let $\widetilde{x}$ be a shifted version of x,

$$\widetilde{x}(t) = x(t - \ell T), \qquad t \in \mathbb{R}, \tag{E5.3-2}$$

for some fixed $\ell \in \mathbb{Z}$. Then, for any $k \in \mathbb{Z}$,

$$\begin{aligned}\langle \widetilde{x}, \varphi_k \rangle &\overset{(a)}{=} \langle x(t-\ell T), \varphi_k(t)\rangle_t \overset{(b)}{=} \langle x(t-\ell T), g(t-kT)\rangle_t \\ &\overset{(c)}{=} \int_{-\infty}^{\infty} x(t-\ell T) g^*(t-kT)\, dt \overset{(d)}{=} \int_{-\infty}^{\infty} x(\tau) g^*(\tau - (k-\ell)T)\, d\tau \\ &\overset{(e)}{=} \langle x(\tau), g(\tau-(k-\ell)T)\rangle_\tau \overset{(f)}{=} \langle x, \varphi_{k-\ell}\rangle \overset{(g)}{=} 0,\end{aligned}$$

where (a) follows from (E5.3-2); (b) and (f) from (5.59); (c) and (e) from the definition of the $\mathcal{L}^2(\mathbb{R})$ inner product; (d) from the change of variable $\tau = t - \ell T$; and (g) from (E5.3-1). This shows that $\widetilde{x}$ is also in $S^\perp$, so $S^\perp$ is a shift-invariant subspace with respect to shift T.

(ii) A function x is in S when

$$x = \sum_{k\in\mathbb{Z}} \alpha_k \varphi_k \tag{E5.3-3}$$

for some coefficient sequence $\alpha \in \ell^2(\mathbb{Z})$. Let $\widetilde{x}$ be a shifted version of x as in (E5.3-2) for some fixed $\ell \in \mathbb{Z}$. Then

$$\begin{aligned}\widetilde{x}(t) &\overset{(a)}{=} \sum_{k\in\mathbb{Z}} \alpha_k \varphi_k(t-\ell T) \overset{(b)}{=} \sum_{k\in\mathbb{Z}} \alpha_k g(t-\ell T - kT) \\ &\overset{(c)}{=} \sum_{k\in\mathbb{Z}} \alpha_k \varphi_{k+\ell}(t) \overset{(d)}{=} \sum_{m\in\mathbb{Z}} \alpha_{m-\ell}\varphi_m(t),\end{aligned}$$

where (a) follows from (E5.3-2) and (E5.3-3); (b) and (c) from (5.59); and (d) from the change of variable $m = k + \ell$. Thus, $\widetilde{x} \in S$; shifting x by ℓT has shifted the coefficient sequence α by ℓ.

5.4. *Interpolation of oversampled signals*

Let $x \in \text{BL}[-\pi, \pi]$. Guaranteeing exact recovery of x from samples at the Nyquist rate $\omega_s = 2\pi$ requires the use of the sinc interpolation postfilter g from (5.68) with $T = 1$, which has a slow decay of the order $1/t$. Sampling faster enables the use of filters with faster decay such as those obtained by frequency-domain convolution of ideal filters as in Example 5.20.

In each of the following, h_1 and h_2 are ideal filters, with cutoff frequencies $\frac{1}{2}\omega_1$ and $\frac{1}{2}\omega_2$, respectively.

(i) Let $\omega_s = 3\pi$. Find ω_1 and ω_2 such that $g_2(t) = h_1(t)\, h_2(t)$ is an appropriate interpolation postfilter and give an expression for g_2.

(ii) Let $\omega_s = 4\pi$. Find ω_1 and ω_2 such that $g_3(t) = h_1(t)\, (h_2(t))^2$ is an appropriate interpolation postfilter, give an expression for g_3, and show that G_3 has a continuous derivative.

(iii) Let $\omega_s = (i+1)\pi$. Find ω_1 and ω_2 such that $g_i(t) = h_1(t)\, (h_2(t))^{(i-1)}$ is an appropriate interpolation postfilter, give an expression for g_i, and show that G_i has continuous derivatives of order up to $(i-2)$.

Solution: All parts of this problem use sampling frequencies of the form $\omega_s = m\pi$. Using (5.72), since $x \in \text{BL}[-\pi, \pi]$, its sampled and interpolated version before postfiltering occupies the following set of frequencies:

$$\cdots \cup [-(m+1)\pi, -(m-1)\pi] \cup [-\pi, \pi] \cup [(m-1)\pi, (m+1)\pi] \cup \cdots. \tag{E5.4-1}$$

(i) With $\omega_s = 3\pi$, (E5.4-1) implies that the sampled and interpolated version before postfiltering occupies the following set of frequencies:

$$\cdots \cup [-4\pi, -2\pi] \cup [-\pi, \pi] \cup [2\pi, 4\pi] \cup \cdots,$$

and thus G_2 must satisfy

$$G_2(\omega) = \begin{cases} 1, & \text{for } |\omega| \le \pi; \\ 0, & \text{for } |\omega| \ge 2\pi. \end{cases} \tag{E5.4-2}$$

Since $G_2 = H_1 * H_2$, we use (E5.4-2) to find the cutoff frequencies ω_1 and ω_2,

$$\tfrac{1}{2}\omega_1 + \tfrac{1}{2}\omega_2 = 2\pi, \qquad \tfrac{1}{2}\omega_1 - \tfrac{1}{2}\omega_2 = \pi,$$

yielding $\omega_1 = 3\pi$ and $\omega_2 = \pi$, or, from (5.68), $g_2(t) = \text{sinc}(\frac{3}{2}\pi t)\,\text{sinc}(\frac{1}{2}\pi t)$; this function clearly decays as $1/t^2$.

(ii) With $\omega_s = 4\pi$, (E5.4-1) implies that the sampled and interpolated version before postfiltering occupies the following set of frequencies:

$$\cdots \cup [-5\pi, -3\pi] \cup [-\pi, \pi] \cup [3\pi, 5\pi] \cup \cdots,$$

and thus G_3 must be of the form

$$G_3(\omega) = \begin{cases} 1, & \text{for } |\omega| \le \pi; \\ 0, & \text{for } |\omega| \ge 3\pi. \end{cases} \tag{E5.4-3}$$

Since $G_3 = H_1 * H_2 * H_2$, we use (E5.4-3) to find the cutoff frequencies ω_1 and ω_2,

$$\tfrac{1}{2}\omega_1 + 2\tfrac{1}{2}\omega_2 = 3\pi, \qquad \tfrac{1}{2}\omega_1 - 2\tfrac{1}{2}\omega_2 = \pi,$$

yielding $\omega_1 = 4\pi$ and $\omega_2 = \pi$, or, from (5.68), $g_3(t) = \text{sinc}(2\pi t)\,\text{sinc}^2(\frac{1}{2}\pi t)$; this function clearly decays as $1/t^3$.

To show that G_3 has a continuous derivative, we can use the convolution-in-frequency property of the Fourier transform, (4.65), and the fact that the differentiation of a convolution can be achieved by differentiating either of the two arguments of the convolution as shown in Example 4.7. The Fourier transform of the sinc function is a box function, whose derivative is a pair of Dirac delta functions. Then, the convolution of this derivative with the triangle function (Fourier transform of the squared sinc function) results in a difference of triangle functions. As this difference is continuous, we conclude that the derivative of G_3 is continuous.

(iii) With $\omega_s = (i+1)\pi$, (E5.4-1) implies that the sampled and interpolated version before postfiltering occupies the following set of frequencies:

$$\cdots \cup [-(i+2)\pi, -i\pi] \cup [-\pi, \pi] \cup [i\pi, (i+2)\pi] \cup \cdots,$$

and thus G_i must be of the form

$$G_i(\omega) = \begin{cases} 1, & \text{for } |\omega| \leq \pi; \\ 0, & \text{for } |\omega| \geq i\pi. \end{cases} \tag{E5.4-4}$$

Since

$$G_i = H_1 * \underbrace{H_2 * \cdots * H_2}_{(i-1)\text{ times}},$$

we use (E5.4-4) to find ω_1 and ω_2,

$$\tfrac{1}{2}\omega_1 + (i-1)\tfrac{1}{2}\omega_2 = i\pi, \qquad \tfrac{1}{2}\omega_1 - (i-1)\tfrac{1}{2}\omega_2 = \pi,$$

yielding $\omega_1 = (i+1)\pi$ and $\omega_2 = \pi$, or, from (5.68),

$$g_i(t) = \operatorname{sinc}\big(\tfrac{1}{2}(i+1)\pi t\big) \left(\operatorname{sinc}\left(\tfrac{1}{2}\pi t\right)\right)^{(i-1)};$$

this function clearly decays as $1/t^i$.

To show that G_i has continuous derivatives of order up to $(i-2)$, we use the same argument as in (ii). We use the convolution-in-frequency property of the Fourier transform, (4.65), and the fact that the differentiation of a convolution can be achieved by differentiating either of the two arguments of the convolution, as shown in Example 4.7. As G_i is the convolution of i box functions, differentiating $(i-2)$ times will lead to $(i-2)$ pairs of Dirac delta functions convolved with a triangle function, leading to $(i-2)$ differences of triangle functions. As the difference of triangle functions is continuous, we conclude that the derivatives of order up to $(i-2)$ of G_i are continuous.

5.5. *Properties of the Dirichlet kernel*
Prove Theorem 5.24.
Solution:

(i) The first expression (5.114a) follows immediately from using (3.286) to combine $k = \ell$ and $k = -\ell$ pairs of terms in (5.113).

To derive the second expression (5.114b), recall the general finite geometric series

$$\sum_{k=a}^{b} t^k = \begin{cases} \dfrac{t^a - t^{b+1}}{1-t}, & \text{for } t \neq 1; \\ b - a + 1, & \text{for } t = 1, \end{cases} \tag{E5.5-1}$$

which can derived from (P2.54-1). Simplifying (5.113) as a geometric series yields

$$\begin{aligned} d(t) &= \sum_{k=-K}^{K} e^{jk2\pi t/T} \overset{(a)}{=} \frac{\exp(-jK2\pi t/T) - \exp(j(K+1)2\pi t/T)}{1 - \exp(j2\pi t/T)} \\ &\overset{(b)}{=} \frac{\exp(j\pi t/T)}{\exp(j\pi t/T)} \cdot \frac{\exp(-j(K+\frac{1}{2})2\pi t/T) - \exp(j(K+\frac{1}{2})2\pi t/T)}{\exp(-j\pi t/T) - \exp(j\pi t/T)} \\ &\overset{(c)}{=} \frac{-2j\sin((K+\frac{1}{2})2\pi t/T)}{-2j\sin(\pi t/T)} = \frac{\sin((K+\frac{1}{2})2\pi t/T)}{\sin(\pi t/T)}, \end{aligned} \tag{E5.5-2}$$

for $t/T \notin \mathbb{Z}$, where (a) follows from (E5.5-1); (b) from factoring both numerator and denominator to be in forms of $\exp(-j\theta) - \exp(j\theta)$; and (c) from canceling the leading factor and applying (3.286) in both numerator and denominator. The $t/T \in \mathbb{Z}$ case in (5.114b) follows directly from the $t = 1$ case in (E5.5-1).

From the definition of the sinc function, (3.9a), $\sin t = t \operatorname{sinc} t$. Thus, the third expression, (5.114c), follows directly from (5.114b).

(ii) Comparing (5.113) with (4.94b) shows that (5.113) is the Fourier series reconstruction of the Dirichlet kernel d from the Fourier series coefficients D specified in (5.115).

(iii) The truncation in (5.116a) is precisely the pointwise multiplication of the Fourier series coefficients X by the Fourier series coefficients of the Dirichlet kernel in (5.115). The desired result thus follows directly from the convolution theorem for the Fourier series, (4.113).

(iv) For each $\ell \in \{0, 1, \ldots, 2K\}$, let $\varphi_\ell(t) = d(t - \tau_\ell)$. The Fourier series coefficients of φ_0 are given by (5.115). Using the shift-in-time property (4.107), the Fourier series coefficients of φ_ℓ are given by

$$\Phi_{\ell,k} = \begin{cases} \exp(-jk2\pi\tau_\ell/T), & \text{for } k = -K, -K+1, \ldots, K; \\ 0, & \text{otherwise.} \end{cases} \tag{E5.5-3}$$

Thus, each φ_ℓ has bandwidth $2K + 1$. Since $\mathrm{BL}\{-K, \ldots, K\}$ is closed, we can conclude that $\overline{\mathrm{span}}(\{\varphi_k\}_{k=0}^{2K}) \subseteq \mathrm{BL}\{-K, \ldots, K\}$.

Now let x be any T-periodic function in $\mathrm{BL}\{-K, \ldots, K\}$. We would like to show that x can be written as a linear combination of $\{\varphi_\ell\}_{\ell=0}^{2K}$. This will show that $x \in \overline{\mathrm{span}}(\{\varphi_k\}_{k=0}^{2K})$, from which we conclude that $\mathrm{BL}\{-K, \ldots, K\} \subseteq \overline{\mathrm{span}}(\{\varphi_k\}_{k=0}^{2K})$. The uniqueness of the expansion will also show the linear independence of $\{\varphi_\ell\}_{\ell=0}^{2K}$, thus completing a proof that $\{\varphi_\ell\}_{\ell=0}^{2K}$ is a basis for $\mathrm{BL}\{-K, \ldots, K\}$ (see Definition 2.34).

Let $x = \sum_{\ell=0}^{2K} \alpha_\ell \varphi_\ell$. As equality of Fourier series, this is $X = \sum_{\ell=0}^{2K} \alpha_\ell \Phi_\ell$. Since X and Φ_ℓ, $\ell = 0, 1, \ldots, 2K$ have common support of $\{-K, \ldots, K\}$, this can be written as a system of linear equations with a square matrix of size $2K + 1$. Because of (E5.5-3), this is a Vandermonde matrix; because of (2.249), it is invertible because each τ_ℓ is distinct.

(v) For $\ell = 0, 1, \ldots, 2K$, let $\tau_\ell = (T/(2K+1))\ell$ and

$$\varphi_\ell(t) = \frac{1}{\sqrt{T(2K+1)}} d(t - \tau_\ell).$$

The previous part of the theorem shows that $\{\varphi_\ell\}_{\ell=0}^{2K}$ is a basis; it remains just to show orthonormality. For any $\ell, m \in \{0, 1, \ldots, 2K\}$,

$$\langle \varphi_\ell, \varphi_m \rangle \overset{(a)}{=} T \langle \Phi_\ell, \Phi_m \rangle \overset{(b)}{=} \delta_{\ell - m},$$

where (a) follows from the generalized Parseval equality for Fourier series, (4.104b); and (b) from separate elementary computations for $\ell = m$ and $\ell \neq m$.

Exercises

5.1. *Approximation by piecewise-constant functions.*(1)
Consider the interval $[0, 1)$ and the linear function $x(t) = t$. For any $k \in \mathbb{N}$, let $\widehat{x}_k$ be the least-squares approximation of x among piecewise-constant functions that are constant over intervals $[n2^{-k}, (n+1)2^{-k})$, $0 \leq n < 2^k$. Compute $\|x - \widehat{x}_k\|_2^2$ and comment on its behavior as $k \to \infty$.

5.2. *Sampling and interpolation for bandlimited vectors.*(1)
A vector $x \in \mathbb{C}^M$ is called *bandlimited* when there exists an odd $k_0 \in \{0, 1, \ldots, M-1\}$ such that its DFT coefficient sequence X satisfies

$$X_k = 0 \quad \text{for all } k \text{ with } \frac{k_0 + 1}{2} \leq k \leq M - \frac{k_0 + 1}{2}. \tag{P5.2-1}$$

The smallest such k_0 is called the *bandwidth* of x. A vector in $\mathbb{C}^M$ that is not bandlimited is called a *full-band* vector. The set of vectors in $\mathbb{C}^M$ with the bandwidth of at most k_0 is a closed subspace. For x in such a bandlimited subspace, find Φ so that the sampling followed by interpolation described by $\Phi\Phi^*$ in Section 5.2.1 achieves perfect recovery, $\widehat{x} = x$.

5.3. *Sampling the DTFT of a finite-length sequence.*(2)
Let x be a sequence with support in $\{-\frac{1}{2}(k_0-1), \ldots, \frac{1}{2}(k_0-1)\}$. Show that if its DTFT $X(e^{j\omega})$ is sampled at k_0 points $\omega_k = 2\pi k/k_0$, with $k = -\frac{1}{2}(k_0-1), \ldots, \frac{1}{2}(k_0-1)$, then the nonzero samples of x can be recovered from $X(e^{j\omega_k})$.

5.4. *Bandlimited space with rational sampling rate changes.*(2)
Consider sampling followed by interpolation in Figure P5.4-1 with input sequence $x \in \text{BL}[-2\pi/K, 2\pi/K]$. For the cases below, what condition on the filter g ensures that $\widehat{x} = x$?

(i) $M = 2$, $N = 3$, and $K = 3$.
(ii) $M = 2$, $N = 3$, and $K = 4$.
(iii) General M, N, and K (with $M < N$).

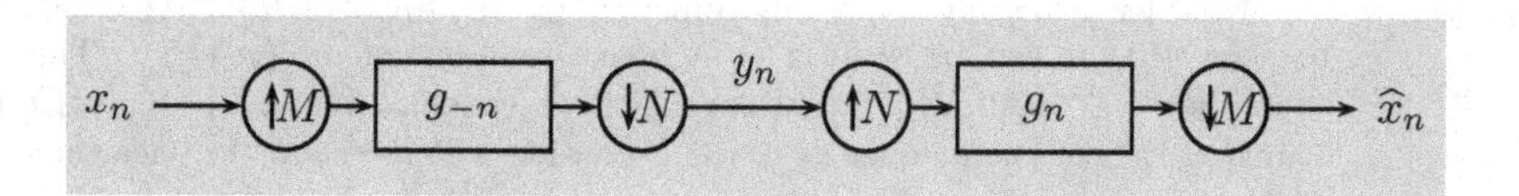

Figure P5.4-1 Sampling followed by interpolation with rational sampling rate change ($M < N$).

5.5. *Bases for shift-invariant subspaces.*(3)
Let $x_n = \delta_n + \delta_{n-1} + \delta_{n-2}$, and let S be the shift-invariant subspace with respect to shift 2 generated by x. Find a generator for S that is orthonormal to its shifts by integer multiples of 2.

5.6. *Ideally matched sampling and interpolation with nonorthogonal sequences.*(3)
Let

$$g = \begin{bmatrix} \ldots & 0 & \frac{1}{2} & \boxed{1} & \frac{1}{2} & 0 & \ldots \end{bmatrix}^{\top}$$

as in (5.53). Find $\widetilde{g}$ such that sampling with rate $\frac{1}{2}$ and prefilter $\widetilde{g}$ is ideally matched with interpolation with spacing 2 and postfilter g. You may assume that the values

$$\widetilde{g}_0 = \frac{1}{\sqrt{2}} \quad \text{and} \quad \widetilde{g}_{-1} = \widetilde{g}_1 = 1 - \frac{1}{\sqrt{2}}$$

in (5.54) are correct.

5.7. *Downsampling by N.*(1)
Let $x_n = q(nT)$ and $y_n = q(nNT) = x_{nN}$ be sampled versions of the same continuous-time signal $q(t)$, with sampling periods T and NT, respectively. Use this to verify the expression for downsampling by N, (3.191).
(*Hint:* Use (5.72) and (5.73).)

5.8. *Multirate system.*(1)
Given is the discrete-time system $y = U_3 G D_2 x$, with G an LSI filter, U_3 the upsampling-by-3 operator, and D_2 the downsampling-by-2 operator.

(i) Express the z-transform of the output sequence y in terms of the z-transform of the input sequence x and the z-transform of the filter g.
(ii) Suppose that the input sequence x is obtained by sampling the function $q \in \text{BL}[-\pi/T, \pi/T]$ with sampling frequency $1/T$ Hz. Write $Y(e^{j\omega})$ as a function of $Q(\omega)$ and $G(e^{j\omega})$. What are the conditions on $Q(\omega)$ to avoid aliasing?

5.9. *Bandpass sampling.*(2)
Refer to Example 5.17 and the spectrum $X(\omega)$ in (5.78).

(i) Show that the solution described in the example can be expressed as an orthonormal expansion for the bandlimited bandpass subspace $\text{BL}([-3\pi, -2\pi] \cup [2\pi, 3\pi])$ and give the basis functions.

(ii) Show that bandpass signals with frequency support $[-\frac{1}{2}(K+1)\omega_0, -\frac{1}{2}K\omega_0] \cup [\frac{1}{2}K\omega_0, \frac{1}{2}(K+1)\omega_0]$ can be sampled with sampling frequency $\omega_s = \omega_0$ and perfectly reconstructed.

5.10. *Sampling movies.*(1)
Classic movies were shot at 24 frames/s, each frame being a snapshot of the scene. Given a carriage with wagon wheels with 16 spokes of radius 1 m, what is/are the speed(s) of the carriage for which the wheels appear motionless?

5.11. *Continuous-time modulation using discrete-time operators.*(2)
In Theorem 5.16, we saw how continuous-time convolution can be implemented in discrete time. Consider two bandlimited functions, $x \in \mathrm{BL}[-\frac{1}{2}\omega_x, \frac{1}{2}\omega_x]$ and $g \in \mathrm{BL}[-\frac{1}{2}\omega_g, \frac{1}{2}\omega_g]$.

(i) Show how the multiplication $g(t)\,x(t)$ can be implemented in discrete time.

(ii) Compute the product $g(t)\,x(t)$ for x in (5.77) and g as in (5.79) with $\omega_0 = 3\pi$.

5.12. *Multichannel sampling.*(2)
Let $x \in \mathrm{BL}[-\pi, \pi]$ and $T = 1$. Consider a generalization of Figure 5.33 to an N-channel system with sampling prefilters $\widetilde{g}_i$, $i = 0, 1, \ldots, N-1$, followed by uniform sampling with period NT.

(i) Let $N = 3$, and let the sampling prefilters be derivative filters: $\widetilde{G}_i(\omega) = (j\omega)^i$, $i = 0, 1, 2$. Show that the determinant of the matrix $\widetilde{G}(\omega)$ is a nonzero constant.

(ii) Let $N \in \mathbb{Z}^+$, and let the sampling prefilters be derivative filters: $\widetilde{G}_i(\omega) = (j\omega)^i$, $i = 0, 1, \ldots, N-1$. Show that the determinant of the matrix $\widetilde{G}(\omega)$ is a nonzero constant.

(iii) Let $N = 3$, and let the sampling prefilters be delay filters: $\widetilde{G}_0(\omega) = 1$, $\widetilde{G}_1(\omega) = e^{-j\omega(1+\alpha)}$, and $\widetilde{G}_2(\omega) = e^{-j\omega(1+\beta)}$, with $\alpha, \beta \in [-1, 1]$. For which values of α and β is the matrix $\widetilde{G}(\omega)$ singular? Numerically compute the condition number $\kappa(\widetilde{G}(\omega))$; you should obtain that it does not depend on ω. Plot $\widetilde{G}(\omega)$ for $\alpha = 1$ and $\beta \in [-\frac{1}{2}, \frac{1}{2})$. Comment on the result.

5.13. *Continuous-time stochastic process leading to i.i.d. samples.*(1)
Let $T \in \mathbb{R}^+$, and let x be a zero-mean WSS continuous-time stochastic process with power spectral density

$$A_{\mathrm{x}}(\omega) = \begin{cases} 1, & \text{for } \omega \in [-\pi/T, \pi/T]; \\ 0, & \text{otherwise.} \end{cases}$$

Show that the discrete-time stochastic process $\mathrm{y}_n = \mathrm{x}(nT)$, $n \in \mathbb{Z}$, is a discrete-time white noise process.

5.14. *Equivalence of shifts for subspaces of periodic functions.*(2)
Let $W \subset \mathcal{L}^2([-\frac{1}{2}T, \frac{1}{2}T))$ be a shift-invariant subspace with respect to shift τ, and suppose that $T/\tau = p/q$, where p and q are coprime. Show that W is a shift-invariant subspace with respect to shift τ/q.

5.15. *Adjoint operators.*(1)
Prove that the sampling operator (5.106) and the interpolation operator (5.109) are adjoints of each other.

5.16. *Dirac comb Fourier series pair.*(2)
Prove that (5.124) is indeed a Fourier-series transform pair.

5.17. *Fourier series with triangle spectrum.*(1)
Let $K \in \mathbb{Z}^+$, and let x be the 1-periodic function in $\mathrm{BL}\{-K, \ldots, K\}$, with Fourier series coefficients

$$X_k = \begin{cases} 1 - |k|/K, & \text{for } |k| \le K; \\ 0, & \text{otherwise.} \end{cases}$$

(i) Find a simple expression for x.

(ii) What is the largest sampling period T_s that enables perfect function recovery?

(iii) With the sampling period T_s determined in the previous part, express the condition for perfect function recovery in terms of a system of linear equations, by expressing each sample of x through its Fourier series reconstruction formula, (4.94b).

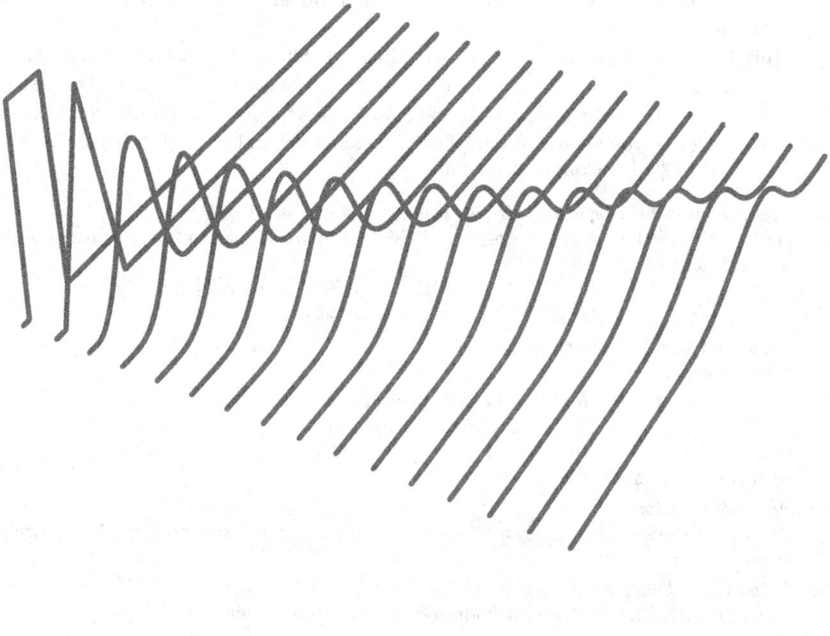

Chapter 6

Approximation and compression

> "Far better an approximate answer to the right question, which is often vague, than an exact answer to the wrong question, which can always be made precise."
>
> — *John Tukey*

Contents

In previous chapters, we saw how to write a sequence or a function as an expansion with a basis (Chapters 2–4) or using sampling and interpolation (Chapter 5). Often, however, we do not have the luxury of representing the sequence or function exactly, necessitating the development of approximate representations.

Truncation methods are used when an exact series expansion of a function or sequence requires too many coefficients. For example, truncating a Fourier series representation results in a bandlimited approximation of a periodic function. Digital storage or transmission of a function or sequence requires both truncation and that all quantities take values in countable sets.

Approximation theory deals with the choice of expansion coefficients to keep; compression theory deals with approximating those coefficients. Some methods are familiar from calculus, such as approximation via Taylor series, while others are likely unfamiliar, such as nonlinear approximation with bases. We review a collection of approaches with a clear message: there is no *one-size-fits-all* technique. Each case requires a careful analysis and a well-engineered solution. A good solution can have high impact, as demonstrated by the billions of cameras, mobile phones, and computers using the MP3 format to represent music, the JPEG format to represent images, and the MPEG format to represent videos; see the *Further reading.*

6.1 Introduction

There are many motivations for forming an approximation of a function.

(i) The function might not be known everywhere, but rather only at some specific points. We may then form an approximation of the function by interpolating between those points. Examples of interpolation on the entire real line were developed in Section 5.4.

(ii) A series expansion of the function might have infinitely many terms. We may then form an approximation with a finite number of terms to reduce the computational complexity, storage, and communication requirements. Examples of infinite series expansions are the Fourier series expansions of periodic functions in Section 4.5 and the sampling expansions in Chapter 5.

(iii) Regardless of the first two motivations, digital computation, storage, and communication require real numbers to be approximated by numbers from countable sets. When this discretization is coarse, it has a significant impact on the approximation quality as well.

To make these points concrete, we provide several approximations of a simple example function. One example is not enough to reveal the relative strengths and weaknesses of the various methods; these will become more apparent from other examples provided throughout the chapter.

Function to be approximated Suppose that x is a piecewise-constant function defined on the unit interval $[0,\,1]$. Assume the number of constant pieces N to be known, while the points of discontinuity $\{t_n\}_{n=1}^{N-1} \subset (0,\,1)$ and the values of the function on each piece $\{\alpha_n\}_{n=0}^{N-1} \subset \mathbb{R}$ are unknown. An example function for $N = 3$ is shown in Figure 6.1. Such a function is called parametric since, given N, the $2N-1$ parameters $\{t_n\}_{n=1}^{N-1}$ and $\{\alpha_n\}_{n=0}^{N-1}$ specify x.

Least-squares polynomial approximation First, we try fitting x with a polynomial of degree K. We should not expect too much, since the function x is discontinuous, while polynomials are smooth. Least-squares approximation p_K is achieved by orthogonal projection to the subspace of polynomials of degree K, as shown in Figure 6.2 for $K = 0,\,1,\,2$.

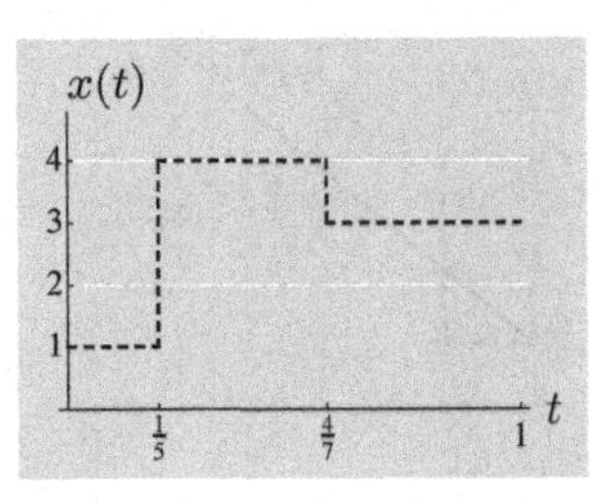

Figure 6.1 Piecewise-constant function on $[0, 1]$ with $N = 3$ pieces, points of discontinuity $\{t_1, t_2\} = \{\frac{1}{5}, \frac{4}{7}\}$, and values $\{\alpha_0, \alpha_1, \alpha_2\} = \{1, 4, 3\}$ on the pieces.

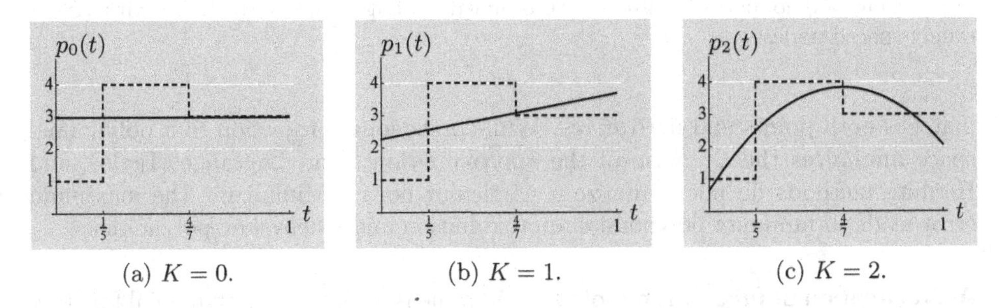

(a) $K = 0$. (b) $K = 1$. (c) $K = 2$.

Figure 6.2 Least-squares polynomial approximations p_K (solid lines) of x from Figure 6.1 (dashed lines) using polynomials of degree K.

Lagrange interpolation: Matching points Instead of using orthogonal projection to a subspace of polynomials, we can try polynomial interpolation where the value of the function is exactly matched at specific points – called *nodes*; this is Lagrange interpolation. An advantage is that the function need not be known everywhere, but only at the nodes, which must be distinct. Figure 6.3 shows the result of approximating the function in Figure 6.1 with one, two, and three nodes. The results are not entirely satisfactory, unsurprisingly so, since polynomials are smooth, unlike the function we are trying to match.

Taylor series expansion: Matching derivatives When an analytical expression of the function to be approximated is available and derivatives exist up to some order, we can match the function and its derivatives using the Taylor series at a point of interest. The advantage is that the representation is exact at that point, while it gradually worsens when moving away. In some sense, one can think of Taylor series as function extrapolation, as opposed to the Lagrange method, which is an interpolation method. Since the function x from Figure 6.1 is piecewise-constant, Taylor series expansions are not interesting to plot.

Other polynomial approximation methods A mixture of Lagrange interpolation and Taylor extrapolation is a hybrid method called Hermite interpolation, which

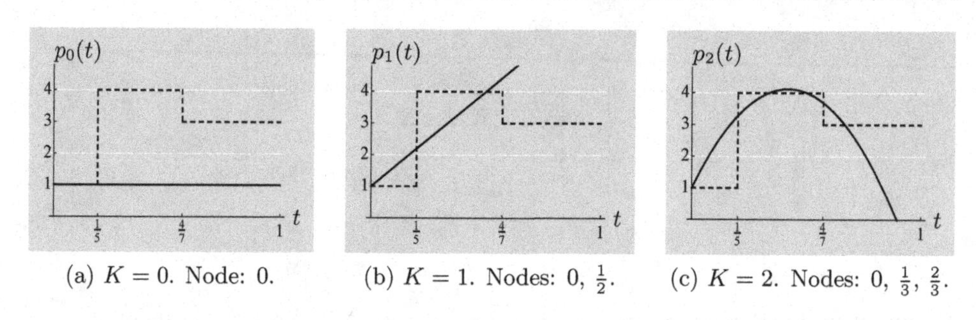

(a) $K = 0$. Node: 0. (b) $K = 1$. Nodes: 0, $\frac{1}{2}$. (c) $K = 2$. Nodes: 0, $\frac{1}{3}$, $\frac{2}{3}$.

Figure 6.3 Polynomial approximations p_K (solid lines) of x from Figure 6.1 (dashed lines) using polynomials of degree K determined by Lagrange interpolation with $K + 1$ evenly spaced nodes.

matches both points and derivatives. While orthogonal projection to a polynomial space minimizes the $\mathcal{L}^2$ norm of the approximation error, Lagrange, Taylor, and Hermite methods do not minimize a particular norm. Minimizing the maximum error leads to minimax polynomial approximation and Chebyshev polynomials.

Approximation of functions by splines A spline is a piecewise-polynomial function of degree K that has continuous derivatives up to order $K - 1$. We consider in particular splines that are uniform, meaning that their pieces are of equal length; uniform splines generate a shift-invariant subspace and are closely related to regular sampling. Functions in spline spaces are another example (besides bandlimited functions) where discrete-time processing implements continuous-time processing in a precise way, and we will show this specifically for derivatives and integrals.

Consider approximating the function in Figure 6.1 using uniform, degree-0 splines. This seems like a good idea because both a degree-0 spline and the function we want to approximate are piecewise-constant. The spline is constant over intervals of length $1/N$ when the interval $[0, 1]$ is split into N pieces of equal length. In Figure 6.4, we show least-squares approximations using 3, 6, and 9 pieces. As the number of pieces N increases, the approximation improves, although, in general, even for a function that is piecewise-constant, there is no value of N such that the function is exactly represented with a uniform, degree-0 spline.[99]

Linear approximation in Fourier bases Since we are considering a function on an interval, it is natural to look at the Fourier series representation. We already know that the Gibbs phenomenon will lead to poor convergence at discontinuities. Figure 6.5 shows Fourier series approximations using the K lowest-frequency[100] terms, with $K = 3$, $K = 9$, and $K = 15$. Since the Fourier series is an orthonormal system, we obtain a least-squares approximation by simple truncation; the

[99]The function in Figure 6.1 is represented exactly when N is any integer multiple of 35; exact representation for certain values of N occurs because the points of discontinuity are rational.

[100]Low refers to low absolute value; we could call these K *central* frequencies, with indices in $\{-\frac{1}{2}(K-1), \ldots, \frac{1}{2}(K-1)\}$, for odd K.

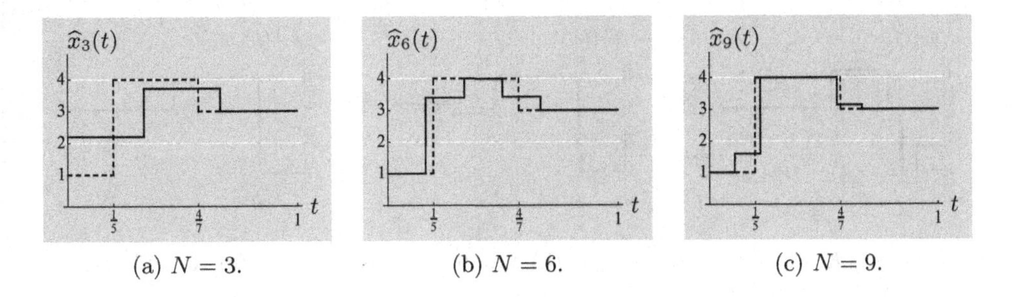

(a) $N = 3$. (b) $N = 6$. (c) $N = 9$.

Figure 6.4 Uniform spline approximations $\widehat{x}_N$ (solid lines) of x from Figure 6.1 (dashed lines) using splines of degree 0 with knots at integer multiples of $1/N$.

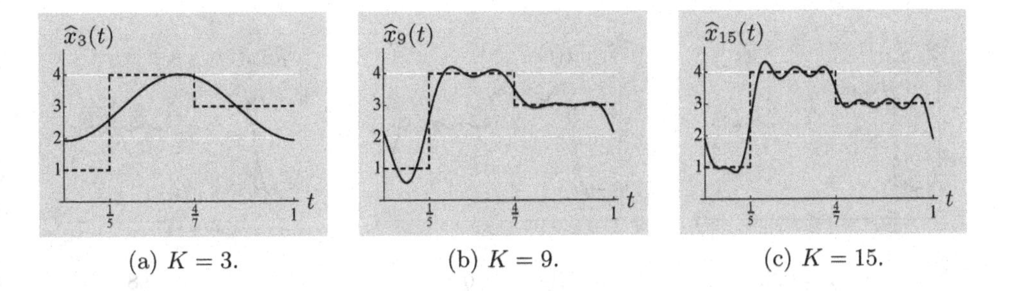

(a) $K = 3$. (b) $K = 9$. (c) $K = 15$.

Figure 6.5 Linear approximations $\widehat{x}_K$ (solid lines) of x from Figure 6.1 (dashed lines) using the K lowest-frequency terms of the Fourier series.

trigonometric polynomials used in the Fourier series are not a good match to the piecewise-constant function and thus convergence is slow.

The approximation we just saw is an example of *linear approximation*. This is because we decided a priori that $\widehat{x}$ would be the orthogonal projection of x onto a (fixed) subspace spanned by the K lowest-frequency Fourier series basis vectors, and the approximation is thus a linear operator applied to x.

Nonlinear approximation An alternative is to choose the largest-magnitude coefficients in the orthonormal basis; the set of basis vectors used now depends on the particular x we wish to approximate. This is an *adaptive subspace approximation*, because we choose the *best* subspace depending on the function to be approximated; it is a nonlinear function of x because the subspace selected to approximate a sum $x + y$ is generally not the same as the subspaces selected to approximate the summands x and y separately. In Figure 6.6, we show this approximation using an orthonormal Haar wavelet basis from (1.6)–(1.7), retaining the largest coefficients in the expansion (1.9). We explore both linear and nonlinear approximations using orthonormal bases in what follows.

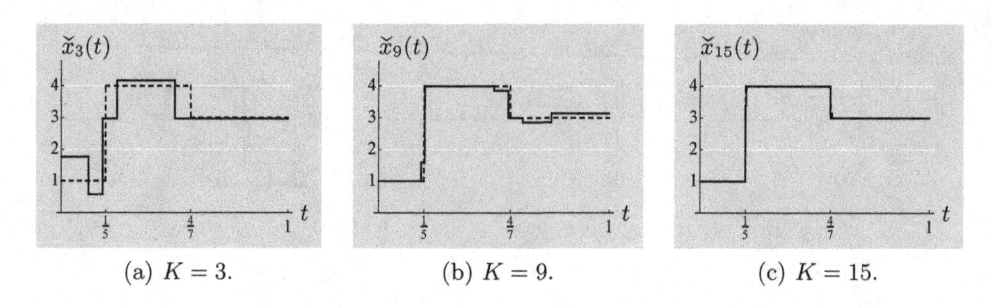

(a) $K = 3$. (b) $K = 9$. (c) $K = 15$.

Figure 6.6 Nonlinear approximations $\check{x}_K$ (solid lines) of x from Figure 6.1 (dashed lines) using the K largest-magnitude coefficients in a Haar wavelet basis expansion.

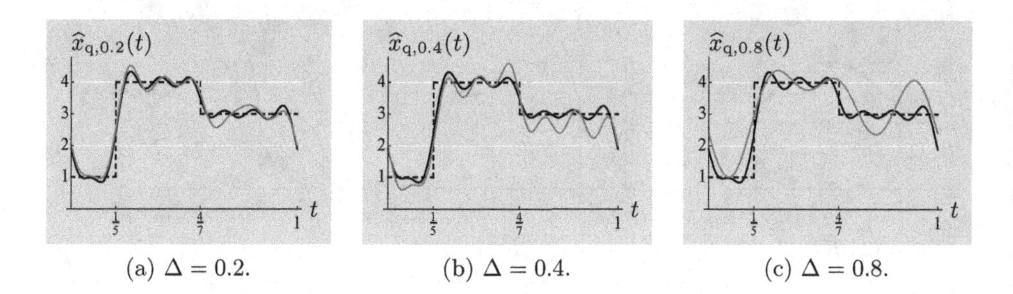

(a) $\Delta = 0.2$. (b) $\Delta = 0.4$. (c) $\Delta = 0.8$.

Figure 6.7 Linear approximations with quantized coefficients $\widehat{x}_{q,\Delta}$ (solid gray lines) of x from Figure 6.1 (dashed lines) using rounding to the nearest multiple of Δ of the 15 lowest-frequency terms of the Fourier series. The linear approximation with unquantized coefficients $\widehat{x}_{15}$ from Figure 6.5 is also shown (solid lines).

Compression Finally, we turn to compression, where we need to quantize the parameters of any representation. This means that when a representation is to be stored on a computer or digitally transmitted, the parameters must be in countable sets, and the number of bits used to describe the signal becomes the currency. Take the truncated Fourier series representation with 15 terms in Figure 6.5(c). One way to quantize the coefficients is to round to the nearest multiple of a quantization step size Δ. Increasing Δ reduces the number of bits needed to represent each quantized coefficient but increases the approximation error. Figure 6.7 shows some results.

The example raises many questions that are at the heart of signal compression: What are the best representations for compression? What are good quantization strategies? How should one allocate bits in a representation? The answers will depend both on the signal models and on the acceptable level of computational complexity. A branch of information theory called rate–distortion theory establishes some fundamental bounds in a somewhat abstract setting where complexity is not an issue. In practical signal compression, transform coding is the most common technique, especially in multimedia compression standards.

Chapter outline

The chapter follows the progression of topics in this brief introduction. Section 6.2 discusses approximation of functions on finite intervals by polynomials, starting with the orthonormal basis given by Legendre polynomials, moving to matching a function at specific points using Lagrange interpolation, and reviewing the classic Taylor expansion that matches a function and its derivatives at a given point. We discuss minimax approximation, near-minimax approximation using Chebyshev polynomials, and filter design. Section 6.3 looks into approximating functions on the real line by splines and the shift-invariant subspaces they generate. We calculate explicit projections on spline spaces, as well as orthogonalizations. We also calculate continuous-time operators such as derivatives and integrals using the discrete sequence of spline coefficients. Section 6.4 considers approximations in bases via truncation of series, both with linear and with nonlinear methods. For stochastic processes, the Karhunen–Loève transform (KLT) is derived as an optimal linear approximation method. In Section 6.5, we present the basic components of compression – entropy coding and quantization – and combine them with a linear change of basis to yield transform coding. Section 6.6 closes with computational aspects.

6.2 Approximation of functions on finite intervals by polynomials

The previous chapter focused on sampling and interpolation operators and both exact and approximate representations of sequences and functions using these operators. The infinite lengths encountered there made it imperative to use LSI filtering in the sampling and interpolation operations; otherwise, the computations could become prohibitively complicated. We now shift our attention to approximating a function on a finite interval using polynomials. Here, we will not have shift-invariance properties; in fact, the behavior near the endpoints of the interval is often quite different from the behavior near the center.

Throughout this section, we denote the set of polynomials of degree at most K defined on the finite interval $[a, b] \subset \mathbb{R}$ as $\mathcal{P}_K[a, b]$ (see (3.296)). We denote the function to be approximated on $[a, b]$ by x, the approximating polynomial in $\mathcal{P}_K[a, b]$ by

$$p_K(t) = \sum_{k=0}^{K} \alpha_k t^k, \qquad t \in [a, b], \tag{6.1a}$$

and the error between the function x and its approximation p_K by

$$e_K(t) = x(t) - p_K(t), \qquad t \in [a, b]. \tag{6.1b}$$

Minimizing the different norms of the error leads to different types of approximations, and we also consider approximations based on enforcing certain equalities – in values and derivatives – between x and p_K. These approximations have various strengths and weaknesses.

We start in Section 6.2.1 with least-squares polynomial approximation, in which series expansions with respect to orthogonal polynomials in general, and

Legendre polynomials in particular, arise naturally. Matching approximating polynomials to a given function at specific points gives Lagrange interpolation in Section 6.2.2, matching derivatives up to a certain order at a single point gives Taylor series expansion in Section 6.2.3, and matching both values and derivatives at specific points gives Hermite interpolation in Section 6.2.4. Minimizing the maximum absolute error (the $\mathcal{L}^\infty$ norm of approximation error) is first studied generally in Section 6.2.5 and then applied to FIR filter design in Section 6.2.6.

6.2.1 Least-squares approximation

Let x be a real-valued function in $\mathcal{L}^2([a, b])$. An approximation $\widehat{x}$ that minimizes

$$\|x - \widehat{x}\|_2^2 = \int_a^b (x(t) - \widehat{x}(t))^2 \, dt$$

over some set of candidates for $\widehat{x}$ is called a *least-squares* approximation. When the set of candidates is a closed subspace, the least-squares approximation is given by the orthogonal projection to the subspace (see Theorem 2.26). Furthermore, when an orthonormal basis for the subspace is available, this orthogonal projection can be computed using Theorem 2.41.

The set $\mathcal{P}_K[a, b]$ is a closed subspace of $\mathcal{L}^2([a, b])$ with dimension $K+1$. With $\{\varphi_0, \varphi_1, \ldots, \varphi_K\}$ an orthonormal basis for $\mathcal{P}_K[a, b]$, the least-squares approximation of x is

$$p_K(t) = \sum_{k=0}^{K} \langle x, \varphi_k \rangle \varphi_k(t).$$

This is a polynomial of degree at most K because each φ_k is a polynomial of degree at most K.

Example 6.1 (Least-squares approximation) Let us find the least-squares degree-1 polynomial approximation of $x(t) = \sin(\frac{1}{2}\pi t)$ on $[0, 1]$. For this, we need an orthonormal basis for $\mathcal{P}_1[0, 1]$; one such basis is $\{1, \sqrt{3}(2t-1)\}$.[101] The least-squares approximation is

$$\begin{aligned} p_1(t) &= \left\langle \sin(\tfrac{1}{2}\pi t), 1 \right\rangle \cdot 1 + \left\langle \sin(\tfrac{1}{2}\pi t), \sqrt{3}(2t-1) \right\rangle \sqrt{3}(2t-1) \\ &= \frac{2}{\pi} + \frac{2\sqrt{3}(4-\pi)}{\pi^2}\sqrt{3}(2t-1) = \frac{12(4-\pi)}{\pi^2}t + \frac{8(\pi-3)}{\pi^2}. \end{aligned}$$

This approximation is illustrated in Figure 6.8.

The method above uses an orthonormal basis for $\mathcal{P}_K[a, b]$. Such an orthonormal basis is not unique. However, if one seeks a single sequence of vectors $\{\varphi_0, \varphi_1, \ldots\}$ such that, for every $K \in \mathbb{N}$,

$$\{\varphi_0, \varphi_1, \ldots, \varphi_K\} \quad \text{is an orthonormal basis for} \quad \mathcal{P}_K[a, b],$$

[101] This particular basis comes from applying the Gram–Schmidt procedure to $(1, t)$.

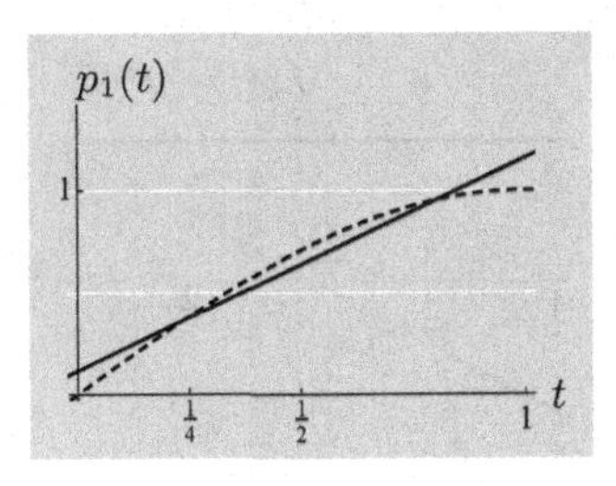

Figure 6.8 Least-squares approximation p_1 (solid line) of $x(t) = \sin(\frac{1}{2}\pi t)$ (dashed line) on $[0, 1]$ among polynomials of degree $K = 1$.

the solution is unique up to multiplications by ± 1 and is obtained by using the Gram–Schmidt procedure on $\{1, t, t^2, \ldots\}$. This construction creates an orthonormal basis for $\mathcal{P}_K[a, b]$ that is the union of an orthonormal basis for $\mathcal{P}_{K-1}[a, b]$ and a degree-K polynomial φ_K. With this nested sequence of bases, the least-squares approximations satisfy the recursion

$$p_K = p_{K-1} + \langle x, \varphi_K \rangle \varphi_K, \tag{6.2}$$

which is a special case of the successive approximation property (2.108).

Legendre polynomials The *Legendre polynomials,*

$$L_k(t) = \frac{1}{2^k k!} \frac{d^k}{dt^k} (t^2 - 1)^k, \qquad k \in \mathbb{N}, \tag{6.3}$$

are orthogonal on $[-1, 1]$. The first few are shown in Figure 6.9 and are listed below:

$$L_0(t) = 1, \qquad L_3(t) = \tfrac{1}{2}(5t^3 - 3t),$$
$$L_1(t) = t, \qquad L_4(t) = \tfrac{1}{8}(35t^4 - 30t^2 + 3),$$
$$L_2(t) = \tfrac{1}{2}(3t^2 - 1), \qquad L_5(t) = \tfrac{1}{8}(63t^5 - 70t^3 + 15t).$$

Legendre polynomials are orthogonal but not orthonormal; an orthonormal set can be obtained by dividing each polynomial by its norm $\|L_k\|_2 = \sqrt{2/(2k+1)}$. The first few in normalized form are

$$\bar{L}_0(t) = \tfrac{1}{\sqrt{2}}, \qquad \bar{L}_3(t) = \tfrac{\sqrt{7}}{2\sqrt{2}}(5t^3 - 3t),$$
$$\bar{L}_1(t) = \tfrac{\sqrt{3}}{\sqrt{2}}t, \qquad \bar{L}_4(t) = \tfrac{3}{8\sqrt{2}}(35t^4 - 30t^2 + 3),$$
$$\bar{L}_2(t) = \tfrac{\sqrt{5}}{2\sqrt{2}}(3t^2 - 1), \qquad \bar{L}_5(t) = \tfrac{\sqrt{11}}{8\sqrt{2}}(63t^5 - 70t^3 + 15t).$$

Once we have the Legendre polynomials, changing the interval of interest does not require a tedious application of the Gram–Schmidt procedure as we have done in Solved exercise 2.5. Instead, polynomials that are orthogonal with respect to the $\mathcal{L}^2$ inner product on $[a, b]$ can be found by shifting and scaling the Legendre polynomials; see Exercise 6.1.

Example 6.2 (Approximation with Legendre polynomials) Let

$$x_1(t) = t \sin 5t. \tag{6.4}$$

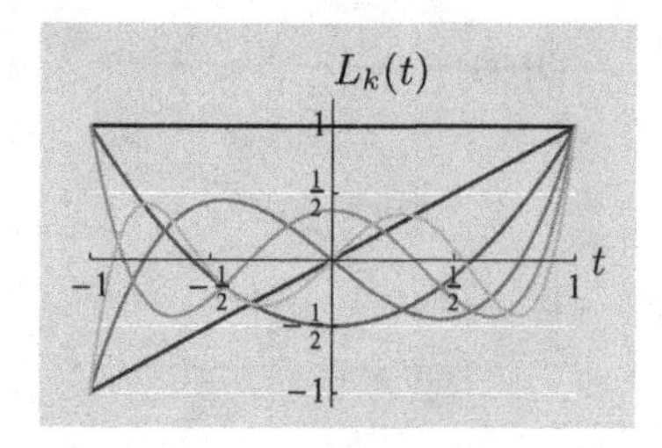

Figure 6.9 The first six Legendre polynomials $\{L_k\}_{k=0}^{5}$ (solid lines, from darkest to lightest); these are orthogonal on the interval $[-1, 1]$.

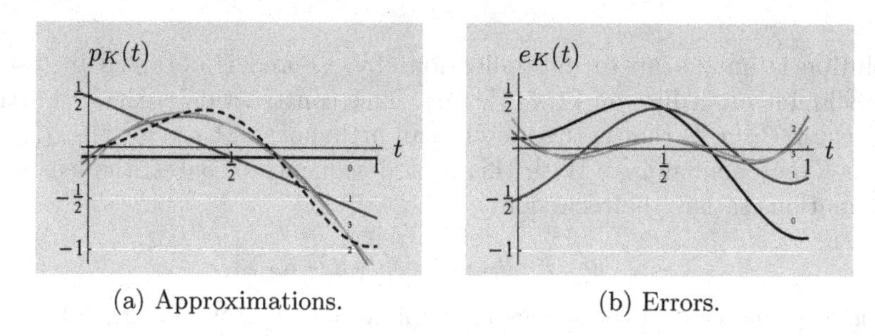

(a) Approximations. (b) Errors.

Figure 6.10 Least-squares approximations p_K and approximation errors e_K, $K = 0, 1, 2, 3$ (solid lines, from darkest to lightest), of x_1 in (6.4) on $[0, 1]$ (dashed line) using an orthogonal polynomial basis of degree K. The curves are labeled by the polynomial degree.

To form least-squares polynomial approximations on $[0, 1]$, we need orthogonal polynomials in $\mathcal{L}^2([0, 1])$; we can obtain these by shifting and scaling the Legendre polynomials. The first few, in normalized form, are

$$\begin{aligned} \varphi_0(t) &= 1, & \varphi_2(t) &= \sqrt{5}(6t^2 - 6t + 1), \\ \varphi_1(t) &= \sqrt{3}(2t - 1), & \varphi_3(t) &= \sqrt{7}(20t^3 - 30t^2 + 12t - 1). \end{aligned}$$

(The first two of these polynomials were used in Example 6.1.) The best constant approximation is $p_0 = \langle x, \varphi_0 \rangle \varphi_0$, and higher-degree least-squares approximations can be found through (6.2). The least-squares approximations up to degree 3 are

$$\begin{aligned} p_0(t) &\approx -0.10, & p_2(t) &\approx -0.11 + 2.42t - 3.59t^2, \\ p_1(t) &\approx 0.49 - 1.17t, & p_3(t) &\approx -0.20 + 3.56t - 6.43t^2 + 1.90t^3. \end{aligned}$$

The resulting approximations and errors are shown in Figure 6.10.

Other orthogonal polynomials While Legendre polynomials and their shifted versions solve our least-squares approximation problems, changing the inner product to

$$\langle x, y \rangle = \int_a^b x(t)\, y(t)\, W(t)\, dt,$$

Family	Weight	Interval	Degree-k polynomial Expression	Recursion
Legendre	1	$[-1, 1]$	$\frac{1}{2^k k!}\frac{d^k}{dt^k}(t^2-1)^k$	$\frac{2k-1}{k}tL_{k-1}(t) - \frac{k-1}{k}L_{k-2}(t)$
Chebyshev	$\frac{1}{\sqrt{1-t^2}}$	$[-1, 1]$	$\cos(k \arccos t)$	$2tT_{k-1}(t) - T_{k-2}(t)$
Laguerre	e^{-t}	$[0, \infty)$	$\frac{e^t}{k!}\frac{d^k}{dt^k}(e^{-t}t^k)$	$\frac{2k-1-t}{k}L_{k-1}(t) - \frac{k-1}{k}L_{k-2}(t)$
Hermite	e^{-t^2}	$(-\infty, \infty)$	$a^{-3k/2}k!t^k \cdot \sum_{\ell=0}^{\lfloor k/2 \rfloor}\frac{(-2t^2/a)^{-\ell}}{\ell!(k-2\ell)!}$	$2tH_{k-1}(t) - 2(k-1)H_{k-2}(t)$

Table 6.1 Families of orthogonal polynomials.

where $W(t)$ is a nonnegative *weight function*, changes the orthogonality relationships among the polynomials. One of the ramifications is that applying the Gram–Schmidt procedure to the ordered monomials $\{1, t, t^2, \ldots\}$ yields a different set of polynomials. Also, if the weight function has adequate decay, inner products between polynomials on $[a, \infty)$ or $(-\infty, \infty)$ can be finite, so we need not consider only finite intervals. Table 6.1 gives families of orthogonal polynomials that arise from a few choices of weight functions and intervals of orthogonality.

Orthogonal polynomials have many applications in approximation theory and numerical analysis; Exercises 6.2 and 6.3 explore some of their properties. We will consider Chebyshev polynomials in Section 6.2.5 because they play an important role in $\mathcal{L}^\infty$ approximation.

6.2.2 Lagrange interpolation: Matching points

In many situations in which we desire a polynomial approximation of a function on an interval, we cannot use inner products of the function with a set of basis functions because we might not know the function on the entire interval, or we might not want to compute the required integrals.[102] However, if we know the values of the function at certain points in the interval, we can find the polynomial that matches the function at these points. We now look at when such matching exists, when it is unique, and how to bound the approximation error.

From (6.1a), a polynomial of degree K has $K+1$ parameters and is thus determined by $K+1$ independent and consistent constraints. Let us look carefully at whether specifying p_K at $K+1$ points provides suitable constraints.

Fix a set of $K+1$ distinct nodes $\{t_k\}_{k=0}^{K} \subset [a, b]$, and assume that the values of the function x at the nodes, $y_k = x(t_k)$, $k = 0, 1, \ldots, K$, are known. Requiring p_K to match x at the nodes gives a system of $K+1$ linear equations with $K+1$

[102]The difficulty of computing integrals is one common reason for forming a polynomial approximation; integrating the polynomial approximant leads to numerical integration techniques such as the trapezoidal rule and Simpson's rule.

unknowns,

$$\begin{bmatrix} 1 & t_0 & t_0^2 & \cdots & t_0^K \\ 1 & t_1 & t_1^2 & \cdots & t_1^K \\ \vdots & \vdots & \vdots & & \vdots \\ 1 & t_K & t_K^2 & \cdots & t_K^K \end{bmatrix} \begin{bmatrix} \alpha_0 \\ \alpha_1 \\ \vdots \\ \alpha_K \end{bmatrix} = \begin{bmatrix} y_0 \\ y_1 \\ \vdots \\ y_K \end{bmatrix}. \qquad (6.5)$$

The matrix in (6.5) is a square Vandermonde matrix, and, as such, it is invertible if and only if the nodes are distinct; see (2.249). Thus, the values of x at $K+1$ distinct nodes uniquely specify an interpolation polynomial of degree at most K. As we know from Appendix 2.B, having more samples (equations) will not lead to a solution: the range of the Vandermonde matrix will be a proper subspace and the vector of samples will (in general) not be in that subspace. On the other hand, having fewer samples always leaves the polynomial unspecified, as there will be infinitely many solutions to (6.5).

Solving the system of linear equations (6.5) is one way to find an interpolating polynomial; however, we use this only for the uniqueness argument. Knowing now that the degree-K polynomial interpolating $K+1$ distinct points $\{(t_k, y_k)\}_{k=0}^K$ is unique, any formula for a degree-K polynomial p_K that satisfies the interpolation condition $p_K(t_k) = y_k$, $k = 0, 1, \ldots, K$, gives the interpolating polynomial. One such formula is as follows:

DEFINITION 6.1 (LAGRANGE INTERPOLATING POLYNOMIAL) Given $K \in \mathbb{N}$ and values of a function x at $K+1$ distinct nodes $\{t_k\}_{k=0}^K$,

$$p_K(t) = \sum_{k=0}^{K} x(t_k) \prod_{\substack{i=0 \\ i \neq k}}^{K} \frac{t - t_i}{t_k - t_i} \qquad (6.6)$$

is called the *Lagrange interpolating polynomial* for $\{(t_k, x(t_k))\}_{k=0}^K$.

The Lagrange polynomial interpolates correctly because

$$\prod_{\substack{i=0 \\ i \neq k}}^{K} \frac{t_\ell - t_i}{t_k - t_i} = \begin{cases} 1, & \text{for } \ell = k \in \{0, 1, \ldots, K\}; \\ 0, & \text{for } \ell \in \{0, 1, \ldots, K\} \setminus \{k\}. \end{cases}$$

The Lagrange interpolating polynomial equation is derived in Exercise 6.4.

EXAMPLE 6.3 (LAGRANGE INTERPOLATION) Let us construct approximations using (6.6) for two functions on $[0, 1]$, one continuous, x_1 from (6.2), and the other not,

$$x_2(t) = \begin{cases} t, & \text{for } 0 \le t < 1/\sqrt{2}; \\ t-1, & \text{for } 1/\sqrt{2} \le t < 1. \end{cases} \qquad (6.7)$$

Let the nodes be k/K for $k = 0, 1, \ldots, K$ (although even spacing of nodes is not a requirement). Figure 6.11 shows the functions (dashed lines) and the

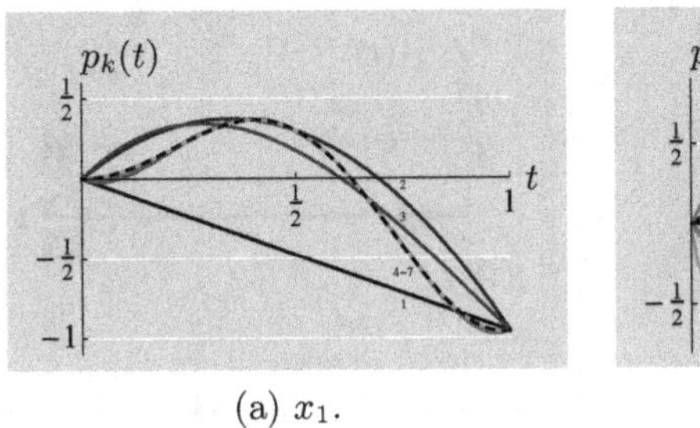

(a) x_1.

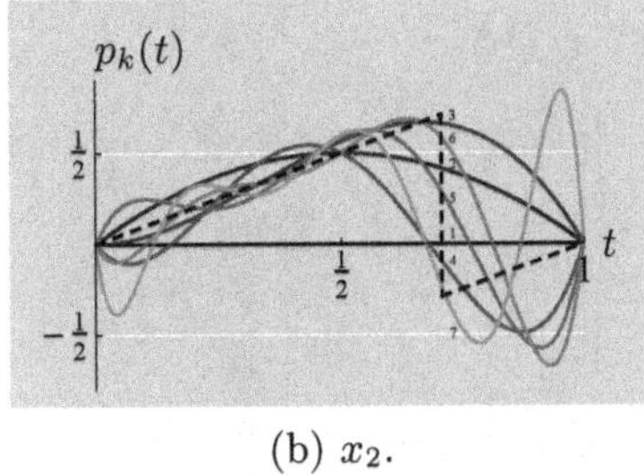

(b) x_2.

Figure 6.11 Lagrange interpolations p_K, $K = 1, 2, \ldots, 7$ (solid lines, from darkest to lightest), of x_1 in (6.4) and x_2 in (6.7) (dashed lines) using $K+1$ nodes evenly spaced over $[0, 1]$: $t_k = k/K$, $k = 0, 1, \ldots, K$. The curves are labeled by the polynomial degree.

interpolating polynomials for $K = 1, 2, \ldots, 7$ (solid lines, from darkest to lightest). The continuous function x_1 is approximated much more closely than the discontinuous function x_2.

This example suggests that, when polynomial interpolation is used, the quality of the approximation is affected by the smoothness of the function. Indeed, for functions that are sufficiently smooth over the range of interest (encompassing the nodes but also wherever the approximation is to be evaluated), the pointwise error can be bounded precisely using the following theorem:

THEOREM 6.2 (ERROR OF LAGRANGE INTERPOLATION) Let $K \in \mathbb{N}$, let $\{t_k\}_{k=0}^{K} \subset [a, b]$ be distinct, and assume that x has $K+1$ continuous derivatives on $[a, b]$. Then, for p_K defined in (6.6) and any $t \in [a, b]$, the error $e_K = x - p_K$ satisfies

$$e_K(t) = \frac{\prod_{k=0}^{K}(t - t_k)}{(K+1)!} x^{(K+1)}(\xi) \tag{6.8a}$$

for some ξ between the minimum and maximum of $\{t, t_0, t_1, \ldots, t_K\}$. Thus,

$$|e_K(t)| \leq \frac{\prod_{k=0}^{K}|t - t_k|}{(K+1)!} \max_{\xi \in [a, b]} \left|x^{(K+1)}(\xi)\right|, \tag{6.8b}$$

for every $t \in [a, b]$.

One immediate consequence of the theorem is that if x is a polynomial of degree K, the interpolant p_K matches everywhere. This follows from (6.8a) or (6.8b) because $x^{(K+1)}$ is identically zero. Of course, this was already obvious from the uniqueness of the interpolating polynomial. In general, the factor $\max_{\xi \in [a, b]}|x^{(K+1)}(\xi)|$ in the error bound is a global smoothness measure; it affects the pointwise error bound identically over the entire interval.

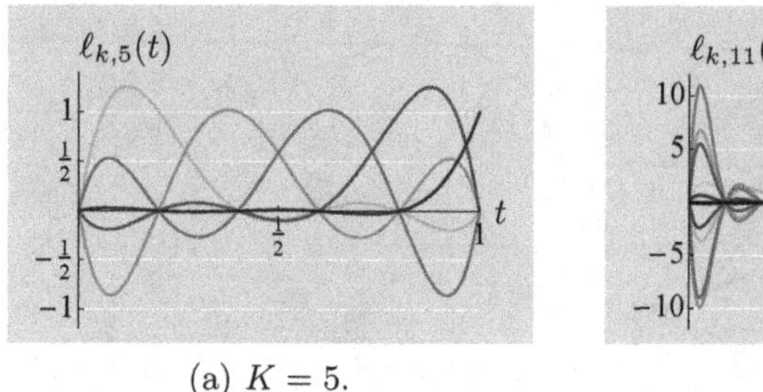

(a) $K = 5$.

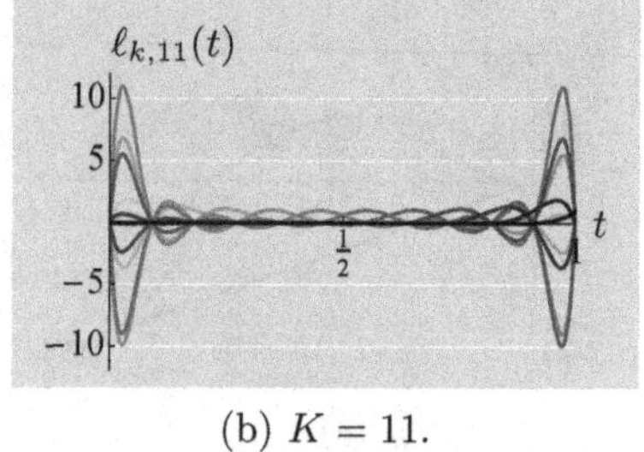

(b) $K = 11$.

Figure 6.12 Bases of degree-K Lagrange interpolating polynomials for $K+1$ nodes evenly spaced over $[0, 1]$: $t_k = k/K$, $k = 0, 1, \ldots, K$ (solid lines, from darkest to lightest).

One interesting aspect of the error bound (6.8b) is that it depends on t through the $\prod_{k=0}^{K} |t - t_k|$ factor. The error is zero at any node because of the interpolating property, but the bound behaves differently in the neighborhoods of different nodes. Moving away from a node by a small amount ε, that is, setting $t = t_k + \varepsilon$, we have

$$\prod_{k=0}^{K} |t - t_k| \;\approx\; |\varepsilon| \prod_{i,\, i \neq k} |t_k - t_i|.$$

If the nodes are evenly spaced, the error bound becomes worse more quickly around an extremal node than around a central node (see also Figure 6.11). The potential for worse behavior at the endpoints can also be seen if we view the Lagrange interpolation formula as a series expansion using $K + 1$ polynomials

$$\ell_{k,K}(t) \;=\; \prod_{\substack{i\,=\,0 \\ i\,\neq\,k}}^{K} \frac{t - t_i}{t_k - t_i}, \qquad k = 0, 1, \ldots, K, \tag{6.9}$$

as a basis. Each basis function depends on all of the nodes; two examples with evenly spaced nodes are shown in Figure 6.12. These illustrate that, at node t_k, the basis function $\ell_{k,K}$ is 1 and the other basis functions are 0. Also, $\ell_{k,K}(t)$ is not necessarily small far away from t_k. This means that the sample $x(t_k)$ affects the interpolation far from t_k. In particular, Figure 6.12(b) illustrates that, unless K is small, the basis functions become large near the ends of the approximation interval. This is problematic for controlling the pointwise error. We will see later that a different spacing of nodes can improve the situation substantially.

6.2.3 Taylor series expansion: Matching derivatives

In the previous section, we used the Vandermonde system (6.5) to establish that matching $K+1$ sample values will uniquely specify a degree-K polynomial approximation of a real-valued function in $\mathcal{L}^2([a, b])$. Other ways of specifying $K + 1$ constraints exist, one of the most familiar being the Taylor series expansion.

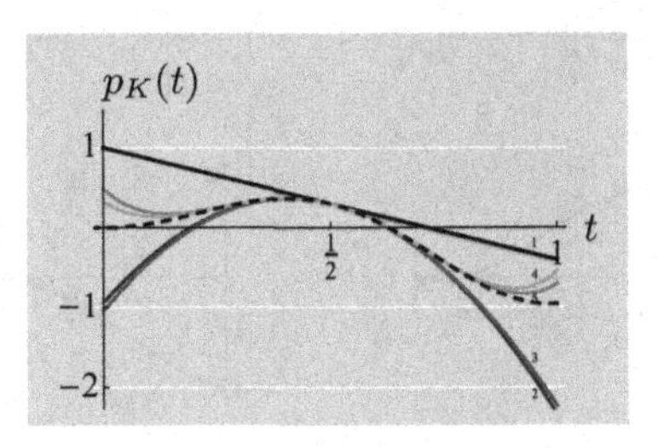

Figure 6.13 Taylor series approximations p_K, $K = 1, 2, \ldots, 5$ (solid lines, from darkest to lightest), of x_1 in (6.4) on $[0, 1]$ (dashed line) using expansion around $\frac{1}{2}$. The curves are labeled by the polynomial degree.

DEFINITION 6.3 (TAYLOR SERIES EXPANSION) Given $K \in \mathbb{N}$ and values of a function x and its first K derivatives at t_0,

$$p_K(t) = \sum_{k=0}^{K} \frac{(t-t_0)^k}{k!} x^{(k)}(t_0) \tag{6.10}$$

is called the *Taylor series expansion* of x around t_0.

The Taylor series expansion has the property that p_K and x have equal derivatives of order $0, 1, \ldots, K$ at t_0. (The zeroth derivative is the function itself.)

EXAMPLE 6.4 (TAYLOR SERIES EXPANSION) Consider Taylor series expansions of the functions x_1 from (6.2) and x_2 from (6.7) around $\frac{1}{2}$. Figure 6.13 shows x_1 and its expansions of degree $K = 1, 2, \ldots, 5$. The Taylor series for x_2 is not interesting to plot because we have $p_K(t) = t$ for all degrees ≥ 1. While this is exact for $t \in [0, 1/\sqrt{2}]$, the error for $t \in (1/\sqrt{2}, 1]$ is not reduced by increasing the degree.

A Taylor series expansion has the peculiar property of getting all its information about x from an infinitesimal interval around t_0. As Example 6.4 illustrated, this means that the approximation can differ from the original function by an arbitrary amount if the function is discontinuous. For functions with sufficiently many continuous derivatives, a precise error bound is given by the following theorem:

THEOREM 6.4 (ERROR OF TAYLOR SERIES EXPANSION) Let $K \in \mathbb{N}$, let $t_0 \in [a, b]$, and assume that x has $K+1$ continuous derivatives on $[a, b]$. Then, for p_K defined in (6.10), the error $e_K = x - p_K$ satisfies

$$e_K(t) = \frac{(t-t_0)^{K+1}}{(K+1)!} x^{(K+1)}(\xi) \tag{6.11a}$$

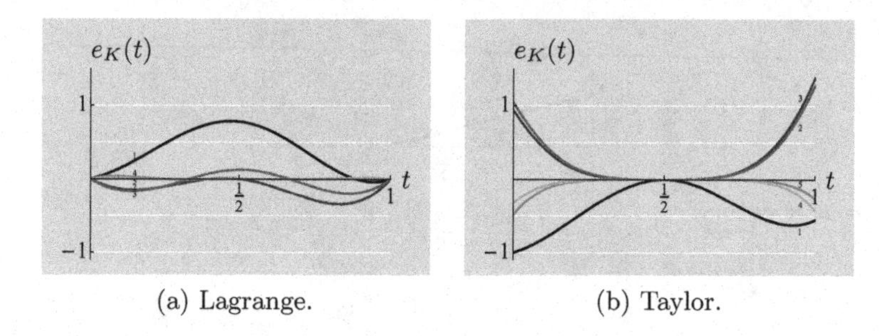

(a) Lagrange. (b) Taylor.

Figure 6.14 Errors of Lagrange and Taylor approximations e_K, $K = 0, 1, \ldots, 5$ (solid lines, from darkest to lightest), of x_1 in (6.4). The Lagrange interpolations are with evenly spaced nodes (see Figure 6.11). The Taylor series expansions are around $\frac{1}{2}$ (see Figure 6.13). The curves are labeled by the polynomial degree.

for some ξ between t and t_0. Thus,

$$|e_K(t)| \leq \frac{|t - t_0|^{K+1}}{(K+1)!} \max_{\xi \in [a, b]} \left| x^{(K+1)}(\xi) \right|, \tag{6.11b}$$

for every $t \in [a, b]$.

The error bounds (6.8b) and (6.11b) are very similar. Both are proportional to

$$\frac{1}{(K+1)!} \max_{\xi \in [a, b]} \left| x^{(K+1)}(\xi) \right|,$$

the global smoothness measure. They differ in the dependence on t, but in a way that is consistent: If every t_k in (6.8b) is replaced by t_0, then the two error bounds are identical. Lagrange interpolation depends on the nodes being distinct, so we cannot literally make all the t_k values equal. However, having $K+1$ nodes distinct but closely clustered is similar to having derivatives up to order K at a single node. This is explored further in Solved exercise 6.1. Figure 6.14 shows a comparison of the errors for Lagrange and Taylor approximations of x_1 in (6.4).

The error bounds (6.8b) and (6.11b) require greater smoothness as the polynomial degree is increased. Furthermore, there exist infinitely differentiable functions for which these bounds do not even decrease as K is increased; see Exercise 6.5.

6.2.4 Hermite interpolation: Matching points and derivatives

A natural combination of the ideas of Lagrange interpolation and Taylor series is to determine a polynomial by fixing some number of derivatives at each of several nodes. This is called *Hermite interpolation.*

EXAMPLE 6.5 (HERMITE INTERPOLATION) Suppose that we are given the values of a function x and its derivative at distinct t_0 and t_1:

$$y_0 = x(t_0), \qquad z_0 = x'(t_0), \qquad y_1 = x(t_1), \qquad z_1 = x'(t_1).$$

We want to confirm that fixing these four values uniquely determines a cubic polynomial. Write the cubic polynomial and its derivative as

$$\begin{aligned} p_3(t) &= \alpha_0 + \alpha_1 t + \alpha_2 t^2 + \alpha_3 t^3, \\ p_3'(t) &= \alpha_1 + 2\alpha_2 t + 3\alpha_3 t^2. \end{aligned}$$

Then finding their values at t_0 and t_1 yields

$$\begin{bmatrix} 1 & t_0 & t_0^2 & t_0^3 \\ 1 & t_1 & t_1^2 & t_1^3 \\ 0 & 1 & 2t_0 & 3t_0^2 \\ 0 & 1 & 2t_1 & 3t_1^2 \end{bmatrix} \begin{bmatrix} \alpha_0 \\ \alpha_1 \\ \alpha_2 \\ \alpha_3 \end{bmatrix} = \begin{bmatrix} y_0 \\ y_1 \\ z_0 \\ z_1 \end{bmatrix}. \tag{6.12}$$

The matrix in this equation is invertible whenever $t_0 \neq t_1$ because then its determinant $-(t_1 - t_0)^4$ is nonzero. Thus, the polynomial is uniquely determined.

More generally, suppose that the derivatives of order $0, 1, \ldots, d_k$ are specified at each distinct node t_k for $k = 0, 1, \ldots, L$. Since the constraints are independent, a polynomial of degree $K = (\sum_{k=0}^{L}(d_k + 1)) - 1$ can be uniquely determined (see Exercise 6.6).

6.2.5 Minimax polynomial approximation

The techniques we have studied thus far determine a polynomial approximation through linear operations: Least-squares approximations – minimizing the $\mathcal{L}^2$ norm of an approximation error – come through the linear operation of orthogonal projection to a subspace, and the interpolation and extrapolation methods use some combination of samples of a function and its derivatives to determine a polynomial approximation through the solution of a system of linear equations. We now turn to the minimization of the maximum pointwise error – the $\mathcal{L}^\infty$ norm of the error. Since $\mathcal{L}^\infty([a, b])$ is not a Hilbert space, we do not have geometric notions such as the orthogonality of the error to the subspace of polynomials $\mathcal{P}_K[a, b]$ to guide the determination of the optimal approximation. We will see that the optimal polynomial approximation is generally more difficult to compute, but interpolation with specially chosen nodes is nearly optimal.

DEFINITION 6.5 (MINIMAX APPROXIMATION) Given $x \in \mathbb{C}^{[a,b]}$, an approximation $\widehat{x}$ that minimizes

$$\|x - \widehat{x}\|_\infty = \max_{t \in [a,b]} |x(t) - \widehat{x}(t)|$$

over some set of candidates for $\widehat{x}$ is called a *minimax approximation.*

The following theorem shows that the set of polynomials is rich enough to enable arbitrarily small $\mathcal{L}^\infty$ error for any continuous function. A constructive proof is discussed in Exercise 6.8.

THEOREM 6.6 (WEIERSTRASS APPROXIMATION THEOREM) Let x be continuous on $[a, b]$ and let $\varepsilon > 0$. Then, there exists a polynomial p for which

$$|e(t)| = |x(t) - p(t)| \leq \varepsilon \qquad \text{for every } t \in [a, b]. \tag{6.13}$$

Denote by p_K the minimax approximation among polynomials of degree at most K, and let $e_{p,K}$ be the error of that approximation with $\mathcal{L}^\infty$ norm $\varepsilon_{p,K} = \|e_{p,K}\|_\infty$. The theorem states that the mere continuity of x is enough to ensure that $\varepsilon_{p,K}$ can be made arbitrarily small by choosing K large enough. This contrasts starkly with the $\mathcal{L}^\infty$ error bounds that we have seen thus far, (6.8b) for Lagrange interpolation and (6.11b) for Taylor series expansion, which require greater smoothness as the polynomial degree is increased.

Sometimes bounds can be unduly pessimistic even when performance is reasonable, so it is useful to examine the performance more closely. With Lagrange interpolation, the difficulty from an $\mathcal{L}^\infty$ perspective is that the maximum errors between pairs of neighboring nodes can differ greatly, so the $\mathcal{L}^\infty$ error can be large even when the function and its approximation are close over most of the approximation interval. When the nodes are evenly spaced, the maximum error tends to be largest near the ends of the approximation interval; we observed this in Figure 6.14(a), connected also to the large peaks shown in Figure 6.12. The large oscillations near the endpoints can be so dramatic that the $\mathcal{L}^\infty$ error diverges as K increases, even for a C^∞ function; see Exercise 6.5. Thus, we must move beyond interpolation with evenly spaced nodes for minimax approximation. Moreover, with an approximating polynomial function q_K, it never pays to have fewer than $K+1$ points at which the error is zero. Both bounding the minimax error and computing a minimax approximation depend on understanding what happens between these points.

Let $\{t_k\}_{k=0}^{K} \subset [a, b]$ be $K+1$ distinct points in increasing order selected to partition $[a, b]$ into $K+2$ subintervals,

$$[a, b] = \bigcup_{k=0}^{K+1} I_k, \qquad \text{where} \quad I_k = \begin{cases} [a, t_0), & \text{for } k = 0; \\ [t_{k-1}, t_k), & \text{for } k = 1, 2, \ldots, K; \\ [t_K, b], & \text{for } k = K+1. \end{cases} \tag{6.14}$$

(While it is natural to think of the specified points as satisfying $e_{q,K}(t_k) = x(t_k) - q_K(t_k) = 0$ for $k = 0, 1, \ldots, K+1$, this is not necessary for the developments that follow.) Since the subintervals cover $[a, b]$, the $\mathcal{L}^\infty$ error is the maximum of the errors on the subintervals,

$$\varepsilon_{q,K} = \|e_{q,K}\|_\infty = \max_{k=0,1,\ldots,K+1} \sup_{t \in I_k} |e_{q,K}(t)|.$$

The following theorem shows that for a minimax approximation, the error $e_{q,K}$ should oscillate in sign from one subinterval I_k to the next and the maximum absolute value of the difference should be the same on every subinterval.

THEOREM 6.7 (DE LA VALLÉE-POUSSIN ALTERNATION THEOREM) Let x be continuous on $[a, b]$, and let $[a, b]$ be partitioned as in (6.14) for some $K \in \mathbb{N}$. Suppose that there exists a polynomial q_K of degree at most K and numbers $s_k \in I_k$ for $k = 0, 1, \ldots, K+1$, such that $e_{q,K}(s_k)$ alternates in sign. Then

$$\min_{k=0,1,\ldots,K+1} |e_{q,K}(s_k)| \leq \varepsilon_{p,K} \leq \varepsilon_{q,K}, \tag{6.15}$$

where $\varepsilon_{p,K}$ is the $\mathcal{L}^\infty$ norm of the error of the minimax approximation among polynomials of degree at most K.

The first of the two inequalities in (6.15) is the interesting one; the second simply states that the minimax approximation p_K is at least as good as q_K.

To use the theorem, we must find an approximating polynomial q_K that creates enough alternations in the sign of the error. Then, the strongest statement is obtained by choosing s_k such that each $|e_{q,K}(s_k)|$ is maximum on the subinterval I_k. The main result of the theorem is that there is no way to change the polynomial to push the error uniformly below the smallest of the local maxima of $|e_{q,K}(t)|$. Intuitively, pushing the worst (largest) of the local maxima down inevitably has the effect of pushing the best (smallest) of the local maxima up. The next theorem makes the stronger statement that equality of the local maxima happens if and only if the minimax approximation has been found.

THEOREM 6.8 (CHEBYSHEV EQUIOSCILLATION THEOREM) Let x be continuous on $[a, b]$, and let $K \in \mathbb{N}$. Denote by $\varepsilon_{p,K}$ the $\mathcal{L}^\infty$ norm of the error of the minimax approximation among polynomials of degree at most K. The minimax approximation p_K is unique and determined by the following property: There are at least $K+2$ points $\{s_k\}_{k=0}^{K+1}$ satisfying

$$a \leq s_0 < s_1 < \cdots < s_{K+1} \leq b$$

for which

$$x(s_k) - p_K(s_k) = \sigma(-1)^k \varepsilon_{p,K}, \qquad k = 0, 1, \ldots, K+1,$$

where $\sigma = \pm 1$, independently of k.

We illustrate both theorems by continuing our previous examples. The numerical calculations of minimax approximations in the following example were performed using the Remez algorithm; see the *Further reading*.

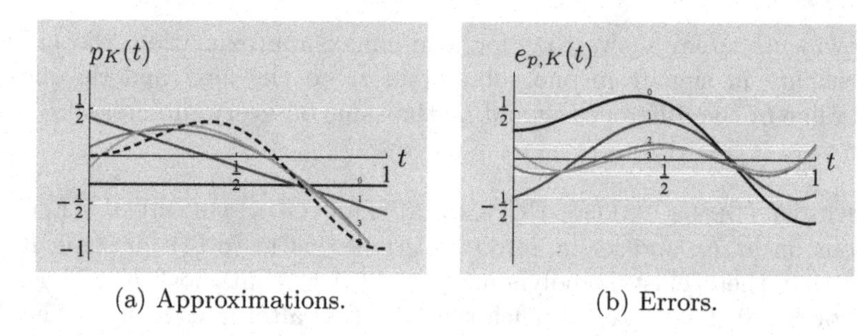

(a) Approximations. (b) Errors.

Figure 6.15 Minimax approximations p_k and approximation errors $e_{p,K}$, $K = 0, 1, 2, 3$ (solid lines, from darkest to lightest), of x_1 in (6.4) on $[0, 1]$ (dashed line). The curves are labeled by the polynomial degree. Horizontal grid lines in the error plot mark the maxima and minima, highlighting that, for each degree, the minimum is the negative of the maximum.

EXAMPLE 6.6 (MINIMAX APPROXIMATION) Consider again approximating $x_1(t) = t \sin 5t$ on $[0, 1]$ (from (6.4)). To draw conclusions from Theorem 6.7, we must find a q_K such that the error $e_{q,K}$ changes sign at least $K+1$ times. This is true for the least-squares approximations computed in Example 6.2, shown in Figure 6.10. For each degree $K \in \{0, 1, 2, 3\}$, the error $e_{q,K}$ does indeed change sign at least $K+1$ times, allowing us to choose $K+2$ values $s_k \in I_k$ to apply Theorem 6.7. (The error of the degree-2 approximation has one more sign change than necessary.) By choosing $\{s_k\}_{k=0}^{K+1}$ to give maxima of $|e_{q,K}(t)|$ in each interval, from (6.15) we get

$$\begin{aligned} 0.46 &\leq \|e_{p,0}\|_\infty, & 0.07 &\leq \|e_{p,2}\|_\infty, \\ 0.33 &\leq \|e_{p,1}\|_\infty, & 0.01 &\leq \|e_{p,3}\|_\infty. \end{aligned}$$

The first four minimax polynomial approximations of $x_1(t) = t \sin 5t$ and the $\mathcal{L}^\infty$ norms of their errors are

$$\begin{aligned} p_0(t) &\approx -0.30, & \|e_{p,0}\|_\infty &\approx 0.66, \\ p_1(t) &\approx 0.40 - 1.00t, & \|e_{p,1}\|_\infty &\approx 0.40, \\ p_2(t) &\approx 0.09 + 1.41t - 2.62t^2, & \|e_{p,2}\|_\infty &\approx 0.16, \\ p_3(t) &\approx -0.12 + 3.22t - 6.31t^2 + 2.13t^3, & \|e_{p,3}\|_\infty &\approx 0.12. \end{aligned}$$

The errors of these approximations $e_{p,K}$ are shown in Figure 6.15. The white lines highlight that the approximation error lies between $\pm\|e_{p,K}\|_\infty$ and reaches these boundaries at least $K+2$ times, satisfying the condition of Theorem 6.8. The intuition behind the theorem is that not reaching the $\pm\|e_{p,K}\|_\infty$ bounds $K+2$ times would waste some of the margin for error.

Computation of minimax approximations is generally difficult because the values of the extrema of the error can depend on the polynomial coefficients in a complicated way.

Near-minimax approximation An alternative to finding an exact minimax approximation is to use an approximation that is simpler to compute but only nearly minimax, such as using Lagrange interpolation with unevenly-spaced nodes chosen based on the approximation interval but not based on the function. One methodical way to keep the $\mathcal{L}^\infty$ norm of the approximation error small is to choose the nodes using roots of Chebyshev polynomials.

THEOREM 6.9 (CHEBYSHEV POLYNOMIALS) The *Chebyshev polynomials*, defined as the unique polynomials for which

$$T_k(t) = \cos(k \arccos t), \qquad t \in [-1, 1], \qquad k \in \mathbb{N}, \tag{6.16}$$

satisfy the following properties:

(i) T_k is a polynomial of degree k.

(ii) For any distinct k and m, T_k and T_m are orthogonal on $[-1, 1]$ with the weight function $(1-t^2)^{-1/2}$; that is,

$$\int_{-1}^{1} T_k(t) T_m(t)\,(1-t^2)^{-1/2}\,dt = 0. \tag{6.17}$$

(iii) The recursion

$$T_{k+1}(t) = 2tT_k(t) - T_{k-1}(t), \qquad k \in \mathbb{Z}^+, \tag{6.18}$$

is satisfied, with $T_0(t) = 1$ and $T_1(t) = t$.

(iv) The roots of T_k are

$$t_m = \cos\left(\frac{m+\frac{1}{2}}{k}\pi\right), \qquad m = 0, 1, \ldots, k-1. \tag{6.19a}$$

(v) The relative maxima and minima of T_k are at

$$\cos\left(\frac{m}{k}\pi\right), \qquad m = 0, 1, \ldots, k. \tag{6.19b}$$

(vi) For any $k \in \mathbb{Z}^+$, the leading coefficient of T_k is 2^{k-1}.

The theorem is proven in Solved exercise 6.2.

The first few Chebyshev polynomials, plotted in Figure 6.16, are

$$\begin{aligned} T_0(t) &= 1, & T_3(t) &= 4t^3 - 3t, \\ T_1(t) &= t, & T_4(t) &= 8t^4 - 8t^2 + 1, \\ T_2(t) &= 2t^2 - 1, & T_5(t) &= 16t^5 - 20t^3 + 5t. \end{aligned}$$

From (6.16), it is clear that $|T_k(t)| \leq 1$ for every $t \in [-1, 1]$ and $k \in \mathbb{N}$. While the

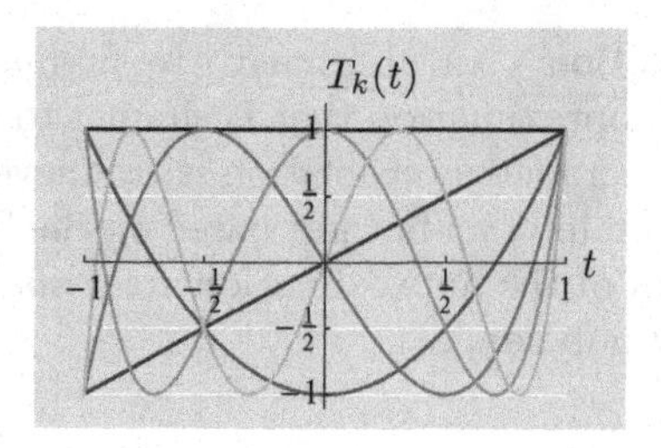

Figure 6.16 The first six Chebyshev polynomials $\{T_k\}_{k=0}^{5}$ (solid lines, from darkest to lightest).

Legendre polynomials also satisfy this bound, they do not cover the interval $[-1, 1]$ evenly; compare Figure 6.16 with Figure 6.9.

Recall our observation regarding the weakness of Lagrange interpolation with evenly spaced nodes from an $\mathcal{L}^\infty$ perspective: the error is large near the ends of the interval of approximation. We can counter this by having more nodes near the ends of the interval at the expense of having fewer near the center. Specifically, we can minimize the maximum of the factor $\prod_{k=0}^{K}|t - t_k|$ from the error bound (6.8b). When the interval of approximation is $[-1, 1]$, the resulting nodes are zeros of the Chebyshev polynomial of degree K.

The factor $\prod_{k=0}^{K}|t - t_k|$ is the absolute value of a polynomial of degree $K+1$ with leading coefficient 1. Among all polynomials with degree $K+1$ and leading coefficient 1, the scaled Chebyshev polynomial $2^{-K}T_{K+1}$ has minimum $\mathcal{L}^\infty([-1, 1])$ norm, which is equal to 2^{-K}. Therefore, choosing $\{t_k\}_{k=0}^{K}$ to be the $K+1$ zeros of T_{K+1} minimizes $\max_{t\in[-1,1]}\prod_{k=0}^{K}|t - t_k|$. The bound (6.8b) then becomes

$$|e_{q,K}(t)| \le \frac{1}{(K+1)!2^K} \max_{\xi\in[-1,1]} \left|x^{(K+1)}(\xi)\right| \tag{6.20}$$

for approximation of x on $[-1, 1]$ with a polynomial q_K of degree at most K. The $\mathcal{L}^\infty([-1, 1])$ norm of the error can be bounded relative to the $\mathcal{L}^\infty([-1, 1])$ norm of the error of the minimax approximation as

$$\varepsilon_{q,K} \le \left(\frac{2}{\pi}\ln(K+1) + 2\right)\varepsilon_{p,K}, \tag{6.21}$$

so it is *near-minimax* in a precise sense.

EXAMPLE 6.7 (APPROXIMATION WITH CHEBYSHEV POLYNOMIALS) Let us return to the approximation of $x_1(t) = t\sin 5t$ on $[0, 1]$ (from (6.4)). If the interval of interest were $[-1, 1]$, we would obtain near-minimax approximations satisfying (6.20) and (6.21) by interpolating with the roots of $T_K(t)$ as the nodes. The only necessary modification is to map the roots from $[-1, 1]$ to $[0, 1]$ with an affine transformation. The errors of the resulting approximations are plotted for $K \in \{0, 1, 2, 3\}$ in Figure 6.17.

Table 6.2 summarizes the $\mathcal{L}^\infty$ error performances of various approximations from this and previous examples. The first five are significantly easier to compute

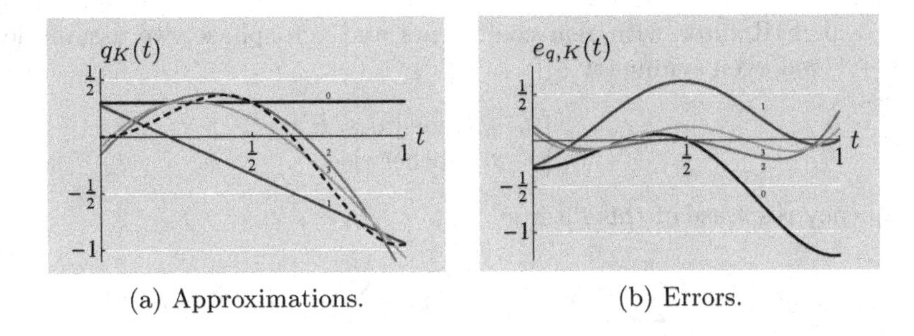

(a) Approximations. (b) Errors.

Figure 6.17 Near-minimax polynomial approximations q_K and approximation errors $e_{q,K}$, $K = 0, 1, 2, 3$ (solid lines, from darkest to lightest), of x_1 in (6.4) on $[0, 1]$ (dashed line) using interpolation with Chebyshev nodes. The curves are labeled by the polynomial degree.

Approximation method	**Polynomial degree**			
	0	1	2	3
Least-squares approximation	0.868	0.489	0.318	0.223
Lagrange interpolation	0.963	0.785	0.346	0.197
Taylor series expansion around $\frac{1}{2}$	1.260	1.000	1.380	1.270
Near-minimax approximation	1.260	0.637	0.298	0.144
Minimax approximation	0.661	0.402	0.149	0.124

Table 6.2 Summary of $\mathcal{L}^\infty$ errors of approximations of $x_1(t) = t \sin 5t$ on $[0, 1]$.

than the minimax approximation. Least-squares approximation is an orthogonal projection to the subspace of polynomials; it is optimal for $\mathcal{L}^2$ error by definition, and its $\mathcal{L}^\infty$ error is not necessarily small. A Taylor series expansion is accurate near the point at which the function is measured (assuming that the function is smooth), but quite poor farther away. Interpolation using uniform nodes is improved upon by the use of Chebyshev nodes; this becomes increasingly important as the polynomial degree is increased. Note also that (6.21) is satisfied.

6.2.6 Filter design

The design of FIR filters with linear phase (see Section 3.4.4) and a given desired magnitude response can be posed as finding a polynomial to approximate a certain function. Therefore, filter design provides case studies for the least-squares and minimax polynomial approximation methods of the preceding sections. It also motivates the extension of minimax methods to weighted minimax approximation.

We consider the case of zero-phase filters, which for any length greater than 1 are noncausal but can be shifted to produce causal, linear-phase filters. Suppose that a desired frequency response H^{d} is given, and it is real (zero phase) and even.

To design an FIR filter with real coefficients and zero phase, we assume length $L = 2K + 1$ and even symmetry,

$$h_n = \begin{cases} h_{-n}, & \text{for } |n| \le K; \\ 0, & \text{otherwise.} \end{cases} \tag{6.22}$$

The frequency response of this filter is

$$\begin{aligned} H(e^{j\omega}) &= \sum_{n=-K}^{K} h_n e^{-j\omega n} \stackrel{(a)}{=} h_0 + \sum_{n=1}^{K} h_n \left(e^{-j\omega n} + e^{j\omega n}\right) \\ &\stackrel{(b)}{=} h_0 + 2\sum_{n=1}^{K} h_n \cos n\omega, \end{aligned}$$

where (a) follows from the symmetry imposed in (6.22); and (b) from (3.286).

Let us now see why the choice of coefficients $\{h_n\}_{n=0}^{K}$ to make the frequency response H approximate the desired frequency response H^{d} is a polynomial approximation problem. By Theorem 6.9(i), $\cos n\omega$ is a polynomial of $t = \cos\omega$ of degree n. Therefore, $H(e^{j\omega})$ is a polynomial of t of degree K. The relation between ω and t provides a one-to-one correspondence between $\omega \in [0, \pi]$ and $t \in [-1, 1]$, and there is no need to consider $\omega \in [-\pi, 0)$ separately because $H(e^{j\omega})$ and $H^{\mathrm{d}}(e^{j\omega})$ are even functions of ω. Thus, we may view our design problem as the approximation of a function by a polynomial of degree K on $[-1, 1]$. While this will be useful when applying the minimax criterion, it is of no benefit when applying the least-squares criterion.

Least-squares approximation A simple criterion to apply is to minimize the squared $\mathcal{L}^2$ norm of the frequency response approximation error,

$$\underset{\{h_0,h_1,\ldots,h_K\}}{\arg\min} \ \|H^{\mathrm{d}}(e^{j\omega}) - H(e^{j\omega})\|_2^2 = \underset{\{h_0,h_1,\ldots,h_K\}}{\arg\min} \int_{-\pi}^{\pi} |H^{\mathrm{d}}(e^{j\omega}) - H(e^{j\omega})|^2 \, d\omega.$$

By the Parseval equality (3.107), this is equivalent to minimizing the squared ℓ^2 norm of the impulse response error,

$$\underset{\{h_0,h_1,\ldots,h_K\}}{\arg\min} \ \|h^{\mathrm{d}} - h\|_2^2 = \underset{\{h_0,h_1,\ldots,h_K\}}{\arg\min} \sum_{n\in\mathbb{Z}} |h_n^{\mathrm{d}} - h_n|^2. \tag{6.23}$$

Since the objective function is a sum of nonnegative terms, the best we can do is have zero contribution from each of the terms $n \in \{-K, -K+1, \ldots, K\}$; thus, the minimum is attained by setting

$$h_n = h_n^{\mathrm{d}}, \qquad \text{for } n = -K, -K+1, \ldots, K. \tag{6.24}$$

In other words, the least-squares approximation is simply the truncation of the desired filter's impulse response to its central $2K + 1$ entries.

While the $\mathcal{L}^2$ norm of the approximation error is minimized, the maximum error can remain large. In particular, if the desired frequency response is discontinuous, the error will be relatively large (at least half the size of the discontinuity)

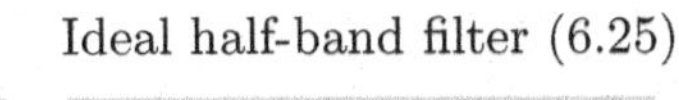

Ideal half-band filter (6.25)

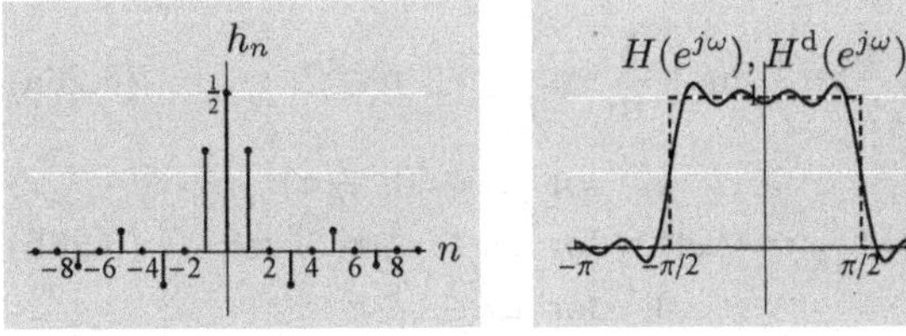

(a) Impulse response.

(b) Frequency response.

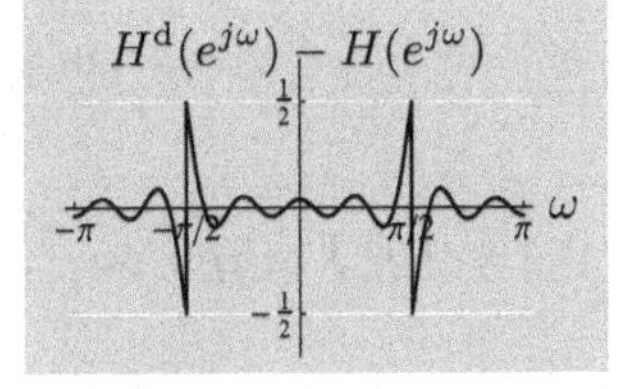

(c) Frequency response error.

Nonideal half-band filter (6.26)

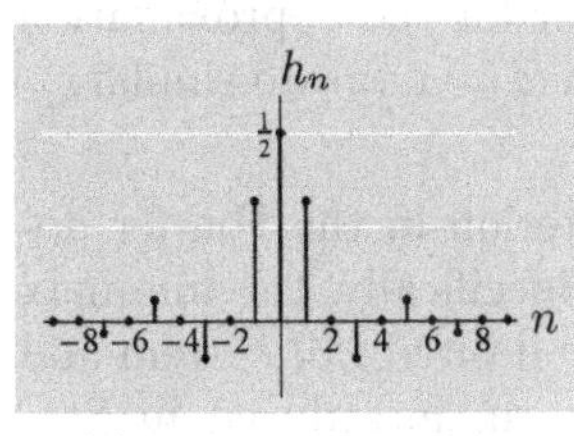

(d) Impulse response.

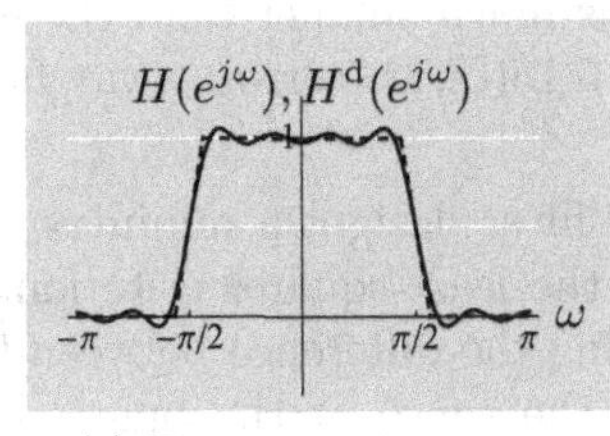

(e) Frequency response.

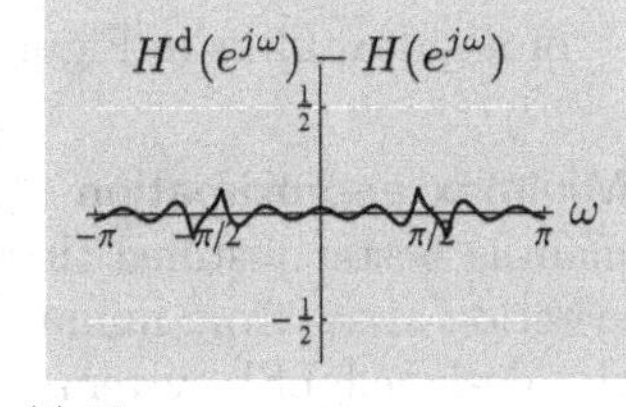

(f) Frequency response error.

Figure 6.18 Least-squares approximations of length 15 to half-band lowpass filters (see Example 6.8).

on at least one side of the discontinuity. Also, for an ideal lowpass filter, the Gibbs phenomenon leads to oscillations that do not diminish in amplitude as the filter length is increased (see Figure 3.9); this causes the error to be about 9% just above the cutoff frequency.

Example 6.8 (Least-squares design of half-band lowpass filters) An ideal half-band lowpass filter with unit passband gain is obtained by rescaling (3.112) with $\omega_0 = \pi$,

$$h_n^{\mathrm{d}} = \frac{1}{2}\,\mathrm{sinc}\left(\frac{1}{2}\pi n\right) \quad \overset{\mathrm{DTFT}}{\longleftrightarrow} \quad H^{\mathrm{d}}(e^{j\omega}) = \begin{cases} 1, & \text{for } \omega \in [0,\, \frac{1}{2}\pi]; \\ 0, & \text{for } \omega \in (\frac{1}{2}\pi,\, \pi]. \end{cases} \tag{6.25}$$

The least-squares approximation of length 15 to this filter is obtained by truncating the impulse response to $n \in \{-7, -6, \ldots, 7\}$, as shown in Figure 6.18(a). Figures 6.18(b) and (c) show its frequency response and the approximation error of the frequency response; the large magnitude of the error at $\omega = \pm\frac{1}{2}\pi$ arises from the discontinuity of $H^{\mathrm{d}}(e^{j\omega})$ at those frequencies.

Continuous-time lowpass filters with continuous frequency responses were developed in Example 5.20. A discrete-time analogue to (5.85) for a half-band lowpass filter is

$$h_n^{\mathrm{d}} = \frac{1}{2}\operatorname{sinc}\left(\frac{1}{2}\pi n\right)\operatorname{sinc}\left(\frac{1}{16}\pi n\right), \qquad n \in \mathbb{Z} \tag{6.26a}$$

$$\overset{\mathrm{DTFT}}{\longleftrightarrow} \quad H^{\mathrm{d}}(e^{j\omega}) = \begin{cases} 1, & \text{for } |\omega| \in [0,\, \frac{7}{16}\pi); \\ (8/\pi)\left(\frac{9}{16}\pi - |\omega|\right), & \text{for } |\omega| \in [\frac{7}{16}\pi,\, \frac{9}{16}\pi); \\ 0, & \text{for } |\omega| \in [\frac{9}{16}\pi,\, \pi]. \end{cases} \tag{6.26b}$$

The least-squares approximation of length 15 to this filter, its frequency response, and the frequency response approximation error are shown in Figures 6.18(d)–(f). Near $\omega = \pm\frac{1}{2}\pi$, the error is much smaller than the error for the approximation of an ideal filter in Figure 6.18(c); the errors away from $\omega = \pm\frac{1}{2}\pi$ are similar.

Minimax approximation In filter design, a minimax criterion in the Fourier domain is better justified than the least-squares criterion. Specifically, the minimax criterion arises from minimizing the difference between the desired LSI system and the designed LSI system in the sense of the operator norm introduced in Section 2.3.3.

THEOREM 6.10 (MINIMAX DESIGN CRITERION) Let H^{d} be a desired frequency response and H its approximation, with corresponding impulse responses h^{d} and h. Define the *error system* $E : \ell^2(\mathbb{Z}) \to \ell^2(\mathbb{Z})$ as the difference between filtering by h^{d} and filtering by h,

$$Ex = h^{\mathrm{d}} * x - h * x.$$

Then the operator norm of E is given by

$$\|E\| = \max_{\omega\in[-\pi,\pi]} |H^{\mathrm{d}}(e^{j\omega}) - H(e^{j\omega})|.$$

Furthermore, assuming that H^{d} is real and even, the zero-phase FIR filter h satisfying (6.22) that minimizes the energy of the difference between filtering by h^{d} and filtering by h over unit-energy inputs is

$$\underset{\{h_0,h_1,\ldots,h_K\}}{\arg\min} \max_{\omega\in[0,\pi]} |H^{\mathrm{d}}(e^{j\omega}) - H(e^{j\omega})|. \tag{6.27}$$

Proof. We leave part of the proof to Solved exercise 6.3. Let $g = h^{\mathrm{d}} - h$; this is the impulse response of the error system. The operator norm of the error system as a mapping from $\ell^2(\mathbb{Z})$ to $\ell^2(\mathbb{Z})$ is given by

$$\|E\| \overset{(a)}{=} \max_{\omega\in[-\pi,\pi]} |G(e^{j\omega})| \overset{(b)}{=} \max_{\omega\in[-\pi,\pi]} |H^{\mathrm{d}}(e^{j\omega}) - H(e^{j\omega})|,$$

where (a) follows from Solved exercise 6.3; and (b) from the definition of g and the linearity of the DTFT.

When H^{d} and H have even symmetry (the latter following from (6.22)), the magnitude of their difference $|H^{\mathrm{d}}(e^{j\omega}) - H(e^{j\omega})|$ has even symmetry, so maximizing over $\omega \in [-\pi, \pi]$ is equivalent to maximizing over $\omega \in [0, \pi]$. The minimax design criterion (6.27) follows.

Letting $t = \cos\omega$ converts the problem (6.27) into a minimax polynomial approximation problem. When the desired frequency response H^{d} is continuous, the change of variable gives a continuous function to approximate with a polynomial. Thus, the minimax solution will satisfy the Chebyshev equioscillation theorem, Theorem 6.8. In particular, Theorem 6.8 indicates a necessary condition for an optimal solution, namely that there will be $K + 2$ points with maximum error. However, when the desired frequency response is discontinuous, the maximum error will inevitably be large near the discontinuity, like in the least-squares case, and Theorem 6.8 does not apply.

The importance of continuity of H^{d} is illustrated by the following example. Like in Example 6.6, the numerical calculations of minimax approximations were performed using the Remez algorithm; see the *Further reading*.

EXAMPLE 6.9 (MINIMAX DESIGN OF HALF-BAND LOWPASS FILTERS) Again consider the ideal half-band lowpass filter (6.25) and the half-band lowpass filter with continuous frequency response (6.26) from Example 6.8. A length-15 minimax approximation of the ideal half-band filter is shown in Figures 6.19(a)–(c). The maximum error in the frequency domain is $\frac{1}{2}$; we could not have expected any better because we are using a continuous function to approximate a discontinuous function with a jump of 1. In fact, aside from requiring $H(e^{-j\pi/2}) = H(e^{j\pi/2}) = \frac{1}{2}$, the set of filters with $\mathcal{L}^\infty$ error in the frequency domain of $\frac{1}{2}$ is relatively unconstrained. In comparison with Figure 6.18(c), there is barely any improvement in maximum error relative to the least-squares design; the squared $\mathcal{L}^2$ error is much worse.

The length-15 minimax approximation of the half-band filter (6.26) is shown in Figures 6.19(d)–(f). Since the desired frequency response is continuous, the maximum error is much smaller and can be driven to zero by increasing the filter length. This optimal design is unique and satisfies the equioscillation property from Theorem 6.8. In comparison with Figure 6.18(f), the maximum error is somewhat smaller than that of the least-squares design.

Weighted minimax approximation The magnitude of the frequency response H^{d} of an ideal filter is constant in the filter's passband and zero in the filter's stopband. Since the frequency response H of an FIR approximation h of such a filter is continuous, the error $H^{\mathrm{d}}(e^{j\omega}) - H(e^{j\omega})$ will inevitably be large for frequencies near the edges of the passband and stopband. The conventional practice is to ignore these frequencies by using a *weighted* error

$$E(\omega) = W(\omega)\left(H^{\mathrm{d}}(e^{j\omega}) - H(e^{j\omega})\right),$$

where W is a nonnegative weight function. The weight function is zero around the discontinuities of H^{d} and also sets the relative importance across ω of how

Ideal half-band filter (6.25)

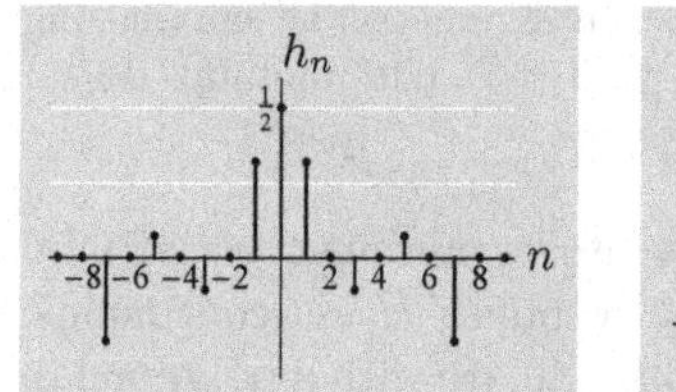

(a) Impulse response.

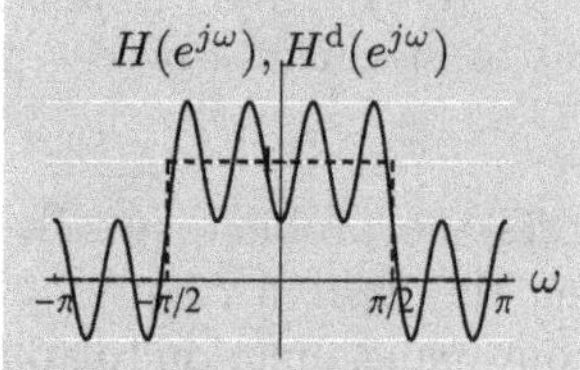

(b) Frequency response.

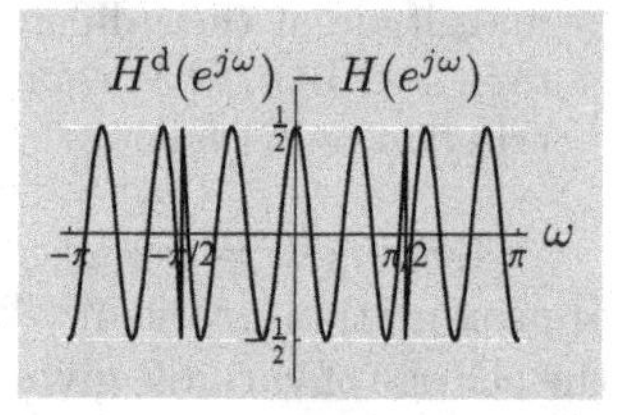

(c) Frequency response error.

Nonideal half-band filter (6.26)

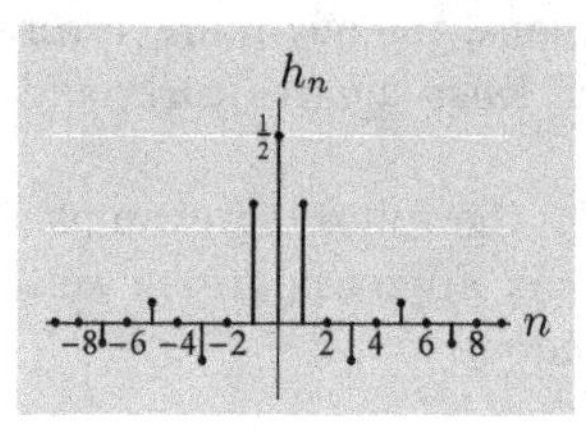

(d) Impulse response.

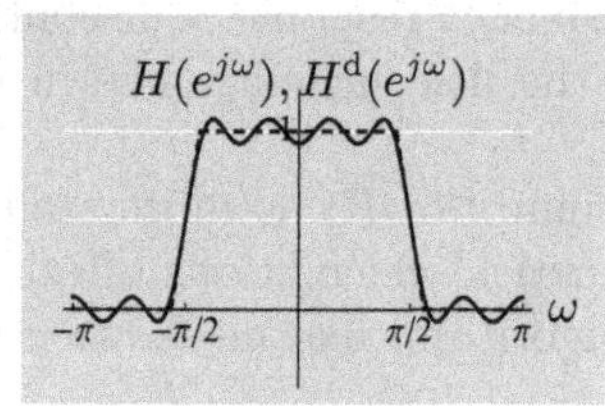

(e) Frequency response.

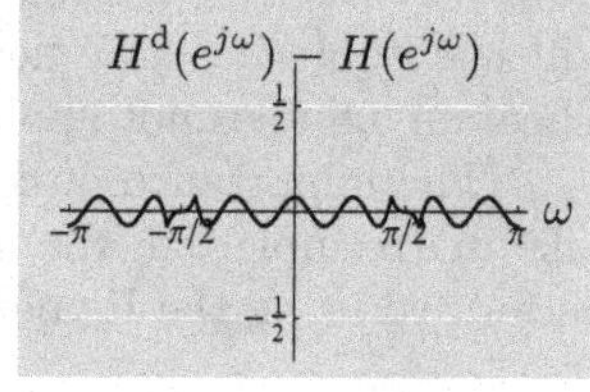

(f) Frequency response error.

Figure 6.19 Minimax approximations of length 15 to half-band lowpass filters (see Example 6.9).

closely $H^{d}(e^{j\omega})$ is approximated by $H(e^{j\omega})$. The weighted minimax approximation minimizes the maximum of this weighted error,

$$\varepsilon \;=\; \max_{\omega\in[0,\pi]} |E(\omega)| \;=\; \max_{\omega\in[0,\pi]} \left|W(\omega)\left(H^{d}(e^{j\omega}) - H(e^{j\omega})\right)\right|. \tag{6.28}$$

Suppose that the goal is to approximate an ideal lowpass filter with cutoff frequency $\frac{1}{2}\omega_0$ and unit gain in the passband,

$$H^{d}(e^{j\omega}) \;=\; \begin{cases} 1, & \text{for } \omega \in [0,\, \frac{1}{2}\omega_0]; \\ 0, & \text{for } \omega \in (\frac{1}{2}\omega_0,\, \pi]. \end{cases}$$

The weight function is typically piecewise-constant of the form

$$W(\omega) \;=\; \begin{cases} 1, & \text{for } \omega \in [0,\, \frac{1}{2}\omega_0 - \omega_t]; \\ 0, & \text{for } \omega \in (\frac{1}{2}\omega_0 - \omega_t,\, \frac{1}{2}\omega_0 + \omega_t); \\ \gamma, & \text{for } \omega \in [\frac{1}{2}\omega_0 + \omega_t,\, \pi], \end{cases} \tag{6.29}$$

for some $\gamma \in \mathbb{R}^{+}$. This establishes a passband of $[0,\, \frac{1}{2}\omega_0 - \omega_t]$, a stopband of $[\frac{1}{2}\omega_0 + \omega_t,\, \pi]$, and a *transition band* of width $2\omega_t$ in between. The constant γ fixes the relative importance of the passband and stopband accuracy; if $\gamma > 1$ then keeping $|H(\omega) - 0|$ small in the stopband is more important than keeping $|H(\omega) - 1|$ small in the passband, and vice versa.

As before, using $t = \cos\omega$, minimization of the weighted maximum error ε in (6.28) over the choice of symmetric filters of length $2K+1$ is equivalent to approximation of a function by a polynomial of degree K; rather than minimize the $\mathcal{L}^\infty$ norm of an approximation error, we are now minimizing a weighted $\mathcal{L}^\infty$ norm. An extension of Theorem 6.8 applies to the minimization of the weighted maximum error. It establishes that the optimal solution is determined uniquely by the existence of $K+2$ frequencies,

$$0 \leq \omega_0 < \omega_1 < \cdots < \omega_{K+1} \leq \pi,$$

such that the weighted error at these frequencies alternates in sign,

$$E(\omega_0) = -E(\omega_1) = E(\omega_2) = -E(\omega_3) = \cdots,$$

while the absolute value of the weighted error reaches the maximum value at these frequencies:

$$|E(\omega_0)| = |E(\omega_1)| = \cdots = |E(\omega_{K+1})| = \varepsilon.$$

This is the basis of the Parks–McClellan algorithm; see the *Further reading*.

EXAMPLE 6.10 (WEIGHTED MINIMAX DESIGN OF A LOWPASS FILTER) We consider the design of a lowpass filter with cutoff frequency at $\frac{1}{2}\omega_0 = \frac{1}{2}\pi$. The ideal impulse response is given in (3.112b), and we want to design symmetric FIR approximations of length 15. The least-squares approximation and (unweighted) minimax approximation were shown in Figures 6.18 and 6.19.

Weighted minimax approximation allows us to introduce a transition band $(\frac{1}{2}\pi - \omega_t, \frac{1}{2}\pi + \omega_t)$. By assigning no importance (zero weight) to the error in the transition band, the deviation from the desired response in the passband and stopband can be reduced. This is illustrated in Figure 6.20, where we see that weighted minimax designs (with $\gamma = 1$ in (6.29)) have reduced passband and stopband approximation error (*ripple*) as the width of the transition band is increased. Varying the relative weighting of the passband and stopband error ($\gamma \in \{\frac{1}{3}, 1, 3\}$ in (6.29)) with $\omega_t = \frac{1}{8}\pi$ held constant allows us to trade off these errors, as shown in Figure 6.21.

The conventional way to specify a frequency-selective filter is through the parameters illustrated in Figure 6.22. A deviation of $\pm\delta_p$ from the ideal passband gain of 1 is allowed, along with a deviation of $\pm\delta_s$ from the ideal stopband gain of 0. These are applied outside of a transition band of width $2\omega_t$ centered at the nominal cutoff frequency of $\frac{1}{2}\omega_0$. A systematic way to design such a filter is to let

$$W(\omega) = \begin{cases} 1, & \text{for } \omega \in [0, \frac{1}{2}\omega_0 - \omega_t]; \\ 0, & \text{for } \omega \in (\frac{1}{2}\omega_0 - \omega_t, \frac{1}{2}\omega_0 + \omega_t); \\ \delta_p/\delta_s, & \text{for } \omega \in [\frac{1}{2}\omega_0 + \omega_t, \pi], \end{cases}$$

and solve the weighted minimax design problem. If the specifications are too stringent for the selected value of K, the weighted minimax error will be greater than δ_p; to design the shortest filter that meets the specifications, one can increase K until the weighted minimax error no longer exceeds δ_p.

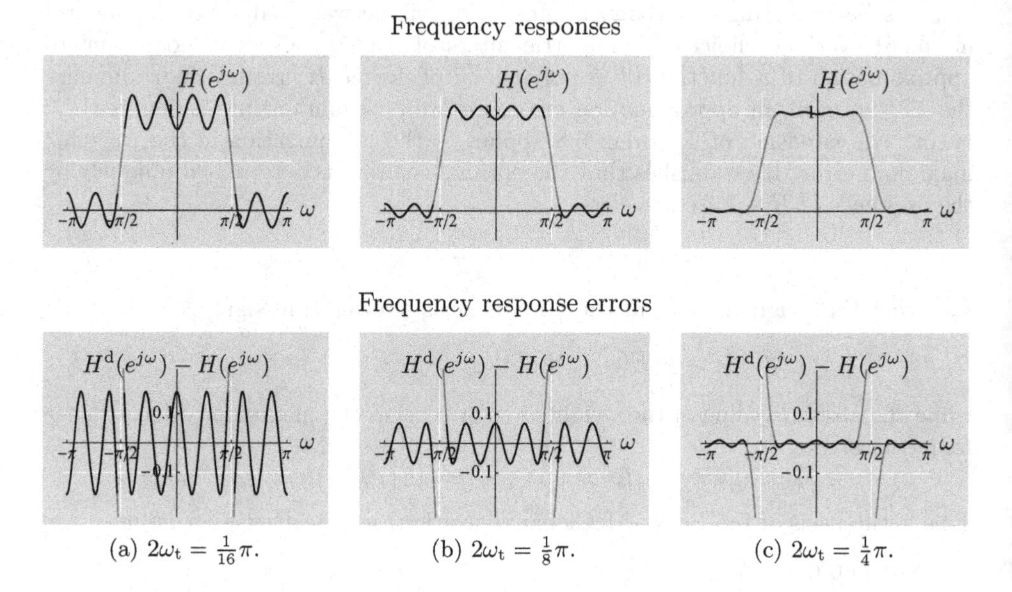

(a) $2\omega_t = \frac{1}{16}\pi$. (b) $2\omega_t = \frac{1}{8}\pi$. (c) $2\omega_t = \frac{1}{4}\pi$.

Figure 6.20 Weighted minimax designs of a half-band lowpass filter of length 15 for various widths of the transition band. The weightings of error in the passband and stopband are equal ($\gamma = 1$).

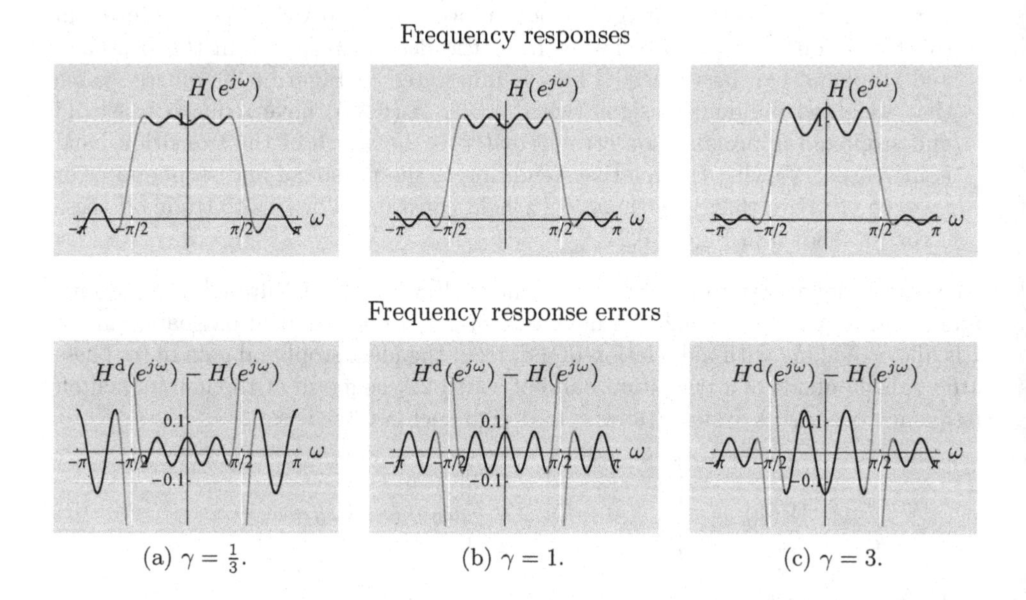

(a) $\gamma = \frac{1}{3}$. (b) $\gamma = 1$. (c) $\gamma = 3$.

Figure 6.21 Weighted minimax designs of a half-band lowpass filter of length 15 for various relative weightings of passband and stopband error. The transition band is $(\frac{3}{8}\pi, \frac{5}{8}\pi)$.

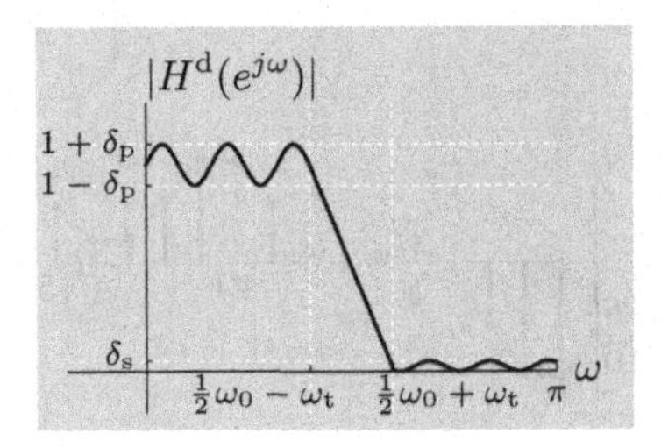

Figure 6.22 Specification and one possible solution for the design of a lowpass filter with cutoff frequency $\frac{1}{2}\omega_0$. The width of the transition band is $2\omega_{\text{t}}$, and δ_{p} and δ_{s} are the error margins for the passband and the stopband.

6.3 Approximation of functions by splines

In the last section, we saw several methods for finding polynomial approximations of functions. Polynomials have the virtue of smoothness: they are infinitely differentiable, with derivatives of order $K+1$ and greater being identically zero for polynomials of degree K. Also, as shown by the Weierstrass approximation theorem, Theorem 6.6, a polynomial approximation can be made arbitrarily close to a given continuous function over a given closed interval by choosing a high-enough degree. While endowed with these benefits, polynomial approximation does suffer from some drawbacks: it cannot approximate a discontinuous function well, and, even when one is trying to approximate a continuous function, increasing the polynomial degree can be problematic (see the discussion of Figure 6.12).

In this section, we develop the approximation of functions by *piecewise* polynomials. By allowing different polynomial approximations on disjoint subintervals of $\mathbb{R}$, we are able to better approximate both continuous and discontinuous functions for any given polynomial degree. Splines are distinguished among piecewise polynomials in that they are as smooth as possible: a spline with pieces of degree K has continuous derivatives up to order $K-1$ and is differentiable up to order K everywhere except at the piece boundaries.

The points at which the polynomial expression for a spline changes are called *knots*. In general, splines can have their knots anywhere; this is emphasized by the term *free-knot spline*. We will limit our attention to *uniform splines*, which have their knots on a uniform grid. The vector spaces of uniform splines are discussed in Section 6.3.1, followed by the development of bases for these spaces in Section 6.3.2. Section 6.3.3 turns this around to discuss the Strang–Fix condition, which describes when a function and its shifts can be used to express a polynomial; this Fourier-domain condition is closely related to error bounds for uniform spline approximations. Uniform spline representations enable discrete-time processing of continuous-time signals, as developed in Section 6.3.4.

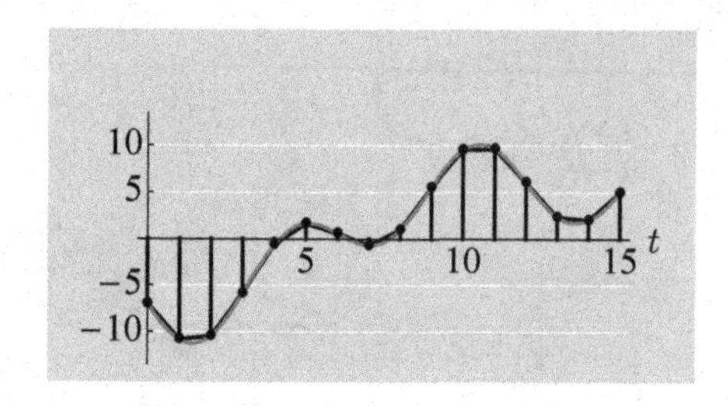

Figure 6.23 Examples of splines of degree 1 (black line) and 3 (gray line) interpolating the same set of points $\{(t_k, y_k)\}_{k=0}^{15}$.

6.3.1 Splines and spline spaces

Let $\ldots, \tau_{-1}, \tau_0, \tau_1, \ldots$ be a strictly increasing sequence of real numbers. A function that is a polynomial of degree K on each interval $[\tau_n, \tau_{n+1})$, possibly different polynomials on different intervals, is determined by $K+1$ parameters per interval. Such a function is generally not continuous at the points $\{\tau_n\}_{n\in\mathbb{Z}}$. A spline arises from adding constraints to make this function as smooth as possible, just shy of specifying that the function is a single polynomial on $\mathbb{R}$ rather than a piecewise polynomial. Specifically, the polynomials on $[\tau_{n-1}, \tau_n)$ and $[\tau_n, \tau_{n+1})$ can be required to match at τ_n for continuity and furthermore match in derivatives of order $1, 2, \ldots, K-1$ at τ_n for continuity of derivatives. The $K+1$ parameters per interval are reduced by these K constraints to a single free parameter per interval. If the pieces were also required to match in the Kth derivative, the two polynomial pieces would be the same polynomial.

In our formal definition, the knot sequence can be finite or infinite. The derivative of order zero is the function itself.

DEFINITION 6.11 (SPLINE, UNIFORM SPLINE, AND SPLINE SPACE) Let $\tau = (\tau_n)_{n\in\mathcal{I}}$, $\tau_n \in \mathbb{R}$, be a strictly increasing sequence, where $\mathcal{I}$ is an ordered index set that might be finite or countably infinite, and let $K \in \mathbb{N}$. A function is called a *spline* of degree K with *knots* τ when it is a polynomial of degree at most K on each interval $[\tau_n, \tau_{n+1})$, $n \in \mathcal{I}$, and its derivatives of order $0, 1, \ldots, K-1$ are continuous. The set of such splines in $\mathcal{L}^2(\mathbb{R})$ is called the *spline space* of degree K with knots τ and denoted $S_{K,\tau}$. When the knot sequence τ is evenly spaced and doubly infinite, the spline and spline space are called *uniform*.

Splines of degree 1 and 3 are especially common. *Spline of degree* 1 is a formal way to express that one uses straight lines to connect the dots given by values at a sequence of knots. Splines of degree 3 are common because their continuity up to the second derivative gives a pleasingly smooth appearance while avoiding the oscillatory behavior of high-order polynomial interpolation. An example is shown in Figure 6.23.

For a fixed degree K and knot sequence τ, the spline space $S_{K,\tau}$ is a subspace of $\mathcal{L}^2(\mathbb{R})$.[103] This follows simply from a sum of polynomials of degree at most K being a polynomial of degree at most K, on each interval $[\tau_n, \tau_{n+1})$, while the linearity of differentiation maintains the continuity of derivatives. Other properties of spline spaces are developed in Exercise 6.11.

Fitting a spline A doubly infinite spline has one degree of freedom per segment $[\tau_n, \tau_{n+1})$. This is exploited in Section 6.3.2 to produce series expansions for splines on $\mathbb{R}$. Restricting a spline to a finite interval leaves extra degrees of freedom, as illustrated in the following example.

EXAMPLE 6.11 (FITTING A SPLINE) A spline on $[0, 1)$ of degree 2 with knots $(0, \frac{1}{2}, 1)$ is given by

$$\widehat{x}(t) = \begin{cases} a_{0,2}t^2 + a_{0,1}t + a_{0,0}, & \text{for } t \in [0, \frac{1}{2}); \\ a_{1,2}t^2 + a_{1,1}t + a_{1,0}, & \text{for } t \in [\frac{1}{2}, 1). \end{cases}$$

Regardless of the function to be approximated with $\widehat{x}$, continuity of $\widehat{x}$ at $\frac{1}{2}$ gives the constraint

$$\frac{1}{4}a_{0,2} + \frac{1}{2}a_{0,1} + a_{0,0} = \frac{1}{4}a_{1,2} + \frac{1}{2}a_{1,1} + a_{1,0}, \tag{6.30a}$$

and continuity of its derivative at $\frac{1}{2}$ gives the constraint

$$a_{0,2} + a_{0,1} = a_{1,2} + a_{1,1}. \tag{6.30b}$$

Four additional constraints will fix $\widehat{x}$. We illustrate this through several approximations of x_1 from (6.4) and x_2 from (6.7), shown in Figure 6.24.

(i) Matching the values of the function at the three knots leaves one additional degree of freedom. This could be used, for example, to match the derivative at 0. The resulting system of equations to solve is

$$\begin{bmatrix} \frac{1}{4} & \frac{1}{2} & 1 & -\frac{1}{4} & -\frac{1}{2} & -1 \\ 1 & 1 & 0 & -1 & -1 & 0 \\ 0 & 0 & 1 & 0 & 0 & 0 \\ 0 & 0 & 0 & \frac{1}{4} & \frac{1}{2} & 1 \\ 0 & 0 & 0 & 1 & 1 & 1 \\ 0 & 1 & 0 & 0 & 0 & 0 \end{bmatrix} \begin{bmatrix} a_{0,2} \\ a_{0,1} \\ a_{0,0} \\ a_{1,2} \\ a_{1,1} \\ a_{1,0} \end{bmatrix} = \begin{bmatrix} 0 \\ 0 \\ x(0) \\ x(\frac{1}{2}) \\ x(1) \\ x'(0) \end{bmatrix}, \tag{6.31}$$

where the first two rows are from (6.30), the next three rows from matching values at the knots, and the last row from matching the derivative at 0. The matrix in (6.31) is nonsingular; thus, any set of values $x(0)$, $x(\frac{1}{2})$, $x(1)$, and $x'(0)$ yields a unique spline approximation.

[103] We have defined spline spaces within $\mathcal{L}^2(\mathbb{R})$, rather than more generally in $\mathbb{C}^{\mathbb{R}}$, even though the defining characteristics of splines are unrelated to having finite $\mathcal{L}^2(\mathbb{R})$ norm. This restriction is included here because most of the tools and techniques of the book are for Hilbert spaces.

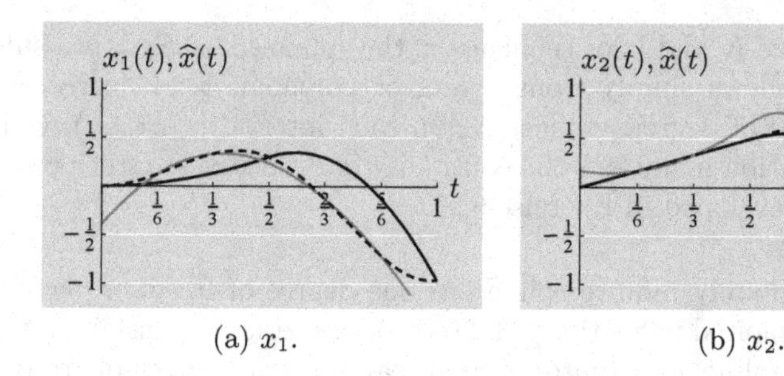

Figure 6.24 Spline approximations of degree 2 with knots $(0, \frac{1}{2}, 1)$ of x_1 in (6.4) and x_2 in (6.7) (dashed lines) using the two methods developed in Example 6.11 (black for part (i) and gray for part (ii)).

(ii) Another possibility is to match values of the function at four points, two on each piece, such as $\frac{1}{6}$, $\frac{1}{3}$, $\frac{2}{3}$, and $\frac{5}{6}$. The resulting system of equations to solve is

$$\begin{bmatrix} \frac{1}{4} & \frac{1}{2} & 1 & -\frac{1}{4} & -\frac{1}{2} & -1 \\ 1 & 1 & 0 & -1 & -1 & 0 \\ \frac{1}{36} & \frac{1}{6} & 1 & 0 & 0 & 0 \\ \frac{1}{9} & \frac{1}{3} & 1 & 0 & 0 & 0 \\ 0 & 0 & 0 & \frac{4}{9} & \frac{2}{3} & 1 \\ 0 & 0 & 0 & \frac{25}{36} & \frac{5}{6} & 1 \end{bmatrix} \begin{bmatrix} a_{0,2} \\ a_{0,1} \\ a_{0,0} \\ a_{1,2} \\ a_{1,1} \\ a_{1,0} \end{bmatrix} = \begin{bmatrix} 0 \\ 0 \\ x(\frac{1}{6}) \\ x(\frac{1}{3}) \\ x(\frac{2}{3}) \\ x(\frac{5}{6}) \end{bmatrix}, \tag{6.32}$$

where the first two rows are from (6.30), and the last four rows are from matching values at four points. Again, the matrix in (6.32) is nonsingular; thus, any set of values $x(\frac{1}{6})$, $x(\frac{1}{3})$, $x(\frac{2}{3})$, and $x(\frac{5}{6})$ yields a unique spline approximation.

(iii) Suppose that instead we match values of the function at four points on the first piece, such as 0, $\frac{1}{8}$, $\frac{1}{4}$, and $\frac{3}{8}$. The resulting system of equations to solve is

$$\begin{bmatrix} \frac{1}{4} & \frac{1}{2} & 1 & -\frac{1}{4} & -\frac{1}{2} & -1 \\ 1 & 1 & 0 & -1 & -1 & 0 \\ 0 & 0 & 1 & 0 & 0 & 0 \\ \frac{1}{64} & \frac{1}{8} & 1 & 0 & 0 & 0 \\ \frac{1}{16} & \frac{1}{4} & 1 & 0 & 0 & 0 \\ \frac{9}{64} & \frac{3}{8} & 1 & 0 & 0 & 0 \end{bmatrix} \begin{bmatrix} a_{0,2} \\ a_{0,1} \\ a_{0,0} \\ a_{1,2} \\ a_{1,1} \\ a_{1,0} \end{bmatrix} = \begin{bmatrix} 0 \\ 0 \\ x(0) \\ x(\frac{1}{8}) \\ x(\frac{1}{4}) \\ x(\frac{3}{8}) \end{bmatrix}, \tag{6.33}$$

where the first two rows are from (6.30) and the last four rows are from matching values at four points. The matrix in (6.33) is singular; thus, a set of values $x(0)$, $x(\frac{1}{8})$, $x(\frac{1}{4})$, and $x(\frac{3}{8})$ either will not be consistent with a spline of the desired form or will not uniquely determine the spline. For the two functions we are considering, the samples of x_1 are not consistent with a spline of the desired form, and the samples of x_2 do not uniquely determine a spline of the desired form.

Fixing values of the spline at the knots, as in part (i) of the example, is conventional. Suppose that we are fitting a spline of degree K with $L+1$ knots. The spline is defined by L polynomial pieces, so there are $L(K+1)$ parameters to determine. Values at the knots give $L+1$ constraints, and continuity of derivatives of order up to $K-1$ at the $L-1$ interior knots gives $(L-1)K$ additional constraints. There are $L(K+1)-(L+1+(L-1)K) = K-1$ excess degrees of freedom (in part (i) of the example above, $K=2$, so there was one additional degree of freedom). The spline is usually determined uniquely by specifying a total of $K-1$ derivatives at the first and/or last knot.

In the common case of $K=3$, two constraints are needed, and these could be obtained by specifying the first derivative at the first and last knots. Another common method is to require the derivative of order 3 to be continuous at the second and penultimate knots. (The degree 3 of the spline ensures continuity of derivatives of order 0, 1, and 2.) This additional continuity requirement effectively removes the second and penultimate knots from the knot sequence and is thus called the *not-a-knot condition.* This method was used to produce the spline of degree 3 in Figure 6.23.

6.3.2 Bases for uniform spline spaces

From now on, we restrict our attention to uniform splines and uniform spline spaces, so the knot sequence is doubly infinite. With the uniform knot spacing $\tau_{n+1}-\tau_n = T$, for all $n \in \mathbb{Z}$, the spline space $S_{K,\tau}$ is a shift-invariant subspace with respect to shift T. These shift-invariant subspaces have generators; that is, they have bases that are shifts of a single function. For the remainder of this section, we consider only $T=1$. We will see how the approximation error varies with T in Section 6.3.3.

Elementary B-splines The centered unit-width box function,

$$\beta^{(0)}(t) = \begin{cases} 1, & \text{for } t \in [-\frac{1}{2}, \frac{1}{2}); \\ 0, & \text{otherwise}, \end{cases} \tag{6.34a}$$

is called the *elementary B-spline of degree* 0. It is a spline of degree 0 with knots $(-\frac{1}{2}, \frac{1}{2})$. Shifts of $\beta^{(0)}$ are called B-splines of degree 0.

The *elementary B-spline of degree* K is defined by repeated convolution of $\beta^{(0)}$ with itself,

$$\beta^{(K)} = \beta^{(K-1)} * \beta^{(0)}, \qquad K = 1, 2, \ldots. \tag{6.34b}$$

Shifts of $\beta^{(K)}$ are called B-splines of degree K. By a simple calculation, the elementary B-spline of degree 1 is given by

$$\beta^{(1)}(t) = \begin{cases} 1+t, & \text{for } t \in [-1, 0); \\ 1-t, & \text{for } t \in [0, 1); \\ 0, & \text{otherwise}, \end{cases} \tag{6.35}$$

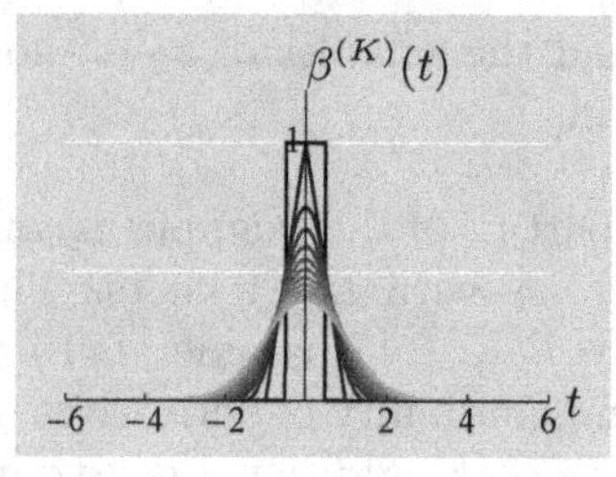

(a) Elementary B-splines.

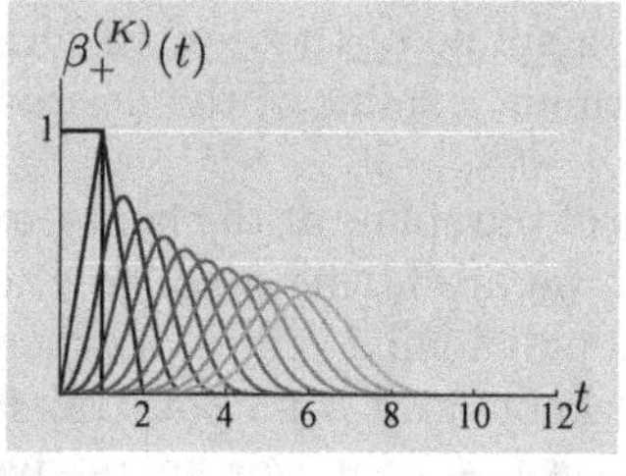

(b) Causal elementary B-splines.

Figure 6.25 Elementary B-splines of degree 0, 1, ..., 11 (solid lines, from darkest to lightest) and causal versions.

which is the triangle function we have seen many times. Convolving with $\beta^{(0)}$ once again gives

$$\beta^{(2)}(t) = \begin{cases} \frac{1}{2}(\frac{3}{2}+t)^2, & \text{for } t \in [-\frac{3}{2}, -\frac{1}{2}); \\ \frac{3}{4} - t^2, & \text{for } t \in [-\frac{1}{2}, \frac{1}{2}); \\ \frac{1}{2}(\frac{3}{2}-t)^2, & \text{for } t \in [\frac{1}{2}, \frac{3}{2}); \\ 0, & \text{otherwise.} \end{cases} \tag{6.36}$$

This process can be repeated to obtain any desired elementary B-spline. The first few elementary B-splines are shown in Figure 6.25(a), and a general expression is developed in Exercise 6.12.

Using the convolution property, (4.62), and the Fourier transform of the box function, (4.75), the Fourier transform of the elementary B-spline of degree K is

$$B^{(K)}(\omega) = \operatorname{sinc}^{K+1}\left(\frac{1}{2}\omega\right). \tag{6.37}$$

The elementary B-spline of degree K is a spline with knots $(-\frac{1}{2}(K+1), \ldots, \frac{1}{2}(K+1))$, the proof of which is left for Exercise 6.13. Note that, for odd K, the knots are at integer values; for even K, the knots are offset by $\frac{1}{2}$ from integer values.

Causal elementary B-splines For any $K \in \mathbb{N}$, shifting $\beta^{(K)}$ by $\frac{1}{2}(K+1)$ gives a spline with support in $[0, K+1]$ and knots $(0, 1, \ldots, K+1)$,

$$\beta_+^{(K)}(t) = \beta^{(K)}\left(t - \frac{1}{2}(K+1)\right), \qquad t \in \mathbb{R}. \tag{6.38}$$

This is called the *causal elementary B-spline of degree* K; the first few of these are shown in Figure 6.25(b). An equivalent definition is as follows:

$$\beta_+^{(0)}(t) = \begin{cases} 1, & \text{for } t \in [0, 1); \\ 0, & \text{otherwise,} \end{cases} \tag{6.39a}$$

$$\beta_+^{(K)} = \beta_+^{(K-1)} * \beta_+^{(0)}, \qquad K = 1, 2, \ldots. \tag{6.39b}$$

B-spline bases B-splines can be used as building blocks for representing functions in uniform spline spaces. Consider first the B-splines of degree 0. The shift by n of $\beta_+^{(0)}$ is a spline of degree 0 with knots $(n, n+1)$. The closure of the span of these functions for $n \in \mathbb{Z}$ gives every piecewise-constant function with points of discontinuity in $\mathbb{Z}$ – in other words, the uniform spline space $S_{0,\mathbb{Z}}$. The recovery of functions in $S_{0,\mathbb{Z}}$ from samples and the approximation of functions by the closest element of $S_{0,\mathbb{Z}}$ were together the central illustration of sampling and interpolation in Section 5.1. Specifically, in sampling followed by interpolation with spacing $T = 1$, using the time-reversed $\beta_+^{(0)}$ as the sampling prefilter and $\beta_+^{(0)}$ as the interpolation postfilter gives a system that computes the orthogonal projection onto $S_{0,\mathbb{Z}}$; see Figure 5.5(b). This shows that $\beta_+^{(0)}$ is a generator of the shift-invariant subspace $S_{0,\mathbb{Z}}$; it is the $K = 0$ case of the following theorem:

THEOREM 6.12 (B-SPLINE BASES FOR UNIFORM SPLINE SPACES) For any $K \in \mathbb{N}$, let $\beta^{(K)}$ be the elementary B-spline of degree K defined in (6.34) and let $\beta_+^{(K)}$ be its causal version defined in (6.38). Then, the following statements hold:

(i) The causal elementary B-spline $\beta_+^{(K)}$ is a generator of the shift-invariant subspace $S_{K,\mathbb{Z}}$ with respect to shift 1.

(ii) $\overline{\text{span}}(\{\beta^{(K)}(t-k)\}_{k\in\mathbb{Z}}) = \begin{cases} S_{K,\mathbb{Z}}, & \text{for odd } K; \\ S_{K,\mathbb{Z}+1/2}, & \text{for even } K. \end{cases}$

(iii) No function with support shorter than that of $\beta_+^{(K)}$ is a generator of $S_{K,\mathbb{Z}}$.

We will first explain why part (i) of the theorem has already been proven for $K = 1$ in the previous chapter. Then, we discuss the steps of a constructive proof of part (i) for general K. Part (ii) follows easily from part (i), and it illustrates why it is slightly more convenient to use the causal elementary B-splines in the remainder of the section – namely we need not distinguish between odd K and even K. We will not prove part (iii); see the *Further reading*.

In part (i) of the theorem, it is clear that any linear combination of $\beta_+^{(K)}$ and its integer shifts is in $S_{K,\mathbb{Z}}$, so $\overline{\text{span}}(\{\beta_+^{(K)}(t-k)\}_{k\in\mathbb{Z}}) \subseteq S_{K,\mathbb{Z}}$. The challenge is to show that $S_{K,\mathbb{Z}} \subseteq \overline{\text{span}}(\{\beta_+^{(K)}(t-k)\}_{k\in\mathbb{Z}})$. Thus, any mechanism to find the coefficient sequence α for the expansion

$$x(t) = \sum_{k\in\mathbb{Z}} \alpha_k \beta_+^{(K)}(t-k), \qquad t \in \mathbb{R}, \tag{6.40}$$

where x is an arbitrary element of $S_{K,\mathbb{Z}}$, will complete the proof. In the language of Section 5.4.3, x in (6.40) is obtained by interpolating the sequence α with spacing 1 and postfilter $\beta_+^{(K)}$. By Theorem 5.19, the existence of a function $\widetilde{\beta}_+^{(K)}$ such that

$$\left\langle \beta_+^{(K)}(t-k), \widetilde{\beta}_+^{(K)}(t-n) \right\rangle_t = \delta_{k-n}, \qquad k, n \in \mathbb{Z},$$

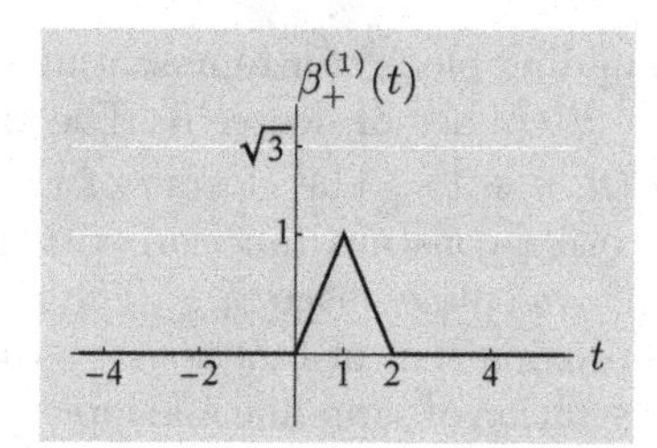

(a) Causal elementary B-spline of degree 1.

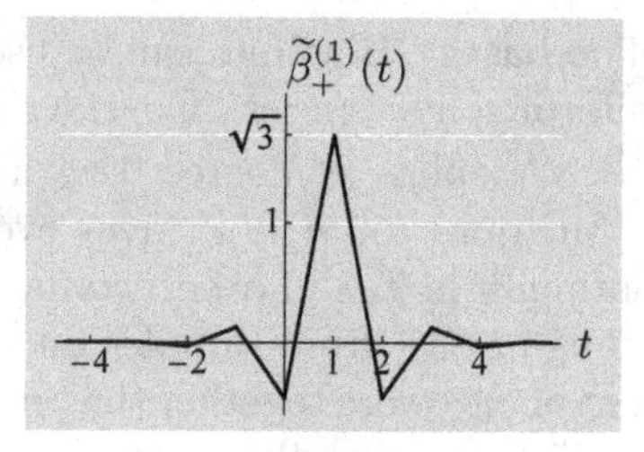

(b) Its canonical dual.

Figure 6.26 The causal elementary B-spline of degree 1, $\beta_+^{(1)}$, and the unique function $\widetilde{\beta}_+^{(1)}$ such that $\{\beta_+^{(1)}(t-k)\}_{k\in\mathbb{Z}}$ and $\{\widetilde{\beta}_+^{(1)}(t-k)\}_{k\in\mathbb{Z}}$ are a biorthogonal pair of bases for $S_{1,\mathbb{Z}}$.

is satisfied implies that a suitable sequence α exists. Specifically, sampling using time-reversed and conjugated $\widetilde{\beta}_+^{(K)}$ as the prefilter and spacing 1 would then give α. All these arguments apply with the elementary B-spline $\beta^{(K)}$ replacing its causal version $\beta_+^{(K)}$ except that, for even values of K, the generated space is $S_{K,\mathbb{Z}+1/2}$ rather than $S_{K,\mathbb{Z}}$.

For the case of $K = 1$, recall that the elementary B-spline $\beta^{(1)}$ is the triangle function used as an interpolation postfilter in Examples 5.24–5.26 in Section 5.4.3. Many duals exist, including the ones in Figures 5.36(b) and 5.37(b), so, after accounting for the shift by 1, part (i) of the theorem is proven for $K = 1$. Specifically, let $\widetilde{\beta}^{(1)}$ denote the time-reversed and conjugated version of any sampling prefilter that is consistent with $\beta^{(1)}$ as an interpolation postfilter, as in Example 5.25. Then, we have

$$x(t) = \sum_{k\in\mathbb{Z}} \langle x(t), \widetilde{\beta}^{(1)}(t-k)\rangle_t \, \beta^{(1)}(t-k), \qquad t \in \mathbb{R},$$

for any $x \in S_{1,\mathbb{Z}}$, showing that $\{\beta^{(1)}(t-k)\}_{k\in\mathbb{Z}}$ is a basis for $S_{1,\mathbb{Z}}$. When, in addition, $\widetilde{\beta}^{(1)} \in S_{1,\mathbb{Z}}$, we have that $\{\beta^{(1)}(t-k)\}_{k\in\mathbb{Z}}$ and $\{\widetilde{\beta}^{(1)}(t-k)\}_{k\in\mathbb{Z}}$ are a biorthogonal pair of bases for $S_{1,\mathbb{Z}}$. The unique function $\widetilde{\beta}^{(1)}$ that gives this property is called the *canonical dual spline*; it was determined in Example 5.26 and was shown in Figure 5.37(b).

Since $\beta_+^{(1)}$ is $\beta^{(1)}$ shifted by 1, its canonical dual is simply $\widetilde{\beta}^{(1)}$ shifted by 1; we denote the resulting function $\widetilde{\beta}_+^{(1)}$. Figure 6.26 shows $\beta_+^{(1)}$ alongside its canonical dual $\widetilde{\beta}_+^{(1)}$. These satisfy

$$x(t) = \sum_{k\in\mathbb{Z}} \langle x(t), \widetilde{\beta}_+^{(1)}(t-k)\rangle_t \, \beta_+^{(1)}(t-k), \qquad t \in \mathbb{R},$$

for any $x \in S_{1,\mathbb{Z}}$. Furthermore, $\{\beta_+^{(1)}(t-k)\}_{k\in\mathbb{Z}}$ and $\{\widetilde{\beta}_+^{(1)}(t-k)\}_{k\in\mathbb{Z}}$ are a biorthogonal pair of bases for $S_{1,\mathbb{Z}}$.

To provide a similar constructive proof for an arbitrary value of $K \in \mathbb{N}$, we can find $\widetilde{\beta}_+^{(K)}$ such that $\{\beta_+^{(K)}(t-k)\}_{k\in\mathbb{Z}}$ and $\{\widetilde{\beta}_+^{(K)}(t-k)\}_{k\in\mathbb{Z}}$ are a biorthogonal pair of bases for $S_{K,\mathbb{Z}}$. To have $\overline{\text{span}}(\{\widetilde{\beta}_+^{(K)}(t-k)\}_{k\in\mathbb{Z}}) \subseteq \overline{\text{span}}(\{\beta_+^{(K)}(t-k)\}_{k\in\mathbb{Z}})$, let

$$\widetilde{\beta}_+^{(K)}(t) \;=\; \sum_{m\in\mathbb{Z}} c_m \beta_+^{(K)}(t-m), \qquad t \in \mathbb{R}, \tag{6.41a}$$

for some sequence $c \in \ell^2(\mathbb{Z})$. Biorthogonality, (2.111), requires

$$\left\langle \widetilde{\beta}_+^{(K)}(t-i),\, \beta_+^{(K)}(t-k) \right\rangle_t \;=\; \delta_{i-k} \qquad \text{for every } i,\, k \in \mathbb{Z},$$

but, since the inner product depends only on $i-k$, it suffices to enforce only

$$\left\langle \widetilde{\beta}_+^{(K)}(t),\, \beta_+^{(K)}(t-k) \right\rangle_t \;=\; \delta_k \qquad \text{for every } k \in \mathbb{Z}. \tag{6.41b}$$

We thus require sequence c to satisfy

$$\begin{aligned}
\delta_k &\overset{(a)}{=} \left\langle \widetilde{\beta}_+^{(K)}(t),\, \beta_+^{(K)}(t-k) \right\rangle_t \\
&\overset{(b)}{=} \left\langle \sum_{m\in\mathbb{Z}} c_m \beta_+^{(K)}(t-m),\, \beta_+^{(K)}(t-k) \right\rangle_t \\
&\overset{(c)}{=} \sum_{m\in\mathbb{Z}} c_m \langle \beta_+^{(K)}(t-m),\, \beta_+^{(K)}(t-k)\rangle_t \\
&\overset{(d)}{=} \sum_{m\in\mathbb{Z}} c_m h_{k-m}^{(K)} \overset{(e)}{=} c * h^{(K)},
\end{aligned} \tag{6.42}$$

where (a) follows from (6.41b); (b) from (6.41a); (c) from the linearity in the first argument of the inner product; (d) from introducing

$$h_n^{(K)} \;=\; \langle \beta_+^{(K)}(t),\, \beta_+^{(K)}(t-n)\rangle_t \;=\; \int_{-\infty}^{\infty} \beta_+^{(K)}(t)\, \beta_+^{(K)*}(t-n)\, dt, \quad n \in \mathbb{Z}, \tag{6.43}$$

which is the deterministic autocorrelation of the function $\beta_+^{(K)}$ evaluated at the integers; and (e) from recognizing the sum as a convolution. What remains is to show why such a sequence $c \in \ell^2(\mathbb{Z})$ exists.

Writing the equivalent of (6.42) in the z-transform domain gives

$$C(z)\, H^{(K)}(z) \;=\; 1. \tag{6.44}$$

We thus seek a stable sequence c with $C(z) = 1/H^{(K)}(z)$; such a c is obtained by inverting the z-transform if there is a region of convergence for $1/H^{(K)}(z)$ that includes the unit circle. Since $h^{(K)}$ is the deterministic autocorrelation of a stable real sequence and has a rational z-transform (see Exercise 6.15), it follows from Theorem 3.13 that the zeros of $H^{(K)}(z)$ satisfy certain symmetries. As long as no zeros of $H^{(K)}(z)$ are on the unit circle, there will be a zero z_0 of $H^{(K)}(z)$ such that $\{z \mid |z_0|^{-1} < |z| < |z_0|\}$ is the unique valid ROC for $C(z)$. This condition on $H^{(K)}(z)$ is established in Exercise 6.15. Since c is uniquely determined, the canonical dual $\widetilde{\beta}_+^{(K)}$ is uniquely determined. This completes the proof.

Orthonormal bases for uniform spline spaces Synthesizing functions using B-splines is very convenient, since the basis functions $\{\beta_+^{(K)}(t-k)\}_{k\in\mathbb{Z}}$ are of finite support. However, the canonical dual splines $\widetilde{\beta}_+^{(K)}$ are of infinite support, albeit with fast decay (exponential decay for any fixed K). Orthonormal bases have other convenient aspects, such as using the same functions in analysis and synthesis, and simple computations of orthogonal projections. Orthonormal bases for uniform spline spaces can be derived from the B-spline bases. We use degree-1 splines to illustrate the method.

EXAMPLE 6.12 (ORTHONORMAL BASIS FOR $S_{1,\mathbb{Z}}$) We want to derive a generator for the uniform spline space $S_{1,\mathbb{Z}}$ that forms an orthonormal set with its integer shifts. Since the causal elementary B-spline of degree 1, $\beta_+^{(1)}$, is a generator of this space, we can express our desired generator $\beta_\perp^{(1)}$ using the B-spline basis,

$$\beta_\perp^{(1)}(t) = \sum_{m\in\mathbb{Z}} d_m \beta_+^{(1)}(t-m), \tag{6.45}$$

for some sequence $d \in \ell^2(\mathbb{Z})$. For the new generator to be orthonormal to its integer shifts, it must satisfy

$$\begin{aligned}
\delta_k &= \langle \beta_\perp^{(1)}(t),\, \beta_\perp^{(1)}(t-k)\rangle_t \\
&\overset{(a)}{=} \left\langle \sum_{m\in\mathbb{Z}} d_m \beta_+^{(1)}(t-m),\, \sum_{\ell\in\mathbb{Z}} d_\ell \beta_+^{(1)}(t-k-\ell)\right\rangle_t \\
&\overset{(b)}{=} \sum_{m\in\mathbb{Z}}\sum_{\ell\in\mathbb{Z}} d_m d_\ell^* \langle \beta_+^{(1)}(t-m),\, \beta_+^{(1)}(t-k-\ell)\rangle_t \\
&\overset{(c)}{=} \sum_{m\in\mathbb{Z}}\sum_{\ell\in\mathbb{Z}} d_m d_\ell^* h_{k+\ell-m}^{(1)} \overset{(d)}{=} \sum_{m\in\mathbb{Z}}\sum_{n\in\mathbb{Z}} d_m d_{m-n}^* h_{k-n}^{(1)} \\
&\overset{(e)}{=} \sum_{n\in\mathbb{Z}}\left(\sum_{m\in\mathbb{Z}} d_m d_{m-n}^*\right) h_{k-n}^{(1)} \overset{(f)}{=} \sum_{n\in\mathbb{Z}} a_n h_{k-n}^{(1)} \overset{(g)}{=} a * h^{(1)},
\end{aligned} \tag{6.46}$$

where (a) follows from (6.45); (b) from the linearity in the first argument and the conjugate linearity in the second argument of the inner product; (c) from (6.43); (d) from the change of variable $n = m - \ell$; (e) from interchanging summations; (f) from denoting the deterministic autocorrelation of d by a; and (g) from recognizing the sum as a convolution. This is a constraint on the sequence d.

Writing the equivalent of (6.46) in the z-transform domain gives

$$A(z)\, H^{(1)}(z) = 1, \tag{6.47}$$

which is nearly the same as (6.44). The distinction is that, rather than wanting the inverse z-transform of $1/H^{(1)}(z)$, we want a sequence with deterministic autocorrelation equal to the inverse z-transform of $1/H^{(1)}(z)$. Using (3.146) and (6.47), we have

$$D(z)\, D_*(z^{-1}) = \frac{1}{H^{(1)}(z)}, \tag{6.48}$$

which indicates that any appropriate spectral factor gives a sequence d through which a desired generator of $S_{1,\mathbb{Z}}$ can be determined.

It is straightforward to calculate the sampled deterministic autocorrelation of $\beta_+^{(1)}$ by computing the required integrals in (6.43):

$$\begin{aligned} h_n^{(1)} &= \begin{cases} \frac{2}{3}, & \text{for } n = 0; \\ \frac{1}{6}, & \text{for } n = \pm 1; \\ 0, & \text{otherwise} \end{cases} \\ &= \frac{1}{6}\delta_{n+1} + \frac{2}{3}\delta_n + \frac{1}{6}\delta_{n-1}. \end{aligned}$$

Substituting the z-transform of this sequence into (6.48) and factoring gives

$$D(z)D_*(z^{-1}) = \frac{6}{z + 4 + z^{-1}} = \frac{6(2-\sqrt{3})}{\left(1 + (2-\sqrt{3})z^{-1}\right)\left(1 + (2-\sqrt{3})z\right)}.$$

Therefore, one possible factorization is

$$D(z) = \frac{\sqrt{6(2-\sqrt{3})}}{1 + (2-\sqrt{3})z^{-1}}, \tag{6.49a}$$

which yields

$$d_n = \begin{cases} \sqrt{6(2-\sqrt{3})}(\sqrt{3}-2)^n, & \text{for } n \geq 0; \\ 0, & \text{otherwise.} \end{cases} \tag{6.49b}$$

The resulting basis function,

$$\beta_\perp^{(1)}(t) = \sum_{k=0}^{\infty} d_m \beta_+^{(1)}(t - m), \tag{6.50}$$

orthonormal to its integer shifts, is shown in Figure 6.27. The choice of $D(z)$ is not unique; for example,

$$D(z) = \frac{\sqrt{6(2-\sqrt{3})}}{1 + (2-\sqrt{3})z}$$

would lead to an anticausal sequence d, and any allpass sequence can be convolved with d from (6.49b) to yield additional solutions.

This orthogonalization method generalizes to any degree K; a variation on it is explored in Exercise 6.16.

One remarkable property satisfied both by $\beta^{(0)}$ and by $\beta^{(1)}$ is the interpolation property; that is, $\beta^{(0)}(n) = \delta_n$ and $\beta^{(1)}(n) = \delta_n$ for any $n \in \mathbb{Z}$. B-splines of higher degree do not have this property; for bases for uniform spline spaces with this property, one needs to use *cardinal splines* instead (see the *Further reading*).

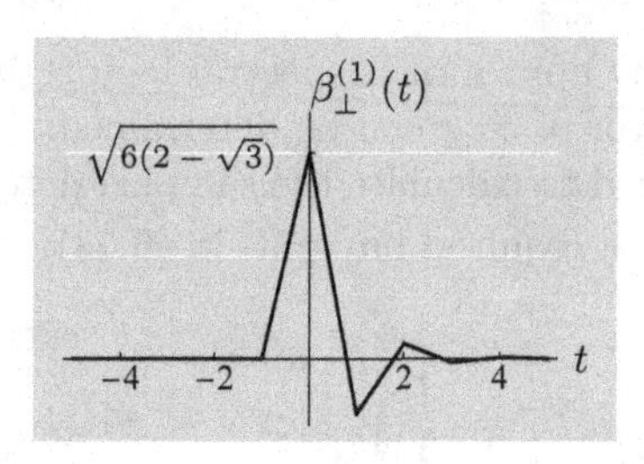

Figure 6.27 A function $\beta_\perp^{(1)}$ that is a generator of the uniform spline space $S_{1,\mathbb{Z}}$ while also forming an orthonormal set with its integer shifts. The function is not unique in having this property.

Best approximation Using an orthonormal basis or biorthogonal pair of bases for a uniform spline space $S_{K,\mathbb{Z}}$, we can obtain an orthogonal projection operator from $\mathcal{L}^2(\mathbb{R})$ onto $S_{K,\mathbb{Z}}$ and hence a way to find the best approximation in $S_{K,\mathbb{Z}}$ of an arbitrary function in $\mathcal{L}^2(\mathbb{R})$.

Let $\beta_\perp^{(K)}$ be a real-valued generator of $S_{K,\mathbb{Z}}$ that is orthogonal to its integer shifts so that $\{\beta_\perp^{(K)}(t-k)\}_{k\in\mathbb{Z}}$ is an orthonormal basis for $S_{K,\mathbb{Z}}$ as in (6.50) for $K=1$. Then, analogously to Theorem 5.11, for any $x \in \mathcal{L}^2(\mathbb{R})$, the spline function

$$\widehat{x}(t) = \sum_{k\in\mathbb{Z}} y_k \beta_\perp^{(K)}(t-k), \qquad t \in \mathbb{R}, \tag{6.51a}$$

computed from the coefficients

$$y_k = \int_{-\infty}^{\infty} x(\tau)\, \beta_\perp^{(K)}(\tau-k)\, d\tau, \qquad k \in \mathbb{Z}, \tag{6.51b}$$

is the orthogonal projection of x onto $S_{K,\mathbb{Z}}$. Similarly, the set of integer shifts $\{\beta_+^{(K)}(t-k)\}_{k\in\mathbb{Z}}$ of the causal elementary B-spline forms a biorthogonal pair of bases for $S_{K,\mathbb{Z}}$ with the set of integer shifts $\{\widetilde{\beta}_+^{(K)}(t-k)\}_{k\in\mathbb{Z}}$ of the canonical dual of the causal elementary B-spline. Thus, analogously to Theorem 5.19, for any $x \in \mathcal{L}^2(\mathbb{R})$, the spline function

$$\widehat{x}(t) = \sum_{k\in\mathbb{Z}} y_k \beta_+^{(K)}(t-k), \qquad t \in \mathbb{R}, \tag{6.52a}$$

computed from the coefficients

$$y_k = \int_{-\infty}^{\infty} x(\tau)\, \widetilde{\beta}_+^{(K)}(\tau-k)\, d\tau, \qquad k \in \mathbb{Z}, \tag{6.52b}$$

is the orthogonal projection of x onto $S_{K,\mathbb{Z}}$.

6.3.3 Strang–Fix condition for polynomial representation

We have seen that $\beta^{(K)}$, the elementary B-spline of degree K, is a generator of the uniform spline space of degree K, so any uniform spline of degree K can be written as a linear combination of $\beta^{(K)}$ and its integer shifts. This result from Theorem 6.12 is not obvious – indeed, it was a major breakthrough of Schoenberg in 1969 – but it is natural since $\beta^{(K)}$ is itself a uniform spline of degree K. As discussed below, any polynomial of degree K can be expressed using $\beta^{(K)}$; this too is unsurprising because a polynomial is the simplest form of uniform spline – even its derivatives of order K are continuous. A surprising fact, however, is that some functions that look nothing like polynomials can also be used to express any polynomial; a simple Fourier-domain test for this property is called the Strang–Fix condition. The approximation error bound we present based on the Strang–Fix condition thus applies more generally, but we will focus our attention on elementary B-spline generators and uniform spline spaces.

Representing polynomials with B-splines Let p_K be a polynomial of degree K. Then, p_K can be expressed as

$$p_K(t) = \sum_{k\in\mathbb{Z}} \alpha_k \beta^{(K)}(t-k), \qquad t \in \mathbb{R}, \tag{6.53}$$

for some coefficient sequence α. Except in the trivial case of $p_K = \mathbf{0}$, the polynomial p_K is not in $\mathcal{L}^2(\mathbb{R})$ since $|p_K(t)|$ will grow without bound for $t \to \pm\infty$. Thus, the equality in (6.53) is pointwise, and $\alpha \notin \ell^2(\mathbb{Z})$.[104] We illustrate the existence of an expression in the form of (6.53) for any polynomial with degree $K = 2$ in the following example.

Example 6.13 (Polynomial reproduction with $\beta^{(2)}$) Consider the elementary B-spline of degree 2 as in (6.36). To show that an arbitrary polynomial of degree 2 can be expressed using $\{\beta^{(2)}(t-k)\}_{k\in\mathbb{Z}}$, as in (6.53), we first find ways to express the monomials 1, t, and t^2.

(i) We start by showing that any constant function can be written as a linear combination of $\{\beta^{(2)}(t-k)\}_{k\in\mathbb{Z}}$. At any fixed t, at most three elements of $\{\beta^{(2)}(t-k)\}_{k\in\mathbb{Z}}$ are nonzero, so the coefficient sequence can be any element of $\ell^\infty(\mathbb{Z})$ without raising convergence issues. To obtain a constant function, the coefficient sequence α would have to be constant. By direct computation, we can easily verify that

$$\sum_{k\in\mathbb{Z}} \beta^{(2)}(t-k) = 1, \tag{6.54a}$$

meaning that $\alpha_k = 1$, for all $k \in \mathbb{Z}$, yields the monomial 1. Specifically, since the sum is a periodic function with period 1, it suffices to verify that $\beta^{(2)}(t+1) + \beta^{(2)}(t) + \beta^{(2)}(t-1) = 1$ on the interval $[-\frac{1}{2}, \frac{1}{2}]$,

$$\frac{1}{2}\left(\frac{1}{2}-t\right)^2 + \left(\frac{3}{4}-t^2\right) + \frac{1}{2}\left(\frac{1}{2}+t\right)^2 = 1.$$

[104] For any $t \in \mathbb{R}$, at most $K+1$ terms of (6.53) are nonzero, so the series converges pointwise.

(ii) Next, we show that the function t can be written as a linear combination of $\{\beta^{(2)}(t-k)\}_{k\in\mathbb{Z}}$ using coefficient sequence $\alpha_k = k$, $k \in \mathbb{Z}$,

$$\sum_{k\in\mathbb{Z}} k\beta^{(2)}(t-k) \;=\; t. \tag{6.54b}$$

To verify this, consider $t \in [n-\frac{1}{2}, n+\frac{1}{2}]$ for some $n \in \mathbb{Z}$. Then, the terms in (6.54b) are nonzero only for $k \in \{n-1, n, n+1\}$, and (6.54b) reduces to

$$(n-1)\frac{1}{2}\left(\frac{1}{2}+n-t\right)^2 + n\left(\frac{3}{4}-(t-n)^2\right) + (n+1)\frac{1}{2}\left(\frac{1}{2}-n+t\right)^2 \;=\; t.$$

(iii) Finally, we show that the function t^2 can be written as a linear combination of $\{\beta^{(2)}(t-k)\}_{k\in\mathbb{Z}}$ using the coefficient sequence $\alpha_k = k^2 - \frac{1}{4}$, $k \in \mathbb{Z}$,

$$\sum_{k\in\mathbb{Z}} \left(k^2 - \frac{1}{4}\right)\beta^{(2)}(t-k) \;=\; t^2. \tag{6.54c}$$

This can be verified similarly to (6.54b).

Using (6.54), an arbitrary polynomial of degree 2 can be written as a linear combination of $\{\beta^{(2)}(t-k)\}_{k\in\mathbb{Z}}$,

$$a_2t^2 + a_1t + a_0 \;=\; \sum_{k\in\mathbb{Z}} \alpha_k \beta^{(2)}(t-k), \tag{6.55}$$

where $\alpha_k = a_2(k^2-\frac{1}{4}) + a_1k + a_0$.

The property (6.54a) of reproducing a constant holds for any spline degree K,

$$\sum_{k\in\mathbb{Z}} \beta^{(K)}(t-k) \;=\; 1. \tag{6.56}$$

This is easy to verify in the time domain for $K=0$ and $K=1$, and we have just verified it for $K=2$ in the previous example. The following theorem gives a general Fourier-domain characterization that is not only applicable to splines; through this characterization, (6.56) holds for any value of K by using (6.37).

THEOREM 6.13 (PARTITION OF UNITY) Let $\varphi \in \mathcal{L}^1(\mathbb{R})$ have Fourier transform Φ. The periodized version of φ with period 1 as in (4.34) satisfies

$$\varphi_1(t) \;=\; \sum_{n\in\mathbb{Z}} \varphi(t-n) \;=\; 1, \qquad t \in \mathbb{R}, \tag{6.57a}$$

if and only if

$$\Phi(2\pi k) \;=\; \delta_k, \qquad k \in \mathbb{Z}. \tag{6.57b}$$

Proof. Since φ_1 is periodic with period 1, it can be represented as a Fourier series (4.94b),

$$\varphi_1(t) = \sum_{k\in\mathbb{Z}} \Phi_{1,k} e^{j2\pi kt}, \tag{6.58a}$$

with coefficients from (4.94a),

$$\begin{aligned}\Phi_{1,k} &= \int_{-1/2}^{1/2} \varphi_1(t)\, e^{-j2\pi kt}\, dt \overset{(a)}{=} \int_{-1/2}^{1/2} \left(\sum_{n\in\mathbb{Z}} \varphi(t-n)\right) e^{-j2\pi kt}\, dt \\ &\overset{(b)}{=} \sum_{n\in\mathbb{Z}} \int_{-1/2}^{1/2} \varphi(t-n)\, e^{-j2\pi kt}\, dt \overset{(c)}{=} \sum_{n\in\mathbb{Z}} \int_{-1/2-n}^{1/2-n} \varphi(\tau)\, e^{-j2\pi k(\tau+n)}\, d\tau \\ &\overset{(d)}{=} \sum_{n\in\mathbb{Z}} \int_{-1/2-n}^{1/2-n} \varphi(\tau)\, e^{-j2\pi k\tau}\, d\tau \overset{(e)}{=} \int_{-\infty}^{\infty} \varphi(\tau)\, e^{-j2\pi k\tau}\, d\tau \overset{(f)}{=} \Phi(2\pi k), \end{aligned} \tag{6.58b}$$

where (a) follows from the definition of φ_1 in (6.57a); (b) from interchanging the sum and integral; (c) from the change of variable $\tau = t - n$; (d) from the periodicity of $e^{-j2\pi kt}$; (e) from combining integrals over unit-length intervals that partition $\mathbb{R}$ into a single integral over $\mathbb{R}$; and (f) from recognizing the Fourier transform evaluated at $\omega = 2\pi k$. Now both implications of the theorem follow from the Fourier series pair

$$1 \quad \overset{\text{FS}}{\longleftrightarrow} \quad \delta_k$$

and the uniqueness of the Fourier series.

Strang–Fix condition The partition of unity is the $K = 0$ case of the following theorem:

THEOREM 6.14 (POLYNOMIAL REPRODUCTION (STRANG–FIX)) Let $K \in \mathbb{N}$, and let φ be a function with Fourier transform Φ. If φ has sufficiently fast decay,

$$\int_{-\infty}^{\infty} (1 + |t|^K)\, |\varphi(t)|\, dt < \infty, \tag{6.59}$$

then the following statements are equivalent:

(i) Any polynomial p_K of degree at most K can be expressed as

$$p_K(t) = \sum_{k\in\mathbb{Z}} \alpha_k \varphi(t-k), \qquad t \in \mathbb{R},$$

for some coefficient sequence α, where the convergence is pointwise.

(ii) The Fourier transform Φ and its first K derivatives satisfy the *Strang–Fix condition of order* $K+1$:

$$\Phi(0) \neq 0, \tag{6.60a}$$

$$\Phi^{(k)}(2\pi\ell) = 0, \qquad k = 1, 2, \ldots, K, \quad \ell \in \mathbb{Z}\setminus\{0\}. \tag{6.60b}$$

A proof of the theorem can be found in the book of Strang and Fix; see the *Further reading.* Solved exercise 6.4 proves a special case of the theorem for when φ is an interpolating function, that is, when

$$\varphi(n) = \delta_n, \qquad n \in \mathbb{Z}. \tag{6.61}$$

In the theorem, the decay condition (6.59) is trivially satisfied by any φ having finite support; for infinitely supported φ, however, sufficient decay might not be satisfied.

The Fourier transform of the elementary B-spline of degree K has a Kth-order zero at nonzero multiples of 2π since it is the product of K sinc functions (see (6.37)). Therefore, the B-splines of degree K reproduce polynomials of degree up to K. This argument is made in detail for $K = 2$ in the following example, providing a different way of coming to the same conclusion as in Example 6.13.

Example 6.14 (Polynomial reproduction with $\beta^{(2)}$ (Strang–Fix)) The elementary B-spline of degree 2 has the Fourier transform (from (6.37))

$$B^{(2)}(\omega) = \operatorname{sinc}^3\left(\frac{\omega}{2}\right).$$

Let us check Condition (ii) in Theorem 6.14. Upon evaluating the Fourier transform at zero, we simply have

$$B^{(2)}(0) = \operatorname{sinc}^3(0) = 1 \neq 0,$$

so (6.60a) is satisfied. For the first derivative, for any $\ell \in \mathbb{Z} \setminus \{0\}$ we get

$$\begin{aligned}\frac{d}{d\omega}\left(B^{(2)}(\omega)\right)\Big|_{\omega=2\pi\ell} &= \frac{d}{d\omega}\operatorname{sinc}^3\left(\frac{\omega}{2}\right)\Big|_{\omega=2\pi\ell} \\ &\overset{(a)}{=} \frac{3}{2}\operatorname{sinc}^2\left(\frac{\omega}{2}\right)\frac{d}{d\omega}\operatorname{sinc}\left(\frac{\omega}{2}\right)\Big|_{\omega=2\pi\ell} \overset{(b)}{=} 0, \end{aligned} \tag{6.62}$$

where (a) follows from the chain rule for differentiation; and (b) from $\operatorname{sinc}(\pi\ell) = \delta_\ell$. Taking one more derivative gives, for any $\ell \in \mathbb{Z} \setminus \{0\}$,

$$\begin{aligned}\frac{d^2}{d\omega^2}\left(B^{(2)}(\omega)\right)\Big|_{\omega=2\pi\ell} &= \frac{d}{d\omega}\left(\frac{3}{2}\operatorname{sinc}^2\left(\frac{\omega}{2}\right)\frac{d}{d\omega}\operatorname{sinc}\left(\frac{\omega}{2}\right)\right)\Big|_{\omega=2\pi\ell} \\ &\overset{(a)}{=} \frac{3}{2}\operatorname{sinc}\left(\frac{\omega}{2}\right)\left(\frac{1}{2}\operatorname{sinc}\left(\frac{\omega}{2}\right)\frac{d^2}{d\omega^2}\operatorname{sinc}\left(\frac{\omega}{2}\right) + \left(\frac{d}{d\omega}\operatorname{sinc}\left(\frac{\omega}{2}\right)\right)^2\right)\Big|_{\omega=2\pi\ell} \\ &\overset{(b)}{=} 0, \end{aligned} \tag{6.63}$$

where (a) follows from the product and chain rules of differentiation; and (b) from $\operatorname{sinc}(\pi\ell) = \delta_\ell$. Together, (6.62) and (6.63) show that (6.60b) holds for $k = 1, 2$. The theorem now implies that polynomials up to degree 2 can be expressed using $\beta^{(2)}$ and its integer shifts, as we had already seen in Example 6.13 using a time-domain argument.

Bounds on approximation error The ability to represent polynomials with φ and its shifts is not very interesting in itself, but it reflects an ability to approximate smooth functions. Specifically, it is desirable for φ to satisfy the Strang–Fix condition for a high order $K+1$ because this ensures that the approximation error – for sufficiently smooth functions – decays quickly as the spacing of shifts of φ is reduced.

Let the function φ satisfy the Strang–Fix condition of order $K+1$, and let $T \in \mathbb{R}^+$. An approximation of $x \in \mathcal{L}^2(\mathbb{R})$ of the form

$$\widehat{x}(t) = \sum_{k\in\mathbb{Z}} \alpha_k \varphi\left(\frac{t-kT}{T}\right), \qquad t \in \mathbb{R},$$

lies in the shift-invariant subspace of $\mathcal{L}^2(\mathbb{R})$ with respect to shift T, generated by $\varphi(t/T)$.[105] If φ is a generator of the uniform spline space $S_{K,\mathbb{Z}}$, such as a causal elementary B-spline, then $\widehat{x}$ is in the uniform spline space $S_{K,T\mathbb{Z}}$. Our interest is in the quality of the best such estimate $\widehat{x}$. Assume, therefore, that the coefficient sequence α is chosen to minimize $\|x - \widehat{x}\|$.

Suppose that derivatives up to order K of x are continuous, and the derivative of order $K+1$ exists almost everywhere, with the resulting function $x^{(K+1)}$ in $\mathcal{L}^2(\mathbb{R})$. Then

$$\|x - \widehat{x}\| \leq c_K T^{K+1} \|x^{(K+1)}\|, \tag{6.64}$$

where c_K is a known constant. We will not prove this result (see the *Further reading*), but we will interpret it in light of polynomial representation and give an example.

Let p_K be an arbitrary polynomial of degree K. Since φ satisfies the Strang–Fix condition of order $K+1$, by Theorem 6.14, p_K can be represented exactly with $\varphi(t)$ and its integer shifts. Dilation by $T \in \mathbb{R}^+$ does not change whether a function is a polynomial, so p_K can also be represented exactly with $\varphi(t/T)$ and its shifts by integer multiples of T. Consistently with this, (6.64) predicts zero error in approximating p_K because the derivative of order $K+1$ of p_K is zero everywhere, implying that $\|p_K^{(K+1)}\| = 0$.[106]

Example 6.15 (Approximation in $S_{0,T\mathbb{Z}}$ and $S_{1,T\mathbb{Z}}$ spaces) Let $T \in \mathbb{R}^+$. The uniform spline space $S_{0,T\mathbb{Z}}$ is the set of piecewise-constant functions with breakpoints at $\{kT\}_{k\in\mathbb{Z}}$. The orthogonal projection of $x \in \mathcal{L}^2(\mathbb{R})$ to $S_{0,T\mathbb{Z}}$ is

$$\widehat{x}(t) = \frac{1}{T}\sum_{k\in\mathbb{Z}} \left\langle x(t), \widetilde{\beta}_+^{(0)}\left(\frac{t-kT}{T}\right)\right\rangle_t \beta_+^{(0)}\left(\frac{t-kT}{T}\right), \qquad t \in \mathbb{R}, \tag{6.65}$$

where $\beta_+^{(0)}$ is its own canonical dual, $\widetilde{\beta}_+^{(0)} = \beta_+^{(0)}$. Note that $\beta_+^{(0)}$ satisfies the Strang–Fix condition of order 1. Similarly, the uniform spline space $S_{1,T\mathbb{Z}}$ is the

[105] Including a normalization factor, $(1/\sqrt{T})\varphi(t/T)$, would keep the norm invariant to T; this is not necessary here.

[106] Technically, (6.64) does not apply because a nonzero polynomial cannot be in $\mathcal{L}^2(\mathbb{R})$. The evaluation of (6.64) is interesting nonetheless.

set of piecewise-linear functions with breakpoints at $\{kT\}_{k\in\mathbb{Z}}$. The orthogonal projection of $x \in \mathcal{L}^2(\mathbb{R})$ to $S_{1,T\mathbb{Z}}$ is

$$\widehat{x}(t) = \frac{1}{T}\sum_{k\in\mathbb{Z}}\left\langle x(t),\, \widetilde{\beta}_+^{(1)}\left(\frac{t-kT}{T}\right)\right\rangle_t \beta_+^{(1)}\left(\frac{t-kT}{T}\right), \qquad t \in \mathbb{R}, \tag{6.66}$$

where the canonical dual $\widetilde{\beta}_+^{(1)}$ is shown in Figure 6.26(b). Note that $\beta_+^{(1)}$ satisfies the Strang–Fix condition of order 2.

For the approximation error bound (6.64) to be applicable for approximation both in $S_{0,T\mathbb{Z}}$ and in $S_{1,T\mathbb{Z}}$, the function x must have at least one continuous derivative, and the second derivative $x^{(2)}$ must exist almost everywhere and be in $\mathcal{L}^2(\mathbb{R})$. The function x_1 in (6.4) satisfies these conditions. Figure 6.28(a) shows approximations of x_1 in $S_{0,T\mathbb{Z}}$ and $S_{1,T\mathbb{Z}}$ for $T = \frac{1}{8}$. (The approximation in $S_{1,T\mathbb{Z}}$ is almost completely obscured by x_1 itself.) The corresponding approximation errors are shown in Figure 6.28(c). Naturally, the piecewise-linear approximation is better than the piecewise-constant approximation. Figure 6.28(e) shows the norm of the approximation error as T is varied. The difference between $\|x - \widehat{x}\| = O(T)$ for piecewise-constant approximations and $\|x - \widehat{x}\| = O(T^2)$ for piecewise-linear approximations, as implied by (6.64), is apparent.

Uniform spline approximations can be useful even when (6.64) is not applicable. For example, the function x_2 in (6.7) is not continuous, so the bound (6.64) does not even apply for $K = 0$. However, the approximation results shown in Figure 6.28(b) show that the method is reasonably effective. Except at the point of discontinuity $1/\sqrt{2}$, x_2 is linear. Thus, if the discontinuity had not been present, we would have expected $O(T)$ error behavior for approximation in $S_{0,T\mathbb{Z}}$ and no error for approximation in $S_{1,T\mathbb{Z}}$. With splines, the effect of the discontinuity is mostly localized to a neighborhood of size $O(T)$ near the discontinuity; more precisely, it decays with distance from the discontinuity as fast as the canonical dual spline, which has an exponential decay. The contribution to $\|x - \widehat{x}\|^2$ from the discontinuity is thus $O(T)$, resulting in $\|x - \widehat{x}\| = O(T^{1/2})$, irrespective of the degree of the spline. The norm of the approximation error as T is varied, shown in Figure 6.28(f), is consistent with this.

Bounds similar to (6.64) apply to $\|x - \widehat{x}\|_p$ for $p \neq 2$, including $p = \infty$, that is, to pointwise error.

6.3.4 Continuous-time operators in spline spaces implemented with discrete-time processing

The present section and Section 5.4 of the previous chapter both study discrete-time representations of continuous-time signals, that is, sequence representations of functions. Various representations differ in terms of which types of processing they facilitate. As illustrated in Examples 2.56 and 2.57, the basic principle is that a linear operator $A : H_0 \to H_1$ has a convenient form when $A\varphi$ has a simple representation in the basis used for H_1, where φ is any element of the basis used

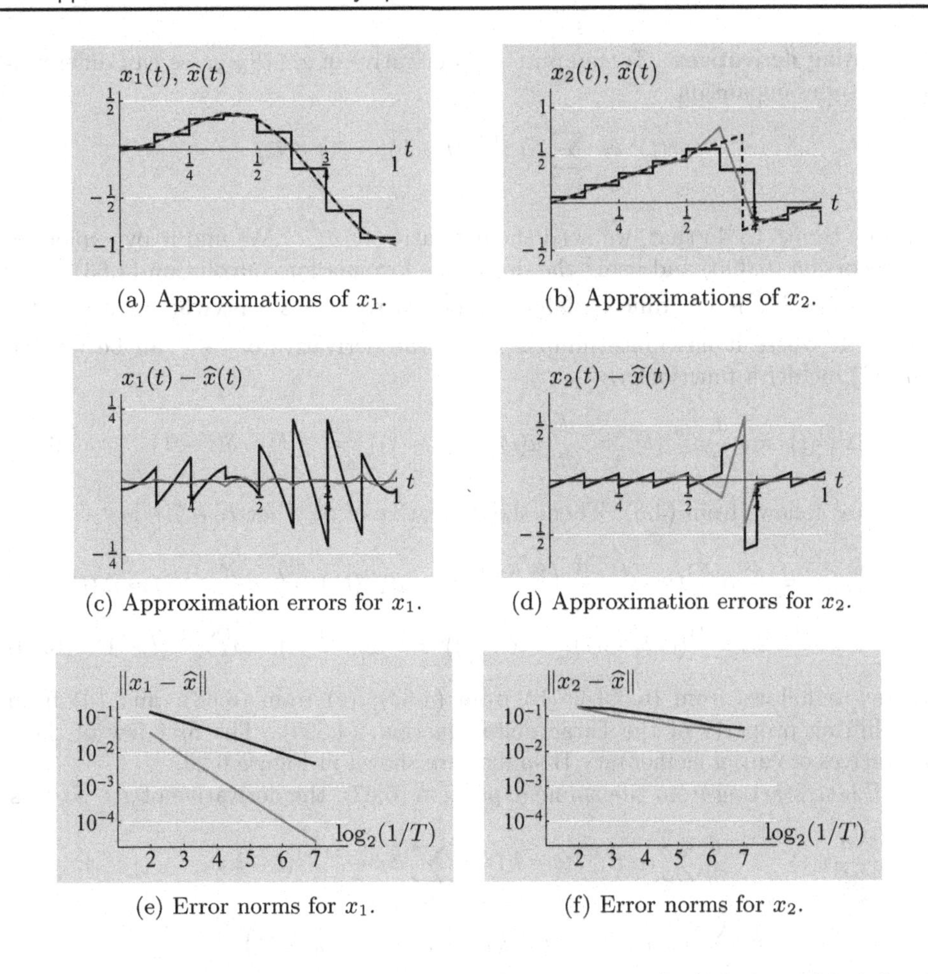

(a) Approximations of x_1. (b) Approximations of x_2.

(c) Approximation errors for x_1. (d) Approximation errors for x_2.

(e) Error norms for x_1. (f) Error norms for x_2.

Figure 6.28 Approximations of x_1 from (6.4) and x_2 from (6.7) in $S_{0,\frac{1}{8}\mathbb{Z}}$ (black lines) and $S_{1,\frac{1}{8}\mathbb{Z}}$ (gray lines) on [0, 1] using uniform splines.

for H_0. For example, with representations based on sinc functions, a continuous-time convolution has a simple equivalent in discrete time that is also a convolution (see Theorem 5.16). For spline representations, integration and differentiation are among the continuous-time operations that have simple forms in discrete time. In fact, we saw this once already in Example 2.57.

Recall that the causal elementary B-spline $\beta_+^{(K)}$ is a generator of the shift-invariant subspace $S_{K,\mathbb{Z}}$ with respect to shift 1, without having to make a distinction between odd and even values of K; see Theorem 6.12. These uniform spline spaces are related by integration and differentiation: if a function is in $S_{K,\mathbb{Z}}$, its derivative will be in $S_{K-1,\mathbb{Z}}$ (assuming that $K \in \mathbb{Z}^+$) and its integral will be in $S_{K+1,\mathbb{Z}}$.

Computing derivatives To compute the derivative of $x \in S_{K,\mathbb{Z}}$, we can differentiate its series expansion

$$x(t) = \sum_{k\in\mathbb{Z}} \alpha_k \beta_+^{(K)}(t-k), \qquad t \in \mathbb{R}, \tag{6.67}$$

term by term. To do that, we need the derivative of $\beta_+^{(K)}$. We find it by exploiting the recursion (6.39b) and using the derivative formula for convolution (4.63).

While $\beta_+^{(0)}$ is not differentiable, its derivative exists everywhere except at 0 and at 1, where it has finite jumps. Thus, the derivative of $\beta_+^{(0)}$ can be written using Dirac delta functions as

$$\Delta^{(0)}(t) = \frac{d}{dt}\beta_+^{(0)}(t) = \frac{d}{dt}(u(t) - u(t-1)) \overset{(a)}{=} \delta(t) - \delta(t-1), \tag{6.68}$$

where (a) follows from (4.6). Then, the derivative of $\beta_+^{(K)}$ for $K \in \mathbb{Z}^+$ is

$$\begin{aligned}\Delta^{(K)}(t) &= \frac{d}{dt}\beta_+^{(K)}(t) \overset{(a)}{=} \frac{d}{dt}\left(\beta_+^{(K-1)}(t) *_t \beta_+^{(0)}(t)\right) \overset{(b)}{=} \beta_+^{(K-1)}(t) *_t \Delta(t) \\ &\overset{(c)}{=} \beta_+^{(K-1)}(t) *_t (\delta(t) - \delta(t-1)) \overset{(d)}{=} \beta_+^{(K-1)}(t) - \beta_+^{(K-1)}(t-1),\end{aligned} \tag{6.69}$$

where (a) follows from (6.39b); (b) from (4.63); (c) from (6.68); and (d) from the shifting property of the Dirac delta function, (4.32e). The first few of these derivatives of causal elementary B-splines are shown in Figure 6.29.

Then, starting from the spline expansion (6.67), the derivative of $x \in S_{K,\mathbb{Z}}$ is

$$\begin{aligned}\frac{d}{dt}x(t) &= \frac{d}{dt}\sum_{k\in\mathbb{Z}} \alpha_k \beta_+^{(K)}(t-k) = \sum_{k\in\mathbb{Z}} \alpha_k \frac{d}{dt}\beta_+^{(K)}(t-k) \\ &\overset{(a)}{=} \sum_{k\in\mathbb{Z}} \alpha_k \left(\beta_+^{(K-1)}(t-k) - \beta_+^{(K-1)}(t-k-1)\right) \\ &= \sum_{k\in\mathbb{Z}} \alpha_k \beta_+^{(K-1)}(t-k) - \sum_{k\in\mathbb{Z}} \alpha_k \beta_+^{(K-1)}(t-k-1) \\ &\overset{(b)}{=} \sum_{k\in\mathbb{Z}} \alpha_k \beta_+^{(K-1)}(t-k) - \sum_{\ell\in\mathbb{Z}} \alpha_{\ell-1} \beta_+^{(K-1)}(t-\ell) \\ &\overset{(c)}{=} \sum_{k\in\mathbb{Z}} (\alpha_k - \alpha_{k-1})\beta_+^{(K-1)}(t-k) \overset{(d)}{=} \sum_{k\in\mathbb{Z}} \alpha'_k \beta_+^{(K-1)}(t-k),\end{aligned} \tag{6.70}$$

where (a) follows from (6.69); (b) from the change of variable $\ell = k+1$ in the second sum; (c) from combining the two sums; and (d) from defining the *first-order backward difference*, or *discrete derivative*, of the sequence α,

$$\alpha'_k = \alpha_k - \alpha_{k-1}, \qquad k \in \mathbb{Z}. \tag{6.71}$$

Thus, to compute the derivative of a function $x \in S_{K,\mathbb{Z}}$, we can apply the discrete derivative to the sequence that represents x; the result is the sequence representing the derivative in $S_{K-1,\mathbb{Z}}$.

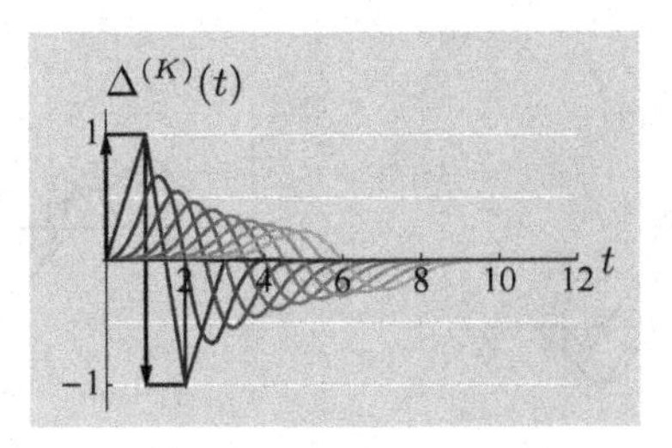

Figure 6.29 Derivatives $\Delta^{(K)}$ of the causal elementary B-splines $\beta_+^{(K)}$ of degree 0, 1, ..., 11 (solid lines, from darkest to lightest).

The discrete derivative operation (6.71) has a simple form using matrix–vector multiplication:

$$\begin{bmatrix} \vdots \\ \alpha'_{-1} \\ \boxed{\alpha'_0} \\ \alpha'_1 \\ \vdots \end{bmatrix} = \begin{bmatrix} & \vdots & \vdots & \vdots & \vdots & \\ \cdots & -1 & 1 & 0 & 0 & \cdots \\ \cdots & 0 & -1 & \boxed{1} & 0 & \cdots \\ \cdots & 0 & 0 & -1 & 1 & \cdots \\ & \vdots & \vdots & \vdots & \vdots & \end{bmatrix} \begin{bmatrix} \vdots \\ \alpha_{-2} \\ \alpha_{-1} \\ \boxed{\alpha_0} \\ \alpha_1 \\ \vdots \end{bmatrix}. \tag{6.72}$$

This shows that the matrix representation of the derivative operator – with respect to using the bases $\{\beta_+^{(K)}(t-k)\}_{k\in\mathbb{Z}}$ for the domain and $\{\beta_+^{(K-1)}(t-k)\}_{k\in\mathbb{Z}}$ for the codomain – is the matrix in (6.72). This matrix representation was previously derived for $K = 1$ in Example 2.57.

EXAMPLE 6.16 (DIFFERENTIATION IN $S_{2,\mathbb{Z}}$) The function x shown in Figure 6.30(a) is a uniform spline of degree 2 with knot sequence $\mathbb{Z}$. Thus, it can be differentiated on each interval $[k, k+1)$, $k \in \mathbb{Z}$, yielding a polynomial of degree 1 on each of these intervals.

An alternative is to use (6.70). The expansion coefficient sequence of x with respect to the causal elementary B-spline basis of degree 2 is

$$\alpha = \begin{bmatrix} \cdots & 0 & \boxed{0} & 1 & 3 & 4 & 5 & 4 & 2 & -2 & -1 & 0 & 0 & \cdots \end{bmatrix}^\top.$$

The discrete derivative of this sequence,

$$\alpha' = \begin{bmatrix} \cdots & 0 & \boxed{0} & 1 & 2 & 1 & 1 & -1 & -2 & -4 & 1 & 1 & 0 & \cdots \end{bmatrix}^\top,$$

is the expansion coefficient sequence of the derivative of x with respect to the causal elementary B-spline basis of degree 1. The derivative is shown in Figure 6.30(b).

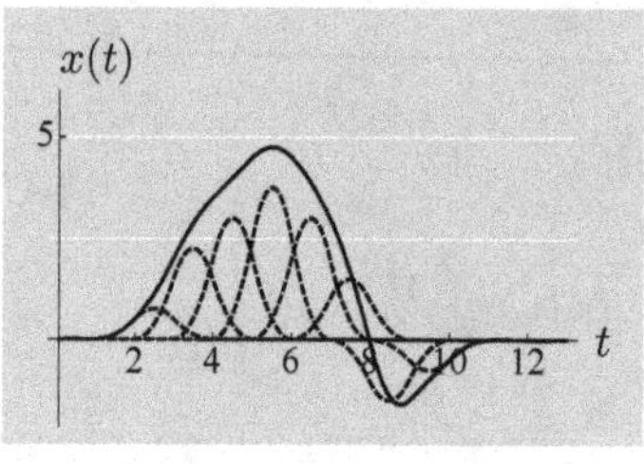

(a) Function in $S_{2,\mathbb{Z}}$.

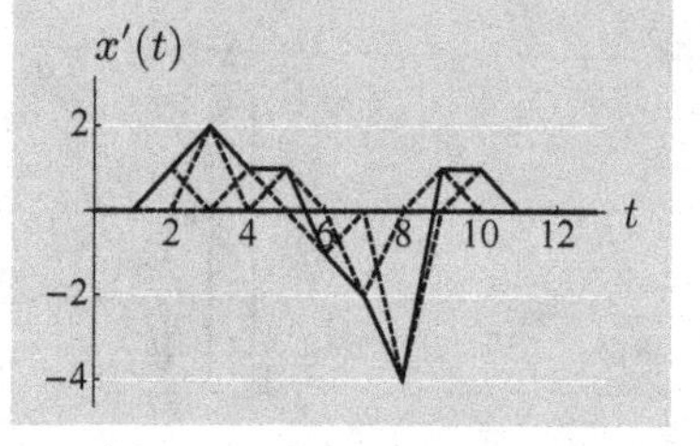

(b) Its derivative in $S_{1,\mathbb{Z}}$.

Figure 6.30 Differentiation in $S_{2,\mathbb{Z}}$. (a) Any function in $S_{2,\mathbb{Z}}$ is a linear combination of causal elementary B-splines of degree 2 (dashed lines) with some coefficient sequence α. (b) Its derivative is in $S_{1,\mathbb{Z}}$ and is a linear combination of causal elementary B-splines of degree 1 (dashed lines) with coefficient sequence α', where α' is the discrete derivative of α (see Example 6.16).

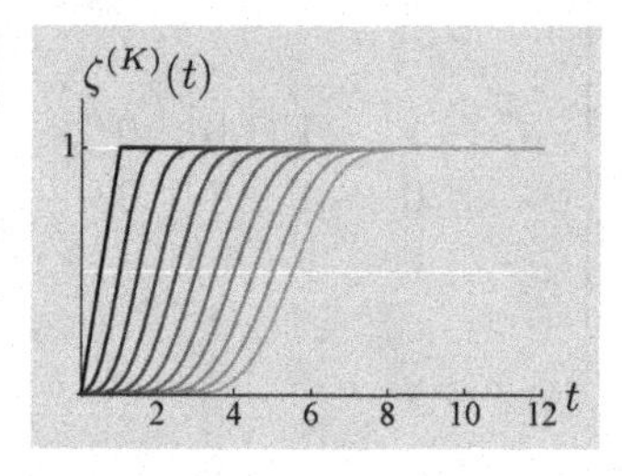

Figure 6.31 Integrals $\zeta^{(K)}$ of the causal elementary B-splines $\beta_+^{(K)}$ of degree 0, 1, ..., 11 (solid lines, from darkest to lightest).

Computing integrals Like differentiation, integration of a series can be done term by term. To parallel the development for differentiation, we find an expression for the integral of $\beta_+^{(K)}$,

$$\begin{aligned} \zeta^{(K)}(t) &= \int_{-\infty}^{t} \beta_+^{(K)}(\tau)\, d\tau \overset{(a)}{=} \sum_{m=0}^{\infty} \int_{t-m-1}^{t-m} \beta_+^{(K)}(\tau)\, d\tau \\ &\overset{(b)}{=} \sum_{m=0}^{\infty} \int_{-\infty}^{\infty} \beta_+^{(K)}(\tau)\, \beta_+^{(0)}(t-m-\tau)\, d\tau \overset{(c)}{=} \sum_{m=0}^{\infty} \beta_+^{(K+1)}(t-m), \end{aligned} \tag{6.73}$$

where (a) follows from breaking the interval of integration into unit-length subintervals; (b) from using the causal elementary B-spline of degree 0 from (6.39a) to restrict the integral to $[t-m-1,\, t-m]$; and (c) from recognizing the integral as a convolution and using (6.39b). The first few of these integrals of causal elementary B-splines are shown in Figure 6.31.

Then, starting from the spline expansion (6.67), the integral of $x \in S_{K,\mathbb{Z}}$ is

$$\begin{aligned}
\int_{-\infty}^{t} x(\tau)\, d\tau &= \int_{-\infty}^{t} \sum_{k\in\mathbb{Z}} \alpha_k \beta_+^{(K)}(\tau - k)\, d\tau \overset{(a)}{=} \sum_{k\in\mathbb{Z}} \alpha_k \int_{-\infty}^{t} \beta_+^{(K)}(\tau - k)\, d\tau \\
&\overset{(b)}{=} \sum_{k\in\mathbb{Z}} \alpha_k \int_{-\infty}^{t-k} \beta_+^{(K)}(s)\, ds \overset{(c)}{=} \sum_{k\in\mathbb{Z}} \alpha_k \sum_{m=0}^{\infty} \beta_+^{(K+1)}(t - k - m) \\
&\overset{(d)}{=} \sum_{k\in\mathbb{Z}} \alpha_k \sum_{n=k}^{\infty} \beta_+^{(K+1)}(t - n) \overset{(e)}{=} \sum_{n\in\mathbb{Z}} \sum_{k=-\infty}^{n} \alpha_k \beta_+^{(K+1)}(t - n) \\
&\overset{(f)}{=} \sum_{k\in\mathbb{Z}} \alpha_k^{(1)} \beta_+^{(K+1)}(t - k),
\end{aligned} \tag{6.74}$$

where (a) follows from interchanging the integral and sum; (b) from the change of variable $s = \tau - k$; (c) from (6.73); (d) from the change of variable $n = k + m$; (e) from interchanging the sums; and (f) from defining the *discrete integral* of the sequence α,

$$\alpha_k^{(1)} = \sum_{m=-\infty}^{k} \alpha_k, \qquad k \in \mathbb{Z}. \tag{6.75}$$

Thus, to compute the integral of a function $x \in S_{K,\mathbb{Z}}$, we can apply the discrete integral to the sequence that represents x; the result is the sequence representing the integral in $S_{K+1,\mathbb{Z}}$.

The discrete integral (6.75) has a simple form using matrix–vector multiplication:

$$\begin{bmatrix} \vdots \\ \alpha_{-1}^{(1)} \\ \boxed{\alpha_0^{(1)}} \\ \alpha_1^{(1)} \\ \vdots \end{bmatrix} = \begin{bmatrix} & \vdots & \vdots & \vdots & \vdots & \\ \cdots & 1 & 1 & 0 & 0 & \cdots \\ \cdots & 1 & 1 & \boxed{1} & 0 & \cdots \\ \cdots & 1 & 1 & 1 & 1 & \cdots \\ & \vdots & \vdots & \vdots & \vdots & \end{bmatrix} \begin{bmatrix} \vdots \\ \alpha_{-2} \\ \alpha_{-1} \\ \boxed{\alpha_0} \\ \alpha_1 \\ \vdots \end{bmatrix}. \tag{6.76}$$

Similarly to differentiation, this shows that the matrix representation of the integral operator – with respect to using the bases $\{\beta_+^{(K)}(t-k)\}_{k\in\mathbb{Z}}$ for the domain and $\{\beta_+^{(K+1)}(t-k)\}_{k\in\mathbb{Z}}$ for the codomain – is the matrix in (6.76).

EXAMPLE 6.17 (INTEGRATION IN $S_{1,\mathbb{Z}}$) To look at Example 6.16 in reverse, let x now be the name of the function in Figure 6.30(b). This function is in the uniform spline space $S_{1,\mathbb{Z}}$, and its expansion coefficient sequence with respect to the causal elementary B-spline basis of degree 1 is

$$\alpha = \begin{bmatrix} \ldots & 0 & \boxed{0} & 1 & 2 & 1 & 1 & -1 & -2 & -4 & 1 & 1 & 0 & \ldots \end{bmatrix}^{\top}.$$

The discrete integral of this sequence,

$$\alpha^{(1)} = \begin{bmatrix} \ldots & 0 & \boxed{0} & 1 & 3 & 4 & 5 & 4 & 2 & -2 & -1 & 0 & 0 & \ldots \end{bmatrix}^{\top},$$

is the expansion coefficient sequence of the integral of x with respect to the causal elementary B-spline basis of degree 2. The integral is shown in Figure 6.30(a).

We have touched on only a few of the remarkable properties of splines. For example, Exercise 6.17 develops the use of B-spline representations for computing inner products. Generalization of the spline concept beyond polynomial pieces with exponential splines is also useful for signal processing; see the *Further reading*.

6.4 Approximation of functions and sequences by series truncation

The approximations in uniform spline spaces in Section 6.3 are infinite series using countable bases (see (6.51) and (6.52)). The approximations obtained by sampling followed by interpolation in Chapter 5 have a similar character. We now shift our attention to approximations formed by truncation of infinite series. Whether an infinite series is an exact representation or merely an approximation, after truncation we will generally have only an approximation. While most series of interest are structured – such as Fourier series, the series studied in Chapter 5 and Section 6.3, and wavelet series [57] – most of the developments in this section do not depend on such structure, so we work with bases abstractly; the examples often use Fourier series. We will see that the approximation quality depends both on the choice of a basis and on the manner of truncation. These choices are also related to the quality of certain estimates of a signal computed from a noisy observation. We develop most results for orthonormal bases; some extend to biorthogonal pairs of bases and frames (see Exercise 6.18).

6.4.1 Linear and nonlinear approximations

Consider a Hilbert space H for which we have an orthonormal basis $\{\varphi_k\}_{k\in\mathbb{N}}$. Given any $x \in H$, we can express it using the expansion (2.94a),

$$x = \sum_{k\in\mathbb{N}} \alpha_k \varphi_k, \qquad \text{where } \alpha_k = \langle x, \varphi_k\rangle \text{ for } k \in \mathbb{N}.$$

For any particular index set $\mathcal{I} \subset \mathbb{N}$, the best approximation of x using $\{\varphi_k\}_{k\in\mathcal{I}}$ is

$$x_{\mathcal{I}} = P_{\mathcal{I}}\, x = \sum_{k\in\mathcal{I}} \alpha_k \varphi_k, \tag{6.77}$$

which is the orthogonal projection onto $S_{\mathcal{I}} = \overline{\operatorname{span}}(\{\varphi_k\}_{k\in\mathcal{I}})$ (see Theorem 2.41); this appears in block diagram form in Figure 6.32.

The approximation error is

$$e_{\mathcal{I}} = x - x_{\mathcal{I}} = \sum_{k\in\mathbb{N}\setminus\mathcal{I}} \alpha_k \varphi_k. \tag{6.78}$$

Since the basis is orthonormal, the approximation and approximation error satisfy $e_{\mathcal{I}} \perp x_{\mathcal{I}}$, and thus,

$$\|x\|^2 = \|x_{\mathcal{I}}\|^2 + \|e_{\mathcal{I}}\|^2, \tag{6.79a}$$

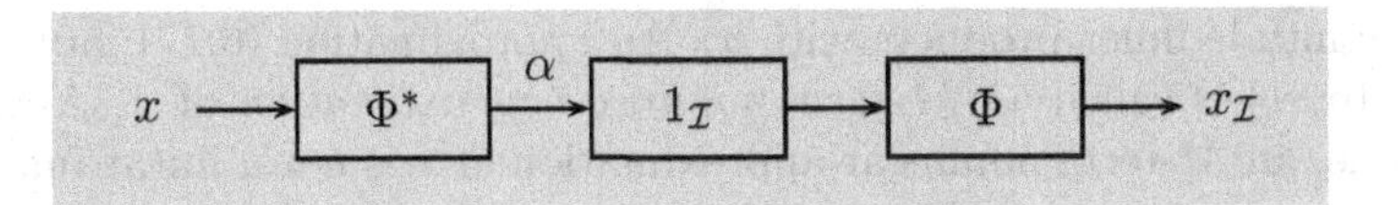

Figure 6.32 Depiction of approximation by series truncation. The analysis operator Φ^* produces the coefficient sequence α. Coefficients with indices in the set $\mathcal{I}$ are retained for use with the synthesis operator Φ; the rest are set to zero. When $|\mathcal{I}| = M$, the output $x_{\mathcal{I}}$ is called an M-term approximation of x. When $\mathcal{I}$ is fixed for a set of signals without depending directly on x, the output is the M-term linear approximation, denoted $\widehat{x}_M$. When $\mathcal{I}$ depends on x through (6.82), the output is the M-term nonlinear approximation, denoted $\breve{x}_M$.

where

$$\|x_{\mathcal{I}}\|^2 = \sum_{k\in\mathcal{I}} |\alpha_k|^2 \tag{6.79b}$$

and

$$\|e_{\mathcal{I}}\|^2 = \sum_{k\in\mathbb{N}\setminus\mathcal{I}} |\alpha_k|^2. \tag{6.79c}$$

Throughout this section, the index set $\mathcal{I}$ has M terms, so $x_{\mathcal{I}}$ is called an *M-term approximation* of x. We are interested in the quality of these approximations for different ways of choosing $\mathcal{I}$.

Linear approximation When $\mathcal{I}$ is fixed (does not depend on x), the approximation $x_{\mathcal{I}}$ is a linear function of x because $P_{\mathcal{I}}$ is a (fixed) linear operator. The approximation is thus called the *M-term linear approximation* of x. For example, when the index set is $\mathcal{I} = \{0, 1, \ldots, M-1\}$, the M-term linear approximation of x is given by

$$\widehat{x}_M = \sum_{k\in\mathcal{I}} \alpha_k\varphi_k = \sum_{k=0}^{M-1} \alpha_k\varphi_k. \tag{6.80a}$$

The resulting approximation error is $x - \widehat{x}_M = \sum_{k=M}^{\infty} \alpha_k\varphi_k$, and the squared norm of the approximation error is

$$\|x - \widehat{x}_M\|^2 = \sum_{k=M}^{\infty} |\alpha_k|^2. \tag{6.80b}$$

Nonlinear approximation Fixing the size $|\mathcal{I}| = M$ but not the index set itself, the best choice of $\mathcal{I}$ is clear from (6.79): the norm of the error $e_{\mathcal{I}}$ is smallest when the norm of the approximation $x_{\mathcal{I}}$ is largest. To get the largest possible terms in (6.79b), the index set $\mathcal{I}$ should satisfy

$$|\alpha_k| \geq |\alpha_m| \qquad \text{for all } k \in \mathcal{I} \text{ and } m \notin \mathcal{I}. \tag{6.81}$$

In other words, $\mathcal{I}$ should contain the indices that correspond to the largest-magnitude coefficients, which are the indices that correspond to the basis elements with largest-magnitude inner products with x. An approximation (6.77) formed with $\mathcal{I}$ satisfying (6.81) is called an *M-term nonlinear approximation* of x. As suggested by the name, an M-term nonlinear approximation of x is not a linear function of x because $\mathcal{I}$ depends on x (see Exercise 6.19).

To find the nonlinear approximation $\check{x}_M$, we can start by creating an ordered sequence $(\alpha_{n_k})_{k\in\mathbb{N}}$ of the expansion coefficients such that[107]

$$|\alpha_{n_k}| \geq |\alpha_{n_{k+1}}| \qquad \text{for all } k \in \mathbb{N}. \tag{6.82a}$$

The first M entries of n are indices of the basis vectors with the M largest-magnitude inner products with x, so we choose

$$\mathcal{I}_x = \{n_0, n_1, \ldots, n_{M-1}\}, \tag{6.82b}$$

where $\mathcal{I}_x$ makes it explicit that this set depends on x. The M-term nonlinear approximation of x is given by

$$\check{x}_M = \sum_{k\in\mathcal{I}_x} \alpha_k \varphi_k = \sum_{k=0}^{M-1} \alpha_{n_k} \varphi_{n_k}. \tag{6.83a}$$

The resulting approximation error is $x - \check{x}_M = \sum_{k=M}^{\infty} \alpha_{n_k}\varphi_{n_k}$, and the squared norm of the approximation error is

$$\|x - \check{x}_M\|^2 = \sum_{k=M}^{\infty} |\alpha_{n_k}|^2. \tag{6.83b}$$

Comparing linear and nonlinear approximations Because of our choice of $\mathcal{I}_x$,

$$\sum_{n\in\mathcal{I}_x} |\alpha_n|^2 \geq \sum_{n\in\mathcal{I}} |\alpha_n|^2, \tag{6.84}$$

where $\mathcal{I}$ is any set of M indices. Thus, using (6.80b) and (6.83b),

$$\|x - \check{x}_M\|^2 \leq \|x - \widehat{x}_M\|^2. \tag{6.85}$$

The fact that nonlinear approximation is at least as good as linear approximation is rather obvious since nonlinear approximation uses the M most useful basis vectors for representing x while linear approximation uses some fixed set of M basis vectors.

The difference between linear and nonlinear approximations is illustrated in Figure 6.33. The Hilbert space $\mathbb{R}^2$ has the orthonormal basis $\{\varphi_0, \varphi_1\}$. The 1-term linear approximation of $x \in \mathbb{R}^2$ is the orthogonal projection of x onto the subspace spanned by φ_0, as shown in Figure 6.33(a). The 1-term nonlinear approximation of $x \in \mathbb{R}^2$ is either the orthogonal projection of x onto the subspace spanned by φ_0 or the orthogonal projection of x onto the subspace spanned by φ_1, depending on which gives a smaller error; clearly it is the latter for the x shown in Figure 6.33(b), because $|\langle x, \varphi_1\rangle| > |\langle x, \varphi_0\rangle|$.

[107] The sequence $(n_k)_{k\in\mathbb{N}}$ is a permutation of $\mathbb{N}$, and, unless the inequality (6.82a) is strict for all $k \in \mathbb{N}$, this permutation is not unique.

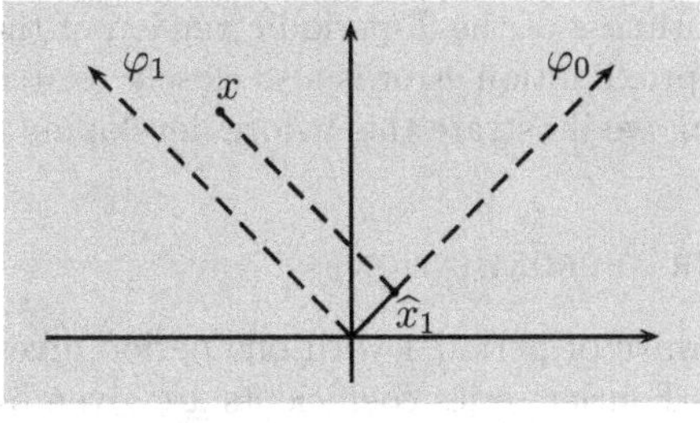

(a) Linear approximation.

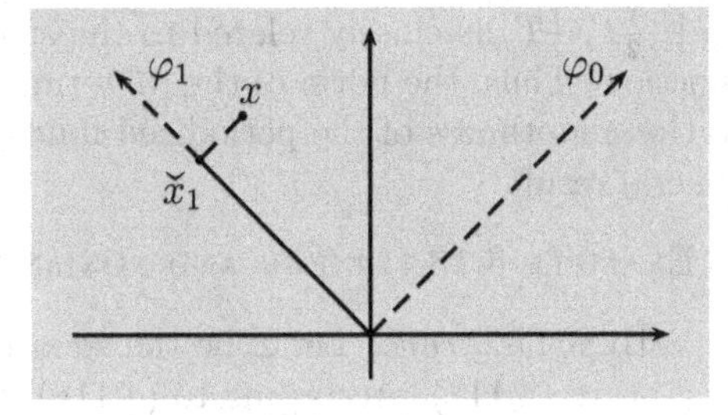

(b) Nonlinear approximation.

Figure 6.33 1-term linear and nonlinear approximation in $\mathbb{R}^2$. The orthonormal basis is $\varphi_0 = \begin{bmatrix} 1 & 1 \end{bmatrix}^\top / \sqrt{2}$ and $\varphi_1 = \begin{bmatrix} -1 & 1 \end{bmatrix}^\top / \sqrt{2}$. (a) Linear approximation keeps only the coefficient of φ_0. (b) Nonlinear approximation keeps the largest-magnitude coefficient.

In this simple two-dimensional setting, because of the orientation of the orthonormal basis, the 1-term linear and nonlinear approximations are identical when x is in the first or third quadrant; the nonlinear approximation is superior when x is in the second or fourth quadrant. This is illustrative of the general principle that the performance difference between linear and nonlinear approximation depends both on the basis and on the signal. Exercises 6.20 and 6.21 explore certain worst-case and average analyses.

The difference between linear and nonlinear approximations can be substantial, depending on the signal and the basis. This was illustrated in Section 6.1, where we saw that linear approximation using Fourier series (Figure 6.5) performs much worse than nonlinear approximation using a Haar basis (Figure 6.6) for a piecewise-constant function. Fourier series bases and Haar bases each have elements that capture broad-scale trends (low frequencies) and fine-scale details (high frequencies). The advantage illustrated in Figure 6.6 depends both on the types of bases and on the use of nonlinear approximation: the short support intervals of the Haar basis functions cause the effects of discontinuities to be limited to a few coefficients, and the selection process of using the largest-magnitude coefficients can then focus on the fine scale only where using these coefficients will be highly advantageous for the quality of approximation.

Fourier series When approximating a function either by linear or by nonlinear approximation, the key to the quality of the approximation is the decay of the magnitudes of the expansion coefficients, either in natural ordering in the linear case, or reordered in decreasing order in (6.82a) in the nonlinear case. Through (6.80b) and (6.83b), fast decay of the coefficients implies fast decay with M of the norm of the M-term approximation error.

For Fourier series, the natural order for linear approximation is from low frequencies to high frequencies since, generally speaking, details (high frequencies) are not useful without first correctly capturing the general trends (low frequencies). The decay with increasing frequency of Fourier series coefficients of a function defined on $[-\frac{1}{2}T, \frac{1}{2}T)$ is closely related to the smoothness of the T-periodic version of the function. Thus, the norm of the M-term approximation error is also closely related to the smoothness of the periodized function; we illustrate this before developing a general result.

EXAMPLE 6.18 (LINEAR AND NONLINEAR APPROXIMATION)

(i) *Square wave:* Let x be the square wave of period 1 with one period given in (4.117). According to (4.118), its Fourier series coefficients are given by

$$X = -\frac{2j}{\pi}\begin{bmatrix} \cdots & -\frac{1}{5} & 0 & -\frac{1}{3} & 0 & -1 & \boxed{0} & 1 & 0 & \frac{1}{3} & 0 & \frac{1}{5} & \cdots \end{bmatrix}^{\top}.$$

The magnitudes of the coefficients are symmetric around the origin, and all even-indexed terms are zero.

Using linear approximation, we keep the central M terms. For $M = 4K-1$ with $K \in \mathbb{Z}^+$, we obtain

$$\begin{aligned} \|x - \widehat{x}_M\|^2 &\stackrel{(a)}{=} \frac{4}{\pi^2}\sum_{|k|\geq K}\frac{1}{(2|k|+1)^2} = \frac{8}{\pi^2}\sum_{k=K}^{\infty}\frac{1}{(2k+1)^2} \\ &\stackrel{(b)}{=} \frac{8}{\pi^2}\int_{K-1}^{\infty}\frac{1}{(2\lceil t\rceil+1)^2}\,dt \stackrel{(c)}{\leq} \frac{8}{\pi^2}\int_{K-1}^{\infty}\frac{1}{(2t+1)^2}\,dt \\ &\stackrel{(d)}{=} \frac{8}{\pi^2}\cdot\frac{1}{2(2K-1)} \stackrel{(e)}{=} \frac{8}{\pi^2(M-1)} = \Theta(M^{-1}), \end{aligned} \tag{6.86a}$$

where (a) follows from summing the squared magnitudes of the omitted coefficients and using the Parseval equality, (4.104a); (b) from representing the sum by an integral of a staircase function; (c) from bounding the staircase function from above; (d) from evaluating the integral; and (e) from $M = 4K-1$.

Choosing nonlinear approximation instead, we can skip all the zero terms. All steps but the last of the computation above hold with $M = 2K$; we thus obtain

$$\|x - \check{x}_M\|^2 \leq \frac{4}{\pi^2(M-1)} = \Theta(M^{-1}). \tag{6.86b}$$

The squared norm of the approximation error is improved by about a factor of 2, but the dependence on M is unchanged.

(ii) *Triangle wave:* Let y be the triangle wave of period 1 with one period given in (4.119). According to (4.120), its Fourier series coefficients are given by

$$Y = \frac{1}{\pi^2}\begin{bmatrix} \cdots & \frac{1}{25} & 0 & \frac{1}{9} & 0 & 1 & \boxed{\frac{1}{4}\pi^2} & 1 & 0 & \frac{1}{9} & 0 & \frac{1}{25} & \cdots \end{bmatrix}^{\top}.$$

Like for the square wave, the magnitudes of the coefficients are symmetric around the origin and all terms with nonzero even indices equal zero. The coefficients decay more quickly than those of the square wave, which is consistent with the fact that the triangle wave is continuous whereas the square wave is not.

One can show (see Exercise 6.22) that the performance of linear and nonlinear approximation satisfy

$$\|y - \widehat{y}_M\|^2 \leq \frac{8}{3\pi^4(M-1)^3} = \Theta(M^{-3}), \tag{6.87a}$$

$$\|y - \check{y}_M\|^2 \leq \frac{1}{3\pi^4(M-2)^3} = \Theta(M^{-3}). \tag{6.87b}$$

Again nonlinear approximation improves upon linear approximation by a constant factor because some Fourier series coefficients are zero. The dependence on M is significantly improved compared with the approximations of the square wave.

Let $x \in \mathcal{L}^2([-\frac{1}{2}T, \frac{1}{2}T))$ have q continuous derivatives. Then, repeating (4.131c), the Fourier series coefficients of x satisfy

$$|X_k| \leq \frac{\gamma}{1+|k|^{q+1}} \qquad \text{for all } k \in \mathbb{Z}, \tag{6.88}$$

for some positive constant γ. Choosing the M central terms, with $M = 2K-1$ for some $K \in \mathbb{Z}^+$,

$$\begin{aligned}
\|x - \widehat{x}_M\|^2 &\overset{(a)}{=} T \sum_{|k| \geq K} |X_k|^2 \overset{(b)}{\leq} T \sum_{|k| \geq K} \frac{\gamma^2}{(1+|k|^{q+1})^2} = 2T \sum_{k=K}^{\infty} \frac{\gamma^2}{(1+k^{q+1})^2} \\
&\leq 2T \sum_{k=K}^{\infty} \frac{\gamma^2}{k^{2(q+1)}} \overset{(c)}{=} 2T\gamma^2 \int_{K-1}^{\infty} \frac{1}{\lceil t \rceil^{2(q+1)}}\, dt \overset{(d)}{\leq} 2T\gamma^2 \int_{K-1}^{\infty} \frac{1}{t^{2(q+1)}}\, dt \\
&\overset{(e)}{=} \frac{2T\gamma^2}{(2q+1)(K-1)^{2q+1}} \overset{(f)}{=} \frac{2^{2q+2}T\gamma^2}{(2q+1)(M-1)^{2q+1}} = \Theta\left(M^{-(2q+1)}\right),
\end{aligned} \tag{6.89}$$

where (a) follows from summing the squared magnitudes of the omitted coefficients and using the Parseval equality, (4.104a); (b) from (6.88); (c) from representing the sum by an integral of a staircase function; (d) from bounding the staircase function from above; (e) from evaluating the integral; and (f) from $M = 2K-1$. This analysis is based only on a bound on $|X_k|$ that is monotonically decreasing with $|k|$, and it gives us no way to predict any improvement from replacing linear approximation by nonlinear approximation. Example 6.18 illustrated that the improvement from nonlinear approximation when using Fourier series is generally only by a constant factor.

The $\Theta(M^{-1})$ dependence on M in (6.86) matches (6.89) for $q = 0$. While the square wave is not continuous, it is continuous almost everywhere, so the match is not too surprising even though (6.89) does not apply to the square wave for $q = 0$. Similarly, the $\Theta(M^{-3})$ dependence on M in (6.87) matches (6.89) for $q = 1$. While the triangle wave is not continuously differentiable, it is continuously differentiable almost everywhere, so the match is again not too surprising even though (6.89) does not apply to the triangle wave for $q = 1$.

6.4.2 Linear approximation of random vectors and stochastic processes

Choosing a basis to obtain the best possible approximation of a class of functions is a difficult problem for which there are few general and useful solutions. One cornerstone result, the optimality of the Karhunen–Loève transform (KLT) for minimizing MSE, has great importance despite its limitation to linear approximation. Here we develop the KLT for finite-dimensional random vectors; the KLT for stochastic processes is more technical, so we instead discuss linear approximation of discrete-time stochastic processes briefly and informally without the use of the KLT.

Linear approximation of random vectors Let $\mathrm{x} = \begin{bmatrix} \mathrm{x}_0 & \mathrm{x}_1 & \ldots & \mathrm{x}_{N-1} \end{bmatrix}^{\top}$ be an N-dimensional random vector for which the mean vector and covariance matrix exist. By subtracting the mean vector if necessary, we may assume each x_k to have mean zero, so the covariance and autocorrelation matrix of x are given by

$$\Sigma_{\mathrm{x}} = \mathrm{E}[\,\mathrm{x}\mathrm{x}^*\,]. \tag{6.90}$$

Using any orthonormal basis $\{\varphi_k\}_{k=0}^{N-1}$ for $\mathbb{C}^N$, one can expand x as

$$\mathrm{x} = \sum_{k=0}^{N-1} \alpha_k \varphi_k, \qquad \text{where } \alpha_k = \langle \mathrm{x}, \varphi_k \rangle \text{ for } k = 0, 1, \ldots, N-1. \tag{6.91}$$

The important distinction from the developments in Section 6.4.1 is that each α_k is now a (scalar) random variable.

The M-term linear approximation

$$\widehat{\mathrm{x}}_M = \sum_{k=0}^{M-1} \alpha_k \varphi_k \tag{6.92}$$

is a random vector; this appears in block diagram form in Figure 6.34. The squared norm of the approximation error is a random variable,

$$\|\mathrm{x} - \widehat{\mathrm{x}}_M\|^2 = \left\| \sum_{k=M}^{N-1} \alpha_k \varphi_k \right\|^2 \overset{(a)}{=} \sum_{k=M}^{N-1} |\alpha_k|^2,$$

where (a) follows from the orthonormality of the basis. The expected squared norm of the approximation error is thus

$$\varepsilon_M = \mathrm{E}\big[\,\|\mathrm{x} - \widehat{\mathrm{x}}_M\|^2\,\big] = \mathrm{E}\left[\sum_{k=M}^{N-1} |\alpha_k|^2\right] \overset{(a)}{=} \sum_{k=M}^{N-1} \mathrm{E}\big[\,|\alpha_k|^2\,\big], \tag{6.93}$$

where (a) follows from the linearity of the expectation operator. The term $\mathrm{E}\big[\,|\alpha_k|^2\,\big]$ is the power of the coefficient α_k, so the MSE ε_M is the sum of the powers of the last $N - M$ coefficients.

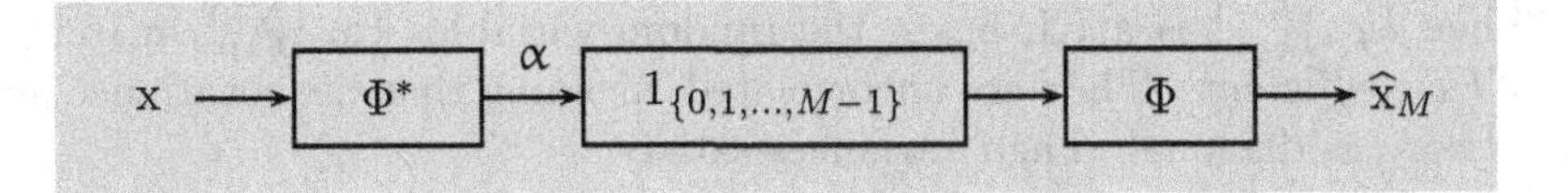

Figure 6.34 Depiction of M-term linear approximation of a random vector x. The analysis operator Φ^* produces the coefficient vector α. Coefficients with indices in the set $\{0, 1, \ldots, M\}$ are retained and the rest are set to zero. Application of the synthesis operator Φ gives the M-term linear approximation $\widehat{x}_M$. It is optimal for Φ^* to be a KLT of x; this depends on the covariance matrix of x but not on the realization of x.

A basis that minimizes ε_M is optimal for linear approximation of x. To understand the impact of choosing the basis, first consider the sum of the powers of all the coefficients:

$$\begin{aligned}\sum_{k=0}^{N-1} \mathrm{E}\big[\,|\alpha_k|^2\,\big] &= \mathrm{E}\big[\,\|\alpha\|^2\,\big] \stackrel{(a)}{=} \mathrm{E}\big[\,\|\mathrm{x}\|^2\,\big] = \sum_{k=0}^{N-1} \mathrm{E}\big[\,|\mathrm{x}_k|^2\,\big] \\ &\stackrel{(b)}{=} \sum_{k=0}^{N-1} (\Sigma_\mathrm{x})_{kk} = \mathrm{tr}(\Sigma_\mathrm{x}), \end{aligned} \tag{6.94}$$

where (a) follows from the Parseval equality and the orthonormality of $\{\varphi_k\}_{k=0}^{N-1}$; and (b) from the definition of the covariance matrix, (2.259). The key observation is that the sum of the powers of all the coefficients is a fixed number that does not depend on the choice of orthonormal basis. To minimize ε_M, this fixed total power should be spread with the least possible power in $\alpha_M, \alpha_{M+1}, \ldots, \alpha_{N-1}$ (and hence also with the most possible power in $\alpha_0, \alpha_1, \ldots, \alpha_{M-1}$). This is achieved with a Karhunen–Loève (KL) basis.

Definition 6.15 (Karhunen–Loève transform in $\mathbb{C}^N$) Let $\{\varphi_k\}_{k=0}^{N-1}$ be an orthonormal basis for $\mathbb{C}^N$ with analysis operator Φ^* and synthesis operator Φ. The operator Φ^* is called a *Karhunen–Loève transform* for an N-dimensional random vector x with covariance matrix Σ_x when $\Phi^*\Sigma_\mathrm{x}\Phi = \Lambda$, where Λ is a diagonal matrix with nonincreasing entries on the diagonal. The basis $\{\varphi_k\}_{k=0}^{N-1}$ is then called a *Karhunen–Loève basis* for x.

Recall that a covariance matrix is Hermitian and that Hermitian matrices have real eigenvalues and are diagonalized by an orthonormal matrix of eigenvectors (see (2.241)). Thus, we can always find a KL basis for x by choosing orthonormal eigenvectors of Σ_x, with the vectors taken in an order that makes the eigenvalues nonincreasing:

$$\Sigma_\mathrm{x}\varphi_k = \lambda_k\varphi_k, \qquad k = 0, 1, \ldots, N-1, \tag{6.95a}$$

$$\lambda_k \geq \lambda_{k+1}, \qquad k = 0, 1, \ldots, N-2. \tag{6.95b}$$

When $\{\varphi_k\}_{k=0}^{N-1}$ is a KL basis, the random variables $\{\alpha_k\}_{k=0}^{N-1}$ in (6.91) are called *KLT coefficients.* They are uncorrelated, meaning that the covariance matrix $\Sigma_\alpha = \mathrm{E}[\,\alpha\alpha^*\,]$ is diagonal. Their variances satisfy

$$\mathrm{E}\big[\,|\alpha_k|^2\,\big] \;\geq\; \mathrm{E}\big[\,|\alpha_{k+1}|^2\,\big], \qquad k = 0, 1, \ldots, N-2. \tag{6.96}$$

Moreover, they satisfy the following property: over any choice of orthonormal basis and for each $M \in \{1, 2, \ldots, N-1\}$, the power of the first M coefficients, $\sum_{k=0}^{M-1} \mathrm{E}\big[\,|\alpha_k|^2\,\big]$, is maximized and the power of the last $N-M$ coefficients, $\sum_{k=M}^{N-1} \mathrm{E}\big[\,|\alpha_k|^2\,\big]$, is minimized. Proofs of these properties are developed in Solved exercise 6.5.

Since the MSE of M-term linear approximation ε_M in (6.93) is minimized, a KL basis is optimal for linear approximation. Note that the same basis is optimal for every value of M. This basis depends on the distribution of x, but it does not depend on the realization of x, so the M-term approximation $\widehat{\mathrm{x}}_M$ is a linear function of the realization of x.

Example 6.19 (Karhunen–Loève transform in $\mathbb{R}^2$) Let x be a Gaussian random vector with mean zero and covariance matrix

$$\Sigma_{\mathrm{x}} \;=\; \begin{bmatrix} 5 & 2 \\ 2 & 2 \end{bmatrix}. \tag{6.97}$$

The PDF of x is depicted in Figure 6.35 by plotting level curves. The eigenvalues of Σ_{x} are $\lambda_0 = 6$ and $\lambda_1 = 1$. The KLT is

$$\Phi^* \;=\; \frac{1}{\sqrt{5}} \begin{bmatrix} 2 & 1 \\ -1 & 2 \end{bmatrix} \;\approx\; \begin{bmatrix} 0.8944 & 0.4472 \\ -0.4472 & 0.8944 \end{bmatrix}, \tag{6.98}$$

and the KL basis is shown in Figure 6.35. The first basis element φ_0 is aligned with the direction in which x has maximum variation, so a 1-term linear approximation is as good as possible.

Linear approximation of WSS processes For WSS discrete-time stochastic processes, we change the notion of series truncation somewhat from the developments in Section 6.4.1, but the spirit remains similar. We will require infinitely many terms to produce an approximation, so truncation will be replaced by a reduction in sampling rate.

Let x be a WSS discrete-time stochastic process, and consider the approximation $\mathrm{x}_{\mathcal{I}}$ generated by the system depicted in Figure 6.36. If we were to have a finite index set $\mathcal{I}$ with $|\mathcal{I}| = M$, the approximation $\mathrm{x}_{\mathcal{I}}$ would depend on M scalar random variables, similarly to (6.92). However, because of the wide-sense stationarity of x, a finite number of scalar random variables cannot adequately describe it; in fact, $\mathrm{x}_{\mathcal{I}}$ would have no better MSE than approximating x by its constant mean, $\mu_{\mathrm{x}} = \mathbf{0}$. Instead, to assign a size to set $\mathcal{I}$ even when it is infinite, let

$$\eta \;=\; \lim_{K\to\infty} \frac{|\mathcal{I} \cap \{-K, -K+1, \ldots, K\}|}{2K+1},$$

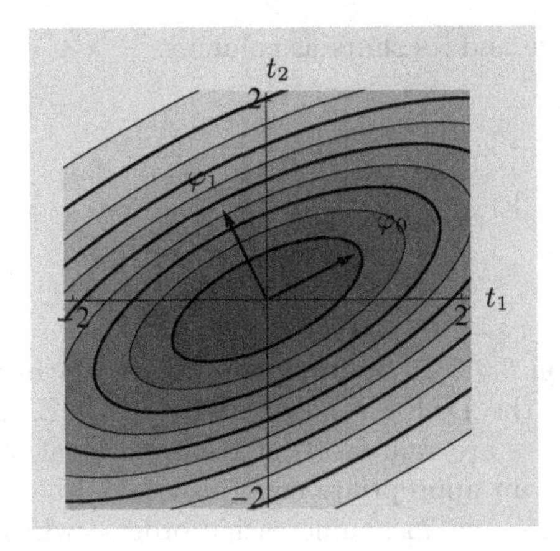

Figure 6.35 Level curves of the PDF of a two-dimensional Gaussian random vector x and a KL basis for x. The first basis element φ_0 is aligned with the major axis of each elliptical level curve.

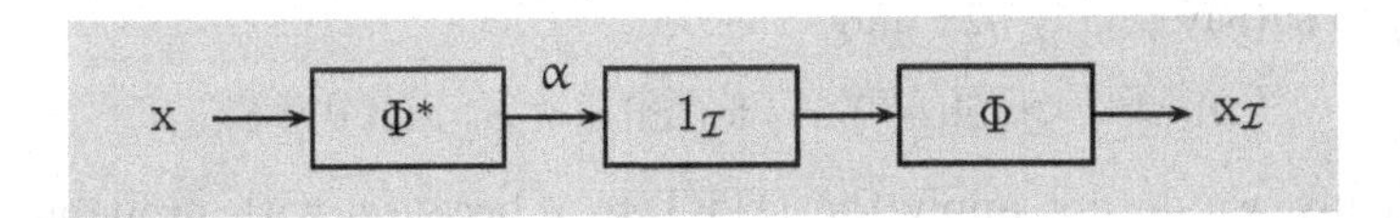

Figure 6.36 Depiction of linear approximation of a WSS discrete-time stochastic process x. The analysis operator Φ^* produces the coefficient stochastic process α. Coefficients with indices in the set $\mathcal{I} \subset \mathbb{Z}$ are retained and the rest are set to zero. Application of the synthesis operator Φ gives the linear approximation $x_{\mathcal{I}}$. The set $\mathcal{I}$ depends on the autocorrelation (or power spectral density) of x but not on the realization of x.

which we assume to exist and interpret as a sampling rate. We will consider only the case when $\mathcal{I}$ depends on the distribution of x but not on a specific realization, so the approximation is linear; we denote that approximation by $\widehat{x}_\eta$.

The two key lessons from the optimality of the KLT for linear approximation of random vectors are as follows:

(i) The analysis operator should produce uncorrelated transform coefficients; that is, it should diagonalize the covariance matrix of x.

(ii) The transform coefficients with largest power (depending on the distribution of x, not on the realization of x) should be retained; that is, their indices should form $\mathcal{I}$.

Let us now apply these lessons. Assume that x has zero mean and autocorrelation sequence a with decay sufficient for the power spectral density $A(e^{j\omega})$ to exist. The autocorrelation matrix Σ_x is then an infinite-dimensional Toeplitz matrix with the autocorrelation sequence and its shifts as columns,

$$\Sigma_{\mathrm{x}} = \mathrm{E}[\,\mathrm{x}\mathrm{x}^*\,] = \begin{bmatrix} & \vdots & \vdots & \vdots & \\ \cdots & a_0 & a_{-1} & a_{-2} & \cdots \\ \cdots & a_1 & \boxed{a_0} & a_{-1} & \cdots \\ \cdots & a_2 & a_1 & a_0 & \cdots \\ & \vdots & \vdots & \vdots & \end{bmatrix}. \tag{6.99}$$

As developed in Section 3.4.3, Toeplitz operators are diagonalized by the DTFT; the eigensequences are the DTFT sequences $e^{j\omega k}$, $k \in \mathbb{Z}$, $\omega \in [0, 2\pi)$, and the corresponding eigenvalues are values of the power spectral density $A(e^{j\omega})$. Thus, the DTFT seems to be an appropriate counterpart for the KLT, except that the eigenvalues have not been put in nonincreasing order – which we will account for shortly.

Having the continuous quantity ω rather than a discrete index changes the reconstruction method from a series to an integral. The sampling rate η translates to integrating over $\omega \in S_\eta \subseteq [0, 2\pi)$ with $|S_\eta|/(2\pi) = \eta$. Since the eigenvalues have not been put in nonincreasing order, instead of simply choosing $S_\eta = [0, \eta 2\pi)$, choose S_η to satisfy $|S_\eta| = \eta 2\pi$ and

$$A(e^{j\omega}) \geq A(e^{j\theta}) \qquad \text{for all } \omega \in S_\eta \text{ and } \theta \notin S_\eta.$$

Formally, we do not apply the DTFT to x because, with probability 1, the DTFT does not converge for a realization of a WSS random process. Instead, note that, when this convergence is not an issue, first applying the DTFT, then setting the resulting spectrum to zero outside of S_η, and finally applying the inverse DTFT is equivalent to LSI filtering with frequency response

$$H_\eta(\omega) = \begin{cases} 1, & \text{for } \omega \in S_\eta; \\ 0, & \text{otherwise.} \end{cases} \tag{6.100a}$$

Thus, avoiding application of the DTFT to x, the linear approximation is

$$\widehat{\mathrm{x}}_\eta = h_\eta * \mathrm{x}, \tag{6.100b}$$

where h_η is the inverse DTFT of H_η.

We can now compute the expected squared error of the linear approximation from the power spectral density. Analogously to (6.93),

$$\varepsilon_\eta = \mathrm{E}\big[\,|\mathrm{x}_n - (\widehat{\mathrm{x}}_\eta)_n|^2\,\big] = \frac{1}{2\pi}\int_{\omega \notin S_\eta} A(e^{j\omega})\,d\omega,$$

since $A(e^{j\omega})$ plays the role of the power of a transform coefficient indexed by ω.

We illustrate linear approximation on an AR-1 process.

Example 6.20 (Linear approximation of an AR-1 process) Let x be the AR-1 process in Example 3.39 with coefficient $a = 0.9$. According to (3.245), the power spectral density of the process is

$$A(e^{j\omega}) = \frac{1}{1.81 - 1.8\cos\omega}, \tag{6.101}$$

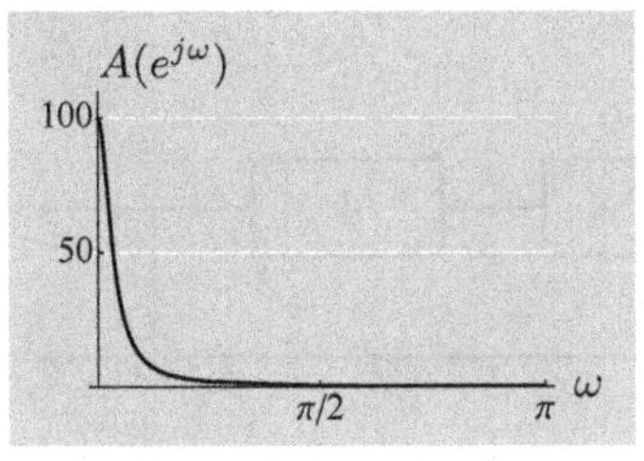

(a) Power spectral density of x.

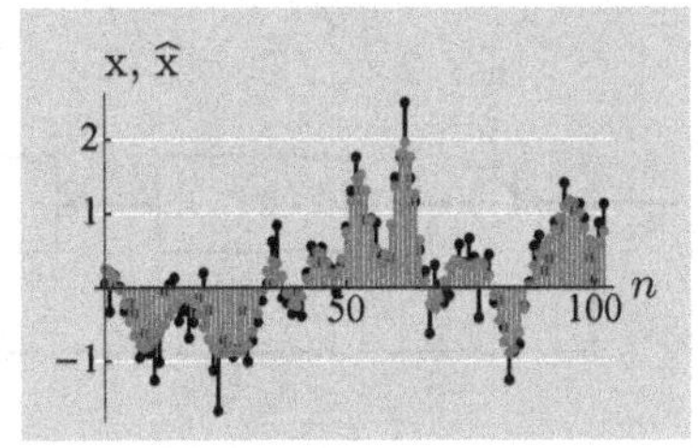

(b) Linear approximation ($\eta = \frac{1}{4}$).

Figure 6.37 Linear approximation of an AR-1 process is the restriction of the monotonically decreasing power spectral density $A(e^{j\omega})$ from (6.101) to $[-\eta\pi, \eta\pi]$.

as shown in Figure 6.37(a). Since the power spectral density is monotonically decreasing with $|\omega|$, the set S_η is simply $[-\eta\pi, \eta\pi]$. For example, for $\eta = \frac{1}{4}$, the linear approximation is the projection of x onto $\mathrm{BL}[-\frac{1}{4}\pi, \frac{1}{4}\pi]$. This can be achieved by ideal lowpass filtering with a quarter-band filter. Figure 6.37(b) shows a portion of a single realization of x and the approximation $\widehat{x}_{1/4}$.

6.4.3 Linear and nonlinear diagonal estimators

There are many ways to estimate a signal from a noisy observation and many ways to assess the performance of an estimator, as reviewed in Appendix 2.C.3. In principle, one can choose an appropriate performance criterion and find the estimator that gives the best possible performance. In practice, the accuracy of any chosen model is limited and optimal estimators under most models are computationally intractable. Thus, estimators are often limited to certain simple forms involving diagonal operators applied to transform coefficient sequences.

Under the constraint of using a diagonal operator, it becomes natural to ask how the choice of basis affects estimator performance. The bases that are best for approximation are generally also best for estimation. We will see this in a few simple settings; see the *Further reading* for pointers to more advanced results.

We will start with linear estimation of random vectors and stochastic processes, where an exact optimality of diagonal estimators arises for certain problems. We then shift to classical estimation settings, where the nonlinear approximation introduced in Section 6.4.1 can be applied as an estimation technique.

Linear estimation of random vectors Let x and w be independent N-dimensional random vectors for which the mean vectors and covariance matrices exist. By subtracting the mean vectors if necessary, we may assume that $\mathrm{E}[\,\mathrm{x}\,] = \mathbf{0}$ and $\mathrm{E}[\,\mathrm{w}\,] = \mathbf{0}$, so the covariance and autocorrelation matrices of x and w are given by $\Sigma_{\mathrm{x}} = \mathrm{E}[\,\mathrm{x}\mathrm{x}^*\,]$ and $\Sigma_{\mathrm{w}} = \mathrm{E}[\,\mathrm{w}\mathrm{w}^*\,]$. Let $\mathrm{y} = \mathrm{x} + \mathrm{w}$ be a noisy observation of x. Then

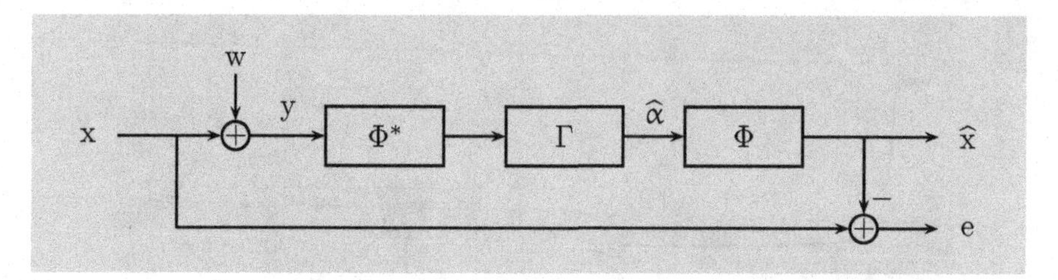

Figure 6.38 Estimation of a random vector x from an observation y that is corrupted by signal-independent additive noise w. The analysis operator Φ^* gives expansion coefficients from which the linear operator Γ gives coefficient estimates $\widehat{\alpha}$. The synthesis operator Φ then gives the estimate $\widehat{x}$. An LMMSE estimator minimizes $E[\,\|e\|^2\,]$ by choosing Γ depending on the distributions of x and w but not on their realizations.

y has mean zero, and the crosscovariance of x and y is

$$\begin{aligned}\Sigma_{x,y} &= E[\,xy^*\,] = E[\,x(x+w)^*\,] \stackrel{(a)}{=} E[\,xx^*\,] + E[\,xw^*\,] \\ &\stackrel{(b)}{=} E[\,xx^*\,] + E[\,x\,]\,E[\,w^*\,] = \Sigma_x + \mathbf{0}\,\mathbf{0}^* = \Sigma_x, \end{aligned} \tag{6.102a}$$

where (a) follows from the linearity of the expectation; and (b) from the independence of x and w. Similarly, the covariance of y is

$$\Sigma_y = \Sigma_x + \Sigma_w. \tag{6.102b}$$

Consider the estimation of x from $y = t \in \mathbb{C}^N$. The linear estimator that minimizes the expected squared norm of the estimation error, $E[\,\|x - \widehat{x}\|^2\,]$, is called the LMMSE estimator. It was shown in Exercise 2.60 that the LMMSE estimator in the present setting – where we have not assumed that x and y are jointly Gaussian – is the same as the MMSE estimator from the setting where x and y are jointly Gaussian. Using results from Example 2.67,

$$\widehat{x} \stackrel{(a)}{=} \Sigma_{x,y}\Sigma_y^{-1} t \stackrel{(b)}{=} \Sigma_x(\Sigma_x + \Sigma_w)^{-1} t, \tag{6.103}$$

where (a) follows from (2.266a) with $\mu_x = \mathbf{0}$ and $\mu_y = \mathbf{0}$; and (b) from (6.102).

Using any orthonormal basis $\{\varphi_k\}_{k=0}^{N-1}$ for $\mathbb{C}^N$, any linear estimator can be implemented as shown in Figure 6.38, where $\Gamma : \mathbb{C}^N \to \mathbb{C}^N$ is a linear operator. Upon combining $\widehat{x} = \Phi\Gamma\Phi^* t$ from the block diagram with (6.103), we find that the optimal choice is

$$\Gamma = \Phi^*\Sigma_x(\Sigma_x + \Sigma_w)^{-1}\Phi.$$

We now look at a special case that arises in particular when the noise is white. Suppose that Φ^* diagonalizes both Σ_x and Σ_w, so

$$\Phi^*\Sigma_x\Phi = \Lambda_x = \operatorname{diag}(\sigma_{x,0}^2, \sigma_{x,1}^2, \ldots, \sigma_{x,N-1}^2), \tag{6.104a}$$

$$\Phi^*\Sigma_w\Phi = \Lambda_w = \operatorname{diag}(\sigma_{w,0}^2, \sigma_{w,1}^2, \ldots, \sigma_{w,N-1}^2). \tag{6.104b}$$

Then

$$\Gamma = \Phi^* \Sigma_{\mathrm{x}}(\Sigma_{\mathrm{x}} + \Sigma_{\mathrm{w}})^{-1}\Phi \overset{(a)}{=} \Phi^*\Phi\Lambda_{\mathrm{x}}\Phi^*(\Phi\Lambda_{\mathrm{x}}\Phi^* + \Phi\Lambda_{\mathrm{w}}\Phi^*)^{-1}\Phi \overset{(b)}{=} \Lambda_{\mathrm{x}}(\Lambda_{\mathrm{x}} + \Lambda_{\mathrm{w}})^{-1}, \tag{6.105}$$

where (a) follows from (6.104); and (b) from $\Phi^*\Phi = I$, which is implied by the orthonormality of the basis. Since Λ_{x} and Λ_{w} are diagonal matrices, Γ is also a diagonal matrix, with

$$\Gamma_{kk} = \frac{\sigma^2_{\mathrm{x},k}}{\sigma^2_{\mathrm{x},k} + \sigma^2_{\mathrm{w},k}}, \qquad k = 0, 1, \ldots, N-1. \tag{6.106}$$

Recall that one can always find a KL basis for x by eigendecomposition of Σ_{x}. The computation above shows that, when such a KL basis is also a KL basis for w, the operator Γ is diagonal. This occurs in particular when w is white, that is, $\Sigma_{\mathrm{w}} = \sigma^2_{\mathrm{w}} I_N$ for a scalar σ^2_{w}; any orthonormal basis is then a KL basis for w.

Linear estimation of WSS processes Let x and w be uncorrelated WSS discrete-time stochastic processes with mean zero, and consider again the system depicted in Figure 6.38. If the overall estimation operator $\Phi\Gamma\Phi^*$ is constrained to be linear and shift-invariant, minimizing the power of the estimation error e is a problem that was solved in Section 3.8.5 (compare Figures 3.36 and 6.38). The answer is the Wiener filter, and this result can be extended to show that allowing a shift-varying linear operator does not provide any improvement.

The Wiener filter is conveniently represented in the DTFT domain; see (3.261) for the general case and (3.263) for the specialization to the uncorrelated additive noise case. This is a diagonal representation of the estimator, with Φ^* representing the DTFT, Γ the pointwise multiplication in the DTFT domain, and Φ the inverse DTFT. The DTFT-domain expression of the filter (3.263) is a direct analogue of (6.106) with $A_{\mathrm{x}}(e^{j\omega})$ playing the role of the variance of the component of x at frequency ω (and similarly for $A_{\mathrm{w}}(e^{j\omega})$). Like in Section 6.4.2, the DTFT is playing the role of the KLT. This works because the DTFT diagonalizes the autocorrelation operator of every WSS process.

Classical estimation: Oracle scaling While Wiener filtering is an important technique, it has two key limitations: linear estimators generally perform worse than nonlinear estimators,[108] and an average over a signal distribution can be less meaningful than a classical performance analysis, which holds for each possible value of the unknown signal. We now shift our attention to nonlinear estimators and classical analysis.[109] We limit our attention to estimators in the form shown in Figure 6.39, where a diagonal operator is applied to the expansion coefficients of the observation with respect to some orthonormal basis. This class of estimators

[108] As discussed in Appendix 2.C.3, MMSE and MAP estimators are linear when a signal of interest and an observation are jointly Gaussian; optimal estimators are rarely linear otherwise.

[109] The distinction between Bayesian and classical analysis of estimators was detailed in Appendix 2.C.3.

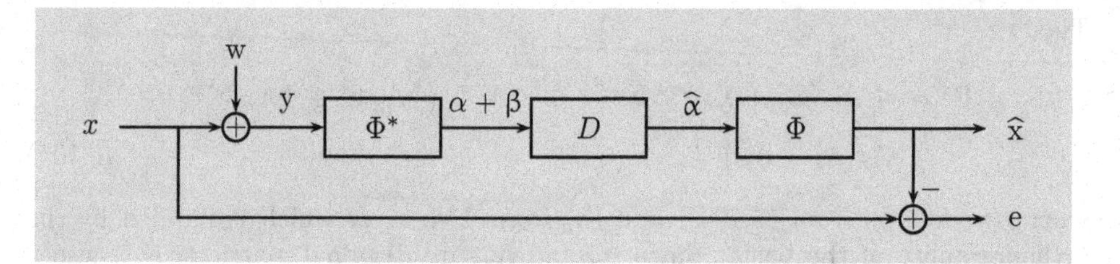

Figure 6.39 Estimation of an unknown deterministic vector x from an observation y that is corrupted by additive noise w. The analysis operator Φ^* gives expansion coefficients $\alpha + \beta$, from which the diagonal operator D gives coefficient estimates $\widehat{\alpha}$. The synthesis operator Φ then gives the estimate $\widehat{x}$. We consider both linear and nonlinear operators D.

has low complexity and potentially good performance – depending on the choice of basis.

Let $\{\varphi_k\}_{k\in\mathbb{N}}$ be an orthonormal basis for H, and suppose a signal $x \in H$ has expansion coefficient sequence $\alpha \in \ell^2(\mathbb{N})$ with respect to this basis. The observation of x is with additive noise $w \in H$, which has expansion coefficient sequence $\beta \in \ell^2(\mathbb{N})$ satisfying

$$\mathrm{E}[\,\beta_k\,] \;=\; 0 \qquad \text{and} \qquad \mathrm{E}[\,\beta_k\beta_m^*\,] \;=\; \sigma^2\delta_{k-m}, \qquad k, m \in \mathbb{N}. \tag{6.107}$$

The expansion coefficients of the noisy observation y are thus $\alpha + \beta$. From this sequence, the diagonal estimator D gives coefficient sequence $\widehat{\alpha} = D(\alpha + \beta)$, from which the estimate of x is

$$\widehat{x} \;=\; \sum_{k\in\mathbb{N}} \widehat{\alpha}_k\varphi_k.$$

The MSE of this estimate is

$$\mathrm{E}[\,\|x - \widehat{x}\|^2\,] \overset{(a)}{=} \mathrm{E}[\,\|\alpha - \widehat{\alpha}\|^2\,] \overset{(b)}{=} \mathrm{E}\left[\sum_{k\in\mathbb{N}} |\alpha_k - \widehat{\alpha}_k|^2\right] \overset{(c)}{=} \sum_{k\in\mathbb{N}} \mathrm{E}[\,|\alpha_k - \widehat{\alpha}_k|^2\,],$$

where (a) follows from the Parseval equality; (b) from the definition of the $\ell^2(\mathbb{N})$ norm; and (c) from the linearity of the expectation.

Since we consider only diagonal estimators,

$$\widehat{\alpha}_k \;=\; \gamma_k(\alpha_k + \beta_k), \qquad k \in \mathbb{N}, \tag{6.108}$$

for some sequence $\gamma \in \mathbb{C}^{\mathbb{N}}$. Thus,

$$\begin{aligned}
\mathrm{E}[\,|\alpha_k - \widehat{\alpha}_k|^2\,] &\overset{(a)}{=} \mathrm{E}[\,|\alpha_k - \gamma_k(\alpha_k + \beta_k)|^2\,] \;=\; \mathrm{E}[\,|(1-\gamma_k)\alpha_k - \gamma_k\beta_k|^2\,] \\
&\overset{(b)}{=} |(1-\gamma_k)\alpha_k|^2 - (1-\gamma_k)\alpha_k\gamma_k^*\mathrm{E}[\,\beta_k^*\,] - (1-\gamma_k^*)\alpha_k^*\gamma_k\mathrm{E}[\,\beta_k\,] + |\gamma_k|^2\mathrm{E}[\,|\beta_k|^2\,], \\
&\overset{(c)}{=} |(1-\gamma_k)\alpha_k|^2 + |\gamma_k|^2\sigma^2,
\end{aligned} \tag{6.109}$$

where (a) follows from (6.108); (b) from expanding the square, using the linearity of the expectation and that α_k and γ_k are deterministic quantities; and (c) from (6.107). This contribution to the MSE is minimized over the choice of γ_k by

$$\gamma_k = \frac{|\alpha_k|^2}{|\alpha_k|^2 + \sigma^2}, \qquad k \in \mathbb{N}, \tag{6.110}$$

resulting in

$$\mathrm{E}\left[\,|\alpha_k - \widehat{\alpha}_k|^2\,\right] = \frac{|\alpha_k|^2\sigma^2}{|\alpha_k|^2 + \sigma^2}, \qquad k \in \mathbb{N}. \tag{6.111}$$

Note the similarity of (6.110) to (6.106); in both cases, the optimal scaling factor is the ratio of the signal energy to the sum of the signal energy and the noise energy. Another way to write the scaling factor is as $\rho_k/(1+\rho_k)$, where $\rho_k = |\alpha_k|^2/\sigma^2$ or $\rho_k = \sigma^2_{\mathrm{x},k}/\sigma^2_{\mathrm{w},k}$ is the *signal-to-noise ratio* (SNR). From this it is clear that the scaling factor goes to 0 in the limit of low SNR and goes to 1 in the limit of high SNR.

The estimator in (6.110) is called an *oracle estimator* because it depends on information that is not ordinarily known – in this case the magnitudes of the coefficient sequence α of the unknown signal x. An oracle estimator provides a benchmark for other estimators. Calling this *oracle scaling*, we denote its MSE by

$$\varepsilon_{\mathrm{OS}} = \sum_{k\in\mathbb{N}} \frac{|\alpha_k|^2\sigma^2}{|\alpha_k|^2 + \sigma^2}. \tag{6.112}$$

Classical estimation: From oracle projectors to nonlinear projectors If γ_k in (6.108) is constrained to be in $\{0, 1\}$ for all $k \in \mathbb{N}$, the estimator is an orthogonal projection operator. Using (6.109), the best choice is

$$\gamma_k = \begin{cases} 1, & \text{if } |\alpha_k|^2 \geq \sigma^2; \\ 0, & \text{otherwise,} \end{cases} \qquad k \in \mathbb{N}, \tag{6.113}$$

resulting in

$$\mathrm{E}\left[\,|\alpha_k - \widehat{\alpha}_k|^2\,\right] = \min\{|\alpha_k|^2, \sigma^2\}, \qquad k \in \mathbb{N}. \tag{6.114}$$

This is an *oracle projector* because the choice of projection operator requires knowledge of whether $|\alpha_k|^2$ is larger than σ^2. Its MSE is given by

$$\varepsilon_{\mathrm{OP}} = \sum_{k\in\mathbb{N}} \min\{|\alpha_k|^2, \sigma^2\}. \tag{6.115}$$

Define an index set by

$$\mathcal{I} = \left\{k \in \mathbb{N} \;\middle|\; |\alpha_k|^2 \geq \sigma^2\right\},$$

and let $M = |\mathcal{I}|$. The estimate $\widehat{\mathrm{x}}$ produced by the oracle projector is the M-term approximation of $\mathrm{y} = \sum_{k\in\mathbb{N}}(\alpha_k + \beta)\varphi_k$ obtained by keeping the terms with $k \in \mathcal{I}$.

Thus, its MSE combines the error from truncating the expansion of x with the error from having additive noise on the M retained coefficients:

$$\varepsilon_{\mathrm{OP}} = \|x - \check{x}_M\|^2 + M\sigma^2, \tag{6.116}$$

where $\check{x}_M$ is the M-term nonlinear approximation of x defined in Section 6.4.1. From this characterization of the error of the oracle projector, the dependence on the choice of orthonormal basis is the same for nonlinear approximation and this form of estimation.

The MSE of oracle scaling is comparable to the MSE of the oracle projector (see Exercise 6.24):

$$\frac{1}{2}\varepsilon_{\mathrm{OP}} \leq \varepsilon_{\mathrm{OS}} \leq \varepsilon_{\mathrm{OP}}. \tag{6.117}$$

Thus, using an orthonormal basis in which the coefficient sequence decays quickly is good for the oracle scaling estimator as well; see Exercise 6.25.

Without oracle knowledge of $|\alpha_k|^2$, one can attempt to guess whether $|\alpha_k|^2$ is larger than σ^2 from the observation $\alpha_k + \beta_k$. Since α_k is unknown and β_k has a symmetric distribution, the reasonable way to guess is to conclude that $|\alpha_k|^2 \geq \sigma^2$ when $|\alpha_k + \beta_k|$ is larger than some threshold value $\nu \in \mathbb{R}^+$. The resulting estimator

$$\widehat{\alpha}_k = \begin{cases} \alpha_k + \beta_k, & \text{for } |\alpha_k + \beta_k| \geq \nu; \\ 0, & \text{otherwise,} \end{cases} \qquad k \in \mathbb{N}, \tag{6.118}$$

is a *nonlinear projector* – nonlinear since the subspace upon which it projects depends on the input to the estimator. It is also called a *hard threshold estimator.* With an appropriate choice of ν, the MSE of the nonlinear projector can be bounded relative to the oracle projector to show that it is approximately minimax for diagonal estimators; see the *Further reading.*

6.5 Compression

In the previous sections, we concentrated on approximating a given function or sequence by using a finite sequence of coefficients – either coefficients of a series expansion or coefficients of a polynomial. In this section, we will focus on approximating with bits rather than real or complex numbers, which is termed *compression* or *source coding.* When the signal can be exactly recovered from the bits, the compression is called *lossless*; otherwise, it is called *lossy.*

Our interest is mostly in lossy compression. The limits of lossy compression are within the purview of rate–distortion theory, which is rather divorced from practice because it suggests methods that have high computational cost for encoding (see the *Further reading*). The practice of lossy compression is dominated by a modular approach with low computational cost, called *transform coding,* in which lossy operations are applied only on one scalar expansion coefficient at a time. Transform coding combines a basis expansion with quantization of expansion coefficients and lossless compression of the resulting discrete values. We discuss lossless compression in Section 6.5.1, then scalar quantization in Section 6.5.2, and finally transform coding in Section 6.5.3.

6.5.1 Lossless compression

Lossless compression is possible only when a signal has been restricted a priori to a countable set. Consider a discrete random variable v that takes values in a countable set $\mathcal{I}$. If $|\mathcal{I}|$ is finite, then each element of $\mathcal{I}$ can be numbered with an index in $\{0, 1, \ldots, |\mathcal{I}| - 1\}$. The binary expansion of this index assigns a unique binary string of length $\lceil \log_2 |\mathcal{I}| \rceil$ to each element of $\mathcal{I}$. The main idea behind lossless compression is that the average length of the binary representation of v can be reduced by assigning shorter strings to more likely values of v even though this requires assigning longer strings to less likely values of v.

DEFINITION 6.16 (LOSSLESS CODE, EXTENSION, UNIQUELY DECODABLE) A *lossless code* γ on a countable set $\mathcal{I}$ assigns a unique binary string, called a *codeword*, to each $i \in \mathcal{I}$. For any $N \in \mathbb{N}$, the *extension* of γ maps the finite sequence $\begin{bmatrix} i_0 & i_1 & \ldots & i_{N-1} \end{bmatrix} \in \mathcal{I}^N$ to the concatenation of the outputs of γ, $\gamma(i_0), \gamma(i_1), \ldots, \gamma(i_{N-1})$. A lossless code is said to be *uniquely decodable* when its extension is one-to-one.

Since the codewords are unique, a lossless code is always invertible. However, we will generally impose the condition of a lossless code being uniquely decodable. This is so that it can be used on a sequence of discrete random variables, with the entire sequence recoverable from the resulting binary string without any punctuation to show where one codeword ends and the next begins. In a *prefix code*, no codeword is a prefix of any other codeword; prefix codes are thus always uniquely decodable. We shall illustrate these concepts shortly.

DEFINITION 6.17 (CODE LENGTH AND OPTIMAL CODE) The *code length* of the lossless code γ applied to discrete random variable v taking values in $\mathcal{I}$ is

$$L_{\mathrm{v}}(\gamma) = \mathrm{E}[\ell(\gamma(\mathrm{v}))] = \sum_{i \in \mathcal{I}} p_{\mathrm{v}}(i)\, \ell(\gamma(i)),$$

where p_{v} is the PMF of v and the *length function* ℓ gives the length of a binary string. A lossless code γ is *optimal* for compression of v when it minimizes $L_{\mathrm{v}}(\gamma)$ over all prefix codes.

EXAMPLE 6.21 (LOSSLESS CODES) Let v be the discrete random variable taking values in $\{0, 1, \ldots, 5\}$ with respective probabilities $\{0.3, 0.26, 0.14, 0.13, 0.09, 0.08\}$.

(i) A simple binary expansion gives the *fixed-rate* lossless code

$$\gamma_{\mathrm{FR}}(i) = \begin{cases} \texttt{000}, & \text{for } i = 0; \\ \texttt{001}, & \text{for } i = 1; \\ \texttt{010}, & \text{for } i = 2; \\ \texttt{011}, & \text{for } i = 3; \\ \texttt{100}, & \text{for } i = 4; \\ \texttt{101}, & \text{for } i = 5. \end{cases}$$

Clearly $L_{\mathrm{v}}(\gamma_{\mathrm{FR}}) = 3$ since $\ell(\gamma_{\mathrm{FR}}(i)) = 3$ for every $i \in \mathcal{I}$. This code is a prefix code and hence is uniquely decodable.

(ii) The lossless code

$$\gamma_{\text{1-to-1}}(i) = \begin{cases} \texttt{0}, & \text{for } i = 0; \\ \texttt{1}, & \text{for } i = 1; \\ \texttt{00}, & \text{for } i = 2; \\ \texttt{01}, & \text{for } i = 3; \\ \texttt{10}, & \text{for } i = 4; \\ \texttt{11}, & \text{for } i = 5 \end{cases}$$

gives a lower code length,

$$L_{\mathrm{v}}(\gamma_{\text{1-to-1}}) = 1 \cdot 0.3 + 1 \cdot 0.26 + 2 \cdot 0.14 + 2 \cdot 0.13 + 2 \cdot 0.09 + 2 \cdot 0.08 = 1.44.$$

The code is a one-to-one mapping, but its extension is not, so the code is not uniquely decodable. For example, the output string `001` could arise from any of the inputs $(0, 0, 1)$, $(0, 3)$, or $(2, 1)$. Codes that are not uniquely decodable are of little value because having to indicate the boundaries between codewords reduces the effectiveness of compression.

(iii) The lossless code

$$\gamma_{\mathrm{SFE}}(i) = \begin{cases} \texttt{00}, & \text{for } i = 0; \\ \texttt{01}, & \text{for } i = 1; \\ \texttt{100}, & \text{for } i = 2; \\ \texttt{101}, & \text{for } i = 3; \\ \texttt{110}, & \text{for } i = 4; \\ \texttt{111}, & \text{for } i = 5 \end{cases}$$

is an example of a *Shannon–Fano–Elias code* (see the *Further reading*). It is a prefix code and hence is uniquely decodable. Its code length is

$$L_{\mathrm{v}}(\gamma_{\mathrm{SFE}}) = 2 \cdot 0.3 + 2 \cdot 0.26 + 3 \cdot 0.14 + 3 \cdot 0.13 + 3 \cdot 0.09 + 3 \cdot 0.08 = 2.44.$$

While a Shannon–Fano–Elias code is not always optimal, in this particular example it is (compare with Example 6.23 in Section 6.6.1).

In Section 6.6.1, we present an algorithm for the design of optimal codes that is due to Huffman. The code length of an optimal code is bounded from below and from above using the concept of entropy; see the *Further reading* for pointers to proofs of these results.

Definition 6.18 (Entropy) The *entropy* of a random variable v that takes values in a countable set $\mathcal{I}$ is

$$H(\mathrm{v}) = \mathrm{E}[-\log_2 p_{\mathrm{v}}(\mathrm{v})] = -\sum_{i\in\mathcal{I}} p_{\mathrm{v}}(i)\log_2 p_{\mathrm{v}}(i), \qquad (6.119)$$

where p_{v} is the PMF of v.

Theorem 6.19 (Entropy bound on optimal code length) Let γ be an optimal lossless code for a discrete random variable v with entropy $H(\mathrm{v})$. Then

$$H(\mathrm{v}) \le L_{\mathrm{v}}(\gamma) < H(\mathrm{v}) + 1. \qquad (6.120)$$

Example 6.22 (Entropy) The entropy of the discrete random variable v defined in Example 6.21 is

$$\begin{aligned} H(\mathrm{v}) &= -0.3\log_2 0.3 - 0.26\log_2 0.26 - 0.14\log_2 0.14 \\ &\quad - 0.13\log_2 0.13 - 0.09\log_2 0.09 - 0.08\log_2 0.08 \\ &\approx 2.41. \end{aligned}$$

The code length of 2.44 from Example 6.21(iii) satisfies (6.120).

We ignore the up-to-1-bit gap in (6.120) for the remainder of the section. If $H(\mathrm{v})$ is large, ignoring that 1 bit introduces only a small relative error. Also, even for low entropies and code rates, $L_{\mathrm{v}}(\gamma) \approx H(\mathrm{v})$ can effectively be attained by coding blocks of symbols together. For example, if v_0 and v_1 are discrete random variables, then the pair $(\mathrm{v}_0, \mathrm{v}_1)$ is itself a discrete random variable, and

$$H((\mathrm{v}_0, \mathrm{v}_1)) \le H(\mathrm{v}_0) + H(\mathrm{v}_1), \qquad (6.121)$$

with equality if and only if v_0 and v_1 are independent (see Exercise 6.26). The code length of an optimal lossless code for $(\mathrm{v}_0, \mathrm{v}_1)$ satisfies

$$\begin{aligned} L_{(\mathrm{v}_0,\mathrm{v}_1)}(\gamma) &\overset{(a)}{<} H((\mathrm{v}_0, \mathrm{v}_1)) + 1 \overset{(b)}{\le} H(\mathrm{v}_0) + H(\mathrm{v}_1) + 1 \\ &= \left(H(\mathrm{v}_0) + \tfrac{1}{2}\right) + \left(H(\mathrm{v}_1) + \tfrac{1}{2}\right), \end{aligned}$$

where (a) follows from (6.120); and (b) from (6.121). Effectively, the excess above the entropy lower bound has been reduced from at most 1 bit to at most $\frac{1}{2}$ bit. By losslessly coding N variables together, the excess is reduced to at most $1/N$ bits.

6.5.2 Scalar quantization

Digital computation and communication use discrete values rather than continuous values such as real numbers. In the discussion of the precision of computation in

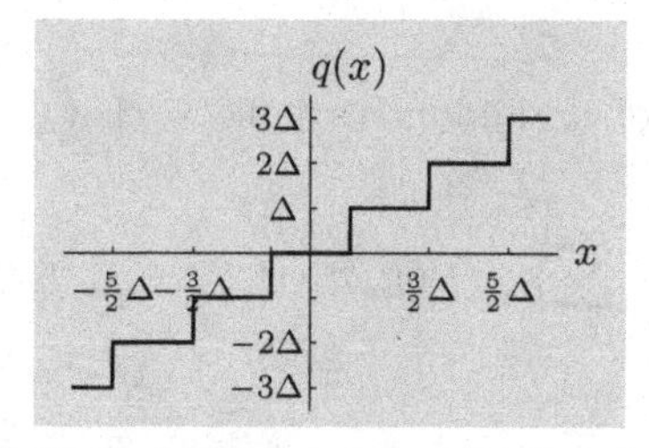

(a) Outputs are integer multiples of Δ.

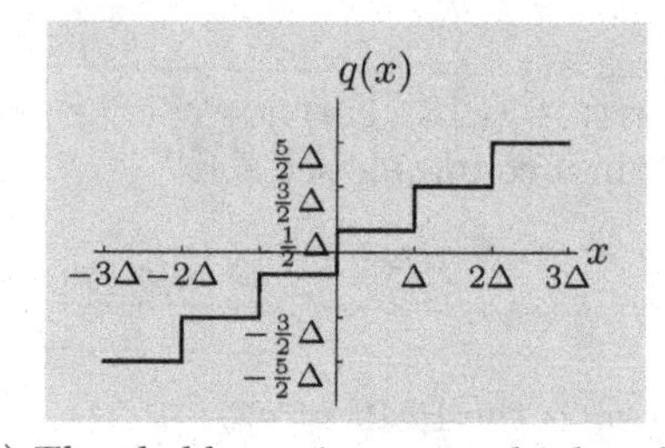

(b) Thresholds are integer multiples of Δ.

Figure 6.40 Uniform quantization.

Section 2.6.2, it was appropriate to think of huge finite sets of values (2^{32} and 2^{64} for the common 32-bit and 64-bit arithmetic). We now examine coarser discretizations of real numbers for compression, where typical average numbers of bits per real number are small (up to 8 but often less than 1).

In lossy compression, the mean approximation error is called the *distortion* and the code length in bits per scalar input is called the *rate*. Just as we have concentrated mostly on measuring the approximation error by MSE, we consider only MSE distortion.

Uniform quantization The simplest and most common form of quantization is *uniform quantization*. Typically, it takes the form of either rounding to the nearest integer multiple of a *step size* Δ, as in Figure 6.40(a), or using the integer multiples of the step size as threshold values at which the output jumps, as in Figure 6.40(b).

While even these simplest forms of quantization can be difficult to analyze precisely, they allow us to see the typical trade-off between rate and distortion that arises in quantizer design. Consider x uniformly distributed on the interval $[0, 1)$. A uniform quantizer as in Figure 6.40(b), with K cells and step size $\Delta = 1/K$, quantizes $x \in [m\Delta, (m+1)\Delta)$ to $(m + \frac{1}{2})\Delta$ for $m = 0, 1, \ldots, K-1$. Since K codewords cover $[0, 1)$, a fixed-rate lossless code γ_{FR} for $q(\mathrm{x})$ will have code length

$$R = L_{q(\mathrm{x})}(\gamma_{\mathrm{FR}}) = \lceil \log_2 K \rceil \approx \log_2 K = -\log_2 \Delta. \tag{6.122a}$$

In this example, since $q(\mathrm{x})$ has a uniform distribution, an optimal lossless code would have the same code length as a fixed-rate lossless code. The MSE distortion is

$$\begin{aligned} D &= \mathrm{E}\left[\,|\mathrm{x} - q(\mathrm{x})|^2\,\right] \overset{(a)}{=} \int_0^1 (s - q(s))^2 f_{\mathrm{x}}(s)\,ds \\ &\overset{(b)}{=} \sum_{m=0}^{K-1} \int_{m\Delta}^{(m+1)\Delta} (s - q(s))^2\,ds \overset{(c)}{=} \sum_{m=0}^{K-1} \int_{m\Delta}^{(m+1)\Delta} \left(s - \left(m + \frac{1}{2}\right)\Delta\right)^2 ds \\ &\overset{(d)}{=} \frac{1}{12}\Delta^2 \overset{(e)}{\approx} \frac{1}{12} 2^{-2R}, \end{aligned} \tag{6.122b}$$

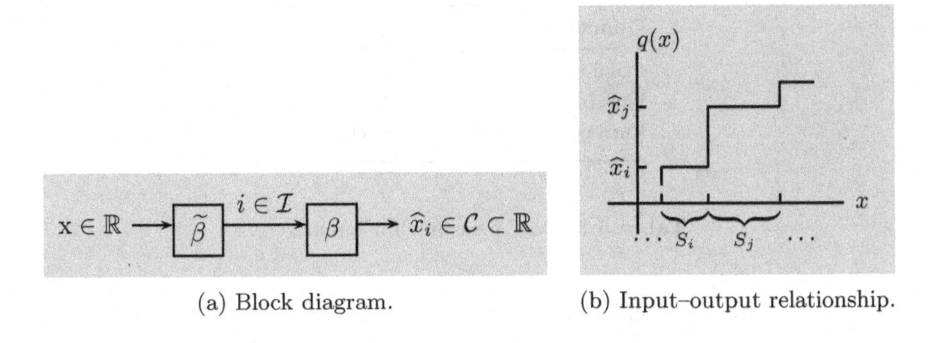

(a) Block diagram. (b) Input–output relationship.

Figure 6.41 Notation for a general scalar quantizer.

where (a) follows from the definition of the expectation; (b) from $f_{\mathrm{x}}(s) = 1$ and breaking the integral into a sum of K integrals over subintervals so that $q(s)$ is constant on each subinterval; (c) from substituting for $q(s)$; (d) from computing the integral; and (e) from (6.122a). We will see that, when using the MSE distortion measure, the 2^{-2R} dependence of distortion on rate will almost always be present.

General scalar quantization We now consider scalar quantization more generally, and we will give simple properties of optimal quantizers and approximations for their performance. Nonuniform scalar quantization provides substantial performance improvements over uniform scalar quantization in some situations. It tends to be most useful at low rates and when scalar quantization is used without lossless compression. We will see shortly that at high rates, there is a bounded performance gap between optimal nonuniform quantization and uniform quantization combined with optimal lossless coding.

DEFINITION 6.20 (SCALAR QUANTIZER) A *scalar quantizer* q is a mapping from $\mathbb{R}$ to a *reproduction codebook* $\mathcal{C} = \{\widehat{x}_i\}_{i\in\mathcal{I}} \subset \mathbb{R}$, where $\mathcal{I}$ is an arbitrary countable index set. Quantization can be decomposed into two operations, β and $\widetilde{\beta}$, with $q = \beta \circ \widetilde{\beta}$. The *lossy encoder* $\widetilde{\beta} : \mathbb{R} \to \mathcal{I}$ is specified by a partition $\{S_i\}_{i\in\mathcal{I}}$ of $\mathbb{R}$ with *partition cells* $S_i = \widetilde{\beta}^{-1}(i) = \{x \in \mathbb{R} \mid \widetilde{\beta}(x) = i\}$, $i \in \mathcal{I}$. The *reproduction decoder* $\beta : \mathcal{I} \to \mathbb{R}$ is specified by the codebook $\mathcal{C}$.

The notation from this definition is illustrated in Figure 6.41.

Returning to uniform quantization as an example, the quantizer in Figure 6.40(a) can now be specified formally as

$$\mathcal{I} = \mathbb{Z}, \quad \mathcal{C} = \Delta\mathbb{Z}, \quad \widetilde{\beta}(x) = \mathrm{round}\Big(\frac{x}{\Delta}\Big), \quad \beta(i) = i\Delta, \tag{6.123a}$$

where round(·) denotes rounding to the nearest integer. Similarly, the quantizer in

Quantizer	Rate R
Fixed-rate	$\lceil \log_2 \lvert \mathcal{I} \rvert \rceil$
Variable-rate	$\mathrm{E}[\,\ell(\gamma(\widetilde{\beta}(\mathrm{x})))\,]$
Entropy-constrained	$H(\widetilde{\beta}(\mathrm{x}))$

Table 6.3 Rate measures for quantizers.

Figure 6.40(b) can be specified as

$$\mathcal{I} = \mathbb{Z}, \quad \mathcal{C} = \Delta\left(\mathbb{Z} - \frac{1}{2}\right), \quad \widetilde{\beta}(x) = \left\lceil \frac{x}{\Delta} \right\rceil, \quad \beta(i) = \left(i - \frac{1}{2}\right)\Delta. \tag{6.123b}$$

The quality of a quantizer is determined by its distortion and rate. The MSE distortion for quantizing the random variable $\mathrm{x} \in \mathbb{R}$ is

$$D = \mathrm{E}\left[\,\lvert \mathrm{x} - q(\mathrm{x})\rvert^2\,\right] = \mathrm{E}\left[\,\lvert \mathrm{x} - \beta(\widetilde{\beta}(\mathrm{x}))\rvert^2\,\right].$$

The rate can be measured in a few ways. The lossy encoder output $\widetilde{\beta}(\mathrm{x})$ is a discrete random variable that is typically entropy coded because the output symbols will have unequal probabilities. Associating an entropy code γ with the lossy encoder $\widetilde{\beta}$ and reproduction decoder β gives a *variable-rate quantizer* specified by $(\widetilde{\beta}, \beta, \gamma)$. The rate of this quantizer is the code length of γ applied to $\widetilde{\beta}(\mathrm{x})$. Not specifying an entropy code (or specifying the use of a fixed-rate lossless code) gives a *fixed-rate quantizer* with rate $R = \lceil \log_2 \lvert \mathcal{I} \rvert \rceil$. Measuring the rate by the entropy lower bound (6.120) gives $R = H(\widetilde{\beta}(\mathrm{x}))$; the quantizer in this case is called *entropy-constrained*. These rates are summarized in Table 6.3. Even though a lossless compression method will not meet the bound (6.120) in general, the entropy-constrained case is important because lossless coding of blocks of quantizer outputs approaches the bound and adds little complexity – especially compared with unconstrained vector quantization (see Section 6.5.3 and the *Further reading*).

Optimal quantization An optimal quantizer minimizes the distortion subject to an upper bound on the rate. Consider the case of fixed-rate quantization. The distortion is given by

$$D = \mathrm{E}\left[\,\lvert \mathrm{x} - q(\mathrm{x})\rvert^2\,\right] = \int_{-\infty}^{\infty} \left(s - \beta(\widetilde{\beta}(s))\right)^2 f_{\mathrm{x}}(s)\, ds. \tag{6.124}$$

With the reproduction decoder β fixed, the choice of lossy encoder $\widetilde{\beta}$ can minimize the distortion by minimizing the integrand of (6.124) pointwise; this has no effect on the rate because the codebook size is unchanged. This implies that the optimal lossy encoder $\widetilde{\beta}$ returns the index of the nearest codeword:

$$i = \widetilde{\beta}(x) = \arg\min_{j \in \mathcal{I}} \lvert x - \widehat{x}_j \rvert. \tag{6.125a}$$

In particular, each partition cell S_i must be a single interval (see Exercise 6.27). Similarly, with the partition $\{S_i\}_{i\in\mathcal{I}}$ (or, equivalently, $\widetilde{\beta}$) fixed, the reproduction decoder β must satisfy

$$\widehat{x}_i = \beta(i) = \mathrm{E}[\,\mathrm{x} \mid \mathrm{x} \in S_i\,], \qquad \text{for all } i \in \mathcal{I}, \tag{6.125b}$$

to be optimal (see Exercise 6.28). The use of (6.125) for quantizer design is discussed in Section 6.6.2.

The variable-rate and entropy-constrained cases are a bit more complicated. Since β has no effect on the rate, the necessary condition (6.125b) remains valid. However, since $\widetilde{\beta}$ might affect the rate, pointwise minimization of the distortion through (6.125a) is no longer optimal. A standard method to optimize $\widetilde{\beta}$ with β held fixed is to minimize a weighted combination of rate and distortion; see the *Further reading*.

Because of simple shifting and scaling properties, an optimal quantizer for a random variable x can be easily deduced from an optimal quantizer for the normalized random variable $(\mathrm{x} - \mu_\mathrm{x})/\sigma_\mathrm{x}$, where μ_x and σ_x are the mean and standard deviation of x, respectively. One consequence of this is that optimal quantizers have performance

$$D = \sigma_\mathrm{x}^2\, g(R), \tag{6.126}$$

where $g(R)$ is the performance of optimal quantizers for the normalized source.

High-resolution analysis For most sources, it is impossible to analytically optimize quantizers or express the performance of optimal quantizers. Fortunately, approximations obtained when it is assumed that the quantization is very fine are reasonably accurate even at low to moderate rates.

High-resolution analysis is based on approximating the PDF f_x on the entire interval S_i by its value at the midpoint. Assuming that f_x is smooth, this approximation is accurate when each S_i is short.[110] Optimization of scalar quantizers then turns into finding how the optimal lengths of the partition cells depend on the PDF f_x. While the details of these optimizations for an arbitrary PDF are beyond our scope, Exercise 6.29 develops optimal fixed-rate and entropy-constrained quantizers for a source with a very simple PDF that are suggestive of the following general results (see the *Further reading*):

(i) For fixed-rate quantization, it is optimal for the cell containing s to have length approximately proportional to $f_\mathrm{x}^{-1/3}(s)$. The resulting distortion satisfies

$$D_{\mathrm{FR}} \approx \frac{1}{12}\left(\int_{-\infty}^{\infty} f_\mathrm{x}^{1/3}(s)\,ds\right)^3 2^{-2R}. \tag{6.127}$$

[110] An infinite partition cell does not have a midpoint. Thus, for a quantizer with a finite codebook applied to a source with infinite support, this approximation does not make sense for the extremal partition cells. It is furthermore assumed that the distortion contribution from the extremal partition cells is not dominant.

(ii) For entropy-constrained quantization, it is optimal for each S_i to have approximately equal length; that is, *uniform quantization is optimal.* The resulting distortion satisfies

$$D_{\text{EC}} \approx \frac{1}{12} 2^{2h(\mathrm{x})} 2^{-2R}, \tag{6.128}$$

where

$$h(\mathrm{x}) = -\int_{-\infty}^{\infty} f_{\mathrm{x}}(s) \log_2 f_{\mathrm{x}}(s)\, ds \tag{6.129}$$

is the *differential entropy* of x.

For a Gaussian random variable with variance σ^2, (6.127) yields

$$D_{\text{FR}} \approx \frac{\sqrt{3}\pi}{2} \sigma^2 2^{-2R} \tag{6.130}$$

for fixed-rate quantization, and (6.128) yields

$$D_{\text{EC}} \approx \frac{\pi e}{6} \sigma^2 2^{-2R}. \tag{6.131}$$

for entropy-constrained quantization.

Summarizing (6.127)–(6.131), we see that high-resolution quantizer performance is described by

$$D \approx c\sigma^2 2^{-2R}, \tag{6.132}$$

where σ^2 is the variance of the source and c is a constant that depends on the normalized density of the source and the type of quantization (fixed-rate, variable-rate, or entropy-constrained). This is consistent with (6.126).

6.5.3 Transform coding

Transform coding is a method for lossy compression that combines the approximation power of a well-chosen basis expansion, as in Section 6.4.2, with the simplicity of scalar quantization and lossless compression. The encoding and decoding are modular, which is advantageous for design, implementation, and understanding.

Motivation Conceptually, it is simple to extend quantization of real numbers to quantization of real vectors; in Definition 6.20, each instance of $\mathbb{R}$ can be replaced by $\mathbb{R}^N$. Computationally, however, this is not so simple. For a fixed number of bits per scalar component, the size of the codebook increases exponentially with N, and the cost of computing the distance to a codeword for nearest-codeword encoding (6.125a) is proportional to N. For scalars, one can perform nearest-codeword encoding (6.125a) by comparing an input to a precomputed list of endpoints of partition cells; for vectors ($N > 1$), partition cells are more complicated N-dimensional sets rather than intervals, so nearest-codeword encoding may require a full search over the codebook. The cost of this search is prohibitive for many applications. Thus, methods for structuring the codebook for efficient search – even at the expense of having a suboptimal codebook – are important; see the *Further reading*.

Dimension	Distortion decrease	Rate decrease
2	0.167 dB	0.028 bits
3	0.257 dB	0.043 bits
4	0.366 dB	0.061 bits
5	0.422 dB	0.070 bits
6	0.496 dB	0.082 bits
7	0.561 dB	0.093 bits
8	0.654 dB	0.109 bits
∞	1.533 dB	0.255 bits

Table 6.4 Gains of vector quantization over scalar quantization for entropy-constrained coding (based on high-resolution analysis).

Most of the benefit of vector quantization over scalar quantization comes from exploiting statistical dependencies between components of a vector. To illustrate that there is some advantage beyond this, consider the vector quantization of an i.i.d. Gaussian vector. The best performance, allowing the dimension N to grow without bound, is

$$D = \sigma^2 2^{-2R}. \tag{6.133}$$

The distortion given by (6.131) is worse only by a factor of $\pi e/6$ (≈ 1.53 dB). Equivalently, the increase in rate to achieve equal distortion is only $\frac{1}{2}\log_2(\pi e/6) \approx 0.255$ bits. These gaps from the best possible performance caused by using scalar (rather than vector) quantization diminish slowly as the dimension of quantization increases (see Table 6.4). Since the computational cost of unstructured vector quantization is exponential in N, the benefit (decrease in distortion or rate) is often not justified if statistical dependencies are exploited in some other way.

Pixels of an image tend to be similar to their neighbors, or differ in partially predictable ways. These tendencies, arising from the continuity, texturing, and boundaries of objects, the similarity of objects in an image, gradual lighting changes, etc., might extend over an entire image with millions of pixels. With the dimension of a signal being so high, it is important to use methods with computational costs that grow slowly with dimension. Thus, state-of-the-art lossy compression methods for images divide the encoding operation into a sequence of three relatively simple steps: the computation of a linear transformation of the data designed primarily to produce uncorrelated coefficients, separate quantization of each scalar coefficient, and entropy coding. Contemporary methods for compression of video and music follow similar steps; standards for speech compression operate rather differently and include vector quantization.

Terminology A transform code is defined as follows (see Figure 6.42).

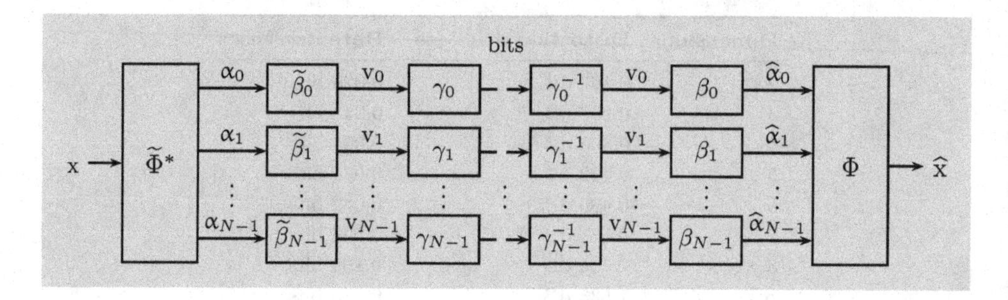

Figure 6.42 The encoder of a transform code applies the analysis operator $\widetilde{\Phi}^*$ to $x \in \mathbb{R}^N$ to compute transform coefficients $\{\alpha_k\}_{k=0}^{N-1}$. The lossy encoders of scalar quantizers $\{\widetilde{\beta}_k\}_{k=0}^{N-1}$ then produce discrete variables $\{v_k\}_{k=0}^{N-1}$ that are compressed with lossless codes $\{\gamma_k\}_{k=0}^{N-1}$. The decoder reverses these steps by first inverting the lossless codes, then applying reproduction decoders of the scalar quantizers $\{\beta_k\}_{k=0}^{N-1}$, and finally applying the synthesis operator Φ. The bases used in analysis and synthesis are a biorthogonal pair of bases, so $\Phi^{-1} = \widetilde{\Phi}^*$.

DEFINITION 6.21 (TRANSFORM CODE) A *transform code* for $x \in \mathbb{R}^N$ uses a biorthogonal pair of bases $\{\varphi_k\}_{k=0}^{N-1}$ and $\{\widetilde{\varphi}_k\}_{k=0}^{N-1}$ for $\mathbb{R}^N$, a set of scalar quantizers $\{(\widetilde{\beta}_k, \beta_k)\}_{k=0}^{N-1}$, and a set of lossless codes $\{\gamma_k\}_{k=0}^{N-1}$.

(i) *Encoding:* Applying the analysis operator $\widetilde{\Phi}^*$ associated with $\{\widetilde{\varphi}_k\}_{k=0}^{N-1}$ to x produces *transform coefficients* $\{\alpha_k\}_{k=0}^{N-1}$. Applying the lossy encoder $\widetilde{\beta}_k$ to α_k produces v_k, and applying the lossless code γ_k to v_k produces a portion of the encoded representation, for $k = 0, 1, \ldots, N-1$.

(ii) *Decoding:* Applying the reproduction decoder β_k to v_k produces $\widehat{\alpha}_k$, for $k = 0, 1, \ldots, N-1$. Applying the synthesis operator Φ associated with $\{\varphi_k\}_{k=0}^{N-1}$ to $\widehat{\alpha}$ produces the reproduction $\widehat{x}$.

(iii) *Distortion:* The distortion is the per-component MSE

$$D = \frac{1}{N} \mathrm{E}\big[\|x - \widehat{x}\|^2\big]. \tag{6.134}$$

(iv) *Rate:* For fixed-rate coding, the rate is

$$R_{\mathrm{FR}} = \frac{1}{N} \sum_{k=0}^{N-1} \log_2 K_k = \frac{1}{N} \log_2 \left(\prod_{k=0}^{N-1} K_k \right), \tag{6.135a}$$

where K_k is the size of the codebook of quantizer $(\widetilde{\beta}_k, \beta_k)$. For entropy-constrained coding, the rate is

$$R_{\mathrm{EC}} = \frac{1}{N} \sum_{k=0}^{N-1} H(\widetilde{\beta}(v_k)). \tag{6.135b}$$

The simplicity of transform coding makes large values of N practical. Computing each of the transforms $\widetilde{\Phi}^*$ and Φ requires at most N^2 multiplications and $N(N-1)$ additions. A structured transform, such as a DFT, a DCT, or a discrete wavelet transform (DWT), is often used to reduce the complexity of these steps further.

Bit allocation Separate quantization and entropy coding of each transform coefficient requires splitting of the total number of bits among the transform coefficients, implying some sort of a *bit allocation* among the components. Bit allocation problems can be stated in a single common form. Given is a set of quantizers described by their distortion–rate performances as

$$D_k = g_k(R_k), \qquad R_k \in \mathcal{R}_k, \qquad k = 0, 1, \ldots, N-1,$$

where each set of available rates $\mathcal{R}_k$ is a subset of the nonnegative real numbers and might be discrete or continuous. The problem is to minimize the distortion

$$D = \frac{1}{N} \sum_{k=0}^{N-1} D_k,$$

subject to the maximum rate

$$R = \frac{1}{N} \sum_{k=0}^{N-1} R_k.$$

If the distortion can be reduced by taking bits away from one component and giving them to another, the initial bit allocation is not optimal. Applying this reasoning with infinitesimal changes in the component rates, a necessary condition for an optimal allocation is that the slope of each g_k at R_k is equal to a common constant value; see the *Further reading* for pointers to more formal arguments.

The approximate performance given by (6.132) leads to a particularly easy bit allocation problem with

$$g_k(R_k) = c_k \sigma_k^2 2^{-2R_k}, \qquad \mathcal{R}_k = [0, \infty), \qquad k = 0, 1, \ldots, N-1. \tag{6.136}$$

Ignoring the fact that each component rate must be nonnegative, an equal-slope argument shows that the optimal bit allocation is

$$R_k = R + \frac{1}{2}\log_2 \frac{c_k}{\left(\prod_{n=0}^{N-1} c_n\right)^{1/N}} + \frac{1}{2}\log_2 \frac{\sigma_k^2}{\left(\prod_{n=0}^{N-1} \sigma_n^2\right)^{1/N}}. \tag{6.137a}$$

With these rates, $D_0 = D_1 = \cdots = D_{N-1} = D$, and this distortion is

$$D = \left(\prod_{n=0}^{N-1} c_n\right)^{1/N} \left(\prod_{n=0}^{N-1} \sigma_n^2\right)^{1/N} 2^{-2R}. \tag{6.137b}$$

This solution is valid when each R_k given above is nonnegative. For lower rates, the components with smallest $c_k \sigma_k^2$ are allocated no bits, and the remaining components have correspondingly higher numbers of bits.

Since (6.137) gives equal distortions for each component, it has a very simple consequence for entropy-constrained quantization. Under high-resolution analysis, uniform quantizers are approximately optimal. Furthermore, the component distortion D_k obtained from uniform quantization with step size Δ_k is $\Delta_k^2/12$, without any dependence on the source distribution. Therefore, using the optimal bit allocation implies having equal step sizes for all N uniform quantizers. See Exercise 6.30 for a related computation.

Visualizing the effect of the transform Beyond two or three dimensions, it is difficult to visualize vectors – let alone the effects of transforms and quantizers on vectors. Fortunately, the most important case for transform coding lends itself to visualization. Assume that the transform coefficient vector α is quantized with uniform scalar quantizers with equal step sizes for each component. The partitioning for α is then into axis-aligned hypercubes, and we can visualize the effect of this partitioning on the original vector of interest x because a linear transform combines rotating, scaling, and shearing such that a hypercube is always mapped to a parallelepiped.

For illustration, consider a two-dimensional random vector x with a zero-mean Gaussian distribution. The level curves of its PDF are ellipses centered at the origin with collinear major axes, as shown earlier in Figure 6.35 and repeated in Figure 6.43(a). Applying the analysis operator $\widetilde{\Phi}^*$ results in a Gaussian vector (see Appendix 2.C.2), so the level curves of the PDF of α are also ellipses centered at the origin with collinear major axes; Figure 6.43(b) shows an example. Uniform scalar quantization of α, with equal step sizes for each component, partitions the transform domain into squares, which are also shown in Figure 6.43(b). Applying the synthesis operator Φ returns to the original coordinates, and the partitioning is deformed into identical parallelograms, as shown in Figure 6.43(c).

The partition in the original coordinates is what is truly relevant. It shows which source vectors are mapped to the same reconstruction vector, thus giving an indication of the average distortion. Looking at the number of cells with appreciable probability gives an indication of the rate.

A singular transform is a degenerate case. As shown in Figure 6.43(d), the transform coefficients have probability mass only along a line. Inverting the transform is not possible, but we may still return to the original coordinates to view the partition induced by quantizing the expansion coefficients. The cells are unbounded in one direction, as shown in Figure 6.43(e). This is undesirable unless the variation of the source in the direction in which the cells are unbounded is very small.

It is inherently suboptimal to have the parallelogram-shaped partition cells that arise from using nonorthogonal bases (see Exercise 6.31). To get rectangular partition cells, the basis vectors must be orthogonal. For square cells, the basis vectors, in addition to being orthogonal, should have equal lengths. A KL basis is an orthonormal basis and thus gives square cells; in addition, the partition is aligned with the axes of the source PDF, as shown in Figures 6.43(f) and (g).

Transform optimization Under various assumptions, one can show that KL bases are optimal for transform coding. We will concentrate on results for orthonormal

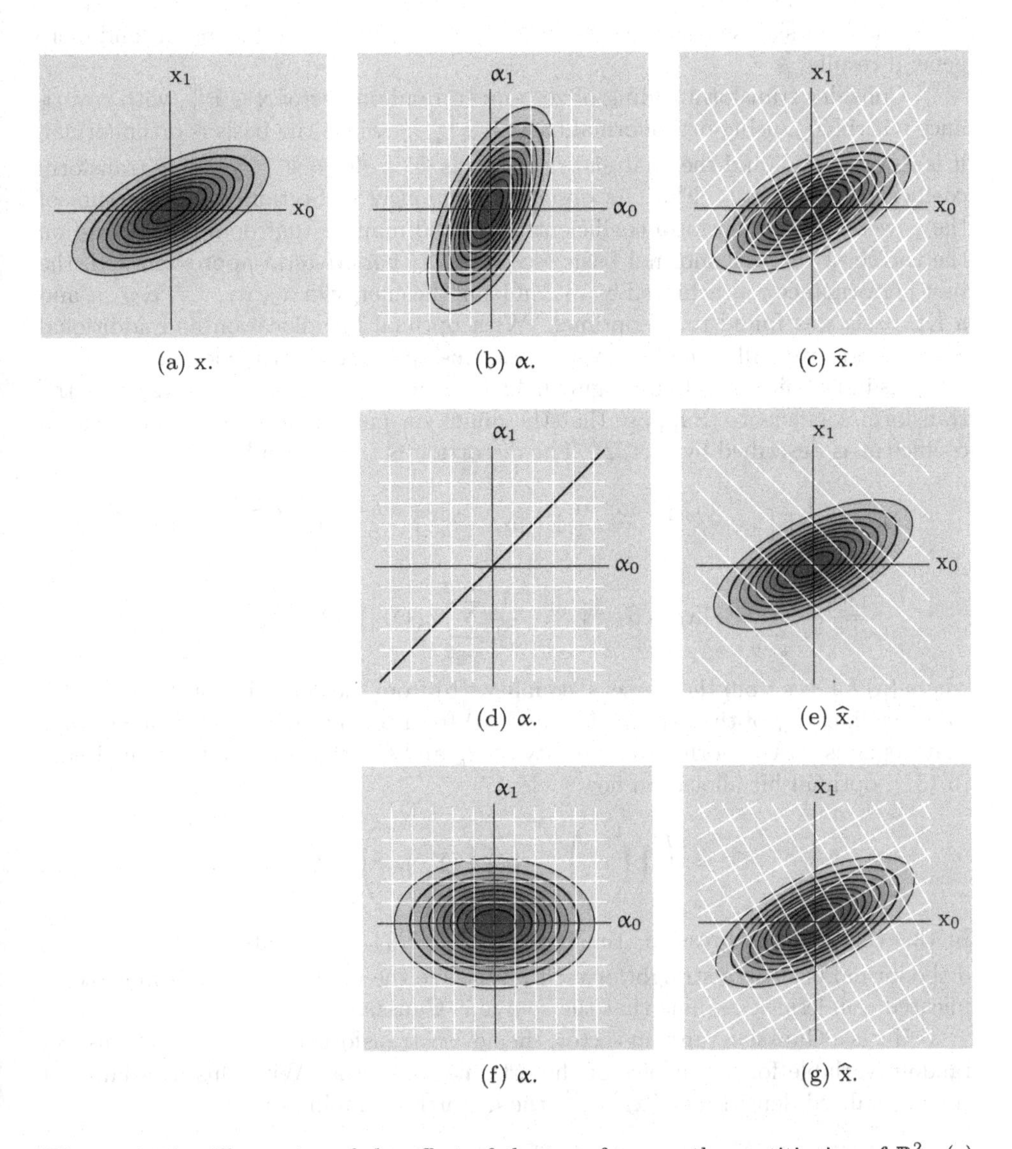

Figure 6.43 Illustration of the effect of the transform on the partitioning of $\mathbb{R}^2$. (a) A Gaussian source x is depicted by level curves of its PDF. (b) An analysis operator $\widetilde{\Phi}^*$ results in Gaussian transform coefficients α, which are in a space partitioned into squares by uniform scalar quantization with equal step sizes in each dimension. (c) Inverting the transform with the synthesis operator Φ returns to the original coordinates, with partitioning into parallelograms. (d) A degenerate case of $\widetilde{\Phi}^*$ having a one-dimensional range results in linearly dependent transform coefficients. (e) In the degenerate case, the partitioning of the original space has unbounded cells. (f) A KLT gives uncorrelated transform coefficients. (g) When the KLT has been used, the partitioning of the original space is aligned with the axes of the source PDF.

bases and Gaussian sources; see the *Further reading* for pointers to proofs and more general results.

Consider transform coding of zero-mean random vector $\mathrm{x} \in \mathbb{R}^N$, with covariance matrix Σ_{x}, using orthonormal basis $\{\varphi_k\}_{k=0}^{N-1}$. Since the basis is orthonormal, it is its own dual, and the analysis transform is $\widetilde{\Phi}^* = \Phi^* = \Phi^{-1}$, yielding transform coefficient vector $\alpha = \Phi^*\mathrm{x}$. As observed previously in Section 6.4.2, the sum of the powers of the transform coefficients is a fixed number that does not depend on the choice of the orthonormal basis (see (6.94)). For M-term approximation, the best performance was achieved by maximizing the energy in $\alpha_0, \alpha_1, \ldots, \alpha_{M-1}$, and a KL basis was found to be optimal. With optimal bit allocation and additional assumptions, we will come to the same conclusion for transform coding.

As in Definition 6.21 and Figure 6.42, let $\widehat{\alpha}$ denote the quantized versions of the transform coefficients. Suppose that the quantizer performance on each transform coefficient is described by (6.132). The distortion is then given by

$$\begin{aligned} D &= \frac{1}{N}\mathrm{E}\left[\,\|\mathrm{x} - \widehat{\mathrm{x}}\|^2\,\right] \stackrel{(a)}{=} \frac{1}{N}\mathrm{E}\left[\,\|\alpha - \widehat{\alpha}\|^2\,\right] \stackrel{(b)}{=} \frac{1}{N}\mathrm{E}\left[\sum_{k=0}^{N-1}|\alpha_k - \widehat{\alpha}_k|^2\right] \\ &\stackrel{(c)}{=} \frac{1}{N}\sum_{k=0}^{N-1}\mathrm{E}\left[\,|\alpha_k - \widehat{\alpha}_k|^2\,\right] \stackrel{(d)}{=} \frac{1}{N}\sum_{k=0}^{N-1} c_k (\Sigma_\alpha)_{kk} 2^{-2R_k}, \end{aligned} \tag{6.138}$$

where (a) follows from the Parseval equality; (b) from the definition of the norm; (c) from the linearity of the expectation; and (d) from (6.132), with each c_k a constant that depends on the normalized density of α_k and the type of quantization. Using (6.137), optimal bit allocation now yields

$$D = \left(\prod_{n=0}^{N-1} c_n\right)^{1/N} \left(\prod_{n=0}^{N-1} (\Sigma_\alpha)_{nn}\right)^{1/N} 2^{-2R}. \tag{6.139}$$

In this distortion expression, the factor $\prod_{n=0}^{N-1}(\Sigma_\alpha)_{nn}$ depends on the choice of orthonormal basis in a straightforward way, while the factor $\prod_{n=0}^{N-1} c_n$ is more complicated – unless we assume that the source is Gaussian.

If x is a Gaussian random vector, then every transform coefficient is a Gaussian random variable for any choice of the orthonormal basis. With this invariance of the normalized densities of $\{\alpha_k\}_{k=0}^{N-1}$, the distortion simplifies to

$$D = c\left(\prod_{n=0}^{N-1} (\Sigma_\alpha)_{nn}\right)^{1/N} 2^{-2R}, \tag{6.140}$$

where c depends on the type of quantization (for example, $c = \sqrt{3}\pi/2$ for fixed-rate quantization and $c = \pi e/6$ for entropy-constrained quantization).

The distortion in (6.140) can be used to define a figure of merit called the *coding gain*:

$$\text{coding gain} = \frac{\left(\prod_{n=0}^{N-1}(\Sigma_{\mathrm{x}})_{nn}\right)^{1/N}}{\left(\prod_{n=0}^{N-1}(\Sigma_\alpha)_{nn}\right)^{1/N}}. \tag{6.141}$$

The coding gain is the factor by which the distortion is reduced because of the transform, assuming high rate and optimal bit allocation.

To maximize the coding gain, and hence minimize the distortion, the choice of orthonormal basis should minimize the geometric mean of the transform coefficient variances. This minimization is achieved by a KL basis. Specifically,

$$\prod_{n=0}^{N-1} (\Sigma_\alpha)_{nn} \overset{(a)}{\geq} \det(\Sigma_\alpha) \overset{(b)}{=} \det(\Phi^* \Sigma_x \Phi) \overset{(c)}{=} \det(\Phi^*) \det(\Sigma_x) \det(\Phi)$$
$$\overset{(d)}{=} \det(\Sigma_x) \overset{(e)}{=} \prod_{n=0}^{N-1} \lambda_n, \qquad (6.142)$$

where (a) follows from Hadamard's inequality; (b) from Solved exercise 6.5(i); (c) from $\det(AB) = \det(A)\det(B)$; (d) from unitary matrices having unit determinant; and (e) from (2.228), where $\{\lambda_n\}_{n=0}^{N-1}$ are the eigenvalues of Σ_x. Thus, $\prod_{n=0}^{N-1}(\Sigma_\alpha)_{nn}$ has a lower bound that is independent of the choice of basis and achieved with equality by a KL basis.

The optimality of the KLT under optimal bit allocation and expressions from high-resolution analysis is a special case of the following more general result:

THEOREM 6.22 (OPTIMALITY OF KARHUNEN–LOÈVE TRANSFORM) Consider transform coding using an orthonormal basis. Suppose that there is a single nonincreasing function g that describes the quantization of each transform coefficient through

$$\mathrm{E}\left[(\alpha_k - \widehat{\alpha}_k)^2 \right] = \sigma_k^2 \, g(R_k), \qquad k = 0, 1, \ldots, N-1,$$

where σ_k^2 is the variance of α_k and R_k is the rate allocated to α_k. Then, a KLT minimizes the distortion for any nonincreasing bit allocation $(R_0, R_1, \ldots, R_{N-1})$.

6.6 Computational aspects

In this section, we present design algorithms for lossless and lossy source codes. We then discuss estimation from quantized samples.

6.6.1 Huffman algorithm for lossless code design

Huffman introduced a simple algorithm for constructing optimal entropy codes. The algorithm starts with a graph with one node for each symbol and no edges. At each step of the algorithm, the probabilities of the disconnected sets of nodes are sorted and the two least probable sets are merged through the addition of a parent node connected with edges to each of the two sets. The edges are assigned labels of 0 and 1. When a tree has been formed, codewords are assigned to each leaf node by concatenating the edge labels on the path from the root to the leaf. We illustrate the algorithm through an example.

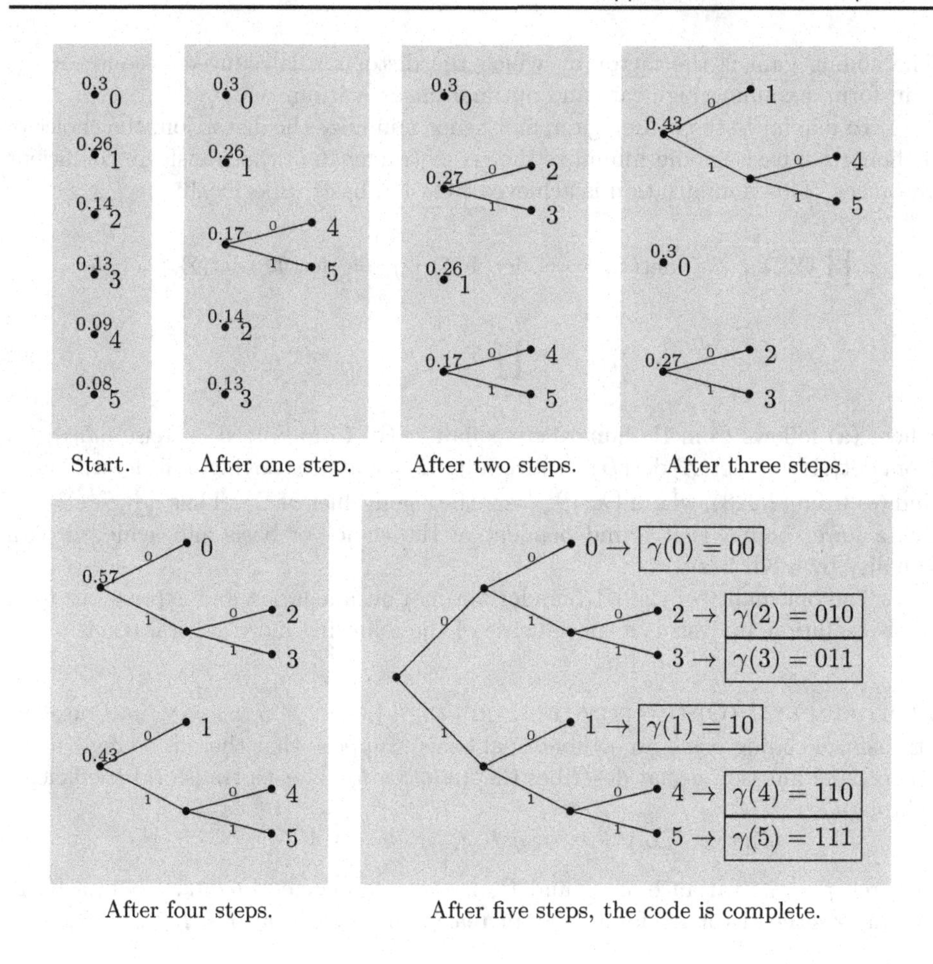

Figure 6.44 A Huffman code construction. One starts with a graph with no edges. At each step, the probabilities are sorted and the two least probable sets are merged by adding a parent node. When a tree has been formed, the codewords can be read off the tree by concatenating the edge labels.

EXAMPLE 6.23 (HUFFMAN CODE DESIGN) Consider the random variable v defined in Example 6.21. Figure 6.44 shows the steps of the Huffman code design algorithm for v. The code length is

$$L_{\mathrm{v}}(\gamma) = 0.3 \cdot 2 + 0.26 \cdot 2 + 0.14 \cdot 3 + 0.13 \cdot 3 + 0.09 \cdot 3 + 0.08 \cdot 3 = 2.44,$$

which is quite close to the entropy of 2.41 bits obtained in Example 6.22.

6.6.2 Iterative design of quantizers

The design of optimal quantizers is a challenging problem. Consider fixed-rate quantization of x with codebook $\mathcal{C} = \{\widehat{x}_i\}_{i=0}^{K-1}$, with

$$\widehat{x}_0 < \widehat{x}_1 < \widehat{x}_2 < \cdots < \widehat{x}_{K-1},$$

and partition cells

$$S_i = [y_i, y_{i+1}), \qquad i = 0, 1, \ldots, K-1,$$

where $y_0 = -\infty$ and $y_K = \infty$.[111] The distortion is

$$D \overset{(a)}{=} \int_{-\infty}^{\infty} \left(s - \beta(\widetilde{\beta}(s))\right)^2 f_{\mathrm{x}}(s)\, ds \overset{(b)}{=} \sum_{i=0}^{K-1} \int_{y_i}^{y_{i+1}} \left(s - \beta(\widetilde{\beta}(s))\right)^2 f_{\mathrm{x}}(s)\, ds$$
$$\overset{(c)}{=} \sum_{i=0}^{K-1} \int_{y_i}^{y_{i+1}} (s - \widehat{x}_i)^2 f_{\mathrm{x}}(s)\, ds, \tag{6.143}$$

where (a) follows from (6.124); (b) from breaking the integral into a sum of K integrals over subintervals so that $\widetilde{\beta}(s)$ is constant on each subinterval; and (c) from substituting the codebook entries. This is a complicated function of $\{\widehat{x}_i\}_{i=0}^{K-1}$ and $\{y_i\}_{i=1}^{K-1}$, and often the best that can be hoped for is to find a local minimum of D.

The optimality conditions (6.125) are the basis of an iterative design algorithm due to Lloyd. The following two steps improve the partition with the codebook fixed and improve the codebook with the partition fixed:

(i) *Partition optimization:* With the codebook $\{\widehat{x}_i\}_{i=0}^{K-1}$ fixed, using (6.125a), each interior partition endpoint should be at the midpoint between codebook entries:

$$y_i = \frac{1}{2}\left(\widehat{x}_{i-1} + \widehat{x}_i\right), \qquad i = 1, 2, \ldots, K-1. \tag{6.144a}$$

(ii) *Codebook optimization:* With the partition endpoints $\{y_i\}_{i=1}^{K-1}$ fixed, using (6.125b), each codebook entry should be at the centroid of the corresponding cell:

$$\widehat{x}_i = \mathrm{E}[\mathrm{x} \mid \mathrm{x} \in [y_i, y_{i+1})] = \frac{\int_{y_i}^{y_{i+1}} s f_{\mathrm{x}}(s)\, ds}{\int_{y_i}^{y_{i+1}} f_{\mathrm{x}}(s)\, ds}, \qquad i = 0, 1, \ldots, K-1. \tag{6.144b}$$

Repeating these steps creates a nonincreasing sequence of distortions; the iteration can be halted when the change of distortion is very small. When $\log f_{\mathrm{x}}$ is concave, the distortion has a single local minimum, so convergence is to the optimal quantizer; see the *Further reading.*

The design of variable-rate and entropy-constrained quantizers is a bit more complicated because (6.125a) is not a necessary condition for optimality of the lossy encoder. This changes the partition optimization step of the iterative design algorithm; see the *Further reading.*

[111] Exercise 6.27 establishes that each partition cell of an optimal quantizer is a single interval.

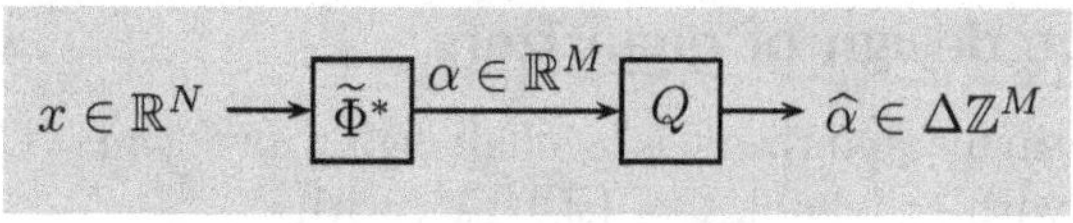

$x \in \mathbb{R}^N \rightarrow \boxed{\widetilde{\Phi}^*} \xrightarrow{\alpha \in \mathbb{R}^M} \boxed{Q} \rightarrow \widehat{\alpha} \in \Delta\mathbb{Z}^M$

Figure 6.45 Quantized expansion coefficients obtained with a frame analysis operator are used to form an estimate of x.

6.6.3 Estimating from quantized samples

The difference between a quantizer's output and its input can be considered as noise that the quantizer has added to a signal. While generic methods like those described in Section 6.4.3 may be employed to mitigate this noise, two peculiarities should inspire caution in analysis:

(i) *Determinism:* The error $q(x) - x$ is a function of the input x. For a random input x, the error $q(\mathrm{x}) - \mathrm{x}$ is a random variable that is a deterministic function of x.

(ii) *Boundedness:* The error $q(x) - x$ is often bounded, unlike the Gaussian model for noise under which many analyses are performed. In particular, if the quantizer is uniform with step size Δ, then $|q(x) - x| \leq \frac{1}{2}\Delta$ for all inputs x. More generally, $q(x) = \widehat{x}_i$ implies that $x \in S_i$, and S_i generally has finite length.

The boundedness property introduces novel geometric aspects to estimation problems and leads to algorithms that perform better than LMMSE estimation. Here we will introduce a finite-dimensional estimation problem, interpret it geometrically, and present a few algorithmic approaches.

Estimation from a quantized frame expansion Let $\{\varphi_k\}_{k=0}^{M-1}$ be a frame in $\mathbb{R}^N$ with $M > N$, and let $\{\widetilde{\varphi}_k\}_{k=0}^{M-1}$ be its canonical dual frame. Also let $Q : \mathbb{R}^M \rightarrow \Delta\mathbb{Z}^M$ be the operator that applies the uniform scalar quantizer with step size Δ in Figure 6.40(a) (rounding to the nearest integer multiple of Δ) separately to each of the M entries of its input. Analysis of $x \in \mathbb{R}^N$ with $\{\widetilde{\varphi}_k\}_{k=0}^{M-1}$ results in coefficient vector $\alpha = \widetilde{\Phi}^* x \in \mathbb{R}^M$. Suppose that we observe its quantized version, $\widehat{\alpha} = Q(\alpha) = Q(\widetilde{\Phi}^* x)$ as in Figure 6.45, and wish to form an estimate for x.

The redundancy from the expansion of a vector in $\mathbb{R}^N$ to a higher-dimensional space $\mathbb{R}^M$ allows us to mitigate some of the error introduced by the quantization. From the canonical dual property, $\widetilde{\Phi}^*\Phi : \mathbb{R}^M \rightarrow \mathbb{R}^M$ is an orthogonal projection operator onto $S = \mathcal{R}(\widetilde{\Phi}^*)$. The noiseless coefficients α must lie in S, so it is sensible to compute the orthogonal projection of $\widehat{\alpha}$ onto S before synthesizing with Φ. This would result in the estimate

$$\widehat{x}_{\mathrm{L}} = \Phi(\widetilde{\Phi}^*\Phi\widehat{\alpha}) = (\Phi\widetilde{\Phi}^*)\Phi\widehat{\alpha} \stackrel{(a)}{=} \Phi\widehat{\alpha},$$

where (a) follows from $\Phi\widetilde{\Phi}^* = I$ for any pair of dual frames. This computation shows an optimality property of canonical dual frames.

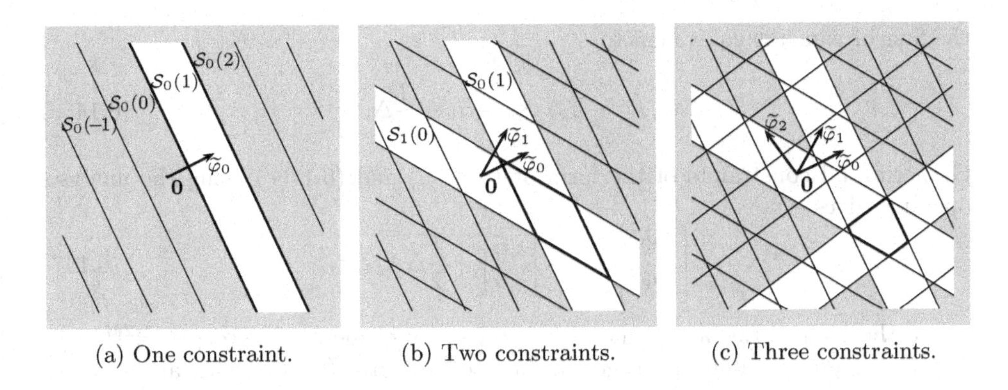

(a) One constraint. (b) Two constraints. (c) Three constraints.

Figure 6.46 Visualizing the information present in a quantized frame expansion of $x \in \mathbb{R}^2$. (a) A single quantized expansion coefficient specifies a strip in $\mathbb{R}^2$ perpendicular to $\widetilde{\varphi}_0$. (b) A second quantized expansion coefficient specifies a second strip in $\mathbb{R}^2$, perpendicular to $\widetilde{\varphi}_1$. Together, they determine a parallelogram as in Figure 6.43. (c) A third quantized expansion coefficient localizes x further, and many shapes for the partition cell containing x are possible.

Under the model that $\beta = Q(\alpha) - \alpha$ is a random vector with mean zero and uncorrelated entries of equal variance, $\widehat{x}_{\mathrm{L}}$ is the LMMSE estimate of x. As M increases with N fixed, the MSE of this estimate is generally a function with $\Theta(M^{-1})$ decay (there is a dependence on the construction of the frames).

Other estimates can give faster decay of MSE with M, often $\Theta(M^{-2})$. The key observation is that the boundedness of quantization error implies that each quantized coefficient $\widehat{\alpha}_k$ specifies hard constraints on x that one might exploit. Figure 6.46 illustrates the geometry for $x \in \mathbb{R}^2$. The first quantized coefficient $\widehat{\alpha}_0$ specifies

$$\langle x, \widetilde{\varphi}_0 \rangle \in \left[\widehat{\alpha}_0 - \frac{1}{2}\Delta, \widehat{\alpha}_0 + \frac{1}{2}\Delta\right] = \mathcal{S}_0(\widehat{\alpha}_0/\Delta),$$

which is the intersection of two half-spaces. This is illustrated in Figure 6.46(a), with one of several sets $\mathcal{S}_0(i)$, $i \in \mathbb{Z}$, shaded. Each additional quantized coefficient specifies an additional constraint

$$\langle x, \widetilde{\varphi}_k \rangle \in \left[\widehat{\alpha}_k - \frac{1}{2}\Delta, \widehat{\alpha}_k + \frac{1}{2}\Delta\right] = \mathcal{S}_k(\widehat{\alpha}_k/\Delta). \tag{6.145}$$

In general, the linear estimate $\widehat{x}_{\mathrm{L}}$ might not satisfy all the constraints.

Linear programming The constraint (6.145) can be written as a pair of constraints

$$\langle x, \widetilde{\varphi}_k \rangle \geq \widehat{\alpha}_k - \frac{1}{2}\Delta, \tag{6.146a}$$

$$\langle x, \widetilde{\varphi}_k \rangle \leq \widehat{\alpha}_k + \frac{1}{2}\Delta, \tag{6.146b}$$

the first of which is equivalent to

$$\langle x, -\widetilde{\varphi}_k \rangle \leq -\widehat{\alpha}_k + \frac{1}{2}\Delta, \tag{6.146c}$$

Gathering all constraints of the form of (6.146b) and (6.146c) using the analysis operator gives

$$\begin{bmatrix} \widetilde{\Phi}^* \\ -\widetilde{\Phi}^* \end{bmatrix} x \leq \begin{bmatrix} \widehat{\alpha} \\ -\widehat{\alpha} \end{bmatrix} + \frac{1}{2}\Delta\, \mathbf{1}_{2M\times 1}, \tag{6.147}$$

where the inequalities are elementwise and every element of $\mathbf{1}_{2M\times 1} \in \mathbb{R}^{2M}$ is 1. Any linear programming package can be used to solve (6.147), with an arbitrary objective function, to find an estimate in $\cap_{k=0}^{M-1} \mathcal{S}_k(\widehat{\alpha}_k/\Delta)$.

Linear programming generally yields a corner of the feasible set, where at least one constraint in (6.147) is active. Typically, the estimate can be improved by maximizing the slack in (6.147), meaning that each constraint is as far from active as possible. To this end, let

$$A = \begin{bmatrix} \widetilde{\Phi}^* & \mathbf{1}_{M\times 1} \\ -\widetilde{\Phi}^* & \mathbf{1}_{M\times 1} \end{bmatrix}, \qquad b = \begin{bmatrix} \widehat{\alpha} \\ -\widehat{\alpha} \end{bmatrix}, \qquad c = \begin{bmatrix} \mathbf{0}_{N\times 1} \\ -1 \end{bmatrix}, \tag{6.148a}$$

and use a linear programming method to

$$\text{minimize} \quad c^\top \begin{bmatrix} x \\ \delta \end{bmatrix} \quad \text{subject to} \quad A \begin{bmatrix} x \\ \delta \end{bmatrix} \leq b \tag{6.148b}$$

and return the first N components of the minimizer as $\widehat{x}$.

Iterative projection Since the set $\mathcal{S}_k(\widehat{\alpha}_k/\Delta)$ in (6.145) is convex, the POCS method of Section 5.6.1 can be applied to the estimation of x. For an iterative method, set $\widehat{x}^{(n+1)}$ to the orthogonal projection of $\widehat{x}^{(n)}$ onto $\mathcal{S}_k(\widehat{\alpha}_k/\Delta)$ for some appropriate choice of k. This update is

$$\widehat{x}^{(n+1)} = \begin{cases} \widehat{x}^{(n)} + (\widehat{\alpha}_k - \frac{1}{2}\Delta - \langle \widehat{x}^{(n)}, \widetilde{\varphi}_k\rangle)\widetilde{\varphi}_k/\|\widetilde{\varphi}_k\|^2, & \text{if } \langle \widehat{x}^{(n)}, \widetilde{\varphi}_k\rangle < \widehat{\alpha}_k - \frac{1}{2}\Delta; \\ \widehat{x}^{(n)} + (\widehat{\alpha}_k + \frac{1}{2}\Delta - \langle \widehat{x}^{(n)}, \widetilde{\varphi}_k\rangle)\widetilde{\varphi}_k/\|\widetilde{\varphi}_k\|^2, & \text{if } \langle \widehat{x}^{(n)}, \widetilde{\varphi}_k\rangle > \widehat{\alpha}_k + \frac{1}{2}\Delta; \\ \widehat{x}^{(n)}, & \text{otherwise.} \end{cases} \tag{6.149}$$

Convergence to the intersection $\cap_{k=0}^{M-1} \mathcal{S}_k(\widehat{\alpha}_k/\Delta)$ extracts all the information from $\widehat{\alpha}$. In a traditional POCS method, this requires the number of orthogonal projections to each $\mathcal{S}_k(\widehat{\alpha}_k/\Delta)$ to be unbounded. The optimal $\Theta(M^{-2})$ decay with M is also possible even when each constraint $x \in \mathcal{S}_k(\widehat{\alpha}_k/\Delta)$ is used only once; see the *Further reading*.

Other formulations and algorithms The estimation problem was posed here in non-Bayesian terms. Given a prior on x, an MMSE or MAP estimate can be computed efficiently with message-passing algorithms; see the *Further reading*.

Chapter at a glance

Method	Approximation criterion	Approximating polynomial $p_K(t)$	Error $e_K(t)$
Least-squares	$\min_{p_K} \\|x - p_K\\|_2^2$	$\sum_{k=0}^{K} \langle x, \varphi_k \rangle \varphi_k(t)$	$x(t) - p_K(t)$
Lagrange	$p_K(t_k) = x(t_k)$, $k = 0, 1, \ldots, K$	$\sum_{k=0}^{K} x(t_k) \prod_{i=0, i \neq k}^{K} \frac{t - t_i}{t_k - t_i}$	$\frac{\prod_{k=0}^{K}(t - t_k)}{(K+1)!} x^{(K+1)}(\xi)$
Taylor series	$p_K^{(k)}(t_0) = x^{(k)}(t_0)$, $k = 0, 1, \ldots, K$	$\sum_{k=0}^{K} \frac{(t - t_0)^k}{k!} x^{(k)}(t_0)$	$\frac{(t - t_0)^{K+1}}{(K+1)!} x^{(K+1)}(\xi)$
Minimax	$\min_{p_K} \\|x - p_K\\|_\infty$		
Near-minimax (on $[-1, 1]$)		Lagrange interp. with $\{t_k\}_{k=0}^{K}$ zeros of T_K	$\leq \frac{1}{(K+1)! 2^K} \\|x^{(K+1)}\\|_\infty$

Table 6.5 Approximation of functions by polynomials of degree K. In the least-squares case, an orthogonal basis (for example, Legendre polynomials) is constructed. Matching a function at several points gives a Lagrange interpolation, and matching a function and several derivatives at one point gives a Taylor series. The minimax case has no closed-form solution, but a near-minimax solution can be expressed simply using the zeros of T_K, the Chebyshev polynomial of degree K.

Method	Approximation
B-splines	$\widehat{x}(t) = \sum_{k \in \mathbb{Z}} \langle x(t), \widetilde{\beta}^{(N)}(t-k) \rangle_t \, \beta^{(N)}(t-k)$ $\beta^{(0)}(t) = \begin{cases} 1, & \lvert t \rvert \leq \frac{1}{2}; \\ 0, & \text{otherwise} \end{cases} \quad \overset{\text{FT}}{\longleftrightarrow} \quad B^{(0)}(\omega) = \operatorname{sinc}(\tfrac{1}{2}\omega)$ $\beta^{(N)}(t) = \beta^{(N-1)}(t) * \beta^{(0)}(t) \quad \overset{\text{FT}}{\longleftrightarrow} \quad B^{(N)}(\omega) = \operatorname{sinc}^{N+1}(\tfrac{1}{2}\omega)$
Orthogonalized splines	$\widehat{x}(t) = \sum_{k \in \mathbb{Z}} \langle x(t), \varphi^{(N)}(t-k) \rangle_t \, \varphi^{(N)}(t-k)$ $\varphi^{(N)}(t) = \sum_{k=0}^{\infty} d_k^{(N)} \beta^{(N)}(t-k)$
Partition of unity	$\sum_{n \in \mathbb{Z}} \varphi(t-n) = 1 \quad \overset{\text{FT}}{\longleftrightarrow} \quad \Phi(2\pi k) = \delta_k$
Polynomial reproduction (Strang–Fix)	$p_K(t) = \sum_{k \in \mathbb{Z}} \alpha_k \varphi(t-k)$ $\Phi(0) \neq 0; \; \Phi^{(k)}(2\pi \ell) = 0$ for $k = 1, 2, \ldots, K$, $\ell \in \mathbb{Z} \setminus \{0\}$

Table 6.6 Approximation of functions by splines.

Method	Approximation $\widehat{x}_M$	Coefficients used	Squared $\mathcal{L}^2$ norm of error ε_M^2
Linear	$\sum_{k=0}^{M-1} \langle x, \varphi_k \rangle \varphi_k$	First M	$\sum_{k=M}^{\infty} \lvert\langle x, \varphi_k \rangle\rvert^2$
Nonlinear	$\sum_{k \in \mathcal{I}_M} \langle x, \varphi_k \rangle \varphi_k$	Largest M	$\sum_{k \notin \mathcal{I}_M} \lvert\langle x, \varphi_k \rangle\rvert^2$

Table 6.7 Approximation of functions and sequences by series truncation.

Historical remarks

One of the names appearing prominently in this chapter is that of **Pafnuty Lvovich Chebyshev (1821–1894)**, who is considered to be the founding father of Russian mathematics. His contributions are many, in fields ranging from probability and statistics to number theory. Chebyshev polynomials were described in this chapter for use in minimax approximation; they are also responsible for Chebyshev's name finding its way into signal processing, through the family of Chebyshev filters. As an interesting aside, a crater on the Moon was named after Chebyshev.

The origins of **splines** are particularly interesting. They date back to ship building; naval engineers needed a method to thread a smooth curve through a given set of points. This resulted in thin wooden strips, *splines*, placed between pairs of points, which were called *ducks*, *rats*, or *dogs*. The method was then used both in the aircraft and in the automobile industry in the late 1950s and early 1960s. Engineers at Citroën, Renault, and General Motors developed the theory further; in particular, Pierre Bézier, a French engineer working at Renault, became a leader in using mathematical and computational tools in design and manufacturing. With the advent of computers, splines took over from polynomials as a tool for interpolating functions.

Transform coding was invented to reduce the bandwidth required to transmit analog speech signals. An early speech transmission system used a ten-channel vocoder (*voice coder*) [28]. The channels carried correlated, continuous-time, continuous-amplitude signals representing estimates, local in time, of the power in ten contiguous frequency bands. By adding modulated versions of these power signals, the synthesis unit resynthesized speech. The vocoder's ancestor, *Pedro, the Voder*, was presented at the 1939 World's Fair. Kramer and Mathews [58] showed that the total bandwidth necessary to transmit the signals with a prescribed fidelity could be reduced by transmitting an appropriate set of linear combinations of the signals instead of the signals themselves. This was not source coding yet because it did not involve discretization. Thus, one could ascribe a later birth to transform coding. Huang and Schultheiss [46] introduced the structure shown in Figure 6.42. They studied the coding of Gaussian sources while assuming independent expansion coefficients and optimal fixed-rate scalar quantization. They first showed that $\Phi^{-1} = \widetilde{\Phi}^*$ is optimal and then that $\widetilde{\Phi}^*$ should have orthogonal rows. Transform coding has since spread through its use in the popular media standards, such as MP3, JPEG, and MPEG.

Further reading

Polynomial approximation An excellent introductory textbook on numerical analysis is the one by Atkinson [1]; it includes details on polynomial approximations and their error bounds. Our statements of Theorems 6.7 and 6.8 are patterned on [1]. We also recommend an earlier textbook by Davis [20] for polynomial approximations, including a full proof of the Chebyshev equioscillation theorem, Theorem 6.8.

Filter design Numerous filter design techniques exist; for an overview, see [72]. They all try to approximate the desired specifications of the filter by a realizable discrete-time filter. For IIR filters, one of the standard methods is to design the discrete-time filters from continuous-time ones using the bilinear transformation. For FIR filters, windowing is often used to approximate the desired response by truncating it with a window, a topic we touched upon in Example 3.4. Many software packages implement minimax polynomial approximation with variants of an algorithm introduced by E. Y. Remez in 1934 and brought into common use for FIR filter design by Parks and McClellan [79]; see also [69] for additional history.

Splines The initial result on B-spline bases for uniform spline spaces is due to Schoenberg [87]. The magazine review article by Unser [102] gives a thorough overview of splines and their use in signal processing, including cardinal splines, error analyses, and a number of references. For more details on error analysis, see [104]. The book by Strang and Fix [95] contains further details on polynomial reproduction and the Strang–Fix theorem, Theorem 6.14.

Approximation and estimation Some important connections between approximation and estimation were developed by Donoho [22], and the near-minimax performance of threshold estimators was developed by Donoho and Johnstone [24]. Both are covered in detail in Chapter 10 of Mallat's book [66].

Lossless compression The entropy bound for optimal code length, Theorem 6.19, is one of the foundational results of lossless source coding theory. It is Theorem 5.4.1 of [17], Theorem 5.1 of [65], or a combination of Theorems 3.5, 3.16, and 3.17 of [110]. Huffman codes were introduced in [47]; the book by Cover and Thomas [17] covers them in Section 5.6 and Shannon–Fano–Elias codes in Section 5.9.

Quantization The review by Gray and Neuhoff [39] is the authoritative source on the history and foundational results in quantization theory. It discusses optimality conditions for variable-rate and entropy-constrained quantizers, high-resolution quantization theory, the complexity of unconstrained vector quantization, the roots of its advantage over scalar quantization, and methods to reduce complexity. For design algorithms and details on types of constrained vector quantizers, see the book by Gersho and Gray [34]

The quantizer design algorithm in Section 6.6.2 was introduced by Lloyd as "Method I" in an unpublished Bell Labs technical memorandum in 1957, and was not published until 25 years later [62]. Lloyd's "Method II" from the same paper was independently discovered by Max [68]. Lloyd's Method I is easily extended to vector quantization, including various types with structural constraints. The sufficient condition for global optimality of the locally optimal scalar quantizer is due to Fleischer [29] and Trushkin [101].

Transform coding The review by Goyal [36] provides detailed coverage of transform coding, including a description of settings where a Karhunen–Loève transform is not optimal. The entries in Table 6.4 are derived from the dimensionless second moments of space-filling polytopes given in a paper by Conway and Sloane [15]. The optimal bit allocation (6.137) can be derived using the method of Lagrange multipliers [88]. The techniques used when the available rates are discrete are quite different [90]. Theorem 6.22 is derived from Theorem 6 of [38].

Estimation from quantized samples The study of improved estimation from quantized samples based on boundedness was initiated by Thao and Vetterli [97–99]. The linear programming formulation is from [37,71]. Iterative projection using each set just once was introduced by Rangan and Goyal [84] and subsequently studied by Powell [81]. Figure 6.46 is taken from Kamilov *et al.* [52], which also provides a message-passing algorithm for the estimation problem; this algorithm allows the use of non-regular quantizers.

Exercises with solutions

6.1. *Lagrange interpolation with coincident nodes*

(i) Write the Lagrange interpolation formula (6.6) for two nodes, one at 0 and the other at $\varepsilon > 0$. Using the definition of the derivative, prove that, in the limit $\varepsilon \to 0$, the Lagrange interpolation yields the first-order Taylor series expansion around 0.

(ii) Generalize (i) to arbitrary order.
(*Hint:* Instead of the Lagrange form of the interpolation polynomial, (6.6), use the Newton form,

$$p_K(t) = \sum_{k=0}^{K} \alpha_k \pi_k(t), \tag{E6.1-1a}$$

with the Newton basis functions

$$\pi_0(t) = 0, \tag{E6.1-1b}$$

$$\pi_k(t) = \prod_{\ell=0}^{k-1} (t - t_\ell), \qquad k = 1, 2, \ldots, K.) \tag{E6.1-1c}$$

Solution:

(i) Writing (6.6) for $K = 1$, $t_0 = 0$, and $t_1 = \varepsilon$ gives

$$\begin{aligned} p_1(t) &= x(t_0)\frac{t-t_1}{t_0-t_1} + x(t_1)\frac{t-t_0}{t_1-t_0} = x(0)\frac{t-\varepsilon}{0-\varepsilon} + x(\varepsilon)\frac{t-0}{\varepsilon-0} \\ &= x(0) + \frac{x(\varepsilon)-x(0)}{\varepsilon}t. \end{aligned}$$

Thus,

$$\lim_{\varepsilon\to 0} p_1(t) = \lim_{\varepsilon\to 0}\left(x(0) + \frac{x(\varepsilon)-x(0)}{\varepsilon}t\right) = x(0) + x'(0)t,$$

and so the first-order interpolating polynomial does indeed yield the first-order Taylor expansion in the limit where the points coincide.

(ii) Since the interpolating polynomial of order K through $K+1$ points is unique, the Lagrange form and the Newton form will yield the same polynomial. With the Newton form, (E6.1-1), we lose the convenience that the expansion coefficients are simply the function values at the knots.

To find the coefficients of the interpolating polynomial with respect to the monomial basis, we could solve a system of linear equations with a Vandermonde matrix, (6.5). To find the coefficients α_k in the Newton form, we again solve a system of linear equations (since we have $K+1$ equations at the interpolation nodes),

$$\begin{bmatrix} \pi_0(t_0) & \cdots & \pi_K(t_0) \\ \vdots & \ddots & \vdots \\ \pi_0(t_K) & \cdots & \pi_K(t_K) \end{bmatrix} \begin{bmatrix} \alpha_0 \\ \vdots \\ \alpha_K \end{bmatrix} = \begin{bmatrix} x(t_0) \\ \vdots \\ x(t_K) \end{bmatrix}.$$

Using (E6.1-1b) and (E6.1-1c), we get

$$\begin{bmatrix} 1 & 0 & 0 & \cdots & 0 \\ 1 & (t_1 - t_0) & 0 & \cdots & 0 \\ 1 & (t_2 - t_0) & (t_2 - t_0)(t_2 - t_1) & \cdots & 0 \\ \vdots & \vdots & \vdots & \ddots & \vdots \\ 1 & (t_K - t_0) & (t_K - t_0)(t_K - t_1) & \cdots & \prod_{k=0}^{K-1}(t_K - t_k) \end{bmatrix} \begin{bmatrix} \alpha_0 \\ \alpha_1 \\ \alpha_2 \\ \vdots \\ \alpha_K \end{bmatrix} = \begin{bmatrix} x(t_0) \\ x(t_1) \\ x(t_2) \\ \vdots \\ x(t_K) \end{bmatrix}.$$

The solution to this system of equations can be easily found in a recursive form, starting from the top, since the matrix is lower-triangular. The coefficients α_k are the *divided differences*, $\alpha_k = x[t_0, t_1, \ldots, t_k]$, $k = 0, 1, \ldots, K$, which are defined by the following recursion:

$$x[\tau_0] \equiv x(\tau_0),$$

$$x[\tau_0, \ldots, \tau_k] \equiv \frac{x[\tau_1, \ldots, \tau_k] - x[\tau_0 - \tau_{k-1}]}{t_k - t_0},$$

for any points $\tau_0, \tau_1, \ldots, \tau_k$.

From the definition of divided difference, it follows that

$$x[t, t_0, \ldots, t_{k-1}] = x[t_0, \ldots, t_k] + (t - t_k)\, x[t, t_0, \ldots, t_k], \tag{E6.1-2}$$

and

$$x[t] = x[t_0] + (t - t_0)x[t, t_0]. \tag{E6.1-3}$$

We can replace $x[t, t_0]$ in (E6.1-3) by (E6.1-2) with $k = 1$,

$$x[t] = x[t_0] + (t - t_0)x[t_0, t_1] + (t - t_0)(t - t_1)x[t, t_0, t_1].$$

The first two terms are actually the interpolating polynomial through points t_0 and t_1, so we write

$$x[t] = p_1(t) + (t - t_0)(t - t_1)x[t, t_0, t_1].$$

We can continue this recursively, to yield

$$x[t] = p_K(t) + x[t, t_0, \ldots, t_K] \prod_{k=0}^{K} (t - t_k). \tag{E6.1-4}$$

Recall now the expression for the error of the polynomial approximation (6.8a),

$$x(t) - p_K(t) = \frac{\prod_{k=0}^{K}(t - t_k)}{(K+1)!} x^{(K+1)}(\xi). \tag{E6.1-5}$$

Comparing (E6.1-4) with (E6.1-5) we conclude that

$$x[t, t_0, \ldots, t_K] = \frac{x^{(K+1)}(\xi)}{(K+1)!},$$

for some $\xi \in [\min\{t, t_0, \ldots, t_K\}, \max\{t, t_0, \ldots, t_K\}]$. Since this holds for any choice of $t, t_0, \ldots, t_K$ and sufficiently smooth x, we can also write $x[t_0, \ldots, t_k] = x^{(k)}(\xi_k)/k!$ for some ξ_k between the smallest and largest values in $\{t_0, t_1, \ldots, t_k\}$.

Going back to the Newton form, and upon substituting for α_k and π_k, we get

$$p_K(t) = \sum_{k=0}^{K} x[t_0, \ldots, t_k] \prod_{\ell=0}^{k-1} (t - t_\ell). \tag{E6.1-6}$$

For the desired result, let $t_0 = 0$ and $t_k \to 0$ for $k = 1, 2, \ldots, K$. Then, $\xi_k \to 0$ for $k = 1, 2, \ldots, K$. Thus, (E6.1-6) becomes the Kth-order Taylor expansion around 0, concluding the proof.

6.2. *Chebyshev polynomials*
Prove Theorem 6.9.
Solution: The recursion in (6.18) aids in proving that T_k is a polynomial of degree k, so we prove part (iii) before part (i).

(iii) Recall the trigonometric identity

$$\cos((k \pm 1)\theta) = \cos(k\theta)\cos\theta \mp \sin(k\theta)\sin\theta. \qquad \text{(E6.2-1)}$$

For any $k \in \mathbb{Z}^+$, with $\theta = \arccos t$,

$$\begin{aligned} T_{k+1}(t) &\overset{(a)}{=} \cos((k+1)\theta) \overset{(b)}{=} \cos(k\theta)\cos\theta - \sin(k\theta)\sin\theta \\ &= \cos(k\theta)\cos\theta - \sin(k\theta)\sin\theta + \cos(k\theta)\cos\theta - \cos(k\theta)\cos\theta \\ &= 2\cos(k\theta)\cos\theta - (\cos(k\theta)\cos\theta + \sin(k\theta)\sin\theta) \\ &\overset{(c)}{=} 2\cos(k\theta)\cos\theta - \cos((k-1)\theta) \overset{(d)}{=} 2tT_k(t) - T_{k-1}(t), \end{aligned}$$

where (a) and (d) follow from (6.16); and (b) and (c) from (E6.2-1).

(i) Having established the recursion (6.18), we can prove that T_k is a polynomial of degree k by induction on k. The base case is that $T_0(t) = 1$ is a polynomial of degree 0 and $T_1(t) = t$ is a polynomial of degree 1. Now assume that T_m is a polynomial for all $m \in \{0, 1, \ldots, k\}$. Then, since

$$T_{k+1}(t) = 2tT_k(t) - T_{k-1}(t),$$

it is clear that T_{k+1} is a polynomial of degree $k+1$: multiplying $T_k(t)$ by $2t$ gives a polynomial of degree $k+1$ since T_k is a polynomial of degree k, and subtracting $T_{k-1}(t)$ cannot change the degree because T_{k-1} is a polynomial of degree $k-1$. The result thus follows by induction.

(ii) Let $k, m \in \mathbb{N}$ be distinct. Then

$$\begin{aligned} &\int_{-1}^{1} T_k(t)T_m(t)\,(1-t^2)^{-1/2}\,dt \\ &\overset{(a)}{=} \int_{-1}^{1} \cos(k \arccos t)\cos(m \arccos t)\,(1-t^2)^{-1/2}\,dt \overset{(b)}{=} \int_0^{\pi} \cos(k\theta)\cos(m\theta)\,d\theta, \end{aligned}$$

where (a) follows from (6.16); and (b) from the change of variable $t = \cos\theta$. (Under this change of variable, $t \in [-1, 1]$ implies that $\theta \in [0, \pi]$, so $\sin\theta$ is positive and

$$(1-t^2)^{-1/2} = (\sin^2\theta)^{-1/2} = 1/(\sin\theta).$$

Also, the $1/(\sin\theta)$ factor is canceled by $dt = -\sin\theta\,d\theta$.) Now

$$\begin{aligned} \int_0^{\pi} \cos(k\theta)\cos(m\theta)\,d\theta &\overset{(a)}{=} \int_0^{\pi} (\cos((k+m)\theta) + \cos((k-m)\theta))\,d\theta \\ &\overset{(b)}{=} \left(\frac{\sin((k+m)\theta)}{k+m} + \frac{\sin((k-m)\theta)}{k-m} \right)\Bigg|_0^{\pi} = 0, \end{aligned}$$

where (a) follows from (E6.2-1); and (b) uses $k+m \neq 0$ and $k-m \neq 0$.

(iv) From the expression $T_k(t) = \cos(k \arccos t)$, we get that the roots of T_k are where

$$k \arccos t = (2m+1)\frac{\pi}{2}, \qquad m \in \{0, 1, \ldots, k-1\}; \qquad \text{(E6.2-2)}$$

the values for m come from the range of the arccos function being $[0, \pi]$. Rearranging (E6.2-2) gives the desired expression (6.19a).

(v) Similarly to part (iv), the extrema of T_k are where

$$k \arccos t = m\pi, \qquad m \in \{0, 1, \ldots, k\}. \qquad \text{(E6.2-3)}$$

Rearranging (E6.2-3) gives the desired expression (6.19b).

(vi) We can prove that the coefficient of t^k in $T_k(t)$ is 2^{k-1} by induction over k. This statement clearly holds for $k = 0$ and $k = 1$. Now assume that the coefficient of t^m in T_m is 2^{m-1} for all $m \in \{0, 1, \ldots, k\}$. Then, $T_k(t) = 2^{k-1}t^k + r_{k-1}(t)$, where the remainder r_{k-1} is a polynomial of degree at most $k-1$. Since

$$\begin{aligned} T_{k+1}(t) &= 2tT_k(t) - T_{k-1}(t) \\ &= 2t\left(2^{k-1}t^k + r_{k-1}(t)\right) - T_{k-1}(t) \\ &= 2^k t^{k+1} + 2tr_{k-1}(t) - T_{k-1}(t) \end{aligned}$$

with r_{k-1} and T_{k-1} polynomials of degree at most $k-1$, the coefficient of t^{k+1} in $T_{k+1}(t)$ is 2^k. The result thus follows by induction.

6.3. *Operator norm of LSI system*
Let $A : \ell^2(\mathbb{Z}) \to \ell^2(\mathbb{Z})$ be an LSI system with impulse response $h \in \ell^1(\mathbb{Z})$ and frequency response H. Prove that the operator norm of A is given by

$$\|A\| = \max_{\omega\in[-\pi,\pi]} |H(e^{j\omega})|.$$

Solution: Since $h \in \ell^1(\mathbb{Z})$, its DTFT H is continuous (see Section 3.4.2). Thus,

$$\lambda_{\max} = \max_{\omega\in[-\pi,\pi]} |H(e^{j\omega})| \quad \text{and} \quad \omega_{\max} = \arg\max_{\omega\in[-\pi,\pi]} |H(e^{j\omega})|$$

are well defined (though $\omega_{\max}$ need not be unique). (The choice of the name $\lambda_{\max}$ emphasizes that, as developed in Section 3.4.1, each value of the frequency response $H(e^{j\omega})$ as ω takes values in $[-\pi, \pi]$ is an eigenvalue of the LSI system A.) Intuitively, the amplification of a complex exponential input sequence $e^{\omega_{\max} n}$ by the system determines the operator norm of the system. The technical complication is that a complex exponential sequence is not in $\ell^2(\mathbb{Z})$ and hence not allowed when computing the operator norm.

We first show that $\|A\| \le \lambda_{\max}$:

$$\begin{aligned}
\|A\| &= \sup_{\|x\|=1} \|Ax\| = \sup_{\|x\|=1} \|h * x\| \overset{(a)}{=} \sup_{\|X\|=\sqrt{2\pi}} \frac{1}{\sqrt{2\pi}} \|H\,X\| \\
&\overset{(b)}{=} \sup_{\|X\|=\sqrt{2\pi}} \frac{1}{\sqrt{2\pi}} \left(\int_{-\pi}^{\pi} |H(e^{j\omega})X(e^{j\omega})|^2 \, d\omega \right)^{1/2} \\
&\overset{(c)}{\le} \sup_{\|X\|=\sqrt{2\pi}} \frac{1}{\sqrt{2\pi}} \left(\lambda_{\max}^2 \int_{-\pi}^{\pi} |X(e^{j\omega})|^2 \, d\omega \right)^{1/2} \\
&\overset{(d)}{=} \sup_{\|X\|=\sqrt{2\pi}} \frac{1}{\sqrt{2\pi}} \lambda_{\max} \|X\| = \lambda_{\max},
\end{aligned}$$

where (a) follows from the Parseval equality for the DTFT, (3.107), and the convolution in time property of the DTFT, (3.96); (b) and (d) from the definition of the $\mathcal{L}^2([-\pi, \pi))$ norm; and (c) from replacing $|H(e^{j\omega})|$ by its upper bound $\lambda_{\max}$.

Now, for any $\varepsilon > 0$, it is also true that $\|A\| > \lambda_{\max} - \varepsilon$, completing the proof. To show this, we would like to construct a unit-norm input sequence x such that $\|Ax\| > \lambda_{\max} - \varepsilon$. Since H is continuous, there is an interval I of length $\varepsilon' > 0$ (centered at $\omega_{\max}$) such that $|\int_I H(e^{j\omega})\, d\omega| > (\lambda_{\max} - \varepsilon)\varepsilon'$. We now choose x by specifying its DTFT X. Since x has unit norm, by the Parseval equality, X has norm $\sqrt{2\pi}$. Thus,

$$X(e^{j\omega}) = \begin{cases} 2\pi/\varepsilon', & \text{for } \omega \in I; \\ 0, & \text{for } \omega \in [-\pi, \pi) \setminus I \end{cases}$$

specifies an input sequence that proves $\|A\| > \lambda_{\max} - \varepsilon$.

The argument for $\|A\| \le \lambda_{\max}$ did not depend on continuity of H, though the maximum would have to be replaced by the supremum in the definition of $\lambda_{\max}$. The argument for $\|A\| > \lambda_{\max} - \varepsilon$ can be extended to any H for which $|H(e^{j\omega})| > \lambda_{\max} - \varepsilon$ on some open interval. For example, it suffices for $H(e^{j\omega})$ to be left continuous or right continuous at a maximum of $|H(e^{j\omega})|$. This allows the result to be extended to some $h \notin \ell^1(\mathbb{Z})$.

6.4. *Strang–Fix theorem for an interpolating $\varphi(t)$*
Assume an interpolating function $\varphi(t)$ as in (6.61) that is sufficiently localized as in (6.59). Prove that (i) and (ii) in Theorem 6.14 are equivalent.

Solution: Condition (i) means that there exist coefficient sequences $\alpha_n^{(k)}$ such that

$$\sum_{n\in\mathbb{Z}} \alpha_n^{(k)} \varphi(t-n) = (t-t_0)^k, \qquad k = 0, 1, \ldots, K. \tag{E6.4-1}$$

Because of the interpolation property ($\varphi(m) = \delta_m$ for $m \in \mathbb{Z}$), when $t = m$,

$$\sum_{n\in\mathbb{Z}} \alpha_n^{(k)} \varphi(m-n) = \alpha_m^{(k)} = (m-t_0)^k.$$

Thus, (E6.4-1) becomes

$$\sum_{n\in\mathbb{Z}} (n-t_0)^k \varphi(t-n) = (t-t_0)^k, \qquad k = 0, 1, \ldots, K,$$

which, upon setting $t_0 = t$, yields

$$\sum_{n\in\mathbb{Z}} (n-t)^k \varphi(t-n) = \begin{cases} 1, & k = 0; \\ 0, & k = 1, 2, \ldots, K. \end{cases} \tag{E6.4-2}$$

The left-hand side is a 1-periodic function, and, because of the localization (6.59), converges to an $\mathcal{L}^1$ function on the interval $[0, 1]$. This function has a Fourier series representation for each $k = 0, 1, \ldots, K$. The coefficients are

$$\begin{aligned} c_\ell^{(k)} &= \int_0^1 \sum_{n\in\mathbb{Z}} (n-t)^k \varphi(t-n) e^{-j2\pi\ell t}\, dt \overset{(a)}{=} \sum_{n\in\mathbb{Z}} \int_0^1 (n-t)^k \varphi(t-n) e^{-j2\pi\ell t}\, dt \\ &\overset{(b)}{=} \int_{-\infty}^{\infty} (-t)^k \varphi(t) e^{-j2\pi\ell t}\, dt \overset{(c)}{=} \frac{1}{(2\pi j)^k} \Phi^{(k)}(2\pi\ell), \end{aligned} \tag{E6.4-3}$$

where (a) follows from (2.209); (b) from merging individual integrals into a single integral over the real line; and (c) from the Fourier transform, (4.40a), as well as the differentiation property in frequency in Table 4.1. For $k = 0$, the function in (E6.4-2) is equal to 1, so the Fourier series coefficients are $c_\ell^{(0)} = \delta_\ell$ and thus

$$\Phi(2\pi\ell) = \delta_\ell, \qquad \ell \in \mathbb{Z}.$$

For $k = 1, 2, \ldots, K$, the function in (E6.4-2) is zero, or

$$c_\ell^{(k)} = 0, \qquad k = 1, 2, \ldots, K,$$

leading to

$$\Phi^{(k)}(2\pi\ell) = 0, \qquad k = 1, 2, \ldots, K, \quad \ell \in \mathbb{Z},$$

hence verifying (6.60).

6.5. *Karhunen–Loève transform*

Let x be an N-dimensional random vector with mean zero and covariance matrix Σ_{x}. Let $\{\varphi_k\}_{k=0}^{N-1}$ be an orthonormal basis for $\mathbb{C}^N$ with analysis operator Φ^* and synthesis operator Φ, and let $\mathrm{y} = \Phi^*\mathrm{x}$ be the vector of transform coefficients of x obtained with this basis.

(i) Show that the covariance of y is $\Phi^*\Sigma_{\mathrm{x}}\Phi$.

(ii) Show that if $\{\varphi_k\}_{k=0}^{N-1}$ is a KL basis, then the random variables $\{\mathrm{y}_k\}_{k=0}^{N-1}$ are uncorrelated and the ordering of variances in (6.96) holds.

(iii) For any Hermitian matrix $A \in \mathbb{C}^{N\times N}$ and any $M \in \{1, 2, \ldots, N-1\}$, the eigenvalues $\lambda_0 \ge \lambda_1 \ge \cdots \ge \lambda_{N-1}$ of A satisfy

$$\max_{\{u_k\}_{k=0}^{M-1}} \sum_{k=0}^{M-1} \langle Au_k, u_k\rangle = \sum_{k=0}^{M-1} \lambda_k, \tag{E6.5-1a}$$

$$\min_{\{v_k\}_{k=M}^{N-1}} \sum_{k=M}^{N-1} \langle Av_k, v_k\rangle = \sum_{k=M}^{N-1} \lambda_k, \tag{E6.5-1b}$$

where the extrema are over orthonormal sets of vectors in $\mathbb{C}^N$. Equivalently,

$$\max_U \operatorname{tr}(UAU^*) = \sum_{k=0}^{M-1} \lambda_k \quad \text{and} \quad \min_V \operatorname{tr}(VAV^*) = \sum_{k=M}^{N-1} \lambda_k, \tag{E6.5-2}$$

where the extrema are over $U \in \mathbb{C}^{M\times N}$ satisfying $UU^* = I_M$ and $V \in \mathbb{C}^{(N-M)\times N}$ satisfying $VV^* = I_{N-M}$. Use any of these variational characterizations to prove that a KL basis is optimal for M-term linear approximation.

Solution:

(i) The mean of y is

$$\mu_{\mathrm{y}} = \mathrm{E}[\Phi^*\mathrm{x}] \overset{(a)}{=} \Phi^*\mathrm{E}[\mathrm{x}] \overset{(b)}{=} \Phi^*\mathbf{0} = \mathbf{0},$$

where (a) follows from the linearity of the expectation; and (b) from x having zero mean. The covariance matrix of $\mathrm{y} = \Phi^*\mathrm{x}$ is

$$\Sigma_{\mathrm{y}} \overset{(a)}{=} \mathrm{E}[(\mathrm{y}-\mu_{\mathrm{y}})(\mathrm{y}-\mu_{\mathrm{y}})^*] \overset{(b)}{=} \mathrm{E}[\mathrm{y}\mathrm{y}^*] \overset{(c)}{=} \mathrm{E}[\Phi^*\mathrm{x}(\Phi^*\mathrm{x})^*]$$
$$= \mathrm{E}[\Phi^*\mathrm{x}\mathrm{x}^*\Phi] \overset{(d)}{=} \Phi^*\mathrm{E}[\mathrm{x}\mathrm{x}^*]\Phi \overset{(e)}{=} \Phi^*\Sigma_{\mathrm{x}}\Phi,$$

where (a) follows from the definition of the covariance matrix; (b) from $\mu_{\mathrm{y}} = \mathbf{0}$; (c) from the definition of y; (d) from the linearity of the expectation; and (e) from the definition of covariance matrix.

(ii) If $\{\varphi_k\}_{k=0}^{N-1}$ is a KL basis, then, by Definition 6.15, $\Sigma_{\mathrm{y}} = \Phi^*\Sigma_{\mathrm{x}}\Phi = \Lambda$ is a diagonal matrix with nonincreasing entries on the diagonal. The covariance between y_k and y_m, where $k \neq m$, is the (k, m) entry of Σ_{y}; thus, since Σ_{y} is diagonal, the random variables $\{\mathrm{y}_k\}_{k=0}^{N-1}$ are uncorrelated. The variance of y_k is the (k, k) entry of Σ_{y}; thus, the nonincreasing order of the entries of the diagonal of Σ_{y} is precisely the ordering of variances (6.96).

(iii) Here, we need to show the correct associations between quantities in (E6.5-1) or (E6.5-2) and quantities relevant to the analysis of M-term linear approximation.

Suppose that one uses the orthonormal basis $\{\varphi_k\}_{k=0}^{N-1}$ for M-term linear approximation of x as in (6.92). The MSE is

$$\begin{aligned}
\varepsilon_M &= \mathrm{E}[\|\mathrm{x}-\widehat{\mathrm{x}}_M\|^2] = \mathrm{E}[\langle \mathrm{x}-\widehat{\mathrm{x}}_M, \mathrm{x}-\widehat{\mathrm{x}}_M\rangle] \\
&\overset{(a)}{=} \mathrm{E}\left[\left\langle \sum_{k=M}^{N-1} \langle \mathrm{x}, \varphi_k\rangle\varphi_k, \sum_{m=M}^{N-1} \langle \mathrm{x}, \varphi_m\rangle\varphi_m \right\rangle\right] \\
&\overset{(b)}{=} \mathrm{E}\left[\sum_{k=M}^{N-1}\sum_{m=M}^{N-1} \langle \langle \mathrm{x}, \varphi_k\rangle\varphi_k, \langle \mathrm{x}, \varphi_m\rangle\varphi_m\rangle\right] \\
&\overset{(c)}{=} \mathrm{E}\left[\sum_{k=M}^{N-1}\sum_{m=M}^{N-1} \langle \mathrm{x}, \varphi_k\rangle\langle \mathrm{x}, \varphi_m\rangle^*\langle \varphi_k, \varphi_m\rangle\right] \\
&\overset{(d)}{=} \mathrm{E}\left[\sum_{k=M}^{N-1}\sum_{m=M}^{N-1} \langle \mathrm{x}, \varphi_k\rangle\langle \mathrm{x}, \varphi_m\rangle^*\delta_{k-m}\right] \overset{(e)}{=} \mathrm{E}\left[\sum_{k=M}^{N-1} \langle \mathrm{x}, \varphi_k\rangle\langle \mathrm{x}, \varphi_k\rangle^*\right] \\
&\overset{(f)}{=} \mathrm{E}\left[\sum_{k=M}^{N-1} \langle \mathrm{x}, \varphi_k\rangle\langle \varphi_k, \mathrm{x}\rangle\right] \overset{(g)}{=} \mathrm{E}\left[\sum_{k=M}^{N-1} \varphi_k^*\mathrm{x}\mathrm{x}^*\varphi_k\right] \\
&\overset{(h)}{=} \sum_{k=M}^{N-1} \varphi_k^*\mathrm{E}[\mathrm{x}\mathrm{x}^*]\varphi_k \overset{(i)}{=} \sum_{k=M}^{N-1} \varphi_k^*\Sigma_{\mathrm{x}}\varphi_k \overset{(j)}{=} \sum_{k=M}^{N-1} \langle \Sigma_{\mathrm{x}}\varphi_k, \varphi_k\rangle,
\end{aligned}$$

where (a) follows from (6.92); (b) from the distributivity of the inner product; (c) from the linearity in the first argument and conjugate linearity in the second argument of the inner product; (d) from the orthonormality of the basis; (e) from the sum over m being nonzero only for $m = k$; (f) from the Hermitian symmetry of the inner product; (g) from writing the $\mathbb{C}^N$ inner product using a vector–vector product as in (2.22a); (h) from the linearity of the expectation; (i) from the definition of the covariance matrix, (6.90); and (j) from recognizing the summand as a $\mathbb{C}^N$ inner product.

Now consider optimization of the orthonormal basis. Since the covariance matrix Σ_{x} is Hermitian, by using (E6.5-1b), the minimum possible value for ε_M is $\sum_{k=M}^{N-1} \lambda_k$, where $\lambda_0 \geq \lambda_1 \geq \cdots \geq \lambda_{N-1}$ are the eigenvalues of Σ_{x}. A KL basis clearly achieves this minimum, proving the desired result.

Use of the trace operator enables an equivalent (but more compact) sequence of steps:

$$\begin{aligned}\varepsilon_M &\overset{(a)}{=} \mathrm{E}\left[\sum_{k=M}^{N-1} |\mathrm{y}_k|^2\right] \overset{(b)}{=} \mathrm{E}[\widetilde{\mathrm{y}}^*\widetilde{\mathrm{y}}] \overset{(c)}{=} \mathrm{E}\left[(\widetilde{\Phi}^*\mathrm{x})^*\widetilde{\Phi}^*\mathrm{x}\right] = \mathrm{E}\left[\mathrm{x}^*\widetilde{\Phi}\widetilde{\Phi}^*\mathrm{x}\right] \\ &\overset{(d)}{=} \mathrm{E}\left[\mathrm{tr}(\mathrm{x}^*\widetilde{\Phi}\widetilde{\Phi}^*\mathrm{x})\right] \overset{(e)}{=} \mathrm{E}\left[\mathrm{tr}(\widetilde{\Phi}^*\mathrm{x}\mathrm{x}^*\widetilde{\Phi})\right] \overset{(f)}{=} \mathrm{tr}(\widetilde{\Phi}^*\mathrm{E}[\,\mathrm{x}\mathrm{x}^*\,]\,\widetilde{\Phi}) \\ &\overset{(g)}{=} \mathrm{tr}\left(\widetilde{\Phi}^*\Sigma_{\mathrm{x}}\widetilde{\Phi}\right),\end{aligned}$$

where (a) follows from (6.93); (b) from introducing $\widetilde{\mathrm{y}} = \begin{bmatrix}\mathrm{y}_M & \mathrm{y}_{M+1} & \cdots & \mathrm{y}_{N-1}\end{bmatrix}^\top$; (c) from introducing $\widetilde{\Phi} = \begin{bmatrix}\varphi_M & \varphi_{M+1} & \cdots & \varphi_{N-1}\end{bmatrix}$; (d) from the trace of a scalar being the scalar value itself; (e) from $\mathrm{tr}(AB) = \mathrm{tr}(BA)$; (f) from the linearity of the expectation and trace; and (g) from the definition of the covariance matrix.

Now note that $\widetilde{\Phi}^*$ plays the role of V in (E6.5-2): it is an $(N-M)\times N$ matrix, and it satisfies $\widetilde{\Phi}^*\widetilde{\Phi} = I_{N-M}$ because $\{\varphi_k\}_{k=0}^{N-1}$ is an orthonormal basis. Thus, the minimum possible value for ε_M is $\sum_{k=M}^{N-1}\lambda_k$, where $\lambda_0 \geq \lambda_1 \geq \cdots \geq \lambda_{N-1}$ are the eigenvalues of Σ_{x}. A KL basis clearly achieves this minimum.

Exercises

6.1. *Basic properties of Legendre polynomials.*(2)

(i) The Legendre polynomials $\{L_k\}_{k\in\mathbb{N}}$ from (6.3) satisfy the differential equation

$$\frac{d}{dt}\left[(t^2-1)L_k'(t)\right] + k(k+1)L_k(t) \;=\; 0. \tag{P6.1-1}$$

Use this to show that L_k and L_m are orthogonal on $[-1,\,1]$ for any distinct $k, m \in \mathbb{N}$.

(ii) Show that $L_k\left((2t-(b+a))/(b-a)\right)$, $k\in\mathbb{N}$, are orthogonal on $[a,\,b]$.

(iii) Let $\{v_0,\, v_1,\, \ldots\}$ be a set of real-valued polynomials that are orthogonal on $[a,\,b]$, where v_k has degree at most k for each $k\in\mathbb{N}$. Express v_k in terms of Legendre polynomials.

(iv) Prove the following recurrence relation for Legendre polynomials:

$$L_{k+1}(t) \;=\; \frac{2k+1}{k+1}tL_k(t) - \frac{k}{k+1}L_{k-1}(t), \qquad k\in\mathbb{Z}^+.$$

(*Hint:* Find the $(k+1)$st derivative of $(t^2-1)^{k+1}$ in terms of the $(k-1)$st derivatives of $(t^2-1)^{k-1}$ and $(t^2-1)^k$, and recognize this as relating L_{k+1}, L_{k-1}, and the $(k-1)$st derivative of $(t^2-1)^k$. Also, express L_{k+1} using the definition (6.3) and the generalized Leibniz rule,

$$(fg)^{(n)} \;=\; \sum_{k=0}^{n}\binom{n}{k} f^{(n)}g^{(n-k)},$$

on $(d^k/dt^k)\left[t(t^2-1)^k\right]$.)

6.2. *Orthogonal polynomials and nesting of polynomial subspaces.*(1)
Let $\{v_0,\, v_1,\, \ldots\}$ be a set of orthogonal polynomials, where v_k has degree at most k for each $k\in\mathbb{N}$, and let p be a polynomial of degree m. Prove that $\langle p,\, v_k\rangle = 0$ for every $k > m$.

6.3. *Roots of orthogonal polynomials.*(3)
Let $\{v_0,\, v_1,\, \ldots\}$ be a set of polynomials orthogonal on $[a,\,b]$, where v_k has degree at most k for each $k\in\mathbb{N}$. Prove that v_k has exactly k real, distinct roots in the open interval $(a,\,b)$, for each $k\in\mathbb{N}$.

6.4. *Derivation of Lagrange interpolating polynomial.*(2)
Let $K \in \mathbb{N}$ and let values of a function x be given at $K+1$ distinct nodes $\{t_k\}_{k=0}^{K}$.

(i) Let p_{K-1} be the interpolating polynomial of degree $K-1$ uniquely specified by $\{(t_k,\, x(t_k))\}_{k=0}^{K-1}$. Find constant c such that

$$p_K(t) \;=\; p_{K-1}(t) + c(t-t_0)(t-t_1)\cdots(t-t_{K-1}) \qquad \text{(P6.4-1)}$$

is the interpolating polynomial specified by $\{(t_k,\, x(t_k))\}_{k=0}^{K}$.

(ii) Using the result of the previous part, justify (6.6).

6.5. *Poor $\mathcal{L}^\infty$ behavior of Lagrange interpolation and Taylor series.*(2)
Let $x(t) = 1/(1+t^2)$. This function is infinitely differentiable but potentially difficult to approximate with a polynomial [1].

(i) Let p_K denote the Lagrange interpolating polynomial for x with $K+1$ evenly spaced nodes over $[-5,\, 5]$,

$$t_k \;=\; -5 + \frac{10}{K}, \qquad k = 0,\, 1,\, \ldots,\, K,$$

and let e_K denote its error. Plot p_K and e_K for a few values of K. Observe empirically that $\|e_K\|_\infty$ grows without bound as K increases, and comment on how the error behavior varies over $[-5,\, 5]$.

(ii) Let p_K denote the Taylor series expansion of x around 0 of degree K, and let e_K denote its error. Plot p_K and e_K for a few values of K. Observe empirically that $\|e_K\|_\infty$ grows without bound as K increases, and comment on how the error behavior varies over $[-5,\, 5]$.

(iii) Show that

$$\frac{1}{1+t^2} \;=\; \sum_{k=0}^{\infty}(-1)^k t^{2k}, \qquad \text{for } |t| < 1.$$

(iv) Show that

$$\left.\frac{d^n}{dt^n}x(t)\right|_{t=0} \;=\; \begin{cases} (-1)^{\lceil n/2 \rceil} n!, & \text{for even } n \in \mathbb{Z}^+; \\ 0, & \text{for odd } n \in \mathbb{Z}^+. \end{cases}$$

(v) Show that, for the Taylor approximations in part (ii),

$$\lim_{K\to\infty} p_K(t) \;=\; x(t), \qquad \text{for } |t| < 1.$$

6.6. *Hermite interpolation.*(2)
For each distinct node $\{t_k\}_{k=0}^{L}$, suppose that the value of a function and its derivatives up to order d_k are given. Prove that these values uniquely determine an approximating polynomial of degree $L + \sum_{k=0}^{L} d_k$.

6.7. *Bernstein polynomials.*(2)
For any positive integer K, the $K+1$ *Bernstein polynomials* of degree K are

$$B_{k,K}(t) \;=\; \binom{K}{k} t^k (1-t)^{K-k}, \qquad k = 0,\, 1,\, \ldots,\, K. \qquad \text{(P6.7-1)}$$

The Bernstein polynomials are nonnegative on $[0,\, 1]$ and satisfy the symmetry property

$$B_{k,K}(t) \;=\; B_{K-k,K}(1-t).$$

(i) Show that the Bernstein polynomials of degree K are a basis for the vector space of polynomials of degree K.

(ii) Show that $B_{k,K}$ has a single maximum over $[0,\, 1]$ that is achieved at k/K.

6.8. *Approximation by Bernstein polynomials.*(2)
Let x be defined on $[0,\, 1]$. The Bernstein polynomial approximation p_K of degree K of x is given by

$$p_K(t) \;=\; \sum_{k=0}^{K} x\left(\frac{k}{K}\right) B_{k,K}(t), \qquad \text{(P6.8-1)}$$

where the Bernstein polynomials are defined in Exercise 6.7. If x is continuous, then p_K converges uniformly to x, so the Bernstein polynomial approximations provide a constructive proof of the Weierstrass approximation theorem (Theorem 6.6).

(i) Show that the Bernstein polynomial approximation p_K for x matches x at 0 and 1 but not necessarily at the other points k/K, $k = 1, 2, \ldots, K-1$, used in forming the approximation.

(ii) For $x(t) = t \sin 5t$ and $K = 4, 8, 16, 32$, plot approximations by Bernstein polynomials. For $K = 4$, plot and compare approximations by Bernstein polynomials and by Lagrange interpolation using the same samples of x.

(iii) For $x(t) = t^2$, show that

$$\lim_{K\to\infty} K e_K(t) = t(1-t),$$

where e_K is the error of p_K. This shows that

$$e_K(t) \asymp \frac{1}{K} t(1-t), \qquad \text{so} \qquad \|e_K(t)\|_\infty \asymp \frac{1}{4K},$$

which is a relatively slow decay even for a very simple function.

6.9. *Near-minimax approximation.*[(3)]

(i) Prove that

$$\max_{t\in[-1,1]} \prod_{k=0}^{K} |t - t_k|$$

is minimized by choosing $\{t_k\}_{k=0}^{K}$ to be the $K+1$ zeros of the Chebyshev polynomial T_{K+1}.

(ii) Show that the interpolation error bound (6.8b) becomes

$$|e_K(t)| \le \frac{1}{(K+1)!2^K} \max_{\xi\in[-1,1]} |x^{(K+1)}(\xi)|$$

for approximation of x on $[-1, 1]$ with a polynomial of degree at most K.

6.10. *Truncation as orthogonal projection.*[(1)]

(i) Using the orthogonality property, Theorem 2.26(ii), show that the truncation operation (6.24) yields the least-squares approximation of the desired filter h^{d}.

(ii) Show that the truncation operation (6.24) is an orthogonal projection operator. Use this to argue that it yields the least-squares approximation of the desired filter h^{d}.

6.11. *Properties of spline spaces.*[(1)]

(i) Let τ be a subsequence of the strictly increasing sequence η. Show $S_{K,\tau} \subset S_{K,\eta}$ for any $K \in \mathbb{N}$.

(ii) Let $x \in S_{K_x,\tau_x}$ and $y \in S_{K_y,\tau_y}$, where K_x, $K_y \in \mathbb{N}$, and τ_x and τ_y are strictly increasing sequences. Does $x + y$ belong to a spline space? Either demonstrate that it does not or describe the spline space that contains $x + y$.

6.12. *Explicit form for elementary B-splines.*[(1)]
For any $K \in \mathbb{N}$, let p_+^K denote the *truncated power function*

$$p_+^K(t) = t^K u(t) = \begin{cases} t^K, & \text{for } t \ge 0; \\ 0, & \text{for } t < 0. \end{cases} \qquad \text{(P6.12-1)}$$

For any $K \in \mathbb{N}$, the elementary B-spline of degree K is given explicitly by

$$\beta^{(K)}(t) = \frac{1}{K!} \sum_{k=0}^{K+1} \binom{K+1}{k} (-1)^k p_+^K\left(t - k + \frac{K+1}{2}\right). \qquad \text{(P6.12-2)}$$

(i) Verify that (P6.12-2) matches the expression (6.34a) for $K = 0$.

(ii) Verify that (P6.12-2) matches the expression (6.35) for $K = 1$.

(iii) Verify that (P6.12-2) matches the expression (6.36) for $K = 2$.

(iv) Show that the Fourier transform P_+^K of p_+^K satisfies

$$(j\omega)^{K+1} P_+^K(\omega) \;=\; K!. \tag{P6.12-3}$$

(*Hint:* Consider the derivative of order $K+1$ of p_+^K.)

(v) Using the result of the previous part, show that (P6.12-2) holds.
(*Hint:* Start from the Fourier domain expression (6.37).)

6.13. *Elementary B-splines are splines.*$^{(2)}$
Show that $\beta^{(K)}$, the elementary B-spline of degree K, is a spline of degree K with knots $(-\frac{1}{2}(K+1), \ldots, \frac{1}{2}(K+1))$.

6.14. *Canonical dual of causal elementary B-spline of degree 2.*$^{(2)}$
Let $\beta_+^{(2)}$ denote the causal elementary B-spline of degree 2, as defined in (6.39) or by using (6.38) to shift $\beta^{(2)}$ given in (6.36).

(i) Let $h^{(2)}$ be the deterministic autocorrelation of $\beta_+^{(2)}$ evaluated at the integers, as in (6.43). Find $h^{(2)}$ explicitly.

(ii) Let $H^{(2)}(z)$ be the z-transform of $h^{(2)}$. Show that

$$H^{(2)}(e^{j\omega}) \;>\; 0, \qquad \text{for all } \omega \in [-\pi, \pi].$$

(iii) Find $c \in \ell^2(\mathbb{Z})$ such that $C(z) = 1/H^{(2)}(z)$. (It might be easier to find an expression with numerically approximated coefficients than an exact algebraic expression.)

(iv) Plot the canonical dual spline

$$\widetilde{\beta}_+^{(2)}(t) \;=\; \sum_{m\in\mathbb{Z}} c_m \beta_+^{(2)}(t-m).$$

6.15. *Existence of unique canonical dual B-splines.*$^{(2)}$

(i) Let $x \in \mathcal{L}^2(\mathbb{R})$, and let

$$h_n \;=\; \langle x(t),\, x(t-n)\rangle_t, \qquad n \in \mathbb{Z},$$

be its deterministic autocorrelation evaluated at the integers. Show that

$$H(e^{j\omega}) \;=\; \sum_{k\in\mathbb{Z}} |X(\omega - 2\pi k)|^2, \tag{P6.15-1}$$

where H is the DTFT of h and X is the Fourier transform of x.

(ii) Let $K \in \mathbb{N}$, and let $x = \beta_+^{(K)}$ be the causal elementary B-spline of degree K. Show that

$$H(e^{j\omega}) \;>\; 0, \qquad \text{for all } \omega \in [-\pi, \pi].$$

(iii) Using the result of part (ii), complete the argument outlined in Section 6.3.2 for the existence of a unique $\widetilde{\beta}_+^{(K)}$ of the form (6.41a) satisfying (6.41b) – a canonical dual of $\beta_+^{(K)}$.

6.16. *Orthogonalizing splines.*$^{(2)}$
Let $B^{(K)}$ be the Fourier transform of $\beta^{(K)}$, the elementary B-spline of degree K.

(i) Prove that

$$\sum_{n\in\mathbb{Z}} |\Phi(\omega + 2n\pi)|^2 \;=\; 1, \tag{P6.16-1}$$

where

$$\Phi(\omega) \;=\; \frac{B^{(K)}(\omega)}{\sqrt{\sum_{k\in\mathbb{Z}} |B^{(K)}(\omega + 2k\pi)|^2}}. \tag{P6.16-2}$$

(ii) Prove that

$$\langle \varphi(t), \varphi(t-n)\rangle_t = \delta_n, \tag{P6.16-3}$$

where φ is the inverse Fourier transform of Φ.

This is a general procedure for orthogonalizing splines; $\{\varphi(t-n)\}_{n\in\mathbb{Z}}$ forms an orthonormal basis for $S_{K,\mathbb{Z}}$ or $S_{K,\mathbb{Z}+1/2}$ (depending on whether K is odd or even). For $K = 2$, the function φ is called a *Battle–Lemarié scaling function* and is used to generate *Battle–Lemarié wavelets*; see the companion volume to this book, [57], for details.

6.17. *Computing inner products with splines.*(2)
Consider a function $x \in \mathcal{L}^2(\mathbb{R})$ that is zero outside of $[0, \infty)$. We want to compute

$$\langle x(t), \beta_+^{(K)}(t-n)\rangle_t, \qquad n \in \mathbb{Z},$$

where $\beta_+^{(K)}$ is the causal elementary B-spline of degree K, using operations on a sequence of samples of the primitive of order $K+1$ of x.

(i) Let $K = 0$. Using the definitions of the inner product and of $\beta_+^{(0)}$ directly, show that

$$\langle x(t), \beta_+^{(0)}(t-n)\rangle_t = \begin{cases} X_{n+1} - X_n, & \text{for } n \in \mathbb{N}; \\ 0, & \text{otherwise,} \end{cases} \tag{P6.17-1a}$$

where

$$X_n = \int_0^n x(t)\, dt \tag{P6.17-1b}$$

is the primitive of x evaluated at integers.

(ii) Show that (P6.17-1) holds by using the derivative of $\beta_+^{(0)}$ as in (6.68).

(iii) Let $\eta^{(K)}$ denote the derivative of order $K+1$ of $\beta_+^{(K)}$, $K \in \mathbb{N}$. Show that

$$\eta^{(K)}(t) = (\delta(t) - \delta(t-1)) *_t (\delta(t) - \delta(t-1)) *_t \cdots *_t (\delta(t) - \delta(t-1)), \tag{P6.17-2}$$

where there are K convolutions.

(iv) For any $K \in \mathbb{N}$, let the sequence $X_n^{(K)} = x^{(K)}(n)$, $n \in \mathbb{Z}$, be the primitive of order K of x evaluated at integers. For any $K \in \mathbb{N}$, find the filter $h^{(K)}$ such that

$$\langle x(t), \beta_+^{(K)}(t-n)\rangle_t = \left(h^{(K)} * X^{(K+1)}\right)_n, \qquad n \in \mathbb{Z}. \tag{P6.17-3}$$

(*Hint:* The result is analogous to part (iii).)

6.18. *Series truncation with a biorthogonal pair of bases.*(2)
Let $\{\varphi_k\}_{k\in\mathbb{N}}$ and $\{\widetilde{\varphi}_k\}_{k\in\mathbb{N}}$ be a biorthogonal pair of basis for a Hilbert space H. Then, for any $x \in H$,

$$x = \sum_{k\in\mathbb{N}} \alpha_k \varphi_k, \qquad \text{where } \alpha_k = \langle x, \widetilde{\varphi}_k\rangle,\ k \in \mathbb{N}.$$

(i) Fix an index set $\mathcal{I} \subset \mathbb{N}$ with $|\mathcal{I}| = M$. What is the orthogonal projection of $x \in H$ onto $S_{\mathcal{I}} = \overline{\operatorname{span}}(\{\varphi_k\}_{k\in\mathcal{I}})$?

(ii) Let $\check{x}_M$ denote the best approximation of x in the form $\sum_{k\in\mathcal{I}} \beta_k \varphi_k$, where $\mathcal{I}$ is any index set of size M and $(\beta_k)_{k\in\mathcal{I}}$ is any coefficient sequence. Prove or disprove that $\check{x}_M$ is determined by the M inner products α_k that are largest in magnitude.

6.19. *Nonlinear approximation operator.*(1)
Specify an orthonormal basis of $\mathbb{R}^2$ and $x, y \in \mathbb{R}^2$ such that the 1-term nonlinear approximation of $x + y$ is not $\check{x}_1 + \check{y}_1$, where $\check{x}_1$ and $\check{y}_1$ are the 1-term nonlinear approximations of x and y, respectively. This shows that nonlinear approximation is not a linear operator.

6.20. *Linear vs. nonlinear approximation: Worst case.*(2)
For $p \in \{1, 2, \infty\}$, let

$$B_p = \{x \in \mathbb{R}^2 \mid \|x\|_p \le 1\},$$

which is the unit ball in $\mathbb{R}^2$ under the p norm as shown in Figure 2.7.

(i) For each $p \in \{1, 2, \infty\}$, find the worst-case approximation errors

$$\sup_{x \in B_p} \|x - \widehat{x}_1\| \qquad \text{and} \qquad \sup_{x \in B_p} \|x - \check{x}_1\|$$

of the 1-term linear approximation and 1-term nonlinear approximation, respectively, using the standard basis.

(ii) For each $p \in \{1, 2, \infty\}$, find the orthonormal basis that minimizes the worst-case approximation error of the 1-term nonlinear approximation.

6.21. *Linear vs. nonlinear approximation: Average case.*$^{(2)}$
Let x be a two-dimensional jointly Gaussian random vector with zero mean and identity covariance matrix. Compare the expected squared norm of the approximation error for 1-term linear and nonlinear approximations using the standard basis.

6.22. *Linear and nonlinear approximation of triangle wave.*$^{(3)}$
Complete Example 6.18(ii) by showing that (6.87a) and (6.87b) hold.

6.23. *Karhunen–Loève basis in* $\mathbb{R}^2$.$^{(2)}$
Let x be a random vector with mean zero and covariance matrix

$$\Sigma_{\mathrm{x}} = \begin{bmatrix} 5 & 2 \\ 2 & 2 \end{bmatrix}$$

as in Example 6.19. Find the expected squared norm of the 1-term linear approximation error for x using every rotation of the standard basis. With this computation, confirm that the KL basis in (6.98) is optimal for the 1-term linear approximation of x.

6.24. *Oracle scaling and oracle projection.*$^{(1)}$
Show that the MSE of oracle scaling and the MSE of the oracle projector satisfy (6.117). (*Hint:* First show that $\frac{1}{2}\min(x, y) \leq xy/(x + y) \leq \min(x, y)$ for all $x, y \in \mathbb{R}^+$.)

6.25. *Dependence of estimation performance on approximation performance.*$^{(3)}$
Consider the estimation of an unknown deterministic vector $x \in H$ from an observation y that is corrupted by additive noise w as in Figure 6.39. Assume that the expansion coefficients of the noise using orthonormal basis $\{\varphi_k\}_{k\in\mathbb{N}}$ for H satisfy (6.107) with $\sigma^2 < 1$. Also assume that the expansion coefficient sequence $\alpha \in \ell^2(\mathbb{N})$ for x satisfies

$$|\alpha_k| \leq \frac{1}{1 + k^s}, \qquad \text{for all } k \in \mathbb{N}, \tag{P6.25-1}$$

for some $s \in (\frac{1}{2}, \infty)$. Find an upper bound for the MSE of the oracle projector that shows improved performance as the noise variance σ^2 shrinks and as s grows.

6.26. *Joint entropy.*$^{(1)}$
Let v_0 and v_1 be discrete random variables.

(i) Show that, when v_0 and v_1 are independent,

$$H((\mathrm{v}_0, \mathrm{v}_1)) = H(\mathrm{v}_0) + H(\mathrm{v}_1).$$

(ii) *Jensen's inequality* states that

$$\mathrm{E}[f(\mathrm{x})] \geq f(\mathrm{E}[\mathrm{x}]) \tag{P6.26-1}$$

for any convex function f. Use Jensen's inequality to show that

$$H((\mathrm{v}_0, \mathrm{v}_1)) \leq H(\mathrm{v}_0) + H(\mathrm{v}_1).$$

6.27. *Nearest-neighbor encoding.*$^{(2)}$

(i) Show that the lossy encoder $\widetilde{\beta}$ of an optimal fixed-rate scalar quantizer satisfies (6.125a). Furthermore, with the codewords given in ascending order

$$\widehat{x}_0 < \widehat{x}_1 < \cdots < \widehat{x}_{K-1}, \tag{P6.27-1}$$

show that, for $i = 1, 2, \ldots, K-2$,

$$\widetilde{\beta}(x) = i \qquad \text{when } x \in \left(\tfrac{1}{2}(\widehat{x}_{i-1} + \widehat{x}_i), \tfrac{1}{2}(\widehat{x}_i + \widehat{x}_{i+1})\right). \tag{P6.27-2}$$

(ii) A scalar quantizer is called *regular* when every partition cell S_i is a single interval and each S_i contains the corresponding codeword $\widehat{x}_i$. Show that an optimal fixed-rate scalar quantizer is always regular.

6.28. *Centroid decoding.*(2)
Show that the reproduction decoder β of an optimal scalar quantizer satisfies (6.125b). (*Hint:* With the lossy encoder $\widetilde{\beta}$ fixed, consider the estimation problem to be solved by the reproduction decoder.)

6.29. *Quantizer optimization.*(2)
Let x have the following simple, piecewise-constant PDF:

$$f_{\mathrm{x}}(s) = \begin{cases} \frac{1}{3}, & \text{for } s \in [0,\ 1); \\ \frac{2}{3}, & \text{for } s \in [2,\ 3); \\ 0, & \text{otherwise.} \end{cases}$$

Consider quantizers for x that have K_1 codewords evenly spaced over $[0,\ 1)$ and K_2 codewords evenly spaced over $[2,\ 3)$:

$$\mathcal{C} = \{(k+\tfrac{1}{2})/K_1\}_{k=0,1,\ldots,K_1-1} \cup \{2+(k+\tfrac{1}{2})/K_2\}_{k=0,1,\ldots,K_2-1}.$$

(i) Find the MSE distortion D as a function of K_1 and K_2.

(ii) Find the value of K_2/K_1 that minimizes D under a constraint on $K_1 + K_2$ and the resulting fixed-rate distortion. Your analysis may treat K_1 and K_2 as arbitrary positive real numbers, ignoring that they should be integers. Compare the result with (6.127).

(iii) Find the output entropy $H(\widetilde{\beta}(\mathrm{x}))$ as a function of K_1 and K_2.

(iv) Find the value of K_2/K_1 that minimizes D under a constraint on $H(\widetilde{\beta}(\mathrm{x}))$ and the resulting entropy-constrained distortion. Your analysis may again treat K_1 and K_2 as arbitrary positive real numbers, ignoring that they should be integers. Compare the result with (6.128).

6.30. *Direct bit allocation calculation.*(1)
Let the $\mathbb{R}^2$-valued random vector x have the uniform distribution on $[0,\ 1]^2$.

(i) Find the rate R and the distortion D for lossy coding of x obtained by uniform scalar quantization of x_1 and x_2 with K_1 and K_2 codewords.

(ii) Find the value of K_2/K_1 that minimizes D under a constraint on the total number of cells K_1K_2. Your analysis may treat K_1 and K_2 as arbitrary positive real numbers, ignoring that they should be integers.

(iii) What partition cell shape results from the optimization in part (ii)?

6.31. *Partition cell shapes in* $\mathbb{R}^2$.(1)
Let the $\mathbb{R}^2$-valued random vector x have the uniform distribution on the parallelogram-shaped region shown in Figure P6.31-1.

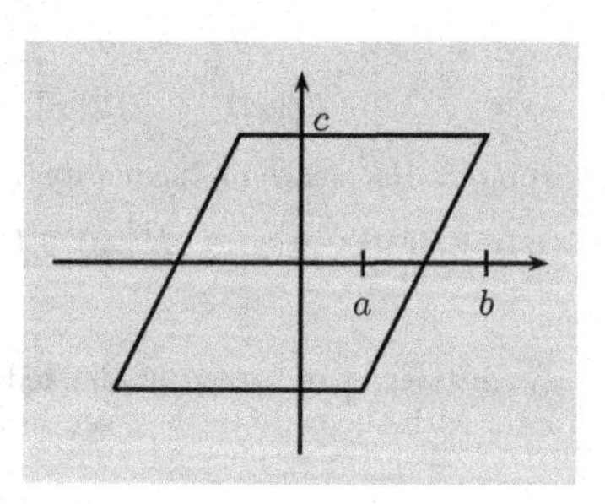

Figure P6.31-1 Support of the random vector x in Exercise 6.31.

(i) Under what condition is the area of the parallelogram 1?

(ii) Under the condition in part (i), what values of a, b, and c minimize $\mathrm{E}[\,\|\mathrm{x}\|^2\,]$?

(iii) Using the result of Part (ii), find conditions on the transform and scalar quantizers for optimal entropy-constrained transform coding of x at high rates.

Chapter 7

Localization and uncertainty

"The more precise the measurement of position, the more imprecise the measurement of momentum, and vice versa."
— *Werner Karl Heisenberg*

Contents

A major theme of this book is the construction of sets of vectors $\{\varphi_k\}$ to use for signal analysis and synthesis. The expansions with respect to bases and frames

$$x = \sum_k \alpha_k \varphi_k$$

we considered were, for the most part, quite general; apart from orthonormality for bases and tightness for frames, we did not impose further requirements on $\{\varphi_k\}$. For many applications, the utility of a representation is tied to time and frequency properties of $\{\varphi_k\}$.

Our primary goal in this chapter is to explore time and frequency properties first of *individual* vectors and then of *sets* of vectors, as these will be our tools

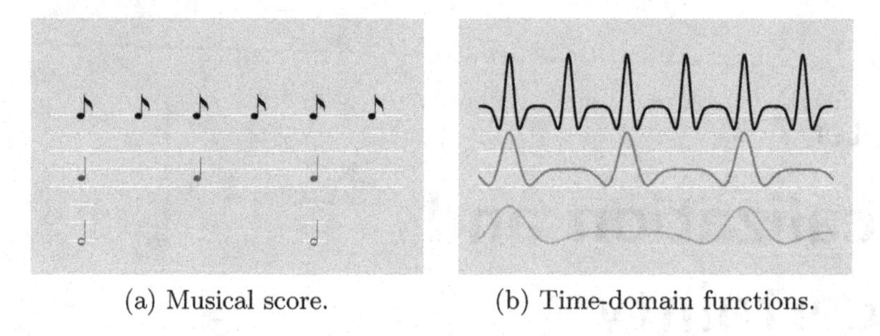

(a) Musical score. (b) Time-domain functions.

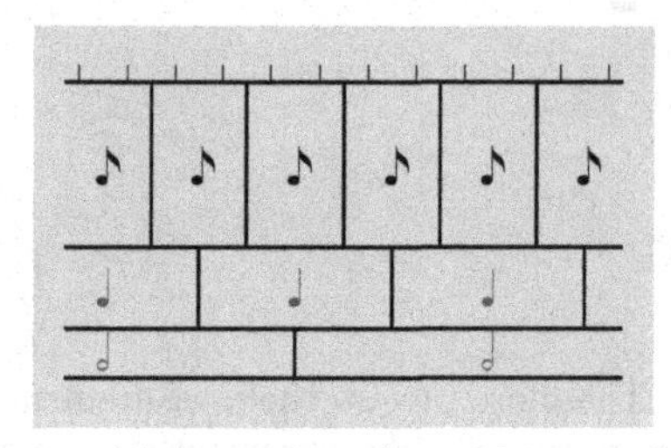

(c) Time–frequency plane.

Figure 7.1 Musical score as an illustration of a time–frequency plane [107].

for extracting information about a given signal (function or sequence). We extract information by computing an inner product of the signal with φ_k, the result of which is the expansion coefficient

$$\alpha_k = \langle x, \varphi_k \rangle \stackrel{(a)}{=} \frac{1}{2\pi} \langle X, \Phi_k \rangle, \tag{7.1}$$

where (a) follows from the generalized Parseval equality. Understanding what we can deduce about the signal from α_k is our goal; one of the main tools we will use is the uncertainty principle, which helps us understand how much local information in time and frequency we can extract using one such inner product.

7.1 Introduction

For certain simple signals, time and frequency properties are quite intuitive. Think of a note on a musical score. It is of a certain frequency (for example, middle A has the frequency of 440 Hz), it has a start time, and its value (𝅗𝅥, ♩, ♪) indicates its relative duration. We can think of the musical score as a time–frequency plane with a logarithmic frequency axis and notes as rectangles in that plane with horizontal extent determined by start and end times, and vertical position related in some way to frequency, as in Figure 7.1.

Localization in time and frequency Time and frequency views of a signal are intertwined in several ways. In this chapter, we consider how various forms of the uncertainty principle determine the trade-off between fine localization in the two domains: signals that are finely localized in time cannot be finely localized in frequency; conversely, signals that are finely localized in frequency cannot be finely localized in time. We can conceptually illustrate this trade-off using extreme cases of two transform pairs involving Dirac delta functions (see Table 4.1),

$$\delta(t) \quad \stackrel{\text{FT}}{\longleftrightarrow} \quad 1,$$
$$1 \quad \stackrel{\text{FT}}{\longleftrightarrow} \quad 2\pi\delta(\omega).$$

The trade-off is precisely specified by the uncertainty principle, which bounds from below the product of what we will define as spreads in time and frequency, with the bound reached by Gaussian functions.

Scale Given a signal, its scale changes when it is contracted or stretched. Signals that are scales of each other might convey the same information. For example, given a portrait of a person, recognizing that person should not depend on whether they occupy one-tenth or one-half of the image; thus, image recognition should be scale-invariant.

We can gain an intuitive understanding of scale by examining maps. In a map at scale 1:100 000, an object of length 1 km is represented by a length of

$$\frac{1 \text{ km}}{100\,000} = 1 \text{ cm};$$

in other words, the scale factor $a = 10^5$ is used as a contraction factor, to map $x(t)$ into its scaled version $y(t) = x(at)$.

Now consider using φ to extract some information about x. The inner product between φ and the function scaled by a is

$$\begin{aligned} \langle \sqrt{a}\, x(at),\, \varphi(t)\rangle_t &= \sqrt{a}\int_{-\infty}^{\infty} x(at)\,\varphi^*(t)\,dt = \frac{1}{\sqrt{a}}\int_{-\infty}^{\infty} x(\tau)\,\varphi^*\!\left(\frac{\tau}{a}\right)d\tau \\ &= \left\langle x(t),\, \frac{1}{\sqrt{a}}\varphi\!\left(\frac{t}{a}\right)\right\rangle_t, \end{aligned} \tag{7.2}$$

where the multiplicative factor of $\sqrt{a}$ in $\sqrt{a}x(at)$ is present to keep the norm of the function unchanged. This shows that examining a contracted function $x(at)$ with $\varphi(t)$ is equivalent to examining the function $x(t)$ with the stretched function $\varphi(t/a)$. If stretched and contracted versions of a single φ are available, large-scale features in x can be extracted using a stretched φ ($a > 1$) and small-scale features (fine details) using a contracted φ ($a < 1$).

This scale invariance is a purely continuous-time property, however, since discrete-time signals cannot be scaled so easily. For example, downsampling by a factor of N is in general a lossy operation, while upsampling by a factor of N is not. We will consider sampling and interpolation operations from Chapter 5 as scaling in the discrete domain: downsampling preceded by ideal lowpass prefiltering makes a sequence shorter, and upsampling followed by ideal lowpass postfiltering makes a sequence longer.

Resolution We often speak of the resolution of a digital photograph in terms of the number of pixels. However, a blurry photograph does not have the resolution of a sharp one, even when the two have the same number of pixels. This is because the resolution of an image usually refers to whether nearby fine details of the underlying scene can be distinguished.

Without a formal mathematical definition, we consider resolution to be related to the information content of a signal, such as the number of degrees of freedom per unit time or space. This is related to bandwidth in that signals in a bandlimited set have an upper bound on the degrees of freedom. Note, however, that sets of full-band signals can also have limited degrees of freedom.

Consider the space of bandlimited functions $\mathrm{BL}[-\frac{1}{2}\omega_0, \frac{1}{2}\omega_0]$ as in Definition 5.13. The sampling theorem, Theorem 5.15, states that samples taken every $T = 2\pi/\omega_0$ seconds, or $x(nT)$, $n \in \mathbb{Z}$, uniquely specify $x \in \mathrm{BL}[-\frac{1}{2}\omega_0, \frac{1}{2}\omega_0]$. This equivalence means that the functions in $\mathrm{BL}[-\frac{1}{2}\omega_0, \frac{1}{2}\omega_0]$ have $\omega_0/(2\pi)$ complex degrees of freedom per unit time. Since a real function has real samples, functions in the subset of $\mathrm{BL}[-\frac{1}{2}\omega_0, \frac{1}{2}\omega_0]$ with real-valued functions have $\omega_0/(2\pi)$ real degrees of freedom per unit time. The set of piecewise-constant functions from (5.1) is an example of a set of functions that, while not bandlimited, still have a finite number of degrees of freedom per unit time. A function from such a space has one degree of freedom per unit time, but an unbounded spectrum since it is discontinuous at every integer.

Interactions Clearly, scale and resolution interact; this is most obvious with images, as illustrated with a drawing by C. Allan Gilbert, *All is Vanity*, in Figure 7.2.[112] It is designed to be perceived as either a woman sitting in front of a mirror (when seen from nearby and at high resolution as in Figure 7.2(a)) or as a skull (when seen from afar or at low resolution as in Figure 7.2(b)). Figure 7.2(c) illustrates what happens when the scale changes; even though the scale has been halved, the resolution remains unchanged and thus our perception of the image remains the same as long as our visual acuity is high enough. Figure 7.2(d) shows a postage-stamp version; the scale has been further reduced, and our perception of the image changes (we see only the skull) since our visual acuity is not high enough to capture details.

Filtering can affect resolution as well. If a function of bandwidth ω_0 is perfectly lowpass-filtered to $|\omega| < \frac{1}{2}\beta\omega_0$, with $0 < \beta < 1$, then its resolution changes from $\omega_0/(2\pi)$ to $\beta\omega_0/(2\pi)$. The same holds for sequences, where an ideal lowpass filter with cutoff frequency $\beta\pi$, $0 < \beta < 1$, reduces the resolution to β samples per unit time.

While high-resolution signals have high bandwidth, the converse is not necessarily true; a signal can have high bandwidth without necessarily being of high resolution. For example, adding noise increases bandwidth without increasing resolution. Figure 7.3 compares reducing bandwidth with adding noise, indicating the similarity of the two effects; too much noise or too little bandwidth both preclude

[112]Another beautiful optical illusion is Salvador Dalí's *Gala Contemplating the Mediterranean Sea which at Twenty Meters Becomes the Portrait of Abraham Lincoln (Homage to Rothko).*

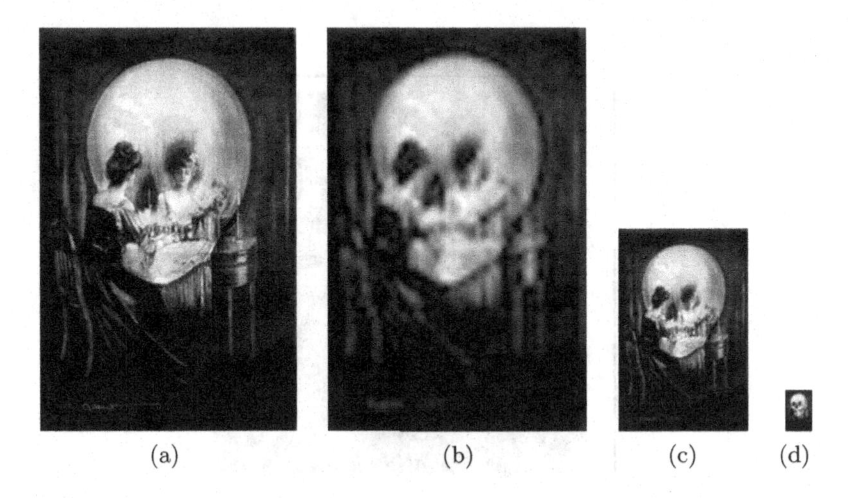

Figure 7.2 Interplay of scale and resolution. (a) The original, high-resolution version. (b) A blurred, lower-resolution version; the resolution is lower and thus our perception of the image has changed. (c) A scaled version of half size in each dimension; the resolution is unchanged and thus our perception of the image remains unchanged as long as our visual acuity is high enough. (d) A version scaled further, to the point that our perception of the image changes (we see only the skull) since our visual acuity is not high enough to capture details.

our ability to extract relevant information.

Chapter outline

The present chapter explores the time and frequency properties both of individual vectors and of sets of vectors as tools for extracting information about a signal. Section 7.2 discusses localization concepts for functions, while Section 7.3 does the same for sequences. Section 7.4 discusses localization concepts for sets of functions and sequences, while Section 7.5 illustrates signal analysis with simple local Fourier and wavelet bases, both for functions and for sequences. Section 7.6 summarizes the main concepts of the book and discusses issues that arise when applying the tools from the book to real-world problems, illustrating them with examples of real-world signals.

7.2 Localization for functions

Consider a function $x \in \mathcal{L}^2(\mathbb{R})$, where the domain of the function is a time index. In Sections 7.4 and 7.5, we will build structured sets of functions from one or more prototype functions by certain basic operations. Those basic operations are as follows:

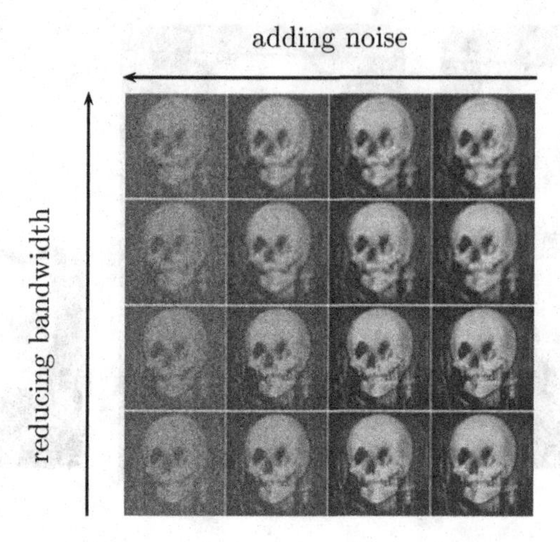

Figure 7.3 Interplay of bandwidth and noise; reducing bandwidth and adding noise both affect our ability to perceive relevant information. The original image is in the lower-right-hand corner; the SNR decreases in increments of 6 dB from right to left, and the bandwidth is progressively halved from bottom to top.

(i) *Shift in time* by $t_0 \in \mathbb{R}$:

$$y(t) = x(t - t_0) \quad \stackrel{\text{FT}}{\longleftrightarrow} \quad Y(\omega) = e^{-j\omega t_0} X(\omega). \tag{7.3a}$$

(ii) *Shift in frequency (modulation)* by $\omega_0 \in \mathbb{R}$:

$$y(t) = e^{j\omega_0 t} x(t) \quad \stackrel{\text{FT}}{\longleftrightarrow} \quad Y(\omega) = X(\omega - \omega_0). \tag{7.3b}$$

(iii) *Scaling in time* by $a \in \mathbb{R}^+$ (which implies scaling in frequency by $1/a$):

$$y(t) = \sqrt{a}\, x(at) \quad \stackrel{\text{FT}}{\longleftrightarrow} \quad Y(\omega) = \frac{1}{\sqrt{a}} X\left(\frac{\omega}{a}\right). \tag{7.3c}$$

7.2.1 Localization in time

We now discuss localization of a function in time. When the function is supported on any finite interval, its Fourier transform is not (see Exercise 7.1); this implies that using support length is too coarse a measure of locality, as all functions with finite support in time have infinite support in frequency.

A concise way to describe locality (or lack thereof) is to introduce a spreading measure akin to standard deviation. The normalized function $|x(t)|^2/\|x\|^2$ can be interpreted as the PDF of a random variable; we call the mean of that random variable the time center of x and the standard deviation of the same random variable the time spread of x.

DEFINITION 7.1 (TIME CENTER AND SPREAD FOR FUNCTIONS) The *time center* μ_t and the *time spread* Δ_t of $x \in \mathcal{L}^2(\mathbb{R})$ are

$$\mu_t = \frac{1}{\|x\|^2} \int_{-\infty}^{\infty} t\, |x(t)|^2\, dt, \tag{7.4a}$$

$$\Delta_t = \left(\frac{1}{\|x\|^2} \int_{-\infty}^{\infty} (t - \mu_t)^2\, |x(t)|^2\, dt \right)^{1/2}. \tag{7.4b}$$

EXAMPLE 7.1 (TIME CENTERS AND SPREADS FOR FUNCTIONS) To illustrate the concepts of time center and time spread, we compute them for three functions.

(i) The sinc function from (4.76) has $\mu_t = 0$ because of its even symmetry. It has infinite Δ_t because the decay of $\operatorname{sinc}^2 t$ is canceled by the t^2 factor in (7.4b), causing the integrand of (7.4b) to have no decay.

(ii) The box function from (4.75) also has $\mu_t = 0$ because of even symmetry. A simple calculation gives $\Delta_t = t_0/(2\sqrt{3})$.

(iii) The Gaussian function from (4.78) also has $\mu_t = 0$ because of even symmetry. Normalizing (4.78) and comparing with (2.261) shows that $\Delta_t = 1/(2\sqrt{\alpha})$.

From the example, we see that the time spread can vary widely; the box and Gaussian functions have finite spreads, while the sinc function has an infinite one, showing that the time spread can be unbounded. As further illustration, Solved Exercise 7.1 computes the time spreads for B-splines of various degrees.

Under the operations in (7.3), the time center and time spread satisfy the following (see also Solved Exercise 7.2 and Table 7.1):

(i) *Shift in time* by $t_0 \in \mathbb{R}$, (7.3a), causes the time center to shift and leaves the time spread unchanged,

$$\mu_t(y) = \mu_t(x) + t_0, \tag{7.5a}$$

$$\Delta_t(y) = \Delta_t(x). \tag{7.5b}$$

(ii) *Shift in frequency (modulation)* by $\omega_0 \in \mathbb{R}$, (7.3b), leaves both the time center and the time spread unchanged,

$$\mu_t(y) = \mu_t(x), \tag{7.5c}$$

$$\Delta_t(y) = \Delta_t(x). \tag{7.5d}$$

(iii) *Scaling in time* by $a \in \mathbb{R}^+$, (7.3c), causes both the time center and the time spread to scale,

$$\mu_t(y) = \frac{1}{a}\mu_t(x), \tag{7.5e}$$

$$\Delta_t(y) = \frac{1}{a}\Delta_t(x). \tag{7.5f}$$

7.2.2 Localization in frequency

We now discuss localization of a function in frequency; the concepts are dual to those for localization in time. The normalized function $|X(\omega)|^2/\|X\|^2$ can be interpreted as the PDF of a random variable; we call the mean of that random variable the frequency center of x and the standard deviation of the same random variable the frequency spread of x.

DEFINITION 7.2 (FREQUENCY CENTER AND SPREAD FOR FUNCTIONS) The *frequency center* μ_f and the *frequency spread* Δ_f of $x \in \mathcal{L}^2(\mathbb{R})$ with Fourier transform X are

$$\mu_f = \frac{1}{2\pi\|x\|^2}\int_{-\infty}^{\infty} \omega\,|X(\omega)|^2\,d\omega, \tag{7.6a}$$

$$\Delta_f = \left(\frac{1}{2\pi\|x\|^2}\int_{-\infty}^{\infty} (\omega-\mu_f)^2\,|X(\omega)|^2\,d\omega\right)^{1/2}. \tag{7.6b}$$

Note that the frequency center will be 0 for all real functions because of the symmetry of the magnitude of the Fourier transform.

EXAMPLE 7.2 (FREQUENCY CENTERS AND SPREADS FOR FUNCTIONS) To illustrate the concepts of frequency center and frequency spread, we compute them for the same functions as in Example 7.1.

(i) The sinc function from (4.76) has $\mu_f = 0$ and $\Delta_f = \omega_0/(2\sqrt{3})$.
(ii) The box function from (4.75) has $\mu_f = 0$ and infinite Δ_f, as $|X(\omega)|^2$ decays only as $|\omega|^{-2}$.
(iii) The Gaussian function from (4.78) has $\mu_f = 0$ and $\Delta_f = \sqrt{\alpha}$.

From the example, we see that the frequency spread can vary widely; the sinc and Gaussian functions have finite spreads, while the box function has an infinite spread, showing that the frequency spread can be unbounded. As further illustration, Solved Exercise 7.1 computes the frequency spreads for B-splines of various degrees.

Under the operations in (7.3), the frequency center and frequency spread satisfy the following (see also Solved Exercise 7.2 and Table 7.1):

(i) *Shift in time* by $t_0 \in \mathbb{R}$, (7.3a), leaves both the frequency center and the frequency spread unchanged,

$$\mu_f(y) = \mu_f(x), \tag{7.7a}$$

$$\Delta_f(y) = \Delta_f(x). \tag{7.7b}$$

(ii) *Shift in frequency (modulation)* by $\omega_0 \in \mathbb{R}$, (7.3b), causes the frequency center to shift and leaves the frequency spread unchanged,

$$\mu_f(y) = \mu_f(x) + \omega_0, \tag{7.7c}$$

$$\Delta_f(y) = \Delta_f(x). \tag{7.7d}$$

(iii) *Scaling in time* by $a \in \mathbb{R}^+$, (7.3c), causes both the frequency center and the frequency spread to scale,

$$\mu_f(y) = a\mu_f(x), \tag{7.7e}$$
$$\Delta_f(y) = a\Delta_f(x). \tag{7.7f}$$

In Sections 7.4 and 7.5, we will show constructions of simple structured sets obtained by applying these operations to a single prototype function. Thus, knowing the time and frequency centers and spreads of the prototype function, we will know the time and frequency centers and spreads of all the other functions in the set as well.

One-sided frequency center and spread We mentioned already that the frequency center of any real function will be 0 (because of the symmetry of the magnitude of the Fourier transform). We might prefer a definition of the frequency center that captures the spectral characteristics of a real bandpass function.

DEFINITION 7.3 (ONE-SIDED FREQUENCY CENTER AND SPREAD) The *one-sided frequency center* μ_f^+ and the *one-sided frequency spread* Δ_f^+ of $x \in \mathcal{L}^2(\mathbb{R})$ with Fourier transform X are

$$\mu_f^+ = \frac{1}{\pi\|x\|^2}\int_0^\infty \omega\,|X(\omega)|^2\,d\omega, \tag{7.8a}$$
$$\Delta_f^+ = \left(\frac{1}{\pi\|x\|^2}\int_0^\infty (\omega-\mu_f^+)^2\,|X(\omega)|^2\,d\omega\right)^{1/2}. \tag{7.8b}$$

EXAMPLE 7.3 (FREQUENCY CENTER FOR A REAL BANDPASS FUNCTION) Let x be the difference between two Gaussian functions,

$$x(t) = \exp\Big(-\frac{\pi}{8}(3+2t)^2\Big) - \exp\Big(-\frac{\pi}{8}(3-2t)^2\Big) \tag{7.9a}$$
$$\stackrel{\text{FT}}{\longleftrightarrow}\quad X(\omega) = \sqrt{2}\exp\Big(-\frac{\omega(\omega+3j\pi)}{2\pi}\Big)\,(e^{3j\omega}-1), \tag{7.9b}$$

as shown in Figure 7.4. Clearly, $\mu_f = 0$ because the magnitude of the spectrum has even symmetry. However, in certain circumstances, we might want to know where the bulk of the spectrum is located on either side of the frequency axis; the frequency center in Definition 7.2 will not provide such information. Using Definition 7.3 to compute the one-sided frequency center of x, we get

$$\mu_f^+ = \frac{1}{\pi\|x\|^2}\int_0^\infty \omega\,|X(\omega)|^2\,d\omega \approx 1.098,$$

which is close to and slightly larger than $\frac{1}{3}\pi$, consistent with what we would guess by looking at Figure 7.4.

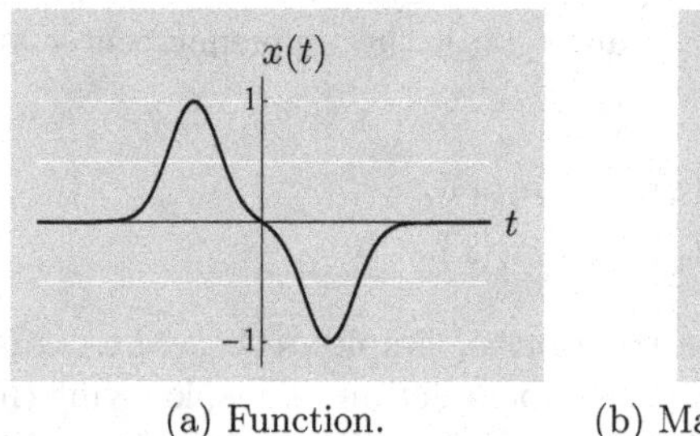

(a) Function.

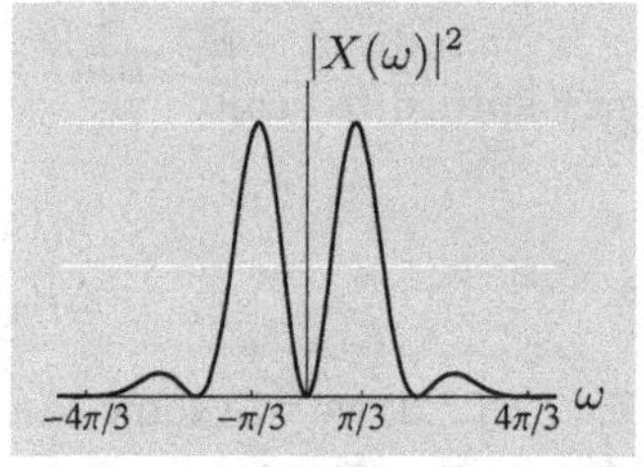

(b) Magnitude squared of the spectrum.

Figure 7.4 A function with bandpass Fourier spectrum.

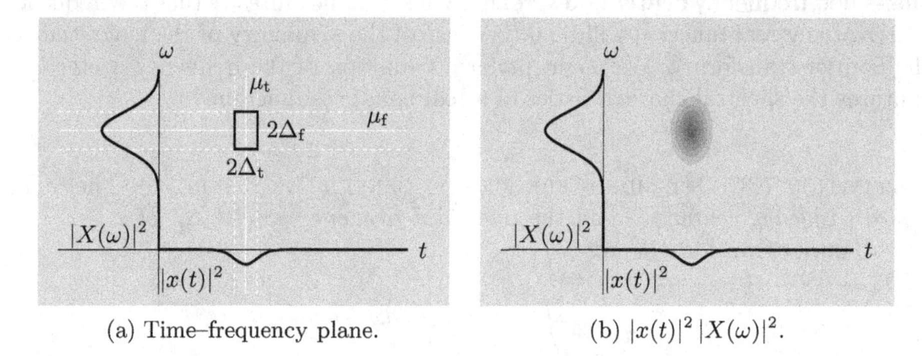

(a) Time–frequency plane. (b) $|x(t)|^2\,|X(\omega)|^2$.

Figure 7.5 The time–frequency plane. The function x with Fourier transform X has an associated Heisenberg box centered at $(\mu_{\mathrm{t}},\,\mu_{\mathrm{f}})$, of width $2\Delta_{\mathrm{t}}$ and height $2\Delta_{\mathrm{f}}$. Interpreting $|x(t)|^2/\|x\|^2 \cdot |X(\omega)|^2/\|X\|^2$ as the joint PDF of two random variables yields $(\mu_{\mathrm{t}},\,\mu_{\mathrm{f}})$ as the means of the random variables and Δ_{t}, Δ_{f} as the standard deviations along the two directions.

7.2.3 Uncertainty principle for functions

Heisenberg box Given a function x and its Fourier transform X, we have just introduced the 4-tuple $(\mu_{\mathrm{t}},\,\Delta_{\mathrm{t}},\,\mu_{\mathrm{f}},\,\Delta_{\mathrm{f}})$ describing the function's center in time and frequency $(\mu_{\mathrm{t}},\mu_{\mathrm{f}})$ and its spread in time and frequency $(\Delta_{\mathrm{t}},\Delta_{\mathrm{f}})$. It is convenient to show this pictorially (see Figure 7.5), as it conveys the idea that there is a center of mass $(\mu_{\mathrm{t}},\mu_{\mathrm{f}})$ around which a rectangular box of width $2\Delta_{\mathrm{t}}$ and height $2\Delta_{\mathrm{f}}$ is located. The plane on which this is drawn is called the *time–frequency plane*, and the box is usually called a *Heisenberg box* or a *time–frequency tile*. We adopt the convention of showing only the first quadrant of the time–frequency plane; we reserve the second quadrant for the squared magnitude of the Fourier transform of the function, $|X|^2$, and the fourth quadrant for the squared absolute value of the function itself, $|x|^2$.

Given that we know the effects of the three operations (7.3) on the time and frequency centers and spreads of a function x, (7.5) and (7.7), we can deduce their effects on the Heisenberg box of the function (see also Table 7.1):

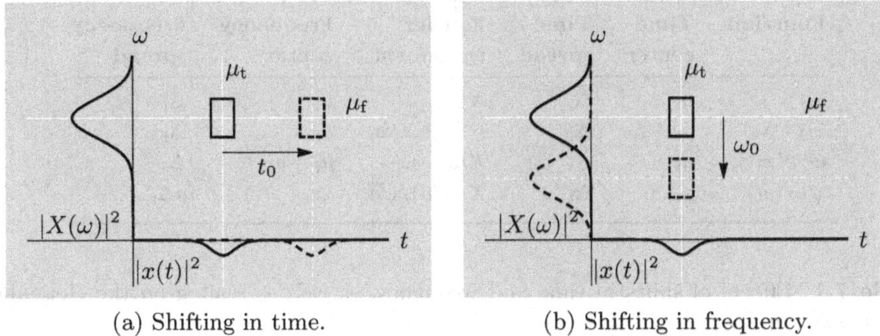

(a) Shifting in time. (b) Shifting in frequency.

Figure 7.6 Shifting of a function with Heisenberg box centered at (μ_t, μ_f), of width $2\Delta_t$ and height $2\Delta_f$. (a) Shifting in time by t_0, $x(t - t_0)$, shifts the Heisenberg box to $(\mu_t + t_0, \mu_f)$; the area remains the same. (b) Shifting in frequency by ω_0 (modulating), $X(\omega - \omega_0)$, shifts the Heisenberg box to $(\mu_t,\ \mu_f + \omega_0)$; the area remains the same. (Note that in the figure we used a negative ω_0.)

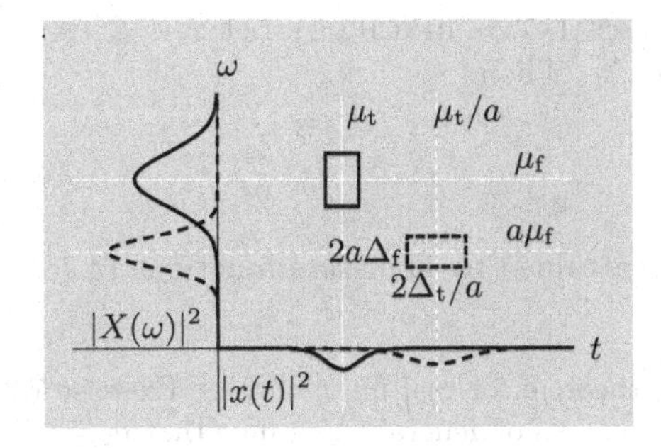

Figure 7.7 Scaling of a function with Heisenberg box centered at $(\mu_t,\ \mu_f)$, of width $2\Delta_t$ and height $2\Delta_f$, shifts the Heisenberg box to $(\mu_t/a,\ a\mu_f)$ and scales its width to $2\Delta_t/a$ and its height to $2a\Delta_f$; the area remains the same. (Illustrated for $a = \frac{4}{7}$.)

(i) *Shift in time* by $t_0 \in \mathbb{R}$, (7.3a), causes the Heisenberg box to shift in time by t_0 (dashed box in Figure 7.6(a)).

(ii) *Shift in frequency (modulation)* by $\omega_0 \in \mathbb{R}$, (7.3b), causes the Heisenberg box to shift in frequency by ω_0 (dashed box in Figure 7.6(b)).

(iii) *Scaling in time* by $a \in \mathbb{R}^+$, (7.3c), causes the Heisenberg box to shift in time to μ_t/a and in frequency to $a\mu_f$, and to scale in time by $1/a$ and in frequency by a (dashed box in Figure 7.7).

Uncertainty principle Neither shifting nor scaling of the function affects the area of the associated Heisenberg box, even when the time and the frequency spreads change. Thus, when manipulating a function through these operations, for one spread to decrease the other must increase. This agrees with what we have seen in

Function	Time center	Time spread	Fourier transform	Frequency center	Frequency spread
$x(t)$	μ_t	Δ_t	$X(\omega)$	μ_f	Δ_f
$x(t-t_0)$	$\mu_t + t_0$	Δ_t	$e^{-j\omega t_0}X(\omega)$	μ_f	Δ_f
$e^{j\omega_0 t}x(t)$	μ_t	Δ_t	$X(\omega-\omega_0)$	$\mu_f + \omega_0$	Δ_f
$\sqrt{a}x(at)$	μ_t/a	Δ_t/a	$X(\omega/a)/\sqrt{a}$	$a\mu_f$	$a\Delta_f$

Table 7.1 Effects of shifts in time and frequency, as well as scaling on the Heisenberg box $(\mu_t, \Delta_t, \mu_f, \Delta_f)$.

Examples 7.1 and 7.2; a function that is narrow in one domain will be wide in the other. It is formalized in the uncertainty principle as the area of the Heisenberg box being lower bounded; no function can be arbitrarily narrow in both time and frequency.

THEOREM 7.4 (UNCERTAINTY PRINCIPLE) Let $x \in \mathcal{L}^2(\mathbb{R})$ have time spread Δ_t and frequency spread Δ_f. Then,

$$\Delta_t \Delta_f \geq \frac{1}{2}, \tag{7.10}$$

with the lower bound attained by Gaussian functions (4.78).

Proof. We prove the theorem for real functions; see Exercise 7.2 for the complex case. Let $x \in \mathcal{L}^2(\mathbb{R})$. Without loss of generality, assume that $\mu_t = 0$ and $\|x\| = 1$; otherwise, we may shift and scale x appropriately. Since x is real, $\mu_f = 0$.

Suppose that the derivative of x exists and is in $\mathcal{L}^2(\mathbb{R})$; if not, the decay of $X(\omega)$ is $|\omega|^{-2}$ or slower (see (4.83b)), so $\Delta_f = \infty$ and the statement holds trivially. Similarly, we may assume that $tx(t)$ is in $\mathcal{L}^2(\mathbb{R})$ because otherwise $\Delta_t = \infty$ and the statement holds trivially.

Consider the following integral:

$$\begin{aligned}\left|\int_{-\infty}^{\infty} tx(t)\,x'(t)\,dt\right|^2 &\overset{(a)}{\leq} \left(\int_{-\infty}^{\infty} |tx(t)|^2\,dt\right)\left(\int_{-\infty}^{\infty} |x'(t)|^2\,dt\right)\\ &\overset{(b)}{=} \left(\int_{-\infty}^{\infty} |tx(t)|^2\,dt\right)\left(\frac{1}{2\pi}\int_{-\infty}^{\infty} |j\omega X(\omega)|^2\,d\omega\right)\\ &= \Delta_t^2\Delta_f^2,\end{aligned} \tag{7.11}$$

where (a) follows from the Cauchy–Schwarz inequality, (2.35); and (b) from the Parseval equality (4.71a) and the differentiation in frequency property of the Fourier transform, (4.59a). Simplifying the integral on the left-hand side gives

$$\begin{aligned}\int_{-\infty}^{\infty} tx(t)\,x'(t)\,dt &\overset{(a)}{=} \frac{1}{2}\int_{-\infty}^{\infty} t\frac{dx^2(t)}{dt}\,dt \overset{(b)}{=} \frac{1}{2}\,tx^2(t)\Big|_{-\infty}^{\infty} - \frac{1}{2}\int_{-\infty}^{\infty} x^2(t)\,dt\\ &\overset{(c)}{=} -\frac{1}{2}\int_{-\infty}^{\infty} x^2(t)\,dt \overset{(d)}{=} -\frac{1}{2},\end{aligned} \tag{7.12}$$

where (a) follows from $(x^2(t))' = 2x'(t)\,x(t)$; (b) from integration by parts; (c) from $\lim_{t\to\pm\infty} tx^2(t) = 0$, which holds because $x \in \mathcal{L}^2(\mathbb{R})$ implies that it decays faster than $1/\sqrt{|t|}$ for $t \to \pm\infty$ (see Section 4.4.2); and (d) from x being real-valued and $\|x\| = 1$. Substituting (7.12) into (7.11) yields (7.10).

To find functions that meet the bound with equality, recall that the Cauchy–Schwarz inequality (2.29) becomes an equality if and only if the two vectors are scalar multiples of each other, which here means that $x'(t) = \beta t x(t)$; solving this differential equation yields a Gaussian function. Indeed, from Examples 7.1 and 7.2, for a Gaussian function, $\Delta_\mathrm{t} = 1/(2\sqrt{\alpha})$ and $\Delta_\mathrm{f} = \sqrt{\alpha}$, yielding the product $\Delta_\mathrm{t}\Delta_\mathrm{f} = \frac{1}{2}$.

Other localization measures While the uncertainty principle uses a spreading measure akin to standard deviation, other measures can be defined. Though they typically lack fundamental bounds of the kind given by the uncertainty principle (7.10), they can be useful as well as intuitive. One such measure finds the centered intervals containing a given fraction β of the energy in time and frequency (and β is typically 0.90 or 0.95).

For a unit-norm function x with time center μ_t and frequency center μ_f, the modified time spread $\Delta_\mathrm{t}^{(\beta)}$ and the modified frequency spread $\Delta_\mathrm{f}^{(\beta)}$ are defined such that

$$\int_{\mu_\mathrm{t}-(1/2)\Delta_\mathrm{t}^{(\beta)}}^{\mu_\mathrm{t}+(1/2)\Delta_\mathrm{t}^{(\beta)}} |x(t)|^2\, dt \;=\; \beta, \tag{7.13a}$$

$$\frac{1}{2\pi}\int_{\mu_\mathrm{f}-(1/2)\Delta_\mathrm{f}^{(\beta)}}^{\mu_\mathrm{f}+(1/2)\Delta_\mathrm{f}^{(\beta)}} |X(\omega)|^2\, d\omega \;=\; \beta. \tag{7.13b}$$

Exercise 7.3 shows that $\Delta_\mathrm{t}^{(\beta)}$ and $\Delta_\mathrm{f}^{(\beta)}$ satisfy the same shift, modulation, and scaling properties as do Δ_t and Δ_f in Table 7.1.

7.3 Localization for sequences

Thus far, we have restricted our attention to the study of localization properties and bounds for functions. Analogous results for sequences are neither as elegant nor do they precisely parallel those for functions, as they require restrictions beyond membership in $\ell^2(\mathbb{Z})$ to avoid complications caused by the periodicity of the frequency axis (see the *Further reading* for alternative approaches).

Consider a sequence $x \in \ell^2(\mathbb{Z})$, where the domain of the sequence is a discrete time index. In Sections 7.4 and 7.5, we will build structured sets of sequences from one or more prototype sequences by certain operations. Those basic operations are as follows:

(i) *Shift in time* by $n_0 \in \mathbb{Z}$:

$$y_n \;=\; x_{n-n_0} \quad \overset{\text{DTFT}}{\longleftrightarrow} \quad Y(e^{j\omega}) \;=\; e^{-j\omega n_0} X(e^{j\omega}). \tag{7.14a}$$

(ii) *Shift in frequency (modulation)* by $\omega_0 \in (-\pi, \pi]$:

$$y_n \;=\; e^{j\omega_0 n} x_n \quad \overset{\text{DTFT}}{\longleftrightarrow} \quad Y(e^{j\omega}) \;=\; X(e^{j(\omega-\omega_0)}). \tag{7.14b}$$

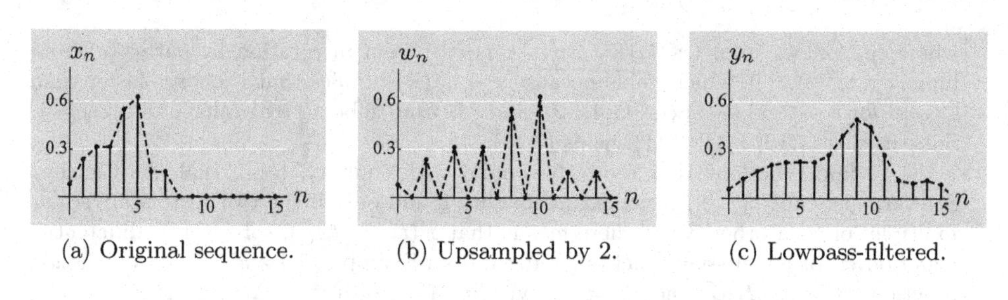

(a) Original sequence. (b) Upsampled by 2. (c) Lowpass-filtered.

Figure 7.8 Upsampling by 2 followed by ideal lowpass postfiltering as in Figure 5.12(b), with g an ideal half-band filter from Table 3.5. The upsampled sequence is denoted by w. Note that g is of unit norm, so $g_0 = 1/\sqrt{2}$.

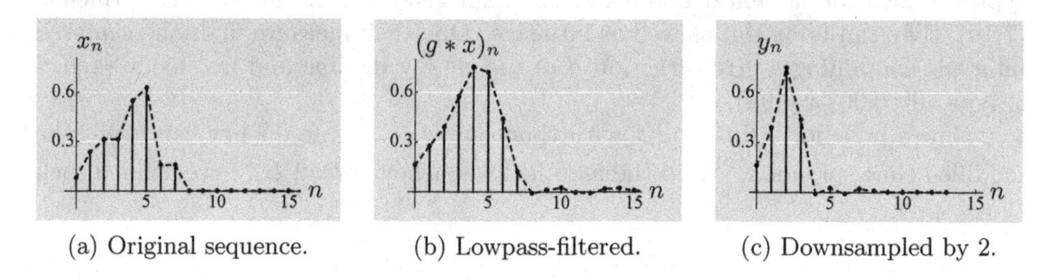

(a) Original sequence. (b) Lowpass-filtered. (c) Downsampled by 2.

Figure 7.9 Downsampling by 2 preceded by ideal lowpass postfiltering as in Figure 5.12(a), with g an ideal half-band filter from Table 3.5. Note that g is of unit norm, so $g_0 = 1/\sqrt{2}$.

(iii) *Upsampling* by $N \in \mathbb{Z}^+$ followed by ideal lowpass postfiltering:

$$\begin{aligned} y_n &= g_n *_n \begin{cases} x_{n/N}, & \text{for } n/N \in \mathbb{Z}; \\ 0, & \text{otherwise} \end{cases} \\ \overset{\text{DTFT}}{\longleftrightarrow} \quad Y(e^{j\omega}) &= \begin{cases} \sqrt{N} X(e^{jN\omega}), & \text{for } \omega \in (-\pi/N, \pi/N]; \\ 0, & \text{otherwise,} \end{cases} \end{aligned} \tag{7.14c}$$

where g is the ideal Nth-band filter (see Table 3.5). Figure 7.8 illustrates this form of discrete scaling for $N = 2$.

(iv) *Downsampling* by $N \in \mathbb{Z}^+$ preceded by ideal lowpass prefiltering:

$$y_n = (g * x)_{Nn} \quad \overset{\text{DTFT}}{\longleftrightarrow} \quad Y(e^{j\omega}) = \frac{1}{\sqrt{N}} X(e^{j\omega/N}), \quad \omega \in [-\pi, \pi), \tag{7.14d}$$

where g is the ideal Nth-band filter. Figure 7.9 illustrates this form of discrete scaling for $N = 2$.

For sequences, scaling has been separated into two cases: upsampling followed by ideal lowpass postfiltering and downsampling preceded by ideal lowpass prefiltering, which are the interpolation and sampling operations defined in Chapter 5.

7.3.1 Localization in time

We now discuss localization of a sequence in time. When the sequence is finitely supported, its DTFT is not (it can only have isolated zeros); see Exercise 7.1. This implies that using support length is too coarse a measure of locality, as all sequences with finite support in time have full-band frequency support.

We again describe locality concisely by introducing a spreading measure akin to standard deviation. The normalized sequence $|x_n|^2/\|x\|^2$ can be interpreted as the probability mass function of a discrete random variable; we call the mean of that random variable the time center of x and the standard deviation of the same random variable the time spread of x. Note that neither the time center nor the time spread is necessarily an integer.

DEFINITION 7.5 (TIME CENTER AND SPREAD FOR SEQUENCES) The *time center* μ_t and the *time spread* Δ_t of $x \in \ell^2(\mathbb{Z})$ are

$$\mu_t = \frac{1}{\|x\|^2} \sum_{n\in\mathbb{Z}} n\,|x_n|^2, \tag{7.15a}$$

$$\Delta_t = \left(\frac{1}{\|x\|^2} \sum_{n\in\mathbb{Z}} (n-\mu_t)^2\,|x_n|^2 \right)^{1/2}. \tag{7.15b}$$

EXAMPLE 7.4 (TIME CENTERS AND SPREADS FOR SEQUENCES) To illustrate the concepts of time center and time spread, we compute them for two sequences.

(i) The sinc sequence from Table 4.5 has $\mu_t = 0$ because of its even symmetry. It has infinite Δ_t because the decay of $\mathrm{sinc}^2(\omega_0 n)$ is canceled by the n^2 factor in (7.15b), causing the summand of (7.15b) to have no decay.
(ii) The right-sided box sequence of length n_0 from (3.13) has $\mu_t = \frac{1}{2}(n_0 - 1)$ and $\Delta_t = \sqrt{n_0^2 - 1}/(2\sqrt{3})$.

From the example, we see that the time spread can vary widely; the box sequence has a finite spread, while the sinc sequence has an infinite one, showing that the time spread can be unbounded.

Under the operations in (7.14), the time center and time spread satisfy the following (see also Exercise 7.5 and Table 7.2):

(i) *Shift in time* by $n_0 \in \mathbb{Z}$, (7.14a), causes the time center to shift and leaves the time spread unchanged,

$$\mu_t(y) = \mu_t(x) + n_0, \tag{7.16a}$$

$$\Delta_t(y) = \Delta_t(x). \tag{7.16b}$$

(ii) *Shift in frequency (modulation)* by $\omega_0 \in (-\pi, \pi]$, (7.14b), leaves both the time center and the time spread unchanged,

$$\mu_t(y) = \mu_t(x), \tag{7.16c}$$
$$\Delta_t(y) = \Delta_t(x). \tag{7.16d}$$

(iii) *Upsampling* by $N \in \mathbb{Z}^+$ followed by ideal lowpass postfiltering, (7.14c), causes both the time center and the time spread to scale,

$$\mu_t(y) = N\mu_t(x), \tag{7.16e}$$
$$\Delta_t(y) = N\Delta_t(x). \tag{7.16f}$$

(iv) *Downsampling* by $N \in \mathbb{Z}^+$ preceded by ideal lowpass prefiltering, (7.14d), causes both the time center and the time spread to scale,

$$\mu_t(y) = \frac{1}{N}\mu_t(x), \tag{7.16g}$$
$$\Delta_t(y) = \frac{1}{N}\Delta_t(x), \tag{7.16h}$$

provided that the original signal is bandlimited, $x \in \mathrm{BL}[-\pi/N, \pi/N]$. Of course, when $x \in \mathrm{BL}[-\pi/N, \pi/N]$, the ideal lowpass prefiltering is simply scalar multiplication by $1/\sqrt{N}$; when x is not bandlimited, the effect of downsampling on the time center and the time spread is more complicated.

7.3.2 Localization in frequency

We now discuss localization of a sequence in frequency; the concepts are dual to those for localization in time and similar to the localization of a function in frequency. The normalized function $|X(e^{j\omega})|^2/\|X\|^2$ can be interpreted as the PDF of a random variable; we call the mean of that random variable the frequency center of x and the standard deviation of the same random variable the frequency spread of x.

Definition 7.6 (Frequency center and spread for sequences) The *frequency center* μ_f and the *frequency spread* Δ_f of $x \in \ell^2(\mathbb{Z})$ with DTFT X are

$$\mu_f = \frac{1}{2\pi\|x\|^2}\int_{-\pi}^{\pi} \omega\,|X(e^{j\omega})|^2\,d\omega, \tag{7.17a}$$
$$\Delta_f = \left(\frac{1}{2\pi\|x\|^2}\int_{-\pi}^{\pi} (\omega-\mu_f)^2\,|X(e^{j\omega})|^2\,d\omega\right)^{1/2}. \tag{7.17b}$$

Note that the frequency center will be 0 for all real sequences because of the symmetry of the magnitude of the DTFT. Unlike for functions, the frequency spread of a sequence is always finite; from the interpretation of $|X(e^{j\omega})|^2/\|X\|^2$ as a PDF, the frequency spread is the standard deviation of a random variable taking values in $[-\pi, \pi]$, so it cannot exceed π.

EXAMPLE 7.5 (FREQUENCY CENTERS AND SPREADS FOR SEQUENCES) To illustrate the concepts of frequency center and frequency spread, we compute them for the same sequences as in Example 7.4.

(i) The sinc sequence from Table 4.5 has $\mu_f = 0$ and $\Delta_f = \omega_0/(2\sqrt{3})$.
(ii) The right-sided box sequence of length n_0 from (3.13) has $\mu_f = 0$ and

$$\Delta_f = \left(\frac{1}{2\pi}\int_{-\pi}^{\pi}\omega^2\left(1+2\sum_{m=1}^{n_0-1}\left(1-\frac{m}{n_0}\right)\cos m\omega\right)d\omega\right)^{1/2}; \tag{7.18}$$

proving this is left for Exercise 7.4.

Under the operations in (7.14), the frequency center and frequency spread satisfy the following (see also Exercise 7.5 and Table 7.2):

(i) *Shift in time* by $n_0 \in \mathbb{Z}$, (7.14a), leaves both the frequency center and the frequency spread unchanged,

$$\mu_f(y) = \mu_f(x), \tag{7.19a}$$

$$\Delta_f(y) = \Delta_f(x). \tag{7.19b}$$

(ii) *Shift in frequency (modulation)* by $\omega_0 \in (-\pi, \pi]$, (7.14b), generally causes the frequency center and the frequency spread to change. These properties are analogous to those for functions only when a shift by ω_0 does not cause any frequency content to cross from $(-\pi, \pi]$ to outside this finite interval, since only this interval is used in defining the frequency center and frequency spread. Specifically, for $\omega_0 \in (0, \pi]$, if $x \in \ell^2(\mathbb{Z})$ has DTFT X satisfying $X(e^{j\omega}) = 0$ for $\omega \in (\pi - \omega_0, \pi]$, then the frequency center shifts and the frequency spread is unchanged,

$$\mu_f(y) = \mu_f(x) + \omega_0, \tag{7.19c}$$

$$\Delta_f(y) = \Delta_f(x). \tag{7.19d}$$

The same conclusion holds for $\omega_0 \in (-\pi, 0]$ when $X(e^{j\omega}) = 0$ for $\omega \in (-\pi, -\pi - \omega_0]$.

(iii) *Upsampling* by $N \in \mathbb{Z}^+$ followed by ideal lowpass postfiltering, (7.14c), causes both the frequency center and the frequency spread to scale,

$$\mu_f(y) = \frac{1}{N}\mu_f(x), \tag{7.19e}$$

$$\Delta_f(y) = \frac{1}{N}\Delta_f(x). \tag{7.19f}$$

(iv) *Downsampling* by $N \in \mathbb{Z}^+$ preceded by ideal lowpass prefiltering, (7.14d), causes both the frequency center and the frequency spread to scale,

$$\mu_f(y) = N\mu_f(x), \tag{7.19g}$$
$$\Delta_f(y) = N\Delta_f(x), \tag{7.19h}$$

provided that the original signal is bandlimited, $x \in \mathrm{BL}[-\pi/N, \pi/N]$. As before, when $x \in \mathrm{BL}[-\pi/N, \pi/N]$, the ideal lowpass prefiltering is simply scalar multiplication by $1/\sqrt{N}$; when x is not bandlimited, the effect of downsampling on the time center and the time spread is more complicated.

In Sections 7.4 and 7.5, we will show constructions of simple structured sets obtained by applying these basic operations to a single prototype sequence. Thus, knowing the time and frequency centers and spreads of the prototype sequence, we will know the time and frequency centers and spreads of all the other sequences in the set as well.

One-sided frequency center and spread We mentioned already that, because of the symmetry of the magnitude of the DTFT, the frequency center of any real sequence is 0. We might prefer a definition of the frequency center that captures the spectral characteristics of a real bandpass sequence.

DEFINITION 7.7 (ONE-SIDED FREQUENCY CENTER AND SPREAD) The *one-sided frequency center* μ_f^+ and the *one-sided frequency spread* Δ_f^+ of $x \in \ell^2(\mathbb{Z})$ with DTFT X are

$$\mu_f^+ = \frac{1}{\pi\|x\|^2} \int_0^\pi \omega\, |X(e^{j\omega})|^2\, d\omega, \tag{7.20a}$$
$$\Delta_f^+ = \left(\frac{1}{\pi\|x\|^2} \int_0^\pi (\omega - \mu_f^+)^2\, |X(e^{j\omega})|^2\, d\omega \right)^{1/2}. \tag{7.20b}$$

EXAMPLE 7.6 (FREQUENCY CENTER FOR A REAL BANDPASS SEQUENCE) Let x be a sequence formed by sampling the function from (7.9a) at $t = \frac{1}{2}n$, for $n = -10, -9, \ldots, 10$,

$$x_n = \exp\Big(-\frac{\pi}{8}(3+n)^2\Big) - \exp\Big(-\frac{\pi}{8}(3-n)^2\Big), \qquad n = -10, -9, \ldots, 10. \tag{7.21a}$$

Its DTFT is

$$X(e^{j\omega}) = \sum_{n=-10}^{10} \Big(\exp\Big(-\frac{\pi}{8}(3+n)^2\Big) - \exp\Big(-\frac{\pi}{8}(3-n)^2\Big) \Big)\, e^{-j\omega n}. \tag{7.21b}$$

Both the sequence and the magnitude of its DTFT are shown in Figure 7.10. Clearly, $\mu_f = 0$ because the magnitude of the spectrum has even symmetry.

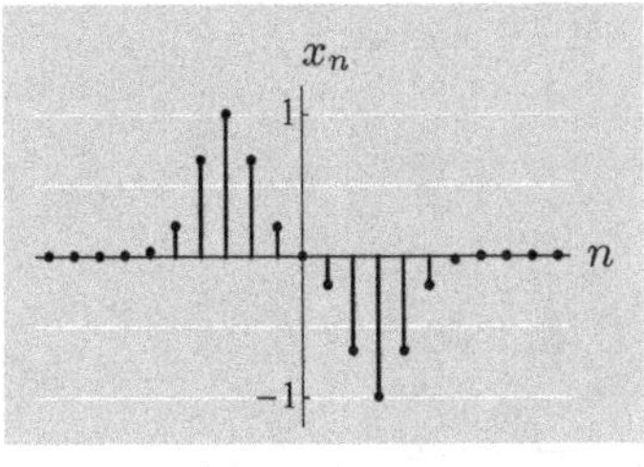

(a) Sequence.

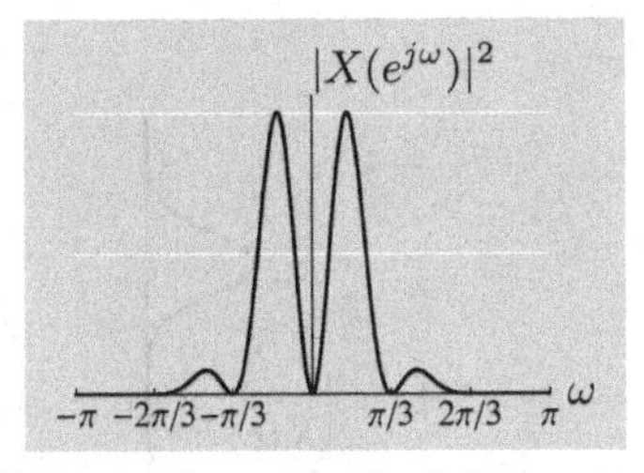

(b) Magnitude squared of the spectrum.

Figure 7.10 A sequence with bandpass DTFT spectrum.

However, in certain circumstances, we might want to know where the bulk of the spectrum is located on either side of the frequency axis; the frequency center in Definition 7.6 will not provide such information. Using Definition 7.7 to compute the one-sided frequency center of x, we get

$$\begin{aligned}\mu_{\mathrm{f}}^{+} &= \frac{1}{\pi\|x\|^2}\int_0^{\pi} \omega\,|X(e^{j\omega})|^2\,d\omega \\ &\approx \frac{1}{3.9967\pi}\int_0^{\pi} \omega\left|\sum_{n=-10}^{10}\left(e^{-\frac{1}{8}\pi(3+n)^2} - e^{-\frac{1}{8}\pi(3-n)^2}\right)e^{-j\omega n}\right|^2 d\omega \\ &\approx 0.5490,\end{aligned}$$

which is close to and slightly larger than $\frac{1}{6}\pi$, consistent with what we would guess by looking at Figure 7.10.

7.3.3 Uncertainty principle for sequences

Heisenberg box Similarly to functions, given a sequence x and its DTFT X, we have just introduced the 4-tuple $(\mu_{\mathrm{t}}, \Delta_{\mathrm{t}}, \mu_{\mathrm{f}}, \Delta_{\mathrm{f}})$ describing the sequence's center in time and frequency $(\mu_{\mathrm{t}}, \mu_{\mathrm{f}})$ and its spread in time and frequency $(\Delta_{\mathrm{t}}, \Delta_{\mathrm{f}})$. As before, we show this pictorially (see Figure 7.11), with a center of mass $(\mu_{\mathrm{t}}, \mu_{\mathrm{f}})$ around which a rectangular box of width $2\Delta_{\mathrm{t}}$ and height $2\Delta_{\mathrm{f}}$ is located, again producing a Heisenberg box, but now for sequences. As before, we show only the first quadrant of the time–frequency plane; we reserve the second quadrant for the squared magnitude response of the DTFT of the sequence, $|X|^2$, and the fourth quadrant for the squared absolute value of the sequence itself, $|x|^2$.

Given that we know the effects of the four basic operations (7.14) on the time and frequency centers and spreads of a sequence x, (7.16) and (7.19), we can deduce their effects on the Heisenberg box of the sequence (see also Table 7.2):

(i) *Shift in time* by $n_0 \in \mathbb{Z}$, (7.14a), causes the Heisenberg box to shift in time by n_0 (dashed box in Figure 7.12(a)).

(ii) *Shift in frequency (modulation)* by $\omega_0 \in (-\pi, \pi]$, (7.14b), causes the Heisenberg box to shift in frequency by ω_0 (dashed box in Figure 7.12(b)), provided

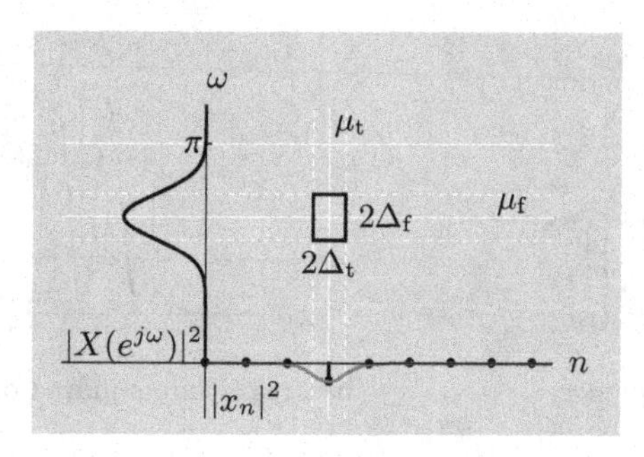

Figure 7.11 The time–frequency plane. The sequence x with DTFT X has an associated Heisenberg box centered at (μ_t, μ_f), of width $2\Delta_t$ and height $2\Delta_f$.

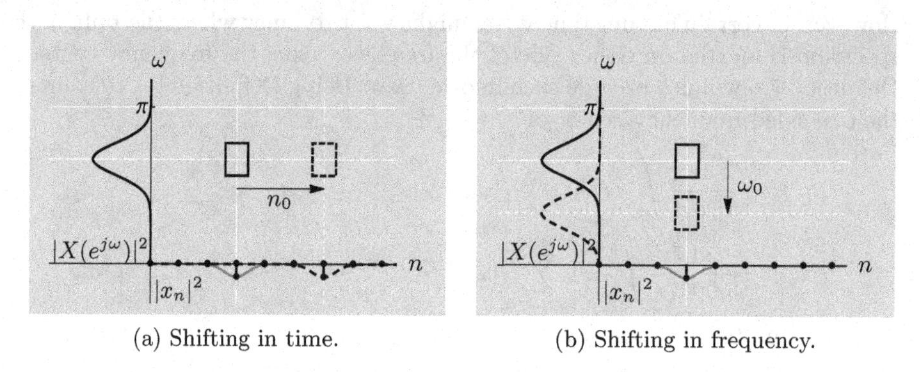

(a) Shifting in time. (b) Shifting in frequency.

Figure 7.12 Shifting of a sequence with Heisenberg box centered at (μ_t, μ_f), of width $2\Delta_t$ and height $2\Delta_f$. (a) Shifting in time by n_0, x_{n-n_0}, shifts the Heisenberg box to $(\mu_t + n_0, \mu_f)$; the area remains the same. (b) When no frequency content crosses from $(-\pi, \pi]$ to outside this interval, shifting in frequency by ω_0 (modulating), $X(e^{j(\omega-\omega_0)})$, shifts the Heisenberg box to $(\mu_t, \mu_f + \omega_0)$; the area remains the same. (Note that in the figure we used a negative ω_0.)

that the shift does not cause any frequency content to cross from $(-\pi, \pi]$ to outside this interval. For $\omega_0 \in (-\pi, 0]$, this requires $X(e^{j\omega}) = 0$ for $\omega \in (-\pi, -\pi-\omega_0]$; for $\omega_0 \in (0, \pi]$, this requires $X(e^{j\omega}) = 0$ for $\omega \in (\pi-\omega_0, \pi]$.

(iii) *Upsampling* by $N \in \mathbb{Z}^+$ followed by ideal lowpass postfiltering, (7.14c), causes the Heisenberg box to shift in time to $N\mu_t$ and in frequency to μ_f/N, and also to scale in time by N and in frequency by $1/N$ (dashed box in Figure 7.13).

(iv) *Downsampling* by $N \in \mathbb{Z}^+$ preceded by ideal lowpass prefiltering, (7.14d), causes the Heisenberg box to shift in time to μ_t/N and in frequency to $N\mu_f$, and also to scale in time by $1/N$ and in frequency by N (similarly to what is shown in Figure 7.13), provided that the original signal is bandlimited, $x \in \mathrm{BL}[-\pi/N, \pi/N]$.

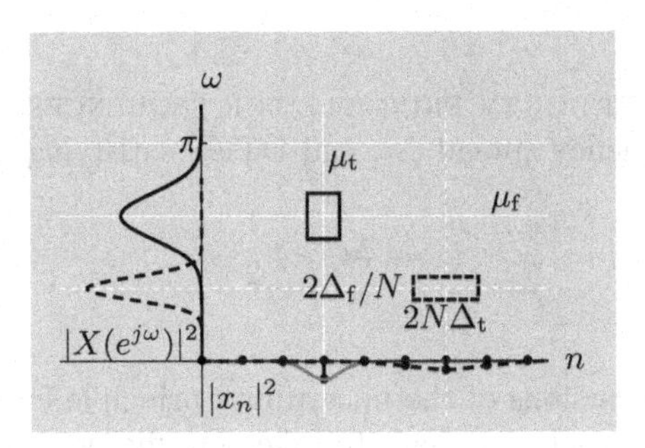

Figure 7.13 Upsampling followed by ideal lowpass postfiltering of a sequence with Heisenberg box centered at (μ_t, μ_f), of width $2\Delta_t$ and height $2\Delta_f$, shifts the Heisenberg box to $(N\mu_t, \mu_f/N)$ and scales its width to $2N\Delta_t$ and its height to $2\Delta_f/N$; the area remains the same. (Illustrated for $N = 2$.)

Sequence	Time center	Time spread	DTFT	Frequency center	Frequency spread
x_n	μ_t	Δ_t	$X(e^{j\omega})$	μ_f	Δ_f
x_{n-n_0}	$\mu_t + n_0$	Δ_t	$e^{-j\omega n_0} X(e^{j\omega})$	μ_f	Δ_f
$e^{j\omega_0 n} x_n$	μ_t	Δ_t	$X(e^{j(\omega-\omega_0)})$	$\mu_f + \omega_0$	Δ_f
upsampled & postfiltered	$N\mu_t$	$N\Delta_t$	$\sqrt{N} X(e^{jN\omega})$	μ_f/N	Δ_f/N
prefiltered & downsampled	μ_t/N	Δ_t/N	$\frac{1}{\sqrt{N}} X(e^{j\omega/N})$	$N\mu_f$	$N\Delta_f$

Table 7.2 Effects of shifts in time and frequency, as well as upsampling followed by postfiltering and downsampling preceded by prefiltering, on the Heisenberg box $(\mu_t, \Delta_t, \mu_f, \Delta_f)$. The properties for frequency center and spread after shift in frequency hold only when a shift by ω_0 does not cause any frequency content to cross from $(-\pi, \pi]$ to outside this interval. For the last row, the DTFT expression holds for $\omega \in (-\pi, \pi]$, and it is assumed that the signal is bandlimited, $x \in \mathrm{BL}[-\pi/N, \pi/N]$.

Uncertainty principle Shifting, upsampling followed by postfiltering, and downsampling preceded by prefiltering (under certain conditions) do not affect the area of the Heisenberg box associated with a sequence, even when the time and frequency spreads change. Thus, when manipulating a sequence through these basic operations, for one spread to decrease the other must increase. This agrees with what we have seen in Examples 7.4 and 7.5; a sequence that is narrow in one domain will be wide in the other. With the definitions paralleling those for continuous-time signals, we can thus obtain a result similar to Theorem 7.4; a proof is given in Exercise 7.6.

THEOREM 7.8 (UNCERTAINTY PRINCIPLE FOR SEQUENCES) Let $x \in \ell^2(\mathbb{Z})$ have time spread Δ_t, frequency spread Δ_f, and DTFT satisfying $X(e^{j\pi}) = 0$. Then,

$$\Delta_t \Delta_f > \frac{1}{2}. \tag{7.22}$$

Note that there exist versions of the uncertainty principle for sequences that use a different measure of frequency spread (to take periodicity into account) and do not require $X(e^{j\pi}) = 0$; see the *Further reading* for details.

EXAMPLE 7.7 (UNCERTAINTY BOUND DOES NOT APPLY) The Kronecker delta sequence from (3.8) is interesting to discuss as it possesses perfect time localization; it is easy to calculate that $\mu_t = 0$ and $\Delta_t = 0$. This means that the Heisenberg box for this sequence has zero width. Theorem 7.8 is not contradicted because it does not apply; the theorem requires $X(e^{j\pi}) = 0$, but in this case $X(e^{j\omega}) = 1$ for all $\omega \in \mathbb{R}$.

EXAMPLE 7.8 (UNCERTAINTY BOUND FOR BOX SEQUENCES) Using results from Examples 7.4 and 7.5 for the right-sided box sequence of length n_0, we can compute the time–frequency product

$$\Delta_t \Delta_f = \left(\frac{n_0^2 - 1}{24\pi} \int_{-\pi}^{\pi} \omega^2 \left(1 + 2 \sum_{m=1}^{n_0 - 1} \left(1 - \frac{m}{n_0} \right) \cos m\omega \right) d\omega \right)^{1/2}.$$

For any even value of n_0, the DTFT of the box sequence at $\omega = \pi$ is zero, so Theorem 7.8 applies to bound $\Delta_t \Delta_f$ from below. Table 7.3 gives this product for several lengths. As we can see, $\Delta_t \Delta_f$ increases monotonically, away from the lower bound, suggesting that the two-point box sequence is the most compact in terms of the time–frequency spread.

n_0	2	4	8	16
$\Delta_t \Delta_f$	0.5679	0.9211	1.3452	1.9176

Table 7.3 Time–frequency products of the box sequence for a few selected lengths.

7.3.4 Uncertainty principle for finite-length sequences

In addition to the uncertainty principle for infinite sequences, there exists a simple and powerful uncertainty principle for finite-length sequences and their DFTs.

THEOREM 7.9 (UNCERTAINTY PRINCIPLE FOR FINITE-LENGTH SEQUENCES) Let a nonzero $x \in \mathbb{C}^N$ have N_t nonzero entries and DFT X with N_f nonzero entries. Then,

$$N_t N_f \geq N. \tag{7.23}$$

Proof. The proof hinges on the fact that x having N_t nonzero entries implies that X cannot have N_t consecutive zeros, where *consecutive* is interpreted mod N. We prove this intermediate result by contradiction.

Suppose that there exist N_t consecutive zero entries in the DFT of x, X_{k+m} for $m = 0, 1, \ldots, N_t - 1$, for some $k \in \{0, 1, \ldots, N-1\}$. Denote by $i_0, i_1, \ldots, i_{N_t-1}$ the indices of the N_t nonzero entries of x, and use these indices to form $y \in \mathbb{C}^{N_t}$ by $y_n = x_{i_n}$ and $z_n = W_N^{i_n}$ for $n = 0, 1, \ldots, N_t - 1$. Observe that

$$X_{k+m} = \sum_{n=0}^{N-1} x_n W_N^{n(k+m)} \stackrel{(a)}{=} \sum_{n=0}^{N_t-1} y_n z_n^{(k+m)}, \qquad m = 0, 1, \ldots, N_t - 1,$$

where (a) follows from summing only over nonzero entries of x. This system of linear equations can be expressed as

$$\widehat{X} = \begin{bmatrix} X_k \\ \vdots \\ X_{k+N_t-1} \end{bmatrix} = Z \begin{bmatrix} y_0 \\ \vdots \\ y_{N_t-1} \end{bmatrix} = Zy,$$

where Z is an $N_t \times N_t$ matrix with elements $Z_{m,n} = z_n^{k+m}$, $0 \leq m, n \leq N_t - 1$.

By assumption, $\widehat{X}$ is a zero vector and the vector y is nonzero; hence, the matrix Z has a nontrivial null space. Since Z is a square matrix, it must be rank-deficient; that is, its rank is smaller than N_t. However, $Z = \widehat{Z}D$, where $\widehat{Z}_{m,n} = z_n^m$, $0 \leq m, n \leq N_t - 1$, and $D = \operatorname{diag}(z_0^k, z_1^k, \ldots, z_{N_t-1}^k)$. The matrix $\widehat{Z}$ is a Vandermonde matrix as in (2.248) constructed from nonzero entries; hence, it is of full rank N_t. The matrix D is a diagonal matrix with nonzero diagonal elements; hence, it is also of full rank N_t. As a product of two square, full-rank matrices, Z must also be a full-rank matrix, contradicting our previous statement. Thus, X cannot have N_t consecutive zero entries.

We now come to the result itself. Arrange the points of X in a circle and choose one nonzero entry to start from. Because of what we just showed, this nonzero entry can be followed by at most $(N_t - 1)$ zero ones. Continuing the argument until we reach the initial point, we will have at least $\lceil N/N_t \rceil$ nonzero entries. Thus, the total number of nonzero entries will be

$$N_f \geq \left\lceil \frac{N}{N_t} \right\rceil \geq \frac{N}{N_t} \quad \Rightarrow \quad N_t N_f \geq N.$$

Exercise 7.8 establishes a set of vectors that achieves equality in (7.23).

7.4 Tiling the time–frequency plane

Having seen the localization properties of *individual* functions and sequences in the previous two sections, a natural next step is to consider the localization properties of *sets* of functions or sequences. We consider sets generated by taking a prototype function and applying basic operations of time shift, frequency shift, and scaling from (7.3), or taking a prototype sequence and applying basic operations of time shift, frequency shift, upsampling followed by lowpass postfiltering, and downsampling preceded by lowpass prefiltering, from (7.14). Compared with arbitrary sets, these structured sets can be described simply because of the presence of a single prototype function or sequence and a few parameters to describe the shifting and scaling operations. This simplicity is essential in applications, in particular because the structure often leads to efficient algorithms for analysis and synthesis. These algorithms and the properties of structured bases and frames are central to the companion volume [57].

Displaying all the Heisenberg boxes for a specific set of functions or sequences is termed the *actual time–frequency plot* of the set; that set of Heisenberg boxes might have overlaps or gaps. In this section, we concentrate on selecting combinations of operations from (7.3) or (7.14) that lead to actual time–frequency plots that, in a loose sense, have Heisenberg boxes that are evenly spaced on the time–frequency plane – even when the boxes have different aspect ratios. We will also introduce *idealized time–frequency tilings* where only the time and frequency centers play a role.

7.4.1 Localization for structured sets of functions

In Figures 7.5–7.7, we illustrated the concepts of the time–frequency plane, localization of a function described by the 4-tuple $(\mu_{\mathrm{t}}, \Delta_{\mathrm{t}}, \mu_{\mathrm{f}}, \Delta_{\mathrm{f}})$, and basic operations of shifting in time, shifting in frequency, and scaling. We use a Heisenberg box for each individual function in a set $\Phi = \{\varphi_i\}_{i\in\mathbb{Z}}$ to visually represent that set.

If the set Φ has no particular structure, the time–frequency plane display of the corresponding Heisenberg boxes will look quite random and is probably of little interest. We thus consider only structured sets, related to desired features such as some invariance to shifts in time/frequency, and to computational efficiency.

Given a prototype function φ, generate a collection of functions by using the following operations from (7.3):

(i) *Shifts in time* by mt_0, $m \in \mathbb{Z}$, $t_0 \in \mathbb{R}^+$,

$$\Phi = \{\varphi(t - mt_0)\}_{m\in\mathbb{Z}}; \tag{7.24a}$$

see Figure 7.14(a).

(ii) *Shifts in frequency (modulation)* by $k\omega_0$, $k \in \mathbb{Z}$, $\omega_0 \in \mathbb{R}^+$,

$$\Phi = \{e^{jk\omega_0 t}\varphi(t)\}_{k\in\mathbb{Z}}; \tag{7.24b}$$

see Figure 7.14(b).

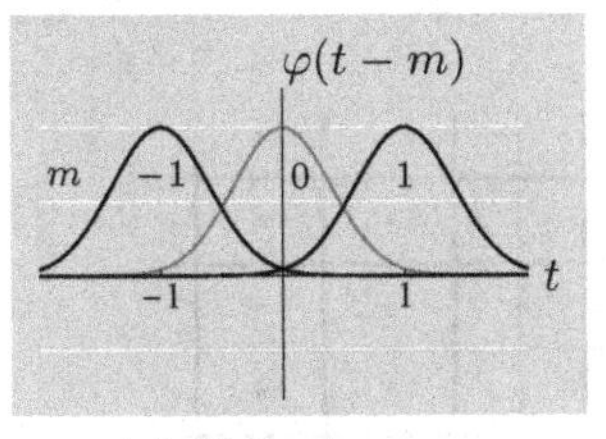

(a) Shifts in time.

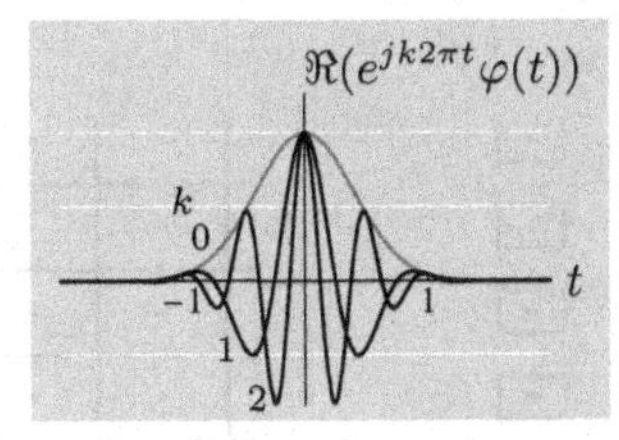

(b) Shifts in frequency.

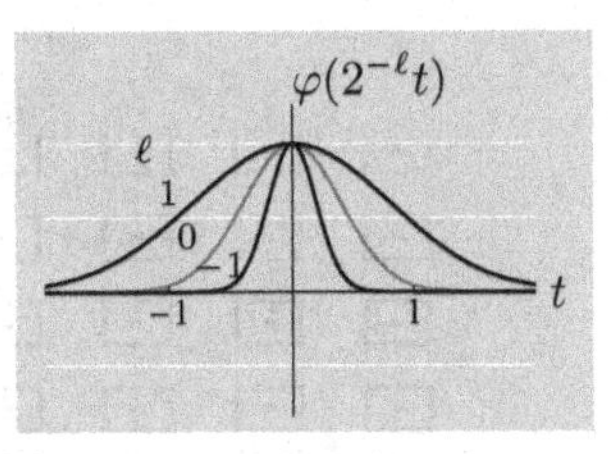

(c) Scales in time/frequency.

Figure 7.14 Basic operations (7.24) on a Gaussian function $\varphi(t) = 2^{1/4}e^{-\pi t^2}$. (Illustrated for $t_0 = 1$, $\omega_0 = 2\pi$, and $a = 2$.)

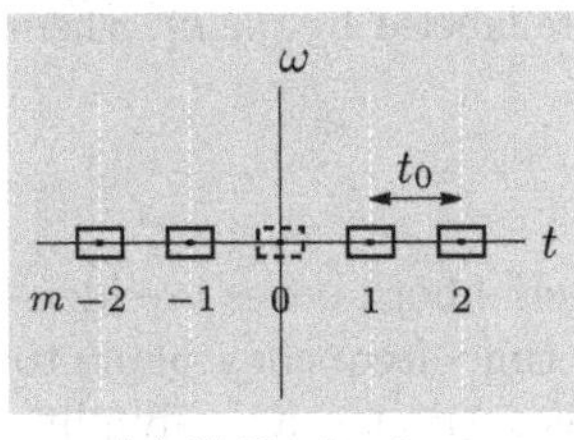

(a) Shifts in time.

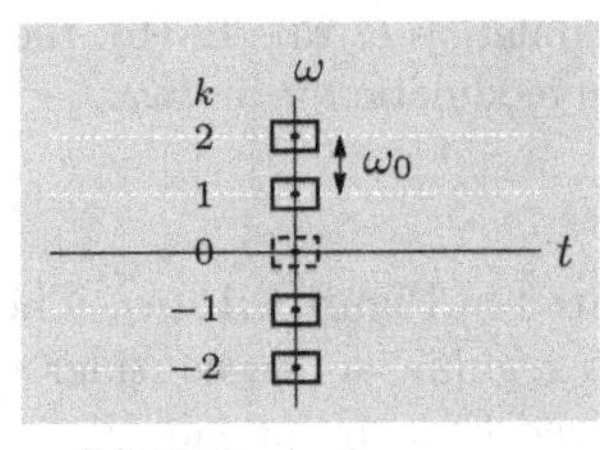

(b) Shifts in frequency.

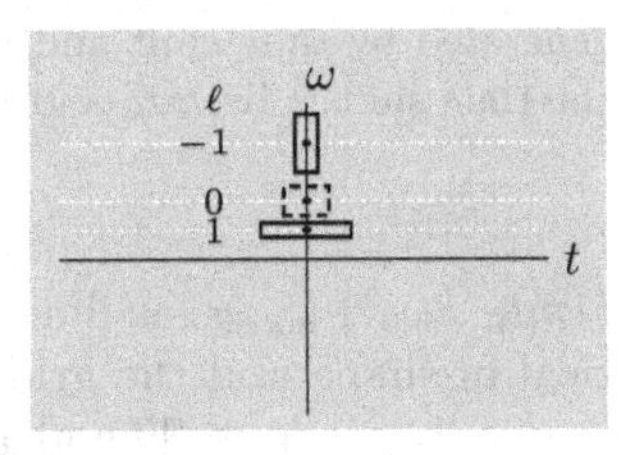

(c) Scales in time/frequency.

Figure 7.15 Actual time–frequency plots resulting from basic operations (7.24) on functions. The Heisenberg box of the prototype function φ is highlighted (dashed lines).

(iii) *Scales in time* by a^ℓ, $a \in (1, \infty)$, $\ell \in \mathbb{Z}$,

$$\Phi = \{\varphi(a^{-\ell}t)\}_{\ell\in\mathbb{Z}}; \tag{7.24c}$$

see Figure 7.14(c). Since $a > 1$, the scale of the function decreases for $\ell < 0$ and increases for $\ell > 0$.

When the operations used are shifts in time and frequency, the prototype function is typically a lowpass function (a function centered around the origin in time and frequency); when the operation used is scaling, the prototype function is typically a bandpass function (a function centered around some nonzero frequency ω_0). Figure 7.15 displays examples of actual time–frequency plots resulting from these operations.

Sets generated by time shift and modulation Given a prototype function φ (typically lowpass) and $t_0, \omega_0 \in \mathbb{R}^+$, generate the set

$$\varphi_{k,m}(t) = e^{jk\omega_0 t}\varphi(t - mt_0), \qquad k, m \in \mathbb{Z}. \tag{7.25}$$

Since the time shifts are on a regular grid $\{mt_0\}_{m\in\mathbb{Z}}$ and the frequency shifts are also on a regular grid $\{k\omega_0\}_{k\in\mathbb{Z}}$, the actual time–frequency plot of the set consists of the Heisenberg box of φ along with its shifts on a two-dimensional grid

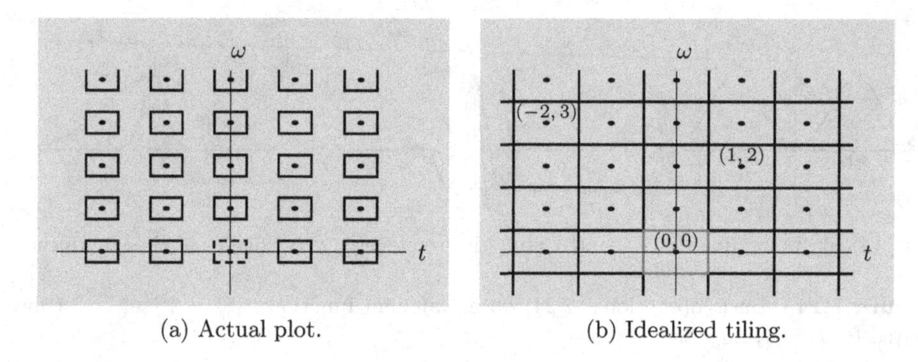

(a) Actual plot. (b) Idealized tiling.

Figure 7.16 Actual time–frequency plot and idealized time–frequency tiling for sets generated by time shift and modulation (7.25). In (b), tiles are labeled by (m, k), where the time shift is by mt_0 and the modulation is by $k\omega_0$.

$\{(mt_0, k\omega_0)\}_{m,k\in\mathbb{Z}}$, as illustrated in Figure 7.16(a). The Heisenberg boxes are identical in shape and the grid is regular, so we consider the time–frequency plane to be evenly covered. The idealized time–frequency tiling is constructed as a simplification of the actual time–frequency plot using just the grid of time and frequency shifts $\{(mt_0, k\omega_0)\}_{m,k\in\mathbb{Z}}$, without using the dimensions of the Heisenberg box of φ, as illustrated in Figure 7.16(b) (we assumed that $\mu_t = \mu_f = 0$). As we will discuss in Section 7.5.1, local Fourier bases have tilings of this form.

Sets generated by time shift and scaling Given a prototype function φ (typically bandpass), the set of scales in (7.24c) gives Heisenberg boxes of differing aspect ratios as shown in Figure 7.15(c) for $\mu_t = 0$ and $\mu_f > 0$. Using equally spaced shifts in time would create a higher density of Heisenberg boxes for low $|\omega|$ and a lower density for high $|\omega|$. Instead, for $t_0 \in \mathbb{R}^+$, $a \in (1, \infty)$, generate the set

$$\varphi_{\ell,m}(t) = \varphi(a^{-\ell}t - mt_0), \qquad \ell, m \in \mathbb{Z}. \tag{7.26}$$

In this set of functions, the scaled version $\varphi(a^{-\ell}t)$ appears along with its shifts by integer multiples of $a^{\ell}t_0$ because

$$\varphi(a^{-\ell}t - mt_0) = \varphi(a^{-\ell}(t - ma^{\ell}t_0)).$$

At scale ℓ, the time spread is proportional to a^{ℓ} and the time shifts are on the grid $\{ma^{\ell}t_0\}_{m\in\mathbb{Z}}$, so the time–frequency plane is again considered to be evenly covered. The actual time–frequency plot is illustrated in Figure 7.17(a), while the idealized time–frequency tiling of this set of functions for $a = 2$ is shown in Figure 7.17(b); we assumed that $\mu_t = 0$. Such a tiling is called *dyadic* because of scaling by 2. As we will discuss in Section 7.5.1, wavelet bases have tilings of this form.

7.4.2 Localization for structured sets of sequences

What we just saw for sets of functions can be carried over to sets of sequences as well. In Figures 7.11–7.13, we illustrated the concepts of the time–frequency

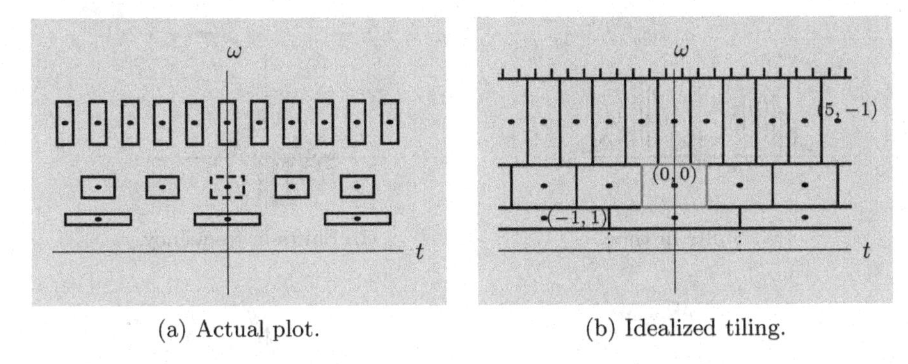

(a) Actual plot. (b) Idealized tiling.

Figure 7.17 Actual time–frequency plot and idealized time–frequency tiling for sets generated by time shift and scaling (7.26). In (b), tiles are labeled by (m, ℓ), where the time shift is by $ma^{\ell}t_0$ and the scaling is by $a^{-\ell}$. (Illustrated for $a = 2$.)

plane, localization of a sequence described by the 4-tuple $(\mu_t, \Delta_t, \mu_f, \Delta_f)$, and basic operations of shifting in time, shifting in frequency, and discrete scaling (upsampling followed by ideal lowpass postfiltering and downsampling preceded by ideal lowpass prefiltering). We use a Heisenberg box for each individual sequence in a set $\Phi = \{\varphi_i\}_{i\in\mathbb{Z}}$ to visually represent that set, and, as before, we distinguish actual time–frequency plots from idealized time–frequency tilings. One distinction from the previous discussion is that we will no longer assume that the filters we use are ideal lowpass filters.

Given a prototype sequence φ, generate a collection of sequences by using the following operations from (7.14):

(i) *Shifts in time* by mn_0, $m \in \mathbb{Z}, n_0 \in \mathbb{Z}^+$,

$$\Phi = \{\varphi_{n-mn_0}\}_{m\in\mathbb{Z}}; \tag{7.27a}$$

see Figure 7.18(a).

(ii) *Shifts in frequency (modulation)* by $k\omega_0$, $k \in \{0, 1, \ldots, N-1\}$, $\omega_0 \in (-\pi, \pi]$,

$$\Phi = \{e^{jk\omega_0 n}\varphi_n\}_{k\in\{0,1,\ldots,N-1\}}; \tag{7.27b}$$

see Figure 7.18(b). The frequency ω_0 is typically chosen as $\omega_0 = 2\pi/N$ for some $N \in \mathbb{Z}^+$; since the DTFT of φ is 2π-periodic, the finite set of k above produces all N distinct frequency shifts, as the rest are identical mod 2π.

(iii) *Upsampling* by $N^{\ell-1}$, $\ell \in \mathbb{Z}^+$, $N \in \mathbb{Z}^+$, $N > 1$, followed by appropriate postfiltering,

$$\Phi = \left\{g^{(\ell-1)} * \varphi^{(N^{\ell-1})}\right\}_{\ell\in\mathbb{Z}^+}, \tag{7.27c}$$

where $\varphi^{(N^{\ell-1})}$ denotes the prototype sequence φ upsampled by $N^{\ell-1}$, $g^{(\ell-1)}$ is an appropriate lowpass filter that depends on scale with $g^{(1)} = g$ and $g^{(0)} = \delta$, and N is typically a small integer; see Figure 7.18(c). Often, the lowpass filter

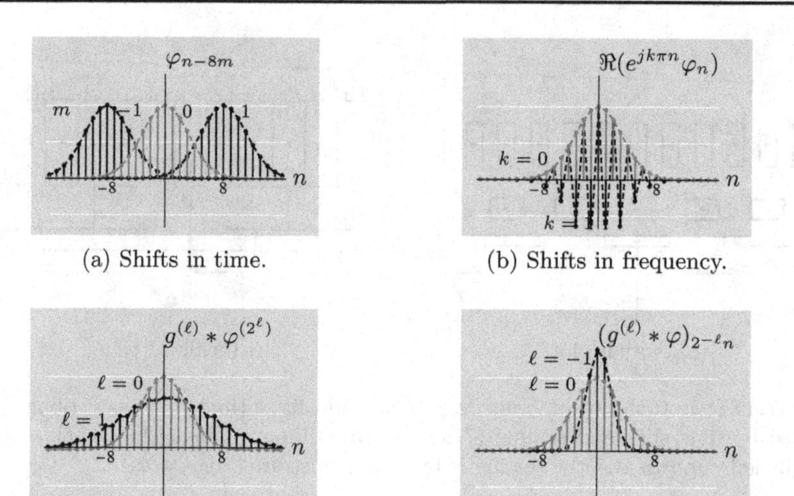

(a) Shifts in time. (b) Shifts in frequency.

(c) Upsampling followed by filtering. (d) Downsampling preceded by filtering.

Figure 7.18 Basic operations (7.27) on a Gaussian sequence $\varphi_n = 2^{1/4}e^{-\pi(n/8)^2}$. (Illustrated for $n_0 = 8$, $\omega_0 = \pi$, $N = 2$, and $g_n = (\delta_n + \delta_{n-1})/\sqrt{2}$.)

$g^{(\ell)}$ is created as a convolution of upsampled versions of a prototype lowpass filter g with itself. Because of upsampling and multiple convolutions, this type of scaling creates long sequences and is thus an increase of scale.

Example 7.9 (Upsampling) Let $N = 2$, the lowpass filter be $g_n = (\delta_n + \delta_{n-1})/\sqrt{2}$, and the prototype sequence be $\varphi_n = \delta_n - \delta_{n-1}$.

The resulting sequence at scale $\ell = 1$ is

$$g^{(0)} * \varphi^{(2^0)} = \delta * \varphi = \varphi,$$

that is, the prototype sequence.

The resulting sequence at scale $\ell = 2$ is the convolution of g and the upsampled-by-2 version of φ,

$$\begin{aligned}(g^{(1)} * \varphi^{(2^1)})_n &= \frac{1}{\sqrt{2}}(\delta_n + \delta_{n-1}) *_n (\delta_n - \delta_{n-2}) \\ &= \frac{1}{\sqrt{2}}(\delta_n + \delta_{n-1} - \delta_{n-2} - \delta_{n-3}).\end{aligned}$$

The resulting sequence at scale $\ell = 3$ is the convolution of $g^{(2)}$ and the upsampled-by-4 version of φ. When $g^{(2)}$ is constructed by convolving

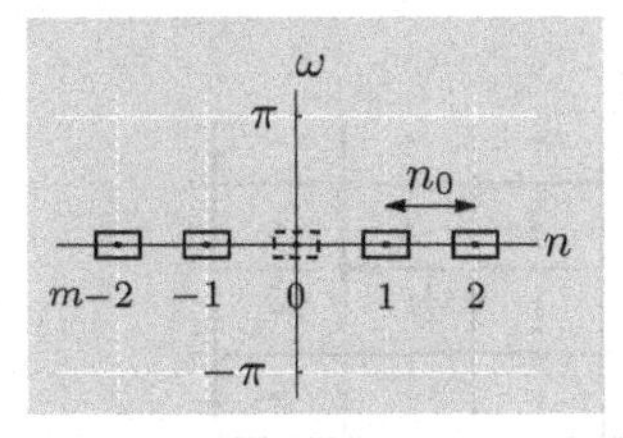

(a) Shifts in time.

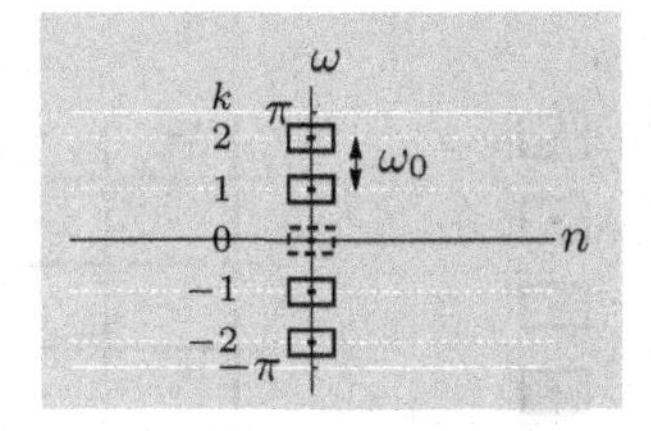

(b) Shifts in frequency (modulation).

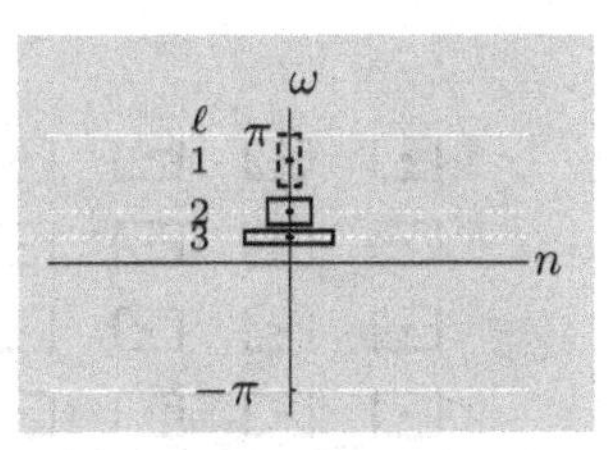

(c) Upsampling followed by filtering.

Figure 7.19 Time–frequency tilings resulting from basic operations (7.27) on sequences. The Heisenberg box of the prototype sequence φ is highlighted (dashed lines).

g with the upsampled-by-2 version of g, the resulting sequence is

$$(g^{(2)} * \varphi^{(2^2)})_n = \frac{1}{2}((\delta_n + \delta_{n-1}) *_n (\delta_n + \delta_{n-2})) *_n (\delta_n - \delta_{n-4})$$
$$= \frac{1}{2}(\delta_n + \delta_{n-1} + \delta_{n-2} + \delta_{n-3}) *_n (\delta_n - \delta_{n-4})$$
$$= \frac{1}{2}(\delta_n + \delta_{n-1} + \delta_{n-2} + \delta_{n-3} - \delta_{n-4} - \delta_{n-5} - \delta_{n-6} - \delta_{n-7}).$$

Note that for illustration, here we have used the construction of the lowpass filter $g^{(\ell)}$ as a convolution of upsampled versions of a prototype lowpass filter g with itself; this is not necessary in general.

We see that, indeed, increasing ℓ increases the length of the sequence.

When the operations used are shifts in time and frequency, the prototype sequence is typically a lowpass sequence (a sequence centered around the origin in time and frequency). Note that when the operations used are shifts in time and discrete-time scaling, we use upsampling only. The prototype φ then determines the lowest available scale (this is a difference between sequences and functions), and is typically a highpass sequence. Figure 7.19 displays the time–frequency tilings resulting from these operations; one period of the frequency axis is shown.

Sets generated by time shift and modulation Given a prototype sequence φ (typically lowpass) and $n_0, N \in \mathbb{Z}^+$, generate the set

$$\varphi_{k,m,n} = e^{jk\omega_0 n}\varphi_{n-mn_0}, \qquad k \in \{0, 1, \ldots, N-1\},\ m \in \mathbb{Z}, \tag{7.28}$$

where $\omega_0 = 2\pi/N$. The actual time–frequency plot is illustrated in Figure 7.20(a), while the idealized time–frequency tiling for $\mu_t = \mu_f = 0$ is shown in Figure 7.20(b). As we will discuss in Section 7.5.2, local Fourier bases have tilings of this form.

Sets generated by time shift and scaling Given a prototype sequence φ (typically highpass), the set of scales in (7.27c) gives Heisenberg boxes of differing aspect ratios

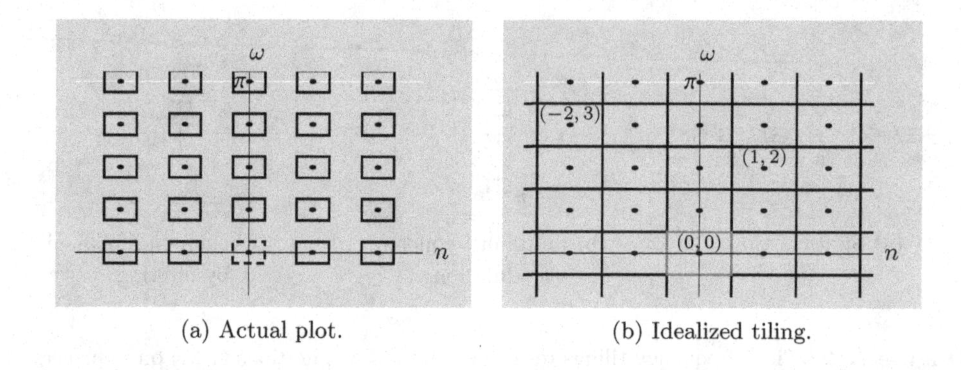

(a) Actual plot. (b) Idealized tiling.

Figure 7.20 Actual time–frequency plot and idealized time–frequency tiling for sets generated by time shift and modulation (7.28). In (b), tiles are labeled by (m, k), where the time shift is by mn_0 and the modulation is by $k\omega_0 = k2\pi/N$. (Illustrated for $N = 4$.)

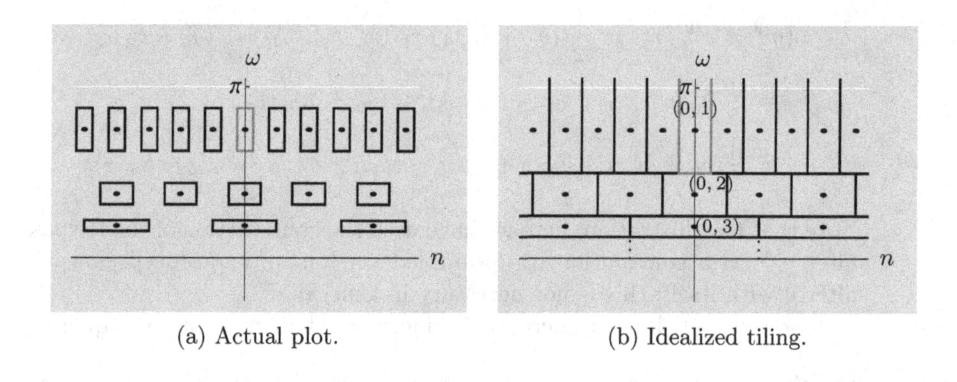

(a) Actual plot. (b) Idealized tiling.

Figure 7.21 Actual time–frequency plot and idealized time–frequency tiling for sets generated by time shift and scaling (7.29). In (b), tiles are labeled by (m, ℓ), where the time shift is by $mN^{\ell-1}n_0$ and the scaling is by $N^{-\ell}$. (Illustrated for $n_0 = N = 2$.)

as shown in Figure 7.19(c) for $\mu_t = 0$ and $\mu_f > 0$. Using equally spaced shifts in time would create a higher density of Heisenberg boxes for low $|\omega|$ and a lower density for high $|\omega|$. Instead, for $N, n_0 \in \mathbb{Z}^+$, $N > 1$, generate the set

$$\varphi_{\ell,m,n} = \left(g^{(\ell-1)} * \varphi^{(N^{\ell-1})}\right)_{n-mN^{\ell-1}n_0}, \qquad \ell \in \mathbb{Z}^+,\ m \in \mathbb{Z}. \tag{7.29}$$

At scale ℓ, the time spread is proportional to $N^{\ell-1}$ and the shifts are on the grid $\{mN^{\ell-1}n_0\}_{m\in\mathbb{Z}}$, so the time–frequency plane is evenly covered. The actual time–frequency plot is illustrated in Figure 7.21(a), while the idealized time–frequency tiling for such sets with $N = n_0 = 2$ and $\mu_t = 0$ is shown in Figure 7.21(b). Such a *dyadic* tiling can be obtained, for example, with the discrete-time wavelet basis we discuss in Section 7.5.2.

7.5 Examples of local Fourier and wavelet bases

The discussion of time–frequency tilings and structured bases thus far has been mostly conceptual. We now show examples of simple orthonormal bases with the structured time–frequency tilings developed in Section 7.4. Local Fourier bases are generated by time shift and modulation of a prototype function/sequence, and wavelet bases are generated by time shift and scaling of a prototype function/sequence. Constructing bases with desirable time–frequency localization and other properties is a major topic of the companion volume, [57].

7.5.1 Local Fourier and wavelet bases for functions

Local Fourier bases for functions

Our first task is to create an orthonormal basis generated from a single prototype function by time shift and modulation as in Section 7.4.1, ideally producing a tiling as in Figure 7.16(b).

Basis Construct an orthonormal set generated from a prototype function φ by shifts mt_0, $m \in \mathbb{Z}$, and modulations $k\omega_0$, $k \in \mathbb{Z}$, as in (7.25), where

$$\varphi(t) \;=\; w(t), \quad t_0 \;=\; 1, \quad \omega_0 \;=\; 2\pi,$$

and w is the box function from (4.8a),

$$w(t) \;=\; \begin{cases} 1, & \text{for } |t| \le \frac{1}{2}; \\ 0, & \text{otherwise.} \end{cases} \tag{7.30a}$$

Together, these yield a *local Fourier basis*

$$\varphi_{k,m}(t) \;=\; e^{jk2\pi t} w(t-m), \qquad k, m \in \mathbb{Z}; \tag{7.30b}$$

a few of the basis functions are shown in Figure 7.22. According to Theorem 4.14, for a fixed shift m, the set $\{\varphi_{k,m}\}_{k\in\mathbb{Z}}$ forms an orthonormal basis for $\mathcal{L}^2([m-\frac{1}{2}, m+\frac{1}{2}))$. For two different shifts $m \neq \ell$, the sets $\{\varphi_{k,m}\}_{k\in\mathbb{Z}}$ and $\{\varphi_{k,\ell}\}_{k\in\mathbb{Z}}$ are orthogonal to each other because the supports of the functions do not overlap. Thus, $\{\varphi_{k,m}\}_{k,m\in\mathbb{Z}}$ is an orthonormal set. Below, we will establish that $\{\varphi_{k,m}\}_{k,m\in\mathbb{Z}}$ is a basis for $\mathcal{L}^2(\mathbb{R})$ by showing the completeness of this set.

Expansion We compute the expansion coefficients of $x \in \mathcal{L}^2(\mathbb{R})$ with respect to the local Fourier basis (7.30) as

$$\begin{aligned} X_{k,m} \;&=\; \langle x, \varphi_{k,m}\rangle \;=\; \int_{-\infty}^{\infty} x(t)\, \varphi_{k,m}^*(t)\, dt \\ &\overset{(a)}{=}\; \int_{-\infty}^{\infty} x(t)\, e^{-jk2\pi t} w(t-m)\, dt \;\overset{(b)}{=}\; \int_{m-1/2}^{m+1/2} x(t)\, e^{-jk2\pi t}\, dt, \end{aligned} \tag{7.31a}$$

where (a) follows from (7.30b); and (b) from the finite support of w. For a fixed m, $\{X_{k,m}\}_{k\in\mathbb{Z}}$ is the Fourier series coefficient sequence of $\widetilde{x}_m(t) = x(t+m)w(t)$.

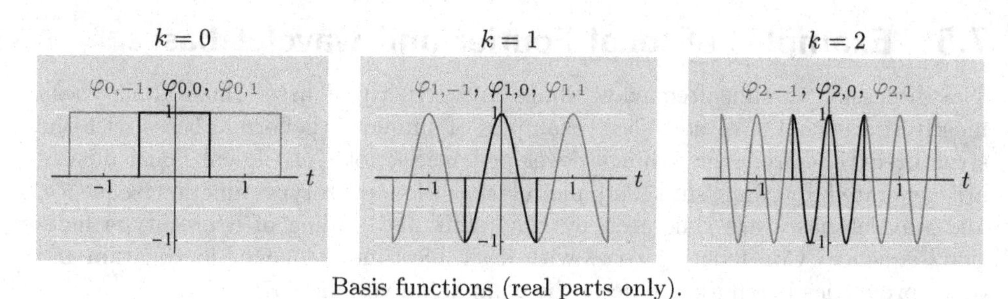

Basis functions (real parts only).

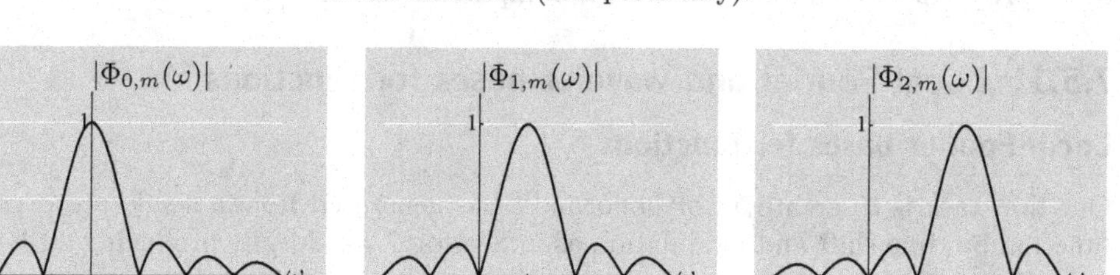

Magnitudes of the Fourier transform.

Figure 7.22 Example local Fourier basis functions (real parts only) from (7.30b) and the magnitudes of their Fourier transforms. For a given k, the absolute values $|\Phi_{k,m}|$ are the same for all $m \in \mathbb{Z}$.

By Theorem 4.15(i), the Fourier series reconstruction $\widehat{x}_m$ obtained from $\{X_{k,m}\}_{k\in\mathbb{Z}}$ satisfies

$$\|x_m - \widehat{x}_m\| = 0$$

using the $\mathcal{L}^2$ norm. Thus, since x can be decomposed into a sum of pieces over unit-length intervals

$$x(t) = \sum_{m\in\mathbb{Z}} \widetilde{x}_m(t-m),$$

we may combine the Fourier series reconstructions, (4.94b), of $\{\widetilde{x}_m\}_{m\in\mathbb{Z}}$, to obtain

$$x(t) = \sum_{m\in\mathbb{Z}} \sum_{k\in\mathbb{Z}} X_{k,m} e^{jk2\pi t} w(t-m), \tag{7.31b}$$

where the equality means that

$$\lim_{M\to\infty} \lim_{K\to\infty} \left\| x(t) - \sum_{m=-M}^{M} \sum_{k=-K}^{K} X_{k,m} e^{jk2\pi t} w(t-m) \right\| = 0,$$

again using the $\mathcal{L}^2$ norm. The expansion (7.31b) generally suffers from the Gibbs phenomenon at interval boundaries, since the 1-periodic version of each $\widetilde{x}_m$ is usually discontinuous at $\{k + \frac{1}{2}\}_{k\in\mathbb{Z}}$ (the periodized version of $\widetilde{x}_m$ is continuous only if x is continuous on $[m - \frac{1}{2}, m + \frac{1}{2}]$ and $x(m - \frac{1}{2}) = x(m + \frac{1}{2})$).

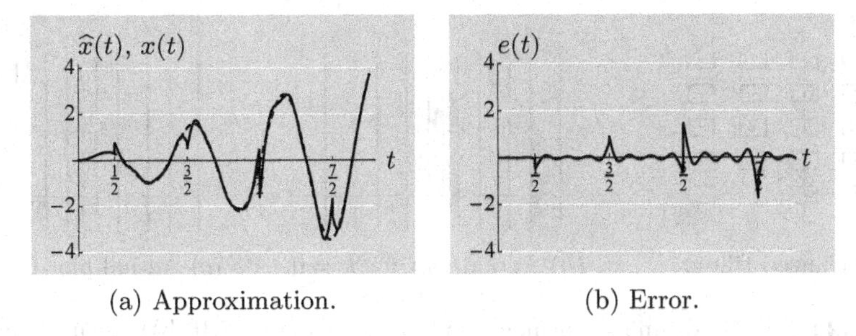

(a) Approximation. (b) Error.

Figure 7.23 Approximation $\widehat{x}$ (solid line) and approximation error $e = x - \widehat{x}$ of x (dashed line) from (7.32) on $[0, 4]$ using the local Fourier basis with $k = -3, -2, \ldots, 3$.

EXAMPLE 7.10 (APPROXIMATION WITH A LOCAL FOURIER BASIS) Let

$$x(t) = \begin{cases} t \sin 5t, & \text{for } -\frac{1}{2} \leq t \leq \frac{9}{2}; \\ 0, & \text{otherwise}, \end{cases} \tag{7.32}$$

that is, the function x_1 from (6.4) restricted to $[-\frac{1}{2}, \frac{9}{2}]$. Figure 7.23 shows an approximation of x on $[0, 4]$ that is obtained by truncating (7.31b) to frequencies $k = -3, -2, \ldots, 3$. Since the approximation uses the seven lowest frequencies per interval and the approximation is shown for an interval of length 4, it is essentially an approximation formed with 28 expansion coefficients. Discontinuities of the approximation at $\frac{1}{2}$, $\frac{3}{2}$, $\frac{5}{2}$, and $\frac{7}{2}$ are apparent. The sizes of the discontinuities would not be diminished by including additional frequencies, but the $\mathcal{L}^2$ norm of the approximation error would vanish.

Time–frequency localization How similar is the actual time–frequency plot of this basis to the idealized tiling from Figure 7.16(b) (repeated in Figure 7.24(a))? From Examples 7.1 and 7.2, we know that, for the prototype box function with $t_0 = 1$, the time spread is $\Delta_t = 1/(2\sqrt{3})$, while the frequency spread Δ_f is infinite, making the time–frequency product unbounded. The Heisenberg boxes for $\varphi_{k,m}$ will be centered at $(m, 2\pi k)$, $m, k \in \mathbb{Z}$, of width $1/\sqrt{3}$; the height is unbounded, however, leaving us far from the desired tiling (see Figures 7.24(b) and (c)).

Haar wavelet basis for functions

Our next task is to create an orthonormal basis generated from a single prototype function by time shift and scaling as in Section 7.4.1, ideally producing a tiling as in Figure 7.17(b).

Basis Construct an orthonormal set generated from a prototype function φ by shifts mt_0, $m \in \mathbb{Z}$, and scalings a^ℓ, $\ell \in \mathbb{Z}$, as in (7.26), where

$$\varphi(t) = \psi(t), \quad t_0 = 1, \quad a = 2,$$

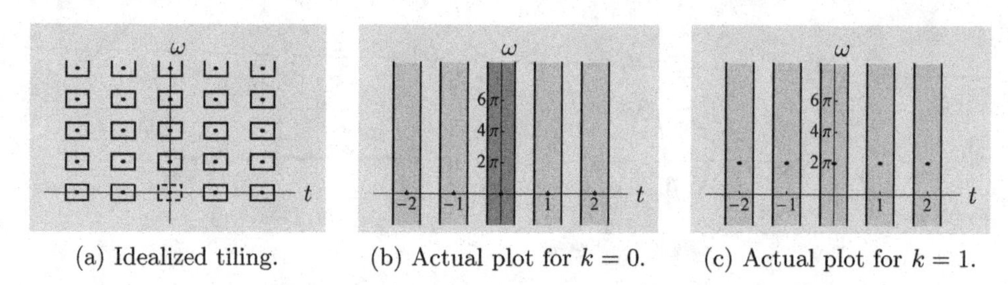

(a) Idealized tiling. (b) Actual plot for $k = 0$. (c) Actual plot for $k = 1$.

Figure 7.24 Idealized time–frequency tiling (same as Figure 7.16(b)) versus actual time–frequency plots for the local Fourier basis in (7.30b). The Heisenberg box of the prototype function, centered at $(0, 0)$, of width $1/\sqrt{3}$ and infinite height, is shown in dark gray, while the Heisenberg boxes of the rest of the basis functions are shown in light gray. Because Δ_f is infinite, for a fixed m and $k \in \mathbb{Z}$, boxes overlap in frequency; we thus show the tilings for $k = 0$ and $k = 1$ separately for clarity.

and ψ is the Haar wavelet function from (1.6),

$$\psi(t) = \begin{cases} 1, & \text{for } 0 \le t < \frac{1}{2}; \\ -1, & \text{for } \frac{1}{2} \le t < 1; \\ 0, & \text{otherwise}. \end{cases} \tag{7.33a}$$

Together, these yield the *dyadic Haar wavelet basis*

$$\varphi_{\ell,m}(t) = 2^{-\ell/2}\psi(2^{-\ell}t - m), \qquad \ell, m \in \mathbb{Z}; \tag{7.33b}$$

a few of the basis functions are shown in Figure 7.25. The prototype function, the Haar wavelet ψ, is bandpass in nature; in particular, $\Psi(0) = 0$. For a fixed scale ℓ, the basis functions are orthogonal to each other since their supports do not overlap; when their supports do overlap, their inner product is zero because one changes sign over the constant span of the other. Moreover, a simple calculation shows that each one is of unit norm; thus, the set is orthonormal. One can show that the set $\{\varphi_{\ell,m}\}_{\ell,m\in\mathbb{Z}}$ is also a basis for $\mathcal{L}^2(\mathbb{R})$ (see the *Further reading*).

Expansion We compute the expansion coefficients of $x \in \mathcal{L}^2(\mathbb{R})$ with respect to the Haar wavelet basis (7.33) as

$$\begin{aligned} X_{\ell,m} &= \langle x, \varphi_{\ell,m}\rangle = \int_{-\infty}^{\infty} x(t)\,\varphi^*_{\ell,m}(t)\,dt \\ &\overset{(a)}{=} 2^{-\ell/2}\int_{-\infty}^{\infty} x(t)\,\psi(2^{-\ell}t - m)\,dt \\ &\overset{(b)}{=} 2^{-\ell/2}\left(\int_{2^\ell m}^{2^\ell(m+1/2)} x(t)\,dt - \int_{2^\ell(m+1/2)}^{2^\ell(m+1)} x(t)\,dt\right), \end{aligned} \tag{7.34a}$$

where (a) follows from (7.33b); and (b) from (7.33a). The expansion is

$$x(t) = \sum_{\ell\in\mathbb{Z}} 2^{-\ell/2}\sum_{m\in\mathbb{Z}} X_{\ell,m}\psi(2^{-\ell}t - m). \tag{7.34b}$$

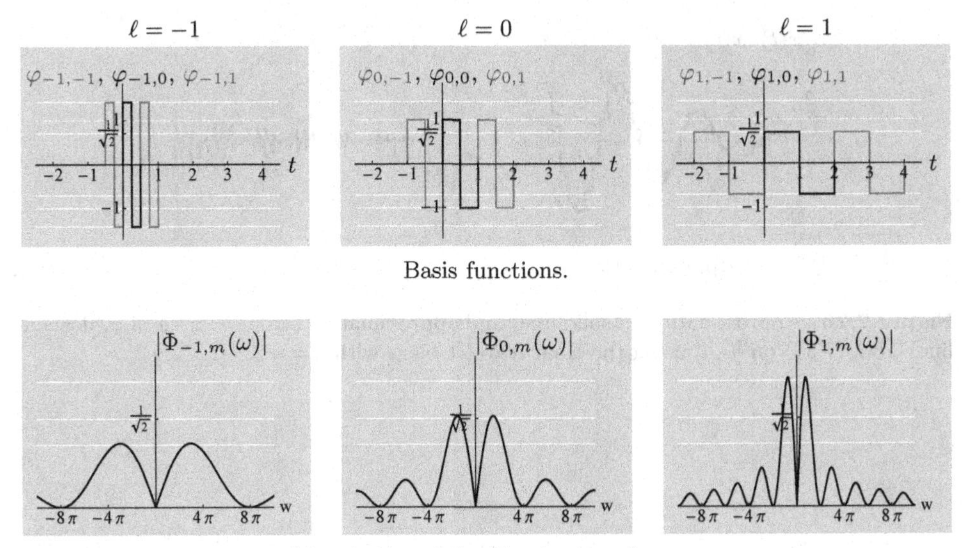

Basis functions.

Magnitudes of the Fourier transform.

Figure 7.25 Example Haar wavelet basis functions from (7.33b) and the magnitudes of their Fourier transforms. For a given ℓ, the absolute values $|\Phi_{\ell,m}|$ are the same for all $m \in \mathbb{Z}$.

In practice, one often fixes a coarsest scale J and uses the wavelets at scales $\ell = -\infty, \ldots, J-1, J$. The contributions from the omitted scales $\ell = J+1, J+2, \ldots, \infty$ are then represented with the *scaling function* (1.8) at scale J together with its shifts, $\{2^{-J/2}\varphi(2^{-J}t - m)\}_{m\in\mathbb{Z}}$. The expansion then becomes

$$x(t) = 2^{-J/2} \sum_{m\in\mathbb{Z}} \widetilde{X}_{J,m} \varphi(2^{-J}t - m) + \sum_{\ell=-\infty}^{J} 2^{-\ell/2} \sum_{m\in\mathbb{Z}} X_{\ell,m} \psi(2^{-\ell}t - m), \quad (7.34c)$$

where $\widetilde{X}_{J,m} = \langle x(t), 2^{-J/2}\varphi(2^{-J}t - m)\rangle_t$ is the expansion coefficient with respect to the scaling function at scale J and shift m. For more details, see the companion volume, [57].

EXAMPLE 7.11 (APPROXIMATION WITH A WAVELET BASIS) Figure 7.26 shows an approximation of the function x from (7.32) on $[0, 4]$ by truncating (7.34c) to 32 terms: one scaling coefficient for scale $\ell = 2$ and 31 wavelet coefficients for scales $\ell = -2, -1, \ldots, 2$ and shifts $m = 0, 1, \ldots, 2^{-(\ell-2)} - 1$. The approximation using this basis looks quite different from that obtained using the local Fourier basis, which was shown in Figure 7.23. Comparing Figure 7.26(b) to Figure 7.23(b), rather than seeing the Gibbs phenomenon, we see that the magnitude of the error is more uniform.

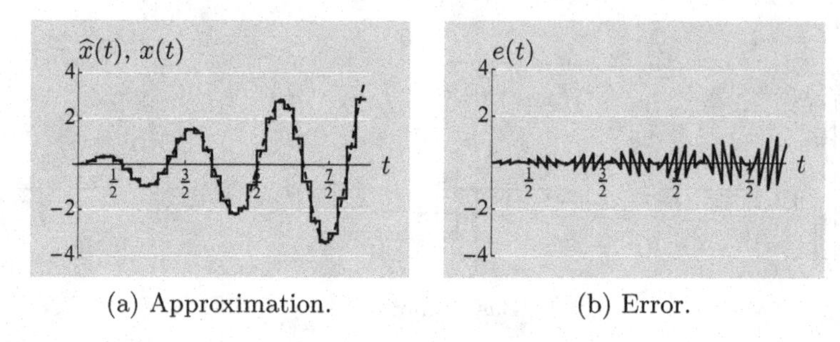

(a) Approximation. (b) Error.

Figure 7.26 Approximation $\widehat{x}$ (solid line) and approximation error $e = x - \widehat{x}$ of x (dashed line) from (7.32) on $[0, 4]$ using the Haar wavelet basis with $\ell = -2, -1, \ldots, 2$

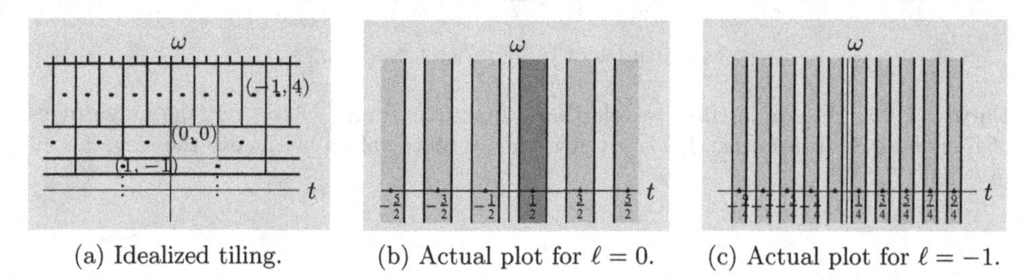

(a) Idealized tiling. (b) Actual plot for $\ell = 0$. (c) Actual plot for $\ell = -1$.

Figure 7.27 Idealized time–frequency tiling (same as Figure 7.17(b) shifted) versus actual time–frequency plots for the dyadic Haar wavelet basis in (7.33b). The Heisenberg box of the prototype function, centered at $(\frac{1}{2}, 0)$, of width $1/\sqrt{3}$ and infinite height, is shown in dark gray, while the Heisenberg boxes of the rest of the basis functions are shown in light gray. Because Δ_{f} is infinite, for a fixed m and $\ell \in \mathbb{Z}$, boxes overlap in frequency; we thus show the tilings for $\ell = 0$ and $\ell = -1$ separately for clarity.

Time–frequency localization How similar is the actual time–frequency plot of this basis to the idealized tiling from Figure 7.17(b) (repeated in Figure 7.27(a), shifted because the prototype basis function has $\mu_{\mathrm{t}} = \frac{1}{2}$)? For the prototype function, we can compute the time spread to be $\Delta_{\mathrm{t}} = 1/(2\sqrt{3})$, and the frequency spread Δ_{f} to be infinite, making the time–frequency product unbounded, just as for the local Fourier basis. The Heisenberg boxes for $\varphi_{\ell,m}$ will be centered at $(m2^{\ell}, 0)$, $\ell, m \in \mathbb{Z}$, of width $1/(2^{-\ell}\sqrt{3})$; the height is unbounded, however, leaving us far from the desired tiling (see Figure 7.27(b) and (c)). In fact, not only are the boxes far from the ideal boxes, but also the time–frequency centers all have $\mu_{\mathrm{f}} = 0$ because the basis functions are real. In this case, not even the one-sided frequency center μ_{f}^{+} from (7.8a) would help since the integral does not converge.

7.5.2 Local Fourier and wavelet bases for sequences

We now consider the discrete-time counterparts of the continuous-time bases we have just seen, namely local Fourier bases and the Haar wavelet basis.

Local Fourier bases for sequences

As in continuous time, our first task is to create an orthonormal basis generated from a single prototype sequence by time shift and modulation as in Section 7.4.2, ideally producing a tiling as in Figure 7.20(b).

Basis Construct an orthonormal set generated from a prototype sequence φ by shifts mn_0, $m \in \mathbb{Z}$, and modulations $k\omega_0$, $k \in \{0, 1, \ldots, N-1\}$, as in (7.28), where

$$\varphi_n = w_n, \quad n_0 = N, \quad \omega_0 = 2\pi/N,$$

and w is the normalized right-sided box sequence from (3.13),

$$w_n = \begin{cases} 1/\sqrt{N}, & \text{for } 0 \leq n \leq N-1; \\ 0, & \text{otherwise.} \end{cases} \tag{7.35a}$$

Together, these yield a *discrete local Fourier basis*

$$\varphi_{k,m,n} = e^{jk(2\pi/N)n} w_{n-mN}, \qquad k \in \{0, 1, \ldots, N-1\},\ m \in \mathbb{Z}; \tag{7.35b}$$

a few of the basis sequences are shown in Figure 7.28. According to (3.165), for a fixed shift m, the set $\{\varphi_{k,m}\}_{k\in\{0,1,\ldots,N-1\}}$ forms a basis for $\ell^2(\{mN, mN+1, \ldots, (m+1)N-1\})$. For two different shifts $m \neq \ell$, the sets $\{\varphi_{k,m}\}_{k\in\{0,1,\ldots,N-1\}}$ and $\{\varphi_{k,\ell}\}_{k\in\{0,1,\ldots,N-1\}}$ are orthogonal to each other because the supports of the sequences do not overlap. Thus, $\{\varphi_{k,m}\}_{k\in\{0,1,\ldots,N-1\}}, m \in \mathbb{Z}$ is an orthonormal set. Below, we will establish that $\{\varphi_{k,m}\}_{k\in\{0,1,\ldots,N-1\},m\in\mathbb{Z}}$ is a basis for $\ell^2(\mathbb{Z})$ by showing the completeness of this set.

Expansion We compute the expansion coefficients of $x \in \ell^2(\mathbb{Z})$ with respect to the discrete local Fourier basis (7.35) as

$$\begin{aligned} X_{k,m} &= \langle x, \varphi_{k,m} \rangle = \sum_{n\in\mathbb{Z}} x_n \varphi^*_{k,m,n} \\ &\overset{(a)}{=} \sum_{n\in\mathbb{Z}} x_n e^{-jk(2\pi/N)n} w_{n-mN} \overset{(b)}{=} \frac{1}{\sqrt{N}} \sum_{n=mN}^{(m+1)N-1} x_n e^{-jk(2\pi/N)n} \\ &= \frac{1}{\sqrt{N}} \sum_{n=mN}^{(m+1)N-1} x_n W_N^{kn}, \end{aligned} \tag{7.36a}$$

where (a) follows from (7.35b); and (b) from the support of w. For a fixed m, $\{X_{k,m}\}_{k\in\{0,1,\ldots,N-1\}}$ is the DFT coefficient sequence of $\widetilde{x}_{m,n} = x_{n+mN} w_n$. By (3.163b), we can obtain the DFT reconstruction of $\widetilde{x}_m$

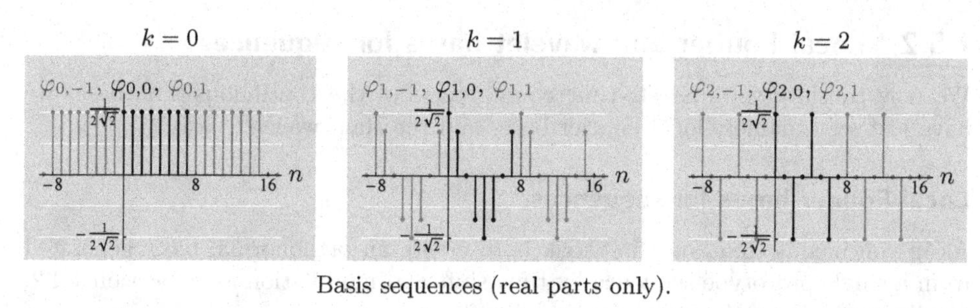

Basis sequences (real parts only).

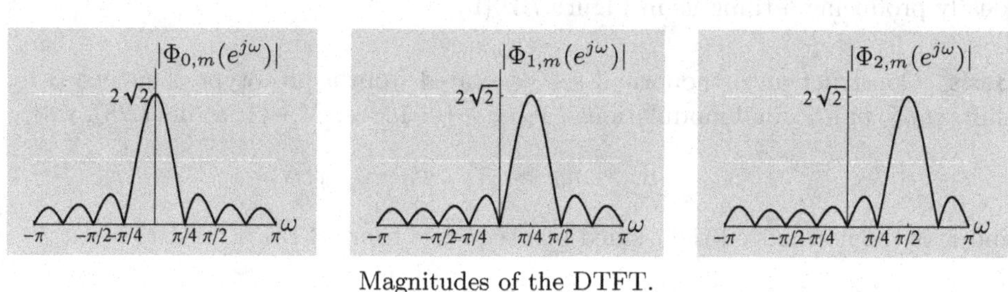

Magnitudes of the DTFT.

Figure 7.28 Example local Fourier basis sequences (real parts only) from (7.35b) and the magnitudes of their DTFTs. For a given k, the absolute values $|\Phi_{k,m}|$ are the same for all $m \in \mathbb{Z}$. (Illustrated for $N = 8$.)

from $\{X_{k,m}\}_{m\in\mathbb{Z},k\in\{0,1,\ldots,N-1\}}$. Thus, since x can be decomposed into a sum of pieces over intervals of length N,

$$x_n = \sqrt{N}\sum_{m\in\mathbb{Z}} \widetilde{x}_{m,n-mN}$$

(where the factor $1/\sqrt{N}$ appears because of (7.35a)), we may combine the DFT reconstructions, (3.163b), of $\{\widetilde{x}_m\}_{m\in\mathbb{Z}}$, to obtain

$$x_n = \frac{1}{\sqrt{N}}\sum_{m\in\mathbb{Z}}\sum_{k=0}^{N-1} X_{k,m}e^{jk(2\pi/N)n}w_{n-mN}. \tag{7.36b}$$

This expansion is often called a *block-by-block transform*, or simply a *block transform*.

Example 7.12 (Approximation with a discrete local Fourier basis) Let

$$x_n = \begin{cases} \frac{1}{8}n\sin\left(\frac{5}{8}n\right), & \text{for } n = 0, 1, \ldots, 31; \\ 0, & \text{otherwise,} \end{cases} \tag{7.37}$$

that is, the function from (7.32) sampled at $t = \frac{1}{8}n$. Figure 7.29 shows an approximation of x on $0, 1, \ldots, 31$ that is obtained with $N = 8$ by truncating (7.36b) to frequencies $k = -2, -1, \ldots, 2$. Since the approximation uses the five lowest frequencies per interval and the approximation is shown on four nonoverlapping intervals, it is essentially an approximation formed with 20 expansion coefficients.

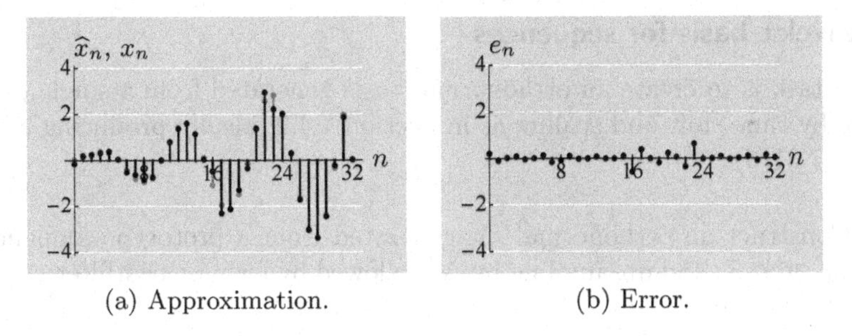

(a) Approximation. (b) Error.

Figure 7.29 Approximation $\widehat{x}$ (black stems) and approximation error $e = x - \widehat{x}$ of x (gray stems) from (7.37) on 0, 1, ..., 31 using the local Fourier basis with $k = -2, -1, \ldots, 2$ and $N = 8$.

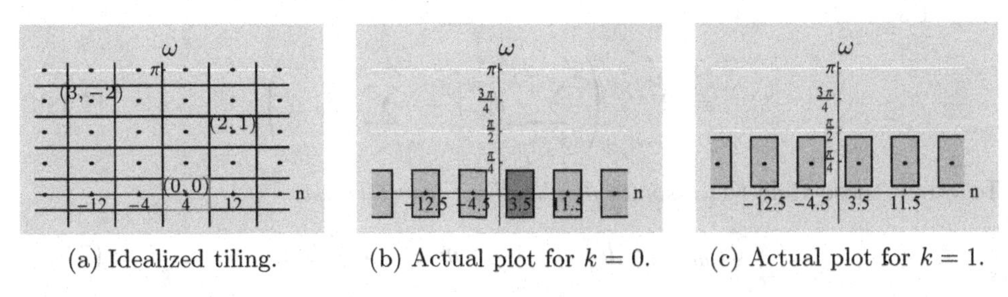

(a) Idealized tiling. (b) Actual plot for $k = 0$. (c) Actual plot for $k = 1$.

Figure 7.30 Idealized time–frequency tiling (same as Figure 7.20(b) shifted) versus actual time–frequency plots for the local Fourier basis in (7.35b). The Heisenberg box of the prototype sequence, centered at $(3.5, 0)$, of width 4.5826 and height 1.1742, is shown in dark gray in (b), while the Heisenberg boxes of the rest of the basis sequences are shown in light gray. Because for a fixed m and $k \in \mathbb{Z}$ boxes overlap in frequency, we show tilings for $k = 0$ and $k = 1$ separately for clarity. Note that the areas of the Heisenberg boxes are not preserved across shifts in frequency, since the prototype sequence is not bandlimited.

Time–frequency localization How similar is the actual time–frequency plot of this basis to the idealized tiling from Figure 7.20(b)? From Examples 7.4 and 7.5, we know that, for the prototype window sequence with $n_0 = N$, the time spread is $\Delta_t = \sqrt{N^2-1}/(2\sqrt{3})$; unlike for functions, the frequency spread Δ_f is no longer unbounded, as we have seen in (7.18). For $k = 0$, the Heisenberg boxes for $\varphi_{0,m}$ will be centered at $(mN + \frac{1}{2}(N-1),\, 0)$, $m \in \mathbb{Z}$, of width $\sqrt{(N^2-1)/3}$; the height is $2\Delta_f$ with Δ_f as in (7.18). Because φ is not bandlimited, the shift in frequency causes the frequency spread to change; the frequency center is in general not shifted by the amount of shift in frequency. Because of this, unlike for functions, the area of the Heisenberg box will change. Figure 7.30 illustrates this for $N = 8$.

Haar wavelet basis for sequences

Our final task is to create an orthonormal basis generated from a single prototype sequence by time shift and scaling as in Section 7.4.2, ideally producing a tiling as in Figure 7.21(b).

Basis Construct an orthonormal set generated from a prototype sequence φ by shifts mn_0, $m \in \mathbb{Z}$, and upsampling by N^ℓ followed by lowpass postfiltering, $\ell \in \mathbb{Z}^+$, as in (7.29), where

$$\varphi_n = h_n, \quad n_0 = 2, \quad N = 2,$$

and h is the Haar wavelet sequence

$$h_n = \begin{cases} 1/\sqrt{2}, & \text{for } n = 0; \\ -1/\sqrt{2}, & \text{for } n = 1; \\ 0, & \text{otherwise}, \end{cases} \tag{7.38a}$$

and, at scale ℓ,

$$h_n^{(\ell)} = 2^{-\ell/2} \left(\sum_{k=0}^{2^{\ell-1}-1} \delta_{n-k} - \sum_{k=2^{\ell-1}}^{2^\ell-1} \delta_{n-k} \right). \tag{7.38b}$$

Together, these yield the *discrete dyadic Haar wavelet basis*

$$\varphi_{\ell,m,n} = h^{(\ell)}_{n-2^\ell m}, \qquad m \in \mathbb{Z},\ \ell \in \mathbb{Z}^+; \tag{7.38c}$$

a few of the basis sequences are shown in Figure 7.31. The prototype sequence, the Haar wavelet h, is highpass in nature; in particular, $H(e^{j\omega})|_{\omega=0} = H(1) = 0$. For a fixed scale ℓ, the basis sequences are orthogonal to each other since their supports do not overlap; when their supports do overlap, their inner product is zero because one changes sign over the constant span of the other. Moreover, a simple calculation shows that each one is of unit norm; thus, the set is orthonormal. One can show that the set $\{\varphi_{\ell,m}\}_{\ell\in\mathbb{Z}^+, m\in\mathbb{Z}}$ is also a basis for $\ell^2(\mathbb{Z})$; for details, see the companion volume [57].

Expansion We compute the expansion coefficients of $x \in \ell^2(\mathbb{Z})$ with respect to the discrete Haar wavelet basis (7.38) as

$$\begin{aligned} X_{\ell,m} &= \langle x, \varphi_{\ell,m} \rangle = \sum_{n\in\mathbb{Z}} x_n \varphi^*_{\ell,m,n} \overset{(a)}{=} \sum_{n\in\mathbb{Z}} x_n h^{(\ell)}_{n-2^\ell m} \\ &\overset{(b)}{=} \sum_{n=2^\ell m}^{2^\ell(m+1)-1} x_n 2^{-\ell/2} \left(\sum_{k=0}^{2^{\ell-1}-1} \delta_{n-2^\ell m-k} - \sum_{k=2^{\ell-1}}^{2^\ell-1} \delta_{n-2^\ell m-k} \right), \end{aligned} \tag{7.39a}$$

where (a) follows from (7.38c); and (b) from (7.38b). The expansion is

$$x_n = \sum_{\ell\in\mathbb{Z}^+} \sum_{m\in\mathbb{Z}} X_{\ell,m} h^{(\ell)}_{n-2^\ell m}. \tag{7.39b}$$

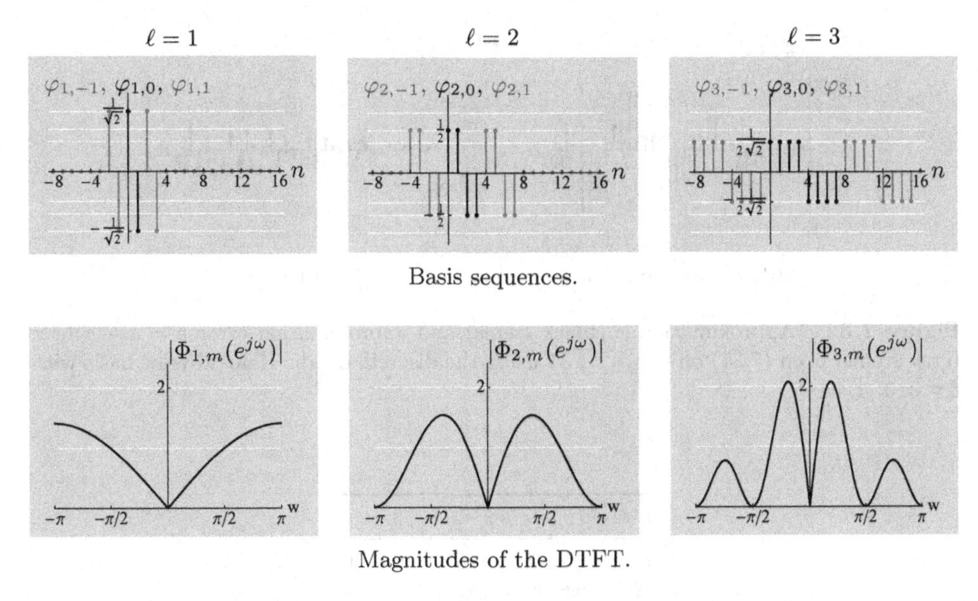

Figure 7.31 Example Haar wavelet basis sequences from (7.38c) and the magnitudes of their DTFTs. For a given ℓ, the absolute values $|\Phi_{\ell,m}|$ are the same for all $m \in \mathbb{Z}$.

In practice, one often fixes a coarsest scale J and uses the wavelet sequences at scales $\ell = 1, 2, \ldots, J$. The contributions from the omitted scales $\ell = J+1, J+2, \ldots, \infty$ are then represented with the *scaling sequence*, a right-sided box sequence g of length 2, at scale J together with its shifts, $\{g^{(J)}_{n-2^J m}\}_{m\in\mathbb{Z}}$, where $g^{(J)}$ is

$$g^{(J)}_n = 2^{-J/2} \sum_{k=0}^{2^J-1} \delta_{n-k}. \tag{7.39c}$$

The expansion then becomes

$$x_n = \sum_{m\in\mathbb{Z}} \widetilde{X}_{J,m} g^{(J)}_{n-2^J m} + \sum_{\ell=1}^{J} \sum_{m\in\mathbb{Z}} X_{\ell,m} h^{(\ell)}_{n-2^\ell m}, \tag{7.39d}$$

where $\widetilde{X}_{J,m} = \langle x_n, g^{(J)}_{n-2^J m}\rangle_n$ is the expansion coefficient with respect to g at scale J and shift m; $\widetilde{X}$ and X are often referred to as the DWT coefficients. For more details, see the companion volume, [57].

Example 7.13 (Approximation with a discrete Haar wavelet basis) Figure 7.32 shows an approximation of the sequence x from (7.37) on $0, 1, \ldots, 31$ by truncating (7.39d) to 16 terms: one scaling coefficient for scale $\ell = 5$ and 15 wavelet coefficients for scales $\ell = 2, 3, 4, 5$ and shifts $m = 0, 1, \ldots, 2^{(5-\ell)} - 1$. The approximation using this basis looks quite different from that using the local Fourier basis shown in Figure 7.29.

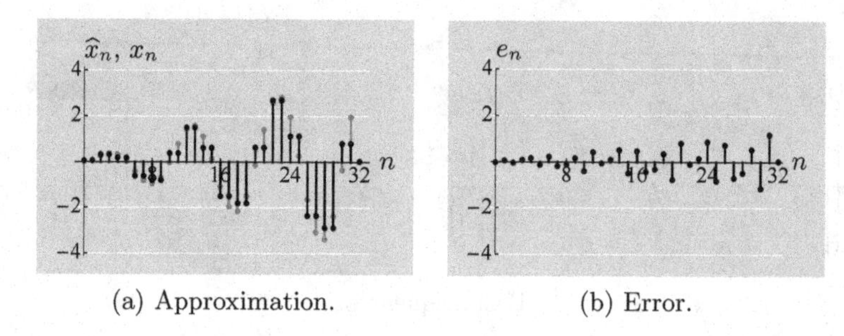

(a) Approximation. (b) Error.

Figure 7.32 Approximation $\widehat{x}$ (black stems) and approximation error $e = x - \widehat{x}$ of x (gray stems) from (7.37) on 0, 1, ..., 31 using the discrete dyadic Haar wavelet basis with $\ell = 2, 3, 4, 5$.

ℓ	1	2	3
μ_f^+	2.2074	1.2879	0.8150
Δ_f^+	0.6459	0.4924	0.5904

Table 7.4 One-sided frequency centers and spreads for Haar basis sequences.

Time–frequency localization How similar is the actual time–frequency plot of this basis to the idealized tiling from Figure 7.21(b) (repeated in Figure 7.33(a) and shifted because the prototype basis sequence is not centered around zero)? For the prototype sequence and all basis sequences for $\ell = 1$, we compute the time spread to be $\Delta_t = \frac{1}{2}$ and the frequency spread to be $\Delta_f = \sqrt{2 + \pi^2/3} \approx 2.3000$. The Heisenberg boxes all have $\mu_f = 0$ because the basis sequences are real; they correspond to the highest-frequency boxes in Figure 7.33(a) and are shown in Figure 7.33(b). For $\ell = 2$, $\Delta_t = \sqrt{5}/2$, and $\Delta_f \approx 1.3788$; they also all have $\mu_f = 0$ and correspond to the next-highest-frequency boxes in Figure 7.33(a) and are shown in Figure 7.33(c).

This is an example where the one-sided concepts come in handy; the one-sided frequency centers and spreads for $\ell = 1, 2, 3$ are given in Table 7.4; the corresponding plots for $\ell = 1$ and $\ell = 2$ are shown in Figures 7.33(d) and (e).

7.6 Recap and a glimpse forward

We are now ready to close this chapter, and the book, with a summary of tools and a discussion of issues that arise when adapting these tools to real-world problems. We end with a few bite-size illustrations of locality and localized bases on real-world signals.

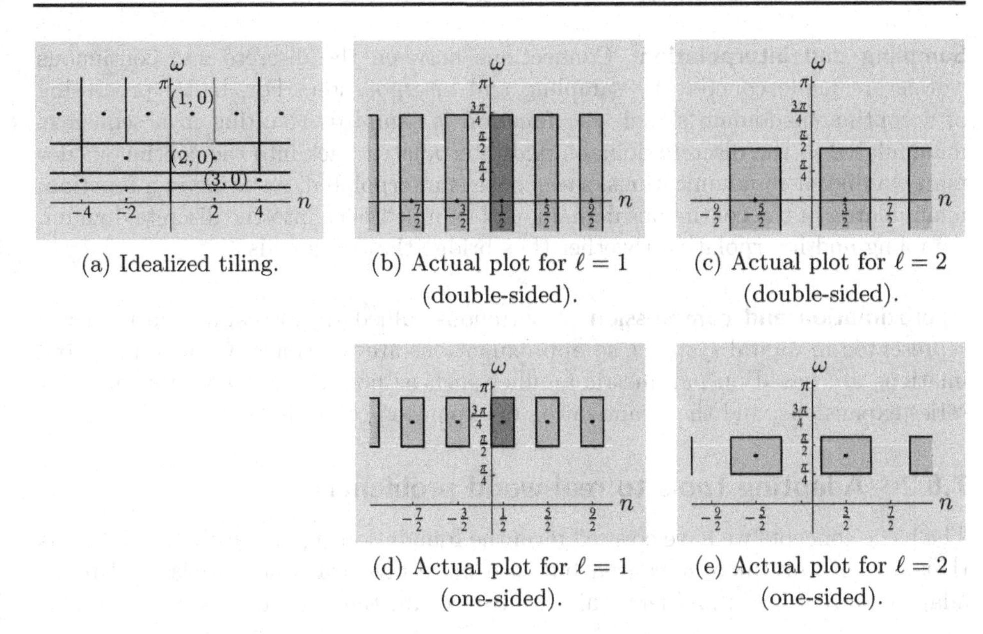

Figure 7.33 Idealized time–frequency tiling (same as Figure 7.20(b) shifted) versus actual (both double-sided and one-sided) time–frequency plots for the dyadic Haar wavelet basis in (7.38c). The Heisenberg box of the prototype sequence, centered at $(\frac{1}{2}, 0)$, of width 1 and height 4.6, is shown in dark gray (in (b) and (d)), while the Heisenberg boxes of the rest of the basis sequences are shown in light gray. Because boxes overlap in frequency, we show tilings for $\ell = 1$ and $\ell = 2$ separately for clarity.

7.6.1 Tools

We have seen a number of basic yet powerful tools and concepts.

Geometry Inner products lead to a powerful geometric view of signals and spaces. This includes orthogonality between signals (vectors), and best approximation in a subspace by orthogonal projection.

Bases and frames The fact that (separable) Hilbert spaces have countable bases leads to natural signal representations using orthonormal bases and biorthogonal pairs of bases. For a linear operator, a representation with respect to a basis of eigenvectors is particularly attractive because it is diagonal. Beyond bases are frames, which are sets of vectors that span a space but are not necessarily linearly independent. Similarly to bases, they also provide signal representations.

Fourier representations Linear, shift-invariant systems operate on signals using a convolution operator. In particular, the eigensequence/eigenfunction property of complex exponentials leads naturally to various forms of the Fourier transform.

Sampling and interpolation Connections between the discrete and continuous worlds are made concrete by sampling and interpolation. For digital processing of a continuous-domain signal, the function is sampled, resulting in a sequence, manipulated in the discrete domain, and interpolated back into the continuous domain; in digital communications, a sequence is interpolated, resulting in a function, manipulated in the continuous domain, and sampled back into the discrete domain. Sampling and interpolation together thus bridge the two worlds.

Approximation and compression Continuous-valued signals cannot be perfectly represented in digital systems, so approximations are necessary. Common approximations are based on parametric families such as polynomials, the truncation of series expansions, and the quantization of expansion coefficients.

7.6.2 Adapting tools to real-world problems

The basic concepts we have covered form the foundations upon which to build tools that are not only more advanced but also more practical; these tools need to be adapted to real-world problems taking into account the following issues.

Finiteness and localization While real-world signals are of finite duration, we use infinite ones as a convenient mathematical abstraction. Moreover, real-world signals often contain both smooth parts and sharp discontinuities. Signal analysis that permits localization in both time and frequency can thus aid in finding characteristics of interest.

Prior knowledge It helps to know what we are looking for: more often than not, we have some prior information about the signal or event we are interested in. Such priors on the signal class or the noise will help shape the solution.

Performance bounds There are limits on what can be done: in the linear measurement case, bounds such as the uncertainty principle set limits to how local an analysis can ever be. In a noisy setting, estimation theory provides lower bounds on the variance of an unbiased estimator. For compression and communications, information theory bounds the performance of any possible scheme, by specifying rate–distortion and capacity regions. Such bounds are useful in at least two fundamental ways: they separate what can be done from what is impossible, and they provide yardsticks against which to compare practical systems.

Computational aspects Whatever the solution, it needs to be computable; for example, some constructive solutions providing bounds on performance lead to hopelessly complex algorithms. Even seemingly simple tasks such as finding the best approximation in a frame require exhaustive search and thus become impractical for real-world problems. Instead, we seek approximate solutions together with performance bounds. Often, the problem is structured and can thus lead to savings in computation. Some computations can be simplified using the FFT. However, since real-world signals can be arbitrarily long, even the $N \log N$ cost of the FFT for a length-N signal is sometimes too great. Furthermore, computations on an entire signal or very long blocks can require too much memory or create too much delay. Thus, time localization is necessary from a computational point of view as well.

7.6.3 Music analysis, communications, and compression

To close this book and motivate some of the constructions in the companion volume [57], we illustrate the importance of locality for a few real-world signals.

Time–frequency analysis of music Ravel's *Boléro* is one of the most famous and popular pieces of Western classical music. It starts with a single flute, whispering the theme, and ends with the full, 120-instrument orchestra thundering the finale. The characteristics of the piece evolve dramatically from beginning to end, illustrated by the 15-minute time-domain display of the acoustical signal in Figure 7.34(a). While the Fourier transform of that sequence, shown in Figure 7.34(b), exhibits a number of spectral peaks corresponding to some key harmonic structures, the signature evolution of the Bolero from the introduction to the grand finale is lost.

To understand the local behavior, we look at two short pieces, at 500 s and 800 s, respectively, each of 0.1 s in duration. Figures 7.34(c) and (e) show the local behavior in time, while Figures 7.34(d) and (f) show the local behavior in frequency. Upon comparing the two segments, we can tell that locally in time the dynamic range has changed, and that locally in frequency, different harmonic components are present, reflecting the numbers and types of instruments involved.

Another way to study locality is by using a spectrogram of a short time segment, as shown in Figure 7.35 for a segment of length 10 s starting at 15 s. A spectrogram computes a DFT on partitions of the sequence, with partitions overlapping by a specified offset. With a shorter partition, as in Figure 7.35(a), we can observe the underlying beat of the piece (time-local events), while with a longer partition, as in Figure 7.35(b), we see that the time information has been smoothed out and the frequency information has become clearer. Figure 7.35(c) zooms in on the lower frequencies in Figure 7.35(b); we can now see the spectral component of the flute playing as well as the rough underlying beat (see the musical score in Figure 7.35(d)).

Digital communications Wireless communication systems operate in allotted portions of the electromagnetic spectrum, as illustrated in Figure 7.36(a). The transmitted and received signals are real-valued, so their Fourier spectra have conjugate symmetry, which implies that the spectra at negative frequencies are completely determined by the spectra at positive frequencies. This is exploited by shifting in frequency to the baseband, resulting in complex-valued signals with bandwidth ω_0, as illustrated in Figure 7.36(b). Typically, the communication channel is approximated as an LSI system with an impulse response h and additive noise w,

$$y = h * x + w,$$

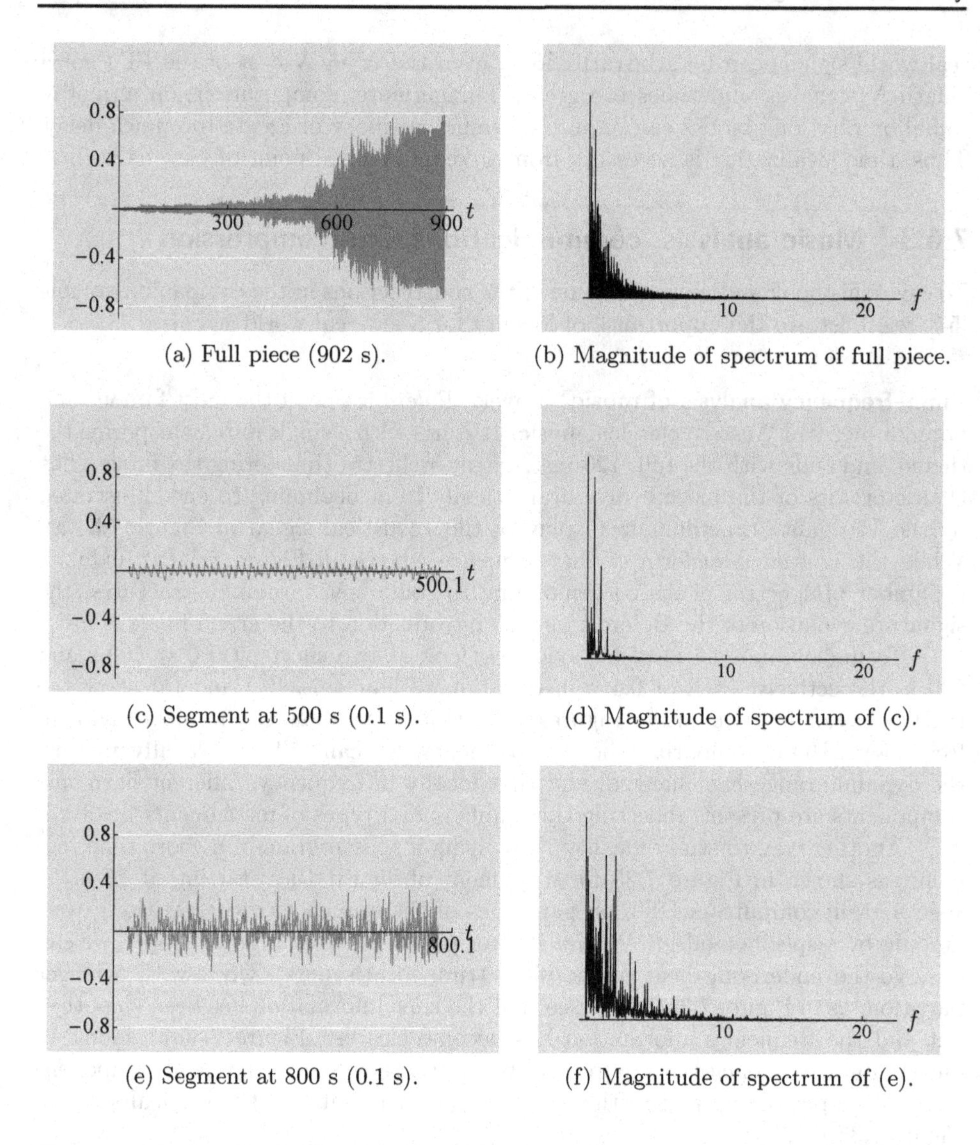

(a) Full piece (902 s).

(b) Magnitude of spectrum of full piece.

(c) Segment at 500 s (0.1 s).

(d) Magnitude of spectrum of (c).

(e) Segment at 800 s (0.1 s).

(f) Magnitude of spectrum of (e).

Figure 7.34 Ravel's *Boléro* in the time and frequency domains. The time axes are in seconds and the frequency axes in kHz. The unusually large dynamic range of the piece is reflected in (a), and little structure is apparent in the global spectrum in (b). At 500 s, a single trumpet plays, and the spectrum locally exhibits simple harmonic structure (see (c) and (d)). At 800 s, many more instruments play; the music is louder and the spectrum is more complicated (see (e) and (f)).

with baseband-equivalent transmitted signal x and baseband-equivalent received signal y. The length of the support of h depends on physical characteristics such as the distances to objects from which there are strong reflections.

From the sampling and interpolation theory in Section 5.4.2, in principle

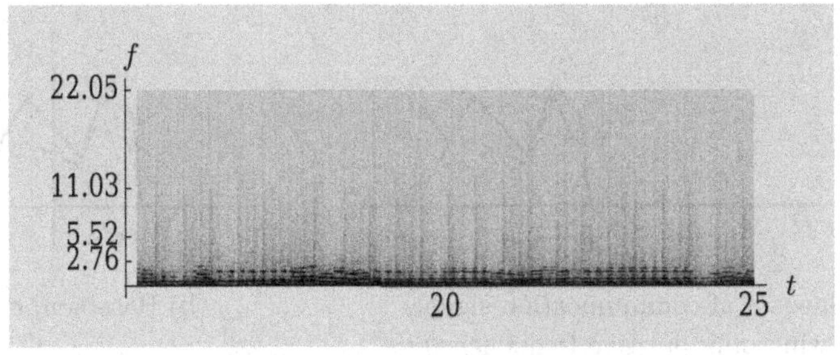

(a) Partition length = 2^9 samples.

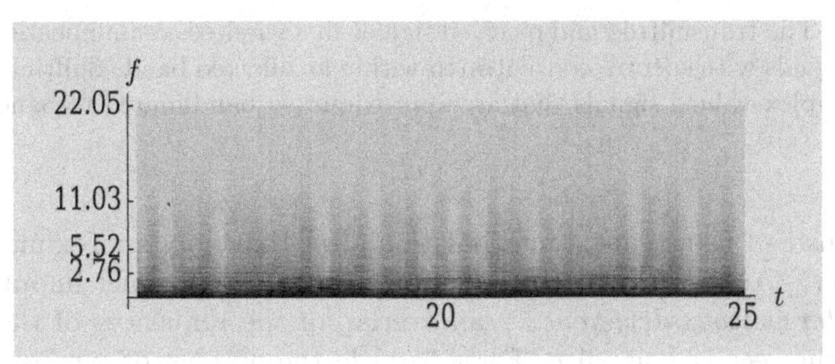

(b) Partition length = 2^{13} samples.

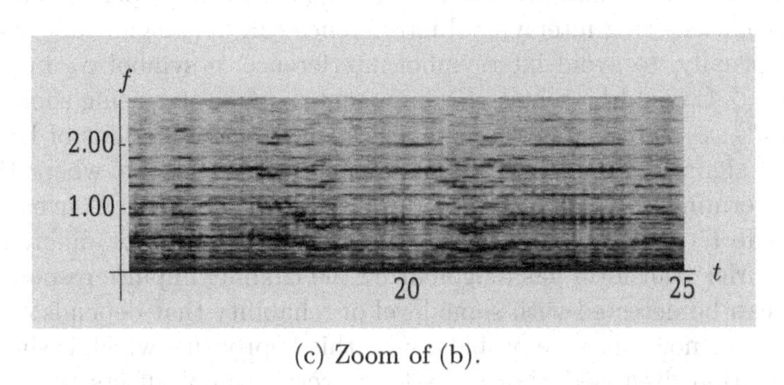

(c) Zoom of (b).

(d) Musical score.

Figure 7.35 Time–frequency analysis of Ravel's *Boléro*. A segment from 15 s to 25 s is used. The offset is set to one third of the partition length. The time axes are in seconds and the frequency axes in kHz.

$x \in \mathrm{BL}[-\frac{1}{2}\omega_0, \frac{1}{2}\omega_0]$ can be generated by sinc interpolation of a sequence with a sampling rate of at most $\omega_0/(2\pi)$ symbols per second; for digital communications, each symbol would be drawn from some finite codebook $\mathcal{C} \subset \mathbb{C}$. Such a function x is impractical because it depends noncausally on the sequence of symbols; thus,

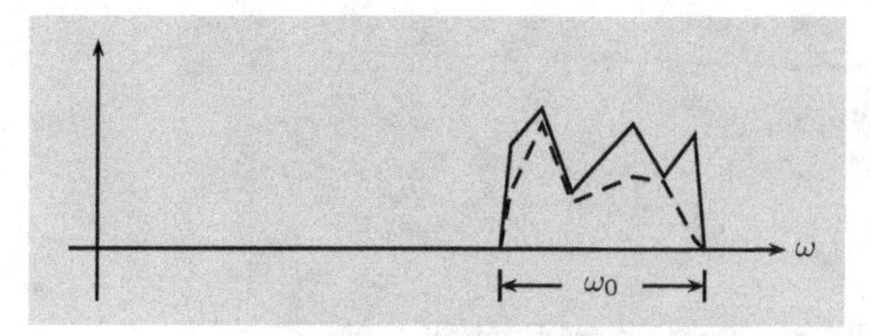

(a) Spectra of communication signals (showing only positive frequencies).

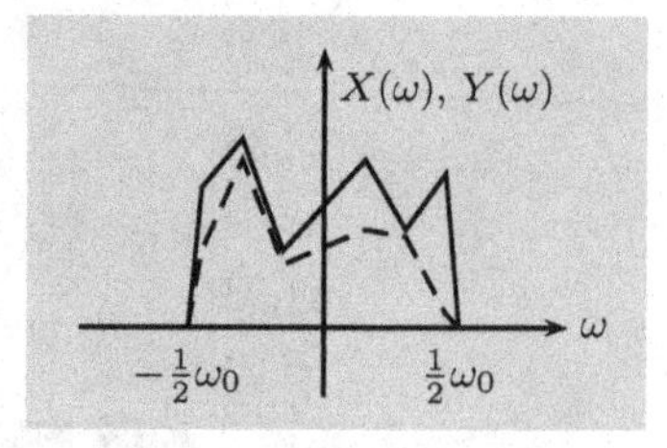

(b) Baseband equivalent.

Figure 7.36 The transmitted and received signals in a wireless communication system are real-valued signals with spectra concentrated within an allotted band. Shifting to baseband results in complex-valued signals that are approximately bandlimited to bandwidth ω_0.

one would instead use a causal interpolation filter that only approximates an ideal lowpass filter. An additional complication is that the channel impulse response h causes *intersymbol interference* – an overlap of the influences of the individual symbols in the received signal y. For a fixed h, the number of symbols that interfere increases as the communication rate of $\omega_0/(2\pi)$ symbols per second increases. Therefore, counteracting intersymbol interference can be a significant challenge.

Conceptually, to avoid intersymbol interference, a symbol α_k in some finite codebook $\mathcal{C}_k \subset \mathbb{C}$ could be sent as a component $\alpha_k e^{j\omega_k t}$ of x using some frequency $\omega_k \in [-\frac{1}{2}\omega_0, \frac{1}{2}\omega_0)$. Since complex exponentials are eigenfunctions of LSI systems, the received signal y would have a component $H(\omega_k)\,\alpha_k e^{j\omega_k t}$, where the scaling $H(\omega_k)$ is determined by the frequency response of the channel. Conveniently, canceling the effect of the channel can be done independently for different symbols since the Fourier transform has diagonalized the channel impulse response h. Each symbol α_k can be detected with some level of reliability that depends on the characteristics of the noise w. One problem with this approach – which is shared by the sinc interpolation discussed above – is that every symbol affects the transmitted signal for all $t \in (-\infty, \infty)$, violating causality at the transmitter and implying that the receiver would have to wait forever to perform optimal detection. Note also that, in this impractical description, the frequencies $\{\omega_k\}$ can be arbitrarily close to each other; with any finite set of frequencies, the number of symbols communicated per unit time is zero because the transmitted and received signals have infinite duration.

Real communication systems have causal transmitters, and it is furthermore desirable for each symbol to affect the transmitted signal for only a short duration so that the receiver may detect the transmitted symbols without great delay and with little intersymbol interference. The model of the channel as LSI is itself accurate only over short time scales, so it is necessary to handle variations in the channel. Being able to achieve these aims can be explained in part using time–frequency tilings.

The use of a sinc-interpolated symbol sequence divides the time–frequency

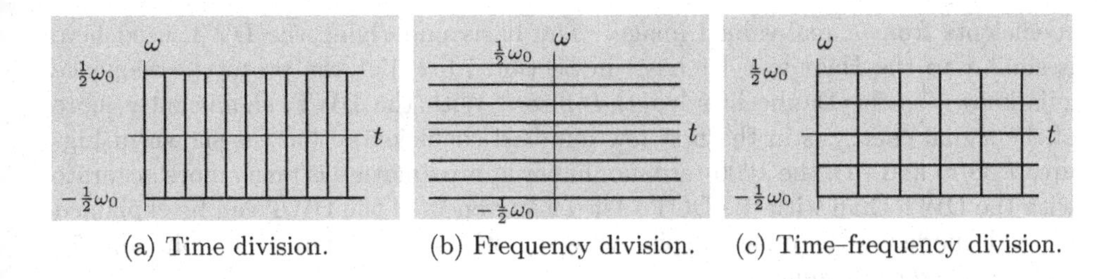

(a) Time division. (b) Frequency division. (c) Time–frequency division.

Figure 7.37 Wireless communication concepts contrasted using idealized time–frequency tilings. Among other possible weaknesses, time division leads to high intersymbol interference and frequency division leads to difficulties with time-varying channels. Practical systems use division in both time and frequency.

plane finely in time but not at all in frequency, as shown in Figure 7.37(a); being wider than the time division, the channel impulse response creates high intersymbol interference. Signaling with complex exponentials divides the time–frequency plane finely in frequency but not at all in time, as shown in Figure 7.37(b); in addition to the reasons for impracticality discussed above, the channel cannot be considered LSI over the symbol duration. Practical approaches – such as orthogonal frequency-division multiplexing in several current standards – divide in time and frequency, as shown in Figure 7.37(c). Time division allows the use of a local LSI approximation h, and when the time increments are similar to the length of the support of h, there is little if any intersymbol interference. Additionally, dividing the time–frequency plane enables the use of different codebook sizes for different symbols, depending on the SNR locally in time and frequency. The tiling in Figure 7.37(c) is achieved by local Fourier bases, and these bases are important in understanding modern communication methods in more detail.

Image compression As a final example, consider the compression of images. The transform coding paradigm introduced in Section 6.5.3 demonstrated that the choice of a basis can affect the compression performance, for example, in terms of the coding gain defined in (6.141). The high-resolution analysis used to arrive at the coding gain gives close approximations to performance when each transform coefficient is allocated at least 1 bit. In this case, the average number of bits per coefficient might be much higher, so the coding gain is not necessarily relevant at typical coding rates for images, which are about 1 bit per pixel. The effect of the choice of a basis can be even more dramatic at lower rates, where the differences between linear and nonlinear approximations developed in Section 6.4.1 become significant.

To illustrate two choices of bases, Figure 7.38(b) shows the magnitudes of the transform coefficients of the image in Figure 7.38(a), sorted in descending order and on a log scale, for a two-dimensional DCT and a two-dimensional DWT. The DCT here is a block-by-block transform with blocks of size 8×8; the underlying basis has locality in space and frequency quite similar to the local Fourier bases in Section 7.5.2, and it is convenient for image processing because it produces real-valued coefficients from a real-valued image. The basis underlying the DWT used here is similar to the Haar wavelet basis in Section 7.5.2, but the prototype sequence φ is known as the Daubechies length-8 filter. With the DWT, significantly more of the signal energy is in the first few hundred coefficients; thus, as shown in Figures 7.38(c) and (d), the 1000-term nonlinear approximation is much more accurate with the DWT than with the DCT. The effectiveness of the DWT can be explained through modeling the image as piecewise-smooth; see the companion volume, [57], and the *Further reading*.

Improvements in nonlinear approximation translate to improvements in compression, but this can be reduced or even overridden by the cost of specifying which coefficients are the largest in magnitude. The impact of the basis choice is thus often smaller in compression. As shown in Figures 7.38(e) and (f), sharp features are more accurately reproduced with JPEG2000 compression (which uses a DWT) than with JPEG compression (which uses an 8×8 block-by-block DCT); see the *Further reading*.

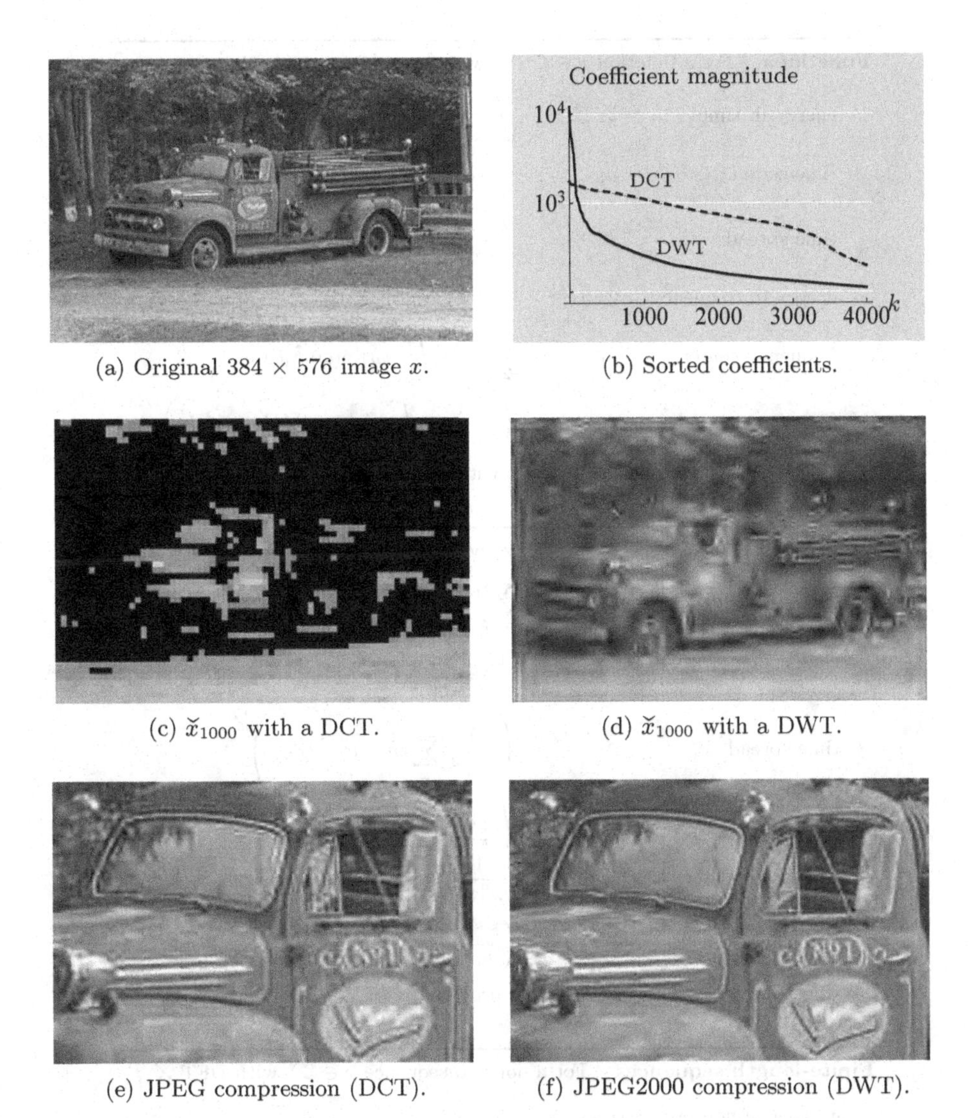

(a) Original 384×576 image x.

(b) Sorted coefficients.

(c) $\check{x}_{1000}$ with a DCT.

(d) $\check{x}_{1000}$ with a DWT.

(e) JPEG compression (DCT).

(f) JPEG2000 compression (DWT).

Figure 7.38 Approximation and compression of Giovanni Pacifici's *Fire Truck* image using a two-dimensional DCT on 8×8 blocks and a two-dimensional DWT. In (d), the nonlinear approximation uses the separable, orthonormal DWT with Daubechies length-8 filters. In (e) and (f), the rate is 0.6 bits per pixel.

Chapter at a glance

Functions For a function $x \in \mathcal{L}^2(\mathbb{R})$ with Fourier transform X:

energy in time	$\|x\|^2$	$\int_{-\infty}^{\infty} \|x(t)\|^2 \, dt$
time center	μ_t	$\frac{1}{\|x\|^2} \int_{-\infty}^{\infty} t \, \|x(t)\|^2 \, dt$
time spread	Δ_t	$\left(\frac{1}{\|x\|^2} \int_{-\infty}^{\infty} (t - \mu_t)^2 \, \|x(t)\|^2 \, dt \right)^{1/2}$
energy in frequency	$2\pi \, \|x\|^2$	$\int_{-\infty}^{\infty} \|X(\omega)\|^2 \, d\omega$
frequency center	μ_f	$\frac{1}{2\pi \|x\|^2} \int_{-\infty}^{\infty} \omega \, \|X(\omega)\|^2 \, d\omega$
frequency spread	Δ_f	$\left(\frac{1}{2\pi \|x\|^2} \int_{-\infty}^{\infty} (\omega - \mu_f)^2 \, \|X(\omega)\|^2 \, d\omega \right)^{1/2}$

Uncertainty principle: $\Delta_t \Delta_f \geq \frac{1}{2}$, with equality achieved by a Gaussian x.

Sequences For a sequence $x \in \ell^2(\mathbb{Z})$ with DTFT X:

energy in time	$\|x\|^2$	$\sum_{n \in \mathbb{Z}} \|x_n\|^2$
time center	μ_t	$\frac{1}{\|x\|^2} \sum_{n \in \mathbb{Z}} n \, \|x_n\|^2$
time spread	Δ_t	$\left(\frac{1}{\|x\|^2} \sum_{n \in \mathbb{Z}} (n - \mu_t)^2 \, \|x_n\|^2 \right)^{1/2}$
energy in frequency	$2\pi \, \|x\|^2$	$\int_{-\pi}^{\pi} \|X(\omega)\|^2 \, d\omega$
frequency center	μ_f	$\frac{1}{2\pi \|x\|^2} \int_{-\pi}^{\pi} \omega \, \|X(\omega)\|^2 \, d\omega$
frequency spread	Δ_f	$\left(\frac{1}{2\pi \|x\|^2} \int_{-\pi}^{\pi} (\omega - \mu_f)^2 \, \|X(\omega)\|^2 \, d\omega \right)^{1/2}$

Uncertainty principle: $\Delta_t \Delta_f > \frac{1}{2}$, provided that $X(e^{j\pi}) = 0$.

Finite-length sequences For a nonzero sequence $x \in \mathbb{C}^N$ with DFT X:

number of nonzero entries of x	N_t
number of nonzero entries of X	N_ω

Uncertainty principle: $N_t N_\omega \geq N$.

Table 7.5 Uncertainty principles.

Historical remarks

Uncertainty principles stemming from the Cauchy–Schwarz inequality have a long and rich history. The best known one is Heisenberg's uncertainty principle in quantum physics, which was first developed in a 1927 essay [44]. **Werner Karl Heisenberg (1901–1976)** was a German physicist, credited as a founder of quantum mechanics, for which he was awarded the Nobel Prize in 1932. He had seven children, one of whom, Martin Heisenberg, was a celebrated geneticist. He collaborated with Bohr, Pauli, and Dirac, among others. While he was initially attacked by the Nazi war machine for promoting Einstein's views, he did head the Nazi nuclear project during the war. His role in the project has been a subject of controversy ever since, with differing views on whether he was deliberately stalling Hitler's efforts.

Kennard is credited with the first mathematically exact formulation of the uncertainty principle, and Robertson and Schrödinger provided generalizations. The uncertainty principle presented in Theorem 7.4 was proven by Weyl and Pauli and introduced to signal processing by **Dennis Gabor (1900–1979)** [31], a Hungarian–British electrical engineer and inventor, winner of the 1971 Nobel Prize in Physics for the invention of holography. By finding a lower bound to $\Delta_t\Delta_f$, Gabor was intending to define an information measure or capacity for signals. Shannon's communication theory [89] proved more fruitful for this purpose, but Gabor's proposal of signal analysis by shifted and modulated Gaussian functions has been a cornerstone of time–frequency analysis ever since. Slepian's survey [92] is enlightening on these topics.

Further reading

Uncertainty principles Many of the uncertainty principles for discrete-time signals are considerably more complicated than Theorem 7.8. We have given only a result that follows papers by Ishii and Furukawa [48] and by Calvez and Vilbé [12]. In [78], Parhizkar, Barbotin, and Vetterli use a definition of frequency spread that avoids the requirement of $X(e^{j\pi}) = 0$, and they construct sequences of minimum time spread for a given frequency spread.

In addition to Theorem 7.9, Donoho and Stark [25] derived several new uncertainty principles. Particularly influential was the demonstration of the significance of uncertainty principles for signal recovery (see Exercise 7.9). Moreover, Donoho and Huo [23] introduced performance guarantees for ℓ^1 minimization-based signal recovery algorithms; this has sparked a large body of work under the name *compressed sensing*.

Bases with time and frequency localization Localized orthonormal bases corresponding to the short-time Fourier transform have been sought ever since Gabor proposed a localized Fourier analysis [31]. Unfortunately, a negative result known as the Balian–Low theorem shows that there are no Fourier-like orthonormal bases with good time and frequency localization [3, 5, 63]. Further results on time–frequency tilings and on Fourier and wavelet constructions can be found in books by Daubechies [19] and by Mallat [66], and in the companion volume, [57].

Image compression The JPEG image compression standard is named after the standards committee that developed it from the late 1980s to 1992 – the Joint Photographic Experts Group. It follows the basic principles for transform coding described in Section 6.5.3, with additions that include selecting quantization step sizes that are consistent with human visual perception and using differential coding for the zero-frequency components of neighboring 8×8 blocks. The JPEG2000 standard deviates further from the basic transform coding paradigm so that it can provide a variety of features, including variable resolution and regions of interest. The book edited by Taubman and Marcellin [96] describes the standard and a range of issues around wavelet-based compression.

Exercises with solutions

7.1. *Uncertainty bounds for B-splines*

(i) Find the time spread Δ_t and frequency spread Δ_f of the elementary B-splines of degree 0, 1, and 2.

(ii) For these functions, compare the product $\Delta_t\Delta_f$ with the lower bound in Theorem 7.4.

Solution:

(i) For any degree $K \in \mathbb{N}$, the elementary B-spline $\beta^{(K)}$ has even symmetry, so $\mu_t(\beta^{(K)}) = 0$. It is real, so $\mu_f(\beta^{(K)}) = 0$.

Using (6.34a) for $\beta^{(0)}$, the squared norm of $\beta^{(0)}$ is

$$\|\beta^{(0)}\|^2 = \int_{-1/2}^{1/2} 1\, dt = 1.$$

Its squared time spread is

$$\Delta_t^2(\beta^{(0)}) = \frac{1}{\|\beta^{(0)}\|^2}\int_{-1/2}^{1/2} t^2\, dt = \frac{1}{12}.$$

Its squared frequency spread is

$$\begin{aligned}\Delta_f^2(\beta^{(0)}) &= \frac{1}{2\pi\,\|\beta^{(0)}\|^2}\int_{-\infty}^{\infty} \omega^2\,|B^{(0)}(\omega)|^2\, d\omega \\ &\overset{(a)}{=} \frac{1}{2\pi}\int_{-\infty}^{\infty} \omega^2 \operatorname{sinc}^2\left(\tfrac{1}{2}\omega\right) d\omega \overset{(b)}{=} \infty,\end{aligned}$$

where (a) follows from (6.37); and (b) from the lack of decay of the integrand.

Using (6.35) for $\beta^{(1)}$, the squared norm of $\beta^{(1)}$ is

$$\|\beta^{(1)}\|^2 = \int_{-1}^{0} (1+t)^2\, dt + \int_{0}^{1} (1-t)^2\, dt = \frac{2}{3}.$$

Its squared time spread is

$$\begin{aligned}\Delta_t^2(\beta^{(1)}) &= \frac{1}{\|\beta^{(1)}\|^2}\left(\int_{-1}^{0} t^2(1+t)^2\, dt + \int_{0}^{1} t^2(1-t)^2\, dt\right) \\ &= \frac{1}{2/3}\left(\frac{1}{30} + \frac{1}{30}\right) = \frac{1}{10}.\end{aligned}$$

Its squared frequency spread is

$$\Delta_f^2(\beta^{(1)}) = \frac{1}{2\pi \|\beta^{(1)}\|^2} \int_{-\infty}^{\infty} \omega^2 |B^{(1)}(\omega)|^2 \, d\omega \stackrel{(a)}{=} \frac{1}{2\pi(2/3)} \int_{-\infty}^{\infty} \omega^2 \operatorname{sinc}^4\left(\tfrac{1}{2}\omega\right) d\omega = 3,$$

where (a) follows from (6.37).

Using (6.36) for $\beta^{(2)}$, the squared norm of $\beta^{(1)}$ is

$$\|\beta^{(2)}\|^2 = \frac{11}{20}.$$

Its squared time spread is

$$\Delta_t^2(\beta^{(2)}) = \frac{1}{\|\beta^{(2)}\|^2} \int_{-3/2}^{3/2} t^2 |\beta^{(2)}(t)|^2 \, dt = \frac{1}{11/20} \cdot \frac{43}{560} = \frac{43}{308}.$$

Its squared frequency spread is

$$\Delta_f^2(\beta^{(2)}) = \frac{1}{2\pi \|\beta^{(2)}\|^2} \int_{-\infty}^{\infty} \omega^2 |B^{(2)}(\omega)|^2 \, d\omega \stackrel{(a)}{=} \frac{1}{2\pi(11/20)} \int_{-\infty}^{\infty} \omega^2 \operatorname{sinc}^6\left(\tfrac{1}{2}\omega\right) d\omega = \frac{20}{11},$$

where (a) follows from (6.37).

(ii) The products of time and frequency spreads are

$$\Delta_t(\beta^{(0)})\, \Delta_f(\beta^{(0)}) = \infty > \frac{1}{2},$$

$$\Delta_t(\beta^{(1)})\, \Delta_f(\beta^{(1)}) = \sqrt{\frac{3}{10}} \approx 0.5477 > \frac{1}{2},$$

$$\Delta_t(\beta^{(2)})\, \Delta_f(\beta^{(2)}) = \sqrt{\frac{215}{847}} \approx 0.5038 > \frac{1}{2}.$$

The sequence of time–frequency products is decreasing and approaching the lower bound of $\frac{1}{2}$.

7.2. *Properties of time and frequency spreads for functions*
The time center μ_t, time spread Δ_t, frequency center μ_f, and frequency spread Δ_f of $x \in \mathcal{L}^2(\mathbb{R})$ are defined in (7.4) and (7.6).

(i) Show that shifting x in time by $t_0 \in \mathbb{R}$, (7.3a), shifts μ_t and leaves μ_f, Δ_t, and Δ_f unchanged; that is, verify (7.5a)–(7.5b) and (7.7a)–(7.7b).

(ii) Show that shifting x in frequency (modulation) by $\omega_0 \in \mathbb{R}$, (7.3b), shifts μ_f and leaves μ_t, Δ_t, and Δ_f unchanged; that is, verify (7.5c)–(7.5d) and (7.7c)–(7.7d).

(iii) Show that scaling x in time by $a \in \mathbb{R}^+$, (7.3c), scales μ_t and Δ_t by $1/a$ and scales μ_f and Δ_f by a; that is, verify (7.5e)–(7.5f) and (7.7e)–(7.7f).

Solution: Without loss of generality assume that $\|x\| = 1$.

(i) The time center of $y(t) = x(t - t_0)$ is

$$\mu_t(y) = \int_{-\infty}^{\infty} t\, |y(t)|^2 \, dt = \int_{-\infty}^{\infty} t\, |x(t - t_0)|^2 \, dt \stackrel{(a)}{=} \int_{-\infty}^{\infty} (\tau + t_0)\, |x(\tau)|^2 \, d\tau = \int_{-\infty}^{\infty} \tau\, |x(\tau)|^2 \, d\tau + t_0 \int_{-\infty}^{\infty} |x(\tau)|^2 \, d\tau \stackrel{(b)}{=} \mu_t(x) + t_0,$$

where (a) follows from the change of variable $\tau = t - t_0$; and (b) from $\|x\| = 1$.

The time spread of y is

$$\begin{aligned}\Delta_t(y) &= \left(\int_{-\infty}^{\infty} (t - \mu_t(y))^2 \, |y(t)|^2 \, dt\right)^{1/2} \\ &= \left(\int_{-\infty}^{\infty} (t - \mu_t(x) - t_0)^2 \, |x(t - t_0)|^2 \, dt\right)^{1/2} \\ &\overset{(a)}{=} \left(\int_{-\infty}^{\infty} (\tau - \mu_t(x))^2 \, |x(\tau)|^2 \, d\tau\right)^{1/2} = \Delta_t(x),\end{aligned}$$

where (a) again follows from the change of variable $\tau = t - t_0$.

The frequency center and frequency spread of y depend on $Y(\omega)$ through its absolute value. Since $|Y(\omega)| = |e^{-j\omega t_0} X(\omega)| = |X(\omega)|$, the frequency center and frequency spread are not changed by the time shift.

(ii) The frequency center of $y(t) = e^{j\omega_0 t} x(t)$ is

$$\begin{aligned}\mu_f(y) &= \frac{1}{2\pi}\int_{-\infty}^{\infty} \omega \, |Y(\omega)|^2 \, d\omega \overset{(a)}{=} \frac{1}{2\pi}\int_{-\infty}^{\infty} \omega \, |X(\omega - \omega_0)|^2 \, d\omega \\ &\overset{(b)}{=} \frac{1}{2\pi}\int_{-\infty}^{\infty} (\theta + \omega_0) \, |X(\theta)|^2 \, d\theta \\ &= \frac{1}{2\pi}\int_{-\infty}^{\infty} \theta \, |X(\theta)|^2 \, d\theta + \omega_0 \frac{1}{2\pi}\int_{-\infty}^{\infty} |X(\theta)|^2 \, d\theta \overset{(c)}{=} \mu_f(x) + \omega_0,\end{aligned}$$

where (a) follows from $Y(\omega) = X(\omega - \omega_0)$; (b) from the change of variable $\theta = \omega - \omega_0$; and (c) from $\|X\|^2 = 2\pi$ by the Parseval equality.

The frequency spread of y is

$$\begin{aligned}\Delta_f(y) &= \left(\frac{1}{2\pi}\int_{-\infty}^{\infty} (\omega - \mu_f(y))^2 \, |Y(\omega)|^2 \, d\omega\right)^{1/2} \\ &= \left(\frac{1}{2\pi}\int_{-\infty}^{\infty} (\omega - \mu_f(x) - \omega_0)^2 \, |X(\omega - \omega_0)|^2 \, d\omega\right)^{1/2} \\ &\overset{(a)}{=} \left(\frac{1}{2\pi}\int_{-\infty}^{\infty} (\theta - \mu_f(x))^2 \, |X(\theta)|^2 \, d\theta\right)^{1/2} = \Delta_f(x),\end{aligned}$$

where (a) again follows from the change of variable $\theta = \omega - \omega_0$.

Since $|y(t)| = |e^{j\omega_0 t} x(t)| = |x(t)|$, the time center and time spread are not changed by the frequency shift.

(iii) The time center of $y(t) = \sqrt{a} x(at)$ is

$$\begin{aligned}\mu_t(y) &= \int_{-\infty}^{\infty} t \, |y(t)|^2 \, dt = \int_{-\infty}^{\infty} t \, |\sqrt{a} \, x(at)|^2 \, dt \\ &\overset{(a)}{=} \frac{1}{a}\int_{-\infty}^{\infty} \tau \, |x(\tau)|^2 \, d\tau = \frac{1}{a}\mu_t(x),\end{aligned}$$

where (a) follows from the change of variable $\tau = at$.

The time spread of y is

$$\begin{aligned}\Delta_t(y) &= \left(\int_{-\infty}^{\infty} (t - \mu_t(y))^2 \, |y(t)|^2 \, dt\right)^{1/2} \\ &= \left(\int_{-\infty}^{\infty} \left(t - \frac{1}{a}\mu_t(x)\right)^2 \, |\sqrt{a} \, x(at)|^2 \, dt\right)^{1/2} \\ &\overset{(a)}{=} \frac{1}{a}\left(\int_{-\infty}^{\infty} (\tau - \mu_t(x))^2 \, |x(\tau)|^2 \, d\tau\right)^{1/2} = \frac{1}{a}\Delta_t(x),\end{aligned}$$

where (a) again follows from the change of variable $\tau = at$.

The frequency center of y is

$$\mu_f(y) = \frac{1}{2\pi}\int_{-\infty}^{\infty} \omega\, |Y(\omega)|^2\, d\omega \stackrel{(a)}{=} \frac{1}{2\pi}\int_{-\infty}^{\infty} \omega \left|\frac{1}{\sqrt{a}} X\left(\frac{\omega}{a}\right)\right|^2 d\omega$$
$$\stackrel{(b)}{=} a\frac{1}{2\pi}\int_{-\infty}^{\infty} \theta\, |X(\theta)|^2\, d\theta = a\mu_f(x),$$

where (a) follows from $Y(\omega) = (1/\sqrt{a})X(\omega/a)$; and (b) from the change of variable $\theta = \omega/a$.

The frequency spread of y is

$$\Delta_f(y) = \left(\frac{1}{2\pi}\int_{-\infty}^{\infty} (\omega - \mu_f(y))^2\, |Y(\omega)|^2\, d\omega\right)^{1/2}$$
$$\stackrel{(a)}{=} \left(\frac{1}{2\pi}\int_{-\infty}^{\infty} (\omega - a\mu_f(x))^2 \left|\frac{1}{\sqrt{a}} X\left(\frac{\omega}{a}\right)\right|^2 d\omega\right)^{1/2}$$
$$\stackrel{(b)}{=} a\left(\frac{1}{2\pi}\int_{-\infty}^{\infty} (\theta - \mu_f(x))^2\, |X(\theta)|^2\, d\theta\right)^{1/2} = a\Delta_f(x),$$

where again (a) follows from $Y(\omega) = (1/\sqrt{a})X(\omega/a)$; and (b) from the change of variable $\theta = \omega/a$.

Exercises

7.1. *Impossibility of finite support in time and frequency.*$^{(2)}$

(i) Show that if a nonzero sequence has a finite number of nonzero entries, then its DTFT cannot be zero over an open interval (that is, it can have only isolated zeros). (*Hint:* Use the fundamental theorem of algebra, Theorem 3.23.)

(ii) Show that if a nonzero function is bandlimited, then the support of the function is infinite.
(*Hint:* Consider sampling the function above its Nyquist rate and use part (i).)

7.2. *Uncertainty principle for complex functions.*$^{(3)}$
Prove Theorem 7.4 without assuming that x is a real-valued function.
(*Hint:* The proof requires more than the Cauchy–Schwarz inequality and integration by parts. Use the product rule of differentiation, $(|x(t)|^2)' = x'(t)x^*(t) + x(t)(x^*(t))'$. Also, use that $|\alpha| \geq |\alpha + \alpha^*|/2$ for any $\alpha \in \mathbb{C}$.)

7.3. *Properties of modified time and frequency spreads for functions.*$^{(1)}$
The modified time spread $\Delta_t^{(\beta)}$ and frequency spread $\Delta_f^{(\beta)}$ of $x \in \mathcal{L}^2(\mathbb{R})$ are defined in (7.13a) and (7.13b).

(i) Show that shifting x in time by $t_0 \in \mathbb{R}$, (7.3a), leaves $\Delta_t^{(\beta)}$ and $\Delta_f^{(\beta)}$ unchanged.

(ii) Show that shifting x in frequency (modulation), (7.3b), leaves $\Delta_t^{(\beta)}$ and $\Delta_f^{(\beta)}$ unchanged.

(iii) Show that scaling x in time by $a \in \mathbb{R}^+$, (7.3c), scales $\Delta_t^{(\beta)}$ by $1/a$ and scales $\Delta_f^{(\beta)}$ by a.

7.4. *Frequency spread of the box sequence.*$^{(2)}$
Show that the expression in (7.18) is the frequency spread of the right-sided box sequence of length n_0 from (3.13).

7.5. *Properties of time and frequency spreads for sequences.*$^{(2)}$
The time center μ_t, time spread Δ_t, frequency center μ_f, and frequency spread Δ_f of $x \in \ell^2(\mathbb{Z})$ are defined in (7.15) and (7.17).

K	$g_n^{(K)}$	$G^{(K)}(z)$
0	$(1/\sqrt{2})(\delta_n + \delta_{n-1})$	$(1/\sqrt{2})(1+z^{-1})$
1	$(1/\sqrt{6})(\delta_n + 2\delta_{n-1} + \delta_{n-2})$	$(1/\sqrt{6})(1+z^{-1})^2$
2	$(1/\sqrt{20})(\delta_n + 3\delta_{n-1} + 3\delta_{n-2} + \delta_{n-3})$	$(1/\sqrt{20})(1+z^{-1})^3$
3	$(1/\sqrt{70})(\delta_n + 4\delta_{n-1} + 6\delta_{n-2} + 6\delta_{n-3} + \delta_{n-4})$	$(1/\sqrt{70})(1+z^{-1})^4$

Table P7.7-1 The first few causal discrete spline sequences.

(i) Show that shifting x in time by $n_0 \in \mathbb{Z}$, (7.14a), shifts μ_t and leaves μ_f, Δ_t, and Δ_f unchanged; that is, verify (7.16a)–(7.16b) and (7.19a)–(7.19b).

(ii) Let $\omega_0 \in (0, \pi]$, and let x have DTFT X satisfying

$$X(e^{j\omega}) = 0 \qquad \text{for } \omega \in (\pi - \omega_0, \pi]. \tag{P7.5-1}$$

Show that shifting x in frequency (modulation) by ω_0, (7.14b), shifts μ_f and leaves μ_t, Δ_t, and Δ_f unchanged; that is, verify (7.16c)–(7.16d) and (7.19c)–(7.19d).

(iii) Show that upsampling x by N followed by ideal lowpass postfiltering, (7.14c), scales μ_t and Δ_t by N and scales μ_f and Δ_f by $1/N$; that is, verify (7.16e)–(7.16f) and (7.19e)–(7.19f).

(iv) Let $x \in \mathrm{BL}[-\pi/N, \pi/N] \subset \ell^2(\mathbb{Z})$. Show that downsampling x by N scales μ_t and Δ_t by $1/N$ and scales μ_f and Δ_f by N; that is, verify (7.16g)–(7.16h) and (7.19g)–(7.19h).

7.6. *Uncertainty principle for sequences.*(3)
Prove Theorem 7.8 for real sequences. Do not forget to provide an argument for the strictness of inequality (7.22).
(*Hint:* Use the Cauchy–Schwarz inequality (2.29) to bound $\left| \int_{-\pi}^{\pi} \omega X(e^{j\omega}) \left[\frac{\partial}{\partial \omega} X(e^{j\omega}) \right] d\omega \right|^2$.)

7.7. *Uncertainty bounds for discrete splines.*(2)
Causal *discrete spline sequences* are obtained by successive convolutions of the length-2 right-sided box sequence; they are also sometimes called *binomial sequences* since their entries are binomial coefficients. In the z-transform domain,

$$G^{(K)}(z) = \binom{2(K+1)}{K+1}^{-1/2} \left(1 + z^{-1}\right)^{K+1}, \qquad K = 0, 1, \ldots,$$

where the leading coefficient ensures that $g^{(K)}$ is of unit norm; see Table P7.7-1.

(i) Numerically compute the time spread Δ_t and frequency spread Δ_f of the causal discrete splines for $K = 0, 1, 2, 3$.

(ii) For these sequences, compare the product $\Delta_t \Delta_f$ with the lower bound in Theorem 7.8.

7.8. *Equality in the uncertainty principle for finite-length sequences.*(3)
Let $N \in \mathbb{Z}^+$ be any composite number. Show that there exists nonzero $x \in \mathbb{C}^N$ such that (7.23) holds with equality.
(*Hint:* The proof of Theorem 7.9 suggests how the extreme case is achieved.)

7.9. *Signal recovery based on the uncertainty principle for finite-length sequences.*(2)
Suppose that the DFT X of a length-N sequence x is known to have only N_f nonzero entries (in positions that are not necessarily known). Show that x is uniquely determined from any M time-domain entries, provided that $2(N - M)N_f < N$.
(*Hint:* Use Theorem 7.9 to show that nonunique recovery leads to a contradiction.)

Image and quote attribution

Images

Source		Permission	
s1	Christiane Grimm, Geneva, Switzerland	p1	Permission of the copyright holder
s2	Martin Vetterli, Grandvaux, Switzerland	p2	Authors own copyright
s3	Wikimedia Commons	p3	Copyright expired
s4	KerryR.net	p4	Public domain in the USA
s5	Alan C. Gilbert	p5	NASA material, not protected by copyright
s6	Giovanni Pacifici, New York, NY, USA	p6	Public domain
s7	Nobelprize.org		

Front Material

Cover photograph.[s1, p1] Experimental set-up by Professor Libero Zuppiroli, Laboratory of Optoelectronics of Molecular Materials, EPFL, Lausanne, Switzerland.

Chapter 1

Rainbow over Grandvaux.[s2, p2]

Rainbow explained.[s3, p4]

Chapter 2

Euclid of Megara.[s3, p3] Panel from the Series "Famous Men," Justus of Ghent, about 1474. Urbino, Galleria Nazionale delle Marche.

David Hilbert.[s3, p4] This photograph was taken in 1912 for postcards of faculty members at the University of Göttingen which were sold to students at the time (see *Hilbert*, Constance Reid, Springer 1970).

Chapter 3

Apollo 8 Earth picture.[s3, p5] Earth visible above the lunar surface, taken by Apollo 8 crew member Bill Anders on December 24, 1968.

Chapter 4

Jean Baptiste Joseph Fourier.[s3, p3] "Portraits et histoire des hommes utiles, collection de cinquante portraits," Société Montyon et Franklin, 1839–1840.

Josiah Willard Gibbs.[s3, p3]

Chapter 5

Claude Shannon.[s4, p6]

Chapter 6

Pafnuty Lvovich Chebyshev.[s3, p2]

Chapter 7

All Is Vanity.[s5, p3]

Fire Truck.[s6, p1] Old fire truck in Hidden Valley Camp, Maine.

Werner Karl Heisenberg.[s3, p3]

Dennis Gabor.[s3, p6]

Quotes

Chapter 1

Samuel L. Clemens (Mark Twain), in an account of Queen Victoria's jubilee that first appeared in the June 23, 1897, issue of the *San Francisco Examiner.* See *Mark Twain: A Tramp Abroad, Following the Equator, Other Travels* (The Library of America, 2010), p. 1052.

Chapter 2

Henri Poincaré, as translated by Francis Maitland from "... la mathématique est l'art de donner le même nom à des choses différentes," *Science et méthode* (Ernst Flammarion, 1920), p. 29. Translated as *Science and Method* (Thomas Nelson and Sons, *c.* 1918), p. 34.

Chapter 3

Julius Caesar, unknown.

Chapter 4

Joseph Fourier, as translated by Alexander Freeman from "L'étude approfondie de la nature est la source la plus féconde des découvertes mathématiques," *Théorie analytique de la chaleur* (Firmin Didot, 1822), p. xiii. Translated as *The Analytical Theory of Heat* (Cambridge University Press, 1878), p. 7.

Chapter 5

Max Planck, in "The meaning and limits of exact science," *Science*, vol. 110, no. 2857, pp. 319–327, September 1949.

Chapter 6

John Tukey, in "The future of data analysis," *Annals of Mathematical Statistics*, vol. 33, no. 1, pp. 1–67, March 1962.

Chapter 7

Werner Karl Heisenberg, in "Über den anschaulichen Inhalt der quantentheoretischen Kinematik und Mechanik," *Zeitschrift für Physik*, vol. 43, no. 3–4, pp. 172–198, 1927.

References

[1] K. E. Atkinson. *An Introduction to Numerical Analysis.* John Wiley & Sons, New York, NY, second edition, 1989.

[2] A. V. Balakrishnan. A note on the sampling principle for continuous signals. *IRE Trans. Inf. Theory*, IT-3(2):143–146, June 1957.

[3] R. Balian. Un principe d'incertitude fort en théorie du signal ou en mécanique quantique. *C. R. Acad. Sci. Paris*, 292:1357–1362, 1981.

[4] Y. K. Belyaev. Anlalitičestkie slučajnie protsessy [Analytical random processes]. *Teor. Veroyatnostei i Ee Primeneniya*, 4(4):437–444, 1959. In Russian.

[5] J. J. Benedetto, C. Heil, and D. F. Walnut. Differentiation and the Balian–Low theorem. *J. Fourier Anal. Appl.*, 1(4):355–402, 1995.

[6] D. P. Bertsekas and J. N. Tsitsiklis. *Introduction to Probability.* Athena Scientific, Belmont, MA, second edition, 2008.

[7] R. E. Blahut. *Fast Algorithms for Digital Signal Processing.* Addison-Wesley, Reading, MA, 1985.

[8] T. Blu, P.-L. Dragotti, M. Vetterli, P. Marziliano, and L. Coulot. Sparse sampling of signal innovations. *IEEE Signal Process. Mag.*, 25(2):31–40, March 2008.

[9] C. B. Boyer. Euclid of Alexandria. In *A History of Mathematics*, pages 111–133. John Wiley & Sons, 1968.

[10] R. N. Bracewell. *The Fourier Transform and Its Applications.* McGraw-Hill, New York, NY, second edition, 1986.

[11] P. Brémaud. *Mathematical Principles of Signal Processing: Fourier and Wavelet Analysis.* Springer, 2002.

[12] L. C. Calvez and P. Vilbé. On the uncertainty principle in discrete signals. *IEEE Trans. Circuits Syst. II, Analog Digit. Signal Process.*, 39(6):394–395, June 1992.

[13] W. Cheney and A. A. Goldstein. Proximity maps for convex sets. *Proc. Am. Math. Soc.*, 10(3):448–450, June 1959.

[14] O. Christensen. *An Introduction to Frames and Riesz Bases.* Birkhäuser, Boston, MA, 2002.

[15] J. H. Conway and N. J. A. Sloane. Voronoi regions of lattices, second moments of polytopes, and quantization. *IEEE Trans. Inf. Theory*, IT-28(2):211–226, March 1982.

[16] J. W. Cooley and J. W. Tukey. An algorithm for the machine calculation of complex Fourier series. *Math. Comput.*, 19:297–301, April 1965.

[17] T. M. Cover and J. A. Thomas. *Elements of Information Theory.* John Wiley & Sons, New York, NY, 1991.

[18] R. E. Crochiere and L. R. Rabiner. *Multirate Digital Signal Processing.* Prentice Hall, Englewood Cliffs, NJ, 1983.

[19] I. Daubechies. *Ten Lectures on Wavelets.* SIAM, Philadelphia, PA, 1992.

[20] P. J. Davis. *Interpolation and Approximation.* Blaisdell, Boston, MA, 1963.

[21] P. A. M. Dirac. *The Principles of Quantum Mechanics.* Oxford University Press, London, fourth edition, 1958.

[22] D. L. Donoho. Unconditional bases are optimal bases for data compression and for statistical estimation. *Appl. Comput. Harmon. Anal.*, 1(1):100–115, December 1993.

[23] D. L. Donoho and X. Huo. Uncertainty principles and ideal atomic decomposition. *IEEE Trans. Inf. Theory*, 47(7):2845–2862, November 2001.

[24] D. L. Donoho and I. M. Johnstone. Ideal spatial adaptation via wavelet shrinkage. *Biometrika*, 81:425–455, 1994.

[25] D. L. Donoho and P. B. Stark. Uncertainty principles and signal recovery. *SIAM J. Appl. Math*, 49(3):906–931, June 1989.

[26] B. Draščić. Sampling reconstruction of stochastic signals – The roots in the fifties. *Austrian J. Stat.*, 36(1):65–72, 2007.

[27] D. E. Dudgeon and R. M. Mersereau. *Multidimensional Digital Signal Processing.* Prentice Hall, Englewood Cliffs, NJ, 1984.

[28] H. W. Dudley. The vocoder. *Bell Lab. Rec.*, 18:122–126, December 1939.

[29] P. E. Fleischer. Sufficient conditions for achieving minimum distortion in a quantizer. In *IEEE Int. Conv. Rec.*, volume 1, pages 104–111, 1964.

[30] G. B. Folland. *A Course in Abstract Harmonic Analysis.* CRC Press, London, 1995.

[31] D. Gabor. Theory of communication. *J. IEE*, 93:429–457, 1946.

[32] R. G. Gallager. *Stochastic Processes: Theory for Applications.* Cambridge University Press, Cambridge, 2013.

[33] R. W. Gerchberg. Super resolution through error energy reduction. *Optica Acta*, 21(9):709–720, 1974.

[34] A. Gersho and R. M. Gray. *Vector Quantization and Signal Compression.* Kluwer Academic Publishers, Norwell, MA, 1992.

[35] I. Gohberg and S. Goldberg. *Basic Operator Theory.* Birkhäuser, Boston, MA, 1981.

[36] V. K. Goyal. Theoretical foundations of transform coding. *IEEE Signal Process. Mag.*, 18(5):9–21, September 2001.

[37] V. K. Goyal, M. Vetterli, and N. T. Thao. Quantized overcomplete expansions in $\mathbb{R}^N$: Analysis, synthesis, and algorithms. *IEEE Trans. Inf. Theory*, 44(1):16–31, January 1998.

[38] V. K. Goyal, J. Zhuang, and M. Vetterli. Transform coding with backward adaptive updates. *IEEE Trans. Inf. Theory*, 46(4):1623–1633, July 2000.

[39] R. M. Gray and D. L. Neuhoff. Quantization. *IEEE Trans. Inf. Theory*, 44(6):2325–2383, October 1998.

[40] G. R. Grimmett and D. R. Stirzaker. *Probability and Random Processes.* Oxford University Press, third edition, 2001.

[41] R. W. Hamming. Mathematics on a distant planet. *Am. Math. Mon.*, 105(7):640–650, August–September 1998.

[42] T. L. Heath and Euclid. *The Thirteen Books of Euclid's Elements.* Dover Publications, New York, NY, 1956.

[43] C. Heil. *A Basis Theory Primer: Expanded Edition.* Birkhäuser, New York, NY, 2010.

[44] W. Heisenberg. Über den anschaulichen inhalt der quantentheoretischen Kinematik und Mechanik. *Zeitschrift für Physik*, 43:172–198, 1927.

[45] R. A. Horn and C. R. Johnson. *Matrix Analysis.* Cambridge University Press, Cambridge, 1985.

[46] J. J. Y. Huang and P. M. Schultheiss. Block quantization of correlated Gaussian random variables. *IEEE Trans. Commun. Syst.*, CS-11(3):289–296, September 1963.

[47] D. A. Huffman. A method for the construction of minimum redundancy codes. *Proc. IRE*, 40:1098–1101, September 1952.

[48] R. Ishii and K. Furukawa. The uncertainty principle in discrete signals. *IEEE Trans. Circuits Syst.*, 33(10):1032–1034, October 1986.

[49] A. J. Jerri. The Shannon sampling theorem – Its various extensions and applications: A tutorial review. *Proc. IEEE*, 65:1565–1596, November 1977.

[50] M. C. Jones. The discrete Gerchberg algorithm. *IEEE Trans. Acoust., Speech, Signal Process.*, 34(3):624–626, June 1986.

[51] T. Kailath. *Linear Systems.* Prentice Hall, Englewood Cliffs, NJ, 1980.

[52] U. Kamilov, V. K. Goyal, and S. Rangan. Message-passing de-quantization with applications to compressed sensing. *IEEE Trans. Signal Process.*, 60(12):6270–6281, December 2012.

[53] R. Konsbruck. *Source Coding in Sensor Networks.* PhD thesis, EPFL, Lausanne, 2009.

[54] J. Kovačević. z-transform. In *Handbook of Circuits and Filters*, pages 119–136. CRC Press, Boca Raton, FL, June 1995.

[55] J. Kovačević and A. Chebira. Life beyond bases: The advent of frames (Part I). *IEEE Signal Process. Mag.*, 24(4):86–104, July 2007.

[56] J. Kovačević and A. Chebira. An introduction to frames. *Found. Trends Signal Process.*, 2(1):1–94, 2008.

[57] J. Kovačević, V. K. Goyal, and M. Vetterli. *Fourier and Wavelet Signal Processing.* Cambridge University Press, Cambridge, 2014. http://www.fourierandwavelets.org/.

[58] H. P. Kramer and M. V. Mathews. A linear coding for transmitting a set of correlated signals. *IRE Trans. Inf. Theory*, 23(3):41–46, September 1956.

[59] E. Kreyszig. *Introductory Functional Analysis with Applications.* John Wiley & Sons, New York, NY, 1978.

[60] D. C. Lindberg. Late medieval optics. In *A Source Book in Medieval Science.* Harvard University Press, Cambridge, MA, 1974.

[61] S. P. Lloyd. A sampling theorem for stationary (wide sense) stochastic processes. *Trans. Am. Math. Soc.*, 92(1):1–12, July 1959.

[62] S. P. Lloyd. Least squares quantization in PCM. *IEEE Trans. Inf. Theory*, IT-28(2):129–137, March 1982. (Originally an unpublished Bell Telephone Laboratories technical memorandum, July 31, 1957.).

[63] F. Low. Complete sets of wave packets. In *A Passion for Physics – Essays in Honor of Geoffrey Chew*, pages 17–22. World Scientific, Singapore, 1985.

[64] D. G. Luenberger. *Optimization by Vector Space Methods.* John Wiley & Sons, New York, NY, 1969.

[65] D. J. C. MacKay. *Information Theory, Inference, and Learning Algorithms.* Cambridge University Press, Cambridge, 2003.

[66] S. Mallat. *A Wavelet Tour of Signal Processing.* Academic Press, New York, NY, third edition, 2009.

[67] R. J. Marks II. *Introduction to Shannon Sampling and Interpolation Theory.* Springer-Verlag, New York, NY, 1991.

[68] J. Max. Quantizing for minimum distortion. *IRE Trans. Inf. Theory*, 6(1):7–12, March 1960.

[69] J. H. McClellan and T. W. Parks. A personal history of the Parks–McClellan algorithm. *IEEE Signal Process. Mag.*, 22(2):82–86, March 2005.

[70] I. Newton. *Opticks or A Treatise of the Reflections, Refractions, Inflections and Colours of Light.* Royal Society, London, 1703.

[71] H. Q. Nguyen, V. K. Goyal, and L. R. Varshney. Frame permutation quantization. *Appl. Comput. Harmon. Anal.*, 31(1):74–97, July 2011.

[72] A. V. Oppenheim and R. W. Schafer. *Discrete-Time Signal Processing.* Prentice Hall, Upper Saddle River, NJ, third edition, 2010.

[73] A. Papoulis. *The Fourier Integral and Its Applications.* McGraw-Hill, New York, NY, 1962.

[74] A. Papoulis. *Probability, Random Variables, and Stochastic Processes.* McGraw-Hill, New York, NY, 1965.

[75] A. Papoulis. A new algorithm in spectral analysis and band-limited extrapolation. *IEEE Trans. Circuits Syst.*, 22(9):735–742, September 1975.

[76] A. Papoulis. Generalized sampling expansion. *IEEE Trans. Circuits Syst.*, 24(11):652–654, November 1977.

[77] A. Papoulis. *Signal Analysis.* McGraw-Hill, New York, NY, 1977.

[78] R. Parhizkar, Y. Barbotin, and M. Vetterli. Sequences with minimal time-frequency uncertainty. *Appl. Comput. Harmon. Anal.*, 2014.

[79] T. W. Parks and J. H. McClellan. Chebyshev approximation for nonrecursive digital filters with linear phase. *IEEE Trans. Circuit Theory*, 19(2):189–194, 1972.

[80] B. Porat. *Digital Processing of Random Signals.* Prentice Hall, Englewood Cliffs, NJ, 1994.

[81] A. M. Powell. Mean squared error bounds for the Rangan–Goyal soft thresholding algorithm. *Appl. Comput. Harmon. Anal.*, 29(3):251–271, November 2010.

[82] P. Prandoni and M. Vetterli. *Signal Processing for Communications.* EPFL Press, Lausanne, 2008.

[83] M. Püschel and J. M. F. Moura. Algebraic signal processing theory: Foundation and 1-D time. *IEEE Trans. Signal Process.*, 56(8):3572–3585, 2008.

[84] S. Rangan and V. K. Goyal. Recursive consistent estimation with bounded noise. *IEEE Trans. Inf. Theory*, 47(1):457–464, January 2001.

[85] K. A. Ross. *Elementary Analysis: The Theory of Calculus.* Springer-Verlag, New York, NY, 1980.

[86] V. P. Sathe and P. P. Vaidyanathan. Effects of multirate systems on the statistical properties of random signals. *IEEE Trans. Signal Process.*, 41(1):131–146, January 1993.

[87] I. J. Schoenberg. Cardinal interpolation and spline functions. *J. Approx. Theory*, 2:167–206, 1969.

[88] A. Segall. Bit allocation and encoding for vector sources. *IEEE Trans. Inf. Theory*, 22(2):162–169, March 1976.

[89] C. E. Shannon. A mathematical theory of communication. *Bell Syst. Tech. J.*, 27:379–423, July and 27:623–656, October 1948.

[90] Y. Shoham and A. Gersho. Efficient bit allocation for an arbitrary set of quantizers. *IEEE Trans. Acoust., Speech, Signal Process.*, 36(9):1445–1453, September 1988.

[91] W. M. Siebert. *Circuits, Signals, and Systems.* MIT Press, Cambridge, MA, 1986.

[92] D. Slepian. Some comments on Fourier analysis, uncertainty and modeling. *SIAM Rev.*, 25(3):379–393, July 1983.

[93] G. Strang. *Linear Algebra and Its Applications.* Brooks/Cole Publishing Co., St. Paul, MN, fourth edition, 2006.

[94] G. Strang. *Introduction to Linear Algebra.* Wellesley–Cambridge Press, Wellesley, MA, fourth edition, 2009.

[95] G. Strang and G. Fix. *An Analysis of the Finite Element Method.* Wellesley–Cambridge Press, Wellesley, MA, second edition, 2008.

[96] D. S. Taubman and M. W. Marcellin, editors. *JPEG2000: Image Compression Fundamentals, Standards and Practice.* Kluwer Academic Publishers, Boston, MA, 2002.

[97] N. T. Thao and M. Vetterli. Deterministic analysis of oversampled A/D conversion and decoding improvement based on consistent estimates. *IEEE Trans. Signal Process.*, 42(3):519–531, March 1994.

[98] N. T. Thao and M. Vetterli. Reduction of the MSE in R-times oversampled A/D conversion from $O(1/R)$ to $O(1/R^2)$. *IEEE Trans. Signal Process.*, 42(1):200–203, January 1994.

[99] N. T. Thao and M. Vetterli. Lower bound on the mean-squared error in oversampled quantization of periodic signals using vector quantization analysis. *IEEE Trans. Inf. Theory*, 42(2):469–479, March 1996.

[100] E. Tolsted. An elementary derivation of Cauchy, Hölder, and Minkowksi inequalities from Young's inequality. *Math. Mag.*, 37(1):2–12, January–February 1964.

[101] A. V. Trushkin. Sufficient conditions for uniqueness of a locally optimal quantizer for a class of convex error weighting functions. *IEEE Trans. Inf. Theory*, IT-28(2):187–198, March 1982.

[102] M. Unser. Splines: A perfect fit for signal and image processing. *IEEE Signal Process. Mag.*, 16(6):22–38, November 1999.

[103] M. Unser. Sampling – 50 years after Shannon. *Proc. IEEE*, 88(4):569–587, April 2000.

[104] M. Unser and I. Daubechies. On the approximation power of convolution-based least squares versus interpolation. *IEEE Trans. Signal Process.*, 45(7):1697–1711, July 1997.

[105] P. P. Vaidyanathan. *Multirate Systems and Filter Banks.* Prentice Hall, Englewood Cliffs, NJ, 1992.

[106] P. P. Vaidyanathan and S. K. Mitra. Polyphase networks, block digital filtering, LPTV systems, and alias-free QMF banks: A unified approach based on pseudo-circulants. *IEEE Trans. Acoust., Speech, Signal Process.*, 36:381–391, March 1988.

[107] M. Vetterli and J. Kovačević. *Wavelets and Subband Coding.* Prentice Hall, Englewood Cliffs, NJ, 1995. http://waveletsandsubbandcoding.org/.

[108] M. Vetterli, P. Marziliano, and T. Blu. Sampling signals with finite rate of innovation. *IEEE Trans. Signal Process.*, 50(6):1417–1428, June 2002.

[109] E. Wong and B. Hajek. *Stochastic Processes in Engineering Systems.* Springer-Verlag, New York, NY, 1985.

[110] R. W. Yeung. *A First Course in Information Theory.* Kluwer Academic Publishers, New York, NY, 2002.

[111] N. Young. *An Introduction to Hilbert Space.* Cambridge University Press, Cambridge, 1988.

Index

Printed in the United States
By Bookmasters

Printed in the United States
By Bookmasters